Food Properties Handbook

Second Edition

Food Properties Handbook

Second Edition

Edited by

M. Shafiur Rahman

CRC Press
Taylor & Francis Group
Boca Raton London New York

CRC Press is an imprint of the
Taylor & Francis Group, an **informa** business

CRC Press
Taylor & Francis Group
6000 Broken Sound Parkway NW, Suite 300
Boca Raton, FL 33487-2742

Printed in the United States of America on acid-free paper
10 9 8 7 6 5 4 3 2 1

International Standard Book Number-13: 978-0-8493-5005-4 (Hardcover)

Library of Congress Cataloging-in-Publication Data

Food properties handbook / edited by M. Shafiur Rahman. -- 2nd ed.
p. cm.
Previous edition has main entry under Rahman, Shafiur.
Includes bibliographical references and index.
ISBN-13: 978-0-8493-5005-4
ISBN-10: 0-8493-5005-0
1. Food--Analysis. 2. Food industry and trade. I. Rahman, Shafiur. II. Title.

TX541.R37 2009
664--dc22 2008019583

Visit the Taylor & Francis Web site at
http://www.taylorandfrancis.com

and the CRC Press Web site at
http://www.crcpress.com

Contents

Preface

A food property is a particular measure of a food's behavior as a matter or its behavior with respect to energy, or its interaction with the human senses, or its efficacy in promoting human health and well-being. An understanding of food properties is essential for scientists and engineers who have to solve the problems in food preservation, processing, storage, marketing, consumption, and even after consumption. Current methods of food processing and preservation require accurate data on food properties; simple, accurate, and low-cost measurement techniques; prediction models based on fundamentals; and links between different properties. The first edition was a well received bestseller, and it received an award. Appreciation from scientists, academics, and industry professionals around the globe encouraged me to produce an updated version. This edition has been expanded with the addition of some new chapters and by updating the contents of the first edition. The seven chapters in the first edition have now been expanded to 24 chapters.

In this edition, the definition of the terminology and measurement techniques are clearly presented. The theory behind the measurement techniques is described with the applications and limitations of the methods. Also, the sources of errors in measurement techniques are compiled. A compilation of the experimental data from the literature is presented in graphical or tabular form, which should be very useful for food engineers and scientists. Models can reduce the number of experiments, thereby reducing time and expenses of measurements. The empirical and theoretical prediction models are compiled for different foods with processing conditions. The applications of the properties are also described, mentioning where and how to use the data and models in food processing.

Chapter 1 provides an overview of food properties, including their definition, classification, and predictions. Chapters 2 through 4 present water activity and sorption isotherm and include terminology, measurement techniques, data for different foods, and prediction models. Chapters 5 through 12 present thermodynamic and structural characteristics including freezing point, glass transition, gelatinization, crystallization, collapse, stickiness, ice content, and state diagram. Chapters 13 through 15 discuss the density, porosity, shrinkage, size, and shape of foods. Chapters 15 through 23 present the thermophysical properties including specific heat, enthalpy, thermal conductivity, thermal diffusivity, and heat transfer coefficient. Chapter 24 provides the acoustic properties of foods.

This second edition will be an invaluable resource for practicing and research food technologists, engineers, and scientists, and a valuable text for upper-level undergraduate and graduate students in food, agriculture/biological science, and engineering. Writing such a book is a challenge, and any comments to assist in future compilations will be appreciated. Any errors that remain are entirely mine. I am confident that this edition will prove to be interesting, informative, and enlightening.

Mohammad Shafiur Rahman
Sultan Qaboos University
Muscat, Sultanate of Oman

Acknowledgments

I would like to thank Almighty Allah for giving me life and blessing to gain knowledge to update this book. I wish to express my sincere gratitude to the Sultan Qaboos University (SQU) for giving me the opportunity and facilities to initiate such an exciting project to develop the second edition, and supporting me toward my research and other intellectual activities. I would also like to thank all my earlier employers, Bangladesh University of Engineering and Technology, University of New South Wales (UNSW), and HortResearch, from whom I built my knowledge and expertise through their encouragement, support, and resources. I wish to express my appreciation to the UNSW, SQU, and HortResearch library staffs, who assisted me patiently with online literature searches and interlibrary loans.

I sincerely acknowledge the sacrifices made by my parents, Asadullah Mondal and Saleha Khatun, during my early education. Appreciation is due to all my teachers, especially Professors Nooruddin Ahmed, Iqbal Mahmud, Khaliqur Rahman, Jasim Zaman, Ken Buckle, Drs. Prakash Lal Potluri and Robert Driscoll, and Habibur Rahman, for their encouragement and help in all aspects of pursuing higher education and research. I would like to express my appreciation to Professor Anton McLachlan, Drs. Saud Al-Jufaily, Yasen Al-Mula, Nadya A-Saadi, and S. Prathapar for their support toward my teaching, research, and extension activities at the SQU. Special thanks to my colleagues Dr. Conrad Perera, Professor Dong Chen, Drs. Nejib Guizani, Ahmed Al-Alawi, Shyam Sablani, Bhesh Bhandar, and Mushtaque Ahmed, and my other research team members, especially Mohd Hamad Al-Ruzeiki, Rashid Hamed Al-Belushi, Salha Al-Maskari, Mohd Khalfan Al-Khusaibi, Nasser Abdulla Al-Habsi, Insaaf Mohd Al-Marhubi, Intisar Mohd Al-Zakwani, and Zahra Sulaiman Al-Kharousi. I owe many thanks to my graduate students for their hard work in their projects related to food properties and building my knowledge base. Special thanks for the contributing authors; it was a great pleasure working with them. I would also like to appreciate the enthusiasm, patience, and support provided by the publisher.

I wish to thank my relatives and friends, especially Professor Md. Mohar Ali and Dr. Md. Moazzem Hossain, Dr. Iqbal Mujtaba, and Arshadul Haque for their continued inspiration. I am grateful to my wife, Sabina Akhter (Shilpi), for her patience and support during this work, and to my daughter, Rubaba Rahman (Deya), and my son, Salman Rahman (Radhin), for allowing me to work at home. It would have been very hard for me to write this book without my family's cooperation and support.

Editor

Mohammad Shafiur Rahman is an associate professor at the Sultan Qaboos University, Sultanate of Oman. He has authored or coauthored more than 200 technical articles including 81 refereed journal papers, 71 conference papers, 40 book chapters, 33 reports, 8 popular articles, and 4 books. He is the editor of the internationally acclaimed 2003 bestseller, *Handbook of Food Preservation* published by CRC Press. He was invited to serve as one of the associate editors for the *Handbook of Food Science, Engineering and Technology*, and he is one of the editors for the *Handbook of Food and Bioprocess Modeling Techniques*, also published by CRC Press. Dr. Rahman has initiated the *International Journal of Food Properties* (Marcel Dekker, Inc.) and has served as its founding editor for more than 10 years. He is a member of the Food Engineering Series editorial board of Springer Science, New York. Presently, he is serving as a section editor for the Sultan Qaboos University journal, *Agricultural Sciences*. In 1998, he was invited to serve as a food science adviser for the International Foundation for Science (IFS) in Sweden.

Dr. Rahman is a professional member of the New Zealand Institute of Food Science and Technology and the Institute of Food Technologists; a member of the American Society of Agricultural Engineers and the American Institute of Chemical Engineers; and a member of the executive committee for International Society of Food Engineering (ISFE). He received his BSc Eng (chemical) (1983) and MSc Eng (chemical) (1984) from Bangladesh University of Engineering and Technology, Dhaka; his MSc (1985) in food engineering from Leeds University, England; and his PhD (1992) in food engineering from the University of New South Wales, Sydney, Australia. Dr. Rahman has received numerous awards and fellowships in recognition of research/teaching achievements, including the HortResearch Chairman's Award, the Bilateral Research Activities Program (BRAP) Award, CAMS Outstanding Researcher Award 2003, SQU Distinction in Research Award 2008, and the British Council Fellowship. The Organization of Islamic Countries has named Rahman as the fourth ranked agroscientist in a survey of the leading scientists and engineers in its 57 member states.

Contributors

Benu P. Adhikari
School of Science and Engineering
The University of Ballarat
Mount Helen, Victoria, Australia

Jasim Ahmed
Polymer Source, Inc.
Dorval, Quebec, Canada

Ghalib Said Al-Saidi
Department of Food Science and Nutrition
Sultan Qaboos University
Muscat, Sultanate of Oman

Bhesh R. Bhandari
School of Land, Crop and Food Sciences
The University of Queensland
Brisbane, Queensland, Australia

G.I. Bisharat
Department of Chemical Engineering
National Technical University of Athens
Athens, Greece

Adriana Delgado
School of Agriculture, Food Science and Veterinary Medicine
University College Dublin
Dublin, Ireland

Ferruh Erdoğdu
Department of Food Engineering
University of Mersin
Mersin, Turkey

Didem Z. Icoz
Department of Food Science
Rutgers, The State University of New Jersey
New Brunswick, New Jersey

Kirsi Jouppila
Department of Food Technology
University of Helsinki
Helsinki, Finland

Stefan Kasapis
Department of Chemistry
National University of Singapore
Singapore

Jozef L. Kokini
Department of Food Science
Rutgers, The State University of New Jersey
New Brunswick, New Jersey

Magdalini K. Krokida
Department of Chemical Engineering
National Technical University of Athens
Athens, Greece

Piotr P. Lewicki
Department of Food Engineering and Process Management
Warsaw University of Life Sciences
Warsaw, Poland

K.M. Machado-Velasco
Chemical and Food Engineering Department
University of the Americas–Puebla
Cholula, Puebla, Mexico

Dimitris Marinos-Kouris
Department of Chemical Engineering
National Technical University of Athens
Athens, Greece

Agata Marzec
Department of Food Engineering and Process Management
Warsaw University of Life Sciences
Warsaw, Poland

Panagiotis A. Michailidis
Department of Chemical Engineering
National Technical University of Athens
Athens, Greece

Mohammad Shafiur Rahman
Department of Food Science and Nutrition
Sultan Qaboos University
Muscat, Sultanate of Oman

Zbigniew Ranachowski
Institute of Fundamental Technological Research
Polish Academy of Sciences
Warsaw, Poland

Yrjö H. Roos
Department of Food Science and Technology
University College Cork
Cork, Ireland

Shyam S. Sablani
Department of Biological Systems Engineering
Washington State University
Pullman, Washington

R. Paul Singh
Department of Biological and Agricultural Engineering
University of California, Davis
Davis, California

M.E. Sosa-Morales
Chemical and Food Engineering Department
University of the Americas–Puebla
Cholula, Puebla, Mexico

Da-Wen Sun
School of Agriculture, Food Science and Veterinary Medicine
University College Dublin
Dublin, Ireland

Jorge F. Velez-Ruiz
Chemical and Food Engineering Department
University of the Americas–Puebla
Cholula, Puebla, Mexico

Liyun Zheng
School of Agriculture, Food Science and Veterinary Medicine
University College Dublin
Dublin, Ireland

CHAPTER 1

Food Properties: An Overview

Mohammad Shafiur Rahman

CONTENTS

1.1 DEFINITION OF FOOD PROPERTY

A property of a system or material is any observable attribute or characteristic of that system or material. The state of a system or material can be defined by listing its properties (ASHRAE, 1993). A food property is a particular measure of the food's behavior as a matter, its behavior with respect to energy, its interaction with the human senses, or its efficacy in promoting human health and well-being (McCarthy, 1997; Rahman and McCarthy, 1999). It is always attempted to preserve product characteristics at a desirable level for as long as possible. Food properties, in turn, define the functionality of foods (Karel, 1999). Food functionality, as defined by Karel, refers to the control of food properties that provides a desired set of organoleptic properties, wholesomeness (including health-related functions), as well as properties related to processing and engineering, in particular, ease of processing, storage stability, and minimum environmental impact.

In general, food preservation and processing affects the properties of foods in a positive or negative manner. During food processing, attempts to achieve the desired characteristics can be grouped as (1) controlling food characteristics by adding of ingredients/preservatives or removing components detrimental to quality, (2) applying different forms of energy, such as heat, light, electricity, and physical forces, and (3) controlling or avoiding recontamination. In the investigation of foods in the temperature range between −80°C and 350°C many effects can be observed during processing, preservation, and storage. These phenomena may either be endothermic (such as melting, denaturation, gelatinization, and evaporation) or exothermic processes (such as freezing, crystallization, and oxidation). Through precise knowledge of such phase transitions, optimum conditions for safe storage or processing of foods can be defined. In addition to thermal energy, other forms of energy, such as electricity, light, electromagnetism, and pressure are also used in food processing.

1.2 CLASSIFICATION OF FOOD PROPERTY

Classifying food properties is a difficult task, and any attempt to do so is likely to be controversial. However, it is necessary to develop a well-defined terminology and classification of food properties (Rahman, 1998). Rahman and McCarthy (1999) attempted to develop a widely accepted classification terminology for food properties. A good classification could facilitate sound interdisciplinary approaches to the understanding of food properties, and of the measurement and use of food property data, leading to better process design and food product characterization. Jowitt (1974) proposed a classification of foodstuffs and their physical properties. Rahman (IJFP, 1998) settled on the list that appears at the end of the first issue of the *International Journal of Food Properties*, after several revisions based on discussions with many academics and scientists around the world (Table 1.1). The classification now proposed contains four major classes (Rahman and

Table 1.1 Food Properties Grouped in the First Issue of *International Journal of Food Properties*

Acoustical properties
Colorimetric properties
Electrical properties
Functional properties
Mass transfer properties
Mass–volume–area-related properties
Mechanical properties
Medical properties
Microbial death–growth-related properties
Morphometric properties
Optical properties
Physico-chemical constants
Radiative properties
Respiratory properties
Rheological properties
Sensory properties
Surface properties
Thermodynamic properties
Textural properties
Thermal properties
Quality kinetics parameters

Source: From *Int. J. Food Prop.*, 1, 78, 1998.

Table 1.2 List of Four Classes of Food Properties

Physical and physicochemical properties
a. Mechanical properties
   1. Acoustic properties
   2. Mass–volume–area-related properties
   3. Morphometric properties
   4. Rheological properties
   5. Structural characteristics
   6. Surface properties
b. Thermal properties
c. Thermodynamic properties
d. Mass transfer properties
e. Electromagnetic properties
f. Physicochemical constants

Kinetic properties
a. Quality kinetic constants
b. Microbial growth, decline, and death kinetic constants

Sensory properties
a. Tactile properties
b. Textural properties
c. Color and appearance
d. Taste
e. Odor
f. Sound

Health properties
a. Positive health properties
   1. Nutritional composition
   2. Medical properties
   3. Functional properties
b. Negative health properties
   1. Toxic at any concentration
   2. Toxic after critical concentration level
   3. Excessive or unbalanced intake

Source: Rahman, M.S. and McCarthy, O.J., *Int. J. Food Prop.*, 2, 1, 1999.

McCarthy, 1999): (1) physical and physicochemical properties, (2) kinetic properties, (3) sensory properties, and (4) health properties (Table 1.2).

1.2.1 Physical and Physicochemical Properties

Physical and physicochemical properties are properties defined, measured, and expressed in physical and physicochemical ways. However, there is no clear dividing line between these two types of properties. Paulus (1989) classified physical properties as mechanical, thermal, transport, and other electrical and optical properties. It is considered misleading to use transport as a subclass of physical properties, since many mechanical, thermal, and electrical properties are considered transport properties, e.g., electrical conductivity and thermal conductivity. Moreover, among thermal properties, specific heat is a constitutive property, whereas thermal conductivity and diffusivity are transport properties. The classification proposed here is similar to the classification of physical properties proposed by Jowitt (1974): first, two new subclasses, thermodynamic and mass transfer properties, replace Jowitt's subclass of diffusion-related properties; most of the properties included in Jowitt's subclass are in fact thermodynamic ones. The new mass transfer properties subclass now proposed includes mass transfer by both diffusion and other mechanisms, and is thus more generic. Second, a new subclass of physicochemical constants has been added.

Mechanical properties are related to food's structure and its behavior when physical force is applied. Structure is the form of building or construction, the arrangement of parts or elements of something constructed or of a natural organism to give an organization of foods. In addition to natural structure, man-made structured foods use assembly or structuring processes to build product microstructure. Examples of essential tools to create microstructure are crystallization, phase inversion, phase transition, glass transition, emulsification (e.g., margarine, ice cream, sauces, and mayonnaise), freeze alignment, foaming (e.g., whipped cream), extrusion, puffing, drying, kneading of dough, and baking. In these products, a complicated multiphase microstructure is held together by binding forces between the various phases. This microstructure leads to acceptance of desired product texture and mouth feel during mastication, which is the key to final product quality and is appreciated by the consumer. The control of the microstructure of man-made structured foods is the key quality-determining factor apart from requirements on microbial stability and safety. A detailed review on structuring processes since the past 25 years as well as the challenges that lie ahead has been presented by Bruin and Jongen (2003). In the past 25 years, substantial progress has been made in the understanding and control of product microstructure, and new ways of achieving them have been developed. The mechanical properties based on structure are further classified into six subclasses: acoustic properties, mass–volume–area-related properties, morphometric properties, rheological properties, structural characteristics, and surface properties.

Thermal properties are related to heat transfer in food, and thermodynamic properties are related to the characteristics indicating phase or state changes in food. Mass transfer properties are related to the transport or flow of components in food. Electromagnetic properties are related to the food's behavior with the interaction of electromagnetic energy (e.g., dielectric constant, dielectric loss, and electrical resistance).

1.2.2 Kinetic Properties

Kinetic properties are kinetic constants characterizing the rates of changes in foods. These can be divided into two groups. The first comprises kinetic constants characterizing the rates of biological, biochemical, chemical, physicochemical, and physical changes in food. It could include respiratory constants, rate constant, decimal reduction time, half-life, Arrhenius equation constants, temperature quotient (Q_{10}), and D and z values. The second comprises kinetic constants characterizing the rates of growth, decline, and death of microorganisms in food. It could include properties such as specific growth rate, the parameters of the logistic and Gompertz equations (mathematical models of microbial growth), generation time, square root (Ratkowsky) equation constants, and decimal-reduction time. It should be noted that these properties are not actually properties of food, but properties of microorganisms as moderated by the food they are in (Rahman and McCarthy, 1999).

1.2.3 Sensory Properties

A sensory property can be defined as the human physiological–psychological perception of a number of physical and other properties of food and their interactions. The physiological apparatus (fingers, mouth, eyes, taste and aroma receptors, and ears) examines the food and reacts to the food's properties. Signals are sent to the brain, which interprets the signals and comes to a decision about the food's sensory quality; this is the psychological bit. Sensory properties are measured subjectively using trained and untrained panels, and individuals or consumers. Sensory properties can be subdivided into tactile properties, textural properties, color and appearance, taste, odor, and sound. Tactile properties are perceived as touch, i.e., by the fingers. For example, the surface roughness and softness of a food can be evaluated by touch. The main difference between texture and other sensory attributes is that texture is perceived mainly by biting and masticating, i.e., by the mouth. Many of the sensory properties are related to physical and physicochemical properties as measured objectively with instruments. However, this does not mean that instrumentally measured characteristics are sensory properties. The following discussion could help to highlight the difference. The rheological nature of a food and the food's texture are two different things. Rheological properties are measured objectively using suitable instruments that allow controlled deformation of the food. Texture, however, has to be measured subjectively. It depends partly, of course, on the food's rheological properties, but also, potentially, on a number of other properties (e.g., shape, size, porosity, and thermal properties) and on the expectations and prior experience of the person(s) assessing the texture. In many cases, texture can be correlated quite well with an instrumentally measured rheological property (often an empirical or imitative one), but texture as such can be measured only by subjective means (Rahman and McCarthy, 1999). Both subjective and objective methods have their own advantages and limitations. However, food properties measured by subjective methods could be correlated with properties measured by objective methods and this could make the quality control process easy during processing, preservation, and storage.

1.2.4 Health Properties

Health properties relate to the efficacy of foods in promoting human health and well-being. Not all foods consumed are safe; thus foods have positive or negative impacts on health. Positive effects can be subdivided into nutritional composition (as defined in nutritional composition tables), medical properties, and functional properties. Functional properties are those that impact on an individual's general health, physical well-being, and mental health, and slow the aging process; medical properties are those that prevent and treat diseases. It is not easy to make a clear-cut distinction between functional and medical properties. For example, the antioxidant character of a food has effects both in controlling heart disease (a medical effect) and in slowing down the aging process (a functional effect). Some components of foods, such as pesticides and fungicides are toxic at any level, or when some critical level is exceeded. It is not safe to consume unlimited quantities of some foods. Some components (e.g., sugar, salt, fat, fat soluble vitamins, and alcohol) have negative effects if intake is excessive, or if the diet as a whole is unbalanced. Thus negative health properties are grouped as toxic at any concentration, toxic above a critical concentration level, and excessive or unbalanced intake.

1.3 APPLICATIONS OF FOOD PROPERTIES

An understanding of food properties is essential for scientists and engineers to solve the problems in food preservation, processing, storage, marketing, consumption, and even after consumption. It would be very difficult to find a branch of food science and engineering that does not need the knowledge of food properties. The application of food properties are discussed in the following sections.

1.3.1 Process Design and Simulation

Processing causes many changes in the biological, chemical, and physical properties of foods. A basic understanding of these properties of food ingredients, products, processes, and packages is essential for the design of efficient processes and minimization of undesirable changes due to processing. The present lack of sufficient data on physical properties (such as rheological, thermal, mass, and surface properties) of basic components under real conditions has limited the application of many well-established engineering principles (IFT, 1993).

1.3.1.1 Process Design

Food properties are used in the engineering design, installation, optimization, and operation of food processing equipments including a complete plant. For example, during canning, foods need to be heated for sterilization. The duration and the temperature at which heating needs to be carried out can be based on quality, safety, nutrition content, and process efficiency. In this case, thermal properties such as thermal conductivity and diffusivity as well as microbial lethality and nutrition loss are required for all heat transfer calculations and to predict the end point of heating process.

1.3.1.2 Process Simulation

Process simulation is an important tool for food engineers to develop concept, design, operation, and improvement of food processes. For example, flow modeling can investigate more alternatives of better products in less time at a lower cost. Dhanasekharan et al. (2004) described examples in which flow modeling was used to overcome challenging design problems in extrusion, mixing, and food safety by incorporating HACCP. Schad (1998) warned not to gamble with physical properties when making the most of process simulation benefits. Physical properties are critical in simulating a process. Thus, it is important to know from where pure-component properties have come, what basic property models are being used, and from where the basic equation has originated. It is important to be careful in interpreting the results of the simulation based on the quality and source of critical physical property data. The missing or inadequate physical properties undermine the accuracy of simulation. The problem is that the simulation software is not likely to tell us whether answer is erroneous. The results may appear to be correct, but they may be totally wrong. It is our responsibility alone to ensure that we are using the right property models and have inputted or accessed correct and sufficient data to describe our physical properties. There are no shortcuts (Schad, 1998).

1.3.1.3 Continuous Need

We may think that food properties are important only for the initial design of a plant or process, thus only those who are building new equipment need food properties. It is misleading to think that after the plant has been commissioned, food properties are not required for process design, process operation, and product development. In many instances, the existing equipment need to be updated for new product lines or when some units are not operating efficiently. In this case, it is very expensive to replace whole processing lines or equipment. Some modifications in the process need the applications of process design. Thus, the use of food properties in process design is necessary during the entire life of a processing plant.

1.3.2 Quality and Safety

Quality is an illusive, ever-changing concept. It is a relative perception and is always pegged to expectations based on past experiences. It may have different dimensions or attributes, which could be rotated based on the types of users. Several authorities have defined quality in various ways, but the term generally appears to be associated with the degree of fitness for use or the satisfaction level of consumers (ITC, 1993); the absence of defects or a degree of excellence (Shewfelt, 1999); the degree of conformance to the desired functionality (Karel, 1999); or the degree of acceptability of a

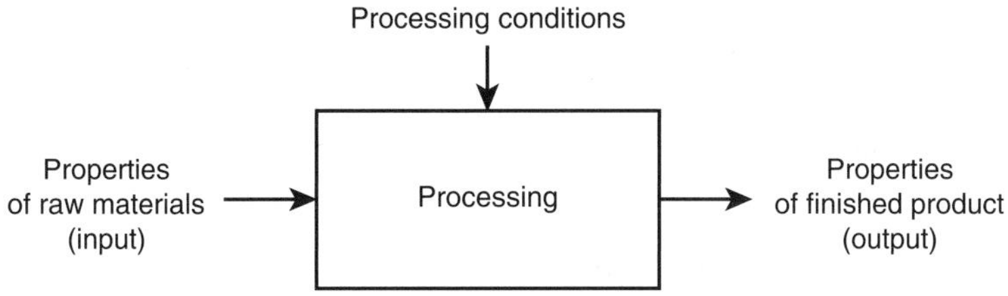

Figure 1.1 Interaction of processing variables with input and output materials variables.

product to users. Every food product has characteristics measurable by sensory evaluation methods or physicochemical tests. Some characteristics or properties are physical and are easily perceived; others are unseen. The applications of food properties can also be described as characterizing the defects in a food product.

Understanding these quality characteristics and emotional factors and familiarity with the appropriate measuring tools are vital to the quality control of food products. Quality loss can be minimized at any stage thus quality retaining depends on the overall control of the processing chain. When preservation fails, the consequences range broadly from being extremely hazardous to the loss of color. Automatic control of food-processing systems helps to improve final product quality, increase process efficiency, and reduce waste of raw materials. Food processes are generally multiple-input, multiple-output systems involving complex interactions between process inputs and outputs (Figure 1.1).

1.3.3 Packaging Design

It is important to know product and packaging characteristics, food-packaging interaction, and stability of packaging during storage and distribution (Petersen et al., 1999). Information on food properties is needed in the selection of packaging materials, and in the design of packages, packaging operations, and packaging machines (McCarthy, 1997). It is important to know how food materials interact with packaging materials, and deterioration kinetics of food during storage and distribution.

1.4 PREDICTION OF FOOD PROPERTIES

The experimental measurement is very costly, labor intensive, and may require specialist knowledge. Computer models can be run very quickly, and in many cases do not require a lot of detailed technical knowledge. They can be used to predict what might happen in the process, handling, storage, and consumption. One of the best features of computer models is that they can be used to explore any number of "what if" scenarios. In many instances simulation refers to "what if" scenarios and optimization refers to "best way to do it." This can be useful as tools, since they can be used to investigate the possible effects before undertaking detailed and time-consuming experimental work. There is a need for models that can predict complicated phenomena such as taste development or the effect of complex food processing events on product properties.

1.5 CONCLUSION

A clear definition of food properties is presented followed by well-defined classifications. The needs of understanding food properties are clearly identified with the different applications in food processing, preservation, storage, and quality control. Food properties can be measured experimentally when needed. This task could be achieved by developing prediction models, which would save money for costly instruments or methods, reduce labor costs, and avoid hiring skilled operators for complex methods. However, prediction models will not be able to replace the needs of developing measurement techniques.

REFERENCES

ASHRAE. 1993. *ASHRAE Fundamentals*. American Society of Heating, Refrigerating and Air-Conditioning Engineers, New York.

Bruin, S. and Jongen, T.R.G. 2003. Food process engineering: The last 25 years and challenges ahead. *Comprehensive Reviews in Food Science and Food Safety*, 2: 42–80.

Dhanasekharan, K.M., Grald, E.W., and Mathur, R. 2004. How flow modeling benefits the food industry. *Food Technology*, 58(3): 32–35.

IFT. 1993. IFT special report: America's food research needs into the 21st century. *Food Technology*, 47(3): 1S–39S.

IJFP. 1998. Instructions for preparation of manuscript. *International Journal of Food Properties*, 1: 95–99.

ITC. 1993. *Quality Control for the Food Industry: An Introductory Handbook*. International Trade Centre UNCTAD/GATT, Geneva.

Jowitt, R. 1974. Classification of foodstuffs and physical properties. *Lebensmittel-Wissenschaft und Technologie*, 7(6): 358–378.

Karel, M. 1999. Food research tasks at the beginning of the new Millennium—a personal vision. In: *Water Management in the Design and Distribution of Quality of Foods*, Roos, Y.H., Leslie, R.B., and Lillford, P.J. (eds.). Technomic Publishing, Lancaster, Pennsylvania, pp. 535–559.

McCarthy, O.J. 1997. Physical properties of foods and packaging materials—an introduction. In: *Food and Packaging Engineering I Course Material*. Department of Food Technology, Massey University, Plmerston North.

Paulus, K. 1989. Nutritional and sensory properties of processed foods. In: *Food Properties and Computer-Aided Engineering of Food Processing Systems*, Singh, R.P. and Medina, A.G. (eds.). Kluwer Academic Publishers, New York, pp. 177–200.

Petersen, K., Nielsen, P., Bertelsen, G., Lawther, M., Olsen, M.B., Nilsson, N.H., and Morthensen, G. 1999. Potential biobased materials for food packaging. *Trends in Food Science and Technology*, 10: 52–68.

Rahman, M.S. 1998. Editorial. *International Journal of Food Properties*, 1(1): v–vi.

Rahman, M.S. and McCarthy, O.J. 1999. Classification of food properties. *International Journal of Food Properties*, 2(2): 1–6.

Schad, R.C. 1998. Make the most of process simulation. *Chemical Engineering Progress*, 94(1): 21–27.

Shewfelt, R.L. 1999. What is quality? *Postharvest Biology and Technology*, 15: 197–200.

CHAPTER 2

Water Activity Measurement Methods of Foods

Mohammad Shafiur Rahman and Shyam S. Sablani

CONTENTS

2.1 INTRODUCTION

Water is an important constituent of all foods. In the middle of the twentieth century, scientists began to discover the existence of a relationship between the water contained in a food and its relative tendency to spoil. They also began to realize that the chemical potential of water is related to its vapor pressure relative to that of pure water was more important. This relative vapor pressure (RVP) is termed as water activity or a_w. Scott (1957) clearly stated that the water activity of a medium correlated with the deterioration of food stability due to the growth of microorganisms. Thus, it is possible to develop generalized rules or limits for the stability of foods using water activity. This was the main reason why food scientists started to emphasize water activity along with water content. Since then, the scientific community has explored the great significance of water activity in determining the physical characteristics, processes, shelf life, and sensory properties of foods. Recently, Rahman and Labuza (2007) have presented a detailed review on this aspect of water activity. Details of the various measurement techniques are presented by Labuza et al. (1976), Rizvi (1995), Rahman (1995), and Bell and Labuza (2000).

Water activity, a thermodynamic property, is defined as the ratio of the vapor pressure of water in a system to the vapor pressure of pure water at the same temperature, or the equilibrium relative humidity (ERH) of the air surrounding the system at the same temperature. Thus, water activity can be expressed as follows:

$$a_{\mathrm{w}} = \frac{\left(P_{\mathrm{w}}^{\mathrm{v}}\right)_{\mathrm{sy}}}{P_{\mathrm{w}}^{\mathrm{v}}} = \mathrm{ERH} \tag{2.1}$$

where

a_{w} is the water activity (fraction) at t (°C)

$(P_{\mathrm{w}}^{\mathrm{v}})_{\mathrm{sy}}$ and $P_{\mathrm{w}}^{\mathrm{v}}$ are the vapor pressures of water in the system and pure water, respectively, at t °C (Pa)

ERH is the equilibrium relative humidity of air at t °C

2.2 WATER ACTIVITY MEASUREMENT

Wiederhold (1987), Labuza et al. (1976), Rizvi (1995), Smith (1971), and Stoloff (1978) studied the accuracy and precision of various water activity measuring devices and found considerable variations. The accuracy of most of the methods lies in the range of 0.01–0.02 water activity units (Rizvi, 1995). The choice of one technique over another depends on the range, accuracy, cost, response time (speed), suitability, portability, simplicity, precision, maintenance and calibration requirements, and types of foods to be measured (Wiederhold, 1987; Rizvi, 1995; Rahman and Al-Belushi, 2006). The required accuracy of the routine and reference methods is given in Table 2.1. More details of the measurement techniques are presented by Rizvi (1995), Wiederhold (1987), Gal (1981), and Smith (1971). The water activity measurement methods can be classified as given in Table 2.2.

2.2.1 Colligative Properties Methods

2.2.1.1 Vapor Pressure Measurement

The water activity of food samples can be estimated by direct measurement of vapor pressure using a manometer (Sood and Heldman, 1974; Lewicki et al., 1978; Lewicki, 1987, 1989). A simple

Table 2.1 Precision Requirements for Temperature, Water Activity, and Moisture Measurement Equipment

Variable	Routine Method	Reference Method
Temperature (°C)	±0.2	±0.02
Relative humidity (%)	1.0	0.10
X_{we} (%)	0.1	0.01

Source: Spiess, W.E.L. and Wolf, W. in *Water Activity: Theory and Applications to Food*, Rockland, L.B. and Beuchat, L.R. (eds.), Marcel Dekker, Inc., New York, 1987.

Table 2.2 Methods for the Determination of Sorption Isotherm

Colligative properties methods
1. Vapor pressure measurement
2. Freezing point measurement
3. Boiling point measurement

Gravimetric methods
1. Methods with discontinuous registration of mass changes
   a. Static systems (isopiestic method)
   b. Evacuated systems
   c. Dynamic systems
2. Methods with continuous registration of mass changes
   a. Static chamber
   b. Dynamic systems
   c. Evacuated system

Hygrometric systems
1. Mechanical hygrometers
2. Wet and dry bulb hygrometers
3. Dew point hygrometers
4. Hygroscopicity of salts
5. Electronic sensor hygrometers

Other methods

schematic diagram is shown in Figure 2.1. A sample of mass 10–50 g of unknown water activity is placed in the sample flask and sealed on to the apparatus. The airspace in the apparatus is evacuated with the sample flask excluded from the system. The sample flask is connected with the evacuated airspace and the space in the sample flask is evacuated to less than 200 μmHg, which is followed by the evacuation of sample for 1–2 min. After isolating the vacuum source and equilibration for 30–50 min the pressure exerted by the sample is recorded (Δh_1). The sample flask is subsequently excluded from the system, and the desiccant flask is opened. Water vapor is removed by sorption onto $CaSO_4$, and the pressures exerted by volatiles and gases are indicated by Δh_2 after equilibrium. The water activity of the sample is calculated as (Labuza et al., 1976):

$$a_w = \frac{[h_1 - h_2]\rho g}{P_w^v} \tag{2.2}$$

where
P_w^v is the vapor pressure of pure water at t °C (Pa)
ρ is the density of manometric fluid (kg/m^3)
h_1 and h_2 are the manometer readings (m)

Rizvi (1995) mentioned that for precise results it is necessary to maintain the following conditions: (1) the whole system should be maintained at a constant temperature, (2) ratio of the sample volume to vapor space volume should be large enough to minimize changes in water activity due to loss of water by vaporization, and (3) a low-density and low-vapor-pressure oil should be used as the manometric fluid. Apiezon B manometric oil (density = 866 kg/m^3) is generally used as manometric fluid. If T_{sa} (sample temperature) and T_{me} (medium temperature) are different, then water activity is corrected as (Rizvi, 1995):

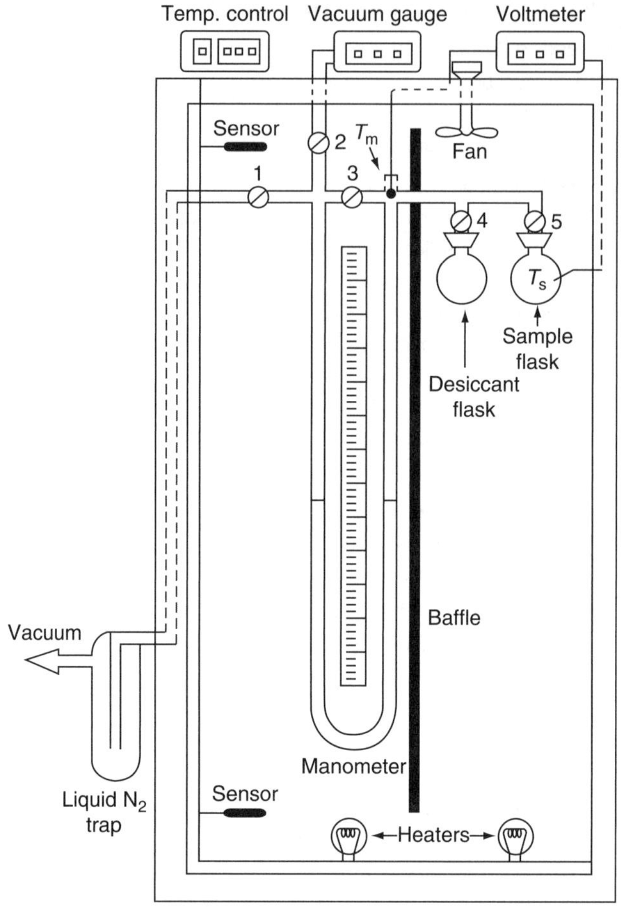

Figure 2.1 Schematic diagram of a thermostatized vapor pressure manometer apparatus. Numbers 1–5 indicate the locations of the stopcocks used for performing an experiment. (From Rizvi, S.S.H., in *Engineering Properties of Foods*, 2nd edn., Rao, M.A., Rizvi, S.S.H., and Datta, A. (eds.), CRC Press, Boca Raton, FL, 1995.)

$$a_w = \left[\frac{\Delta h_1 - \Delta h_2}{P_w^v}\right] \times \left[\frac{T_{sa}}{T_{me}}\right]\rho g \tag{2.3}$$

The capacitance manometer can be used for more compactness of the large setup and better temperature control (Troller, 1983). In order to incorporate the change in volume that occurs when water vapor is eliminated from the air–water mixture during desiccation, Nunes et al. (1985) presented the following corrections:

$$a_w = \frac{[h_1 - Ch_2]\rho g}{P_w^v} \quad \text{and} \quad C = \left[1 + \frac{V_d}{V_s}\right] \tag{2.4}$$

where

C is the correction factor

V_d and V_s are the volumes of vapor space and sample, respectively

The additional step performed for the correction of volume requires initially placing 1 g of P_2O_5 in both sample and desiccant flasks. With stopcocks 1, 3, and 5 in open position and stopcock 4 in close position the sample flask is evacuated. Manometric reading (h_1) is taken by closing stopcock 3, and the manometric reading (h_2) is obtained with stopcock 5 in close position and stopcock 4 in open position. The void volumes V_s and V_d corresponding to the sample and the desiccant flasks, respectively are also measured. The details were provided by Rahman et al. (2001). Although vapour pressure manometer (VPM) is considered a standard method, it is not suitable for materials either containing large amounts of volatiles and bacteria or mold, or undergoing respiration processes. This method can be used only in the laboratory and is limited for field applications. Stamp et al. (1984) measured the water activity of salt solutions and foods by several electronic methods as compared to direct vapor pressure measurement. An error of approximately 0.01 water activity units was found using the 1 h data but no significantly better regression line was found using the 24 h data. Measurement of the a_w of five foods, however, gave values differing by an average of 0.051 a_w units as compared to the VPM readings. Their study demonstrates justification of the food and drug administration (FDA) cutoff a_w values of 0.85 for low-acid foods as a margin of safety.

2.2.1.2 *Water Activity above Boiling*

Loncin (1988) proposed a method to determine the water activity above 100°C. If any substance initially containing free water is heated in a closed vessel at a temperature above 100°C (say 110°C) and if the pressure is released in order to reach the atmospheric pressure (1.0133×10^5 Pa), then water activity is only a function of temperature (Loncin, 1988). The vapor pressure of water in the product is 1.0133×10^5 Pa because it is in equilibrium with the atmosphere. The vapor pressure of pure water is 1.43×10^5 Pa at 110°C (Table B.1 in Appendix B). Thus, water activity of the product at 110°C is

$$a_w = \frac{1.0133 \times 10^5}{1.43 \times 10^5} = 0.70 \tag{2.5}$$

In this case, water activity does not depend on the binding forces between water and solutes or solids and composition. This fact is very important for extrusion, where water activity at the outlet is a function of the temperature only (Loncin, 1988). Bassal et al. (1993) proposed a method based on the equilibration of food samples with an atmosphere of pure water vapor at constant pressure. The equilibration cell consisted of 100 mL glass bottle with a Teflon stopper through which a capillary tube (1.3 mm diameter and 20 cm long) was inserted. The bottle with the sample was placed in a temperature-regulated oven (air circulated) and the total pressure inside the bottle was measured by a barometer. An in situ weight-measuring device was also attached to the system. Boiling equilibrium (T_B) was assumed to be reached when the variation of the sample mass was less than 0.01 g for 1 h. The test duration depends on the set temperature and air circulation rate in the oven. At equilibrium, the vapor pressure of the sample must be equal to the steam surrounding it and the water activity can be written as

$$a_w = \frac{P_{SS}}{P_{ST}^{v}} \tag{2.6}$$

where P_{SS} is the pressure of the surrounding steam, and P_{ST}^{v} is the vapor pressure of water at temperature T_B from steam tables. The moisture content of the equilibrated sample can be determined by air drying of the equilibrated sample in a conventional dryer. At the end of the test, the capillary tube was damped to prevent any loss of steam from the bottle during cooling. A correction for the partial condensation of water vapor on the sample can be estimated from the known temperature and volume. Bassal et al. (1993) used the above procedure for measuring desorption isotherms of microcrystalline cellulose (MCC) and potato starch at temperatures from 100°C to

150°C at 1 atm and recommended the suitability of this method to measure water activity of foods at higher pressure or vacuum. Boiling point elevation of solution can also be used to predict the water activity by the equation given by Fontan and Chirife (1981) as

$$\ln a_{\mathrm{w}} = 1.1195 \times 10^{-4}(T_{\mathrm{bs}} - T_{\mathrm{bw}})^2 - 35.127 \times 10^{-3}(T_{\mathrm{bs}} - T_{\mathrm{bw}}) \tag{2.7}$$

where T_{bs} and T_{bw} are the boiling points of sample and pure water, respectively.

2.2.1.3 *Water Activity by Freezing Point Measurements*

The determination of water activity by cryoscopy or freezing point depression is very accurate at water activity above 0.85 as mentioned by Wodzinski and Frazier (1960), Strong et al. (1970), Fontan and Chirife (1981), Rey and Labuza (1981), and Lerici et al. (1983). This method is applicable only to liquid foods and provides the water activity at freezing point instead of at room temperature. In the case of solution, the difference is not larger than 0.01 water activity unit (Fontan and Chirife, 1981). Rahman (1991) measured the water activity and freezing point of fresh seafood independently, and found that water activity prediction from freezing point data was 0.02–0.03 units higher than the actual water activity data. This method has advantages at high water activity and for the materials having large quantities of volatile substances which may create error in vapor pressure measurement and in electric hygrometer due to contamination of the sensor.

In a two-phase system (ice and solution) at equilibrium, the vapor pressure of solid water as ice crystals and the interstitial concentrated solution are identical; thus water activity depends only on the temperature, and not on the nature and initial concentration of solutes, present in the third or fourth phase (i.e., with respective kind of food). This creates a basis to estimate the water activity of foods below the freezing point using the equation:

$$a_{\mathrm{w}} = \frac{\text{Vapor pressure of solid water (ice)}}{\text{Vapor pressure of liquid water}} \tag{2.8}$$

At −10°C, water activity in an aqueous system at equilibrium containing ice crystals is equal to $260.0/286.6 = 0.907$ (data Table A.3) and is independent of nature and initial concentration of solutes, presence of third or fourth phase as in the case of ice cream (Loncin, 1988). Fennema (1981) concluded that changes in properties could occur below freezing point without any change in water activity. These include changes in diffusion properties, addition of additives or preservatives, and disruption of cellular systems. The water activity data of ice from 0°C to −50°C are correlated with an exponential function as (Rahman and Labuza, 2007):

$$a_{\mathrm{w}} = 8.727\left[\exp\left(-\frac{595.1}{T}\right)\right] \tag{2.9}$$

where T is measured in kelvin. The maximum error in prediction is 0.012 unit water activity and the average is 0.0066. Other colligative properties such as osmotic pressure and boiling point elevation have not yet been used for food systems (Rizvi, 1995).

2.2.2 Gavimetric Methods Based on Equilibrium Sorption Rate

The gravimetric method is based on the equilibration of samples with its atmosphere of known humidity. In this method, it is important to achieve both hygroscopic and thermal equilibrium (Gal, 1981).

2.2.2.1 Discontinuous Registration of Mass Changes

In this method, the sample in the controlled atmosphere needs to be taken out for weighing and is then placed back in the atmosphere chamber for equilibration. The balance is not a fixed part of the apparatus and samples must be conditioned to different ERH values and conditioning can be carried out in a static or dynamic way. With these methods it is possible to visually examine the samples to detect immediate physical changes, like caking, shrinkage, discoloration, and loss of free-flowing properties (Gal, 1975).

2.2.2.1.1 Static Systems (Isopiestic Method)

The static method is the most simple and common method of measuring water activity of food. This method is also known as isopiestic method. In this method, a weighted sample of known mass (around 2–3 g) is stored in an enclosure and allowed to reach equilibrium with an atmosphere of known ERH (or a_w), for example, by a saturated salt solution, and reweighed at regular intervals until constant weight is established. The condition of equilibrium is thus determined in this manner. The moisture content of the sample is then determined, either directly or by calculation from the original moisture content and the known change in weight. A desiccator is commonly used as a chamber to generate controlled atmosphere (Figure 2.2). The details of measuring water activity using isopiestic methods are presented in Rahman and Al-Belushi (2006), Lewicki and Pomaranska-Lazuka (2003), and Sablani et al. (2001). Several days, or even weeks, may be required to establish equilibrium under static air conditions, but results can be obtained for all relative humidity values simultaneously with little effort if the apparatus is replaced with different salt solutions (Smith, 1971).

The main advantages of this method are its simplicity, low cost, ability to handle many samples simultaneously, and easy operability (Lewicki and Pomaranska-Lazuka, 2003; Rahman and Al-Belushi, 2006). The main disadvantages of this simple method are

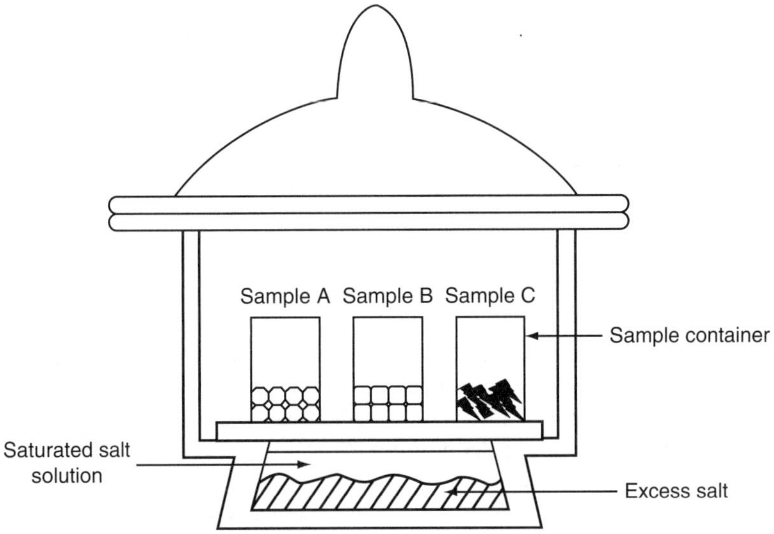

Figure 2.2 Humidity control chamber using desiccator.

- Slowness of the equilibrium process, which usually takes from 3 to 6 weeks. In certain instances, it could take a few months to equilibrate. It is therefore doubtful whether the microbial and physicochemical stability remain valid in the sample during long experimental periods, especially at higher water activity.
- At high relative humidity values, the delay in equilibration can lead to mould or bacterial growth on the samples and consequent invalidation of the results. Although it is recommended to place toluene or thymol in the chamber for slowing the microbial growth, there is no option available to ensure physical and chemical stability in the course of the equilibration period. In addition, care should be taken not to inhale these toxic chemicals from the desiccator chamber while preparing the sample and performing the weighing process.
- Condition of equilibrium is determined by reweighing the sample at regular intervals until constant weight is established. The equilibrium can also be hastened by evacuating the conditioning chamber. The loss of conditioned atmosphere each time the chamber is opened to remove the sample for weighing delays the equilibrium process. Lewicki and Pomaranska-Lazuka (2003) studied the effects of individual operations on this process such as opening the desiccator, and transferring samples to the balance for checking mass. It was shown that opening the desiccator, taking the sample, and closing it again caused the most disturbance. The error depends on the water activity and number of times the desiccator was opened. At low water activity ($a_w < 0.6$) a maximum of about 20% overestimation in sample mass was observed, while at high water activity ($0.6 < a_w < 0.8$) an underestimation of 20% was observed. The process of equilibration can be enhanced in desiccators with or without vacuum. It was shown that equilibration of samples with or without vacuum yields different water contents (Laaksonen et al., 2001). All materials showed clearly that water contents were higher at the high water activities (>0.60) and lower at the low water activities (<0.40) using vacuum desiccators, because of a probable difference in humidity between the external atmosphere and the interior of the desiccators with or without vacuum. Equilibration of water contents was achieved after 2–3 days of storage using vacuum desiccators, while in desiccators without vacuum it took 2–3 weeks to achieve the same over the whole range of relative humidity for all 3 g samples of different materials tested.
- In addition, it is always impossible to find salts, which could be used for each and every water activity ranging from 0 to 1.
- Using the static isopiestic method it is also difficult to measure adsorption and desorption isotherm for the same sample.

Improvement to enhance the equilibrium time can be done by (1) circulating the atmosphere with a fan or pump, and (2) increasing the surface area of the sample by slicing or breaking into pieces and spreading the sample in a thin layer over a large surface area. In order to prevent mould growth at high relative humidity levels of 0.70 and above, small glass bottles containing toluene can be placed inside the sealed container or jar (Labuza, 1984).

The use of a proximity equilibrium cell (PEC) to rapidly equilibrate a sample to an atmosphere of known relative humidity was developed by Lang et al. (1981). This method is better than others because it is rapid, precise, and simple. The rapid equilibration is due to the increase in surface area of saturated salt solution and sample per unit vapor volume as well as shorter mean free path for water vapor. Design of the sample holder suggested by Lang et al. (1981) was found unsuitable for sugar and hygroscopic samples as these samples dissolved and dripped in the saturated salt solution. Kanade and Pai (1988) also proposed a simple and efficient system requiring a balance and a sample equilibration chamber to accommodate hygroscopic samples in PEC under partial vacuum (Figure 2.3). This method can avoid sample dripping in the salt solution due to the improvement in the sample holder. This type of method can decrease the equilibration time from more than 30 days to about 7 days. The desiccator method is also a fairly rapid equilibration method; however, cost of materials and labor is higher. The water activities of the saturated salt solutions used in equilibration cell are presented as a function of temperature in Tables 2.3 through 2.5 for easy reference.

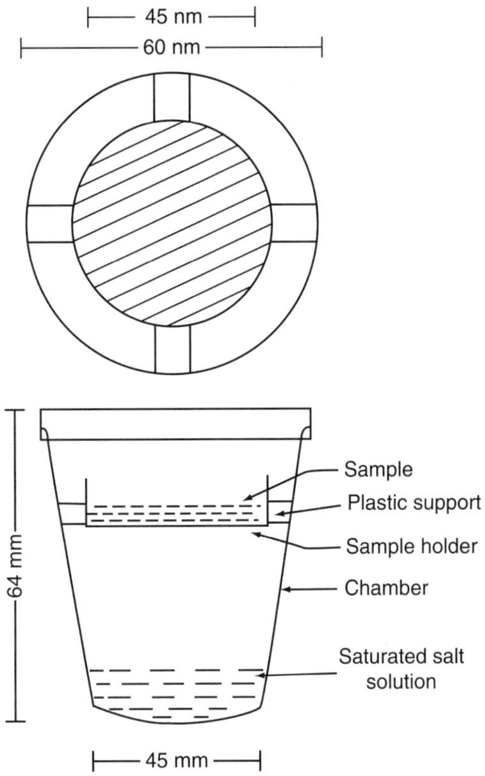

Figure 2.3 PEC with modified sample holder for sugar and hygroscopic samples. (From Kanade, P.B. and Pai, J.S., *J. Food Sci.*, 53, 1218, 1988.)

Spiess and Wolf (1987) mentioned that a group of European research laboratories developed a reference system. This reference system comprises (1) a reference material, (2) simple standard equipment with handling procedure, and (3) evaluation of the results. MCC Avicel PH 101 manufactured by the FMC Company was selected as reference material. This material was selected due to the following reasons (Spiess and Wolf, 1987): (1) It is stable in its crystalline structure in a temperature range −18°C to 80°C, with minimal changes in its sorption characteristics. However, when exposed to temperatures above 100°C the sorption properties may change and thermal degradation is expected to begin at temperatures above 120°C. (2) It is stable in its sorption properties after two to three repeated adsorption and desorption cycles. However, when adsorption and desorption cycles are repeated more than three times, significant changes in the sorption behavior are to be expected. (3) All published sorption data in the pertinent literature resulted in a sigmoid shape of the sorption isotherm. Mean adsorption isotherms of MCC from a collaborative study with 32 participating laboratories within the framework of COST 90 project are given by GAB model. (4) MCC is available as a biochemical analytical agent in constant quality.

The detailed description of the standard equipment and handling procedure is given by Spiess and Wolf (1987). The sorption apparatus is shown in Figures 2.4 and 2.5. The sorption device consists of sorption containers, petri dishes on trivets, and weighing bottles. Samples in the weighing bottles are exposed to the humid atmosphere in the containers. Simple preserving jars (1 L volume) are used as sorption containers which can be tightly sealed by means of rubber seal rings and glass covers to prevent water vapor transport (Figure 2.4). Five weighing bottles on a petri dish (Duran, meltable glass) are placed in the sorption container over the surface of the salt solution. Weighing bottles

Table 2.3 Relative Humidity Variations with Temperature of Saturated Salt Solutions

	Percent Relative Humidity							
Salt	**5°C**	**10°C**	**15°C**	**20°C**	**25°C**	**30°C**	**35°C**	**40°C**
Group A								
Lithium chloride	16	14	13	12	11	11	11	11
Potassium acetate	25	24	24	23	23	23	23	23
Magnesium bromide	32	31	31	31	31	30	30	30
Magnesium chloride	33	33	33	33	33	32	32	31
Potassium carbonate	—	47	45	44	43	42	41	40
Magnesium nitrate	54	53	53	52	52	52	51	51
Sodium bromide	69	59	53	57	57	57	57	57
Cupric chloride	65	63	68	63	67	67	67	67
Lithium acetate	72	72	71	79	68	66	65	64
Strontium chloride	77	77	75	73	71	69	68	66
Sodium chloride	76	75	75	75	75	75	75	75
Ammonium sulfate	81	80	79	79	79	79	79	79
Cadmium chloride	83	83	83	82	82	82	79	79
Potassium bromide	—	86	85	84	83	82	81	80
Lithium sulfate	84	84	84	85	85	85	85	81
Potassium chloride	88	87	87	86	86	84	84	83
Potassium chromate	89	89	88	88	87	85	84	82
Sodium benzoate	88	88	88	89	88	88	85	83
Barium chloride	93	93	92	91	90	89	83	87
Potassium nitrate	96	95	95	94	93	92	91	89
Potassium sulfate	98	97	97	97	97	97	96	95
Disodium phosphate	98	98	98	98	97	96	96	94
Lead nitrate	99	99	98	98	97	96	96	95
Group B								
Zinc nitrate	43	43	41	38	31	24	21	18
Lithium nitrate	—	66	60	58	54	51	48	45
Cobalt chloride	—	—	73	67	64	62	59	57
Zinc sulfate	95	93	92	90	88	85	85	84

Source: Smith, P.R., *The Determination of Equilibrium Relative Humidity or Water Activity in Foods—A Literature Review*, The British Food Manufacturing Industries Research Association, England, 1971.

(DIN 12 605; 25 mm diameter and 25 mm height) with ground-in stopper are satisfactory. Sorption containers must be placed in a temperature-controlled thermostat cabinet. The thermostat should be covered with a plastic foam lid to prevent heat loss. The salt solution shown in Table 2.6 should be prepared with cold distilled water by stirring and be allowed to stand 1 week at closed condition. The solution should be stirred once a day for a brief cohile. To avoid the clumping of salts a metal spatula can be used. The microbial growth at high water activity can be avoided by using sorbent in a special support, i.e., phenyl mercury acetate (highly toxic) or thymol (for nonfatty products).

2.2.2.1.2 Evacuated Systems

Hygroscopic and thermal equilibrium between samples and environment is generally a very slow process. The apparatus can be evacuated in gravimetric techniques. High vacuum hastens the mass transfer but acts as thermal insulation around the samples, so that the rate-controlling

Table 2.4 Water Activities of Selected Salt Solutions at Various Temperatures

	Water Activity						
	Temperature (°C)						
Salt	**5**	**10**	**20**	**25**	**30**	**40**	**50**
Lithium chloride	0.113	0.113	0.113	0.113	0.113	0.112	0.111
Potassium acetate	—	0.234	0.231	0.225	0.216	—	—
Magnesium chloride	0.336	0.335	0.331	0.328	0.324	0.316	0.305
Potassium carbonate	0.431	0.431	0.431	0.432	0.432	—	—
Magnesium nitrate	0.589	0.574	0.544	0.529	0.514	0.484	0.454
Potassium iodide	0.733	0.721	0.699	0.689	0.679	0.661	0.645
Sodium chloride	0.757	0.757	0.755	0.753	0.751	0.747	0.744
Ammonium sulfate	0.824	0.821	0.831	0.810	0.806	0.799	0.792
Potassium chloride	0.877	0.868	0.851	0.843	0.836	0.823	0.812
Potassium nitrate	0.963	0.960	0.946	0.936	0.923	0.891	0.848
Potassium sulfate	0.985	0.982	0.976	0.973	0.970	0.964	0.958

Source: Greenspan, L., *J. Res. Nat. Bur. Stand: Phys Chem*, 81A, 89, 1977.

process in such cases is usually the heat transfer. The vacuum systems offer better precision but require more expensive construction and instrumentation (Gal, 1981).

2.2.2.1.3 Dynamic Systems

In this method two airstreams are mixed: one of them is kept permanently dry while the other is saturated with water vapor. This is the preferred method in dynamic systems allowing a convenient and continuous humidity control at moderate precision. In dynamic systems two systems are used: (1) a two-temperature system and (2) a two-pressure system (Gal, 1981).

In the two-temperature system, water or ice is kept in a thermostated container at a lower temperature than the sample. Thus, the temperature of the vapor source defines its partial pressure

Table 2.5 Regression Equations for Water Activity of Selected Saturated Salt Solutions at Different Temperatures

Salt	**Regression Equation**	r^2
Lithium chloride [LiCl]	$\ln a_w = (500.95 \times 1/T) - 3.85$	0.976
Potassium acetate [CH_3COOK]	$\ln a_w = (861.39 \times 1/T) - 4.33$	0.965
Magnesium chloride [$MgCl_2$]	$\ln a_w = (303.35 \times 1/T) - 2.13$	0.995
Potassium carbonate [K_2CO_3]	$\ln a_w = (145.00 \times 1/T) - 1.30$	0.967
Magnesium nitrate [$Mg(NO_3)_2$]	$\ln a_w = (356.60 \times 1/T) - 1.82$	0.987
Sodium nitrate [$NaNO_3$]	$\ln a_w = (435.96 \times 1/T) - 1.88$	0.974
Sodium chloride [NaCl]	$\ln a_w = (228.92 \times 1/T) - 1.04$	0.961
Potassium chloride [KCl]	$\ln a_w = (367.58 \times 1/T) - 1.39$	0.967
Ammonium sulfate [NH_3SO_4]	$\ln a_w = (154.75 \times 1/T) - 0.689$	0.999
Potassium chloride [KCl]	$\ln a_w = (79.43 \times 1/T) - 0.478$	0.999
Potassium nitrate [KNO_3]	$\ln a_w = (244.37 \times 1/T) - 0.900$	0.999
Potassium sulfate [KSO_4]	$\ln a_w = (55.51 \times 1/T) - 0.214$	0.999
Sodium hydroxide [NaOH]	$\ln a_w = (2634.27 \times 1/T) - 11.28$	0.999

Source: Labuza, T.P., *Moisture Sorptions: Practical Aspects of Isotherm Measurement and Use*, American Association of Cereal Chemists, St. Paul, MN, 1984.

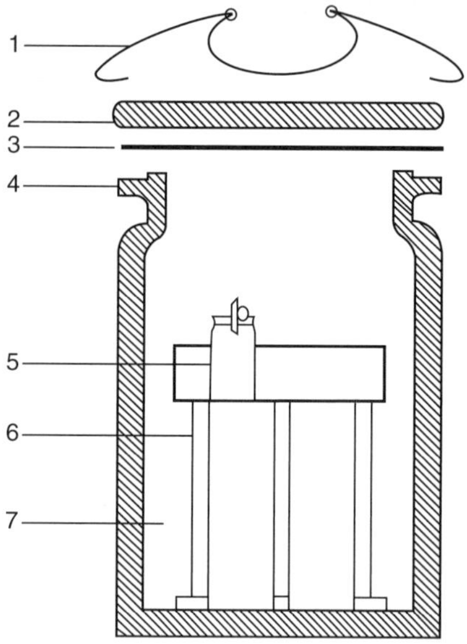

Figure 2.4 Standardized sorption apparatus: (1) locking clamp, (2) lid glass, (3) rubber seal ring, (4) sorption container (glass), (5) weighing bottle with ground-in stopper, (6) petri dish on trivet, (7) saturated salt solution. (From Spiess, W.E.L. and Wolf, W. in *Water Activity: Theory and Applications to Food*, Rockland, L.B. and Beuchat, L.R. (eds.), Marcel Dekker, Inc., New York, 1987.)

within the apparatus. This is the most commonly used method for maintaining constant vapor pressures in research work because of the ease and precision of controlling temperature instead of humidity, and the continuous relative humidity scale available. Near saturation, this method becomes less precise due to the very large dependence of the saturated vapor pressure on temperature. Consequently, the highest precision can be achieved only at the low relative pressure range. The two-temperature system can be combined with the two-pressure system and with saturated solutions to extend its range of applicability beyond that under isothermal conditions (Gal, 1981).

The two-pressure system is a comparatively new way to maintain constant vapor pressures in dynamic sorption measurements. Air or any other inert gas is saturated with water vapor at the same temperature at a higher pressure than that existing in the experimental space. The gas–vapor mixture expands consecutively to atmospheric pressure and becomes unsaturated, as expressed by Dalton's law. The control of relative humidity is thus replaced by pressure control, which can be performed more precisely especially in the high RVP range (Gal, 1981). A modification of this principle was developed by Lowe et al. (1974) for humidity control in testing chambers. In this method, a measured amount of liquid water is continuously evaporated into an airstream saturated previously at 0°C and 5 psig.

2.2.2.2 Methods with Continuous Registration of Mass Changes

The loss of atmosphere during weighing can be avoided by weighing the sample in situ by a scale attached to the sample holder.

2.2.2.2.1 *Static Chamber*

An apparatus consists of a magnetic stirrer unit, equilibrium chamber in a constant temperature bath, and a sensitive weighing balance as shown in Figure 2.6.

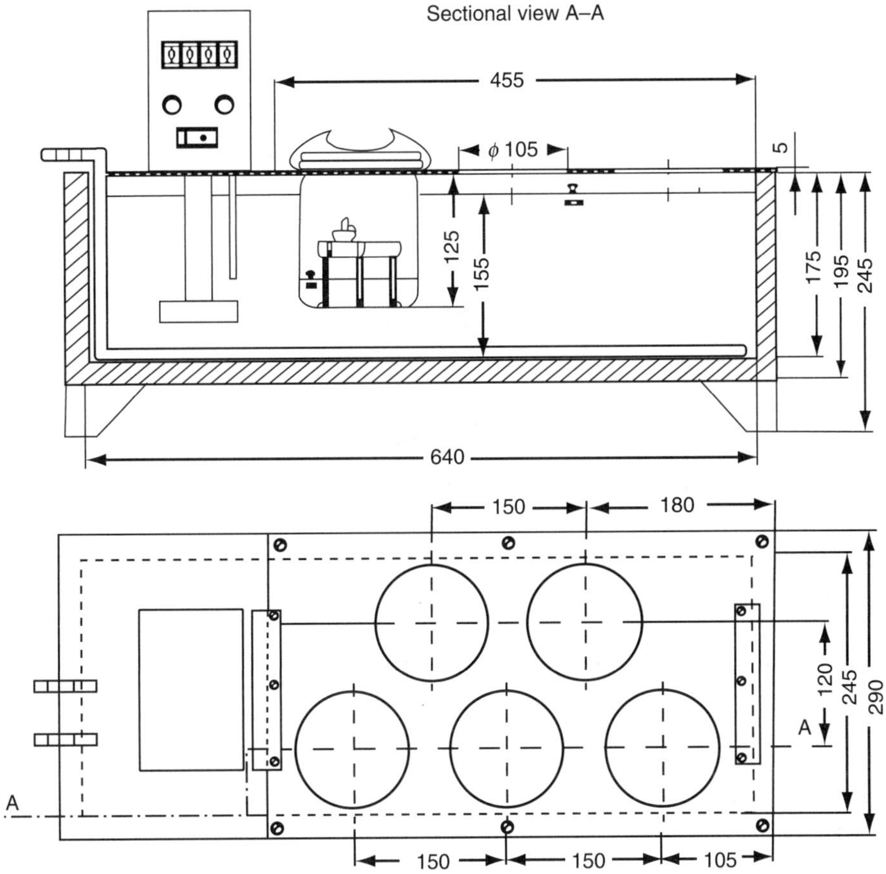

Figure 2.5 Standardized sorption apparatus (rectangular thermostat prepared for the accommodation of five sorption containers). (From Spiess, W.E.L. and Wolf, W. in *Water Activity: Theory and Applications to Food*, Rockland, L.B. and Beuchat, L.R. (eds.), Marcel Dekker, Inc., New York, 1987.)

Table 2.6 Preparation of Recommended Saturated Salt Solutions at 25°C

Salt	Percent Relative Humidity (%)	Salt (g)	Water (mL)
Lithium chloride [LiCl]	11.15	150	85
Potassium acetate [CH_3COOK]	22.60	200	65
Magnesium chloride [$MgCl_2$]	32.73	200	25
Potassium carbonate [K_2CO_3]	43.80	200	90
Magnesium nitrate [$Mg(NO_3)_2$]	52.86	200	30
NaBr	57.70	200	80
$SrCl_2$	70.83	200	50
Sodium chloride [NaCl]	75.32	200	60
Sodium chloride [KCl]	84.32	200	80
Barium chloride [$BaCl_2$]	90.26	250	70

Source: Spiess, W.E.L. and Wolf, W. in *Water Activity: Theory and Applications to Food*, Rockland, L.B. and Beuchat, L.R. (eds.), Marcel Dekker, Inc., New York, 1987.

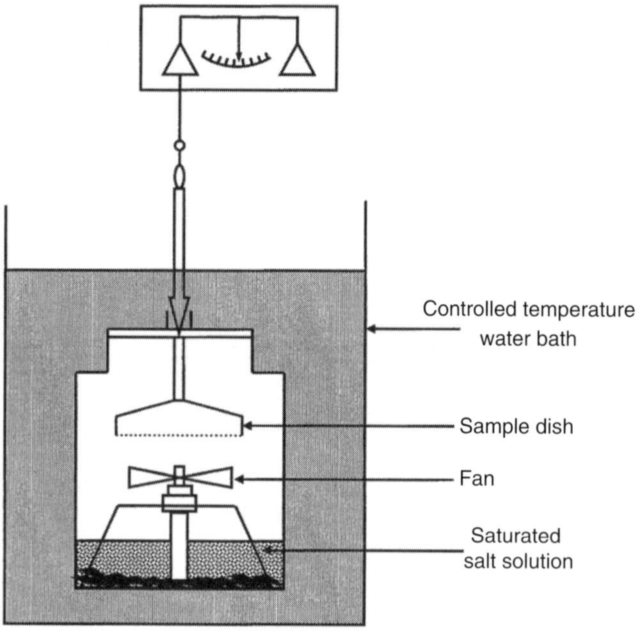

Figure 2.6 Circulated air chamber with in situ weighing system (From Igbeka, J.C. and Blaisdell, J.L., *J. Food Technol.*, 17, 37, 1982.)

2.2.2.2.2 Dynamic Vapor Sorption

The dynamic vapor sorption (DVS) method could overcome the disadvantages of the static isopiestic method. Different types of commercial equipments based on dynamic method with automatic control and data analysis options are available. This includes built-in electro-balance and humidity-controlled process. The main advantages of the dynamic method are its ability to (1) equilibrate the sample rapidly, (2) use the same sample for the entire isotherm, (3) retain the sample in the controlled humidity chamber during the entire isotherm measurement, (4) easily measure the adsorption and desorption isotherms for the same sample, (5) use microlevel sample in the order of milligrams, and (6) measure steps and oscillation mode at any increasing or decreasing water activity. The reasons why this process gives fast equilibration (main advantage) could be due to the small sample size, the continuous circulation of gas, the surface area, and the small chamber or cell volume. This DVS method is also defined as dynamic isopiestic method (Rahman and Al-Belushi, 2006).

Figure 2.7 shows the Symmetrical Gravimetric Analyzer Model 100 (SGA-100) from VTI Corporation, Florida which could be used to generate the sorption and drying data. It is an instrument designed for obtaining water sorption data on solid samples at relative humidity values between 2% and 98% and temperatures between 5°C and 60°C. The drying of samples could be done up to 80°C. This apparatus was designed in such a way as to insert an entire microbalance in a forced nitrogen circulation system, provided with a dynamic nitrogen humidification system, electronically controlled by an accurate mass flow controller. Identical conditions of temperature and humidity for sample and a reference were achieved by using a symmetrical two-chamber aluminum block. The nitrogen gas with a specific humidity was passed over the sample, as well as through the microbalance reference. The relative humidity was determined by relative humidity measuring probe. In order to ensure a constant atmospheric composition during the test and to optimize the nitrogen flux, the difference between the values read by two detectors was controlled. Sample mass changes were recorded using electronics microbalance (CI Electronics, Salisbury,

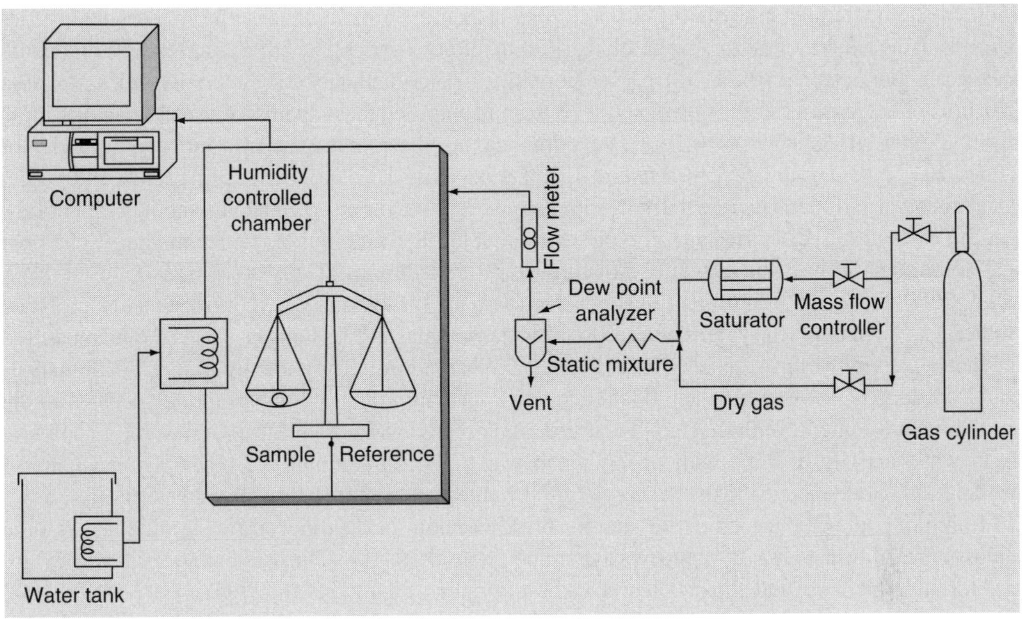

Figure 2.7 Schematic flow diagram of a dynamic isopiestic method. (From Rahman, M.S. and Al-Belushi, R.H., *Int. J. Food Prop.*, 9, 421, 2006.)

Wilts, U.K.) sensitive to 0.1 μm and with a capacity of 5 g. The entire apparatus was maintained at specific temperatures by circulating cold or warm water from a water bath (stability of 0.01°C) around the chamber. The equipment was tested and calibrated with sodium chloride and providone, *N*-vinyl pyrrolidone (PVP) within 10%–80% relative humidity with drying cycle at 60°C. The microbalance was calibrated with 100 mg weight before each isotherm measurement. Spring balance is very popular and widely used for registering the mass of the sample in gravimetric methods. Similar DVS systems are also available from Surface Measurement Systems, London.

Recently surface measurement systems (SMS) has released a new DVS system coupled with a Raman spectroscopy and in addition the system can also accomodate a video option. This instrument can provide three separate streams of data: gravimetric, visual, and spectroscopy. Rahman and Al-Belushi (2006) showed the adsorption isotherm of freeze-dried garlic powder measured by static and DVS methods. Both methods showed similar isotherm within water activity values of 0.35–0.70. At the low water activity range ($a_w < 0.35$), static isopiestic method showed overestimation of the moisture content, while at high water activity range ($a_w > 0.70$) it showed underestimation of the moisture content. One of the reasons could be due to the higher relative humidity of the room atmosphere, which caused condensation or adsorption of humidity resulting in overestimation at low water activity. In the case of desorption, evaporation could occur for the high water activity sample. In addition, high water activity equilibration for long periods of time may also cause continuous physical and chemical deterioration of the sample.

2.2.2.2.3 *Other Applications of DVS*

Rahman and Al-Belushi (2006) discussed other applications of the dynamic method. Dynamic method has high potential to be used for microlevel drying studies. Microlevel drying kinetics could be performed as a function of air composition (modified atmosphere drying with varied oxygen, carbon dioxide, and nitrogen levels), relative humidity, temperature, and varied sample geometry and structure. Roques et al. (1983) used dynamic systems to study sorption and diffusion in rapeseed within the temperature range of 40°C–105°C. May et al. (1997) studied the isothermal drying kinetics of foods using a thermogravimetric analyzer (Perkin-Elmer 7 series), which was automated and could handle samples between 10 and 100 mg. It recorded mass automatically at set intervals with a precision of 0.1 mg. They used a sample size of about 30 mg and recorded mass against time at a set temperature of 40°C with an accuracy of 0.1°C. The purge gas used was nitrogen at a constant velocity of 1.25×10^{-3} m/s. Drying curves of five different foods (apple, potato, carrot, asparagus, and garlic) were analyzed and found different drying characteristics with varied constant and falling rate periods. Lin and Chen (2005) discussed a prototype setup for controlled air humidity and temperature chamber for isothermal drying kinetics and moisture sorption isotherm. Teoh et al. (2001) used DVS (DVS-2000, Surface Measurement Systems, London) analysis to measure cornmeal snack moisture isotherm and compared the values by equilibrating the sample in PEC (Lang et al., 1981) and measured the water activity using Aqualab CX-2 (Pullman, Washington). They identified that DVS analysis could produce rapid isotherms and could accurately condition the individual model systems to the same moisture content. With such rapid equilibration times, it was possible to obtain isotherm points at up to water activity of 0.95, without having to worry about microbial and other physicochemical degradation, while increasing the accuracy of the isotherm modeling at higher water activities.

Rahman et al. (2005) studied the quality of dried lamb meat produced by simulating modified atmosphere (nitrogen gas) drying using dynamic system SGA-100. The dried lamb meat was evaluated for their microbial and physicochemical characteristics. Sannino et al. (2005) used the DVS-1000 system to study the drying process of lasagna pasta at controlled humidity and temperature with a sensing device to measure the electrical conductivity of pasta during the drying process. An anomalous diffusion mechanism has been observed, typical of the formation of a layered sample structure: a glassy shell on the surface of the pasta slice, which inhibits a fast diffusion from the humidity, rubbery internal portion. Internal stresses at the interface of the glassy–rubbery surfaces are responsible for the formation and propagation of cracks and thus lasagna sample delamination and breakage. The kinetics of adsorption and desorption isotherms could be used to explore other structural characteristics, such as glass transition. A break in the mass transfer rate constant or moisture diffusivity at the glass–rubber transition is expected to occur. The mass transfer rate constant or diffusivity could be estimated from the adsorption or desorption kinetics of samples having the same initial moisture content and can be plotted as a function of temperature in order to identify a break in the plot at glass transition. Rahman et al. (2007) measured moisture diffusivity of spaghetti (moisture content: 9.77 kg water/100 kg spaghetti) within the temperature range 10°C–80°C and plotted as a function of temperature. They found a clear break at 50°C, which is close to the glass–rubber transition of the spaghetti measured by differential scanning calorimetry (DSC). Garcia and Pilosof (2000) attempted to correlate water sorption kinetics with the glass transition temperature. They measured adsorption and desorption kinetics by placing samples in a desiccator at 30°C with different relative humidity values maintained by saturated salts. They plotted rate constant as a function of relative humidity and observed a change in slope at the relative humidity when the sample was transformed to a glassy state. In general, the dynamic method used at microlevel has high potential to be used in foods.

Del-Nobile et al. (2004) developed a new approach based on the use of oscillatory sorption tests, which was proposed to determine the water-transport properties of chitosan-based edible films. Oscillatory sorption tests as well as stepwise sorption tests were conducted at 25°C on chitosan films. Two different models were fitted to the experimental data to determine the relationship between the water-diffusion coefficient and the local water concentration. One of the two tested models accounted only for stochastic diffusion while the other accounted also for the superposition of polymer relaxation to stochastic diffusion. A comparison between experimental and predicted water permeability indicated that stepwise sorption tests cannot be used to determine the dependence of water-diffusion coefficient on local water concentration when the diffusion process has characteristic time much smaller than that of polymer relaxation. In fact, in these cases the diffusion process controls only the very early stage of sorption kinetics, whereas the remaining part of the transient is controlled by polymer relaxation (Del-Nobile et al., 2004). In the future, more other potential applications of the dynamic methods could emerge from the literature.

2.2.2.2.4 *Evacuated Systems*

Evacuated or vacuum systems are also used in the dynamic system with continuous registering of the mass of the sample during the equilibration period.

2.2.3 Hygrometric Methods

One of the most commonly used methods of measuring ERH is to equilibrate the sample with air in a closed vessel and then to determine the relative humidity of the air with a hygrometer. Any instrument capable of measuring the humidity or psychrometric state of air is a hygrometer. Hygrometric devices are based on many different scientific principles such as dew point, frost point, wet and dry bulb temperature, expansion of a material, and electric resistance and capacitance of salt. The details of measurement steps are provided in Rahman and Sablani (2001). In a hygrometer there are three zones: testing enclosure, sample environment, and sensor environment. The temperatures within the three zones need to be in equilibrium. Unless the temperature of the sample is known, the relative humidity in this region cannot be determined. Unless the temperature of the sensor is known, the relative humidity indicated the sensor cannot be converted into a reliable estimate of the vapor pressure throughout the system. Without careful control and measurement of temperatures in the sample and the sensor, no meaningful data can be collected (Reid, 2001).

2.2.3.1 *Mechanical Hygrometer*

This method is based on the dimensional changes in natural or synthetic materials suitably amplified by mechanical linkage to indicate atmospheric humidity. The most common one is the hair type hygrometer, in which relative humidity is indicated on a dial by the change in length of a bundle of other fibers. Smith (1971) mentioned that these instruments are comparatively slow to react to changes in the ambient atmosphere and are difficult to maintain in calibration. They are not suitable for use below 0.25 relative humidity and above 50°C, as the element may undergo a permanent change in length under these conditions. Prolonged exposure to high humidity involves the risk of deterioration due to mould growth on the hairs. So hair hygrometers have limited use in ERH measurement but may be useful for approximate ERH.

2.2.3.2 *Wet and Dry Bulb Hygrometer*

The wet bulb temperature depends on the amount of moisture present in the air–vapor mixture. The atmospheric humidity can be estimated from the water vapor pressures at saturation (dew point) and dry bulb temperatures. The vapor pressures can be related to wet and dry bulb temperatures as (Smith, 1971)

$$P_{\mathrm{w}}^{\mathrm{v}} = P_{\mathrm{ws}}^{\mathrm{v}} - \Pi P(T_{\mathrm{sy}} - T_{\mathrm{wb}}) \tag{2.10}$$

where

$P_{\mathrm{w}}^{\mathrm{v}}$ is the vapor pressure of water
$P_{\mathrm{ws}}^{\mathrm{v}}$ is the vapor pressure of water on saturation at the wet bulb temperature
P is the barometric pressure
T_{sy} and T_{wb} are the ambient and wet bulb temperatures
Π is a constant that depends on the instruments and certain conditions

The constant Π is predominantly affected by the airflow rate over the wet bulb and to a lesser extent by the dimensions of the thermometer. Results become independent of flow rate at speeds above 3 m/s for all practical purposes. This can be achieved by physically shifting the instrument or more commonly by circulating the air by means of a fan. The dimensions of the thermometer affect constant Π mainly through conduction of heat along the uncooled stem to the cooled mercury bulb. This effect is considerably reduced by substituting smaller temperature sensors such as platinum resistance thermometers, thermocouples, or thermistors. These have added advantages of greater sensitivity and the ability to provide remote reading facilities (Smith, 1971). This method is not suitable for small volumes of air and has been primarily used to determine the relative humidity of large storage atmospheres and commercial dehydrators (Rizvi, 1995). Small hygrometers especially designed for use in foods are now commercially available. Major limitations of this method are condensation of volatile materials, heat transfer by conduction and radiation, and the minimum wind velocity requirement of at least 3 m/s (Rizvi, 1995).

2.2.3.3 Dew Point Hygrometer

The dew point is the temperature at which the air under investigation will reach saturation point due to the moisture conent and below which a dew or frost will form. A typical simple dew point apparatus consists of a polished silver thimble projecting into the enclosed atmosphere. The thimble is cooled by the forced evaporation of a solvent in the thimble. The temperatures at which dew is just formed and disappears on rewarming are averaged to give the actual dew point. The visual detection of the deposition and disappearance of the dew requires considerable care, and various attempts have been made to improve accuracy by substituting the human eye with detection devices (Smith, 1971). Modern instruments, based on Peltier effect cooling of mirrors and the photoelectric determination of condensation on the reflecting surface via a null-point type of circuit, give very precise values of dew point temperatures (Rizvi, 1995). Dew point measuring devices are reported to have an accuracy of 0.003 water activity unit in the range of 0.75–0.99 (Prior, 1979). At lower water activity levels there is not sufficient vapor in the headspace to cover the reflecting surface and the accuracy of these instruments is therefore diminished. Thus, the measurement based on psychrometry has the highest accuracy near relative humidity 1.0 (Wiederhold, 1987).

2.2.3.4 Hygroscopicity of Salts

The micromethod based on the hygroscopicity of salt does not yield an accurate result but gives a useful indication of the ERH of a sample between fairly narrow limits within about an hour (Smith, 1971). It is based on the fact that water vapor will condense on a salt crystal only from an atmosphere which has a higher relative humidity than the critical relative humidity of the salt. Each salt has its own characteristic or critical relative humidity transition point. A salt crystal will remain dry if the surrounding air relative humidity is lower than the critical relative humidity and will show a wet short line when inspected under a lens if air relative humidity is higher than the critical relative humidity.

An approximate estimate of ERH can be obtained by observing the change in color of a salt enclosed in the headspace over the sample as silica gel which changes color from blue at low humidities, through shades of lilac, to pink at high humidities. Solomon (1945) used papers impregnated with cobalt chloride, potassium thiocyanate, or sodium thiosulfate which is blue at low and pale at high humidities, with a series of lilac colors in between. The error in relative humidity measurement may be up to $\pm 5\%$ from color matching against a suitably calibrated series of standard colors. Solomon (1957) mentioned that the measuring of relative humidity by matching the colors of tissue paper impregnated with cobalt thiocyanate becomes convenient due to the commercial availability of impregnated paper and colored glass standards (commercial kits). However, color correction for temperature and up to 2 h exposure for equilibration is necessary. This method is suitable for general use when elaborate equipment cannot be employed.

2.2.3.5 Electronic Sensor Hygrometer

The hygrometers are more sensitive compared to the static and dynamic methods at water activity above 0.9 (Pezzutti and Crapiste, 1997). Several hygrometers are commercially available for direct determination of water activity. These types of hygrometers are based on the measurement of the conductivity of salt solution (usually LiCl) which is in equilibriums with the air (Figure 2.8). The electrolytic sensor consists of a small hollow cylinder covered with a glass fiber tape, which is impregnated with saturated lithium chloride solution. A spiral bifilar electrode is wound over the tape and a temperature sensor is mounted at the center of the cylinder. An alternating voltage is applied to the electrodes and a current is allowed to pass through the inverting electrolyte. The resulting rise in the temperature opposes the absorption of moisture by the lithium chloride and the sensor rapidly reaches an equilibrium temperature at which the vapor pressure of the salt solution equals that of the air. Temperature is determined by a sensor at the core of the cylinder and a calibration chart is used to convert this relative humidity. Hygroscopic organic polymer films are also used instead of lithium chloride salt. Another type is the anodized aluminum sensor. This sensor consists of an aluminum strip that is anodized by a process that forms a porous oxide layer. A very thin coating of gold is then evaporated over this structure. The aluminum base and the gold layer form the two electrodes of what is essentially an aluminum oxide capacitor (Smith, 1971; ASHRAE, 1993; Rahman, 1995).

Different variations in sensor construction are available. The disadvantages due to the nonlinear response to humidity, large temperature coefficient for conductivity values, and contamination of the lithium chloride can cause erroneous conductivity readings. The range of operation is restricted to 0.15–0.90 ERH and each sensor exhibits individual characteristics and must be calibrated against reference atmospheres (Smith, 1971). The performance of the sensor is subject to change on aging

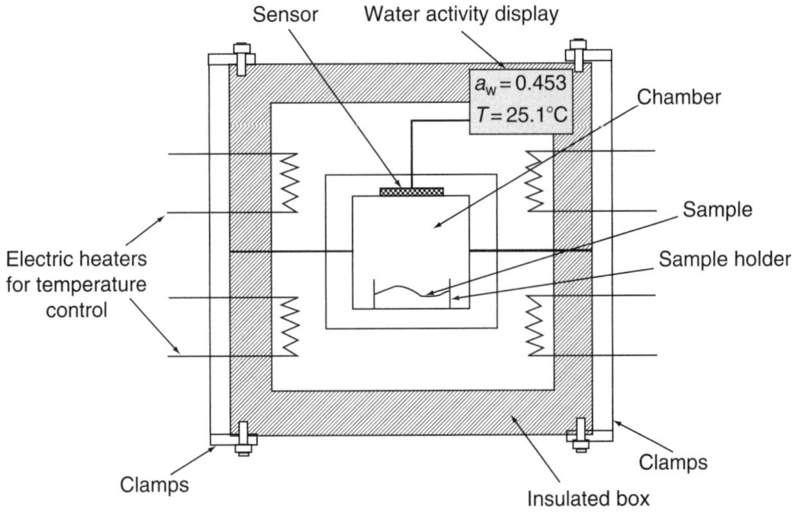

Figure 2.8 Schematic diagram of setup for water activity measurement by electronic sensor. (From Rahman, M.S. and Sablani, S.S., in *Current Protocols in Food Analytical Chemistry*, John-Wiley & Sons, Inc., New York, 2001, A2.5.1–A2.5.4.)

and with contamination of the element by foreign particles, but the probe can be readily calibrated with saturated salt solutions. Equilibrium times obtained in measurement on various food products ranged from a few seconds to a minute or more. Readings can be obtained in small areas and the sensors' physical shape make them particularly suitable for insertion through the small packs and pouches. A small probe-type hygrometer is based upon the effect of humidity on the capacitance and resistance of a condenser. Different commercial probes are available.

This type of instrument provides rapid and reliable means of measuring water activity provided some precautions are taken. The specific problems that arise from the use of an electric hygrometer for measurement of water activity are the equilibration period, the calibration of sensors, and the influence of temperature on the values measured in saturated salt solutions and food products. Stekelenburg and Labots (1991) found that the equilibration time required for various products to reach a constant water activity value increased at higher water activity and mentioned that the following precautions must be taken for reliable measurement: (1) the water activity value should be taken when the reading (0.001 unit) has been constant for 10 min, (2) the humidity sensors should be calibrated regularly to compensate for drift, (3) a separate calibration curve should be made for each sensor, and (4) sensors should be calibrated at the same temperature at which the samples are measured and differences in temperature between sample and sensor should be avoided because of possible formation of condensate on the sensor. Labuza et al. (1976) mentioned that the equilibration time varied from 20 min to 24 h depending on the humidity range and food materials. A mechanical or chemical filter is available to protect the sensor from contamination. The Sina-scope sensor was equipped with a mechanical filter to protect it from dust, oil, and water vapor condensation. A chemical filter can also be used to protect the sensor from chlorine, formaldehyde, ammonia, sulfur dioxide, hydrogen sulfite, amino acids, hydrocarbons, and oil droplets (Labuza et al., 1976). It is difficult to find a sensor that will operate at very low temperature and above 100°C (Wiederhold, 1987).

The sensor consists of a small hollow cylinder covered with a glass fiber tape which is impregnated with saturated lithium chloride solution. A spiral bifilar electrode is wound over the tape and a temperature sensor is mounted at the center of the cylinder. An alternating voltage is applied to the electrodes and a current is allowed to pass through the intervening electrolyte. The resulting rise in temperature opposes the absorption of moisture by the lithium chloride and the sensor rapidly reaches an equilibrium temperature at which the vapor pressure of the salt solution equals that of the air. This temperature is determined by the sensor at the core of the cylinder and a calibration chart is used to convert this relative humidity.

Stoloff (1978) presented collaborative studies of using instruments with immobilized salt solution sensors in different laboratories. A multilaboratory and multi-instrument study of the method, applied to a cross section of commodities and reference standards, demonstrated that measurements of water activity by the described method with instruments using immobilized salt solution sensors can be made with an accuracy and precision within ± 0.01, provided there are no commodity–instrument interactions. Their study showed that commodity–instrument interactions did exist with lupine beans in brine, soy sauce, and cheese spread, but not with fudge sauce. A sampling error is distinctly evident with walnuts and is probably a component of variance with rice.

The sensitivity of the sample or the sensor to vapor transfer must also be considered. Here, the quantity of material represented by the sample or by the sensor is important. Vapor pressure is established by the presence of a particular number of molecules in a defined volume of space. The transfer of water molecules into the vapor phase may cause a measurable change in the gravimetric water contents of the sample and sensor. It is necessary that the sample water content (or the initial sample weight) be known. It is also necessary to know the sensor water content, depending upon the operating principle of the sensor. The amount of moisture transfer from the sample does not significantly change its moisture content (Reid, 2001).

2.2.4 Other Methods

Another method involves preparation of a standard curve by equilibration of a specific amount of dry standard material over different saturated salt solutions. The standard must be stable during reuse of the material. The standard curve is a plot of water activity versus water content of the standard material. The standards that could be used are soy albumin, MCC, glycerol, and sulfuric acid solution (Stoloff, 1978). The jar or desiccator used should be exactly the same as will be used later. For the measurement, the sample size of the standard should be in a controlled narrow range, e.g., $\sim 1.6 \pm 0.1$ g in duplicate or triplicate. Once the standard is made, dry samples of the same weight as in the standard curve are equilibrated over a large quantity of the food material ($\sim$10–20 g). Once equilibrated, the moisture content of the standard material is measured (e.g., by weight gain) and then a_w is estimated from the standard curve (Vos and Labuza, 1974). This method avoids preparation or storing of saturated salts for each determination. Moreover, use of a standard shortens the equilibration, thus requiring less time for measurement. The sensitivity of this method is highly dependent on the accuracy of the calibration curve (Rizvi, 1995). This technique is not accurate below 0.50 and above 0.90 water activity (Troller, 1983).

Gilbert (1993) described the application of inverse gas chromatography (IGC) to determine sorption isotherms in different sorbents. IGC is used to measure sorption as a function of retention volume, but the short transit time of normal chromatography which requires near ideal conditions hindered the application of IGC in complex food systems exhibiting slow or complex changes with water sorption resulting in hysteresis. Chromatography offers another dimension—time, removing the requirement for equilibrium. This makes possible the measurement of limited rates of change approaching zero as in slow reactions or restricted diffusivity in the so-called lag phase.

2.3 SELECTION OF A SUITABLE METHOD

The selection of an appropriate method or instrument for a specific food material and the purpose of measurement is very important for more valuable and meaningful results. It is clear that many of the methods discussed above fail to meet the present requirements of portability, speed, cost, and simplicity. The manometric, volumetric, sorption isotherm, and sorption rate methods are all excluded by their dependence on fixed laboratory equipment. The methods based upon the wetting of hygroscopic salts or the color changes of cobalt salts are free from this limitation and might be adapted to give cheap, semirapid, and approximate procedure suitable for some applications (Smith, 1971). For more accurate measurements, the various electrical hygrometers now commercially available appear to meet the present requirements. Among these, the most promising instruments are anodized aluminum sensors (Smith, 1971). The characteristics of a sensor depend upon the conditions of manufacture, particularly on the nature of the anodized film, and therefore each manufacturer's instrument must be calibrated separately. The anodized sensors are advantageous because of their ruggedness and small dimensions, fast response and freedom from large temperature coefficients, and susceptibility to contamination of lithium chloride conductivity sensors (Smith, 1971). However, there is no single humidity instrument which is suitable for every application (Wiederhold, 1987). It is necessary to develop instruments to measure water activity of food at higher temperatures and pressures in the future.

2.4 CONCLUSION

Different types of water activity measurement equipments are presented in the literature. However, the static isopiestic method is the most widely used due to its simplicity and low cost;

electronic sensor types are also popular due to their simplicity, speed, and portability. In recent times, DVS systems are also being used due to their multidimensional applications for sorption, drying rate, controlled atmosphere, and other structural changes.

REFERENCES

ASHRAE. 1993. Measurement and instruments. In: *ASHRAE Handbook Fundamentals*. American Society of Heating, Refrigeration and Air-Conditioning Engineers, Atlanta, GA, pp. 13.1–13.23.

Bassal, A., Vasseur, J., and Lebert, A. 1993. Measurement of water activity above 100°C. *Journal of Food Science*, 58(2): 449–452.

Bell, L.N. and Labuza, T.P. 2000. *Moisture Sorptions: Practical Aspects of Isotherm Measurement and Use*. Egan Press AACC, Egan, MN.

Del-Nobile, M.A., Buonocore, G.G., and Conte, A. 2004. Oscillatory sorption tests for determining the water-transport properties of chitosan-based edible films. *Journal of Food Science*, 69(1): 44–49.

Fennema, O. 1981. Water activity at subfreezing temperatures. In: *Water Activity: Influences on Food Quality*, Rockland, L.B. and Stewart, G.F. (eds.). Academic Press, New York, pp. 713–732.

Fontan, C.F. and Chirife, J. 1981. The evaluation of water activity in aqueous solutions from freezing point depression. *Journal of Food Technology*, 16: 21–30.

Gal, S. 1975. Recent advances in techniques for the determination of sorption isotherms. In: *Water Relations of Foods*, Duckkworth, R.B. (ed.). Academic Press, London, pp. 139–154.

Gal, S. 1981. Recent developments in techniques for obtaining complete sorption isotherms. In: *Water Activity Influences and Food Quality*, Rockland, L.B., and Stewart, G.F. (eds.). Academic Press, London, pp. 89–111.

Garcia, L.H. and Pilosof, A.M.R. 2000. Kinetics of water sorption in okara and its relationship to the glass transition temperature. *Drying Technology*, 18(9): 2105–2116.

Gilbert, S.G. 1993. Applications of IGC for research in kinetic and thermodynamic problems in food science. In: *Shelf Lie Studies of Foods and Beverages*. Charalambous, G. (ed.). Elsevier Science Publishers B.V., London, pp. 1071–1079.

Greenspan, L. 1977. Humidity fixed points of binary saturated aqueous solutions. *Journal of Research and National Bureau of Standards [A]: Physics and Chemistry*, 81A(1): 89–96.

Igbeka, J.C. and Blaisdell, J.L. 1982. Moisture isotherms of a processed meat product—bologna. *Journal of Food Technology*, 17:37–46.

Kanade, P.B. and Pai, J.S. 1988. Moisture sorption method for hygroscopic samples using a modified proximity equilibrium cell. *Journal of Food Science*, 53(4): 1218–1219.

Laaksonen, T.J., Roos, V.H., and Labuza, T.P. 2001. Comparisons of the use of desiccators with or without vacuum for water sorption and glass transition studies. *International Journal of Food Properties*, 4(3): 545–563.

Labuza, T.P. 1984. *Moisture Sorptions: Practical Aspects of Isotherm Measurement and Use*. American Association of Cereal Chemists, St. Paul, MN.

Labuza, T.P., Acott, K., Tatini, S.R., Lee, R.Y., Flink, J., and McCall, W. 1976. Water activity determination: A collaborative study of different methods. *Journal of Food Science*, 41: 910–917.

Lang, K.W., McCune, T.D., and Steinberg, M.P. 1981. A proximity equilibration cell for rapid determination of sorption isotherms. *Journal of Food Science*, 46: 936–938.

Lerici, C.R., Piva, M., and Rosa, M.D. 1983. Water activity and freezing point depression of aqueous solutions and liquid foods. *Journal of Food Science*, 48: 1667–1669.

Lewicki, P.P. 1987. Design of water activity vapor pressure manometer. *Journal of Food Engineering*, 6: 405–422.

Lewicki, P.P. 1989. Measurement of water activity of saturated salt solutions with the vapor pressure manometer. *Journal of Food Engineering*, 10: 39–55.

Lewicki, P.P. and Pomaranska-Lazuka, W. 2003. Errors in static desiccator method of water sorption isotherms estimation. *International Journal of Food Properties*, 6(3): 557–563.

Lewicki, P.P., Busk, G.C., Peterson, P.L., and Labuza, T.P. 1978. Determination of factors controlling accurate measurement of a_w by the vapor pressure manometric technique. *Journal of Food Science*, 43: 244–246.

Lin, S.X.Q. and Chen, X.D. 2005. An effective laboratory air humidity generator for drying research. *Journal of Food Engineering*, 68: 125–131.

Loncin, M. 1988. Activity of water and its importance in preconcentration and drying of foods. In: *Preconcentration and Drying of Food Materials*, Bruin, S. (ed.). Elsevier Science Publishers B.V., Amsterdam, pp. 15–34.

Lowe, E., Durkee, E.L., and Farkas, D.F. and Silverman, G.J. 1974. Anidea for precisely controlling the water activity in testing chambers. *Journal of Food Science*, 39: 1072–1073.

May, B.K., Shanks, R.A., Sinclair, A.J., Halmos, A.L., and Tran, V.N. 1997. A study of drying characteristics of foods using thermogravimetric analyzer. *Food Australia*, 49(5): 218–220.

Nunes, R.V., Urbicain, M.J., and Rotstein, E. 1985. Improving accuracy and precision of water activity measurements with a water vapor pressure manometer. *Journal of Food Science*, 50: 148–149.

Pezzutti, A. and Crapiste, G.H. 1997. Sorption equilibrium and drying characteristics of garlic. *Journal of Food Engineering*, 31: 113–123.

Prior, B.A. 1979. Measurement of water activity in foods: a review. *Journal of Food Protection*, 42, 668–674.

Rahman, M.S. 1991. Thermophysical properties of seafoods. PhD thesis, University of New South Wales, Sydney.

Rahman, M.S. 1995. *Handbook of Food Properties*. CRC Press, Boca Raton, FL.

Rahman, M.S. and Al-Belushi, R.H. 2006. Dynamic isopiestic method (DIM): Measuring moisture sorption isotherm of freeze-dried garlic powder and other potential uses of DIM. *International Journal of Food Properties*, 9(3): 421–437.

Rahman, M.S. and Labuza, T.P. 2007. Water activity and food preservation. In: *Handbook of Food Preservation*, 2nd edn., Rahman, M.S. (ed.), CRC Press, Boca Raton, FL, pp. 447–476.

Rahman, M.S. and Sablani, S.S. 2001. Measurement of water activity using electronic sensors. In *Current Protocols in Food Analytical Chemistry*, Eds, Wrolstad, R.E., Acree, T.E., Am, H., Decker, E.A., Penner, M.H., Reid, D.S., Schwartz, S.J., Shoemaker, C.F., and Sporms, P., John-Wiley & Sons, Inc., New York, pp. A2.5.1–A2.5.4.

Rahman, M.S., Sablani, S.S., Guizani, N., Labuza, T.P., and Lewicki, P.P. 2001. Direct manometric measurement of vapor pressure. In: *Current Protocols in Food Analytical Chemistry*, John-Wiley & Sons, Inc., New York, pp. A2.4.1–A2.4.7.

Rahamn, M.S., Salman, Z., Kadim, I.T., Mothershaw, A., Al-Riziqi, M.H., Guizani, N., Mahgoub, O., and Ali, A. 2005. Microbial and physico-chemical characteristics of dried meat processed by different methods. *International Journal of Food Engineering*, 1(2): 1–14.

Rahman, M.S., Al-Marhubi, I.M., and Al-Mahrouqi, A. 2007. Measurement of glass transition temperature by mechanical (DMTA), thermal (DSC and MDSC), water diffusion and density methods: A comparison study. *Chemical Physics Letters*, 440: 372–377.

Reid, D. 2001. Factors to consider when estimating water vapor pressure. In: *Current Protocols in Food Analytical Chemistry*, Eds, Wrolstad, R.E., Acree, T.E., Am, H., Decker, E.A., Penner, M.H., Reid, D.S., Schwartz, S.J., Shoemaker, C.F., and Sporms, P., John Wiley and Sons, New York, p. A2.1.1–A2.1.3.

Rey, D.K. and Labuza, T.P. 1981. Characterization of the effect of solute in water-binding and gel strength properties of carrageenan. *Journal of Food Science*, 46: 786.

Rizvi, S.S.H. 1995. Thermodynamic properties of foods in dehydration. In: *Engineering Properties of Foods*, 2nd edn., Rao, M.A., Rizvi, S.S.H., and Datta, A. (eds.). CRC Press, Boca Raton, FL.

Roques, M., Naiha, M., and Briffaud, J. 1983. Hexane sorption and diffusion in rapeseed meals. In: *Food Process Engineering*, Vol. 1, *Food Processing Systems*, Linko, P., Malkki, Y., Olkku, J., and Larinkari, J. (eds.). Applied Science Publishers, London, pp. 13–21.

Sannino, A., Capone, S., Siciliano, P., Ficarella, A., Vasanelli, L., Meffezzoli, A. 2005. Monitoring the drying process of lasagna pasta through a novel sensing divice-based method. *Journal of Food Engineering*, 69: 51–59.

Sablani, S.S., Rahman, M.S., and Labuza, T.P. 2001. Measurement of water activity using isopiestic method. In *Current Protocols in Food Analytical Chemistry*, Eds, Wrolstad, R.E., Acree, T.E., Am, H., Decker, E.A., Penner, M.H., Reid, D.S., Schwartz, S.J., Shoemaker, C.F., and Sporms, P., John-Wiley & Sons, Inc., New York, pp. A2.3.1–A2.3.10.

Scott, W.J. 1957. Water relations of food spoilage microorganisms. *Advances in Food Research*, 72: 83.

Smith, P.R. 1971. *The Determination of Equilibrium Relative Humidity or Water Activity in Foods—A Literature Review*. The British Food Manufacturing Industries Research Association, England.

Solomon, M.E. 1945. The use of cobalt salts as indicators of humidity and moisture. *Annals of Applied Biology*, 32: 75–85.

Solomon, M.E. 1957. Estimation of humidity with cobalt thiocyanate papers and permanent colour standards. *Bulletin Entomological Research*, 48(3): 489–507.

Sood, V.C. and Heldman, D.R. 1974. Analysis of a vapor pressure manometer for measurement of water activity in nofat dry milk. *Journal of Food Science*, 39: 1011–1013.

Spiess, W.E.L. and Wolf, W. 1987. Critical evaluation of methods to determine moisture sorption isotherms, In: *Water Activity: Theory and Applications to Food*, Rockland, L.B. and Beuchat, L.R. (eds.). Marcel Dekker, New York, pp. 215–233.

Stamp, J.A., Linscott, S., Lomauro, C., and Labuza, T.P. 1984. Measurement of water activity of salt solutions and foods by several electronic methods as compared to direct vapor pressure measurement. *Journal of Food Science*, 49: 1139–1142.

Stekelenburg, F.K. and Labots, H. 1991. Measurement of water activity with an electric hygrometer. *International Journal of Food Science and Technology*, 26(1): 111–116.

Stoloff, L. 1978. Calibration of water activity measuring instruments and devices: Collaborative study. *Journal of the Association of Official Analytical Chemists*, 61(5): 1166–1178.

Strong, D.H., Foster, E.M., and Duncan, C.L. 1970. Influence of water activity on the growth of Clostridium perfrigens. *Applied Microbiology*, 19: 980.

Teoh, H.M., Schmidt, S.J., Day, G.A., and Faller, J.F. 2001. Investigation of cornmeal components using dynamic vapor sorption and differential scanning calorimetry. *Journal of Food Science*, 66(3): 434–440.

Troller, J.A. 1983. Methods to measure water activity. *Journal of Food Protection*, 46: 129.

Vos, P. and Labuza, T.P. 1974. A technique for measurement of water activity in the high a_w range. *Journal of Agricultural Food Chemistry*, 22: 326–327.

Wiederhold, P. 1987. Humidity measurements. In: *Handbook of Industrial Drying*, Mujumdar, A.S. (ed.). Marcel Dekker, New York.

Wodzinski, R.J. and Frazier, W.C. 1960. Moisture requirement of bacteria. 1. Influence of temperature and pH on requirements of *Pseudomonas fluorescens*. *Journal of Bacteriology*, 79: 572.

CHAPTER 3

Data and Models of Water Activity. I: Solutions and Liquid Foods

Piotr P. Lewicki

CONTENTS

3.1 INTRODUCTION

Food is a multicomponent and multiphase system that is usually not in a thermodynamic equilibrium. This lack of equilibrium causes many chemical and physical changes, which are passed within the material during storage. Processing creates domains within the food with higher and lower concentration of constituents. Concentration gradients result in diffusion, which in turn enable contact between substrates and can facilitate chemical reactions. Redistribution of constituents, especially water, affects rheological properties of the material, hence its texture, structure, ability to creep and relax all change during storage. Interactions between polymers strongly influence properties of the material and depend on rotational and translational diffusion of polymer chains. In materials with interphase boundaries, mass and surface forces cause changes of structure and constituents, spatial concentration. Destabilization of foams and coalescence of emulsions is a macroscopic result of those changes. All the above-mentioned examples show that food is a dynamic system, far from the equilibrium state, which undergoes many changes during storage. Research conducted over the years prove that the course and the dynamics of all those changes are related to the thermodynamic state of water in the material (Lewicki, 2004).

The thermodynamic state of water in food arises firstly from unusual properties of water and its ability to form strong hydrogen bonds. Interactions with hydrophilic solute lead to the formation of structure water and hydration water. On the other hand, an interaction with hydrophobic solute affects the structure of the solvent water. A clathrate-like structure is formed. Water molecule mobility in both structure and hydration state as well as in clathrate-like structure is reduced. In the presence of electrolytes, ionic interactions also occur and induce some structuring of water molecules. Finally, porosity of the material, which is quite common for foods, also affects the thermodynamic state of water. Curvature of the solvent surface in capillaries, in drops or interstices, reduces the ability of water molecules to do the work. The thermodynamic state of water is expressed by its Gibbs free energy, which is a criterion of feasibility of chemical or physical transformation. The Gibbs free energy is quantitatively expressed by the equation:

$$G = H - TS \tag{3.1}$$

where

H is the enthalpy (J)
T is the temperature (K)
S is the entropy (J/K) and can be understood as total energy (H) diminished by unavailable energy (TS)

Gibbs free energy in differential form:

$$\mathrm{d}G = \mathrm{d}H - T\mathrm{d}S - S\mathrm{d}T \tag{3.2}$$

Substituting $\mathrm{d}H = \mathrm{d}E + P\mathrm{d}V + V\mathrm{d}P$ and $\mathrm{d}E = T\mathrm{d}S - P\mathrm{d}V$ yield the differential change in the Gibbs free energy:

$$\mathrm{d}G = V\mathrm{d}P - S\mathrm{d}T \tag{3.3}$$

where

E is the internal energy (J)
P is the pressure (Pa)
V is the volume (m^3)

The above equation applies to a homogeneous system of constant composition in which only work of expansion takes place. In an open multicomponent system, the Gibbs free energy will depend not only on the temperature and pressure, but also on the amount of each component present in the system. Hence

$$G = f(T, P, n_1, n_2, \ldots, n_n) \tag{3.4}$$

where $n_1, n_2, \ldots, n_n$ is the number of moles of component 1, 2, ..., n. Under this situation, change of Gibbs free energy is described by the equation:

$$\mathrm{d}G = \left.\frac{\partial G}{\partial P}\right)_{T,n_j} \mathrm{d}P + \left.\frac{\partial G}{\partial T}\right)_{P,n_j} \mathrm{d}T + \sum_{n_i}^{n_j} \left.\frac{\partial G}{\partial n}\right)_{T,P,n_j} \mathrm{d}n_i \tag{3.5}$$

in which i denotes that component where concentration changes and j denotes all those components where concentration remains constant. The partial molar Gibbs free energy

$$\left.\frac{\partial G}{\partial n_i}\right)_{T,P,n_j} \tag{3.6}$$

expressed by Equation 3.6 is called chemical potential of component i and is denoted as μ_i. It represents the change in Gibbs free energy of the system caused by addition of one mole of component i keeping temperature, total pressure, and number of moles of all other components constant. Then Equation 3.3 can be written as follows:

$$dG = VdP - SdT + \sum \mu_i dn_i \tag{3.7}$$

Gibbs has shown that for any system the necessary and sufficient condition for equilibrium is the equality of chemical potential of component i in all phases. That is

$$\mu_i^{I} = \mu_i^{II} = \mu_i^{III} \tag{3.8}$$

where the superscripts refer to different phases. Dividing Equation 3.3 by the number of moles of component i, the following can be written:

$$\frac{dG}{dn_i} = \frac{V}{dn_i}dP - \frac{S}{dn_i}dT \tag{3.9}$$

Denoting $V/dn_i = v_i$ and $S/dn_i = s_i$, Equation 3.9 becomes

$$d\mu_i = v_i dP - s_i dT \tag{3.10}$$

where
v_i is the molar volume (m^3/mol)
s_i is the molar entropy J/(mol K)

At constant temperature

$$d\mu_i = v_i dP \tag{3.11}$$

Taking the substance as an ideal gas, Lewis obtained chemical potential in terms of easily measurable properties.

$$v_i = \frac{RT}{P} \quad \text{and} \quad d\mu_i = \frac{RT}{P}dP \tag{3.12}$$

where R is the gas constant (J/mol K). Integrating Equation 3.12 the following formula is obtained:

$$\mu_i - \mu_i^\circ = RT \ln \frac{p}{p_\circ} \tag{3.13}$$

where
μ_i° is the chemical potential of component i at standard conditions
$p_\circ$ is the initial pressure in the system

In a real system properties of gases deviate from the ideal one and to account for the deviation Lewis proposed a new function called fugacity, f. Hence, Equation 3.13 for a real gas takes the following form:

$$\mu_i - \mu_i^\circ = RT \ln \frac{f_i}{f_i^\circ} \tag{3.14}$$

In a pure ideal gas $f_i = p_i$, the partial pressure of the gas. The no ideality of the gas follows from the existence of intermolecular forces (Prausnitz, 1969). The ratio between fugacities is called activity, a. Hence, Equation 3.14 becomes

$$\mu_i - \mu_i^\circ = RT \ln a_i \tag{3.15}$$

Table 3.1 Fugacity and Activity of Water Vapor in Equilibrium with the Liquid at Saturation and at Pressure 0.01 MPa

Temperature (°C)	Pressure (kPa)	Fugacity (kPa)	Activity
0.01	0.611	0.611	0.9995
10	1.227	1.226	0.9992
20	2.337	2.334	0.9988
40	7.376	7.357	0.9974
60	19.920	19.821	0.9950
80	47.362	46.945	0.9912
100	101.325	99.856	0.9855
120	198.53	194.07	0.9775
140	361.35	349.43	0.9670
160	618.04	589.40	0.9537
180	1002.7	939.93	0.9374

Source: Adapted from Hass, J.L., *Geochim. Cosmochim. Acta*, 34, 929, 1970.

Analysis of data presented in Table 3.1 shows that in the range of temperatures important for food processing, the water vapor deviates from the ideal gas by not more than by 6%, and at ambient temperature and pressure, the deviation is less than 0.2%. Thus, under the conditions experienced in food processing and storage activity, water vapor at saturation can be assumed to be equal to 1 and it can be written as

$$\frac{f_w}{f_w^\circ} = \frac{p_w}{p_w^\circ} \tag{3.16}$$

where p_w and p_w° are the vapor pressure of water in the system and of pure water at the same temperature and total pressure, respectively. Equation 3.16 is used to calculate water activity in food when partial pressure of water vapor over that food is known. Assuming that food is in equilibrium with the gas phase, the activity of gas phase calculated from Equation 3.16 is taken as the activity of water in solid or liquid food according to Equation 3.8. Taking a system consisting of an ideal solution and an ideal gas, the equilibrium state can be described as follows:

$$\mu_i^{\text{solution}} = \mu_i^{\text{vapor}} \tag{3.17}$$

An ideal solution behaves analogously to an ideal gas and follows Raoult's law. Fugacity of a component i in an ideal solution is proportional to the concentration of that component. Thus

$$f_i = kx_i \tag{3.18}$$

where

k is the constant

x is the mole fraction of component i

For $x_i = 1$ $f_i = k = f_i^\circ$, hence

$$\frac{f_i^{\text{solution}}}{f_i^\circ} = x_i = a_i^{\text{solution}} \tag{3.19}$$

In liquids, deviations from ideality are strong and arise from electrostatic forces, induction forces between permanent dipoles, forces of attraction, and repulsion between nonpolar molecules and specific interactions such as hydrogen bonds. To account for these interactions, an activity coefficient γ was introduced. Thus

$$a_i^{\text{solution}} = \gamma_i x_i \tag{3.20}$$

and

$$\mu_i^{\text{solution}} - \mu_i^{\circ\ \text{solution}} = RT \ln \gamma_i x_i \tag{3.21}$$

Activity of water in solution can be calculated from Equation 3.20 but the activity coefficient γ_w must be known. However, this is a function of temperature and composition and is usually undetermined.

3.2 WATER ACTIVITY OF SOLUTIONS AND LIQUID FOODS

Water activity of solutions and liquid foods depends on concentration, chemical nature of solutes, and temperature. In most fresh foods, it is close to 1 and its measurement presents some difficulties. Water activity of some solutions and liquid foods is presented in Tables 3.2 through 3.6.

Table 3.2 Water Activity of Some Liquid Foods

Product	Concentration (%)	a_w	Temperature (°C)	Reference
Apple juice		0.986		Chirife and Ferro Fontan (1982)
Apple juice concentrate	40–42	0.91–0.93		Chen (1987a)
Apple juice concentrate	66.0	0.792	24.2	Jarczyk et al. (1995)
	65.4	0.798	24.3	Jarczyk et al. (1995)
	66.2	0.795	23.7	Jarczyk et al. (1995)
	66.2	0.791	23.5	Jarczyk et al. (1995)
	70.1	0.739	23.4	Jarczyk et al. (1995)
	62.4	0.821	24.1	Jarczyk et al. (1995)
	71.2	0.734	23.8	Jarczyk et al. (1995)
Aronia juice concentrate	60.7	0.823	23.7	Jarczyk et al. (1995)
Black currant juice concentrate	64.5	0.810	24.4	Jarczyk et al. (1995)
	65.0	0.812	24.3	Jarczyk et al. (1995)
	65.0	0.822	23.3	Jarczyk et al. (1995)
	60.6	0.846	23.8	Jarczyk et al. (1995)
	63.9	0.803	24.2	Jarczyk et al. (1995)
	66.6	0.815	23.3	Jarczyk et al. (1995)
Cherry juice		0.986		Chirife and Ferro Fontan (1982)
Coffee beverage, freeze-dried	5	0.996		Chen (1987b)
	10	0.991		Chen (1987a)
	20	0.990		Chen (1987b)
	30	0.978		Chen (1987a)
	40	0.964		Chen (1987a)

(continued)

Table 3.2 (continued) Water Activity of Some Liquid Foods

Product	Concentration (%)	a_w	Temperature (°C)	Reference
Corn syrup	17% water	0.62		
Cream, 40% fat		0.979		Chirife and Ferro Fontan (1982)
Flavored milks		0.9899 ± 0.0007		Esteban et al. (1990a)
Fruit juices		0.9901 ± 0.0017		Esteban et al. (1990a)
Glucose, 66%		0.78		Richardson (1986)
Grape juice		0.983		Chirife and Ferro Fontan (1982)
Grape juice	10	0.986		Chen (1987a)
Grape juice concentrate	20	0.976		Chen (1987b)
	30	0.962		Chen (1987b)
Grapefruit juice concentrate	59	0.844 – 0.860		Chen (1987b)
Honey	17% water	0.54		
Honey, buckwheat	17.6%–20.9% water	$a_w = 0.203 + 0.02x$		Bakier (2006)
Honey, rape	17.4%–19.5% water	$a_w = 0.060 + 0.026x$		Bakier (2006)
Lemon juice concentrate	40–42	0.90 – 0.932		Chen (1987a)
Maracuja juice concentrate	54	0.88		Chen (1987a)
Milk, 1.5% fat		0.995		Chirife and Ferro Fontan (1982)
Milk, pasteurized		0.995 ± 0.001		Esteban et al. (1989)
Milk, whole		0.995		Chirife and Ferro Fontan (1982)
Milk, whole		0.994 – 0.995		Chirife and Ferro Fontan (1982)
Milk condensed sweetened		0.833 ± 0.0017		Favetto et al. (1983)
Molasses	26% water	0.76		
NaCl solution	2.18% salt	0.977		Chuang and Toledo (1976)
	10.16% salt	0.927		Chuang and Toledo (1976)
	15.45% salt	0.877		Chuang and Toledo (1976)
	23.41% salt	0.799		Chuang and Toledo (1976)
Orange juice		0.988		Chirife and Ferro Fontan (1982)
Orange juice concentrate	15	0.982		Chen (1987a)
	40	0.908		Chen (1987a)
	50	0.892 – 0.905		Chen (1987a)
	55	0.88 – 0.90		Chen (1987a)
	60	0.84 – 0.87		Chen (1987a)
	65.0	0.824	23.2	Jarczyk et al. (1995)
	65	0.802 – 0.835		Chen (1987a)
	68	0.79		
	72	0.735		Chen (1987a)
Pineapple juice concentrate	61	0.84		Chen (1987a)

Table 3.2 (continued) Water Activity of Some Liquid Foods

Product	Concentration (%)	a_w	Temperature (°C)	Reference
Raspberry juice		0.988		Chirife and Ferro Fontan (1982)
Raspberry juice concentrate	53.6	0.903	23.5	Jarczyk et al. (1995)
Red currant juice concentrate	61.0	0.828	23.2	Jarczyk et al. (1995)
	65.6	0.811	23.6	Jarczyk et al. (1995)
	62.3	0.825	23.3	Jarczyk et al. (1995)
Skim milk solution	10	0.997		Chen (1987b)
	20	0.990		Chen (1987b)
	30	0.982		Chen (1987b)
	40	0.971		Chen (1987b)
Sour cherry juice concentrate	67.0	0.788	24.8	Jarczyk et al. (1995)
	64.2	0.796	24.0	Jarczyk et al. (1995)
	66.2	0.756	24.1	Jarczyk et al. (1995)
	67.4	0.796	23.4	Jarczyk et al. (1995)
	64.3	0.816	24.3	Jarczyk et al. (1995)
	65.3	0.793	24.0	Jarczyk et al. (1995)
Soy sauce		0.810 ± 0.018		Stoloff (1978)
Sucrose	5.87	0.993		Chuang and Toledo (1976)
	15.29	0.982		Chuang and Toledo (1976)
	40.78	0.944		Chuang and Toledo (1976)
	59	0.90		Favetto et al. (1983)
	66	0.86		Richardson (1986)
	66.29	0.839		Chuang and Toledo (1976)
Strawberry juice		0.991		Chirife and Ferro Fontan (1982)
Strawberry juice concentrate	56.0	0.838	22.4	Jarczyk et al. (1995)
Tomato ketchup		0.959 ± 0.012	20	Jakobsen (1983)
Tomato paste, triple concentrated		0.934 ± 0.0013		Favetto et al. (1983)
Tomato purée		0.978 ± 0.029	20	Jakobsen (1983)
Whey cheese		0.996 ± 0.000		Esteban et al. (1989)
Yogurt beverages		0.9876 ± 0.0008		Esteban et al. (1990a)
Culture media used in food microbiology				
Nutrient broth (Difco)		0.997	20	Esteban et al. (1990b)
Brain heart infusion broth (Oxoid)		0.995	20	Esteban et al. (1990b)
Staphyloccocus medium no. 110 (Difco)		0.936	20	Esteban et al. (1990b)
Malt yeast extract 50% glucose broth		0.876	20	Esteban et al. (1990b)
Halophylic broth (25% NaCl)		0.816	20	Esteban et al. (1990b)
Malt yeast extract 70% glucose fructose broth		0.740	20	Esteban et al. (1990b)

Table 3.3 Water Activity of Solutions of Low Molecular Weight Compounds

Concentration (%)	Lactose	Fructose	Sucrose	Glycerol	Glucose	Reference
1	0.999					Chen (1987b)
3	0.997					Chen (1987b)
5	0.995		0.996			Chen (1987b)
5				0.96		Grover and Nicol (1940)
5					0.98	Chen and Karmas (1980)
5.87			0.993			Chuang and Toledo (1976)
8	0.992					Chen (1987b)
10		0.991	0.992			Chen (1987b)
15		0.985	0.988			Chen (1987b)
15				0.93		Grover and Nicol (1940)
15					0.96	Chen and Karmas (1980)
15.29			0.982			Chuang and Toledo (1976)
20			0.983			Chen (1987b)
25.0				0.923		Grover and Nicol (1940)
30		0.961	0.971			Chen (1987b)
35.0				0.887		Grover and Nicol (1940)
40		0.935	0.955			Chen (1987b)
40.78			0.944			Chuang and Toledo (1976)
50.0				0.814		Grover and Nicol (1940)
60.0				0.737		Grover and Nicol (1940)
66.29			0.839			Chuang and Toledo (1976)
75.0				0.587		Grover and Nicol (1940)
83.0				0.446		Grover and Nicol (1940)
0.92				0.275		Grover and Nicol (1940)

	Concentration (%)		
Solute	5	15	
L-Alanine	0.97		Chen and Karmas (1980)
Arabinose	0.98	0.96	Chen and Karmas (1980)
Galactose	0.98	0.96	Chen and Karmas (1980)
Gluconic acid	0.97	0.96	Chen and Karmas (1980)
Glucuronic acid	0.97	0.96	Chen and Karmas (1980)
Glycine	0.97	0.94	Chen and Karmas (1980)
Lactic acid	0.96	0.94	Chen and Karmas (1980)
Malonic acid	0.98	0.95	Chen and Karmas (1980)
Maltose	0.98	0.96	Chen and Karmas (1980)
Mannitol	0.97	0.96	Chen and Karmas (1980)
Mannose	0.98	0.96	Chen and Karmas (1980)
Oxalic acid	0.98	0.96	Chen and Karmas (1980)
Ribose	0.98	0.96	Chen and Karmas (1980)
Sorbitol	0.96	0.95	Chen and Karmas (1980)
Succinic acid	0.99		Chen and Karmas (1980)
Tartaric acid	0.97	0.96	Chen and Karmas (1980)
Xylose	0.98	0.95	Chen and Karmas (1980)

Table 3.4 Water Activity of Glucose Solutions

	Water Activity at Temperature (°C)			
Concentration (%)	**20**	**25**	**30**	**35**
Glucose				
30.0	0.954	0.955	0.956	0.957
35.0	0.942	0.943	0.944	0.945
40.0	0.927	0.929	0.930	0.930
45.0	0.910	0.912	0.913	0.914
50.0	0.891	0.891	0.893	0.895
55.0	0.865	0.867	0.869	0.870
60.0	0.835	0.837	0.839	0.841

Source: Adapted from Viet Bui, A., Minh Nguyen, H., and Muller, J., *J. Food Eng.*, 57, 243, 2003.

Nonideality of the liquid solution can be characterized by the thermodynamic excess function. The excess functions are thermodynamic properties of solutions, which are in excess of those of an ideal solution at the same temperature, pressure, and concentration. Thus

$$\mu_i^{\text{solution}} - \mu_i^{\text{ideal}} = RT\ \ln\frac{a_i}{x_i} = RT\ \ln\gamma_i \tag{3.22}$$

Hence, the excess function, the deviation of properties of a real solution from those of an ideal solution, is expressed by the activity coefficient. The problem resolves to finding the activity coefficient of component i in the solution. Activity coefficients for some sugars are given in Table 3.7. Several approaches have been used to calculate activity coefficients of liquid solutions. This includes empirical equations based on solution composition, equations derived by the thermodynamic approach, and equations based on solution theories (Sereno et al., 2001).

Table 3.5 Water Activity of Sugar Solutions

Glucose		Fructose		Sucrose		Maltose	
W	a_w	*W*	a_w	*W*	a_w	*W*	a_w
0.049	0.995	0.049	0.995	0.049	0.998	0.046	0.998
0.096	0.990	0.096	0.989	0.096	0.995	0.091	0.995
0.142	0.983	0.142	0.985	0.142	0.991	0.134	0.990
0.186	0.977	0.186	0.978	0.186	0.989	0.176	0.988
0.229	0.970	0.229	0.971	0.228	0.985	0.217	0.984
0.272	0.962	0.274	0.965	0.270	0.980	0.258	0.979
				0.292	0.974		
				0.349	0.967		
				0.399	0.958		
				0.445	0.948		

Source: Adapted from Valezmoro, C., Meirelles, A.J., and Vitali, A., in *Engineering & Food at ICEF 7*, Jowitt, R. (Ed.), Sheffield Academic Press, Sheffield, UK, 1997, pp. A145–A148.

Note: *W*, weight fraction for solute.

Table 3.6 Water Activity of Sugars Solutions

	a_w					
	Sucrose (°C)		Glucose (°C)		Fructose (°C)	
Molality	**25**	**50**	**25**	**50**	**25**	**Reference**
0.1	0.9982					Robinson and Stokes (1965)
0.5	0.9907					Robinson and Stokes (1965)
0.617			0.9735	0.9851		Ross (1975)
0.7513					0.985 ± 0.005	Correa et al. (1994b)
1.0	0.9806					Robinson and Stokes (1965)
1.0727					0.979 ± 0.005	Correa et al. (1994b)
1.4	0.9719					Robinson and Stokes (1965)
1.5212					0.971 ± 0.005	Correa et al. (1994b)
1.926	0.9599	0.9617				Ross (1975)
1.9976					0.963 ± 0.005	Correa et al. (1994b)
2.0	0.9581					Robinson and Stokes (1965)
2.379			0.9400	0.9517		Ross (1975)
2.3865					0.954 ± 0.005	Correa et al. (1994b)
2.5	0.9457					Robinson and Stokes (1965)
2.563	0.9442	0.9477				Ross (1975)
2.7466					0.947 ± 0.005	Correa et al. (1994b)
3.0	0.9328					Robinson and Stokes (1965)
3.1491					0.938 ± 0.005	Correa et al. (1994b)
3.5	0.9193					Robinson and Stokes (1965)
3.9199					0.924 ± 0.005	Correa et al. (1994b)
4.0	0.9057					Robinson and Stokes (1965)
4.5	0.8917					Robinson and Stokes (1965)
5.0	0.8776					Robinson and Stokes (1965)
5.0339					0.897 ± 0.005	Correa et al. (1994b)
5.5	0.8634					Robinson and Stokes (1965)
5.508	0.8631	0.8758				Ross (1975)
5.551			0.8722	0.8878		Ross (1975)
6.0	0.8493					Robinson and Stokes (1965)
6.0778					0.870 ± 0.005	Correa et al. (1994b)
7.2271					0.841 ± 0.005	Correa et al. (1994b)

For a simple two-component mixture, the Wohl equations for both components are obtained by combining the total excess Gibbs energy with the Gibbs–Duhem equation.

$$\begin{aligned} \ln \gamma_w &= z_s^2 \left[A + 2z_w \left(B\frac{q_w}{q_s} - A \right) \right] \\ \ln \gamma_s &= z_w^2 \left[B + 2z_s \left(A\frac{q_s}{q_w} - B \right) \right] \end{aligned} \tag{3.23}$$

where

z is the effective volume fraction of component in solution

q is the effective molar volume of component in solution

A, B are constants

subscripts w and s denote water and solute, respectively

Table 3.7 Activity Coefficients for Saccharides at 25°C

Molality	Maltose (Miyajima et al., 1983a)	Maltotriose (Miyajima et al., 1983a)	D-Glucose (Miyajima et al., 1983b)	D-Mannose (Miyajima et al., 1983b)	D-Galactose (Miyajima et al., 1983b)
0.1	1.003	1.003	1.002	1.000	1.001
0.2	1.007	1.010	1.004	1.000	1.001
0.3	1.010	1.016	1.006	1.000	1.002
0.4	1.014	1.023	1.009	1.000	1.002
0.5	1.020	1.030	1.013	1.001	1.003
0.6	1.027	1.040	1.017	1.001	1.004
0.7	1.036	1.049	1.021	1.002	1.005
0.8	1.044	1.061	1.026	1.003	1.006
0.9	1.052	1.074	1.031	1.005	1.007
1.0	1.061	1.088	1.036	1.006	1.009
1.1	1.070	1.103			
1.2	1.081	1.120	1.047	1.010	1.013
1.3	1.091	1.138			
1.4	1.103	1.158	1.059	1.014	1.018
1.5	1.114	1.179			
1.6	1.126	1.199	1.071	1.019	1.025
1.7	1.138	1.220			
1.8	1.150	1.242	1.084	1.025	1.032
1.9	1.163	1.265			
2.0	1.176	1.288	1.097	1.031	1.039
2.5			1.131	1.048	1.061
3.0			1.166	1.068	1.083
3.5			1.200	1.088	1.104
4.0			1.233	1.109	
4.5			1.266	1.130	
5.0			1.298	1.150	
5.5			1.329	1.169	
6.0			1.306	1.189	
Activity coefficients of xylose in water solution		$\ln \gamma_s = 0.02814m + 0.003432m^2 - 0.0006427m^3$, where m is the molality			Uedaira and Uedaira (1969)
Activity coefficients of water in xylose solution		$\ln \gamma_w = -1.282x^2 - 9.093x^3 + 65.61x^4$, where x is the mole fraction of the solute			Uedaira and Uedaira (1969)
Activity coefficients of maltose in water solution		$\ln \gamma_s = 0.007270m + 0.02727m^2 - 0.003207m^3$, where m is the molality			Uedaira and Uedaira (1969)
Activity coefficients of water in maltose solution		$\ln \gamma_w = -2.518x^2 - 60.71x^3 + 302.5x^4$, where x is the mole fraction of the solute			Uedaira and Uedaira (1969)

The Wohl equation for activity coefficient of water is simplified to the following form:

$$\ln \gamma_w = x_s^2(A + kx_w) \tag{3.24}$$

Assuming that molecules are similar in size, shape and chemical nature ($q_1 = q_2$) and that A and B are equal, the two-suffix Margules equations are obtained (Prausnitz et al., 1999)

$$\ln \gamma_w = \frac{A}{RT} x_s^2$$
$$\ln \gamma_s = \frac{A}{RT} x_w^2 \tag{3.25}$$

Further expansions of these derivations are as follows (Poling et al., 2001).

Three-suffix Margules equations:

$$\ln \gamma_w = (2B - A)\, x_s^2 + 2(A - B)\, x_s^3 \tag{3.26}$$

van Laar equations:

$$\ln \gamma_w = A\left(1 + \frac{A}{B}\frac{x_w}{x_s}\right)^{-2}$$
$$\ln \gamma_s = B\left(1 + \frac{B}{A}\frac{x_s}{x_w}\right)^{-2} \tag{3.27}$$

The selected parameters for the above equations for sugar solutions are collected in Table 3.8.

Solution theories were used to develop molar excess Gibbs energy and activity coefficients. Wilson (1964) applied local-composition models and developed the following equation, which can be written for a binary mixture as

$$\ln \gamma_w = -\ln (x_w + \Lambda_{ws} x_s) + x_s \left(\frac{\Lambda_{ws}}{x_w + \Lambda_{ws} x_s} - \frac{\Lambda_{sw}}{x_s + \Lambda_{sw} x_w} \right)$$
$$\Lambda_{12} = \frac{v_s}{v_w} e^{(-E_{ws}/RT)}; \quad \Lambda_{21} = \frac{v_w}{v_s} e^{(-E_{sw}/RT)} \tag{3.28}$$

The molar volume v_w and v_s refer to pure liquids with an activity of 1. A pure liquid phase does not exist for dissolved components, hence the ratio of their molar volumes is used and the equation is simplified to the following form:

$$\ln \gamma_w = -\ln\left(\sum_{k=1}^{n} x_k \Lambda_{wk}\right) + 1 - \sum_{k=1}^{n} \frac{x_k \Lambda_{wk}}{\sum_{i=1}^{n} x_i \Lambda_{ik}} \tag{3.29}$$

Table 3.8 Parameters of Equations 3.24 through 3.27 for Sugar Solutions (de Cindio et al., 1995)

Equation	A	B	k
Glucose			
Margules (Equation 3.25)	−2.217		
Margules (Equation 3.26)	−4.240	−2.977	
van Laar (Equation 3.27)	−8.494	−3.866	
Wohl (Equation 3.24)	−4.240		2.526
Sucrose			
Margules (Equation 3.25)	−5.998		
Margules (Equation 3.26)	2.1819	−2.064	
van Laar (Equation 3.27)	−2.064	−3.948	
Wohl (Equation 3.24)	2.585		−9.916

Source: Adapted from de Cindio, B., Correra, S., and Hoff, V., *J. Food Eng.*, 24, 405, 1995.

Renon and Prausnitz's (1968) equations known as the non-random two liquid equations (NRTL) are also based on the concept of local composition. The NRTL equation for a binary mixture is expressed as

$$\ln \gamma_w = x_s^2 \left[\tau_{sw} \left(\frac{G_{sw}}{x_w + x_s G_{sw}} \right)^2 + \frac{\tau_{ws} G_{ws}}{(x_s + x_w G_{ws})^2} \right]$$
$$G_{ws} = \exp(-\alpha_{12}\tau_{ws}) \ \tau_{ws} = \frac{A_{ws}}{RT} \tag{3.30}$$

Parameters of the NRTL equation are presented for some food volatiles in Table 3.9. The NRTL local composition model was extended to electrolyte solutions (Chen et al., 1982). Recently, attempts have been made to develop equations of state for polymer systems. The segment-based polymer NRTL model was developed by Chen (1993) and the perturbed hard-sphere-chain model was proposed by Song et al. (1994).

Mixing of high molecular weight polymer with solvent causes the entropy change. This reduction in the entropy of mixing was first formulated independently by Flory (1942) and Huggins (1942) using a lattice theory. The developed equation describes a free energy change during mixing of a solvent and amorphous polymer.

$$\mu = \mu_0 + RT\left[\ln(1 - \nu_2) + \left(1 - \frac{1}{b}\right)\nu_2 + \chi\nu_2^2\right]$$
$$b = \frac{\nu_2}{\nu_1} \tag{3.31}$$

where

ν_1 is the molar volume of solvent
ν_2 is the molar volume of polymer
χ is the Flory interaction parameter denoting intermolecular interactions

The above equation is valid for such polymer concentrations at which polymer chains interlace in the solution. When the solution is sufficiently diluted and the chains are not interacting each with other, Equation 3.31 simplifies to the form:

$$\mu = \mu_0 + RT\left[\ln(1 - \nu_2) + \nu_2 + \chi\nu_2^2\right] \tag{3.32}$$

Table 3.9 Parameters of the NRTL Equation for Binary Mixtures

Substance in Water	Temperature (°C)	A_{ws}	A_{sw}	α_{12}	Reference
Acetic acid	100–119	183.82	−47.032	0.39688	Faúndez and Valderrama (2004)
Ethanol	40	1443.6	−391.8	0.1803	Lee and Kim (1995)
Ethyl acetate	50	2241.3	618.1	0.2834	Lee and Kim (1995)
Ethyl acetate	71–76	1881.15	727.01	0.39304	Faúndez and Valderrama (2004)
Methanol	25	699.9	−280.6	0.2442	Lee and Kim (1995)
Methanol	65–100	573.90	−208.12	0.29426	Faúndez and Valderrama (2004)
1-Propanol	88–95	1094.4	−12.175	0.30282	Faúndez and Valderrama (2004)

Equation 3.32 rewritten for water activity in the polymer solution is as follows:

$$\ln a_\mathrm{w} = \ln \phi_\mathrm{w} + \left(1 - \frac{\nu_1}{\nu_2}\right)\phi_\mathrm{p} + \chi\phi_\mathrm{p}^2 \tag{3.33}$$

where
ϕ is the volume fraction
subscripts w and p are for water and polymer, respectively

The Flory–Huggins model was applied to solutions of simple molecules, and the results of calculation are presented in Table 3.10.

The method, based on the group contribution, views molecules as made of a certain number of standard functional groups. The Analytical Solution of Groups (ASOG) was developed by Wilson and Deal (1962), Derr and Deal (1969), and Kojima and Tochigi (1979). According to this method the activity coefficient of component i in a liquid mixture is a result of differences both in molecular size and shape and in intermolecular forces. The ASOG method was used by Correa et al. (1994a) to calculate water activities of solutions of sugars and urea at 25°C. An average deviation of 0.4% was obtained between experimental and predicted values of a_w. In another publication, Correa et al. (1994b) showed that water activity of binary and ternary solutions containing sugars, glycerol, or urea can be predicted with the ASOG method with the deviation between the measured and predicted values not exceeding 1%. The method was applied to electrolytes solutions (Kawaguchi et al., 1981) and it was shown by Correa et al. (1997) that calculated values of water activity for binary, ternary, and quaternary systems deviated from measured values by 0.21%, 0.28%, and 0.20%, respectively.

The concept of group contribution developed by Abrams and Prausnitz (1975) known as Universal Quasi Chemical method was used by Peres and Macedo (1996) to predict thermodynamic properties of sugar aqueous solutions. Predicted water activities of binary systems water/D-glucose, water/D-fructose, and water/sucrose differed from the measured values by 0.28%, 0.44%, and 0.96%, respectively.

Another approach based on the group contribution concept known as universal functional activity coefficient (UNIFAC) was proposed by Fredenslund et al. (1975). This has received more attention and a continued development of this method has been observed. In this method, a molecule is pictured as an aggregate of functional groups, and its physical properties are the sum of

Table 3.10 Parameters of the Flory–Huggins Equation for Glucose Solution

Molality	ϕ_s	χ
0.14	0.011	−77.0
0.28	0.021	−27.6
0.56	0.041	−7.1
0.83	0.060	−4.3
1.11	0.079	−2.6
1.39	0.096	−1.55
2.22	0.146	−0.40
2.56	0.176	−0.07
3.33	0.204	+0.11

Source: Adapted from Napierala, D., Popenda, M., Surma, S., and Plenzler, G., in *Properties of Water in Foods*, Lewicki, P.P. (Ed.), Warsaw Agricultural University Press, Warsaw, 1998, pp. 7–13.

contributions made by the groups. The contributions are independent and additive. Contributions are due to differences in molecular size and shape as well as molecular interactions.

A solution is treated as a mixture of groups, which contribute to the partial molal excess free energy independently and additively. The contributions are associated with differences in molecular size and with interactions of the structural groups (Wilson and Deal, 1962). The UNIFAC model assumes that water activity results from the combinatorial and residual contributions. Hence

$$\ln a_{\mathrm{w}} = \ln a_{\mathrm{w}})^{c} + \ln a_{\mathrm{w}})^{r} \tag{3.34}$$

The combinatorial part is derived from the pure component properties such as group volume and area constants. This part is described by equations:

$$\begin{gathered}\ln a_{\mathrm{w}})^{c} = \ln \Phi_{\mathrm{w}} + \frac{z}{2} q_{\mathrm{w}} \ln \frac{\Theta_{\mathrm{w}}}{\Phi_{\mathrm{w}}} + l_{\mathrm{w}} - \frac{\Phi_{\mathrm{w}}}{x_{\mathrm{w}}}\left(\sum_{j=i}^{n} x_j l_j\right) \\ \Theta_{\mathrm{w}} = \frac{q_{\mathrm{w}} x_{\mathrm{w}}}{\sum_j^n q_j x_j}; \quad \Phi_{\mathrm{w}} = \frac{r_{\mathrm{w}} x_{\mathrm{w}}}{\sum_j^n r_j x_j}; \quad r_i = \sum_k v_k^i R_k; \quad q_i = \sum_k v_k^i Q_k\end{gathered} \tag{3.35}$$

The residual part of the activity is a function of group area fractions and their interactions in pure components and in mixtures. It is given by

$$\ln a_{\mathrm{w}})^{r} = \sum_k v_k^{\mathrm{w}} \left[\ln \Gamma_k - \ln \Gamma_k^{\mathrm{w}}\right] \tag{3.36}$$

where Γ_k is a group residual activity and Γ_k^{w} is the group residual activity of group k in a reference solution containing only molecules of water (w):

$$\begin{gathered}\ln \Gamma_k = Q_k\left[1 - \ln\left(\sum_m \Omega_m \Psi_{mk}\right) - \sum_m \left(\frac{\Omega_m \Psi_{km}}{\sum_n \Omega_n \Psi_{nm}}\right)\right] \\ \Omega_m = \frac{Q_m X_m}{\sum_n Q_n X_n}; \quad X_m = \frac{\sum_j v_m^j x_j}{\sum_j x_j \sum_m v_m^j}; \quad \Psi_{mn} = \exp\left(-\frac{a_{mn}}{T}\right)\end{gathered} \tag{3.37}$$

where a_{mn} is the group interaction constant; $l_i = z/2(r_i - q_i) - (r_i - 1)$; A_k is the van der Waals area of group k; Q_k is the group area constant, $Q_k = (A_k/2.5 \times 10^9)$; R_k is the group volume constant, $R_k = (V_k/15.17)$; v_k is the number of groups of type k in molecule i; V_k is the van der Waals volume of group k; x_i are mole fraction of component i; z is coordination number, $z = 10$. Θ_i is the component area fraction, Φ_i is the component volume fraction, Ω_m is the area fraction of group m, X_m is the mole fraction of group m. Subscripts denote i component i; k, m, n are groups, w is water.

The UNIFAC model was used by Choundry and Le Maguer (1986) to predict water activities of glucose solutions. Catté et al. (1995) used the UNIFAC equations to model aqueous solutions of sugars. In this model, conformational equilibrium was taken into account, thus isomers and anomers can be distinguished. Spiliotis and Tassios (2000) modified a UNIFAC model in order to predict phase equilibriums in aqueous and nonaqueous sugar solutions. The model allowed satisfactory prediction of water activity in sugar solutions. Further modification of a UNIFAC model was done by Peres and Macedo (1997) who introduced a new group in a sugar molecule. The "OH-ring" group behaves in a different way from the alcohol OH-group; hence, the two cyclic structures "PYR" and "FUR" can be treated as single groups. In systems, D-glucose–sucrose–water, water

Table 3.11 Water Activities for the System D-Glucose–Sucrose–Water at 25°C

Molality		a_w	
Sucrose	D-Glucose	Predicted from UNIFAC Model (Peres and Macedo, 1997)	Measured (Lilley and Sutton, 1991)
2.8039	0.5542	0.92098	0.9260
2.3504	1.0996	0.92189	0.9260
1.2166	2.4489	0.92385	0.9260
1.7391	1.8737	0.92207	0.9251
0.8286	2.9503	0.92350	0.9251
0.3648	3.4947	0.92406	0.9251

activity was predicted with relative deviation lower than 0.6%. Some results of the UNIFAC model application are presented in Table 3.11.

The modified UNIFAC model proposed by Peres and Macedo (1997) was used to predict the water activities of fruit juice concentrates and synthetic honey (Peres and Macedo, 1999) assuming that they are mixtures of sugars and water. For a total sugar concentration between 10% and 60% the water activities were predicted with an average deviation of 0.34% for apple juice and 0.37% for grape juice. The UNIFAC model with a new assignment of groups was used to describe the activity coefficients in binary systems of amino acids and peptides in water and other biochemicals. The new assignment of groups was needed because biochemicals have several asymmetric carbons and form electrolyte complexes with ions. Activity coefficients for solutions of DL-proline, xylose, D-glucose, D-mannose, D-galactose, maltose, sucrose, raffinose, and KCl were predicted with root mean square deviation (RMSD) lower than 1%. For L-hydroxyproline, sodium and potassium glutamate, L-agrinine · HCl, L-histidine · HCl, NaCl, and sodium glucuronate the RMSD was between 1% and 2.69%. For other studied compounds, the RMSD was larger than 5% (Kuramochi et al., 1997). Attempts to predict thermodynamic properties of solution of complex molecules such as proteins have been made (Chen et al., 1995; Curtis et al., 2001).

New methods such as cosmo-Rs (Eckert and Klamt, 2002) and group and segment contribution solvation models (Lin and Sandler, 1999, 2002) using quantum chemistry and molecular modeling are under development. The predictions based on group contribution models are good although they consider binary and in some cases ternary solutions. However, liquid foods are multicomponent mixtures and the group contribution models are not very suitable unless there are some dominant components, and the solution can be treated as binary or ternary mixture. For practical application empirical and semiempirical equations are developed.

3.3 SEMIEMPIRICAL EQUATIONS

The semiempirical equations are based on Raoult's law for an ideal solution. Water activity is described by the following equation:

$$a_w = \frac{n_w}{n_w + n_s} \tag{3.38}$$

where

n is the number of moles

subscripts w and s denote water and solid, respectively

Interaction of water molecules with solute molecule leads to formation of hydrated moieties. At solvation equilibrium, average hydration number denotes the average number of molecules of bound solvent per solute molecule. For real solutions the hydration number is given by the following equation:

$$\bar{h} = \frac{55.51}{m} - \frac{a_w}{1 - a_w} \tag{3.39}$$

where

m is the molality

a_w is the water activity

The above approach was taken up by Stokes and Robinson (1966) and equation predicting activity of solution of several solutes based on known data for single solutes was developed

$$\frac{55.51}{m} = \frac{a_w}{1 - a_w} + \frac{\sigma}{\Sigma} \tag{3.40}$$

where

$\Sigma = 1 + Ka_w + (Ka_w)^2 + \cdots + (Ka_w)^n$

$\sigma = Ka_w + \cdots + n(Ka_w)^n$ and K is the solvation equilibrium constant

For sucrose solutions, $n = 11$ and $K = 0.994$. For glucose, $n = 6$ and $K = 0.786$ and for glycerol $n = 3$ and $K = 0.720$. The n value was assumed to be equal to oxygen sites in a molecule able to interact with water molecules. For a mixture of two solutes the equation is

$$\frac{55.51}{m_A + m_B} = \frac{a_w}{1 - a_w} + \frac{m_A h_A + m_B h_B}{m_A + m_B} \tag{3.41}$$

with the assumption that there is no interaction between solutes A and B. For sucrose and sorbitol a very good agreement was obtained for measured and calculated values. Sucrose and glucose, sucrose and arabinose, and sucrose and glycerol deviation not exceeding 1.5% was noticed.

Poliszko and coworkers (2001) developed an equation accounting for hydration water and its effect on water activity of sugar solutions. The equation is as follows:

$$\begin{aligned} a_w &= \exp\left(\frac{\Delta\mu}{RT}\frac{hc}{(1 - c)}\right) \\ \frac{\Delta\mu h}{RT} &= \frac{\overline{\Delta\mu}}{RT}\bar{h}\left(1 + A\cos\left(\frac{2\pi c}{D}\right)\right) \\ h &= \frac{1 - c_s}{c_s} \end{aligned} \tag{3.42}$$

where

$\overline{\Delta\mu}$ is the average excess chemical potential of hydration water

$\bar{h}$ is the average hydration degree

A is the fluctuation amplitude

c is the concentration

c_s is the saturation concentration

Table 3.12 Parameters of Equation 3.42

Sugar	c_s (g/g)	A	$\bar{h}$	$\overline{\Delta\mu}$ (J/mol)
Fructose	0.824	0.0979	0.214	−1226
Glucose	0.470	0.1755	1.128	−285
Maltose	0.380	0.2520	1.632	−144
Xilitol	0.600	0.1680	0.670	−406

Source: From Baranowska, H.M., Klimek-Poliszko, D., and Poliszko, S., in *Properties of Water in Foods*, Lewicki, P.P. (Ed.), Warsaw Agricultural University Press, Warsaw, 2001, pp. 12–20.

Parameters of Equation 3.42 for some sugars are presented in Table 3.12.

Raoult's law written in mass fractions yields the following equation:

$$a_w = \frac{1-x}{1-x+Ex} \tag{3.43}$$

where

x is the solute content in the solution (kg/kg solution)

E is the ratio of the molecular weight of water to molecular weight of the solute

$$E = \frac{M_w}{M_s} \tag{3.44}$$

Assuming that some amount of water is bound with the solute, Schwartzberg (1976) proposed to modify Raoult's law and gave the expression:

$$a_w = \frac{1-x-bx}{1-x-bx+Ex} \tag{3.45}$$

where b is the amount of water bound by unit weight of solid (kg/kg solids). Modification of Raoult's law was presented by Palnitkar and Heldman (1970) in which effective molecular weight of the solute was introduced

$$a_w = \frac{(x_w/M)}{(x_w/M)+(x_s/\mathrm{EMW})} \tag{3.46}$$

where

x_w and x_s are the mass of water and solute, respectively (kg/kg solution)

M is the molecular weight of water (kg/mol)

EMW is the effective molecular weight of solute (kg/mol)

The above equation was used to calculate activity coefficients for sugars, polyols, amines and amino acids, organic acids, inorganic salts, and isolated soy proteins (Chen and Karmas, 1980). It was found that EMW for Promine (Central Soya, Chicago) was between 821.2 and 950.1, depending on its composition.

Caurie (1983) derived an equation based on Raoult's law in which water activity of a solution is expressed as the difference between the values calculated for an ideal solution and the amount of overestimation arising from interactions between solvent and solutes

$$a_w = \frac{55.5}{m_s+55.5} - \frac{m_s}{m_s+55.5}(1-a_w) \tag{3.47}$$

Table 3.13 Parameters of Equation 3.48

Sugar	*E*	*b*
D-Fructose	0.100	0.10–0.18
D-Glucose	0.100	0.15–0.25
Lactose	0.053	0.21–0.46
Maltose	0.053	0.21–0.26
Sucrose	0.053	0.26–0.30
Freeze-dried skim milk (10%–40% concentration)	0.054	0.11–0.13
Freeze-dried coffee beverage (5%–40% concentration)	0.052	0.00–0.02
Concentrated orange juice	0.075	0.15–0.20
Fruit juice (average)	0.089	0.20

Source: Adapted from Chen, C.S., *LWT*, 20, 64, 1987a.

where m_s is the molal concentration of the solute. It was concluded that water activity is a measure of free, available, and unbound fraction of water in a solution.

Chen (1987a) proposed to derive activity coefficient by comparison of ideal and real solution activities. Taking Equations 3.44 and 3.45 the following is obtained:

$$\gamma = 1 - \frac{Eby^2}{1 + y(E - b)} \tag{3.48}$$

where $y = x/(1 - x)$.

Calculated water activities from the above equation up to sugar concentration of 40% agreed with the experimental data within 0.01 unit of a_w. Equation parameters for sugars are collected in Table 3.13.

Chen (1989) developed a simple equation to predict water activity of single solutions. The equation predicts water activity of a wide range of solutes with an accuracy of ± 0.001 a_w. The equation is as follows:

$$a_w = \frac{1}{1 + 0.018(\beta + Bm^n)m} \tag{3.49}$$

where
m is the molality
β and n are constants

Values of constants for some solutes are given in Table 3.14. The above equation in combination with the Ross (1975) equation was used to calculate water activity in ternary and quaternary electrolyte and nonelectrolyte solutions (Chen, 1990). In electrolyte solutions such as NaCl + KCl, $NaCl + KNO_3$, NaCl + KCl + LiCl, water activity was predicted with an accuracy higher than ± 0.01 a_w. For sucrose+glucose and sucrose + glycerol, water activity can be predicted with an accuracy of at least ± 0.005 a_w. The mixture of sucrose and NaCl yielded water activities, which differed from the predicted values by some 0.002 a_w. Water activities of some multicomponent mixtures are presented in Table 3.15.

Prediction of water activity in a mixed solution of interacting components is of particular interest, since equilibrium measurement would be difficult if not impossible. The development is based on the experimental fact of linear or near linear relationship between the molalities of most

Table 3.14 Parameters of Equation 3.49

Solution	β	B	n	Molality Range
Glucose	1	0.0424	0.926	<7.5
Glycerol	1	0.0250	0.855	<14
Sucrose	1	0.1136	0.955	<6
NaCl	1.868	0.0582	1.618	<6

Source: From Chen, C.S., *J. Food Sci.*, 54, 1318, 1989.

isopiestic solutions. Based on the Zdanorskii rule and the Stokes and Robinson (1966) relation the following equation was proposed (Chen et al., 1973):

$$\frac{m_{o1}}{m_{o2}} = \frac{((a_w/1 - a_w) + h_2)}{((a_w/1 - a_w) + h_1)} \tag{3.50}$$

where

m_{o1} is the molality of binary solution of solute 1
m_{o2} is the molality of binary solution of solute 2
h_1 is the average hydration number of solute 1
h_2 is the average hydration number of solute 2 at a given value of a_w in a ternary mixture

Predicted water activities for 51 tested mixtures deviated from the measured values by less than 0.5%.

Table 3.15 Water Activity of Multicomponent Systems at 25°C

System	Concentration (%)		a_w	Reference
Sucrose:NaCl:water	20:20:60		0.744	Chuang and Toledo (1976)
	25:15:60		0.794	Chuang and Toledo (1976)
	30:10:60		0.846	Chuang and Toledo (1976)
	15:15:70		0.827	Chuang and Toledo (1976)
	20:10:70		0.879	Chuang and Toledo (1976)
	10:10:80		0.890	Chuang and Toledo (1976)
	15:5:80		0.963	Chuang and Toledo (1976)
	6:4:90		0.968	Chuang and Toledo (1976)
	7:3:90		0.973	Chuang and Toledo (1976)
	8:2:90		0.976	Chuang and Toledo (1976)
	Molalities			
Mixture	m_1	m_2	a_w	
Sucrose (1) + NaCl (2)	0.2209	0.2943	0.986	Chen (1990)
	0.5082	1.2240	0.950	Chen (1990)
	0.6830	1.4285	0.939	Chen (1990)
	3.419	0.2510	0.914	Chen (1990)
	4.32	0.2666	0.890	Chen (1990)
	4.357	0.2287	0.890	Chen (1990)
	4.467	0.6689	0.874	Chen (1990)
	5.5290	0.2424	0.857	Chen (1990)
Sucrose (1) + glucose (2)	2.8039	0.5542	0.926	Chen (1990)
	0.3648	3.4947	0.925	Chen (1990)
Sucrose (1) + glycerol (2)	2.0405	1.3615	0.930	Chen (1990)
	0.3278	3.4537	0.930	Chen (1990)

Change in activity of solution due to changes in its composition in a rigorous form is given by the Gibbs–Duhem equation:

$$n_w \mathrm{d}\ln a_w + n_1 \mathrm{d}\ln a_1 + n_2 \mathrm{d}\ln a_{w2}, \ldots, n_n \mathrm{d}\ln a_n = 0 \tag{3.51}$$

where
n is the mole fraction of a component
a is the activity
subscripts w, 1, 2, ..., n denote water, component 1, component 2, ..., component n

By integrating the Gibbs–Duhem equation for one solute and substituting it into integral form of Equation 3.51, Ross (1975) developed an equation describing water activity of a mixture of different solutes

$$a_w = a_{w1} a_{w2} a_{w3}, \ldots, a_{wn} \tag{3.52}$$

where
a_w is the water activity
subscripts 1, 2, 3, ..., n denote solutes at the same concentration as in the complex solution

It was shown that the errors in water activities calculated by Equation 3.52 are relatively small and do not exceed 1% in the concentration ranges usual for food products. In derivation of the above equation it was assumed that the interactions solute–solute cancel on the average in the mixture. This was criticized by Caurie (1985) and the Ross equation was corrected. For a three-component mixture the corrected equation is

$$a_w = a_{w1} a_{w2} a_{w3} - \left[\frac{n(m_1 m_2 + m_1 m_3 + m_2 m_3)}{(55.5)^2} + \frac{(n+1) m_1 m_2 m_3}{(55.5)^3}\right] \tag{3.53}$$

where
m is the molal concentration of a component
n is the number of components

The correction shows that the Ross equation overestimates water activity of a complex solution. Corrections made by Caurie (1985) to the Ross equation were criticized by Kitic and Chirife (1988). In conclusion, it may be stated that the development made by Caurie had so many deficiencies that the equation was useless in predicting a_w of simple and multicomponent mixtures. Chen (1990) expressed a similar opinion.

Water activity of strong electrolytes can be calculated from the osmotic coefficient defined as follows:

$$\phi = -\frac{55.51 \ln a_w}{v_i m_i} \tag{3.54}$$

where
v_i is the number of ions into which each solute dissociates
m_i is the molality

$$a_w = \exp(-\phi \cdot 0.018 m_i v_i) \tag{3.55}$$

The osmotic coefficient is given by Pitzer (1973) as

$$\phi - 1 = |z_M z_X| \times f + m\left(\frac{2\upsilon_M \upsilon_X}{\upsilon}\right) B_{MX} + m^2 2\frac{\sqrt{(\upsilon_M \upsilon_X)^3}}{\upsilon} C_{MX} \tag{3.56}$$

where
$f = -A_\phi|\sqrt{I}/1 + b\sqrt{I}|$
$\upsilon = \upsilon_M + \upsilon_X$
$B_{MX} = \beta^\circ_{MX} + \beta^I_{MX}\exp(-\alpha\sqrt{I})$
$I = 1/2\Sigma m_i z_i^2$ and υ_M and υ_X are the number of M and X ions
z_M and z_X are respective charges in electronic units
A_ϕ is the Debye–Hückel coefficient for the osmotic function equal to 0.392 at 25°C
I is the ionic strength
b is the constant equal to 1.2 for all solutions
α is the constant equal to 2 for all solutions β°_{MX}, β^I_{MX}, and C_{MX} constants

Testing 30 different electrolytes as aqueous solutions, Benmergui et al. (1979) showed that Equations 3.55 and 3.56 can be successfully used to predict water activity of those solutions. Using the above presented approach, Kitic et al. (1986) predicted water activities of several saturated salt solutions as a function of temperature and found that the difference between predicted and measured values was of the order of 0.002 a_w. Only in one case, the difference was 0.009.

3.4 EMPIRICAL EQUATIONS

Grover (1947) used actual data for relative vapor pressures of sugar solutions to develop an empirical equation of the form:

$$a_w = 1.04 - 0.10E_s + 0.0045E_s^2 \tag{3.57}$$

where $E_s = s + 1.3i + 0.8g$ and s, i, and g are the concentrations of sucrose, invert sugar, and glucose (g/g water). Relative pressure is predicted by the above equation with an accuracy of ±0.5%. Experiments showed that the relative vapor pressure of investigated sugar solutions was not dependent on temperature. It was shown that soluble nonsugar substances added to sugar solution lower relative vapor pressure of the mixture. Equivalent conversion factors are listed in Table 3.16. Finally, the equivalent sucrose concentration is calculated from the equation:

$$E_s = \sum cf \tag{3.58}$$

where
c is the concentration (g/g water)
f is the conversion factor

If the solution contains crystalline sugar then the sucrose concentration in the solution saturated with sucrose at 20°C can be calculated from the equation:

$$s = 1.994 - 0.339(g + i) + 0.038(g + i)^2 \tag{3.59}$$

Table 3.16 Conversion Factors for the Grover Equation 3.58

Substance	*f*	Reference
Sucrose, lactose	1.0	Grover (1947)
Invert sugar	1.3	Grover (1947)
Gelatin, casein	1.3	Grover (1947)
Confectioners' glucose solids	0.8	Grover (1947)
Starch	0.8	Grover (1947)
Gums, pectin, etc.	0.8	Grover (1947)
Tartaric acid, citric acid, and their salts	2.5	Grover (1947)
Glycerol	4.0	Grover (1947)
	2.49	Munsch et al. (1987)
Sodium chloride	9.0	Grover (1947)
	9.07	Munsch et al. (1987)
Fat	0	Grover (1947)
Sorbitol	1.56	Munsch et al. (1987)
Hexaglycerol	1.70	Munsch et al. (1987)
Glycine	3.60	Munsch et al. (1987)
Corn syrup	0.91	Munsch et al. (1987)
Supro 620 (soya protein with 1.1% salts)	1.11	Munsch et al. (1987)

An empirical equation to calculate water vapor pressure over sucrose solutions from 45% to 85% concentration between the temperatures of 60°C and 95°C was derived by Dunning et al. (1951)

$$\log a_w = 0.4343m_s + \left(0.4721 + \frac{713}{T}\right)m_s^2 - 1.32m_s^3 \quad (3.60)$$

where
T is the temperature (K)
m_s is the mole fraction of sucrose

There is, however, no information on the precision of the above equation.

Norrish (1966) derived an equation to predict water activity of nonelectrolyte solutions on thermodynamic grounds. The equation is as follows:

$$a_w = m_w \exp(km_s^2) \quad (3.61)$$

where
m_w and m_s are the mole fractions of water and solute, respectively
k is the empirical constant

For electrolytes, the deviation from linearity is observed, and it can be eliminated by introduction of an intercept.

$$\log \frac{a_w}{m_w} = km_s^2 + b \quad (3.62)$$

In the case when molecular weight of the solute is not known, Equation 3.61 was modified by Chuang and Toledo (1976) to the following:

$$\log \frac{a_w}{n_w} = k\left(\frac{s_s}{M_s}\right)^2 + b$$
$$\log \frac{a_w}{n_w} = Ks_s^2 + b \qquad (3.63)$$

where
n_w is the moles of water in 100 g of solution
s_s is the grams of solute in 100 g solution
M_s is the molecular weight of solute

Another way to use the Norrish equation, when molecular weight of the solute is not known, was proposed by Rahman and Perera (1997) deriving the following equation:

$$a_w = \frac{X_w}{X_w + EX_s}\left[\exp\left[k\left(1 - \frac{X_w}{X_w + EX_s}\right)^2\right]\right] \qquad (3.64)$$

where
$E = M_w/M_s$ is the ratio of molecular weight of water to molecular weight of solute
X_w and X_s are the masses of water and solute, respectively

Parameters in Equations 3.60, 3.61, and 3.63 are collected in Table 3.17. Using values from Table 3.17 one should be careful because some authors use natural logarithms and others use decimal logarithms. Then the values of k differ by 2.302585.

Table 3.17 Coefficients for the Norrish Equation

	Equation 3.61	
Solute	***k***	**Reference**
β-Alanine	-2.52 ± 0.37	Chirife et al. (1980)
α-Amino-*n*-butyric acid	-2.59 ± 0.14	Chirife et al. (1980)
1,3-Butylene glycol	−0.20	Norrish (1966)
Citric acid	−6.17	Labuza (1984)
Corn syrup (42 DE)	−2.31	Norrish (1966)
Dextrose	−0.70	Norrish (1966)
Fructose	−0.70	Norrish (1966)
Fructose	−2.82	Chirife et al. (1982)
Galactose	−2.24	Leiras et al. (1990)
Glucose	−2.25	Labuza (1984)
Glucose	−2.92	Chirife et al. (1982)
Glycerol	−0.38	Norrish (1966)
Glycerol	−1.16	Labuza (1984)
Glycerol (0.9%–56.3%)	−0.505	Chuang and Toledo (1976)

Table 3.17 (continued) Coefficients for the Norrish Equation

Equation 3.61

Solute	k	Reference
Glycine	+0.87	Labuza (1984)
Glycine	+2.02 ± 0.33	Chirife et al. (1980)
Lactic acid	−1.59	Labuza (1984)
Lactose	−10.20	Labuza (1984)
Lactulose	−8.00	Labuza (1984)
Lysine	−9.3 ± 0.3	Chirife et al. (1980)
Malic acid	−1.82	Labuza (1984)
Maltose	−4.54	Labuza (1984)
Mannitol	−0.91	Labuza (1984)
NaCl	−7.60	Norrish (1966)
NaCl	−17.48	Labuza (1984)
L-Ornithine	−6.4 ± 0.4	Chirife et al. (1980)
L-Proline	−3.9 ± 0.1	Chirife et al. (1980)
Propylene glycol	−0.20	Norrish (1966)
Sorbitol	−0.85	Norrish (1966)
Sorbitol	−1.65	Labuza (1984)
Sucrose	−2.60	Norrish (1966)
Sucrose	−6.47	Labuza (1984)
Sucrose (3.3%–68.9%)	−2.735	Chuang and Toledo (1976)
Tartaric acid	−4.68	Labuza (1984)
Xylose	−1.54	Labuza (1984)

Equation 3.62

Solute	k	B	Reference
NaCl (0.5%–25.9%)	−7.578	−0.0045	Chuang and Toledo (1976)

Equation 3.64

Product	a_w Range	Temperature (°C)	E	K	RSE	Reference
Ethylene glycol	0.113–0.954	15	0.290	−0.3472	0.011	Rahman and Perera (1997)
	0.111–0.927	25	0.290	−0.2642	0.006	Rahman and Perera (1997)
	0.113–0.908	35	0.290	−0.0016	0.006	Rahman and Perera (1997)
Glycerol	0.113–0.954	15	0.195	−0.6549	0.012	Rahman and Perera (1997)
	0.111–0.927	25	0.195	−0.6534	0.009	Rahman and Perera (1997)
	0.113–0.908	35	0.195	−0.4242	0.012	Rahman and Perera (1997)
Propylene glycol	0.113–0.954	15	0.237	+0.0531	0.014	Rahman and Perera (1997)
	0.111–0.927	25	0.237	+0.0639	0.014	Rahman and Perera (1997)
	0.113–0.908	35	0.237	+0.5152	0.027	Rahman and Perera (1997)

Empirical polynomial equations for predicting water activity of some food humectants were derived by Munsch et al. (1987), and they are collected in Table 3.18. Some attempts made to calculate water activity of the solution based on the freezing point depression. Chen (1987)

Table 3.18 Regression Equations for Food Humectants at 24°C ± 0.1°C

Humectant	Equation
Sucrose	$a_w = 1.0001 - 0.077625x + 0.002187x^2$
Corn syrup	$a_w = 1.0092 - 0.080314x + 0.002582x^2$
Sorbitol	$a_w = 1.0010 - 0.12325x + 0.00538x^2$
Glycerol	$a_w = 0.9862 - 0.17637x + 0.001109x^2$
NaCl	$a_w = 0.997 - 0.520408x - 0.400727x^2$

Source: Adapted from Munsch, M.H., Cormier, A., and Chiasson, S., *Labensm. Wiss.-u Technol.*, 20, 319, 1987.
Note: x is g solute/g water.

derived the following equation based on Raoult's law and effective molecular weight of dissolved solids:

$$a_w = \frac{1}{1 + 0.0097\Delta t + C\Delta t^2} \quad (3.65)$$

where
Δt is the freezing point depression (K)
$C = 5 \times 10^{-5}\ K^{-2}$ for ice–water system

Water activity calculated from the above equation is that measured at freezing point temperature. The equation provides a simple and accurate way to calculate water activity of a solution within the temperature range from 0°C to −40°C. The difference between measured and calculated water activity was less than 0.01 a_w. The accuracy of freezing point measurement is important and errors in temperature ±0.2°C result in water activity variation ±0.022 a_w.

Lerici et al. (1983) used the Clapeyron–Clausius equation for solid–vapor and solid–liquid systems and combined it with the Robinson and Stokes equation for the relationship between freezing point and water activity, thus deriving the following equation:

$$-\ln a_w = 27.622 - 528.373(1/T) - 4.579 \ln T \quad (3.66)$$

where T is the freezing temperature (K). Measurements done with electric hygrometer showed that differences between measured and calculated water activities were less than 0.01 a_w. It was stressed that freezing point depression should be properly measured and continuous agitation of the solution during cooling was essential for the accurate temperature measurement. Ferro-Fontan and Chirife (1981) proposed the following equation to calculate water activity of solution knowing its freezing point depression:

$$-\ln a_w = 9.6934 \times 10^{-3}\Delta t + 4.761 \times 10^{-6}\Delta t^2 \quad (3.67)$$

It was shown that the difference between measured and calculated water activities did not exceed 0.01 a_w.

Water activity of the solution is related to its osmotic pressure. Hence, some attempts were made to calculate osmotic coefficient from osmotic pressure and then to estimate water activity. The following relationships are used:

$$\phi = \frac{\pi m_w V_w}{RT\nu m_s}$$
$$-\ln a_w = \frac{\phi \nu m_s}{m_w} \tag{3.68}$$

where

π is the osmotic pressure (Pa)

V_w is partial molal volume of water (m^3/mol)

m is molar fraction

subscripts s and w denote solute and water, respectively

Some data for sucrose solutions are presented in Table 3.19.

Table 3.19 Osmotic Pressure of Water Solutions of Some Solutes

Concentration (mol/L)			Osmotic Pressure at 20°C (kPa)					Reference
0.098			262					Moore (1962)
0.192			513					Moore (1962)
0.282			771					Moore (1962)
0.370			1030					Moore (1962)
0.453			1290					Moore (1962)
0.533			1560					Moore (1962)
0.610			1840					Moore (1962)
0.685			2120					Moore (1962)
0.757			2400					Moore (1962)
0.825			2700					Moore (1962)
	Osmotic Pressure at 25°C (kPa)							
Molality	**NaCl**	**Glucose**	**Fructose**	**Sucrose**	**Maltose**	**Lactose**	**α-Casein**	
2×10^{-4}							0.5	Matsuura and Sourirajan (1986)
4×10^{-4}							1.0	Matsuura and Sourirajan (1986)
6×10^{-4}							1.6	Matsuura and Sourirajan (1986)
6×10^{-4}							2.2	Matsuura and Sourirajan (1986)
1×10^{-3}							2.9	Matsuura and Sourirajan (1986)
1.5×10^{-3}							4.0	Matsuura and Sourirajan (1986)
2×10^{-3}							6.0	Matsuura and Sourirajan (1986)
0.1	462	259	253	248	214	214		Matsuura and Sourirajan (1986)
0.2	917	517	496	503	455	455		Matsuura and Sourirajan (1986)
0.3	1,372	776	790	758	724	724		Matsuura and Sourirajan (1986)
0.4	1,820	1,034	1,013	1,020	933			Matsuura and Sourirajan (1986)

(continued)

Table 3.19 (continued) Osmotic Pressure of Water Solutions of Some Solutes

	Osmotic Pressure at 25°C (kPa)							
Molality	**NaCl**	**Glucose**	**Fructose**	**Sucrose**	**Maltose**	**Lactose**	**α-Casein**	
0.5	2,282	1,293	1,307	1,282				Matsuura and Sourirajan (1986)
0.6	2,744	1,517	1,611	1,551				Matsuura and Sourirajan (1986)
0.7	3,213	1,744	1,824	1,827				Matsuura and Sourirajan (1986)
0.8	3,682		2,067	2,103				Matsuura and Sourirajan (1986)
0.9	4,158		2,310	2,379				Matsuura and Sourirajan (1986)
1.0	4,640		2,564	2,668				Matsuura and Sourirajan (1986)
1.2	5,612		3,101	3,241				Matsuura and Sourirajan (1986)
1.4	6,612		3,587	3,840				Matsuura and Sourirajan (1986)
1.6	7,646			4,447				Matsuura and Sourirajan (1986)
1.8	8,701			5,061				Matsuura and Sourirajan (1986)
2.0	9,784			5,695				Matsuura and Sourirajan (1986)
3.0	15,651			9,128				Matsuura and Sourirajan (1986)
4.0	22,326			12,866				Matsuura and Sourirajan (1986)
5.0	29,875							Matsuura and Sourirajan (1986)
6.0	38,335							Matsuura and Sourirajan (1986)

The above-discussed models to predict water activity of solutions are related to real solutions of small molecules. In some experiments, high molecular weight substances were also investigated and their influence on water activity of the solution was noticed. However, the only way to account for this influence was to calculate equivalent molecular weight as proposed by Chen and Karman (1980), to account for bond water (Schwartzberg, 1976; Chen, 1987) or to use Ross's equation.

Polymers in the solution can exist in two states. The first state is such that the chains of the polymer do not interpenetrate each other. The solution is diluted. In the second state, the polymer is cross-linked and forms a gel network. Osmotic pressure of an ideal solution is described by van't Hoff's law, which states that:

$$\pi = mRT \tag{3.69}$$

where

m is the molality of the solution
R is the gas constant (J/(mol K))
T is the temperature (K)

For a real solution the scaling approach was used according to which osmotic pressure of uncharged polymer solution is expressed by the following equation:

$$\pi = Ac^n \tag{3.70}$$

where
- A is the constant
- c is the concentration of polymer (g/cm^3)
- n is the scaling exponent

Taking the scaling approach as a basis water activity of a diluted uncharged polymer, the solution is (Mizrahi et al., 1997)

$$-\ln a_{\mathrm{w}} = \frac{AV_{\mathrm{w}}c^n}{RT} \tag{3.71}$$

where V_{w} is molar volume of water. For dextran solutions n was found to be 2.1. In a gel, a network formed by cross-linked polymer chains exerts mechanical pressure on liquid embedded in the matrix. The observed osmotic pressure is smaller than that of a polymer solution, hence (Horkay and Zrinyi, 1982)

$$-\ln a_{\mathrm{w}} = \frac{V_{\mathrm{w}}}{RT}(\pi_{\mathrm{os}} - \pi_{\mathrm{net}}) \tag{3.72}$$

Mizrahi et al. (1997) showed that charged and uncharged gels also follow a power law behavior and proposed to substitute $\pi_{\mathrm{os}} - \pi_{\mathrm{net}} = Bc^n$ into Equation 3.72. Hence, a new equation is obtained

$$-\ln a_{\mathrm{w}} = \frac{BV_{\mathrm{w}}c^n}{RT} \tag{3.73}$$

It was also shown that the greater the degree of cross-linking, the larger the pressure network and the higher the value of the scaling exponent. The scaling approach was extended on hydrogels containing small molecular weight constituents (Tesch et al., 1999). Water activity of such a gel can be expressed by the following equation:

$$-\ln a_{\mathrm{w}} = \frac{V_{\mathrm{w}}}{RT}\left[\left(\pi_{\mathrm{os}} + \pi_{\mathrm{os}}^{s}\right) - \pi_{\mathrm{net}}\right] \tag{3.74}$$

where π_{os}^{s} is osmotic pressure exerted by small molecules. The analysis of the above equation showed that low molecular weight solutes exert relatively high osmotic pressure compared to that of the polymeric matrix. Hence, in systems with relatively high concentration of solutes in comparison to concentration of polymers and low network pressure the water activity is determined by solutes concentration. At high network pressures, water activity of the gel below the syneresis point is determined by the original gel osmotic pressure. Above that point the gel composition changes and water activity is practically that of the solute solution.

Some foods possess native plant structure and cells are intact in full turgor. Under these conditions, the state of water arises from the following phenomena:

- Osmotic pressure due to solutes
- Interactions of water with insoluble matrix
- Turgor pressure

On the analogy to chemical potential, the term "water potential" was introduced to express the state of water in intact plant tissue. Hence (Wilson and Rose, 1967)

Water potential = osmotic potential + matrix potential + pressure potential

$$\Psi = \pi + \tau + P \tag{3.75}$$

The matrix potential originates in forces of capillarity, adsorption, and hydration. All these forces depend on the surface of the tissue matrix. The effect of hydrophilic gels in cytoplasm on the water equilibrium can be significant. When the tissue is entirely killed, the pressure potential $P = 0$ and Equation 3.75 reduces to

$$\Psi = \pi + \tau \tag{3.76}$$

The osmotic and matrix potentials are interrelated and usually the sum of both potentials measured separately is not equal to the measured water potential.

$$\pi + \tau = -\frac{RT}{V_w} \ln a_w \tag{3.77}$$

Gołacki (1994) measured water potential for three apple varieties grown in Poland and showed that it was in the range of 0.72–0.92 MPa in October and increased to 1.05–1.20 MPa during 40 weeks of storage at 2°C.

REFERENCES

Abrams, D.S. and Prausnitz, J.M. Statistical thermodynamics of liquid mixtures: A new expression for the excess Gibbs free energy of partly or completely miscible systems. *American Institute of Chemical Engineering Journal*, 21, 116–128, 1975.

Bakier, S. Characteristics of water state in some chosen types of honey found in Poland. *Acta Agrophysica*, 7(1), 1–9, 2006.

Baranowska, H.M., Klimek-Poliszko, D., and Poliszko, S. Effect of hydration water on the properties of saccharide solutions, in *Properties of Water in Foods*, Lewicki, P.P. Ed., Warsaw Agricultural University Press, Warsaw 2001, pp. 12–20.

Benmergui, E.A., Ferro Fontan, C., and Chirife, J. The prediction of water activity in aqueous solutions in connection with intermediate moisture foods. I. a_w Prediction in single aqueous electrolyte solutions. *Journal of Food Technology*, 14, 625–637, 1979.

Catté, M., Dussap, G.-G., and Gros, J.-B. A physical chemical UNIFAC model for aqueous solutions of sugars. *Fluid Phase Equilibria*, 105, 1–25, 1995.

Caurie, M. A research note. Raoult's law, water activity and moisture availability in solutions. *Journal of Food Science*, 48, 648–649, 1983.

Caurie, M. A corrected Ross equation. *Journal of Food Science*, 50, 1445–1447, 1985.

Chen, C.S. Calculation of water activity and activity coefficient of sugar solutions and some liquid foods. *Lebensmittel Weissenschaft und Technologie*, 20, 64–67, 1987a.

Chen, C.S. Relationship between water activity and freezing point depression of food systems. *Journal of Food Science*, 52, 433–435, 1987b.

Chen, C.S. Water activity-concentration models for solutions of sugars, salts and acids. *Journal of Food Science*, 54, 1318–1321, 1989.

Chen, C.S. Predicting water activity in solutions of mixed solutes. *Journal of Food Science*, 55, 494–497, 515, 1990.

Chen, C.C. A segment-based local composition model for the Gibbs energy of polymer solutions. *Fluid Phase Equilibria*, 83, 301–312, 1993.

Chen, A.C.C. and Karmas, E. Solute activity effect on water activity. *Lebensmittel Wissenschaft und Technologie*, 13, 101–104, 1980.

Chen, H., Sangster, J., Teng, T.T., and Lenzi, F. A general method of predicting the water activity of ternary aqueous solutions from binary data. *The Canadian Journal of Chemical Engineering*, 51, 234–241, 1973.

Chen, C.C., Britt, H.I., Boston, J.F., and Evans, L.B. Local composition model Gibbs energy of electrolyte systems. Part I. Single solvent, single completely dissociated electrolyte system. *American Institute of Chemical Engineering Journal*, 28, 588–596, 1982.

Chen, C.C., King, J., and Wang, D.I.C. A molecular thermodynamic model for helix–helix docing and protein aggregation. *American Institute of Chemical Engineering Journal*, 41, 1015–1024, 1995.

Chirife, J. and Ferro Fontan, C. Water activity of fresh foods. *Journal of Food Science*, 47, 661–663, 1982.

Chirife, J., Ferro Fontan, C., and Scorza, O.C. A study of the water activity lowering behaviour of some amino acids. *Journal of Food Technology*, 15, 383–387, 1980.

Chirife, J., Favetto, G., and Ferro Fontan, C. The water activity of fructose solutions in the intermediate moisture range. *Lebensmittel Wissenschaft and Technologie*, 15, 159–160, 1982.

Chuang, L. and Toledo, R.T. Predicting the water activity of multicomponent systems from water sorption isotherms of individual components. *Journal of Food Science*, 41, 922–927, 1976.

Correa, A., Comesaña, J.F., and Sereno, A. Measurement of water activity in water–urea–"sugar" and water–urea–"polyol" systems, and its prediction by the ASOG group contribution method. *Fluid Phase Equilibria*, 98, 189–199, 1994a.

Correa, A., Comesaña, J.F., and Sereno, A. Use of analytical solution of groups (ASOG) contribution method to predict water activity in solutions of sugars, polyols and urea. *International Journal of Food Science and Technology*, 29. 331–338, 1994b.

Correa, A., Comesaña, J.F., Correa, J.M., and Sereno, A. Measurement and prediction of water activity in electrolyte solutions by a modified ASOG group contribution method. *Fluid Phase Equilibria*, 129, 267–283, 1997.

Curtis, R.A., Blanch, H.W., and Prausnitz, J.M. Calculation of phase diagrams for aqueous protein solutions. *Journal of Physical Chemistry B*, 105, 2445–2452, 2001.

de Cindio, B., Correra, S., and Hoff, V. Low temperature sugar–water equilibrium curve by a rapid calorimetric method. *Journal of Food Engineering*, 24, 405–415, 1995.

Derr, E.L. and Deal, C.H. Analytical solution of groups: Correlation of activity coefficients through structural group parameters. *Institution of Chemical Engineers Symposium Series*, 32, 44–51, 1969.

Dunning, W.J., Evans, H.C., and Taulor, M. The vapour pressures of concentrated aqueous sucrose solutions up to the pressure 760 mm. *Journal of Chemical Society*, 2363–2372, 1951.

Eckert, F. and Klamt, A. Fast solvent screening via quantum chemistry: Cosmo-Rs approach. *American Institute of Chemical Engineering Journal*, 48, 369–385, 2002.

Esteban, M.A., Marcos, A., Fernández-Salguero, J., and Alcalá, M. An improved simple gravimetric method for measurement of high water activities. *International Journal of Food Science and Technology*, 24, 139–146, 1989.

Esteban, M.A., Alcalá, M., Marcos, A., and Fernández-Salguero, J. Comparison of methods for determination of high water activities. Application to dairy products and juices. *Food Chemistry*, 35, 153–158, 1990a.

Esteban, M.A., Alcalá, M., Marcos, A., Fernández-Salguero, J., Garcia de Fernando, G.D., Ordoñez, J.A., and Sanz, B. Water activity of culture media used in food microbiology. *International Journal of Food Science and Technology*, 25, 464–468, 1990b.

Faúndez, C.A. and Valderrama, J.O. Phase equilibrium modeling in binary mixtures found in wine and must distillation. *Journal of Food Engineering*, 65, 577–583, 2004.

Favetto, G., Resnik, S., Chirife, J., and Ferro Fontán, C. Statistical evaluation of water activity measurements obtained with the Vaisala Humicup humidity meter. *Journal of Food Science*, 48, 534–538, 1983.

Ferro-Fontan, C. and Chirife, J. The evaluation of water activity in aqueous solutions from freezing point depression. *Journal of Food Technology*, 16, 21–30, 1981.

Flory, P.J. Thermodynamics of high polymer solutions. *Journal of Chemistry and Physics*, 10, 51–61, 1942.

Fredenslund, A., Jones, R.L., and Prausnitz, J.M. Group-contribution estimation of activity coefficient in nonideal mixtures. *American Institute of Chemical Engineering Journal*, 21, 1086–1099, 1975.

Gołacki, H. Water potential of apple tissue. *Journal of Food Physics*, 2, 38–40, 1994.

Grover, D.W. The keeping properties of confectionary as influenced by its water vapour pressure. *Journal of the Society of Chemical Industry*, 66, 201–205, 1947.

Grover, D.W. and Nicol, J.M. The vapour pressure of glycerin solutions at 20°C. *Journal of the Society of Chemical Industry*, 59, 175–177, 1940.

Hass, J.L. Fugacity of H_2O from 0°C to 350°C at liquid–vapor equilibrium and at 1 atmosphere. *Geochimica et Cosmochimica Acta*, 34, 929–934, 1970.

Horkay, F. and Zrinyi, M. Studies on the mechanical and swelling behaviour of polymer networks based on scaling concept. Extention of the scaling approach to equilibrium in diluent of arbitrary activity. *Macromolecules*, 15, 1306–1310, 1982.

Huggins, M.L. Theory of solutions of high polymers. *Journal of Chemical Society*, 64, 1712–1719, 1942.

Jakobsen, M. Filament hygrometer for water activity measurement: interlaboratory evaluation. *Journal of the Association of Official Analytical Chemists*, 66, 1106–1111, 1983.

Jarczyk, A., Fedejko, E., and Krolewicz, P. Water activity in concentrated fruit juices. *Przemysl Fermentacyjny i Owocowo-Warzywny*, (5), 19–21, 1995 (in Polish).

Kawaguchi, Y., Kanai, H., Kajiwara, H., and Arai, Y. Correlation for activities of water in aqueous electrolyte solutions using ASOG model. *Journal of Chemical Engineering of Japan*, 14(3), 243–246, 1981.

Kitic, D. and Chirife, J. Technical note: Criticism of a method for predicting the water activity of simple and multicomponent mixtures of solubles and non-solutes. *International Journal of Food Science and Technology*, 23, 199–201, 1988.

Kitic, D., Pereira Jardim, D.C., Favetto, G.J., Resnik, S.L., and Chirife, J. Theoretical prediction of the water activity of standard saturated salt solutions at various temperatures. *Journal of Food Science*, 51(4), 1037–1041, 1986.

Kuramochi, H., Noritomi, H., Hoshimo, D., and Nagahama, K. Representation of activity coefficients of fundamental biochemicals in water by the UNIFAC model. *Fluid Phase Equilibria*, 130, 117–132, 1997.

Labuza, T.P. *Moisture Sorption: Practical Aspects of Isotherm Measurement and Use*. American Association of Cereal Chemists, St. Paul, MN, 1984.

Lee, S.B. and Kim, K.J. Effect of water activity on enzyme hydration and enzyme reaction rate in organic solvents. *Journal of Fermentation and Bioengineering*, 79, 473–478, 1995.

Leiras, M.C., Alzamora, S.M., and Chirife, J. Water activity of galactose solutions. *Journal of Food Science*, 55, 1174–1174, 1990.

Lerici, C.R., Piva, M., and Dalla Rosa, M. Water activity and freezing point depression of aqueous solutions and liquid foods. *Journal of Food Science*, 48, 1667–1669, 1983.

Lewicki, P.P. Water as the determinant of food engineering properties. A review. *Journal of Food Engineering*, 61, 483–495, 2004.

Lilley, T.H. and Sutton, R.L. The prediction of water activities in multicomponent systems. *Advances in Experimental Medicine and Biology*, 302, 291–304, 1991.

Lin, S.T. and Sandler, S.I. Infinite dilution activity coefficients from Ab initio solvation calculations. *American Institute of Chemical Engineering Journal*, 45, 2606–2618, 1999.

Lin, S.T. and Sandler, S.I. A priori phase equilibrium prediction from a segment contribution solvation model. *Industrial Engineering Chemistry Research*, 41, 899–913, 2002.

Matsuura, T. and Sourirajan, S. Physicochemical and engineering properties of food in reverse osmosis and ultrafiltration, in *Engineering Properties of Foods*, Rao, M.A. and Rizvi, S.S.H., Eds., Marcel Dekker Inc., New York, 1986, pp. 255–327.

Miyajima, K., Sawada, M., and Nakagaki, M. Studies on aqueous solutions of saccharides. I. Activity coefficients of monosaccharides in aqueous solutions at 25°C. *Bulletin of the Chemical Society of Japan*, 56, 1620–1623, 1983a.

Miyajima, K., Sawada, M., and Nakagaki, M. Studies on aqueous solutions of saccharides. II. Viscosity B-coefficients, apparent molar volumes, and activity coefficients of D-glucose, maltose, and maltotriose in aqueous solutions. *Bulletin of the Chemical Society of Japan*, 56, 1954–1957, 1983b.

Mizrahi, S., Ramon, O., Silberger-Bouhnik, M., Eichler, S., and Cohen, Y. Scaling approach to water sorption isotherms of hydrogels and foods. *International Journal of Food Science and Technology*, 32, 95–105, 1997.

Moore, W.J. *Physical Chemistry*. Prentice-Hall Inc., Englewood Cliffs, NJ, 1962.

Munsch, M.H., Cormier, A., and Chiasson, S. Equations for sorption curves of some food humectants. *Lebensmittel Wissenschaft und Technologie*, 20, 319–324, 1987.

Napierala, D., Popenda, M., Surma, S., and Plenzler, G. The study of water activity in glucose solutions, in *Properties of Water in Foods*, Lewicki, P.P., Ed., Warsaw Agricultural University Press, Warsaw, 1998, pp. 7–13.

Norrish, R.S. An equation for the activity coefficients and equilibrium relative humidities of water in confectionery syrups. *Journal of Food Technology*, 1, 25–39, 1966.

Peres, A.M. and Macedo, E.A. Thermodynamic properties of sugars in aqueous solutions: Correlation and prediction using a modified UNIQUAC model. *Fluid Phase Equilibria*, 123, 71–95, 1996.

Peres, A.M. and Macedo, E.A. A modified UNIFAC model for the calculation of thermodynamic properties of aqueous and non-aqueous solutions containing sugars. *Fluid Phase Equilibria*, 139, 47–74, 1997.

Peres, A.M. and Macedo, E.A. Prediction of thermodynamic properties using a modified UNIFAC model: application to sugar industrial systems. *Fluid Phase Equilibria*, 156–160, 391–399, 1999.

Pitzer, K.S. Thermodynamics of electrolytes. I. Theoretical basis and general equations. *Journal of Physical Chemistry*, 77, 268–277, 1973.

Poling, B.E., Prausnitz, J.M., and O'Connell, J.P. *The Properties of Gases and Liquids*. McGraw-Hill Co. Inc., Boston, 2001.

Prausnitz, J.M. *Molecular Thermodynamics of Fluid-Phase Equilibria*. Prentice-Hall, Englewood Cliffs, NJ, 1969.

Prausnitz, J.M., Lichtenthaler, R.N., and Azevedo, E.G. *Molecular Thermodynamics of Fluid Phase Equilibria*. Prentice-Hall, Englewood Cliffs, NJ, 1999.

Rahman, S.M. and Perera, C.O. Evaluation of the GAB and Norrish models to predict the water sorption isotherms in foods, in *Engineering & Food at ICEF 7*, R. Jowitt, Ed., Sheffield Academic Press, Sheffield, U.K., 1997, pp. A101–A104.

Renon, H. and Prausnitz, J.M. Local compositions in thermodynamic excess functions for liquid mixtures. *American Institute of Chemical Engineering Journal*, 14, 135–144, 1968.

Richardson, T. ERH of confectionary food products. *Manufacturing Confectionary*, 66(12), 85–89, 1986.

Robinson, R.A. and Stokes, R.H. *Electrolyte Solutions*. Butterworths Publ., Ltd., London, 1965.

Ross, K.D. Estimation of water activity in intermediate moisture foods. *Food Technology*, 29(3), 26–34, 1975.

Schwartzberg, H.G. Effective heat capacities for the freezing and thawing of food. *Journal of Food Science*, 41, 151–156, 1976.

Sereno, A.M., Hubinger, M.D., Comesaña, J.F., and Correa, A. Prediction of water activity of osmotic solutions. *Journal of Food Engineering*, 49, 103–114, 2001.

Song, Y.S., Lambert, S.M., and Prausnitz, J.M. A perturbed hand-sphere-chain equation of state for normal fluids and polymers. *Industrial and Engineering Chemistry Research*, 33, 1047–1057, 1994.

Spiliotis, N. and Tassios, D. A UNIFAC model for phase equilibrium calculations in aqueous and nonaqueous sugar solution. *Fluid Phase Equilibria*, 173, 39–55, 2000.

Stokes, R.H. and Robinson, R.A. Interactions in aqueous nonelectrolyte solutions. I. Solut-solvent equilibria. *The Journal of Physical Chemistry*, 70, 2126–2130, 1966.

Stoloff, L. Calibration of water activity measuring instruments and devices: Collaborative study. *Journal of the Association of Official Analytical Chemists*, 61, 1166–1178, 1978.

Tesch, R., Ramon, O., Ladyzhinski, I., Cohen, Y., and Mizrahi, S. Water sorption isotherm of solution containing hydrogels at high water activity. *International Journal of Food Science and Technology*, 34, 235–243, 1999.

Uedaira, H. and Uedaira, H. Activity coefficients of aqueous xylose and maltose solutions. *Bulletin of the Chemical Society of Japan*, 42, 2137–2140, 1969.

Valezmoro, C., Meirelles, A.J., and Vitali, A. Prediction of water activity of solutions containing food additives, in *Engineering & Food at ICEF 7*, R. Jowitt, Ed., Sheffield Academic Press, Sheffield, U. K., 1997, pp. A145–A148.

Viet Bui, A., Minh Nguyen, H., and Muller, J. Prediction of water activity of glucose and calcium chloride solutions. *Journal of Food Engineering*, 57, 243–248, 2003.

Wilson, G.M. Vapour–liquid equilibrium. A new expression for the excess free energy of mixing. *Journal of the American Chemical Society*, 86(2), 127–130, 1964.

Wilson, G.M. and Deal, C.H. Activity coefficients and molecular structure. *Industrial and Engineering Chemistry Fundamentals*, 1, 20–23, 1962.

Wilson, W. and Rose, C.W. The components of leaf water potential. I. Osmotic and matric potentials. *Australian Journal of Biological Science*, 20, 329–347, 1967.

CHAPTER 4

Data and Models of Water Activity. II: Solid Foods

Piotr P. Lewicki

CONTENTS

4.1 INTRODUCTION

Solid foods are multicomponent and multiphase bodies, whose properties and structure are dependent on their biological nature, used raw materials, and applied processing. Because of biochemical activity and mechanical, thermal, and concentration stresses during processing, solid foods are not in thermodynamic equilibrium with their surroundings and undergo relaxation during storage. The thermodynamic state of water in solid foods is characterized by activity of the component (Chapter 3), and that state also changes to reach equilibrium with the surrounding environment. Water activity of selected solid foods is shown in Table 4.1.

In solid bodies, surface molecules and atoms are subjected to unbalanced forces of attraction, which are perpendicular to the surface. As a consequence, the surface of the solid body tends to bind molecules from the surroundings to lower the surface energy and to attain a state of equilibrium. Surface forces can be of two kinds: physical and chemical. For solid foods, forces of physical nature are important and they arise from the following phenomena:

- Dispersion forces caused by rapid resonant fluctuations of electron densities in neighboring atoms. Resonant fluctuations described by Lennard–Jones cause dipole moments, which tend to be in the same phase. Dispersion forces are additive. Dispersion forces between free molecules are different from those between surface and a molecule in its vicinity.
- Repulsion forces, which arise from the interpenetration of the electronic clouds of the atoms.
- Electrostatic forces occur when the solid is ionic in nature and the gas is polar. The H-atom of a dipole, lying at the periphery of the molecule, approaching surface forms comparatively strong hydrogen bond. This takes place in water, alcohols, amines, etc.
- Induced electrostatic forces occur when the solid produces electrostatic fields and the adsorbed molecule has no permanent dipole.

Surface forces cause molecules of surrounding gas or vapor to migrate to the solid and are adsorbed on its surface. The process is called adsorption and the term was introduced by Kayser in 1881 to distinguish condensation of gases on free surfaces from absorption—the process in which gas molecules penetrate into the mass of solid. In 1909, McBain proposed to combine both types of phenomena and termed the process as sorption (Gregg and Sing, 1967). The solid that attracts molecules of the gas is called adsorbent, and the gas or vapor molecules, which are attracted are called adsorbate.

Under normal conditions, the surface of solid is struck by surrounding gas molecules, and for air, the flux density is 3×10^{27} mol/(m^2 s) (Atkins, 1998). This means that each point of molecular size on the surface is struck by the gas or vapor molecules several million times every second, and a portion of molecules undergoes adsorption, while the other returns to the gas phase. The process proceeds until an equilibrium is reached, wherein the number of molecules absorbed is equivalent to

Table 4.1 Water Activity of Solid Foods

Product	Water Content %	Temperature (°C)	Water Activity	References
Fresh foods				
Apples			0.975–0.988	Chirife and Ferro Fontan (1982)
Apricots			0.977–0.987	Chirife and Ferro Fontan (1982)
Artichokes			0.976–0.987	Chirife and Ferro Fontan (1982)
Asparagus			0.992–0.994	Chirife and Ferro Fontan (1982)
Avocado			0.989	Chirife and Ferro Fontan (1982)
Bananas			0.964–0.971	Chirife and Ferro Fontan (1982)
Beans, green			0.990–0.996	Chirife and Ferro Fontan (1982)
Beans, lima			0.994	Chirife and Ferro Fontan (1982)
Beef			0.980–0.990	Chirife and Ferro Fontan (1982)
Beets			0.979–0.988	Chirife and Ferro Fontan (1982)
Bilberries			0.989	Chirife and Ferro Fontan (1982)
Blackberries			0.986–0.989	Chirife and Ferro Fontan (1982)
Blueberries			0.982	Chirife and Ferro Fontan (1982)
Broccoli, sprouting			0.991	Chirife and Ferro Fontan (1982)
Brussels sprouts			0.990	Chirife and Ferro Fontan (1982)
Cabbage			0.990–0.992	Chirife and Ferro Fontan (1982)
Carrots			0.983–0.993	Chirife and Ferro Fontan (1982)
Cauliflower			0.984–0.990	Chirife and Ferro Fontan (1982)
Celeriac			0.990	Chirife and Ferro Fontan (1982)
Celery			0.987–0.994	Chirife and Ferro Fontan (1982)
Celery leaves			0.992–0.997	Chirife and Ferro Fontan (1982)
Cherries			0.959–0.986	Chirife and Ferro Fontan (1982)
Chub mackerel	77.9		0.997–0.998	Doe and Heruwati (1988)
Cod			0.990–0.994	Chirife and Ferro Fontan (1982)
Corn sweet			0.994	Chirife and Ferro Fontan (1982)
Cranberries			0.989	Chirife and Ferro Fontan (1982)
Cucumbers			0.992–0.998	Chirife and Ferro Fontan (1982)
Currants			0.990	Chirife and Ferro Fontan (1982)
Dates			0.974	Chirife and Ferro Fontan (1982)
Dewberries			0.985	Chirife and Ferro Fontan (1982)
Eggplant			0.987–0.993	Chirife and Ferro Fontan (1982)
Endive			0.995	Chirife and Ferro Fontan (1982)
Figs			0.974	Chirife and Ferro Fontan (1982)
Gooseberries			0.989	Chirife and Ferro Fontan (1982)
Grapefruit			0.980–0.985	Chirife and Ferro Fontan (1982)
Grapes			0.974–0.986	Chirife and Ferro Fontan (1982)
Lamb carcasses			0.990	Chirife and Ferro Fontan (1982)
Leeks			0.976–0.991	Chirife and Ferro Fontan (1982)
Lemons			0.982–0.989	Chirife and Ferro Fontan (1982)
Lettuce			0.996	Chirife and Ferro Fontan (1982)
Limes			0.980	Chirife and Ferro Fontan (1982)
Lupine beans			0.945 ± 0.016	Stoloff (1978)
Mangoes			0.986	Chirife and Ferro Fontan (1982)
Meat, chopped			0.978–0.989	Wojciechowski (1978)
Meat, portioned			0.981–0.990	Wojciechowski (1978)

(continued)

Table 4.1 (continued) Water Activity of Solid Foods

Product	Water Content %	Temperature (°C)	Water Activity	References
Meats			0.989–0.992	Chirife and Ferro Fontan (1982)
Melons			0.970–0.991	Chirife and Ferro Fontan (1982)
Mushrooms			0.989–0.995	Chirife and Ferro Fontan (1982)
Nectarines			0.984	Chirife and Ferro Fontan (1982)
Onion, green			0.992–0.996	Chirife and Ferro Fontan (1982)
Onions			0.974–0.990	Chirife and Ferro Fontan (1982)
Oranges			0.979–0.987	Chirife and Ferro Fontan (1982)
Papaya			0.990	Chirife and Ferro Fontan (1982)
Parsnips			0.988	Chirife and Ferro Fontan (1982)
Peaches			0.979–0.989	Chirife and Ferro Fontan (1982)
Pears			0.969–0.989	Chirife and Ferro Fontan (1982)
Peas, green			0.980–0.990	Chirife and Ferro Fontan (1982)
Peppers			0.992–0.997	Chirife and Ferro Fontan (1982)
Persimmons			0.976	Chirife and Ferro Fontan (1982)
Pineapple			0.985–0.988	Chirife and Ferro Fontan (1982)
Plums			0.969–0.982	Chirife and Ferro Fontan (1982)
Pork			0.990	Chirife and Ferro Fontan (1982)
Potatoes			0.988–0.997	Chirife and Ferro Fontan (1982)
Potatoes, sweet			0.985	Chirife and Ferro Fontan (1982)
Pumpkins			0.984–0.992	Chirife and Ferro Fontan (1982)
Quinces			0.972–0.981	Chirife and Ferro Fontan (1982)
Radishes			0.980–0.990	Chirife and Ferro Fontan (1982)
Radishes, small			0.994–0.996	Chirife and Ferro Fontan (1982)
Raspberries			0.984–0.994	Chirife and Ferro Fontan (1982)
Rhubarb			0.989	Chirife and Ferro Fontan (1982)
Rice			0.591 ± 0.018	Stoloff (1978)
Rutabagas			0.988	Chirife and Ferro Fontan (1982)
Salsify			0.987	Chirife and Ferro Fontan (1982)
Sea water fish			0.989	Chirife and Ferro Fontan (1982)
Sour cherries			0.971–0.983	Chirife and Ferro Fontan (1982)
Spinach			0.988–0.996	Chirife and Ferro Fontan (1982)
Squash			0.994–0.998	Chirife and Ferro Fontan (1982)
Strawberries			0.986–0.997	Chirife and Ferro Fontan (1982)
Sweet cherries			0.975	Chirife and Ferro Fontan (1982)
Tangerines			0.987	Chirife and Ferro Fontan (1982)
Tomatoes			0.991–0.998	Chirife and Ferro Fontan (1982)
Turnips			0.988	Chirife and Ferro Fontan (1982)
Walnut			0.690 ± 0.034	Stoloff (1978)
Watermelon			0.992	Chirife and Ferro Fontan (1982)
Processed foods				
Bacon		20 ± 1	0.963 ± 0.021	Jakobsen (1983)
Beef, salted			0.887	Favetto et al. (1983)
Blackcurrant jam		20 ± 1	0.804 ± 0.017	Jakobsen (1983)
Blood sausage			0.960–0.970	Wojciechowski (1978)
Bread, sandwich type			0.97	Labuza et al. (1976)
Butter, salted		25	0.91–0.93	Rüegg (1985)
Butter, unsalted		25	>0.99	Rüegg (1985)

Table 4.1 (continued) Water Activity of Solid Foods

Product	Water Content %	Temperature (°C)	Water Activity	References
Canned beef			0.985	Wojciechowski (1978)
Canned pork meat			0.984	Wojciechowski (1978)
Cheese spread			0.965 ± 0.001	Scott and Bernard (1983)
Cheese spread			0.952 ± 0.0023	Scott and Bernard (1983)
Cheese spread			0.946 ± 0.012	Stoloff (1978)
Cheese, fresh		25	0.98–0.99	Rüegg (1985)
Cheese, hard		25	0.86–0.97	Rüegg (1985)
Cheese, soft		25	0.96–0.98	Rüegg (1985)
Chinese dates, dried			0.72	Hocking (1988)
Chocolate frosting			0.816 ± 0.005	Scott and Bernard (1983)
Chub mackerel salted and dried, 32.5% NaCl	56.2		0.821	Doe and Heruwati (1988)
Chub mackerel salted and dried, 33.4% NaCl	59.0		0.862	Doe and Heruwati (1988)
Chub mackerel salted and dried, 35.6% NaCl	55.4		0.728	Doe and Heruwati (1988)
Corned beef			0.982	Wojciechowski (1978)
Currants, dried			0.66–0.67	Hocking (1988)
Dry soup mix, vegetable			0.21	Labuza et al. (1976)
Edam cheese			0.952 ± 0.001	Esteban et al. (1989)
Figs, dry	15		0.62	
Figs, dry	25		0.75	
Frosting			0.793 ± 0.0044	Scott and Bernard (1983)
Frozen desserts		25	0.97–0.98	Rüegg (1985)
Grape jelly		30	0.802 ± 0.009	Stamp et al. (1984)
Grated Parmesan cheese			0.670 ± 0.001	Esteban et al. (1989)
Guava, banana, quinces, cashew pastes	18		0.70	
Guava, banana, quinces, cashew pastes	30		0.80	
Ham		20 ± 1	0.939 ± 0.019	Jakobsen (1983)
Ham with bone, smoked			0.878–0.926	Wojciechowski (1978)
Ham, canned			0.981	Wojciechowski (1978)
Instant tea		30	0.130 ± 0.007	Stamp et al. (1984)
Jams and marmalades			075–0.78	Hocking (1988)
Liver sausage			0.949–0.964	Wojciechowski (1978)
Luncheon meat			0.977	Wojciechowski (1978)
Manchego cheese			0.913 ± 0.003	Esteban et al. (1989)
Marzipan		20 ± 1	0.809 ± 0.010	Jakobsen (1983)
Meat paste			0.871	Wojciechowski (1978)
Milk candy	13		0.75	
Milk jam			0.842 ± 0.0015	Favetto et al. (1983)
Milk, nonfat dry		30	0.137 ± 0.004	Stamp et al. (1984)
Milk products, dried		25	0.1–0.3	Rüegg (1985)
Parmesan cheese		30	0.693 ± 0.007	Stamp et al. (1984)
Pâté			0.976	Wojciechowski (1978)
Preserves			0.826 ± 0.0040	Scott and Bernard (1983)
Processed American cheese			0.97	Labuza et al. (1976)

(continued)

Table 4.1 (continued) Water Activity of Solid Foods

Product	Water Content %	Temperature (°C)	Water Activity	References
Raisins, dried			0.66	Hocking (1988)
Raspberry jam		20 ± 1	0.939 ± 0.029	Jakobsen (1983)
Raspberry preserve			0.835 ± 0.002	Scott and Bernard (1983)
Salami			0.724–0.944	Wojciechowski (1978)
Salami		20 ± 1	0.840 ± 0.019	Jakobsen (1983)
Salami		20 ± 1	0.744 ± 0.013	Jakobsen (1983)
Salami, ham, pickled meat			0.80–0.95	Hocking (1988)
Salami "farmer"		20 ± 1	0.746 ± 0.015	Jakobsen (1983)
Sausage, cervelat			0.820–0.930	Wojciechowski (1978)
Sausage, mettwurst			0.862–0.967	Wojciechowski (1978)
Sausage, scalded			0.949–0.984	Wojciechowski (1978)
Semolina dough + 5.5% whole egg solids			0.93	Labuza et al. (1976)
Sirloin, smoked			0.896	Wojciechowski (1978)
Skipjack salted and dried, 26.4% NaCl	55.2		0.864	Doe and Heruwati (1988)
Skipjack salted and dried, 31.5% NaCl	54.5		0.810	Doe and Heruwati (1988)
Skipjack salted and dried, 37.1% NaCl	50.5		0.770	Doe and Heruwati (1988)
Skipjack salted and dried, 42.3% NaCl	53.2		0.767	Doe and Heruwati (1988)
Thuringer sausage			0.95	Labuza et al. (1976)
Mechanical mixtures				
Starch:wheat flour = 50:50	Starch 14.9, wheat flour 11.44% water		0.444	Chuang and Toledo (1976)
Starch:wheat flour = 30:70	Starch 14.9, wheat flour 11.44% water		0.428	Chuang and Toledo (1976)
Starch:wheat flour = 70:30	Starch 14.9, wheat flour 11.44% water		0.460	Chuang and Toledo (1976)
Starch:wheat flour = 50:50	Starch 22.45, wheat flour 18.76% water		0.879	Chuang and Toledo (1976)
Starch:wheat flour = 50:50	Starch 28.6, wheat flour 26.5% water		0.965	Chuang and Toledo (1976)
Starch:wheat flour = 20:80	Starch 28.6, wheat flour 26.5% water		0.941	Chuang and Toledo (1976)
Microbiological plating media				
Baird–Parker agar (Oxoid)			0.987	Esteban et al. (1990)
Brain heart infusion agar (Oxoid)			0.996	Esteban et al. (1990)
MacConkey agar (Difco)			0.995	Esteban et al. (1990)
Malt agar (Difco)			0.999	Esteban et al. (1990)
Mannitol salt agar (Oxoid)			0.936	Esteban et al. (1990)
Nutrient agar (Difco)			0.998	Esteban et al. (1990)
Plate count agar (Oxoid)			0.998	Esteban et al. (1990)
Violet red bile agar (Oxoid)			0.995	Esteban et al. (1990)
Violet red glucose agar (Oxoid)			0.993	Esteban et al. (1990)

the number of molecules leaving the surface. The process of detachment of adsorbed molecules from the surface is called desorption. Hence, the state of equilibrium is the balance between adsorption and desorption processes.

Adsorption is a spontaneous process accompanied by the decrease of free energy of the system. Adsorption also decreases the degrees of freedom of adsorbate; hence, the entropy of the system is lowered. It can therefore be inferred that adsorption is an exothermic process. The amount of gas or vapor adsorbed by the solid depends on pressure, temperature, and the nature of both gas and solid. For a given gas and known solid, adsorption observed at a constant temperature yields the relationship between the amount adsorbed and partial pressure of adsorbate. This relationship is called adsorption (sorption) isotherm. In foods, sorption of water molecules from the surrounding environment is most important (Spiess and Wolf, 1987; Rahman and Labuza, 1999). Sorption isotherm, in fact, presents changes of free energy caused by binding of gas molecules.

The majority of isotherms resulting from physical adsorption are grouped into five classes (Brunauer et al., 1940). The types of isotherms are presented in Figure 4.1. The first type presents adsorption limited to the monomolecular layer. Other types of isotherms describe multilayer adsorption. Types II and III show asymptotic approaching the saturation pressure, which means

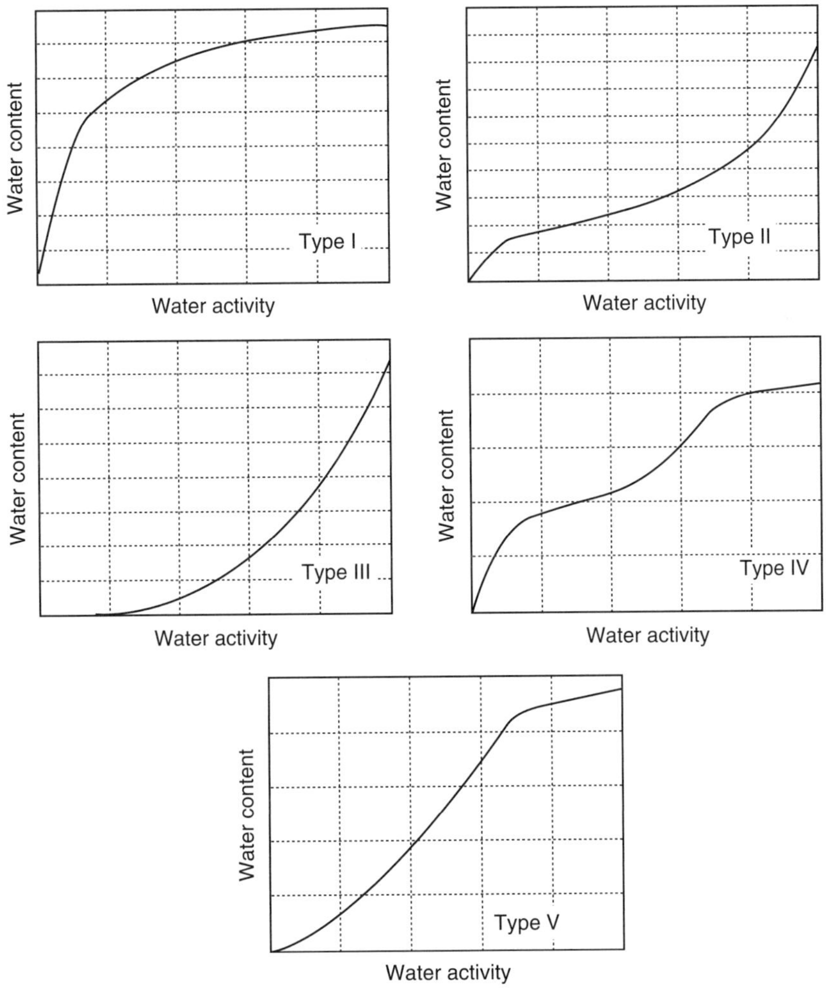

Figure 4.1 Types of sorption isotherm according to Brunauer et al. (1940).

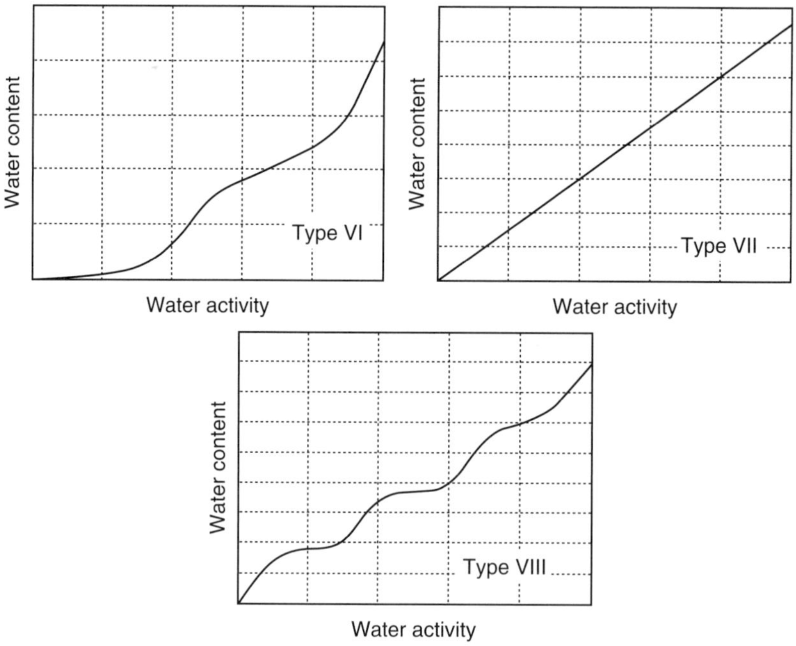

Figure 4.2 Additional types of isotherms proposed by Jovanović (1969).

that equilibrium is attained at infinite dilution. Types IV and V are like types II and III, respectively, but the saturation pressure is reached at finite amount of adsorbed gas or vapor.

It is generally accepted that five types of isotherms exist, but Jovanović (1969) discusses eight types. Type I represents monolayer adsorption, types II and III correspond to the Brunauer et al. (1940) classification. The analogs for capillary condensation in porous adsorbents are represented by types IV and V. Type VI shows two flexion points and a slow adsorption till relatively high pressures. Type VII is a linear isotherm, and type VIII represents stepwise isotherm (Figure 4.2). The irregularities observed as steps are caused by condensation phenomena.

In food, another classification was proposed by Heiss and Eichner (1971). Isotherm of type 1 (Figure 4.3) is typical to very hygroscopic materials. Type 2 presents materials with

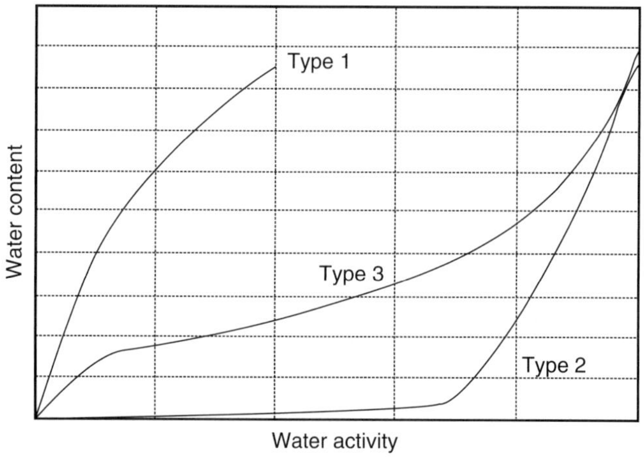

Figure 4.3 Types of sorption isotherms proposed by Heiss and Eichner (1971).

low hygroscopicity, and type 3 is typical for many food products. Equations describing sorption isotherms have been developed on the basis of physical phenomena accompanying adsorption, empirically or in a semitheoretical manner. In theoretically developed equations, two approaches were mainly used. The first is based on the assumption that the surface is homogenous and is adsorption proceeds in a monolayer or multilayer manner. The second takes into account the heterogeneity of the adsorption surface.

4.2 EQUATIONS DESCRIBING PHYSICAL PHENOMENA

Homogenous surface has active centers at which adsorbate molecules are bound, hence the adsorption has a localized nature. Such an assumption was the basis for development of Langmuir's equation, which describes isotherm of type I according to Brunauer et al.'s (1940) classification. Langmuir's equation is

$$u = \frac{u_{\mathrm{m}} k a_{\mathrm{w}}}{1 + k a_{\mathrm{w}}} \tag{4.1}$$

where
- u is the amount adsorbed (kg/kg dry solids)
- u_{m} is the amount of adsorbate filling monolayer (kg/kg dry solids)
- k is the model parameter
- a_{w} is the water activity

Equation 4.1 can be easily linearized, and u_{m} and k can be calculated from the sorption isotherm. Langmuir's equation in linear form is

$$\frac{a_{\mathrm{w}}}{u} = \frac{1}{u_{\mathrm{m}} k} + \frac{1}{u_{\mathrm{m}}} a_{\mathrm{w}} \tag{4.2}$$

In Langmuir's theory of adsorption the process is localized and proceeds in monolayer. Parallel to that theory a multilayer adsorption was developed. Polanyi (1920) developed the potential theory according to which the surface of the solid is surrounded by layers of the same adsorption potential. The adsorption potential decreases with increasing distance from the surface of the solid. The adsorption potential expresses the work of transferring the gas or vapor molecules from a gas phase to the site of adsorption. Adsorption potential is expressed by the following equation:

$$\varepsilon = RT \ln\left(\frac{p_0}{p}\right) \tag{4.3}$$

where
- R is the gas constant (J/(mol K))
- T is the temperature (K)
- p_0 is the saturation water vapor pressure at temperature T (Pa)
- p is the water vapor pressure over the surface of the solid (Pa)

It has been assumed by Polanyi that adsorption potential is not dependent on temperature in a broad range of temperatures. Hence:

$$\varepsilon = RT \ln\left(\frac{p_0}{p}\right) = RT_1 \ln\left(\frac{p_{01}}{p_1}\right) \tag{4.4}$$

where subscript 1 refers to temperature T_1. Drawing adsorption potential in relation to the volume of adsorbate, a characteristic curve for adsorption is obtained, which is not dependent on the temperature of the process according to Equation 4.4. From the potential theory of Polanyi results that the surface of the solid is surrounded by equipotential spaces, which adsorption potential decreases moderately with distance from the surface. Under this situation adsorption proceeds in multilayers, and this phenomena was investigated by many researchers.

4.2.1 BET Model

In 1938, Brunauer, Emmett, and Teller overcame mathematical difficulties and developed equations, which found wide use in food technology. It was assumed that formation of multilayers occurs on a homogeneous surface. Moreover, heat of adsorption occurs only in the first layer and there are no interactions between adsorbed molecules. Adsorption is not limited.

The so-called Brunauer–Emmett–Teller (BET) equation has the following form:

$$u = \frac{u_m c a_w}{(1 - a_w)[1 + (c - 1)a_w]} \tag{4.5}$$

where c is the parameter related to the net heat of sorption. Parameters of the BET equation for numerous foods are shown in Table 4.2. The BET equation can be linearized as follows:

$$\frac{a_w}{u(1 - a_w)} = \frac{1}{u_m c} + \frac{c - 1}{u_m c} a_w \tag{4.6}$$

Isotherm described by the BET equation has an S-like or sigmoid shape with an inflection point, which is assigned to the capacity of monolayer. Water activity at which monolayer is filled can be calculated from the following equation:

$$a_w)_{u=u_m} = \frac{1}{\sqrt{c} + 1} \tag{4.7}$$

Mathematical analysis of the BET equation (Gregg and Sing, 1967) showed that the location of the inflection point depends very much on the value of c. At $c = 9$ and $c \Rightarrow \infty$ $u = u_m$. At values $9 < c < \infty$ the calculated monolayer capacity can differ by as much as 20% from the real value. At c lower than 9 the deviation between the two quantities is greater the lower the value of c. At $c = 2$ the inflection point coincides with the origin of the isotherm. Values of c lower than 2 move the inflection point to imaginary negative values of water activity. Hence, at values $c \leq 2$ the isotherm is no longer of type II, but clearly changes to type III. However, it was also stated that inflection point on the BET isotherm is accurately located when c is somewhat larger than 9 (White, 1947).

The BET equation describes most of the type II isotherms in the water activity range from 0.05 to 0.30. In some cases application to $a_w \approx 0.45$ was observed (Labuza, 1968). Due to strong and highly specific adsorption of water vapor on adsorbents, Gregg and Sing (1967) questioned the validity of the BET equation. According to them, results indicate that the formation of second and higher layers of adsorbed water molecules commences while the first layer is still incomplete. Hence, use of water vapor adsorption for the determination of specific surface of the solid is very doubtful. For adsorption limited in porous body, the following equation was developed (Brunauer et al., 1938):

$$u = u_m \frac{c a_w}{(1 - a_w)} \left[\frac{1 - (n - 1)a_w^n + n a_w^{n+1}}{1 + (c - 1)a_w - c a_w^{n+1}} \right] \tag{4.8}$$

Table 4.2 Parameters of the BET Equation for Adsorption Isotherms

Product	Temperature (°C)	a_w Range	u_m	c	Error	Reference
Agar			9.33	21.4		Masuzawa and Sterling (1968)
Anis	25	0.05–0.40	3.9	20.21		Iglesias and Chirife (1976a)[a]
Apple, freeze-dried	30	0.05–0.40	8.2	3.46		Iglesias and Chirife (1976b)
Avocado	25	0.05–0.40	2.2	7.34		Iglesias and Chirife (1976a)[a]
Bacillus subtilis spores	25	0.05–0.40	10.0	10.50		Rubel (1997)
Barley grain			7.44	4.76		Lewicki (1977)
Beet root	25	0.05–0.40	5.4	11.00		Iglesias and Chirife (1976a)[a]
Beet root, freeze-dried	5	0.05–0.40	8.1	2.89		Iglesias and Chirife (1976b)
Cardamom	25	0.05–0.40	4.8	5.51		Iglesias and Chirife (1976a)[a]
Cayenne red pepper (*Capsicum annum* L. v. Dabok)	15	0–0.43	7.38	78.4	0.414	Kim et al. (1991)
	25	0–0.43	6.91	72.4	0.469	Kim et al. (1991)
	40	0–0.43	6.06	69.9	0.288	Kim et al. (1991)
Celery	25	0.05–0.40	4.3	15.96		Iglesias and Chirife (1976a)[a]
Chamomile	25	0.05–0.40	5.1	6.41		Iglesias and Chirife (1976a)[a]
Chicken, cooked	25	0.05–0.40	4.0	14.18		Iglesias and Chirife (1976a)[a]
Chicken, raw	25	0.05–0.40	4.4	13.48		Iglesias and Chirife (1976a)[a]
Cinnamon	25	0.05–0.40	5.4	13.94		Iglesias and Chirife (1976a)[a]
Cloves	25	0.05–0.40	3.7	27.86		Iglesias and Chirife (1976a)[a]
Carboxymethylcellulose			9.5	21.0		Masuzawa and Sterling (1968)
Coriander	25	0.05–0.40	4.9	14.66		Iglesias and Chirife (1976a)[a]
Corn starch (26% amylase)			7.5	18.6		Masuzawa and Sterling (1968)
Dried malt			4.52	8.10		Lewicki (1977)
Egg plant	25	0.05–0.40	4.6	3.80		Iglesias and Chirife (1976a)[a]
	45	0.05–0.40	5.3	2.20		Iglesias and Chirife (1976b)
	60	0.05–0.40	1.8	3.62		Iglesias and Chirife (1976b)
Emmental cheese	25	0.05–0.40	2.8	6.63		Iglesias and Chirife (1976a)[a]
Fish protein concentrate	25	0.05–0.40	5.0	17.66		Iglesias and Chirife (1976a)[a]
Gelatin			7.4	16.6		Masuzawa and Sterling (1968)
Ginger	25	0.05–0.40	5.9	17.07		Iglesias and Chirife (1976a)[a]
Grapefruit, freeze-dried	25	0.05–0.40	3.4	2.82		Iglesias and Chirife (1976b)
Laurel	25	0.05–0.40	3.4	15.96		Iglesias and Chirife (1976a)[a]
Leek			3.66	7.25		Lewicki (1977)
Lentil	25	0.05–0.40	6.0	12.38		Iglesias and Chirife (1976a)[a]
Mushroom, (*Boletus edulis*)	25	0.05–0.40	3.6	4.65		Iglesias and Chirife (1976a)[a]
Nutmeg	25	0.05–0.40	4.1	29.80		Iglesias and Chirife (1976a)[a]
Oat grain			6.84	7.27		Lewicki (1977)
Onion	10		6.67	3.95		Mazza and LeMaguer (1978)
	30		6.20	3.28		Mazza and LeMaguer (1978)
	45		4.71	2.89		Mazza and LeMaguer (1978)
Onion, flakes			8.78	1.33		Lewicki (1977)
Paranut	25	0.05–0.40	1.7	15.17		Iglesias and Chirife (1976a)[a]
Pear	25	0.05–0.40	17.4	0.75		Iglesias and Chirife (1976b)
Pear, freeze-dried	25	0.05–0.40	9.1	2.84		Iglesias and Chirife (1976b)
Peppermint	25	0.05–0.40	4.9	8.99		Iglesias and Chirife (1976a)[a]

(continued)

Table 4.2 (continued) Parameters of the BET Equation for Adsorption Isotherms

Product	Temperature (°C)	a_w Range	u_m	C	Error	Reference
Pineapple, freeze-dried	5	0.05–0.40	23.0	0.48		Iglesias and Chirife (1976b)
	45	0.05–0.40	8.7	1.32		Iglesias and Chirife (1976b)
	60	0.05–0.40	5.4	2.14		Iglesias and Chirife (1976b)
Prune	23.9	0.05–0.40	12.1	1.39		Iglesias and Chirife (1976b)
Red beet root, ground			4.77	3.97		Lewicki (1977)
Rye grain			6.86	5.92		Lewicki (1977)
Salsify	25	0.05–0.40	4.7	7.21		Iglesias and Chirife (1976a)[a]
Savoy			9.50	0.95		Lewicki (1977)
Sugar beet root	20	0.05–0.35	5.55	12.02	7.1	Iglesias and Chirife (1975b)
Sugar beet root water insoluble components	35	0.05–0.83	6.93	16.21	5.8	Iglesias and Chirife (1975b)
Sweet marjoram	25	0.05–0.40	3.8	17.66		Iglesias and Chirife (1976a)[a]
Tapioca	25	0.05–0.40	6.1	18.57		Iglesias and Chirife (1976a)[a]
Thyme	25	0.05–0.40	4.1	15.69		Iglesias and Chirife (1976a)[a]
Trout, cooked	25	0.05–0.40	3.8	22.75		Iglesias and Chirife (1976a)[a]
Trout, raw	25	0.05–0.40	3.9	22.75		Iglesias and Chirife (1976a)[a]
Wheat grain			6.51	10.17		Lewicki (1977)
Whey protein, denatured, air dried	25		6.37	5.42		Greig (1979)
Whey protein, denatured, drum dried	12		7.18	3.37		Greig (1979)
	25		5.51	11.04		Greig (1979)
	40		4.52	9.40		Greig (1979)
Whey protein, denatured, freeze-dried	12		9.05	3.75		Greig (1979)
	25		8.73	4.48		Greig (1979)
	40		5.58	8.07		Greig (1979)
Whey protein, denatured, spray dried	12		9.25	3.75		Greig (1979)
	25		8.62	4.84		Greig (1979)
	40		7.51	4.30		Greig (1979)
Whey protein, denatured, vacuum dried	25		6.94	7.53		Greig (1979)
Whole egg powder			2.557	5.77		Lai et al. (1986)
Whole egg powder + 1% silica			3.830	19.08		Lai et al. (1986)
Winter savory	25	0.05–0.40	5.1	18.89		Iglesias and Chirife (1976a)[a]
Yogurt	25	0.05–0.40	3.5	51.17		Iglesias and Chirife (1976a)[a]

[a] Recalculated.

where n is the parameter associated with the number of adsorbed layers. Application of Equation 4.8 to calculate water sorption isotherms is presented in Table 4.3. In the BET equation, the assumption is made that $b_2/a_2 = b_3/a_3 = \cdots = b_n/a_n = g$, that is, the condensation (a) and evaporation (b) processes are independent of the number of adsorbed layers and proceed as on the surface of the liquid. Pickett (1945) questioned this assumption and stated that there can be no evaporation from an area covered with n layers when all adjacent areas are also covered with n layers. It was assumed that when $a_w \to 1$ then $n \to u/u_m$ and the following equation was obtained:

$$u = u_m \frac{c a_w (1 - a_w^n)}{(1 - a_w)[1 + (c - 1) a_w]} \tag{4.9}$$

Table 4.3 Parameters of the Modified BET Equation

Product	Temperature (°C)	Parameters			Reference
		u_m	c	n	
Chitosan 10		7.1	31.2	8	Gocho et al. (2000b)
Chitosan 30		6.8	28.4	8	Gocho et al. (2000b)
Chitosan 50		6.7	13.9	8	Gocho et al. (2000b)
Chitosan 90		6.6	13.2	8	Gocho et al. (2000b)
Cellulose acetate CA32[a]	20 ± 0.2	2.5	13.4	9	Gocho et al. (2000a)
Cellulose acetate CA38	20 ± 0.2	2.1	7.5	9	Gocho et al. (2000a)
Cellulose acetate CA44	20 ± 0.2	1.6	3.6	9	Gocho et al. (2000a)

[a] Degree of acetylation (% of acetyl groups): CA32%–32.2%; CA38%–38.4%, and CA44%–44.3%.

For isotherms types IV and V the following equation was developed:

$$u = u_m \frac{c\left[a_w(1 - a_w^n) + (g - 1/2)na_w^n(1 - a_w)\right]}{(1 - a_w)\left[1 - a_w + c(a_w + (g - 1/2)a_w^n)\right]} \tag{4.10}$$

The BET equation was modified in numerous ways. Aguerre et al. (1989a) assumed that heat of sorption for the second and higher adsorbed layers increases or decreases with the layer order number. When the heat of sorption is higher than the heat of liquefaction in layers over the monolayer, then the derived isotherm yields values larger than those obtained from the BET equation. In the opposite case, the isotherm is below that predicted by the BET theory. Equations for the former and latter cases are as follows:

$$u = \frac{u_m c a_w (1 + a_w)}{(1 - a_w)[(1 - a_w)^2 + c a_w]} \tag{4.11}$$

$$u = \frac{u_m c a_w}{(1 - a_w)[1 - c \ln(1 - a_w)]} \tag{4.12}$$

Isotherms with defined inflection point were well described in Equation 4.12, while those containing principally sugars were approximated in Equation 4.11. Depending on the kind of food, the isotherm could be described in the range of water activity from 0.05 to 0.92 (Aguerre et al., 1989b). The parameters of Equations 4.11 and 4.12 are presented in Tables 4.4 and 4.5, respectively. In other experiments, it was confirmed that Equation 4.12 described sorption isotherms up to water activity of 0.75 (Kim et al., 1994).

4.2.2 GAB Model

Anderson (1946) modified the BET equation assuming that the heat of adsorption in the second to ninth layers is lower than the heat of liquefaction. Anderson's equation was later kinetically and statistically derived by de Boer (1953) and Guggenheim (1966), respectively. The equation is as follows and is called the GAB equation:

$$u = \frac{u_m c k a_w}{(1 - k a_w)[1 + (c - 1) k a_w]} \tag{4.13}$$

Table 4.4 Parameters of the Modified BET Equation 4.11

Product	Temperature (°C)	u_m	c	a_w Range	Error (%)	Reference
Ajedrea	25	8.6	16.2	0.05–0.80	3.4	Aguerre et al. (1989b)[a]
Almonds, Moroccan sweet	25	3.7	6.5	0.29–0.94	2.2	Aguerre et al. (1989b)[a]
Beans, field split	27	7.6	10.1	0.06–0.86	4.9	Aguerre et al. (1989b)[a]
Beef, minced	25	9.2	4.4	0.10–0.80	5.4	Aguerre et al. (1989b)[a]
Beef, minced	10	14.3	2.3	0.11–0.87	8.4	Aguerre et al. (1989b)[a]
Cardamom	25	7.9	3.3	0.05–0.80	1.9	Aguerre et al. (1989b)[a]
Cayenne red pepper (*Capsicum annum* L. v. Dabok)	15	10.23	19.5	0–0.75	0.683[b]	Kim et al. (1991)
	25	9.77	18.3	0–0.75	0.734[b]	Kim et al. (1991)
	40	8.86	13.9	0–0.75	0.496[b]	Kim et al. (1991)
Cinnamon	25	7.7	16.2	0.05–0.80	5.3	Aguerre et al. (1989b)[a]
Clove	25	5.9	13.3	0.05–0.80	2.8	Aguerre et al. (1989b)[a]
Cod, unsalted, freeze-dried	25	12.2	5.6	0.07–0.90	5.8	Aguerre et al. (1989b)[a]
Collagen	25	13.4	11.2	0.05–0.90	4.5	Aguerre et al. (1989b)[a]
Coriander	25	6.7	11.1	0.05–0.80	4.2	Aguerre et al. (1989b)[a]
Egg albumin	25	8.1	7.9	0.05–0.90	4.8	Aguerre et al. (1989b)[a]
Fish flour	25	6.8	4.2	0.11–0.85	5.3	Aguerre et al. (1989b)[a]
Gelatin	25	11.7	12.7	0.05–0.90	5.5	Aguerre et al. (1989b)[a]
Ginger	25	8.2	15.3	0.05–0.80	4.6	Aguerre et al. (1989b)[a]
Grams, fried dried	27	6.4	7.4	0.06–0.92	9.4	Aguerre et al. (1989b)[a]
Grams, green split	27	7.9	21.9	0.06–0.86	3.1	Aguerre et al. (1989b)[a]
Grams, kesari split	27	6.7	8.7	0.06–0.86	7.2	Aguerre et al. (1989b)[a]
Grams, red split	27	7.5	25.4	0.06–0.86	3.2	Aguerre et al. (1989b)[a]
Lactoglobulin	25	8.7	6.4	0.05–0.90	4.5	Aguerre et al. (1989b)[a]
Laurel	25	7.5	5.3	0.05–0.80	6.3	Aguerre et al. (1989b)[a]
Lentils split	27	7.4	23.5	0.06–0.86	4.6	Aguerre et al. (1989b)[a]
Marjoram	25	8.3	5.4	0.05–0.80	9.0	Aguerre et al. (1989b)[a]
Mullet roe, unsalted	25	4.8	5.3	0.15–0.84	1.4	Aguerre et al. (1989b)[a]
Mullet, white muscle	25	10.1	6.1	0.14–0.83	3.7	Aguerre et al. (1989b)[a]
Nutmeg	25	5.6	19.3	0.05–0.80	3.4	Aguerre et al. (1989b)[a]
Para nut	25	2.7	9.4	0.05–0.80	4.7	Aguerre et al. (1989b)[a]
Peas, dried	25	8.3	75.1	0.25–0.86	3.3	Aguerre et al. (1989b)[a]
Pecan nut	25	2.9	6.8	0.05–0.80	4.7	Aguerre et al. (1989b)[a]
Potato starch	25	11.5	11.9	0.06–0.75	3.1	Aguerre et al. (1989b)[a]
γ-Pseudoglobulin	25	8.7	13.5	0.05–0.90	4.3	Aguerre et al. (1989b)[a]
Rapeseed, tower	25	5.0	8.4	0.34–0.92	2.3	Aguerre et al. (1989b)[a]
Salmin	25	22.4	1.4	0.05–0.90	8.6	Aguerre et al. (1989b)[a]
Serum albumin	25	8.1	13.2	0.05–0.90	3.7	Aguerre et al. (1989b)[a]
Sesame, whole	27	3.3	28.2	0.05–0.80	7.0	Aguerre et al. (1989b)[a]
Sorghum	38	8.4	19.0	0.04–0.80	6.1	Aguerre et al. (1989b)[a]
Soybean seed	15	8.0	5.6	0.10–0.90	5.4	Aguerre et al. (1989b)[a]
Thyme	25	7.6	7.2	0.05–0.80	6.3	Aguerre et al. (1989b)[a]
Wheat flour	30.1	6.8	57.7	0.13–0.90	9.1	Aguerre et al. (1989b)[a]
Wheat starch	30.1	9.4	9.6	0.10–0.75	1.6	Aguerre et al. (1989b)[a]

[a] MRE.
[b] SE.

Table 4.5 Parameters of the Modified BET Equation 4.1 2

Product	Temperature (°C)	u_m	c	a_w Range	MRE
Apricots	25	3.9	1.7	0.21–0.90	10.4
Banana	25	4.8	1.5	0.05–0.80	18.1
Carrots, v. Chatenay	37	4.5	2.9	0.10–0.70	5.8
Edam cheese	25	2.1	27.6	0.05–0.80	4.7
Emmental cheese	25	1.9	77.7	0.05–0.80	10.3
Figs	25	4.9	1.7	0.11–0.84	5.8
Onion	25	4.1	64.7	0.05–0.80	5.0
Pear	25	5.2	6.1	0.10–0.80	6.2
Pineapple	25	4.7	4.7	0.05–0.80	25.0
Plums	25	4.9	1.9	0.11–0.84	10.8
Radish	25	4.6	9.7	0.10–0.70	8.4
Spinach	37	3.2	64.3	0.10–0.70	1.2
Sultana raisins	25	4.5	48.4	0.11–0.86	8.3
Yogurt	25	2.8	25.2	0.05–0.80	2.2

Source: Adapted from Aguerre, R.J., Suarez, C., and Viollaz, P.E., *Lebensm. Wiss.-u Technol.*, 22, 192, 1989b.

In this equation:

$$k = \exp\left(\frac{d}{RT}\right) \tag{4.14}$$

where

- d is the parameter accounting for energy of new surface formation (J/mol)
- k is the parameter, which corrects the properties of multilayer adsorbed water with respect to the bulk liquid

Parameter $k > 1$ means that sorption will become infinite at a value of a_w less than unity and it is physically unsound. For tested proteins, k falls into a narrow range of 0.82–0.88 with an average value of 0.84 ± 0.03. For starchy materials a narrow range is 0.70–0.77 with an average value of 0.74 ± 0.03. This was obtained in a_w range to 0.90 (Chirife et al. 1992). For large values of c, Equation 4.13 simplifies to the following form:

$$u = \frac{u_m}{1 - ka_w} \tag{4.15}$$

On the other hand, for $k = 1$, Equation 4.13 reduces to the BET equation. The requirements of the GAB model are shown in Table 4.6. It has been shown by Lewicki (1997a) that parameters k and c in the GAB equation must be kept in a certain range and beyond that range the isotherm either moves into unrealistic water activities larger than 1, or the estimated monolayer capacity is loaded with large error. Hence, to have a relatively good description of the sigmoid type of the isotherm and to fulfill the requirements of the BET model the parameters should be kept in the following ranges:

$$0.24 < k \leq 1 \quad \text{and} \quad 5.67 \leq c < \infty$$

Table 4.6 Parameters of the GAB Equation for Adsorption Isotherms

Product	Temperature (°C)	u_m	k	c	Error (%)	Reference
Agar	22	13.30	0.736	47.05		Rahman (1995)
Amaranth starch	25	0.102 ± 0.003	0.81 ± 0.03	16.8 ± 0.5	4.7	Calzetta Resio et al. (1999)[a]
Amylose	20	9.90	0.724	16.16		Rahman (1995)
Anise	25	4.7	0.948	12.91		Rahman (1995)
Apple	10	16.5	0.935	0.790		Rahman (1995)
Apple fiber	23	8.7	0.882	1.707	0.0098	Rahman and Perera (1997)[b]
Apple juice	10	1.8	2.044	0.315		Rahman (1995)
Apple pectin	25	12.48	0.733	9.96	2.6	Weisser (1985)[c]
	40	11.30	0.748	8.13	2.2	Weisser (1985)[c]
	60	9.66	0.777	7.60	3.2	Weisser (1985)[c]
	80	8.24	0.822	7.05	4.0	Weisser (1985)[c]
Apricot	15	3.9	1.119	11.599		Rahman (1995)
	20	10.05	0.9433	2.443		Ayranci et al. (1990)
	30	3.6	1.156	5.844		Rahman (1995)
	36	9.87	0.9849	1.908		Ayranci et al. (1990)
Avicel PH 101	4	4.42 ± 0.09	0.768 ± 0.007	10.1 ± 0.7	1.2	Cadden (1988)[c]
	25	3.65 ± 0.08	0.798 ± 0.007	14.2 ± 1.5	1.3	Cadden (1988)[c]
	37	3.55 ± 0.04	0.751 ± 0.005	18.6 ± 1.2	0.8	Cadden (1988)[c]
Avicel PH 101		3.90	0.7895	9.47	<10.0	Lewicki (1998)[c]
Avicel PH 101	25	4.064	0.772	8.776		Wolf et al. (1984)
	35	0.094 ± 0.002	0.76 ± 0.02	12.9 ± 0.0	4.9	Calzetta Resio et al. (1999)[a]
	50	0.090 ± 0.002	0.80 ± 0.04	9.7 ± 0.4	5.2	Calzetta Resio et al. (1999)[a]
Avicel PH 105	4	5.04 ± 0.41	0.717 ± 0.030	7.1 ± 1.4	3.3	Cadden (1988)[c]
	25	3.21 ± 0.13	0.830 ± 0.012	21.7 ± 6.4	3.2	Cadden (1988)[c]
	37	3.61 ± 0.03	0.751 ± 0.003	14.6 ± 0.5	2.3	Cadden (1988)[c]
Avicel RC 591	23	4.3	0.813	7.339	0.0020	Rahman and Perera (1997)[b]
Barley	25	7.0	0.827	30.14		Rahman (1995)
Beef		3.77	0.9812	14.47	<10.0	Lewicki (1998)[c]
Bread, flat wheat extruded	25	5.321 ± 0.361	0.860 ± 0.046	42.012 ± 4.115		Marzec and Lewicki (2006)

Bread, flat rye extruded	25	4.748 ± 0.047	0.888 ± 0.035	70.952 ± 1.479		Marzec and Lewicki (2006)
Cardamom	25	6.3	0.834	28.67		Rahman (1995)
Carrageenan	22	10.9	0.799	33.971	0.0153	Rahman and Perera (1997)[b]
Carrageenan		11.62	0.9035	42.12	<5.0	Lewicki (1998)[c]
Carrot	30	4.4	1.146	4.377		Rahman (1995)
	45	8.3	1.074	0.978		Rahman (1995)
Casein		8.58	0.6540	7.12	<10.0	Lewicki (1998)[c]
Casein, high micellar, powder	4	7.80	0.75	9.89	0.0953	Foster et al. (2005)[d]
Caseinate, potassium salt	40	7.90	0.780	5.30		Rahman (1995)
Caseinate sodium salt		7.80	0.7541	7.63	<5.0	Lewicki (1998)[c]
Caseinate, sodium salt	25	7.26	0.887	14.27	1.6	Weisser (1985)[c]
Cellulose, microcrystalline	20	5.10	0.806	16.60		Rahman (1995)
Cellulose powder		2.85	0.7294	12.30	<5.0	Lewicki (1998)[c]
Cheese, Gruyere	4	5.6	0.941	11.5		Rahman (1995)
	22	5.1	0.859	126.65		Rahman (1995)
Cinnamon	25	7.8	0.703	17.70		Rahman (1995)
Citrus pulp powder		4.35	0.9422	19.35	<10.0	Lewicki (1998)[c]
	40	6.57	0.891	16.02	1.6	Weisser (1985)[c]
	60	5.70	0.918	12.64	1.3	Weisser (1985)[c]
	80	4.90	0.950	10.10	1.1	Weisser (1985)[c]
Clove	25	4.6	0.853	29.25		Rahman (1995)
Cocoa bean	28	3.3	0.882	903.66		Rahman (1995)
	30	16.1	0.512	11.87		Rahman (1995)
Coffee decaffeinated	20	5.0	1.048	12.69		Rahman (1995)
Coffee extract	20	6.2	1.023	2.85		Rahman (1995)
Coffee freeze-dried	20	3.9	1.182	19.16		Rahman (1995)
Coffee ground	20	3.5	0.963	16.65		Rahman (1995)
Coffee green bean	28	4.0	0.966	4.86		Rahman (1995)
Coffee roasted	25	4.203	0.941	4.186	4.93	Cepeda et al. (1999)[e]
	30	3.923	0.940	3.520	4.77	Cepeda et al. (1999)[e]
	40	3.441	0.937	2.531	2.23	Cepeda et al. (1999)[e]

(*continued*)

Table 4.6 (continued) Parameters of the GAB Equation for Adsorption Isotherms

Product	Temperature (°C)	u_m	k	c	Error (%)	Reference
Coffee roasted and ground	20	3.3	0.940	61.61		Rahman (1995)
	20	3.5	0.963	16.65		Rahman (1995)
	25	3.2	0.993	15.90		Rahman (1995)
	40	3.2	0.987	12.45		Rahman (1995)
	60	3.0	0.993	8.59		Rahman (1995)
	80	2.8	1.029	5.22		Rahman (1995)
Coffee roasted with 15% sugar	25	3.445	0.994	11.70	1.61	Cepeda et al. (1999)[e]
	30	3.288	0.990	9.088	0.74	Cepeda et al. (1999)[e]
	40	3.008	0.982	5.610	2.26	Cepeda et al. (1999)[e]
Coffee short-time roasted and ground (Jacobs Krönung)	20	3.49	0.963	16.65	1.32	Weisser (1985)[c]
Collagen		11.5 ± 0.5	0.80 ± 0.09	17.3 ± 4.4	3.01	Timmermann et al. (2001)[f]
	25	3.22	0.993	15.87	1.10	Weisser (1985)[c]
	40	3.12	0.987	12.45	1.03	Weisser (1985)[c]
	60	3.10	0.993	8.59	3.95	Weisser (1985)[c]
	80	2.74	1.029	5.22	3.09	Weisser (1985)[c]
Cookie	20	3.4	0.982	5.866		Kim et al. (1998)
	30	3.2	0.997	4.204		Kim et al. (1998)
	40	3.0	1.010	2.979		Kim et al. (1998)
Cookie/jam/cookie	20	4.5	0.953	57.978		Kim et al. (1998)
	30	3.8	0.992	27.254		Kim et al. (1998)
	40	3.4	1.022	10.018		Kim et al. (1998)
Coriander	25	6.2	0.753	14.30		Rahman (1995)
Corn bran	5	6.20	0.845	22.046	6.41	Dural and Hines (1993a)[a]
	15	6.20	0.853	15.104	6.44	Dural and Hines (1993a)[a]
	25	6.20	0.832	14.610	4.69	Dural and Hines (1993a)[a]
	37	6.20	0.857	12.333	3.23	Dural and Hines (1993a)[a]
Corn bran	25	7.21 ± 0.75	0.76 ± 0.20	9.8 ± 4.5	3.38	Timmermann et al. (2001)[f]
Corn bran flour		4.62	0.9237	56.18	<5.0	Lewicki (1998)[c]

Corn, continental	25	11.5	0.58	14.3	0.9	Aguerre et al. (1996)[a]
	20	7.85	0.68	8.36	0.2785	Foster et al. (2005)[d]
	37	7.20	0.74	13.17	0.3321	Foster et al. (2005)[d]
	50	4.86	0.86	10.92	0.5310	Foster et al. (2005)[d]
Corn flour, degermed	25	10.27 ± 0.16	0.59 ± 0.03	42.4 ± 5.4	1.07	Timmermann et al. (2001)[f]
Corn starch	25	10.1	0.69	24.3	4.5	Aguerre et al. (1996)[a]
Cotton seed	25	4.6	0.872	61.31		Rahman (1995)
Cracker	20	5.0	0.957	9.012		Kim et al. (1998)
	30	4.4	0.974	5.051		Kim et al. (1998)
	40	4.0	0.987	3.378		Kim et al. (1998)
Cracker/jam/cracker	20	6.0	0.972	67.696		Kim et al. (1998)
	30	5.2	0.976	38.647		Kim et al. (1998)
	40	4.4	0.981	20.085		Kim et al. (1998)
Date pastes cv. Ruziz	5	17.3	0.560	13.980		Alhamdan and Hassan (1999)
	25	15.4	0.840	2.790		Alhamdan and Hassan (1999)
	40	13.7	0.950	4.760		Alhamdan and Hassan (1999)
Egg albumin, coagulated		6.30 ± 0.26	0.78 ± 0.07	11.8 ± 2.3	2.70	Timmermann et al. (2001)[f]
Egg albumin, freeze-dried		6.88 ± 0.47	0.80 ± 0.12	11.6 ± 3.7	3.06	Timmermann et al. (2001)[f]
Elastin		7.61 ± 1.3	0.77 ± 0.32	15.5 ± 13.8	9.2	Timmermann et al. (2001)[f]
Ethylcellulose	9	1.1593	0.9082	5.3597		Velázquez de la Cruz et al. (2001)
	15	1.0867	0.9035	4.6522		Velázquez de la Cruz et al. (2001)
	20	0.9068	0.9109	4.4003		Velázquez de la Cruz et al. (2001)
	25	0.7885	0.9067	3.7338		Velázquez de la Cruz et al. (2001)
	35	0.7505	0.9224	3.0674		Velázquez de la Cruz et al. (2001)
Fig	15	5.9	1.067	22.957		Rahman (1995)
	20	11.00	0.9235	1.425		Ayranci et al. (1990)
	30	5.1	1.129	4.733		Rahman (1995)
	36	10.29	0.9771	1.338		Saravacos et al. (1986)[c]
	45	5.4	1.121	2.690		Rahman (1995)
Fish flour	25	5.80 ± 1.23	0.81 ± 0.33	5.1 ± 3.7	5.58	Timmermann et al. (2001)[f]
Fish rahu (*Labio rohita*), myosin from dorsal portion	10	3.462	0.924	12.962	0.382	Das and Das (2002)[g]
Gelatin		10.3 ± 0.8	0.78 ± 0.14	18.7 ± 7.8	4.31	Timmermann et al. (2001)[f]

(*continued*)

Table 4.6 (continued) Parameters of the GAB Equation for Adsorption Isotherms

Product	Temperature (°C)	u_m	*k*	c	Error (%)	Reference
Gelatin	22	8.6	0.855	84.452	0.0190	Rahman and Perera (1997)[b]
Gelatin		9.23	0.9380	114.06	<5.0	Lewicki (1998)[c]
Ginger	25	7.8	0.738	19.17		Rahman (1995)
Gluten, wheat cv. Spring	25	7.65	0.76	10.10	3.34	De Jong et al. (1996)[c]
	40	6.92	0.71	7.80	2.58	De Jong et al. (1996)[c]
Gluten, wheat cv. Taurus	25	6.93	0.83	13.92	3.71	De Jong et al. (1996)[c]
	40	6.56	0.75	9.60	3.89	De Jong et al. (1996)[c]
	27	4.906	0.897	5.333	0.404	Das and Das (2002)[g]
	45	7.300	0.851	2.417	0.595	Das and Das (2002)[g]
Guar gum, coars	4	9.15 ± 0.10	0.880 ± 0.003	26.9 ± 2.7	1.3	Cadden (1988)[c]
	25	8.98 ± 0.38	0.839 ± 0.013	27.7 ± 10.6	2.5	Cadden (1988)[c]
	37	7.19 ± 0.16	0.882 ± 0.006	25.2 ± 5.5	2.3	Cadden (1988)[c]
Guar gum, fine	4	8.66 ± 0.28	0.887 ± 0.009	23.3 ± 6.5	2.1	Cadden (1988)[c]
	25	8.11 ± 0.36	0.877 ± 0.012	30.8 ± 15.4	2.6	Cadden (1988)[c]
	37	7.05 ± 0.15	0.890 ± 0.005	20.7 ± 3.9	2.2	Cadden (1988)[c]
Hazelnuts v. Negret, kernel	3	2.0	0.914	144.2	0.022	Lopez et al. (1995)[h]
	10	1.8	0.939	37.0	0.054	Lopez et al. (1995)[h]
	30	1.8	0.947	7.4	0.044	Lopez et al. (1995)[h]
Hazelnuts v. Pauetet, kernel	3	2.0	0.892	108.0	0.057	Lopez et al. (1995)[h]
	10	1.7	0.908	29.0	0.006	Lopez et al. (1995)[h]
	30	1.8	0.913	8.1	0.014	Lopez et al. (1995)[h]
Hazelnuts v. Tonda Romana, kernel	3	2.0	0.930	76.5	0.015	Lopez et al. (1995)[h]
	10	2.0	0.938	20.8	0.006	Lopez et al. (1995)[h]
	30	1.9	0.931	6.3	0.010	Lopez et al. (1995)[h]
Horseradish	25	5.8	0.936	4.850		Rahman (1995)
Jam	20	13.5	0.990	75.773		Kim et al. (1998)
	30	8.0	1.009	42.034		Kim et al. (1998)
	40	5.0	1.031	25.980		Kim et al. (1998)
Keratin from wool	24.6	9.20	0.696	8.83		Rahman (1995)
Lactobacillus plantarum	25	8.4	1.01	21.3		Linders et al. (1997)

Lactoglobulin, crystallized		7.72 ± 0.45	0.81 ± 0.10	9.5 ± 2.4	1.93	Timmermann et al. (2001)[f]
Lactoglobulin, freeze-dried		7.35 ± 0.52	0.80 ± 0.13	9.3 ± 2.8	3.39	Timmermann et al. (2001)[f]
Lactose amorphous	12–38	6.27	1.01	2.81	0.5459	Foster et al. (2005)[d]
Laurel	25	5.0	0.891	14.11		Rahman (1995)
Macaroni		4.26	0.9071	10.61	<10.0	Lewicki (1998)[c]
Malt six-row barley v. Plaisant	15	51.030	0.804	24.558		Barreiro et al. (2003)
	25	50.770	0.850	23.205		Barreiro et al. (2003)
	35	46.030	0.904	25.181		Barreiro et al. (2003)
Mango (*Irvingia gabonensis*) osmosed to 40.09% d.m. then oven dried	20	15.79	0.9948	68.77		Falade and Aworh (2004)
	40	12.35	0.9656	19.903		Falade and Aworh (2004)
Manioc, native starch	25	9.3	0.72	26.0	2.2	Aguerre et al. (1996)[a]
Marjoram	25	5.0	0.932	22.62		Rahman (1995)
Methylocarboxycellulose	25	4.06	0.772	8.78		Rahman (1995)
Methylocarboxycellulose-agar gel freeze-dried (MCC:agar = 5:1.5 w/w)	20	7.789	0.706	10.062		Biquet and Labuza (1988)
Methylcellulose	9	3.9284	0.9743	11.0592		Velázquez de la Cruz et al. (2001)
	15	4.0573	0.9564	8.6868		Velázquez de la Cruz et al. (2001)
	20	3.5236	0.9642	11.0973		Velázquez de la Cruz et al. (2001)
	25	3.6727	0.9259	5.0327		Velázquez de la Cruz et al. (2001)
	35	4.8565	0.8773	4.3591		Velázquez de la Cruz et al. (2001)
Milk protein concentrate	4	7.28	0.79	9.85	0.2317	Foster et al. (2005)[d]
	20	7.83	0.75	9.12	0.0900	Foster et al. (2005)[d]
	37	7.68	0.71	11.47	0.3274	Foster et al. (2005)[d]
	50	7.01	0.74	7.48	0.3006	Foster et al. (2005)[d]
Nutmeg	25	5.1	0.766	28.09		Rahman (1995)
Oat bran	4	6.38 ± 0.21	0.793 ± 0.013	23.8 ± 5.2	2.0	Cadden (1988)[c]
	25	6.18 ± 0.16	0.822 ± 0.008	39.4 ± 11.4	2.0	Cadden (1988)[c]
	37	6.52 ± 0.23	0.734 ± 0.014	17.6 ± 2.9	1.5	Cadden (1988)[c]
Oat bran, ground	4	7.31 ± 0.34	0.753 ± 0.019	16.4 ± 3.2	1.9	Cadden (1988)[c]
	25	6.83 ± 0.16	0.771 ± 0.008	25.3 ± 4.1	1.4	Cadden (1988)[c]
	37	7.32 ± 0.31	0.708 ± 0.020	13.3 ± 1.7	1.1	Cadden (1988)[c]

(continued)

Table 4.6 (continued) Parameters of the GAB Equation for Adsorption Isotherms

Product	Temperature (°C)	u_m	*k*	c	Error (%)	Reference
Oat bran flour		3.54	0.9288	43.20	<5.0	Lewicki (1998)[c]
Oat fiber	5	5.66	0.850	19.851	7.99	Dural and Hines (1993a)[a]
	15	5.66	0.843	20.071	6.88	Dural and Hines (1993a)[a]
	25	5.66	0.844	18.463	4.66	Dural and Hines (1993a)[a]
	37	5.66	0.857	13.743	3.45	Dural and Hines (1993a)[a]
Onion	30	4.6	0.948	2.860	0.0049	Rahman and Perera (1997)[b]
	45	9.6	0.923	1.646	0.0040	Rahman and Perera (1997)[b]
	55	11.1	0.866	1.262	0.0036	Rahman and Perera (1997)[b]
	70	7.6	0.950	1.294	0.0039	Rahman and Perera (1997)[b]
	80	4.7	0.948	2.860	0.0049	Rahman and Perera (1997)[b]
Pea	25	3.3	0.744	35.490		Rahman (1995)
Pea flour		5.39	0.8489	53.18	<10.0	Lewicki (1998)[c]
Pecan nut	5	2.4	0.832	10.42		Rahman (1995)
Pectin HM	22	4.9	0.905	150.140	0.0108	Rahman and Perera (1997)[b]
Pectin LM	22	7.7	0.983	9.136	0.0131	Rahman and Perera (1997)[b]
Pectin LM		8.28	0.9541	7.26	<5.0	Lewicki (1998)[c]
Pimento	20	5.8	0.868	38.89		Rahman (1995)
Pistachio nuts cv. Aeginas, kernel	15	6.560	0.938	9.747	1.1	Yanniotis and Zarmboutis (1996)
	25	6.206	0.940	8.457	2.8	Yanniotis and Zarmboutis (1996)
	40	5.584	0.970	9.609	2.1	Yanniotis and Zarmboutis (1996)
Pistachio nuts cv. Aeginas, shell	15	5.795	0.862	9.070	1.0	Yanniotis and Zarmboutis (1996)
	25	5.715	0.835	7.781	1.3	Yanniotis and Zarmboutis (1996)
	40	5.381	0.874	5.689	2.1	Yanniotis and Zarmboutis (1996)
Polyglicine		8.83	0.6268	5.55	<10.0	Lewicki (1998)[c]
Potato	40	5.2	0.83	13.73	5.62	Wang and Brennan (1991)[a]
	50	4.8	0.82	13.41	4.08	Wang and Brennan (1991)[a]
	60	3.6	0.86	21.18	3.41	Wang and Brennan (1991)[a]
	70	2.9	0.90	17.75	5.69	Wang and Brennan (1991)[a]

Potato	30	6.16	0.894	12.9	3.86	McMinn and Magee (2003)[a]
	45	5.26	0.853	10.9	6.79	McMinn and Magee (2003)[a]
	60	3.66	0.750	5.67	7.13	McMinn and Magee (2003)[a]
Potato cv. Pentland Dell	30	8.93	0.78	14.45	3.70	McLaughlin and Magee (1998)[a]
Potato, freeze-dried	20	6.6	0.849	19.10		Rahman (1995)
Potato, vacuum dried	25	13.0	0.700	1.39		Rahman (1995)
Potato flakes		5.45	0.8937	43.78	<10.0	Lewicki (1998)[c]
Potato starch		9.86	0.7467	8.37	<10.0	Lewicki (1998)[c]
Potato starch	30	3.5	0.907	17.6	10.3	Al-Muhtaseb et al. (2004)[a]
	45	2.7	0.905	11.8	10.8	Al-Muhtaseb et al. (2004)[a]
	60	2.1	0.889	8.19	15.0	Al-Muhtaseb et al. (2004)[a]
Potato starch	25	14.6	0.64	12.4		Linders et al. (1997)
	45	7.35	0.81	15.68	4.34	McLaughlin and Magee (1998)[a]
	60	4.62	0.84	19.99	3.13	McLaughlin and Magee (1998)[a]
Potato starch	25	10.312	0.734	7.976		Wolf et al. (1984)
Potato starch	25	8.5	0.80	10.9	3.3	Aguerre et al. (1996)[a]
Potato starch	20	10.10	0.740	17.60		Rahman (1995)
Potato starch, high amylose powder	30	3.1	0.913	11.1	5.54	Al-Muhtaseb et al. (2004)[a]
	45	2.8	0.893	8.52	17.8	Al-Muhtaseb et al. (2004)[a]
	60	2.8	0.91	6.04	7.18	Al-Muhtaseb et al. (2004)[a]
Potato starch, high amylopectin powder	30	3.2	0.888	22.2	6.19	Al-Muhtaseb et al. (2004)[a]
	45	3.2	0.882	15.3	9.24	Al-Muhtaseb et al. (2004)[a]
	60	2.7	0.887	9.86	6.21	Al-Muhtaseb et al. (2004)[a]
Potato starch, native	20	9.79 ± 0.27	0.75 ± 0.05	20.4 ± 3.4	3.11	Timmermann et al. (2001)[f]
Prune	15	5.5	1.079	24.023		Rahman (1995)
	30	5.3	1.121	4.817		Rahman (1995)
	45	20.0	0.961	0.321		Rahman (1995)
Quince jam, 63.6% sugar	25	5.3	1.060	17.700		Rahman (1995)
	35	9.1	1.030	1.050		Rahman (1995)
	45	13.5	1.040	0.493		Rahman (1995)
Quinoa (*Chenopodium quinoa* Willd.) grains	20	8.67	0.70	15.30	0.43	Tolaba et al. (2004)[a]
	30	8.51	0.68	11.91	0.21	Tolaba et al. (2004)[a]
	40	5.90	0.80	9.72	0.29	Tolaba et al. (2004)[a]

(*continued*)

Table 4.6 (continued) Parameters of the GAB Equation for Adsorption Isotherms

Product	Temperature (°C)	u_m	k	c	Error (%)	Reference
Raisin	20	9.98	0.9466	3.076		Ayranci et al. (1990)
	36	9.34	0.9920	6.061		Ayranci et al. (1990)
Raisins, sultana	15	12.5	0.933	1.966		Rahman (1995)
	15	6.1	1.074	16.344		Rahman (1995)
	20	10.2	1.196	0.781	1.768	Saravacos et al. (1986)[c]
	25	7.7	1.091	1.454	6.645	Saravacos et al. (1986)[c]
	30	5.4	1.196	2.442	6.916	Saravacos et al. (1986)[c]
	30	5.6	1.137	3.671		Rahman (1995)
	30	12.5	0.963	1.237		Rahman (1995)
	35	4.2	1.281	3.016	9.520	Saravacos et al. (1986)[c]
	45	6.8	1.098	2.297		Rahman (1995)
	45	12.5	0.991	0.831		Rahman (1995)
	60	12.0	1.017	0.555		Rahman (1995)
Rapeseed	25	3.6	0.876	10.63		Rahman (1995)
Rapeseed	25	3.2	0.923	18.01		Rahman (1995)
Raspberry	10	4.8	1.084	3.008		Rahman (1995)
Raspberry juice	10	4.3	1.850	0.208		Rahman (1995)
Red pepper	15	8.16	0.830	70.4	0.480	Kim et al. (1991)[d]
	25	7.82	0.827	61.3	0.511	Kim et al. (1991)[d]
	40	7.41	0.793	29.3	0.388	Kim et al. (1991)[d]
Rice	4	7.2	0.735	5.49		Rahman (1995)
Rice	25	11.0 ± 0.53	0.58 ± 0.08	19.2 ± 5.3	2.04	Timmermann et al. (2001)[f]
Rice, brown	27	9.7	0.654	16.67		Rahman (1995)
Rice bran flour		3.78	0.9843	42.89	<5.0	Lewicki (1998)[c]
Rice fiber	5	3.70	0.918	7.960	1.91	Dural and Hines (1993a)[a]
	15	3.70	0.922	8.976	4.98	Dural and Hines (1993a)[a]
	25	3.70	0.925	10.308	7.62	Dural and Hines (1993a)[a]
	37	3.70	0.891	8.778	4.00	Dural and Hines (1993a)[a]
Rice, rough	25	7.9	0.75	44.0	4.8	Aguerre et al. (1996)[a]

Savory	25	7.5	0.791	26.22		Rahman (1995)
Skim milk	34	4.27 ± 0.11	0.876 ± 0.05	38.0 ± 10.2	2.98	Timmermann et al. (2001)[f]
Skim milk	25	4.3	0.929	56.42		Rahman (1995)
Skim milk	34	4.0	0.942	69.23		Rahman (1995)
Sorghum	37.8	8.2	0.72	23.4	1.9	Aguerre et al. (1996)[a]
Soya bran flour		3.92	0.9488	123.96	<5.0	Lewicki (1998)[c]
Soy protein	25	6.32	0.873	20.25	3.5	Weisser (1985)[c]
	40	5.94	0.876	16.67	2.1	Weisser (1985)[c]
	60	5.10	0.909	13.36	1.7	Weisser (1985)[c]
	80	4.54	0.942	10.33	1.9	Weisser (1985)[c]
Star apple (*Chrysophyllum albidum*) osmosed to 40.09% d.m. then oven dried	20	15.79	0.9948	68.77		Falade and Aworh (2004)
	40	12.35	0.9565	19.903		Falade and Aworh (2004)
Starch gel		6.33	0.8976	42.27	<5.0	Lewicki (1998)[c]
Starch gel		6.43	0.8484	29.13	<5.0	Lewicki (1998)[c]
Strawberries	25	10.9	0.953	1.530		Rahman (1995)
Strawberries (*Fragaria ananassa* v. Camarosa) whole, freeze-dried	30	5.1	1.16	3.5		Moraga et al. (2004)
Strawberries (*F. ananassa* v. Camarosa) homogenized, freeze-dried	30	3.6	1.28	7.0		Moraga et al. (2004)
Sunflower nutmeat	10	3.8	0.763	34.95		Rahman (1995)
	20	3.5	0.786	36.25		Rahman (1995)
	30	3.3	0.821	37.48		Rahman (1995)
Tapioca starch	25	10.1	0.71	23.5	5.2	Aguerre et al. (1996)[a]
Tea, BOP grade	21	4.4	0.879	7.49		Rahman (1995)
Tea, chamomile	25	6.4	0.931	16.76		Rahman (1995)
Tea, Fennel	25	4.3	0.950	3.11		Rahman (1995)
Tea, Pekoe	28	6.5	0.729	26.35		Rahman (1995)
Thyme	25	5.2	0.891	20.95		Rahman (1995)
Tomato	30	16.6 ± 0.6	0.83 ± 0.06	31.4 ± 9.3	2.38	Timmermann et al. (2001)[f]

(*continued*)

Table 4.6 (continued) Parameters of the GAB Equation for Adsorption Isotherms

Product	Temperature (°C)	u_m	*k*	c	Error (%)	Reference
Tomato insoluble solids, freeze-dried	20	4.5	0.942	26.83		Giovanelli et al. (2002)
Tomato pulp, freeze-dried	20	11.7	1.013	5.86		Giovanelli et al. (2002)
Turkey, cooked	22	6.29 ± 0.26	0.82 ± 0.07	7.41 ± 1.22	3.87	Timmermann et al. (2001)[f]
Wheat	25	8.4	0.743	23.58		Rahman (1995)
Wheat bran	4	6.67 ± 0.19	0.842 ± 0.009	17.8 ± 3.2	2.1	Cadden (1988)[c]
	25	6.28 ± 0.19	0.864 ± 0.009	34.9 ± 12.8	2.0	Cadden (1988)[c]
	37	5.75 ± 0.10	0.855 ± 0.005	20.6 ± 2.9	1.4	Cadden (1988)[c]
Wheat bran	5	5.66	0.895	10.198	3.96	Dural and Hines (1993a)[a]
	15	5.66	0.887	9.425	2.65	Dural and Hines (1993a)[a]
	25	5.66	0.863	10.782	4.54	Dural and Hines (1993a)[a]
	37	5.66	0.849	11.602	2.63	Dural and Hines (1993a)[a]
Wheat bran, ground	4	5.95 ± 0.15	0.893 ± 0.007	14.9 ± 2.4	1.6	Cadden (1988)[c]
	25	5.47 ± 0.18	0.822 ± 0.008	48.5 ± 27.9	1.7	Cadden (1988)[c]
	37	5.43 ± 0.16	0.869 ± 0.008	16.8 ± 3.4	1.6	Cadden (1988)[c]
Wheat flour		8.10	0.7358	17.40	<5.0	Lewicki (1998)[c]
	25	6.44	0.91	22.23		Rahman (1995)
	27	6.7	0.82	31.7	6.6	Aguerre et al. (1996)[a]
	30	6.31	0.90	20.28		Rahman (1995)
	35	6.22	0.87	18.74		Rahman (1995)
	45	6.11	0.84	17.65		Rahman (1995)
	55	6.04	0.81	17.38		36
Wheat gluten	3	6.38 ± 0.54	0.78 ± 0.15	16.4 ± 7.6	4.40	Timmermann et al. (2001)[f]
Wheat semolina	27	7.7	0.76	30.5	3.9	Aguerre et al. (1996)[a]
Wheat starch, native	20	9.89 ± 0.21	0.68 ± 0.04	26.7 ± 3.8	1.94	Timmermann et al. (2001)[f]
Whey protein	25	8.68	0.731	10.25	2.6	Weisser (1985)[c]
	40	6.95	0.802	13.25	3.0	Weisser (1985)[c]
	60	5.54	0.847	12.59	2.6	Weisser (1985)[c]
	80	4.85	0.867	12.08	4.2	Weisser (1985)[c]

Whey protein isolate	4	8.63	0.79	8.96	0.5295	Foster et al. (2005)[d]
	20	6.82	0.94	13.25	0.1190	Foster et al. (2005)[d]
	37	8.33	0.75	9.58	0.5770	Foster et al. (2005)[d]
	50	6.56	0.84	7.63	0.4432	Foster et al. (2005)[d]
Wool		7.33	0.7819	16.13	<10.0	Lewicki (1998)[c]
Yogurt	20	7.59	0.9998	22.5778	3.756	Kim et al. (1994)[c]
	35	6.47	1.0098	14.7562	2.513	Kim et al. (1994)[c]
	50	5.59	1.0189	10.1582	2.010	Kim et al. (1994)[c]
Yogurt concentrated	20	4.14	0.9826	13.6667	4.019	Kim et al. (1994)[c]
	35	3.12	1.0242	9.0933	3.186	Kim et al. (1994)[c]
	50	2.41	1.0644	6.3535	4.034	Kim et al. (1994)[c]
Yogurt concentrated freeze-dried	20	3.81	0.9919	12.2764	2.785	Kim et al. (1994)[c]
	35	2.47	1.0220	10.5030	4.072	Kim et al. (1994)[c]
	50	1.67	1.0506	9.2415	3.471	Kim et al. (1994)[c]
Yogurt concentrated spray dried	20	3.51	1.0011	10.5919	2.667	Kim et al. (1994)[c]
	35	2.35	1.0236	9.3567	3.529	Kim et al. (1994)[c]
	50	1.62	1.0451	8.5129	2.454	Kim et al. (1994)[c]
Yogurt freeze dried	20	7.05	0.9435	23.7095	2.560	Kim et al. (1994)[c]
	35	3.90	0.9644	20.1481	3.006	Kim et al. (1994)[c]
	50	2.27	0.9842	17.7277	4.448	Kim et al. (1994)[c]
Yogurt microwave vacuum dried	20	3.32	1.0109	9.6757	4.530	Kim et al. (1994)[c]
	35	2.21	1.0590	9.7508	3.427	Kim et al. (1994)[c]
	50	1.54	1.1022	9.3503	4.896	Kim et al. (1994)[c]
Yogurt spray dried	20	5.25	0.9933	44.7394	2.572	Kim et al. (1994)[c]
	35	2.75	1.0284	26.2381	2.409	Kim et al. (1994)[c]
	50	1.57	1.0541	11.5234	4.988	Kim et al. (1994)[c]
c-Zeine		4.37 ± 0.21	0.83 ± 0.09	12.8 ± 3.0	4.42	Timmermann et al. (2001)[f]

[a] MRE.
[b] RSE.
[c] RMS.
[d] SE.
[e] Mean point deviation.
[f] Normalized error.
[g] Residual variance
[h] Mean square error 10^4.

Keeping k and c within these ranges assures that calculated monolayer values differ by not more than $\pm 15.5\%$ from the true monolayer capacity. Anderson (1946) also assumed that after r layers are adsorbed the value of d is zero and the isotherm is described by the following equation:

$$u = u_{\mathrm{m}} \frac{cka_{\mathrm{w}}}{[1+(c-1)ka_{\mathrm{w}}]} \left[\frac{1-(ka_{\mathrm{w}})^r}{1-ka_{\mathrm{w}}} + \frac{k^{r-1}a_{\mathrm{w}}^r}{1-a_{\mathrm{w}}} \right] \tag{4.16}$$

Adsorption limited to n layers is described by the equation identical to Equation 4.9 developed by Pickett:

$$u = u_{\mathrm{m}} \frac{ca_{\mathrm{w}}(1-a_{\mathrm{w}}^n)}{(1-a_{\mathrm{w}})[1+(c-1)a_{\mathrm{w}}]} \tag{4.17}$$

When the adsorption surface decreases with the increasing number of adsorbate layers, as it happens in cylindrical capillaries, then (Anderson and Hall, 1948):

$$\begin{aligned} j &= \frac{A_{n+1}}{A_n} \\ u &= u_{\mathrm{m}} \frac{cka_{\mathrm{w}}}{(1-jka_{\mathrm{w}})[1+(c-j)ka_{\mathrm{w}}]} \end{aligned} \tag{4.18}$$

where A is the surface occupied by the layer of adsorbed molecules. The equation is applicable to type IV isotherms. The average pore diameter is related to the parameter j in Anderson's equation:

$$\bar{d} = 6.56 \frac{M}{\rho S} \left[\frac{1}{1-j} \right] \tag{4.19}$$

where

M is the molecular weight
ρ is the density of adsorbate
S is the cross-sectional area of the adsorbate molecule

Yanniotis (1994) developed a method to predict water activity of a substance if only two experimental points are known. Considering the GAB model at high water activities the following relationships can be derived:

$$\begin{aligned} a_{\mathrm{w}} &= \frac{k_{\mathrm{ref}}}{k} a_{\mathrm{w,ref}} \\ u &= \frac{u_{\mathrm{m}}}{u_{\mathrm{m,ref}}} u_{\mathrm{ref}} \end{aligned} \tag{4.20}$$

where subscripts ref are denoting reference material. Reference substances proposed were potato starch and microcrystalline cellulose (MCC) at 25°C. The GAB parameters for those substances are as follows:

$$u_{\mathrm{m}} = 10.312,\ k = 0.734,\ c = 7.976 \quad \text{for potato starch}$$
$$u_{\mathrm{m}} = 4.064,\ k = 0.772,\ c = 8.776 \quad \text{for MCC}$$

The plot of a_{w} versus $a_{\mathrm{w,ref}}$ is made at the same number of monolayers, that is at the same $u/u_{\mathrm{m}} = n$. Hence, for values of (m_{m}, n) water activity is calculated and drawn against the reference water activity calculated for $u = u_{m,ref} \cdot n$. The plot should yield a straight line above one monolayer. Equation 4.13 can also be derived assuming entropy of adsorption smaller than that assumed in the BET model. Assuming nonporous solid and introducing k as a measure of the strength of attractive force field of the adsorbent, Brunauer et al. (1969) derived an equation identical to that derived by Anderson. In Brunauer's derivation:

$$k = \frac{p_0}{g} \exp\left(\frac{E'}{RT}\right) \tag{4.21}$$

where $E' > E_L$ and

$$c = \frac{a_1 g}{b_1} \exp\left(\frac{E_1 - E'}{RT}\right) \tag{4.22}$$

where E is heat of adsorption and subscripts 1 and L refer to the first layer and liquefaction. It was also assumed that at a_w close to one the number of adsorbed layers is far from infinity, as it was presupposed in the BET model. The GAB equation was modified by Jayas and Mazza (1993) by dividing the parameter c by temperature:

$$u = \frac{u_m k a_w (c/t)}{(1 - k a_w)[1 - k a_w + k a_w (c/t)]} \tag{4.23}$$

Another way to account for the effect of temperature on the course of sorption isotherm was by explaining the parameters c and k:

$$c = c_0 \exp\left(\frac{\Delta H_c}{RT}\right) \tag{4.24}$$

$$k = k_0 \exp\left(\frac{\Delta H_k}{RT}\right) \tag{4.25}$$

where
- c_0 and k_0 are the parameters
- ΔH_c is the difference between enthalpies of monolayer and multilayers
- ΔH_k is the difference between enthalpies of bulk liquid and multilayers
- R is the gas constant
- T is the absolute temperature

This way a five-parameter GAB equation is obtained. Assuming that capacity of monolayer is also temperature dependent a six-parameter GAB equation is obtained

$$u_m = u_{m0} \exp\left(\frac{u_{m1}}{RT}\right) \tag{4.26}$$

where u_{m0}, u_{m1} are parameters. The GAB model was empirically modified by Viollaz and Rovedo (1999) to obtain good fitting at high water activities:

$$u = \frac{u_m c k a_w}{(1 - k a_w)[1 + (c - 1) k a_w]} + \frac{u_m c k k_2 a_w^2}{(1 - k a_w)(1 - a_w)} \tag{4.27}$$

Table 4.7 Parameters of Equation 4.27 for Potato Starch

Temperature (°C)	u_m	k	c	k_2	RMS (%)
2	0.10729	0.75207	25.6484	0.00014	1.63
20	0.10709	0.72015	17.1920	0.00055	1.855
40	0.09770	0.67030	16.1790	0.00186	2.450
67	0.20002	0.07340	31.2587	0.03041	1.57

Source: Adapted from Viollaz, P.E. and Rovedo, C.O., *J. Food Eng.*, 40, 287, 1999.

where k_2 is a parameter. An example of application of Equation 4.27 is presented in Table 4.7. Water adsorbed in the first layer, according to the GAB model can be calculated from the following equation:

$$u_1 = \frac{u_m c k a_w}{1 - k a_w + c k a_w} \tag{4.28}$$

where u_1 corresponds to "bound water" (Velazquez et al., 2003). Third sorption stage isotherm developed by Timmermann (1989) and based on the BET model is expressed by the following equation:

$$\begin{aligned} u &= \frac{u_m k c a_w H H'}{(1 - k a_w)[1 + (cH - 1) k a_w]} \\ H &= 1 + \left(\frac{1-k}{k}\right)\frac{(k a_w)^h}{(1 - a_w)} \\ H' &= 1 + \left(\frac{H-1}{H}\right)\left(\frac{1 - k a_w}{1 - a_w}\right)[h + (1 - h) a_w] \end{aligned} \tag{4.29}$$

where h is a parameter. Assuming heterogeneous surface Dural and Hines (1993b) developed a local isotherm for a specific site, given by the GAB equation, expressed by

$$\begin{aligned} u &= \int_{e_{min}}^{e_{max}} \frac{u_m c k a_w}{(1 - k a_w)[1 + (c - 1) k a_w]} E(e) \mathrm{d}e \\ E(e) &= \sum K_j \exp(-L_j e) \\ e &= \exp\left(\frac{q}{RT}\right) - 1 \end{aligned} \tag{4.30}$$

where

e is the site surface energy parameter
$E(e)$ is the function of the energy distribution on the surface
K and L are parameters in the energy distribution function

The final equation is as follows:

$$\frac{u}{u_m} = \frac{k a_w}{1 - k a_w} + \alpha k a_w + 2\beta (k a_w)^2 + 6\gamma (k a_w)^3$$

where

$$\alpha = \frac{1}{L_1}\left[1 - \sum_{j=2}^{m} \frac{K_j}{L_j}\right] + \sum_{j=2}^{m} \frac{K_j}{(-L_j)^2}$$
$$\beta = -\frac{1}{L_1^2}\left[1 - \sum_{j=2}^{m} \frac{K_j}{L_j}\right] + \sum_{j=2}^{m} \frac{K_j}{(-L_j)^3} \qquad (4.31)$$
$$\gamma = \frac{1}{L_1^3}\left[1 - \sum_{j=2}^{m} \frac{K_j}{L_j}\right] + \sum_{j=2}^{m} \frac{K_j}{(-L_j)^4}$$

where α, β, and γ are energy distribution parameters. Equation 4.31 tested for dietary fibers yielded error within 1.04%–5.67%, which was lower than that for the GAB equation (1.91%–7.99%) (Dural and Hines, 1993a). Vullioud et al. (2004) modified the GAB equation introducing a new parameter and temperature. The equation is as follows:

$$u = \frac{C_1}{T^3} a_w \frac{1}{(1 - C_2 a_w)T} \frac{1}{(1 - (C_2 a_w/T) + (C_3 a_w/T^2))} \qquad (4.32)$$

Table 4.8 contains parameters of Equation 4.32. Halsey's model assumed that, in adsorbed layers far away from the surface, the properties of adsorbate are those of liquid in bulk. A theory of film condensation in multilayers was developed. The concept is similar to the potential theory of Polanyi. Frenkel, Halsey, and Hill (Young and Crowell, 1962) considered adsorption at $u/u_m \geq 2$ and assumed that at that distance all periodicity of the surface of adsorbent disappears. Hence, adsorption could be considered as formation of liquid film with properties of bulk liquid. Halsey (1948) assumed that potential energy of adsorbate molecules changes proportionally to r-power of reciprocal distance from the surface. The developed equation is as follows:

$$\ln a_w = -\frac{A}{RT(u/u_m)^r} \qquad (4.33)$$

where A and r are parameters. The power r characterizes interactions between adsorbate and adsorbent. At low values of r the interaction forces are of the van der Waals type, while at large values of r the attraction forces become very specific and react on short distances. Model of surface film can only be used for adsorbates with spherical shape. However, it has been shown that Halsey's equation can be used to describe sorption isotherms of numerous foods in the range of water activities 0.05–0.90 (Iglesias et al., 1975a). In Halsey's original equation there are three unknown parameters, hence the equation was simplified to two unknowns as:

$$\frac{A u_m^r}{RT} = B \qquad (4.34)$$

Table 4.8 Desorption Isotherms of Sweet (v. Napolitana) and Sour Cherry (v. Montmorency) at 0.11 < a_w < 0.85 and 20°C < t < 60°C

SE (%)	C_1	C_2	C_3
6.9	$1.3 \cdot 10^7$	$1.05 \cdot 10^2$	0.9

Source: Adapted from Vullioud, M., Márquez, C.A., and De Michelis, A., *J. Food Eng.*, 63, 15, 2004.

Note: No difference for both fruits was found.

Equation 4.33 can be written as

$$\ln a_w = -Bu^{-r} \tag{4.35}$$

Sorption isotherms of foods described by Equation 4.35 are collected in Table 4.9. It was observed by Iglesias et al. (1975a) that the *RT* term in Halsey's equation does not account for the dependence of *A* and *r* on temperature. Halsey's equation was modified by Iglesias and Chirife (1976c) to account for the effect of temperature. In Equation 4.33 the following substitution was done:

$$\frac{A}{RT} = b \quad \text{and} \quad \ln b + r \ln u_m = \ln C \tag{4.36}$$

and an equation of the following form was obtained:

$$\ln(-\ln a_w) = -r \ln u + \ln C \tag{4.37}$$

where C is a function of temperature

$$\ln C = bt + d \tag{4.38}$$

and

$$\ln a_w = -\frac{\exp(bt + d)}{u^r} \tag{4.39}$$

Table 4.9 Parameters of Halsey's Equation for Adsorption Isotherms

Product	Temperature (°C)	a_w Range	*B*	*r*	Error (%)	Reference
Almond nutmeat	25	0.30–0.90	11.859	1.718		King et al. (1983)
Almonds, bitter Moroccan	15	0.40–0.93	9.4275	1.6548	1.37	Iglesias and Chirife (1982)[a]
	25	0.42–0.93	8.3748	1.6101	1.21	Iglesias and Chirife (1982)[a]
	35	0.45–0.93	7.3376	1.5672	0.86	Iglesias and Chirife (1982)[a]
Almonds, sweet Moroccan	15	0.26–0.94	8.2333	1.6586	2.30	Iglesias and Chirife (1982)[a]
	25	0.29–0.94	7.2802	1.6180	2.20	Iglesias and Chirife (1982)[a]
	35	0.31–0.94	6.5617	1.5807	2.45	Iglesias and Chirife (1982)[a]
Anise	25	0.10–0.80	11.1189	1.3308	2.15	Iglesias and Chirife (1982)[a]
Apricot	20	0.06–0.98	11.588	0.666		Ayranci et al. (1990)
	36	0.06–0.98	5.666	1.171		Ayranci et al. (1990)
Asparagus	10	0.10–0.80	8.8163	1.0807	4.27	Iglesias and Chirife (1982)[a]
Avocado	25	0.10–0.80	7.0060	1.2531	4.34	Iglesias and Chirife (1982)[a]
Bean, freeze-dried	25	0.05–0.80	11.8112	1.2461		Iglesias et al. (1975a)[b]
Beef, freeze-dried	Room	0.10–0.85	13.1025	1.3015		Iglesias et al. (1975a)[b]
Beans v. Rosinha	25	0.10–0.80	77.6447	1.8879	2.10	Iglesias and Chirife (1982)[a]
Beans v. Great Northern	25	0.10–0.80	13.4769	1.2952	2.03	Iglesias and Chirife (1982)[a]
Beef	20	0.07–0.85	7.3131	1.2243	3.74	Iglesias and Chirife (1982)[a]
Beef, cooked, vacuum dried	30	0.10–0.80	15.9212	1.4400	1.85	Iglesias and Chirife (1982)[a]

Table 4.9 (continued) Parameters of Halsey's Equation for Adsorption Isotherms

Product	Temperature (°C)	a_w Range	*B*	*r*	Error (%)	Reference
Beef, cooked, air dried at 55°C	50	0.10–0.75	9.7529	1.2442	2.68	Iglesias and Chirife (1982)[a]
Beef, cooked, air dried at 30°C	30	0.10–0.80	19.8960	1.4767	2.33	Iglesias and Chirife (1982)[a]
Beef, cooked, freeze-dried	21.1	0.10–0.80	21.2136	1.4605	2.62	Iglesias and Chirife (1982)[a]
Beef dried at 30°C	30	0.10–0.80	18.50	1.448	2.6	Iglesias and Chirife (1976d)[c]
Beef dried at 55°C	30	0.10–0.80	15.08	1.392	1.8	Iglesias and Chirife (1976d)[c]
Beef dried at 70°C	30	0.10–0.80	10.70	1.301	3.7	Iglesias and Chirife (1976d)[c]
Beef dried at 55°C	50	0.10–0.80	9.91	1.250	2.6	Iglesias and Chirife (1976d)[c]
Beef muscle frozen at −30°C, freeze-dried		0.10–0.85	9.4235	1.1948	5.14	Iglesias and Chirife (1982)[a]
Beef muscle frozen at −150°C, freeze-dried		0.10–0.85	14.1236	1.3205	3.06	Iglesias and Chirife (1982)[a]
Cabbage, freeze-dried	37	0.05–0.60	3.4968	0.68467		Iglesias et al. (1975a)[b]
Cabbage, v. Savoy, freeze-dried	37	0.10–0.60	3.2302	0.6869	2.38	Iglesias and Chirife (1982)[a]
Cabbage, v. Savoy, air dried	10	0.10–0.80	7.5091	1.0471	1.47	Iglesias and Chirife (1982)[a]
Cardamom	25	0.10–0.80	51.4065	1.8561	1.85	Iglesias and Chirife (1982)[a]
	45	0.10–0.80	14.9946	1.4406	2.37	Iglesias and Chirife (1982)[a]
Carrot seeds	10	0.10–0.70	133.5877	2.3878	2.82	Iglesias and Chirife (1982)[a]
Cashew nuts	25	0.10–0.75	2.3340	1.0652	6.66	Iglesias and Chirife (1982)[a]
Celery, freeze-dried	45	0.10–0.80	4.4222	0.8800	3.05	Iglesias and Chirife (1982)[a]
Champignon, freeze-dried	20	0.05–0.80	6.6167	0.95166		Iglesias et al. (1975a)[b]
Chamomile flower (*Anthemis nobilis* L.)	5	0.05–0.80	18.812	1.367	0.235	Soysal and Öztekin (2001)[d]
	25	0.05–0.80	18.812	1.367	0.235	Soysal and Öztekin (2001)[d]
	45	0.05–0.80	6.021	1.025	0.055	Soysal and Öztekin (2001)[d]
	60	0.05–0.80	4.853	1.083	0.434	Soysal and Öztekin (2001)[d]
Chamomile tea	5	0.10–0.80	21.2757	1.4204	2.21	Iglesias and Chirife (1982)[a]
	25	0.10–0.80	21.2757	1.4204	2.21	Iglesias and Chirife (1982)[a]
	45	0.10–0.80	6.1615	1.0367	1.07	Iglesias and Chirife (1982)[a]
Cheese, Edam	25	0.10–0.80	4.9692	1.0668	3.52	Iglesias and Chirife (1982)[a]
Cheese, Emmental	25	0.10–0.80	5.9967	1.1889	2.36	Iglesias and Chirife (1982)[a]
Chicken sausage, smoked	5	0.113–0.877	5.1461	1.8385	2.85	Singh et al. (2001)[e]
	25	0.113–0.843	2.8224	1.3619	6.57	Singh et al. (2001)[e]
	50	0.111–0.812	0.5112	0.6626	9.57	Singh et al. (2001)[e]
Chives	25	0.10–0.80	11.8931	1.1146	1.99	Iglesias and Chirife (1982)[a]
Cinnamon	60	0.30–0.80	22.9855	1.8015	1.20	Iglesias and Chirife (1982)[a]
Citrus juice, 10% pulp, spray-dried	25	0.11–0.75	12.7186	1.1355	3.23	Iglesias and Chirife (1982)[a]
Citrus juice, 15% pulp, spray-dried	25	0.11–0.75	10.7161	1.1078	3.21	Iglesias and Chirife (1982)[a]
Citrus juice, 20% pulp, spray-dried	25	0.11–0.75	9.6176	1.0939	4.20	Iglesias and Chirife (1982)[a]

(continued)

Table 4.9 (continued) Parameters of Halsey's Equation for Adsorption Isotherms

Product	Temperature (°C)	a_w Range	*B*	*r*	Error (%)	Reference
Citrus juice, 25% pulp, spray-dried	25	0.11–0.75	1.6778	2.8727	3.29	Iglesias and Chirife (1982)[a]
Cloves	25	0.10–0.80	25.6953	1.7926	2.04	Iglesias and Chirife (1982)[a]
	45	0.10–0.80	9.9545	1.4351	2.20	Iglesias and Chirife (1982)[a]
Cloves flower (*Caryophyllus aromaticus* L.)	25	0.05–0.80	23.036	1.737	0.214	Soysal and Öztekin (2001)[d]
	45	0.05–0.80	8.529	1.356	0.114	Soysal and Öztekin (2001)[d]
	60	0.05–0.80	3.002	1.262	1.025	Soysal and Öztekin (2001)[d]
Coffee extract, decaffeinated, freeze-dried	30	0.10–0.60	5.0591	0.9143	2.68	Iglesias and Chirife (1982)[a]
Coffee extract, freeze-dried	30	0.10–0.60	4.0906	0.8036	6.46	Iglesias and Chirife (1982)[a]
Coffee extract, agglomerated, spray-dried	30	0.10–0.60	3.4670	0.7887	6.29	Iglesias and Chirife (1982)[a]
Coffee, soluble	28	0.10–0.77	4.6227	0.9134	2.59	Iglesias and Chirife (1982)[a]
Crackers, Criollitas	20	0.11–0.80	52.5	1.91	3.1	Tubert and Iglesias (1986)[f]
Crackers, Livianitas	20	0.11–0.80	27.0	1.60	1.7	Tubert and Iglesias (1986)[f]
Dextrin from corn starch	28.2	0.09–0.62	32.4438	1.6214	3.97	Iglesias and Chirife (1982)[a]
	10.7	0.10–0.84	52.6899	1.7632	3.66	Iglesias and Chirife (1982)[a]
Egg albumin, heat coagulated	25	0.20–0.80	18.2337	1.4549		Iglesias et al. (1975a)[b]
Egg, spray-dried	60	0.10–0.70	6.4464	1.2875	2.26	Iglesias and Chirife (1982)[a]
Egg white, spray-dried	10	0.10–0.80	33.4609	1.4833	2.00	Iglesias and Chirife (1982)[a]
Egg white, freeze-dried	10	0.10–0.80	30.0316	1.4462	3.33	Iglesias and Chirife (1982)[a]
Figs	20	0.06–0.98	8.758	0.726		Ayranci et al. (1990)
	36	0.06–0.98	5.754	1.130		Ayranci et al. (1990)
Fish protein concentrate	25	0.10–0.80	47.9368	1.9604		Iglesias et al. (1975a)[b]
	35	0.10–0.80	36.9085	1.9043		Iglesias et al. (1975a)[b]
	42	0.10–0.80	25.3188	1.7776		Iglesias et al. (1975a)[b]
Gelatin	25	0.20–0.90	93.8034	1.8227		Iglesias et al. (1975a)[b]
Groundnuts	25	0.40–0.90	10.9835	1.6968	1.05	Iglesias and Chirife (1982)[a]
Groundnut kernel	20	0.37–0.89	9.9880	1.6464	0.50	Iglesias and Chirife (1982)[a]
	30	0.38–0.89	8.9566	1.5996	0.54	Iglesias and Chirife (1982)[a]
Guava: taro flakes 1:2	22	0.17–0.75	23.3065	1.5476	1.86	Iglesias and Chirife (1982)[a]
Guava: taro flakes 3:2	22	0.17–0.75	28.1557	1.4958	1.61	Iglesias and Chirife (1982)[a]
Halibut (*Hippoglossus stenolepis*), freeze-dried	25	0.10–0.80	19.4920	1.3384	1.52	Iglesias and Chirife (1982)[a]
Hops	25	0.08–0.74	16.2079	1.5752	1.67	Iglesias and Chirife (1982)[a]
Horse radish	45	0.10–0.80	6.3707	1.0340	2.67	Iglesias and Chirife (1982)[a]
Laurel	45	0.10–0.80	6.0798	1.1959	1.54	Iglesias and Chirife (1982)[a]
Linseed seed	25	0.25–0.85	21.1685	1.8218	2.26	Iglesias and Chirife (1982)[a]
Maltose, freeze-dried	23	0.10–0.70	6.3405	0.92244		Iglesias et al. (1975a)[b]

Table 4.9 (continued) Parameters of Halsey's Equation for Adsorption Isotherms

Product	Temperature (°C)	a_w Range	*B*	*r*	Error (%)	Reference
Marjoram (*Origanum majorana* L.)	25	0.05–0.80	16.374	1.440	0.209	Soysal and Öztekin (2001)[d]
	45	0.05–0.80	5.227	1.092	0.140	Soysal and Öztekin (2001)[d]
	60	0.05–0.80	3.318	1.073	0.333	Soysal and Öztekin (2001)[d]
Marjoram, sweet	5	0.10–0.80	68.8515	1.9237	2.54	Iglesias and Chirife (1982)[a]
	25	0.10–0.80	17.2126	1.4701	2.22	Iglesias and Chirife (1982)[a]
	45	0.10–0.80	5.5094	1.1235	3.64	Iglesias and Chirife (1982)[a]
	25	0.10–0.80	26.1584	1.5945	1.73	Iglesias and Chirife (1982)[a]
	45	0.10–0.80	5.5094	1.1235	3.64	Iglesias and Chirife (1982)[a]
Papaya: taro flakes 1:2	22	0.17–0.75	34.9202	1.6401	2.85	Iglesias and Chirife (1982)[a]
Papaya: taro flakes 3:2	22	0.17–0.75	35.3323	1.4797	1.71	Iglesias and Chirife (1982)[a]
Paracasein	20	0.10–0.80	53.3476	1.8508	3.41	Iglesias and Chirife (1982)[a]
Para nut	25	0.10–0.80	5.1712	1.6822	1.65	Iglesias and Chirife (1982)[a]
Pea flour	20	0.07–0.85	49.6628	1.9305	4.44	Iglesias and Chirife (1982)[a]
Pea, freeze-dried	25	0.05–0.80	11.2420	1.2366		Iglesias et al. (1975a)[b]
Peanut v. Runner	25	0.11–0.86	31.9494	1.9237	3.10	Iglesias and Chirife (1982)[a]
Peanut v. NC-5, kernel	15	0.10–0.80	4.3079	1.2540	5.13	Iglesias and Chirife (1982)[a]
Peas, freeze-dried	10	0.10–0.80	8.0019	1.0963	4.25	Iglesias and Chirife (1982)[a]
	25	0.10–0.80	10.5023	1.2113	2.89	Iglesias and Chirife (1982)[a]
Pepper, green cv. Charliston	30		9.7555	1.183	6.750	Kaymak-Ertekin and Sultanoğlu (2001)[g]
	45		7.5174	1.166	7.201	Kaymak-Ertekin and Sultanoğlu (2001)[g]
	60		4.5284	1.129	8.579	Kaymak-Ertekin and Sultanoğlu (2001)[g]
Pepper, red cv. Bursa	30		12.7101	1.163	3.716	Kaymak-Ertekin and Sultanoğlu (2001)[g]
	5		8.4985	1.065	4.712	Kaymak-Ertekin and Sultanoğlu (2001)[g]
	60		7.4827	1.136	10.144	Kaymak-Ertekin and Sultanoğlu (2001)[g]
Peppermint (*Mentha piperita* L.)	60	0.05–0.80	7.041	1.317	0.682	Soysal and Öztekin (2001)[d]
Peppermint tea	45	0.10–0.80	10.8261	1.3011	3.04	Iglesias and Chirife (1982)[a]
Potato	25	0.10–0.70	7.9389	1.0912		Iglesias et al. (1975a)[b]
Potato, air dried	10	0.10–0.80	27.4606	1.5068	2.59	Iglesias and Chirife (1982)[a]
	37	0.10–0.80	33.2730	1.6622	3.33	Iglesias and Chirife (1982)[a]
	80	0.10–0.80	17.1200	1.5638	3.86	Iglesias and Chirife (1982)[a]
Potato, roller dried	15	0.10–0.80	22.4896	1.5158	3.33	Iglesias and Chirife (1982)[a]
Potato, cooked, air dried	15	0.10–0.80	58.3943	1.8259	1.99	Iglesias and Chirife (1982)[a]
	37	0.10–0.80	41.6712	1.7666	2.51	Iglesias and Chirife (1982)[a]
	28	0.10–0.80	74.0611	2.0135	1.88	Iglesias and Chirife (1982)[a]
	50	0.10–0.80	30.4792	1.6608	2.00	Iglesias and Chirife (1982)[a]

(continued)

Table 4.9 (continued) Parameters of Halsey's Equation for Adsorption Isotherms

Product	Temperature (°C)	a_w Range	*B*	*r*	Error (%)	Reference
Potato flakes	20	0.07–0.85	37.6598	1.7490	2.25	Iglesias and Chirife (1982)[a]
	25	0.12–0.76	18.1936	1.4714	4.20	Iglesias and Chirife (1982)[a]
Raisins	20	0.06–0.98	10.644	0.709		Ayranci et al. (1990)
	36	0.06–0.98	6.707	1.111		Ayranci et al. (1990)
Rapeseed v. Hector	5	0.40–0.90	14.2157	1.7082	0.56	Iglesias and Chirife (1982)[a]
	15	0.40–0.90	13.9606	1.7105	0.77	Iglesias and Chirife (1982)[a]
	25	0.40–0.90	12.6126	1.7052	0.45	Iglesias and Chirife (1982)[a]
Rapeseed v. Tower	5	0.30–0.90	19.4672	1.7702	0.63	Iglesias and Chirife (1982)[a]
	15	0.30–0.90	16.8157	1.7269	0.58	Iglesias and Chirife (1982)[a]
	25	0.30–0.90	15.6599	1.7187	0.67	Iglesias and Chirife (1982)[a]
Salmon, freeze-dried	37	0.10–0.80	10.4990	1.2150		Iglesias et al. (1975a)[b]
Salsify, freeze-dried	45	0.10–0.80	15.1924	1.3293	1.18	Iglesias and Chirife (1982)[a]
	60	0.10–0.80	6.8883	1.0917	2.51	Iglesias and Chirife (1982)[a]
Skim milk powder	34		24.8439	1.7764	1.90	
Sorghum	21.1	0.04–0.84	127.4496	2.1063		Iglesias et al. (1975a)[b]
Soybean, defatted	30	0.10–0.80	8.8080	1.2603	2.53	Iglesias and Chirife (1982)[a]
Soybean	30	0.10–0.80	3.5132	1.0406		Iglesias et al. (1975a)[b]
Soybean	25	0.25–0.80	18.1424	1.5385	0.74	Iglesias and Chirife (1982)[a]
Soya meal, 1.9% oil db	25	0.40–0.90	13.4399	1.3269	1.82	Iglesias and Chirife (1982)[a]
Soya meal, 20.7% oil db	25	0.40–0.80	17.4814	1.5286	0.45	Iglesias and Chirife (1982)[a]
Soybean seed v. Brown C 1935	15	0.10–0.80	20.9621	1.5817	4.04	Iglesias and Chirife (1982)[a]
Spinach	37	0.05–0.75	8.2524	1.0947		Iglesias et al. (1975a)[b]
Spinach, freeze-dried	10	0.10–0.80	20.5440	1.4265	1.95	Iglesias and Chirife (1982)[a]
Sucrose, freeze-dried	47	0.20–0.80	5.3311	0.85983		Iglesias et al. (1975a)[a]
Sugar beet root, raw	20	0.05–0.70	1.4359	0.9558	3.91	Iglesias et al. (1975b)
	35	0.05–0.70	1.3335	0.89772	5.76	Iglesias et al. (1975b)
	47	0.05–0.70	1.2219	0.76711	8.77	Iglesias et al. (1975b)
	65	0.05–0.70	1.1597	0.63305	6.55	Iglesias et al. (1975b)
Sugar beet root, insoluble components	35	0.11–0.80	1.8309	1.9403	7.17	Iglesias et al. (1975b)
	47	0.11–0.80	1.5045	1.7446	7.65	Iglesias et al. (1975b)
Sugar, icing	10	0.10–0.80	3.9626	1.0095	1.64	Iglesias and Chirife (1982)[a]
	20	0.10–0.80	4.3225	1.0258	5.83	Iglesias and Chirife (1982)[a]
Sultanas	5–35	0.41–0.72	5.4512	0.8127	1.58	Iglesias and Chirife (1982)[a]
Sunflower seed	25	0.25–0.85	14.2094	1.7036	2.56	Iglesias and Chirife (1982)[a]
Tea, black Bop grade	21	0.10–0.80	7.1343	1.2831	4.23	Iglesias and Chirife (1982)[a]
	32	0.10–0.80	9.6222	1.3851	3.45	Iglesias and Chirife (1982)[a]
Tea, black, dust	32	0.10–0.80	13.4015	1.4985	2.85	Iglesias and Chirife (1982)[a]
Tea v. Dar Jeeling, infusion, freeze-dried	10	0.10–0.80	7.8124	1.1094	1.76	Iglesias and Chirife (1982)[a]

Table 4.9 (continued) Parameters of Halsey's Equation for Adsorption Isotherms

Product	Temperature (°C)	a_w Range	B	r	Error (%)	Reference
Thyme (*Thymus vulgaris* L.)	5	0.05–0.80	30.605	0.655	0.345	Soysal and Öztekin (2001)[d]
	25	0.05–0.80	17.583	1.487	0.166	Soysal and Öztekin (2001)[d]
	45	0.05–0.80	9.042	1.290	0.037	Soysal and Öztekin (2001)[d]
	60	0.05–0.80	9.007	1.404	0.538	Soysal and Öztekin (2001)[d]
Thyme	25	0.10–0.80	21.5361	1.5840	2.65	Iglesias and Chirife (1982)[a]
	45	0.10–0.80	9.4088	1.3091	1.02	Iglesias and Chirife (1982)[a]
	45	0.10–0.80	9.4088	1.3091	1.02	Iglesias and Chirife (1982)[a]
Tomato, freeze-dried	17	0.10–0.80	10.0587	0.9704	3.02	Iglesias and Chirife (1982)[a]
Trout, raw, freeze-dried	45	0.10–0.80	8.0277	1.1329	4.36	Iglesias and Chirife (1982)[a]
Thuna, Big-eye (*Thunnus obesus*), freeze-dried	25	0.10–0.80	11.3017	1.1702	1.34	Iglesias and Chirife (1982)[a]
Walnut, kernel	25	0.11–0.81	9.6452	2.1303	3.42	Iglesias and Chirife (1982)[a]
Walnut, kernel	22.5	0.10–0.90	19.1203	2.6815		Iglesias et al. (1975a)[b]
Walnut, shelled	7	0.10–0.80	10.2130	2.0981	4.61	Iglesias and Chirife (1982)[a]
	22.5	0.10–0.80	20.0611	2.7222	2.76	Iglesias and Chirife (1982)[a]
Wheat flour	20.2	0.12–089	171.2978	2.2566		Iglesias et al. (1975a)[b]
	30.1	0.13–0.90	150.6560	2.2561		Iglesias et al. (1975a)[b]
	40.8	0.13–0.90	106.0755	2.1543		Iglesias et al. (1975a)[b]
	50.2	0.15–0.90	66.1388	2.0168		Iglesias et al. (1975a)[b]
Wheat flour	20.2	0.12–0.89	200.5532	2.3122	1.66	Iglesias and Chirife (1982)[a]
Whey protein concentrate, 78.0% protein 1.3% lactose	24	0.12–0.86	17.6265	1.4806	4.23	Iglesias and Chirife (1982)[a]
Whey protein concentrate, 72.8% protein 7.2% lactose	24	0.10–0.80	8.4317	1.0367	4.36	Iglesias and Chirife (1982)[a]
Whey protein concentrate, 87.0% protein, 4.0% lactose	24	0.10–0.80	15.9825	1.3573	1.43	Iglesias and Chirife (1982)[a]
Whey protein concentrate, 83.0% protein, 5.5% lactose	24	0.10–0.80	17.3629	1.3893	1.90	Iglesias and Chirife (1982)[a]
Winter savory	25	0.10–0.80	89.5484	1.9929	2.74	Iglesias and Chirife (1982)[a]
Yam	37	0.10–0.70	10.6590	1.2423	1.60	Iglesias and Chirife (1982)[a]
Yeast, freeze-dried	10	0.10–0.80	16.2487	1.3126	2.04	Iglesias and Chirife (1982)[a]
Yogurt	25	0.10–0.80	6.4806	1.0529	2.63	Iglesias and Chirife (1982)[a]

[a] MRE.
[b] Average error varies between 1.62% and 7.78% depending on the kind of product (recalculated).
[c] *u*-water content on fat free basis. Cooked beef was subjected to drying to final water content less than 0.09.
[d] RSS $\times 10^{-2}$.
[e] RMS.
[f] Criollitas contain wheat flour, fat, salt, and flavor. Livianitas contain wheat flour, wheat bran, fat, salt, and flavor.
[g] Recalculated.

Table 4.10 Parameters of Modified Halsey's Equation 4.39

Product	Temperature Range (°C)	a_w Range	r	d	b	Error (%)
Corn cobs	20–80	0.11–0.87	1.6887	3.88771	$-1.4623 \cdot 10^{-2}$	
Beans, red	21–38	0.11–0.90	1.6933	4.2669	$-1.3382 \cdot 10^{-2}$	
Beans, baby lima	21–38	0.11–0.90	1.7270	4.3867	$-1.2080 \cdot 10^{-2}$	
Beans, pinto	21–38	0.11–0.90	1.7571	4.4181	$-1.1875 \cdot 10^{-2}$	
Beans black	10–38	0.30–0.90	1.9856	5.2003	$-2.2685 \cdot 10^{-2}$	
Pea, dried (Pixton and Henderson, 1979)	5–50	0.10–0.95	2.21	5.482	$-8.764 \cdot 10^{-3}$	1.47
Pea seeds (v. Tainun no. 1) dried at 25°C (Chen, 2003)	5–50	0.10–0.95	1.2201	2.778	$-5.379 \cdot 10^{-3}$	3.58
Peanut, kernel	10–32	0.21–0.97	2.2375	3.9916	$-1.7856 \cdot 10^{-2}$	
Pigeon pea (Shepherd and Bhardwaj, 1986)	5–50	0.10–0.95	1.150	2.455	$-8.136 \cdot 10^{-3}$	8.22
Pumpkin seed flour (Menkov and Durakova, 2005)	10–40		2.144373	4.789093	$-1.911 \cdot 10^{-2}$	1.13
Rapeseed, v. Tower	5–35	0.27–0.90	1.7007	2.8748	$-7.4848 \cdot 10^{-3}$	
Rapeseed, v. Candle	5–35	0.21–0.93	1.7607	3.0026	$-4.8967 \cdot 10^{-3}$	
Rapeseed (canola)	5–25	0.18–0.90	1.8600	3.4890	$-1.0553 \cdot 10^{-2}$	
Soybeans, USA	5–55	0.11–0.90	1.5431	3.3109	$-1.1635 \cdot 10^{-2}$	
Soybeans, England	15–35	0.18–0.96	1.5245	3.0446	$-5.4321 \cdot 10^{-3}$	
Sunflower seeds	5–45	0.11–0.90	2.1201	4.4308	$-1.5532 \cdot 10^{-2}$	

Source: Adapted from Iglesias, H.A., and Chirife, J., *J. Food Technol.*, 11, 109, 1976c.

The parameters of Equation 4.39 are collected in Table 4.10. Finally, the equation is as follows:

$$\ln a_\mathrm{w} = -Cu^{-r} \tag{4.40}$$

Parameters C and r are presented in Table 4.11. Saravacos et al. (1986) modified Equation 4.35 as follows:

$$\ln a_\mathrm{w} = \ln c - Bu^{-r} \tag{4.41}$$

They found that Equation 4.41 describes sorption isotherms of raisins with fairly good precision (Table 4.12).

4.2.3 Bradley Model

De Boer and Zwikker (1929) developed a theory of polarized layers while trying to explain sigmoid shape of sorption isotherms. According to that theory, adsorbent surface induces dipoles in the first layer of adsorbate. Dipoles induced in the first adsorbed layer induce dipoles in the next layer and so on until several layers are built up. Developed equation of sorption isotherm is as follows:

$$\ln\left(\frac{a_\mathrm{w}}{K_3}\right) = K_2 K_1^{(u/u_\mathrm{m})} \tag{4.42}$$

where

K_1, K_2, and K_3 are parameters
u/u_m is the number of adsorbed layers

Table 4.11 Parameters of Modified Halsey's Equation 4.40

Product	Temperature (°C)	a_w Range	r	C
Chicken, freeze-dried	5	0.10–0.70	1.072	11.03
	45		1.072	7.461
	60		1.072	5.914
Fish protein concentrate	25	0.20–0.80	2.132	72.04
	35		2.132	62.05
	42		2.132	55.56
Laurel	25	0.05–0.70	1.318	10.97
	45		1.318	7.50
	60		1.318	5.01
Nutmeg	5	0.10–0.70	1.892	44.03
	25		1.892	32.37
	60		1.892	12.97
Paranut	5	0.10–0.70	1.581	6.01
	25		1.581	4.61
	60		1.581	2.05
Thyme	5	0.10–0.70	1.496	21.43
	25		1.496	17.89
	45		1.496	13.66
	60		1.496	10.60
Wheat flour	20.2	0.12–0.89	2.259	173.7
	30.1		2.259	152.9
	40.8		2.259	141.1
	50.2		2.259	126.1

Source: Adapted from Iglesias, H.A. and Chirife, J., *J. Food Technol.*, 11, 109, 1976c.

Parameter K_3 is usually close to 1, and Equation 4.42 describes isotherm in the relative pressure range from 0.1 to 1.0. Bradley (1936a), assuming a strongly polar surface and dipole induction, developed an equation identical to that of de Boer and Zwikker. The equation is written as

$$T \ln\left(\frac{1}{a_w}\right) = K_1 K_3^u \tag{4.43}$$

or

$$\ln\left(\frac{1}{a_w}\right) = \frac{K_1}{T} K_3^u \tag{4.44}$$

Table 4.12 Parameters of Modified Halsey's Equation 4.41 for Sorption Isotherms of Raisins

Temperature (°C)	a_w Range	r	B	c	Error (%)
20	0.25–0.85	0.702	5.004	1.121	6.622
25	0.25–0.80	0.721	5.198	1.036	10.135
30	0.20–0.83	0.576	3.290	1.156	5.028
35	0.20–0.80	0.523	3.001	1.196	7.768

Source: Adapted from Saravacos, G.D., Tsiourvas, D.A., and Tsami, E., *J. Food Sci.*, 51, 381, 1986.

where K_1 and K_3 are functions of the field of the sportive polar group, the dipole moment of the sorbed gas, and the temperature. K_3 in addition includes a term, which is characteristic of the distribution of the sorbed molecules on the sorptive sites. Parameters of Equation 4.44 for many foods are collected in Table 4.13. Later (Bradley, 1936b) the equation was modified to extend its application to finely divided solids. In such materials, adsorption on convex surfaces forms meniscus different from that formed during capillary condensation. Thus

$$\ln\left(\frac{1}{a_w}\right) = K_2 K_3^u + K_4 \qquad (4.45)$$

Table 4.13 Parameters of the Bradley Equation for Adsorption Isotherms

Product	Temperature (°C)	a_w Range	K_1/T	K_3	MRE
Beef, raw, freeze-dried	21.1	0.10–0.80	2.9867	0.8911	2.22
	40	0.10–0.80	4.4020	0.8226	2.62
Casein (Hoover and Mellon, 1950)[a]	30	0.05–0.95	1.8621	0.8337	3.2
Collagen		0.10–0.80	5.3509	0.9146	1.89
Corn	22	0.10–0.80	6.3766	0.8325	0.76
Corn flour	25.5	0.10–0.80	8.7432	0.8226	1.54
Corn flour, degermed	22.5	0.10–0.80	12.7420	0.8100	0.93
	50	0.10–0.80	11.8322	0.7932	1.38
Cotton (Hoover and Mellon, 1950)[a]	30	0.05–0.95	1.9953	0.7379	3.8
Cross-linked starch, potato	25	0.09–0.81	3.2396	0.8668	2.19
Egg albumin, coagulated, freeze-dried	25	0.10–0.80	4.0081	0.8310	2.82
	40	0.10–0.80	4.2013	0.8140	1.40
Egg white, desalted	20	0.10–0.80	4.3260	0.8339	2.51
Egg, spray-dried	17.1	0.05–0.40	5.8136	0.7268	3.8
Fish protein concentrate, hake	25	0.10–0.80	6.6567	0.7687	0.92
Gelatin, freeze-dried	40	0.10–0.80	4.6337	0.8843	2.32
Linseed seed v. Dutch H 1935	15	0.10–0.80	5.3102	0.7538	4.35
Oleic acid	30	0.20–0.89	2.7871	0.0203	1.24
	80	0.20–0.80	2.9265	0.0699	1.97
Ovalbumin (Hoover and Mellon, 1950)[a]	30	0.05–0.95	1.5136	0.8590	3.0
Polyglycine (Hoover and Mellon, 1950)[a]	30	0.05–0.95	2.1380	0.8017	4.2
Potato, freeze-dried	70	0.10–0.70	4.3115	0.8051	1.58
Rice, freeze-dried	4.4	0.10–0.80	3.3439	0.8390	1.77
Sorghum v. Blockhull Kajr	4.4	0.10–0.80	7.4799	0.8450	0.59
	21.1	0.10–0.80	7.4660	0.8336	0.28
Sorghum v. Bukura Mahemba	35	0.25–0.86	7.3452	0.8219	1.05
Starch, freeze-dried	35	0.10–0.80	4.0844	0.8545	1.10
Starch, maize	30	0.33–0.92	8.6705	0.8275	1.48
	50	0.32–0.87	5.8369	0.8388	0.71
Starch, wheat	25	0.34–0.87	14.2468	0.7886	1.55
	27	0.10–0.80	5.7637	0.8468	1.31
	50	0.32–0.87	5.9465	0.8363	0.52
Trout, cooked, freeze-dried	5	0.10–0.80	3.4626	0.8842	2.16
Walnut kernels	25	0.11–0.81	4.8961	0.5644	3.33
Wheat v. Capelle	25	0.25–0.85	5.1552	0.8551	0.54
Wool (Hoover and Mellon, 1950)[a]	30	0.05–0.95	1.9055	0.8166	5.1

Source: Adapted from Iglesias, H.A. and Chirife, J., in *Handbook of Food Isotherms: Water Sorption Parameters for Food and Food Components*, Academic Press, New York, 1982.

[a] Recalculated.

where $K_2 = K_1/T$ and K_4 account for the difference between the heat of evaporation from the polarized surface and the bulk liquid. Moreover, K_4 is a parameter that is dependent on temperature.

4.2.4 Smith Model

Smith (1947) developed a sorption isotherm for nonswelling gel assuming localized sorption and solution model and dividing water into two portions. One portion is condensed with a normal heat of condensation (*A*), and fraction that is bound by excessive forces (*B*). The portion *A* must increase progressively toward the saturation vapor pressure, and portion *B* reaches its peak at a vapor pressure well short of saturation. Hence, the isotherm for *A* will be convex and that for *B* will be concave. The summation of the isotherms results in a sigmoid isotherm:

$$u = B - A' \ln(1 - a_w) \tag{4.46}$$

where A' is the weight of adsorbate condensed on 100 g of adsorbent required to saturate the first layer of the *A* fraction. Tables 4.14 and 4.15 present the parameters of Smith's equation for many foods. For a nonswelling gel, the plot should be linear. For swelling material, the equation was modified to the following form:

$$\frac{u}{1+u} = B - A' \ln(1 - a_w) \tag{4.47}$$

4.2.5 Kühn Model

Kühn (1964) developed an equation of sorption isotherm assuming capillary condensation and adsorption potential theories. Distribution of capillary radiuses was assumed and adsorption potential related to those radiuses. The equation is as follows:

$$u = k\left(\frac{1}{a_w}\right)^{-z} - B \tag{4.48}$$

where *k*, *z*, and *B* are parameters. Depending on *z*, I, II, or III type of isotherm is obtained. Moreover, *z* is related to the capillary distribution function. Kühn's equation was simplified to the two-parameter equation of the form:

$$u = \frac{K_1}{\ln a_w} + K_2 \tag{4.49}$$

Kühn's equation is used to describe sorption isotherms of food in Table 4.16.

4.2.6 Harkins–Jura Model

Analyzing analogies between adsorbed layers on the solid and insoluble liquid films on the surface of liquids, a theory of condensed phase was proposed by Harkins and Jura (1944) and was discussed further by Jura and Harkins (1946). Using the Gibbs adsorption isotherm a new equation was proposed to describe adsorption on the solid surface:

$$\log a_w = B - \frac{A}{u^2} \tag{4.50}$$

where
A is the parameter related to the specific surface of the adsorbent
B is a parameter

Table 4.14 Water Sorption Isotherms Described by the Smith Equation

Product	Temperature (°C)	a_w Range	Fat (%)	Protein (%)	*B*	*A′*	Reference
Casein	7.2	0.33–0.95			3.993	7.366	Lang et al. (1982)[a]
Casein	20	0.33–0.95			4.698	6.197	Lang et al. (1982)[a]
Casein	30	0.33–0.95			4.154	6.037	Lang et al. (1982)[a]
Cellulose	20	0.30–0.95			1.71	4.934	Lang and Steinberg (1981b)[a]
Fructose	20	0.30–0.95			−20.31	45.770	Lang and Steinberg (1981b)[a]
Glucose	20	0.30–0.95			−142.86	103.983	Lang and Steinberg (1981b)[a]
Paracasein	20 ± 1	0.76–0.95			11.30	4.951	Hardy and Steinberg (1984)[a]
Potato v. Huinkul	20	02–0.8			4.8	11.5	Crapiste and Rotstein (1982)
Rape seed press-cake	20 ± 0.5	0.11–0.97	10.1	30.3	5.06 ± 0.814	8.23 ± 0.504	Sadowska et al. (1995)
Rape seed press-cake	20 ± 0.5	0.11–0.97	9.9	28.5	5.45 ± 0.848	7.73 ± 0.514	Sadowska et al. (1995)
Rape seed raw flakes	20 ± 0.5	0.11–0.97	46.6		3.78 ± 0.691	5.57 ± 0.413	Sadowska et al. (1995)
Rape seed raw flakes	20 ± 0.5	0.11–0.97	47.4		4.62 ± 0.584	4.41 ± 0.349	Sadowska et al. (1995)
Rape seed roasted flakes	20 ± 0.5	0.11–0.97	47.9		2.57 ± 0.427	5.35 ± 0.255	Sadowska et al. (1995)
Rape seed roasted flakes	20 ± 0.5	0.11–0.97	46.7		3.52 ± 0.474	4.31 ± 0.283	Sadowska et al. (1995)
Rape seeds	20 ± 0.5	0.11–0.97	44.6		3.98 ± 0.463	5.24 ± 0.276	Sadowska et al. (1995)
Rape seeds	20 ± 0.5	0.11–0.97	45.6		4.18 ± 0.309	4.82 ± 0.184	Sadowska et al. (1995)

Salt	20	0.30–0.95		−511.18	546.59	Lang et al. (1981a)[a]
Soy flour	20	0.30–0.95		−0.03	13.811	Lang et al. (1981a)[a]
Spaghetti, thin	30			3.0	7.0	Dinçer and Esin (1996)[a]
Spaghetti, thin	40			2.5	6.3	Dinçer and Esin (1996)[a]
Spaghetti, thin	50			1.9	6.2	Dinçer and Esin (1996)[a]
Spaghetti, thin	60			1.7	5.2	Dinçer and Esin (1996)[a]
Spaghetti, thin	70			2.3	4.3	Dinçer and Esin (1996)[a]
Starch	7.2	0.33–0.95		9.708	7.405	Lang et al. (1982)[a]
Starch	20	0.33–0.95		9.890	6.450	Lang et al. (1982)[a]
Starch	30	0.33–0.95		9.127	6.297	Lang et al. (1982)[a]
Sucrose	7.2	0.33–0.95	−72.38		62.564	Lang et al. (1982)[a]
Sucrose	20	0.33–0.95	−60.52		55.082	Lang et al. (1982)[a]
Sucrose	30	0.33–0.95	−51.06		49.414	Lang et al. (1982)[a]
Sweet potato v. Camote	28	0.06–0.81	5.85		9.68	Diamante and Munro (1990)
Sweet potato v. Kumara	25	0.06–0.81	4.03		9.31	Diamante and Munro (1990)
Sweet potato v. Kumara	40	0.06–0.81	3.56		8.20	Diamante and Munro (1990)
Sweet potato v. Kumara	55	0.06–0.81	3.24		8.05	Diamante and Munro (1990)

[a] Recalculated.

Table 4.15 Parameters of the Smith Equation for Amino Acids and Related Compounds

Product	Temperature (°C)	a_w Range	B	A'
Alanine	20	0.33–0.95	0	0.04
γ-Aminobutyric acid	20	0.33–0.95	−44.2	92.98
Arginine	20	0.33–0.95	−65.6	35.00
Asparagine	20	0.33–0.95	0.2	0.13
Aspartate hemimagnesium	20	0.33–0.95	−4.6	48.38
Aspartate monopotassium	20	0.33–0.95	−186.9	155.48
Aspartate monosodium	20	0.33–0.95	−65.6	137.02
Aspartic acid	20	0.33–0.95	0	0.04
Cysteine	20	0.33–0.95	0.2	0.26
Cystine	20	0.33–0.95	−0.4	0.39
Glutamate monopotassium	20	0.33–0.95	−125.0	121.08
Glutamate monosodium	20	0.33–0.95	−143.4	80.04
Glutamic acid	20	0.33–0.95	−1.4	0.78
Glutamine	20	0.33–0.95	0.1	0.17
Histidine	20	0.33–0.95	−2.1	1.09
Isoleucine	20	0.33–0.95	0.1	0.00
Leucine	20	0.33–0.95	−2.1	1.13
Lysine · HCl	20	0.33–0.95	−236.8	130.16
Methionine	20	0.33–0.95	−0.3	0.17
Ornithine · HCl	20	0.33–0.95	−336.8	153.26
Phenylalanine	20	0.33–0.95	0.1	0.04
Proline	20	0.33–0.95	−73.4	102.62
Serine	20	0.33–0.95	−24.8	13.90
Threonine	20	0.33–0.95	−2.1	1.13
Tryptophan	20	0.33–0.95	−0.1	0.09
Tyrosine	20	0.33–0.95	0	0.00
Valine	20	0.33–0.95	−0.3	0.48

Source: Adapted from Anderson, C.B. and Witter, L.D., *J. Food Sci.*, 47, 1952, 1982.
Note: To fulfill Equation 4.46, natural algorithm is used throughout, but in some publications decimal logrithm is used.

Harkins–Jura's equation is based on the two-dimensional gas theory and it applies to regions of water activity at which a condensed film is formed. It applies to a_w of about 0.4–0.5. Sandoval and Barreiro (2002) obtained a good approximation of desorption isotherm of cocoa beans at $a_w > 0.5$ and in the temperature range 25°C–35°C.

4.2.7 D'Arcy–Watt Model

Assuming that adsorption surface contains patches with homogeneous centers of adsorption, the free energy is

$$G_A = G_x + G_y + G_{(A-x-y)} + G_{\text{int}} \tag{4.51}$$

where
subscripts x and y are adsorbate adsorbed on different centers of adsorption
A is the total adsorption
int refers to interactions

Table 4.16 Parameters of Kühn's Equation

Product	Temperature (°C)	a_w Range	K_1	K_2	MRE
Apricot	25	0.40–0.80	−9.2133	0.2514	0.76
Chicken, cooked, freeze-dried	60	0.10–0.80	−3.6178	0.9087	3.23
Cocoa	37	0.10–0.70	−2.8564	2.3419	3.19
Coffee beans, Arabica	25	0.42–0.90	−2.5927	6.5013	2.59
	35	0.45–0.90	−2.6219	6.2315	2.46
Coffee extract, freeze-dried	10	0.10–0.80	−7.3726	−1.2415	3.00
Orange juice	37	0.07–0.71	−11.7633	−3.2407	6.29
Peanut oil	30	0.21–0.76	−0.0359	0.0468	2.58
	80	0.21–0.85	−0.0416	0.0774	5.56
Potato, air dried	50	0.10–0.70	−4.7444	3.7985	1.20
Rice bran, unextracted	15	0.55–0.89	−1.4700	6.2673	2.43
	25	0.59–0.90	−1.4188	5.9644	2.12
Skim milk	20	0.10–0.80	−3.0113	1.3983	2.28
Soya meal	15	0.30–0.86	−4.0861	4.0649	0.67
	35	0.36–0.87	−3.7916	3.9532	2.14
Tea, black, dust grade	21	0.10–0.80	−3.2246	0.6756	5.71
Trout, cooked, freeze-dried	60	0.10–0.80	−3.7359	1.0946	1.76
Yeast (*Saccharomyces cerevisiae*) spray-dried	20	0.10–0.67	−5.3564	1.4603	2.64
Yeast (*S. cerevisiae*), freeze-dried	20	0.10–0.67	−5.0269	0.8324	1.82
Yogurt, freeze-dried	45	0.10–0.80	−3.6752	0.2732	3.84

Source: Adapted from Iglesias, H.A. and Chirife, J., in *Handbook of Food Isotherms: Water Sorption Parameters for Food and Food Components*, Academic Press, New York, 1982.

Hence, it can be said that monolayer is adsorbed both by strongly and weakly binding sites, and then formation of multilayer begins the extent of which is limited by the properties of the substrate. Assuming $G_{\text{int}}=0$, the following equation was derived (D'Arcy and Watt, 1970):

$$u=\frac{K_1K_2a_w}{1+K_2a_w}+K_3a_w+\frac{K_4K_5a_w}{1-K_5a_w}$$
$$K_1=\frac{Mn_i}{N};\quad K_2=\alpha\beta_ip_0;\quad K_3=\frac{MK_5}{Np_0};\quad K_4=\frac{MD}{N};\quad K_5=\alpha\beta_m p_0 \tag{4.52}$$

where

K_2 is an equilibrium constant for binding water to the charged or highly polar groups on the solvent-accessible surface
K_3 is an equilibrium constant for binding water to the polypeptide backbone and the less polar sites
K_5 is an equilibrium constant for binding water to the outer multilayers and the bulk
K_1 and K_4 are parameters
M is the molar mass of sorbate
n_i is the number of adsorption centers of type i
N is Avogadro's number
$\alpha=\exp(\mu_0/RT)$
$\beta=Q_i\exp(E_i/RT)$
Q_i is the molecular partition function for molecules adsorbed on i centers
E_i is the heat of sorption on i centers

Table 4.17 Parameters of the D'Arcy–Watt Equation

Product		K_1	K_2	K_3	K_4	K_5
Collagen		107.6×10^{-3}	704	2.9	80.13×10^{-3}	84.766
Cotton		14.6×10^{-3}	5563	6.42	8.67×10^{-3}	94.512
Egg albumin		34.0×10^{-3}	2402	11.28	17.15×10^{-3}	96.214
Egg albumin, coagulated		23.9×10^{-3}	3342	11.36	8.27×10^{-3}	98.056
Egg albumin, freeze-dried		31.2×10^{-3}	2008	10.56	17.19×10^{-3}	95.365
Gelatin		68.3×10^{-3}	1724	13.96	20.73×10^{-3}	98.004
Keratin		39.0×10^{-3}	2024	12.05	20.98×10^{-3}	90.588
β-Lactoglobulin, crystal		25.7×10^{-3}	2479	13.60	16.58×10^{-3}	99.122
β-Lactoglobulin, freeze-dried		17.3×10^{-3}	5471	13.98	16.58×10^{-3}	96.566
Lysozyme (Smith et al., 2002)		36.0×10^{-3}	3091	13.36	21.5×10^{-3}	93.92
γ-Pseudoglobulin		37.3×10^{-3}	2687	14.30	15.07×10^{-3}	98.504
Serum albumin		46.5×10^{-3}	1448	10.84	17.54×10^{-3}	96.478
Sultana raisins (Saravacos et al., 1986)	20°C	50.01	5.58	−30.85	0.86	30.24
	25°C	6.11	16.24	−66.14	0.83	44.09
	30°C	4.32	3.37	−25.01	0.93	22.54
	35°C	4.43	3.32	−27.76	0.94	23.51
c-Zeine		38.9×10^{-3}	945	2.98	20.65×10^{-3}	90.998
b-Zeine		38.0×10^{-3}	1058	2.78	20.84×10^{-3}	89.662

Source: Adapted from D'Arcy, R.L. and Watt, I.C., *Trans. Faraday Soc.*, 66, 1236, 1970.

p_0 is the saturation pressure of adsorbate
D is the number of centers forming multilayers
$\beta_m = Q_m \exp(E_m/RT)$
Q_m is the molecular partition function for molecules adsorbed in multilayers
E_m is the heat of sorption in multilayers

Table 4.17 presents the application of D'Arcy–Watt's equation.

4.2.8 Kiselev Model

Kiselev (1958) derived the sorption isotherm equation assuming two types of interactions. The first type accounts for interactions between adsorbate molecules and adsorbent, and the second type considers the interactions between adsorbate molecules. Hence, vertical and horizontal interactions are taken into account. The following equation was derived for a monomolecular adsorption:

$$a_w = \frac{(u/u_m)}{K_1(1-(u/u_m))(1+K_2(u/u_m))} \tag{4.53}$$

where
K_1 is the parameter accounting for adsorbate–adsorbent interactions
K_2 is the parameter related to adsorbate–adsorbate interactions

In case of multilayer adsorption, the equation is as follows:

$$a_w = \frac{(u/u_m)(1-K_n a_w)^2}{K_1(1-(u/u_m)(1-K_n a_w))(1+K_2(u/u_m)(1-K_n a_w))} \tag{4.54}$$

where the parameter K_n is assumed to be independent of the number of adsorbed layers and considering vertical interactions between already adsorbed layers and adsorbate molecules. When $a_w = 1$ then $u/u_m \to \infty$ and $K_n = 1$, thus

$$a_w = \frac{(u/u_m)(1 - a_w)^2}{K_1(1 - (u/u_m)(1 - a_w))(1 + K_2(u/u_m)(1 - a_w))} \quad (4.55)$$

4.2.9 Dubinin Model

Dubinin et al. (1947) developed a sorption isotherm assuming Polanyi's potential theory and Gaussian distribution of corresponding adsorption potential. The equation is

$$\ln u = \ln\left(\frac{W}{v}\right) - B\left(\ln\frac{1}{a_w}\right)^2 \quad (4.56)$$

where
- W is the volume of adsorption space
- v is the molar volume of adsorbed liquid
- $B = k/\beta^2\,(RT)^2$
- k is the parameter characterizing the pore size distribution
- β is the Dubinin affinity coefficient being the ratio of adsorption potentials of two different vapors adsorbed on the same surface

According to Equation 4.56, a plot of ln u against $(\ln 1/a_w)^2$ should give a straight line of slope B and intercept ln (W/v). Dubinin et al. (1947) equation that was applied to porous bodies gives the volume of micropores (W) in the material.

4.2.10 Jovanović Model

Jovanović (1969) assumed that the surface of adsorbent is impinged by adsorbate molecules, and the number of impinging molecules follows the classical equation:

$$n = \frac{p}{\sqrt{2\pi mkT}} \quad (4.57)$$

The surface is characterized by a potential ε and the average time of settling is

$$\tau = \tau_0 \exp\left(\frac{\varepsilon}{kT}\right) \quad (4.58)$$

For multilayer adsorption, the following equation was derived:

$$\begin{aligned} u &= u_m[1 - \exp(-ba_w)]\exp(da_w) \\ b &= \frac{\sigma\tau p_0}{\sqrt{2\pi mkT}} \\ d &= \frac{\sigma\tau_l p_0}{\sqrt{2\pi mkT}} \end{aligned} \quad (4.59)$$

where

σ is the area ascribed to one molecule
τ is the average settling time for a molecule in the first layer
τ_l is the average settling time for molecules in the second and higher layers
m is the mass of one molecule
p_0 is the saturation pressure
k is Boltzmann's constant
T is the absolute temperature

Parameters b and d describe adsorption in the first and in the second and higher layers, respectively.

4.2.11 Hailwood and Horrobin Model

Hailwood and Horrobin (1946) developed a sorption isotherm assuming three-phase model based on vapor–dry adsorbent–wetted adsorbent. Moreover, two kinds of adsorption sites were assumed. The first kind forms hydrates on certain polar groups and the second kind is associated with solid solution of water in the polymer. A three-parameter equation was developed

$$\frac{u}{u_{\mathrm{m}}} = \frac{K_1 a_{\mathrm{w}}}{1 + K_1 a_{\mathrm{w}}} + \frac{K_2 a_{\mathrm{w}}}{1 - K_2 a_{\mathrm{w}}} \tag{4.60}$$

Rearranging the above equation, the following is obtained:

$$\frac{a_{\mathrm{w}}}{u} = A + B a_{\mathrm{w}} - C a_{\mathrm{w}}^2$$

where

$$A = \frac{1}{u_{\mathrm{m}}(K_1 + K_2)}; \quad B = \frac{K_1 - K_2}{u_{\mathrm{m}}(K_1 + K_2)}; \quad C = \frac{K_1 K_2}{u_{\mathrm{m}}(K_1 + K_2)}$$

and finally

$$u = \left[\frac{A}{a_{\mathrm{w}}} + B - C a_{\mathrm{w}}\right]^{-1} \tag{4.61}$$

where A, B, and C are parameters. Ferro Fontan et al. (1982) analyzed Hailwood and Horrobin's equation and indicated that thermodynamically it is satisfied only for proteins and starchy foods. Hailwood and Horrobin's equation was tested mainly for proteins. Bettelheim and Volman (1957) applied this equation to water sorption isotherms of pectic substances up to water activity 0.3 and King et al. (1983) described almond nutmeat isotherm at 25°C and in the water activity range $0.3 < a_{\mathrm{w}} < 0.9$.

4.2.12 Lewicki Model 1

Lewicki (2000) developed a sorption isotherm based on Raoult's law assuming that water occurs in food in two states, as a free water and hydration water. The equation is similar to the empirical

Oswin equation but the grounds and the way the equation is used are different. Developed equation is a two-parameter formula of the following form:

$$u = A\left(\frac{1}{a_w} - 1\right)^{b-1} \tag{4.62}$$

in which

$$A = \frac{18.016 m_c}{(1-\chi)M_c}\left(\frac{1}{a_w} - 1\right)^{b}$$

where

m_c is the mass of a hydrated complex
M_c is the molecular weight of the hydrated complex
$M_c = M_s + 18.06n$
χ is the ratio between mass of water in hydration shell (m_h) to the mass of the total water, $\chi = m_h/(m_f + m_h)$
m_f is the free water
n is the number of moles of water forming hydration shell

The equation is used in the following way:

$\alpha = u\,((1/a_w) - 1)$ is calculated and drawn in relation to water activity. The point $\alpha = 0$, $a_w = 1$ is included in the relationship (Figure 4.4). Then the parameters of equation describing the following relationship are sought:

$$\alpha = A\left(\frac{1}{a_w} - 1\right)^{b}$$

and parameters A and b are found.

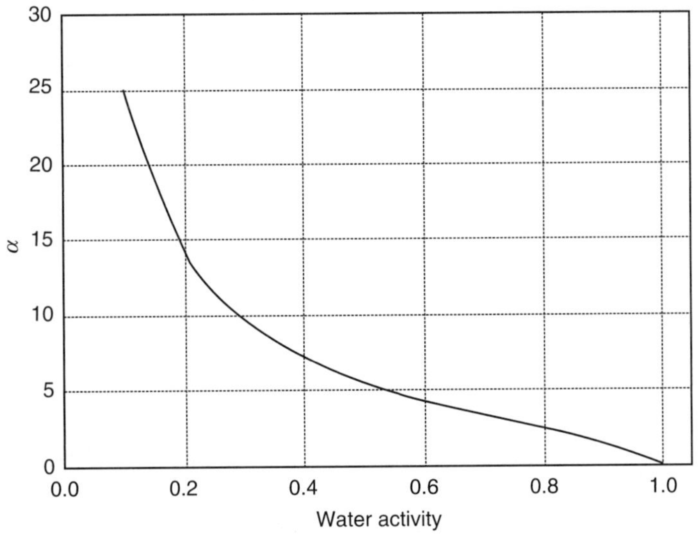

Figure 4.4 Sorption isotherm drawn in coordinates proposed by Lewicki (2000).

Out of 69 analyzed sorption isotherms, 68 could be described by the developed equation with RMS $< \pm 25\%$. Out of this 42.6% were loaded with RMS $< \pm 5\%$, and 70.5% with RMS $< \pm 10\%$. The most frequent RMS was equal to $\pm 4.66\%$ (Lewicki, 2000). Table 4.18 presents parameters of Equation 4.62 for some selected foods.

Table 4.18 Parameters of Raoult-Based Isotherm Equation 4.62

Product	a_w Range	*A*	*b*
Apple cellular fiber	0.11–0.98	9.55	0.2611
Avicel PH 101 MCC	0.11–0.98	5.4	0.5657
Beef	0.03–0.85	7.58	0.4942
Carrageenan	0.11–0.98	22.2	0.5838
Carrot	0.022–0.90	9.71	0.1955
Casein	0.20–0.90	9.58	0.6060
Caseinate sodium salt	0.022–0.865	9.59	0.4908
Caseinate sodium salt	0.015–0.813	8.07	0.5850
Caseinate sodium salt	0.022–0.865	8.73	0.5048
Cellulose powder	0.022–0.865	5.49	0.5826
Cellulose powder	0.11–0.98	3.73	0.5830
Cellulose powder	0.015–0.813	3.11	0.5739
Citrus pulp fiber	0.11–0.98	9.02	0.5252
Coffee	0.05–0.95	4.4	0.2284
Corn bran flour	0.11–0.98	9.33	0.5701
Gelatin	0.11–0.98	19.58	0.5456
Macaroni	0.015–0.813	7.99	0.4802
Mushroom (*Boletus edulis*)	0.03–0.85	8.54	0.5421
Oat bran flour	0.11–0.98	9.82	0.6070
Onion	0.03–0.85	9.78	0.2990
Pea flour	0.03–0.85	14.27	0.6679
Pectin low methoxyl	0.11–0.98	7.15	0.4377
Polyglicine	0.1–0.9	8.35	0.5745
Potato flakes	0.03–0.85	10.42	0.6308
Rice	0.015–0.813	10.85	0.5001
Rice bran flour	0.11–0.98	9.6	0.4696
Soy bran flour	0.11–0.98	8.81	0.5461
Starch gel	0.033–0.980	12.63	0.6020
Starch gel	0.055–0.973	11.77	0.6299
Wheat bran flour	0.11–0.98	9.6	0.4696
Wheat flour	0.015–0.851	11.44	0.6054
Wool	0–0.95	10.56	0.6157
Yeast (*Saccharomyces cerevisiae*)	0.03–0.748	8.78	0.5851
Sodium caseinate:cellulose = 1:1	0.022–0.865	5.56	0.5832
Sodium caseinate:potato starch = 1:1	0.022–0.865	10.89	0.6405
Potato starch:cellulose = 1:1	0.022–0.865	7.47	0.5762
Potato starch:glucose = 1:1	0.022–0.865	6.02	0.6170
Potato starch:NaCl = 4:1	0.022–0.865	10.17	0.5687

Source: Adapted from Lewicki, P.P., *J. Food Eng.*, 43, 31, 2000.
Note: u expressed per 100 g d.m. In some publications, u is given in g/g d.m. In this chapter all data are recalculated to give u in g/100 g d.m.

4.3 SEMIEMPIRICAL MODELS

4.3.1 Chung and Pfost Model

Chung and Pfost (1967) developed an equation describing sorption isotherms assuming the way the free energy of sorption changes with the moisture content. This two-parameter equation is of the form:

$$\ln a_{\mathrm{w}} = -\frac{A}{RT}\exp(-Bu) \tag{4.63}$$

where
R is the universal gas constant
T is the absolute temperature
A and B are parameters (collected in Table 4.19)

The use of temperature term does not eliminate the temperature influence on parameters A and B. Hence, the equation was modified by Pfost et al. (1976) by adding new parameters related to the temperature:

$$u = E - F\ln[-(t + C)\ln a_{\mathrm{w}}] \tag{4.64}$$

where
t is the temperature (°C)
C, E, and F are adjustable parameters

The equation was accepted by the American Society of Agricultural Engineers (1996) as the second standard equation. Parameters of Equation 4.64 for some foods are presented in Table 4.20.

4.3.2 Chen-1 Model

Chen (1971) developed a desorption isotherm, which relates moisture content to drying parameters:

$$a_{\mathrm{w}} = \exp(k + m\exp(nu)) \tag{4.65}$$

Table 4.19 Parameters of the Chung–Pfost Equation at 20°C ± 0.5°C and Water Activity Range 0.11–0.97

Product	Fat Content (%)	Protein Content (%)	*A*	*B*
Rape seed press-cake	10.1	30.3	540.82 ± 126.43	$(6.252 \pm 0.5227) \cdot 10^{-2}$
Rape seed press-cake	9.9	28.5	511.59 ± 127.17	$(6.010 \pm 0.5256) \cdot 10^{-2}$
Rape seeds	44.6		499.40 ± 127.17	$(4.214 \pm 0.3748) \cdot 10^{-2}$
Rape seeds	45.6		448.25 ± 60.17	$(3.915 \pm 0.1778) \cdot 10^{-2}$
Rape seeds raw flakes	46.6		538.38 ± 184.90	$(4.436 \pm 0.5442) \cdot 10^{-2}$
Rape seeds raw flakes	47.4		350.80 ± 118.40	$(3.547 \pm 0.3776) \cdot 10^{-2}$
Rape seeds roasted flakes	47.9		694.30 ± 186.85	$(4.202 \pm 0.4395) \cdot 10^{-2}$
Rape seeds roasted flakes	46.7		472.61 ± 113.52	$(3.4893 \pm 0.2878) \cdot 10^{-2}$

Source: Adapted from Sadowska, J., Ostaszyk, A., and Fornal, J., *Zesz. Probl. Post. Nauk Rol., 427, 137, 1995.* Recalculated.

Table 4.20 Parameters of the Modified Chung and Pfost Equation 4.64

Product	Temperature Range (°C)	a_w Range	*E*	*F*	*C*	Error
Barley	0–50	0.20–0.95	33.363	5.0279	91.323	
Barley (*Hordium vulgare*) (Basunia and Abe, 2005)[a]	5.7–46.3	0.482–0.886	52.694	5.908	1160.72	0.292
Beans	0–50	0.20–0.95	43.001	6.2596	160.626	
Corn, yellow dent	0–50	0.20–0.95	33.876	5.8970	30.205	
Ear corn	20–80	0.11–0.87	32.560	5.3419	30.445	
Oat	25–65	0.15–0.92	28.704	4.7108	35.803	
Peanuts, hull	10–32	0.21–0.97	38.723	6.9896	39.836	
Peanuts, kernel	0–50	0.20–0.95	18.948	3.4196	33.892	
Pumpkin seed flour (Menkov and Durakova, 2005)[b]	10–40		21.849	3.5634	27.7416	1.27
Rice hull	20–40	0.11–0.85	28.645	5.0664	27.733	
Rice rough	0–50	0.20–0.95	29.394	4.6015	35.703	
Rice rough long, Australia	10–38	0.19–0.97	34.351	5.7052	39.016	
Rice rough short, Australia	10–38	0.15–0.95	37.220	6.0006	57.383	
Rice rough short, California	10–40	0.11–0.88	32.675	5.5432	26.674	
Rice rough (Basunia and Abe, 1999)[a]	17.8–45	0.56–0.893	31.05	4.85848	48.19	0.1940
Rice rough, Japan	20–40	0.11–0.85	36.019	5.9312	48.282	
Sorghum	0–50	0.20–0.95	35.649	5.0907	102.849	
Soybeans	0–50	0.20–0.95	41.631	7.1853	100.288	
Sunflower, hulls	10–55	0.11–0.94	24.971	4.676	19.690	
Wheat, durum	0–50	0.20–0.95	37.761	5.5318	112.350	
Wheat, hard	0–50	0.20–0.95	35.616	5.6788	50.998	
Wheat, hard red	5–45	0.11–0.93	0.36058	6.0768	35.590	
Wheat, soft	0–50	0.20–0.95	27.908	4.2360	35.662	

Source: Adapted from Pabis, S., Jayas, D.S., and Cenkowski, S., *Grain Drying*, John Wiley & Sons, New York, 1998.
[a] Standard error of estimation (SEE).
[b] Recalculated SEM.

where k, m, and n are parameters. The equation was derived taking theory of drying as a basis and the steady-state period in which diffusion is a principal mode of water transport. The equation was tested for barley and sorghum and a good agreement was reported between observed and calculated results (Chen, 1971). Parameters in equation are temperature dependent. Chen and Clayton (1971) applied Chen's equation to a number of materials and found that parameter k is close to unity. Chen's equation where simplified becomes equivalent to Bradley's equation (Chirife and Iglesias, 1978).

4.3.3 Peleg Model

Peleg (1993) developed a semiempirical four-parameter equation to describe sigmoid sorption isotherms assuming the mode in which the water is adsorbed by the solid. Hence, it was assumed that sigmoid shape of the line arises from summation of two parabolas (with horizontal and vertical axes, respectively). Hence, the following equation was proposed:

$$u = Aa_w^B + Ca_w^D \tag{4.66}$$

where A, B, C, and D are parameters, and $B < 1$ and $D > 1$. It was shown by Peleg (1993) and Lewicki (1998) that Equation 4.66 gives a better fit than the GAB model. Analysis showed (Lewicki, 1998) that the GAB equation yielded 26.3% while Peleg's equation gave 50.3% fits with RMS < 5%, for the same set of data. Parameters of Peleg's equation are collected in Table 4.21.

Table 4.21 Parameters of Peleg's Equation

Product	Temperature (°C)	a_w Range	A	B	C	D	Error (%)
Agar (Rahman, 1995)	Ambient		31.9	0.53	25.4	14.7	0.65
Animalitos, cookies (Palou et al., 1997)	25	0.11–0.90	9.86	0.77	54.67	5.43	4.69
Carbohydrates 79.80%	35	0.11–0.90	10.31	0.91	76.22	6.30	9.87
Protein 6.10%, fat 4.87%	45	0.11–0.90	10.09	0.96	71.23	6.08	10.60
Avicel PH 101 MCC (Lewicki, 1998)[a]	Ambient	0.11–0.90	8.54	0.6651	8.85	5.49	<10
Beef (Lewicki, 1998)[a]	Ambient	0.03–0.85	7.24	0.4427	84.06	4.87	<10
Carrageenan (Rahman, 1995)	Ambient		34.3	0.61	119.0	12.0	0.74
Carrgeenan (Lewicki, 1998)[a]	Ambient	0.11–0.98	33.18	0.5837	82.91	9.57	<10
Carrot (Lewicki, 1998)[a]	Ambient	0.02–0.90	16.66	1.0602	48.65	4.76	<5
Casein (Rahman, 1995)	Ambient		9.2	0.36	12.7	2.3	0.03
Casein (Lewicki, 1998)[a]	Ambient	0.20–0.90	9.30	0.3651	12.71	2.33	<10
Caseinate, sodium salt (Lewicki, 1998)[a]	Ambient	0.02–0.86	11.73	0.5526	14.60	2.74	<10
Cellulose, powdered (Lewicki, 1998)[a]	Ambient	0.11–0.98	7.34	0.5715	14.20	5.52	<10
Citrus pulp fiber (Lewicki, 1998)[a]	Ambient	0.11–0.98	19.73	1.0185	52.90	16.72	<5
Coffee (Rahman, 1995)	Ambient		12.8	1.50	42.4	10.1	0.11
Coffee (Lewicki, 1998)[a]	Ambient	0.05–0.95	12.86	1.4630	42.43	10.12	<5
Corn bran flour (Lewicki, 1998)[a]	Ambient	0.11–0.98	16.05	0.7164	42.45	12.37	<10
Dextrin (Rahman, 1995)	Ambient		17.9	0.38	4.4	8.1	0.18
Doritos, corn snack (Palou et al., 1997)	25	0.11–0.90	7.94	0.59	45.85	6.52	7.50
Carbohydrates 63.93%	35	0.11–0.90	9.25	0.71	140.25	9.98	7.82
Protein 6.07%, fat 22.77%	45	0.11–0.90	7.45	0.66	47.60	6.18	7.21
Gelatin (Rahman, 1995)	Ambient		28.8	0.58	105.0	12.1	0.90
Gelatin (Lewicki, 1998)[a]	Ambient	0.11–0.98	27.66	0.5474	108.59	11.37	<10
Habaneras, cookies (Palou et al., 1997)	25	0.11–0.90	10.16	0.49	35.72	5.44	2.81
Carbohydrates 77.68%	35	0.11–0.90	10.05	0.51	32.75	5.36	3.21
Protein 6.53%, fat 9.92%	45	0.11–0.90	10.47	0.60	53.15	6.82	5.18
Macaroni (Lewicki, 1998)[a]	Ambient	0.01–0.81	11.25	0.6970	26.60	7.15	<10
Mushroom (*Boletus edulis*) (Lewicki, 1998)[a]	Ambient	0.03–0.85	4.44	0.2178	25.86	2.62	<10
Oat bran flour (Lewicki, 1998)[a]	Ambient	0.11–0.98	14.30	0.6721	35.24	14.23	<10
Pea flour (Lewicki, 1998)[a]	Ambient	0.03–0.85	11.33	0.3735	23.49	6.06	<10
Pectin, low metoxyl (Rahman, 1995)	Ambient		14.0	0.39	90.0	6.0	0.25
Polyglicine (Lewicki, 1998)[a]	Ambient	0.1–0.9	12.89	0.6672	7.68	2.90	<10
Potato flakes (Lewicki, 1998)[a]	Ambient	0.03–0.85	11.41	0.3821	27.66	5.24	<10
Potato starch gel (Lewicki, 1998)[a]	Ambient	0.03–0.98	24.13	0.7029	46.19	20.36	<10
Raisins (Rahman, 1995)	Ambient		11.0	0.34	97.2	4.9	0.63
Ricanelas, cookies (Palou et al., 1997)	25	0.11–0.90	9.32	0.82	55.56	5.63	3.95
Carbohydrates 74.26%	35	0.11–0.90	10.16	0.91	53.10	6.08	5.31
Protein 5.08%, fat 15.59%	45	0.11–0.90	8.81	0.87	40.36	5.07	4.70

(continued)

Table 4.21 (continued) Parameters of Peleg's Equation

Product	Temperature (°C)	a_w Range	*A*	*B*	*C*	*D*	Error (%)
Rice bran flour (Lewicki, 1998)[a]	Ambient	0.11–0.98	16.10	0.7497	133.52	20.04	<10
Soya bran flour (Lewicki, 1998)[a]	Ambient	0.11–0.98	14.51	0.6954	54.83	13.57	<10
Starch, potato (Rahman, 1995)	Ambient		20.9	0.70	16.3	5.5	0.05
Starch, potato (Lewicki, 1998)[a]	Ambient	0.11–0.90	20.79	0.6937	16.32	5.51	<10
Tostitos, corn snack (Palou et al., 1997)	25	0.11–0.90	7.46	0.60	24.91	4.63	6.06
Carbohydrates 65.26% Protein 5.40%, fat 24.42%	35	0.11–0.90	7.02	0.62	21.80	4.09	5.42
	45	0.11–0.90	7.75	0.68	27.28	5.13	6.12
Wheat bran flour (Lewicki, 1998)[a]	Ambient	0.11–0.98	20.14	1.0562	133.62	20.29	<10
Wheat bran (Rahman, 1995)	Ambient		16.4	0.74	62.8	13.8	0.14
Wheat flour (Lewicki, 1998)[a]	Ambient	0.01–0.85	13.43	0.4449	12.35	2.64	<10
Wool (Lewicki, 1998)[a]	Ambient	0–0.95	19.25	0.6906	20.78	13.44	<10
Yeasts (*Saccharomycetes cerevisiae*) (Lewicki, 1998)[a]	Ambient	0.03–0.85	7.55	0.3519	43.31	4.20	<10

[a] RMS.

4.3.4 Lewicki Model 2

The equation developed by Lewicki (1998) was based on the assumption that the sigmoid shape is a result of a difference between two hyperbolae. One of them is characterized by a decreasing slope and predominates at low water activities, and the second one shows monotonous ever-increasing slope, and prevails at high water activities. The following boundary conditions were assumed: $u = 0 \rightarrow a_w = 0$, and $a_w = 1 \rightarrow u \rightarrow \infty$. The equation is as follows:

$$u = F\left(\frac{1}{(1 - a_w)^G} - \frac{1}{1 + a_w^H}\right) \tag{4.67}$$

Parameter G affects the slope of the isotherm at high water activities and its increase moves the inflection point toward the origin of axes. Parameter H affects the slope of the isotherm at low water activities. Increasing H begins to influence the isotherm slope at higher water activities. Analysis of 31 isotherms showed that 71% of them could be successfully described by the developed equation with RMS $< 10\%$. The most frequent RMS was $\pm 2.89\%$ (Lewicki, 1998). Parameters of Equation 4.67 for sorption isotherms of selected foods are collected in Table 4.22.

4.3.5 Norrish Model

Norrish (1966) derived an equation to predict water activity of nonelectrolyte solutions on thermodynamic grounds. Rahman and Perera (1997) proposed to use the Norrish's equation to describe sorption isotherms of solids. Since molecular weight of the solid is not known, the following equation was proposed:

$$a_w = \frac{X_w}{X_w + EX_s}\left[\exp\left[k\left(1 - \frac{X_w}{X_w + EX_s}\right)^2\right]\right] \tag{4.68}$$

where

$E = M_w/M_s$ is the ratio of molecular weight of water to molecular weight of solute

X_w and X_s are the mass fraction of water and solute, respectively

Table 4.22 Parameters of Lewicki's Equation 4.67

Product	a_w Range	F	G	H	RMS (%)
Apple cellular fiber	0.11–0.98	19.89	0.5149	4.3814	<5
Avicel PH 101 MCC	0.11–0.98	9.75	0.2698	0.7333	<10
Beef	0.03–0.85	6.05	0.7576	0.3169	<10
Carrageenan	0.11–0.98	38.5	0.2985	0.7215	<10
Carrot	0.022–0.90	21.15	0.4483	3.0198	<5
Casein	0.20–0.90	22.76	0.1345	0.9299	<10
Caseinate, sodium salt	0.022–0.865	18.2	0.2622	0.7571	<10
Cellulose powder	0.022–0.865	7.15	0.2234	0.6476	<5
Cellulose powder	0.11–0.98	10.57	0.2236	0.8127	<10
Citrus pulp fiber	0.11–0.98	12.38	0.4170	0.6027	<10
Coffee	0.05–0.95	15.09	0.3776	9.6308	<5
Corn bran flour	0.11–0.98	13.38	0.3642	0.5131	<10
Gelatin	0.11–0.98	27.52	0.3981	0.5763	<10
Macaroni	0.015–0.813	8.59	0.5098	0.5628	<10
Mushroom (*Boletus edulis*)	0.03–0.85	6.53	0.8038	0.2979	<10
Oat bran flour	0.11–0.98	11.56	0.3507	0.3852	<10
Pea flour	0.03–0.85	12.23	0.3869	0.2825	<10
Polyglicine	0.1–0.9	21.53	0.1328	0.9649	<10
Potato flakes	0.03–0.85	11.63	0.4712	0.2971	<10
Rice bran flour	0.11–0.98	7.24	0.7012	0.1515	<5
Soy bran flour	0.11–0.98	10.13	0.4592	0.3120	<10
Starch gel	0.033–0.980	20.67	0.2894	0.5150	<10
Starch gel	0.055–0.973	20.79	0.2351	0.6069	<10
Starch, potato	0.112–0.903	25.21	0.2100	0.7940	<10
Wheat bran flour	0.11–0.98	8.01	0.6860	0.2209	<10
Wheat flour	0.015–0.851	20.41	0.2322	0.5634	<10
Wool	0–0.95	20.87	0.2121	0.6456	<10
Yeast (*Saccharomyces cerevisiae*)	0.03–0.85	5.92	0.9699	0.1393	<5

Source: Adapted from Lewicki, P.P., *J. Food Process Eng.*, 21, 127, 1998.
Note: *u* expressed per 100 g d.m. In some publications, *u* is given in g/g d.m. In this chapter all data are recalculated to give *u* in g/100 g d.m.

Taking that $u = X_w/X_s$, Equation 4.67 can be rearranged to the following form:

$$a_w = \frac{u}{u+E}\left[\exp\left[k\left(1-\frac{u}{u+E}\right)^2\right]\right] \tag{4.69}$$

At $k = 0$, the equation is equivalent to Raoult's law for ideal solutions. When m_s tends to 0 and m_w tends to 1, a_w tends to 1. This makes the Norrish model conceptually sound over the entire range of water activity, whereas the GAB model is valid up to water activity of 0.9 (Rahman et al. 1998). Parameters of Equation 4.6 for some foods are collected in Table 4.23.

4.3.6 Ferro Fontan, Chirife, Sancho, and Iglesias Model

Ferro Fontan et al. (1982), analyzing Hailwood and Horrobin's equation and verifying its thermodynamic grounds, showed that isosteric heat of sorption can be related to water content

Table 4.23 Parameters of Norrish's Equation

Product	Temperature (°C)	a_w Range	E	k	RSE
Apple fiber	23	0.110–0.980	6.30	−1.134	0.030
Avicel RC 591	23	0.110–0.980	1.50	−12.266	0.039
Carrageenan	22	0.110–0.980	0.93	−399.955	0.045
Gelatin	22	0.110–0.980	0.47	−1143.900	0.046
HM pectin	22	0.110–0.980	1.50	−35.040	0.019
LM pectin	22	0.110–0.980	3.30	−11.419	0.035
Onion	30	0.113–0.836	4.80	−1.016	0.033
	45	0.112–0.817	9.50	−0.461	0.010
	55	0.110–0.807	8.90	−0.363	0.012
	70	0.108–0.795	7.90	−0.176	0.017
	80	0.105–0.789	4.80	−1.016	0.033

Source: Adapted from Rahman, S.M. and Perera, C.O., in *Engineering & Food at ICEF 7*, Jowitt, R. (ed.), Sheffield Academic Press, Sheffield, UK, 1997, pp. A101–A104. With permission.

Note: u expressed per 100 g d.m. In some publications, u is given in g/g d.m. In this chapter all data are recalculated to give u in g/100 g d.m.

as a power function. Using the Clausius–Clapeyron equation the following relationship was obtained:

$$\ln\left(\frac{\gamma}{a_w}\right) = \frac{Q_1}{Q}\left(\frac{u_1}{u}\right)^r \tag{4.70}$$

where

γ is the activity coefficient
Q_1 is the isosteric heat of sorption at water content u_1
Q is the isosteric heat of sorption at water content
u and r are parameters

Equation 4.70 was simplified (Chirife et al., 1983) to the following form:

$$\ln\left(\frac{\gamma}{a_w}\right) = \alpha u^{-r} \tag{4.71}$$

where

γ is the parameter accounting for the structure of sorbed water
α is a parameter

4.3.7 Ratti, Crapiste, and Rotstein Model

Taking a solution thermodynamics approach and assuming that adsorption energy decreases exponentially with increasing amount of adsorbed adsorbate Ratti et al. (1989) developed the following equation:

$$u = c_1 \exp(-c_2 u) \cdot u^{c_3} \tag{4.72}$$

where c_1, c_2, and c_3 are parameters

4.4 EMPIRICAL MODELS

4.4.1 Henderson Model

Empirical equation developed by Henderson (1952) is written as

$$\ln(1 - a_w) = -kTu^n \tag{4.73}$$

where
- k and n are parameters
- T is the absolute temperature

Parameters of Henderson's equation are presented in Table 4.24. It has been shown that use of temperature in Equation 4.73 does not eliminate the effect of temperature on parameters k and n (Singh and Ojha, 1974). Therefore, Equation 4.73 was modified by Thompson et al. (1968) by adding constants related to temperature:

$$u = \left[\frac{\ln(1 - a_w)}{-K(t + C)}\right]^{\frac{1}{N}} \tag{4.74}$$

where t is the temperature (°C). This equation was adopted by the American Society of Agricultural Engineers (1996) as one of two standard equations used to describe sorption isotherms of agricultural grains. Table 4.25 presents parameters of Henderson's modified equation.

Sometimes Equation 4.73 was modified in a specific way to fit experimental data of some food products. Such a modification was done by Kouhila et al. (2001) by adjusting temperature-related parameters.

$$u = \left[\frac{-\ln(1 - a_w)}{K(1.8T + 492)}\right]^{(1/N)} \tag{4.75}$$

Equation describes well sorption isotherms of sage and verbena (Table 4.26). It has been shown that Henderson's equation yields two or three straight lines when data is plotted in appropriate coordinates (Rockland, 1957). The point of intersection of the straight lines is calculated from the following equation:

$$\ln u = 2.303\frac{Y_{m+1} - Y_m}{n_m - n_{m+1}} \tag{4.76}$$

where $Y = \log kT$ and m is the number of section described by a straight line. According to that finding, the isotherm would be characterized by two or three pairs of constants. Examples of calculations done by Rockland (1957) are presented in Table 4.27.

4.4.2 Chen and Clayton Model

The equation developed by Chen (1971) was modified to account for the effect of temperature on the sorption process (Chen and Clayton, 1971). A four-parameter empirical equation is as follows:

$$\ln a_w = -K_1T^{K_2}\exp(-K_3T^{K_4}u) \tag{4.77}$$

where
- K_1, K_2, K_3, and K_4 are the parameters
- T is the absolute temperature

Table 4.24 Parameters of Henderson's Equation

Product	Temperature (°C)	a_w Range	*K*	*n*	SE	Reference
Corn, shelled	25–28.3		1.59×10^{-6}	2.68		Henderson (1952)
Eggs, spray dried	30		2.95×10^{-5}	2.00		Henderson (1952)
Flaxseed	25–28.3		4.00×10^{-6}	2.81		Henderson (1952)
Lima bean (*Canavalia ensiformis*) roasted, flour	18	0.20–0.973	37.2330×10^{-3}	1.27	0.01	Corzo and Fuentes (2004)[a]
	28	0.20–0.973	34.9650×10^{-3}	1.30	0.02	Corzo and Fuentes (2004)[a]
	38	0.20–0.973	72.5880×10^{-3}	1.07	0.01	Corzo and Fuentes (2004)[a]
	48	0.20–0.973	66.3560×10^{-3}	1.14	0.02	Corzo and Fuentes (2004)[a]
Peaches, dried	23.9		4.11×10^{-4}	0.564		Henderson (1952)
Pigeon pea (*Cajanus cajanas* L. millsp. v. Tovar) roasted, flour	18	0.20–0.973	17.6300×10^{-3}	1.53	0.01	Corzo and Fuentes (2004)[a]
	28	0.20–0.973	7.8448×10^{-3}	1.86	0.01	Corzo and Fuentes (2004)[a]
	38	0.20–0.973	18.2700×10^{-3}	1.57	0.02	Corzo and Fuentes (2004)[a]
	48	0.20–0.973	27.5790×10^{-3}	1.47	0.03	Corzo and Fuentes (2004)[a]
Prunes, dried	23.9		1.25×10^{-4}	0.865		Henderson (1952)
Raisins	23.9		7.13×10^{-5}	1.02		Henderson (1952)
Sorghum	21.1		3.40×10^{-6}	2.31		Henderson (1952)
Soybeans	25		3.20×10^{-5}	1.52		Henderson (1952)
Sweet potato v. Camote	28	0.06–0.81	4.7508×10^{-6}	2.37		Diamante and Munro (1990)[a]
Sweet potato v. Kumara	25	0.06–0.81	2.0772×10^{-5}	1.96		Diamante and Munro (1990)[a]
	40	0.06–0.81	2.0160×10^{-5}	2.04		Diamante and Munro (1990)[a]
	55	0.06–0.81	3.2317×10^{-5}	1.84		Diamante and Munro (1990)[a]
Tomato seeds	30	0.10–0.85	4.06475×10^{-4}	2.291	0.058	Sogi et al. (2003)[a]
	40	0.10–0.85	3.57131×10^{-4}	2.148	0.147	Sogi et al. (2003)[a]
	50	0.10–0.85	1.50759×10^{-4}	2.345	0.120	Sogi et al. (2003)[a]
	60	0.10–0.85	2.46945×10^{-4}	1.807	0.100	Sogi et al. (2003)[a]
	70	0.10–0.85	3.66447×10^{-5}	2.140	0.129	Sogi et al. (2003)[a]
Wheat	32.2		5.59×10^{-7}	3.03		Henderson (1952)

[a] Recalculated.

Table 4.25 Parameters of Modified Henderson's Equation 4.74

Product	Temperature Range (°C)	a_w Range	K	N	C
Barley	0–50	0.20–0.95	2.2919×10^{-5}	2.0123	195.267
Beans	0–50	0.20–0.95	2.0899×10^{-5}	1.8812	254.23
Corn, yellow dent	0–50	0.20–0.95	8.6541×10^{-5}	1.8634	49.810
Cowpea (*Vigna unguiculata*) (Ajibola et al., 2003)	40–80	0.05–0.95	5.69×10^{-5}	2.06	72.1
Ear corn	20–80	0.11–0.87	6.4424×10^{-5}	2.0855	22.150
Oat	25–65	0.15–0.92	8.5511×10^{-5}	2.0087	37.811
Peanuts, hull	10–32	0.21–0.97	1.132×10^{-4}	1.8075	42.154
Peanuts, kernel	0–50	0.20–0.95	65.0413×10^{-5}	1.4984	50.561
Pumpkin seed flour (Menkov and Durakova, 2005)	10–40		1.28×10^{-5}	2.433004	2.071307
Rice rough	0.50	0.20–0.95	1.9187×10^{-5}	2.4451	51.161
Rice rough short, California	10–40	0.11–0.88	3.5502×10^{-5}	2.3100	27.396
Rice rough short, Australia	10–38	0.15–0.95	3.4382×10^{-5}	2.1305	59.535
Rice rough, Japan	20–40	0.11–0.85	4.852×10^{-5}	2.0794	45.646
Rice rough long, Australia	10–38	0.19–0.97	4.1276×10^{-5}	2.1191	49.828
Rice brown	20–40	0.11–0.85	3.2301×10^{-5}	2.2482	34.267
Rice hull	20–40	0.11–0.85	1.4449×10^{-5}	1.9467	24.264
Sorghum	0–50	0.20–0.95	0.8532×10^{-5}	2.4757	113.725
Soybeans	0–50	0.20–0.95	30.5327×10^{-5}	1.2164	134.136
Wheat, durum	0–50	0.20–0.95	2.5738×10^{-5}	2.2110	70.318
Wheat, hard	0–50	0.20–0.95	2.3007×10^{-5}	2.2857	55.815
Wheat, hard red	5–45	0.11–0.93	4.3295×10^{-5}	2.1119	41.565
Wheat, soft	0–50	0.20–0.95	1.2299×10^{-5}	2.5558	64.346
Winged beans	40–70	0.10–0.80	1.64×10^{-4}	1.277	0

Source: Adapted from Pabis, S., Jayas, D.S., and Cenkowski, S., *Grain Drying*, John Wiley & Sons, New York, 1998.

The equation describes adequately corn isotherms in the temperature range from 4.4°C to 60°C. An example of application of that equation is given in Table 4.28.

4.4.3 Chen-2 Model

Chen (1997) developed a sorption isotherm considering evaporation of water in the period of constant rate of drying. The equation is as follows:

$$a_w = \exp\left[-\frac{a}{RT^m}\exp(-bu^j)\right] \tag{4.78}$$

Table 4.26 Sorption Characteristics of Sage and Verbena at 25°C < *t* < 50°C and 0.05 < a_w < 0.91. Equation 4.75

Material	N	K
Sage	1.916	23.5×10^{-6}
Verbena	2.410	14×10^{-6}

Source: Adapted from Kouhila, M., Belghit, A., Daguenet, M., and Boutaleb, B.C., *J. Food Eng.*, 47, 281, 2001.

Table 4.27 Adsorption Isotherms of Selected Foods at $0.10 < a_w < 0.99$ Described by Henderson's Equation

		Part I		Part II		Part III	
Material	**Temperature (°C)**	*n*	Y	*n*	Y	*n*	Y
Beans, v. Great Northern	25.0	3.12	−3.71	2.47	−3.08	1.55	−2.10
Beans, v. Michelite	25.0	3.11	−3.17	2.41	−3.03	1.50	−2.05
Beans, v. Pinto	25.0	3.37	−3.96	2.62	−3.24	1.51	−2.07
Cabbage, v. Savoy	37	1.45	−1.77	0.70	−1.23		
Carrot, v. Chantenay	37	1.00	−1.42	0.74	−1.31		
Coconut, shreded	0.83	−0.84	0.46	−0.62			
Gelatin	25.0	2.96	−3.80	1.88	−2.80	0.81	−1.31
	40.0			1.91	−2.79	0.94	−1.48
Oats, flaked		4.76	−5.39	2.68	−3.34		
Peanut v. Runner, kernel	25.0	3.16	−2.89	1.62	−1.75	0.69	−0.60
Peanut v. Runner, skins	25.0	2.94	−3.90	1.01	−1.37		
Potato white, v. Russet	37	2.25	−2.73	1.65	−2.24		
Rice		1.73	−2.17				
Rice brown, v. Caloro	25	2.73	−3.48				
Rice rough, v. Rexora	25	2.27	−3.28				
Rice white, v. Rexora.	25	2.56	−3.28				
Rice white, v. Rexora. Parboiled	25	2.60	−3.35	1.94	−2.48		
Rice, quick-cooking, long-grain	25	2.90	−3.51	1.35	−1.76		
Salmine	25.0	3.98	−4.24	0.99	−1.70	0.59	−1.09
	40.0	4.20	−4.29	0.95	−1.64	0.66	−1.20
Spinach, v. Prickly winter	37	2.42	−2.53	0.87	−1.35	0.78	−1.25
Starch		3.60	−4.38	2.24	−3.03		
Walnut v. Placentia kernel	7	3.28	−2.34	1.48	−1.30		
	22.5	3.96	−2.64	1.35	−1.13		
Yams, v. Puerto Rico	37	1.98	−2.35	0.83	−1.28		

Source: Adapted from Rockland, R.B., *Food Res.* 22, 604, 1957.

where a, b, j, and m are parameters. The isotherm at different temperatures can be reduced to a single relationship between water content and water activity. Parameters of Equation 4.78 are collected in Table 4.29.

4.4.4 Iglesias and Chirife Model

Iglesias and Chirife (1978) proposed an empirical equation to describe sorption isotherms of high-sugar foods. The theoretical description of these types of isotherms is difficult because during adsorption of water, sugars dissolve and material changes from solid to solid-concentrated solution system. The equation has a form:

$$\ln\left(u + \sqrt{u^2 + u_{0.5}}\right) = ba_w + p \tag{4.79}$$

where

b and p are parameters

$u_{0.5}$ is the moisture content at $a_w = 0.5$

The equation describes adequately sorption isotherms of several fruits.

Table 4.28 Parameters of the Chen and Clayton Equation 4.77

	Parameters					
Product	K_1	K_2	K_3	K_4	**RMS (%)**	**Reference**
Barley	2.475×10^{4}	−1.4245	1.0677×10^{-3}	0.89693		Chen and Clayton (1971)[a]
Corn	1.436×10^{6}	−2.1113	4.9490×10^{-3}	0.64259		Chen and Clayton (1971)[a]
Millet	4.469×10^{8}	−3.1176	70.0016×10^{-3}	0.11792		Chen and Clayton (1971)[a]
Oats	0.967×10^{3}	−0.9247	1.6885×10^{-3}	0.80043		Chen and Clayton (1971)[a]
Rice, hulled	8.704×10^{-2}	0.8393	0.0209×10^{-3}	1.61610		Chen and Clayton (1971)[a]
Rice, rough	0.901×10^{3}	−0.8094	0.2678×10^{-3}	1.16970		Chen and Clayton (1971)[a]
Rye	5.910×10^{4}	−1.6216	7.1188×10^{-3}	0.54004		Chen and Clayton (1971)[a]
Soya	5.466×10^{3}	−1.3386	12.8765×10^{-3}	0.40061		Chen and Clayton (1971)[a]
Sunflower ground nutmeat, 49.66% oil, 33.1% protein d.m.	3780956	−2.391	5.58129	−0.492	5.092	Mok and Hettiarachchy (1990)[a]
Sunflower meal, 63.7% protein d.m.	381586	−1.991	0.93590	−0.299	6.153	Mok and Hettiarachchy (1990)[a]
Sunflower protein concentrate, 72.5% protein d.m.	608.6	−0.846	0.00340	0.700	2.284	Mok and Hettiarachchy (1990)[a]
Sunflower protein isolate, 99.4% protein d.m.	1.534	0.183	$8.3\ 10^{-7}$	2.170	4.516	Mok and Hettiarachchy (1990)[a]
Wheat	1.540×10^{3}	−0.9648	0.9887×10^{-3}	0.90068		Chen and Clayton (1971)[a]

[a] Recalculated.

4.4.5 Oswin Model

Oswin (1946) developed an equation of sigmoid-shaped curves as a mathematical series. The equation is as follows:

$$u = B\left[\frac{a_{\mathrm{w}}}{1 - a_{\mathrm{w}}}\right]^{A} \tag{4.80}$$

where A and B are parameters (collected in Table 4.30). Boquet et al. (1978) stated that Oswin's equation was applicable to protein-based and starchy foods and described reasonably good

Table 4.29 Parameters of Adsorption Isotherm Described by Equation 4.78

Product	a	B	j	m	Error (%)
Bambara groundnut (*Voandzeia subterranea* L. Thouars)	5.104985×10^{21}	1.01	0.5	8	1.4
Paddy rice v. Inga	4.092513×10^{10}	0.26	1	4.5	0.39

Source: Chen, X.D., *Food Res. Int.*, 30, 755, 1997.

Table 4.30 Parameters of the Oswin's Equation

Product	Temperature (°C)	a_w Range	*A*	*B*	Error	Reference
Almond nutmeat	25 ± 1	0.3–0.9	0.494	5.126		King et al. (1983)
Amylose, corn	28.2	0.07–0.61	0.4946	13.8888	1.78	Iglesias and Chirife (1982)[a]
Anise	5	0.10–0.80	0.5155	9.0421	3.57	Iglesias and Chirife (1982)[a]
	45	0.10–0.80	0.6573	7.5054	4.10	Iglesias and Chirife (1982)[a]
Anise (*Pimpinella anisum* L.)	5	0.05–0.80	0.497	8.998	0.135[b]	
	25	0.05–0.80	0.526	8.374	0.265[b]	
	45	0.05–0.80	0.635	7.486	0.161[b]	
Apple juice, spray dried	20	0.10–0.80	0.9745	10.0357	3.27	Iglesias and Chirife (1982)[a]
Avocado, freeze-dried	45	0.10–0.80	0.7386	5.2391	3.57	Iglesias and Chirife (1982)[a]
Barley	25	0.28–0.95	0.3577	11.4189	2.46	Iglesias and Chirife (1982)[a]
Beef, raw, freeze-dried	30	0.10–0.80	0.4292	10.8829	2.48	Iglesias and Chirife (1982)[a]
Bean, black Argentinean			0.610 ± 0.006	8.821 ± 0.115	1.75	Castillo et al. (2003)[c]
Beans v. Rosinha	25	0.60–0.85	0.3337	13.3695	0.52	Iglesias and Chirife (1982)[a]
Beans, runner, freeze-dried	10	0.10–0.80	0.7702	10.4525	3.59	Iglesias and Chirife (1982)[a]
Broccoli, freeze-dried	10	0.10–0.80	0.7423	11.0749	3.59	Iglesias and Chirife (1982)[a]
Cabbage, raw	10	0.10–0.80	0.8369	12.7157	4.27	Iglesias and Chirife (1982)[a]
Cabbage, blanched	10	0.10–0.80	0.7444	12.1523	5.55	Iglesias and Chirife (1982)[a]
Cabbage v. Savoy	25	0.10–0.80	0.8352	9.5873	4.34	Iglesias and Chirife (1982)[a]
	37	0.10–0.80	0.8007	9.7378	1.81	Iglesias and Chirife (1982)[a]
Carboxylmethylcellulose, sodium salt	24	0.10–0.80	0.5888	20.0598	2.70	Iglesias and Chirife (1982)[a]
Cardamom (*Elletaria cardamomum* K.)	5	0.05–0.80	0.337	12.171	0.116[b]	Soysal and Öztekin (2001)
	25	0.05–0.80	0.373	10.512	0.174[b]	Soysal and Öztekin (2001)
	45	0.05–0.80	0.474	8.774	0.290[b]	Soysal and Öztekin (2001)
	60	0.05–0.80	0.430	7.251	1.591[b]	Soysal and Öztekin (2001)
Carrots, freeze-dried	10	0.10–0.80	0.8269	12.8169	2.70	Iglesias and Chirife (1982)[a]
Carrot, blanched	22	0.09–0.80	0.9090	10.0346	5.63	Iglesias and Chirife (1982)[a]
Carrot, frozen	22	0.09–0.80	0.9398	9.7108	4.52	Iglesias and Chirife (1982)[a]
Carrot, DBD	22	0.09–0.80	0.9079	9.4012	3.73	Iglesias and Chirife (1982)[a]
Celery, freeze-dried	5	0.10–0.80	0.6203	12.6920	1.33	Iglesias and Chirife (1982)[a]
	25	0.10–0.80	0.6581	12.0212	1.59	Iglesias and Chirife (1982)[a]

Chicken, cooked, freeze-dried	45	0.10–0.80	0.5386	8.5158	0.55	Iglesias and Chirife (1982)[a]
Chicken, raw, freeze-dried	45	0.10–0.80	0.6503	9.6230	3.02	Iglesias and Chirife (1982)[a]
Chicken sausage, smoked	5	0.113–0.877	0.3855	3.0654	5.79[d]	Singh et al. (2001)
	25	0.113–0.843	0.4968	2.4197	11.24[d]	Singh et al. (2001)
	50	0.111–0.812	1.0251	1.4843	15.06[d]	Singh et al. (2001)
Cinnamon (*Cinnamonum cassia*)	5	0.05–0.80	0.318	13.590	0.655[b]	Soysal and Öztekin (2001)
	25	0.05–0.80	0.359	10.784	0.107[b]	Soysal and Öztekin (2001)
	45	0.05–0.80	0.448	8.129	0.058[b]	Soysal and Öztekin (2001)
	60	0.05–0.80	0.405	7.117	0.127[b]	Soysal and Öztekin (2001)
Cinnamon, vacuum dried	25	0.10–0.80	0.3426	10.7387	2.48	Iglesias and Chirife (1982)[a]
	45	0.10–0.80	0.4337	8.1014	2.18	Iglesias and Chirife (1982)[a]
Cloves (*Caryophyllus aromaticus* L.)	5	0.05–0.80	0.304	8.880	0.162[b]	Soysal and Öztekin (2001)
Cloves, vacuum dried	5	0.10–0.80	0.3155	8.9129	2.31	Iglesias and Chirife (1982)[a]
	5	0.10–0.80	0.2746	10.2268	2.82	Iglesias and Chirife (1982)[a]
Cod, raw, freeze-dried	19.5	0.10–0.64	0.4303	14.3641	1.00	Iglesias and Chirife (1982)[a]
	30	0.10–0.75	0.5256	14.0298	2.55	Iglesias and Chirife (1982)[a]
Coriander (*Coriandrum sativum* L.)	5	0.05–0.80	0.383	8.910	0.120[b]	Soysal and Öztekin (2001)
	25	0.05–0.80	0.383	8.910	0.120[b]	Soysal and Öztekin (2001)
	45	0.05–0.80	0.545	6.584	0.602[b]	Soysal and Öztekin (2001)
Corn flour, whole	50	0.10–0.80	0.2908	11.0956	1.75	Iglesias and Chirife (1982)[a]
Daphne (*Laurus nobilis* L.)	5	0.05–0.80	0.342	10.963	0.443[b]	Soysal and Öztekin (2001)
	25	0.05–0.80	1.389	12.805	0.327[b]	Soysal and Öztekin (2001)
	45	0.05–0.80	0.572	6.434	0.278[b]	Soysal and Öztekin (2001)
	60	0.05–0.80	0.505	4.703	2.422[b]	Soysal and Öztekin (2001)
Egg albumin, freeze-dried	25	0.10–0.80	0.4222	10.0696	3.17	Iglesias and Chirife (1982)[a]
Eggs, spray dried	10	0.10–0.70	0.4805	8.0481	1.45	Iglesias and Chirife (1982)[a]
	37	0.10–0.70	0.5097	7.1133	2.47	Iglesias and Chirife (1982)[a]
	80	0.10–0.70	0.5893	4.6593	1.84	Iglesias and Chirife (1982)[a]
	30	0.06–0.42	0.4286	6.6885	0.54	Iglesias and Chirife (1982)[a]
	40	0.07–0.44	0.4496	6.5016	0.38	Iglesias and Chirife (1982)[a]

(*continued*)

Table 4.30 (continued) Parameters of the Oswin's Equation

Product	Temperature (°C)	a_w Range	*A*	*B*	Error	Reference
	50	0.09–0.45	0.4929	6.4023	0.51	Iglesias and Chirife (1982)[a]
	60	0.10–0.48	0.5040	6.0626	0.42	Iglesias and Chirife (1982)[a]
	70	0.13–0.53	0.5263	5.5450	2.09	Iglesias and Chirife (1982)[a]
Egg white	20	0.10–0.80	0.4374	11.8432	1.39	Iglesias and Chirife (1982)[a]
Egg white, cooked, freeze-dried	10	0.10–0.80	0.4906	12.9140	1.01	Iglesias and Chirife (1982)[a]
Egg yolk, raw, freeze-dried	10	0.10–0.80	0.5238	14.5079	2.19	Iglesias and Chirife (1982)[a]
Egg yolk, cooked freeze-dried	10	0.10–0.80	0.4926	13.4118	1.93	Iglesias and Chirife (1982)[a]
Fennel (*Foeniculum vulgare* Mill.)	5	0.05–0.80	0.665	5.595	0.083[b]	Soysal and Öztekin (2001)
	25	0.05–0.80	0.665	5.595	0.083[b]	Soysal and Öztekin (2001)
	45	0.05–0.80	0.701	5.030	0.044[b]	Soysal and Öztekin (2001)
Fennel tea	5	0.10–0.80	0.6610	5.6321	3.20	Iglesias and Chirife (1982)[a]
	25	0.10–0.80	0.6610	5.6321	3.20	Iglesias and Chirife (1982)[a]
	45	0.10–0.80	0.7066	5.0395	2.35	Iglesias and Chirife (1982)[a]
Fish protein concentrate, whole red hake	25	0.10–0.80	0.3313	9.1425	1.80	Iglesias and Chirife (1982)[a]
	35	0.10–0.80	0.3389	8.5182	1.73	Iglesias and Chirife (1982)[a]
	42	0.10–0.80	0.3654	8.0828	2.99	Iglesias and Chirife (1982)[a]
Gel, sugar-pectin, freeze-dried	25		0.104	13.1	0.015[e]	Tsami et al. (1999)[a,f]
Gel, sugar-pectin, vacuum dried	25		0.909	12.5	0.014[e]	Tsami et al. (1999)[a,e]
Gel, sugar-pectin, microwave dried	25		1.169	13.7	0.012[e]	Tsami et al. (1999)[a,f]
Gel, sugar-pectin, microwave/convective dried	25		1.137	8.0	0.010[e]	Tsami et al. (1999)[a,f]
Gelatin, freeze-dried	25	0.10–0.80	0.3840	15.7591	1.92	Iglesias and Chirife (1982)[a]
Ginger (*Zingiber officinale*, Roscoe)	5	0.05–0.80	0.341	12.147	0.131[b]	Soysal and Öztekin (2001)
	25	0.05–0.80	0.349	11.422	0.108[b]	Soysal and Öztekin (2001)
	45	0.05–0.80	0.458	8.324	0.019[b]	Soysal and Öztekin (2001)
Ginger, vacuum dried	25	0.10–0.80	0.3506	11.4074	1.76	Iglesias and Chirife (1982)[a]
	45	0.10–0.80	0.4628	8.3052	0.98	Iglesias and Chirife (1982)[a]
Gluten, wheat	20.2	0.10–0.80	0.3667	10.334	1.87	Iglesias and Chirife (1982)[a]
	30.1	0.10–0.80	0.3965	9.6526	3.28	Iglesias and Chirife (1982)[a]

Horse radish	5	0.10–0.80	0.4823	13.8900	1.14	Iglesias and Chirife (1982)[a]
	25	0.10–0.80	0.5039	12.9501	3.35	Iglesias and Chirife (1982)[a]
Laurel, vacuum dried	5	0.10–0.80	0.3395	10.9260	1.29	Iglesias and Chirife (1982)[a]
	25	0.10–0.80	0.4905	8.4951	2.94	Iglesias and Chirife (1982)[a]
	45	0.10–0.80	0.5502	7.3283	6.27	Iglesias and Chirife (1982)[a]
Leek, white part	22	0.09–0.80	0.7970	9.1081	5.23	Iglesias and Chirife (1982)[a]
Leek, white part, freeze-dried	22	0.09–0.80	0.7355	9.6308	3.94	Iglesias and Chirife (1982)[a]
Lentil, vacuum dried	25	0.10–0.80	0.3962	12.2191	0.74	Iglesias and Chirife (1982)[a]
	45	0.10–0.80	0.5246	9.2343	2.91	Iglesias and Chirife (1982)[a]
Mango bar, 22.9% water (db)	18 ± 4	0.111–0.929	0.5688	14.9212	31.44	Mir and Nath (1995)
Mango + 2% coconut powder, 20.5% water	18 ± 4	0.111–0.929	0.8964	6.9831	22.65	Mir and Nath (1995)
Mango + 4.5% soy protein concentrate, 19.5% water	18 ± 4	0.111–0.929	1.2097	4.2423	31.44	Mir and Nath (1995)
Marrow, freeze-dried	10	0.10–0.80	0.7212	12.3433	2.13	Iglesias and Chirife (1982)[a]
Muscat (*Myristica fragrans* Houtt.)	5	0.05–0.80	0.285	9.117	0.816[b]	Soysal and Öztekin (2001)
	25	0.05–0.80	0.343	7.849	0.009[b]	Soysal and Öztekin (2001)
	45	0.05–0.80	0.486	6.406	0.024[b]	Soysal and Öztekin (2001)
	60	0.05–0.80	0.472	4.963	0.598[b]	Soysal and Öztekin (2001)
Mushroom (*Boletus edulis*)	45	0.10–0.80	0.8702	8.0644	2.60	Iglesias and Chirife (1982)[a]
Noodle, Japanese udon	20	0.10–0.80	0.3792	11.1561	2.63	Iglesias and Chirife (1982)[a]
Nutmeg, vacuum dried	25	0.10–0.80	0.3473	7.8596	0.67	Iglesias and Chirife (1982)[a]
	45	0.10–0.80	0.4841	6.3978	1.25	Iglesias and Chirife (1982)[a]
Onion, freeze-dried	17	0.10–0.80	0.7070	16.1993	1.37	Iglesias and Chirife (1982)[a]
	27	0.10–0.80	0.7923	13.5043	1.48	Iglesias and Chirife (1982)[a]
Orgeat, freeze-dried	25	0.10–0.80	0.4785	7.9278	2.45	Iglesias and Chirife (1982)[a]
Paranut, vacuum dried	5	0.10–0.80	0.3745	4.0105	2.26	Iglesias and Chirife (1982)[a]
Pepper, green	22.2	0.10–0.75	0.7567	8.4360	4.89	Iglesias and Chirife (1982)[a]
Pekanut, vacuum dried	5	0.10–0.80	0.4457	3.5025	1.58	Iglesias and Chirife (1982)[a]
	25	0.10–0.80	0.4457	3.5025	1.58	Iglesias and Chirife (1982)[a]
	45	0.10–0.80	0.5689	2.7634	2.44	Iglesias and Chirife (1982)[a]

(*continued*)

Table 4.30 (continued) Parameters of the Oswin's Equation

Product	Temperature (°C)	a_w Range	*A*	*B*	Error	Reference
Peppermint (*Mentha piperita* L.)	5	0.05–0.80	0.398	11.523	0.110	Soysal and Öztekin (2001)
	25	0.05–0.80	0.398	11.523	0.110	Soysal and Öztekin (2001)
	45	0.05–0.80	0.510	8.848	0.653	Soysal and Öztekin (2001)
Peppermint tea	5	0.10–0.80	0.4127	11.5537	2.34	Iglesias and Chirife (1982)[a]
	25	0.10–0.80	0.4127	11.5537	2.34	Iglesias and Chirife (1982)[a]
Potato	40	0.054–0.884	0.43	8.2	4.13	Wang and Brennan (1991)
	50	0.054–0.884	0.42	7.4	3.31	Wang and Brennan (1991)
	60	0.054–0.884	0.43	6.2	4.07	Wang and Brennan (1991)
	70	0.054–0.884	0.49	5.1	5.86	Wang and Brennan (1991)
Potato cv. Pentland Dell dried at 40°C	30	0.110–0.970	0.37	13.42	4.54	McLaughlin and Magee (1998)
	45	0.110–0.970	0.40	11.42	4.42	McLaughlin and Magee (1998)
	60	0.110–0.970	0.43	7.51	3.48	McLaughlin and Magee (1998)
Potato, freeze-dried	25	0.10–0.80	0.5975	9.8173	3.23	Iglesias and Chirife (1982)[a]
	37	0.10–0.80	0.4430	10.9452	1.33	Iglesias and Chirife (1982)[a]
Rhubarb	10	0.10–0.80	0.7160	9.1082	1.90	Iglesias and Chirife (1982)[a]
Serum albumin, horse blood	25	0.10–0.80	0.4279	11.3829	1.36	Iglesias and Chirife (1982)[a]
Soy flour, full fat	25	0.10–0.80	0.6406	3.1498	3.36	Iglesias and Chirife (1982)[a]
Soy protein concentrate	1	0.07–0.80	0.3791	12.9405	2.58	Iglesias and Chirife (1982)[a]
	21	0.08–0.80	0.4437	11.5443	3.16	Iglesias and Chirife (1982)[a]
	37	0.09–0.80	0.4870	10.6769	3.49	Iglesias and Chirife (1982)[a]
Thyme, vacuum dried	5	0.10–0.80	0.4043	10.2073	1.85	Iglesias and Chirife (1982)[a]
Turkey, cooked, freeze-dried	22	0.10–0.80	0.4948	8.8714	2.18	Iglesias and Chirife (1982)[a]
Winter savory, vacuum dried	5	0.10–0.80	0.3282	12.7681	1.76	Iglesias and Chirife (1982)[a]

[a] MRE.
[b] RSS $\times 10^{-2}$.
[c] Recalculated.
[d] RMS.
[e] SE.
[f] Gel contained 62% d.m., i.e., 57.4% sugars, 3.3% pectin, and 1.5% citric acid.

Table 4.31 Parameters of the Modified Oswin's Equation 4.81

Product	Temperature Range (°C)	a_w Range	*B*	*C*	*A*
Canola	5–25	0.18–0.90	8.1234	-4.5390×10^{-2}	0.41719
Corn cobs	20–80	0.11–0.87	12.628	-8.8889×10^{-2}	0.46926
Ear corn	20–80	0.11–0.87	15.306	-8.4674×10^{-2}	0.33598
Oats	25–65	0.15–0.92	12.412	-6.0707×10^{-2}	0.34017
Peanut, kernels	10–32	0.21–0.97	6.9812	-4.3870×10^{-2}	0.27012
Popcorn	10–50	0.19–0.93	13.814	-8.2312×10^{-2}	0.38184
Pumpkin seed flour (Menkov and Durakova, 2005)[a]	10–40		11.11582	-4.543×10^{-2}	0.29566
Rice, rough long	10–38	0.19–0.97	14.431	-7.8660×10^{-2}	0.31878
Rice, rough short	20–40	0.11–0.85	14.816	-8.7027×10^{-2}	0.35251
Wheat, durum	5–25	0.18–0.92	14.736	-5.4590×10^{-2}	0.29979
Wheat, hard red	5–25	0.25–0.95	13.101	-5.2626×10^{-2}	0.33348

Source: Adapted from Pabis, S., Jayas, D.S., and Cenkowski, S., *Grain Drying*, John Wiley & Sons, New York, 1998.
[a] SEM = 0.67.

isotherms of meat and vegetables. Chen and Morey (1989) modified Oswin's equation accounting for the influence of temperature on sorption of water vapor by foods. The modified equation is

$$u = (B + Ct)\left(\frac{a_w}{1 - a_w}\right)^A \tag{4.81}$$

where
- A, B, and C are parameters
- t is the temperature (°C)

The parameters of Oswin's modified equation (Equation 4.81) for selected foods are listed in Table 4.31.

4.4.6 Schuchman–Roy–Peleg Model

An equation for sorption isotherm at high water activities was developed by Schuchmann et al. (1990) by taking the GAB equation as a basis for mathematical transformation. The equation is

$$\begin{aligned} u &= \frac{C_1 x}{(1 + C_2 x)(C_3 - x)} \\ x &= \ln\left(\frac{1}{1 - a_w}\right) \end{aligned} \tag{4.82}$$

where C_1, C_2, and C_3 are parameters. An example of application of Equation 4.82 is given in Table 4.32.

4.4.7 Fink and Jackson Model

The whole sigmoid isotherm can be described by the empirical equation developed by Fink and Jackson (1973). The equation is

$$\ln u = A + B \ln\left(a_w^{-c} - 1\right) \tag{4.83}$$

where A, B, and c are parameters.

Table 4.32 Parameters of the Schuchmann et al. (1990) Equation

Product	C_1	C_2	C_3	Mean Square Error
Agar	892	4.20	7.80	0.06
Apple fiber	93	0.00	6.50	0.12
Gelatin	392	0.00	17.4	0.38
Pectin	166	0.01	9.00	0.01
Soy bran	127	1.75	5.04	0.07
Starch	440	1.87	8.99	0.06

Source: Adapted from Schuchman, H., Roy, I., and Peleg, M., *J. Food Sci.*, 55, 759, 1990.

4.4.8 Crapiste and Rotstein Model

Crapiste and Rotstein (1985) proposed the following equation to describe sigmoid isotherms of foods:

$$\ln a_{\mathrm{w}} = -Au^{-b} \exp(-cu)$$
$$A = \frac{a}{R}\left(\frac{1}{T} - \frac{1}{T_1}\right) \tag{4.84}$$

where a, b, and c are parameters.

4.4.9 Mazur and Karpienko Model

Sigmoid type of relationship can be described by an empirical equation of the form (Mazur and Karpienko, 1972):

$$\ln\left(\frac{20\phi}{\ln(100 - \phi)}\right) = \frac{u}{a + bu} \tag{4.85}$$

where ϕ is the relative humidity.

4.4.10 Polynomial Model

Polynomial equation describing sorption isotherms of grains was proposed by Alam and Shove (1973).

$$u = A + Ba_{\mathrm{w}} + Ca_{\mathrm{w}}^2 + Da_{\mathrm{w}}^3 \tag{4.86}$$

where A, B, C, and D are parameters.

4.4.11 Konstance, Craig, and Panzer Model

Analyzing sorption isotherms of bacon, Konstance et al. (1983) found that the isotherms could be adequately described by the equation:

$$u = a\exp(ba_{\mathrm{w}}) \tag{4.87}$$

where a and b are parameters.

4.4.12 Fractal Models

In the classical BET model extended to fractal surfaces the possible number of adsorption sites in the *i*th layer is $i^{-(2-df)}$. Fractal generalization of the Frenkel–Halsey–Hill equation of a polymolecular adsorption allows also for capillary condensation on a fractal surface. The surface fractal dimension is derived as follows:

$$df = 3 + \frac{d[\ln u(a_{\mathrm{w}})]}{d[\ln(-\ln a_{\mathrm{w}})]}$$

and it is valid for $0.2 < a_{\mathrm{w}} < 0.8$. Fripiat et al. (1986) obtained the following equation for the polymolecular adsorption isotherm:

$$u = u_{\mathrm{m}} \frac{c\sum_{i=1}^{n} i^{2-df} \sum_{j=i}^{n} a_{\mathrm{w}}^{j}}{1 + c\sum_{i=1}^{n} a_{\mathrm{w}}^{i}} \sum_{i=1}^{n} i^{2-df} \sum_{j=i}^{n} a_{\mathrm{w}}^{j}$$

$$= \left[\left(a_{\mathrm{w}} + a_{\mathrm{w}}^{2} + a_{\mathrm{w}}^{3} + \cdots + a_{\mathrm{w}}^{n}\right)1^{2-df} + \left(a_{\mathrm{w}}^{2} + a_{\mathrm{w}}^{3} + \cdots + a_{\mathrm{w}}^{n}\right)2^{2-df} + \cdots + a_{\mathrm{w}}^{n} n^{2-df}\right]$$

$$u = u_{\mathrm{m}} \frac{c a_{\mathrm{w}}}{1 + (c-1)a_{\mathrm{w}}} \sum_{i=1}^{\infty} \mathrm{e}^{-(df-2)} a_{\mathrm{w}}^{i} \qquad (4.88)$$

where *df* is the fractal dimension. Sorption isotherm with fractal dimension was also derived by Aguerre et al. (1996):

$$u = u_{\mathrm{m}} \frac{c\sum_{i=1}^{\infty} a_{\mathrm{w}}(2i-1)^{D-2} \sum_{j=1}^{i} (2j-i)^{2-D}}{1 + c\sum_{i=1}^{\infty} a_{\mathrm{w}}^{i}(2i-1)^{D-2}} \qquad (4.89)$$

where i is number of adsorbed layers. Parameters of the fractal isotherms for some foods are presented in Table 4.33. Kutarov and Kats (1993) modified BET equation and improved the accuracy of description of sorption isotherm. The modified equation is

$$u = u_{\mathrm{m}} \frac{c a_{\mathrm{w}}\left(1 - a_{\mathrm{w}}^{n}\right)\left(1 + a_{\mathrm{w}}^{n}\right)^{1/3}}{(1 - a_{\mathrm{w}})[1 + (c-1)a_{\mathrm{w}}]} \qquad (4.90)$$

where n, the number of polylayers, should be positive to have a physical meaning. It was assumed that n reflects the fractal nature of adsorption. Hence, $df = 2 + nf$, where nf is a fractional part of n. Further analysis of the BET equation leads to the following expression:

$$u = u_{\mathrm{m}} \frac{c a_{\mathrm{w}}\left(1 - a_{\mathrm{w}}^{n}\right)}{(1 - a_{\mathrm{w}})(1 - a_{\mathrm{w}} + c a_{\mathrm{w}})} \qquad (4.91)$$

where

- n is $n_{\mathrm{c}} + n_{\mathrm{f}}$
- n_{c} is the real number of polylayers
- n_{f} is the fractional part defined by fractality $D = 2 + n_{\mathrm{f}}$

For potato starch granule, the value of D is between 2.39 and 2.55 (Czepirski et al., 2002).

Table 4.33 Fractal Sorption Isotherms of Some Foods

Product	Mode	Temperature (°C)	a_w Range	u_m	c	D	MRE
Corn continental	Desorption	20	0.25–0.91	9.5	7.5	2.9	1.0
		40	0.25–0.91	8.7	4.6	2.9	2.1
		50	0.25–0.91	7.9	4.0	2.8	1.9
Corn starch	Desorption	25	0.08–0.75	9.8	12.5	2.9	2.3
Native manioc starch	Adsorption	25	0.05–0.90	9.3	15.8	2.8	1.7
Potato starch	Adsorption	25	0.08–0.93	9.5	4.0	2.8	3.3
Rourh rice	Desorption	25	0.10–0.90	9.0	10.7	2.9	2.6
		30	0.11–0.85	9.6	7.9	3.0	1.1
		40	0.11–0.85	8.9	8.3	2.9	1.3
		50	0.11–0.85	8.3	8.3	2.9	1.5
Semolina	Not specified	27	0.06–0.92	7.9	17.9	2.8	3.1
Sorghum	Desorption	20	0.14–0.92	9.6	16.4	2.9	0.9
		40	0.14–0.92	8.9	8.1	2.9	1.1
		50	0.14–0.92	8.8	6.9	2.9	2.1
Tapioca starch	Desorption	25	0.08–0.93	10.2	14.3	2.9	2.4
Wheat durum	Desorption	25	0.12–0.76	8.7	12.6	2.8	0.9
Wheat flour	Not specified	27	0.06–0.92	6.4	18.7	2.6	5.9
Wheat starch	Adsorption	30	0.13–0.90	8.1	9.1	2.8	4.3

Source: Adapted from Aguerre, R.J., Viollaz, P.E., and Suárez, C., *J. Food Eng.*, 30, 227, 1996.

4.5 HYSTERESIS IN SORPTION ISOTHERMS

Adsorption is a process in which adsorbate molecules are attracted by the surface of the solid, which results in an increase of the mass of the material. When activity of water in the solid is higher than that in surroundings, then detachment of molecules from the surface proceeds and a decrease of mass is observed. Hence, process of desorption occurs and relationship between water content and water activity under desorption conditions is called desorption isotherm.

Comparing adsorption and desorption isotherms for the same material, a hysteresis loop is observed (Figure 4.5). Water content on the desorption isotherm is higher than that on the adsorption side at the same water activity. A hysteresis loop was observed for many foods and its shape and extent are dependent on the composition of the material, its structure, temperature, storage time, and number of successive adsorption–desorption cycles.

Many attempts have been made to explain the phenomenon of hysteresis. At present, there is no theory, which would give a complete insight into the processes responsible for that phenomenon. The earliest attempt to explain hysteresis was made by Zsigmondy (1911) and was based on a difference in contact angle during adsorption (dry surface) and desorption (liquid surface and $\theta = 0$). Another explanation proposed by McBain (1935) was based on the "ink bottle" hypothesis. The idea was further developed by Rao (1941). The "ink bottle" has a narrow neck and a wide body. During adsorption, condensation of adsorbate occurs in the wide body (small curvature of meniscus), while during desorption, evaporation takes place in the neck (large curvature of meniscus). Barrer et al. (1956) examined other shapes of pores and emphasized that hysteresis arises also from structural changes in the adsorbent caused by wetting or desiccation.

Cohan (1938) explained hysteresis as the effect of different shape of the meniscus during adsorption and desorption processes. During desorption, capillaries are filled with liquid and the Kelvin mechanism is assumed. The meniscus is hemispherical. During adsorption, capillaries are empty and adsorbed layers form cylindrical meniscus, and the pores are open at both ends. Hence

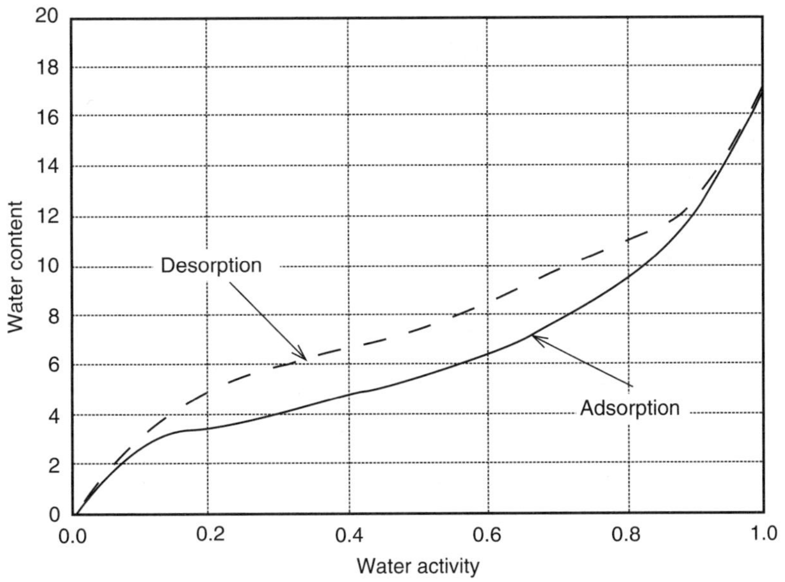

Figure 4.5 Hysteresis loop.

$$\begin{aligned} \ln a_w &= -\frac{2V\sigma}{rRT} \quad \text{desorption} \\ \ln a_w &= -\frac{V\sigma}{rRT} \quad \text{adsorption} \end{aligned} \tag{4.92}$$

where

V is the molar volume
σ is the surface tension

Another explanation of hysteresis was proposed by Foster (1952) who suggested that desorption proceeds from already existing meniscus, while during adsorption time is needed before layers deposited on the surface reach the middle of the capillary and form the meniscus. Hence, the time delay in formation of meniscus is responsible for hysteresis. Gregg and Sing (1967) postulated that hysteresis is due to the differences in molecular structure of adsorbate during adsorption and desorption processes. The analysis was done in relation to the Zsigmondy (1911) hypothesis. Assumption that during desorption contact angle $\theta = 0$ suggests that adsorbed film is indistinguishable in structure from the bulk liquid. In addition, it is expected that layers of adsorbate left on the walls retain, wholly or in part, the liquid-like structure. During adsorption $\theta > 0$, hence, adsorbed film deviates in molecular arrangement from that of bulk liquid. In other words, adsorption leads to the relatively highly ordered structure, while desorption proceeds from the less ordered, more random structure of bulk liquid. On these grounds the following expression was developed:

$$\frac{\cos\theta_a}{\cos\theta_d} = \frac{\ln a_w)_a}{\ln a_w)_d} \tag{4.93}$$

Taking $\theta_d = 0$, Equation 4.92 simplifies to

$$\cos\theta_a = \frac{\ln a_w)_a}{\ln a_w)_d} \tag{4.94}$$

where subscripts a and d denote adsorption and desorption, respectively, and water activities are taken at the same water content. Water polymer interactions in nonrigid materials were suggested as the cause of hysteresis (Wolf et al., 1972) as well as the presence of solutes. Structural changes of polymers can also be responsible for hysteresis (Benson and Richardson, 1955). Chinachoti and Steinberg (1986b) showed that the presence of amorphous sugar causes hysteresis in the starch–sucrose model system. Moreover, the degree of hysteresis was linearly related to the amount of amorphous sucrose present. In natural products such as freeze-dried horseradish root, the changes between glassy and rubbery states were probably one of the main reasons for hysteresis in water sorption (Pääkkönen and Roos, 1990). The extent of water sorption hysteresis on starch was related to the volume of mesopores present in the material (Boki and Ohno, 1991).

The explanation of hysteresis proposed by Yang et al. (1997) is based on the redistribution of sorbate during the sorption process. The phase change occurring during sorption causes temperature gradients, which induce heat transfer. This in turn influences phase changes and mass transfer leading to redistribution of the sorbate. Sorption is either exothermic or endothermic. This can alter the local temperature significantly. Adsorption would proceed at raised local temperatures and desorption at lowered local temperatures. Temperature gradients would induce evaporation or condensation of sorbate. Since $\mu = RTlna_w$ and the system is at equilibrium, the change in temperature must cause change in water activity.

According to Yang et al. (1997), the hysteresis loop is described by the following equations:

BET

$$H = \frac{ACa_w}{(1-a_w)[1+(C-1)a_w]} erfc\left[-\frac{B}{\ln a_w}\right] \tag{4.95}$$

GAB

$$H = \frac{ACDa_w}{(1-Ca_w)[1+Ca_w+CDa_w]} erfc\left[-\frac{B}{\ln a_w}\right] \tag{4.96}$$

Chung–Pfost

$$H = A\left\{C - \frac{1}{D}\ln[-\ln a_w]\right\} erfc\left[-\frac{B}{\ln a_w}\right] \tag{4.97}$$

Halsey

$$H = A\left[-\frac{C}{\ln a_w}\right]^{\frac{1}{D}} erfc\left[-\frac{B}{\ln a_w}\right] \tag{4.98}$$

Henderson

$$H = A\left[-\frac{\ln(1-a_w)}{C}\right]^{\frac{1}{D}} erfc\left[-\frac{B}{\ln a_w}\right] \tag{4.99}$$

Kühn

$$H = A\left[\frac{C}{\ln a_w} + D\right] erfc\left[-\frac{B}{\ln a_w}\right] \tag{4.100}$$

Table 4.34 Parameters of Equations 4.95 through 4.102 for Calculation Hysteresis Loop

Product	Equation	*A*	*B*	*C*	*D*	Coefficient of Variation (%)
Air dried apple	Kühn	3.3318	0.8029	−5.6859	0.4245	3.53
	Smith	5.0166	0.8548	41.1403		4.01
	Halsey	16.1592	0.7775	1.2214	1.2087	4.70
Garnet wheat	Chung-Pfost	5.9466	0.4475	1.4641	1.7852	9.44
	Oswin	9.5347	0.4359	4.6398		9.61
Flour	BET	1.5813	0.2963	43.9486		5.69
	Oswin	2.7691	0.2483	3.0605		5.41
	Henderson	3.6096	0.2320	1.5841	2.7108	5.54
Freeze-dried gluten	Henderson	4.3967	0.3185	1.3366	2.1155	5.62
	Oswin	3.3875	0.3371	2.4126		6.25
Rice	Halsey	8.6658	0.2677	0.4047	1.0156	6.39
	Oswin	4.6372	0.2246	1.9044		9.07

Source: Adapted from Yang, W., Sokhansanj, S., Cenkowski, S., Tang, J., and Wu, Y., *J. Food Eng.*, 33, 421, 1997.

Oswin

$$H = A\left(\frac{1 - a_w}{a_w}\right)^{-\frac{1}{C}} erfc\left[-\frac{B}{\ln a_w}\right] \tag{4.101}$$

Smith

$$H = [A - C\ln(1 - a_w)]erfc\left[-\frac{B}{\ln a_w}\right] \tag{4.102}$$

Parameters of the above equations for some foods are collected in Table 4.34.

4.6 MULTICOMPONENT MIXTURES

Many foods are formulated by mixing dry desired ingredients and packaging in appropriate package. In such a mechanical mixture, equilibration process proceeds and some items lose water in favor of others. At the equilibrium state, all ingredients attain the same water activity, but their water contents are different and do not necessarily assure storage stability. Hence, there is a problem of predicting water activity of a formulated food at the equilibrium state.

According to Briggs (1932), water binding by nonreactive mixed system is additive. Berlin et al. (1973) observed additive water binding by the components in a number of milk protein-containing mixtures. Salwin and Slawson (1959) used sorption isotherms of ingredients to predict water activity and moisture content of a mixture at equilibrium. Lang and Steinberg (1980) developed a mass balance for water in a mixture and showed, in binary and ternary mixtures, that the components sorbed water independently of each other. Applying the Smith equation, Lang and Steinberg (1981b) developed an equation to calculate water activity of a mixture of known composition at given water content. An excellent agreement between the calculated and measured water activities was found in a_w range 0.30–0.95, which proofed the additive rule.

The findings presented above are contradicted by other publications. Palnitkar and Heldman (1971) showed that all predicted water activities of freeze-dried beef were lower than the experimental ones. Iglesias et al. (1975b) had difficulties in predicting water sorption isotherm of sugar beet, and San Jose et al. (1977) showed that water sorption by components is not additive in dried lactose hydrolyzed milk. Gal (1975) has shown that sodium chloride is bound by casein during water adsorption by the mixture. Lang et al. (1981) investigating mixtures of starch with either casein or sucrose found that starch–casein mixture behaved according to the additive rule, while the presence of sucrose in the mixture influenced the enthalpy of water binding by the starch. Interaction of both sodium chloride and sucrose with starch, in desorption as well as adsorption modes, was shown by Chinachoti and Steinberg (1986a). It was explained by a competition for solute between starch and water. Interaction between salt and paracasein in the water activity range 0.76–0.95 was evidenced by Hardy and Steinberg (1984). The system behaved less like salt and more like paracasein, which indicated binding of salt by protein.

Some experiments showed that behavior of the mixture depends on the water activity range and the way the mixture is prepared. Carrillo et al. (1989) prepared a mechanical mixture of starch and sucrose, and dissolved half of it in water and then freeze-dried it. Sorption behavior of mechanical mixture and freeze-dried mixture was different showing that drying process exposed additional sites for water binding. Research done by Bakhit and Schmidt (1992) showed that mechanical mixtures of NaCl/casein at water activities below 0.755 behaved according to the additive rule. At $a_w > 0.755$, experimental values were consistently less than the calculated ones. The freeze-dried mixtures did not follow the additive in the whole investigated a_w range. At $a_w < 0.755$ the mixture adsorbed more water, while at $a_w > 0.755$ the adsorption was less than predicted. It is worth noticing that experiments reported by Hardy and Steinberg (1984) and Chinachoti and Steinberg (1986a) were done on mechanical mixtures wetted and then freeze-dried.

Lewicki and Pomaranska-Lazuka (1994) analyzed sorption behavior of mechanical mixtures of potato starch, sodium caseinate, cellulose, and inorganic salts. It was shown that the sorption properties of mixtures were influenced by composition and water activity range. Mixtures of polymers sorbed more water than predicted. For mixture of potato starch with inorganic salts, the experimental isotherm was below the calculated one in the whole investigated water activity range. Further research done by Lewicki (1997b) showed that mechanical mixtures of macromolecules and mixtures of biopolymers and simple solutes yielded isotherms that were super imposable with those predicted at low water activity range. In the multilayer adsorption range, experimental isotherms were, in most cases, below those predicted. The difference between experimental and predicted isotherms is explained by multiple effects, such as ordering of water on the macromolecular surface, swelling, conformational changes caused by water adsorption, inter-polymer hydrogen bonds, binding of ions, cross linking, competition for water, and plasticization of amorphous regions. The changes are time dependent, and some of them can need long time-scale experiments to be observed. It was suggested that lack of validity in additive rule of sorption processes is rather a rule, not an exception in the case of multicomponent mixtures.

4.7 STATISTICAL ESTIMATION OF GOODNESS OF FIT

Goodness of fit of the isotherm to experimental data can be measured by different methods. The mean percent relative error, standard error of the estimate, and the graph of residuals are most commonly used (Andrieu et al., 1985; Chen and Morey, 1989; Osborn et al., 1989; Mazza and Jayas, 1991). The following equations are used:

1. Root mean square deviation

$$\mathrm{RMSD} = \frac{1}{n}\sqrt{\sum_{1}^{n}(u_e - u_p)^2} \tag{4.103}$$

2. Mean relative error (mean relative percent deviation)

$$\mathrm{MRE} = \frac{100}{n}\sum_{1}^{n}\frac{|u_\mathrm{e} - u_\mathrm{p}|}{u_\mathrm{e}} \tag{4.104}$$

3. Standard error (standard error of the estimated value)

$$\mathrm{SE} = \sqrt{\frac{\sum (u_\mathrm{e} - u_\mathrm{p})^2}{df}} \tag{4.105}$$

4. Root mean square error

$$\mathrm{RMSE} = 100\sqrt{\frac{\sum_{1}^{n}\left(\frac{u_\mathrm{e}-u_\mathrm{p}}{u_\mathrm{e}}\right)^2}{n}} \tag{4.106}$$

5. Variance of regression

$$v = \sum_{1}^{n}\frac{(u_\mathrm{e} - u_\mathrm{p})^2}{n-1} \qquad v = (\mathrm{SE})^2 \tag{4.107}$$

4.8 PHYSICAL EXPLANATIONS OF MODEL PARAMETERS

Although all the theoretical or semitheoretical models are based on the theoretical principles, at the end the parameters are estimated from the regression analysis using experimental data. The reason is that the model parameters cannot be estimated from the theoretical principles used to develop the model. Rahman and Al-Belushi (2006) discussed these issues in detail and their discussions are presented below. At present, the BET and GAB models are widely used and acceptable. The BET model was used mainly because it provides the value of monolayer, which has wide applications in food processing and preservation although it is only valid up to water activity 0.45.

The parameters of BET can be estimated graphically by plotting $[a_\mathrm{w}/(1 - a_\mathrm{w})u_\mathrm{m}]$ versus a_w, which should give a straight line. From the slope $[(c - 1)/u_\mathrm{b}\, c]$ and intercept $(1/u_\mathrm{b}\, c)$ of the line, the BET-monolayer and parameter c can be estimated. The GAB isotherm equation is an extension of the BET model taking into account the modified properties of the sorbate in the multilayer region and the bulk liquid properties through the introduction of a third constant k. The GAB parameters can be estimated by linear multiple regression after parabolic transformation of nonlinear Equation 4.13. Another option is to use nonlinear optimization techniques.

The physical meaning of c in BET may not be valid in many cases. Iglesias and Chirife (1976a) showed that there is limited validity of the relation of c with heat of sorption due to the number of assumptions involved in BET theory. The value of c constant cannot be taken as more than a very rough guide to the magnitude related to heat of sorption. They identified many instances when estimated heat of binding of water from temperature dependency of c was completely wrong although in many instances it was valid. In addition, it is more difficult to explain when monolayer decreases significantly with increasing temperature (Iglesias and Chirife, 1976b). Selected situations are presented when the values of c are very difficult to explain. Maskan and Gogus (1997) reported that negative values of c were an impossibility and were not reported in the literature. Comino (2005) found negative values of c for Limonene-β-cyclodextrin powders. In case of all desorption data, the values of c were negative, whereas in case of adsorption only 9.6% oil content showed negative values. In case of Arabian sweet isotherm in the temperature range of 20°C–50°C, Ahmed et al. (2004) also found negative values of c for BET and GAB models. Similarly, Dinçer and Esin (1996) also observed negative values of c for macaroni in the temperature range 30°C–70°C. Rahman and Al-Belushi (2006) found negative values for desorption isotherm of garlic; in addition it was very difficult to converge in nonlinear regression procedure. Young (1976) also found infinity values of c in the case of peanuts. In case of hazelnut, the values of c were found in the order of 10^6 at 3°C, whereas at 10°C and 30°C the values were in the order of 100 and 10 (Lopez et al., 1995). However, the lack of reliability of the energy parameter c does not preclude the use of BET equation for the determination of monolayer value (Iglesias and Chirife, 1976b). It is well documented for its physical meaning and validity for its monolayer value.

In selecting isotherm models, two points need to be considered. One is the accuracy of the model and another is the physical meaning of the model's parameters (Rahman and Al-Belushi, 2006). Accuracy indicates the mathematical representation of the experimental data, and physical meaning indicates more explanation of the physicochemical process. The whole range of isotherms up to water activity of 0.9 could be predicted using GAB. The GAB model was found most accurate up to water activity 0.9. It is now well accepted that GAB is one of the most accurate models for predicting the isotherm of foods. In many cases, GAB was found accurate and residual was observed at random (Menkov, 2000). Chen and Jayas (1998) evaluated the GAB equation for the isotherms of agricultural products. They found uniform scattered in residual plots of some high protein and high oil materials. Clear patterns in residual plots were found in high starch and some high protein products. The lack of fit for starchy foods was also observed by Bizot (1984). A clear pattern of residual plots was also found for garlic slices (Madamba et al. 1994) and sweet potato slices (Chen and Tsao, 1997). The physical meaning of GAB parameters may not be valid in all cases although this has been emphasized in the literature. These aspects are explained below from the reported literature. Gogus et al. (1998) found completely unreliable monolayer values (very high 40% through 138%) for Turkish delight estimated from GAB model although it gave the most accurate prediction. Theoretically, the values of k should be less than unity (Chirife et al. 1992). However, in the literature, a huge number of papers presented the values of k higher than unity, which was unsound. Lewicki (1997) showed that the GAB model described well sigmoidal type isotherms when parameters are kept in the following regions: $0.24 < k < 1$ and $5.67 \leq C \leq \propto$. Outside these regions, the isotherm is either no longer sigmoid or the monolayer capacity is estimated with the error larger than $\pm 15.5\%$. Keeping constants in the above regions fulfills the requirements of the GAB model.

It is common to use an Arrhenius-type equation for three parameters of the GAB, which transforms six parameters temperature-dependent GAB equation. Chen and Jayas (1998) identified that Arrhenius function did not show a strong relationship between GAB parameters and temperature. In this case, the basis of c and k based on the heat of sorption for mono- and multilayers adsorption could be questioned for its physical meaning. Different regression methods and optimization techniques could be used to estimate the parameters, which may result in different values or many local optimization regions. When it happens, the parameters could vary significantly based on the initial values in the iteration process as well as the variation of location of optimum region. When computing the estimated parameters of the GAB equation for some sorption isotherms of rapeseed, it was very difficult to obtain convergence for the c value. Neither the Gauss–Newton method nor the steepest descent method could obtain the convergence value of c. The value of c always approached infinity no matter what numerical technique was used (Chen and Jayas, 1998). Isotherm data of MCC and cake provided negative values of c and positive value for starch when the temperature was varied from 100°C to 130°C (Bassal and Vasseur, 1992; Bassal et al. 1993a). When isotherm data of MCC cellulose and potato starch from 92°C to 135°C were fitted with the GAB model, Bassal et al. (1993b) found the values of c were positive. The expected values of c should be negative at temperatures above 100°C. Based on the points discussed above, GAB could lose its theoretical basis and transform into just a three-parameter regression model (Rahman, 2005). These are the generic problems when theoretical-based models are extended to fit the experimental data and attempt to relate with physics (Rahman, 2004). In this situation, Rahman and Al-Belushi (2006) strongly believed that only the BET model having two parameters (which could be estimated by graphical methods always giving same values from the same set of data) could be related with the physics for only monolayer value. Computer software could be used to do regression or nonlinear optimization for BET, but it could also be checked with the graphical solution or use the graphical values as the initial values for optimization.

REFERENCES

Aguerre, R.J., Suarez, C., and Viollaz, P.E. New BET type multilayer sorption isotherms. Part I: Theoretical derivation of the model. *Lebensmittel Wissenschaft und Technologie*, 22, 188–191, 1989a.

Aguerre, R.J., Suarez, C., and Viollaz, P.E. New BET type multilayer sorption isotherms. Part II: Modelling water sorption in foods. *Lebensmittel Wissenschaft und Technologie*, 22, 192–195, 1989b.

Aguerre, R.J., Viollaz, P.E., and Suárez, C. A fractal isotherm for multilayer adsorption in foods. *Journal of Food Engineering*, 30, 227–238, 1996.

Ahmed, J., Khan, A.R., and Hanan, A.S. Moisture adsorption of an Arabian sweet (basbusa) at different temperatures. *Journal of Food Engineering*, 64, 187–192, 2004.

Ajibola, O.O., Aviara, N.A., and Ajetumobi, O.E. Sorption equilibrium and thermodynamic properties of cowpea (*Vigna unguiculata*). *Journal of Food Engineering*, 58, 317–324, 2003.

Al-Muhtaseb, A.H., McMeen, W.A.M., and Magee, T.R.A. Water sorption isotherms of starch powders. Part 1. Mathematical description of experimental data. *Journal of Food Engineering*, 61, 297–307, 2004.

Alam, A. and Shove, G.C. Hygroscopicity and thermal properties of soybeans. *Transactions of the American Society of Agricultural Engineers*, 16, 707–709, 1973.

Alhamdan, A.M. and Hassan, B.H. Water sorption isotherms of date pastes as influenced by date cultivar and storage temperature. *Journal of Food Engineering*, 39, 301–306, 1999.

American Society of Agricultural Engineers. Standards 1996. Standard D245.4, 1996.

Anderson, R.B. Modification of the Brunauer, Emmett and Teller equation. *Journal of the American Chemical Society*, 68, 686–691, 1946.

Anderson, R.B. and Hall, K.W. Modification of the Brunauer, Emmett and Teller equation II. *Journal of the American Chemical Society*, 70, 1727–1734, 1948.

Anderson, C.B. and Witter, L.D. Water binding capacity of 22 L-amino acids from water activity 0.33 to 0.95. *Journal of Food Science*, 47, 1952–1954, 1982.

Andrieu, J., Stamatopoulos, A., and Zafiropoulos, M. Equation for fitting desorption isotherms of durum wheat pasta. *Journal of Food Technology*, 20, 651–658, 1985.

Atkins, P. *Physical Chemistry*. Oxford University Press, Oxford, 1998.

Ayranci, E., Ayranci, G., and Doğantan, Z. Moisture sorption isotherms of dried apricot, fig and raisin at 20°C and 36°C. *Journal of Food Science*, 55, 1591–1593, 1625, 1990.

Bakhit, R.M. and Schmidt, S.J. Sorption behaviour of mechanically mixed and freeze-dried NaNcl/Casein mixtures. *Journal of Food Science*, 57, 493–496, 1992.

Barreiro, J.A., Fernández, S., and Sandoval, A.J. Water sorption characteristics of six row barley malt (*Hordeum vulgare*). *Lebensmittel Wissenschaft und Technologia*, 36, 37–42, 2003.

Barrer, R.M., McKenzie, N., and Reay, J.S.S. Capillary condensation in single pores. *Journal of Colloid Science*, 11, 479–495, 1956.

Bassal, A. and Vasseur, J. Measurement of water activity at high temperature, in *Drying'92*, Mujumdar, A.S., Ed., Elsevier Science Publishers, London, pp. 312–321, 1992.

Bassal, A., Vasseur, J., and Loncin, M. Sorption isotherms of food materials above 100°C. *Food Science and Technology*, 26, 505–511, 1993a.

Bassal, A., Vasseur, J., and Lebert, A. Measurement of water activity above 100°C. *Journal of Food Science*, 58(2), 449–452, 1993b.

Basunia, M.A. and Abe, T. Moisture sorption isotherms of rough rice. *Journal of Food Engineering*, 42, 235–242, 1999.

Basunia, M.A. and Abe, T. Adsorption isotherms of barley at low and high temperatures. *Journal of Food Engineering*, 66, 129–136, 2005.

Benson, S.W. and Richardson, R.L. A study of hysteresis in the sorption of polar gases by native and denatured proteins. *Journal of American Chemical Society*, 77, 2585–2590, 1955.

Berlin, E., Anderson, B.A., and Pallansch, M.J. Water sorption by dried dairy products stabilized with carboxymethyl cellulose. *Journal of Dairy Science*, 56, 685–689, 1973.

Bettelheim, F.A. and Volman, D.H. Pectic substances–water. II. Thermodynamics of water vapor sorption. *Journal of Polymer Science*, 24, 445–454, 1957.

Biquet, B. and Labuza, T.P. New model gel system for studying water activity of foods. *Journal of Food Processing and Preservation*, 12, 151–161, 1988.

Bizot, H. Using the G.A.B model to construct sorption isotherms, in *Physical Properties of Foods*, Jowitt, J., Ed., Elsevier Applied Sciences, New York, pp. 43–54, 1984.

Boki, K. and Ohno, S. Moisture sorption hysteresis in kudzu starch and sweet potato starch. *Journal of Food Science*, 56, 125–127, 1991.

Boquet, R., Chirife, J., and Iglesias, H.A. Equations for fitting water sorption isotherms of foods. II. Evaluation of various two-parameter models. *Journal of Food Technology*, 13, 319–327, 1978.

Bradley, S.R. Polymolecular adsorbed films. Part I. The adsorption of argon on salt crystals at low temperatures, and the determination of surface fields. *Journal of Chemical Society*, 1467–1474, 1936a.

Bradley, S.R. Polymolecular adsorbed films. Part II. The general theory of the condensation of vapours on finely divided solids. *Journal of Chemical Society*, 1799–1804, 1936b.

Briggs, D.R. Water relationship in colloids. II. Bound water in colloids. *Journal of Physical Chemistry*, 36, 367–386, 1932.

Brunauer, S., Emmett, P.H., and Teller, E. Adsorption of gases in multimolecular layers. *Journal of the American Chemical Society*, 60, 309–319, 1938.

Brunauer, S., Deming, L.S., Deming, W.S., and Teller, E. On a theory of the van der Waals adsorption of gases. *Journal of the American Chemical Society*, 62, 1723–1732, 1940.

Brunauer, S., Skalny, J., and Bodor, E.E. Adsorption on nonporous solids. *Journal of Colloid and Interface Science*, 30, 546–552, 1969.

Cadden, A.-M. Moisture sorption characteristics of several food fibers. *Journal of Food Science*, 53, 1150–1155, 1988.

Calzetta Resio, A., Aguerre, R.J., and Suárez, C. Analysis of the sorptional characteristics of amaranth starch. *Journal of Food Engineering*, 42, 51–57, 1999 (MRE).

Carrillo, P.J., Gilbert, S.G., and Daun, H. Starch/sucrose interactions by organic probe analysis: An inverse gas chromatography study. *Journal of Food Science*, 54, 162–165, 1989.

Castillo, M.D., Martinez, E.J., González, H.H.L., Pacin, A.M., and Resnik, S.L. Study of mathematical models applied to sorption isotherms of Argentinean black bean varieties. *Journal of Food Engineering*, 60, 343–348, 2003.

Cepeda, E., de Latierro, R.O., San José, M.J., and Olazar, M. Water sorption isotherms of roasted coffee and coffee roasted with sugar. *International Journal of Food Science and Technology*, 34, 287–290, 1999.

Chen, C.S. Equilibrium moisture curves for biological materials. *Transactions of the American Society of Agricultural Engineers*, 14, 924–926, 1971.

Chen, X.D. A new water sorption equilibrium isotherm model. *Food Research International*, 30, 755–759, 1997.

Chen, C. Moisture sorption isotherms of pea seeds. *Journal of Food Engineering*, 58, 45–51, 2003.

Chen, C.S. and Clayton, J.T. The effect of temperature on sorption isotherms of biological materials. *Transactions of the American Society of Agricultural Engineers*, 14, 927–929, 1971.

Chen, C. and Jayas, D.S. Evaluation of the GAB equation for the isotherms of agricultural products. *Transactions of the American Society of Agricultural Engineers*, 41(6), 1755–1760, 1998.

Chen, C.C. and Morey, R.V. Comparison of four EMC/ERH equations. *Transactions of the American Society of Agricultural Engineers*, 32, 983–990, 1989.

Chen, C. and Tsao, T. Equilibrium relative humidity characteristics of sweet potato slices. *Journal of Agricultural Machinery*, 7(1), 99–113, 1997.

Chinachoti, P. and Steinberg, M.P. Interaction of solutes with raw starch during desorption as shown by water retention. *Journal of Food Science*, 51, 450–452, 1986a.

Chinachoti, P. and Steinberg, M.P. Moisture hysteresis is due to amorphous sugar. *Journal of Food Science*, 51, 453–455, 1986b.

Chirife, J. and Ferro Fontan, C. Water activity of fresh foods. *Journal of Food Science*, 47, 661–663, 1982.

Chirife, J. and Iglesias, H.A. Equations for fitting water sorption isotherms of foods: Part 1—a review. *Journal of Food Technology*, 13, 159–174, 1978.

Chirife, J., Boquet, R., Ferro Fontan, C., and Iglesias, H.A. A new model to describe the water sorption isotherms of foods. *Journal of Food Science*, 48, 1382–1383, 1983.

Chirife, J., Timmermann, E.O., Iglesias, H.A., and Boquet, R. Some features of the parameter *k* of the GAB equation as applied to sorption isotherms of selected food products. *Journal of Food Engineering*, 15, 75–82, 1992.

Chuang, L. and Toledo, R.T. Predicting the water activity of multicomponent systems from water sorption isotherms of individual components. *Journal of Food Science*, 41, 922–927, 1976.

Chung, D.S. and Pfost, H.B. Adsorption and desorption of water vapor by cereal grains and their products. *Transactions of the American Society of Agricultural Engineers*, 10, 552–555, 1967.

Cohan, L.H. Sorption hysteresis in the vapor pressure of concave surfaces. *Journal of the American Chemical Society*, 60, 433–435, 1938.

Comino, P.R. The sorption isotherm properties of limonene-β-cyclodextrin complex powder. Master of Philosophy Thesis. The University of Queensland, Brisbane, 2005.

Corzo, O. and Fuentes, A. Moisture sorption isotherms and modeling for pre-cooked flours of pigeon pea (*Cajanus cajanas* L. millsp.) and lima bean (*Canavalia ensiformis*). *Journal of Food Engineering*, 65, 443–448, 2004.

Crapiste, G.H. and Rotstein, E. Prediction of sorptional equilibria data for starch-containing foodstuffs. *Journal of Food Science*, 47, 1501–1507, 1982.

Crapiste, G.H. and Rotstein, E. Sorption equilibrium at changing temperatures, in *Drying of Solids. Recent International Developments*, Mujumdar, A.S., Ed., John Wiley, New York, pp. 41–45, 1985.

Czepirski, L., Komorowska-Czepirska, E., and Szymonska, J. Fitting of different models for water vapour sorption on potato starch granules. *Applied Surface Science*, 196, 150–153, 2002.

D'Arcy, R.L. and Watt, I.C. Analysis of sorption isotherms of non-homogeneous sorbents. *Transactions of the Faraday Society*, 66, 1236–1245, 1970.

Das, M. and Das, S.K. Analysis of moisture sorption characteristics of fish protein myosin. *International Journal of Food Science and Technology*, 37, 223–227, 2002.

de Boer, J.H. and Zwikker, C. Adsorption als Folge von polarisation; die Adsorptionisotherme. *Annalen der Physic und Chemie*, B3, 407–418, 1929.

de Boer, J.H. *The Dynamic Character of Adsorption*. Clarendon Press, Oxford, pp. 61–81, 1953.

de Jong, G.I.W., van den Berg, C., and Kokelaar, A.J. Water vapour sorption behaviour of original and defatted wheat gluten. *International Journal of Food Science and Technology*, 31, 519–526, 1996.

Diamante, L.M. and Munro, P.A. Water desorption isotherms of two varieties of sweet potato. *International Journal of Food Science and Technology*, 25, 140–147, 1990.

Dinçer, T.D. and Esin, A. Sorption isotherms of macaroni. *Journal of Food Engineering*, 27, 211–228, 1996.

Doe, P.D. and Heruwati, E.S. Drying and storage of tropical fish—a model for the prediction of microbial spoilage, in *Food Preservation by Moisture Control*, Seow, C.C., Ed., Elsevier Applied Science, London, pp. 117–135, 1988.

Dubinin, M.M., Zawierina, E.D., and Raduszkiewicz, L.W. Sorption and structure of active coals. I. Research on adsorption of organic vapors. *Zurnal Fiziczeskoj Chimii*, 21, 1351–1362, 1947 (in Russian).

Dural, N.H. and Hines, A.L. Adsorption of water on cereal-bread type dietary fibers. *Journal of Food Engineering*, 20, 17–43, 1993a.

Dural, N.H. and Hines, A.L. A new theoretical isotherm equation for water vapor–food systems: Multilayer adsorption on heterogeneous surfaces. *Journal of Food Engineering*, 20, 75–96, 1993b.

Esteban, M.A., Marcos, A., Fernández-Salguero, J., and Alcalá, M. An improved simple gravimetric method for measurement of high water activities. *International Journal of Food Science and Technology*, 24, 139–146, 1989.

Esteban, M.A., Alcalá, M., Marcos, A., Fernández-Salguero, J., Garcia de Fernando, G.D., Ordoñez, J.A., and Sanz, B. Water activity of culture media used in food microbiology. *International Journal of Food Science and Technology*, 25, 464–468, 1990.

Falade, K.O. and Aworh, O.C. Adsorption isotherms of osmo-oven dried African star apple (*Chrysophyllum albidum*) and African mango (*Irvingia gabonensis*) slices. *European Food Research and Technology*, 218, 278–283, 2004.

Favetto, G., Resnik, S., Chirife, J., and Ferro Fontan, C. Statistical evaluation of water activity measurements obtained with the Vaisala Humicamp humidity meter. *Journal of Food Science*, 48, 534–538, 1983.

Ferro Fontan, C., Chirife, J., Sancho, E., and Iglesias, H.A. Analysis of a model for water sorption phenomena in foods. *Journal of Food Science*, 47, 1590–1594, 1982.

Fink, D.H. and Jackson, R.D. An equation for describing water vapor adsorption isotherms of soils. *Soil Science*, 116, 256–261, 1973.

Foster, A.G. Sorption hysteresis. Part II. The role of the cylindrical meniscus effect. *Journal of the Chemical Society* (London), 1806–1812, 1952.

Foster, K.D., Bronlund, J.E., and (Tony) Paterson, A.H.J. The prediction of moisture sorption isotherms for dairy powders. *International Dairy Journal*, 15, 411–418, 2005 (SE).

Fripiat, J.J., Gatineau, L., and Van Damme, H. Multilayer physical adsorption on fractal surfaces. *Langmuir*, 2, 562–567, 1986.

Gal, S. Solvent versus non-solvent water in casein—sodium chloride—water system, in *Water Relations of Foods*, Duchworth, R.B., Ed., Academic Press, London, pp. 183–191, 1975.

Giovanelli, G., Zanoni, B., Lavelli, V., and Nani, R. Water sorption, drying and antioxidant properties of dried tomato products. *Journal of Food Engineering*, 52, 135–141, 2002.

Gocho, H., Shimizu, H., Tanioka, A., Chou, T.-J., and Nakajima, T. Effect of acetyl content on the sorption isotherm of water by cellulose acetate: comparison with the thermal analysis results. *Carbohydrate Polymers*, 41, 83–86, 2000a.

Gocho, H., Shimizu, H., Tanioka, A., Chou, T.-J., and Nakajima, T. Effect of polymer chain end on sorption isotherm of water by chitosan. *Carbohydrate Polymers*, 41, 87–90, 2000b.

Gogus, F., Maskan, M., and Kaya, A. Sorption isotherms of Turkish delight. *Journal of Food Processing and Preservation*, 22, 345–357, 1998.

Gregg, S.J. and Sing, K.S. *Adsorption, Surface Area and Porosity*. Academic Press, New York, 1967.

Greig, R.I.W. Sorption properties of heat denetured cheese whey protein. Part 1. Moisture sorption isotherms. *Dairy Industries International*, (5), 18–22, 1979.

Guggenheim, E.A. *Application of Statistical Mechanics*. Clarendon Press, Oxford, pp. 186–206, 1966.

Hailwood, A.J. and Horrobin, S. Adsorption of water by polymers: Analysis in terms of a simple model. *Transactions of the Faraday Society*, 42B, 84–92, 1946.

Halsey, G. Physical adsorption on non-uniform surfaces. *Journal of Chemical Physics*, 16, 931–937, 1948.

Hardy, J.J. and Steinberg, M.P. Interaction between sodium chloride and paracasein as determined by water sorption. *Journal of Food Science*, 49, 127–131, 136, 1984.

Harkins, W.D. and Jura, G. Surfaces of solids. XIII. A vapor adsorption method for the determination of the area of a solid without the assumption of a molecular area, and the areas occupied by nitrogen and other molecules on the surface of a solid. *Journal of the American Chemical Society*, 66, 1366–1373, 1944.

Heiss, R. and Eichner, K. Die Haltbarkeit von Lebensmitteln mit niedrigen und mittleren Wassergehalten. *Chemie, Mikrobiologie, Technologie der Lebensmittel*, 1, 33–40, 1971.

Henderson, S.M. A basic concept of equilibrium moisture. *Agricultural Engineering*, 33, 29–31, 1952.

Hocking, A.D. Moulds and yeasts associated with foods of reduced water activity: Ecological interactions, in *Food Preservation by Moisture Control*, Seow, C.C., Ed., Elsevier Applied Science, London, pp. 57–72, 1988.

Hoover, S.R. and Mellon, E.F. Application of polarization theory to sorption of water vapor by high polymers. *Journal of the American Chemical Society*, 72, 2562–2566, 1950.

Iglesias, H.A. and Chirife, J. Isosteric heats of water vapor sorption on dehydrated foods. Part II. Hysteresis and heat of sorption comparison with BET theory. *Lebensmittel Wissenschaft und Technologie*, 9, 123–127, 1976a.

Iglesias, H.A. and Chirife, J. BET monolayer values in dehydrated foods and food components. *Lebensmittel Wissenschaft und Technologie*, 9, 107–113, 1976b.

Iglesias, H.A. and Chirife, J. Prediction of the effect of temperature on water sorption isotherms of food materials. *Journal of Food Technology*, 11, 109–116, 1976c.

Iglesias, H.A. and Chirife, J. Equilibrium moisture contents of air dried beef. Dependence on drying temperature. *Journal of Food Technology*, 11, 565–773, 1976d.

Iglesias, H.A. and Chirife, J. An empirical equation for fitting water sorption isotherms of fruits and related products. *Canadian Institute of Food Science and Technology Journal*, 11, 12–15, 1978.

Iglesias, H.A. and Chirife, J. *Handbook of Food Isotherms: Water Sorption Parameters for Food and Food Components*. Academic Press, New York, 1982.

Iglesias, H.A., Chirife, J., and Lombardi, J.L. An equation for correlating equilibrium moisture content in foods. *Journal of Food Technology*, 10, 289–297, 1975a.

Iglesias, H.A., Chirife, J., and Lombardi, J.L. Water sorption isotherms in sugar beet root. *Journal of Food Technology*, 10, 299–308, 1975b.

Jakobsen, M. Filament hygrometer for water activity measurement: Interlaboratory evaluation. *Journal of Association of Official Analytical Chemists*, 66, 1106–1111, 1983.

Jayas, D.S. and Mazza, G. Comparison of five, three parameter equations for the description of adsorption data of oats. *Transactions of the American Society of Agricultural Engineers*, 36, 119–125, 1993.

Jovanović, D.S. Physical adsorption of gases. I. Isotherms for monolayer and multilayer adsorption. *Kolloid-Zeitschrift und Zeitschrift für Polymere*, 235, 1203–1213, 1969.

Jura, G. and Harkins, W.D. Surfaces of solids. XIV. A unitary thermodynamic theory of the adsorption of vapors on solids and of insoluble films on liquid subphases. *Journal of the American Chemical Society*, 68, 1941–1952, 1946.

Kaymak-Ertekin, F. and Sultanoğlu, M. Moisture sorption isotherm characteristics of peppers. *Journal of Food Engineering*, 47, 225–231, 2001 (recalculated).

Kim, S.S. and Bhowmik, S.R. Moisture sorption isotherms of concentrated yogurt and microwave vacuum dried yogurt powder. *Journal of Food Engineering*, 21, 157–175, 1994.

Kim, H.K., Song, Y., and Yam, K.L. Water sorption characteristics of dried red peppers (*Capsicum annum* L.). *International Journal of Food Science and Technology*, 29, 339–345, 1991.

Kim, S.S., Kim, S.Y., Kim, D.W., Shin, S.G., and Chang, K.S. Moisture sorption characteristics of composite foods filled with strawberry jam. *Lebensmittel Wissenschaft und Technologie*, 31, 397–401, 1998.

King, A.D., Halbrook, W.U., Fuller, G., and Whitehand, L.C. Almond nutmeat moisture and water activity and its influence on fungal flora and seed composition. *Journal of Food Science*, 48, 615–617, 1983.

Kiselev, A.V. Adsorbate–adsorbate interactions in the adsorption of vapor on graphitized carbon blacks. The equation for the adsorption isotherm with adsorbate–adsorbate interactions. *Kolloidnyj Zurnal*, 20, 338–348, 1958 (in Russian).

Konstance, R.P., Craig, J.C., and Panzer, C.C. Moisture sorption isotherms of bacon slices. *Journal of Food Science*, 48, 127–130, 1983.

Kouhila, M., Belghit, A., Daguenet, M., and Boutaleb, B.C. Experimental determination of the sorption isotherms of mint (*Mentha viridis*), sage (*Salvia officinalis*) and verbena (*Lippia citriodora*). *Journal of Food Engineering*, 47, 281–287, 2001.

Kühn, J. A new theoretical analysis of adsorption phenomena. Introductory part: The characteristic expression of the main regular types of adsorption isotherms by a single simple equation. *Journal of Colloid Science*, 19, 685–698, 1964.

Kutarov, V.V. and Kats, B.M. Determination of the fractal dimension of ion-exchange fibers from adsorption data. *Russian Journal of Physical Chemistry*, 67, 1854–1856, 1993.

Lai, C.C., Gilbert, S.G., and Mannheim, C.H. Effect of flow conditioners on water sorption and flow properties of egg powder. *Journal of Food Engineering*, 5, 321–333, 1986.

Labuza, T.P. Sorption phenomena in foods. *Food Technology*, 22, 263–272, 1968.

Labuza, T.P., Acott, K., Tatini, S.R., Lee, R.Y., Flink, J., and McCall, W. Water activity determination: A collaborative study of different methods. *Journal of Food Science*, 41, 910–917, 1976.

Lang, K.W. and Steinberg, M.P. Calculation of moisture content of a formulated food system for any given water activity. *Journal of Food Science*, 45, 1228–1230, 1980.

Lang, K.W. and Steinberg, M.P. Predicting water activity from 0.30 to 0.95 of a multicomponent food formulation. *Journal of Food Science*, 46, 670–672, 680, 1981a.

Lang, K.W. and Steinberg, M.P. Linearization of the water sorption isotherm for homogeneous ingredients over a_w 0.30–0.95. *Journal of Food Science*, 46, 1450–1452, 1981b.

Lang, K.W., McCune, T.D., and Steinberg, M.P. A proximity equilibrium cell for rapid determination of sorption isotherms. *Journal of Food Science*, 46, 936–938, 1981.

Lang, K.W., Whitney, R.McL., and Steinberg, M.P. Mass balance model for enthalpy of water binding by a mixture. *Journal of Food Science*, 47, 110–113, 1982.

Lewicki, P.P. A method to calculate constants in the BET equation applicable to the type III isotherms of the BET classification. *Acta Alimentaria Polonica*, 3/27(1), 67–77, 1977.

Lewicki, P.P. The applicability of the GAB model to food water sorption isotherms. *International Journal of Food Science and Technology*, 32, 553–557, 1997a.

Lewicki, P.P. Water sorption isotherms and their estimation in food mechanical mixtures. *Journal of Food Engineering*, 32, 47–68, 1997b.

Lewicki, P.P. A three parameter equation for food moisture sorption isotherms. *Journal of Food Process Engineering*, 21, 127–144, 1998.

Lewicki, P.P. Raoult's law based food water sorption isotherm. *Journal of Food Engineering*, 43, 31–40, 2000.

Lewicki, P.P. and Pomaranska-Lazuka, W. Water sorption isotherms of food model mixtures, in *Developments in Food Engineering*, Yano, T., Matsuno, R., and Nakamura, K., Eds., Blackie Academic and Professional, London, pp. 185–187, 1994.

Linders, L.J.M., de Jong, G.I.W., Meerdink, G., and van't Riet, K. Carbohydrates and dehydration inactivation of *Lactobacillus plantarum*: the role of moisture distribution and water activity. *Journal of Food Engineering*, 31, 237–250, 1997.

Lopez, A., Pique, M.T., Clop, M., Tasias, J., Romero, A., Boatella, J., and Garcia, J. The hygroscopic behaviour of the hazelnut. *Journal of Food Engineering*, 25, 197–208, 1995.

Madamba, P.S., Driscoll, R.H., and Buckle, K.A. Predicting the sorption behavior of garlic slices. *Drying Technology*, 12(3), 669–683, 1994.

Marzec, A. and Lewicki, P.P. Antiplasticization of cereal-based products by water. Part I. Extruded flat bread. *Journal of Food Engineering*, 73, 1–8, 2006.

Maskan, M. and Gogus, F. The fitting of various models to water sorption isotherms of Pistachio nut paste. *Journal of Food Engineering*, 33, 227–237, 1997.

Masuzawa, M. and Sterling, C. Gel–water relationships in hydrophilic polymers: thermodynamics of sorption of water vapor. *Journal of Applied Polymer Science*, 12, 2023–2032, 1968.

Mazur, P.J. and Karpienko, W.I. Influence of additives on water binding by material. *Piszczewaja Tiechnologia*, (4), 79–81, 1972 (in Russian).

Mazza, G. and Jayas, D.S. Equilibrium moisture characteristics of sunflower seed, hulls and kernels. *Transactions of the American Society of Agricultural Engineers*, 34, 534–538, 1991.

Mazza, G. and LeMaguer, M. Water sorption properties of yellow globe onion (*Allium ceps* L.). *Canadian Institute of Food Science and Technology Journal*, 11(4), 189–193, 1978.

McBain, J.W. An explanation of hysteresis in the hydration and dehydration of gels. *Journal of the American Chemical Society*, 57, 699–700, 1935.

McLaughlin, C.P. and Magee, T.R.A. The determination of sorption isotherm and the isosteric heats of sorption for potatoes. *Journal of Food Engineering*, 35, 267–280, 1998.

McMinn, W.A.M. and Magee, T.R.A. Thermodynamic properties of moisture sorption of potato. *Journal of Food Engineering*, 60, 157–165, 2003.

Menkov, N.D. Moisture sorption isotherms of lentil seeds at several temperatures. *Journal of Food Engineering*, 44, 205–211, 2000.

Menkov, N.D. and Durakova, A.G. Equilibrium moisture content of semi-defatted pumpkin seed flour. *International Journal of Food Engineering*, 1, 1–7, 2005.

Mir, M.A. and Nath, N. Sorption isotherms of fortified mango bars. *Journal of Food Engineering*, 25, 141–150, 1995.

Mok, C. and Hettiarachchy, N.S. Moisture sorption characteristics of ground sunflower nutmeat and its products. *Journal of Food Science*, 55, 786–789, 1990.

Moraga, G., Martinez-Navarrete, N., and Chiralt, A. Water sorption isotherms and glass transition in strawberries: Influence of pretreatment. *Journal of Food Engineering*, 62, 315–321, 2004.

Norrish, R.S. An equation for the activity coefficients and equilibrium relative humidities of water in confectionery syrups. *Journal of Food Technology*, 1, 25–39, 1966.

Osborn, G.S., White, G.M., Sulaiman, A.H., and Walton, L.R. Predicting equilibrium moisture properties of soybeans. *Transactions of the American Society of Agricultural Engineers*, 32, 2109–2113, 1989.

Oswin, C.R. The kinetics of package life. III. Isotherm. *Journal of the Society of Chemical Industry* (London), 65, 419–421, 1946.

Pääkkönen, K. and Roos, Y.H. Effect of drying conditions on water sorption and phase transitions of freeze-dried horseradish roots. *Journal of Food Science*, 55, 206–209, 1990.

Pabis, S., Jayas, D.S., and Cenkowski, S. *Grain Drying*. John Wiley and Sons, New York, 1998.

Palnitkar, M.P. and Heldman, D.R. Equilibrium moisture characteristics of freeze-dried beef components. *Journal of Food Science*, 36, 1015–1018, 1971.

Palou, E., López-Malo, A., and Argaiz, A. Effect of temperature on the moisture sorption isotherms of some cookies and corn snacks. *Journal of Food Engineering*, 31, 85–93, 1997.

Peleg, M. Assessment of a semi-empirical four parameter general model for sigmoid moisture sorption isotherms. *Journal of Food Process Engineering*, 16, 21–37, 1993.

Pfost, H.B., Mourer, S.G., Chung, D.S., and Milliken, G.A. Summarizing and reporting equilibrium moisture data for grains. Technical Report 76–3520. American Society of Agricultural Engineers, St. Joseph, MI., 1976.

Pickett, G. Modification of the Brunauer–Emmeett–Teller theory of multilayer adsorption. *Journal of the American Chemical Society*, 67, 1958–1962, 1945.

Pixton, S.W. and Henderson, S. Moisture relations of dried peas, shelled almonds and lupines. *Journal of Stored Product Research*, 15, 59–63, 1979.

Polanyi, M. Neueres über Adsorption und Ursache der Adsorptionskräfte. *Zeitschrift für Elektrochemie*, 26, 370–374, 1916.

Rahman, S. *Food Properties Handbook*. CRC Press, Boca Raton, FL, pp. 1–86, 1995.

Rahman, M.S. State diagram of date flesh using differential scanning calorimetry (DSC). *International Journal of Food Properties*, 7(3), 407–428, 2004.

Rahman, M.S. Dried food properties: challenges ahead. *Drying Technology*, 23(4), 695–715, 2005.

Rahman, M.S. and Al-Belushi, R.H. Dynamic isopiestic method (DIM): Measuring moisture sorption isotherm of freeze-dried garlic powder and other potential uses of DIM. *International Journal of Food Properties*, 9(3), 421–437, 2006.

Rahman, M.S. and Labuza, T.P. Water activity and food preservation, in *Handbook of Food Preservation*, Rahman, M.S., Ed., Marcel Dekker, New York, pp. 339–382, 1999.

Rahman, S.M. and Perera, C.O. Evaluation of the GAB and Norrish models to predict the water sorption isotherms in foods, in *Engineering & Food at ICEF 7*, Jowitt, R., Ed., Sheffield Academic Press, Sheffield, U.K., pp. A101–A104, 1997.

Rahman, M.S., Perera, C.O., and Thebaud, C. Desorption isotherm and heat pump drying kinetics of peas. *Food Research International*, 30(7), 485–491, 1998.

Rao, K.S. Hysteresis in sorption. *Journal of Physical Chemistry*, 45, 500–539, 1941.

Ratti, C., Crapiste, G.H., and Rotstein, E. A new water sorption equilibrium expression for solid foods based on thermodynamic considerations. *Journal of Food Science*, 54, 738–742, 747, 1989.

Rockland, R.B. A new treatment of hygroscopic equilibria: Application to walnuts (*Juglans regia*) and other foods. *Food Research*, 22, 604–628, 1957.

Rubel, G.O. A non-intrusive method for the measurement of water vapour sorption by bacterial spores. *Journal of Applied Microbiology*, 83, 243–247, 1997.

Rüegg, M. Water in dairy products related to quality, with special reference to cheese, in *Properties of Water in Foods in Relation to Quality and Stability*, Simatos, D. and Multon J.L., Eds., Martinus Nijhoff Publishers, Dordrecht, pp. 603–625, 1985.

Sadowska, J., Ostaszyk, A., and Fornal, J. Water sorption ability of seeds, flakes and press-cakes of rapeseed. *Zeszyty Problemowe Postepow Nauk Rolniczych*, 427, 137–143, 1995.

Salwin, H. and Slawson, V. Moisture transfer in combinations of dehydrated foods. *Food Technology*, 13, 715–718, 1959.

San Jose, C., Asp, N., Burvall, A., Dahlgvist, A., and Logetko, V.P. Water sorption in lactose hydrolyzed dry milk. *Journal of Dairy Science*, 60, 1539–1543, 1977.

Sandoval, A.J. and Barreiro, J.A. Water sorption isotherms of non-fermented cocoa beans (*Theobroma cacao*). *Journal of Food Engineering*, 51, 119–123, 2002.

Saravacos, G.D., Tsiourvas, D.A., and Tsami, E. Effect of temperature on the water adsorption isotherms of sultana raisins. *Journal of Food Science*, 51, 381–383, 387, 1986.

Schuchmann, H., Roy, I., and Peleg, M. Empirical models for moisture sorption isotherms at very high water activities. *Journal of Food Science*, 55, 759–762, 1990.

Scott, V.N. and Bernard, D.T. Influence of temperature on the measurement of water activity of food and salt systems. *Journal of Food Science*, 48, 552–554, 1983.

Shepherd, H. and Bhardwaj, R.K. A study of the desorption isotherms of rewet pigeon pea type-17. *Journal of Food Science*, 51, 595–598, 1986.

Singh, R.S. and Ojha, T.P. Equilibrium moisture content of groundnut and chillies. *Journal of the Science of Food and Agriculture*, 25, 451–459, 1974.

Singh, R.R.B., Rao, K.H., Anjaneyulu, A.S.R., and Patil, G.R. Moisture sorption properties of smoked chicken sausages from spent hen meat. *Food Research International*, 34, 143–148, 2001.

Smith, S.E. The sorption of water vapor by high polymers. *Journal of the American Chemical Society*, 69, 646–651, 1947.

Smith, A.L., Shirazi, H.M., and Mulligan, R.S. Water sorption isotherms and enthalpies of water sorption by lysozyme using the quartz crystal microbalance/heat conduction calorimeter. *Biochimica et Biophysica Acta*, 1594, 150–159, 2002.

Sogi, D.S., Shivhare, U.S., Garg, S.K., and Bawa, A.S. Water sorption isotherm and drying characteristics of tomato seeds. *Biosystems Engineering*, 84, 297–301, 2003.

Soysal, Y. and Öztekin, S. Sorption isosteric heat for some medicinal and aromatic plants. *Journal of Agricultural Engineering Research*, 78(2), 159–166, 2001.

Spiess, W.E.L. and Wolf, W. Water activity, in *Theory and Applications to Food*, Rockland, L.B. and Beuchat, L.R., Eds., Marcel Dekker, New York, pp. 215–233, 1987.

Stamp, J.A., Linscott, S., Lomauro, C., and Labuza, T.P. Measurement of water activity of salt solutions and foods by several electronic methods as compared to direct vapor pressure measurement. *Journal of Food Science*, 49, 1139–1142, 1984.

Stoloff, L. Calibration of water activity measuring instruments and devices: collaborative study. *Journal of the Association of Official Analytical Chemists*, 61(5), 1166–1178, 1978.

Thompson, T.L., Peart, R.M., and Foster, G.H. Mathematical simulation of corn drying—A new model. *Transactions of the American Society of Agricultural Engineers*, 24, 582–586, 1968.

Timmermann, E.O. A BET-like three sorption stage isotherm. *Journal of Chemical Society, Faraday Transactions*, 85, 1631–1645, 1989.

Timmermann, E.O., Chirife, J., and Iglesias, H.A. Water sorption isotherms of foods and foodstuffs: BET or GAB parameters? *Journal of Food Engineering*, 48, 19–31, 2001.

Tolaba, M.P., Peltzer, M., Enriquez, N., and Pollio, M.L. Grain sorption equilibria of quinoa grains. *Journal of Food Engineering*, 61, 365–371, 2004 (MRE).

Tsami, E., Krokida, M.K., and Drouzas, A.E. Effect of drying method on the sorption characteristics of model fruit powders. *Journal of Food Engineering*, 38, 381–392, 1999.

Tubert, A.H. and Iglesias, H.A. Water sorption isotherms and prediction of moisture gain during storage of packed cereal crackers. *Lebensmittel Wissenschaft und Technologie*, 19, 365–368, 1986.

Velázquez de la Cruz, G., Torres, J.A., and Martin-Polo, M.O. Temperature effect on the moisture sorption isotherms for methylcellulose and ethylcellulose films. *Journal of Food Engineering*, 48, 91–94, 2001.

Velazquez, G., Herrera-Gómez., A., and Martin-Polo, M.O. Theoretical determination of first adsorbed layer of water in methylcellulose. *Journal of Food Engineering*, 59, 45–50, 2003.

Viollaz, P.E. and Rovedo, C.O. Equilibrium sorption isotherms and thermodynamic properties of starch and gluten. *Journal of Food Engineering*, 40, 287–292, 1999.

Vullioud, M., Márquez, C.A., and De Michelis, A. Desorption isotherms for sweet and sour cherry. *Journal of Food Engineering*, 63, 15–19, 2004.

Wang, G. and Brennan, J.G. Moisture sorption isotherm characteristics of potatoes at four temperatures. *Journal of Food Engineering*, 14, 269–287, 1991.

Weisser, H. Influence of temperature on sorption equilibria, in *Properties of Water in Foods*, Simatos, D. and Multon, J.L., Eds., Martinus Nijhoff Publishers, Dordrecht, pp. 95–118, 1985.

White, L. A limitation of the determination of surface area by the "point B" method. *Journal of Physical Chemistry*, 51, 644–647, 1947.

Wojciechowski, J. Water activity. Importance of water activity in production of meat products. Part I. *Gospodarka Miesna*, (3), 20–22, 1978 (in Polish).

Wolf, M., Walker, J.E., and Kapsalis, J.G. Water vapor sorption hysteresis in dehydrated foods *Journal of Agricultural and Food Chemistry*, 20, 1073–1077, 1972.

Wolf, W., Spiess, W.E.L., Jung, G., Weisser, H., Bizot, H., and Duckworth, R.B. The water-vapour sorption isotherms of microcristalline cellulose (MCC) and purified potato starch. Results of collaborative study. *Journal of Food Engineering*, 3, 51–73, 1984.

Yang, W., Sokhansanj, S., Cenkowski, S., Tang, J., and Wu, Y. A general model for sorption hysteresis in food materials. *Journal of Food Engineering*, 33, 421–444, 1997.

Yanniotis, S. A new method for interpolating and extrapolating water activity data. *Journal of Food Engineering*, 21, 81–96, 1994.

Yanniotis, S. and Zarmboutis, I. Water sorption isotherms of pistachio nuts. *Food Science and Technology*, 29(4), 372–375, 1996.

Young, J.H. Evaluation of models to describe sorption and desorption equilibrium moisture content isotherms of Virginia-type peanuts. *Transactions of the American Society of Agricultural Engineers*, 19, 146–155, 1976.

Young, D.M. and Crowell, A.D. *Physical Adsorption of Gases*. Butterworth & Co., London, 1962.

Zsigmondy, R. Über die Struktur des Gels der Kieselsäure. Theorie der Entwässerung. *Zeitschrift für Anorganische und Allgemaine Chemie*, 71, 356–377, 1911.

CHAPTER 5

Freezing Point: Measurement, Data, and Prediction

Mohammad Shafiur Rahman, K.M. Machado-Velasco, M.E. Sosa-Morales, and Jorge F. Velez-Ruiz

CONTENTS

5.1 INTRODUCTION

5.1.1 Definition

Freezing of foods is done for long-term preservation and to extend shelf life and retain quality attributes. Freezing modifies the physical state of a substance by changing water into ice as energy is removed by cooling below freezing temperature and this process concentrates the unfrozen phase. Freezing of foods can be carried out by different methods, where the freezing rate determines the physical characteristics and quality of the food (Rahman and Vélez-Ruiz, 2007).

To design and simulate the process equipment, there are several aspects to be considered by food processors, such as the amount of energy to be eliminated (both enthalpies: sensible and latent), food composition and concentration, shape of food (i.e., homogeneous or not) and dimensions (area, height, length, radius, thickness, and width), physical properties (density, freezing point, freezing time, latent heat, specific heat, and thermal conductivity), and initial or equilibrium freezing temperature.

The freezing temperature is the temperature at which the first ice crystals appear at a point where both phases, liquid (water) and solid (ice crystal), coexist in equilibrium. It should be the same as the melting point of ice if the freezing and melting processes are considered completely reversible from a thermodynamic viewpoint. Foods are multicomponent systems, in which there are minerals and organic compounds, including acids, fats, gases, proteins, salts, and sugars, dispersed in water, which is the main component of food. The freezing point of a food is lower than that of pure water. A depression of the freezing point is observed as a consequence of the food constituents (Figure 5.1). Due to high water content in a majority of food items, most raw foods have a freezing point between 0°C and −3.9°C. Other low moisture food items reach lower freezing points as a consequence of their high solid content (Table 5.1). The temperature of the food undergoing the freezing process remains relatively constant (plateau) until most of the water is frozen; after this the temperature decreases to reach the storage temperature or that of the freezing medium (Desrosier, 1970).

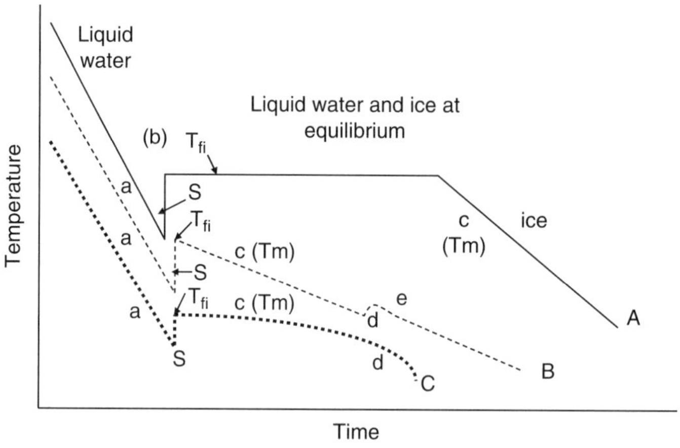

Figure 5.1 Typical cooling curves. A: Water, B: solution, C: foods, S: supercooling. a: Ice crystallization temperature, b: equilibrium or initial freezing point, c: end of freezing, d: formation of first solute crystal, e: eutectic point. (From Rahman, S., *Food Properties Handbook*, CRC Press, Boca Raton, FL, 1995. With permission.)

Table 5.1 Freezing Point of Some Food Items[a]

Food Item	T_f (°C)	X_w^o (Watt and Merril, 1963)	Reference
Cabbage, cauliflower, lettuce	−0.56 to 0	0.924, 0.91, 0.95	Desrosier (1970)
Poblano chili, nopal	−0.1	0.917, 0.934	Machado et al. (2006)
Egg white, milk	−0.5	0.876, 0.874	Ibarz and Barbosa (2001)
Asparagus, pea, spinach, tomato	−1.11 to −0.55	0.917, 0.923, 0.907, 0.935	Desrosier (1970); Heldman (1975)
Corn dough, tortilla, and similar products	−1.1	0.597, 0.448, 0.447	Machado et al. (2006)
Carrots, onion, raspberries	−1.67 to −1.11	0.882, 0.891, 0.867	Desrosier (1970); Heldman (1975)
Apricot, pear, strawberry	−1.67 to −1.11	0.853, 0.862, 0.899	Ibarz and Barbosa (2001)
Apple juice	−1.44	0.878	Heldman (1975)
Beef, fish, potato, sweet cherry	−2.78 to −1.67	0.674, 0.775, 0.798, 0.804	Desrosier (1970); Geankoplis (1998); Singh and Heldman (2001)
Ice cream	−1.86	0.621 to 0.632	Heldman (1975)
Oaxaca cheese	−2.80	0.566	Machado et al. (2006)
Lamb, veal	−3.33 to −2.22	0.625, 0.620	Desrosier (1970)
Banana, coconut, garlic	−4.44 to −3.33	0.757, 0.657, 0.613	Desrosier (1970)
Walnut	−6.67	0.035	Desrosier (1970)
Peanut	−8.33	0.054	Desrosier (1970)
Apple juice concentrate	−11.3	—	Heldman (1975)

[a] Variation in the freezing temperature is due to food composition.

5.1.2 Factors Affecting Food Freezing

From an engineering viewpoint, the freezing process is an unsteady-state heat transfer phenomenon in which the food loses heat by convection through its surface and by conduction at its interior. Heat transfer in foods defined as bulk takes place by a combination of conduction and convection, in which some factors, both internal and external, affect the freezing process. The most important factor affecting the freezing process is food composition, in which the water content (bound and free); the characteristics of other food components, soluble and insoluble solids; and other factors such as specific heat, enthalpy, and thermal and mass diffusivities (Jie et al., 2003) are important. Other influencing factors are food microstructure, particle size, porosity, and certain biological aspects (species, age, and maturity) (Devine et al., 1996; Hamdami et al., 2004). Skrede (1996) indicated variety and maturity stage as two important biological factors that influence the freezing process and the final quality of fruits. For dough and baked products, Inoue and Bushuk (1996) consider the type and concentration of ingredients, flour characteristics, fermentation performance, and gas retention as major factors affecting freezing.

External factors are related to physical conditions in which the freezing process occurs, such as the origin and processing of foods, freezing rate, supercooling, heat and mass transfer coefficients at the surface, and type of freezing equipment used (Lind, 1991). The most important external factor is convective heat transfer, expressed by the magnitude of surface heat transfer coefficient, with values from 5 to 10 W/m^2 K for natural convection by means of still air in a cold room, and from 500 to 1200 W/m^2 K for forced convection as in a liquid immersion process (Rahman and Vélez-Ruiz, 2007).

5.1.3 Freezing Point Depression

When the freezing of pure or aqueous systems is considered, thermodynamic and kinetic factors are involved, affecting each other during the freezing process. Before a crystallization stage can occur at the initial freezing point, a significant amount of energy must be removed. This withdrawal of energy causes sensible heat to decrease below 0°C without initiating a change of phase. Thus, undercooling or supercooling results in a thermodynamically unstable state, which initiates the formation of submicroscopic aqueous aggregates leading to a proper interface (seed) necessary for the transformation of a liquid to a solid state (Figure 5.1).

The presence of a solute in the aqueous medium increases the complexity of crystallization, in which mass transfer plays a dominant role at certain points of the process. Figure 5.1 shows typical cooling responses during the freezing process for three different systems. Homogeneous nucleation occurs in a pure system (at point a), whereas heterogeneous nucleation occurs in nonpure systems (points b and c). Due to the release of latent heat during ice formation (nucleation and crystal growth), the temperature of the food is lower than the corresponding temperature of water, a result of freezing point depression of the solution or food.

The freezing point represents the chemical potential in equilibrium between liquid and solid phases; at 0°C the vapor pressure of pure water and ice reaches equilibrium or equal potential. Therefore, the addition of solutes reduces the partial vapor pressure of water, and hence equilibrium between the two phases can be achieved only through a reduction in temperature, causing freezing point depression due to the difference between the freezing point of pure water and that of the solution or food (Sahagian and Goff, 1996).

5.1.4 Food Properties

As a consequence of freezing, properties of food are affected. This process mainly influences physical or thermal properties such as density, enthalpy, specific heat, thermal conductivity, and thermal diffusivity, besides inducing chemical, nutritional, physicochemical, textural, and sensorial changes.

The freezing point of foods is one of the most important properties, and it is needed in the prediction of thermophysical properties because of the discontinuity exhibited at that point (Rahman, 1994). Thermophysical properties are essential to model and predict some engineering aspects involved in the freezing process, for example, equipment dimensions, freezing times, and heat loads (Mannapperuma and Singh, 1989; Lind, 1991; Valentas et al., 1997; Singh and Heldman, 2001; Vélez, 2007). Freezing also affects other characteristics of foods, such as bound, free, and frozen water; effective molecular weight; enthalpy; and water activity (Rahman, 1994).

5.2 MEASUREMENT METHODS

Although there are several reports on the measurement of thermal properties and freezing point of foods in general, the great variation in composition, origin, and processing of foodstuff requires experimental determination for specific foods (Lind, 1991). Different types of apparatus and equipment are used to measure the freezing point of foods. The most commonly used methods for foods are the cooling curve method and differential scanning calorimetry (DSC). Figure 5.2 shows a commercial cryoscope used to measure the freezing point.

5.2.1 Freezing Curve or Thermistor Cryoscope Method

5.2.1.1 Cooling Curve Method

The cooling curve is one of the most simple, accurate, and widely used methods to measure the freezing point of foods. The wide application of this method is due to its accuracy and simplicity.

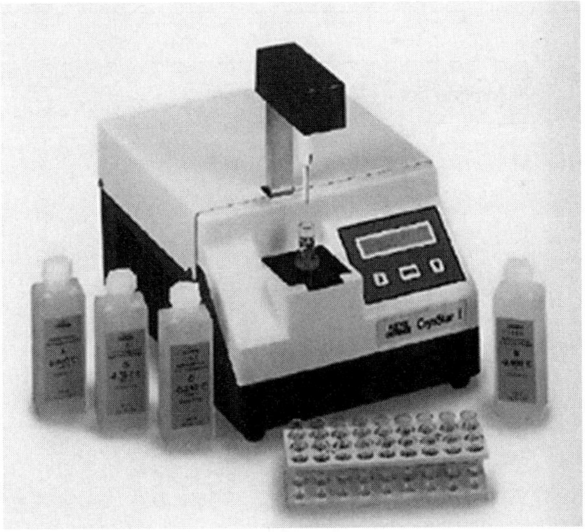

Figure 5.2 Advanced® model 4250 single-sample cryoscope, © Advanced Instruments, Inc., Norwood, MA. All rights reserved.

In the freezing curve method, the temperature–time of a sample is recorded by using a temperature-indicating device (glass thermometer, thermocouple, thermistor, or resistance thermometer). The freezing point is derived from a relatively long temperature plateau, which follows the supercooling stage, on a plot of temperature as a function of time (Figure 5.1). In this determination, the supercooling phenomenon plays an important role; therefore, supercooling must be controlled and compensated to obtain accurate results (Fennema et al., 1973).

There are many factors involved in the tendency of a system to supercool, including temperature, rate of cooling, volume, type of container, particles in the liquid, etc. (Otero and Sanz, 2006). Figure 5.3 shows different extensions of supercooling for a simple solution, in which four levels of supercooling (S^0 to S^3) determine some variations in the freezing temperature (Fennema et al., 1973). Yamada et al. (1993) studied the supercooling of molten salts and found that the supercooling increased as both the sample mass and cooling rate were decreased. Figure 5.4 shows the increase in supercooling temperature with the decrease in cooling rate of calcium chloride.

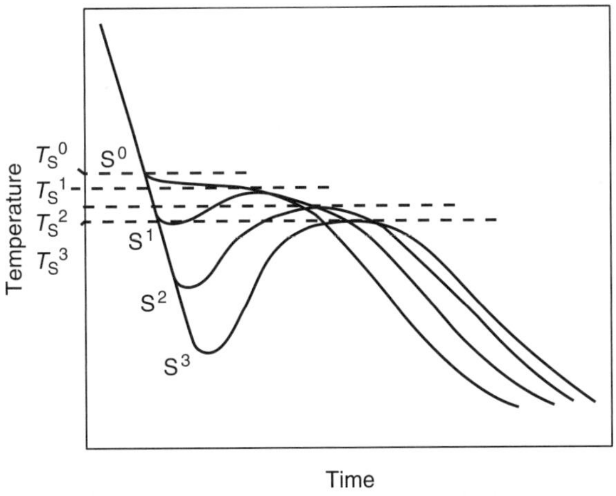

Figure 5.3 Freezing curves for the same solution taking four levels of supercooling: S^0, S^1, S^2, and S^3. (From Rahman, S., *Food Properties Handbook*, CRC Press, Boca Raton, FL, 1995. With permission.)

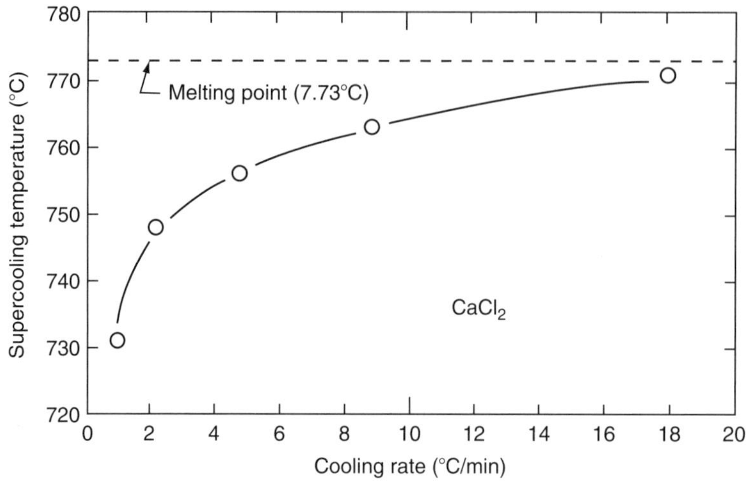

Figure 5.4 Effect of cooling rate on supercooling temperature of calcium chloride. (From Yamada, M., Tago, M., Fukusako, S., and Horibe, A., *Thermochim. Acta*, 218, 401, 1993. With permission.)

The degree of supercooling increases as the sample mass decreases due to a decrease in the number of nuclei. It is evident that the portion of the curve following supercooling is not perfectly horizontal as it is for pure solvent and its position varies with the extent of supercooling. This is because of the concentration effect on freezing.

Typical cooling curves during freezing are shown in Figure 5.1. The abrupt rise in temperature due to the liberation of the heat of fusion after initial supercooling represents the onset of ice crystallization. Supercooling is defined as cooling below the initial freezing point of a sample without ice being formed. Pure water can be undercooled by several degrees before the nucleation phenomenon begins. Once the crystal embryos exceed the critical radius for nucleation, the system nucleates at point a in Figure 5.1 and releases its latent heat faster than the heat removed from the system. The temperature then increases instantly to the initial freezing temperature at point b (T_f). In aqueous solutions, point a is not as low as pure water, since the added solute promotes heterogeneous nucleation, thereby accelerating the nucleation process. The solute greatly decreases the amount of supercooling for two reasons: faster nucleation and lowered freezing point. In the case of water as shown in Figure 5.1, curve A, point c (T_m'), is defined as the temperature drop due to the release of sensible heat after ice crystallization is completed. In the case of a solution as shown in Figure 5.1, curve B, point c (T_m'), is defined as the change in cooling curve slope after solute crystallization. In solutions, supersaturation continues due to the freezing of water, and solute crystals may form by releasing latent heat of solute crystallization, causing a slight shift in temperature from d to e (Figure 5.1, curve B). This point is known as eutectic point. Solute crystallization can occur in foods and solutions. It is not easy to identify the eutectic point in the cooling curve due to the small change in enthalpy. In solutions with multiple solutes or foods it is difficult to determine the eutectic points. Many different eutectic points might be expected, but each plateau would be quite short if small quantities of solutes were involved. When a material is heated from a frozen or glass state, the onset of ice melting is called the end point of the freezing curve. The freezing point of water or end point of ice melting is considered as a thermodynamic equilibrium process, that is, neither cooling or heating rate affects the phase transition point of ice crystallization (Rahman and Driscoll, 1994; Rahman, 1999).

Chen and Chen (1996) proposed a device consisting of two major parts: a freezing vessel and a data acquisition system as shown in Figure 5.5. The freezing vessel is a thermal flask containing liquid nitrogen and has an inverted aluminum cone supported by a polystyrene cap. The flask sits

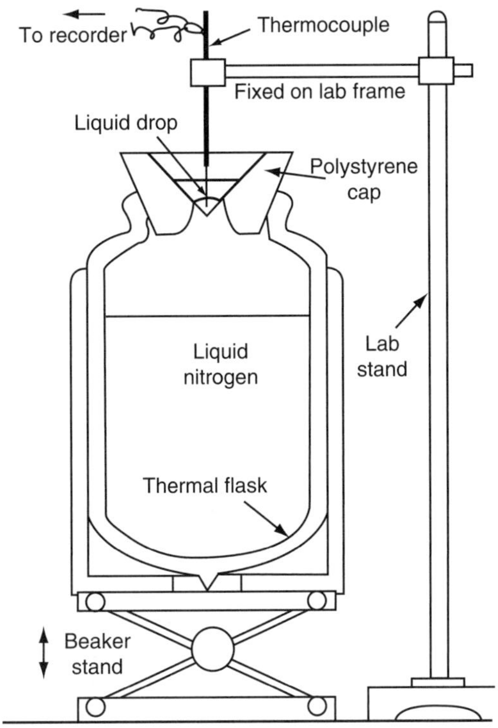

Figure 5.5 Schematic diagram of the experimental setup (−70°C). (From Chen, X.D. and Chen, P., *Food Res. Int.*, 29, 723, 1996. With permission.)

on an adjustable beaker stand. The aluminum cone has an angle of 90°C, a wall thickness of 0.4 mm, and the largest diameter of 2 cm. During each measurement, an aqueous liquid drop is supplied by a syringe to the bottom of the inverted cone and is cooled by the nitrogen vapor in the thermal flask extracting heat from the metal wall. The temperature of the sample in the cone was measured by a thermocouple as a function of time. The location of the measuring point of the temperature can be adjusted. It was observed that the reproducibility of the freezing point of distilled water was within ±0.05°C. They observed that supercooling could be determined by vibrating the thermocouple. It was observed that the shape of the freezing curve changed and large errors (nearly 1°C lower) occurred if supercooling was too large. The cooling curve measured at the lowest location (nearest to the bottom of the cone) has the largest supercooling. Chen and Chen (1996) pointed out that freezing point measurement can be carried out using this device if the following precautions are taken, that is, sample size cannot be too small (preferably not smaller than 0.3 mL), the supercooling cannot be too high (preferably within 5°C), the measuring point cannot be placed too close to the cone metal surface, and the sensitivity of the thermocouple should be high enough to measure the temperature changes within the timescale of the nucleation and crystallization (this means that the finest thermocouple is most beneficial). The supercooling correction can be made by the extrapolation method and by using standard solutions.

5.2.1.1.1 Extrapolation Method

A simple extrapolation method is illustrated in Figure 5.6 where line DE is visually extrapolated back to the cooling curve, giving an approximate freezing point (T_B), point G being the true freezing point (T_G). The difference between G and B increases with the increasing of concentration or

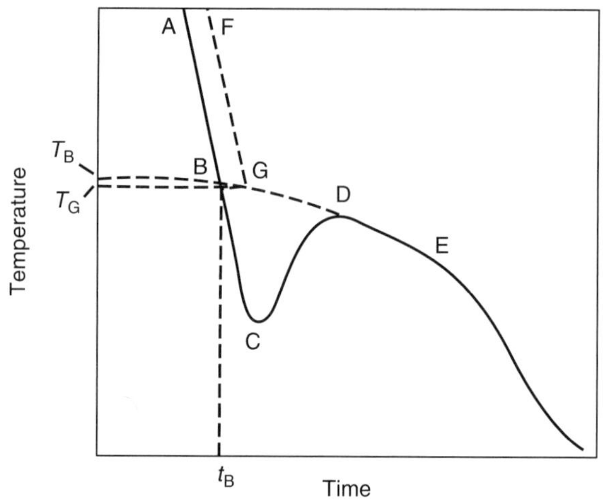

Figure 5.6 Sample extrapolation method for correcting the observed freezing point of a solution for error caused by supercooling. AB: cooling, BC: supercooling, DE: freezing, FG: corrected cooling. (From Fennema, O.R., Powrie, W.D., and Marth, E.H., *Low-Temperature Preservation of Foods and Living Matter*, Marcel Dekker, New York, 1973. With permission.

supercooling (beyond 1°C) and with the decreasing rate of heat removal. More accurate geometrical or optical methods are now available and necessary to obtain lower differences (±0.01°C).

5.2.1.1.2 Calibration with Standard Solution

Supercooling errors can also be corrected with standard solutions of known freezing points. Two standards are generally used, one slightly above and the other below the concerned temperature. Differences between actual and observed freezing points of the standards are used as a basis for correcting subsequent freezing points of the unknown. The same extent of supercooling and the same crystallization temperature and rate are necessary for accurate results (Rahman, 1995). However, this process is complicated as two standards are required close to the sample's freezing point.

5.2.1.2 Cryoscope

Two basic types of freezing point instruments are commonly used (1) conventional cryoscopes and (2) thermistor cryoscopes. The essential features of the cryoscopes are schematically shown in Figure 5.7a and b. Conventional cryoscopes are specified by the ASTM, whereas either type is permitted by the AOAC and the APHA (Fennema et al., 1973). The sample (100 mL) is cooled in a cell (at least 30°C below the freezing point) and constantly agitated with a magnetic stirrer; cell and solution temperatures are recorded. From the experimental graph the freezing temperature (plateau) is obtained (Lerici et al., 1983).

5.2.1.3 Cooling Curve for Freezing Point and End Point of Freezing

Rahman et al. (2002) presented details of the cooling curve method to measure the equilibrium freezing point and also proposed this method to measure the end point of freezing. The schematic diagram of the experimental setup is shown in Figure 5.8. They used sugar solution and starch gel to test their method. Sugar solution or starch gel (15–20 g) was poured into a stainless steel cylinder (internal diameter, 2.2 cm; height, 4.5 cm; wall thickness, 1 mm) insulated at the top, bottom, and side with a polystyrene foam as shown in Figure 5.8. The cylinder was kept around three-quarter full

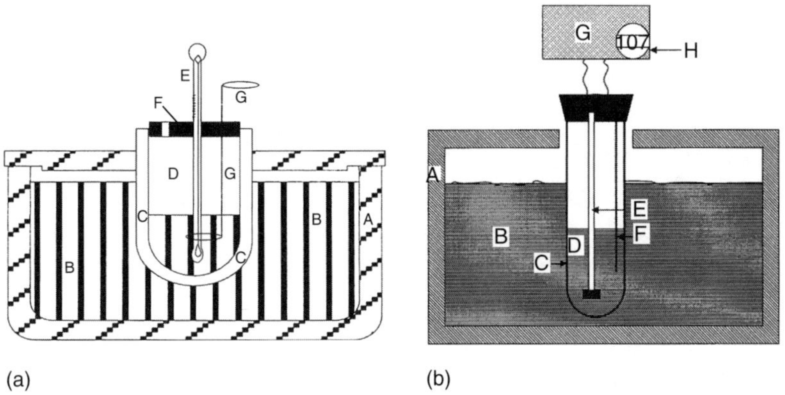

Figure 5.7 (a) Schematic diagram of a conventional cryoscope for freezing point determination. A: insulated container, B: coolant, C: double-walled sample container, D: sample, E: mercury or platinum resistance thermometer, F: cap, G: stiring device. (b) Schematic diagram of a thermistor cryoscope for freezing point determination. A: insulated chamber for coolant, B: stirred coolant, C: glass sample tube, D: sample, E: thermistor, F: vibrating wire for stirring to induce crystallization, G: electronic circuit for thermoelectric cooling vibrating wire, thermistor, and calibration, H: temperature indicator. (From Fennema, O.R., Powrie, W.D., and Marth, E.H., *Low-Temperature Preservation of Foods and Living Matter*, Marcel Dekker, New York, 1973. With permission.)

for all experiments. The cylinder with syrup was then placed into a chest freezer, which can be controlled at constant temperature from −40°C to −85°C. The temperature of the freezer was set at −70°C for all experimental measurements. The temperature at the sample center was measured at every minute interval using a digital thermometer from Barnant Tri-Sense, Illinois. The thermometer was calibrated by measuring the freezing point of distilled water and freezing point of sugar solution (0.40 kg sucrose/kg syrup) and was accurate to ±0.1°C. The equilibrium or initial freezing point was considered at the temperature where the slowest cooling rate was observed as noted by Rahman and Driscoll (1994). The cooling curve was analyzed further to determine other phase and structural changes by calculating the slope. The cooling rate could be varied by placing the cylinder into polystyrene foam blocks of different dimensions. Sucrose syrup (0.40 kg sucrose/kg sample) and starch gel (0.10 kg starch/kg sample) were mainly used in determining the sensitivity and

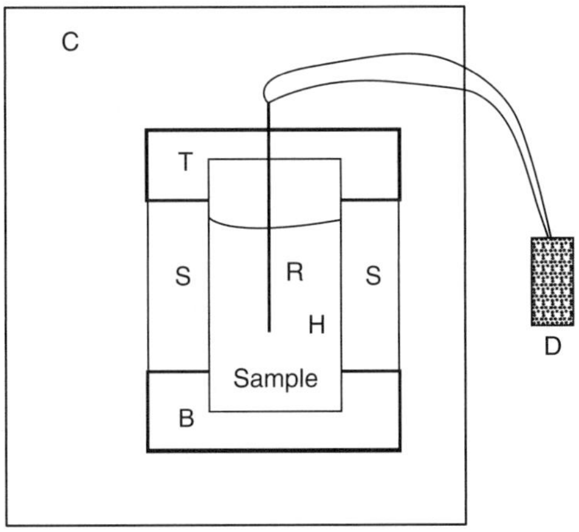

Figure 5.8 Schematic diagram of an experimental setup for sugar solution. T: Top insulated block, B: bottom insulated block, S: side insulated block, R: thermocouple probe, H: stainless steel sample holder, D: temperature recorder, C: chest freezer. (From Rahman, M.S., Guizani, N., Al-Khaseibi, M., Al-Hinai, S., Al-Maskri, S.S., and Al-Hamhami, K., *Food Hydrocolloids*, 16, 653, 2002. With permission.)

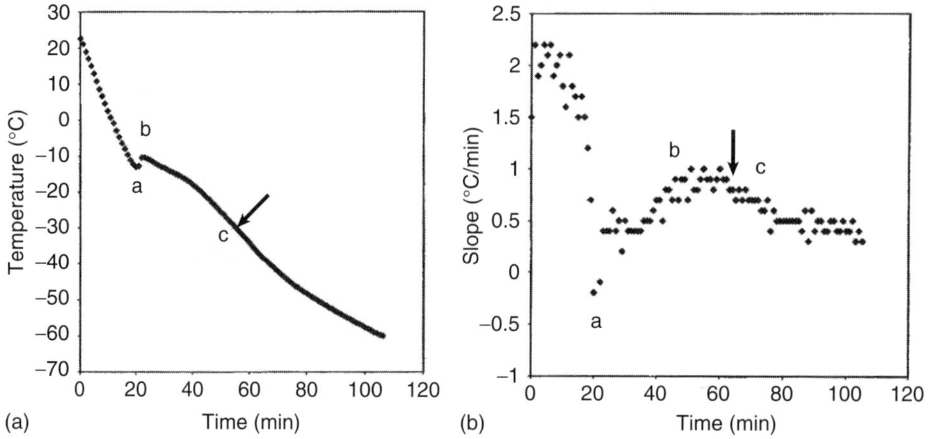

Figure 5.9 Cooling curve of sucrose syrup showing supercooling (point a), initial freezing point (point b), and end point of freezing (point c). (From Rahman, M.S., Guizani, N., Al-Khaseibi, M., Al-Hinai, S., Al-Maskri, S.S., and Al-Hamhami, K., *Food Hydrocolloids*, 16, 653, 2002. With permission.)

validity of the proposed method. In the second set, the concentration was varied to identify its effect on freezing point and the end point of freezing.

Typical cooling curve of sucrose syrup (0.40 kg sucrose/kg sample) is shown in Figure 5.9a, which shows the ice nucleation and initial or equilibrium freezing point. The initial freezing point (T_f, point b in Figure 5.9a) of 0.40 kg sucrose/kg sample was found to be −4.8°C. The literature value of the freezing point of 0.40 kg sucrose/kg sample is −4.7°C (Chen, 1987a,b). Figure 5.9b shows the slope of the cooling curve for sugar syrup as a function of cooling time. The initial slope was used to determine the cooling rate, r_i (°C/min). The slope is decreased to a minimum, when ice crystals form (point a in Figure 5.9b). The plateau in slope indicates the initial freezing point (point b) due to the formation of ice. When the amount of ice formation decreases, the slope starts to increase and reaches a highest value or plateau. The end point of freezing is defined as the point when the cooling rate is highest or the end of the plateau after freezing started since this indicates that freezing is completed. Thus, T'_m is the point at the highest slope value or the end of the plateau after which the slope starts to decrease again. In the case of sucrose, it is easy to find the end point of the highest value of the slope (Figure 5.9b). The average value of T'_m was −30.4°C for sucrose solution. The values of T'_g from DSC for sucrose have been reported as −32°C (Slade and Levine, 1988; Hatley et al., 1991; Schenz et al., 1991) and −35°C (Izzard et al., 1991). These researchers took the midpoint of the change in specific heat as T'_g. Roos and Karel (1991a,b) proposed a value of −46°C for the onset of T'_g, −40°C for the midpoint, and −34°C for the end point of the transition (T'_m). It is more widely agreed that the melting of ice or end of freezing starts around −32°C for frozen sucrose solution (Le Meste and Huang, 1992). This indicates that the values found here for sucrose are close to the values found in the literature measured by DSC and differential mechanical thermal analysis (DMTA) methods.

A typical cooling curve of starch gel is shown in Figure 5.10, which shows the ice nucleation and initial or equilibrium freezing point (point b). Supercooling was not observed in the case of starch gel, thus point a was missing. The initial freezing point of 0.10 kg starch/kg sample was found to be −0.3°C. In the case of starch gel, there is a sharp fall of the slope after the highest value, and the highest value in the slope was considered as T'_m (point c, Figure 5.10b). The average value of T'_m was −7.1°C. Slade and Levine (1995) found $T'_g = -5$°C from the midpoint rather than from the onset of transition for anhydrous starch. Roos and Karel (1991c) found $T'_m = -6$°C for starch and assumed that $T'_m = T'_g$. Jang and Pyun (1997) measured the T'_m and T'_g for wheat starch as −7°C and −13°C, respectively. The value found by this method is close to the values found in the literature. Thus, the initial results show that the cooling curve method for the determination of T'_m has high potential for foods.

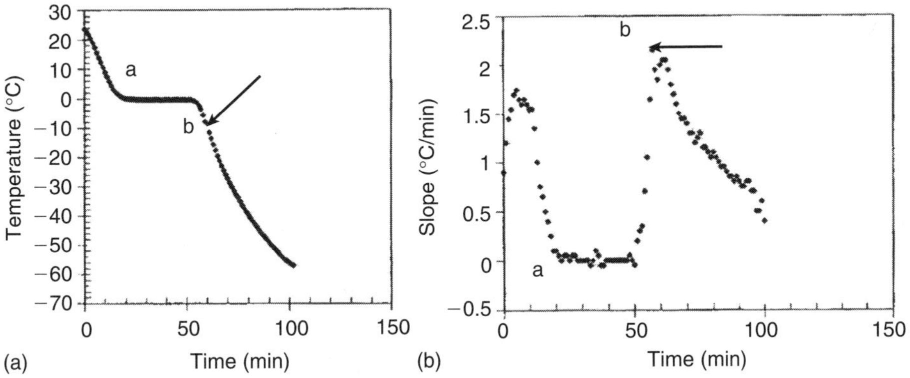

Figure 5.10 Slope of the cooling curve for sucrose syrup showing initial freezing point (point a) and end point of freezing (point b). (From Rahman, M.S., Guizani, N., Al-Khaseibi, M., Al-Hinai, S., Al-Maskri, S.S., and Al-Hamhami, K., *Food Hydrocolloids*, 16, 653, 2002. With permission.)

The hypothesis was tested further varying the solute concentration of sugar and starch. The values of T'_m should be constant when solute concentration is varied if the hypothesis is valid. The average values of T'_m for sugar and starch are −31.9°C ± 2.0°C and −7.4°C ± 1.4°C, respectively. Most of the earlier references did not present information on the standard deviation of the DSC method. However, results from a few recent references are presented on the DSC method. The standard deviations varied from 2.6°C to 6.4°C for lactose (Lloyd et al., 1996), from 1.0°C to 8.0°C for lactose, sucrose, sucrose/starch (Roos and Karel, 1990), from 0.1°C to 2°C for collagen and sucrose (Brake and Fennema, 1999), and from 2.4°C to 3.4°C for extruded starch (Ross et al., 2002). Thus, the variability in the cooling curve method is similar to the DSC method.

The advantages of this method are that it can handle bigger samples and the samples could be in the form of solids, liquids, gels, or powder; measurement and data analysis are simple and require low-cost instruments. Further work needs to be done to identify the effect of sample size, wide variation of r_c with different extent of insulation, and variation of freezer or cooling medium temperature. This method was used to measure the freezing point and end point of freezing for tuna meat (Rahman et al., 2003) and abalone (Sablani et al., 2004). Rahman et al. (2005) used this method to measure the freezing point and end point of freezing for garlic and compared the results with the DSC method. They found that both the DSC and cooling curve gave decreasing T'_m values with increasing solid content. The cooling curve method was more sensitive to the solid contents than the DSC method. This is evident from the slope of T'_m as a function of solid content. The line intersection (around 40 g/100 g sample) indicates that the same values were measured both by the DSC method and the cooling curve method. In addition, spreadability of the data point was wider in the case of the cooling curve than the DSC curve.

5.2.2 DSC Method

The onset, peak, and end of freezing can be determined from the heat exotherm by DSC. Cooling and heating rates of 2°C/min–10°C/min are commonly applied to small samples hermetically closed in aluminum pans (Hamdami et al., 2004). DSC may be used either for scanning at a wide temperature profile or for stepwise (isothermal) experiments (Lind, 1991). The calorimeter should be previously calibrated with distilled water or ice and indium, and small quantities of sample (5–10 mg) are placed in the pan (Sheard et al., 1990; De Qian and Kolbe, 1991). At the end of the experimental determination, the moisture content of the sample should be verified to ensure that no water has escaped from the pan.

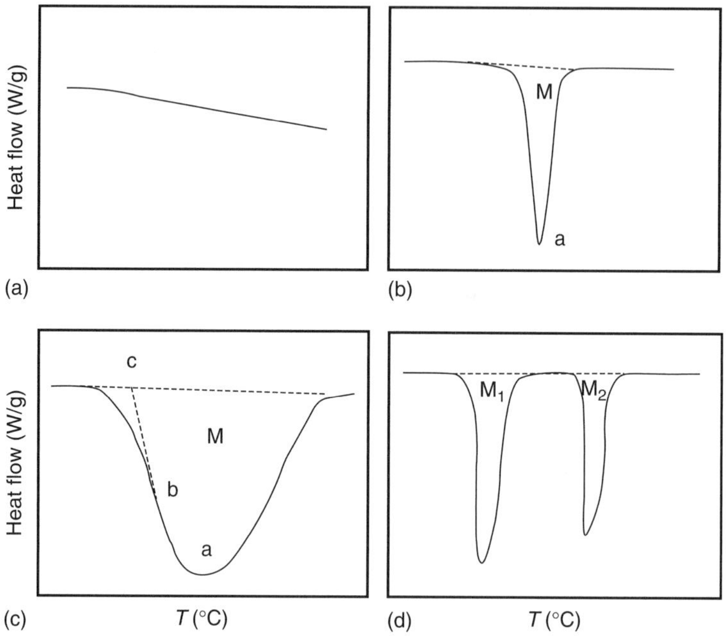

Figure 5.11 Typical melting curves by a DSC instrument.

The advantages of DSC are (1) it works rapidly and simply, and valuable information may be deducted from the thermogram, in which a small quantity of the food item generates relatively accurate data (De Qian and Kolbe, 1991); (2) when inhomogeneities of the food sample are presented, it may imply a major number of replicates (Lind, 1991). A typical melting curve of food by a DSC instrument is shown in Figure 5.11.

The main disadvantage of the DSC method is to exactly locate the freezing point as determined by the cooling curve. In addition, the increased cooling rate shifts the peak to a higher temperature. Although DSC is popular, it has a number of limitations. The equipment is costly and needs to be handled skillfully to obtain meaningful results for foods. DSC uses a small sample size (3–30 mg) and for a multicomponent mixture, such as food, it is difficult to make a representative sample when such small quantities are used (Rahman et al., 2002).

The freezing point is usually taken from the DSC melting endotherm rather than from the freezing exotherm. Typical melting endotherms for melting of ice are shown in Figure 5.11 (Rahman, 2006). The sample containing nonfreezing water shows no first-order transition (Figure 5.11a). The ice melting or freezing point is commonly characterized from the endothermic peak during melting (point a in Figure 5.11b) (Goff et al., 1993; Rahman, 2004). This method provides very accurate determination when a sharp peak is observed. In the case of a wider peak, identifying only peak may not represent the real freezing point and other points need to be located (Figure 5.11c). The wider peak appears because of the wide variation in the water content in foods; in this case, maximum slope of the endotherm (point b in Figure 5.11c) or the extra-plotted peak onset temperature of the ice melting (point c in Figure 5.11b) (Goff et al., 1993; Fonseca et al., 2001; Rahman, 2004). When the sample contains mainly free water, it shows a sharp endothermic peak on melting similar to pure water (Figure 5.11b). Multipeak natures of the DSC curves are found for the metastable states of water in gum from *Acacia senegal* (Phillips et al., 1996) and gellan (Hatakeyama et al., 1996) (Figure 5.11d).

The freezing point of surimi by a cooling curve and DSC thermogram as a function of cryoprotectant concentration is given in Table 5.2. Delgado et al. (1990) measured the latent

Table 5.2 Initial or Equilibrium Freezing Point of Surimi[a] by DSC Thermogram (2°C/min) and Cooling Curve as a Function of Cryoprotectant Concentration

		T_f (°C)			
X_{cp}^{o}	a_w	DSC Thermogram	Cooling Curve	Y_{uw}^{b}(kg Water/kg Solids)	L_w (kJ/kg)
0.01	0.9911	−0.22	−0.35	0.54	231.6
0.04	0.9892	−1.24	−0.75	0.46	237.8
0.06	0.9882	−1.33	−1.01	0.42	240.2
0.08	0.9849	−1.49	−1.16	0.46	237.3
0.12	0.9853	−1.65	−1.63	0.54	231.8

Source: Adapted from Wang, D.Q. and Kolbe, E., *J. Food Sci.*, 55, 302, 1991.
[a] Initial water 80.3%.
[b] Experimental unfreezable water.

heats of strawberries for heating and cooling at different temperature ranges (Table 5.3). The lower values for heating and cooling (−20°C to 10°C) were due to the incomplete phase change in the temperature range studied and due to the difficulty in detecting the clear separation between curve and base line to define the limits of peak integration.

5.3 FREEZING POINT DATA OF FOODS

Freezing and melting points of different foods are given in Tables 5.1 through 5.15. Table 5.15 shows the freezing points of different foods as a function of total solids. The eutectic temperatures of various solutions and foods are given in Table 5.16, whereas the melting points of salts are given in Table 5.17.

5.4 FREEZING POINT PREDICTION MODELS

The freezing point prediction models can be broadly divided into three groups: (1) theoretical, (2) empirical curve fitting, and (3) semiempirical models. The theoretical models are generally based on the assumption that food materials are ideal binary solutions. On the other hand, the other models are based on experimental measurements of the food behavior through the freezing process, in which the determination has been achieved only for unfrozen foods or frozen foods at or near equilibrium conditions (Saad and Scott, 1996).

Table 5.3 DSC Latent Heat (2°C/min) of Strawberry Cultivars Measured for Different Temperature Ranges of Integration

	Composition			Latent Heat (kJ/kg)		
Cultivar and Sample	X_w^o	X_{ss}^o	X_F^o	Heating (−20°C to 10°C)	Cooling (−5°C to 30°C)	Heating (−40°C to 10°C)
Florida (pulp)				270.26	270.72	280.00
Florida (whole)	0.89	0.05	0.013	256.40	267.19	272.50
Tioga (pulp)				237.35	240.61	279.50
Tioga (whole)	0.87	0.05	0.015	227.39	234.29	260.80

Source: Adapted from Delgado, A.E., Rubiolo, A.C., and Gribaudo, L.M., *J. Food Process. Pres.*, 14, 231, 1990.

Table 5.4 Freezing Points of Liquid Foods as a Function of Water Content

Material	X_w^o	T_f (°C)	Reference	Material	X_w^o	T_f (°C)	Reference
Apple juice	0.872	−1.44	Heldman and Singh (1981)	Milk	0.875	−0.60	Polley et al. (1980)
	0.498	−11.33	Heldman (1974)	Orange juice	0.920	−1.10	Lerici et al. (1983)
Berry juice	0.895	−0.96	Dickerson (1968)		0.900	−1.30	Lerici et al. (1983)
Cherry juice	0.867	−1.29	Dickerson (1968)		0.890	−1.17	Heldman (1974)
Fructose	0.900	−1.17	Weast (1982)		0.890	−1.02	Dickerson (1968)
	0.900	−1.12	Lerici et al. (1983)		0.800	−2.30	Lerici et al. (1983)
	0.850	−1.88	Weast (1982)		0.500	−11.50	Chen (1985)
	0.850	−1.90	Lerici et al. (1983)	Peach juice	0.920	−1.10	Lerici et al. (1983)
	0.700	−4.70	Lerici et al. (1983)		0.900	−1.30	Lerici et al. (1983)
	0.600	−7.70	Lerici et al. (1983)		0.800	−2.30	Lerici et al. (1983)
Grape juice	0.900	−1.20	Lerici et al. (1983)	Raspberry juice	0.885	−1.07	Dickerson (1968)
	0.850	−1.78	Heldman and Singh (1981)	Strawberry juice	0.917	−0.74	Dickerson (1968)
	0.850	−1.92	Dickerson (1968)	Sucrose	0.950	−0.29	Weast (1982)
	0.800	−2.90	Lerici et al. (1983)		0.950	−0.30	Lerici et al. (1983)
	0.700	−4.40	Lerici et al. (1983)		0.900	−0.63	Weast (1982)
Juice[a]	0.960	−0.39	Succar and Hayakawa (1983)		0.900	−0.60	Lerici et al. (1983)
	0.870	−1.38	Succar and Hayakawa (1983)		0.850	−1.01	Weast (1982)
	0.750	−3.19	Succar and Hayakawa (1983)		0.850	−1.00	Lerici et al. (1983)
	0.610	−6.98	Succar and Hayakawa (1983)		0.800	−1.46	Weast (1982)
Lactose	0.990	−0.05	Weast (1982)		0.800	−1.40	Lerici et al. (1983)
	0.990	−0.10	Lerici et al. (1983)		0.700	−2.65	Weast (1982)
	0.970	−0.17	Weast (1982)		0.700	−2.80	Lerici et al. (1983)
	0.970	−0.30	Lerici et al. (1983)		0.600	−4.45	Weast (1982)
	0.950	−0.29	Weast (1982)		0.600	−4.70	Lerici et al. (1983)
	0.950	−0.40	Lerici et al. (1983)	Tomato pulp	0.929	−0.72	Heldman (1974)
	0.920	−0.49	Weast (1982)		0.929	−0.57	Dickerson (1968)
	0.920	−0.60	Lerici et al. (1983)	Yogurt	0.802	−2.30	Rocha-Santiago et al. (2005)

[a] Fruits and vegetables.

Table 5.5 Freezing Point of Liquid Foods as a Function of Solute Fraction

NaCl		Fructose		Sucrose		Lactose		Glycerol		Juice		Skim Milk[a]		Coffee[a]		Grape Juice		Tomato Juice	
X_{ss}^{o}	T_f (°C)	X_{ss}^{o}	T_f (°C)	X_{ss}^{o}	T_f (°C)	X_{ss}^{o}	T_f (°C)	X_{ss}^{o}	T_f (°C)	X_{ss}^{o}	T_f (°C)	X_{ss}^{o}	T_f (°C)	X_{ss}^{o}	T_f (°C)	X_{ss}^{o}	T_f (°C)	X_{ss}^{o}	T_f (°C)
0.01	−0.6	0.10	−1.2	0.05	−0.3	0.01	−0.1	0.05	−1.0	0.08	−1.1	0.10	−0.6	0.05	−0.3	0.10	−1.2	0.05	−0.5
0.05	−3.1	0.15	−1.9	0.10	−0.6	0.03	−0.3	0.10	−2.5	0.10	−1.3	0.20	−1.6	0.10	−0.6	0.20	−2.9	0.10	−0.9
0.10	−6.6	0.30	−4.7	0.15	−1.0	0.05	−0.4	0.20	−5.4	0.20	−2.3	0.30	−2.3	0.20	−1.3	0.30	−4.4	0.15	−1.6
0.15	−10.9	0.40	−7.7	0.20	−1.4	0.08	−0.6	0.30	−9.8			0.40	−4.1	0.30	−2.3			0.20	−2.0
0.20	−16.5			0.30	−2.8			0.40	−15.5					0.40	−3.7				
				0.40	−4.7														

Source: Adapted from Chen, C.S., *J. Food Sci.*, 52, 433, 1987.
[a] Freeze dried.

Table 5.6 Freezing Point of Foods as a Function of Water Content

Egg White (Succar and Hayakawa, 1990)		Egg Yolk (Succar and Hayakawa, 1990)		Fruit–Vegetable Juice (Chen and Nagy, 1987; Succar and Hayakawa, 1990)		NaCl (Chen and Nagy, 1987; Succar and Hayakawa, 1990)		Sucrose (Chen and Nagy, 1987; Succar and Hayakawa, 1990)		Squid (Rahman and Driscoll, 1994)	
X_w^o	T_f (°C)	X_w^o	T_f (°C)	X_w^o	T_f (°C)	X_w^o	T_f (°C)	X_w^o	T_f (°C)	X_w^o	T_f (°C)
0.90	−0.01	0.50	−0.12	0.96	−0.40	0.98	−1.19	0.90	−0.63	0.82	−1.03
0.80	−0.16	0.40	−0.41	0.96	−0.39	0.98	−1.22	0.90	−0.60	0.81	−0.99
0.70	−1.00	0.30	−1.03	0.87	−1.41	0.96	−2.42	0.85	−1.00	0.77	−1.28
0.60	−1.48	0.20	−2.65	0.87	−1.38	0.96	−2.52	0.85	−1.02	0.74	−1.69
0.50	−3.14			0.75	−3.28	0.94	−3.70	0.80	−1.45	0.69	−2.39
0.40	−5.28			0.75	−3.19	0.94	−3.71	0.80	−1.50	0.65	−2.56
				0.61	−6.86	0.90	−6.55	0.75	−1.99	0.55	−7.23
				0.61	−6.98	0.90	−6.67	0.75	−2.02	0.43	−11.74
						0.86	−9.94	0.70	−2.66	0.33	−15.52
						0.86	−9.93	0.70	−2.67		
								0.65	−3.48		
								0.65	−3.50		

Table 5.7 Freezing Point of Milk (Determined by Cryoscope) According to the Treatment

Product and Treatment	pH	T_f (°C)
Raw	6.67	−0.526
Raw	6.68	−0.524
Direct ultrahigh temperature (145°C)		−0.517
Indirect ultrahigh temperature (145°C)		−0.523
Heating 74°C, 30 s		−0.524
Heating 85°C, 2.8 s		−0.522
Heating 95°C, 303 s		−0.523
Heating 78°C, 33.6 s		−0.524
Heating 95°C, 50.3 s		−0.521
Raw (without gas)	6.62	−0.525
Saturated with O_2	6.22	−0.529
Saturated with N_2	6.64	−0.529
Saturated with CO_2	5.90	−0.596
Saturated with N_2, O_2, and CO_2 (treatment with gas during 15 min at 10°C)	5.90	−0.590
Stored at 4°C during 24 h	6.58	−0.534
Stored at 4°C during 48 h	6.33	−0.539

Source: Adapted from Kessler, H.G., *Milchwissenschaft*, 39, 339. Cited by Kirk, R.S., Sawyer, R., and Egan, H. (eds.), in *Pearson's Food Composition and Analysis*, 2nd edn., CECSA, Mexico, 1984, p. 600.

Table 5.8 Freezing Point of Fruits and Vegetables as a Function of Water Content

Material	X_w^o	T_f (°C)	Reference	Material	X_w^o	T_f (°C)	Reference
Apple	0.841	−2.00	ASHRAE (1967)	Lime	0.860	−1.70	ASHRAE (1967)
Apple sauce	0.828	−1.52	Dickerson (1968)	Melon	0.926	−1.70	ASHRAE (1967)
Apricot	0.854	−2.20	Polley et al. (1980)	Muskmelon	0.927	−1.70	ASHRAE (1967)
Asparagus	0.926	−0.52	Dickerson (1968)	Nectarine	0.829	−1.70	ASHRAE (1967)
Asparagus	0.930	−1.20	ASHRAE (1967)	Olive	0.752	−1.90	ASHRAE (1967)
Avocado	0.940	−2.70	ASHRAE (1967)	Onion	0.855	−1.29	Dickerson (1968)
Banana	0.748	−2.20	ASHRAE (1967)	Onion	0.855	−1.44	Heldman and Singh (1981)
Bean	0.900	−1.80	ASHRAE (1967)	Orange	0.872	−2.20	ASHRAE (1967)
Beet	0.876	−2.80	ASHRAE (1967)	Parsnip	0.786	−1.70	ASHRAE (1967)
Bilberry	0.851	−1.10	Heldman and Singh (1981)	Pea (green)	0.743	−1.10	ASHRAE (1967)
Blackberry	0.829	−1.40	ASHRAE (1967)	Peach	0.851	−1.41	Dickerson (1968)
Blueberry	0.823	−1.90	ASHRAE (1967)	Peach	0.851	−1.56	Heldman and Singh (1981)
Broccoli	0.899	−1.60	ASHRAE (1967)	Peach	0.869	−1.40	ASHRAE (1967)
Brussel	0.849	−0.60	ASHRAE (1967)	Pear	0.838	−1.46	Dickerson (1968)
Cabbage	0.924	−0.50	ASHRAE (1967)	Pear (bartlet)	0.835	−1.90	ASHRAE (1967)
Cantaloupe	0.927	−1.70	ASHRAE (1967)	Pear	0.838	−1.61	Heldman and Singh (1981)
Carrot	0.875	−0.96	Dickerson (1968)	Plum	0.803	−2.28	Heldman and Singh (1981)
Carrot	0.882	−1.30	ASHRAE (1967)	Potato (white)	0.778	−1.70	ASHRAE (1967)
Carrot	0.875	−1.11	Heldman and Singh (1981)	Potato (sweet)	0.685	−1.90	ASHRAE (1967)
Cauliflower	0.917	−1.10	ASHRAE (1967)	Quince	0.853	−2.20	ASHRAE (1967)
Celery	0.937	−1.30	ASHRAE (1967)	Raspberry	0.820	−1.10	ASHRAE (1967)
Cherry (sweet)	0.770	−2.61	Heldman and Singh (1981)	Raspberry	0.827	−1.22	Heldman and Singh (1981)
Corn (green)	0.739	−1.70	ASHRAE (1967)	Rutabaga	0.891	−1.40	ASHRAE (1967)
Cranberry	0.874	−2.60	ASHRAE (1967)	Spinach	0.902	−0.41	Dickerson (1968)
Date (dried)	0.200	−20.10	ASHRAE (1967)	Spinach	0.902	−0.56	Heldman and Singh (1981)
Dewberry	—	−1.60	ASHRAE (1967)	Spinach	0.927	−0.94	ASHRAE (1967)
Eggplant	0.927	−0.90	ASHRAE (1967)	Strawberry	0.893	−0.74	Dickerson (1968)
Garlic (dry)	0.742	−3.70	ASHRAE (1967)	Strawberry	0.893	−0.89	Heldman and Singh (1981)
Globe	0.837	−1.60	ASHRAE (1967)	Strawberry	0.900	−1.20	ASHRAE (1967)
Gooseberry	0.883	−1.70	ASHRAE (1967)	Strawberry[a]	0.890	−0.83	Delgado et al. (1990)
Grape	0.888	−2.00	ASHRAE (1967)	Strawberry[b]	0.870	−1.41	Delgado et al. (1990)
Grape (American)	0.819	−2.50	ASHRAE (1967)	Tall pea	0.758	−1.68	Dickerson (1968)
Huitlacoche	0.906	−1.60	Sanchez-Rechy (2006)	Tangerine	0.730	−2.20	ASHRAE (1967)
Jerusalem artichoke	0.795	−2.50	ASHRAE (1967)	Tomato (mature)	0.947	−0.90	ASHRAE (1967)

(continued)

Table 5.8 (continued) Freezing Point of Fruits and Vegetables as a Function of Water Content

Material	X_w^o	T_f (°C)	Reference	Material	X_w^o	T_f (°C)	Reference
Kohlrabi	0.901	−1.10	ASHRAE (1967)	Tomato (ripe)	0.941	−0.90	ASHRAE (1967)
Leek	0.882	−1.60	ASHRAE (1967)	Turnip	0.909	−0.80	ASHRAE (1967)
Lemon	0.893	−2.20	ASHRAE (1967)	Watermelon	0.921	−1.60	ASHRAE (1967)
Lettuce	0.948	−0.40	ASHRAE (1967)				

[a] Florida X_{ss}: 0.05.
[b] Tioga X_{ss}: 0.05.

5.4.1 Theoretical Models

5.4.1.1 Ideal Solution

Heldman (1974) derived an equation for ideal solutions, taking the relationship derived by Moore in 1962 (Saad and Scott, 1996), which is valid for $T < T_F$:

$$\frac{L_w}{R}\left(\frac{1}{T_w} - \frac{1}{T_f}\right) = \ln(x_w^o) \tag{5.1}$$

Based on Raoult's law, Chen (1986) applied the concept of equilibrium freezing curve (EFC) to get an equation for the freezing point depression of solutions (Chen, 1988):

$$\Delta T = T_w - T_f = \frac{K_f X_s^o}{\left(1 - X_s^o - BX_s^o\right)M_s} \tag{5.2}$$

Table 5.9 Freezing Point of Several Breeds of Fruits and Vegetables

	T_f (°C)				
Material	**Source 1**	**Source 2**	**Source 3**	**Source 4**	**Jie et al. (2003)**
Apple	−1.10	−1.10	−1.50	−1.50	—
Apple (Jonathan)	—	—	—	—	−2.32
Apple (Starkrimson)	—	—	—	—	−2.20
Banana	−0.80	−0.80	−0.77	−0.77	—
Cherry (sour)	−1.70	−1.70	—	—	—
Cherry (sweet)	−1.80	−1.80	−1.77	—	—
Fig	−2.40	—	—	−2.44	—
Grape (American breed)	−1.60	—	−1.17	−1.17	—
Grape (European breed)	−2.10	—	−1.16	−1.16	—
Grape (Jufeng)	—	—	—	—	−2.75
Lemon	−1.40	−1.40	−1.33	−1.33	—
Lizao	—	—	—	—	−2.83
Orange	−0.80	−0.80	−1.05	−1.05	−1.81
Peach	−0.90	−0.90	−0.88	−0.88	—
Pear	−1.60	−1.60	—	−1.50	—
Pear (snow)	—	—	—	—	−1.96
Pear (ya)	—	—	—	—	−1.83
Persimmon	−2.20	—	−2.16	−2.16	—
Plum	−0.80	—	−0.82	−0.82	—
Strawberry	−0.80	−0.80	−0.77	−0.77	—

Source: Compiled from Jie, W., Lite, L., and Yang, D., *J. Food Eng.*, 60, 481, 2003.

Table 5.10 Initial or Equilibrium Freezing Point of Meat

Material	X_w^o	T_f (°C)	Reference	Material	X_w^o	T_f (°C)	Reference
Beef				Loin	0.524	−0.84	Pham (1987)
Brine	0.740	−0.99	Succar and Hayakawa (1983)	Loin	0.444	−1.10	Mellor (1983)
Carcass	0.450	−2.20	ASHRAE (1981)	Loin	0.444	−0.84	Pham (1987)
Carcass	0.490	−1.70	ASHRAE (1981)	Muscle[a]	0.745	−1.74	Murakami and Okos (1989)
Fat	—	−2.20	ASHRAE (1967)	Muscle[a]	0.740	−1.00	Murakami and Okos (1989)
Flank	0.745	−1.75	Murakami and Okos (1989)	Muscle[a]	0.650	−0.60	Levy (1979)
Liver	0.700	−1.70	ASHRAE (1981)	Muscle[a]	0.610	−1.90	ASHRAE (1981)
Muscle[a]	0.261	−13.46	Pham (1987)	Muscle[a]	0.580	−1.70	ASHRAE (1967)
Muscle[a]	0.450	−4.09	Succar and Hayakawa (1983)	Muscle[a]	0.530	−0.70	Levy (1979)
Muscle[a]	0.500	−2.80	Chen (1986)	Muscle[a]	0.440	−0.80	Levy (1979)
Muscle[a]	0.500	−3.00	Levy (1979)	*Pork*			
Muscle[a]	0.500	−3.63	Pham (1987)	Bacon	0.680	−1.70	ASHRAE (1967)
Muscle[a]	0.570	−2.02	Succar and Hayakawa (1983)	Ham	0.560	−1.70	ASHRAE (1981)
Muscle[a]	0.600	−2.00	Levy (1979)	Muscle	0.745	−1.75	Murakami and Okos (1989)
Muscle[a]	0.602	−1.20	Mannapperuma and Singh (1989)	Muscle	0.740	−1.00	Murakami and Okos (1989)
Muscle[a]	0.630	−1.76	Succar and Hayakawa (1983)	Muscle	0.725	−0.90	Levy (1979)
Muscle[a]	0.683	−1.00	Mannapperuma and Singh (1989)	Muscle	0.600	−2.20	ASHRAE (1967)
Muscle[a]	0.700	−1.01	Succar and Hayakawa (1983)	Muscle	0.560	−0.90	Levy (1979)
Muscle[a]	0.734	−1.10	Mascheroni and Calvelo (1980)	Muscle	0.402	−0.90	Levy (1979)
Muscle[a]	0.740	−0.75	Murakami and Okos (1989)	Shoulder	0.490	−2.20	ASHRAE (1981)
Muscle[a]	0.740	−0.99	Succar and Hayakawa (1983)	*Poultry*			
Muscle[a]	0.740	−0.99	Pham (1987)	Chicken	0.760	−0.79	Dickerson (1968)
Muscle[a]	0.740	−1.00	Chen (1986)	Poultry	0.740	−2.80	ASHRAE (1967)
Muscle[a]	0.745	−0.63	Dickerson (1968)	Turkey	0.740	−2.80	Murakami and Okos (1989)
Muscle[a]	0.745	−1.77	Heldman (1974)	*Others*			
Muscle[a]	0.750	−0.82	Sheard et al. (1990)	Liver	0.655	−1.70	ASHRAE (1967)
Muscle[a]	0.760	−1.11	Mellor (1983)	Sausage (frank)	0.600	−1.70	ASHRAE (1967)
Muscle[a]	0.800	−0.73	Pham (1987)	Sausage (frank)	0.560	−1.70	ASHRAE (1981)
Muscle[a]	0.800	−0.80	Levy (1979)	Sausage (fresh)	0.650	−3.30	ASHRAE (1967)
Round	0.745	−1.75	Murakami and Okos (1989)	Sausage (smoked)	0.600	−3.90	Polley et al. (1980)
Sirloin	0.740	−1.00	Murakami and Okos (1989)	Sausage (smoked)	0.500	−3.90	ASHRAE (1981)

(continued)

Table 5.10 (continued) Initial or Equilibrium Freezing Point of Meat

Material	X_w^o	T_f (°C)	Reference	Material	X_w^o	T_f (°C)	Reference
Sirloin	0.750	−1.20	Mellor (1983)	Tylose MH1000	0.770	−0.60	Succar and Hayakawa (1983)
Veal	0.765	−0.74	Dickerson (1968)	Tylose + NaCl	0.770	−0.95	Succar and Hayakawa (1983)
Veal	0.745	−1.75	Murakami and Okos (1989)	Tylose	0.770	−0.60	Mannapperuma and Singh (1989)
Veal (calf)	0.775	−0.68	Pham (1987)	Tylose	0.770	−0.68	Pham (1987)
Lamb				Tylose (salted)	0.770	−1.19	Pham (1987)
Kidney	0.798	−0.96	Pham (1987)	Tylose	0.750	−0.77	Pham (1987)
Leg	0.710	−1.47	Mellor (1983)	Venison	0.730	−0.74	Dickerson (1968)
Loin	0.649	−0.90	Pham (1987)				

[a] Muscle means that specification was not mentioned.

Table 5.11 Initial or Equilibrium Freezing Point of Fish

Material	X_w^o	T_f (°C)	Reference	Material	X_w^o	T_f (°C)	Reference
Carp	0.778	−0.80	Mannapperuma and Singh (1989)	Haddock	0.836	−0.57	Dickerson (1968)
Catfish	0.803	−1.00	Murakami and Okos (1989)	Haddock	0.836	−0.89	Pham (1987)
Cod	0.820	−0.90	Chen (1986)	Haddock	0.803	−1.00	Murakami and Okos (1989)
Cod	0.820	−0.90	Pham (1987)	Haddock	0.803	−2.94	Charm and Moody (1966)
Cod	0.812	−1.00	Mannapperuma and Singh (1989)	Haddock	0.800	−2.20	ASHRAE (1981)
Cod	0.805	−1.08	Mellor (1983)	Haddock-cod	0.780	−2.20	ASHRAE (1981)
Cod	0.803	−0.63	Dickerson (1968)	Halibut	0.750	−2.20	ASHRAE (1981)
Cod	0.803	−0.91	Pham (1987)	Herring	0.700	−2.20	ASHRAE (1981)
Cod	0.780	−2.20	Murakami and Okos (1989)	Herring (smoked)	0.640	−2.20	ASHRAE (1981)
Cod	0.745	−1.00	Long (1955)	Hake-whiting	0.820	−2.20	ASHRAE (1981)
Cod	0.700	−2.20	ASHRAE (1967)	Menhaden	0.620	−2.20	ASHRAE (1981)
Cod	0.500	−3.50	Chen (1986)	Mackerel	0.570	−2.20	ASHRAE (1981)
Cod	0.500	−3.57	Pham (1987)	Perch	0.803	−1.00	Murakami and Okos (1989)
Fish	0.820	−0.80	Succar and Hayakawa (1983)	Perch	0.791	−0.86	Pham (1987)
Fish	0.800	−0.80	Levy (1979)	Pollock	0.790	−2.20	ASHRAE (1981)
Fish	0.750	−1.00	Succar and Hayakawa (1983)	Redfish	0.803	−1.00	Murakami and Okos (1989)
Fish	0.660	−1.95	Succar and Hayakawa (1983)	Salmon	0.670	−2.20	Murakami and Okos (1989)
Fish	0.600	−2.00	Levy (1979)	Salmon	0.670	−2.20	Murakami and Okos (1989)
Fish	0.570	−2.96	Succar and Hayakawa (1983)	Salmon	0.640	−2.20	ASHRAE (1981)
Fish	0.500	−3.00	Levy (1979)	Tuna	0.700	−2.20	ASHRAE (1981)
Trout	0.803	−1.00	Murakami and Okos (1989)				
Tilapia	0.803	−1.03	Chen and Pan (1995)				

Note: Fish means that specification was not mentioned.

Table 5.12 Initial or Equilibrium Freezing Point of Seafood

Seafood	X_w^o	X_{ss}^o	T_f (°C)	Reference	Seafood	X_w^o	X_{ss}^o	T_f (°C)	Reference
Calamari[a]	0.8367	0.0197	−0.50	Rahman and Driscoll (1991)	Octopus[a]	0.8572	0.0414	−1.47	Rahman and Driscoll (1991)
Calamari[a]	0.8391	0.0231	−0.50	Rahman and Driscoll (1991)	Octopus[b]	0.7863	0.0817	−1.70	Rahman and Driscoll (1991)
Calamari[a]	0.8411	0.0277	−0.63	Rahman and Driscoll (1991)	Octopus[b]	0.8032	0.0349	−1.27	Rahman and Driscoll (1991)
Calamari[b]	0.7944	0.0482	−1.23	Rahman and Driscoll (1991)	Octopus[b]	0.8205	0.0477	−1.00	Rahman and Driscoll (1991)
Calamari[b]	0.7988	0.0599	−1.50	Rahman and Driscoll (1991)	Octopus[b]	0.8223	0.0449	−1.40	Rahman and Driscoll (1991)
Calamari[b]	0.8002	0.0352	−1.23	Rahman and Driscoll (1991)	Octopus[b]	0.8407	0.0495	−1.43	Rahman and Driscoll (1991)
Calamari[b]	0.8013	0.0481	−0.53	Rahman and Driscoll (1991)	Octopus[b]	0.8415	0.0428	−1.30	Rahman and Driscoll (1991)
Calamari[b]	0.8034	0.0303	−1.00	Rahman and Driscoll (1991)	Octopus[b]	0.8437	0.0399	−1.40	Rahman and Driscoll (1991)
Calamari[b]	0.8163	0.0369	−0.97	Rahman and Driscoll (1991)	Octopus[b]	0.8449	0.0483	−1.40	Rahman and Driscoll (1991)
Calamari[c]	0.8274	0.0373	−0.50	Rahman and Driscoll (1991)	Octopus[b]	0.8531	0.0344	−1.10	Rahman and Driscoll (1991)
Calamari[c]	0.8312	0.0296	−1.07	Rahman and Driscoll (1991)	Oyster	0.7732	0.0493	−1.60	Rahman and Driscoll (1991)
Calamari[c]	0.8338	0.0234	−0.60	Rahman and Driscoll (1991)	Oyster (shell)	0.8040	—	−2.80	ASHRAE (1967)
Calamari[c]	0.8361	0.0383	−0.40	Rahman and Driscoll (1991)	Oyster (tube)	0.8700	—	−2.80	ASHRAE (1967)
Cuttle[a]	0.7830	0.0614	−1.13	Rahman and Driscoll (1991)	Scallop	0.8563	0.0120	−0.80	Rahman and Driscoll (1991)
Cuttle[a]	0.7979	0.0499	−0.73	Rahman and Driscoll (1991)	Scallop	0.8030	—	−2.20	ASHRAE (1967)
Cuttle[a]	0.8117	0.0493	−1.20	Rahman and Driscoll (1991)	Shrimp	0.7080	—	−2.20	ASHRAE (1967)
Cuttle[a]	0.8221	0.0474	−0.72	Rahman and Driscoll (1991)	Squid[a]	0.7815	0.0700	−1.58	Rahman and Driscoll (1991)
Crab[d]	0.8017	0.0742	−2.00	Rahman and Driscoll (1991)	Squid[a]	0.7908	0.0554	−1.63	Rahman and Driscoll (1991)
King prawn[e]	0.7390	0.0817	−1.87	Rahman and Driscoll (1991)	Squid[a]	0.7968	0.0523	−1.67	Rahman and Driscoll (1991)
King prawn[e]	0.7563	0.0916	−1.67	Rahman and Driscoll (1991)	Squid[a]	0.8173	0.0514	−1.13	Rahman and Driscoll (1991)
King prawn[e]	0.7568	0.0906	−2.07	Rahman and Driscoll (1991)	Squid[a]	0.8377	0.0246	−1.10	Rahman and Driscoll (1991)
King prawn[f]	0.7649	0.0968	−1.37	Rahman and Driscoll (1991)	Squid[a]	0.8399	0.0313	−1.00	Rahman and Driscoll (1991)
King prawn[f]	0.7652	0.0909	−1.70	Rahman and Driscoll (1991)	Squid[b]	0.7984	0.0559	−1.67	Rahman and Driscoll (1991)
King prawn[f]	0.7706	0.0834	−1.53	Rahman and Driscoll (1991)	Squid[b]	0.8046	0.0321	−0.93	Rahman and Driscoll (1991)
King prawn[f]	0.7730	0.0303	−1.10	Rahman and Driscoll (1991)	Squid[b]	0.8421	0.0318	−0.50	Rahman and Driscoll (1991)
Mussel	0.7744	0.0407	−1.20	Rahman and Driscoll (1991)	Squid[g]	0.8137	0.0440	−1.50	Rahman and Driscoll (1991)
Octopus[a]	0.7744	0.0888	−1.70	Rahman and Driscoll (1991)	Squid[g]	0.8224	0.0329	−0.60	Rahman and Driscoll (1991)
Octopus[a]	0.8061	0.0535	−1.70	Rahman and Driscoll (1991)	Squid[g]	0.8397	0.0362	−0.57	Rahman and Driscoll (1991)

[a] Mantle.
[b] Tentacle.
[c] Wing.
[d] Blue swimmer.
[e] Green.
[f] Tiger.
[g] Tail.

Table 5.13 Experimental Freezing Point of Bakery Products

Material	X_w^o	T_f (°C)	Reference
Deep-fat fried churros	0.62	−2.56	Guevara-Lara (2004)
Stuffed with cajeta	0.58	−3.21	Guevara-Lara (2004)
Stuffed with chocolate	0.55	−3.60	Guevara-Lara (2004)
Precooked biscuits	0.24	−6.73	Sosa-Morales et al. (2004)
Raw croissants	0.30	−2.90	Encinas-Banda et al. (2005)
Sponge cake dough	0.31	−6.80	Morales-de la Peña (2005)

Singh and Mannapperuma (1990) assumed the application of Raoult's law and Clausius–Clapeyron equation to propose an equation (Equation 5.3) between the freezing point and mass fractions of a food model (Valentas et al., 1997):

$$\frac{1}{T_\mathrm{f}} = \frac{1}{T_\mathrm{w}} - \frac{R}{M_\mathrm{w}L} \ln\left(\frac{(X_\mathrm{w}^\mathrm{o} - X_\mathrm{I} - X_\mathrm{b})/M_\mathrm{w}}{((X_\mathrm{w}^\mathrm{o} - X_\mathrm{I} - X_\mathrm{b})/M_\mathrm{w}) + ((X_\mathrm{SNF}^\mathrm{o} + X_\mathrm{F}^\mathrm{o})/M_\mathrm{E})}\right) \tag{5.3}$$

At the beginning of the freezing process, the aqueous solution is diluted; then Raoult's law may be used to calculate the freezing point (Ibarz and Barbosa, 2001).

$$T_\mathrm{f} = K_\mathrm{f}\frac{X_\mathrm{s}}{M_\mathrm{s}} \tag{5.4}$$

Most of the proposed equations to quantify the freezing point of foods are based on the difference between the freezing point of water and the freezing point of the specific food item, that is to say, the freezing point depression (ΔT). Furthermore, all theoretical freezing point depression equations have been derived from thermodynamic relationships in equilibrium and particularly on the basis of the Clausius–Clapeyron equation generated from the first and second law of thermodynamics. The magnitude of the freezing point depression becomes a direct function of the molecular weight, type, and concentration of the food item (Heldman, 1975). A proposal is that the freezing point depression can also be established from concentration and the effective molecular weight of dissolved solids (Chen, 1986):

$$\begin{aligned} \Delta T &= \frac{1000K_\mathrm{f}X_1}{(1 - X_1)M_\mathrm{s}} \\ X_1 &= \frac{W_\mathrm{so}}{W_\mathrm{T}} \end{aligned} \tag{5.5}$$

where

W_so is the weight of dissolved solids

$W_\mathrm{T} = W_\mathrm{w} + W_\mathrm{so}$, being W_w the weight of water.

Table 5.14 Freezing Temperature of Carrot, Reindeer Meat, and White Bread

Material	X_w^o	T_f (°C)	L_w (kJ/kg)	ΔH_w (kJ/kg)	T_{if} (°C)	T_{iif} (°C)	X_{uw}[a]	X_{uw}[b]
Carrot	0.880	−3.4	259.3	432.9	−33	−13.3	0.083	0.034
Reindeer meat	0.749	−3.1	194.5	350.3	−39	−11.8	0.151	0.064
White bread	0.367	−12.2	46.2	172.9	−40	−17.8	0.225	0.029

Source: Roos, Y.H., *J. Food Sci.*, 51, 684, 1986. With permission.

T_f: melting point; T_{if}: incipient freezing; T_{iif}: incipient intensive freezing point.

[a] Unfrozen moisture calculated from latent heat of melting.

[b] Unfrozen moisture calculated from change of enthalpy.

Table 5.15 Freezing Point of Different Foods as a Function of Solid Content

King Fish[1]		King Fish[2]		Garlic[3]		Garlic[4]		Abalone[5]		Date Flesh[6]			Tuna Meat[7]		Apple[8]		Date[9]	
X_s^o	T_f (°C)	X_s^o	T_f (°C)	X_s^{mo}	T_f (°C)	X_s^o	T_f (°C)	X_s^o	T_f (°C)	X_s^o	T_f^a (°C)	T_f^b (°C)	X_s^o	T_f (°C)	X_s^o	T_f^c (°C)	X_s^o	T_f (°C)
0.20	−0.68	0.30	0.78	0.20	−0.90	0.05	−1.10	0.25	−0.90	0.22	−1.80	−4.10	0.27	−1.40	0.14	−4.7	0.12	−1.90
0.30	−1.78	0.35	0.31	0.30	−1.80	0.10	−1.20	0.30	−1.20	0.30	−3.20	−5.50	0.27	−1.30	0.14	−4.6	0.24	−3.60
0.40	−2.56	0.40	−1.06	0.50	−3.90	0.20	−1.80	0.35	−1.90	0.36	−4.90	−7.20	0.29	−1.10	0.27	−8.0	0.29	−5.10
0.50	−3.47	0.45	−1.44	0.60	−8.90	0.30	−2.80	0.40	−2.80	0.41	−7.80	−10.80	0.36	−2.40	0.42	−22.0	0.38	−9.40
0.60	−7.60	0.50	−3.54	0.65	−12.70	0.40	−4.10	0.45	−3.20	0.51	−12.60	−15.20	0.40	−2.90	0.74[d]	−50.3	0.44	−13.60
0.69[d]	−17.40			0.70	−18.60	0.50	−6.10	0.50	−4.30	0.54	−18.30	−20.20	0.44	−3.80			0.54	−24.50
				0.72	−19.20	0.60	−11.10	0.55	−7.10	0.59	−20.30	−23.40	0.45	−3.40			0.59	−31.70
				0.74	−20.70	0.67	−29.50	0.60	−8.90	0.61	−22.00	−25.60	0.45	−3.90			0.64	−47.00
				0.76	−20.50	0.68	−27.00	0.62	−10.10	0.76[d]	—	−43.60	0.49	−4.90				
				0.78	−21.80	0.69	−23.80	0.64	−14.10				0.50	−5.30				
				0.80	−20.70	0.70	−31.10	0.66	−15.90				0.55	−7.60				
				0.82[d]	−26.00			0.68[d]	−18.10				0.55	−7.80				
													0.60	−11.50				
													0.60	−11.40				
													0.62	−15.10				
													0.63	−15.30				
													0.61[d]	−13.30				

1, cooling curve method (Sablani et al., 2007); 2, DSC method, T_F as maximum slope (cooling rate: 2°C–5°C/min) (Sablani et al., 2007); 3, DSC method, T_F as maximum slope (cooling rate: 5°C/min) (Rahman et al., 2005); 4, cooling curve method (Rahman et al., 2005); 5, cooling curve method (Sablani et al., 2004); 6, DSC method (cooling rate: 1°C/min–5°C/min) (Rahman, 2004); 7, cooling curve method (Rahman et al., 2003); 8, DSC method (cooling rate: 10°C/min) (Bai et al., 2001); 9, cooling curve method (Kasapis et al., 2007).

[a] Maximum peak.

[b] Maximum slope.

[c] Extension of maximum slope curve to the base line.

[d] T_m' (maximal-freeze-concentration condition).

Table 5.16 Eutectic Temperature of Various Aqueous Solutions and Foods

Material	T_u (°C)	Material	T_u (°C)
Calcium chloride	−55.0	White bread	−70.0
Potassium chloride	−11.1	Lactose (beta)	−2.3
Sodium chloride	−21.1	Glycerol	−46.5
Sucrose	−9.5	Egg white	−55.0
Glucose	−5.0	Beef	−52.0
Lactose	−0.3	Raspberry juice	−55.0

Source: Fennema, O.R., Powrie, W.D., and Marth, E.H., *Low-Temperature Preservation of Foods and Living Matter*, Marcel Dekker, New York, 1973. With permission.

The Clausius–Clapeyron equation can be written for dilute solution as (Chen and Nagy, 1987)

$$\Delta T = \left(\frac{-RT_f T_w}{L_w}\right) \ln\left(\frac{1 - X_s^o}{1 - X_s^o + EX_s^o}\right) \tag{5.6}$$

For a small change in temperature, the last equation can be written as

$$\Delta T = \frac{K_f}{M_w} \ln\left(\frac{1 - X_w^o}{1 - X_w^o + EX_w^o}\right) \tag{5.7}$$

The assumptions in this equation are (1) all water is freezable, (2) soluble and insoluble solids have a similar effect on the water activity, and (3) Raoult's law is valid. Empirical values for E are given in Table 5.18. Another equation for ideal solutions is the one proposed by Chen (1987a), which shows the relationship between water activity and the ΔT of food systems:

Table 5.17 Melting Point of Salts from Different Sources

	T_m (°C)			
Salt	Janz et al. (1979)	JSME (1986)	IP (1987)	Yamada et al. (1993)
$NaNO_2$	—	285	271	281
$LiNO_3$	253	254	—	252
$NaNO_3$	307	310	307	305
KNO_3	337	337	333	334
LiCl	610	610	605	604
NaCl	800	800	801	800
KCl	770	768	770	768
Li_2CO_3	723	735	725	725
Na_2CO_3	858	854	851	860
K_2CO_3	893	895	891	903
KSCN	—	—	173	175
$CaCl_2$	782	—	772	773
$BaCl_2$	962	—	—	962
LiBr	552	—	—	548
Li_2SO_4	859	—	—	854
Na_2SO_4	884	—	—	889

Source: Yamada, M., Tago, M., Fukusako, S., and Horibe, A., *Thermochim. Acta*, 218, 401, 1993. With permission.

Table 5.18 Model Parameters of Equations 5.7 and 5.9

	Equation 5.7		Equation 5.9		
Material	**X_s^o Range**	***E***	***B* (kg/kg Dry Solids)**	***E***	**Reference**
King fish	0.20–0.69	0.008	0.303	0.028	Sablani et al. (2007)
Garlic	0.20–0.82	0.068	−0.062	0.080	Rahman et al. (2005)
Abalone	0.25–0.66	0.071	0.301	0.034	Sablani et al. (2004)
Date flesh	0.22–0.76	0.147	0.053	0.129	Rahman (2004)
Tuna meat	0.27–0.63	0.071	0.383	0.033	Rahman et al. (2003)
Apple	0.14–0.76	0.238	−0.156	0.320	Bai et al. (2002)
Beef (raw)	—	—	0.185	0.023	Rahman (1994)
Meat[a] (raw)	—	—	0.192	0.021	Rahman (1994)
Seafood[b] (raw)	0.14–0.24	0.049	0.650	0.041	Rahman and Driscoll (1994)
Squid	0.18–0.65	0.082	0.120	0.067	Rahman and Driscoll (1994)

[a] Beef, lamb, pork, poultry, venison, and fish.
[b] Cuttle, octopus, calamari, king prawn, squid, mussel, oyster, scallop, and crab.

$$a_w = \left(\frac{1}{1 + 0.0097\Delta T}\right) \tag{5.8a}$$

Yet this equation is applicable to an ideal dilute solution. For compensating the deviations caused by the use of more concentrated solutions, the equation was modified as

$$a_w = \left(\frac{1}{1 + 0.0097\Delta T + c\Delta T^2}\right) \tag{5.8b}$$

Fennema and Berny (1974, cited by Chen, 1987a) used data for ice–water system, and the value of $c = 5 \times 10^{-5}\ \mathrm{K}^{-2}$ was determined (Chen, 1987a). Peralta et al. (2007) used UNIQUAC model to predict the freezing point of solutions containing solutes such as sodium chloride, calcium chloride, potassium chloride, ethanol, and glucose.

5.4.1.2 Nonideal Systems

For most real solutions, the depression of the freezing point does not obey the ideal equation, except at very dilute concentrations of dissolved solids. In real solution, some water in solution combines with the solute and acts as a nonsolvent water and some water is amorphous in nature. Based on the concept of bound water, Schwartzberg (1976) proposed the following equation:

$$\Delta T = \frac{K_f}{M_w} \ln\left[\frac{X_w^o - BX_s^o}{(X_w^o - BX_s^o) + EX_s^o}\right] \tag{5.9}$$

The estimated values of *B* and *E*, as empirical terms of this equation, are given in Table 5.18 for different foods. Chen (1986) suggested that two sets of data X_s and T_f (which need to be well separated, i.e., X_s at 0.10 and 0.30) were necessary. Miles et al. (1997) proposed the following equation to predict the freezing point of foods from composition data:

$$\Delta T = \frac{K_f}{M_w}\left[\frac{\sum_{i=1}^{n}\left(\frac{X_i^o}{M_w}\right)}{\left(\frac{X_w^o - X_b}{M_w}\right) + \sum_{i=1}^{n}\frac{\tau_i X_i^o}{M_i}}\right] \tag{5.10}$$

where τ_i is the molecular dissociation and other nonideal behavior, and X_b is the fraction of bound water to components of the food, such as protein or starch, and can be estimated from the following equation:

$$X_b = \sum_{i=1}^{n} \sigma_i X_i^o \tag{5.11}$$

where the coefficients σ_i are the component-specific constants. Miles et al. (1997) used only two components: carbohydrate and protein. A review of the literature suggested that σ_i coefficient values of 0.3 and 0.45, respectively, could be used. They also used τ_i equal to 1 for all food components. The components and their molecular mass used by Miles et al. (1997) are presented in Table 5.19. Similarly, Boonsupthip and Heldman (2007) used the food composition data (Table 5.19) to predict the freezing point of foods. Van der Sman and Boer (2005) used this approach to predict the freezing point of meat products. They found from the data fitting that σ_i coefficients of protein and carbohydrate are 0.299 and 0.10, respectively. They also used NaCl, sodium polyphosphate (NaPP), and the remaining ash. The molecular weight of sodium chloride, NaPP ($Na_5P_3O_{10}$), and ash are 58.46, 376, and 72, respectively. The dissociation numbers of sodium chloride, NaPP ($Na_5P_3O_{10}$), and ash used were 2, 8, and 2, respectively.

Mellor (1983) used a different approach to predict the freezing point. He assumed that at −40°C there was no free water present in the system, so that the enthalpy of the product could be defined as zero at that temperature. The enthalpy balance gave the equation as

$$T_f = \frac{(C_p)_r B\gamma\left(1 - X_w^o\right)}{C_p X_w^o} \tag{5.12}$$

Mellor (1983) considered γ as the temperature where the enthalpy becomes small (approximately 1% change per degree Kelvin). He measured the freezing point T_f and B for a number of food

Table 5.19 Components and Their Molecular Masses Used for Predicting the Initial Freezing Point of Food

Components used by Miles et al. (1997)		Components used by Boonsupthip and Heldman (2007)	
Component	**M_i**	**Component**	**M_i**
Water	18	*Minerals and element*	
Monosaccharide	180	Sodium (Na)	22.99
Disaccharide	342	Magnesium (Mg)	24.31
Lactic acid	90	Phosphorus (P)	30.97
Malic acid	134	Chloride (Cl)	35.45
Citric acid and isocitric acid	192	Potassium (K)	39.10
Acetic acid	60	Calcium (Ca)	40.08
Tartaric acid	150	*Carbohydrates*	
Oxalic acid	90	Monosaccharides ($C_6H_{12}O_6$)	180.07
Alcohol	46	Disaccharides ($C_{12}H_{22}O_{11}$)	342.11
Na	23	*Acid and bases*	
K	39	Nitrite (NO_3)	62.00
Ca	40	Oxalic acid (HOOCCOOH)	90.08
Mg	24	Lactic acid [$CH_3CH(OH)COOH$]	134.10
Fe	56	Ascorbic acid ($C_6H_6O_6$)	176.10
P	31	Citric and isocitric acid [$(HO(COOH)(CH_2COOH)_2$]	192.10
Cl	35.5		

materials. The mean percent deviations of the models were less than 7% for sucrose, apple, grapes, beans, peas, cod, beef, and egg (white and whole). But in the case of carrot and milk, the above model gave an error around 26%. The values of B obtained from the experimental data can be used to characterize the effect of dissociation.

Chen (1987b) used an empirical constant, which was a linear coefficient of concentration, in the following equation:

$$\Delta T = -\frac{K_\mathrm{f}}{M_\mathrm{w}} \ln\left(\frac{1 - X_\mathrm{s}^\mathrm{o}}{1 - X_\mathrm{s}^\mathrm{o} + EX_\mathrm{s}^\mathrm{o}\left(1 + cX_\mathrm{s}^\mathrm{o}\right)}\right) \quad (5.13)$$

According to Chen (1987b) c can either be positive or negative depending on the food systems. Later, Chen and Nagy (1987) modified the equation and proposed the following for nonideal systems:

$$\Delta T = -\frac{K_\mathrm{f}}{M_\mathrm{w}} \ln\left(\frac{1 - X_\mathrm{s}^\mathrm{o}}{1 - X_\mathrm{s}^\mathrm{o} + EX_\mathrm{s}^\mathrm{o}\left(f + cX_\mathrm{s}^\mathrm{o}\right)}\right) \quad (5.14)$$

where c and f (empirical constants) can be estimated by regression. The theoretical models described here were used only for liquid food where all solids are soluble. For solid and semisolid foods, the situation is different due to the presence of insoluble solids in the system, which also has interaction with the water and soluble solids. Heldman (1974, 1982), Schwartzberg (1976), and Murakami and Okos (1989) assumed X_s as the concentration of all solutes in the system (soluble and insoluble). The amount of combined water for any given substance is a function of concentration and temperature (Chen and Nagy, 1987).

Mannapperuma and Singh (1989) used a similar model to estimate the freezing point of foods by a computer-aided method for freezing and thawing processes.

$$\frac{1}{T_\mathrm{f}} = \frac{1}{T_\mathrm{w}} - \frac{R}{L_\mathrm{w}} \ln\left(\frac{x_\mathrm{w}^\mathrm{o}}{x_\mathrm{w}^\mathrm{o} + \sum_i x_i^\mathrm{o}}\right) \quad (5.15)$$

Jie et al. (2003) improved Hoo and McLellan's method to measure the freezing point, according to the content of soluble solids of fruits like apple, pear, grape, and orange in several varieties. By recording the freezing point curves with temperature measured every 10 s, and then using a statistical package (SAS) to analyze them, a mathematical model was obtained ($R = -0.994$):

$$T_\mathrm{f} = 1.15 - 0.20X_\mathrm{s}^\mathrm{o} \quad (5.16)$$

Succar and Hayakawa (1990) obtained an equation to predict the freezing point, taking enthalpies at different temperatures. They validated their model with aqueous solutions of sodium chloride and sucrose at different concentrations. They also found good agreement between reported and predicted values for fresh fruits, lean beef meat, lean fish meat, and fresh vegetables. The equation from Succar and Hayakawa (1990) is an iterative model, which is an interesting and different approach in comparison with the other equations:

$$T_\mathrm{f} - \frac{\left|T_\mathrm{f}^{(1-m)}\right|}{Z_1} - Z_2 = 0 \quad (5.17a)$$

$$Z_1 = (m-1)[(C_p)_{\text{un}} - A]/D \tag{5.17b}$$

$$Z_2 = \frac{1}{A - (C_p)_{\text{un}}}\left[A - T_r + \frac{D}{(m-1)|T_r|^{m-1}} - G + H_0\right] \tag{5.17c}$$

m, A, D, and G are determined from the enthalpy data; the food enthalpies were estimated for temperatures above the initial freezing point, taking the following relationship as basis:

$$H = (C_p)_{\text{un}}T + H_{\text{o}} \quad \text{for } T > T_{\text{f}} \tag{5.18a}$$

For temperatures below T_{f} the following equation may be used for the enthalpy as a function of temperature:

$$H = A(T - T_r) + \frac{D}{m-1}\left[\frac{1}{(T_w - T)^{m-1}} - \frac{1}{(T_w - T_r)^{m-1}}\right] + G \tag{5.18b}$$

The first researchers to develop an enthalpy data analysis method for estimating T_{f} were Jason and Lang (1955, cited by Rahman, 1995). They assumed continuity in the derivate at temperatures close to T_{f}, which caused them to obtain much larger specific heats (Succar and Hayakawa, 1990). They considered the discontinuity of the phenomenon to make the predictions more accurate.

5.4.2 Empirical Curve-Fitting Models (ECFM)

Different models have been proposed to fit the experimental freezing point values obtained with groups of foods. Chang and Tao (1981) provided equations to predict the freezing point of food materials, based on statistical correlation techniques. They divided the food materials into three groups: meat/fish, juice, and fruits/vegetables. Using data for three food types tabulated by Dickerson (1968), the correlation between initial freezing point and water content was expressed as linear and quadratic expressions:

$$\text{For the meat group:} \quad \Delta T = 1.9 + 1.47X_{\text{w}}^{\text{o}} \tag{5.19}$$

$$\text{For the vegetables and fruits group:} \quad \Delta T = -14.46 + 49.19X_{\text{w}}^{\text{o}} - 37.07\left(X_{\text{w}}^{\text{o}}\right)^2 \tag{5.20}$$

$$\text{For the juice group:} \quad \Delta T = 152.63 - 327.35X_{\text{w}}^{\text{o}} - 176.49\left(X_{\text{w}}^{\text{o}}\right)^2 \tag{5.21}$$

Another approach considering the solutes present in the food was proposed by Hoo and McLellan (1987) in their study about the effect of apple pectin on the freezing point depression (ΔT) of apple juice concentrates. The results of this investigation indicated that the freezing point depression was affected by the pectin amount, being greater as concentration increased. Juice that was depectinized had a higher freezing point than that of the untreated juice with intact pectin. The degree of ripening in the apples, directly proportional to their soluble pectin content, also affected the ΔT.

It was found that adding both soluble solids and pectin to the juice would depress the freezing point further than only added pectin or added solids, possibly because of interactions formed between soluble solids and pectin. According to the content of pectin intact in the juice, a regression equation was derived for the freezing point in degree Celsius:

$$T_{\text{f}} = -1.44 + 0.1214(^\circ\text{B}) - 0.028(\text{PE}) - 0.00524(^\circ\text{B})^2 - 0.00469(^\circ\text{B})(\text{PE}) \tag{5.22}$$

The predicted freezing point calculated from Equation 5.22 had a variation of ±1.5°C of the actual freezing point. Yet this discrepancy was explained by the varieties and growing season differences of the fruit, and that adding sucrose was the way to adjust degree Brix for the predicted model, whereas vacuum evaporation was used for actual studies.

A third-degree polynomial has been proposed by Chen and Nagy (1987) for correlating the freezing point depression of foods:

$$\Delta T = IX_s^o + J\left(X_s^o\right)^2 + F\left(X_s^o\right)^3 \quad (5.23)$$

The values of I, J, and F (empirical constants) for various aqueous solutions are given in Tables 5.20 through 5.22. Sanz et al. (1989) developed an empirical correlation to predict the freezing point of meat products as a function of moisture content as

$$\Delta T = \frac{\left(X_w^o - 1\right)}{\left(0.069 - 0.439X_w^o\right)} \quad (5.24)$$

Table 5.20 Parameters *I*, *J*, and *F* of Equation 5.23 for Aqueous Solutions

Component	X_s^o Range	I	J	F	Reference
Acetic acid	0–0.36	30.578	16.411	5.556	Chen and Nagy (1987)
Acetone	0–0.10	32.063	9.263	−11.462	Chen and Nagy (1987)
Ammonium chloride	0–0.13	62.084	41.733	316.961	Chen and Nagy (1987)
Ammonium sulfate	0–0.16	31.37	−43.53	191.767	Chen and Nagy (1987)
Calcium chloride	0–0.32	51.409	−24.328	1076.521	Chen and Nagy (1987)
Citric acid	0–0.30	9.302	20.116	−12.978	Chen and Nagy (1987)
Citric acid	0–0.30	10.29	9.475	12.749	Chen (1986)
Ethanol	0–0.68	32.496	155.515	140.76	Chen and Nagy (1987)
Ethanol	0–0.68	31.808	157.439	−141.85	Chen (1986)
Formic acid	0–0.64	42.75	−5.749	66.584	Chen and Nagy (1987)
Formic acid	0–0.64	43.548	−9.573	71.586	Chen (1986)
D-Fructose	0–0.40	10.282	10.693	22.902	Chen and Nagy (1987)
Fructose	0–0.28	10.114	12.738	17.017	Chen (1986)
D-Glucose	0–0.30	10.547	7.991	34.2	Chen and Nagy (1987)
Glucose	0–0.30	10.398	9.709	29.823	Chen and Nagy (1987)
Glycerol	0–0.40	18.702	38.635	26.429	Chen (1986)
Glycerol	0–0.40	20.374	23.188	56.754	Chen (1986)
Hydrochloric acid	0–0.12	98.025	309.176	2491.089	Chen and Nagy (1987)
Inulin	0–0.10	0.528	−2.341	30.605	Chen (1986)
Lactic acid	0–0.2	18.262	19.129	8.384	Chen and Nagy (1987)
Lactic acid	0–0.2	17.781	28.045	−27.53	Chen and Nagy (1987)
Lactose	0–0.08	5.639	−4.588	140.566	Chen and Nagy (1987)
Lactose	0–0.08	5.936	−14.85	225.03	Chen and Nagy (1987)
Magnesium chloride	0–0.05	0.669	80.479	2217.858	Chen (1986)
Magnesium sulfate	0–0.16	18.471	−46.614	462.922	Chen and Nagy (1987)
Manganous sulfate	0–0.20	14.387	−21.948	225.507	Chen and Nagy (1987)
Maltose	0–0.44	5.661	1.443	29.939	Chen and Nagy (1987)
Manitol	0–0.15	10.236	10.41	19.976	Chen and Nagy (1987)
Methanol	0–0.68	57.976	63.938	86.395	Chen and Nagy (1987)

(*continued*)

Table 5.20 (continued) Parameters *I*, *J*, and *F* of Equation 5.21 for Aqueous Solutions

Component	X_s^o Range	*I*	*J*	*F*	Reference
Oxalic acid	0–0.04	30.186	−88.962	442.205	Chen and Nagy (1987)
Phosphoric acid	0–0.40	20.736	15.549	201.265	Chen and Nagy (1987)
Potassium bicarbonate	0–0.14	33.747	−41.176	182.855	Chen and Nagy (1987)
Potassium carbonate	0–0.40	38.636	−87.502	560.779	Chen and Nagy (1987)
Potassium chloride	0–0.13	45.56	9.653	162.337	Chen and Nagy (1987)
Potassium hydroxide	0–0.07	54.987	276.536	−784.533	Chen and Nagy (1987)
Potassium iodide	0–0.40	20.924	10.142	62.058	Chen and Nagy (1987)
Potassium nitrate	0–0.10	33.75	−89.108	272.467	Chen and Nagy (1987)
Potassium phosphate[a]	0–0.10	24.654	−11.848	−115.329	Chen and Nagy (1987)
Potassium phosphate[b]	0–0.08	24.678	−9.916	69.436	Chen and Nagy (1987)

[a] Dihydrogen.
[b] Monohydrogen.

Chen et al. (1990), based on Riedel's equation, proposed a generalized empirical model for fruits, vegetables, and juices:

$$\Delta T = 10X_s^o + 50\left(X_s^o\right)^3 \tag{5.25}$$

Riedel's equation assumes that all juices are completely frozen at −60°C. Rahman and Driscoll (1991) similarly derived an equation for fresh seafood as

$$\Delta T = \frac{\left(X_w^o - 1\right)}{\left(0.0078 - 0.140X_w^o\right)} \tag{5.26}$$

Rahman (1994) studied the accuracy of prediction of freezing point of meat from general (empirical, theoretical, and semiempirical) models and developed a general relationship:

$$\Delta T = \frac{\left(X_w^o - 1\right)}{\left(0.072 - 0.488X_w^o\right)} \tag{5.27}$$

Table 5.21 Parameters *I*, *J*, and *F* of Equation 5.23 for Aqueous Solutions

Component	X_s^o Range	*I*	*J*	*F*	Reference
Potassium sulfate	0–0.05	27.614	−160.628	1557.138	Chen and Nagy (1987)
Propylene glycol	0–0.16	26.551	−40.012	537.229	Chen and Nagy (1987)
Sodium acetate	0–0.09	42.346	68.696	277.82	Chen and Nagy (1987)
Sodium bicarbonate	0–0.06	40.069	−44.198	49.982	Chen and Nagy (1987)
Sodium carbonate	0–0.06	41.43	−247.256	2469.03	Chen and Nagy (1987)
Sodium chloride	0–0.23	59.278	7.332	544.427	Chen and Nagy (1987)
Sodium citrate	0–0.18	22.562	−84.531	499.225	Chen and Nagy (1987)
Sodium hidroxide	0–0.14	84.963	61.028	1241.118	Chen and Nagy (1987)
Sodium nitrate	0–0.07	40.109	−35.006	195.415	Chen and Nagy (1987)
Sucrose	0–0.42	6.001	0.419	33.292	Chen and Nagy (1987)
Sucrose	0–0.42	5.629	2.743	27.415	Chen (1986)
Sodium chloride	0–0.23	59.593	2.473	559.546	Chen (1986)
Tartaric acid	0–0.40	13.743	−0.517	74.335	Chen and Nagy (1987)

Table 5.22 Parameters *I*, *J*, and *F* of Equation 5.23 for Foods

Food	X_s^o Range	*A*	G	*F*	Reference
Skim milk (freeze-dried)	0–0.40	5.604	5.125	14.583	Chen (1986)
Coffee (freeze-dried)	0–0.40	6.079	−3.446	28.555	Chen (1986)
Grape juice	0–0.30	11.056	13.333	0	Chen (1986)
Tomato juice (dried)	0–0.20	9.5	3.33	0	Chen (1986)
Fresh seafood	0.739–0.853	29.55	−25.74	−1038	Rahman and Driscoll (1991)

Errors of 46% or more were obtained when 101 published data points for meat were used to develop this equation. The error can be reduced to 24% by considering only beef. The high error indicated that generalized equations should include more measured variables such as composition or should include more specific information such as meat source and location of the meat in the animal. This also indicates that the presentation of freezing point of foods in terms of only water content has less value since chemical composition of food materials varies with the variety, growing conditions and environment, and product formulation. Jie et al. (2003) measured the freezing points of 11 breeds of fruits and correlated the decreasing effect as a function of soluble solids and obtained a mathematical model as

$$T_f = 0.15 - 0.196X_s^o \tag{5.28}$$

They confirmed that the freezing point of fruits is different from their corresponding juice. Pham (1996) developed the freezing point prediction equation for meat, fish, fruit, and nonfat dairy foods as

$$\Delta T = 4.66\left(\frac{X_{ca}^o}{X_w^o}\right) + 46.40\left(\frac{X_{as}^o}{X_w^o}\right) \tag{5.29}$$

In the above equation, carbohydrate (X_{ca}) represents mainly the sugar content, while ash (X_{as}) represents the salt content. Pham (1996) pointed out that a limited number of data were used and the values of sugar content were low; thus not too much significance should be attached to Equation 5.29 as a general model.

5.4.3 Semiempirical Models

Heldman (1974) estimated the T_f of food through thermodynamical analysis, thus developing semitheoretical formulas. He considered that the Clausius–Clapeyron equation could apply to the liquid, which was also assumed to be a binary solution. Chen (1985) refined his approach and developed the following equation for enthalpy of food at the initial freezing point:

$$H_f = (T_f - T_r)\left(0.37 + 0.3X_s^o + \frac{X_s^o}{M_s}\cdot\frac{RT_w^2}{T_f T_r}\right) \tag{5.30a}$$

where both T_r and M_s are used to determine the food enthalpy. M_s is dependent on food and its moisture content, and its values are obtained by

$$M_s = \frac{d}{(1 + eX_s^o)} \tag{5.30b}$$

Here, *d* and *e* are used for estimating effective molecular weight. For example, for lean beef and cod muscle, $e = -1$, and for apple and orange juices $e = 0.25$. The *d* values are 5.354 and 404.9 for lean beef and cod muscle, respectively, and 200 for juices.

Lerici et al. (1983), after several assumptions, correlated the water activity and the freezing point of aqueous solutions and liquid foods in the following equation:

$$-\ln a_{\mathrm{w}} = \frac{1}{R}(L_{\mathrm{w}} - \Delta C_p T_{\mathrm{w}})\left(\frac{1}{T_{\mathrm{f}}} - \frac{1}{T_{\mathrm{w}}}\right) + \frac{\Delta C_p}{R}\ln\frac{T_{\mathrm{w}}}{T_{\mathrm{f}}} \quad (5.31a)$$

They derived the following model:

$$\ln a_{\mathrm{w}} = 27.622 - 528.373\left(\frac{1}{T_{\mathrm{f}}}\right) - 4.579\ln T_{\mathrm{f}} \quad (5.31b)$$

Chen and Nagy (1987) used an equation for binary systems as an alternative model for calculating ΔT:

$$\Delta T = \frac{K_{\mathrm{f}} X_{\mathrm{s}}^{\mathrm{o}}(1 - gX_{\mathrm{s}}^{\mathrm{o}})}{(1 - X_{\mathrm{s}}^{\mathrm{o}})M_{\mathrm{s}}} \quad (5.32)$$

K values of 0.25–0.27 for citrus juice and medium inverted sugar solution and 0.56 ($\pm$0.13) for model systems are cited by Chen et al. (1990). This equation has no discontinuity of the variable X, and it is useful for weak acids and nonelectrolyte solutions. Chen (1988) proposed an equation for ΔT to characterize the EFC, which is a plot of the freezing point as a function of concentration:

$$\Delta T = \frac{K_{\mathrm{f}} X_{\mathrm{s}}^{\mathrm{o}}}{(1 - X_{\mathrm{s}}^{\mathrm{o}} - BX_{\mathrm{s}}^{\mathrm{o}})M_{\mathrm{s}}} \quad (5.33a)$$

The mass of bound water per unit mass of solids is represented by the constant B (Schwartzberg, 1976 as cited in Chen, 1988). The values of this constant and for M_{s} may be calculated as follows:

$$B = \frac{X_{\mathrm{s1}}^{\mathrm{o}}\Delta T_2 - X_{\mathrm{s2}}^{\mathrm{o}}\Delta T_1}{X_{\mathrm{s1}}^{\mathrm{o}} X_{\mathrm{s2}}^{\mathrm{o}}(\Delta T_2 - \Delta T_1)} \quad (5.33b)$$

$$M_{\mathrm{s}} = \frac{K_{\mathrm{f}} X_{\mathrm{s1}}^{\mathrm{o}}}{(1 - X_{\mathrm{s1}}^{\mathrm{o}} - BX_{\mathrm{s1}}^{\mathrm{o}}\Delta T)} \quad (5.33c)$$

The data are obtained from two sets, 1 and 2, which need to be well separated in concentration. The value of B is a function of concentration, but M_{s} has only one true value. Therefore, with an average value of M_{s}, the value of B is recalculated with the following equation:

$$B = \frac{1}{X_{\mathrm{s1}}^{\mathrm{o}}}\left(1 - X_{\mathrm{s1}}^{\mathrm{o}} - \frac{K_{\mathrm{f}} X_{\mathrm{s1}}^{\mathrm{o}}}{M_{\mathrm{s}}\Delta T_1}\right) \quad (5.33d)$$

Equation 5.33a through d is subject to the following limitation:

$$X_{\mathrm{s}}^{\mathrm{o}} < \frac{1}{1 + B} \quad (5.33e)$$

Chen et al. (1990) compared Equation 5.32 and Equation 5.33a, as well as Riedel's equation (Equation 5.24). Equation 5.31a assumes that at any temperature there exists a definite amount of unfreezable water, while Equation 5.30 assumes solute–water interaction. Riedel also used an EFC model to develop his formula, but Equations 5.31 and 5.32a are semiempirical models that require an empirical constant at different values of molecular weights to predict the ΔT. Another difference from Riedel's approach is that Equations 5.32 and 5.33a imply that juices do not have a finite temperature for being completely frozen. The EFC obtained by Riedel's model gave values close to those obtained by Equation 5.32a with values of $M_s = 202$ and $B = 0.20$, values close to those obtained by Equation 5.32 when $K_f = 0.25$.

5.5 HIGH-PRESSURE FREEZING

High-pressure processing has now emerged as a new method of food preservation and processing; thus, it is now important to know the characteristics of solid water at high pressure. The phase diagram of solid phases of water is shown in Figure 5.12 (Hobbs, 1974). Fuchigami and Teramoto (1997) reviewed different solid phases of water. When pressure was raised to 2400 MPa at −200°C to 80°C, several kinds of high-pressure ices (ice I–ice IX) with different structures and properties were formed (Fletcher, 1970; Hobbs, 1974; Maeno, 1981; Franks, 1989). High-pressure forms of ice have included ices II and III and continued in classic studies reporting ices V, VI, and VII including metastable ice IV (Fletcher, 1970). Ice VIII and ice IX were reported by Whalley and Heath (1966). Ice at atmospheric pressure is denoted as ice I (density 992 kg/m^3). From the coordination number of 4, the structure of ice I is open with much empty space (Fletcher, 1970). It is the only ice less dense than liquid water and thus floats. The density of high-pressure ice is higher, and its crystal structure is very complex. The density values of ice II, ice III, and ice V are 1170, 1140, and 1230 kg/m^3, respectively. These high-pressure ices have crystals with bent structures

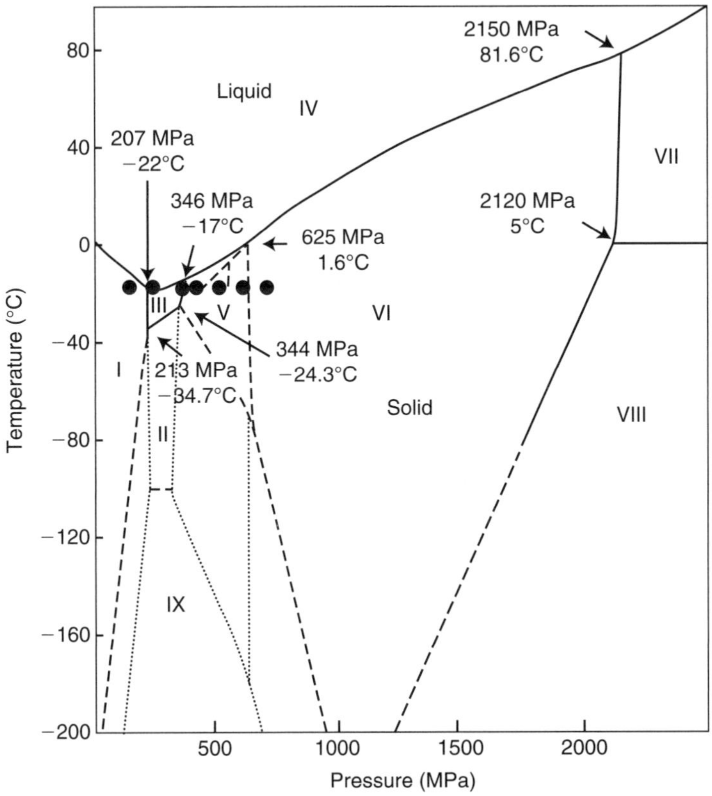

Figure 5.12 Phase diagram of water–solid phases as a function of pressure and temperature. (From Hobbs, P.V. *Ice Physics*, Oxford University Press, London, U.K., 1974. With permission.)

(much denser than ice I); the length (ice I: 1.76 Å, ice II: 2.75–2.84 Å) and angle (ice I: 109°, ice II: 80°–129°) of hydrogen bonds grow, shrink, or bend (Hobbs, 1974; Maeno, 1981). Ice VI (1310 kg/m^3) has a dual structure with two sets of crystals formed in one, and density of ice VII and ice VIII is highest (1500 kg/m^3). When water is pressurized above 2200 MPa at temperatures above 82°C, hot ice is formed. A region below 0°C (liquid phase) has enabled the nonfreezing preservation of foods (Deuchi and Hayashi, 1990) when ice was pressurized to 200 MPa, because this resulted in pressure fusion (melting). Zhu et al. (2006) measured the freezing point depression of different foods (water, potato, tylose, salmon, pork) as a function of pressure for ice I (0–200 MPa). A general correlation was developed as (Zhu et al., 2006)

$$T_f = -0.92 - 6.66 \times 10^{-2}P - 18.3 \times 10^{-4}P^2 \tag{5.34}$$

where P is the pressure in MPa. They identified that pressure-dependent phase transition temperature in real food products can be accurately determined through isothermal P-scan technique using a high-pressure DSC system.

5.6 CONCLUSIONS

The freezing point is one of the most important properties of food because of the discontinuity exhibited at this point. It can be measured mainly by cooling curve, cryoscope, and DSC. Each method has its advantages and disadvantages compared to other methods. Experimental data could be used when they are available in the literature for the specific foods. Freezing points of foods are mainly predicted by semiempirical models and empirical curve-fitting correlations. The empirical models are limited to the experimental region based on which data were used to develop the models. However, a semiempirical model based on Clausius–Clapeyron equation based on the two parameters (amount of bound water and molecular weight ratio of water and solutes) is widely used for its accuracy and theoretical basis.

NOMENCLATURE

A	constant of Equation 5.17 and 5.18
a	activity
B	bound or unfreezable water (kg/kg solids)
°B	degree Brix
C	specific heat (kJ/kg K)
c	constant of Equation 5.8b or Equation 5.13
D	constant of Equation 5.17 and 5.18
d	constant of Equation 5.30b
E	molecular weight ratio (M_w/M_s)
e	constant of Equation 5.30b
F	constant of Equation 5.23
f	constant of Equation 5.14
G	constant of Equation 5.17 and 5.18
g	constant of Equation 5.32
H	enthalpy (kJ/kg)
I	constant of Equation 5.23

J	constant of Equation 5.23
K	molar freezing or cryogenic constant for water (1.86 kg K/kg mol)
L	latent heat (J/kg)
M	molecular weight
m	constant of Equations 5.17 and 5.18
n	number of components
P	pressure (Pa or MPa)
PE	pectin
R	ideal gas constant
r	cooling rate (°C/min)
T	temperature (°C or K)
X	mass fraction (kg/kg sample)
x	mole fraction
Y	mass fraction (kg/kg solids)
Z	empirical function of Equation 5.17

Greek Symbols

τ	dissociation constant
γ	temperature when enthalpy becomes negligible
σ	coefficient of Equation 5.11
Δ	difference

Subscript

as	ash
B	apparent freezing
b	bound water
c	cooling
ca	carbohydrate
cp	cryoprotectant
E	equivalent
F	fat
f	freezing point
G	true freezing
g	glass transition
I	ice
i	ith component
if	incipient freezing
iif	intensive incipient freezing
m	end point of freezing or melting
o	at 0°C
p	constant pressure
r	reference
SNF	solids nonfat
SS	soluble solids
s	solids
s1	solids at condition 1
s2	solids at condition 2

u eutectic
un unfrozen
uw unfrozen water
w water

Superscript

′ maximal-freeze-concentration condition
o initial before freezing

REFERENCES

ASHRAE. 1967. Thermal properties of foods. In *Fundamentals Handbook*. American Society of Heating, Refrigeration and Air conditioning Engineers, New York.

ASHRAE. 1981. Thermal properties of foods. In *Fundamentals Handbook*. American Society of Heating, Refrigeration and Air conditioning Engineers, New York.

Bai, Y., Rahman, M.S., Perera, C.O., Smith, B., and Melton, L.D. 2001. State diagram of apple slices: glass transition and freezing curves. *Food Research International*, 34(2–3): 89–95.

Boonsupthip, W. and Heldman, D.R. 2007. Prediction of frozen food properties during freezing using product composition. *Journal of Food Science*, 72(5): E254–E262.

Brake, N.C. and Fennema, O.R. 1999. Glass transition values of muscle tissue. *Journal of Food Science*, 64(1): 25–32.

Chang, H.D. and Tao, L. 1981. Correlations of enthalpies of food systems. *Journal of Food Science*, 46: 1493–1497.

Charm, S.E. and Moody, P. 1966. Bound water in haddock muscle. *ASHRAE Journal*, 8(4): 39–42.

Chen, C.S. 1985. Thermodynamic analysis of freezing and thawing of foods: Enthalpy and apparent specific heat. *Journal of Food Science*, 50(4): 1158–1162.

Chen, C.S. 1986. Effective molecular weight of aqueous solutions and liquid foods calculated from the freezing point depression. *Journal of Food Science*, 51(6): 1537–1539.

Chen, C.S. 1987a. Relationship between water activity and freezing point depression of food systems. *Journal of Food Science*, 52(2): 433–435.

Chen, C.S. 1987b. Sorption isotherm and freezing point depression equations for glycerol solutions. *Transactions of the ASAE*, 30(1): 279–282.

Chen, C.S. 1988. Bound water and freezing point depression of concentrated orange juices. *Journal of Food Science*, 53(3): 983–984.

Chen, X.D. and Chen, P. 1996. Freezing of aqueous solution in a simple apparatus designed for measuring freezing point. *Food Research International*, 29(8): 723–729.

Chen, C.S. and Nagy, S. 1987. Prediction and correlation of freezing point depression of aqueous solutions. *Transactions of the ASAE*, 30(4): 1176–1180.

Chen, Y. and Pan, B.S. 1995. Freezing tilapia by airblast and liquid nitrogen-freezing point and freezing rate. *International Journal of Food Science and Technology*, 30: 167–173.

Chen, C.S., Nuguyen, T.K., and Braddock, R.J. 1990. Relationship between freezing point depression and solute composition of fruit juice systems. *Journal of Food Science*, 55(2): 566–569.

Delgado, A.E., Rubiolo, A.C., and Gribaudo, L.M. 1990. Characteristic temperatures determination for strawberry freezing and thawing. *Journal of Food Processing and Preservation*, 14(3): 231–240.

Desrosier, N.W. 1970. *The Technology of Food Preservation*. The AVI Publishing Company, Inc., Westport, CN, pp. 92–122.

Deuchi, T. and Hayashi, R. 1990. A new approach for food preservation: use of non-freezing conditions at subzero temperature generated under moderate high pressure (Japanese). In *Kaatsu Shokuhin—Kenkyu to Kaihatsu* (*Pressure-Processed Food—Research and Development*), Hayashi, R. (ed.), Chapter 3. San'ei Press, Kyoto, Japan.

Devine, C.E., Bell, R., Lovatt, S., Chrystall, B.B., and Jeremiah, L.E. 1996. Red meats. In *Freezing Effects on Food Quality*, Jeremiah, L.E. (ed.). Marcel Dekker, New York, pp. 51–84.

De Qian, W. and Kolbe, E. 1991. Thermal properties of surimi analyzed using DSC. *Journal of Food Science*, 56(2): 302–308.

Dickerson, R.W. 1969. Thermal properties of foods. In *The Freezing Preservation of Food*, Tressler, D.K., Van Arsdel, W.B., and Copley, M.R. (eds.). AVI Publishing, Westport, CT, pp. 26–51.

Encinas-Banda, D., Gonzalez-Loo, H., Garcia, M., Velez-Ruiz, J.F., and Sosa-Morales, M.E. 2005. Effect of different additives on quality characteristics of frozen bread. Poster presentation 36G-2, IFT Annual Meeting, New Orleans, LO, July 15–20.

Fennema, O.R., Powrie, W.D., and Marth, E.H. 1973. *Low-Temperature Preservation of Foods and Living Matter*. Marcel Dekker, New York.

Fletcher, N.H. 1970. *The Chemical Physics of Ice*. Cambridge University Press, Cambridge.

Fonseca, F., Obert, J.P., Beal, C., and Marin, M. 2001. State diagrams and sorption isotherms of bacterial suspensions and fermented medium. *Thermochimica Acta*, 366: 167–182.

Franks, F. 1989. *Biophysics and Biochemistry at Low Temperatures* (Japanese translation by N. Murase and C. Katagiri). (Cambridge University Press), Hokkaido University Press, Sapporo, p. 1320.

Fuchigami, M. and Teramoto, A. 1997. Structure and textural changes in kinu-tofu due to high-pressure-freezing. *Journal of Food Science*, 62(4): 828–832, 837.

Geankoplis, C.J. 1998. *Transport Processes and Unit Operations*. CECSA, Mexico.

Goff, H., Caldwell, K.B., Stanley, D.W., and Maurice, T.P. 1993. The influence of polysaccharides on the glass transition in frozen sucrose solutions and ice cream. *Journal of Dairy Science*, 76: 1268–1277.

Guevara-Lara, J. 2004. Development of a semi-industrial process to elaborate frozen deep-fat fried churros with different stuffs. Undergraduate Thesis, Universidad de las Americas, Puebla, Mexico.

Hamdami, N., Monteau, J.Y., and Le Bail, A. 2004. Transport properties of a high porosity model food at above and sub-freezing temperatures. Part 1. Thermophysical properties and water activity. *Journal of Food Engineering*, 62: 373–383.

Hatakeyama, T., Quinn, F.X., and Hatakeyama, H. 1996. Changes in freezing bound water in water–gellan systems with structure formation. *Carbohydrate Polymers*, 30: 155–160.

Hatley, R.H.M., Van Den Berg, C., and Franks, F. 1991. The unfrozen water content of maximally freeze concentrated carbohydrate solutions. *Cryo-Letters*, 12: 113–126.

Heldman, D.R. 1974. Predicting the relationship between unfrozenwater fraction and temperature during food freezing using freezing point depression. *Transactions of the ASAE*, 17(1): 63–66.

Heldman, D.R. 1975. *Food Process Engineering*. The AVI Pub. Co., Westport, CT.

Heldman, D.R. 1982. Food properties during freezing. *Food Technology*. 36: 92–96.

Heldman, D.R. and Singh, R.P. 1981. *Food Process Engineering*, 2nd edn., AVI Publishing Co., Westport, CT.

Hobbs, P.V. 1974. *Ice Physics*. Oxford University Press, London, UK.

Hoo, A.F. and McLellan, M.R. 1987. The contributing effect of apple pectin on the freezing point depression of apple juice concentrates. *Journal of Food Science*, 52(2): 372–374, 377.

Ibarz, A. and Barbosa, G. 2001. *Unit Operations in Food Engineering*. Ed. Acribia, Spain.

Inoue, Y. and Bushuk, W. 1996. Effects of freezing, frozen storage, and thawing on dough and baked goods. In *Freezing Effects on Food Quality*, Jeremiah, L.E. (ed.). Marcel Dekker, Inc., New York, pp. 367–400.

IP, Iwanami Publish. 1987. *Rikagaku Jiten*, 4th edn.

Izzard, M.J., Ablett, S., and Lillford, P.J. 1991. Calorimetric study of the glass transition occurring in sucrose solutions. In *Food Polymers, Gels and Colloids*, Dickinson, E. (ed.). The Royal Society of Chemistry, London, pp. 289–300.

Jang, J.K. and Pyun, Y.R. 1997. Effect of moisture level on the crystallinity of wheat starch aged at different temperatures. *Starch*, 49(5): 272–277.

Janz, G.J., Allen, C.B., Donwey, J.R., Tomkins, R.P.T. 1979. Physical Properties Data Compilation Relevant to Energy Storage, NSRDS-NBS 61.

JSME, Japan Society Mechanical Engineering. 1986. *Technical Data Book. Heat Transfer*, 4th edn.

Jie, W., Lite, L., and Yang, D. 2003. The correlation between freezing point and soluble solids of fruits. *Journal of Food Engineering*, 60: 481–484.

Kasapis, S., Sablani, S.S., Rahman, M.S., Al-Marhoobi, I.M., and Al-Amri, I.S. 2007. Porosity and the effect of structural changes on the mechanical glass transition temperature. *Journal of Agricultural and Food Chemistry*, 55: 2459–2466.

Kessler, H.G. 1984. *Milchwissenschaft*, 39, 339. Cited by Kirk, R.S., Sawyer, R., and Egan, H. (eds.) *Pearson's Food Composition and Analysis*, 2nd edn., CECSA, Mexico, p. 600.

Le Meste, M. and Huang, V. 1992. Thermomechanical properties of frozen sucrose solutions. *Journal of Food Science*, 57(5): 1230–1233.

Lerici, C.R., Piva, M. and Dalla Rosa, M. 1983. Water activity and freezing point depression of aqueous solutions and liquid foods. *Journal of Food Science*, 48: 1667–1669.

Levy, F.L. 1979. Enthalpy and specific heat of meat and fish in the freezing range. *Food Technology*, 14: 549–560.

Lind, I. 1991. The measurement and prediction of thermal properties of food during freezing and thawing—a review with particular reference to meat and dough. *Journal of Food Engineering*, 13: 285–319.

Lloyd, R.J., Chen, X.D., and Hargreaves, J.B. 1996. Glass transition and caking of spray-dried lactose. *International Journal of Food Science and Technology*, 31: 305–311.

Long, A.K. 1955. Some thermodynamic properties of fish and their effect on the rate of freezing. *Journal of the Science and Food Agriculture*, 6: 621–632.

Machado, K., Sosa, M.E., and Velez, J. 2006. Physical properties of some frozen Mexican foods. Written for presentation at the 2006 CIGR, Section VI. *International Symposium on Future of Food Engineering*, Warsaw, Poland, April 26–28, p. 6.

Maeno, N. 1981. *Kori no Kagaku* (*Science of Ice*). Hokkaido University Press, Sapporo.

Mannapperuma, J.D. and Singh, R.P. 1989. A computer-aided method for the prediction of properties and freezing/thawing times of foods. *Journal of Food Engineering*, 9: 275–304.

Mascheroni, R.H. and Calvelo, A. 1980. Relationship between heat transfer parameters and the characteristic damage variables for the freezing of beef. *Meat Science*, 4: 267–285.

Mellor, J.D. 1983. Critical evaluation of thermophysical properties of foodstuffs and outline of future development. In *Physical Properties of Foods*, Jowitt, R., Escher, F., Hallstrom, B., Meffert, H.F.T., Spiess, W.E.L., and Vos, G. (eds.). Applied Science Publishers, New York.

Miles, C.A., Mayer, Z., Morley, M.J., and Houska, M. 1997. Estimating the initial freezing point of foods from composition data. *International Journal of Food Science and Technology*, 32: 389–400.

Morales-de la Peña, M. 2005. Reformulation of sponge cake dough to increase its shelf-life in frozen storage. Undergraduate Thesis, Universidad de las Americas, Puebla, Mexico.

Murakami, E.G. and Okos, M.R. 1989. Measurement and prediction of thermal properties of foods. In *Food Properties and Computer Aided Engineering of Food Processing Systems*, Singh, R.P. and Medina, A.G. (eds.). Kluwer Academic Publishers, New York.

Otero, L. and Sanz, P.D. 2006. High-pressure-shift freezing: main factors implied in the phase transition time. *Journal of Food Freezing*, 72: 354–363.

Peralta, J.M., Rubiolo, A.C., and Zorrilla, S.E. 2007. Prediction of heat capacity, density and freezing point of liquid refrigerant solutions using an excess Gibbs energy model. *Journal of Food Engineering*, 82: 548–558.

Pham, Q.T. 1987. Calculation of bound water in frozen food. *Journal of Food Science*, 52(1): 210–212.

Pham, Q.T. 1996. Prediction of calorimetric properties and freezing time of foods from composition data. *Journal of Food Engineering*, 30: 95–107.

Phillips, G.O., Takigami, S., and Takigami, M. 1996. Hydration characteristics of the gum exudate from *Acacia senegal. Food Hydrocolloids*, 10(1): 11–19.

Polley, S.L., Snyder, O.P., and Kotnour, P. 1980. A compilation of thermal properties of foods. *Food Technology*, 34(11): 76–94.

Rahman, S. 1994. The accuracy of prediction of the freezing point of meat from general methods. *Journal of Food Engineering*, 21: 127–136.

Rahman, S. 1995. *Food Properties Handbook*. CRC Press, Boca Raton, FL.

Rahman, M.S. 1999. Glass transition and other structural changes in foods. In *Handbook of Food Preservation*, 1st edn., Rahman, M.S. (ed.). Marcel Dekker, Inc., New York, pp. 75–94.

Rahman, M.S. 2004. State diagram of date flesh using differential scanning calorimetry (DSC). *International Journal of Food Properties*, 7(3): 407–428.

Rahman, M.S. 2006. State diagram of foods: its potential use in food processing and product stability. *Trends in Food Science & Technology*, 17: 129–141.

Rahman, M.S. and Driscoll, R.H. 1991. Thermal conductivity of seafoods: calamari, octopus and prawn. *Food Australia*, 43(8): 356–360.

Rahman, M.S. and Driscoll, R.H. 1994. Freezing points of selected seafoods (invertebrates). *International Journal of Food Science and Technology*, 29(1): 51–61.

Rahman, S. and Vélez-Ruiz, J.F. 2007. Food preservation by freezing. In *Handbook of Food Preservation*, 2nd edn., Rahman, M.S. (ed.). CRC Press, Boca Raton, FL, pp. 635–665.

Rahman, M.S., Guizani, N., Al-Khaseibi, M., Al-Hinai, S., Al-Maskri, S.S., and Al-Hamhami, K. 2002. Analysis of cooling curve to determine the end point of freezing. *Food Hydrocolloids*, 16(6): 653–659.

Rahman, S., Kasapis, S., Guizani, N., and Saud Al-Amri, O. 2003. State diagram of tuna meat: freezing curve and glass transition. *Journal of Food Engineering*, 57: 321–326.

Rahman, M.S., Sablani, S.S., Al-Habsi, N., Al-Maskri, S., and Al-Belushi, R. 2005. State diagram of freeze-dried garlic powder by differential scanning calorimetry and cooling curve methods. *Journal of Food Science*, 70(2): E135–E141.

Rocha-Santiago, L.M., Sosa-Morales, M.E., and Velez-Ruiz, J.F. 2005. Physical properties of carbonated yoghurt prepared with a freeze-dried base, *Inter American Drying Conference (IADC) D-10*, Montreal, Canada, August 21–23, p. 7.

Roos, Y.H. 1986. Phase transitions and unfreezable water content of carrots, reindeer meat and white bread studied using differential scanning calorimetry. *Journal of Food Science*, 51(3): 684–686.

Roos, Y. and Karel, M. 1990. Differential scanning calorimetry study of phase transitions affecting the quality of dehydrated materials. *Biotechnology Progress*, 6: 159–163.

Roos, Y. and Karel, M. 1991a. Phase transitions of amorphous sucrose and frozen sucrose solutions. *Journal of Food Science*, 56(1): 266–267.

Roos, Y. and Karel, M. 1991b. Amorphous state and delayed ice formation in sucrose solutions. *International Journal of Food Science and Technology*, 26: 553–566.

Roos, Y. and Karel, M. 1991c. Water and molecular weight effects on glass transitions in amorphous carbohydrates and carbohydrate solutions. *Journal of Food Science*, 56(6): 1676–1681.

Ross, K.A., Campanella, O.H., and Okos, M.R. 2002. The effect of porosity on glass transition measurement. *International Journal of Food Properties*, 5(3): 611–628.

Saad, Z. and Scott, E.P. 1996. Estimation of temperature dependent thermal properties of basic food solutions during freezing. *Journal of Food Engineering*, 28: 1–19.

Sablani, S., Kasapis, S., Rahman, M.S., Al-Jabri, A., and Al-Habsi, N. 2004. Sorption isotherms and the state diagram for evaluating stability criteria of abalone. *Food Research International*, 37(10): 915–924.

Sablani, S.S., Rahman, M.S., Al-Busaidi, S., Guizani, N., Al-Habsi, N., Al-Belushi, R., and Soussi, B. 2007. Thermal transitions of king fish whole muscle, fat and fat-free muscle by differential scanning calorimetry. *Thermochimica Acta*, 462: 56–63.

Sahagian, M.E. and Goff, H.D. 1996. Fundamentals aspects of freezing process. In *Freezing Effects on Food Quality*, Jeremiah, L.E. (ed.). Marcel Dekker, Inc., New York, pp. 1–50.

Sanchez-Rechy, M.G. 2006. Air and freeze drying of huitlacoche. Undergraduate Thesis, Universidad de las Americas, Puebla, Mexico.

Sanz, P.D., Dominguez, M., and Mascheroni, R.H. 1989. Equations for the prediction of thermophysical properties of meat products. *Latin American Applied Research*, 19: 155–160.

Schenz, T.W., Israel, B., and Rosolen, M.A. 1991. Thermal analysis of water-containing systems. In *Water Relationships in Food*, Levine, H. and Slade, L. (eds.). Plenum Press, New York, pp. 199–214.

Schwartzberg, H.G. 1976. Effective heat capacities for freezing and thawing of food. *Journal of Food Science*, 41(1): 152–156.

Sheard, P.R., Jolley, P.D., Katib, A.M.A., Robinson, J.M., and Morley, M.J. 1990. Influence of sodium tripolyphosphate on the quality of UK-style grillsteaks: relationship to freezing point depression. *International Journal of Food Science & Technology*, 25: 643–656.

Singh, R.P. and Heldman, D.R. 2001. *Introduction to Food Engineering*. Academic Press, Glascow, Great Britain, pp. 410–446.

Singh, R.P. and Mannapperuma, J.D. 1990. Development in food freezing. In *Biotechnology Food Process Engineering*, Schwartzberg, H.G. and Rao, M.A. (eds.). Marcel Dekker, New York.
Skrede, G. 1996. Fruits. In *Freezing Effects on Food Quality*, Jeremiah, L.E. (ed.). Marcel Dekker, Inc., New York, pp. 183–245.
Slade, L. and Levine, H. 1988. Structural stability of intermediate moisture foods—a new understanding? In *Food Structure—Its Creation and Evaluation*, Blanshard, J.M.V. and Mitchell, J.R. (eds.). Butterworths, London, pp. 115–147.
Slade, L. and Levine, H. 1995. Glass transitions and water–food structure interactions. *Advances in Food and Nutrition Research*, 38: 103–209.
Sosa-Morales, M.E., Guerrero-Cruz, G., Gonzalez-Loo, H., Velez-Ruiz, J.E. 2004. Modeling of heat and mass transfer during baking of biscuits. *Journal of Food Processing and Research*, 28(6): 417–432.
Succar, J. and Hayakawa, L. 1983. Empirical formulae for prediction thermal physical properties of food at freezing or defrosting temperatures. *Food Science & Technology*, 16(6): 326–331.
Succar, J. and Hayakawa, K. 1990. A method to determine initial freezing point of foods. *Journal of Food Science*, 55(6): 1711–1713.
Valentas, K.J., Rotstein, E., and Singh, R.P. 1997. *Food Engineering Practice*. CRC Press, New York, pp. 71–123.
Van der Sman, R.G.M. and Boer, E. 2005. Predicting the initial freezing point and water activity of meat products from composition data. *Journal of Food Engineering*, 66: 469–475.
Vélez, J. 2007. Notes of Food Engineering II. Universidad de las Americas, Puebla (unpublished document).
Wang, D.Q. and Kolbe, E. 1991. Thermal properties of surimi analyzed using DSC. *Journal of Food Science*, 55(6): 302–308.
Watt, B.K. and Merril, A.L. 1963. Composition of foods. In *Agricultural Handbook No. 8*. U.S. Department of Agriculture, Washington, DC.
Weast, R.C. 1982. *Handbook of Chemistry and Physics*, 63rd edn., CRC Press, West Palm Beach, FL.
Whalley, E. and Heath, J.B.R. 1966. Dielectric properties of ice VII; ice VIII: a new phase of ice. *Journal of Chemical Physics*, 45: 3976–3982. Cited by Fuchigami, M. and Teramoto, A. 1997. Structure and textural changes in kinu-tofu due to high-pressure-freezing. *Journal of Food Science*, 62(4): 828–832, 837.
Yamada, M., Tago, M., Fukusako, S., and Horibe, A. 1993. Melting point and supercooling characteristics of molten salt. *Thermochimica Acta*, 218: 401–411.
Zhu, S., Le Bail, A., and Ramaswamy, H.S. 2006. High-pressure differential scanning calorimetry: Comparison of pressure-dependent phase transition in food materials. *Journal of Food Engineering*, 75: 215–222.

CHAPTER 6

Prediction of Ice Content in Frozen Foods

Mohammad Shafiur Rahman

CONTENTS

6.1 INTRODUCTION

The changes in thermal properties during freezing are dominated by the change in phase of the water from liquid to ice. Good knowledge and accurate prediction of the ice fraction-temperature dependence have significant importance for reliable determination of the thermophysical characteristics and enthalpy variation during freezing or thawing of foodstuffs, as well as for proper selection of the temperature regimes during refrigeration processing, storage, and distribution in the frozen state (Fikiin, 1998; Rahman, 2005). The amount of different types of water in foods is also important to determine the stability and quality of frozen foods during storage.

6.2 TYPES OF WATER IN FROZEN FOODS

Different types of water are found in foods. In the case of frozen foods, different types of water are usually defined as total water (X_w^o), ice (X_I), unfreezable water (X_w'), and unfrozen water (X_w^u). Total water can be written as

$$X_w^o = X_w^u + X_I + X_w^{'} \tag{6.1}$$

At any temperature in the frozen state the total water before freezing consists of three components. Unfreezable water ($X_w^{'}$) is defined as the water that could not be formed as ice even at low temperature (i.e., −40°C or below). The frozen water is the ice content (X_I) that increased with the decrease in temperature. At any temperature below freezing point, the amount of ice increased with the decrease of unfrozen water.

6.3 PREDICTION OF ICE CONTENTS

The methods of predicting ice content can be grouped as (Sakai and Hosokawa, 1984): (1) freezing point method, (2) freezing curve method, (3) using Clausius–Clapeyron equation, and (4) using enthalpy value.

6.3.1 Freezing Point

Nakaide (1968) proposed a method to estimate the ice content from the freezing point data. The ice content at any temperature below freezing can be simply calculated as

$$X_I = X_w^o\left(1 - \frac{t_F}{t}\right) \tag{6.2}$$

Equation 6.2 is derived based on Raoult's law and is valid for very dilute solutions. The experimental data show that Equation 6.2 slightly underestimates the real ice contents for temperatures near the initial freezing point, and with the lowering of temperature, Equation 6.2 begins to overestimate them (Tchigeov, 1979). As all water in the sample is not freezable, better prediction could be achieved based on the unfreezable water as

$$X_I = (X_w^o - X_w^{'}) \times \left(1 - \frac{t_F}{t}\right) \tag{6.3}$$

where $(X_w^o - X_w^{'})$ is the total freezable water. Unfreezable water is usually measured and expressed in terms of kilogram of unfreezable water per kilogram of sample or total dry solids:

$$X_w^{'} = BX_s^o = B(1 - X_w^o) \tag{6.4}$$

where B is the unfreezable or bound water as kilogram per kilogram dry solids. Other forms of equations are available to predict the ice content when the freezing point is known. Jadan (1992) proposed the following equation:

$$X_I = X_w^o\left[1 - t_F\left(\frac{1.12 - 0.05t}{t}\right)\right] \tag{6.5}$$

Equation 6.5 is applicable with satisfactory precision only for $t < -10$°C and with most accurate results when $-1°\mathrm{C} \le t_F \le -0.4°\mathrm{C}$ (Nakaide, 1968; Pham et al., 1994). Wang and Kolbe (1991) used data in Table 6.1 to develop an empirical correlation in cubic form in a function of temperature below the freezing point. Tchigeov (1956, 1979) and Fikiin (1980) proposed an empirical relationship as

$$X_I = \frac{1.105X_w^o}{[1 + (0.7138/\ln(t_F - t + 1))]} \tag{6.6}$$

Table 6.1 Measured Unfrozen Water Content of Surimi at Various Temperature and Cryoprotectant Concentration When Initial Water Is 80.3% (Wet Basis)

	Unfrozen Water Fraction (Wet Basis)											
	Temperature (°C)											
X_{cp}	−40	−35	−30	−25	−20	−15	−10	−8	−6	−4	−2	t_F
0.00	0.106	0.107	0.108	0.111	0.115	0.121	0.137	0.151	0.174	0.223	0.399	0.803
0.04	0.091	0.091	0.091	0.092	0.096	0.107	0.133	0.155	0.195	0.298	0.591	0.803
0.06	0.083	0.083	0.084	0.086	0.093	0.107	0.136	0.159	0.200	0.296	0.577	0.803
0.08	0.091	0.091	0.091	0.096	0.106	0.124	0.162	0.193	0.253	0.385	0.715	0.803
0.12	0.106	0.107	0.112	0.121	0.135	0.160	0.211	0.252	0.325	0.467	0.752	0.803

Source: Adapted from Wang, D.Q. and Kolbe, E., *J. Food Sci.*, 56, 302, 1991.
Note: Cryoprotectant: half sucrose and half sorbitol. Scanned rate: 2°C/min from −80°C to 40°C.

The above equation represents a very good fit of the experimental data for various products (meat, fish, milk, eggs, fruits, and vegetables) and provides fully satisfactory precision when $-45°C \leq t \leq t_F$ and $-2 \leq t_F \leq -0.4°C$. The calculations of various thermal properties by means of Equations 6.5 and 6.6 usually give results very close to those of Riedel (1951) and other established authors of classical reference works (Fikiin, 1998). The relationship proposed by Latyshev (1992)

$$X_I = X_w^o - \left[1 - \frac{\delta}{(1 + ((\delta - 1)\ln(t_F - t_e + \gamma)/\ln(t_F - t + \gamma)))}\right] \tag{6.7}$$

where t_e is the product cryohydric (eutectic) temperature, and δ and γ are food-specific constants, given in Table 6.2. Latyshev (1992), assumed that in all cases, t_e is approximately equal to the evaporation temperature of liquid nitrogen at atmospheric pressure, that is, $t_e = -196°C$.

Table 6.2 Values of δ and γ for Different Foods

Product	δ	γ
Pork	1.10	1.14
Veal and beef	1.10	1.13
Boiled beef	1.10	1.34
Beef liver	1.20	1.21
Pancreases of cattle	1.20	1.25
Rabbit meat	1.10	1.13
Milk, curds, yogurt, kefir, cream, butter, whey	1.10	1.08
Condensed milk with sugar	1.10	1.15
Processed cheese	1.30	1.94
Ice cream		
Milk	1.28	1.49
Typical cream	1.30	1.55
Plumber milk	1.28	1.58
Berry	1.30	1.70
Green tea leaves	1.10	1.13

Source: Latyshev, V., Scientific thermophysical fundamentals of refrigerated processing and storage of foods, DSc Thesis, SPTIHP, St. Petersburg, Russia, 1992.

6.3.2 Freezing Curve

The freezing curve can be used to determine the ice content in frozen foods. The following sections discuss different uses of the freezing curves.

6.3.2.1 *Using Lever Arm Rule*

A typical freezing curve is shown in Figure 6.1. At any temperature below freezing (point R in the freezing curve), the ice content can be estimated from the equilibrium solute concentration using the lever rule as

$$X_I = \left(\frac{\text{Length of line QR}}{\text{Length of line PR}}\right) = \frac{X_s^e - X_s^o}{X_s^e} \tag{6.8}$$

6.3.2.2 *Using Clausius–Clapeyron Equation*

Barlett (1944) proposed the following equation based on Clapeyron's equation and Raoult's law as

$$X_I = X_w^o\left[1 - \left\{\frac{\exp\,[k_1(T_w - T_F) + k_2(T_w - T_F)^2] - 1}{\exp\,[k_1(T_w - T_F) + k_2(T_w - T_F)^2] - 1}\right\}\right] \tag{6.9}$$

where k_1 is 9.703×10^{-3} 1/K and k_2 is 4.794×10^{-6} $1/K^2$. Similarly, Heldman (1977) proposed the following equation:

$$X_I = X_w^o\left[1 - \left\{\frac{[1 - \exp\,(ab)] \times [\exp\,(ac)]}{[1 - \exp\,(ac)] \times [\exp\,(ab)]}\right\}\right] \tag{6.10}$$

where $a = (L_w \lambda_w / R)$, $b = [(1/T_w) - (1/T_F)]$, $c = [(1/T_w) - (1/T)]$

Because Raoult's law is valid only for ideal (highly diluted) solutions, the predictive equations based on Raoult's law are comparatively precise for cryoscopic temperatures close to the freezing

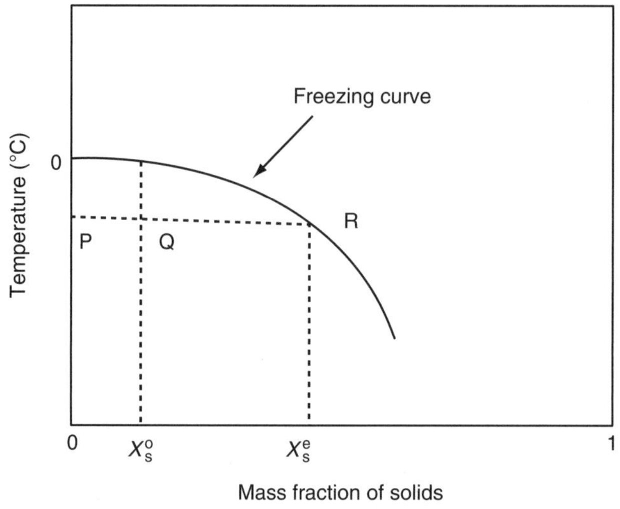

Figure 6.1 Freezing curve to determine ice content.

temperature of the pure solvent (i.e., water). The mass fraction of water below the freezing point can be estimated as (Singh and Mannapperuma, 1990)

$$X_{\mathrm{W}} = X_{\mathrm{W}}^{\mathrm{o}}\left(\frac{F_{\mathrm{W}} - F_{\mathrm{F}}}{F - F_{\mathrm{F}}}\right) \quad \text{where } F = \exp\left(\frac{\lambda_{\mathrm{W}} L_{\mathrm{W}}}{RT}\right) \tag{6.11}$$

Since the sum of ice and water fractions remains constant during the freezing process, the ice fraction can also be calculated as

$$X_{\mathrm{I}} = X_{\mathrm{W}}^{\mathrm{o}} - X_{\mathrm{s}}^{\mathrm{o}} - X_{\mathrm{W}} \tag{6.12}$$

where X_{w} is equal to $X_{\mathrm{W}}^{\mathrm{u}} + X_{\mathrm{W}}'$. A comparison of predicted and experimental values indicates that the accuracy of the above equation is poor in foods containing significant amounts of unfreezable water. The effect of unfreezable water on the mass fraction of unfrozen water at any temperature can be accounted for by the following modification (Singh and Mannapperuma, 1990):

$$X_{\mathrm{W}} = X_{\mathrm{W}}' + (X_{\mathrm{W}}^{\mathrm{o}} - X_{\mathrm{W}}') \times \left(\frac{F_{\mathrm{W}} - F_{\mathrm{F}}}{F - F_{\mathrm{F}}}\right) \tag{6.13}$$

This method is equivalent to the methods proposed by Schwartzberg (1976) and Chen (1985). Since the latent heat of fusion of ice decreases as temperature decreases, Mannapperuma and Singh (1989) allowed for this variation by assuming that the latent heat is a linear function of temperature. Thus, a more complete scheme for more accurate prediction of the composition of frozen foods is given by

$$L_{\mathrm{W}} = g + hT \tag{6.14}$$

The latent heat of fusion of ice as a function of temperature can be expressed as (Nakaide, 1968)

$$L_{\mathrm{W}} = -3.893 \times 10^{-3}T^2 + 4.245T - 535.1 \tag{6.15}$$

where L_{w} is in kJ/kg and T in K. The linear relation was given by Randall (1930) as

$$L_{\mathrm{W}} = 333.88 + 2.11(T - 273.1) \tag{6.16}$$

The latent heat of water in food at temperature T is proposed by Schwartzberg (1976) as

$$L_{\mathrm{W}} = L_{\mathrm{W}}^{\mathrm{o}} - (C_{\mathrm{W}} - C_{\mathrm{I}})(T_{\mathrm{F}} - T) \tag{6.17}$$

Riedel (1978) gave the following equation for the temperature dependence of latent heat of fusion of ice in foods:

$$L_{\mathrm{W}} = 334.1 + 2.05T - 4.19 \times 10^{-3}T^2 \tag{6.18}$$

Lind (1991) mentioned that none of the above equations is suitable to predict the latent heat of fusion of water in foods below −20°C. Using the above equations for latent heat, the water content can be predicted:

$$X_{\mathrm{W}} = X_{\mathrm{W}}' + (X_{\mathrm{W}}^{\mathrm{o}} - X_{\mathrm{W}}') \times \left(\frac{F_{\mathrm{W}}' - F_{\mathrm{W}}'}{F' - F_{\mathrm{W}}'}\right), \quad \text{where } F' = T^{-(\lambda_{\mathrm{w}}h/R)} \exp\left(\frac{\lambda_{\mathrm{W}} g}{RT}\right) \tag{6.19}$$

6.3.2.3 Using Prediction Model of Freezing Curve

The theoretical Clausius–Clapeyron equation was used to estimate the freezing point; the equation can be written as

$$\Delta = -\frac{\beta}{\lambda_w} \ln\left[\frac{X_w^o}{X_w^o + EX_s^o}\right] = -\frac{\beta}{\lambda_w} \ln\left[\frac{1 - X_s^o}{1 - X_s^o + EX_s^o}\right] \quad (6.20)$$

where
- Δ is the freezing point depression $(T_w - T_F)$
- T_F is the freezing point of food (K)
- T_w is the freezing point of water (K)
- β is the molar freezing point constant of water (1860 kg K/kg mol)
- λ_w is the molecular weight of water
- X_s^o is the initial solids mass fraction
- E is the molecular weight ratio of water and solids (λ_w/λ_s)

The values of E can be estimated by nonlinear regression using freezing point data as a function of initial water or solids. The values of E were found to be 0.147 for dates flesh (Rahman, 2004), 0.071 for tuna (Rahman et al., 2003), 0.238 for apple (Bai et al., 2001), 0.068 for garlic (Rahman et al., 2005), 0.082 for squid (Rahman and Driscoll, 1994), and 0.049 for fresh seafood (Rahman and Driscoll, 1994).

The Clausius–Clapeyron equation is limited to the ideal solution (i.e., for a very dilute solution). Theoretical models can be improved by introducing parameters for nonideal behavior when the fraction of total water is unavailable for forming ice. The unfreezable water content B can be defined as the ratio of unfrozen water even at very low temperature to the total solids. Equation 6.20 can be modified based on this concept (Schwartzberg, 1976):

$$\Delta = -\frac{\beta}{\lambda_w} \ln\left[\frac{X_w^o - BX_s^o}{X_w^o + EX_s^o}\right] = -\frac{\beta}{\lambda_w} \ln\left[\frac{1 - X_s^o - BX_s^o}{1 - X_s^o + EX_s^o}\right] \quad (6.21)$$

The model parameters B and E can be estimated by nonlinear regression. The values of B and E were found to be 0.053 and 0.129 for date flesh (Rahman, 2004), 0.383 and 0.033 for tuna (Rahman et al., 2003), 0.303 and 0.027 for king fish (Sablani et al., 2007), 0.12 and 0.067 for squid mantle (Rahman and Driscoll, 1994), 0.65 and 0.041 for fresh seafood (Rahman and Driscoll, 1994), and −0.062 and 0.080 for garlic (Rahman et al., 2005). More values of B are compiled in Table 6.3, and Murakami and Okos (1996) also compiled B for different foods.

Although the values of unfreezable water can be estimated from the values of B from the above equation, it is not recommended for a number of reasons (Rahman, 2004; Rahman et al., 2005). These are explained below. In many instances the values of B were found negative. The negative value of B indicated that a fraction of solids behave as solvent. It could be an inadequately accepted argument. Chen (1986) indicated that freezing point depression is caused by the complex interaction between solutes and water, and it can be expressed by the equivalent an increase in free water. Thus, values of B can either be positive or negative depending on the behavior of freezing point data. In the case of numbers of solutes, they found negative values of B. Murakami and Okos (1996) used Equation 6.21 to estimate bound or unfreezable water for fish and their values of B varied from 0.10 (X_w' : 0.022) to 0.3 (X_w' : 0.045), which is also relatively low. It is unrealistic to have such variations (by 3 factors) even within the different types of fish.

Table 6.3 Unfreezable Water Content in Foods

Product	X_w^o	t_F (°C)	B (kg Water/kg Solids)	Reference
Beef meat	0.740	−0.95	0.257	Fikiin (1998)
Fish				
Haddock	0.836	−0.83	0.270	Fikiin (1998)
Cod	0.803	−0.91	0.278	Fikiin (1998)
Sea perch	0.791	−0.83	0.280	Fikiin (1998)
Glucose	—	—	0.150–0.200	Schwartzberg (1976)
Meat and fish	—	—	0.240–0.270	Schwartzberg (1976)
Meat and fish	—	—	0.140–0.320	Pham (1996)
Egg white	0.864	−0.45	0.275	Fikiin (1998)
Egg	—	—	0.103–0.109	Pham (1996)
Bread	—	—	0.111–0.143	Pham (1996)
Orange juice	—	—	0.000	Heldman (1974)
Grape juice	0.800	−2.90	0.000	Lerici et al. (1983)
Tomato juice	0.850	−1.60	0.180	Lerici et al. (1983)
Yeast	0.730	−1.37	0.167	Fikiin (1998)
Green peas	0.760	−1.74	0.080	Fikiin (1998)
Spinach	0.800	−0.55	0.117	Fikiin (1998)
Sucrose	—	—	0.300	Schwartzberg (1976)
Fructose	—	—	0.150–0.200	Schwartzberg (1976)
Vegetables	—	—	0.180–0.250	Schwartzberg (1976)

In addition, the unfreezable water content measured by other techniques is shown in Table 6.4. The values determined by enthalpy and state diagram (discussed in Sections 6.3.3 and 6.3.4) do not relate with the values determined by this method. For this reason, Rahman (2004) indicated that although Equation 6.21 can predict the freezing point with accuracy, this does not explain the physics when experimental data on freezing point are used to fit the equation (although it is originally based on the physics) (Rahman et al., 2005). The reasons could be as follows: when the parameters are estimated by nonlinear regression, different local optimized regions could be found based on the initial values used in optimization. However, this is one of the generic problems that exist when a theoretical-based model with more than two parameters is used to fit the experimental data (Rahman, 2005; Rahman and Al-Belushi, 2006). In addition, when data are fitted by regression, they provide the mathematics rather then physics. The accuracy of Equation 6.21 was also demonstrated by Rahman (1994) and Rahman and Driscoll (1994) without extensive evidence that its parameters represent real physical meaning. It is recommended by the author that the determination of unfreezable water is most acceptable and accurate when T_m' in the state diagram is used since this is the real point when all freezable water forms ice and is experimentally evident by achieving maximal freeze concentration condition, which could be achieved with slow cooling and annealing (i.e. holding sample at a specific temperature for predetermined duration).

6.3.3 Enthalpy Diagram

The ice content can be determined from the enthalpy balance by considering the water phase, ice phase, solid phase, and latent heat. Sakai and Hosokawa (1984) compared the ice content during freezing of sodium chloride and sugar by the above methods and found that the enthalpy balance method was more accurate than others when the phase diagram method was considered as reference. The main disadvantage of this method is the necessity of the enthalpy diagram of the foods, which is

Table 6.4 Unfreezable Water Determined by Different Methods

				Unfreezable Water (X'_w)					
				From Equation 6.22					
Sample	t_F Range (°C)	t'_m (°C)	X^o_w Range	Range	Average	Equation 6.24	Equation 6.21	State Diagram[a]	Reference
Apple	−4.6 to −22.0	−57.8	0.58–0.86	—	—	—	—	0.264	Bai et al. (2001)
Abalone	−0.9 to −15.9	−18.0	0.34–0.75	—	—	—	0.072	0.317	Sablani et al. (2004)
Date flesh	−1.8 to −22.0	−43.6	0.39–0.78	0.247–0.308	0.290 (0.026)	0.320	0.050	0.238	Rahman (2004)
Date pits[b]	—	—	0.25–0.60	0.221–0.229	0.226	0.239	—	—	Rahman et al. (2007)
Date pits[c]	—	—	0.25–0.80	0.166–0.273	0.212	0.294	—	—	Rahman et al. (2007)
Date pits[d]	—	—	0.24–0.50	0.188–0.238	0.239	0.239	—	—	Rahman et al. (2007)
Date pits[e]	—	—	0.24–0.50	0.122–0.219	0.184	0.222	—	—	Rahman et al. (2007)
Garlic	−0.9 to −20.7	−26.0	0.20–0.80	—	—	0.200	−0.062	0.180	Rahman et al. (2005)
Tuna	−1.4 to −15.3	−13.3	0.37–0.73	—	—	—	0.281	0.390	Rahman et al. (2003)
King fish	−1.5 to −6.3	−17.4	0.50–0.70	—	—	0.367	0.071	0.312	Sablani et al. (2007)
Sucrose	—	−34.0	0.32–0.80	0.227–0.388	0.292 (0.058)	0.233	—	0.200	Roos and Karel (1991)
Gooseberry	−2.2 to −19.7	−41.8	0.31–0.80	0.335–0.745	0.564 (0.200)	0.198	0.024	0.153	Wang et al. (2008)

[a] Determined the moisture content at t'_m.
[b] Roasted-defatted.
[c] Roasted with fat.
[d] Unroasted-defatted.
[e] Roasted-defatted.

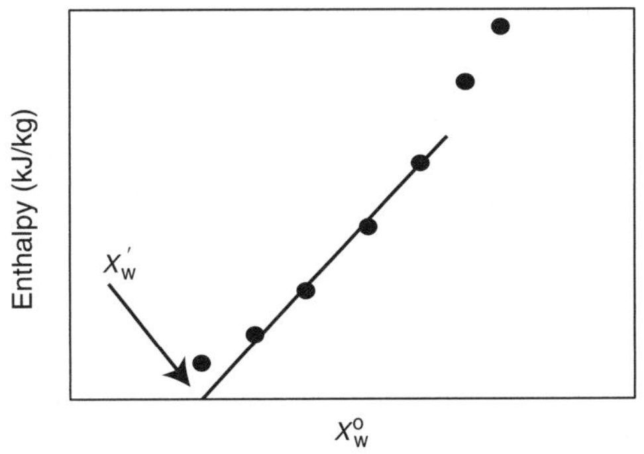

Figure 6.2 Plot of enthalpy change during freezing versus initial moisture content before freezing.

not readily available in the literature. Levine and Slade (1986) proposed a method based on the determination of ΔH_m for 20% solution, which reportedly provides a fast method to determine the unfreezable water content approximately. The unfreezable water content can be calculated from the amount of ice:

$$X'_w = X^o_w - \frac{(\Delta H_m)_{sample}}{(\Delta H_m)_w} \tag{6.22}$$

$(\Delta H_m)_{sample}$ is the enthalpy change during melting of ice in the sample, and $(\Delta H_m)_w$ is the latent heat of melting of ice (334 kJ/kg). The use of the constant latent heat of pure water for the calculation of unfrozen water content from the latent heat of melting significantly below the melting point of pure water showed highly erroneous values of unfreezable water (Roos and Karel, 1991; Rahman, 2004). The measured unfrozen water fractions of surimi at various temperatures are given in Table 6.1. Pham et al. (1994) developed an unfreezable water prediction model based on enthalpy balance during freezing by considering meat, fish, fruit, and nonfat dairy foods as

$$X'_w = X^o_w \left[0.342(1 - X^o_w) + 4.51X^o_{as} + 0.167X^o_p\right] \tag{6.23}$$

Simatos et al. (1975) plotted $(\Delta H_m)_{sample}$ as a function of water content (Figure 6.2), and the unfreezable water content was calculated from the linear relationship extending it to zero $(\Delta H_m)_{sample}$. Alternatively, a linear regression for date samples can be developed as

$$(\Delta H_m)_{sample} = dX^o_w + e \tag{6.24}$$

The value of $(\Delta H_m)_{sample}$ is zero when X^o_w is equal to X'_w; then the above equation can be written as

$$X'_w = -\frac{d}{e} \tag{6.25}$$

It is recommended that higher moisture content values should not be used since nonlinearity exists at high water content.

6.3.4 State Diagram

The unfreezable water fraction can also be determined from the state diagram at T'_m. Rahman (2004) proposed the following method to estimate X'_s. In Figure 6.3, ab is the freezing curve based

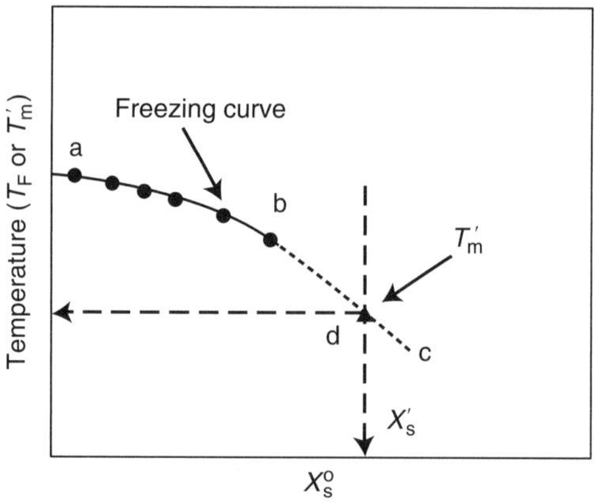

Figure 6.3 Extension of freezing curve to determine T_m' and X_s'.

on the experimental freezing points. The freezing curve ab could be extended to bc by maintaining the same curvature of the freezing curve. A horizontal line could be drawn from the *y*-axis at the value of T_m', which intersects curve bc at point d. The value of X_s' can be estimated from a vertical line intersecting the *x*-axis. Alternatively, Equation 6.21 could also be used to estimate X_s' by using an iterative process considering $\Delta = T_m' - T_w$ and $X_s = X_s'$. The values of unfreezable water determined by different methods are given in Table 6.4.

6.4 HEAT LOAD CALCULATION DURING FREEZING

The following two examples provide the procedures to estimate the heat load during the freezing or melting process (Rahman, 1995).

Example 1

When pure water is frozen to ice 334 kJ/kg must be removed at 0°C. Similarly, the heat of vaporization for water is 2256 kJ/kg at 100°C and 1 atm. How much heat should be removed when 5 kg ice at −20°C is transformed to water at 80°C at atmosphere pressure? Given: specific heat values of water and ice are 4.20 and 2.12 kJ/kg K, respectively. Assume that specific heat is not a function of temperature.

Solution

(a) Latent heat of melting	
5×334	= 1670 kJ
(b) Sensible heat to raise ice temperature	
$(5 \times 2.12) \times (0 + 20)$	= 212 kJ
(c) Sensible heat to raise water temperature	
$(5 \times 4.20) \times (80 - 0)$	= 1680 kJ
	3562 kJ

Thus, 3562 kJ heat needs to be added to ice at −20°C to raise the temperature to 80°C. Similarly, 3562 kJ heat needs to be extracted or removed to cool water at 80°C to ice at −20°C.

Example 2

Beef meat with 77% water content (wet basis) is to be frozen from 25°C to −30°C. Calculate how much heat must be removed if unfrozen water content is 0.40 kg water per kg total solid. Given: latent heat of fusion of water is 334 kJ/kg, specific heat of ice is 2.10 kJ/kg K, specific heat of water is 4.23 kJ/kg K, specific heat of solids is 1.20 kJ/kg K, and freezing point is −1.2°C. Assume that specific heat and latent heat are not a function of temperature, all freezable water is transformed to ice at the freezing point, and unfrozen water has the same specific heat as the bulk water.

SOLUTION

(a) Latent heat of freezing	
(334)[0.77 − (1 − 0.77)(0.40)]	= 226.5 kJ/kg
(b) Sensible heat to cool ice	
[0.77 − (1 − 0.77)(0.40)](2.10)(30 − 1.2)	= 41.0 kJ/kg
(c) Sensible heat to cool freezable water	
[0.77 − (1 − 0.77)(0.40)](4.23)(25 + 1.2)	= 75.1 kJ/kg
(d) Sensible heat to cool unfreezable water	
[(1 − 0.77)(0.40)](4.23)(25 + 30)	= 21.4 kJ/kg
(e) Sensible heat to solids	
(1 − 0.77)(1.2)(25 + 30)	= 15.2 kJ/kg
	379.2 kJ/kg

Thus, 379.2 kJ/kg heat must be removed from beef meat to freeze the meat at 25°C to −30°C.

6.5 CONCLUSION

Different methods and procedures are available in the literature to estimate the ice fraction during freezing. Most of them are complicated and do not provide significant improvement in prediction. The author recommends Equation 6.3 to estimate the fraction of ice during freezing considering unfreezable X'_w from the state diagram. If the state diagram is not available, then it should be predicted from the enthalpy values of foods and water using Equation 6.22.

NOMENCLATURE

a model parameter (Equation 6.10)
B bound water (kg water/kg dry solids)
b model parameter (Equation 6.10)
C specific heat at constant pressure (J/kg K)
c model parameter (Equation 6.10)
d model parameter (Equation 6.24)
E molecular weight ratio of water and solutes (λ_w/λ_s)
e model parameter (Equation 6.24)
F function defined in Equation 6.11
g model parameter (Equation 6.14)
H enthalpy (J/kg)

h model parameter (Equation 6.14)
k model parameter (Equation 6.9)
L latent heat of fusion (J/kg)
R ideal gas constant
T temperature (K)
t temperature (°C)
X mass fraction (wet basis, kg/kg sample)

Greek Symbol

δ model parameter (Equation 6.7)
γ model parameter (Equation 6.7)
λ molecular weight
β molar freezing point constant of water (1860 kg K/kg mol)
Δ freezing point depression ($T_w - T_F$) or change

Subscript

as ash
e eutectic
F freezing
I ice
m maximal freeze concentration condition in T or melting in H
p protein
s solids
w water
1, 2 parameters 1, 2

Superscript

e equilibrium
o before processing or freezing
u unfrozen
′ unfreezable in X, or modified function in F, or maximal-freeze-concentration condition in T

REFERENCES

Bai, Y., Rahman, M.S., Perera, C.O., Smith, B., and Melton, L.D., State diagram of apple slices: Glass transition and freezing curves, *Food Research International*, 34, 89–95, 2001.
Bartlett, L.H., A thermodynamic examination of the latent heat of food, *Refrigeration Engineering*, 47(5), 377–380, 1944.
Chen, C.S., Thermodynamic analysis of freezing and thawing of foods: Enthalpy and apparent specific heat, *Journal of Food Science*, 50(4), 1158–1162, 1985.
Chen, C.S., Effective molecular weight of aqueous solutions and liquid foods calculated from the freezing point depression, *Journal of Food Science*, 51(6), 1537–1543, 1986.
Fikiin, K.A., *Refrigeration Technological Process and Systems*. Technica, Sofia, 1980.
Fikiin, K.A., Ice content prediction methods during food freezing: A survey of the Eastern European literature, *Journal of Food Engineering*, 38, 331–339, 1998.

Heldman, D.R., Predicting the relationship between unfrozen water fraction and temperature during food freezing using freezing point depression, *Transactions of the ASAE*, 17(1), 63–66, 1974.

Heldman, D.R., *Food Process Engineering*. The AVI Publishing Company, Westport, CT, 1977.

Jadan, V., Calculation of frozen water quality, *Refrigeration Engineering*, 6, 12–13, 1992 (in Russian).

Latyshev, V., Scientific thermophysical fundamentals of refrigerated processing and storage of foods. D.Sc. Thesis, SPTIHP, St. Petersburg, Russia, 1992.

Lerici, C.R., Piva, M., and Rosa, M.D., Water activity and freezing point depression of aqueous solutions and liquid foods, *Journal of Food Science*, 48, 1667–1669, 1983.

Levine, H. and Slade, L., A polymer physico–chemical approach to the study of commercial starch hydrolysis products (SHPs), *Carbohydrate Polymer*, 6, 213–244, 1986.

Lind, I., The measurement and prediction of thermal properties of food during freezing and thawing—a review with particular reference to meat and dough, *Journal of Food Engineering*, 13(4), 285–319, 1991.

Mannapperuma, J.D. and Singh, R.P., A computer-aided method for the prediction of properties and freezing/thawing times of foods, *Journal of Food Engineering*, 9(4), 275–304, 1989.

Murakami, E.G. and Okos, M.R., Calculation of initial freezing point, effective molecular weight and unfreezable water of food materials from composition and thermal conductivity data, *Journal of Food Process Engineering*, 19, 301–320, 1996.

Nakaide, M., *Shokuhinkoogyo-no-reitoo (Refrigeration in Food Engineering)*, Koorin Shoin, Tokoyo, 1968.

Pham, Q.T., Prediction of calorimetric properties and freezing time of foods from composition data, *Journal of Food Engineering*, 30, 95–107, 1996.

Pham, Q.T., Wee, H.K., Kemp, R.M., and Lindsay, D.T., Determination of the enthalpy of foods by an adiabatic calorimeter, *Journal of Food Engineering*, 21(2), 137–156, 1994.

Rahman, M.S., The accuracy of prediction of the freezing point of meat from general models, *Journal of Food Engineering*, 21, 127–136, 1994.

Rahman, M.S., *Food Properties Handbook*, 1st edn., CRC Press, Boca Raton, FL, 1995.

Rahman, M.S., State diagram of date flesh using differential scanning calorimetry (DSC), *International Journal of Food Properties*, 7(3), 407–428, 2004.

Rahman, M.S., Dried food properties: Challenges ahead, *Drying Technology*, 23(4), 695–715, 2005.

Rahman, M.S. and Al-Belushi, R.H., Dynamic isopiestic method (DIM): measuring moisture sorption isotherm of freeze-dried garlic powder and other potential uses of DIM, *International Journal of Food Properties*, 9(3), 421–437, 2006.

Rahman, M.S. and Driscoll, R.H., Freezing points of selected seafoods (invertebrates), *International Journal of Food Science and Technology*, 29(1), 51–61, 1994.

Rahman, M.S., Kasapis, S., Guizani, N., and Al-Amri, O., State diagram of tuna meat: freezing curve and glass transition, *Journal of Food Engineering*, 57(4), 321–326, 2003.

Rahman, M.S., Sablani, S.S., Al-Habsi, N., Al-Maskri, S., and Al-Belushi, R., State diagram of freeze-dried garlic powder by differential scanning calorimetry and cooling curve methods, *Journal of Food Science*, 70(2), E135–E141, 2005.

Rahman, M.S., Kasapis, S., Al-Kharusi, N.S.Z., Al-Marhubi, I.M., and Khan, A.J., Composition characterization and thermal transition of date pits powders, *Journal of Food Engineering*, 80(1), 1–10, 2007.

Randell, M., *International Critical Tables*, Vol. VII. McGraw Hill Book, New York, 1930.

Riedel, L., The refrigeration required to freeze fruits and vegetables, *Refrigeration Engineering*, 59, 670–673, 1951.

Riedel, L., Eine formed zur berechnung der enthalpie fettarmer lebensmitteln in abhangigkeit von wassergehalt und temperature, *Chemie, Mikrobiologie und Technologie von Lebensmitteln*, 5, 129, 1978.

Roos, Y. and Karel, M., Amorphous state and delayed ice formation in sucrose solutions, *International Journal of Food Science and Technology*, 26, 553–566, 1991.

Sablani, S.S., Kasapis, S., Rahman, M.S., Al-Jabri, A., and Al-Habsi, N., Sorption isotherms and the state diagram for evaluating stability criteria of abalone, *Food Research International*, 37, 915–924, 2004.

Sablani, S.S., Rahman, M.S., Al-Busaidi, S., Guizani, N., Al-Habsi, N., Al-Belushi, R., and Soussi, B., Thermal transitions of king fish whole muscle, fat, fat-free muscle by differential scanning calorimetry, *Thermochimica Acta*, 462, 56–63, 2007.

Sakai, N. and Hosokawa, A., Comparison of several methods for calculating the ice content of foods, *Journal of Food Engineering*, 3, 13, 1984.

Schwartzberg, H.G., Effective heat capacities for the freezing and thawing of food, *Journal of Food Science*, 41(1), 152–156, 1976.

Simatos, D., Faure, M., Bonjour, E., and Couach, M., The physical state of water at low temperature in plasma with different water contents as studied by differential thermal analysis and differential scanning calorimetry, *Cryobiology*, 12, 202–208, 1975.

Singh, R.P. and Mannapperuma, J.D., Development in food freezing, in *Biotechnology Food Process Engineering*, Schwartzberg, H.G. and Rao, M.A. (eds.). Marcel Dekker, New York, 1990.

Tchigeov, G., *Problems of the Theory of Food Engineering*. Food Industry, Moscow, 1956.

Tchigeov, G., *Thermophysical Process in Food Refrigeration Technology*. Food Industry, Moscow, 1979.

Wang, D.Q. and Kolbe, E., Thermal properties of surimi analyzed using DSC, *Journal of Food Science*, 56(2), 302–308, 1991.

Wang, H., Zhang, S., and Chen, G., Glass transition and state diagram for fresh and freeze-dried Chinese gooseberry, *Journal of Food Engineering*, 84, 307–312, 2008.

CHAPTER 7

Glass Transitions in Foodstuffs and Biomaterials: Theory and Measurements

Stefan Kasapis

CONTENTS

7.1 HISTORIC ITINERARY OF THE PHENOMENON OF GLASS TRANSITION

In the last century, rigorous scientific investigation facilitated the development of a clear concept regarding the crystalline solid state. This was based on the space lattice theory, in terms of atom groups arranged in regular periodic ordering, thus being unable to implement translatory motions relative to the molecular assembly (Sheppard and Houck, 1930). Similarly, an equally clear concept was obtained for the gaseous state, which according to the molecular kinetic theory, comprises an ensemble of particles in random translatory motion (Atkins, 1984). The nature of the liquid state has been the subject of less definite conceptions but, nonetheless, thermodynamic reasoning recognized the existence of sharp discontinuities separating the three states of matter (Haase, 1971). Kinetic considerations emphasize the unorganized movement of liquid molecules but modified by temperature-dependent interactions, leading to molecular groupings over a microscopic timescale of observation.

Studies documented that organic liquids subjected to rapid cooling possess temperature ranges in which they change to vitreous solids or glasses. It was suggested that hardening takes place rather sharply within a narrow temperature interval and with a maximum in heat capacity change (Parks and Huffman, 1926). That would make it reasonable to regard glass as a fourth state of matter, distinct from the liquid and crystalline states, and yet showing to some extent characteristics of both states. Nowadays, it is well understood that glass is a supercooled liquid with the thermal transition entirely distinct from crystallization, and the study of glassy phenomena has evolved to a highly specialized subject, especially one that cuts across several conventional disciplines (Ngai, 2000a).

In the glassy state, characteristics exhibited by materials are those of brittleness, high strength, clarity, and ultimately low molecular mobility (Rahman, 1995). Thus, the basic description of vitrification phenomena maintains that glass is a liquid that has lost its ability to flow and instead of taking the shape of its container, glass itself can serve as the container for liquids. On heating, materials soften progressively and macromolecular matrices achieve properties related to the rubbery state (Roos, 1995). Everyone knows that a glassy material is hard and fragile, but that rubber bands splinter like dropped goblets when impacted at liquid nitrogen temperature usually comes as a surprise. Examples of partial or total glassy behavior include hair, dry cotton shirts, biscuits, coffee granules, pasta, spaghetti, ice cream, as well as inorganic oxide systems, organic and inorganic polymers, and carbohydrate or protein matrices in an aqueous environment or in a mixture with high levels of sugars (Slade and Franks, 2002).

In the past 80 years or so that the scientific understanding of glassy systems has evolved, a fundamental question relates to the thermodynamic or kinetic nature of vitrification (Ferry, 1991). Glasses are formed when a liquid or a rubbery system is cooled so rapidly that there is no time for the molecules to rearrange themselves and pack into crystalline domains (Hutchinson, 1995). At the glass transition region there is no discontinuity in the primary thermodynamical variables of volume, enthalpy, and free energy. On the other hand, the first-derivative variables of coefficient of expansion (α_p), heat capacity (C_p), etc., undergo marked changes in the course of a few degrees. Furthermore, the spike in α_p and C_p observed at the crystallization temperature has no counterpart during vitrification. The process is considered to be a second-order thermodynamic transition in which the material undergoes a change in state but not in phase.

Experimentally, this is supported by the following: (1) x-ray data that show that no sudden structural change is involved; (2) calorimetric studies on supercooled glycerol, which produce a step change in heat capacity as a function of temperature at 190 K (Allen, 1993); (3) devitrification of polydisperse food materials on heating, which shows that softening does not occur at a fixed point with the absorption of a characteristic latent heat. Instead, networks soften over quite a large range of temperatures centered about a mean value called the glass transition temperature (T_g; Binder et al., 2003).

It is preferable, however, to refer to the molecular process as a glass transition rather than as a second-order transition to avoid implying a thermodynamic state at which equilibrium conditions are achieved. This is because increasing rates of cooling shift the T_g at higher temperatures and produce a less dense glass, thus arguing that the equilibrium glass condition lies below the experimentally accessible values (Christensen, 1977).

In foods, the glass phenomenon has been introduced rather recently following the development by the food processing industry of standards of quality control based on the concept of water activity (Slade and Levine, 1991). Thus, the importance of the glassy state was first highlighted by White and Cakebread (1966). They discussed the importance of the glassy and rubbery states in relation to the quality control of a number of high solids systems. Vitrification was considered as a reference point for the development of a new branch of technology and, generally, the discussion of properties has been in terms of temperatures above or below T_g (Zimeri and Kokini, 2002; Lazaridou and Biliaderis, 2002). Thus, a low T_g means that at room or mouth temperature the food is soft and relatively elastic, and at higher temperatures it may even flow. In contrast, a food with a high T_g will be hard and brittle at ambient temperature.

An important consideration in the discussion of the behavior of these foodstuffs is the concept of plasticization and its effect on the glass transition temperature (Perry and Donald, 2002). A plasticizer is defined as a substance incorporated in a material to increase the material's workability, flexibility, or extensibility. For example, proteins or polysaccharides are plasticized by low molecular weight diluents. Here, the T_g of the biopolymer is considerably higher than that of the diluent, so as the concentration of the diluent increases the T_g of the molecularly miscible mixture decreases. Water is a very effective plasticizer. Sugars also plasticize proteins and polysaccharides but in relation to water, for example, in a starch–water system they act as antiplasticizers in that they raise the T_g of the system (Kumagai et al., 2002; Momany and Willett, 2002).

The following section of the chapter is written with the view to exposing food scientists to the theoretical advances in the field and in particular the synthetic polymer approach, which constitutes the cornerstone of molecular understanding of biorubbers and glasses.

7.2 THEORIES OF GLASS TRANSITION

There is as yet no fully satisfactory theory of rubber to glass transition in biomaterials but there are several valuable ways of describing it in a quantitative fashion and there are statistical treatments, which aim at interpreting the phenomenon in terms of molecular processes. Synthetic science and, in particular, the subject of the viscoelasticity of polymers has reached a notable level of development, with the majority of the scientific community adopting certain schemes and a prevailing degree of order. Thus, the phenomenological theory of linear viscoelasticity is essentially complete. The molecular origin of the viscoelastic behavior peculiar to polymers is semiquantitatively understood, as are their dependences on temperature, pressure, molecular weight, concentration, and other variables (Ferry, 1980; Montserrat and Hutchinson, 2002).

Moreover, the relationships are understood well to permit predictions of behavior in practical situations and, of course, it would make a world of difference if the sophisticated synthetic polymer approach could be extended to the structural conundrums of biomaterials and foodstuffs. In recent years, this way of thinking has been further supported by the advent of microcomputing and system software, which allowed the rapid development of rheological techniques, with computer-driven rheometers becoming commonplace in the laboratory. These are now established as the most productive line of attack for the development of function–structure–texture relationships in food preparations (Richardson and Kasapis, 1998).

7.2.1 Combined Framework of Williams, Landel, and Ferry/Free Volume Theory

In view of the nature of the ensuing discussion on the applicability of the synthetic polymer approach to biomaterials, a concise account of the methodology as historically exploited to great effect in synthetics is appropriate. In 1954, Marvin assembled the most complete set of rheological data available for a single polymer, a sample of polyisobutylene, and constructed the master curve of Figure 7.1. Data covered a frequency window of 20 decades, which is well in excess of the range available on a dynamic oscillatory rheometer (three to four decades), and it was achieved by transposing mechanical spectra of storage (G') and loss (G'') modulus taken at different temperatures along the log frequency axis. The damping factor, tan δ, is taken to be the ratio of viscous to elastic modulus (G''/G'). In the absence of a phase transition, the span of moduli at each experimental temperature reflects a change in state, which can be associated with a reference temperature if data are superposed horizontally by a shift factor, thus implementing the method of reduced variables (otherwise known as the time–temperature superposition [TTS] principle; Tobolsky, 1956).

Clearly the viscoelastic spectrum of Figure 7.1 is divided into four frequency areas. Very low experimental frequencies or very long timescales of measurement ($\omega = 1/t$) allow complete energy dissipation and chain relaxation, with the values of storage modulus on shear becoming effectively zero. Probing at higher frequencies generates a rapid increase in both G' and G'' proportional to ω^2 and ω (terminal zone; part I), respectively, and eventually the viscoelastic parameters crossover into the plateau zone (tan $\delta = 1$; Takahashi et al., 1994). As shown in part II, the elastic component remains flat and the viscous response goes through a minimum, which is the result of more segments of the network supporting the applied stress. This can be the outcome of extensive topological

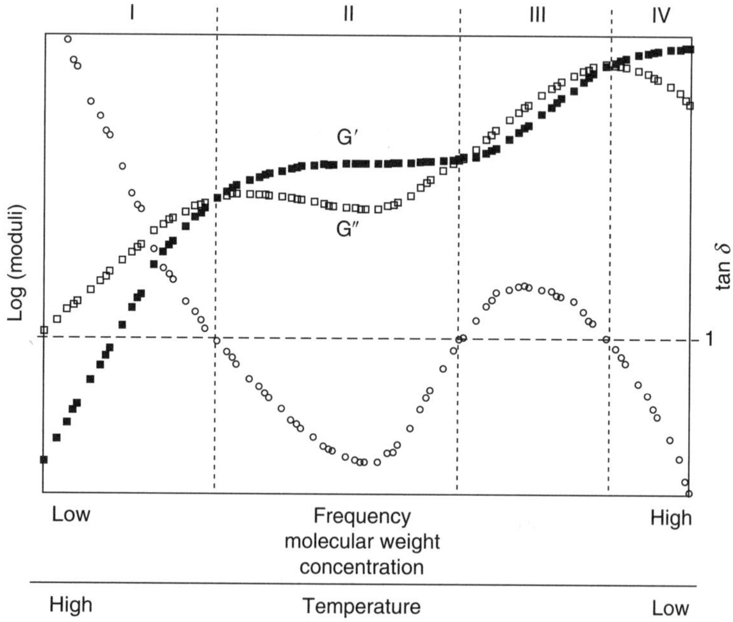

Figure 7.1 Master curve for the oscillatory mechanical response transcending the relaxation, rubbery, glass transition, and glassy states with increasing frequency, polymer molecular weight, concentration (diluted preparations), and reducing temperature.

entanglement or covalent cross-linking with the latter, of course, not exhibiting a relaxation zone (Huang et al., 2002).

Further on, the frequency dependence of viscoelasticity accelerates for a second time and the loss modulus becomes once more dominant. Part III is known as the glass transition zone where the long-range movements of the main chain are restricted (contributing to elasticity) but local relaxation processes that dissipate energy, for example of pendant groups, can occur. Eventually, at extremely short timescales of measurement the moduli crossover for a third time ($\tan \delta = 1$) and enter the glassy state (part IV), where secondary transitions and stretching or bending of chemical bonds are allowed (Frick and Richter, 1995). Values of G' around $10^{8.5}$ Pa have been reported for the glassy state of poly(*n*-octyl methacrylate) at −14.3°C (Dannhauser et al., 1958).

The approach used extensively by polymer scientists to develop a mechanistic understanding of rubber to glass transition, that is, area II to area IV in Figure 7.1, is based on the concept of macromolecular free volume. According to Ferry (1980) holes between the packing irregularities of long chain segments or the space required for their string-like movements accounts for free volume (u_f). Adding to that the space occupied by the van der Waals radii of polymeric contours and the thermal vibrations of individual residues, that is, the occupied volume (u_o), we obtain the total volume per unit mass (u) of a macromolecule. In polymer melts, the proportion of free volume is usually 30% of the total volume (Cangialosi et al., 2003), and the theory predicts that it collapses to about 3% at the glass transition temperature. At this point, the thermal expansion coefficient of free volume (α_f) undergoes a discontinuity, which reflects a change in slope in the graph of the linear dependence of total volume with temperature (Kovacs, 1964). A schematic representation of the concept of free volume is given in Figure 7.2.

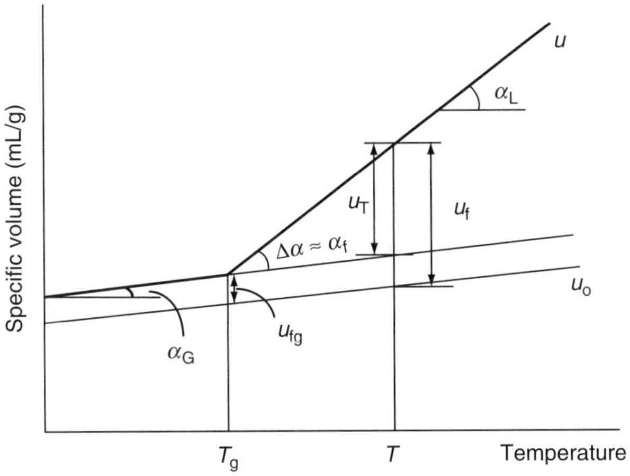

Figure 7.2 (1) If the occupied volume u_o is a constant fraction of the total volume below T_g, then a line can be drawn nearly parallel to the total specific volume (u) below T_g, with the difference a small constant fraction of u.
(2) Above the knee temperature T_g, the expansion of u_o does not match the overall expansion, leaving an increasing volume difference that is termed u_f (free volume).
(3) At and below T_g there is a certain small fraction of free volume u_{fg} that is assumed to be constant.
(4) Difference between the expansivity below T_g (α_G) and that above T_g (α_L) leads to a free volume component increasing with temperature (u_T) according to the relation $u_T = (\alpha_L - \alpha_G)(T - T_g)$ if T is the temperature of observation.
(5) Free volume total is $u_f = u_{fg} + u_T$, or $u_f = u_{fg} + (\alpha_L - \alpha_G)(T - T_g)$ as a function of temperature. The difference ($\Delta\alpha$) between α_L and α_G is written as α_f, the thermal expansion coefficient of the free volume.

Adopting this way of thinking, Doolittle and Doolittle (1957) followed the viscosity variation of liquid alkanes over a wide temperature range and found that the normally used equation of Arrhenius was inferior to the following mathematical expression:

$$\ln \eta = \ln A + B(u - u_{\mathrm{f}})/u_{\mathrm{f}} \tag{7.1}$$

where A and B are empirical constants, and for several polymers the value of B is more or less equal to unity (Neway et al., 2003). The next step was to define the fractional increase in free volume (f) as the dimensionless ratio u_{f}/u and work out from Equation 7.1 the shift factor (a_{T}) that integrates two sets of temperature data:

$$\log a_{\mathrm{T}} = \frac{B}{2.303}\left(\frac{1}{f} - \frac{1}{f_{\mathrm{o}}}\right) \tag{7.2}$$

The assumption of a rapid and linear development of the fractional free volume at temperatures above the glass transition can be considered in terms of the expansion coefficient α_{f} at a reference temperature T_0 as follows:

$$f = f_{\mathrm{o}} + \alpha_{\mathrm{f}}(T - T_0) \tag{7.3}$$

Use of the above relation for the fractional free volume in Equation 7.2 gives

$$\log a_{\mathrm{T}} = -\frac{(B/2.303 f_{\mathrm{o}})(T - T_0)}{(f_{\mathrm{o}}/\alpha_{\mathrm{f}}) + T - T_0} \tag{7.4}$$

which in the form of $C_1^{\mathrm{o}} = B/2.303 f_{\mathrm{o}}$ and $C_2^{\mathrm{o}} = f_{\mathrm{o}}/\alpha_{\mathrm{f}}$ is the equation proposed by Williams, Landel, and Ferry (WLF) to describe the temperature dependence of viscoelasticity at the glass transition (Williams et al., 1955). The WLF theory becomes inappropriate at temperatures below T_{g}, or higher than $T_{\mathrm{g}} + 100°\mathrm{C}$ when the temperature dependence of relaxation processes is heavily controlled by specific features, for example, the chemical structure of molecules in the melt. Within these limits, however, Equation 7.4 holds for any reference temperature (including the T_{g}), and a straightforward algorithm of simultaneous equations can be devised to determine parameters (C_1^{g}, C_2^{g}, f_{g}, α_{f}) related to the glassy state of a polymer:

$$C_1^{\mathrm{o}} = C_1^{\mathrm{g}} C_2^{\mathrm{g}} / \left(C_2^{\mathrm{g}} + T_0 - T_{\mathrm{g}}\right), \quad C_2^{\mathrm{o}} = C_2^{\mathrm{g}} + T_0 - T_{\mathrm{g}} \tag{7.5}$$

It is quite striking to see that the combined WLF/free volume theoretical framework is not only applicable to the vitrification of synthetic polymers but also to low molecular weight organic liquids and inorganic materials (Wang et al., 2003; Dlubek et al., 2003). This has prompted calls for the universality of the approach in glass-forming systems where changes in the free volume appear to be independent of chemical features.

7.2.2 Stretched Exponential Function of Kohlrausch, Williams, and Watts

The WLF/free volume approach is also known as thermorheological simplicity (TS) since it implies that all relaxation processes have the same temperature dependence, that is, a change in temperature shifts the time or frequency scale of all of them by the same amount. However, this is not a universal observation and, instead, thermorheological complexity (TC) has been reported on the superposition of viscoelastic functions in a number of amorphous synthetic polymers and epoxy resins (Ngai and Plazek, 1995). In addition, there is certain opposition in the use of free volume, because, in physics, intermolecular interactions are more fundamental and the ultimate determining factor of molecular dynamics in densely packed polymers (Ngai and Rendell, 1993).

Going back to TC, this refers to the lack of reduction of mechanical data along the logarithmic frequency axis. TC is seen, for example, in the temperature dependence of the shape of the tan δ peak covering the frequency range that corresponds to compliance values in the softening region of polystyrene (Cavaille et al., 1987). A subtle form of TC involves the superposition of data onto a single curve, which exhibits two tan δ peaks because of the presence of different temperature dependences of molecular processes, for example, polyisobutylene in Plazek et al. (1995). The lack of data superposition was attributed to the significant contribution of the local segmental motions of the polymer, which reflect a molecular mechanism triggered only in the short-time portion of the glass transition region. Dynamic oscillatory data of $G'(\omega)$ and $G''(\omega)$ monitoring these motions were first converted by Ngai (2000b) to stress relaxation moduli, $G(t)$:

$$G'(\omega) = G_e + \omega \int_0^{\infty} [G(t) - G_e] \sin \omega t \, dt$$
$$G''(\omega) = \omega \int_0^{\infty} [G(t) - G_e] \cos \omega t \, dt \tag{7.6}$$

and then fitted using the stretched exponential function of Kohlrausch, Williams, and Watts (KWW), as follows:

$$G(t) = (G_g - G_e) \, \exp[-(t/\tau^*)^{1-n}] + G_e \tag{7.7}$$

where

G_g is the glassy modulus
G_e is the equilibrium modulus of the local segmental motions
t is the time after the application of a fixed strain
τ^* is a measured relaxation time
n is the coupling constant, which ranges from 0 to 1.0

It was found that strongly coupled (interacting) systems have high values of n and an apparently broad distribution of relaxation times, a result that is the cornerstone of the coupling theory (Roland et al., 1999). The parameter n need not be constant during a relaxation, hence the theory is not thermorheologically simple, which is a basic assumption of the WLF/free volume theory discussed earlier. This allows the development of a connection between the chemical structure of the monomer and the viscoelastic properties of the polymer, which explains anomalous experimental facts observed in the literature. See, for example, the rationalization of the apparently anomalous finding that at rather slow cooling rates, all the cooling curves of a certain epoxy resin appear to reach an asymptotic glassy state at about the same temperature (Ngai et al., 2000). It is anticipated that much attention will be focused on the area of the coupling theory in the future. Further work on polymer–solvent mixtures, polymer blends, and block copolymers will provide additional tests as to a more general applicability of the model.

Admittedly, TC is more pronounced on low molecular weight materials and the degree of TC decreases with increasing chain length, with high molecular weight products exhibiting TS (Ngai, 1999). This may reduce the utility of the approach in high molecular weight fractions of biomaterials, an area in which the theory remains untested and which continues to offer tremendous challenges and opportunities to researchers in the field. Thus, recent work on four molecular gelatin fractions unveiled excellent superposition of viscoelastic data with a single broad peak of tan δ, which indicates TS (Kasapis et al., 2003a). Furthermore, the framework of the coupling theory is under development, with the current correlation between the shift factor of the local segmental motions and the temperature range of the glass transition creating the following mathematical expression (Plazek and Ngai, 1991):

$$(1 - n)\log a_{\mathrm{T}} = -\frac{C_1(T - T_{\mathrm{g}})}{(C_2 T_{\mathrm{g}}) + T - T_{\mathrm{g}}} \tag{7.8}$$

Clearly, the above formula offers no additional insights because it is comparable to the WLF equation.

7.3 RHEOLOGICAL MEASUREMENTS OF BIOGLASSES

7.3.1 Viscosity

Viscosity is the simplest rheological parameter to measure and was considered as a factor in determining whether a liquid will crystallize or form a glass on cooling (Richardson et al., 1998). Normally, samples are liquefied by heating crystalline materials, for example, α-glucose, in a flask immersed in a paraffin bath. Glucose crystals melt at about 145°C and then maintained under a vacuum at 155°C–160°C for a short time, usually about 5 min, to eliminate bubbles of air or water vapor. When the liquid becomes fairly clear, air is admitted and the sample is allowed to cool. Glucose is usually straw-colored. Care should be taken, since considerably longer periods of heating at 155°C or above lead to noticeable decomposition and a product that is dark brown in color (Bhandari and Roos, 2003).

Angell (1988) reproduced viscosity data for inorganic melts and organic liquids achieving 10^{12} Pa s at the glassy state, a value that is now considered as an indication of the glass transition temperature. This approach is valid provided that Newtonian liquids are analyzed or the shear-thinning properties of the material are well mapped out. However, the paper on the vitrification behavior of maltose–water mixtures demonstrates the difficulty of experimenting at conditions of extreme sample rigidity (Noel et al., 1991). Undercooled liquid sugars assume all the superficial aspects of a solid and at low temperatures it is found impractical to carry on viscosity determinations by the standard method of concentric cylinders.

In general, readings do not exceed $10^{6.5}$ Pa s and a long extrapolation to 10^{12} Pa s is implemented in an attempt to predict the value of T_g. However, in the absence of concrete evidence of an exponential (Arrhenius) temperature dependence of viscosity or a WLF function of molecular processes the arbitrary treatment of results is fundamentally flawed. Similarly, the temperature dependence of viscosity for honey was recorded with a view to predict its glass transition temperature from relatively low viscosity values (below 2.4 kPa s in Bhandari et al., 1999). It was also stated that the T_g can be a useful tool to predict crystallization behavior of honey but, in our view, these are fundamentally different molecular processes and should not be referred to interchangeably.

Further difficulties in developing a viscosity-related T_g become apparent from work on the textural properties of high sugar preparations in the presence of protein or polysaccharide at levels of normal industrial use (below 10% and 1%, respectively). These are recently enjoying increasing attention from researchers and product developers alike, and there is a need to develop useful theoretical and empirical relationships. Figure 7.3 illustrates a typical mechanical spectrum of the glassy state covering a frequency range from 0.1 to 100 rad/s (Kasapis et al., 1999). The bioglass is a mixture of 5% acid pigskin gelatin ($M_n = 68{,}000$), 30% sucrose, and 50% glucose syrup with a dextrose equivalent of 42. Within the accessible range of observation, the values of G' lie above those of G'' with both moduli being relatively independent of the frequency of oscillation in the

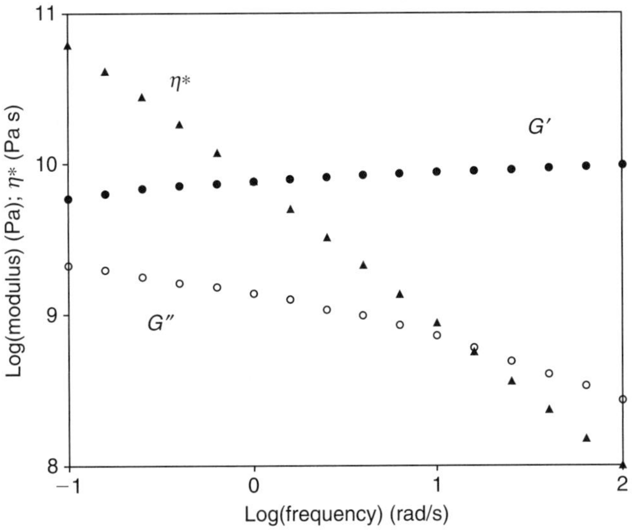

Figure 7.3 Frequency sweep of storage modulus, loss modulus, and complex dynamic viscosity for 5% acid pigskin gelatin in the presence of 30% sucrose and 50% glucose syrup obtained at −55°C and a strain of 0.001%.

abscissa, features that are characteristic of a solid-like behavior. The complex dynamic viscosity (η^*) is given by the following equation (Steffe, 1996):

$$\eta^* = \left[(G')^2 + (G'')^2\right]^{1/2}/\omega \tag{7.9}$$

Clearly, the log of the complex viscosity descends steeply from almost 10^{11} to 10^8 Pa s as a function of log ω at the experimental temperature of observation (−55°C). Thus, the absence of a plateau in the trace of η^* makes predictions of T_g from viscosity readings rather tenuous.

7.3.2 Dynamic Mechanical Thermal Analysis

To successfully address the complexities and correlations of the structure–function properties, additional insights into viscosity should be obtained using viscoelastic measurements (Walton, 2000). Dynamic mechanical thermal analysis (DMTA) has been used as a means of extracting that information from samples with a low water content (typically between 10% and 20%) and a relatively high ratio of biopolymer to cosolute (>1:1). Biopolymers are mainly starch hydrolysates of a range of molecular sizes, casein and gluten (Kalichevsky and Blanshard, 1992). Measurements are carried out in extension or bending mode on sample bars that are formed by pressing the material at about 10^8 Pa (Mizuno et al., 1998). At this low range of water content, samples are brittle and particular care must be taken to establish the strain range of the linear viscoelastic regime (Brent et al., 1997). It is a curious fact that the literature is largely devoid of data on strain sweeps of synthetics and biomaterials, which would have established unambiguously the appropriate range of deformation for conducting small deformation experiments.

In addition, experimentalists should be aware of limitations regarding the application of the necessary force (stress) to the sample under investigation. High forces may interfere with the compliance of the instrument whereas low forces may fall below the stiffness of the clamp suspension. Heating runs are usually recorded within the range of −50°C to 150°C but since it is difficult to seal hermetically the sample-clamp measuring arrangement, water loss might affect the reliability of results at the high-temperature end and diminish the value of a subsequent cooling scan

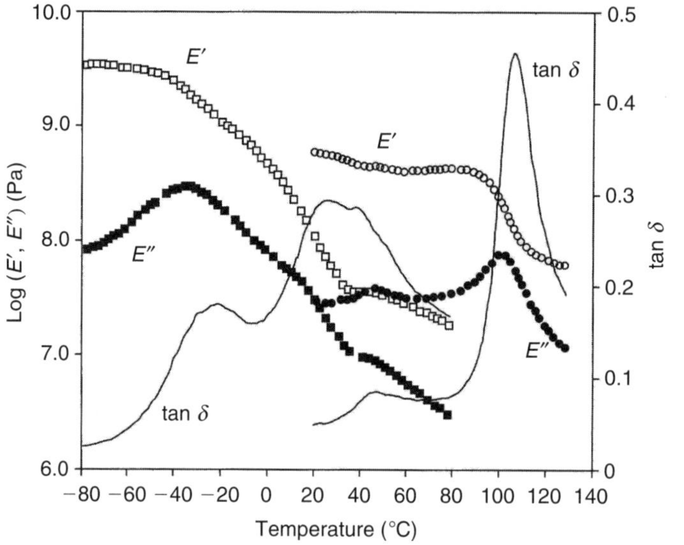

Figure 7.4 DMTA heating of amylopectin with 10.5% water monitored within the temperature range of 20°C–130°C, and a second heating run from −80°C to 80°C for amylopectin with 25% fructose and ≈19% water.

(Liu et al., 1997). If these issues are addressed successfully, additional insights into the mechanism of temperature-induced structural changes will be obtained from (1) aspects of thermal reversibility of a network due to its entropic rubber to glass transition and (2) the development of thermal hysteresis signifying enthalpic aggregation in the system.

The use of DMTA can be demonstrated on the effect of different moisture levels and type of sugar on the structural properties of waxy maize starch. Figure 7.4 reproduces the heating profiles of storage (E') and loss (E'') Young modulus on longitudinal deformation and their viscoelastic ratio ($\tan\delta = E''/E'$) for amylopectin-based preparations. There is an order of magnitude variation in E' of the amylopectin in the presence of 10.5% water, which has been attributed to the glass transition of the sample (Kalichevsky et al., 1992). The peak of the tan δ was used as an empirical pointer of the glass transition temperature (T_g) occurring at 106°C. There is another smaller increment in the values of E' at temperatures below 40°C perhaps due to short-range local motions, whereas the glass transition has been attributed to freezing of the translational and rotational processes of the main chain.

The plasticizing effect of water was confirmed by increasing its content to ≈25%, which moved the T_g of the amylopectin samples to 29°C (tan δ peak). Incorporation of sugar (fructose, glucose, sucrose, xylose) affected the vitrification of samples as expected by the change in the molecular weight distribution of the mixture, that is, sugar acted as an antiplasticizer in comparison to water but replacing the polymer with the cosolute at the same water content results in a fall in the T_g—the cosolute is more plasticizing than the polymer (Ollett et al., 1991; Kalichevsky et al., 1993a). At cosolute concentrations below 9% the decrease in T_g values was taken from thermal spectra exhibiting a single wave of structure alteration, as seen for sugar-free waxy starch preparations.

As illustrated in Figure 7.4, however, a more complex behavior of viscoelasticity develops at higher levels of sugar (Kalichevsky and Blanshard, 1993). Thus, there are two steps of structure in the presence of 25% fructose and inevitably two clear maxima in tan δ values. The first one is attributed to amylopectin with the second peak reflecting the vitrification of cosolute (T_g's are 25°C and −23°C, respectively). The strong network of the amylopectin–fructose mixture, log (E'/Pa) = 9.5 at −80°C, strongly argues that the cosolute viscoelasticity becomes dominant at the low-temperature range. Furthermore, the appearance of a second tan δ peak (glass transition) indicates that some of the cosolute forms a separate phase, which is not mixed on a molecular level with amylopectin and not acting as the polymer's plasticizer.

7.3.3 Dynamic Oscillation on Shear

Oscillatory rheology has been employed to put the qualitative findings of DMTA on a sound theoretical basis. The free volume concept is particularly popular, in this respect, partly because it is intuitively appealing. Often (but not invariably), it is able to explain observed trends correctly and is easy for workers in materials science coming from many different backgrounds. Food scientists have mainly worked on gelatin and polysaccharides, which comprise an important class of confectionery products. The issue is as follows: gelatin, which for almost a century has been produced on an industrial scale, is the most frequently used structuring agent in confectionery products, but is increasingly falling out of fashion with consumers and producers alike (Poppe, 1992). Reasons for the change in attitudes include the need to circumvent diet and health problems or perceptions such as the BSE scare, vegetarianism, and religious dietary laws (e.g., Muslim and Hindu). Furthermore, high sugar–gelatin confectioneries tend to become sticky during handling, or when stored in warehouses at high ambient temperatures, resulting in partial structural collapse, welding, and crystallization (caking) of the product (Shim, 1985; Wolf et al., 1989). For these reasons there is an incentive to understand the behavior of polysaccharides in high sugar environments since this class of biopolymers may provide an alternative to gelatin (DeMars and Ziegler, 2001).

7.3.3.1 Sample Preparation Methods

7.3.3.1.1 Open Boiling Method

Materials are dispersed in preheated distilled water from 60°C to 90°C with constant stirring until they are fully dissolved, after which time the appropriate amounts of sugars (mainly glucose syrup and sucrose) are added. For most purposes, volumes of 250–500 mL are sufficient for structural, textural, or sensory analysis. Mixtures are kept hot up to 60 min to provide formulations with the required level of solids. Final moisture contents should be checked by vacuum oven drying at 80°C. The method is appropriate for samples containing between 60% and 85% solids (Tsoga, 2001).

7.3.3.1.2 Reduced Pressure Method

When the solid content is 85% and above, boiling under vacuum is generally used in an attempt to minimize the temperature and decrease the length of time necessary to produce samples of known composition. This method reduces the risk of chemical changes such as carameliza-tion and macromolecular degradation. A typical apparatus used is illustrated in Figure 7.5 (Deszczynski, 2003).

Samples are prepared in a 250 mL round bottom Quickfit glass flask. The flask, containing a magnetic stirrer, is first weighed. The appropriate amounts of biopolymer, water, and salts (if any) are dispersed in distilled water at 85°C with constant stirring until the sample is fully dissolved. The cosolute is then added and the flask is reweighed. When all the components have been dissolved, the flask is transferred to an aqueous glycerol bath. The head (H) is placed on the flask (A) and the pump starts with the taps (S) and (T) being open. The tube (V) acts as a bypass to prevent material being sucked out of the flask when the pump is started. Tap (S) is then closed and clip (T) tightened, to produce a rapid boil in flask (A) at a temperature of approximately 70°C. When sufficient water

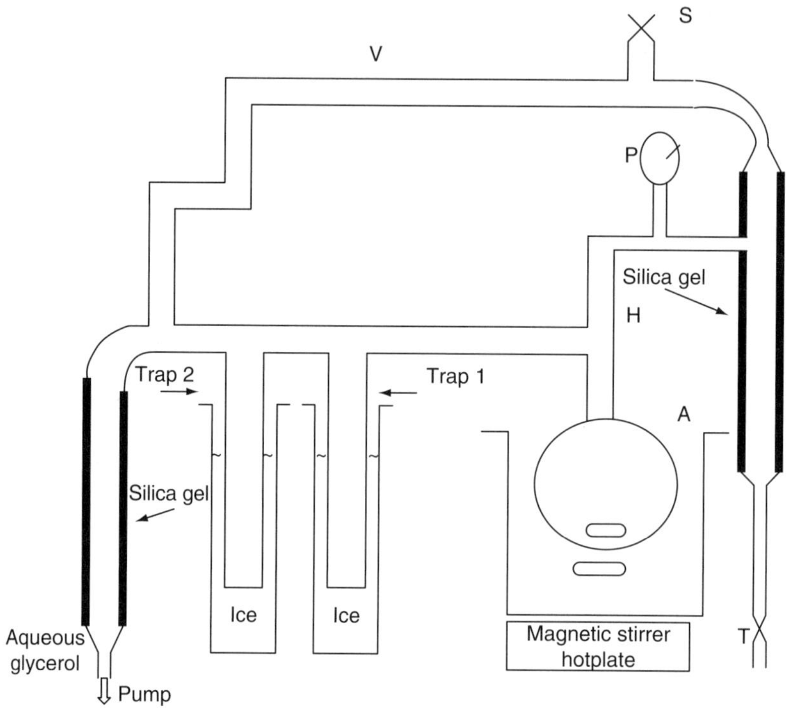

FIGURE 7.5 Apparatus for the production of confectionery products under reduced pressure.

has evaporated, flask (A) is removed, stoppered, washed with acetone, and weighed. While the sample is still hot it should be used for the ensuing analysis, for example, poured onto the preheated rheometer plates. This method is suited for all industrially important polysaccharides but not for gelatin-based formulations due to problems with foaming of the protein.

7.3.3.1.3 Commercial Products

Whereas the above two methods are employed for research and development of new formulations (Sworn, 1998), production of commercial confectioneries involves deposition in starch moulds. Gummy sweets are based on mixtures of gelatin and polysaccharide in the presence of sucrose and glucose syrups with a dextrose equivalent of 42–70 and typically contain 10%–25% moisture in the final product (Carr et al., 1995). At least 50% of the cosolute is glucose syrup whose polydispersity prevents crystallization at ambient temperatures. Products are made by first preparing the liquor, which is a mixture with all the ingredients at 30% moisture content. Liquors are deposited hot (e.g., 70°C) into dry-powdered starch moulds and the excess moisture is extracted by stoving the sweets in these moulds for the required period of time at about 50°C (Ong et al., 1998).

7.3.3.2 Application of the Synthetic Polymer Approach to High Solid Bioglasses

Delving in the synthetic polymer literature along the lines described in Section 7.2.1 provides insights into the mechanism of vitrification of biomaterials. The first step of analysis would be to examine G' and G'' as a function of temperature at a constant frequency or time of measurement. This is illustrated in Figure 7.6 for a gelatin–sugar preparation. As indicated earlier, these find

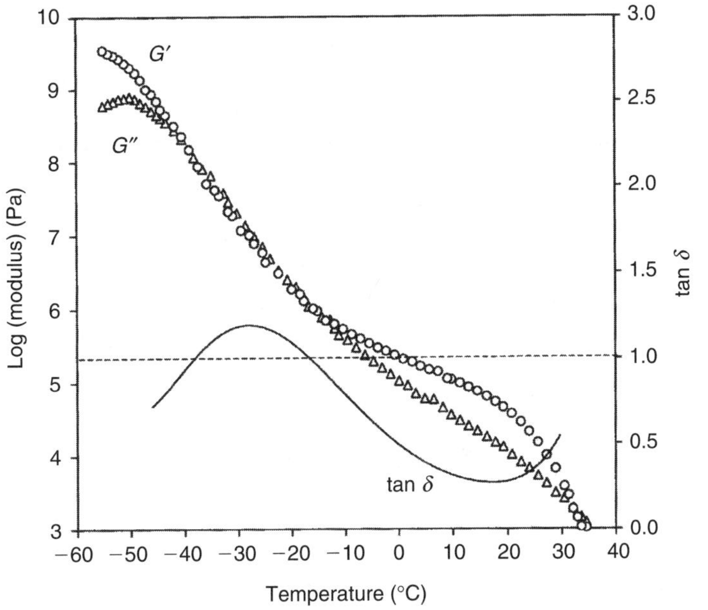

Figure 7.6 Cooling profiles of G' (○), G'' (△), and tan δ (—) for a mixture of 25% acid pigskin gelatin with 40% sucrose + 15% glucose syrup obtained at a scan rate of 1°C/min and frequency of 1 rad/s. Applied strain was varied in steps during the thermal run from 1% in the rubbery region to 0.00072% in the glassy state.

application in confections (gummy bears, wine gums, etc.) and, additionally, in flavor encapsulation and the preservation of bioactive molecules in glassy matrices (Norton and Foster, 2002). The sample was made by mixing gelatin with sugar at intermediate levels of solids and then allowing gentle dehydration at 60°C to yield a composition of 25% gelatin, 40% sucrose, and 15% glucose syrup.

Using the idealized profile in Figure 7.1, it is clear that a considerable part of the master curve of viscoelasticity of gelatin has been captured within the experimentally accessible temperature range in Figure 7.6. To cope with the spectacular changes in the rigidity of the sample (from 10^3 to almost 10^{10} Pa with cooling) a controlled strain rheometer with an air-lubricated force rebalance transducer is required (Windhab, 1996/7). Nevertheless, it should be checked that any inherent machine compliance is insufficient to significantly offset measured values from the high modulus glass systems. This should be achieved by progressive adjustment of geometry settings while measuring samples of known intermediate and high modulus, for example, polydimethylsiloxene (PDMS) at 30°C ($G_c = 2.5 \times 10^4$ Pa) and ice at −5°C ($G' = 10^9$ Pa). For the measuring geometry of parallel plates, maximum plate diameter and minimum measuring gap between the two plates consistent with accurate results can be established (Al-Ruqaie et al., 1997).

Capture of the rubber-to-glass transition requires precise control of the scan rate over a wide temperature range. An air convection oven with a dual element heater–cooler and with counter-rotating airflow is recommended. Depending on the moisture content of the samples, the glassy state is unveiled at temperatures as low as −70°C at a scan rate of 0.1°C/min–2°C/min (Kasapis et al., 2001). Silicone fluids (Dow Corning 100–200 cs) are used to cover the exposed edges between the parallel plates. This prevents the formation of skin on the surface of the sample at the high-temperature end of the experimental routine at which the rubbery plateau is manifest. The fluid should be of sufficient low molecular weight to remain liquid-like at subzero temperatures. Nowadays, the measuring software is flexible enough to interrupt the cooling–heating runs at constant temperature intervals of a few degrees centigrade to carry out frequency sweeps between 0.01 and 100 rad/s (Simon and Ploehn, 1997).

Implementation of the above protocol unveils a classic viscoelastic behavior for the gelatin–sugar mixture in Figure 7.6, as pioneered by research in synthetic polymers. At the upper range of temperature, the gelatin–cosolute mixture is a melt with the loss modulus dominating the storage modulus. Cooling sees a development in G' with the two traces crossing over at a gelling point (about 31°C), which demarcates the onset of the rubbery region (Sablani et al., 2002). Mixtures are true gels with elastic consistency that can support their shape against gravity at ambient temperature. On further cooling (below −10°C), the shear moduli develop rapidly and we consider values up to $10^{8.5}$ Pa, which define the rubber-to-glass transition region. Main features of the transition include a dominant viscous component ($G'' > G'$) and a spectacular dependence of viscoelastic functions on temperature. In Figure 7.6, the ratio of G'' to G' ($\tan \delta$) between −15°C and −40°C acquires values higher than 1 and, within the same temperature range, the increase in shear modulus is three orders of magnitude. Finally, at the lowest temperature (below −42°C), there is yet another development and a hard solid response is obtained (G' values approach 10^{10} Pa), which according to the idealized profile in Figure 7.1, demarcates the fourth region of the master curve, the so-called glassy state (Li and Yee, 2003).

The thermal profiles of Figures 7.1 and 7.6 allow a first insight into the structural properties of materials (Pak et al., 2003). A more valuable find, however, allows complete dissociation of the contributions of temperature and frequency to the overall mechanical behavior according to the postulates of TTS discussed in Section 7.2.1. The approach affords a device for changes in temperature to be seen as shifts in the frequency–timescale of mechanical spectra taken in a sequence of temperature intervals (Normand et al., 2000). Figure 7.7a and b demonstrates the application of TTS in the gelatin–sugar mixture by citing frequency sweeps of the real and imaginary parts of the complex shear modulus.

Measurements were taken at 11 different temperatures between 14°C and −55°C and at 16 different frequencies from 0.1 to 100 rad/s. The modulus data covering the long timescales of the upper range of temperature remain relatively flat within the rubbery region, whereas further cooling sees a rapid reinforcement of viscoelasticity, which unveils the glass transition region. At the

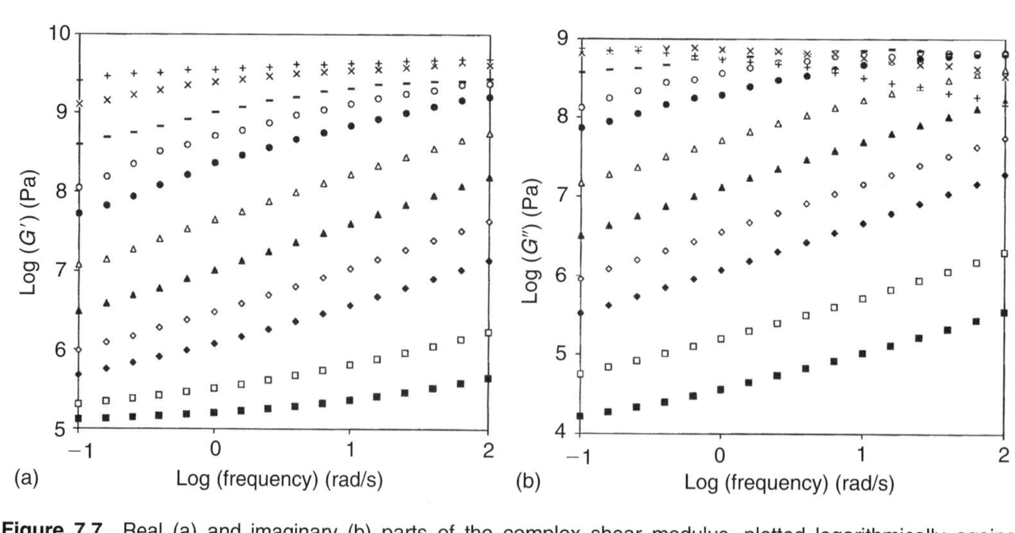

Figure 7.7 Real (a) and imaginary (b) parts of the complex shear modulus, plotted logarithmically against frequency of oscillation for a sample of 25% acid pigskin gelatin with 40% sucrose and 15% glucose syrup. Bottom curve is taken at 14°C (■); other curves successively upward, −1°C (□), −16°C (◆), −22°C (◇), −28°C (▲), −34°C (△), −40°C (●), −43°C (○), −46°C (–), −52°C (×), and −55°C (+)°C, respectively.

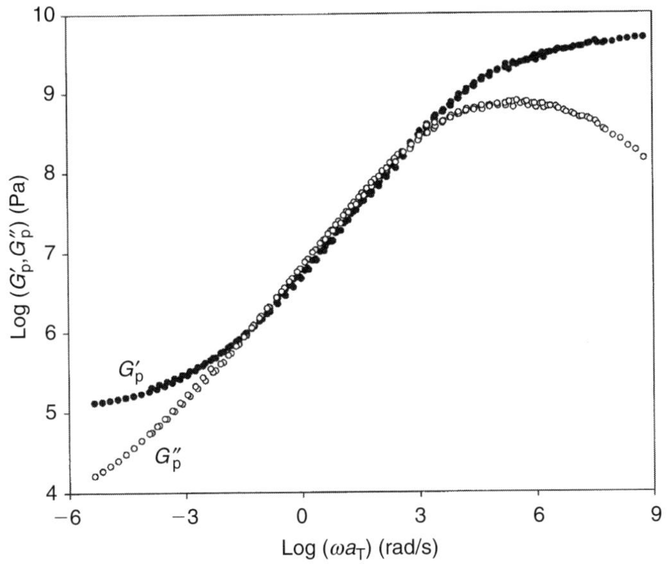

Figure 7.8 Master curve of reduced shear moduli (G'_p and G''_p) as a function of reduced frequency of oscillation (ωa_T) for a sample of 25% acid pigskin gelatin with 40% sucrose and 15% glucose syrup (reference temperature = −25°C).

low-temperature end of the experimental range, the values of G' level off once more with a quantifiable endpoint (Figure 7.7a) thus unveiling the glassy state.

Application of TTS to the data of Figure 7.7a and b verified that shear modulus measured at a certain frequency of oscillation and temperature T is equivalent to a modulus measured as the product of the same frequency multiplied with a scaling factor a_T, at the reference temperature T_0 (Ward and Hadley, 1993). The scaling factor is identical to the shift factor in Equation 7.4, and T_0 is arbitrarily chosen but within the temperature range of the glass transition region. Figure 7.8 reproduces the modulus traces of Figure 7.7a and b, which are superposed horizontally along the abscissa at the reference temperature of −25°C. This creates master curves of viscoelasticity, with the reduced variables, G'_p and G''_p, being plotted logarithmically.

7.3.3.3 *Definition of a Fundamental Indicator of the Glass Transition Temperature*

The pictorial manifestation of the rubber-to-glass transition in gelatin–sugar mixtures is not unique but has been reproduced for polysaccharide preparations, a result which begs the question of universality of the viscoelasticity in these systems (Kasapis, 2001a). To pinpoint the glass transition temperature, which is one of the main concerns of the food processing industry, we use the premises of the free volume theory. As discussed in Section 7.2.1, the theory has been widely employed in a quantitative fashion to interpret glassy phenomena in terms of molecular processes (Dlubek et al., 2000). According to the theory, T_g should be located at the end of the glass transition region where the free volume declines to insignificant levels, that is, about 3% of the total volume of the material. For the thermal profile in Figure 7.6, the conjunction of the glass transition region with the glassy state is clearly discernable, thus making the experimental T_g of the gelatin–sugar mixture equal to −39°C.

The aforementioned thermal profile enables pinpointing of the experimental T_g value. This should be compared with the prediction of the WLF equation to check the ability of the concept of

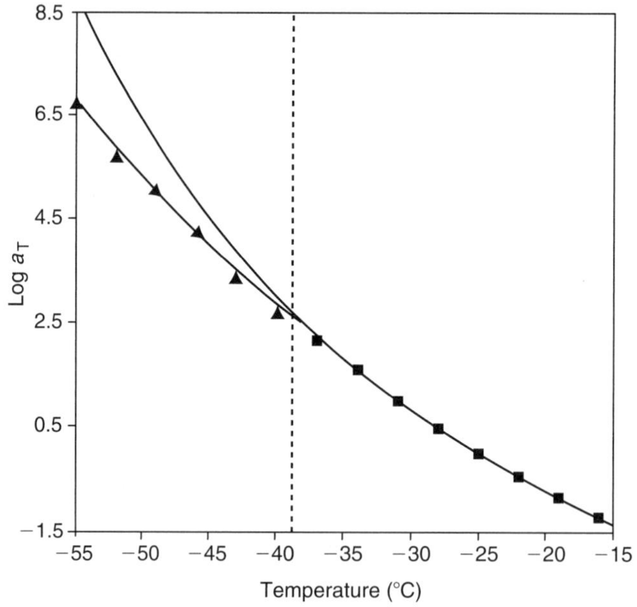

Figure 7.9 Temperature variation of the factor a_T within the glass transition region (■) and the glassy state (▲) for the mixture in Figure 7.8, with the solid lines reflecting the WLF and Andrade fits of the shift factors in the glass transition region and the glassy state, respectively; dashed line pinpoints the T_g prediction.

free volume to follow the effect of temperature on viscoelastic functions throughout the glass transition region (Singh and Eftekhari, 1992). In doing so, logarithmic plots of the shift factors used in the construction of the master curve in Figure 7.8 are plotted against the experimental temperature of each frequency sweep in Figure 7.9. Clearly, the WLF equation provides a good fit of the empirically derived shift factors in the glass transition region, which determine how much the frequency scale shifts with temperature.

Fitting of the shift factors to the WLF framework can be achieved by plotting $1/\log a_T$ against $1/(T - T_0)$ and obtaining the two parameters C_1^o and C_2^o from the slope and intercept of the linear fit, respectively (Peleg, 1992). Assuming that the WLF equation holds for any temperature within the glass transition region including T_g, calculation of the free volume parameters at T_g is feasible through Equation 7.5. The fractional free volume ($f_g = 0.035$) and the thermal expansion coefficient ($\alpha_f = 7.0 \times 10^{-4}$ deg^{-1}) at T_g have been calculated and found to correlate well with parameters obtained for polysaccharide–sugar mixtures and from synthetic polymer research (Kasapis, 2001b).

In addition to the glass transition region, the master curves in Figure 7.8 cover the glassy state where the values of G' approach a maximum and those of G'' decrease rapidly. As shown in Figure 7.9, the shift factors of mechanical spectra in the glassy state unveil a pattern of behavior that cannot be followed by the WLF equation (Slade and Levine, 1993). Instead, progress in mechanical properties at the region of the lowest temperatures is better described by the mathematical expression of Andrade:

$$\log a_T = \frac{E_a}{2.303R}\left(\frac{1}{T} - \frac{1}{T_0}\right) \qquad (7.10)$$

This yields the concept of activation energy (E_a) for an elementary flow process in the glassy state, which is independent of temperature. Within the glassy state, the factor a_T is an exponential function of the reciprocal absolute temperature, so the logarithmic form with a constant energy of activation for an elementary flow process can be used for calculating numerical values (Gunning et al., 2000). For the gelatin–sugar mixture, E_a was found to be in the order of 230 ± 30 kJ/mol.

The energy of vitrification (E_v) within the glass transition region is associated with the difficulty for transverse string-like vibrations over several molecules to occur. Instead of being temperature independent as the reaction-rate theory within the glassy state has predicted, it increases rapidly with cooling of the sample from the reference to the glass transition temperature. This can be derived by differentiating the WLF equation (D'Haene and Van Liederkerke, 1996):

$$E_v = Rd \ln a_T / d(1/T) = 2.303RC_1^o C_2^o T^2 / (C_2^o + T - T_0)^2 \tag{7.11}$$

where R is the gas constant. For amorphous synthetic polymers it is of the order of 260 kJ/mol if $T_g = 200$ K and 1047 kJ/mol if $T_g = 400$ K. Application of Equation 7.11 to biosolids generates an energetic cost of vitrification within the synthetic polymer range, hence being in accordance with experience (Plazek, 1996).

As discussed above, the predictions of the rheological T_g, obtained by treating the master curves in Figure 7.8 and the shift factors in Figure 7.9 with the WLF and Andrade equations, are congruent with the experimental observations shown in the thermal profile of Figure 7.6. Mechanistically, this coincides with the transformation from free-volume derived effects in the glass transition region to the process of an energetic barrier to rotation in the solid-like environment of the glassy state. Work has evaluated this approach in polysaccharide–cosolute mixtures and its apparent widespread application assigns physical significance to the rheological T_g as the threshold of the two distinct molecular processes (Kasapis, 1998).

The need to identify an objective way to assess the temperature dependence of molecular processes during vitrification has led to the development of a glass transition temperature using small deformation dynamic oscillation. This is compared advantageously with early attempts to demarcate glassy phenomena, which produced empirical definitions of the mechanical T_g. Thus, DMTA in extension or bending mode attempted to identify the right glass transition temperature (Rieger, 2001). This has been taken as the initial drop in the values of storage modulus (E') on heating or the maxima in loss modulus (E'') and tan δ traces within the realm of the rubber-to-glass transition. As has been pointed out, however, in the absence of a distinct molecular process associated with these selections, the approach denotes merely an empirical index of convenience (Peleg, 1995; Mitchell, 2000).

7.3.3.4 Partially Vitrified Foodstuffs

The high sugar–gelatin or sugar–polysaccharide mixtures utilized in confectionery formulations exhibit classic viscoelastic behavior, which facilitates application of a basic approach for pinpointing the glass transition temperature. In terms of food processing and storage, interesting examples related to vitrification phenomena include the dehydration and preservation of fruits, vegetables, and meat or fish in relation to the product quality. Food dehydration dramatically increases the concentration of solids in the remaining water thus bringing the system close to its rubber-to-glass transition (Matveev et al., 2000; Bai et al., 2001). This, of course, imparts heavily to the textural properties of dried foodstuffs, and the accurate identification of a T_g value should offer insights into the quality control and the ultimate acceptability of these products.

It appears, however, that the mechanical transition of dried fish and fruits is not that clearly defined. This is illustrated in Figure 7.10 for the thermal profile of tuna meat dried to 70% solids. To start with, the magnitude of the plateau at temperatures above 0°C is unusually high reaching values in excess of 10^8 Pa for the storage modulus; compare, for example, with the corresponding values in Figure 7.6 between 0°C and 30°C. Second, the change in moduli at subzero temperatures is

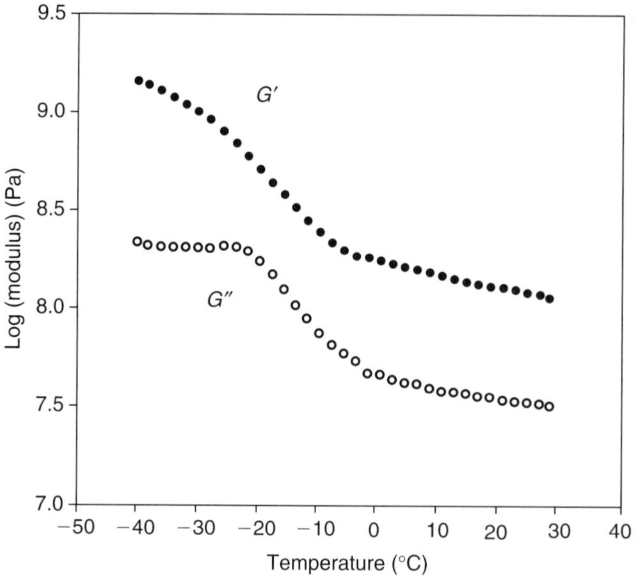

Figure 7.10 Heating profiles of storage (G') and loss (G'') modulus for dried tuna at 70% solids; scan rate: 1°C/min, frequency: 1 rad/s, strain: 0.005%.

less than an order of magnitude from one step to another. Third, the liquid-like response, which is the primary indication of glassy relaxation processes (Farhat et al., 2003), is substantially diminished with the values of tan δ throughout the experimental range remaining well below that of 1. In contrast, the rubber-to-glass transition in Figure 7.6 develops from $10^{5.5}$ to $10^{8.5}$ Pa and, within this range, the values of tan δ are higher than 1.

It appears that the work by Favier et al. (1995) can bridge the gap between the two extrema of mechanical behavior. These researchers prepared suspensions of ordered cellulose whiskers at a nanoscale size and mixed them homogeneously with a lattice of copolymerized styrene and butyl acrylate. Figure 7.11 shows temperature profiles for the synthetic rubbery matrix at different contents of the rodlike cellulosic inclusions. Clearly, there is a dramatic enhancement of network rigidity at the plateau region of the high-temperature end with a concomitant distortion of the magnitude, shape, and temperature band of the following transition.

A similar experiment was carried out on polyethylene terephthalate, a material that is amorphous when quenched from the melt but crystallizes partially during slow cooling (Thompson and Woods, 1956). The effect of crystalline regions was a much more gradual transition of one order of magnitude, with the tan δ trace collapsing from a value of 3 to about 0.2 as a function of the reduced cooling rate. On the basis of this evidence, one envisages the mechanical profile in Figure 7.6 as the outcome of vitrification of a largely amorphous system. In contrast, the curtailed vitrification in Figure 7.10 should be because of a complex transformation occurring in a composite of amorphous localities and low-mobility ordered expanses (Cruz et al., 2001).

Clearly, drying of tuna meat produces tightly packed aggregates of reduced molecular mobility, an outcome that diminishes the opportunity for a full manifestation of the glass transition in these systems. In support of this, it is difficult to obtain discernable glass transition spectra using the standard technique of modulated differential scanning calorimetry (MDSC; see Section 7.4) at various scan rates and annealing temperatures or times for dried foodstuffs. On the other hand, one can obtain and contrast T_g values from MDSC with those of the rheological T_g for mixtures

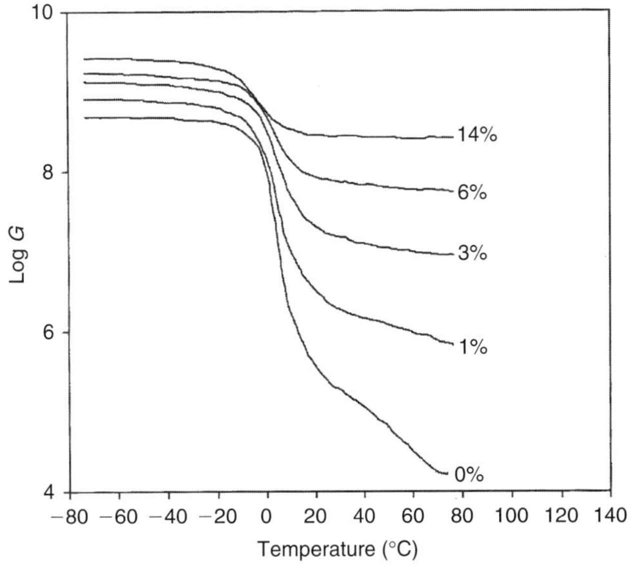

FIGURE 7.11 DMTA measurements during heating of a 35% styrene and 65% butyl acrylate latex suspending various concentrations (0%–14%) of cellulose whiskers shown by the individual traces.

of high sugar in the presence of low to medium levels of gelatin or polysaccharide (Deszczynski et al., 2002).

To circumvent the problem of a diminished glass transition in dehydrated fish and fruits and to identify an objective way to assess the temperature dependence of molecular processes in these systems, one can put forward a reasoning based on first principles. As discussed, the thermal effect on the structural properties of dried fish and fruits is less pronounced, thus yielding a partial glass transition in terms of temperature band and viscoelastic functions. Nevertheless, the fundamentals of a fully developed glass transition in amorphous polymers originate from the same dynamics of molecular chains as in curtailed vitrification owing to partially ordered or crystalline matrices (Maltini and Anese, 1995). Based on this postulate, a potentially useful device can be conceived for the estimation of the extent of vitrification in partially glassy materials. This entails plotting of the first derivative of shear modulus as a function of the sample temperature versus the sample temperature during vitrification. An example is shown for the gelatin–cosolute mixture in Figure 7.12, which is an expanded version of Figure 7.9 (Kasapis, 2004).

The storage modulus becomes the appropriate parameter for consideration because its trace is reduced to a minimum at the conjunction of the WLF/free volume and Andrade theories (Peleg, 1992). This reproduces the T_g value of −39°C for the completion of the softening process in the gelatin–cosolute mixture. Based on this principle, similar treatments were pursued for the viscoelastic behavior of partially amorphous tuna, abalone, apple, apricot, pear, plum, and dates. As illustrated in Figure 7.13, these produced smooth first-derivative curves, with the relaxations spectrum of the viscous component of the network unraveling rapidly compared with the solid element for all foodstuffs.

The minimum of the storage modulus trace clearly demarcates the mechanical glass transition temperature for 70% tuna (−18°C), 70% abalone (−22°C), 67.2% apple (−43°C), 67% apricot (−47°C), 70.6% pear (−35°C), 60.9% plum (−52°C), 80.2% dates (−24°C), and 83.8% dates solids (−20°C), respectively. The approach should be extended to other partially amorphous foodstuffs, thus identifying the function of storage modulus as the relevant indicator of molecular mobility during vitrification (Pomeranz, 1991).

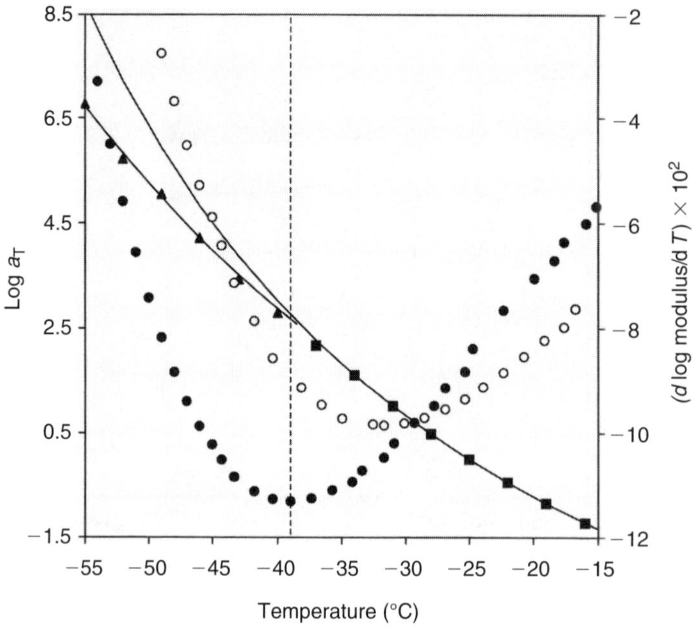

FIGURE 7.12 Temperature variation of the factor a_T within the glass transition region (■) and the glassy state (▲) for the mixture in Figure 7.8, with the solid lines reflecting the WLF and Andrade fits of the shift factors in the glass transition region and the glassy state, respectively (left-hand side *y*-axis); and first-derivative plot of log G' (●) and log G'' (○) as a function of sample temperature, with the dashed line pinpointing the T_g prediction (right-hand side *y*-axis).

7.4 DIFFERENTIAL SCANNING CALORIMETRY

7.4.1 Conventional DSC

For almost half a century, differential scanning calorimetry (DSC) has been used to measure, as a function of temperature, the difference in energy inputs between a substance and its reference, with both materials being subjected to a controlled temperature program (Biliaderis, 1983). The most common instrumental design for making DSC measurements is the heat flux design shown in Figure 7.14. In this design, a metallic disk (made of constantan alloy) is the primary means of heat transfer to and from the sample and reference. The sample, contained in a metal pan and the reference (an empty pan) sit on raised platforms formed in the constantan disk. As heat is transferred through the disk, the differential heat flow to the sample and the reference is measured by area thermocouples formed by the junction of the constantan disk and chromel wafers, which cover the underside of the platforms (Verdonck, 1999).

The thermocouples are connected in series to measure the differential heat flow using the thermal equivalent of Ohm's law: $dQ/dt = \Delta T/R_D$, where dQ/dt is the heat flow, ΔT the temperance difference between the reference and sample, and R_D is the thermal resistance of the constantan disk. Chromel and alumel wires attached to the chromel wafers form thermocouples that directly measure the sample temperature. Purge gas is admitted to the sample chamber through an orifice in the heating block before entering the sample chamber. The result is a uniform, stable thermal environment, which assures good baseline flatness and sensitivity (low signal-to-noise ratio; Noel et al., 2003). The temperature regime seen by the sample and reference is linear heating or cooling at rates as fast as 100°C/min to rates as slow as 0°C/min (isothermal).

The most common DSC application is the precise measurement of a transition temperature whether melting of a polymer or the polymorphic process of a neutraceutical (Dierckx and Huyghebaert, 2002). Thus, the formation of cold setting polysaccharide networks is arrested with DSC, as reproduced in Figure 7.15 for κ-carrageenan with increasing levels of cosolute. These span the range of solids from table jellies (30%–40%) to confectioneries (80%–90%) that may vitrify upon cooling (Izzo et al., 1995). Drawing a baseline underneath the exothermic peak and then

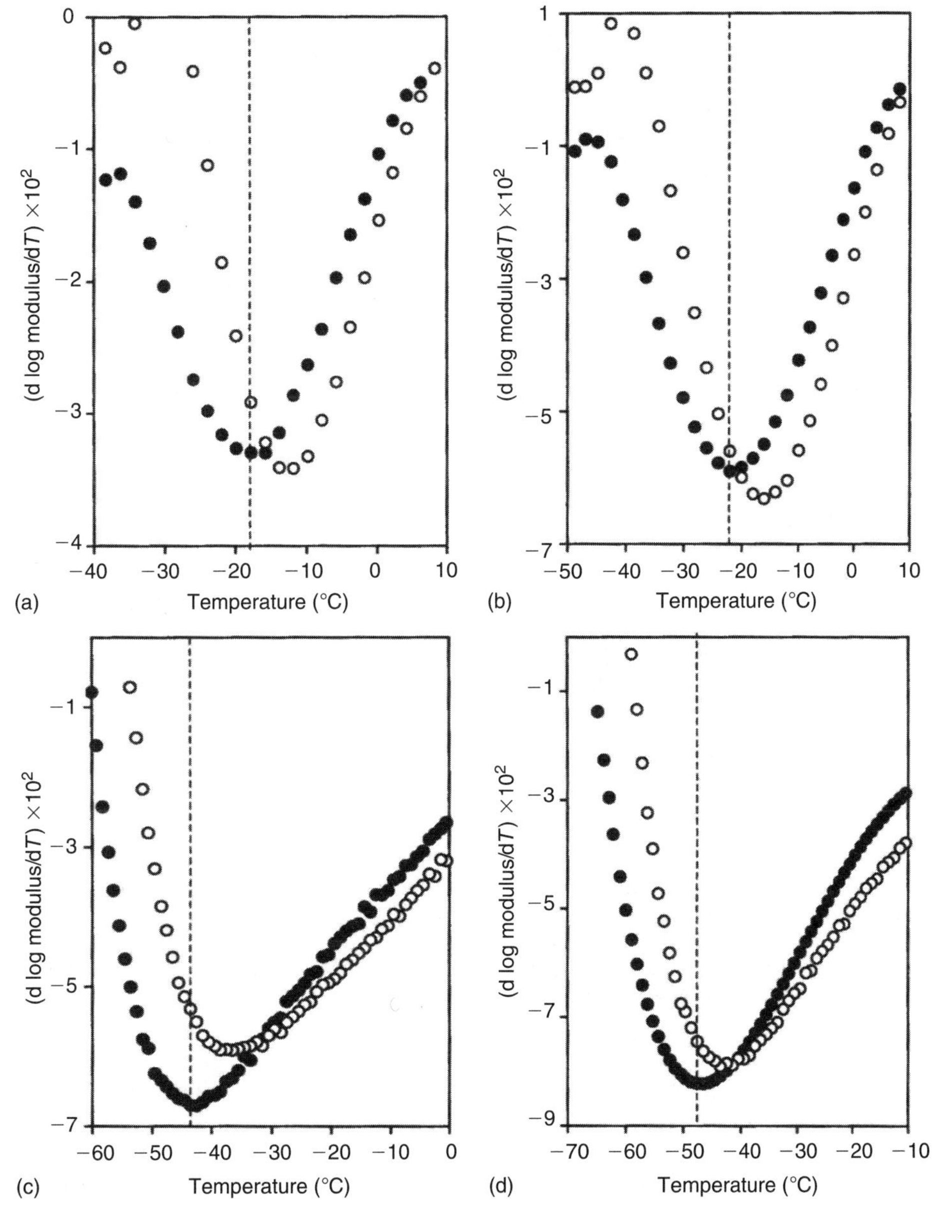

Figure 7.13 First-derivative plot of log G' (•) and log G'' (○) as a function of sample temperature plotted against temperature for (a) 70% tuna, (b) 70% abalone, (c) 67.2% apple, (d) 67% apricot

(continued)

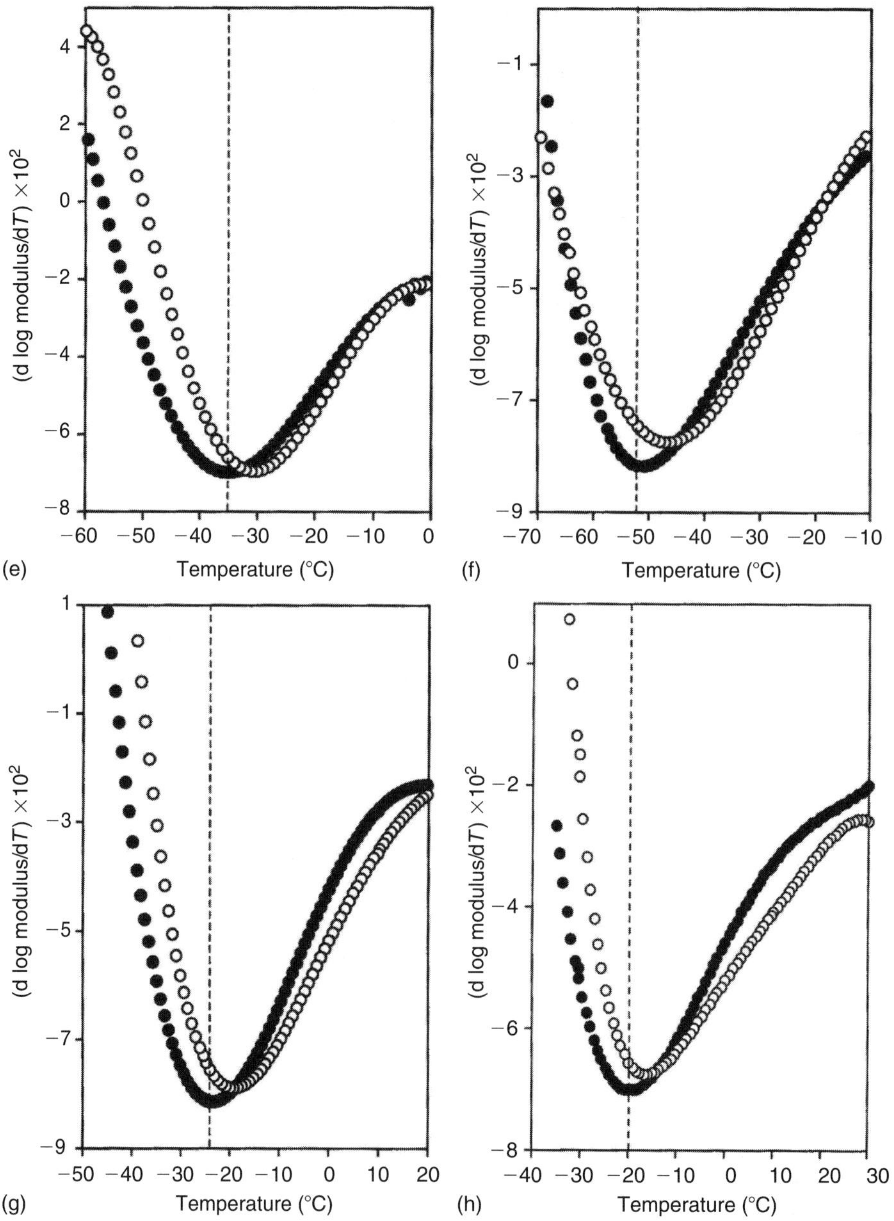

Figure 7.13 (continued) (e) 70.6% pear, (f) 60.9% plum, (g) 80.2% dates, and (h) 83.8% dates solids, respectively, with the dashed line pinpointing the T_g prediction.

subtracting this baseline from the experimental trace allows accurate estimation of the enthalpy (ΔH) and the midpoint temperature (T_m) of the gelation process.

There is a positive development in enthalpy from ≈26 to 34 J/g at 10% and 50% glucose syrup, respectively. The beginning of the transition is sharp (especially at 10% and 30% cosolute), thus suggesting a cooperative process of coil-to-aggregate formation (Kara et al., 2003). In accordance with optical rotation profiles, the end of the transition is quite broad and forms a tail, which is attributed to polydispersity and the formation of a heterogeneous network encapsulating a range of

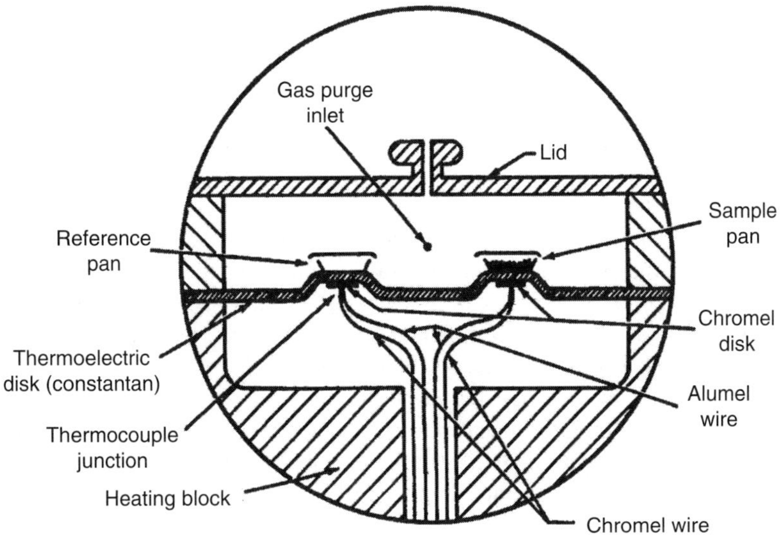

Figure 7.14 Heat flux schematic of DSC.

temperature-induced relaxation processes. At higher levels of solids, however, there is a drop in the enthalpy of the thermal transition, which is reduced to 13.8 J/g at 85% glucose syrup. Furthermore, a gradual ordering process develops on the high-temperature side of the peak, which argues for reduction in cooperativity, the exotherms now being perfectly symmetrical (80% and 85% cosolute). In spite of the diminishing order at the upper range of the glucose syrup, there is no abatement

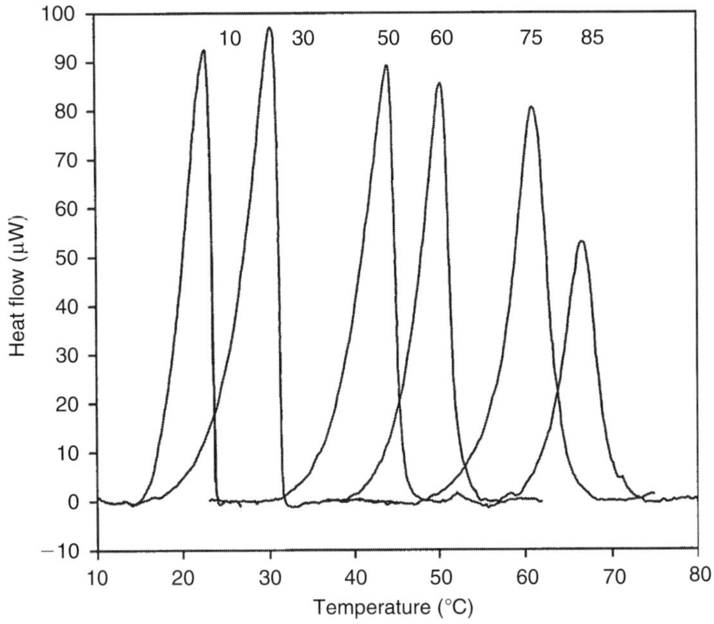

Figure 7.15 Cooling exotherms of 0.5% κ-carrageenan samples in the presence of added 0.01 M KCl and at levels of glucose syrup shown by the individual traces (scan rate: 0.1°C/min).

in the shift of exotherms to higher temperatures, with values following a linear dependence from ≈21.3°C to 66.3°C at 10% and 85% cosolute, respectively (Evageliou, 1998).

The decline in the magnitude of enthalpic transitions is accompanied by a drop in the mechanical strength of polysaccharide networks at the same levels of cosolute monitored by small deformation rheology. It appears, therefore, that the saturating levels of cosolute limit the availability of water molecules to the polysaccharide, molecules that are required for structure development (Chandrasekaran and Radha, 1995). This should prevent excessive aggregation, with the chain segments remaining largely in the disordered form. Large deformation compression further corroborates this hypothesis because addition of cosolute transforms the brittle polysaccharide networks into elastic structures that can stretch further before relaxing (Deszczynski et al., 2003a). Yield strain is recorded to be ∼0.3 and 1.0 units of strain for aqueous and confectionery formulations, respectively.

The transformation from enthalpic to entropic, lightly cross-linked networks with increasing additions of sugar allows monitoring at subzero temperatures of the rubber-to-glass transition described in Section 7.3. DSC can also trace vitrification processes by providing a direct, continuous measurement of a sample's heat capacity. Figure 7.16 illustrates the thermograms obtained during heating of glucose syrup preparations with a dextrose equivalent of 42 at 75% and 80% levels of solids. Small amounts of gelling polysaccharide, that is, 1% κ-carrageenan (30 mM added KCl), 1% high methoxy pectin (degree of esterification [DE] 92), 1% agarose, and 1% deacylated gellan (7 mM added $CaCl_2$) were also included in the formulations. Clearly, there is little change in the pattern of heat capacity in the presence of the polysaccharide, a result that is contrasted with the effect of the macromolecule on the rheological manifestation of the glass transition temperature (see the ensuing discussion on comparing calorimetric with rheological studies). On average, the values of DSC T_g in Figure 7.16 are about −47.9°C and −42°C at 75% and 80% solids, respectively.

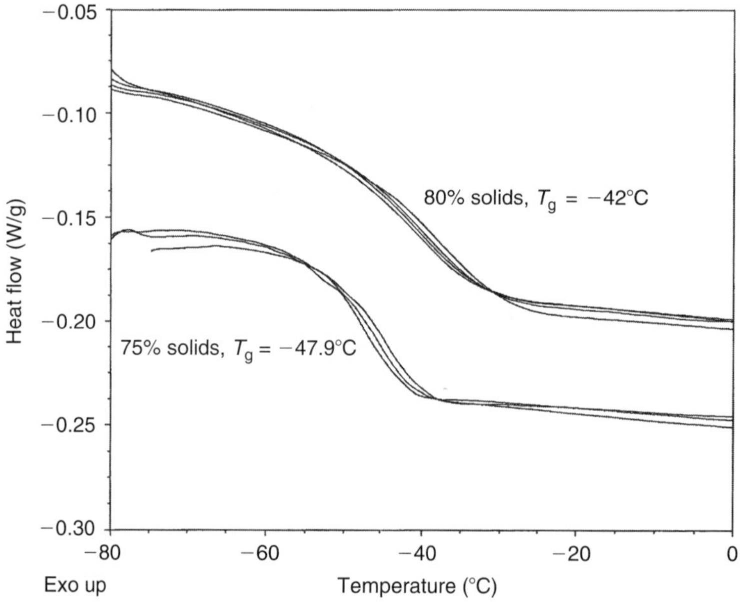

Figure 7.16 Devitrification recorded calorimetrically for samples of 75% solids (sugar, sugar + κ-carrageenan, sugar + high methoxy pectin) and 80% solids (sugar, sugar + κ-carrageenan, sugar + agarose, sugar + deacylated gellan) during heating at a scan rate of 1°C/min.

Clearly, water plasticizes food materials and DSC has been used widely to estimate the T_g values for sugars, oligosaccharides, proteins, maltodextrins, and starches at various levels of water content (Roos et al., 1996; Noel et al., 1999). In this respect, Gordon and Taylor proposed an empirical equation to predict the glass transition temperature of mixtures comprising amorphous synthetics. Today, this is commonly used to assess the vitrification properties of multicomponent biomaterials:

$$T_{gm} = \frac{X_s T_{gs} + k X_w T_{gw}}{X_s + k X_w} \quad (7.12)$$

where

T_{gm}, T_{gs}, and T_{gw} are the glass transition temperatures of the mixture, solids, and water, respectively

X_s and X_w are the mass fractions of solids and water

k is the Gordon–Taylor parameter, which from the thermodynamic standpoint is equivalent to the ratio of change of the component specific heat at their T_g (Couchman and Karasz, 1978)

Equation 7.12 has proven to be particularly useful in fitting experimental data on T_g of foodstuffs and model systems, thus greatly advancing the concept of a state diagram. This is useful in evaluating the effects of food composition on glass transition-related properties that affect the shelf life and quality (Roos and Karel, 1991; Rahman, 1999).

7.4.2 MDSC

Despite its utility, DSC does have some important limitations. In pure systems, different types of transitions such as melting and recrystallization in a semicrystalline material may overlap. In multicomponent systems, transitions of different compounds may partially overlap. In order to increase the sensitivity and resolution of thermal analysis, provide the heat capacity and heat flow in a single experiment, and measure the thermal conductivity, 10 years ago, MDSC was developed and commercialized. As a result, complex transitions can be separated into molecular processes with examples including the enthalpic relaxation that occurs at the glass transition region and changes in heat capacity during the exothermic cure reaction of a thermoset (Aubuchon et al., 1998).

Specific applications in synthetics and pharmaceuticals include the determination of the percentage of polymer crystallinity following processing (e.g., polyethylene terephthalate), the effect of plasticizer and water on vitrification temperature (e.g., polyvinyl chloride/nylon), the feasibility of shipping solid petroleum pitch at elevated temperatures, and the glass transition of amorphous drugs, etc. (Gmelin, 1997). More recently, the technique found a variety of applications in the research of biomaterials such as the glass transition of lactose, the subambient transitions of frozen sucrose, and the isothermal crystallization of concentrated starch systems (Truong et al., 2002).

MDSC is a technique that also measures the difference in heat flow between a sample and an inert reference as a function of time and temperature. In addition, the same heat flux cell design is used. However, in MDSC a different heating profile is applied to the sample and reference. Specifically, a sinusoidal modulation (oscillation) is overlaid on the conventional linear heating or cooling ramp to yield a profile in which the average sample temperature continuously changes with time but not in a linear fashion (Schawe, 1996). The solid line in Figure 7.17 shows the overall profile for an MDSC heating experiment. This is the net effect of imposing a complex heating profile on the sample and it can be analyzed in two simultaneously running experiments: one experiment at the traditional linear (average) heating rate (dashed line in Figure 7.17) and another at a sinusoidal (instantaneous) heating range (dashed-dot line in Figure 7.17). The actual rates for these two simultaneous experiments are dependent on three operator-selectable variables, which are applied mostly at the following ranges:

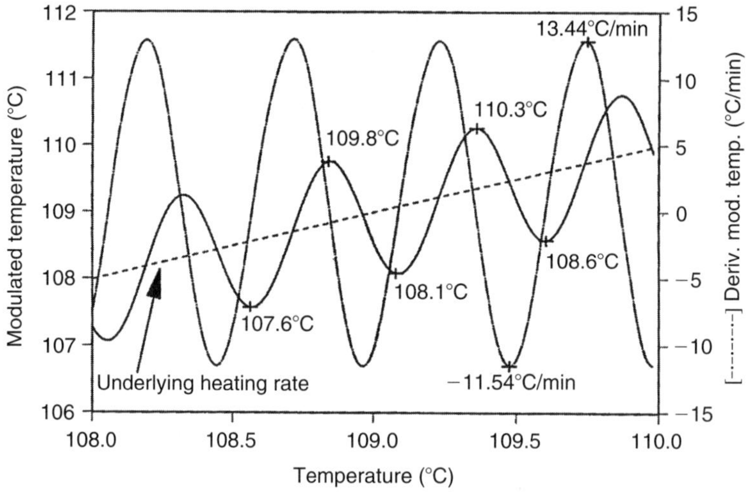

Figure 7.17 Typical MDSC heating profile.

1. Underlying heating rate (range 0°C/min–10°C/min)
2. Period of modulation (range 10–60 s)
3. Temperature amplitude of modulation (range ±0.01°C–2°C)

Although the actual sample temperature changes in a sinusoidal fashion during this process, the analyzed signals are ultimately plotted versus a linear temperature, which is calculated from the average value as measured by the sample thermocouple (essentially the dashed line in Figure 7.17).

The general equation that describes the resultant heat flow at any point of the experiment is

$$dQ/dt = C_p\beta + f(T,t) \tag{7.13}$$

where

dQ/dt is the total heat flow
C_p is the heat capacity
β is the heating rate
$f(T,t)$ is the heat flow from the kinetic processes (absolute temperature and time dependence)

From Equation 7.13, the total heat flow, measured also by conventional DSC, is composed of two components: (1) the sample's heat capacity and rate of temperature change and (2) absolute temperature and time. In the literature, the heat capacity component, $C_p\beta$, is known as reversing heat flow, whereas the kinetic component is referred to as nonreversing heat flow (Murase et al., 2002).

Due to the modulated heating rate, there exists a resultant modulated heat flow curve, and the total heat flow recorded as the final quantitative result is continuously calculated as the moving average of the raw modulated heat flow signal. The reversing component of the total heat flow is calculated by multiplying the measured heat capacity with the average (underlying) heating rate used in the experiment (Boller et al., 1995). The kinetic (nonreversing) component of the total heat flow is determined as the arithmetic difference between the total heat flow and the heat capacity component. Phenomena such as glass transitions and melting are reversing or heat capacity events. Nonreversing signals contain kinetic events such as crystallization, crystal perfection, and reorganization, cure, and decomposition (Sopade et al., 2002). The advantage MDSC has over standard DSC in resolving the two components of total heat flow thus better characterizing material properties is demonstrated in the determination of glass transition temperature of small polyhydric compounds like lactose. Often the event is masked by the evolution of water and enthalpic relaxation. MDSC is capable of resolving the glass transition in the reversing heat flow signal whereas the evolution of water and the enthalpic relaxation being kinetic phenomena are separated into the nonreversing heat flow signal (Goff et al., 2002).

7.4.3 Comparing Calorimetric with Rheological Studies

Earlier in Section 7.3, an index of physical significance was defined, which can be associated with the rheological glass transition temperature as the threshold of the free volume and reaction-rate theories. The fundamental index of T_g is, of course, frequency dependent. Ngai and Roland (2002) hold the view that the T_g of materials should be defined as the temperature at which the apparent relaxation times of the local segmental motions should be in the order of 100 s, that is, at least 10 orders of magnitude higher than the unrestrained time period for the relaxation of these primitive species (about 10^{-10} s). This school of thought was also implemented in the discussion surrounding Figures 7.6 through 7.9, 7.12, and 7.13, because results were obtained or extrapolated to an oscillatory frequency of 0.1 rad/s ($t = 62.8$ s).

It was also argued that this definition of T_g is more reliable than the values obtained from calorimetric measurements. It is true that there is no clear-cut relationship between molecular mobility and thermal events in calorimetric experiments, which forces researchers to resort to limiting factors in the form of T_{g1}, T_{g2}, and T_{g3} for the onset, middle, and completion of a particular case, respectively (Prolongo et al., 2002). MDSC thermograms reported in Figure 7.16 report glass transition temperatures at 75% and 80% solids at the middle on the pattern of heat capacity change (T_{g2}). Nevertheless, glass formation is in the nature of a second-order thermodynamic transition, which is accompanied by a heat capacity change and detected readily by calorimetry. Conventional DSC involves a constant scan rate, which imposes no frequency effects on the collected data. Normally, those effects are of no concern in MDSC provided the modulation period, which determines frequency, is constant when comparing materials (Schawe, 1995). Even in the situation where the period varies, frequency effects are insignificant because of the narrow range of useable periods of modulation indicated above (10–60 s).

However, calorimetrically determined glass transition temperatures are affected by the heating rate, which should be reported (Mazzobre et al., 2003). This is demonstrated in Figure 7.18 for the midpoint glass transition temperatures of gelatin films. These were hydrated at 65% relative humidity and analyzed using DSC at the heating rates of 4°C/min–32°C/min. In addition, Figure 7.18 includes data on the fundamental derivation of the rheological glass transition temperature (T_{gr}) discussed here and an empirical marker of T_g based on the tan δ peak observed in the glass transition region (Kasapis et al., 2003b). The mechanical data were recorded by cooling or heating a mixture of 1% κ-carrageenan (30 mM KCl added) with 81% glucose syrup at a rate of 1°C/min. Thermal profiles stretched from the rubbery plateau to the glassy state and, accordingly, the applied strain dropped from 1% to 0.0008%). Thermal runs were repeated seven times at the following experimental frequencies of oscillation: 0.1, 0.3, 1, 3, 10, 30, and 100 rad/s.

Gelatin data have been used in the last decade or so to advance a popular misconception on the comparison between the mechanical and thermal glass transitions temperatures. A linear fit of the DSC glass transition temperatures of gelatin as a function of heating rate was attempted as illustrated in Figure 7.18, and it was stated incorrectly that the zero heating rate T_g was approximately 46.5°C (Kalichevsky et al., 1993b). However, it is clear from Figure 7.18 that experimentation at lower heating rates would have followed an exponential T_g trend seen for the T_{gr} and tan δ_{max} values of κ-carrageenan–sugar mixtures at the low range of circular frequencies.

It was further argued, and it is today a popular belief among some practitioners in the field, that in synthetic polymers the rubber-to-glass transition obtained by dynamic mechanical techniques at

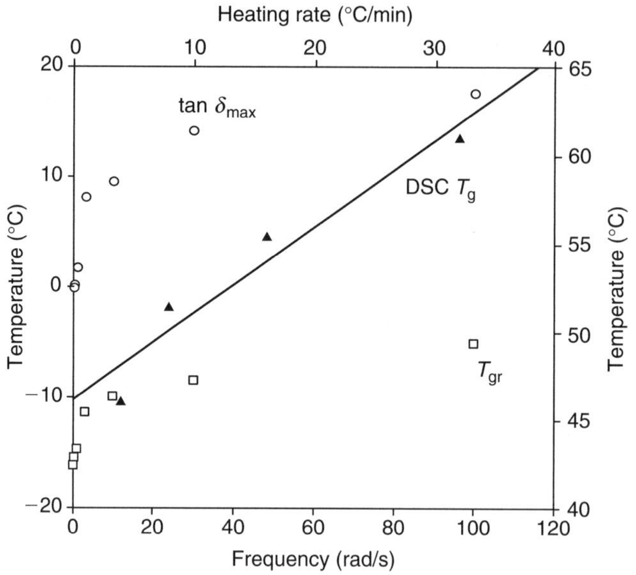

Figure 7.18 Effect of frequency of oscillation on T_{gr} (□) and tan δ_{max} (○) of a mixture of 1% κ-carrageenan (30 mM KCl added) with 81% glucose syrup (left-hand side *y*-axis), and the effect of heating rate on the DSC T_g for gelatin films (▲) hydrated at 65% relative humidity (right-hand side *y*-axis).

0.001 Hz corresponds to the DSC zero heating rate T_g (Braga da Cruz, 2002). Nevertheless, one is unable to find in the literature real evidence in support of this claim. Instead, Figure 7.19 reproduces the isothermal cure of an epoxy resin monitored with MDSC and DMTA at 1 Hz (Thomas and

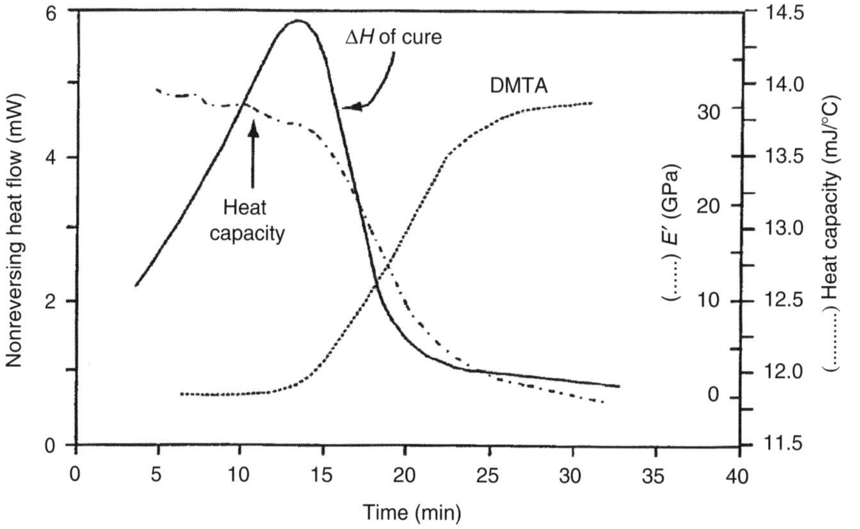

Figure 7.19 Isothermal cure of an epoxy thermoset at 80°C showing changes in heat capacity and nonreversing heat flow obtained at ±0.5°C amplitude and 60 s period of modulation, and in storage Young's modulus recorded with DMTA at a frequency of 1 Hz.

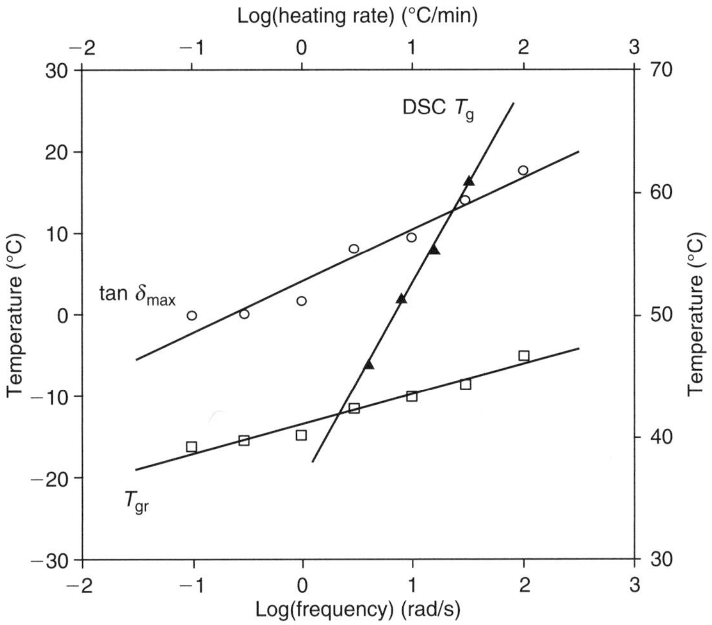

Figure 7.20 Semilogarithmic effects of frequency on T_{gr} and tan δ_{max} for the sugar–κ-carrageenan sample in Figure 7.18 (left-hand side *y*-axis), and of heating rate on the DSC T_g for gelatin films in Figure 7.18 (right-hand side *y*-axis).

Aubuchon, 1999). The exothermic peak of curing, that is, linear polymerization followed by cross-linking of the epoxy, coincides with a decrease in heat capacity due to reduction in diffusion mobility and free volume in the system. Evaluation by DMTA demonstrates that the heat capacity changes at exactly the same time with the increase in Young's modulus measured at 1 Hz, an outcome that invalidates the preceding hypothesis. Failure to recognize this feature has led to some erroneous conclusions about the applicability of comparison between thermal and mechanical vitrification phenomena.

In order to make meaningful allowances for the measuring principles used in each technique, the dependence of T_{gr} and tan δ_{max} on frequency and DSC T_g on heating rate have been plotted semilogarithmically in Figure 7.20. It should be remarked that the gradients of linear-log fits of mechanical and thermal events show the ongoing effect of the experimental parameters, a result that makes nonkinetic determinations of T_g no longer appropriate. Vitrification phenomena can be further manipulated by annealing at a temperature above T_g shown for DSC thermograms of fructose solutions with a level of solids between 60% and 77% (Ablett et al., 1993).

As a final demonstration of the care that must be taken in interpreting rheological and calorimetric data, a comparison is made between the T_g values of sugar and polysaccharide–sugar mixtures obtained at the same level of solids using both techniques. In Figure 7.21, the pattern of T_g variation of glucose syrup was established calorimetrically and then the effect of small additions of polysaccharide on the mobility of glucose syrup molecules was investigated. There is a 90°C increase in the DSC measured T_g of glucose syrup because the level of solids increases from 82% to 99%. Inclusions of 0.7% agarose or κ-carrageenan to glucose syrup preparations do not alter the values of T_g, which are dominated by the sugar spectrum (Shamblin et al., 1996).

This is in direct contrast to the increase in the values of the rheological T_g due to the addition of biopolymers to glucose syrup, shown for a range of solids from 80% to 85% (Kasapis et al., 2003c). Thus, addition of gelatin fractions of increased molecular weight (M_n), pectin of increased DE,

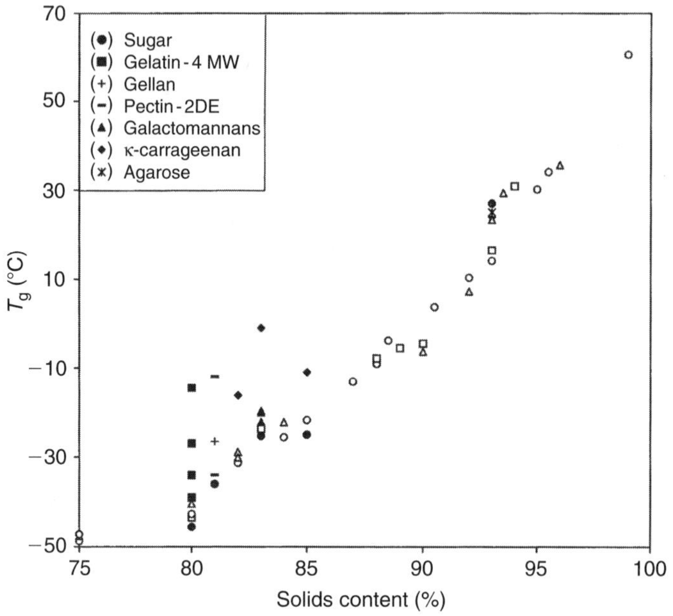

Figure 7.21 Variation of the calorimetric glass transition temperature with solids content of glucose syrup (○), 0.7% agarose + glucose syrup (□), and 0.7% κ-carrageenan + glucose syrup (△); and of the rheological glass transition temperature for single sugar samples (●) and in mixture with 25% gelatin (four M_n of 68, 55.8, 39.8, and 29.2 kDa) (■), 0.5% deacylated gellan (+), 1%–1.3% pectin (two DEs of 92 and 22) (—), 0.5%–1% galactomannans (▲), 0.5%–1% κ-carrageenan (◆), and 0.7% agarose (*) as indicated in the inset.

deacylated gellan with added sodium, and κ-carrageenan with added potassium, that is, characteristics that enhance gelation, accelerates the mechanical manifestation of vitrification in these systems. This is not the case for the agarose and κ-carrageenan samples with a solid content in excess of 90%, where there is no significant difference between rheology and calorimetry. Similarly, addition of nongelling polysaccharides such as guar or locust bean gum has little effect on the conventional trend of vitrification phenomena (galactomannans in the inset of Figure 7.21).

In conclusion, unlike the DSC T_g, the rheological T_g is affected by the nature of the biopolymer and cannot be predicted by the basic mathematical expressions for mixed systems, such as the Couchman–Karasz equation. It does appear that the apparent increase in the T_{gr} is related to the ability of the biopolymer to form a network (Deszczynski et al., 2003b). Thus increasing the removal of water from the polysaccharide network reduces the capacity for development of intermolecular associations to such an extent that at extremely high levels of solids (>90%) formation of a three-dimensional structure is abandoned. Network formation is a process that rheology is extremely well qualified to follow. In contrast, calorimetry provides information primarily on the mobility of the sugar molecules and the small addition of a biopolymer is a mere cross-contamination (Kasapis et al., 2004).

7.5 OTHER METHODS

It is quite remarkable that the free energy, volume, or enthalpy relaxations associated with glassy systems are universal phenomena and are not associated with specific details of chemical structure (Perez et al., 1992). This allows utilization of a wide range of techniques, beyond mechanical and thermal analysis, with a micro- or macrostructural nature. Thus, it is common to make reference to quantities such as free volume or configurational entropy as discussed earlier and obtain results, which have physical meaning from several methods (Angell, 1991). The following is not intended to be an exhaustive account of such methods but rather to convey briefly the typical information that emerges from such studies.

7.5.1 Dielectric Relaxation

Dynamic electric analyzers are used to measure the dielectric permitivity (ε') and the dielectric loss factor (ε'') with the complex dielectric constant (ε^*) being expressed by the equation: $\varepsilon^* = \varepsilon' - i\varepsilon''$ (Montserrat et al., 2003). The permitivity depends on the orientation polarization, which is the outcome of (1) changes in the dipole moment due to the chemical reaction and (2) the concentration of the dipoles due to the volume contraction during the process of vitrification. The loss factor is a measure of the energy loss. In synthetic polymer research, a common objective of these studies is to relate the kinetics of chemical and structural events with dielectric parameters (Nandan et al., 2003). Dielectric relaxation during vitrification has been examined in terms of the diffusion of functional groups and quantitative interpretation was attempted with modified Cole's, WLF-type, or other empirical equations.

This technique can offer insights beyond rheology and it has been ascertained that the orientation dynamics of pendant groups is governed by the network characteristics of the polymeric matrix (Ribierre et al., 2003). In addition, dielectric relaxation employs a wide frequency range (e.g., 10 Hz–100 kHz), compared with that of rheology, thus being able to monitor the relaxation of small dipolar groups and to correlate it with secondary vitrification processes recorded by MDSC. Work culminates in the calculation of activation energies for motions of molecular groups, which are influenced by specific interactions or partial segmental mixing between the components of a blend (Lechuga-Ballesteros et al., 2002). In the case of freeze-dried sucrose, spectra indicate two frequency-dependent groups believed to measure the cooperative motions of whole sucrose molecules.

7.5.2 Nuclear Magnetic Resonance

The growing recognition of the importance of mono-, oligo-, and polysaccharides to neutraceuticals, drugs, and biological systems initiated a drive for the identification of a technique that can cope with the micromolecular aspects of the enormous complexity of these compounds. Nuclear magnetic resonance (NMR) appears to be a technique of promise, and sugars are mainly investigated with ^{1}H NMR. These are usually deuterium-exchanged preparations; therefore, only C-linked protons are detected. In synthetic science, NMR is used in the characterization of compounds such as polyindene and its hydrogenation product (polyolefins), and then a relationship can be established between the molecular weight of the product and glass transition temperature (Hahn and Hillmyer, 2003).

Experiments have been performed in binary mixtures of amylopectin at various levels of fructose, glucose, sucrose, or xylose with the concentration of the polymeric component being kept higher than that of the cosolute (Kalichevsky et al., 1993a). It was found that small additions of sugar reduced the glass transition temperature of the polysaccharide according to the postulates of the Couchman–Karasz equation. Nevertheless, mixtures containing the highest amount of cosolute (amylopectin/sugar ratio = 2:1) exhibited less plasticization than expected because of partial phase separation between the two components. Changing dramatically the type of solids content generates systems at which the biopolymer-to-sugar ratio is well below 1, for example, 1% κ-carrageenan with 80%–95% sugar. Contrary to the acceleration of vitrification reported rheologically in the presence of a gelling agent, the addition of a biopolymer has no effect on the T_g or the molecular mobility in the glassy state of the sugar sample (Kumagai et al., 2002). It should be emphasized that it remains difficult to accurately deconvolute the spectrum of a mixture of carbohydrates or a multicomponent system comprising protein, carbohydrate, and water (Reid, 2002).

7.5.3 Thermogravimetric Analysis

Thermogravimetric analysis (TGA) is used to follow the weight loss of a sample as a function of temperature under vacuum or pressure and in a variety of gaseous atmospheres (Ma et al., 1990). Weight loss can be recorded as percentile within the temperature region of 0°C–500°C with a linear heating rate of up to 10°C/min. Samples can be as small as 2–5 mg. A weight loss process is defined as the amount of material lost from a sample between two zero points around a peak in the first derivative of the weight loss versus temperature curve. Thus, TGA can help elucidate various transitions due to loss of surface water, loss of hydration water, the occurrence of chemical processes like the demethanolation of aspartame, and the in vitro fermentation characteristics of maize (Marvin et al., 1996). Application of the technique to soy protein isolates suggests that protein films exhibit considerable thermal degradation above 180°C, thus limiting the temperature of product development to 150°C. Introduction of various levels of glycerol to the soy films indicates that the higher the initial level of plasticizer the lower the final weight percent remaining. TGA can be used in combination with mechanical studies and microscopy to pinpoint the glass transition temperature in biomaterials (Ogale et al., 2000).

7.5.4 X-Ray Diffractometry

X-ray diffractometers are used mainly to measure the relative crystallinity of starch samples. This technique is particularly useful in the assessment of the loss of crystallinity of extruded starch for pasta, snack, or breakfast-cereal processing (Donald et al., 1993). Typical diffractograms show a series of peaks and each separated point of minimum intensity at the beginning and the end of the peak is joined by a straight line. The area between the baseline and the peak is calculated and for a particular type of starch the following is verified for the level of crystallinity: native starch > retrograded starch > gelatinized starch (Mizuno et al., 1998). Best results are recorded for anhydrous crystals suspended in an amorphous matrix that retains the available water. Reliable quantitative results are obtained at levels of crystallization between 10% and 90% in the system.

7.5.5 FTIR Microspectroscopy

There is relatively little information on the infrared applications to vitrification phenomena. Farhat et al. (2003) studied the degree of heterogeneity in extrudates of amylopectin in the presence of sucrose. Thin films (5–10 μm) were cut by cryomicrotoming and transmission FTIR measurements were obtained keeping the samples in a humidified air environment to avoid drying of the films. Despite the thorough mixing conveyed by the twin-screw extrusion process, the technique revealed the extent of heterogeneity in the form of fluctuations in the sucrose content across the samples. The heterogeneous character of the mixtures was further verified by the appearance of multiple transitions in the DMTA thermograms.

FTIR was also employed in the study of the chain conformation in wafers, which are low moisture (about 6%) starch-based systems (Livings et al., 1993). Work focused on wafers aged for 1 year at room temperature due to the detrimental loss in texture. It was documented that the glass transition temperature lies well above the room temperature and thus the textural changes during aging occur below T_g. There was further confirmation of the DSC and x-ray scattering results that on heating an aged wafer to 100°C there is an irreversible thermal change. The technique can be developed further by improving the spatial resolution and devising a normal mode of analysis.

ACKNOWLEDGMENTS

The author would like to express his gratitude to Drs. Shafiur Rahman, SQU and Marcin Deszczynski, University of Nottingham for encouragement and critical evaluation of this manuscript.

REFERENCES

Ablett, S., Izzard, M.J., Lillford, P.J., Arvanitoyannis, I., and Blanshard, J.M.V. Calorimetric study of the glass transition occurring in fructose solutions. *Carbohydrate Research*, 246, 13–22, 1993.

Allen, G. A history of the glassy state, in *The Glassy State in Foods*, J.M.V. Blanshard and P.J. Lillford (eds.), Nottingham University Press, Nottingham, 1993, pp. 1–12.

Al-Ruqaie, I.M., Kasapis, S., Richardson, R.K., and Mitchell, G. The glass transition zone in high solids pectin and gellan preparations. *Polymer*, 38, 5685–5694, 1997.

Angell, C.A. Perspective on the glass transition. *Journal of Physics and Chemistry of Solids*, 9, 863–871, 1988.

Angell, C.A. Relaxation in liquids, polymers and plastic crystals—strong/fragile patterns and problems. *Journal of Non-Crystalline Solids*, 131–133, 1991.

Atkins, P.W. Molecules in motion: The kinetic theory of gases, in *Physical Chemistry*, Oxford University Press, Oxford, 1984, pp. 859–887.

Aubuchon, S.R., Thomas, L.C., Theuerl, W., and Renner, H. Investigations of the sub-ambient transitions in frozen sucrose by modulated differential scanning calorimetry (MDSC®). *Journal of Thermal Analysis*, 52, 53–64, 1998.

Bai, Y., Rahman, M.S., Perera, C.O., Smith, B., and Melton, L.D. State diagram of apple slices: Glass transition and freezing curves. *Food Research International*, 34, 89–95, 2001.

Bhandari, B.R. and Roos, Y.H. Dissolution of sucrose crystals in the anhydrous sorbitol melt. *Carbohydrate Research*, 338, 361–367, 2003.

Bhandari, B., D'Arcy, B., and Kelly, C. Rheology and crystallization kinetics of honey: Present status. *International Journal of Food Properties*, 2, 217–226, 1999.

Biliaderis, C.G. Differential scanning calorimetry in food research—a review. *Food Chemistry*, 10, 239–265, 1983.

Binder, K., Baschnagel, J., and Paul, W. Glass transition of polymer melts: Test of theoretical concepts by computer simulation. *Progress in Polymer Science*, 28, 115–172, 2003.

Boller, A., Schick, C., and Wunderlich, B. Modulated differential scanning calorimetry in the glass transition region. *Thermochimica Acta*, 266, 97–111, 1995.

Braga da Cruz, I., MacInnes, W.M., Oliveira, J.C., and Malcata, F.X. Supplemented state diagram for sucrose from dynamic mechanical thermal analysis, in *Amorphous Food and Pharmaceutical Systems*, H. Levine (ed.), The Royal Society of Chemistry, Cambridge, 2002, pp. 59–70.

Brent, Jr. J.L., Mulvaney, S.J., Cohen, C., and Bartsch, J.A. Thermomechanical glass transition of extruded cereal melts. *Journal of Cereal Science*, 26, 301–312, 1997.

Cangialosi, D., Schut, H., van Veen, A., and Picken, S.J. Positron annihilation lifetime spectroscopy for measuring free volume during physical aging of polycarbonate. *Macromolecules*, 36, 142–147, 2003.

Carr, J.M., Sufferling, K., and Poppe, J. Hydrocolloids and their use in the confectionery industry. *Food Technology*, 41–42, 44, 1995.

Cavaille, J.Y., Jordan, C., Perez, J., Monnerie, L., and Johari, G.P. Time-temperature superposition and dynamic mechanical behaviour of atactic polystyrene. *Journal of Polymer Science: Part B: Polymer Physics*, 25, 1235–1251, 1987.

Chandrasekaran, R. and Radha, A. Molecular architectures and functional properties of gellan gum and related polysaccharides. *Trends in Food Science & Technology*, 6, 143–148, 1995.

Christensen, R.M. A thermodynamical criterion for the glass-transition temperature. *Transactions of the Society of Rheology*, 21(2), 163–181, 1977.

Couchman, P.R. and Karasz, F.E. A classical thermodynamic discussion of the effect of composition on glass-transition temperatures. *Macromolecules*, 11, 117–119, 1978.

Cruz, I.B., Oliveira, J.C., and MacInnes, W.M. Dynamic mechanical thermal analysis of aqueous sugar solutions containing fructose, glucose, sucrose, maltose and lactose. *International Journal of Food Science and Technology*, 36, 539–550, 2001.

Dannhauser, W., Child, Jr. W.C., and Ferry, J.D. Dynamic mechanical properties of poly-*n*-octyl-methacrylate. *Journal of Colloid Science*, 13, 103–113, 1958.

DeMars, L.L. and Ziegler, G.R. Texture and structure of gelatin/pectin-based gummy confections. *Food Hydrocolloids*, 15, 643–653, 2001.

Deszczynski, M. Fundamental and technological aspects of high sugar/biopolymer mixtures. PhD thesis, University of Nottingham, Sutton Bonington, 2003.

Deszczynski, M., Kasapis, S., MacNaughton, W., and Mitchell, J.R. High sugar/polysaccharide glasses: resolving the role of water molecules in structure formation. *International Journal of Biological Macromolecules*, 30, 279–282, 2002.

Deszczynski, M., Kasapis, S., and Mitchell, J.R. Rheological investigation of the structural properties and aging effects in the agarose/co-solute mixture. *Carbohydrate Polymers*, 53, 85–93, 2003a.

Deszczynski, M., Kasapis, S., MacNaughtan, W., and Mitchell, J.R. Effect of sugars on the mechanical and thermal properties of agarose gels. *Food Hydrocolloids*, 17, 793–799, 2003b.

D'Haene, P. and Van Liederkerke, B. Viscosity prediction of starch hydrolysates from single point measurements. *Starch*, 48, 327–334, 1996.

Dierckx, S. and Huyghebaert, A. Effects of sucrose and sorbitol on the gel formation of a whey protein isolate. *Food Hydrocolloids*, 16, 489–497, 2002.

Dlubek, G., Fretwell, H.M., and Alam, M.A. Positron/positronium annihilation as a probe for the chemical environment of free volume holes in polymers. *Macromolecules*, 33, 187–192, 2000.

Dlubek, G., Bondarenko, V., Pionteck, J., Supej, M., Wutzler, A., and Krause-Rehberg, R. Free volume in two differently plasticized poly(vinyl chloride)s: a positron lifetime and PVT study. *Polymer*, 44, 1921–1926, 2003.

Donald, A.M., Warburton, S.C., and Smith, A.C. Physical changes consequent of the extrusion of starch, in *The Glassy State in Foods*, J.M.V. Blanshard and P.J. Lillford (eds.), Nottingham University Press, Nottingham, 1993, pp. 375–393.

Doolittle, A.K. and Doolittle, D.B. Studies in Newtonian flow. V. Further verification of the free-space viscosity equation. *Journal of Applied Physics*, 28, 901–905, 1957.

Evageliou, V., Kasapis, S., and Hember, M.W.N. Vitrification of κ-carrageenan in the presence of high levels of glucose syrup. *Polymer*, 39, 3909–3917, 1998.

Farhat, I.A., Mousia, Z., and Mitchell, J.R. Structure and thermomechanical properties of extruded amylopectin-sucrose systems. *Carbohydrate Polymers*, 52, 29–37, 2003.

Favier, V., Chanzy, H., and Cavaillé, J.Y. Polymer nanocomposites reinforced by cellulose whiskers. *Macromolecules*, 28, 6365–6367, 1995.

Ferry, J.D. *Viscoelastic Properties of Polymers*, John Wiley, New York, 1980.

Ferry, J.D. Some reflections on the early development of polymer dynamics: viscoelasticity, dielectric dispersion, and self-diffusion. *Macromolecules*, 24, 5237–5245, 1991.

Frick, B. and Richter, D. The microscopic basis of the glass transition in polymers from neutron scattering studies. *Science*, 267, 1939–1947, 1995.

Gmelin, E. Classical temperature-modulated calorimetry: A review. *Thermochimica Acta*, 304/305, 1–26, 1997.

Goff, H.D., Montoya, K., and Sahagian, M.E. The effect of microstructure on the complex glass transition occurring in frozen sucrose model systems and foods, in *Amorphous Food and Pharmaceutical Systems*, H. Levine (ed.), The Royal Society of Chemistry, Cambridge, 2002, pp. 145–157.

Gunning, Y.M., Parker, R., and Ring, S.G. Diffusion of short chain alcohols from amorphous maltose-water mixtures above and below their glass transition temperature. *Carbohydrate Research*, 329, 377–385, 2000.

Haase, R. Thermodynamic properties of gases, liquids, and solids, in *Physical Chemistry—An Advanced Treatise: Thermodynamics*, H. Eyring, D. Henderson, and W. Jost (eds.), Academic Press, New York, 1971, pp. 293–365.

Hahn, S.F. and Hillmyer, M.A. High glass transition temperature polyolefins obtained by the catalytic hydrogenation of polyindene. *Macromolecules*, 36, 71–76, 2003.

Huang, Y., Szleifer, I., and Peppas, N.A. A molecular theory of polymer gels. *Macromolecules*, 35, 1373–1380, 2002.

Hutchinson, J.M. Physical aging of polymers. *Progress in Polymer Science*, 20, 703–760, 1995.

Izzo, M., Stahl, C., and Tuazon, M. Using cellulose gel and carrageenan to lower fat and calories in confections. *Food Technology*, 49(7), 45–49, 1995.

Kalichevsky, M.T. and Blanshard, J.M.V. A study of the effect of water on the glass transition of 1:1 mixtures of amylopectin, casein and gluten using DSC and DMTA. *Carbohydrate Polymers*, 19, 271–278, 1992.

Kalichevsky, M.T. and Blanshard, J.M.V. The effect of fructose and water on the glass transition of amylopectin. *Carbohydrate Polymers*, 20, 107–113, 1993.

Kalichevsky, M.T., Jaroszkiewicz, E.M., Ablett, S., Blanshard, J.M.V., and Lillford, P.J. The glass transition of amylopectin measured by DSC, DMTA and NMR. *Carbohydrate Polymers*, 18, 77–88, 1992.

Kalichevsky, M.T., Jaroszkiewicz, E.M., and Blanshard, J.M.V. A study of the glass transition of amylopectin-sugar mixtures. *Polymer*, 34, 346–358, 1993a.

Kalichevsky, M.T., Blanshard, J.M.V., and Marsh, R.D.L. Applications of mechanical spectroscopy to the study of glassy biopolymers and related systems, in *The Glassy State in Foods*, J.M.V. Blanshard and P.J. Lillford (eds.), Nottingham University Press, Nottingham, 1993b, pp. 133–156.

Kara, S., Tamerler, C., Bermek, H., and Pekcan, O. Cation effects on sol-gel and gel-sol phase transitions of κ-carrageenan-water system. *International Journal of Biological Macromolecules*, 31, 177–185, 2003.

Kasapis, S. Structural properties of high solids biopolymer systems, in *Functional Properties of Food Macromolecules*, S.E. Hill, D.A. Ledward, and J.R. Mitchell (eds.), Aspen, Gaithersburg, 1998, pp. 227–251.

Kasapis, S. Critical assessment of the application of the WLF/free volume theory to the structural properties of high solids systems: A review. *International Journal of Food Properties*, 4, 59–79, 2001a.

Kasapis, S. Advanced topics in the application of the WLF/free volume theory to high sugar/biopolymer mixtures: a review. *Food Hydrocolloids*, 15, 631–641, 2001b.

Kasapis, S. Definition of a mechanical glass transition temperature for dehydrated foods. *Journal of Agricultural and Food Chemistry*, 52, 2262–2268, 2004.

Kasapis, S., Al-Marhoobi, I.M.A., and Giannouli, P. Molecular order versus vitrification in high-sugar blends of gelatin and κ-carrageenan. *Journal of Agricultural and Food Chemistry*, 47, 4944–4949, 1999.

Kasapis, S., Al-Marhoobi, I.M.A., and Sworn, G. α and β mechanical dispersions in high sugar/acyl gellan mixtures. *International Journal of Biological Macromolecules*, 29, 151–160, 2001.

Kasapis, S., Al-Marhoobi, I.M., and Mitchell, J.R. Molecular weight effects on the glass transition of gelatin/co-solute mixtures. *Biopolymers*, 70, 169–185, 2003a.

Kasapis, S., Al-Marhoobi, I.M., and Mitchell, J.R. Testing the validity of comparisons between the rheological and the calorimetric glass transition temperatures. *Carbohydrate Research*, 338, 787–794, 2003b.

Kasapis, S., Al-Marhoobi, I.M., Deszczynski, M., Mitchell, J.R., and Abeysekera, R. Gelatin vs. polysaccharide in mixture with sugar. *Biomacromolecules*, 4, 1142–1149, 2003c.

Kasapis, S., Mitchell, J., Abeysekera, R., and MacNaughtan, W. Rubber-to-glass transitions in high sugar/biopolymer mixtures. *Trends in Food Science & Technology*, 15, 298–304, 2004.

Kovacs, A.J. Transition vitreuse dans les polymères amorphes. Etude phènomènologique. *Advances in Polymer Science*, 3, 394–507, 1964.

Kumagai, H., MacNaughtan, W., Farhat, I.A., and Mitchell, J.R. The influence of carrageenan on molecular mobility in low moisture amorphous sugars. *Carbohydrate Polymers*, 48, 341–349, 2002.

Lazaridou, A. and Biliaderis, C.G. Thermophysical properties of chitosan, chitosan-starch and chitosan-pullulan films near the glass transition. *Carbohydrate Polymers*, 48, 179–190, 2002.

Lechuga-Ballesteros, D., Miller, D.P., and Zhang, J. Residual water in amorphous solids: measurement and effects on stability, in *Amorphous Food and Pharmaceutical Systems*, H. Levine (ed.), The Royal Society of Chemistry, Cambridge, 2002, pp. 275–316.

Li, L. and Yee, A.F. Effect of the scale of local segmental motion on nanovoid growth in polyester copolymer glasses. *Macromolecules*, 36, 2793–2801, 2003.

Liu, H., Qi, J., and Hayakawa, K. Rheological properties including tensile fracture stress of semolina extrudates influenced by moisture content. *Journal of Food Science*, 62, 813–815, 820, 1997.

Livings, S.J., Donald, A.M., and Smith, A.C. Ageing in confectionery wafers, in *The Glassy State in Foods*, J.M.V. Blanshard and P.J. Lillford (eds.), Nottingham University Press, Nottingham, 1993, pp. 507–511.

Ma, C.-Y., Harwalkar, V.R., and Maurice, T.J. Instrumentation and techniques of thermal analysis in food research, in *Thermal Analysis of Foods*, V.R. Harwalkar and C.-Y. Ma (eds.), Elsevier, London, 1990, pp. 1–15.

Maltini, E. and Anese, M. Evaluation of viscosities of amorphous phases in partially frozen systems by WLF kinetics and glass transition temperatures. *Food Research International*, 28, 367–372, 1995.

Marvin, R.S. The dynamic mechanical properties of polyisobutylene, in *Proceedings of the Second International Congress Rheology*, V.G. Harrison (ed.), Butterworth, London, 1954, pp. 156–163.

Marvin, H.J.P., Krechting, C.F., van Loo, E.N., Snijders, C.H.A., Nelissen, L.N.I.H., and Dolstra, O. Potential of thermal analysis to estimate chemical composition and in vitro fermentation characteristics of maize. *Journal of Agricultural and Food Chemistry*, 44, 3467–3473, 1996.

Matveev, Y.I., Grinberg, V.Y., and Tolstoguzov, V.B. The plasticizing effect of water on proteins, polysaccharides and their mixtures. Glassy state of biopolymers, food and seeds. *Food Hydrocolloids*, 14, 425–437, 2000.

Mazzobre, M.F., Aguilera, J.M., and Buera, M.P. Microscopy and calorimetry as complementary techniques to analyze sugar crystallisation from amorphous systems. *Carbohydrate Research*, 338, 541–548, 2003.

Mitchell, J.R. Hydrocolloids in low water and high sugar environments, in *Gums and Stabilisers for the Food Industry 10*, P.A. Williams and G.O. Phillips (eds.), The Royal Society of Chemistry, Cambridge, 2000, pp. 243–254.

Mizuno, A., Mitsuiki, M., and Motoki, M. Effect of crystallinity on the glass transition temperature of starch. *Journal of Agricultural and Food Chemistry*, 46, 98–103, 1998.

Momany, F.A. and Willett, J.L. Molecular dynamics calculations on amylose fragments. I. Glass transition temperatures of maltodecaose at 1, 5, 10, and 15.8% hydration. *Biopolymers*, 63, 99–110, 2002.

Montserrat, S. and Hutchinson, J.M. On the measurement of the width of the distribution of relaxation times in polymer glasses. *Polymer*, 43, 351–355, 2002.

Montserrat, S., Roman, F., and Colomer, P. Vitrification and dielectric relaxation during the isothermal curing of an epoxy-amine resin. *Polymer*, 44, 101–114, 2003.

Murase, N., Ruike, M., Yoshioka, S., Katagiri, C., and Takahashi, H. Glass transition and ice crystallisation of water in polymer gels, studied by Oscillation DSC, XRD-DSC simultaneous measurements, and Raman spectroscopy, in *Amorphous Food and Pharmaceutical Systems*, H. Levine (ed.), The Royal Society of Chemistry, Cambridge, 2002, pp. 339–346.

Nandan, B., Kandpal, L.D., and Mathur, G.N. Glass transition behaviour of poly(ether ether ketone)/poly(aryl ether sulphone) blends: dynamic mechanical and dielectric relaxation studies. *Polymer*, 44, 1267–1279, 2003.

Neway, B., Hedenqvist, M.S., and Gedde, U.W. Effect of thermal history on free volume and transport properties of high molar mass polyethylene. *Polymer*, 44, 4003–4009, 2003.

Ngai, K.L. Synergy of entropy and intermolecular coupling in supercooling liquids. *Journal of Chemical Physics*, 111, 3639–3643, 1999.

Ngai, K.L. Dynamic and thermodynamic properties of glass-forming substances. *Journal of Non-Crystalline Solids*, 275, 7–51, 2000a.

Ngai, K.L. Short-time and long-time relaxation dynamics of glass-forming substances: A coupling model perspective. *Journal of Physics: Condensed Matter*, 12, 6437–6451, 2000b.

Ngai, K.L. and Plazek, D.J. Identification of different modes of molecular motion in polymers that cause thermorheological complexity. *Rubber Chemistry and Technology*, 68, 376–434, 1995.

Ngai, K.L. and Rendell, R.W. Cooperative dynamics in relaxation: A coupling model perspective. *Journal of Molecular Liquids*, 56, 199–214, 1993.

Ngai, K.L. and Roland, C.M. Development of cooperativity in the local segmental dynamics of poly(vinylacetate): Synergy of thermodynamics and intermolecular coupling. *Polymer*, 43, 567–573, 2002.

Ngai, K.L., Magill, J.H., and Plazek, D.J. Flow, diffusion and crystallization of supercooled liquids: Revisited. *Journal of Chemical Physics*, 112, 1887–1892, 2000.

Noel, T.R., Ring, S.G., and Whittam, M.A. Kinetic aspects of the glass-transition behaviour of maltose-water mixtures. *Carbohydrate Research*, 212, 109–117, 1991.

Noel, T.R., Parker, R., Ring, S.M., and Ring, S.G. A calorimetric study of structural relaxation in a maltose glass. *Carbohydrate Research*, 319, 166–171, 1999.

Noel, T.R., Parker, R., and Ring, S.G. Effect of molecular structure on the conductivity of amorphous carbohydrate-water-KCl mixtures in the supercooled liquid state. *Carbohydrate Research*, 338, 433–438, 2003.

Normand, V., Lootens, D.L., Amici, E., Plucknett, K.P., and Aymard, P. New insight into agarose gel mechanical properties. *Biomacromolecules*, 1, 730–738, 2000.

Norton, I.T. and Foster, T.J. Hydrocolloids in real food systems, in *Gums and Stabilisers for the Food Industry 11*, P.A. Williams and G.O. Phillips (eds.), The Royal Society of Chemistry, Cambridge, 2002, pp. 187–200.

Ogale, A.A., Cunningham, P., Dawson, P.L., and Acton, J.C. Viscoelastic, thermal and microstructural characterisation of soy protein isolate films. *Journal of Food Science*, 65, 672–679, 2000.

Ollett, A.-L., Parker, R., and Smith, A.C. Mechanical Properties of wheat starch plasticized with glucose and water, in *Food Polymers, Gels, and Colloids*, E. Dickinson (ed.), The Royal Society of Chemistry, Cambridge, 1991, pp. 537–541.

Ong, M.H., Whitehouse, A.S., Abeysekera, R., Al-Ruqaie, I.M., and Kasapis, S. Glass transition-related or crystalline forms in the structural properties of gelatin/oxidised starch/glucose syrup mixtures. *Food Hydrocolloids*, 12, 273–281, 1998.

Pak, J., Pyda, M., and Wunderlich, B. Rigid amorphous fractions and glass transitions in poly(oxy-2,6-dimethyl-1,4-phenylene). *Macromolecules*, 36, 495–499, 2003.

Parks, G.S. and Huffman, H.M. Glass as a fourth state of matter. *Science*, LXIV, 363–364, 1926.

Peleg, M. On the use of the WLF model in polymers and foods. *Critical Reviews in Food Science and Nutrition*, 32, 59–66, 1992.

Peleg, M. A note on the tan δ (T) peak as a glass transition indicator in biosolids. *Rheological Acta*, 34, 215–220, 1995.

Perez, J., Muzeau, E., and Cavaille, J.Y. α and β mechanical relaxations in amorphous polymers: Rubber-glass transition and physical aging. *Plastics, Rubber and Composites Processing and Applications*, 18, 139–148, 1992.

Perry, P.A. and Donald, A.M. The effect of sugars on the gelatinisation of starch. *Carbohydrate Polymers*, 49, 155–165, 2002.

Pomeranz, Y. Part II: Engineering foods, in *Functional Properties of Food Components*, Academic Press, San Diego, 1991, pp. 331–525.

Poppe, J. Gelatin, in *Thickening and Gelling Agents for Food*, A. Imeson (ed.), Chapman and Hall, Glasgow, 1992, pp. 98–132.

Plazek, D.J. 1995 Bingham medal address: Oh, thermorheological simplicity, wherefore art thou? *Journal of Rheology*, 40, 987–1015, 1996.

Plazek, D.J. and Ngai, K.L. Correlation of polymer segmental chain dynamics with temperature-dependent time-scale shifts. *Macromolecules*, 24, 1222–1224, 1991.

Plazek, D.J., Chay, I.-C., Ngai, K.L., and Roland, C.M. Viscoelastic properties of polymers. 4. Thermorheological complexity of the softening dispersion in polyisobutylene. *Macromolecules*, 28, 6432–6436, 1995.

Prolongo, M.G., Salom, C., and Masegosa, R.M. Glass transitions and interactions in polymer blends containing poly(4-hydroxystyrene) brominated. *Polymer*, 43, 93–102, 2002.

Rahman, S. Phase transitions in foods, in *Food Properties Handbook*, CRC Press, Boca Raton, 1995, pp. 87–177.

Rahman, S. Glass transition and other structural changes in foods, in *Handbook of Food Preservation*, Marcel Dekker, New York, 1999, pp. 75–93.

Reid, D.S. Use, misuse and abuse of experimental approaches to studies of amorphous aqueous systems, in *Amorphous Food and Pharmaceutical Systems*, H. Levine (ed.), The Royal Society of Chemistry, Cambridge, 2002, pp. 325–338.

Ribierre, J.-C., Mager, L., Fort, A., and Mery, S. Effects of viscoelastic properties on the dielectric and electrooptic responses of low-T_g guest-host polymers. *Macromolecules*, 36, 2516–2525, 2003.

Richardson, R.K. and Kasapis, S. Rheological methods in the characterization of food biopolymers, in *Instrumental Methods in Food and Beverage Analysis*, D.L.B. Wetzel and G. Charalambous (eds.), Elsevier, Amsterdam, 1998, pp. 1–48.

Richardson, P.H., Willmer, J., and Foster, T.J. Dilute solution properties of guar and locust bean gum in sucrose solutions. *Food Hydrocolloids*, 12, 339–348, 1998.

Rieger, J. The glass transition temperature T_g of polymers—comparison of the values from differential thermal analysis (DTA, DSC) and dynamic mechanical measurements (torsion pendulum). *Polymer Testing*, 20, 199–204, 2001.

Roland, C.M., Santangelo, P.G., and Ngai, K.L. The application of the energy landscape model to polymers. *Journal of Chemical Physics*, 111, 5593–5598, 1999.

Roos, Y.H. Prediction of the physical state, in *Phase Transitions in Foods*, Academic Press, San Diego, 1995, pp. 157–192.

Roos, Y.H. and Karel, M. Applying state diagrams to food processing and development. *Food Technology*, 45, 66, 68–71, 107, 1991.

Roos, Y.H., Karel, M., and Kokini, J.L. Glass transitions in low moisture and frozen foods: Effects on shelf life and quality. *Food Technology*, 50(11), 95–108, 1996.

Sablani, S.S., Kasapis, S., Al-Rahbi, Y., and Al-Mugheiry, M. Water sorption isotherms and glass transition properties of gelatin. *Drying Technology*, 20, 2081–2092, 2002.

Schawe, J.E.K. Principles for the interpretation of modulated temperature DSC measurements. Part 1. Glass transition. *Thermochimica Acta*, 261, 183–194, 1995.

Schawe, J.E.K. Modulated temperature DSC measurements: The influence of the experimental conditions. *Thermochimica Acta*, 271, 127–140, 1996.

Shamblin, S.L., Huang, E.Y., and Zografi, G. The effects of co-lyophilised polymeric additives on the glass transition temperature and crystallisation of amorphous sucrose. *Journal of Thermal Analysis*, 47, 1567–1579, 1996.

Sheppard, S.E. and Houck, R.C. The fluidity of liquids. I. The relation of fluidity to temperature. *Journal of Rheology*, 1, 349–371, 1930.

Shim, J.L. Gellan gum/gelatin blends, U.S. patent 4,517,216, May 14, 1985.

Simon, P.P. and Ploehn, H.J. Molecular-level modeling of the viscoelasticity of crosslinked polymers: Effect of time and temperature. *Journal of Rheology*, 41, 641–670, 1997.

Singh, J.J. and Eftekhari, A. Free volume model for molecular weights of polymers. *Nuclear Instruments and Methods in Physics Research*, B63, 477–483, 1992.

Slade, L. and Franks, F. Appendix I: Summary report of the discussion symposium on chemistry and application technology of amorphous carbohydrates, in *Amorphous Food and Pharmaceutical Systems*, H. Levine (ed.), The Royal Society of Chemistry, Cambridge, 2002, pp. x–xxvi.

Slade, L. and Levine, H. Beyond water activity: Recent advances based on an alternative approach to the assessment of food quality and safety, *Critical Reviews in Food Science and Nutrition*, F.M. Clydesdale (ed.), 30(2–3), 115–360, 1991.

Slade, L. and Levine, H. The glassy state phenomenon in food molecules, in *The Glassy State in Foods*, J.M.V. Blanshard and P.J. Lillford (eds.), Nottingham University Press, Nottingham, 1993, pp. 35–101.

Sopade, P.A., Bhandari, B., D'Arcy, B., Halley, P., and Caffin, N. A study of vitrification of Australian honeys at different moisture contents, in *Amorphous Food and Pharmaceutical Systems*, H. Levine (ed.), The Royal Society of Chemistry, Cambridge, 2002, pp. 169–183.

Steffe, J.F. *Rheological Methods in Food Process Engineering*, Freeman Press, East Lansing, 1996.

Sworn, G. Novel gellan gum gels: Scientific and technological aspects. PhD thesis, Cranfield University, Silsoe, 1998.

Takahashi, Y., Hase, H., Yamaguchi, M., and Noda, I. Viscoelastic properties of polyelectrolyte solutions. III. Dynamic moduli from terminal to plateau regions. *Journal of Non-Crystalline Solids*, 172–174, 911–916, 1994.

Thomas, L. and Aubuchon, S. Heat capacity measurements using quasi-isothermal MDSC. *TA Instruments Technical Bulletin*, No. 230, 1–5, 1999.

Thompson, A.B. and Woods, D.W. The transitions of polyethylene terephthalate. *Transactions of the Faraday Society*, 52, 1383–1397, 1956.

Tobolsky, A.V. Stress relaxation studies of the viscoelastic properties of polymers. *Journal of Applied Physics,* 27, 673–685, 1956.

Truong, V., Bhandari, B.R., Howes, T., and Adhikari, B. Analytical models for the prediction of glass transition temperature of food systems, in *Amorphous Food and Pharmaceutical Systems*, H. Levine (ed.), The Royal Society of Chemistry, Cambridge, 2002, pp. 31–58.

Tsoga, A.K. Effect of co-solutes on polysaccharides gelation. PhD thesis, Cranfield University, Silsoe, 2001.

Verdonck, E., Schaap, K., and Thomas, L.C. A discussion of the principles and applications of modulated temperature DSC (MTDSC). *International Journal of Pharmaceuticals*, 192, 3–20, 1999.

Walton, A. Modern rheometry in characterising the behaviour of foods. *Food Science and Technology Today*, 14, 144–146, 2000.

Wang, B., Gong, W., Liu, W.H., Wang, Z.F., Qi, N., Li, X.W., Liu, M.J., and Li, S.J. Influence of physical aging and side group on the free volume of epoxy resins probed by positron. *Polymer*, 44, 4047–4052, 2003.

Ward, I.M. and Hadley, D.W. Experimental studies of linear viscoelastic behaviour as a function of frequency and temperature: Time-temperature equivalence, in *An Introduction to the Mechanical Properties of Solid Polymers*, John Wiley & Sons, Chichester, 1993, pp. 84–108.

White, G.W. and Cakebread, S.H. The glassy state in certain sugar-containing food products. *Journal of Food Technology*, 1, 73–82, 1966.

Williams, M.L., Landel, R.F., and Ferry, J.D. The temperature dependence of relaxation mechanisms in amorphous polymers and other glass-forming liquids. *Journal of the American Chemical Society*, 77, 3701–3707, 1955.

Windhab, E.J. Recent developments in the area of rheometry. *European Food and Drink Directory*, 65–69, 1996/7.

Wolf, C.L., Lavelle, W.M., and Clark, R.C. Gellan gum/gelatin blends, US patent 4,876,105, October 24, 1989.

Zimeri, J.E. and Kokini, J.L. The effect of moisture content on the crystallinity and glass transition temperature of inulin. *Carbohydrate Polymers*, 48, 299–304, 2002.

CHAPTER 8

Glass Transition Data and Models of Foods

Mohammad Shafiur Rahman

CONTENTS

8.1 TERMINOLOGY

The concepts of state diagram, glass transition, and freezing point are also presented in other chapters. It is important to clearly define the terminology before presenting the data related to the glass transition. This will avoid confusion and help in appropriate use of the data in a meaningful way. An updated state diagram from Rahman (2006) is presented in Figure 8.1 and the terminologies are discussed below.

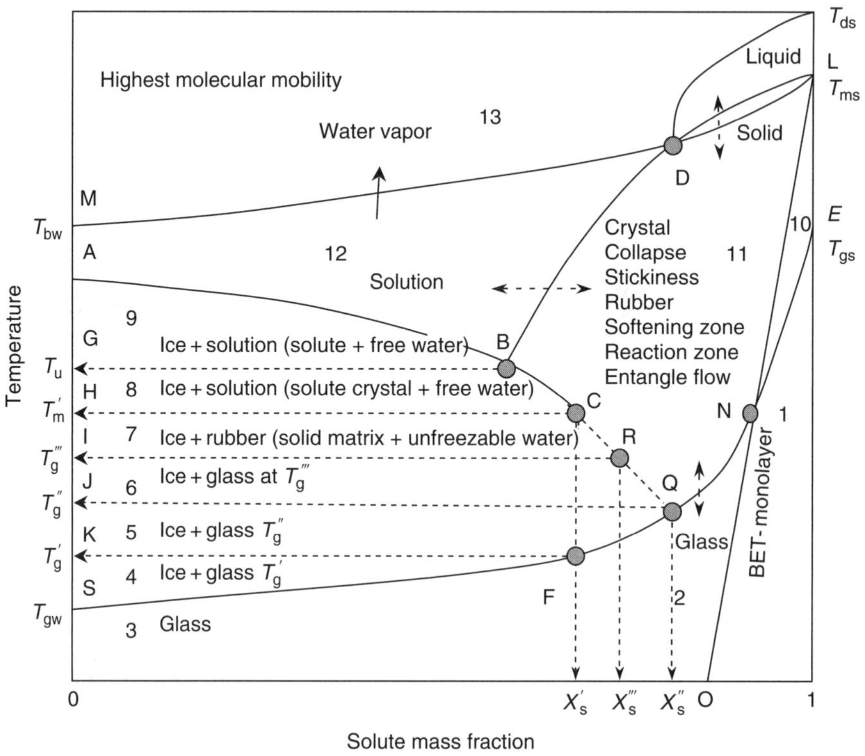

Figure 8.1 State diagram showing different regions and state of foods. (Adapted from Rahman, M.S., *Trends Food Sci. Technol.*, 17, 129, 2006.)

8.2 STATE DIAGRAM AND ITS COMPONENTS

8.2.1 State Diagram

A state diagram is the map of different states of a food as a function of water or solid content and temperature (Rahman, 2004). The main advantages of drawing a map are to help in understanding the complex changes when food's water content and temperature are changed. It also assists in identifying food's stability during storage and selecting a suitable condition of temperature and moisture content for processing. In addition the structural characteristics of foods are determined from different regions of the state diagram. Figure 8.1 shows a state diagram indicating different states as a function of temperature and solids mass fraction.

8.2.2 Components of State Diagram

Earlier state diagram was constructed with only a freezing curve and glass transition line. Recently attempts have been made to add other structural changes with glass line, freezing curve, and solubility line in the state diagram. Numbers of microregions and new terminologies are being included in constructing the state diagram. The state diagram presented in Figure 8.1 is updated from Rahman (2004, 2006). In Figure 8.1, the freezing line (ABC) and solubility line (BD) are shown in relation to the glass transition line (EFS). The point F (X_s' and T_g') lower than T_m' (point C) is a characteristic transition (maximal-freeze-concentration condition) in the state diagram defined as the intersection of the vertical line from T_m' to the glass line EFS (Rahman, 2006). The water content at point F or C is considered as the unfreezable water ($1 - X_s'$). Unfreezable water mass fraction is the amount of water remaining unfrozen even at very low temperature. It includes both uncrystallized free water and bound water attached to the solids matrix. The point Q is defined as T_g'' and X_s'' as the intersection of the freezing curve to the glass line by maintaining a similar curvature of the freezing curve. Matveev (2004) proposed a method to estimate the T_g'' and X_s'' intersection point in the state diagram of frozen solution using the glass transition temperature of the solute.

Point R is defined as T_g''' the glass transition of the solids matrix in the frozen sample, which is determined by differential scanning calorimetry (DSC) below T_m'. This is owing to the formation of same solid matrix associated unfreezable water and transformation of all free water into ice although the sample contains different levels of total water before the start of DSC scanning (Rahman et al., 2005). The values of T_g''' decreased with the increase in solids contents. The point R is the minimum value of T_g'''. In the region AGB, the phases present are ice and solution (solute and free water). Below point B, first crystallization of solute occurs and the transforming GBCH region transforms to three states: ice, solution, and solute crystal. There is no free water (i.e., able to form ice) existing right side of point C (T_m', end point of freezing with maximal freeze-concentration-condition) and then the concentrated solution is transformed to the rubber state. The maximal-freeze-concentration condition could be achieved using optimum conditions by slow cooling and annealing of the samples. The region HCRI contains ice, rubber, and solid matrix. The point F is the T_g', below this point a portion of the rubber state is transformed to glass state. The rate of cooling can shift the points B, C, R, Q, and F. The change in point for different rates of cooling is shown in Figure 8.2. More detailed effects of cooling on the shift are discussed by Rahman (2004).

The region BQEL is important in food processing and preservation; many characteristics such as crystallization, stickiness, and collapse are observed in this region (Roos and Karel, 1991a; Roos, 1995). In case of cereal proteins using G' and G'', Kokini et al. (1994) determined

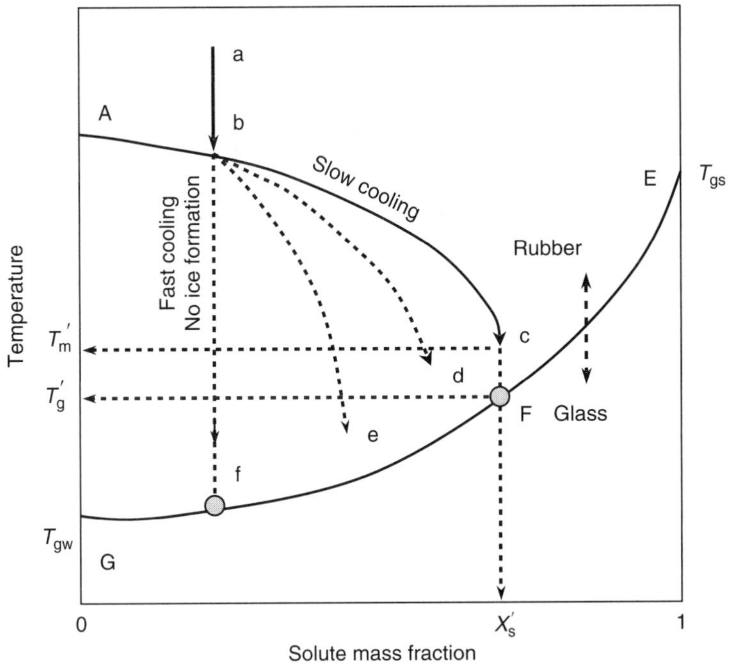

Figure 8.2 Effect of cooling rate on the freezing curve and end point of freezing. (From Rahman, M.S., *Int. J. Food Prop.*, 7, 407, 2004.)

the entangled polymer flow region when both G' and G'' decreased with the increase in temperature. A reaction zone was defined when both G' and G'' increased from a minimum value and started separating each other and then decreased again starting the softening region. All these transitions are observed in the region BQEL. The line BDL is the melting line, which is important when products go to high temperatures during processing, such as frying, baking, roasting, and extrusion cooking.

In the case of a multicomponent mixture, such as food, a clear melting is difficult to observe at high temperature because of the reactions between components. In this case, Rahman (2004) defined it as the decomposition temperature. Line MDL is the boiling line for water evaporation from the liquid phase (line MD) and solid matrix (line DL). T_{ds} is the deterioration temperature when liquid is converted to vapor or combustion compounds.

Recently many papers presented data on water activity and glass transition as a function of water content. However, it was not identified where the link is between them to determine the stability. Karel et al. (1994) attempted to relate water activity and glass transition by plotting equilibrium water content and glass transition into two y-axes as a function of water activity. By drawing a vertical line on the graph, the stability criterion could be determined from the isotherm curve and glass transition line. At any temperature (say 25°C), the stability moisture content from the glass transition line was much higher than the stability moisture from the isotherm. The question is how to use both. At present, it is a real challenge to link them. As a first attempt, Rahman (2006) plotted BET-monolayer value as LO line in the state diagram shown in Figure 8.1. It intersects at point N with the glass line EQS, which shows that at least in one location (point N) glass and water activity concepts provide the same stability criterion. This approach forms more microregions, which could give different stability in the state diagram. More studies regarding stability need to be done on the left (above and below glass) and right sides (above and below) of the line LO. A successful combination of water activity and glass transition could open more in-depth knowledge on stability criteria. In addition how other factors, such as pH preservatives, be linked with these concepts. We are far away could developing a unified theoretical basis. The region of the drying and freezing process can be easily visualized in the diagram, and product stability could be assessed based on moisture content and temperature. Most of the transitions defined in the state diagram are commonly measured by the DSC method using appropriate protocol. The thermomechanical analysis (TMA) and oscillation methods are less commonly used; however, these methods are more sensitive. More details of measurement methods are presented by Rahman (1995, 2006).

It is evident from the review by Rahman (2006) that the variation of stability below glass transition indicating only the glass transition temperature is not enough for developing the stability of foods. The types or characteristics of the glassy state form in different types of foods with variations of composition and water content should be used to characterize the stability criterion. In addition the effect of temperature below T'_m, T'''_g, T''_g, and T'_g should also be explored. Samples with freezable water are more complex and four temperatures are defined as $T'_m > T'''_g > T''_g > T'_g$ (Rahman et al., 2005). There are only few references available including all four characteristic temperatures with their moisture content. It is important to know how these temperatures affect the stability of foods. It would be interesting to explore the differences that exist in the stability in the product within these different ranges (Rahman, 2006).

8.2.3 Equilibrium and Nonequilibrium State

Complex foods exist in states of either unstable nonequilibrium or metastable equilibrium, but never in true thermodynamic equilibrium (Fennema, 1995). Fennema (1995) defined the terminology as follows.

8.2.3.1 *Thermodynamic Equilibrium*

Any food consisting of only one phase requires the minimization of free energy to attain thermodynamic equilibrium. For foods containing two or more phases, thermodynamic equilibrium requires that the chemical potential be equal in every part of the system for each substance present. The chemical potential determines whether a substance will undergo a chemical reaction or diffuse from one part of a system to another. An equilibrium state can be attained through many possible paths, that is, the same properties must be obtainable at a given temperature regardless of whether the temperature is approached by cooling or warming.

8.2.3.2 *Metastable Equilibrium*

Metastable equilibrium refers to a state of pseudoequilibrium, or apparent equilibrium, which is stable over practical time periods but is not the most stable state possible. A metastable state can exist (i.e., conversion to a more stable equilibrium state will not occur) when the activation energy for conversion to a more stable equilibrium state is so high that the rate of conversion is of no practical importance.

8.2.3.3 *Nonequilibrium*

Nonequilibrium refers to a state that is inherently unstable, that is, change to a more stable state is likely to occur at a rate of practical importance. The exact rate at which destabilization occurs depends on the particular system and the conditions to which it is exposed.

8.2.3.4 *Cooling Rate and Thermodynamic Equilibrium*

Fennema's (1995) schematic depiction of a binary system is used to simplify the presentation. The basic format of the figures is shown in Figure 8.3, a plot of sample temperature versus equilibrium state. The columns represent different rates of cooling: equilibrium cooling, moderate cooling, and rapid cooling. The left column (equilibrium cooling) represents cooling at an exceedingly slow rate, the middle column (moderate cooling) represents cooling at a moderate rate consistent with commercial practice, and the right column (rapid cooling) represents cooling at a rate that is exceedingly rapid. Cooling will follow a downward path in the columns.

8.2.3.4.1 *Equilibrium Cooling*

Cooling in the first column in Figure 8.3 is in accord with thermodynamic equilibrium, which is neither possible under practical circumstances nor desirable. Thermodynamic freezing is, however, worthy of consideration for conceptual reasons. The path begins at point S, which represents an aqueous solution containing one solute at thermodynamic equilibrium (Figure 8.4). Cooling must occur at an exceedingly slow rate to preserve equilibrium conditions (Zhao and Notis, 1993), which will eventually bring the solution to its initial freezing point T. At this point, an ice crystal must be added to avoid undercooling and subsequent nucleation of ice, both of which are nonequilibrium events. With the ice crystal in place, further cooling will result in the formation of additional pure ice crystals and a decline in the freezing point of the unfrozen phase. Eventually the temperature decreases to the saturation or eutectic point U. Again at this point, a small crystal of solute must be added to avoid supersaturation and subsequent nucleation of solute, both of which are nonequilibnum events. Fennema (1995) noted that solute crystallization is mandatory to sustain thermodynamic equilibrium, and it is common even in the presence of seed crystals of

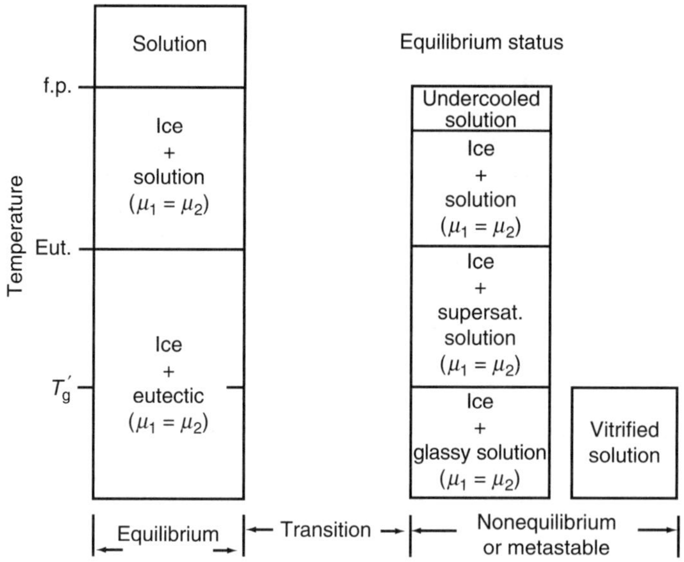

Figure 8.3 Equilibrium status based on rate of cooling. (From Fennema, O., in *Food Preservation by Moisture Control: Fundamentals and Applications*, Barbosa-Canovas, G.V. and Welti-Chanes, J. (eds.), Technomic Publishing, Lancaster, PA, 1995, p. 243. With permission.)

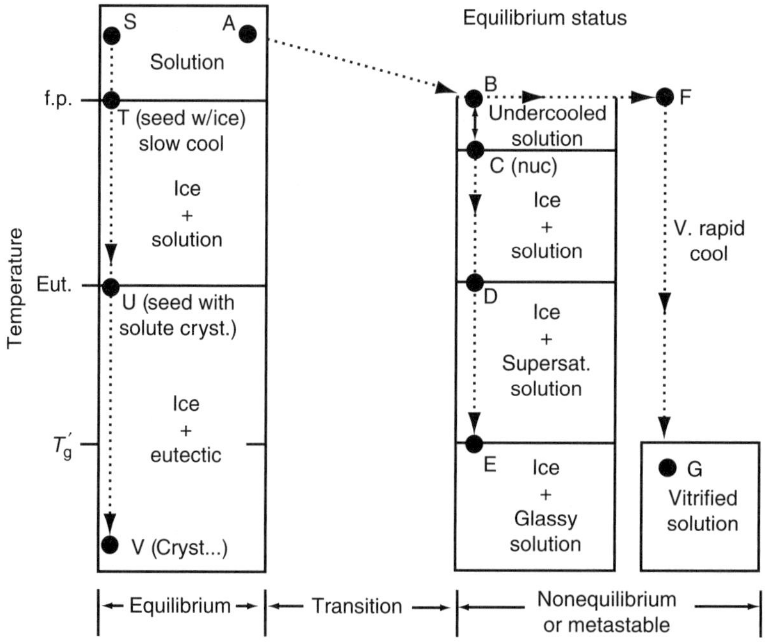

Figure 8.4 Equilibrium status and path of cooling. (From Fennema, O., in *Food Preservation by Moisture Control: Fundamentals and Applications*, Barbosa-Canovas, G.V. and Welti-Chanes, J. (eds.), Technomic Publishing, Lancaster, PA, 1995, p. 243. With permission.)

solute that the solute will not crystallize at a subeutectic temperature. However, it is assumed here to form solute crystallization. Further cooling results in crystallization of ice and solute in constant proportion, leaving the unfrozen phase unchanged in composition and freezing point. This dual crystallization process continues at constant temperature until crystallization of water and solute is as complete as possible. Further cooling will simply lower the sample temperature with no further change in physical state.

8.2.3.4.2 Moderate Cooling

The metastable or nonequilibrium pathways are quite different from the equilibrium cooling. Both paths start in the solution at point A, and cooling brings the sample to its initial freezing point B. At this point, further cooling at a moderate rate, consistent with commercial practices, results in undercooling to point C. This nonequilibrium supercooling eventually results in nucleation, release of latent heat of crystallization, and if cooling is relatively slow, a reversal in temperature almost to the initial freezing point B. As ice forms with further cooling, the freezing point of the solution phase declines. With further cooling, more of the original water converts to ice and the solute eventually attains its saturation concentration D (eutectic point). Further cooling does not result typically in nucleation of solute crystals, rather the solution becomes increasingly supersaturated with solute, and this condition is normally metastable. Continued cooling to point E will cause the supersaturated unfrozen phase to convert to a metastable, amorphous solid (a glass) with a very high viscosity (about 10^{12} Pa s). This temperature is the glass transition temperature (T_g), which is usually determined by the composition of the sample and the rate of cooling. If cooling has been slow by commercial standards, the unfrozen solution can be maximally freeze-concentrated, and T_g, under this circumstance, will assume quasi-invariant value known as T_g', which is dependent only on the solute composition of the sample. In practice, maximum freeze concentration is usually not obtained, and the observed T_g differs from T_g'. Numerical data for different food components have been compiled by Rahman (1995).

8.2.3.4.3 Very Fast Cooling

A third possible cooling pathway (Figure 8.4) involves very rapid removal of heat from very small samples. This path has no commercial significance for foods. Thus, the ABFG path results in vitrification of the entire sample. The glass temperature is not an equilibrium process, and the cooling rate also affects the glass temperature. With very slow cooling during freezing, there is a possibility of heterogeneous nucleation of ice and less possibility of ice formation. In the extreme case (fast cooling), the material may transform from liquid to glass at the melting point without further ice formation.

8.2.4 State of Water in Foods

Different states of water such as bound, free, capillary, mobile, nonsolvent, and unfreezable are defined in the literature (Rahman, 1995). The state of water can be measured with different techniques or methods. The water sorption isotherm is based on the three types of water: monolayer, multiplayer, and mobile or free water (Rockland, 1969). The BET-monolayer is estimated from water sorption isotherm and is commonly presented in the literature. It could be mentioned that only the BET-monolayer has a strong theoretical basis and should be used in stability determination. The BET-monolayer for large numbers of foods and its components have been compiled by Rahman and Labuza (1999). It is not recommended to use the GAB-monolayer value because of the number of defects in estimating its real value although it is popular for its validity up to a water activity of 0.9 (Rahman, 2005; Rahman and Al-Belushi, 2006).

Unfreezable water content can be estimated comparing DSC endotherms of samples with freezable water. Paakkonen and Plit (1991a) measured the unfreezable water of cabbage by this method. Unfreezable water can be estimated from the plot of meting enthalpy as a function of the water content. This procedure was used for model crackers (Given, 1991), strawberry (Roos, 1987), dates (Rahman, 2004), sucrose (Ablett et al., 1992), and garlic (Rahman et al., 2005). Usually unfreezable water is independent of the total water present in the system. In the case of chitin, unfreezable water increased with the increase in total water and the amount of freezable water is relatively low compared with the unfreezable water (Paakkonen and Plit, 1991b). In the case of water–gellan systems, unfreezable water increased with the size of the junction zone (Hatakeyama et al., 1996). Using nuclear magnetic resonance (NMR) technique, Li et al. (1998) studied the mobility of freezable and unfreezable water in waxy corn starch determined by DSC. Water was found to be isotropically mobile for samples over a range of water contents (6.3%–47%) at room temperature. Mobility increased with increasing water content and temperature. A large fraction of unfreezable water was relatively mobile comparable to a liquid state even down to −32°C. The decreasing fraction of mobile water with decreasing temperature suggested that only some of the so-called unfreezable water could be progressively immobilized as temperature decreased. Much of the water remained high in mobility, regardless of the relatively rigid starch molecules in the glassy solid state. This means that water in the glassy state of starch can greatly influence reactions at both ambient and freezing temperatures. At least in this example, the glassy state of the solid materials is not an appropriate term to imply or to predict the molecular dynamics of water and its influence on food stability. Bell et al. (2002) determined water mobility in the PVP system as determined through NMR and found that water mobility was not affected by the glass transition. Providone n-vinyl pyrolidone (PVP) systems at constant water activities and water contents, but different physical states (glassy and rubbery), had the same water mobility. An evaluation of four chemical reactions showed no relation between water mobility and kinetic data. The effect of water on chemical reactions is multidimensional and cannot be reduced to a single physicochemical parameter.

From the state diagram shown in Figure 8.1, unfreezable water can be estimated from points C and F. A comparison of determining unfreezable water using different methods was presented for dates (Rahman, 2004) and garlic (Rahman et al., 2005). It is always found that BET-monolayer values are much lower than the unfreezable water (Duckworth and Smith, 1963). Other techniques used to determine the state of water are dielectric spectroscopy, Fourier transformation infrared spectroscopy (FTIR), x-ray scattering, NMR, magnetic resonance imaging, electrical resistance, and self-diffusion (Kaatze, 1990; Abbott et al., 1996; Hardman, 1986; Labuza and Hyman, 1998). Three states of water (polymer, capillary, and free) were identified in whey by NMR (Padua et al., 1991). The NMR and dielectric measurements of starch paste showed one type of water while agar gels contain two types of water when samples contain less than 55% moisture (Padua, 1993). Lang and Steinberg (1983) studied the types of water in corn starch, sugar, sodium chloride, and mixture of starch and sugar by NMR. It was found that sucrose is a structure-former, while sodium chloride is a structure-breaker. Three types of water mobility were observed in sucrose solution by NMR techniques (Richardson et al., 1987). Solute–solvent and solute–solute interactions by way of hydrogen bonding are suggested as the mechanism to explain the observed decrease in water mobility. Lai et al. (1993) studied the water mobility in starch-based food products when fat is replaced by fat mimetic components. The active water in starch–sucrose system was strongly dependent on sucrose content (Chinachoti and Stengle, 1990). Molecular mobility of starch and water in starch–water mixtures was studied with the NMR technique and related with the water sorption isotherm (Choi and Kerr, 2003). The effect of bound water on glycinin was studied by FTIR spectra (Abbott et al., 1996). In addition to the above techniques mentioned, many techniques are used to determine the mobility and state of water and solutes available for chemical reactions, but their interpretation is far from straightforward (Hardman, 1986).

8.3 MEASUREMENT TECHNIQUES

8.3.1 DSC

In many cases, the glass transition temperature is difficult to determine in real food systems due to their complexity, heterogeneity, multidomain region, and little change in specific heat at the transition. The most common and popular method used to determine glass transition is the DSC, which detects the change in heat capacity occurring over the transition temperature range. For almost half a century, most of the work in these systems has been carried out by DSC, which measures as a function of temperature the difference in energy inputs into a substance and its reference, with both materials being subjected to a control temperature program. In the 1990s, modulated DSC was commercialized to increase the sensitivity and resolution of thermal analysis and to provide the heat capacity and heat flow in a single experiment (Verdonck et al., 1999; Kasapis, 2005). It is common to use mainly the heating DSC curve instead of the cooling curve to study the characteristic transitions, and usually the heating rate at 5°C/min–20°C/min is used. However, the heating rate affects the values of glass transition. The experimental conditions such as the cooling rate, sample size, and annealing conditions used should be always reported with glass transition values. Calorimetric or spectroscopic techniques have some limitations in terms of sample size and shape and water content control. In some complex food materials, for example in the case of starch and fish muscle, it is less sensitive (Sablani et al., 2007).

The typical DSC curves shown in Figures 8.5 through 8.9 are based on the level of moisture contents and types of the samples. Figure 8.5 shows DSC graphs for low moisture content (i.e., high solids) when there is no freezable water in the sample. In the case of concentrated samples with

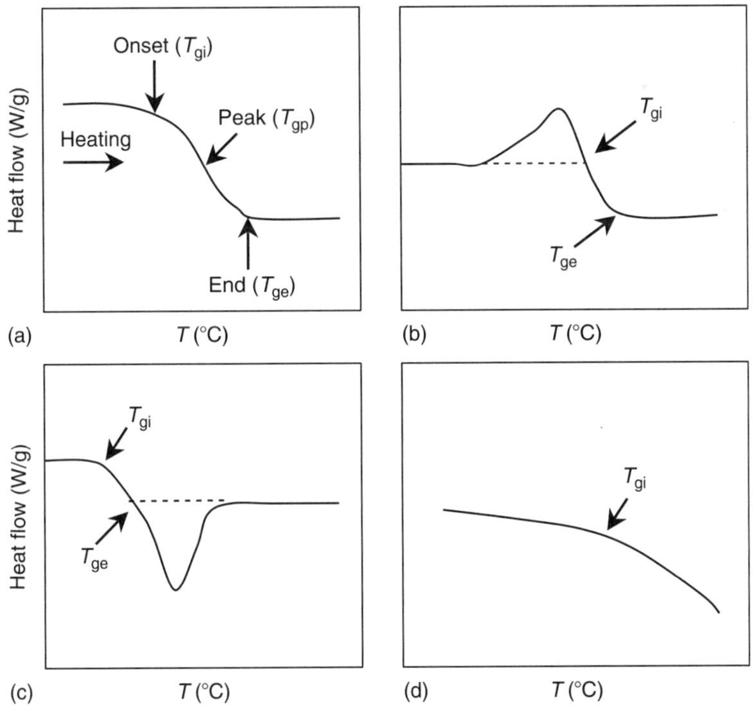

Figure 8.5 Typical DSC thermograms for glass transition of samples containing unfreezable water. (a) Shift, (b) exothermic peak before shift, (c) endothermic peak after shift, and (d) change in slope. (From Rahman, M.S., *Trends Food Sci. Technol.*, 17, 129, 2006. With permission.)

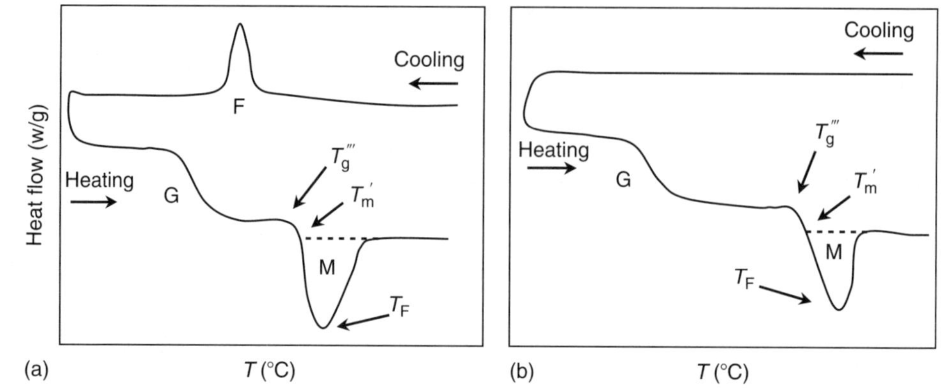

Figure 8.6 Typical DSC thermograms showing glass transition, freezing, and melting endotherms for sample containing freezable water. (a) Exothermic peak during cooling and (b) no peak observed during cooling. (From Rahman, M.S., *Trends Food Sci. Technol.*, 17, 129, 2006. With permission.)

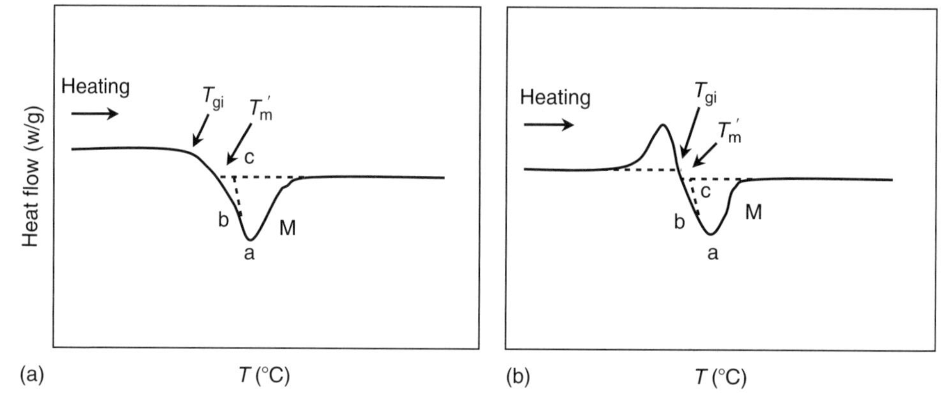

Figure 8.7 Typical DSC thermograms showing melting of ice and no glass transition shift in the thermogram line. (a) No exothermic before peak for melting of ice and (b) exothermic peak before peak for melting of ice.

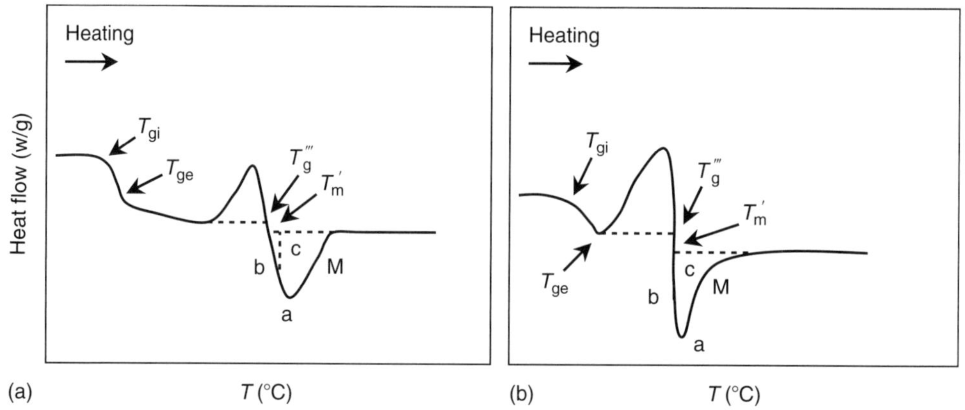

Figure 8.8 Typical DSC thermograms showing glass transition by a thermogram line shift and exothermic peak before melting of ice. (a) Exothermic peak before melting of ice and (b) exothermic peak after glass transition.

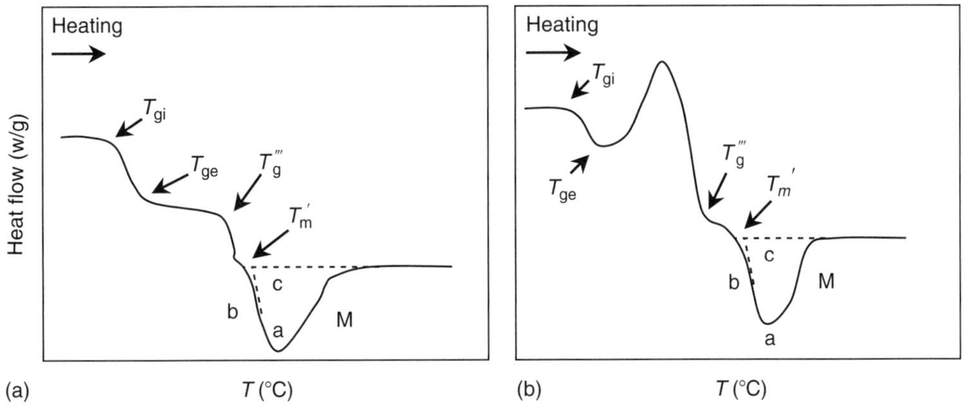

Figure 8.9 Typical DSC thermograms showing two glass transitions by thermogram line shift; with and without exothermic peak before melting of ice. (a) Noexothermic peak and (b) exothermic peak after glass transition.

freezable water, no ice crystallization is observed during cooling, indicating that vitrification was accomplished. Many foods or food components showed an exothermic (Figure 8.5B) or endothermic (Figure 8.5C) peak. Kasapis (2005) pointed that (at a low moisture content <30%) regardless of the polysaccharide types an endothermic peak appeared consistently within the 45°C–80°C temperature band during the first heating scan on a calorimeter. The position of the peak remained constant and independent of the temperature shifts of the glass transition (when it appeared), but the associated enthalpy increased with the moisture content (Appelqvist et al., 1993). This was interpreted on the basis of stabilizing enthalpic associations between water molecules and ordered macromolecular sequences. The endothermic overshoot may be deliberately used to help in detecting the glass transition, with materials for which the heat capacity jump is particularly small and smeared out over a broad temperature range. There is no consensus for the definition of the glass transition point on a DSC curve among the various points that may be chosen as onset (T_{gi}), mid (T_{gp}), and end (T_{ge}). In Figure 8.5, the location of onset and end when there is an exothermic or endothermic peak was also pointed out. Earlier data presented in the literature mainly presented the mid, peak, or initial point; however, the recent trend is to present onset, mid, and end points. Champion et al. (2000) pointed out that a glass transition should be characterized by at least two parameters indicating its onset or mid and the width of the transition. However, the onset (T_{gi}) is more important to know for determining the product stability during storage because molecular mobility starts at the onset. Figure 8.5d shows the change of slope in the thermogram line instead of a shift in the thermogram line. In the case of very complex systems, it was difficult to observe the baseline shift.

Figures 8.6 through 8.9 show DSC cooling and heating curves for samples containing freezable water. If the moisture content is high and the cooling rate is relatively slow, freezing of water is observed during cooling as shown by F in Figure 8.6a. If the sample contains relatively low freezable water and the cooling rate is relatively fast, the freezing exotherm does not appear during cooling (Figure 8.6b). During heating, Figure 8.6 shows a glass transition at G followed by the ice melting endotherm marked as M and the location of T_m' and T_g'''.

Other different types of heating thermograms are shown in Figures 8.7 through 8.9. In Figure 8.7, no glass transition (i.e., shift in thermogram line) is observed before the ice melting endotherm M and T_{gi} could be considered as T_g'''. In addition, Figure 8.7b shows an exothermic endotherm before the ice melting thermogram. The locations of glass transition (T_{gi} and T_{ge}), apparent maximal-freeze-concentration condition (T_g''' and T_m'), and melting are shown in Figures 8.8 and 8.9. To determine T_m', optimum annealing needs to be done to maximize the ice formation at a temperature between T_g''' and apparent T_m' (without annealing). It is common to perform annealing at $T'_m - 1$. Because of kinetic constraints, solutions with high initial solutes (60%–80%) may require several days or even weeks of annealing at $T'''_g < T < T'_m$ until the maximally freeze-concentrated state is achieved (Karel et al., 1994). In many samples, the exothermic peak is observed after glass transition and melting endotherm in the heating DSC curve (Figures 8.8 and 8.9).

The observed exothermic enthalpy relaxation peak during rewarming, between the glass transition and the melting endotherm, may disappear after annealing or rescanning (Baroni et al., 2003; Rahman, 2004). However, this procedure changes the original state of the samples. This process, generally called devitrification, corresponds to ice crystallization. Freezable water that had remained unfrozen due to hindered crystallization during a fast cooling, and then freezable water crystallizes into ice at lower temperature during heating cycle.

In many cases, two glass transitions are observed before the ice melting endotherm even with annealing (Fonseca et al., 2001; Rahman, 2004). In extreme cases, a second glass transition could be observed with the melting endotherm (Figure 8.9). Different hypotheses are proposed for two or more glass transitions. The two transitions occurred, one due to the backbone of a large polymer or less mobile component, and the other due to the less mobile or side chains (Rahman, 2004). Another reason could be the incompatibility of different solutes in the mixture (Morales-Diaz and Kokini, 1998; Li and Chen, 2001). Another proposed concept for solution even with a single solute is that it is the result of formation of a solute-crystal-rich, unequilibrated phase trapped around or within the rapidly nucleated ice crystals, or solute inclusion within the ice crystals itself (Goff et al., 2002). Li and Chen (2001) used these two glass transitions to identify the degree of compatibility of rice starch–hydrocolloid mixtures. In the case of a compatible sample, such as rice starch–high methoxyl pectin mixture, the sample exhibited a new single glass transition, which is between the transitions of two individual components. In contrast, incompatible rice starch–low methoxyl pectin and rice starch–locust bean gum showed two transitions corresponding to the two individual components.

8.3.2 Dynamic Mechanical Thermal Analysis

Other useful and sometimes more sensitive methods include TMA, dynamic mechanical analysis (DMA), dynamic mechanical thermal analysis (DMTA), and dynamic oscillation method. In this method, structural properties are examined with G' and G'' as a function of temperature at a constant frequency or time of measurement. Figure 8.10 shows typical curves. Again it is necessary to specify how the transition temperature (T_r) is defined from the experimental curves. The temperature is commonly taken from the maximum of the loss factor (tan δ), which is easily determined ($T_{\delta p}$). The maximum of the loss modulus (E'' or G″) is much better for the transition

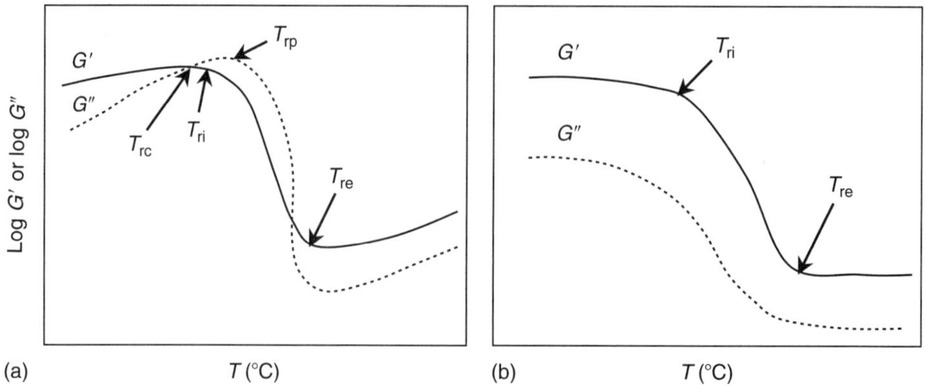

Figure 8.10 Typical plots of log G' or log G'' showing glass transition. (From Rahman, M.S., *Trends Food Sci. Technol.*, 17, 129, 2006. With permission.)

determined from the point of view of its physical meaning (Champion et al., 2000). Moreover, no tan δ peak is expected with small molecular weight systems. In Figure 8.9a, it is proposed that the rheological glass transition is the point between the glass transition and the glassy state (Kasapis et al., 2001). The transition from T_g (DSC glass) and T_r (mechanical or rheological glass) should not be considered as fully equivalent. Shalaev and Kanev (1994) mentioned that mechanical glass transition occurs above DSC glass because of the sample's ability to keep its form. In DSC and DMTA, the sample is subjected to stresses of different physical nature (change of temperature in DSC, shearing or compression in DMTA). The experimental time may also be different (depending on cooling–heating rates and annealing in DSC on measurement frequency in DMTA). Blond (1994) and Kasapis et al. (2003) studied the comparisons of glass transition from DSC and mechanical methods. A different coupling of the imposed perturbations with the structural units (with particular relaxation times) may be responsible for discrepancies in the data obtained with different techniques. In Figure 8.10b, the crossover between G′ and G″ is not observed. In general, it could be recommended to present T_{ri}, T_{re}, T_{rc}, and T_{rp} in comparison to glass transition by DSC.

8.4 GLASS TRANSITION DATA

Tables 8.1 and 8.2 show the effects of the annealing, heating, and cooling rate on the glass transition temperature of sucrose, fructose, and date flesh. In general, the glass transition shifted to higher temperature with the increase in cooling or heating rate and annealing time. The change in specific heat at its glass transition decreased with the increase in scanning rate (Table 8.3). In the case of the samples containing no freezable water, the glass transition temperature of different foods with varied water or solid content is presented in Tables 8.4 through 8.13. Sobral and Habitante (2001) measured the phase transitions of pigskin gelatin by the DSC method and observed its complexity (Table 8.10). The first scan showed different characteristics based on the water content in gelatin. At a_w 0.11, the glass transition was merged with its melting endotherm and at a_w 0.33 and 0.43 an exothermic peak was observed just followed by glass transition. At a_w 0.52, 0.64, 0.75, and 0.88, an endothermic peak was observed followed by glass transition; however, a_w 0.88 showed a separation of the glass transition thermogram shift and endothermic peak. In case of heating cycle sample remained at 105°C and annealed for 24 h, and then second scan removed the endothermic and exothermic peaks from the thermograms. The DSC traces obtained in the first scan corresponds to that of partially crystalline polymers, while those of the second scan are characteristics of completely amorphous polymers. It was due to the elimination of the crystals (junctions).

Table 8.1 Effect of DSC Scan Rate on Glass Transition Temperature

Fructose[a]		Sucrose[b]				Date[c]				
r_c (°C/min)	t_g (°C)	r_h (°C/min)	t_g (°C)	r_c (°C/min)	t_g (°C)	r_h (°C/min)	θ_{an} (h)	t_{gi} (°C)	t_{gp} (°C)	t_{ge} (°C)
2.5	−82.8	2.5	−67.1	5.0	−69.2	5	0.0	−71.5	−68.3	−65.4
5.0	−83.7	5.0	−65.9	10.0	−69.6	5	0.5	−71.9	−67.8	−64.8
10.0	−81.5	10.0	−64.5	20.0	−71.0	5	1.0	−73.1	−69.4	−66.9
20.0	−74.8	20.0	−62.5			1	0.0	−72.9	−69.3	−66.9
40.0	−70.6	30.0	−59.3			1	0.5	−73.8	−70.0	−67.6
		40.0	−57.8							
		50.0	−56.0							

[a] X_s: 0.600, sample mass: 10–35 mg (Ablett et al., 1993).
[b] X_s: 0.701, sample mass: 10–35 mg (Whittam et al., 1991).
[c] X_s: 0.675 and a_w: 0.755 (Rahman, 2004).

Table 8.2 Effect of Annealing Time on the Glass Transition Temperature of Devitrifying Solution (Rapid Cooling)

Sucrose[a]		Fructose[b]		Sucrose[c]	
Time (min)	t_{ge} (°C)	Time (min)	t_{ge} (°C)	Time (min)	t_{ge} (°C)
0.0	−73.1	0.0	−84.0	0.0	−74.0
0.1	−71.7	4.0	−83.0	15.0	−74.0
1.0	−71.3	15.0	−82.0	30.0	−72.0
2.0	−65.0	45.0	−80.0	60.0	−67.0
4.0	−52.5	60.0	−53.0	120.0	−54.0
15.0	−43.7			300.0	−46.0
33.0	−42.1				

[a] Annealing time at −35°C (X_s^0 : 0.65, DSC rapid cooling) (Izzard et al., 1991).
[b] Annealing time at −50°C (X_s^0 : 0.60, DSC 5°C/min, sample: 10–35 mg) (Ablett et al., 1993).
[c] Annealing time at −35°C (X_s^0 : 0.68, DSC heating at 5°C/min from −100°C to 0°C) (Roos and Karel, 1991a).

The disappearance of the endothermic or exothermic peaks in the second scan is an expected fact, considering that the helix-coil transition is a first-order transition and of kinetic character. The heating of the sample after the glass transition provokes disruption of the crystalline region and the gelatin behaves as a molecularly disrupted molten plastic (Sobral and Habitante, 2001).

Table 8.14 shows the maximal-freezing condition, glass transition, and melting temperatures of pure solutes of different polyhydroxy compounds. The values of T_g' for commercial starch hydrolysis products (SHP) are compiled in Table 8.15. The freezing and melting points of diethyl sulphoxide (cryoprotectant) are presented in Table 8.16. Freezing temperature, enthalpy, and maximal-freeze concentration conditions of different foods (containing freezable water) are compiled in Tables 8.17 and 8.18.

8.5 MODELING OF GLASS TRANSITION

The free volume theory treats the glassy state as an iso-free volume state in which 1/40 of the volume of a liquid is free volume at the glass transition temperature. In this case

$$V_v = \frac{1}{40} + (\alpha_l - \alpha_g)(T - T_g) \tag{8.1}$$

Table 8.3 Effect of DSC Scan Rate on the $(\Delta C_p)_g$ Values of Sucrose Glasses

	$(\Delta C_p)_g$ (kJ/kg K)			
X_s^0	10°C/min	20°C/min	30°C/min	40°C/min
0.650	0.670	0.720	0.680	0.630
0.681	0.804	0.748	0.730	0.700
0.701	0.755	0.712	0.660	0.670
0.774	0.733	0.737	0.742	0.724
0.817	0.828	0.786	0.731	0.715

Source: Izzard, M.J., Ablett, S., and Lillford, P.J., in *Food Polymers, Gels and Colloids*, Dickinson, E. (ed.), The Royal Society of Chemistry, London, 1991. With permission.

Table 8.4 Glass Transition Temperature of Food Components and Foods as a Function of Water Content

Maltose[a]		Sucrose[b]		Sucrose[c]		Maltose[c]			Fructose[d]		Maltotriose[a]		Maltohexaose[a]		Gelatin[e]		Gelatin[f]		Gelatin[g,h]	
X_w	t_g (°C)	X_w	t_g (°C)	X_w	t_g (°C)	a_w	X_w	t_g (°C)	X_s	t_g (°C)	X_w	t_g (°C)	X_w	t_g (°C)	X_s	t_g (°C)	X_s	t_g (°C)	X_s	t_{rc} (°C)
0.000	93.9	0.033	42	0.000	62	0.00	0.000	87.0	0.000	10	0.000	130.5	0.000	173.3	1.00	195.0	0.986	107.4	1.000	160.5
0.026	52.5	0.035	38	0.035	29	0.11	0.011	59.0	0.050	4	0.018	112.6	0.028	116.4	1.00	199.1	0.981	92.9	0.978	70.8
0.042	30.9	0.049	30	0.200	−46	0.23	0.027	39.0	0.100	−17	0.041	67.6	0.052	76.9	1.00	118.4	0.968	80.9	0.922	47.3
0.062	26.9	0.061	23	0.250	−59	0.33	0.043	2.9	0.180	−38	0.060	47.3	0.058	70.0	1.00	172.5	0.953	90.5	0.862	20.8
0.080	13.9	0.180	−33	0.300	−68	0.43	0.073	11.1	0.099	−46	0.079	31.7	0.064	76.9	1.00	177.5	0.941	78.4	0.805	16.3
0.124	−2.5	0.196	−39			0.52	0.092	−4.0	0.215	−50	0.106	23.6	0.097	52.0	1.00	93.4	0.932	78.5	0.754	6.9
0.135	−6.5	0.212	−43						0.240	−57	0.125	10.8	0.132	32.0	1.00	189.0	0.929	67.4	0.750	7.0
		0.249	−53						0.260	−63			0.213	−5.2			0.927	60.9	0.800	15.0
		0.270	−60						0.300	−69							0.920	57.4	0.858	21.0
		0.299	−64						0.350	−78							0.908	56.4	0.970	35.0
		0.315	−68						0.398	−85							0.891	53.0		
		0.350	−75														0.869	57.0		
																	0.861	40.5		
																	0.842	21.0		
																	0.825	22.5		
																	0.801	21.0		
																	0.772	−13.5		
																	0.763	−11.0		
																	0.735	−11.0		
																	0.728	−19.0		
																	0.719	−9.0		
																	0.705	−13.5		
																	0.680	−16.5		
																	0.656	−9.4		

[a] Orford et al. (1989).
[b] DSC heating rate: 5°C/min (Izzard et al., 1991).
[c] Heating rate: 5°C/min (Roos, 1992).
[d] DSC heating rate: 5°C/min, sample mass: 10–35 mg (Ablett et al., 1993).
[e] Rheological method (Marshall and Petrie, 1980; Koleske and Faucher, 1965).
[f] DSC method (Slade and Levine, 1991; Mitchell, 2000; Marshall and Petrie, 1980; Tseretely and Smirnova, 1992).
[g] Rheological method (Kasapis and Sablani, 2005).
[h] Rheological method (Sablani et al., 2002).

Table 8.5 Glass Transition Temperature of Maltodextrin as a Function of Water Content

	Maltrin M040 (λ_s: 3600, DE 5)		Maltrin M100 (λ_s: 1800, DE 10)		Maltrin M150 (λ_s: 1200, DE 15)		Maltrin M200 (λ_s: 900, DE 20)		Maltrin M250 (λ_s: 720, DE 25)		Maltrin M365 (λ_s: 720, DE 36)	
a_w	X_w	t_{gi} (°C)	X_w	t_{gi} (°C)	X_w	t_{gi} (°C)	X_w	t_{gi} (°C)	X_w	t_{gi} (°C)	X_w	t_{gi} (°C)
0.00	0.000	188	0.000	160	0.000	—	0.000	141	0.000	121	0.000	100
0.11	0.019	135	0.020	103	0.020	99	0.024	86	0.021	83	0.017	67
0.23	0.038	102	0.047	84	0.046	83	0.052	73	0.044	60	0.038	45
0.33	0.043	90	0.051	66	0.057	65	0.054	42	0.049	36	0.049	31
0.43	0.059	87	0.065	60	0.066	57	0.058	40	0.059	34	0.054	27
0.52	0.082	58	0.076	38	0.074	40	0.084	37	0.081	29	0.098	6
0.75	0.096	44	0.095	30	0.102	8	0.139	−9	0.148	−18	0.170	−35
0.85	0.150	23	0.158	−6	0.166	−15	0.208	−32	0.215	−39	0.238	−52

Source: Roos, Y. and Karel, M., *Biotechnol. Prog.*, 7, 49, 1991b. With permission.

Table 8.6 Glass Transition Temperature of Food Components and Foods as a Function of Water Content

Corn Embryos[a]		Sucrose[b]		Sucrose[c]		Maltose[c]			Fructose[d]		Wafer[e]			Pineapple[f]				
X_w	t_g (°C)	X_w	t_g (°C)	X_w	t_g (°C)	a_w	X_w	t_{gi} (°C)	X_w	t_{gp} (°C)	X_w	t_{gi} (°C)	t_{gp} (°C)	X_w	a_w	t_{gi} (°C)	t_{gp} (°C)	t_{ge} (°C)
0.099	50.1	0.033	42	0.000	62	0.00	0.000	87.0	0.000	10	0.047	65.1	68.1	0.037	0.11	34.49	38.39	45.40
0.109	28.8	0.035	38	0.035	29	0.11	0.011	59.0	0.050	4	0.062	55.0	61.0	0.069	0.33	1.64	8.89	13.84
0.109	22.5	0.049	30	0.200	−46	0.23	0.027	39.0	0.100	−17	0.073	50.0	55.9	0.083	0.43	−18.52	−9.88	−2.62
0.120	1.3	0.061	23	0.250	−59	0.33	0.043	2.9	0.180	−38	0.088	30.1	32.3	0.059	0.53	−28.20	−20.94	−14.59
0.121	−1.9	0.180	−33	0.300	−68	0.43	0.073	11.1	0.099	−46	0.104	29.5	33.0	0.103	0.64	−41.96	−34.70	−29.28
0.122	−3.8	0.196	−39			0.52	0.092	−4.0	0.215	−50	0.133	8.4	10.4	0.261	0.75	−69.42	−62.21	−57.72
0.131	−20.0	0.212	−43						0.240	−57	0.153	−8.1	−5.3	0.307	0.80	−79.51	−70.01	−62.79
0.137	−22.5	0.249	−53						0.260	−63	0.192	−26.2	−21.2					
0.177	−47.5	0.270	−60						0.300	−69								
0.191	−55.6	0.299	−64						0.350	−78								
0.205	−60.6	0.315	−68						0.398	−85								
0.206	−61.9	0.350	−75															
0.222	−63.8																	

[a] Two melting points at about −15°C (with oil); glass temperature was measured after extracting oil. X_{uw}: 0.20; solids: 20%; sucrose: 17%; raffinose: 3%; DSC heating rate: 20°C/min (Williams and Leopold, 1989).
[b] DSC heating rate: 5°C/min (Izzard et al., 1991).
[c] Heating rate: 5°C/min (Roos, 1992).
[d] DSC heating rate: 5°C/min, sample mass: 10–35 mg (Ablett et al., 1993).
[e] Carbohydrate: 0.795, protein: 0.109, lipids: 0.029 (wet fraction), DSC heating rate: 5°C/min, sample mass: 10 mg (Martinez-Navarrete et al., 2004).
[f] DSC method, heating rate: 10°C/min, sample size: 10 mg (Telis and Sobral, 2001).

Table 8.7 Glass Transition Temperature above Unfrozen Water Content When There Is No Formation of Ice

Date Flesh[a]					Garlic[b]				Kiwifruit[c]		Apple[d,e]				
a_w	X_s	t_{gi} (°C)	t_{gp} (°C)	t_{ge} (°C)	X_s	T_{gi} (°C)	T_{gp} (°C)	T_{ge} (°C)	X_s	T_{gp} (°C)	a_w	X_w	T_{gi} (°C)	T_{gp} (°C)	T_{ge} (°C)
0.755	0.675	−71.5	−68.3	−65.4	0.820	−49.1	−39.7	−25.2	1.00	65.31	0.00	0.000	9.3	16.2	21.2
0.717	0.716	−51.3	−54.7	−58.1	0.830	−34.7	−26.8	−20.0	1.00	71.12	0.03	0.008	−8.3	−1.8	4.6
0.570	0.782	−43.7	−46.4	−49.9	0.850	−25.1	−17.2	−12.3	0.969	21.67	0.11	0.027	−19.4	−3.9	−6.3
0.544	0.802	−34.6	−38.2	−42.6	0.830	−19.3	−15.6	−6.1	0.965	17.06	0.23	0.055	−22.3	−14.9	−10.4
0.432	0.821	−26.3	−29.8	−34.3	0.880	−20.4	−12.9	−4.5	0.958	12.43	0.33	0.080	−23.3	−16.4	−12.4
—	0.832	−23.7	−29.3	−35.6	0.900	−10.3	−6.8	−2.9	0.956	9.05	0.43	0.107	−27.5	−22.8	−18.9
0.331	0.880	−13.8	−17.2	−32.6	0.950	3.5	13.2	20.7	0.949	1.97	0.52	0.137	−39.4	−30.9	−22.3
0.231	0.851	−13.9	−20.3	−24.9	0.995	41.2	45.5	52.8	0.945	−5.40	0.79	0.288	−74.0	−65.0	−60.1
0.113	0.862	−19.6	−24.3	−27.8					0.939	−4.51	0.00	0.027	0.1	3.6	6.6
—	0.889	−17.2	−23.3	−28.4					0.922	−17.77	0.12	0.052	−1.9	2.4	5.3
—	0.927	29.2	32.7	35.6					0.920	−14.73	0.23	0.077	−13.6	−8.2	−1.9
—	0.932	29.8	33.2	35.9					0.916	−17.20	0.33	0.108	−22.1	−15.4	−9.7
—	0.969	33.9	36.4	38.5					0.897	−28.94	0.44	0.140	−30.3	−25.6	−20.6
0.000	1.000	62.8	69.2	76.2					0.896	−31.09	0.53	0.161	−43.7	−37.0	−29.6
									0.895	−26.20	0.58	0.176	−50.5	−44.8	−38.6
									0.887	−33.59	0.61	0.277	−55.8	−50.6	−44.0
									0.820	−46.23	0.76	0.331	−72.9	−70.0	−65.3
									0.816	−50.85					
									0.809	−54.56					
									0.808	−51.82					

[a] Rahman (2004) (without annealing and cooling rate 5°C/min).
[b] Rahman et al. (2005) (without annealing and cooling rate 5°C/min).
[c] Moraga et al. (2006) (heating rate: 5°C/min, 30 min annealing at −35°C, t'_m: −40.4°C, t'_g: −52.0, X'_s: 0.814).
[d] Del Valle et al. (1998) (heating rate: 10°C/min).
[e] Sa et al. (1999) (heating rate: 5°C/min).

Table 8.8 Glass Transition Temperature above Unfrozen Water Content When There Is No Formation of Ice

	Honey[a]				Strawberry[b–d]					Onion (Freeze-Dried Powder)[c]				Onion (Freeze-Dried Flake)[c]				Grape[c]			
y	X_s	t_{gi} (°C)	t_{gp} (°C)	t_{ge} (°C)	a_w	X_s	t_{gi} (°C)	t_{gp} (°C)	t_{ge} (°C)	a_w	X_s	t_{gi} (°C)	t_{ge} (°C)	a_w	X_s	t_{gi} (°C)	t_{ge} (°C)	a_w	X_s	t_{gi} (°C)	t_{ge} (°C)
100	0.793	−45.7	−38.2	−35.7	0.11	0.978	23.9	29.5	34.9	0.12	0.965	15.0	26.5	0.12	0.979	30.8	46.8	0.12	0.929	−15.8	−7.0
98	0.777	−50.3	−44.0	−41.1	0.23	0.964	8.0	13.1	17.7	0.23	0.945	—	—	0.23	0.964	14.8	25.0	0.23	0.922	−22.3	−9.0
90	0.714	−64.2	−58.7	−55.8	0.32	0.952	−5.4	−0.8	4.9	0.33	0.925	−11.0	−1.5	0.33	0.941	−1.5	10.8	0.33	0.911	−22.6	−14.8
					0.43	0.933	−15.2	−8.5	−1.2	0.44	0.891	−21.1	−8.7	0.44	0.909	−12.0	−0.5	0.44	0.871	−35.2	−26.4
					0.50	0.893	−33.7	−26.4	−19.0	0.53	0.855	−26.1	−15.4	0.53	0.885	−26.5	−13.4	0.53	0.832	−44.5	−37.1
					0.75	0.528	−69.0	−65.0	−59.3	0.61	0.812	−46.6	−34.7	0.61	0.828	−40.0	−23.9	0.61	0.782	−59.3	−51.7
					0.12	0.992	20.1	—	25.1	0.76	0.728	−68.8	−57.6	0.76	0.761	−59.3	−45.9	0.76	0.710	−75.5	−67.7
					0.33	0.948	−11.5	—	0.0	0.85	0.629	−87.5	−78.0	0.85	0.659	−83.9	−69.5	0.85	0.618	−86.3	−80.9
					0.44	0.900	−24.7	—	−14.7												
					0.53	0.856	−38.3	—	−26.8												
					0.76	0.730	−73.6	—	−64.7												
					0.85	0.615	−90.8	—	−83.0												
					0.00	1.000	36.3	—	52.8												
					0.00	1.000	39.5	—	53.9												
					0.00	1.000	47.8	—	57.6												
					0.12	0.990	21.4	—	46.7												
					0.23	0.974	2.3	—	42.0												
					0.33	0.939	−10.3	—	4.5												
					0.44	0.909	−23.3	—	−9.1												
					0.52	0.885	−34.4	—	−20.2												
					0.75	0.767	−65.3	—	−54.3												

y: Percent honey.
[a] Kantor et al. (1999) (DSC heating rate: 10°C/min).
[b] Moraga et al. (2004) (DSC heating rate: 5°C/min).
[c] Sa and Sereno (1994) (DSC heating rate: 5°C/min).
[d] Roos (1987) (DSC heating rate: 5°C/min).

Table 8.9 Glass Transition Temperature above Unfrozen Water Content When There Is No Formation of Ice

Pear[a]			Apple (Air-Dried at 30°C)[b]			Apple (Air-Dried at 60°C)[b]			Apple (Freeze-Dried)[b]			Apple Juice (Freeze-Dried)[b]			Cabbage[c]			
a_w	X_w	t_{gp} (°C)	a_w	X_w	t_{gi} (°C)	a_w	X_w	t_{gi} (°C)	a_w	X_w	t_{gi} (°C)	a_w	X_w	t_{gi} (°C)	a_w	X_w	t_{gi} (°C)	t_{ge} (°C)
ND	0.877	ND	0.231	0.030	−20.6	0.000	0.000	−2.2	0.000	0.000	4.5	0.231	0.038	−25.5	—	0.043	38.6	72.7
0.885	0.350	−45.0	0.326	0.057	−23.6	0.000	0.020	−8.5	0.000	0.003	−12.5	0.326	0.045	−26.5	—	0.049	10.7	31.1
0.602	0.185	−40.5	0.436	0.097	−39.4	0.231	0.026	−13.2	0.231	0.025	−21.6	0.436	0.057	−33.9	—	0.099	−9.2	8.4
0.419	0.110	−23.2	0.539	0.137	−51.8	0.326	0.034	−18.6	0.326	0.049	−31.2	0.539	0.130	−52.6	—	0.150	−31.0	−11.3
0.388	0.100	−20.7	0.640	0.188	−65.8	0.436	0.092	−43.0	0.436	0.074	−38.7	0.640	0.187	−67.8	—	0.215	−58.4	−44.2
0.290	0.055	−7.1	0.756	0.244	−78.5	0.539	0.135	−54.8	0.539	0.113	−42.1	0.756	0.257	−81.4	0.26	0.266	−66.3	−47.5
0.236	0.043	−1.4				0.640	0.188	−70.3	0.640	0.167	−63.2				0.29	0.300	−68.4	−54.8
						0.756	0.251	−79.4	0.756	0.233	−79.3				—	0.377	−82.9	−69.5

[a] Ling et al. 2005 (Cooling rate: 10°C/min).
[b] Welti-Chanes et al. (1999).
[c] Paakkonen and Plit (1991a).

Table 8.10 Phase Transitions of Pigskin Gelatin at Different Moisture Contents

		First Scan						Second Scan		
		Glass Transition			Melting			Glass Transition		
a_w	X_w	t_{gi} (°C)	t_{gp} (°C)	t_{ge} (°C)	t_{mi} (°C)	t_{mp} (°C)	t_{me} (°C)	t_{gi} (°C)	t_{gp} (°C)	t_{ge} (°C)
Samples prepared by absorption										
0.11	0.063	118.9	121.8	126.4	135.7	138.5	141.9	105.6	114.2	118.9
0.33	0.099	83.0	87.0	91.1	103.8	106.4	109.5	71.4	77.2	84.1
0.43	0.115	70.8	73.7	77.7	95.0	96.1	100.7	59.8	65.0	71.9
0.52	0.130	53.9	57.4	58.5	88.6	89.7	93.7	51.0	57.9	63.7
0.64	0.154	49.8	52.1	53.8	82.7	83.3	86.7	41.1	46.3	54.4
0.75	0.184	35.3	38.2	41.0	72.2	72.8	77.4	24.9	30.6	37.5
0.84	0.217	19.1	23.1	32.2	48.5	49.6	52.5	9.8	16.1	25.9
Samples prepared by desorption										
0.11	0.063	103.4	104.7	107.8	110.3	110.4	113.5	78.3	86.3	94.9
0.33	0.099	69.1	73.4	75.9	80.2	82.1	84.0	45.7	53.7	59.9
0.43	0.115	56.2	61.8	66.1	75.3	76.4	78.5	39.0	46.4	53.8
0.52	0.130	55.1	58.8	59.4	70.5	71.2	74.3	35.4	41.6	47.1
0.64	0.154	47.8	52.1	54.0	66.3	66.9	69.4	27.5	34.3	42.3
0.75	0.184	39.8	44.8	51.6	63.3	63.3	65.8	20.8	29.4	34.9
0.84	0.217	14.7	19.0	27.6	51.0	54.1	57.8	7.3	12.9	20.3

Source: Adapted from Sobral, P.J.A. and Habitante, A.M.Q.B., *Food Hydrocolloid.*, 15, 377, 2001.
Note: DSC heating at 10°C/min until 105°C and annealed for 24 h. A second scan at 10°C/min until glass transition was observed.

where V_v is the free volume at temperature T, and $(\alpha_l - \alpha_g)$ is the volume expansion coefficient of the liquid minus the volume expansion coefficient of the glass. If free volumes are assumed to be additive, the effect of a diluent on the glass transition temperature of polymer–diluents system can be written as (Bueche, 1979)

Table 8.11 Glass Transition of Food Components as a Function of Water Content

Sorbitol[a]		Lignin[b,1]		Lignin[b,2]		Lignin[b,3]		Wheat Starch[c,4]		Wheat Starch[c,5]	
X_w	t_{gp} (°C)	X_w	t_{gp} (°C)	X_w	t_{gp} (°C)	X_w	t_{gp} (°C)	X_w	t_{gp} (°C)	X_w	t_{gp} (°C)
0.000	−2.7	0.000	166.6	0.000	156.4	0.000	124.0	0.072	127.5	0.130	88.3
0.044	−16.3	0.041	144.5	0.036	119.2	0.021	94.0	0.075	120.0	0.131	84.2
0.044	−20.0	0.129	113.6	0.071	109.6	0.050	86.8	0.094	107.2	0.136	81.9
0.060	−17.7	0.217	101.6	0.100	104.8	0.068	79.6	0.096	103.3	0.141	75.4
0.060	−20.0	0.234	98.3	0.150	100.0	0.086	72.4	0.108	92.3	0.142	65.8
0.060	−23.8	0.316	94.9	0.204	96.4	0.121	70.0	0.118	74.9	0.144	64.7
0.108	−35.0	0.387	93.3			0.135	67.7	0.120	74.9	0.147	63.2
0.123	−37.5					0.200	65.2	0.122	66.8	0.149	62.0
0.150	−43.9					0.300	65.2	0.130	61.7	0.151	57.2
0.150	−46.0							0.142	54.2	0.157	55.5
0.164	−47.3							0.152	40.0	0.159	53.9
0.206	−51.7							0.182	28.7	0.163	53.9
0.220	−56.5							0.197	24.7	0.163	50.0

(*continued*)

Table 8.11 (continued) Glass Transition of Food Components as a Function of Water Content

Sorbitol[a]		Lignin[b,1]		Lignin[b,2]		Lignin[b,3]		Wheat Starch[c,4]		Wheat Starch[c,5]	
X_w	t_{gp} (°C)	X_w	t_{gp} (°C)	X_w	t_{gp} (°C)	X_w	t_{gp} (°C)	X_w	t_{gp} (°C)	X_w	t_{gp} (°C)
0.262	−58.4									0.170	50.0
0.301	−63.9									0.170	48.4
0.343	−65.9									0.170	47.5
0.371	−68.8									0.171	46.3
0.439	−71.8									0.172	47.5
0.500	−77.7									0.176	40.9
										0.179	42.9
										0.180	41.7
										0.182	37.5
										0.187	34.3
										0.188	36.1
										0.190	34.5
										0.191	33.4
										0.193	32.2
										0.200	30.0
										0.201	26.9
										0.203	24.7
										0.210	26.3
										0.213	24.2
										0.219	24.2

[a] Quinquenet et al. (1988).
[b] Kelley et al. (1987).
[c] Zeleznak and Hoseney (1987).

$$T_g = \frac{(\alpha_{pl} - \alpha_{pg})V_p T_{gp} + (\alpha_{pl} - \alpha_{pg})(1 - V_p)T_{dg}}{(\alpha_{pl} - \alpha_{pg})V_p + (\alpha_{dl} - \alpha_{dg})(1 - V_d)} \quad (8.2)$$

where V_p and V_d are the volume of polymer and diluent (m^3), $(\alpha_{pl} - \alpha_{pg})$ and $(\alpha_{dl} - \alpha_{dg})$ are the volume expansion coefficient difference of polymer and diluent, and T_{pg} and T_{dg} are the glass

Table 8.12 Glass Transition, Crystallization, and Solute Melting of Sucrose and Lactose

			Glass Transition		Crystallization		Melting	
Material	a_w	X_w	t_{gi} (°C)	t_{ge} (°C)	T_{ci} (°C)	T_{ce} (°C)	T_{mi} (°C)	T_{me} (°C)
Sucrose	0.00	0.005	58.6	63.9	101.1	106.6	157.0	181.9
	0.11	0.033	33.2	39.0	74.6	80.3	140.6	169.7
	0.22	0.047	21.8	27.5	61.9	72.2	133.0	163.2
Lactose	0.00	0.004	93.2	97.3	154.0	162.6	ND	ND
	0.11	0.044	49.4	54.2	97.7	98.5	ND	ND
	0.22	0.057	32.3	39.7	76.5	83.4	ND	ND
	0.33	0.063	26.8	33.8	72.3	78.8	ND	ND

Source: Ottenhof, M., MacNaughtan, W., and Farhat, I.A., *Carbohydr. Res.*, 338, 2198, 2003. With permission.
Note: Heating: 3°C/min.

Table 8.13 Calorimetric Glass Transition Temperature and α- and β-Dielectric Relaxation Peak Parameters for Dry Amorphous Carbohydrates and Their 10% Water Mixtures

						α-Relaxation		β-Relaxation	
Carbohydrate	t_{gp} (°C)	$t_{\delta p}$ (°C)	Tan δ_{max}	$t_{\beta p}$ (°C)	Tan β_{max}	Log a	*Ea* (kJ/mol)	Log a	*Ea* (kJ/mol)
Dry carbohydrate									
D-Glucose	39	60	0.24	−62	0.037	51.7	320	13.5	42
D-Fructose	11	42	0.26	−50	0.045	51.0	290	13.1	43
D-Mannose	38	60	0.28	−63	0.036	49.5	295	14.4	46
L-Rhamnose	37	58	0.32	−70	0.017[a]	53.9	320	A	A
D-Xylose	10	31	0.34	B	B	45.1	250	A	C
Glucitol	−2	16	0.27	−30	0.082[b]	52.5	275	14.0	55[c]
Maltose	97	103	0.11	−50	0.062	59.5	405	13.8	45
Hydrated carbohydrate (X_s: 0.10)									
D-Glucose	−17	5	0.24	−43	0.064	44.2	220	16.1	59
D-Fructose	−24	−3	0.25	−47	0.070	46.7	225	15.0	49
D-Mannose	−15	7	0.27	−46	0.058	47.4	240	15.6	52
L-Rhamnose	−11	9	0.31	−60	0.042	46.9	235	8.5	17[d]
D-Xylose	−29	−10	0.33	B	B	44.1	205	A	A
Glucitol	−33	−16	0.27	B	B	69.6	330	A	A
Maltose	7	32	0.17	−55	0.098	53.5	295	18.0	62
Maltotriose	19	B	B	−65	0.125	A	A	16.0	52

Source: Adapted from Noel, T., Parker, R., and Ring, S., *Carbohydr. Res.*, 329, 839, 2000. With permission.
Note: Spectroscopic transitions were measured at 1 kHz; DSC cooling rate: 10°C/min, and DSC glass was considered as peak.
[a] Measured at 100 kHz.
[b] Measured at 400 Hz.
[c] Data only 100 Hz–1 kHz.
[d] Data only 400 Hz–kHz.
A Activation energy not determined.
B Peak not observed.

transition temperature of the polymer and diluent. The value of α_{dl} is difficult to estimate, but it can be considered as 10^{-3} K^{-1} and for most polymers α_p is approximately 4.8×10^{-4} K^{-1} (Bueche, 1979). For starch, the semiempirical rule is (Whittam et al., 1991)

$$T_g = (2/3)T_m \tag{8.3}$$

Gordon and Taylor (1952) proposed the following equation to predict the glass transition temperature of polymer blends:

$$t_g = \frac{X_w t_{gw} + \sum_{i=1}^{n} X_i[(t_g)_i/k_i]}{X_w + \sum_{i=1}^{n} (X_i/k_i)} \tag{8.4}$$

where i represents the solutes and k_i represents the Gordon–Taylor parameter for ith components. For a two-component mixture, such as water and solids, the above equation can be written as

$$T_g = \frac{X_w T_{gw} + kX_s T_{gs}}{X_w + kX_s} \tag{8.5}$$

Table 8.14 Values of t'_g and t_{gs} for Low Molecular Weight Polyhydroxy Compounds

Compound	λ_s	t'_g (°C)	t_{gs} (°C)	M'_w	t_m (°C)	Reference
β-1-*o*-Methyl glucoside	194.2	−47.0	—	1.29	—	Levine and Slade (1992)
1,3-Butanediol	90.1	−63.5	—	1.41	—	Levine and Slade (1992)
1-*o*-Ethyl galactoside	208.2	−45.0	—	1.26	—	Levine and Slade (1992)
1-*o*-Ethyl glucoside	208.2	−46.5	—	1.35	—	Levine and Slade (1992)
1-*o*-Ethyl mannoside	208.2	−43.5	—	1.21	—	Levine and Slade (1992)
1-*o*-Methyl galactoside	194.2	−44.5	—	0.86	—	Levine and Slade (1992)
1-*o*-Methyl mannoside	194.2	−43.5	—	1.43	—	Levine and Slade (1992)
1-*o*-Propyl galactoside	222.2	−42.0	—	1.05	—	Levine and Slade (1992)
1-*o*-Propyl glucoside	222.2	−43.0	—	1.22	—	Levine and Slade (1992)
1-*o*-Propyl mannoside	222.2	−40.5	—	0.95	—	Levine and Slade (1992)
2,3,4,6-*o*-Methyl glucoside	236.2	−45.5	—	1.41	—	Levine and Slade (1992)
2-*o*-Ethyl fructoside	208.2	−46.5	—	1.15	—	Levine and Slade (1992)
2-*o*-Methyl fructose	194.2	−51.5	—	1.61	—	Levine and Slade (1992)
3-*o*-Methyl glucoside	194.2	−45.5	—	1.34	—	Levine and Slade (1992)
6-*o*-Methyl galactoside	194.2	−45.5	—	0.98	—	Levine and Slade (1992)
Allose	180.2	−41.5	—	—	—	Levine and Slade (1992)
Altrose	180.2	−43.5	10.5	—	107.0	Levine and Slade (1992); Roos (1992)
Arabinose	150.1	−47.5 to −48.5	—	1.23	87.0	Levine and Slade (1992); Franks (1985)
Arabitol	152.1	−47.0 to −48.0	—	0.89	—	Levine and Slade (1992)
Cellobiose	342.3	−29.0	77.0	—	249.0	Levine and Slade (1992)
Cellobiulose	342.3	−32.5	—	—	—	Levine and Slade (1992)
Erythritol	122.1	−53.5	—	Eutectic	—	Levine and Slade (1992)
Erythrose	120.1	−50.0	—	1.39	—	Levine and Slade (1992)
Ethyl alcohol	—	—	−177.0 to −183.0	—	—	White and Cakebread (1966)
Ethylene glycol	62.1	−85.0	—	1.90	—	Levine and Slade (1992)
Fructose (deoxygalactose)	164.2	−43.0	—	1.11	—	Levine and Slade (1992)
Fructose	180.2	−42.0	7.0–17.6	0.95–0.96	124.0	Levine and Slade (1992); Franks (1985); Roos (1992); Finegold et al. (1989)
Fructose:glucose (1:1)	180.2	−41.5 to −42.5	13.0–21.6	—	—	Levine and Slade (1992); Orford et al. (1990); Finegold et al. (1989)

Fructose:glucose (1:3)	—	—	28.6	—	—	Finegold et al. (1989)
Fructose:glucose (1:0.33)	—	—	16.9	—	—	Finegold et al. (1989)
Fructose:sucrose (1:15.67)	—	—	55.8	—	—	Finegold et al. (1989)
Fructose:sucrose (1:1.94)	—	—	46.4	—	—	Finegold et al. (1989)
Fructose:sucrose (1:0.64)	—	—	35.1	—	—	Finegold et al. (1989)
Galactitol	182.2	−39.0	—	Eutectic	—	Levine and Slade (1992)
Galactose	180.2	−40.5 to −42.0	110.0	0.75–0.77	170.0	Levine and Slade (1992); Franks (1985); Roos (1992)
Gentiobiose	342.3	−31.5	—	0.26	—	Levine and Slade (1992)
Glucitol	—	—	0.0	—	—	Orford et al. (1990)
Glucoheptose	210.2	−37.5	—	—	—	Levine and Slade (1992)
Glucoheptulose	210.2	−36.5	—	0.77	—	Levine and Slade (1992)
Glucose	180.2	−36.5 to −43.0	20.0–36.8	0.40–0.41	150.0–158.0	Levine and Slade (1992); Franks (1985); White and Cakebread (1966); Orford et al. (1990); Finegold et al. (1989)
Glucose:galactose (1:1)	—	—	32.0	—	—	Orford et al. (1990)
Gluose	180.2	−42.5	—	—	—	Levine and Slade (1992)
Glycerol	92.1	−65.0 to −95.0	−83.0 to −93.0	0.85	18.0	Levine and Slade (1992); Franks (1985); Roos (1992); White and Cakebread (1966); Ollett et al. (1991)
Idose	180.2	−44.0	—	—	—	Levine and Slade (1992)
Inositol	180.2	−35.5 to −36.0	—	0.31	—	Levine and Slade (1992); Franks (1985)
Isomaltose	342.3	−32.0 to −32.5	78.0	0.70	—	Levine and Slade (1992)
Isomaltotroise	504.5	−30.5	—	0.50	—	Levine and Slade (1992)
Isomaltulose (palatinose)	342.3	−35.5	—	—	—	Levine and Slade (1992)
Isopropyl alcohol	—	—	−153.0	—	—	White and Cakebread (1966)
Lactose	342.3	−28.0	101.0–103.0	0.69–0.70	241.0	Levine and Slade (1992); Franks (1985); Roos (1992)
Lactulose	342.3	−30.0	—	0.72	—	Levine and Slade (1992)
Laminaribiose	342.3	−31.5	—	—	—	Levine and Slade (1992)
Lyxose	150.1	−47.5	8.0	—	115	Levine and Slade (1992); Roos (1992)
Maltitol	344.3	−34.5	—	0.59	—	Levine and Slade (1992)
Maltoheptaose	1153.0	−13.5	138.5	0.27	—	Levine and Slade (1992)

(continued)

Table 8.14 (continued) Values of t_g' and t_{gs} for Low Molecular Weight Polyhydroxy Compounds

Compound	λ_s	t_g' (°C)	t_{gs} (°C)	M_w'	t_m (°C)	Reference
Maltohexaose	990.9	−14.5	134.0–175.0	0.50	—	Levine and Slade (1992)
Maltopentaose	829.9	−16.5	125.0–165.0	0.47	—	Levine and Slade (1992)
Maltose	342.3	−29.5 to −41.0	43.0–87.0	0.25	129.0	Levine and Slade (1992); Franks (1985); Roos (1992)
Maltotetratose	666.6	−19.5	111.5–147.0	0.55	—	Levine and Slade (1992)
Maltotriose	504.5	−23.0 to −24.0	76.0–135.0	0.45	133.5	Levine and Slade (1992)
Maltotriose:glucitol (1:1)	—	—	20.0	—	—	Orford et al. (1990)
Maltotriose:xylose (1:1)	—	—	43.0	—	—	Orford et al. (1990)
Maltotriose:xylitol (1:1)	—	—	0.0	—	—	Orford et al. (1990)
Maltrin M040	—	—	188.0	—	—	Roos (1993)
Maltrin M100	—	—	160.0	—	—	Roos (1993)
Maltrin M200	—	—	141.0	—	—	Roos (1993)
Maltrin M250	—	—	121.0	—	—	Roos (1993)
Maltrin M365	—	—	100.0	—	—	Roos (1993)
Maltulose	342.3	−29.5	—	—	—	Levine and Slade (1992)
Mannitol	182.2	−40.0	—	Eutectic	—	Levine and Slade (1992)
Mannobiose	342.3	−30.5	90.0	0.91	205.0	Levine and Slade (1992)
Mannoheptose	210.2	−36.5	—	—	—	Levine and Slade (1992)
Mannose	180.2	−41.0	30.0	0.35	139.5–140.0	Levine and Slade (1992); Roos (1992)
Melibiose	342.3	−30.5	95.0	—	—	Levine and Slade (1992)
Methyl riboside	164.2	−53.0	—	0.96	—	Levine and Slade (1992)
Methyl xyloside	164.2	−49.0	—	1.01	—	Levine and Slade (1992)
Nigerose	342.3	−35.5	—	—	—	Levine and Slade (1992)
Nystose	666.6	−26.5	77.0	—	—	Levine and Slade (1992)
Panose	504.5	−28.0	—	0.59	—	Levine and Slade (1992)
Perseitol (mannoheptitol)	212.2	−32.5	—	Eutectic	—	Levine and Slade (1992)
n-Propyl alcohol	—	—	−172.0	—	—	White and Cakebread (1966)
Propylene glycol	76.1	−67.5	—	1.28	—	Levine and Slade (1992)

Psicose	180.2	−44.0	—	—	—	Levine and Slade (1992)
Quinovose (deoxyglucose)	164.2	−43.5	—	1.11	—	Levine and Slade (1992)
Raffinose	504.5	−26.5	—	0.70–0.71	—	Levine and Slade (1992); Franks (1985)
Rhamnose (deoxymannose)	164.2	−43.0	—	0.90	—	Levine and Slade (1992)
Ribitol	152.1	−47.0	—	0.82	—	Levine and Slade (1992)
Ribose	150.1	−47.0 to −49.0	−10.0	0.49	87.0	Levine and Slade (1992); Roos (1992)
Ribulose	150.1	−50.0	—	—	—	Levine and Slade (1992)
Sorbitol	182.2	−43.0 to −44.0	−2.0	0.23	111.0	Levine and Slade (1992); Franks (1985)
Sorbose	180.2	−41.0	—	—	—	Levine and Slade (1992)
Starch	—	−6 to −2	—	0.39	151.0–227.0	Levine and Slade (1987); Ollett et al. (1991); Van den Berg (1981)
Starchyose	666.6	−23.5	—	1.12	—	Levine and Slade (1992)
Sucrose	342.3	−32.0 to −46.0	52.0–70.0	0.56	192.0	Levine and Slade (1992); Franks (1985); Roos (1992); White and Cakebread (1966); Finegold et al. (1989)
Sucrose:fructose (1:1)	—	—	25.0	—	—	Orford et al. (1990)
Sucrose:glucose (1:1)	—	—	47.0	—	—	Orford et al. (1990)
Tagatose	180.2	−40.5	—	—	—	Levine and Slade (1992)
Talose	180.2	−44.0	11.5	—	140.0	Levine and Slade (1992)
Thyminose (deoxyribose)	134.1	−52.0	—	1.32	—	Levine and Slade (1992)
Trehalose	342.3	−29.5 to −30.0	77.0–79.0	0.20–0.21	203.0	Levine and Slade (1992); Franks (1985); Roos (1992)
Treose	120.1	−45.5	—	—	—	Levine and Slade (1992)
Turanose	342.3	−31.0	52.0	0.64	177.0	Levine and Slade (1992),
Water	—	—	−125.0 to −150.0	—	—	White and Cakebread (1966); Ollett et al. (1991); and Orford et al. (1990)
Xylitol	152.1	−46.5	−18.0 to −18.5	0.75	94.0	Levine and Slade (1992); Roos (1992)
Xylose	150.1	−47.0 to −48.0	9.5	0.45	153.0	Levine and Slade (1992); Franks (1985)
α-1-*o*-Methyl glucoside	194.2	−44.5	—	1.32	—	Levine and Slade (1992)
α-Cyclodextrin	972.9	−9.0	—	—	—	Levine and Slade (1992)

Table 8.15 Maximally Freeze-Concentrated Conditions for Commercial SHP

SHP	Manufacturer	Source	DE	t'_g (°C)
AB 7436	Anheuser Busch	Waxy maize	0.5	−4.0
Paselli SA-2	AVEBE	Potato	2.0	−4.5
Stadex 9	Staley	Dent corn	3.4	−4.5
78NN128	Staley	Potato	0.6	−5.0
78NN122	Staley	Potato	2.0	−5.0
ARD2326	Amaizo	Dent corn	0.4	−5.5
ARD2308	Amaizo	Dent corn	0.3	−6.0
AB7435	Anheuser Bush	Waxy/dent blend	0.5	−6.0
Star Dri 1	Staley	Dent corn	1.0	−6.0
Maltrin M050	GPC	Dent corn	6.0	−6.0
Paselli MD-6	AVEBE	Potato	6.0	−6.5
Dextrin 11	Staley	Tapioca	1.0	−7.5
MD-6-12	V-labs	—	2.8	−7.5
Stadez 27	Staley	Dent corn	10.0	−7.5
MD-6-40	V-Labs	—	0.7	−8.0
Star Dri 5	Staley	Dent corn	5.0	−8.0
Paselli MD-10	AVEBE	Potato	10.0	−8.0
Morrex 1910	CPC	Dent corn	10.0	−9.5
Star Dri 10	Staley	Dent corn	10.0	−10.0
Maltrin M040	GPC	Dent corn	5.0	−10.5
Frodex 5	Amaizo	Waxy maize	5.0	−11.0
Morrex 1918	CPC	Waxy maize	10.0	−11.5
Maltrin M100	GPC	Dent corn	10.0	−11.5
Lodex 5	Amaizo	Waxy maize	7.0	−12.0
Maltrin M500	GPC	Dent corn	10.0	−12.5
Lodex 10	Amaizo	Waxy maize	12.0	−12.5
Maltrin M150	GPC	Dent corn	15.0	−13.5
MD-6-1	V-Labs	—	20.5	−13.5
Frodex 15	Amaizo	Waxy maize	18.0	−14.0
Frodex 10	Amaizo	Waxy maize	10.0	−15.5
Lodex 15	Amaizo	Waxy maize	18.0	−15.5
Maltohexaose	V-Labs	—	18.2	−15.5
Maltrin M200	GPC	Dent corn	20.0	−15.5
Maltrin M250	GPC	Dent corn	25.0	−17.5
Stalet 200	Staley	Corn	26.0	−19.5
Frodex 24	Amaizo	Waxy maize	28.0	−20.5
Dri Sweet 36	Hubinger	Corn	36.0	−22.0
Maltrin M365	GPC	Dent corn	36.0	−22.5
Staley 300	Staley	Corn	35.0	−23.5
Globe 1052	CPC	Corn	37.0	−23.5
Maltotriose	V-Labs	—	35.7	−23.5
Frodex 42	Amaizo	Waxy maize	42.0	−25.5
Neto 7300	Staley	Corn	42.0	−26.5
Globe 1132	CPC	Corn	43.0	−27.5
Staley 1300	Staley	Corn	43.0	−27.5
Neto 7350	Staley	Corn	50.0	−27.5

Table 8.15 (continued) Maximally Freeze-Concentrated Conditions for Commercial SHP

SHP	Manufacturer	Source	DE	t'_g (°C)
Maltose	Sigma	—	52.6	−29.5
Globe 1232	CPC	Corn	54.5	−30.5
Staley 2300	Staley	Corn	54.0	−31.0
Sweerose 4400	Staley	Corn	64.0	−33.5
Sweetose 4300	Staley	Corn	64.0	−34.0
Globe 1642	CPC	Corn	63.0	−35.0
Globe 1632	CPC	Corn	64.0	−35.0
Royal 2626	CPC	Corn	95.0	−42.0
Glucose	Sigma	Corn	100.0	−43.0

Source: Levine, H. and Slade, L., *Carbohydr. Polym.*, 6, 213, 1986. With permission.

Couchman (1978) and Couchman and Karasz (1978) showed from the thermodynamic point that the k value in the above equation was equivalent to the ratio of the change in component mixture specific heats at their T_g as

$$k = \frac{(\Delta C_p)_{gs}}{(\Delta C_p)_{gd}} \tag{8.6}$$

Kwei (1984, cited by Couchman, 1978) proposed another equation, which is more accurate in the case of polymeric compounds:

$$T_g = \frac{X_p T_{pg} + k X_d T_{dg}}{X_p + k X_d} + q X_d X_p \tag{8.7}$$

Table 8.16 Freezing Point (Ice Forming) and Melting (Solids) Point of Diethyl Sulphoxide (Cryoprotectant)

Freezing Point					Melting Point of Solutes				
X_s^0	t_{Fi} (°C)	t_{Fp} (°C)	t_{Fe} (°C)	ΔH (kJ/kg)	X_s^0	T_{mi} (°C)	T_{mp} (°C)	T_{me} (°C)	ΔH (kJ/kg)
0	0.0	0.3	2.4	334.0	0.45–0.70	N	N	N	N
0.05	−3.5	−1.0	0.8	270.2	0.80a	−39.7	−23.1	−20.0	52.1
0.10	−5.7	−2.2	−1.0	231.6	0.85	−31.1	−14.8	−11.3	61.2
0.15	−7.9	−3.9	−2.7	174.3	0.90	−19.8	−5.6	−2.1	113.9
0.20	−11.0	−6.0	−4.9	147.1	0.95	−6.2	4.1	7.0	142.9
0.25	−14.7	−8.4	−7.0	117.5	1.00	20.2	24.5	25.4	197.8
0.30	−19.3	−11.3	−10.0	86.5					
0.35	−24.5	−14.9	−13.6	67.3					
0.40	−30.8	−19.8	−18.3	46.8					
0.45–0.70	n	N	n	N					

Source: Markarian, S.A., Bonora, S., Bagramyan, K.A., and Arakelyan, V.B., *Cryobiology*, 49, 1, 2004. With permission.

Note: t'_m: −30.8°C.

n, no transition was observed. N, glass transition observed at −114°C and −65.5°C.

Table 8.17 Freezing Point and Glass Transition of Pineapple as a Function of Moisture Content

Sample	X_s^0	a_w	t_F[a] (°C)	t_F[b] (°C)	t_F[c] (°C)	$(\Delta H)_{sample}$ (kJ/kg)	t'_m (°C)	Transition 1 t_{gi} (°C)	Transition 1 t_{gp} (°C)	Transition 1 t_{ge} (°C)	Transition 2 (with Melting of Ice) t_{gi} (T''_g) (°C)	Transition 2 (with Melting of Ice) t_{ge} (t'_m) (°C)
Pineapple (Telis and Sobral, 2001; without annealing, DSC method, cooling rate: 10°C/min, sample size: 10 mg)	0.645	0.82	−19.84	−22.53	−22.08	25.07	−24.33	−81.31	−75.49	−69.66	−24.78	−24.33
	0.642	0.81	−20.72	−22.52	−22.97	28.65	−23.87	—	—	—	−26.13	−23.87
	0.635	0.84	−20.29	−25.67	−25.67	17.30	−28.37	−85.82	−78.64	−73.25	−31.96	−28.37
	0.635	0.84	−25.35	−26.90	−25.35	—	−31.47	−88.37	−79.79	−72.57	−32.34	−31.47
	0.607	0.85	−17.59	−24.78	−26.57	53.78	−31.51	−84.47	−80.88	−74.60	−33.30	−31.51
	0.578	0.87	−14.45	−20.29	−25.22	—	−32.85	−89.86	−83.57	−76.84	−35.39	−32.85
	0.570	0.88	−16.15	−22.08	−27.92	—	−35.10	−93.00	−87.61	−81.78	−44.97	−35.10
	0.90	—	−13.33	−21.61	−26.24	65.92	−34.49	−50.75	−48.49	−43.51	−38.06	−34.49
	0.516	—	−7.46	−12.69	−19.78	77.70	−27.99	−52.24	−49.27	−46.27	−38.43	−27.99
	0.451	—	−4.48	−8.21	−12.69	108.88	−25.37	−51.49	−48.13	−45.42	−36.57	−25.37
	0.403	—	−4.48	−7.84	−12.31	127.78	−24.63	—	—	—	−37.31	−24.63
	0.300	—	−0.75	−4.85	−9.33	167.63	−17.54	—	—	—	−36.57	−17.54
	0.199	—	0.37	−2.99	−6.72	222.91	−14.93	—	—	—	−36.19	−14.93
	0.098	—	4.10	0.37	−2.61	268.42	−8.21	—	—	—	−37.31	−8.21
Pineapple (Telis and Sobral, 2001; with annealing, DSC method, cooling rate: 10°C/min; sample size: 10 mg)	0.642	0.81	−19.37	−23.42	−24.32	—	−28.83	−68.47	−64.41	−58.56	−39.98	−28.83
	0.645	0.82	−19.39	−25.22	−25.67	—	−28.82	−71.01	−63.82	−55.75	−34.65	−28.82
	0.635	0.84	−20.74	−25.22	−25.67	—	−29.26	−72.35	−64.72	−58.54	−33.75	−29.26
	0.607	0.85	−18.04	−25.22	−29.26	—	−37.79	−59.39	−53.95	−50.81	−42.73	−37.79
	0.578	0.87	−14.90	−19.84	−21.18	—	−22.98	−89.86	−83.57	−83.12	−41.83	−22.98
	0.570	0.88	−15.35	−19.84	−27.47	—	−35.13	−69.66	−66.07	−62.03	−43.18	−35.13
Honey (Kantor et al., 1999; DSC method, cooling rate: 10°C/min)	0.634[d]	—	−24.8 (equilibrium)	−16.2 (end)	−36.5 (initial)	—	−36.5	−80.1	−75.0	−71.7	—	—
	0.555	—	−15.9	−22.9	−29.7	45.2	−38.2	−74.9	−72.6	−68.6	−43.6	−38.2
	0.476	—	−12.4	−17.5	−25.1	77.5	−39.1	−61.8	−57.9	−53.5	−42.3	−39.1
	0.397	—	−8.1	−13.7	−21.6	112.3	−36.2	−61.0	−58.8	−53.3	−41.0	−36.2
	0.238	—	−4.2	−7.0	−12.2	194.0	−21.2	−66.1	−61.2	−55.0	−42.2	−21.2

Apple (Del Valle et al., 1998; heating rate: 10°C/min)	0.669	0.81	−19.7	−23.3	−23.7	—	−25.6	−77.9	−73.9	−69.4	−27.2	−25.6
	0.614	0.85	−16.7	−22.0	−26.5	—	−31.7	−83.0	−80.9	−77.7	−40.0	−31.7
	0.580	0.87	−14.4	−19.1	−25.5	—	−33.4	−87.6	−84.1	−80.6	−42.9	−33.4
	0.518	0.90	−9.0	−13.5	−22.0	—	−30.4	−58.4	−56.2	−52.5	−41.1	−30.4
	0.439	0.93	−7.6	−11.1	−20.2	—	−28.0	−62.4	−57.3	−53.0	−41.1	−28.0
Strawberry (Sa et al., 1999; heating rate: 5°C/min)	Fresh	—	—	−3.6	−5.5	—	−7.8	−56.9	−52.9	−47.0	−31.4	−7.8
Grape (Sa and Sereno, 1994; heating rate: 5°C/min)	0.618	0.850	−37.7	−44.7	−47.1		−47.1	−67.6	−64.2	−62.0	−59.0	−47.1
Garlic (Rahman et al., 2005; heating rate: 5°C/min)	0.200	—	—	−0.9	—	176	—	—	—	—	—	—
	0.300	—	—	−1.8	—	162	—	—	—	—	—	—
	0.500	—	—	−3.9	—	102	—	—	—	—	—	—
	0.600	—	—	−8.9	—	57	−19.1	−43.2	−36.7	−36.7	−26.0	−19.1
	0.650	—	—	−12.7	—	58	−23.6	−48.6	−36.6	−32.2	−24.6	−23.6
	0.700	—	—	−18.6	—	26	−22.9	−49.5	−39.5	−34.7	−24.9	−22.9
	0.720	—	—	−19.2	—	23	−22.6	−50.9	−42.6	−35.8	−26.0	−22.6
	0.740	—	—	−20.7	—	15	−22.2	−50.7	−41.6	−35.3	−26.6	−22.2
	0.760	—	—	−20.5	—	9	−21.6	−52.6	−42.1	−35.8	−25.9	−21.6
	0.780	—	—	−21.8	—	4	−22.0	−53.2	−42.5	−36.8	−26.2	−22.0
	0.800	—	—	−20.7	—	5	−22.1	−48.6	−36.6	−32.2	−26.5	−22.1
Gooseberry (Wang et al., 2008; heating rate: 10°C/min)	0.200	—		−2.2	—	172	−29.6	—	—	—	—	−29.6
	0.300	—		−2.7	—	171	−33.2	—	—	—	—	−33.2
	0.530	—		−7.7	—	102	−39.6	−61.1	−55.2	−51.6	—	−39.6
	0.650	—		−19.0	—	35	−33.6	−59.8	−54.9	−52.0	—	−33.6
	0.670	—		−21.1	—	33	−41.3	−48.1	−41.6	−39.1	—	−41.3
	0.690	—		−19.7	—	40	−41.8	−60.5	−55.0	−51.3	—	−41.8

(continued)

Table 8.17 (continued) Freezing Point and Glass Transition of Pineapple as a Function of Moisture Content

								Transition 1			Transition 2 (with Melting of Ice)	
Sample	X_s^0	a_w	t_F[a] (°C)	t_F[b] (°C)	t_F[c] (°C)	$(\Delta H)_{sample}$ (kJ/kg)	t'_m (°C)	t_{gi} (°C)	t_{gp} (°C)	t_{ge} (°C)	t_{gi} (T'''_g) (°C)	t_{ge} (t'_m) (°C)
Apple (Cornillon, 2000; osmosed, heating rate: 5°C/min)	—	—	−0.65	−4.8	−7.69	148	−22.7	−35.3	−33.1	−32.7	−32.7	−22.7
Sucrose (Roos and Karel, 1991c)	0.200	—	—	—	—	225	−34.0	−46.0	−41.0	−36.0	−36.0	−34.0
	0.300	—	—	—	—	185	−34.0	−46.0	−40.0	−35.0	−35.0	−34.0
	0.400	—	—	—	—	145	−34.0	−46.0	−41.0	−36.0	−36.0	−34.0
	0.500	—	—	—	—	104	—	−47.0	—	—	—	—
	0.650	—	—	—	—	46	−34.0	−46.0	−42.0	−36.0	−36.0	−34.0
	0.680	—	—	—	—	37	−34.0	−46.0	−41.0	−36.0	−36.0	−34.0

[a] Maximum peak.
[b] Maximum slope.
[c] Extension of maximum slope curve to the base line.

TABLE 8.18 Freezing Point and Glass Transition of Foods as a Function of Moisture Content

							Transition 1			Transition 2 (Close to Ice Melting)		
Sample	r_c	X_s	t_F[a] (°C)	t_F[b] (°C)	$(\Delta H)_{sample}$ (kJ/kg)	(t'_m)[c] (°C)	t_{gi} (°C)	t_{gp} (°C)	t_{ge} (°C)	t_{gi} (T'''_g) (°C)	t_{gp} (°C)	t_{ge} (t'_m) (°C)
Date	5	0.220	−1.8	−4.1	172	−26.1	−61.4	−57.6	−54.5	−44.2	−42.6	−41.5
	5	0.298	−3.2	−5.5	151	−34.9	−60.6	−58.4	−56.0	−44.3	−42.2	−40.0
	5	0.360	−4.9	−7.2	122	−31.0	−60.7	−58.4	−54.9	−44.7	−42.9	−40.7
	5	0.414	−7.8	−10.8	87	−40.9	−60.9	−57.1	−54.8	−46.3	−43.3	−41.4
	1	0.511	−12.6	−15.2	59	−37.1	−59.7	−56.9	−54.4	−46.4	−44.3	−42.4
	1	0.541	−18.3	−20.2	55	−38.9	−58.5	−55.3	−54.0	−44.1	−39.7	−39.4
	1	0.589	−20.3	−23.4	41	−37.9	−62.6	−58.6	−56.1	−44.8	−42.6	−39.9
	1	0.611	−22.0	−25.6	27	−43.6	−63.4	−59.3	−57.0	−46.4	−45.4	−43.8

Source: Rahman, M.S., *Int. J. Food Prop.*, 7, 407, 2004. With permission.
Note: DSC method, r_c (cooling rate, °C/min) and AT (annealing time) are the determined optimum conditions.
[a] Maximum peak.
[b] Maximum slope.
[c] At optimum annealing condition (30 min at $t'_m - 1$).

8.6 CONCLUSION

The state diagram is the map of different states of a food as a function of water or solid content and temperature. It assists in identifying food's stability during storage and in selecting a suitable condition of temperature and moisture content for processing. Different parameters are needed to develop the state diagram, such as glass transition, freezing point, and maximal-freeze-concentration conditions. The most common and popular method used to determine glass transition is the DSC. Other methods include TMA and DMA. The values of these parameters for different foods have been compiled and presented in tabular form. The terminologies have been clearly defined and explained before presenting the data.

NOMENCLATURE

A	Parameter in Table 8.13
a	Activity
C	Specific heat (kJ/kg K)
DE	Dextrose equivalent
E	Modulus for longitudinal mode
E_a	Activation energy (kJ/mol)
G	Modulus for oscillation mode (Pa)
H	Enthalpy (kJ/kg)
k	Gordon–Taylor parameter
M	Mass fraction (kg/kg dry solids)
n	Number of constituents in a mixture
q	Kwei model parameter
r	Rate (°C/min)
T	Temperature (K)
t	Temperature (°C)
V	Volume (m^3)
X	Mass fraction (kg/kg sample)
y	Percent honey

Greek Symbol

α	Volume expansion coefficient
δ	Loss factor
λ	Molecular weight
θ	Time (h)
Δ	Difference

Subscript

an	Annealing
c	Cooling
ce	Crystallization end
ci	Crystallization initial
d	Diluent
dg	Diluent glass
dl	Diluent liquid

ds Deterioration
F Freezing
Fe End freezing
Fi Initial freezing
Fp Peak freezing
g Glass transition
gd Glass diluent
ge End of glass transition (i.e., end of glass–rubber transition)
gi Onset in glass transition (i.e., start of glass–rubber transition)
gp Peak in glass transition (i.e., mid of glass–rubber transition)
gs Glass solids
gw Glass water
h Heating
i *i*th constituent
l Liquid
m Melting
me Melting end
mi Melting initial
mp Melting peak
ms Melting solids
p Polymer or constant pressure in *C*
pg Polymer glass
pl Polymer liquid
R Rheological glass transition
rc Rheological glass–rubber transition from crossover (Figure 8.10)
re End of rheological glass–rubber transition (Figure 8.10)
ri Onset of rheological glass–rubber transition (Figure 8.10)
rp Rheological glass from E'' or G'' peak (Figure 8.10)
s Solids or solutes
v Free volume or void
w Water
δp Glass transition from maximum peak of loss factor
βp Peak in β transition

Superscript

0 Before freezing
$'$ Maximal-freeze-concentration condition (defined in Figure 8.1) in *T* or *t* or storage in *G*
$''$ Maximal-freeze-concentration condition (defined in Figure 8.1) in *T* or *t* or loss in *G*
$'''$ Maximal-freeze-concentration condition (defined in Figure 8.1)

REFERENCES

Abbott, T.P., Nabetani, H., Sessa, D.J., Wolf, W.J., Liebman, M.N., and Dukor, R.K. 1996. Effects of bound water on FTIR spectra of glycinin. *Journal of Agriculture and Food Chemistry*. 44: 2220–2224.

Ablett, S., Izzard, M.J., and Lillford, P.J. 1992. Differential scanning calorimetry study of frozen sucrose and glycerol solutions. *Journal of Chemical Society: Faraday Transactions*. 88(6): 789–794.

Ablett, S., Izzard, M.J., Lillford, P.J., Arvanitoyannis, I., and Blanshard, J.M.V. 1993. Calorimetric study of the glass transition occurring in fructose solutions. *Carbohydrate Research*. 246: 13–22.

Appelqvist, I.A.M., Cooke, D., Gidley, M.J., and Lane, S.J. 1993. Thermal properties of polysaccharides at lowmoisture: 1. An endothermic melting process and water–carbohydrate interactions. *Carbohydrate Polymer*. 20: 291–299.

Baroni, A.F., Sereno, A.M., and Hubinger, M.D. 2003. Thermal transitions of osmotically dehydrated tomato by modulated temperature differential scanning calorimetry. *Thermochimica Acta*. 395: 237–249.

Bell, L.N., Bell, H.M., and Glass, T.E. 2002. Water mobility in glassy and rubbery solids as determined by oxygen-17 Nuclear Magnetic Resonance: Impact on chemical stability. *Food Science and Technology*. 35: 108–113.

Blond, G. 1994. Mechanical properties of frozen model solutions. *Journal of Food Engineering*. 22: 253–269.

Bueche, F. 1979. *Physical Properties of Polymers*. R.E. Krieger Publishing, Huntington, New York.

Champion, D., Meste, M. Le., and Simatos, D. 2000. Towards an improved understanding of glass transition and relaxations in foods: Molecular mobility in the glass transition range. *Trends in Food Science & Technology*. 11: 41–55.

Chinachoti, P. and Stengle, T.R. 1990. Water mobility in starch/sucrose systems: An oxygen-17 NMR. *Journal of Food Science*. 55(6): 1732–1734.

Choi, S. and Kerr, W.L. 2003. 1H NMR studies of molecular mobility in wheat starch. *Food Research International*. 36(4): 341–348.

Cornillon, P. 2000. Characterization of osmotic dehydrated apple by NMR and DSC. *Food Science and Technology*. 33: 261–267.

Couchman, P.R. 1978. Compositional variation of glass-transition temperatures. 2. Application of the thermodynamic theory to compatible polymer blends. *Macromolecules*. 11(6): 1156–1161.

Couchman, P.R. and Karasz, F.E. 1978. A classical thermodynamic discussion of the effect of composition on glass-transition temperatures. *Macromolecules*. 11(1): 117–119.

Del Valle, J.M., Cuadros, T.R.M., and Aguilera, J.M. 1998. Glass transitions and shrinkage during drying and storage of osomosed apple pieces. *Food Research International*. 31(3): 191–204.

Duckworth, R.B. and Smith, G.M. 1963. The environment for chemical change in dried and frozen foods. *Proceedings of the Nutrition Society*. 22: 182–189.

Fennema, O. 1995. Metastable and nonequilibrium states in frozen food and their stabilization. In: *Food Preservation by Moisture Control: Fundamentals and Applications*, Barbosa-Canovas, G.V. and Welti-Chanes, J. Eds. Technomic Publishing, Lancaster, PA, p. 243.

Finegold, L., Franks, F., and Hatley, R.H.M. 1989. Glass/rubber transitions and heat capacities of binary sugar blends. *Journal of Chemical Society: Faraday Transactions 1*. 85(9): 2945–2951.

Fonseca, F., Obert, J.P., Beal, C., and Marin, M. 2001. State diagrams and sorption isotherms of bacterial suspensions and fermented medium. *Thermochimica Acta*. 366: 167–182.

Franks, F. 1985. Complex aqueous systems at subzero temperatures. In: *Properties of Water in Foods*, Simatos, D. and Multon, J.L. Eds., Martinus Nijhoff Publishers, Dordrecht, pp. 497–509.

Given, P.S. 1991. Molecular behavior of water in a flour-water baked model system. In: *Water Relationships in Food*, Levine, H. and Slade, L. Eds. Plenum Press, New York, pp. 465–483.

Goff, H.D., Montoya, K., and Sahagian, M.E. 2002. The effect of microstructure on the complex glass transition occurring in frozen sucrose model systems and foods. In: *Amorphous Food and Pharmaceutical Systems*, Levine, H. Ed., The Royal Society of Chemistry, Cambridge, pp. 145–157.

Gordon, M. and Taylor, J.S. 1952. Ideal copolymers and the second order transitions of synthetic rubbers. I. Non-crystalline copolymers. *Journal of Applied Chemistry*. 2: 493–500.

Hardman, T.M. 1986. Interaction of water with food components. In: *Interactions of Food Components*, Birch, G.G. and Lindley, M.G. Eds. Elsevier Applied Science Publishers, London, pp. 19–30.

Hatakeyama, T., Quinn, F.X., and Hatakeyama, H. 1996. Changes in freezing bound water in water-gellan systems with structure formation. *Carbohydrate Polymers*. 30: 155–160.

Izzard, M.J., Ablett, S., and Lillford, P.J. 1991. Calorimetric study of the glass transition occurring in sucrose solutions. In: *Food Polymers, Gels and Colloids*, Dickinson, E. Ed., The Royal Society of Chemistry, London, pp. 289–300.

Kaatze, U. 1990. On the existence of bound water in biological systems as probed by dielectric spectroscopy. *Physics in Medicine and Biology*. 35(12): 1663–1681.

Kantor, Z., Pitsi, G., and Thoen, J. 1999. Glass transition temperature of honey as a function of water content as determined by differential scanning calorimetry. *Journal of Agricultural Food Chemistry*. 47: 2327–2330.

Karel, M., Anglea, S., Buera, P., Karmas, R., Levi, G., and Roos, Y. 1994. Stability-related transitions of amorphous foods. *Thermochimica Acta*. 246: 249–269.

Kasapis, S. 2005. Glass transition phenomena in dehydrated model systems and foods: A review. *Drying Technology*. 23(4): 731–758.

Kasapis, S. and Sablani, S.S. 2005. A fundamental approach for the estimation of the mechanical glass transition temperature in gelatin. *International Journal of Biological Macromolecules*. 36: 71–78.

Kasapis, S., Al-Marhobi, I.M.A., and Sworn, G. 2001. α and β mechanical dispersions in high sugar/acylgellan mixtures. *International Journal of Biological Macromolecules*. 29: 151–160.

Kasapis, S., Al-Marhoobi, I.M., and Mitchell, J.R. 2003. Testing the validity of comparisons between the rheological and calorimetric glass transition temperatures. *Carbohydrate Research*. 338: 787–794.

Kelley, S.S., Rials, T.G., and Glasser, W.G. 1987. Relaxation behaviour of the amorphous components of wood. *Journal of Materials Science*. 22: 617–624.

Kokini, J.L., Cocero, A.M., Madeka, H., and De Graaf, E. 1994. The development of state diagrams for cereal proteins. *Trends in Food Science and Technology*. 5: 281–288.

Koleske, J.V. and Faucher, J.A. 1965. Transitions in gelatin nitrified gelatin–water systems. *Journal of Physical Chemistry*. 69: 4040–4042.

Labuza, T.P. and Hyman, C.R. 1998. Moisture migration and control in multi-domain foods. *Trends in Food Science and Technology*. 9: 47–55.

Lai, H.M., Schmidt, S.J., Chiou, R.G., Slowinski, L.A., and Day, G.A. 1993. Mobility of water in a starch-based fat replacer by ^{17}O NMR spectroscopy. *Journal of Food Science*. 58(5): 1103–1106.

Lang, K.W. and Steinberg, M.P. 1983. Characterization of polymer and solute bound by pulsed NMR. *Journal of Food Science*. 48: 517–520, 533, 538.

Levine, H. and Slade, L. 1986. A polymer physico-chemical approach to the study of commercial starch hydrolysis products (SHPs). *Carbohydrate Polymer*. 6: 213–244.

Levine, H. and Slade, L. 1992. Glass transitions in foods. In: *Physical Chemistry of Foods*, Schwartzberg, H.G. and Hartel, R.W. Eds., Marcel Dekker, New York, pp. 83–221.

Li, T.Y. and Chen, J.T. 2001. Evaluation of rice starch-hydrocolloid compatibility at low-moisture content by glass transitions. *Journal of Food Science*. 66(5): 698–704.

Li, S., Dickinson, L.C., and Chinachoti, P. 1998. Mobility of "unfreezable" and freezable water in waxy corn starch by 2H and 1H NMR. *Journal of Agricultural Food Chemistry*. 46: 62–71.

Ling, H., Birch, J., and Lim, M. 2005. The glass transition approach to determination of drying protocols for colour stability in dehydrated pear slices. *International Journal of Food Science and Technology*. 40: 921–927.

Markarian, S.A., Bonora, S., Bagramyan, K.A., and Arakelyan, V.B. 2004. Glass-forming property of the system diethyl sulphoxide/water and its cryoprotective action on *Escherichia coli* survival. *Cryobiology*. 49: 1–9.

Marshall, A.S. and Petrie, S.E.B. 1980. Thermal transitions in gelatin and aqueous gelatin solutions. *The Journal of Photographic Science*, 28: 128–134.

Martinez-Navarrete, N., Moraga, G., Talens, P., and Chiralt, A. 2004. Water sorption and plasticization effect in wafers. *International Journal of Food Science and Technology*. 39: 555–562.

Matveev, Y.I. 2004. Modification of the method for calculation of the Cg' and Tg' intersection point in state diagrams of frozen solutions. *Food Hydrocolloids*. 18: 363–366.

Mitchell, J.R. 2000. Hydrocolloids in low water and high sugar environments. In: *Gums and Stabilisers for Food Industry*, Williams, P.A. and Phillips, G.O. Eds., The Royal Society of Chemistry, Cambridge, p. 243.

Moraga, G., Martinez-Navarrete, N., and Chiralt, A. 2004. Water sorption isotherms and glass transition in strawberries: Influence of pretreatment. *Journal of Food Engineering*. 62: 315–321.

Moraga, G., Martinez-Navarrete, N., and Chiralt, A. 2006. Water sorption isotherms and phase transitions in kiwifruit. *Journal of Food Engineering*. 72: 147–156.

Morales-Diaz, A. and Kokini, J.L. 1998. Phase transitions of soy globulins and the development of state diagrams. In: *New Techniques in the Analysis of Foods*, Tunick, M.H., Palumbo, S.A., and Fratamico, P.M. Eds., Kluwer Academic/Plenum Press, New York. pp. 69–77.

Noel, T., Parker, R., and Ring, S. 2000. Effect of molecular structure and water content on the dielectric relaxation behaviour of amorphous low molecular weight carbohydrates above and below their glass transition. *Carbohydrate Research*. 329: 839–845.

Ollett, A., Parker, R., and Smith, A.C. 1991. Mechanical properties of wheat starch plasticized with glucose and water. In: *Food Polymers, Gels, and Colloids*, Dickinson, E.. Ed., The Royal Society of Chemistry, London, pp. 537–541.

Orford, P.D., Parker, R., Ring, S.G., and Smith, A.C. 1989. Effect of water as a diluent on the glass transition behaviour of malto-oligosaccharides, amylose and amylopectin. *International Journal of Biological Macromolecule*. 11: 91–96.

Orford, P.D., Parker, R., and Ring, S.G. 1990. Aspects of the glass transition behaviour of mixtures of carbohydrates of low molecular weight. *Carbohydrate Research*. 196: 11–18.

Ottenhof, M., MacNaughtan, W., and Farhat, I.A. 2003. FTIR study of state and phase transitions of low moisture sucrose and lactose. *Carbohydrate Research*. 338: 2198–2202.

Paakkonen, K. and Plit, L. 1991a. Equilibrium water content and the state of water in dehydrated white cabbage. *Journal of Food Science*. 56(6): 1597–1599.

Paakkonen, K. and Plit, L. 1991b. Equilibrium moisture content and state of water in chitin. *Food Science and Technology*. 24(3): 259–262.

Padua, G.W. 1993. Proton NMR and dielectric measurements on sucrose filled agar gels and starch pastes. *Journal of Food Science*. 58(3): 603–626.

Padua, G.W., Richardson, S.J., and Steinberg, M.P. 1991. Water associated with whey protein investigated by pulser NMR. *Journal of Food Science*. 56(6): 1557–1561.

Quinquenet, S., Grabielle-Madelmont, C., and Ollivon, M. 1988. Influence of water on pure sorbitol polymorphism. *Journal of Chemical Society: Faraday Transactions 1*. 84(8): 2609–2618.

Rahman, M.S. 1995. *Food Properties Handbook*, 1st edn., CRC Press, Boca Raton, FL.

Rahman, M.S. 2004. State diagram of date flesh using differential scanning calorimetry (DSC). *International Journal of Food Properties*. 7(3): 407–428.

Rahman, M.S. 2005. Dried food properties: Challenges ahead. *Drying Technology*. 23(4): 695–715.

Rahman, M.S. 2006. State diagram of foods: Its potential use in food processing and product stability. *Trends in Food Science and Technology*. 17: 129–141.

Rahman, M.S. and Al-Belushi, R.H. 2006. Dynamic isopiestic method (DIM): Measuring moisture sorption isotherm of freeze-dried garlic powder and other potential uses of DIM. *International Journal of Food Properties*. 9(3): 421–437.

Rahman, M.S. and Labuza, T.P. 1999. Water activity and food preservation. In: *Handbook of Food Preservation*, Rahman, M.S. Ed., 1st edn., Marcel Dekker, Inc. New York, pp. 339–382.

Rahman, M.S., Sablani, S.S., Al-Habsi, N., Al-Maskri, S., and Al-Belushi, R. 2005. State diagram of freeze-dried garlic powder by differential scanning calorimetry and cooling curve methods. *Journal of Food Science*. 70(2): E135–E141.

Richardson, S.J., Baianu, I.C., and Steinberg, M.P. 1987. Mobility of water in sucrose solutions determining by deuterium and oxygen-17 nuclear magnetic resonance measurements. *Journal of Food Science*. 52(3): 806–812.

Rockland, L.B. 1969. The practical approach to better low-moisture foods, water activity and storage stability. *Food Technology*. 23: 11–18, 21.

Roos, Y.H. 1987. Effect of moisture on the thermal behavior of strawberries studied using differential scanning calorimetry. *Journal of Food Science*. 52(1): 146–149.

Roos, Y.H. 1992. Phase transitions and transformations in food systems. In: *Handbook of Food Engineering*, Heldman, R. and Lund, D.B. Eds., Marcel Dekker, New York, pp. 145–197.

Roos, Y. 1993. Melting and glass transitions of low molecular weight carbohydrates. *Carbohydrate Research*. 238: 39–48.

Roos, Y. 1995. Characterization of food polymers using state diagrams. *Journal of Food Engineering*. 24: 339–360.

Roos, Y. and Karel, M. 1991a. Phase transitions of amorphous sucrose and frozen sucrose solutions. *Journal of Food Science*. 56(1): 266–267.

Roos, Y. and Karel, M. 1991b. Phase transitions of mixtures of amorphous polysaccharides and sugars. *Biotechnology Progress*. 7: 49–53.

Roos, Y. and Karel, M. 1991c. Amorphous state and delayed ice formation in sucrose solutions. *International Journal of Food Science and Technology*. 26: 553–566.

Sa, M.M. and Sereno, A.M. 1994. Glass transitions and state diagrams for typical natural fruits and vegetables. *Thermochimica Acta*. 246: 285–297.

Sa, M.M., Figueiredo, A.M., and Sereno, A.M. 1999. Glass transitions and state diagrams for fresh and processed apple. *Thermochimica Acta*. 329: 31–38.

Sablani, S.S., Kasapis, S., Al-Rahbi, Y., and Al-Mugheiry, M. 2002. Water sorption isotherms and glass transition properties of gelatin. *Drying Technology*. 20(10): 2081–2092.

Sablani, S.S., Rahman, M.S., Al-Busaidi, S., Guizani, N., Al-Habsi, N., Al-Belushi, R., and Soussi, B. 2007. Thermal transitions of king fish whole muscle, fat and fat-free muscle by differential scanning calorimetry. *Thermichimica Acta*. 462: 56–63.

Shalaev, E.Y. and Kanev, A.N. 1994. Study of the solid–liquid state diagram of the water–glycine–sucrose system. *Cryobiology*. 31: 374–382.

Slade, L. and Levine, H. 1991. Beyond water activity. *Critical Reviews in Food Science and Nutrition*. 30(2–3): 115–360.

Sobral, P.J.A. and Habitante, A.M.Q.B. 2001. Phase transitions of pigskin gelatin. *Food Hydrocolloids*. 15: 377–382.

Telis, V.R.N. and Sobral, P.J.A. 2001. Glass transition and state diagram for freeze-dried pineapple. *Food Science and Technology*. 34: 199–205.

Tseretely, G.I. and Smirnova, O.I. 1992. DSC study of melting and glass transition in gelatins. *Journal of Thermal Analysis*. 38: 1189–1201.

Van den Berg, C. 1981. Vapour sorption equilibria and other water–starch interactions: A physico–chemical approach. Ph.D thesis, Wageninger, the Netherlands.

Verdonck, E., Schaap, K., and Thomas, L.C. 1999. A discussion of the principles and applications of modulated temperature DSC (MTDSC). *International Journal of Pharmaceutics*. 1992: 3–20.

Wang, H., Zhang, S., and Chen, G. 2008. Glass transition and state diagram for fresh and freeze-dried Chinese gooseberry. *Journal of Food Engineering*. 84: 307–312.

Welti-Chanes, J., Guerrero, J.A., Barcenas, M.E., Aguilera, J.M., Vergara, F., and Barbosa-Canovas, G.V. 1999. Glass transition temperature (T_g) and water activity (a_w) of dehydrated apple products. *Journal of Food Process Engineering*. 22: 91–101.

White, G.W. and Cakebread, S.H. 1966. The glassy state in certain sugar-containing food products. *Journal of Food Technology*. 1: 73–82.

Whittam, M.A., Noel, T.R., and Ring, S.G. 1991. Melting and glass/rubber transitions of starch polysaccharides. In: *Food Polymers, Gels and Colloids*, Dickinson, E. Ed., The Royal Society of Chemistry, London, pp. 277–300.

Williams, R.J. and Leopold, A.C. 1989. Glassy states in dormant corn embryos. *Thermochimica Acta*. 155: 109–114.

Zeleznak, K.J. and Hoseney, R.C. 1987. The glass transition in starch. *Cereal Chemistry*. 64(2): 121–123.

Zhao, J. and Notis, M.R. 1993. Phase transition kinetics and the assessment of equilibrium and metastable states. *Journal of Phase Equilibrium*. 14: 303.

CHAPTER 9

Gelatinization of Starch

Shyam S. Sablani

CONTENTS

9.1 INTRODUCTION

Starch is the second most abundant biomaterial available in nature, next to cellulose. Starch is a reserve carbohydrate in the plant kingdom and is generally deposited in the form of minute granules or cells ranging from 1 to 100 μm or more in diameter (Zobel and Stephen, 1995). It is produced by photosynthesis in the amyloplast of higher plants. Starch can be found in leaves, stems, fruits, and roots. Starch is a polymeric carbohydrate consisting of anhydroglucose units linked primarily through α-D-(1 → 4) glucosidic bonds. Starch mainly consists of two major components, amylose and amylopectin. Amylose is essentially a linear polymer in which the anhydroglucose units are linked through α-D-(1 → 4) glucosidic bonds whereas amylopectin is a branched polymer with α-D-(1 → 4) glucosidic bonds, with periodic branches at the *O*-6 position (Buleon et al., 1998). Starch granules are semicrystalline, composed of alternating crystalline and amorphous lamella. It is found that amylose is more concentrated at the periphery of the granule.

In addition to providing a major source of energy in food products, starch plays a crucial role in textural modification on heating in the presence of water. Commercial starches are obtained from seeds (corn, wheat, and rice) and from tubers and roots, mainly potato, sweet potato, and cassava (Whistler and BeMiller, 1997). Native starch granules are mostly indigestible. Starch granules are insoluble in cold water due to closer and more orderly packing of the starch granules at the surface of a granule than its interior. If a suspension of starch in excess water is heated, diffusion of water into granules causes swelling (i.e., increases the granule volume) and disrupts hydrogen bonding and subsequently water molecules become attached to hydroxyl groups in starch. This phase transformation is termed as gelatinization. Gelatinized starch is easily digestible. Water in the amorphous parts of starch acts as a plasticizer and decreases the glass transition temperature. Gelatinization temperatures are mainly influenced by moisture, solute (ionic and nonionic), presence of other biomaterials, and processing and pretreatment conditions. However, gelatinization is a melting process by heat in the presence of water.

Olkku and Rha (1978, cited by Lund, 1984) summarized the steps of gelatinization as follows: (1) granules hydrate and swell to several times their original size, (2) granules lose their birefringence, (3) the clarity of the mixture increases, (4) rapid increase in consistency occurs and reaches a maximum, (5) linear molecules dissolve and diffuse from ruptured granules, and (6) the uniformly dispersed matrix forms a gel or paste-like mass.

To optimize processing conditions and obtain desired quality of starch-based foods, a thorough understanding of starch gelatinization is required. Although there have been numerous studies on starch gelatinization, there is still little information on the interactions between starch and water, which play an important role in the gelatinization mechanism (Tananuwong and Reid, 2004).

9.2 GELATINIZATION TEMPERATURES AND ENTHALPY

Gelatinization temperature is one of the most important parameters during gelatinization. Our understanding of the process of gelatinization has greatly increased over the past two decades. This is due in part to greater knowledge of the structure of the granule, and also due to numerous investigations of the gelatinization process using applications of several techniques. Various methods have been used to probe into the physics of gelatinization including differential scanning calorimetry (DSC), x-ray diffraction, nuclear magnetic resonance (NMR) spectroscopy, Fourier transform infrared (FTIR) spectroscopy, optical microscopy, electron microscopy, and rheometry (Liu et al., 2002).

9.2.1 Measurement Techniques

The gelatinization process influences several physicochemical properties of food materials and the gelatinization temperature is measured by monitoring the changes in these properties as affected by the gelatinization process.

9.2.1.1 *Birefringence Method*

This is one of the most sensitive methods of measurement of starch gelatinization. In some cases, birefringence is difficult to apply because starch granules are not only difficult to count in heterogeneous mixtures but also difficult to separate from other components in cooked material (Fang and Chinnan, 2004). In this method, polarized light microscopy with a heating stage is mainly used in the microscopic analysis. When native starch granules are heated in the presence of water, and the temperature rises to a critical value, the birefringence of the native starch granules is lost. This critical value is referred to as the gelatinization temperature. The loss of birefringence is measured using electrically heated microscopic hot stage. A small drop of 0.1%–0.2% aqueous suspensions is spotted on a microscopic slide and surrounded by a continuous ring of high viscosity mineral oil. A cover slip is placed on the aqueous drop such that no air bubbles are present. This oil barrier prevents the escape of steam that fogs the glass plate and prevents air penetration under the cover plate. The loss of birefringence of the granules in the field is observed by the uniform heating rate (2°C/min). Spies and Hoseney (1982) studied the effect of sugars on starch gelatinization with a Koefler hot-stage microscope. They designated the temperature at which 50% of the granules had lost birefringence as the gelatinization temperature. However, Bryant and Hamaker (1997) referred to the temperature at which the first granule was noted to lose birefringence as the onset temperature and that at which more than 95% of the granules had lost birefringence as the final temperature. Thus, when it is employed for determining gelatinization temperature, the result is reported subjectively by different researchers (Li et al., 2004).

9.2.1.2 *Viscosity Method*

The swelling and gelatinization of starch increases the viscosity of suspension. The most important practical method is to use a continuous automatic recording viscometer, and brabender amylograph is most widely used in starch industries (Goto, 1969). The use of brabender amylograph is normally restricted to less than 10%, which is too low to be compared with the concentration routinely used in food industries. The slurry is heated at a rate of 1.5°C/min and the sample holder is rotated at 30–150 rpm (Lund, 1984). Varriano-Marston et al. (1980) studied different methods to determine starch gelatinization in bakery foods and found that the amylographic method was less reliable than x-ray diffraction, polarization microscope, and enzymatic methods.

9.2.1.3 *X-Ray Diffraction Method*

The x-ray diffraction pattern can be used to determine the area of amorphous and crystalline phase of starch. X-rays are electromagnetic radiation with wavelength from 0.1 to 1.0 nm, which is in the order of molecular spacing in a crystal. The recorded diffracted beams give the information of crystal and molecular structure within the crystal. Diffraction is due to the interaction of the incident beam with an obstacle. The samples should be less than 200 mesh (74 μm) in size and packed as densely as possible in a sample holder to get high-intense patterns (Lund, 1984). Recently Liu et al. (2002) used a rotating anode x-ray diffraction instrument with copper, nickel filtered $K\alpha$ radiation (0.154178 nm) by using an RU-200 BH, CN4148 H_2 to measure the area of crystalline amorphous phase of potato starch. They operated the instrument at 55 kV and 190 mA. The scan was made from a diffraction angle (2θ) of 2°–50° and slit width of 0.2 mm, and scan speed of 0.6°/min. They prepared the sample in thin-walled (0.01 mm) glass capillary tubes (1.0 or 1.5 mm diameter). Weighted starch samples were loaded into a tube and distilled water was injected through a needle to the desired level. To pack the starch–water suspension at the bottom, the capillary was centrifuged and sealed using a flame. The sample temperature was controlled using an oven surrounding the sample. The crystallinity was calculated from the areas of crystalline and amorphous phases. The intensities of diffracted peaks were observed at $2\theta = 5.3°$ and 16.9°. The loss of crystallinity was considered as a manifestation of starch gelatinization (Liu et al., 2002). Starch gelatinization studied with bakery foods has shown that the crystallographic (x-ray diffraction) method was not sensitive to small change in the degree of swelling (Varrino-Marston et al., 1980). Lugay and Juliano (1965) observed five characteristic diffraction peaks of starch during the gelatinization process and the degree of crystallinity of starch granules depends on the amylose content. Liu et al. (2002) observed that x-ray diffraction gives a higher end transition temperature than other techniques (microscopy, calorimetry, and FTIR spectroscopy) used.

9.2.1.4 Amylose–Iodine Blue Value Method

The linear fraction or amylose gives a deep blue complex with iodine, and this characteristic has been used as an analytical tool to measure the amylose content. The amylose content in solution can be determined by (1) potentiometric titration of the dissolved starch with standard iodine, (2) similar amperometric titration, (3) spectrophotometric determination of the intensity of blue coloration with iodine, (4) sorption of Congo red (Schoch, 1964). The absorbance of blue color is usually measured in a spectrophotometer at 600 nm. The iodine blue value method provides a rapid determination of amylose content, but incomplete solubilization of amylose may give lower blue value. The maximum solubilization can be achieved by a vigorous mixing (Lund, 1984). This method is not applicable to more highly hydrolyzed products, enzymatic dextrin, oxidized starches, and starch derivatives (Schoch, 1964). In this case, the iodine titration curve does not give a sharp inflection point and a linear segment that can be extrapolated. Thus, the calcium chloride method has been used extensively for starch derivatives due to the more complete adsorption of iodine by starch in calcium chloride solution than in potassium iodide–potassium chloride used in the standard method (Colburn and Schoch, 1964). Again this method is limited to parboiled rice since gelatinized rice flour is insoluble in water. So McCready and Hassid (1943) modified this method by dispersing the starch in alkali solution for dissolving amylose and then subsequently neutralizing it with acid solution. A critical concentration of alkali was found: 0.2 N KOH for gelatinized starch whereas 0.5 N KOH for raw starch (Birch and Priestly, 1973). The optimum or critical concentration could also be used to differentiate raw and gelatinized starch (Lund, 1984). This method has been used for bread, cake, crackers, cookies, potato chips, biscuit, wafers, etc. (Leach, 1979; Wootton and Chaudhary, 1980; Wootton et al., 1971). Wootton et al. (1971) used this method to estimate the degree of gelatinization (DG) of different food products and found it most suitable for estimation of the degree of gelatinization. Bayram (2005, 2006) used the amylose/iodine method to determine the degree of starch gelatinization in wheat during cooking for the production of bulgur. He noticed that the cooking degree obtained from amylase/iodine method was higher than the center cutting and light scattering methods to 40 min due to differences in measuring procedures of each method.

9.2.1.5 Nuclear Magnetic Resonance

Lelievre and Mitchell (1975) used pulsed NMR to investigate the gelatinization process in foods. The variation of spin–spin relaxation times of protons in starch suspension with temperature and starch concentration indicated the gelatinization process. Prior to gelatinization, the relaxation time decreased with increasing temperature and passed a minimum and then again increased to a value similar to that of an untreated suspension. Similar to DSC measurements, the rapid decrease in spin–spin relaxation time corresponds to the onset of the melting of crystalline polymer of starch granules. Lelievre and Mitchell (1975) also suggested that gelatinization is similar to a melting process with the transition of starch from a partly crystalline to an amorphous state. At this stage, the mobility of water decreased as the temperature is increased from 20°C to 60°C and this is due to the formation of starch–water complex. In last two decades, several studies have used NMR to explore the starch–water interaction and the mechanism of starch gelatinization (Gidley and Bociek, 1985; Hill et al., 1990; Cheetam and Tao, 1998; Waigh et al., 2000; Chatakanonda et al., 2003; Tananuwong and Reid, 2004).

9.2.1.6 Differential Scanning Calorimetry

DSC is the most common and widely used analytical method to study the starch gelatinization phenomena. The purpose of DSC is to record the difference between an enthalpy change due to the gelatinization process in a sample and in an inert reference material when both are heated. The instrument is programmed to heat at a constant rate and then the temperature difference between the sample and reference is a function of (1) the enthalpy change, (2) the heat capacity, and (3) total thermal resistance to heat flow. The thermal resistance to heat flow depends on the nature of the sample and the packing and the extent of thermal contact (Lund, 1984). The temperature sensors attached to the pan can reduce the thermal resistance due to the sample itself. A quantity of 5–20 mg of sample is commonly used in the case of DSC. The contact surface between the pan and sample can be maximized by having the sample as thin disks, films, or fine granules (Lund, 1984). In the case of heterogeneous food materials, the sample must be homogenized. The aluminum pan must be hermetically sealed to prevent evaporation of water from water-containing sample. High-pressure DSC cells (up to 30 atm) are also available. Calibration of the instrument is generally carried out with a high purity metal of known enthalpy of fusion and melting point. Lund (1984) mentioned that the most commonly used material for calibration is indium (ΔH: 28.45 kJ/kg; T_m: 156.4°C). The rate of heating (scan rate) can affect onset-, mid-, and end value of gelatinization temperature and enthalpy of gelatinization (Tables 9.1 and 9.2).

DSC studies have also been useful in explaining the mechanism of the gelatinization process. According to DSC studies, during heating of a starch–water mixture, two endothermic peaks (namely G and M1) related to the gelatinization process are observed (Donavan, 1979). The enthalpy of both peaks and the position of the M1 peak is water dependent (Donavan, 1979; Hoseney et al., 1986; Rolee and LeMeste, 1999). Tananuwong and Reid (2004) summarized three classical models advanced in the literature based on DSC studies. The first model proposed by Donavan (1979) suggested that the first gelatinization peak (G endotherm) is a result of swelling-driven crystalline disruption, in which the swelling of the amorphous regions is considered to "strip" polymer chains from the surface of crystalline, while the second peak (M1 endotherm) represents the melting of the remaining less hydrated crystallites. A second model proposed by Evans and Haisman (1982) considers the system as a population of starch granules with a range of properties. The successive gelatinization peaks have been suggested to reflect the melting of the crystallites with different stabilities due to a gradient of water within the sample. The first peak G represents the highly cooperative melting of less stable crystallites. The excess water present facilitates the gelatinization by lowering the melting point of crystallites. As water molecules are absorbed by the disordered polysaccharides chains, resulting in remaining water insufficient to

Table 9.1 Effect of Heating Rate on Endotherm Temperatures and ΔH_G of Wheat Starch[a]

	Gelatinization Temperature (°C)			
Heating Rate (°C)	T_{GI}	T_{GP}	T_{GE}	ΔH_G (kJ/kg Starch)
8	52	67	78	21.76
16	50	68	86	19.66
32	46	65	85	14.64

Source: Wootton, M. and Bamunuarachchi, A., *Starch*, 31, 264, 1979.
Note: T_{GI}, T_{GP}, and T_{GE} are the initial, peak, and end of gelatinization temperature; ΔH_G is the enthalpy of gelatinization.
[a] Starch water ratio 1:2.

Table 9.2 Effect of Heating Rate on Gelatinization Peak Temperature of Potato Starch

Heating Rate (°C)	Potato (Pravisani et al., 1985)[a]		ND 860-2 (Leszkowiat et al., 1990)[b]			Norchip (Leszkowiat et al., 1990)[b]	
	T_{GP} (°C)	T_{GE} (°C)	T_{GP} (°C)	ΔH_G (kJ/kg)	T_{G1} (°C)	T_{GP} (°C)	ΔH_G (kJ/kg)
1	67.0	71.9	73.8	14.04	69.1	71.3	14.12
3	68.8	72.4	74.3	16.90	70.0	72.9	15.89
5	69.3	72.5	74.9	17.60	71.0	73.4	15.39
7	69.5	72.7	75.3	14.00	70.9	73.6	14.87
10	71.0	73.2	76.2	13.96	71.9	74.9	13.17
15	72.8	73.7	76.8	12.33	72.9	76.1	12.23
20	73.8	74.4	78.2	15.07	73.5	77.3	12.32
25	74.7	75.1	79.0	14.42	73.4	77.7	14.74

[a] Reference material, indium and 8–10 replicates for each sample.
[b] Sample weight, 10 mg and six replicates for each sample; reference material, 8 and 2 mg distilled water.

facilitates the melting of more stable crystallites. Hence, a higher temperature is needed to melt these remaining crystallites, causing the M1 endotherm to appear. A third model, proposed by Slade and Levine (1988), applies the glass transition concept to explain the gelatinization phenomena. This model suggests that G endotherm is considered to primarily reflect plasticization in the amorphous region, which is required before the start of melting of crystallites, and the M1 endotherm reflects nonequilibrium melting of crystallites. Tananuwong and Reid (2004) also noted that a more detailed comparison of all aspects of the three models indicates that none gives an entirely satisfactory overview of the gelatinization mechanism. These models do not provide a detailed description of the structural changes during gelatinization. Recently, Waigh et al. (2000) have proposed another mechanism, focusing on the change in the crystalline structure during gelatinization. This model considers amylopectin molecules to be a side-chain liquid–crystalline polymer. Gelatinization is described as the coupling between self-assembly, described as the dissociation of amylopectin double helices side by side (helix–helix dissociation), and the breakdown of the overall crystalline structure during heating.

Recently, modulated differential scanning calorimeter (MDSC) has been used to characterize thermal transition in food materials (Lai and Lii, 1999; Baik et al., 1999; Tan et al., 2004; Srikaei et al., 2005). In this technique, the sample is subjected to a sinusoidal modulation (oscillation) overlaid on the conventional linear heating or cooling ramp. The total, reversing, and nonreversing heat-flow changes, as well as complex heat capacity, during transition of the sample can then be quantified (Gallaghar, 1997). This technique can separate reversible and irreversible thermal events, which can result in an accurate value of heat capacity. In addition, the reversing and nonreversing signals disclose the thermodynamic and kinetic characteristics, respectively. Examples of events associated with nonreversible signals are the endothermic relaxation of amorphous materials, gelatinization, recrystallization, and protein denaturation. Reversible events include glass transition and simultaneous crystallization. Lai and Lii (1999) studied the effects of MDSC variables on thermodynamic and kinetic characteristics during gelatinization of waxy rice starch. Using MDSC, Srikaei et al. (2005) were able to detect a possible glass transition that occurred during starch gelatinization in soft and hard wheat, even though there were relatively small changes in heat capacity (0.02–0.06 J/g°C) and a broad range of glass transition (48°C–67°C).

9.2.1.7 Microscopy Analysis

Microscopy examination of the starch granules can be used to monitor starch gelatinization. It is not a quantitative method, but it is the most appropriate approach to conclusively determine whether the instrumental or chemical method reflected the physical changes taking place (Owusu-Ansah et al., 1982). Recently, a variety of microscopes have been used for the examination of cereals, their components, and derived products (Fulcher et al., 1994; Kaláb et al., 1995). This includes conventional direct light microscopes (LM) of various types for example, bright field, polarizing, and fluorescence LMs (Autio and Salmenkallio-Marttila, 2001). Transmission and scanning electron including environmental scanning electron microscopes (TEM, SEM, and ESEM), which provide high resolution and magnification have also been used (Hoseney et al., 1977; Parades-Lopez and Bushuk, 1983; Liu and Zhao, 1990; Freeman and Shelton, 1991; Koksel et al., 1998; McDonough and Rooney, 1999; Roman-Gutierrez et al., 2002, Blaszczak et al., 2005a,b).

9.2.1.8 Enzymatic Method

In situ starch gelatinization phenomena in seed/cereal have been studied by a method based on the enzymatic digestion susceptibility (Lund, 1984; Haruhito et al., 1990). This is one of the most sensitive methods of measurement of starch gelatinization (Shetty et al., 1974). Fang and Chinnan (2004) used the enzymatic method to determine the degree of starch gelatinization in cowpea seed using mixed enzyme system: beta-amylase-pullulanase (0.8 IU amylase and 3.14 IU pullulanase/mL). One unit of β-amylase liberates 1.0 mg of maltose from starch in 3 min at pH 4.8 and 20°C. One unit of pullulanase liberates 1.0 μmol of maltotriose from pullulan per minute at pH 5.0 and at 25°C. They used a tissue grinder to thoroughly homogenize the samples. Reducing sugar content in the sample was estimated using the standard curve for maltose solution. A standard for a completely gelatinized sample corresponds to 100% degree of gelatinization and raw cowpea sample as a standard for 0% degree of gelatinization. Absorbance values of the same sample treated first by alkali and then by enzyme were obtained. The absorbance values were converted to reducing sugar contents on the basis of the standard curve. Before conversion of absorbance values to reducing sugar contents, absorbance values were corrected by deducting the value of the blank test.

9.2.1.9 Fourier Transform Infrared Spectroscopy

This method has been used to probe starch gelatinization and retrogradation. However, this technique is limited to low starch–water ratio and temperature over which transition phenomena can be studied (Liu et al., 2002). The infrared (IR) transmission of the sample of starch–water suspensions at different moisture levels is monitored. The measurements are made as a function of temperature. The spectra are analyzed using the software. Liu et al. (2002) used a Mattson Surius 100 FTIR equipped with tri-glycine sulfate detector, operating at 4 cm^{-1} resolution to study potato starch gelatinization. They observed that during starch gelatinization infrared spectra were sensitive to change in the range 900–1200 cm^{-1}. They noticed that at temperature below 65°C, there were no changes in the absorption bands but above 65°C, a shoulder peak appeared at 1020 cm^{-1}. The intensity of this shoulder increased with the temperature while the intensity of the absorbance at 1640 cm^{-1} decreased. At 71°C, the intensity of both bands reached maximum value but decreased above 71°C. The change in spectra corresponds to O—H stretching in starch and absorbed water (1640 cm^{-1}) and vibration of C—O—H deformation (1020 cm^{-1}) and it is related to a change in the amorphous and crystalline parts of the granule (Van Soest et al., 1995). The changes at 1020 and 1640 cm^{-1} indicate that gelatinization is primarily a hydration process. The rupture of the most inter- and intrahydrogen bonds and hydrophobic bonds occurs near 65°C. The increase in absorbance from 65°C to 71°C suggests exchange of hydrogen bonds between starch and water, indicative of sol formation. Above 71°C, the intensity of absorbance at these bands decreased due to further rupturing of intermolecular H-bonds and hydrophobic bonds in starch, resulting in increased solution of starch molecules at these temperatures (Liu et al., 2002).

9.2.1.10 *Ohmic Heating Method*

A method for measuring starch gelatinization temperature and degree of gelatinization, determined from a change in electrical conductivity (σ), has been used recently (Wang and Sastry, 1997; Karapantsios et al., 2000; Li et al., 2004). The method involved suspension of starch–water solution to Ohmic heating and measuring the change in electric conductivity of the suspension as a function of temperature. The electrical conductivity and the change in electrical conductivity with temperature ($d\sigma/dT$) are plotted against the temperature. As the temperature of the suspension initially increases, the electric conductivity increases almost linearly with time up to the onset of gelatinization temperature about 62°C. After this point, electric conductivity decreases and attains a minimum value, which corresponds to the end of gelatinization. Therefore during the gelatinization period, electrical conductivity decreases gradually. After the suspension is gelatinized, the electrical conductivity increases linearly. Because a linear relationship exists between the conductivity and the temperature, $d\sigma/dT$ is constant before and after the starch gelatinization, but the value of $d\sigma/dT$ is different. The shape of the $d\sigma/dT$ and T curve is more like the endothermic peak on a DSC thermogram in the gelatinization temperature range. Li et al. (2004) named the section of the curve with a shape similar to a DSC endothermic peak, on the $d\sigma/dT$–T curve, as "block peak." They observed that the position and area of the peak were influenced by starches of different origin and concentration of starch suspension.

Traditional methods such as the center cutting method and light scattering method have been used to determine the degree of cooking of the wheat kernel (Smith et al., 1964; Singh and Dodda, 1979; Bayram, 2005). In the center cutting method, 100 cooked kernels are cut with a razor blade and the endosperms are examined for opaque white centers. The percentages of kernels containing white centers are designated as "white kernel count" percentage. Bayram (2005) used a light scattering method where the opaque core in the center of the kernel during cooking was determined for 100 cooked kernels over a glass plate using a high-intensity light source (150 W) in a dark cabinet.

9.2.2 Gelatinization Temperature and Enthalpy Data

The gelatinization temperature and enthalpy of different starches are compiled in Table 9.3. There are several factors that affect the gelatinization process (gelatinization temperature range and enthalpy of gelatinization): (1) origin of starch and water–starch ratio, (2) presence of other solutes, and (3) pretreatments.

Table 9.3 Gelatinization Temperature of Different Starches

Type	Method	ΔH_G (kJ/kg)	Gelatinization Temperature (°C) T_{GI}	T_{GP}	T_{GE}	Reference
Amylomaize (70%)	DSC	2.92	65.0	71.0	—	Eberstein et al. (1980)
Corn	Plastograph		62.0		74.0	Goto (1969)
Hafer	DSC	9.20	52.0	58.3	64.0	Eberstein et al. (1980)
Hafer	Microscopic	—	54.0	58.0	61.0	Eberstein et al. (1980)
Wheat	DSC	10.04	54.0	69.0	86.0	Stevens and Elton (1971)
Wheat	DSC	10.46	54.0	69.0	85.0	Stevens and Elton (1971)
Wheat	DSC	11.72	54.0	69.0	86.0	Stevens and Elton (1971)

Table 9.3 (continued) Gelatinization Temperature of Different Starches

Type	Method	ΔH_G (kJ/kg)	Gelatinization Temperature (°C)			Reference
			T_{GI}	T_{GP}	T_{GE}	
Wheat	DSC	12.13	54.0	66.0	87.0	Stevens and Elton (1971)
Wheat	DSC	10.04	52.0	59.0	65.0	Eberstein et al. (1980)
Wheat	Microscopic	—	55.0	61.0	66.0	Eberstein et al. (1980)
Wheat	Plastograph	—	54.0	—	67.0	Goto (1969)
Wheat	Turbidity	—	55.0	—	95–100	Banks and Greenwood (1975)
Rice	DSC	14.23	66.0	82.0	100.0	Stevens and Elton (1971)
Rice	DSC	16.32	68.0	79.0	108.0	Stevens and Elton (1971)
Rice	DSC	12.97	70.0	76.3	82.0	Eberstein et al. (1980)
Rice	Microscopic	—	72.0	75.0	79.0	Eberstein et al. (1980)
Rice	Plastograph	—	66.0	—	80.0	Goto (1969)
Rice	DSC	12.55	68.0	74.0	79.0	Wirakartakusumah (1981)
Roggen	DSC	10.04	49.0	54.0	61.0	Eberstein et al. (1980)
Roggen	Microscopic	—	51.0	54.0	58.0	Eberstein et al. (1980)
Maize 1	DSC	15.48	67.0	78.0	95.0	Stevens and Elton (1971)
Maize 2	DSC	14.64	68.0	79.0	97.0	Stevens and Elton (1971)
Maize 3	DSC	18.83	70.0	79.0	108.0	Stevens and Elton (1971)
Maize	DSC	13.81	65.0	70.6	77.0	Eberstein et al. (1980)
Maize	Microscopic	—	65.0	69.0	76.0	Eberstein et al. (1980)
Meranta	DSC	16.74	67.0	74.6	85.0	Eberstein et al. (1980)
Meranta	Microscopic	—	69.0	76.0	84.0	Eberstein et al. (1980)
Tapioca	DSC	16.64	66.0	78.0	100.0	Stevens and Elton (1971)
Tapioca	DSC	15.06	63.0	68.3	79.0	Eberstein et al. (1980)
Tapioca	Microscopic	—	64.0	69.0	80.0	Eberstein et al. (1980)
Tapioca	Plastograph	—	66.0	—	75.0	Goto (1969)
Arrowroot	DSC	19.25	73.0	84.0	106.0	Stevens and Elton (1971)
Potato	DSC	21.34	59.0	71.0	95.0	Stevens and Elton (1971)
Potato	DSC	23.01	57.0	71.0	94.0	Stevens and Elton (1971)
Potato	DSC	17.99	58.6	64.9	71.2	Donavan and Mapes (1980)
Potato	DSC	17.57	61.0	65.1	71.0	Eberstein et al. (1980)
Potato	Microscopic	—	58.0	64.0	68.0	Eberstein et al. (1980)
Potato	Plastograph	—	56.0	—	70.0	Goto (1969)
Potato	Microscopic	—	59.0	63.0	68.0	Schoch and Maywald (1956)
Amylodextrin (3 days)	DSC	18.41	63.0	69.5	76	Donavan and Mapes (1980)
Amylodextrin (8 weeks)	DSC	17.99	54.7	73.4	92.1	Donavan and Mapes (1980)
Sweet potato	Plastograph	—	62.0	—	80.0	Goto (1969)
Waxy corn	Plastograph	—	62.0	—	73.0	Goto (1969)
Waxy maize	DSC	16.74	65.0	72.3	80.0	Eberstein et al. (1980)
Waxy maize	Microscopic	—	64.0	70.0	78.0	Eberstein et al. (1980)
Waxy rice	DSC	14.23	61.0	68.6	79.0	Eberstein et al. (1980)
Waxy rice	Microscopic	—	59.0	69.0	76.0	Eberstein et al. (1980)
Korean ginseng (grade 1)	DSC	15.0	43.5	55.7	77.3	Koo et al. (2005)
Korean ginseng (grade 1)	DSC	14.7	44.5	55.6	72	Koo et al. (2005)
Korean ginseng (grade 1)	DSC	14.2	45.4	56.9	76.5	Koo et al. (2005)

9.2.2.1 Origin of Starch and Water–Starch Ratio

Gelatinization behavior varies with the type or origin of starch because of differences in composition and structure. Within a given type of starch, such as potato starch, the composition and functional properties can vary depending on the cultivar, cultivation conditions, and the method of isolation of starch (Lisinska and Leszcynski, 1989). Liu et al. (2002) reported that the gelatinization transition temperature of potato starch was influenced by different cultivar, storage, and genetic transformation of potato tubers. The water–starch ratio can influence the onset of gelatinization as well as enthalpy of gelatinization. The effect of excess water on gelatinization enthalpy is given in Table 9.4. Kugimiya et al. (1980) studied the irreversible endothermic gelatinization of starch (maize, potato, and wheat) with an excess of water and found the temperature range from 65°C to 75°C (Table 9.5). A second reversible endothermic transition near 100°C was observed for maize and wheat starch but not for potato starch or waxy maize starch (Table 9.5). The second transition is because both maize and wheat starches contain significant amounts of lipid and amylose. Kugimiya et al. (1980) also observed second and third transitions when lipid is removed by methanol extraction. Fornal et al. (1998) and Freitas et al. (2004) reported that the starches with high amylase content require more energy during gelatinization.

Eliasson (1980) studied the influence of water on the gelatinization of wheat starch when heated at 140°C with 30%–80% (dry basis) water by DSC. Three endothermic transitions were observed when wheat starch was heated with water in the interval 35%–80% (dry basis). The first endothermic transition did not change with the water content, whereas second and third transitions were shifted toward higher temperatures with the decrease in water content (Table 9.6). Enough water made the second transition peak disappear. Ghiasi et al. (1982a,b) also observed a single endotherm at high water–starch ratio (2:1). The minimum water necessary for starch gelatinization can be calculated from a plot of gelatinization enthalpy and moisture content. The values are 33% (dry basis) (Eliasson, 1980) and 45% (dry basis) (Wootton and Bamunuarachchi, 1979) for wheat starch. The enthalpy of the second transition reached a maximum at a water content of 45% (dry basis) for wheat starch (Eliasson, 1980) and 31% (dry basis) water for potato starch (Donavan, 1979). The effect of excess water on gelatinization is also given in Table 9.7. The minimum moisture for gelatinization for different starches is given in Table 9.8.

Wang et al. (1991) studied the effect of water on phase transition and gelatinization of Amioca (waxy corn starch) in detail. The peak temperature decreases as the moisture content of starch samples increases until a threshold value is reached where water is in excess (Donavan, 1979; Billiaderis et al., 1980; Wang et al., 1991). Wang et al. (1991) studied the DSC peak temperatures (scan rate: 5°C/min) for Amioca–water mixtures (amylose content: 3.3%) by varying the water content from 0% to 99% (wet basis) and found that samples with water content higher than 60% (wet basis) had constant peak temperature of 71°C. As the water content decreased from 60%, the peak temperature increased linearly from 71°C to 230°C for bone dry sample and the correlation was proposed as

$$T_{GP} = 227.92 - 267.4X_{wo} \tag{9.1}$$

where temperature is in degree Celsius. Wang et al. (1991) mentioned that these two regions of peak temperature indicated the presence of two physicochemical mechanisms in starch granules

Table 9.4 Effect of Water Level on Gelatinization Enthalpy for Wheat Starch[a]

M_w^o:	0.50	0.75	1.00	1.25	1.50	1.75	2.00
ΔH_G	0.84	2.93	8.37	10.04	15.06	17.15	19.66

Source: Wootton, M. and Bamunuarachchi, A., *Starch*, 31, 264, 1979.
[a] DSC heating rate: 16°C/min.

Table 9.5 Gelatinization Temperature of Starches with Excess Water

Source	ε_w^o	T_{GP} (°C)	$(\Delta H)_G$ (kJ/kg)	T_{III} (°C)	$(\Delta H)_G$ (kJ/kg)
Potato starch	0.94	65	18.83	m	m
Waxy maize starch	0.94	73	15.90	m	m
Maize starch	0.91	72	13.39	97	1.67
Wheat starch	0.93	65	10.88	96	2.09

Source: Kugimiya, M., Donavan, J.W., and Wong, R.Y., *Starch*, 32, 265, 1980.
Note: ε_w^o is initial volume fraction of water and T_m is melting.
m, the absence of second endotherm.

Table 9.6 Gelatinization Temperature of Wheat Starch as a Function of Water[a]

M_w^o	T_{GP} (°C)	T_{II} (°C)	T_{III} (°C)
0.35	61	70	100
0.40	60	70	105
0.45	61	72	108
0.50	61	81	118
0.60	61	87	120
0.70	61	96	125
0.80	61	106	139

Source: Adapted from Eliasson, A.C., *Starch*, 32, 270, 1980.
Note: M_w^o is the initial mass fraction of water (g/g dry solids) and T_{II} and T_{III} are the first and second melting temperature.
[a] Scanning rate: 10°C/min.

Table 9.7 Effect of Water on Gelatinization Temperature[a]

(kg Water/kg Starch)	T_I (°C)	T_{GP} (°C)	T_{GE} (°C)
0.75	74.3	81.5	111.3
1.00	72.3	80.0	103.8
1.50	72.1	77.3	94.3
2.00	72.6	77.3	87.8
3.00	71.3	77.2	85.3

Source: Adapted from Wirakartakusumah, M.A., Kinetics of starch gelatinization and water absorption in rice, PhD thesis, Department of Food Science, University of Wisconsin, Madison, 1981.
[a] DSC rate: 10°C/min.

Table 9.8 Minimum Water for Gelatinization, Bound Water, and Difference for Some Starches

Starch	Minimum Moisture for Gelatinization (kg Water/kg Starch)	Bound Water (kg Water/kg Starch)	Excess Water (kg Water/kg Starch)
Wheat	0.45	0.33	0.12
Maize	0.45	0.30	0.15
Waxy Maize	0.47	0.32	0.15
Amylomaize	0.52	0.29	0.23

Source: Wootton, M. and Bamunuarachchi, A., *Starch*, 31, 264, 1979.

gelatinization or melting process and the extent of each of these mechanisms in starch granules depends on the available water in the sample. When peak temperatures remained at 71°C, excess water was present and complete gelatinization occurred (i.e., DG: 1.0). This is the critical water content (60%, wet basis) at 71°C. Starch granules melt at 230°C in the absence of water. At water contents up to 60% (wet basis), starch is partly bounded with water molecule through gelatinization process converting Amioca dough as a semisolid. The remaining starch could be melted as a solid when enough heat is supplied. A single endotherm occurred in starch–water mixtures containing more than 60% water and less than 60% a prominent second transition occurred at higher temperatures. Similar results were also observed for potato starch (Donavan, 1979), azuki bean starch, smooth pea starch, lentil starch (Billiaderis et al., 1980), wheat starch (Burt and Russell, 1983), rice starch (Billiaderis et al., 1986), and waxy corn starch (Amioca) (Wang et al., 1989). Wang et al. (1991) showed by simulation that a minimum ratio of 14 water molecules to one anhydrous glucose unit was required for complete gelatinization considering the critical water content 60% (wet basis). The unreacted water content and water conversion were calculated using the following equations to estimate the critical stoichiometric ratio of water to glucose (Wang et al., 1991):

$$X_{\mathrm{urw}} = X_{\mathrm{wo}} - \frac{18(X_{\mathrm{STO}})(\mathrm{SC})(\psi)}{162} \tag{9.2}$$

$$\mathrm{DC} = \frac{X_{\mathrm{urw}}}{X_{\mathrm{urw}} + X_{\mathrm{STO}}(1 - \mathrm{SC})} \tag{9.3}$$

where

- X_{wo} is the initial water content
- X_{STO} is the initial starch content
- SC is the fraction converted starch
- ψ is the stoichiometric ratio of water to glucose
- X_{urw} is the unreacted water after a specific degree of conversion and a specific stoichiometric ratio

Since the critical water content was 0.60, considering the initial water content of 0.60 and SC setting to 1.0 (i.e., 100% conversion), the stoichiometric ratio (ψ) was 14.

9.2.2.2 *Presence of Other Solutes*

Starch gelatinization is known to be affected by various solutes (electrolyte and nonelectrolyte). The effect of solutes on gelatinization is given in Tables 9.9 through 9.12. The effect of neutral salts on starch gelatinization followed the order of Hofmeister (lyotropic) series (Evans and Haisman, 1982; Sandstedt et al., 1960). Anion has structure-making and -breaking effect on starch gelatinization. Sodium sulfate increases gelatinization temperature, and sodium thiocyanade and potassium iodide decrease starch gelatinization temperature (Evans and Haisman, 1982; Jane, 1993). Potassium thiocyanade (>2 molal) and calcium chloride (>3 molal) solutions are both known to prompt starch gelatinization at room temperature (Evans and Haisman, 1982; Sandstedt et al., 1960; Jane, 1993). Other salts with chloride ion have no specific order of increase or decrease (Tables 9.10 and 9.11). Starch in the presence of concentrated calcium chloride (>4 molal) gelatinized with an exothermic enthalpy (Evans and Haisman, 1982; Jane, 1993). Slade and Levine (1987) studied the starch gelatinization of a three-component mixture of native wheat starch, glucose polymer, and water and plotted the gelatinization temperature as a function of inverse cosolvent (water and glucose polymer) molecular weight.

Table 9.9 Effect of Solute on the Gelatinization Temperature of Corn Starch

Na-Carbonate		Na-Sulfate		Na-Polyacrylic Acid	Na-Chloride	Na-Acrylate
x^a	T_{GP} (°C)	x^a	T_{GP} (°C)	T_{GP} (°C)	T_{GP} (°C)	T_{GP} (°C)
25	60.0	0	67.7	67.7	67.7	67.7
53	57.7	20	73.8	72.5	72.7	71.0
78	57.1	40	77.5	74.9	75.0	72.9
132	56.4	60	80.9	76.2	76.2	73.9
184	61.2	80	84.1	77.6	77.1	74.1
239	64.9	100	86.6	78.6	77.9	73.7
		120	89.0	79.6	78.2	73.9
		140	91.0	80.6	78.4	73.7
		160	—	81.5	78.0	73.6
		180	—	82.1	77.3	72.4
		200	—	82.7	76.2	—
		220	—	—	75.8	—

Source: Roosendaal, B.J.O., *Starch*, 34, 233, 1982; Samec, M., *In Die kollodchemie der starke*, Verlag Theodor Steinkopff, Leipzig, 1927.
[a] *x*, g salt/kg water.

Table 9.10 Effect of Solute on Gelatinization Temperature and Enthalpy Change of Starch

	Na_2SO_4[a]			KSCN[a]			KI[a]			KCl[a]	
x	T_{GP} (°C)	ΔH_G (kJ/kg)	x	T_{GP} (°C)	ΔH_G (kJ/kg)	x	T_{GP} (°C)	ΔH_G (kJ/kg)	x	T_{GP} (°C)	ΔH_G (kJ/kg)
0	61.8	11.7	0	61.8	11.7	0	61.8	11.7	0	61.8	11.7
18	68.6	13.7	49	60.3	9.72	80	61.8	—	59	70.9	11.9
68	77.5	14.1	87	51.3	8.03	142	56.7	7.97	130	70.4	11.7
128	87.0	15.4	127	40.8	4.97	200	48.7	6.16	224	70.0	10.6
201	113	21.8	163	6.6	0.00	249	37.2	3.09	271	67.1	9.29

Source: Jane, J., *Starch*, 45, 161, 1993.
[a] *x*, g salt/kg water; corn starch:water, 1:2; DSC method, heating rate 10°C/min.

Table 9.11 Effect of Solute on Gelatinization Temperature and Enthalpy Change of Starch (Effect of Chloride Ion and Sucrose)

	$CaCl_2$[a]			NaCl[a]			RbCl[a]			LiCl[a]		Sucrose[b]	
x	T_{GP} (°C)	ΔH_G (kJ/kg)	x	T_{GP} (°C)	ΔH_G (kJ/kg)	x	T_{GP} (°C)	ΔH_G (kJ/kg)	x	T_{GP} (°C)	ΔH_G (kJ/kg)	x	T_{GP} (°C)
0	61.8	11.7	0	61.8	11.74	0	61.8	11.74	0	61.8	11.74	0	62.2
86	74.4	11.4	47	74.0	12.64	93	70.0	11.86	63	75.1	14.02	100	64.9
171	64.1	7.95	105	73.1	12.39	195	70.0	11.57	126	73.0	13.14	500	77.9
216	47.4	3.09	185	71.1	11.05	319	70.0	10.56	228	0.0	0.26	1000	93.8
307	34.0	0.00	260	60.3	8.37	421	65.3	8.37	298	0.0	0.26		
358	53.0	16.3	367	58.3	—	492	57.5	5.71	373	37.1	—		

Source: Jane, J., *Starch*, 45, 161, 1993.
[a] *x*, g salt/kg water; corn starch:water, 1:2; DSC method, heating rate 10°C/min.
[b] *x*, g salt/kg water; native wheat starch; starch:water, 1:1; DSC method, Slade and Levine (1987).

Table 9.12 Effect of Sucrose and Sodium Chloride on Gelatinization Temperature of Wheat Starch[a]

Sucrose[b]						Sodium Chloride[b]					
x	ΔH_G (kJ/kg)	DG	T_{GI} (°C)	T_{GP} (°C)	T_{GE} (°C)	x	ΔH_G (kJ/kg)	DG	T_{GI} (°C)	T_{GP} (°C)	T_{GE} (°C)
0	19.66	1.00	50	68	86	0	19.66	1.00	50	68	86
176	13.39	0.68	50	70	86	31	11.30	0.57	58	71	88
429	11.72	0.60	50	73	86	64	10.46	0.53	64	75	88
818	9.62	0.49	50	75	86	99	10.88	0.55	68	78	88
						136	11.30	0.57	65	77	88
						176	11.30	0.57	65	77	88
						266	11.72	0.60	61	80	90
						429	13.81	0.70	59	79	91

Source: Wootton, M. and Bamunuarachchi, A., *Starch*, 32, 126, 1980.
[a] Starch (water, 13.6%; protein, 0.29%; fat, 0.46%; ash, 0.21%).
[b] *x*, g salt/kg water; wheat starch:water, 1:2; DSC method, heating rate 16°C/min.

The effect of sucrose on the thermal gelatinization of several starches has been studied, and it has been found that the gelatinization temperature increased with increasing sucrose concentration (Evans and Haisman, 1982; Spies and Hoseney, 1982, Chinachoti et al., 1991; Ahmed and William, 1999; Jang et al., 2001, Maaurf et al., 2001, Hirashima et al., 2005). Other low-molecular sugars such as fructose, glucose, maltose, etc., exhibited equal effects on thermal starch gelatinization but influenced the gelatinization characteristics to a different extent (Evans and Haisman, 1982; Ahmed and William, 1999; Beleia et al., 1996). This inhibitory effect of sugars on starch gelatinization has been attributed to the reduction of mobility of the solvent and decrease in the water activity values thereby impeding the penetration of water into the granule. Spies and Hoseney (1982) suggested that sugar–starch interactions stabilize amorphous regions by sugar molecules forming bridges between starch chains and hence increase the energy required for starch gelatinization. Tomasik et al. (1995) observed the formation of complexes of low-molecular sugars with starch by polarimetric measurements and concluded that these inclusion complexes are developed by penetration of sugar molecules into the interior of the starch molecule opened by starch swelling. Marcotte et al. (2004) reported that the wheat starch gelatinization peak temperature increased to 95°C in the presence of cake ingredients such as sugar, cocoa, shortening, eggs, and baking powder. Hirashima et al. (2005) observed that a lower content of sucrose enhanced the swelling of starch granules, while a higher content of sucrose decreased the rate of swelling of starch granules. The starch gelatinization temperature was shifted to higher temperatures with increasing sucrose concentration and gelatinization was not completed in the presence of excessive sucrose (Hirashima et al., 2005).

The effects of salts and sugars on high hydrostatic pressure gelatinization of starch were comparable to thermal gelatinization. However, the extent of influence of salts on pressure-induced starch gelatinization seems to be dependent on the type of starch. Potato starch, although the most pressure-resistant, was the most susceptible to salts and gelatinized at lower salt concentrations under pressure than wheat and tapioca starches (Rumpold and Knorr, 2005). The gelatinization pressure was lowered by sugars whereas the degree of gelatinization was linearly correlated with the number of equatorial hydroxyl groups. Since pressure-induced starch gelatinization is strongly dependent on the water content and is decreased in the presence of sugars and water structure making ions (SO_4^{2-}) and increased by structure breaking ions as SCN^-, it was suggested that the availability of free water is crucial for starch gelatinization under pressure (Rumpold and Knorr, 2005).

Lipids are used as food additives in starch products to modify the textural properties. Addition of lipids in starch can inhibit the starch gelatinization process (Kugimiya et al., 1980; Eliasson et al.,

Table 9.13 Lipid Content and Characteristics of Rice Starch

Starch	Lipid Content (%)	Iodine Affinity (%)	Iodine Color Intensity at 625 nm	Gelatinization Temperature (°C)		
				0.1[a]	0.5[a]	1.0[a]
Native	0.62	2.24	0.425	53.5	59.3	65.0
Defatted	0.39	2.94	0.445	53.0	59.3	65.0
Defatted	0.36	3.15	0.457	53.1	59.2	64.7
Defatted	—	3.09	0.452	53.1	59.2	64.7
Defatted	0.34	3.14	0.459	53.2	59.1	64.8
Defatted	0.12	3.76	0.466	51.5	58.1	64.3
RES[b]	0.41	3.21	0.445	54.1	58.5	63.8
RES[c]	0.73	2.64	0.429	53.3	58.5	64.7

Source: Adapted from Ohashi, K., Goshima, G., Kusuda, H., and Tsuge, H., *Starch*, 32, 54, 1980.
[a] Degree of gelatinization.
[b] Starch re-embraced with 0.5% palmitate.
[c] Starch re-embraced with 1.0% palmitate.

1981a,b). Larsson (1980) mentioned that amylose leaching can be blocked by the formation of an amylose–lipid complex on the granule surface. The degree of inhibition by lipids is related to the physical state of lipids, lipid monomer concentration, and amylose leaching rate from the starch surface (Larsson, 1980). These factors are important when lipids are added as antistaling agent to bread. Maningat and Juliano (1980) also observed that defatting reduced the gelatinization temperature and gel viscosity of rice starch. The characteristics of native and defatted rice starch depend on the trace amount of embraced lipid and the saturated amount is 1% or less deduced from amylograph (Ohashi et al., 1980).

Ghiasi et al. (1982a,b) studied the effect of surfactant (monoglycerides or sodium stearoyl Iactylate) on gelatinization of wheat starch by heating at 60°C, 70°C, 80°C, and 95°C in the presence of excess water. At low temperatures, surfactants enter into the starch granules and form a surfactant–amylose complex. At 95°C, the dissociation of the starch–amylose complex was observed and confirmed by the formation of an amylose–iodine complex. Hoover and Hadziyev (1981) studied the complexing ability of commercially available monoglycerides (palmitic and stearic acid) with the amylose the of potato granules process. X-ray diffraction and IR spectroscopy also indicated the formation of α- and β-crystals. The effects of lipid on gelatinization are shown in Tables 9.13 and 9.14. Eliasson (1986) studied the effect of surface active agents on the gelatinization of starch and observed the second melting endotherm due to the amylose–lipid complexes (Table 9.15).

Table 9.14 Effect of Lipids Removal on Gelatinization of Wheat Starch

Starch	Native[a]					Lipid Removed[a]				
	M_w^o	T_{GI} (°C)	T_{GP} (°C)	T_{II} (°C)	T_{III} (°C)	M_w^o (°C)	T_{GI} (°C)	T_{GP} (°C)	T_{II} (°C)	T_{III} (°C)
Amy	0.35	50	60	106	139	0.35	52	60	105	140
Maris Huntasman	0.35	49	58	108	140	0.35	53	60	107	—
Amy	0.40	51	60	98	128	0.40	48	61	98	129
Maris Huntasman	0.40	49	61	89	—	0.40	50	60	94	128
Amy	0.45	50	62	88	121	0.45	50	63	91	122
Maris Huntasman	0.45	46	62	87	122	0.45	44	62	88	122

Source: Adapted from Eliasson, A.C., Larsson, K., and Miezis, Y., *Starch*, 33, 231, 1981.
[a] T_{II} and T_{III} are second and third endothermic peaks.

Table 9.15 Gelatinization Temperature of Wheat and Potato Starches in the Presence of Surface Active Agents

	Wheat Starch					Rice Starch				
Additive[a]	T_{GI} (°C)	T_{GP} (°C)	ΔH_G (kJ/kg)	T_{II} (°C)	ΔH_{II} (kJ/kg)	T_{GI} (°C)	T_{GP} (°C)	ΔH_G (kJ/kg)	T_{II} (°C)	ΔH_{II} (kJ/kg)
Native	57.0	61.3	12.7	101	1.33	58.2	63.7	17.0	—	—
SDS	54.7	60.1	9.5	94.2	4.40	55.4	60.8	13.9	88.6	1.84
CTAB	57.6	61.7	8.5	92.8	4.41	58.6	63.9	14.1	91.9	2.05
SSL	58.4	62.3	10.3	100	2.72	59.5	64.2	17.9	102	0.42
SMG	56.7	60.5	12.8	100	2.26	58.8	63.3	17.8	105	0.34
Lysolecithin	55.7	60.6	7.5	105	6.36	57.8	63.4	15.8	110	5.91
Lecithin	56.7	60.8	12.2	96.9	0.88	59.1	63.7	18.1	—	—

Source: Eliasson, A.C., *Carbohyd. Polym.*, 6, 463, 1986.

[a] SDS, sodium dodecyl sulfate; SMG, saturated monoglycerides; CTAB, cetyltrimethyl ammonium bromide; SSL, soduim stearoyl-2-lactylate; additives are 5% of the dry starch; starch water ratio is 1:3; DSC heating rate (10°C/min).

Eliasson (1983) studied a starch–gluten system and found that the gelatinization peak temperature of the starch increased and the enthalpy decreased in the presence of gluten protein. Mohamed and Rayas-Duarte (2003) reported that a higher amount of protein in the wheat starch–protein/gluten blend increased the onset and peak temperatures and decreased the enthalpy of the starch gelatinization. However, the results of Srikaei et al. (2005) about onset and peak temperatures did not clearly agree with the findings from Eliasson (1983) and Mohamed and Rayas-Duarte (2003). The presence of sugars and salts has been found to inhibit gelatinization reaction and increase the gelatinization temperature. Tan et al. (2004) investigated the starch gelatinization in a water–glycerol system and reported that addition of glycerol increased the gelatinization onset temperature with an extent that depended on the water content in the system. They also noticed that glycerol promoted starch gelatinization at low water content. Chaisawang and Suphantharika (2005) reported that the presence of guar and xanthan gums influenced the gelatinization characteristics of tapioca starch significantly by increasing the onset gelatinization temperature and decreasing the gelatinization enthalpy (Table 9.16).

9.2.2.3 *Pretreatments*

Thermal studies have shown that the peak gelatinization temperature of a thermally treated waxy maize starch (WMS) solution occurs at about 60°C at high moisture levels (>70%). As moisture content is decreased about 30%, the gelatinization peak is observed at a higher temperature (Zanoni et al., 1995). Donavan et al. (1983) studied the gelatinization temperature of native potato and wheat starches heated at 100°C for 16 h at moisture contents 18% and 27% by DSC and found a broadening of the gelatinization temperature range and a shift of the endotherm transition toward higher

Table 9.16 Gelatinization Temperature of Tapioca Starch in the Presence of Guar and Xanthan Gums

Sample	T_{GI} (°C)	T_{GP} (°C)	T_{GE} (°C)	ΔH (kJ/kg)	$T_{GE} - T_{GI}$ (°C)
Starch	53.9c	65.7a	73.2a	11.1a	19.3a
Starch + guar gum	55.9b	65.3a	72.0a	8.33b	16.1b
Starch + xanthan gum	58.3a	65.5a	71.9a	5.2c	13.7c

Source: Chaisawang, M. and Suphantharika, M., *Carbohyd. Polym.*, 61, 288, 2005.
Same letter in each column is not significantly different ($P > 0.05$).

temperature compared with untreated starches. The shift was greater for potato starch than for wheat starch. The endotherms of treated starches (potato and wheat) were biphasic indicating two types of granule structure. Defatted starches after and before treatment gave different endotherms. Again in the presence of small water, treated wheat starch gave three endothermic peaks whereas treated potato starch gave two endothermic peaks. Donavan et al. (1983) suggested that these transitions were due to the melting of (1) amylopectin crystals, (2) amylose–lipid complex, and (3) unknown structure. Gough and Pybus (1971) studied the effect of the gelatinization temperature of wheat starch granules treated in water for 72 h and found that gelatinization occurred at a higher but much more sharply defined temperature (Table 9.17). The effect of heat treatment on enthalpy is shown in Table 9.18. Marcotte et al. (2004) reported the peak gelatinization temperature of waxy maize starch in the range of 64°C–67°C depending on the thermal treatment. The peak gelatinization temperature (T_p) shifted to a higher value as the temperature of thermal treatment increased. The T_p of untreated WMS solution sample was 62.3°C. Recently Sablani et al. (2007) also reported the shifts in peak gelatinization temperature with increasing time of thermal and high-pressure treatments (Figure 9.1).

Stevens and Elton (1971) observed a 3°C shift in the endothermic peak for small granules than large granules. Ghiasi et al. (1982a,b) also observed a similar difference in the case of peak, but insignificant difference in the case of onset temperature. Thus, according to Ghiasi et al. (1982a,b) the difference in peak temperature may be due to the sample size and heating rate. The effect of soaking and steaming on the gelatinization of parboiled rice is given in Table 9.19, and the effect of freezing and drying on gelatinization is given in Table 9.20.

Palav and Seetharaman (2006) used microwave heating to investigate the process of gelatinization of wheat starch suspensions. They observed granule swelling and leaching of polymers after complete loss of granule birefringence, suggesting an asynchronous process of gelatinization compared with conduction modes of heating. The kinetics of polymer leaching was strongly dependent on the initial starch concentration. The loss of crystalline arrangement in a microwave-heated sample occurred at a lower temperature than that observed for a conduction-heated sample. They suggested that the vibrational motion of the polar molecules during microwave heating had a direct impact on the crystalline lamella of the amylopectin thus disrupting the radial arrangement of the amylopectin lamellae. Therefore, the crystalline arrangement is destroyed before the glass transition of the amorphous region of the granule, thus resulting in no swelling before loss of birefringence.

Table 9.17 Effect of Heat Treatment on Gelatinization of Starch

Rice Starch[a]				Wheat Starch[b]			
Treatment	T_{GI} (°C)	T_{GP} (°C)	T_{GE} (°C)	Treatment	T_{GI} (°C)	T_{GP} (°C)	T_{GE} (°C)
Untreated	71.0	76.8	82.5	Untreated[b]	57.2	60.5	62.6
1 h at 60°C	73.8	77.0	82.3	24 h at 25°C	58.0	61.4	63.8
24 h at 60°C	79.1	82.0	85.5	72 h at 25°C	58.8	62.0	63.5
48 h at 60°C	79.5	82.5	86.1	24 h at 40°C	59.9	62.4	64.4
72 h at 60°C	78.7	81.8	85.3	72 h at 40°C	63.0	64.5	65.5
				24 h at 50°C	66.0	68.4	69.5
				72 h at 50°C	67.0	69.0	69.5
				Untreated[c]	52.0		61.0
				72 h at 50°C	65.4		65.8
				Untreated[c]	56.0		61.0
				72 h at 50°C	65.4		65.6

[a] Wirakartakusumah (1981).
[b] Lorenz and Kulp (1980.)
[c] Gough and Pybus (1971).
ΔH_G is 12.55 KJ/kg in all cases.
DSC heating rate: 10°C/min; water/starch: 10.

Table 9.18 Effect of Heat Treatment and Water Content on the Enthalpy of Waxy Maize Starch

	Treatment		
M_w^o	Temperature (°C)	Time (min)	ΔH_G (kJ/kg)
0.65	UT[a]	0	16.0
	62	15	11.0
	70	15	1.7
	80	15	0.0
0.55	UT[a]	0	16.0
	65	30	13.0
	70	10	8.0
	80	15	0.7
	90	15	0.0
0.45	UT[a]	0	16.0
	65	15	14.0
	75	15	12.0
	80	15	10.0
	95	15	7.0
	110	15	0.0

Source: Adapted from Maurice, T.J., Slade, L., Sirett, R.R., and Page, C.M., in *Properties of Water in Foods*, Simatos, D. and Multon, J.L. (eds.), Martinus Nijhoff Publishers, Dordrecht, 1985.

[a] Untreated sample.

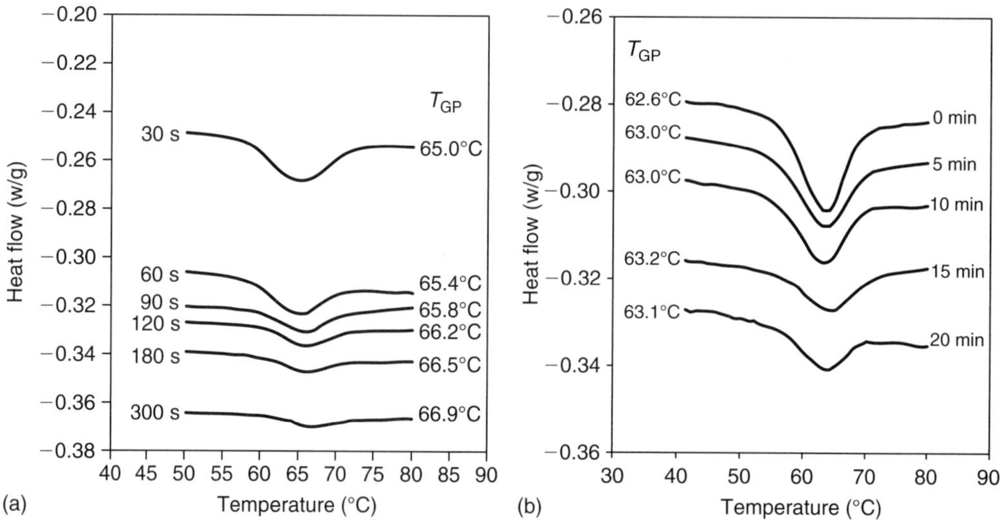

Figure 9.1 Shifts in peak of gelatinization temperature of 25% waxy maize starch at a heating rate of 5°C/min with (a) samples being pretreated at 60°C and (b) samples being pretreated at 450 MPa. (From Sablani, S.S., Kasapis, S., Al-Tarqe, Z., Al-Marhubi, I., Al-Khuseibi, M. and Al-Khabori, T., *J. Food Eng.*, 82, 443, 2007.)

Table 9.19 Degree of Gelatinization of Parboiled Rice[a] at Various Process Conditions

Soaking Temperature (°C)	Steaming Temperature (°C)	Time (min)	ΔH_G (kJ/kg)	DG	T_{GI} (°C)	T_{GP} (°C)	T_{GE} (°C)
63	112	10	5.44	0.291	52.2	58.1	65.5
71	112	10	5.02	0.352	53.5	60.0	66.0
63	121	10	3.35	0.582	49.7	55.9	62.0
71	121	10	2.93	0.632	49.0	56.4	61.9
63	112	15	5.02	0.368	54.1	58.1	66.5
71	112	15	3.77	0.537	49.2	55.5	61.6
63	121	15	2.93	0.646	49.3	55.8	61.5
71	121	15	2.09	0.741	49.7	56.5	62.0
UT	UT	UT	7.95	0.00	69.0	75.0	80.5

Source: Wirakartakusumah, M.A., Kinetics of Starch Gelatinization and Water Absorption in Rice, PhD Thesis, Department of Food Science, University of Wisconsin, Madison, 1981.

[a] Labonnet variety rice.

High hydrostatic pressure can facilitate gelatinization of starch at room temperature in the presence of water. The high pressure causes reversible hydration of the amorphous phase followed by irreversible distortion of the crystalline region, which in turn leads to the destruction of the granular structure (Blaszczak et al., 2005a). The effect of high pressure on the mechanism of starch gelatinization and granule structure is not similar to the heat–water treatment (Stute et al., 1996; Rubens and Heremans, 2000). Waxy maize starch disintegrates under high pressure, whereas amylomaize starch does not change its granular structure even at 900 MPa (Stolt et al., 2001). The high pressure-treated starch showed two distinctly differentiated zones (Blaszczak et al., 2005a). The outer zone of the granules remained unchanged and it corresponded to the more organized part of the granule, whereas the interior part of the granules was completely destroyed and formed gel-like structures. The SEM observation was also confirmed by ^{13}C NMR studies on pressurized potato starch conducted by Blaszczak and coworkers. The influence of high hydrostatic pressure on starch–water suspension also depends on the types of starch. Microscopy analysis showed that pressurization of waxy corn starch resulted in a complete breakdown of the granules, whereas Hylon VII retained a granular structure after 9 min of treatment (Blaszczak et al., 2005b). Katopo et al. (2002) reported that high-pressure treatment of A-type starches in the presence of water resulted in partial conversion to the B-type crystallinity. They also observed that the B-type x-ray pattern was more resistant to the high-pressure treatment in a water suspension. DSC thermogram of the A-type crystalline starch displayed an additional peak (between 41°C and 46°C) just after the pressure treatment. This indicated that

Table 9.20 Effect of Freezing and Drying Methods on Gelatinization Temperature of Potato Starch Measured by DSC

Starch	T_{GI} (°C)
Potato starch	66
Freeze-dried potato starch	60
Potato starch frozen with water[a]	66
Potato starch frozen without extra water	66
Potato starch dried at 135°C for 3 h	61

Source: Adapted from Eliasson, A.C., Larsson, K., and Miezis, Y., *Starch*, 33, 231, 1981.

[a] Water:starch ratio 5:1.

the crystallinity transformation was a result of pressure-induced rearrangements of double helices in the A-type starch.

It is known that high pressure favors the formation of hydrogen bonds and therefore the retarding effect of high pressure on the gelatinization of carbohydrate systems has been hypothesized due to the result of the stabilization of hydrogen bonds, which maintain the carbohydrate granule in its original state (Suzuki and Taniguchi, 1972; Fujii et al., 1980). The extent of pressure-facilitated gelatinization depends on the pressure applied, moisture content, treatment time and temperature, and types of starch. Muhr and Blanshard (1982) reported that for wheat starch the peak gelatinization temperature shifted upward with high-pressure treatment up to 300 MPa; however, they observed a subsequent fall in the gelatinization temperature as the pressure increased to 1500 MPa. One of the possible explanations given was the change in the water structure at very high pressure, which can result in changing stabilities with the pressure of interbiopolymer hydrogen bonds and water–biopolymer hydrogen bonds.

Douzals et al. (1996) found that wheat starch gelatinization starts at pressure over 300 MPa. A complete gelatinization was induced at the level of 600 MPa. It is also reported that the B-type starches were more pressure resistant (Stute et al., 1996), that is, potato starch needed the pressure of 800–1000 MPa to reach total gelatinization (Kudla and Tomasik, 1992). The excessive pressurization of 10% waxy maize starch suspension, however, formed a weak gel structure (Stolt et al., 1999). Ahromrit et al. (2007) studied the starch gelatinization in rice and reported that gelatinization was not observed below 300 MPa. They also noted that the extent of gelatinization at any time, temperature, and pressure could be correlated with the grain moisture content. Sablani et al. (2007) observed that the high hydrostatic pressure treatment of 25% waxy maize starch suspension resulted in an upward shift in the peak gelatinization temperature.

Acid modification is widely used in the starch industry to prepare thin boiling starches for use in food, paper, textile, and other industries (Rohwer and Klem, 1984). The inherent structures of starch before acid modification play an important role in determining its functionality. The starch gelation is strongly affected by its amylose content, molecular size of amylose and amylopectin, and short and long branch chains in amylopectin (Wang et al., 2003). The voscoelastic and thermal properties of starches are affected by acid hydrolysis. Wang et al. (2003) observed a slightly lower onset (from 69.7°C to 68.4°C) and peak (from 73.5°C to 72.4°C) temperature of gelatinization of 1.0 N HCl-treated starch.

9.3 PREDICTION

9.3.1 Gelatinization Temperature

The gelatinization temperature can be estimated from the theory of melting in polymers. The melting point (loss of crystallinity) of polymer diluents can be predicted by the equation as (Flory, 1953)

$$\frac{1}{T_{\mathrm{G}}} - \frac{1}{T_{\mathrm{GS}}} = \left(\frac{R}{\Delta H_{\mathrm{ru}}}\right)\left(\frac{\nu_{\mathrm{ru}}}{\nu_{\mathrm{D}}}\right)\left(\varepsilon_{\mathrm{D}} - \chi_{12}\varepsilon_{\mathrm{D}}^{2}\right) \tag{9.4}$$

where

R is the gas constant
ΔH_{ru} is the enthalpy of fusion per repeating unit
$(\nu_{\mathrm{ru}}/\nu_{\mathrm{D}})$ is the ratio of molar volume of the repeating unit to the molar volume of the diluent
χ_{12} is the Flory–Huggins polymer–diluent interaction parameter
ε_{D} is the volume fraction of diluent
T_{GS} is the melting point of perfect crystal in the absence of diluent due to gelatinization (K)

The values of ΔH_{ru} and χ_{12} can be estimated from the above equation using T_G at various volume fraction ε_D if T_{GS} is known. Naturally occurring polymers decompose before melting and T_{GS} cannot be determined directly (Lelievre, 1976). In this case, the above equation can be rearranged as (Flory et al., 1954)

$$\left(\frac{1 - T_G/T_{GS}}{\varepsilon_D^2}\right) = \left(\frac{R\nu_{\text{ru}}}{\Delta H_{\text{ru}}\nu_1}\right)\frac{T_G}{\varepsilon_D} - \left(\frac{R}{\Delta H_{\text{ru}}}\right)\left(\frac{\nu_{\text{ru}}}{\nu_D}\right)\chi_{12}T_G \tag{9.5}$$

If ΔH_{ru} and the product χ_{12} T_G are assumed constant over the temperature range, then the plot of $[1 - (T_G/T_{GS})]/\varepsilon_D^2$ versus T_G/ε_D should be linear. From the best fitted plot, the values of T_{GS}, (ΔH_{ru}), and χ_{12} can be estimated. Donavan (1979) calculated values of $\Delta H_{\text{ru}} = 56.48 \pm 1.67$ kJ/mol D-glucose unit, $T_{GS} = 441$ K (168°C) and χ_{12} may be assigned a value of 0.0875. Zobel et al. (1965) found $\Delta H_{\text{ru}} = 62.34$ kJ/mol glucose residue and $T_{GS} = 153$°C. Lelievre (1973) employed the loss of birefringence of wheat starch using a polarizing microscope with a hot stage at a heating rate of 1°C/30 min as a measure of T_G and found a value of $\Delta H_{\text{ru}} = 25.10$ kJ/mol, $\chi_{12} = 0.5$, and $T_{GS} = 495$ K (Marchant and Blanshard, 1980).

Lelievre (1976) extended Equation 9.4 (Flory, 1953) in the case of diluent (1), polymer (2), and solute (3) from the thermodynamic point of view as

$$\frac{1}{T_G} - \frac{1}{T_{GS}} = \left(\frac{R\nu_{\text{ru}}}{\Delta H_{\text{ru}}\nu_1}\right)\left[\varepsilon_1 + \frac{\varepsilon_3}{x_3} + \chi_{13}\varepsilon_1\varepsilon_3 - \frac{\chi_{12}\varepsilon_1 + \chi_{32}\varepsilon_3}{x_3}(\varepsilon_1 + \varepsilon_3)\right] \tag{9.6}$$

where

x_1, x_2, and x_3 are the number of segment units per molecule in the respective species (i.e., number of monomers)
χ_{13} is the pair interaction parameter
ε_1, ε_2, and ε_3 are the volume fractions of segments 1, 2, and 3

The assumptions in the above equations are (1) x_2 is large and ε_2 is comparatively large, (2) χ_{21} is equal to χ_{12}. Lelievre (1976) estimated the interaction parameters χ_{12} and χ_{13} for starch, water, and sucrose by assuming χ_{32} is zero and χ_{12} is equal to χ_{13}. The accurate estimation of the interaction constants makes the above equation limited to practical purposes (Evans and Haisman, 1982). Evans and Haisman (1982) plotted the gelatinization temperature as a function of water activity instead of concentration, and there is no universal relationship and each solute followed its own curve. Evans and Haisman (1982) derived an equation by transforming the chemical potential into water activity and neglecting the product of interaction terms:

$$\frac{1}{T_G} - \frac{1}{T_{GS}} \sim \frac{T_{GS} - T_G}{(T_{GS})^2} \sim K\left[1 - \ln\frac{a}{\varepsilon_1}\right] \tag{9.7}$$

where a is the water activity.

Evans and Haisman (1982) plotted the initial gelatinization temperature against $\ln(a/\varepsilon_1)$ for a large number of electrolyte and nonelectrolyte solutes and found a linear increasing relation. Thus, it accounts not only for raising of gelatinization temperature by sugars and polyhydroxy compounds, but also for lowering of the gelatinization temperature by simple salts. This indicated that the main variable affecting the gelatinization temperature is the ratio (a/ε_1).

9.3.2 Enthalpy

Our understanding of the process of gelatinization has greatly increased over the past two decades. Several new techniques have been used to probe into the physics and mechanism of gelatinization. Most studies have reported the gelatinization temperature range and only limited studies reported the enthalpy data and modeling kinetics of starch gelatinization. Wirakartakusumah (1981) plotted the fraction of ungelatinized rice starch as a function of time at various heating temperatures and observed two stages of kinetics. One is equilibrium period and the other is nonequilibrium period. The equilibrium fraction of ungelatinized starch decreased with the increase in temperature and the rate of gelatinization process increased with the increase in temperature. Thus, both time and temperature affect the gelatinization process kinetics or degree of gelatinization. The presence of nonequilibrium period is due to the diffusion process of water into starch granules, which causes swelling and hydration (Blanshard, 1979). A first-order reaction model is used in most of the cases to study the kinetics of the gelatinization process (Kubota et al., 1979; Bakshi and Singh, 1980; Wirakartakusumah, 1981; Lund, 1984; Kokini et al., 1992; Zanoni et al., 1995; Okechukwu and Rao, 1996; Marcotte et al., 2004; Sablani et al., 2007):

$$(1-\alpha) = \exp(-Kt) \tag{9.8}$$

where
- α is the degree of gelatinization
- t is the gelatinization time (s)
- K is the reaction rate constant (1/s)

The degree of gelatinization (α) is defined, as a function of time, t, as

$$\alpha(t) = 1 - \left[\frac{Q(t)}{Q_{\text{max}}}\right] \tag{9.9}$$

where $Q(t)$ is the heat uptake for partially gelatinized starch and Q_{max} is the heat uptake for the gelatinization of raw starch suspension (Figure 9.2). The rate constant can be determined by measuring the extent of gelatinization as a function of time at a constant temperature (isothermal condition) and can be related with temperature by an Arrhenius type of equation (Lund, 1984).

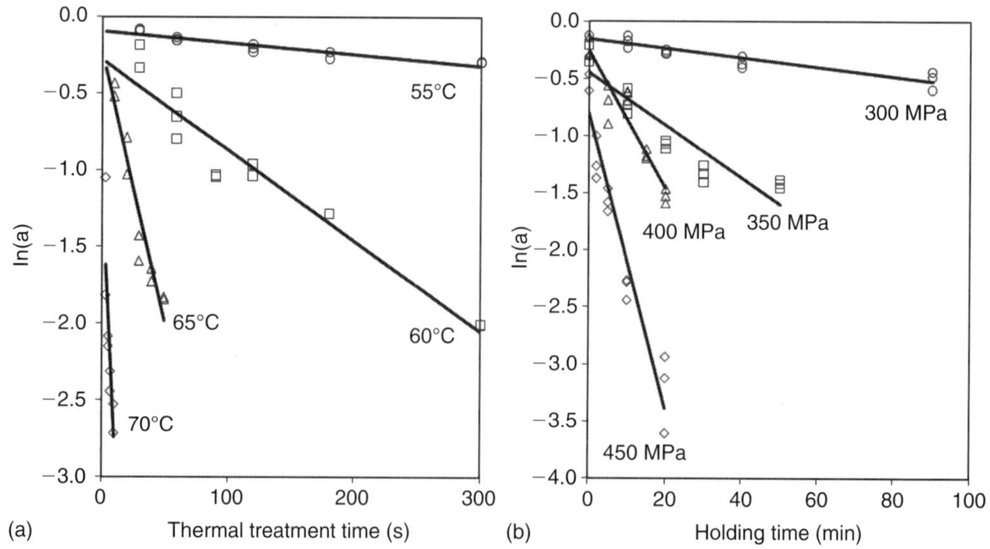

Figure 9.2 First-order kinetics of 25% waxy maize starch gelatinization following (a) thermal treatment (b) high hydrostatic pressure treatment. (From Sablani, S.S., Kasapis, S., Al-Tarqe, Z., Al-Marhubi, I., Al-Khuseibi, M., and Al-Khabori, T., *J. Food Eng.*, 82, 443, 2007.)

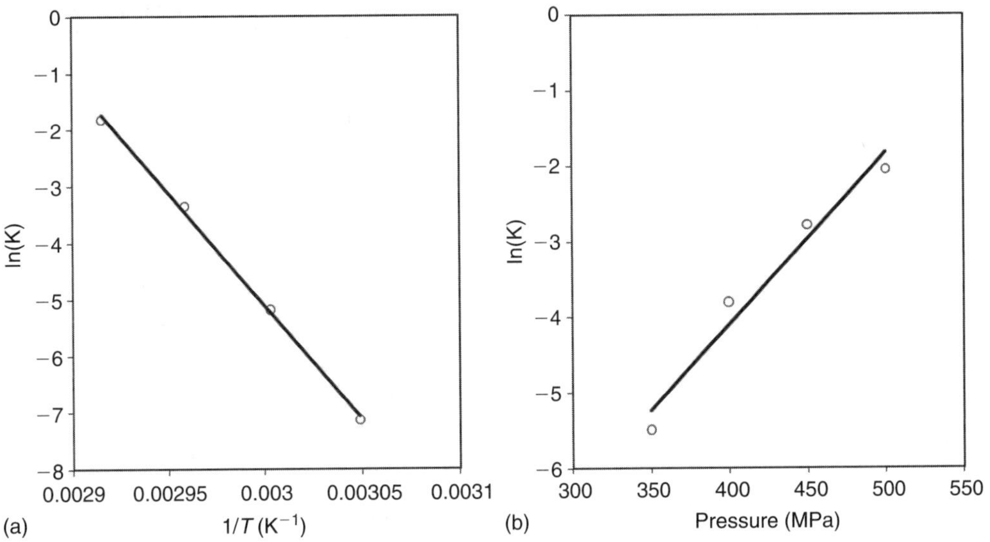

Fligure 9.3 Temperature (a) and pressure (b) effects on the reaction rate constants of 25% waxy maize starch gelatinization. (From Sablani, S.S., Kasapis, S., Al-Tarqe, Z., Al-Marhubi, I., Al-Khuseibi, M. and Al-Khabori, T., *J. Food Eng.*, 82, 443, 2007.)

The temperature dependence of constants can be described by an Arrhenius equation:

$$K = K_o \exp\left(-\frac{E_a}{RT}\right) \tag{9.10}$$

where
K_o is the reaction frequency factor
E_a is the activation energy of gelatinization (J/mol)
R is the gas constant (8.314 J/K mol)
T is the absolute temperature (K) (Figure 9.3)

The activation energy values for thermal gelatinization of different starches are given in Table 9.21. In addition, the nonisothermal kinetics approach of Kissinger (1957) for solid-state reaction has also been used to calculate kinetic parameters:

$$\ln\left(\frac{\beta}{T_{GP}^2}\right) = -\frac{E_a}{RT} + \ln\left(K\frac{R}{E_a}\right) \tag{9.11}$$

where
β is the heating rate (dT/dt, in K/min)
T_{GP} is the peak gelatinization temperature (in K)

The activation energy obtained by this method is more accurate than that obtained from the isothermal method (Starink, 1996; Lai and Lii, 1999). Other dependencies between T_p, β, and E_a can be derived according to many other kinetic equations that take into account the mechanism of the process: for example, the Mehl–Johnson–Avrami equation, two-, and three-dimensional diffusion models

Table 9.21 Activation Energy (E_a) Values for Thermal Gelatinization of Starch at Atmospheric Pressure Obtained by Fitting First-Order Rate Constant at Various Temperatures to the Arrhenius Equation

Materials	Temperature (°C)	E_a (kJ/mol)	Reference
White rice	75–110	79.4	Suzuki et al. (1976)
	110–150	36.8	
White rice (Taiwan rice)	48–63.5	62.9	Lin (1993)
	Above 63	11.1	
White rice (*Basmati*)	60–75	38	Ramesh (2001)
Brown rice (short grain: S6)	50–85	77.4	Bakshi and Singh (1980)
	85–120	43.8	
Rough rice	50–85	103.3	
	85–120	40.1	
Rice flour (long grain: Irgo)	<70	287	Ojeda et al. (2000)
	>70	30	
Rice starch	<85	187	Birch and Priestly (1973)
	>85	99	
Rice starch	73–92	42	Kubota et al. (1979)
Potato starch	60–63	961.4	
Potato starch	60–67	748.2	Donavan (1979)[a]
Potato	<67.5	819.3	Pravisani et al. (1985)
	>67.5	243.3	
Rice starch	<76	306	Yeh and Li (1996)
	>76	43	
Waxy rice starch	59–72	385	Lai and Lii (1999)
Cowpea starch	67–86	233.6	Okechukwu and Rao (1996)
Rough rice	30–60	289.3	Bello et al. (2007)
	60–90	16.6	
Hard wheat starch (in situ)	60–75	133	Turhan and Gunasekaran (2002)
	75–100	76	
Hard wheat starch (in vivo)	60–100	73	
Soft wheat starch (in situ)	60–75	140	
	75–100	82	
Soft wheat starch (in vivo)	60–100	78	
Soybean starch (in situ)	<60	40	Kubota (1979)
	>60	20	
Corn starch (in situ)	<80	111	Cabrera et al. (1984)
	>80	45	
Potato starch	53–65	110	Shiotsubo (1983)
Tarhana starch	60–120	3325	Ibanoglu and Ainsworth (1997)
Cake mix (wheat flour in the presence of sugar, cacao, shortening, egg, and baking powder)	67–97	204.7	Marcotte et al. (2004)
Bread dough (wheat flour and salt)	60–90	139	Zanoni et al. (1995a)

[a] Calculated by Pravisani et al. (1985) from the data reported by Donavan (1979).

Mehl–Johnson–Avrami equation:

$$1.5 \ln \beta = -\frac{E_a}{R} \frac{1}{T_{GP}} + \text{const} \tag{9.12}$$

Two-dimensional diffusion model:

$$\ln \frac{\beta^3}{T_{GP}^2} = -2\frac{E_a}{R} \frac{1}{T_{GP}} + \text{const} \tag{9.13}$$

Three-dimensional diffusion model:

$$\ln \frac{\beta^4}{T_{GP}^2} = -3\frac{E_a}{R} \frac{1}{T_{GP}} + \text{const} \tag{9.14}$$

Spigno and De Faveri (2004) used these three models and the Kissinger model to determine the energy of activation of the gelatinization of rice starch (Table 9.22). They also estimated the influence of starch water ratio on activation energy (Table 9.23). They used the following iso-conversional method proposed earlier by Calzetta Resio and Suarez (2001) to evaluate the kinetic parameters of both rate constants and reaction orders:

$$\ln \frac{d\alpha}{dt} = \ln K - n \ln (1 - \alpha) - \frac{E_a}{R} \frac{1}{T} \tag{9.15}$$

The values of α and $d\alpha/dt$ corresponding to the preselected temperature of the DSC scans can be used to simultaneously estimate K, n, and E_a using the multilinear regression procedure (Table 9.23). Spigno and De Faveri (2004), however, cautioned that nonisothermal DSC should be used to characterize the gelatinization process only if the kinetics of the process was already known because different models can give different values of E_a. Ozawa (1965) proposed nonisothermal approach where the plot of ln of the heating rates, in place of the rate constant, as a function of the reciprocal of the absolute temperature at different gelatinization percentages. This is obtained by the slicing of the endothermic spectra and representing as degree of gelatinization. The angular fit of this plot gives the $-E_a/R$, and the linear fit the pre-exponential factor. Later, this approach was used to calculate E_a of the gelatinization of potato, yam, and cassava starches (Pielichowski et al., 1998; Freitas et al., 2004).

The pressure primarily affects the volume of a system (Hermans, 1995). The Le Chatelier principle expresses that pressure favors reactions that result in a decrease in volume and inhibits reactions with an increase in volume (Masson, 1992). The influence of pressure on the reaction rate may be described by the transition state theory; the rate constant of a reaction in a liquid phase is proportional to the equilibrium constant for the formation of the active reactants (Tauscher, 1995; Knorr et al., 2006). Based on this assumption, Hermans (1995) and Tauscher (1995) reported that at constant temperature, the pressure dependence of the reaction rate constant (K) is due to the activation volume of the reaction (ΔV^*):

$$\left(\frac{\delta \ln K}{\delta P}\right)_T = -\frac{\Delta V^*}{RT} \tag{9.16}$$

where

P is the pressure (MPa)
R is the gas constant (8.314 cm^3/MPa/K/mol)
T is the temperature (K)

Table 9.22 Energy of Activation of Starch Gelatinization Obtained from DSC Nonisothermal Traces according to Different Kinetic Models

	E_a (kJ/mol) and r^2									
	20% Water		40% Water		60% Water		80% Water		90% Water	
Kinetic Model	**Two Lines**	**One**	**Two Lines**	**One**	**Two Lines**	**One**	**Two Lines**	**One**	**Two Lines**	**One**
Kissinger	<79.5°C		<82.4°C		<83.8°C		<81.7°C		<82.8°C	
	144 (0.99)	26 (0.76)	74 (0.99)	46.6 (0.91)	88 (0.99)	68.4 (0.98)	125 (0.99)	69 (0.90)	97 (0.99)	76.5 (0.97)
	>79.5°C		>82.4°C		>83.4°C		>81.7°C		>82.8°C	
	14 (0.91)		30 (0.93)		56 (0.99)		41 (0.99)		67 (0.89)	
Mehl–Johnson–Avrami	<79.5°C		<82.4°C		<83.8°C		<81.7°C		<82.8°C	
	220 (0.99)	48 (0.78)	120 (0.99)	79 (0.91)	141 (0.99)	111.5 (0.98)	197 (0.99)	112 (0.92)	154 (0.99)	124 (0.97)
	>79.5°C		>82.4°C		>83.8°C		>81.7°C		>82.8°C	
	30 (0.95)		54 (0.95)		92 (0.99)		70 (0.99)		109 (0.90)	
Two-dimensional diffusion	<79.5°C		<82.4°C		<83.8°C		<81.7°C		<82.8°C	
	218 (0.99)	45 (0.76)	117 (0.99)	76 (0.91)	138 (0.99)	109 (0.98)	194 (0.99)	110 (0.92)	151 (0.99)	120 (0.97)
	>79.5°C		>82.4°C		>83.8°C		>81.7°C		>82.8°C	
	27 (0.95)		51(0.94)		89 (0.99		67 (0.99)		106 (0.90)	
Three-dimensional diffusion	<79.5°C		<82.4°C		<83.8°C		<81.7°C		<82.8°C	
	195 (0.99)	41 (0.76)	104 (0.99)	68 (0.92)	124 (0.99)	97 (0.98)	173 (0.99)	98 (0.91)	135 (0.99)	108 (0.97)
	>79.5°C		>82.4°C		>83.8°C		>81.7°C		>82.8°C	
	24 (0.95)		46 (0.94)		80 (0.99)		60 (0.99)		95 (0.90)	

Source: Spigno, G. and De Faveri, D.M., *J. Food Eng.*, 62, 337, 2004.

Table 9.23 Kinetic Constants of Rice Starch at Different Water: Starch Ratios and Heating Rates

Heating Rate (°C/min)	20% H_2O		40% H_2O		60% H_2O		80% H_2O		90% H_2O	
	n	E_a (r^2)	*n*	E_a (r^2)	*N*	E_a (r^2)	*n*	E_a (r^2)	*n*	E_a (r^2)
3	0.37	139 (0.93)	1.5	406 (0.90)	1.13	276 (0.97)	0.47	233 (0.91)	0.98	313 (0.91)
5	0.86	170 (0.73)	1	328 (0.98)	0.88	219 (0.93)	1.04	213 (0.92)	1	214 (0.92)
7	0.2	69 (0.93)	0.92	262 (0.97)	0.96	202 (0.96)	0.94	259 (0.96)	0.65	278 (0.99)
10	0.45	88 (0.93)	0.75	199 (0.95)	0.73	164 (0.98)	0.67	200 (0.97)	0.89	236 (0.96)
15	0.58	155 (0.99)	0.4	91 (0.98)	0.71	179 (0.98)	0.59	194 (0.98)	0.76	160 (0.93)
Average E_a		124		257		208		220		240

Source: Spigno, G. and De Faveri, D.M., *J. Food Eng.*, 62, 337, 2004.

The pressure and temperature dependence of the reaction rate constant k can be expressed combining the equations of Arrhenius and Eyring (Knorr et al., 2006):

$$\ln(K) = \ln(K_o) + \frac{E_a}{RT} + \left(\frac{-\Delta V^*}{RT}\right) \tag{9.17}$$

Activation volume values for pressure gelatinization of different starches are given in Table 9.24.

9.4 RETROGRADATION

During the gelatinization process, starch granules swell and gradually lose their molecular order; the amylase chains solubilize and a starch gel is formed. Upon cooling, the gel undergoes transformation leading to a partially crystalline structure; both amylase and amylopectin taking part in this process result in the formation of retrograded starch. Retrograded starch, as native starch granules, is also indigestible (Colonna et al., 1992; Garcia-Alonso et al., 1999). Retrogradation of starch in the processed product during storage is the major reason for the deterioration of various desirable sensory qualities. Several methods used for retarding starch retrogradation include enzyme treatment, adding oligosaccharides, and adding lipids. Mixtures of different starches have also shown retardation in retrogradation behavior (Yao et al., 2003). Retrogradation in starch has been quantified using DSC (Liu and Thompson, 1998; Jane et al., 1999; Stolt et al., 2001), dynamic viscoelasticity (Morikawa and Nishinari, 2000; Tako and Hizukuri, 2000), enzymatic analysis (Tsuge et al., 1992; Kim et al., 1997), x-ray analysis (Jagannath et al., 1998; Jouppila et al., 1998), infrared spectroscopy (Ogawa et al., 1998; Smits et al., 1998), and pulsed NMR (Wursch and Gumy, 1994; Farhat et al., 2000; Teo et al., 2000; Yao et al., 2003).

Table 9.24 Activation Volume (ΔV) Values for Pressure Gelatinization of Starch Obtained at Given Temperature by Fitting First-Order Rate Constant at Various Pressures to the Eyring Equation

Materials	Temperature (°C)	ΔV (cm^3/mol)	Reference
Thai glutinous rice	20	−7.064	Ahromrit et al. (2007)
	50	−11.816	
	60	−9.690	
	70	−11.122	
Waxy maize starch	25	−57.4	Sablani et al. (2007)

9.5 IMPORTANCE OF GELATINIZATION

Gelatinized starch plays an important role in determining the structural and textural properties of many foods. The proportion of raw and gelatinized starch in ready-to-serve starchy products may be critical in determining the acceptability. Texture of many foods such as breakfast cereals, beverages, rice, noodles, pasta, and dried soups depends on the fraction of gelatinized starch in the product (Guraya and Toledo, 1993). Gelatinization of starch is important in the processes such as baking of bread, gelling of pie fillings, formulation of pasta products, and thickening of sauces to produce a desirable texture or consistency (Olkku and Rha, 1978). The degree of starch gelatinization affects the final oil content of the chips and the oil distribution in the chips after frying (Kawas and Moreira, 2001).

Parboiling of rice is an ancient process in which paddy rice is soaked in warm water for several hours, the soaked rice is soaked usually under pressure to gelatinize the rice in the endosperm, and then dried and milled. This facilitates removal of the hulls and less damage or breakage of rice. The nutritional value of milled parboiled rice is higher than white rice. In addition, parboiled rice has the advantage that the grains show little or no tendency to clump or become pasty when prepared for eating, and when cooked it is firmer and less cohesive. For this reason, many consumers prefer this type of rice. The changes in the process or the intensity of heat treatment also affects the amount of soluble starch in the milled rice, which is probably related to the degree of gelatinization (Roberts et al., 1954). The presence of an insoluble amylose complex appears to be responsible for the above characteristics of parboiled rice (Priestley, 1976). Gelatinized rice is easy to digest. Gelatinization temperature is important in case of rice and rice products in terms of consumer acceptance. More than one transition in starch occur during bread making when dough is heated under baking conditions. The stickiness of rice can be controlled by controlling the gelatinization process. Low-molecular weight amylose in mashed potatoes gives a gluey, sticky, or gummy texture. Gelatinization appeares to be an obvious choice as index of the cooking process of rice (Birch and Priestly, 1973). Different sources of starch have different gelatinization temperatures and enthalpy. The starch gelatinization kinetics is important in engineering design of a starch cooking system and to achieve a desired level of gelatinized starch in formulated products with acceptable quality. Therefore, engineers must take this into account when designing thermal heat transfer processes for cooking starch systems that undergo gelatinization (Lund, 1984).

Pressure-induced gelatinization is very sensitive to changes in temperature, pressure, and treatment time. Hence, this characteristic can be exploited as an extrinsic pressure, temperature, and time integrator (Bauer and Knorr, 2005). The major benefit of pressure-induced starch gelatinization is the variety of pressure sensitivity depending on the type of starch. Starches can be systematically selected that gelatinize over a pressure range at the temperature and treatment time applied in the high-pressure range requested for ensuring process efficiency and safety. These can easily be incorporate in packaging material and in conjunction with a color marker to provide a tool for process monitoring.

REFERENCES

Ahmed, F.B. and William, P.A. 1999. Effect of sugars on the thermal and rheological properties of sago starch, *Biopolymers* 50: 401–412.

Ahromrit, A., Ledward, D.A., and Niranjan, K. 2007. Kinetics of high pressure facilitated starch gelatinization in Thai glutinous rice, *Journal of Food Engineering* 79: 834–841.

Autio, K. and Salmenkallio-Marttila, M. 2001. Light microscopic investigations of cereal grains, doughs and breads, *Food Science and Technology* 34: 18–22.

Baik, O.D., Sablani, S.S., Mrcotte, M., and Castaigne, F. 1999. Modeling the thermal properties of a cup cake during baking, *Journal of Food Science* 64: 295–299.

Bakshi, A.S. and Singh, R.P. 1980. Kinetics of water diffusion and starch gelatinization during rice parboiling, *Journal of Food Science* 45: 1387–1392.

Banks, W. and Greenwood, C.T. 1975. *Starch and Its Composition*, John-Wiley & Sons, New York.

Bauer, B.A. and Knorr, D. 2005. The impact of pressure, temperature and treatment time on starches: Pressure-induced starch gelatinization as pressure time temperature indicator for high hydrostatic pressure processing, *Journal of Food Engineering* 68: 329–334.

Bayram, M. 2005. Modeling of cooking of wheat to produce bulgur, *Journal of Food Engineering* 71: 179–186.

Bayram, M. 2006. Determination of the cooking degree for bulgur production using amylose/iodine, centre cutting and light scattering methods, *Food Control* 17: 331–335.

Beleia, A., Miller, R.A., and Hoseney, R.C. 1996. Starch gelatinization in sugar solutions, *Starch* 48: 259–262.

Bello, M.O., Tolaba, M.P., and Suarez, C. 2007. Water absorption and starch gelatinization in whole rice grain during soaking, *LWT-Food Science and Technology* 40: 313–318.

Billiaderis, C.G., Maurice, T.J., and Vose, J.R. 1980. Starch gelatinization phenomena studied by differential scanning calorimetry, *Journal of Food Science* 45: 1669–1674.

Billiaderis, C.G., Page, C.M., Maurice, T.J., and Juliano, B.O. 1986. Thermal characterization of rice starch: A polymeric approach to phase transition of granular starch, *Journal of Agricultural and Food Chemistry* 34: 6–14.

Birch, G.G. and Priestly, R.J. 1973. Degree of gelatinization of cooked rice, *Starch* 25: 98–100.

Blanshard, J.M.V. 1979. Physicochemical aspects of starch gelatinization. In: *Polysaccharides in Food*, Blanshard, J.M.V. and Mitchell, J.R. (Eds.), Butterworths, London, pp. 139–152.

Blaszczak, W., Valverde, S., and Fornal, J. 2005a. Effect of high pressure on the structure of potato starch, *Carbohydrate Polymers* 59: 377–383.

Blaszczak, W., Fornal, J., Valverde, S., and Garrido, L. 2005b. Pressure-induced changes in the structure of corn starches with different amylose content, *Carbohydrate Polymers* 61: 132–140.

Bryant, C.M. and Hamaker, B.R. 1997. Effect of lime on gelatinization of corn flour and starch, *Cereal Chemistry* 74(2): 171–175.

Buleon, A., Colonna, P., Planchot, V., and Ball, S. 1998. Starch granules: Structure and biosynthesis, *International Journal of Biological Macromolecules* 23: 85–112.

Burt, D.J. and Russell, P.L. 1983. Gelatinization of low water content wheat starch–water mixtures, *Starch* 35: 354–360.

Cabrera, E., Pineda, J.C., Duran De Bazua, C., Segurajauregeui, J.S., and Vernon, E.J. 1984. Kinetics of water diffusion and starch gelatinization during corn nixtamalization. In: *Engineering and Food*, Vol. 1, McKenna, B.M. (Ed.), London, UK, Elsevier, pp. 117–125.

Calzetta Resio, A. and Suarez, C. 2001. Gelatinization kinetics of amaranth starch, *International Journal of Food Science and Technology* 36: 441–448.

Chaisawang, M. and Suphantharika, M. 2005. Effects of guar gum and xanthan gum additions on physical and rheological properties of cationic tapioca starch, *Carbohydrate Polymers* 61: 288–295.

Chatakanonda, P., Dickinson, L.C., and Chinachoti, P. 2003. Mobility and distribution of water in cassava and potato starches by ^1H and ^2H NMR, *Journal of Agricultural and Food Chemistry* 51: 7445–7449.

Cheetam, N.W.H. and Tao, L. 1998. Oxygen-17 NMR relaxation studies on gelatinization temperature and water mobility in maize starch, *Carbohydrate Polymers* 35: 279–286.

Chinachoti, P., Steinberg, M.P., and Villota, R. 1990. A model for quantitating energy and degree of starch gelatinization based on water, sugar and salt contents, *Journal of Food Science* 55(2): 543–546.

Chinachoti, P., Kim-Shin, M.-S., Mari, F., and Lo, L. 1991. Gelatinization of wheat starch in the presence of sucrose and sodium chloride: Correlation between gelatinization temperature and water mobility as determined by oxygen-17 nuclear magnetic resonance, *Cereal Chemistry* 68: 245–248.

Colburn, C.R. and Schoch, T.J. 1964. Iodimetric determination of amylose. In: *Methods in Carbohydrate Chemistry*, Vol. 4, Whistler, R.L., Smith, R.J., Be Miller, J.N., and Wolform, M.L. (Eds.), Academic Press, New York, pp. 161–164.

Colonna, P., Leloup, V., and Buleon, A. 1992. Limiting factors of starch hydrolysis, *European Journal of Clinical Nutrition* 46(Suppl.): S17–S32.

Donavan, J.W. 1979. Phase transitions of the starch–water system, *Biopolymers* 18: 263–275.

Donavan, J.W. and Mapes, C.J. 1980. Multiple phase transition of starches and nageli amylodextrins, *Starch* 32: 190–193.

Donavan, J.W., Lorenz, K., and Kulp, K. 1983. Differential scanning calorimetry of heat–moisture treated wheat and potato starches, *Cereal Chemistry* 60: 381–387.

Douzals, J.P., Marechal, P.A., Coquille, J.C., and Gervais, P. 1996. Microscopic study of starch gelatinization under high hydrostatic pressure, *Journal of Agriculture and Food Chemistry* 44: 1403–1408.

Eberstein, V.K., Hamburg, Hopcke, R., Kleve, Konieczny-Janda, G., and Stute, R. 1980. DSC-Untersuchungen an starken, *Starch* 32(12): 397–400.

Eliasson, A.C. 1980. Effect of water content on the gelatinization of wheat starch, *Starch* 32(8): 270–272.

Eliasson, A.C. 1983. Differential scanning calorimetry studies on wheat starch–gluten mixture I: Effect of gluten on the gelatinization of wheat starch, *Journal of Cereal Science* 1: 199–205.

Eliasson, A.C. 1986. On the effects of surface active agents on the gelatinization of starch—a calorimetric investigation, *Carbohydrate Polymers* 6: 463–476.

Evans, I.D. and Haisman, D.R. 1982. The effect of solutes on the gelatinization temperature range of potato starch, *Starch* 34(7): 224–231.

Eliasson, A.C., Carlson, T.L.G., Larsson, K., and Miezis, Y. 1981a. Some effect of starch lipids on the thermal and rheological properties of wheat starch, *Starch* 33: 130–134.

Eliasson, A.C., Larsson, K., and Miezis, Y. 1981b. On the possibility of modifying the gelatinization properties of starch by lipid surface coating, *Starch* 33: 231–235.

Fang, C. and Chinnan, M.S. 2004. Kinetic of cowpea starch gelatinization and modeling of starch gelatinization during steaming of intact cowpea, *Food Science and Technology (lwt)* 37: 345–354.

Farhat, I.A., Blanshard, J.M.V., and Mitchell, J.R. 2000. The retrogradation of waxy maize starch extrudates: Effects of storage temperature and water content, *Biopolymers* 53(5): 411–422.

Flory, P.J. (1953) *Principles of Polymer Chemistry*, Cornell University Press, Ithaca, NY.

Flory, P.J., Garett, R.R., Newman, S., and Mandelkern, L. 1954. Thermodynamics of crystallization in high polymers cellulose trinitrate, *Journal of Polymer Science* 12: 97–107.

Fornal, J., Blaszczak, W., and Lewandowicz, G. 1998. Microstructure of starch acetates from different botanical sources, *Polish Journal of Food and Nutrition Sciences, Supplement* 3(7/48): 86–95.

Freeman, T.P. and Shelton, D.R. 1991. Microstructure of wheat starch: From kernel to bread, *Food Technology* 45: 164–168.

Freitas, R.A., Paula, R.C., Feitosa, J.P.A., Rocha, S., and Sierakowski, M.R. 2004. Amylose contents, rheological properties and gelatinization kinetics of yam (*Dioscorea alata*) and cassava (*Manihot utilissima*) starches, *Carbohydrate Polymers* 55: 3–8.

Fujii, S., Miyagawa, K., and Watanbe, T. 1980. The influence of applied hydrostatic pressure on gel formation of a (1→3)-β-glucan, *Carbohydrate Research* 84: 265–272.

Fulcher, R.G., Faubion, J.M., Ruan, R., and Miller, S.S. 1994. Quantitative microscopy in carbohydrate analysis, *Carbohydrate Polymers* 25: 285–293.

Gallaghar, P.K. 1997. Thermoanalytical instrumentation, technique and methodology. In: *Thermal Characterization of Polymeric Materials*, Vol. 1, 2nd ed., Turi, E.A. (Ed.), Academic Press, New York, pp. 2–203.

Garcia-Alonso, A., Jimenez-Escrig, A., Martin-Carron, N., Bravo, L., and Saura-Calixto, F. 1999. Assessment of some parameters involved in the gelatinization and retrogradation of starch, *Food Chemistry* 66: 181–187.

Ghiasi, K., Varriano-Marston, E., and Hoseney, R.C. 1982a. Gelatinization of wheat starch II. Starch–surfactant interaction, *Cereal Chemistry* 59: 86–88.

Ghiasi, K., Hoseney, R.C., and Varriano-Marston, E. 1982b. Gelatinization of wheat starch III. Comparison by different scanning calorimetry and light microscopy, *Cereal Chemistry* 59: 258–262.

Gidley, M.J. and Bociek, S.M. 1985. Molecular organization in starches: A ^{13}C CP/MAS NMR study, *Journal of the American Chemical Society* 107: 7040–7044.

Goto, F. 1969. Determination of gelatinization property of highly concentrated starch suspension by brabender plastograph, *Starch* 21: 128–132.

Gough, B.M. and Pybus, J.N. 1971. Effect on gelatinization temperature of wheat starch granules of prolonged treatment with water at 50°C, *Starch* 23(6): 210–213.

Guraya, H.S. and Toledo, R.T. 1993. Determining gelatinized starch in a dry starchy product, *Journal of Food Science* 58(4): 888–889.

Haruhito, T., Mayumi, H., Hiroaki, I., Satoshi, W., and Gisho, G. 1990. Enzymatic evaluation for the degree of starch retrogradation in foods and foodstuffs, *Starch/Starke* 42: 213–216.

Hermans, K. 1995. High pressure effects on biomolecules. In: *High Pressure Processing of Foods*, Ledward, D.A., Johnston, D.E., and Earnshaw, R.G. (Eds.), Nottingham University Press, Nottingham, pp. 81–98.

Hill, B.P., Takacs, S.F., and Belton, P.S. 1990. A new interpretation of proton NMR relaxation time measurements of water in food. *Food Chemistry* 37: 95–111.

Hirashima, M., Takahashi, R., and Nishinari, K. 2005. Changes in the viscoelasticity of maize starch paste by adding sucrose at different stages, *Food Hydrocolloids* 19: 777–784.

Hoover, R. and Hadziyev, D. 1981. The effect of monoglycerides on amylose complexing during a potato granule process, *Starch* 33(10): 346–355.

Hoseney, R.C., Atwell, W.A., and Lineback, D.R. 1977. Scanning electron microscopy of starch isolated from baked products, *Cereal Foods World* 22: 56–60.

Hoseney, R.C., Zeleznak, K.J., and Yost, D.A. 1986. A note on the gelatinization of starch, *Starch* 38: 407–409.

Ibanoglu, S. and Ainsworth, P. 1997. Kinetics of starch gelatinization during extrusion of tarhana, a traditional Turkish wheat flour–yoghurt mixture, *International Journal of Food Sciences and Nutrition* 48: 201–204.

Jagannath, J.H., Jayaraman, K.S., Arya, S.S., and Somashekar, R. 1998. Differential scanning calorimetry and wide-angle X-ray scattering studies of bread staling, *Journal of Applied Polymer Science* 67(9): 1597–1603.

Jane, J. 1993. Mechanism of starch gelatinization in neutral salt solutions, *Starch* 45(6): 161–166.

Jane, J., Chen, Y.Y., Lee, L.E., McPherson, A.E., Wong, K.S., Radosavijevic, M., and Kasemsuwan, T. 1999. Effect of amylopectin branch chain length and amylose content on the gelatinization and pasting properties of starch, *Cereal Chemistry* 76(5): 629–637.

Jang, J.K., Lee, S.H., Cho, S.C., and Pyun, Y.R. 2001. Effect of sucrose on glass transition, gelatinization, and retrogradation of wheat starch, *Cereal Chemistry* 78(2): 186–192.

Jouppila, K., Kansikas, J., and Roos, Y.H. 1998. Factors affecting crystallization and crystallization kinetics in amorphous corn starch, *Carbohydrate Polymers* 36(203): 143–149.

Kaláb, M., Allan-Wojtas, P., and Miller, S.S. 1995. Microscopy and other imaging techniques in food structure analysis, *Trends in Food Science & Technology* 6: 177–186.

Karapantsios, T.D., Sakonidou, E.P. and Raphaelides, S.N. 2000. Electric conductance study of fluid motion and heat transport during starch gelatinization, *Journal of Food Science* 65: 144–150.

Katopo, H., Song, Y., and Jane, J.-L. 2002. Effect and mechanism of ultrahigh hydrostatic pressure on the structure and properties of starches, *Carbohydrate Polymers* 47: 233–244.

Kawas, M.L. and Moreira, R.G. 2001. Effect of degree of starch gelatinization on quality attributes of fried tortilla chips, *Journal of Food Science* 66(2): 300–306.

Kim, J.O., Kim, W.S., and Shin, M.S. 1997. A comparative study on retrogradation of rice starch gels by DSC, X-ray and α-amylase methods, *Starch* 49(2): 71–75.

Kissinger, H.E. 1957. Reaction kinetics in differential thermal analysis, *Analytical Chemistry* 29: 1702–1706.

Knorr, D., Heinz, V., and Buckow, R. 2006. High pressure application for food biopolymers, *Biochemica et Biophysica Acta* 1764: 619–631.

Kokini, J.K., Lai, L.S., and Chedid, L.L. 1992. Effect of starch structure on starch rheological properties, *Food Technology* 46(6): 124–139.

Koksel, H., Sivri, D., Scanlon, M.G., and Bushuk, W. 1998. Comparison of physical properties of raw and roasted chickpeas (leblebi), *Food Research International* 31: 659–665.

Koo, H.-J., Park, S.-H., Jo, J., Kim, B.-Y., and Baik, M.-Y. 2005. Gelatinization and retrogradation of 6 year old Korean ginseng starches studied by DSC, *Food Science and Technology (lwt)* 38: 59–65.

Kubota, K. 1979. Studies on the soaking and cooking rate equations of soybean, *Journal of Faculty of Applied and Biological Sciences* 18: 1–9.

Kubota, K., Hosokawa, Y., Suzuki, S., and Hosaka, H. 1979. Studies on the gelatinization rate of rice and potato starches, *Journal of Food Science* 44: 1394–1397.

Kudla, E. and Tomasik, P. 1992. Modification of starch by high pressure, Part II: Comparison of starch with additives, *Starch* 44: 253–259.

Kugimiya, M., Donavan, J.W., and Wong, R.Y. 1980. Phase transitions of amylose–lipid complexes in starch: A calorimetric study, *Starch* 32: 265–270.

Lai, V.M.F. and Lii, C.Y. 1999. Effects of modulated differential scanning calorimetry (MDSC) variables on thermodynamic and kinetic characteristics during gelatinization of waxy rice starch, *Cereal Chemistry* 76: 519–525.

Larsson, K. 1980. Inhibition of starch gelatinization by amylose–lipid complex formation, *Starch* 32(4): 125–126.

Leach, H.W. 1979. Gelatinization of starch. In: *Starch Chemistry and Technology*, Chapter XII, Whistler, R.L. and Paschall, E.F. (Eds.), Academic Press, New York, pp. 287–306.

Lelievre, J. 1973. Starch gelatinization, *Journal of Applied Polymer Science* 18: 293–296.

Lelievre, J. 1976. Theory of gelatinization in a starch–water–solute system, *Polymer* 17: 854–858.

Lelievre, J. and Mitchell, J. 1975. A pulsed NMR study of some aspects of starch gelatinization, *Starch* 27: 113–115.

Leszkowiat, M.J., Yada, R.Y., Coffin, R.H., and Stanley, D.W. 1990. Starch gelatinization in cold temperature sweetening resistant potatoes, *Journal of Food Science* 55(3): 1338–1340.

Li, F.-D., Li, L.-T., Li, Z., and Tatsumi, E. 2004. Determination of starch gelatinization by Ohmic heating, *Journal of Food Engineering* 62: 113–120.

Lin, S.H. 1993. Water uptake and gelatinization of white rice, *Lebensmittel-Wissenchaft und Technologie* 26(3): 276–278.

Lisinska, G. and Leszcynski, W. 1989. *Potato Science and Technology*, Elsevier Applied Science, London.

Liu, J.M. and Zhao, S.L. 1990. Scanning electron microscope study on gelatinization of starch granules in excess water, *Starch/Stärke* 42: 96–98.

Liu, Q. and Thompson, D.B. 1998. Effects of moisture content and different gelatinization heating temperatures on retrogradation of waxy-type maize starches, *Carbohydrate Research* 314(3–4): 221–235.

Liu, Q., Charlet, G., Yelle, S., and Arul, J. 2002. Phase transition in potato starch–water system. I. Starch gelatinization at high moisture level, *Food Research International* 35: 397–407.

Lorenz, K. and Kulp, K. 1980. Steeping of starch at various temperatures-effect on functional properties, *Starch* 32: 181–186.

Lugay, J.C. and Juliano, B.O. 1965. Crystallinity of rice starch and its fraction in relation to gelatinization and pasting characteristics, *Journal of Applied Polymer Science* 9: 3775–3790.

Lund, D. 1984. Influence of time, temperature, moisture, ingredients, and processing conditions on starch gelatinization, *CRC Critical Reviews in Food Science and Nutrition* 20: 249–273.

Maaurf, A.G., Che Man, Y.B., Asbi, B.A., Junainah, A.H., and Kennedy, J.F. 2001. Gelatinization of sago starch in the presence of sucrose and sodium chloride as assessed by differential scanning calorimetry, *Carbohydrate Polymers* 45: 335–345.

Maningat, C.C. and Juliano, B.O. 1980. Starch lipids and their effect on rice starch properties, *Starch* 32(3): 76–82.

Marchant, J.L. and Blanshard, J.M.V. 1980. Changes in the birefringence characteristics of cereal starch granules at different temperatures and water activities, *Starch* 32(7): 223–226.

Marcotte, M., Sablani, S.S., Kasapis, S., Baik, O.D., and Fustier, P. 2004. The thermal kinetics of starch gelatinization in the presence of other cake ingredients, *International Journal of Food Science and Technology* 39: 807–810.

Masson, P. 1992. Pressure denaturation of protein. In: *High Pressure and Biotechnology*, Balny, C. Hayashi, R. Harmans, K. and Masson, P. (Eds.), Colloque Inserm/John Libbey Eurotex, Ltd., Montrouge, pp. 89–98.

Maurice, T.J., Slade, L., Sirett, R.R., and Page, C.M. 1985. Polysaccharides–water interactions—Thermal behavior of rice starch. In: *Properties of Water in Foods*, Simatos, D. and Multon, J.L. (Eds.), Martinus Nijhoff Publishers, Dordrecht, pp. 211–227.

McCready, R.M. and Hassid, W.Z. 1943. The separation and quantitative estimation of amylose and amylopectin in potato starch, *Journal of American Chemical Society* 65: 1154–1157.

McDonough, C.M. and Rooney, L.W. 1999. Use of the environmental scanning electron microscope in the study of cereal-based foods, *Cereal Foods World* 44: 342–348.

Mohamed, A.A. and Rayas-Duarte, P. 2003. The effect of mixing and wheat protein/gluten on the gelatinization of wheat starch, *Food Chemistry* 81: 533–545.

Morikawa, K. and Nishinari, K. 2000. Effects of concentration dependence of retrogradation behavior of dispersions for native and chemically modified potato starch, *Food Hydrocolloid* 14(4): 395–401.

Muhr, A.H. and Blanshard, J.M.V. 1982. Effect of hydrostatic pressure on starch gelatinization, *Carbohydrate Polymers* 2: 61–74.

Ogawa, K., Yamazaki, I., Yoshimura, T., Ono, S., Rengakuji, S., Nakamura, Y., and Shimasaki, C. 1998. Studies on the retrogradation and structural properties of waxy corn starch, *Bulletin of Chemical Society of Japan* 71(5): 1095–1100.

Ohashi, K., Goshima, G., Kusuda, H., and Tsuge, H. 1980. Effect of embraced lipid on the gelatinization of rice starch, *Starch* 32(2): 54–58.

Olkku, J. and Rha, C. 1978. Gelatinization of starch and wheat flour starch—a review, *Food Chemistry* 3: 293–311.

Ojeda, C.A., Tolaba, M.P., and Suarez, C. 2000. Modeling starch gelatinization kinetics of milled rice flour, *Cereal Chemistry* 77(2): 145–147.

Okechukwu, P.E. and Rao, M.A. 1996. Kinetics of cowpea starch gelatinization based on granule swelling, *Starch/Staerke* 48: 43–47.

Owusu-Ansah, J., Van de Voort, F.R., and Stanley, D.W. 1982. Determination of starch gelatinization by X-ray diffractometry, *Cereal Chemistry* 59: 167–171.

Ozawa, T. 1965. A new method of analysing thermo gravimetric data, *Bulletin of the Chemical Society of Japan* 38: 1881–1886.

Palav, T. and Seetharaman, K. 2006. Mechanism of starch gelatinization and polymer leaching during microwave heating, *Carbohydrate Polymers* 65: 364–370.

Parades-Lopez, O. and Bushuk, W. 1983. Development and "underdevelopment" of wheat dough by mixing: Microscopic structure and its relation to bread making quality, *Cereal Chemistry* 60: 24–27.

Pielichowski, K., Tomasik, P., and Sikora, M. 1998. Kinetics of gelatinization of potato starch studied by non-isothermal DSC, *Carbohydrate Polymers* 35: 49–54.

Pravisani, C.I., Califano, A.N., and Calvelo, A. 1985. Kinetics of starch gelatinization in potato, *Journal of Food Science* 50: 657–660.

Priestley, R.J. 1976. Studies on parboiled rice: Part-I. Comparison of the characteristics of raw and parboiled rice, *Food Chemistry* 1: 5–14.

Ramesh, M. 2001. An application of image analysis for the study of kinetics of hydration of milled rice in hot water, *International Journal of Food Properties* 4(2): 271–284.

Robert, R.L., Potter, A.L., Kester, E.B., and Keneaster, K.K. 1954. Effect of processing conditions on the expanded volume, color and soluble starch of parboiled rice, *Cereal Chemistry* 31: 121–129.

Rohwer, R.G. and Klem, R.E. 1984. Acid-modified starch: Production and use. In: *Starch: Chemistry and Technology*, BeMiller, J.N. and Paschall, E.F. (Eds.), Academic Press, Orlando, FL, pp. 529–541.

Rolee, A. and LeMeste, M. 1999. Effect of moisture content on thermomechanical behavior of concentrated wheat starch–water preparations, *Cereal Chemistry* 76: 452–458.

Roosendaal, B.J.O. 1982. Tentative hypothesis to explain how electrolytes affect gelatinization temperature of starches in water, *Starch* 34(7): 233–239.

Roman-Gutierrez, A.D., Guilbert, S., and Cuq, B. 2002. Description of microstructural changes in wheat flour and flour components during hydration by using environmental scanning electron microscopy, *Food Science and Technology* 35: 730–740.

Rubens, P. and Heremans, K. 2000. Pressure–temperature gelatinization phase diagram of starch: An in situ Fourier transform infrared study, *Biopolymers* 54: 524–530.

Rumpold, B.A. and Knorr, D. 2005. Effect of salts and sugars on pressure-induced gelatinization of wheat, tapioca, and potato starches, *Starch* 57: 370–377.

Sablani, S.S., Kasapis, S., Al-Tarqe, Z., Al-Marhubi, I., Al-Khuseibi, M., and Al-Khabori, T. 2007. Isobaric and isothermal kinetics of gelatinization of waxy maize starch, *Journal of Food Engineering* 82: 443–449.

Samec, M. 1927. *In Die kollodchemie der starke*, Verlag Theodor Steinkopff, Leipzig.

Sandstedt, R.M., Kempf, W., and Abbott, R.C. 1960. The effects of salts on the gelatinization of wheat starch, *Starch* 12(11): 333–336.

Schoch, J.T. 1964. Iodimetric determination of amylose. In: *Methods in Carbohydrate Chemistry*, Vol. 4, Whistler, R.L., Smith, R.J., Be Miller, J.N., and Wolform, M.L. (Eds.), Academic Press, New York, pp. 157–160.

Schoch, J.T. and Maywald, E.C. 1956. Microscopic examination of modified starches, *Analytical Chemistry* 28: 382–387.

Shetty, R.M., Lineback, D.R., and Seib, P.A. 1974. Determining the degree of starch gelatinization, *Cereal Chemistry* 51: 364–375.

Shiotsubo, T. 1983. Starch gelatinization at different temperatures as measured by enzymatic digestion method, *Agricultural Biology and Chemistry* 47: 2421–2425.

Singh, B. and Dodda, L.M. 1979. Studies on the preparation and nutrient composition of bulgur from triticale, *Journal of Food Science* 44: 449–452.

Slade, L. and Levine, H. 1987. Recent advantages in starch retrogradation. In: *Industrial Polysaccharides*, Stivala, S.S., Crescenzi, V., and Dea, M. (Eds.), Gordon and Breach Science Publishers, New York, pp. 387–430.

Smith, G.S., Barta, E.J., and Lazar, M.E. 1964. Bulgur production by continuous atmospheric pressure process, *Food Technology* 18: 89–92.

Smits, A.L.M., Ruhnau, F.C., Vtiegenthart, J.F.G., and van Soest, J.J.G. 1998. Aging of starch based systems as observed with FT-IR and solid state NMR spectroscopy, *Starch* 50(11–12): 478–483.

Spigno, G. and De Faveri, D.M. 2004. Gelatinization kinetics of rice starch studied by non-isothermal calorimetric technique: influence of extraction method, water concentration and heating rate, *Journal of Food Engineering* 62: 337–344.

Spies, R.D. and Hoseney, R.C. 1982. Effect of sugar on starch gelatinization, *Cereal Chemistry* 59(2): 128–131.

Srikaei, K., Furst, J.E., Ashton, J.F., Hosken, R.W., and Sopade, P.A. 2005. Wheat grain cooking process as investigated by modulated temperature differential scanning calorimetry, *Carbohydrate Polymers* 61: 203–210.

Starink, M.J. 1996. A new method for the derivation of activation energies from the experiments performed at constant heating rate, *Thermochimica Acta* 288: 97–104.

Stevens, D.J. and Elton, G.A. 1971. Thermal properties of the starch/water system. I. Measurement of heat of gelatinization by differential scanning calorimetry, *Starch* 23: 8–11.

Stolt, M., Oinonen, S., and Autio, K. 2001. Effect of high pressure on the physical properties of barley starch, *Innovative Food Sciences & Emerging Technologies* 1: 167–175.

Stolt, M., Stoforos, N.G., Taoukis, P.S., and Autio, K. 1999. Evaluation and modeling of rheological properties of high pressure treated waxy maize starch dispersions, *Journal of Food Engineering* 40: 293–298.

Stute, R., Klingler, R.W., Boguslawski, S., Eshtiaghi, M.N., and Knorr, D. 1996. Effects of high pressures treatment on starches, *Starch/Starke* 48: 399–408.

Stute, R. 1997. High pressure treated starch. European Patent Application EP 0 804 884 A2.

Suzuki, K. and Taniguchi, Y. 1972. The effects of pressure on organisms. In: *Effects of Pressure on Organisms*, Sleigh, M.A. and MacDonald, A.G. (Eds.), Cambridge University Press, UK, pp. 103–124.

Suzuki, K., Kubota, K., Omichi, M., and Hosaka, H. 1976. Kinetics studies on cooking of rice, *Journal of Food Science* 41: 1180–1183.

Tako, M. and Hizukuri, S. 2000. Retrogradation mechanism of rice starch, *Cereal Chemistry* 77(4): 473–477.

Tan, I., Wee, C.C., Sopade, P.A., and Halley, P.J. 2004. Investigation of the starch gelatinization phenomena in water–glycerol systems: Application of modulated temperature differential scanning calorimetry, *Carbohydrate Polymers* 58: 191–204.

Tananuwong, K. and Reid, D.S. 2004. DSC and NMR relaxation studies of starch–water interactions during gelatinization, *Carbohydrate Polymers* 58: 345–358.

Tauscher, B. 1995. Pasteurization of food by hydrostatic pressure: Chemical aspects, *Z. Lebensmittel. Unters. Forsch.* 200: 3–13.

Teo, C.H., Abd Cheah, P.B., Norziah, M.H., and Seow, C.C. 2000. On the roles of protein and starch in the aging of non-waxy rice flour, *Food Chemistry* 69(3): 229–236.

Tomasik, P., Wang, Y.-J., and Jane, J.L. 1995. Complexes of starch with low-molecular saccharides, *Starch* 47: 185–191.

Tsuge, H., Tatsumi, E., Ohtani, N., and Nakazima, A. 1992. Screening of alpha-amylase suitable for evaluating the degree of starch retrogradation, *Starch* 44(1): 29–32.

Turhan, M. and Gunasekaran, S. 2002. Kinetics of in situ and in vitro gelatinization of hard and soft wheat starches during cooking in water, *Journal of Food Engineering* 52: 1–7.

Van Soest, J.J.G., Tournois, H., De Wit, D., and Vliegenthat, J.J.G. 1995. Short-range structure in (partially) crystalline potato starch determined with attenuated total reflectance Fourier-transform IR spectroscopy, *Carbohydrate Research* 279: 201–214.

Varrino-Marston, E., Ke, V., Huang, G., and Ponte, J. 1980. Comparison of methods to determine starch gelatinization in bakery foods, *Cereal Chemistry* 57: 242–248.

Waigh, T.A., Gidley, M.J., Komanshek, B.U., and Donald, A.M. 2000. The phase transformations in starch during gelatinization: A liquid crystalline approach, *Carbohydrate Research* 328: 165–176.

Wang, W.C. and Sastry, S.K. 1997. Starch gelatinization in ohmic heating, *Journal of Food Engineering* 34: 225–242.

Wang, Y.-J., Truong, V.-D., and Wang, L. 2003. Structural and rheological properties of corn starch as affected by acid hydrolysis, *Carbohydrate Polymers* 52: 327–333.

Wang, S.S., Chiang, W.C., Yeh, A.-I., Zhao, B., and Kim, I.H. 1989. Kinetics of phase transition of waxy corn starch at extrusion temperatures and moisture contents, *Journal of Food Science* 54: 1298–1301.

Wang, S.S., Chiang, W.C., Zhao, B., Zheng, X.G., and Kim, I.H. 1991. Experimental analysis and computer simulation of starch water interactions during phase transition, *Journal of Food Science* 56: 121–124.

Whistler, R.L. and BeMiller, J.N. 1997. *Carbohydrate Chemistry for Food Scientists*, Eagan Press, St. Paul.

Wirakartakusumah, M.A. 1981., Kinetics of starch gelatinization and water absorption in rice, Ph.D. thesis, Department of Food Science, University of Wisconsin, Madison.

Wootton, M. and Bamunuarachchi, A. 1979. Application of differential scanning calorimetry to starch gelatinization, *Starch* 31: 264–269.

Wootton, M. and Bamunuarachchi, A. 1980. Application of differential scanning calorimetry to starch gelatinization, III. Effect of sucrose and sodium chloride, *Starch* 32(4): 126–129.

Wootton, M. and Chaudhary, M.A. 1980. Gelatinization and in vitro digestibility of starch in baked products, *Journal of Food Science* 45: 1783–1784.

Wootton, M., Weeden, D., and Munk, N. 1971. A rapid method for the estimation of starch gelatinization in processed foods, *Food Technology Australia* 23: 612–615.

Wursch, P. and Gumy, D. 1994. Inhibition of amylopectin retrogradation by partial beta-amylolysis, *Carbohydrate Research* 256(1): 129–137.

Yao, Y., Zhang, J., and Ding, X. 2003. Retrogradation of starch mixtures containing rice, *Journal of Food Science* 68: 260–265.

Yeh, A. and Li, J.Y. 1996. A continuous measurement of swelling of rice starch during heating, *Journal of Cereal Science* 23: 277–283.

Zanoni, B., Schiraldi, A., and Simoneta, R. 1995a. A naive model of starch gelatinization kinetics, *Journal of Food Engineering* 24: 25–33.

Zobel, H.F. and Stephen, A.M. 1995. Starch: Structure, analysis, and application. In: *Food Polysaccharides and Their Applications*, Stephen, A.M. (Ed.), Marcel Dekker, New York, pp. 19–65.

Zobel, H.F., Senti, F.R., and Brown, D.S. 1965. Paper presented at the 50th Annual Meeting of AACC, Kansas.

CHAPTER 10

Crystallization: Measurements, Data, and Prediction

Kirsi Jouppila and Yrjö H. Roos

CONTENTS

10.1 INTRODUCTION

Crystallization and melting are first-order phase transitions of the liquid to solid and solid to liquid states, respectively. Crystallization involves the arrangement of disordered molecules to a highly ordered crystalline structure. This requires an initial arrangement of a few molecules to form stable nuclei for crystal growth. The driving force for the process can be described by Gibbs energy, which has the lowest value for an equilibrium state. Hence, a driving force can be defined for supersaturated solutions or supercooled liquids below the equilibrium melting temperature of the substance. The driving force for nucleation increases with increasing level of supersaturation and decreasing temperature below the equilibrium melting temperature of the crystalline state.

Nucleation requires formation of stable nuclei, which can take place at and above critical values of extents of supersaturation (solution) and supercooling (melt). Stable nuclei will subsequently grow to crystals through propagation and maturation steps (e.g., Hartel and Shastry [1], Roos [2], Hartel [3]). Homogeneous nucleation refers to formation of nuclei by small clusters of the crystallizing molecules while heterogeneous nucleation involves impurities, foreign particles, or a surface (e.g., dust particles or microscopic structures in the vessel wall) at which molecules can adhere to initiate nucleation. Secondary nucleation is also possible when existing crystals enhance nucleation and crystal growth. Addition of such seed crystals is often used in crystallization processes to control crystallization and formation of desired crystal forms. This process is known as seeding.

Propagation or growth of crystals may occur with the following steps depending on the crystallizing compound: (1) mutarotation to correct anomeric form, (2) diffusion of crystallizing molecules to the crystal interface, (3) removal of hydration water, (4) counterdiffusion of noncrystallizing molecules from the crystal interface, (5) orientation of molecules at the crystal–liquid interface, (6) incorporation of molecules into the crystal lattice, and (7) removal of latent heat (e.g., Hartel and Shastry [1], Roos [2], Hartel [3]). Any of these steps can limit the rate of crystallization. Maturation includes recrystallization (or ripening) resulting in formation of crystals with lower energy state and crystal perfection.

Crystallization is a common unit operation in the food industry and it may often occur in food storage. Sugar and salt industries are based on crystallization processes, crystallization is essential in the fats and oils industry as well as in manufacturing of confectionary, and ice formation is an important part of the frozen foods industry. Crystallization into several different polymorphic crystal forms with various crystal structures and physicochemical properties is often problematic and requires particular understanding to obtain desired, stable crystal structures. Recrystallization and crystallization of noncrystalline amorphous solids in food systems may also take place in manufacturing and storage. Uncontrolled crystallization of amorphous solids is often detrimental. A typical example is crystallization of noncrystalline sugars in dehydrated foods, such as lactose crystallization in dairy powders or frozen dairy desserts. Furthermore, recrystallization of ice in frozen foods and blooming of chocolate are well-known recrystallization phenomena resulting in quality deterioration.

Understanding crystallization in food systems is important in the control of crystallization processes and preventing crystallization and recrystallization in food manufacturing and storage. Information on water sorption, glass transition, and kinetics of crystallization under various conditions is often needed to allow the control of both recrystallization processes and crystallization of amorphous food components.

There are several analytical techniques that can be used to follow and measure crystallization and crystallinity. However, there are differences in how these techniques observe crystals. The measurement of crystallization using qualitative and quantitative techniques is discussed. Qualitative measurements include observation of the presence of crystals and identification of crystal forms. Quantitative measurements include determination of the degree of crystallinity and kinetics of crystallization, that is, crystal growth or changes in crystallinity as a function of time.

10.2 MEASUREMENT OF CRYSTALLIZATION

Analytical techniques used in the measurement of crystallization and crystallinity include thermoanalytical, spectroscopic, x-ray diffraction (XRD), and gravimetric techniques. Most of these techniques can be used in both qualitative and quantitative analysis of crystallization and crystallinity. Also, several microscopic techniques are available for visualization of the shape and size of crystals and the formation of crystals.

10.2.1 Qualitative Measurement of Crystallization

Crystallization and presence of crystals can be observed and often quantified using XRD; spectroscopic techniques such as infrared (IR) spectroscopy, FTIR spectroscopy, near infrared (NIR) spectroscopy, and Raman spectroscopy; thermoanalytical techniques such as differential scanning calorimetry (DSC); and gravimetric techniques. The presence of crystals may also be observed using microscopic techniques such as polarizing light microscopy and scanning electron microscopy (SEM). For example, the presence of semicrystalline native starch granules can be

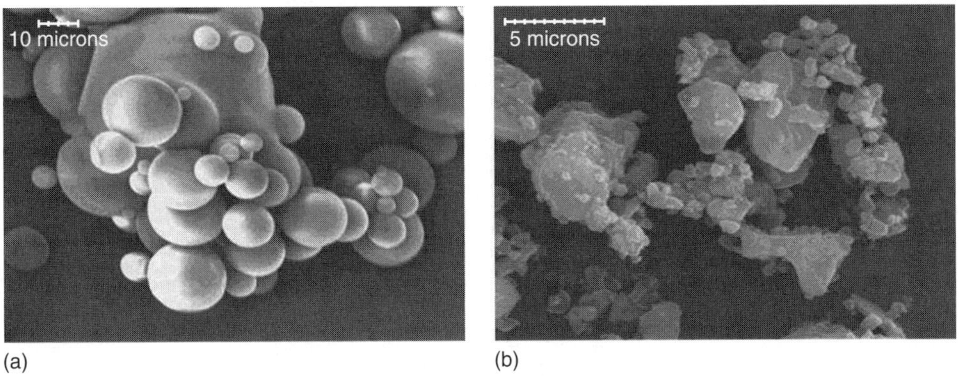

Figure 10.1 Scanning electron micrographs for amorphous spray-dried lactose (a) and crystalline spray-dried lactose obtained by storing over RVP of 76% for 144 h (b). (From Haque, M.K. and Roos, Y.H., *Innovat. Food Sci. Emerg. Technol.*, 7, 62, 2006. With permission.)

detected using polarizing light microscopy from a typical dark cross, the so-called Maltese cross, indicating crystalline structures (e.g., French [4]). The presence of crystals may also be observed from scanning electron micrographs, as shown in Figures 10.1 and 10.2. Amorphous spray-dried lactose particles had smooth surfaces and they were round-shaped whereas crystalline spray-dried lactose particles were tomahawk-shaped [5]. Amorphous freeze-dried lactose particles were glass-like membranes whereas crystalline freeze-dried lactose appeared as rodlike or needlelike structures [5]. Formation of crystals has also been followed visually using polarized light video-microscopy (e.g., Mazzobre et al. [6]).

The presence of crystals in various materials can be detected using XRD, because crystals reflect x-rays at very specific angles, which give sharp peaks in XRD patterns of crystalline materials while amorphous materials produce a flat and blunt XRD pattern, as shown in Figure 10.3 (e.g., Jouppila and Roos [8], Savolainen et al. [9]). However, the presence of tiny crystals and small amounts of crystals is not always detected using XRD (e.g., Hartel [3]).

Crystals can also be observed from the appearance of the peaks in Raman spectra. Amorphous materials exhibit a spectrum with flattened peaks whereas crystalline material exhibit a spectrum with sharp peaks resulting from a more narrow energy distribution of bond vibrations for crystalline materials than for amorphous material (e.g., Söderholm et al. [10], Jørgensen et al. [11]). Moreover, crystallization of amorphous materials (e.g., spray-dried lactose) could be detected from changes in NIR spectra [12] and in mid-IR spectra [13].

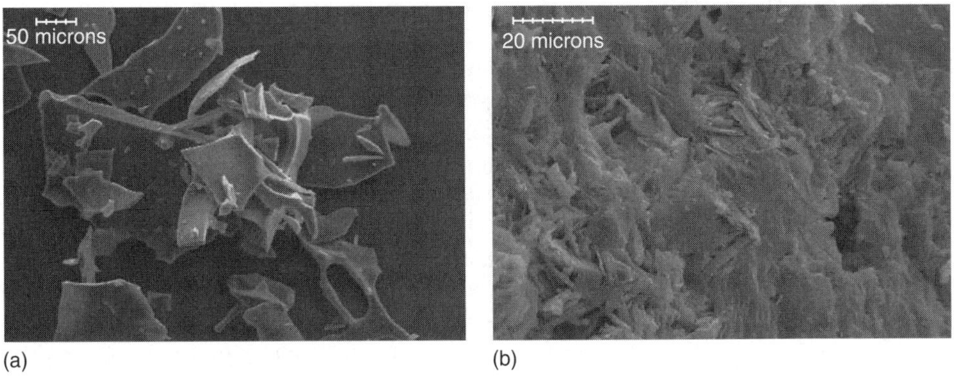

Figure 10.2 Scanning electron micrographs for amorphous freeze-dried lactose (a) and crystalline freeze-dried lactose obtained by storing over RVP of 76% for 72 h (b). (From Haque, M.K. and Roos, Y.H., *Innovat. Food Sci. Emerg. Technol.*, 7, 62, 2006. With permission.)

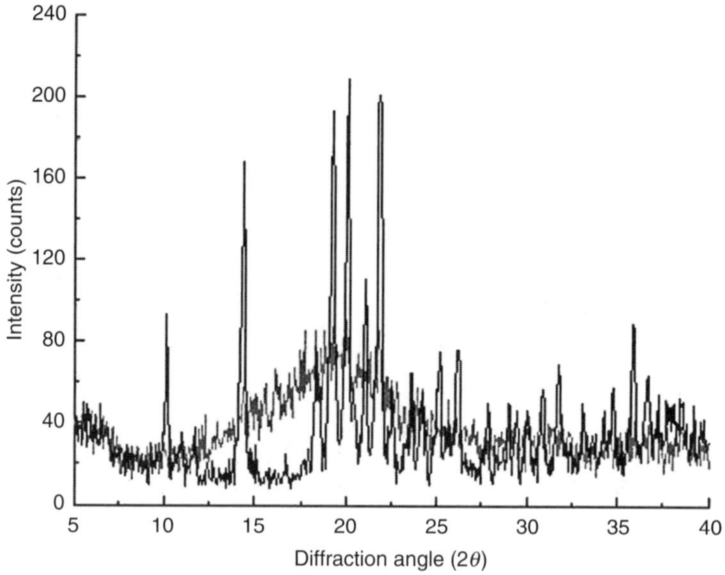

Figure 10.3 XRD pattern of freeze-dried maltose stored for 16 days at RVP of 54% at 25°C, which shows XRD pattern typical of amorphous materials (gray curve) and XRD pattern of freeze-dried maltose stored for 16 days at RVP of 85% at 25°C, which crystallized as β-maltose monohydrate. (Data from Jouppila, K., Lähdesmäki, M., Laine, P., Savolainen, M., and Talja, R.A., Comparison of water sorption and crystallization behaviour of freeze-dried lactose, lactitol, maltose and maltitol, ISOPOW 10, September 2–7, 2007, Bangkok, Thailand, Poster #PP I–14, 2007.)

The use of thermoanalytical techniques, such as DSC, in studies of crystallinity is based on the melting properties of crystals (e.g., Roos and Karel [14]). Materials containing crystals often show a characteristic melting endotherm. However, detection of melting may be difficult if other simultaneous thermal processes occur and interfere with crystallization. Such thermal changes have included nonenzymatic browning in milk powders [15].

There are clear differences in water sorption properties of amorphous and crystalline materials: water sorption of an amorphous, hygroscopic material is substantially higher than that of a crystalline material at the same relative humidity (RH) (e.g., Roos [2]). This difference may be used in the estimation of the presence of crystallinity using gravimetric techniques (e.g., Saleki-Gerhardt et al. [16], Buckton and Darcy [17]). Also, the occurrence and propagation of crystallization can be determined using gravimetric techniques. These techniques are based on the determination of water contents during crystallization. Gravimetric techniques have most often been used in studies of crystallization of amorphous sugars or materials containing amorphous sugars (e.g., Makower and Dye [18], Jouppila and Roos [19], Haque and Roos [20]). Many of these studies have used samples stored in vials in evacuated desiccators under given relative vapor pressure (RVP) conditions. The weight of the samples was monitored at intervals (e.g., Jouppila and Roos [19], Haque and Roos [20], Miao and Roos [21], Omar and Roos [22]). Dynamic vapor sorption (DVS) techniques have become popular in continuous and rapid observation of changes in water contents and they also can be used in gravimetric studies of crystallization. DVS techniques record weight changes in situ over dynamically changing RH conditions. For example, Al-Hadithi et al. [23] studied the crystallization of spray-dried trehalose at RH of 53% and 75% at 25°C using DVS.

Crystal forms are most often identified using XRD and spectroscopic techniques. Packing and orientation of molecules are different in various crystal forms resulting in differences in physical properties such as melting temperature (e.g., Hartel [3]). Hence, crystal forms can also be identified from melting temperatures using thermoanalytical techniques, for example, DSC. This requires that

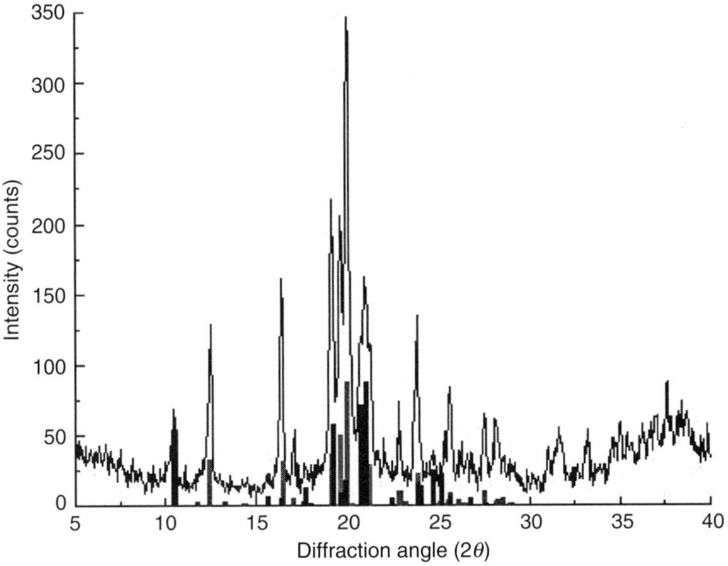

Figure 10.4 XRD pattern recorded for freeze-dried lactose stored for 11 days at RVP of 54% at 25°C; gray bars show typical peaks of α-lactose monohydrate and black peaks typical peaks of anhydrous β-lactose. (Data from Jouppila, K., Lähdesmäki, M., Laine, P., Savolainen, M., and Talja, R.A., Comparison of water sorption and crystallization behaviour of freeze-dried lactose, lactitol, maltose and maltitol, ISOPOW 10, September 2–7, 2007, Bangkok, Thailand, Poster #PP I–14, 2007.)

samples contain only crystal forms with the same melting temperatures or crystals that differ sufficiently in their melting temperatures. Crystal forms may also be identified visually from the shape of the crystals. For example, α-lactose monohydrate crystals often appear as tomahawk-shaped crystals whereas needle- or rod-shaped crystals are typical of anhydrous α-lactose and β-lactose crystals [5]. However, crystallization conditions may affect the shape of crystals [3].

Various crystal forms can be identified from peaks occurring at given diffraction angles of XRD patterns, which also show the intensities of diffracted x-ray beams against diffraction angle values. For example, identification of lactose crystal forms can be performed on the basis of their unique peaks at given diffraction angles, for example, 10.5° for anhydrous β-lactose [24], 16.4° for α-lactose monohydrate [25], and 22.1° for anhydrous crystals with α- and β-lactose in a molar ratio of 5:3 [26], as shown in Figure 10.4. Also, ratios of intensities of various peaks can be used to identify crystal forms. For example, XRD patterns of various crystal forms of lactose occurred at almost the same diffraction angles and the peaks were quite broad and overlapping (e.g., Jouppila et al. [27,28]). In these studies, the ratios of intensities of various peaks allowed accurate identification of crystal forms and the intensities were useful in the estimation of proportions of various crystal forms in the material.

Spectroscopic techniques, including IR reflectance, NIR reflectance, and nuclear magnetic resonance (NMR) spectroscopy, can be successfully used to study crystal forms of lactose. For example, Earl and Parrish [29] identified five crystal forms of lactose from differences in their NMR spectra. Vuataz [30] and Buckton et al. [31] could distinguish α-lactose monohydrate and β-lactose from NIR spectra. Drapier-Beche et al. [32] identified α-lactose monohydrate, anhydrous α-lactose, anhydrous β-lactose, and anhydrous crystals containing both α- and β-lactose based on their IR spectra.

10.2.2 Quantitative Measurement of Crystallization

Determination of crystallinity and crystallization kinetics, that is, the velocity of crystallization, is quantitative measurement of crystallization, which may be carried out by DSC and XRD techniques. Other techniques, such as isothermal microcalorimetry, gravimetry, NIR spectroscopy, and solid-state NMR, have also been applied. Crystallization kinetics is often studied using XRD, thermoanalytical, and spectroscopic techniques. Crystallization from amorphous biological solids is often recorded gravimetrically from changes in weight corresponding to changes in sorbed water content. XRD, DSC, and gravimetric methods have provided quantitative crystallization data in numerous studies of crystallization kinetics in food systems (e.g., Roos and Karel [14], Jouppila et al. [27], Kedward et al. [33]) although spectroscopic techniques, such as NIR and Raman spectroscopy, have also been used in crystallization studies of pharmaceutical materials (e.g., Buckton et al. [31], Jørgensen et al. [11]).

The degree of crystallinity can be determined using XRD by comparing intensities of diffraction peaks in XRD pattern recorded for a partially crystalline material with intensities of peaks recorded for a totally crystalline material. Traditional XRD techniques are applicable in detection of crystallinity above detection limits ranging from 5% to 10%. For example, the detection limit of 10% was found for the mixture containing freeze-dried sucrose and crystalline sucrose [16] and the detection limit of 5% for the mixture containing spray-dried lactose and α-lactose monohydrate [34]. Chen et al. [35] used the full diffraction pattern to quantify amorphous lactose in the mixture containing α-lactose monohydrate and freeze-dried lactose and suggested that the detection limit for amorphous lactose based on comparison of the full diffraction patterns was 0.37%.

DSC is also useful in analyzing crystallization kinetics. Roos and Karel [14] followed crystallization heats of crystallizing samples and compared values at various lengths of crystallization to the heat of crystallization of totally amorphous material. Heat of crystallization may also be determined by isothermal microcalorimetry. The relationship between the heat of crystallization and amorphous sugar content in the mixture at constant RH was found to be linear [23]. Water sorption measurements [36], solid-state NMR [37], and NIR spectroscopy [34] have also been used to detect amorphous sugar contents in predominantly crystalline sugars.

In XRD measurements, the development of crystallinity can be followed from increasing intensities of specific peaks in XRD patterns (e.g., Palmer et al. [38], Jouppila et al. [27], Haque and Roos [39]). Crystallization of a compound into a particular crystal form is studied by following the intensity of the unique peak or peaks corresponding to the presence of the crystal form at intervals. For example, crystallization of lactose in freeze-dried skim milk as anhydrous crystals with α- and β-lactose in a molar ratio of 5:3 can be detected from increasing intensity of its unique peak at a diffraction angle of 22° [27]. Barham et al. [40] studied crystallization of lactose and lactose–protein mixtures using in situ XRD analysis. They followed appearance and disappearance of crystals of anhydrous β-lactose, α-lactose monohydrate, and α–β mixed phases from increasing and decreasing intensities at diffraction angles of the unique peaks of those crystal forms. XRD studies of overall crystallization of a compound with several crystal forms may be difficult. In the case of lactose, overall crystallization has been studied by detecting increases in intensities of the peaks at diffraction angles of 19° and 20° (20° [28,41]; 19° [42]; both [39,21]). These peaks were chosen as several crystal forms of lactose show diffraction at the same diffraction angle, which allows quantification of these forms from the peak intensities.

Thermoanalytical techniques (DSC and isothermal microcalorimetry) can also be used in the determination of kinetics of crystallization. For example, the induction time for crystallization can be derived from DSC data obtained from isothermal measurements (e.g., Kedward et al. [43], Kedward et al. [33], Mazzobre et al. [6]). Similarly, crystallization time (time at the onset or peak of the crystallization exotherm) can be derived from DSC data (e.g., Roos and Karel [44]). Time to crystallization, that is, time to occurrence of crystallization exotherm, is often determined from isothermal microcalorimetry data (e.g., Briggner et al. [45], Sebhatu et al. [46], Lechuga-Ballesteros et al. [47], Al-Hadithi et al. [23]). Isothermal microcalorimetry measurements are most often carried out at 25°C under various RH conditions. In isothermal measurements, temperatures are often chosen so that crystallization occurs within minutes or in a few hours. Kinetics of crystallization can also be followed from melting enthalpies of crystals formed during storage. Melting enthalpy is determined especially in such studies where crystallization occurs slowly in the course of days or weeks, for example, studies of crystallization in starch at low water contents (e.g., Jouppila and Roos [8]). Roos and Karel [14] followed the kinetics of lactose crystallization by determining the heat released during crystallization of remaining amorphous lactose at various storage times. Kinetics of crystallization of amorphous lactose has also been monitored using Raman spectroscopy [11].

Kinetics of crystallization of amorphous carbohydrates is often modeled using the Avrami equation (e.g., Jouppila et al. [27], Biliaderis et al. [42], Haque and Roos [39], Miao and Roos [21]). This modeling needs the values of crystallinity at the beginning (value in amorphous state) and at the end of crystallization (leveling-off value) as well as the levels of crystallinity at various times during crystallization. Avrami modeling provides the rate constant, r, which refers to the rate of crystallization. The Avrami exponent, n, may estimate the direction of propagation of crystallization [48], which may enable the prediction of the crystal form produced. These parameters can also be used to obtain the half-life for crystallization, that is, time for 50% crystallinity.

10.3 CRYSTALLIZATION IN FOOD SYSTEMS

Amorphous food materials and food components may be in the glassy state with high viscosity and low-molecular mobility or in the more liquid-like state with decreased viscosity, increased molecular mobility, and time-dependent flow (e.g., Roos [2]). Glass transition is the solid–liquid transition of the amorphous, supercooled, nonequilibrium phase and it occurs over a temperature range. The temperature and broadness of glass transition are often dependent on molecular size, composition, and water content. Glass transition temperature (T_g) refers to the state transition and its value decreases with decreasing molecular size of homopolymers and increasing water content. Above the glass transition, the increasing translational mobility of crystallizing species may lead to crystallization.

10.3.1 Dehydrated Food Systems

Water content and temperature affect molecular mobility, crystallization behavior, and kinetics of crystallization of amorphous components in dehydrated food systems. Crystallization in storage of foods may occur when the water content is sufficient to decrease the T_g to below storage temperature (T). The rate of crystallization is likely to increase with increasing $T - T_g$.

Crystallization of amorphous sugars in dehydrated products has been studied in numerous investigations. Supplee [49] found a gradual loss of sorbed water in skim milk powder during storage, which was reported to occur as a result of lactose crystallization at RH higher than 50% at 25°C. Studies on the effects of storage RH on crystallization of sugars have been reported by several authors (e.g., Makower and Dye [18], Jouppila et al. [27], Haque and Roos [20], Haque and Roos [39], Miao and Roos [21]). Most of these studies were carried out at room temperature to follow crystallization under typical storage conditions. Experimental data for the time to complete crystallization of various sugars are listed in Table 10.1.

Crystallization may occur at above a critical storage RH with an increasing rate with increasing RH at a constant temperature. At room temperature, crystallization of disaccharides requires a higher RH than crystallization of monosaccharides. Crystallization often occurs to varying extents at different RH values, that is, to a RH-dependent leveling-off extent. Maximum leveling-off extent of crystallization in dehydrated lactose and lactose-containing products was reported to occur at RH ranging from 60% to 70% when studied using XRD [27,28,42]. Lower values at lower RH were probably due to lower molecular mobility and diffusion as well as steric hindrance of molecules caused by the crystals formed at the beginning of crystallization. At higher RH, increasing solubility

Table 10.1 Time to Complete Crystallization of Various Sugars, Sugar Alcohols, and a Mixture of Sugars as Observed from Loss of Sorbed Water during Storage at Various Temperatures and RHs

Material	*T* (°C)	RH (%)	Time to Complete Crystallization
Sugars			
Glucose, rapidly cooled melt that was powdered at −25°C [18]	25	8.6	500 days
		11.8	150 days
		16.2	25 days
		24.0	5 days
		28.2	4 days
		33.6	3 days
Sucrose, spray-dried [18]	25	16.2	>850 days
		24.0	>400 days
		28.2	50 days
		33.6	<2 days
Lactose, spray-dried [51]	25	55	25 h (1.0 day)
Lactose, spray-dried [52]	Room temperature	57	14 h (0.6 day)
Lactose, spray-dried [53]	21	57	13 h (0.5 day)
		75	10 h (0.4 day)
		84	8 h (0.3 day)
Lactose, spray-dried [17]	25	75	6 h (0.2 day)
Lactose, spray-dried [12]	25	75	2 h (0.1 day)
Lactose, spray-dried [54]	22–23	54.5	21 h (0.9 day)
		65.6	20 h (0.8 day)
		76.1	20 h (0.8 day)
Lactose, freeze-dried [51]	25	55	8 h (0.3 day)
Lactose, freeze-dried [55]	20	53	27 h (1.1 days)
Lactose, freeze-dried [56]	24	44	32 days
		54	8 days
		66	2 days
		76	36 h (1.5 days)
Lactose, freeze-dried [21]	22–23	54.5	48 h (2.0 days)
		65.6	21 h (0.9 day)
		76.1	10 h (0.4 day)
Lactose, freeze-dried [5]	22–23	54.5	48 h (2.0 days)
		76.1	21 h (0.9 day)
Lactose, freeze-dried [22]	22–23	54.4	70 h (2.9 days)
		65.6	20 h (0.8 day)
		76.1	20 h (0.8 day)
Lactose, freeze-dried [7]	25	54	80 h (3.3 days)
		66	50 h (2.1 days)
		76	70 h (2.9 days)
Maltose, freeze-dried [7]	25	66	150 h (6.2 days)
		76	150 h (6.2 days)
		85	~120 h (~5.0 days)
Sucrose, spray-dried [18]	25	16.2	>850 days
		24.0	>400 days
		28.2	50 days
		33.6	<2 days

Table 10.1 (continued) Time to Complete Crystallization of Various Sugars, Sugar Alcohols, and a Mixture of Sugars as Observed from Loss of Sorbed Water during Storage at Various Temperatures and RHs

Material	*T* (°C)	RH (%)	Time to Complete Crystallization
Sucrose, freeze-dried [57]	30	32.4	10 h (0.4 day)
Trehalose, spray-dried [23]	25	53	7 h (0.3 day)
		75	~2 h (~0.1 day)
Trehalose, freeze-dried [21]	22–23	54.5	24 h (1.0 day)
		65.6	~10 h (~0.4 day)
		76.1	~10 h (~0.4 day)
Mixture of sugars			
Lactose–trehalose mixture (1:1), freeze-dried [21]	22–23	65.6	100 h (4.2 days)
		76.1	90 h (3.8 days)
Sugar alcohols			
Lactitol, freeze-dried [7]	25	44	350 h (15 days)
		54	350 h (15 days)
		66	250 h (10 days)
Maltitol, freeze-dried [7]	25	54	250 h (10 days)
		66	120 h (5.0 days)

Source: Updated from Jouppila, K., in *Carbohydrates in Food*, 2nd edn., Eliasson, A.-C. (ed.), CRC Press, Boca Raton, FL, 2006, 41–88.

of lactose crystals in sorbed water was assumed to decrease the rate of crystallization and total crystallinity. The presence of other substances, such as polysaccharides, often causes delays in sugar crystallization probably as a result of molecular interactions and increased viscosity of the system [58].

Crystallization in dehydrated food systems may cause defects in product quality. For example, lactose crystals have low solubility resulting in dispersibility problems of milk powders. Amorphous lactose is present in most dairy powders. Amorphous lactose may crystallize into various crystal forms depending on RH conditions at room temperature. The crystalline forms were also found to depend on dehydration method and the presence of other compounds (Table 10.2). Freeze- and spray-dried lactose has been reported to crystallize in different crystal forms but in a study by Haque and Roos [39] both freeze- and spray-dried lactose crystallized into the same crystal form at the same storage conditions. However, Haque and Roos [5] reported differences in crystal morphology for the freeze- and spray-dried lactose. In crystallized spray-dried lactose, tomahawk-shaped crystals typical of α-lactose monohydrate were formed. Freeze-dried lactose crystallized to needle- or rod-shaped crystals typical of anhydrous α- and β-lactose. Also, the presence of other compounds in lactose-containing materials, such as other carbohydrates and milk solids, has been shown to affect crystallization behavior of amorphous lactose during storage.

Crystallization may occur when the T_g of a material is lower than its storage temperature. T_g values for various anhydrous carbohydrates and carbohydrate-containing products are given in Table 10.3. Water decreases the glass transition and T_g at various levels of water plasticization can be predicted using the Gordon–Taylor equation with k values based on experimental T_g data for each material (Table 10.3). An increase in water content decreases the T_g. For example, crystallization in freeze-dried lactose during dynamic heating was found to occur at lower temperatures at increasing water contents [33]. At a constant storage temperature, a decrease in T_g results in an

Table 10.2 Crystal Forms of Lactose Present in Freeze- and Spray-Dried Lactose and Lactose-Containing Products Stored at Various Temperatures and RHs Detected Using XRD

Material	*T* (°C)	RH (%)	Crystal Forms of Lactose Present
Lactose			
Freeze-dried lactose [51]	25	55	A mixture of α-lactose monohydrate and anhydrous β-lactose
Freeze-dried lactose [55]	20	53	A mixture of anhydrous β-lactose and α-lactose monohydrate
Freeze-dried lactose [28,59]	24	44	A mixture of α-lactose monohydrate and anhydrous crystals of α- and β-lactose in a molar ratio of 5:3; traces of anhydrous β-lactose and anhydrous crystals of α- and β-lactose in a molar ratio of 4:1
	24	54, 66, 76	A mixture of α-lactose monohydrate and anhydrous crystals of α- and β-lactose in a molar ratio of 5:3; traces of anhydrous crystals of α- and β-lactose in a molar ratio of 4:1 and unstable α-lactose
Freeze-dried lactose [42]	25	54, 64, 75, 84	A mixture of anhydrous β-lactose, α-lactose monohydrate, and anhydrous crystals of α- and β-lactose in a molar ratio of 5:3
Freeze-dried lactose [39]	22–23	54, 66, 76	A mixture of α-lactose monohydrate and anhydrous β-lactose
Freeze-dried lactose [21]	22–23	54, 66, 76	A mixture of anhydrous β-lactose and α-lactose monohydrate
Spray-dried lactose [51]	25	55	A mixture of α-lactose monohydrate and anhydrous β-lactose
Spray-dried lactose [60]	Room temperature	75	A mixture of α-lactose monohydrate and anhydrous β-lactose
Spray-dried lactose [45]	25	53, 65, 75, 85	A mixture of α-lactose monohydrate and anhydrous β-lactose
Spray-dried lactose [46]	25	57, 75, 84, 100	A mixture of α-lactose monohydrate and anhydrous β-lactose
Spray-dried lactose [39]	22–23	54, 66, 76	A mixture of α-lactose monohydrate and anhydrous β-lactose
Spray-dried lactose [40]	25	88	A mixture of anhydrous β-lactose and α-lactose monohydrate
Spray-dried lactose [9]	Room temperature	85	A mixture of α-lactose monohydrate and anhydrous β-lactose
Lactose in model systems			
Freeze-dried lactose–pullulan (3:1) [42]	25	54, 64	A mixture of α-lactose monohydrate and anhydrous crystals of α- and β-lactose in a molar ratio of 5:3
		75	A mixture of α-lactose monohydrate and anhydrous crystals of α- and β-lactose in a molar ratio of 5:3; traces of anhydrous β-lactose
		84	A mixture of α-lactose monohydrate, anhydrous β-lactose, and anhydrous crystals of α- and β-lactose in a molar ratio of 5:3
Freeze-dried lactose–pullulan (2:1) [42]	25	54, 64, 75, 84	A mixture of α-lactose monohydrate and anhydrous crystals of α- and β-lactose in a molar ratio of 5:3

Freeze-dried lactose–pullulan (1:1) [42]	25	64, 75, 84	A mixture of α-lactose monohydrate and anhydrous crystals of α- and β-lactose in a molar ratio of 5:3
Freeze-dried lactose–sucrose–invertase [61]	24	54, 66, 76	α-Lactose monohydrate; traces of anhydrous crystals of α- and β-lactose in a molar ratio of 5:3
Freeze-dried lactose–sucrose–invertase–carrageenan [61]	24	66, 76	α-Lactose monohydrate; traces of anhydrous crystals of α- and β-lactose in a molar ratio of 5:3
Freeze-dried lactose–gelatin (3:1) [41]	22–23	76.1	α-Lactose monohydrate; traces of anhydrous β-lactose
Freeze-dried lactose–gelatin (5:1) [41]	22–23	76.1	A mixture of α-lactose monohydrate and anhydrous β-lactose
Freeze-dried lactose–Na-caseinate (3:1 and 5:1) [41]	22–23	76.1	α-Lactose monohydrate
Freeze-dried lactose–whey protein isolate (WPI) (3:1) [41]	22–23	65.6	α-Lactose monohydrate; traces of anhydrous β-lactose
		76.1	α-Lactose monohydrate
Freeze-dried lactose–WPI (5:1) [41]	22–23	65.6	A mixture of α-lactose monohydrate and anhydrous β-lactose
		76.1	α-Lactose monohydrate; traces of anhydrous β-lactose
Spray-dried lactose–gelatin (3:1 and 5:1) [41]	22–23	76.1	α-Lactose monohydrate
Spray-dried lactose–Na-caseinate (3:1 and 5:1) [41]	22–23	76.1	α-Lactose monohydrate
Spray-dried lactose–WPI (3:1 and 5:1) [41]	22–23	65.6, 76.1	α-Lactose monohydrate
Spray-dried lactose–albumin (3:1) [40]	25	88	α-Lactose monohydrate
Spray-dried lactose–Na-caseinate (3:1) [40]	25	88	α-Lactose monohydrate
Spray-dried lactose–WPI (3:1) [40]	25	88	A mixture of anhydrous crystals of α- and β-lactose in a molar ratio of 5:3 and α-lactose monohydrate
Lactose in milk products			
Freeze-dried solution made from spray-dried skim milk [51]	25	55	α-Lactose monohydrate
Freeze-dried skim milk [51]	25	55	Anhydrous crystals of α- and β-lactose in a molar ratio of 5:3
Freeze-dried skim milk [27]	24	54, 66	Anhydrous crystals of α- and β-lactose in a molar ratio of 5:3
	24	76	Anhydrous crystals of α- and β-lactose in a molar ratio of 5:3 (mainly) and 4:1, stable anhydrous α-lactose
	24	86	Anhydrous crystals of α- and β-lactose in a molar ratio of 5:3 (mainly) and 4:1, stable anhydrous α-lactose, traces of α-lactose monohydrate
Spray-dried skim milk [62]	0–37	75	α-Lactose monohydrate

(*continued*)

Table 10.2 (continued) Crystal Forms of Lactose Present in Freeze- and Spray-Dried Lactose and Lactose-Containing Products Stored at Various Temperatures and RHs Detected Using XRD

Material	*T* (°C)	RH (%)	Crystal Forms of Lactose Present
Skim milk powder [55]	20	43	Anhydrous β-lactose
	20	53	A mixture of anhydrous β-lactose and α-lactose monchydrate
	20	59	α-Lactose monohydrate
	20	75	α-Lactose monohydrate
Spray-dried whole milk [63]	60		Anhydrous β-lactose
Spray-dried whole milk [62]	37	$<$20	Anhydrous β-lactose
	0–37	75	α-Lactose monohydrate
Spray-dried whole milk containing 28%, 47%, and 68% lactose [64]	20–22	77	α-Lactose monohydrate
Spray-dried whey and partially demineralized whey [60]	Room temperature	75	α-Lactose monohydrate

Source: Updated from Jouppila, K., in *Carbohydrates in Food*, 2nd edn., Eliasson, A.-C. (ed.), CRC Press, Boca Raton, FL, 2006, 41–88.

Table 10.3 Glass Transition Temperature (T_g, Onset Value) Determined Using DSC at a Scanning Rate of 5°C min^{-1} and k Values for Various Anhydrous Amorphous Carbohydrates and Products Containing Amorphous Carbohydrates

Material (Reference)	T_g (°C)	k^a	k^b
Sugars—pentose-monosaccharides			
Arabinose, melt [65]	−2		3.55
Ribose, melt [65]	−20		3.02
Xylose, melt [65]	6		3.79
Sugars—hexose-monosaccharides			
Fructose, melt [65]	5		3.76
Fucose, melt [65]	26		4.37
Galactose, melt [65]	30		4.49
Glucose, melt [66]	37[c]		4.69
Glucose, melt [65]	31		4.52
Mannose, melt [65]	25		4.34
Rhamnose, melt [65]	−7		3.40
Rhamnose, melt [67]	37[c]		4.69
Sorbose, melt [65]	19		4.17
Sugars—disaccharides			
Lactose, freeze-dried [68]	101		6.57
Lactose, freeze-dried [69]	97	6.7	6.45
Lactose, freeze-dried [5]	105	6.6	6.69
Lactose, spray-dried [53]	104[c]	3.8	6.66
Lactose, spray-dried [70]	101	6.2	6.57
Lactose, spray-dried [20]	105	6.9	6.69
Lactulose, melt [59]	79		5.92
Maltose, freeze-dried [71]	87	6	6.16
Maltose, melt [66]	91[c]		6.28
Melibiose, melt [65]	85		6.10
Sucrose, freeze-dried [72]	62	4.7	5.43
Sucrose, freeze-dried [57]	74[d]	6.7	5.78
Sucrose, spray-dried [53]	77[c]	9.1	5.87
α,α-Trehalose, melt [65]	100		6.54
Trehalose, freeze-dried [73]	85		6.10
Sugar-containing model systems			
Lactose–albumin (3:1), spray-dried [20]	108	8.0	6.77
Lactose–gelatin (3:1), spray-dried [20]	113	7.9	6.92
Lactose–Na-caseinate (3:1), spray-dried [20]	104	7.2	6.66
Lactose–sucrose (1.86:1) with invertase, freeze-dried [61]	60	5.8	5.37
Lactose–sucrose–carrageenan (1.8:1:0.026) with invertase, freeze-dried [61]	63	6.3	5.46
Lactose–WPI (3:1), spray-dried [20]	112	8.1	6.89
Sugar-containing products			
Horseradish, freeze-dried [74]	58	5.3	5.31
Skim milk, freeze-dried [69]	92	5.7	6.31

(continued)

Table 10.3 (continued) Glass Transition Temperature (T_g, Onset Value) Determined Using DSC at a Scanning Rate of 5°C min^{-1} and k Values for Various Anhydrous Amorphous Carbohydrates and Products Containing Amorphous Carbohydrates

Material (Reference)	T_g (°C)	k^a	k^b
Skim milk with hydrolyzed lactose, freeze-dried [69]	49	8.0	5.05
Strawberry, freeze-dried [74]	36	4.7	4.66
Sugar alcohols			
Erytritol, melt [75]	−45		2.29
Xylitol, melt [65]	−29		2.76
Xylitol, melt [75]	−24	2.8	2.91
Sorbitol, melt [65]	−9		3.35
Sorbitol, melt [75]	−6	3.0	3.43
Lactitol, freeze-dried [7]	50	8.1	5.08
Maltitol, melt [65]	39		4.75
Maltitol, freeze-dried [7]	34	7.1	4.61
Polysaccharides			
Maltodextrin M365, freeze-dried [71]	100	6	
Maltodextrin M200, freeze-dried [71]	141	6.5	
Maltodextrin M100, freeze-dried [71]	160	7	
Maltodextrin M040, freeze-dried [71]	188	7.7	

Source: Updated from Jouppila, K., in *Carbohydrates in Food*, 2nd edn., Eliasson, A.-C. (ed.), CRC Press, Boca Raton, FL, 2006, 41–88.

[a] The k values were calculated using the Gordon–Taylor equation.

[b] The k values were calculated from experimental T_g values for anhydrous materials using the equation $k = 0.0293 \times T_g + 3.61$ [65].

[c] The scanning rate was 10°C min^{-1}. The midpoint temperature of glass transition was taken as T_g.

[d] The scanning rate was 10°C min^{-1}. It was not reported which temperature of glass transition was taken as T_g.

increase in the temperature difference of storage temperature and T_g (i.e., $T - T_g$). At a constant water content an increase in storage temperature similarly increases the $T - T_g$. The rate of crystallization has been found to increase with increasing $T - T_g$ in several studies (e.g., Roos and Karel [14], Jouppila and Roos [8], Jouppila et al. [27], Jouppila et al. [28]).

The rate of crystallization in partially crystalline synthetic polymers has been shown to have a maximum at a temperature between the T_g and equilibrium melting temperature of crystals, T_m. This follows the rate of nucleation, which is the highest at temperatures above T_g (translational mobility and nucleation ceases around T_g) and the rate of crystal growth is the highest at temperatures below but approaching T_m (e.g., Slade and Levine [76]). Kedward et al. [43] found a relationship for the rate of crystallization of freeze-dried lactose and sucrose as determined using isothermal DSC scanning at temperatures ranging from T_g to T_m, as shown in Figure 10.5.

Roos and Karel [14] studied the time to crystallization of amorphous sugars and found that the time to crystallization decreased with increasing $T - T_g$. Isothermal data for crystallization at various RH conditions were also used to show that the Williams–Landel–Ferry (WLF)-type relationship (Equation 10.1) could be used to model time to crystallization of amorphous lactose at temperatures above T_g. This suggested that crystallization was related to viscosity and diffusional

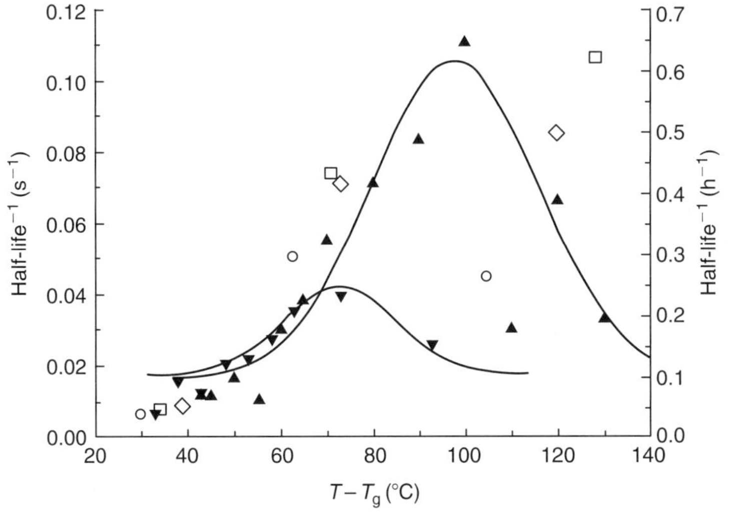

Figure 10.5 Rate of crystallization, given as half-life^{-1}, in amorphous lactose and sucrose as a function of $T - T_g$: data obtained using isothermal DSC scanning for freeze-dried lactose (▲) and sucrose (▼) by Kedward et al. [43]; data obtained using XRD for freeze-dried lactose (○) by Jouppila et al. [28], and for freeze-dried lactose (□) and spray-dried lactose (◇) by Haque and Roos [39]. (Updated from Jouppila, K., in *Carbohydrates in Food*, 2nd edn., Eliasson, A.-C. (ed.), CRC Press, Boca Raton, FL, 2006.)

properties of amorphous lactose above the glass transition. Hence, crystallization of amorphous sugars can be time-dependent with an exponential and dramatic decrease in crystallization times at small increases in thermal or water plasticization increasing the storage $T - T_g$.

$$\log a_T = \frac{-C_1(T - T_S)}{C_2 + (T - T_S)} \tag{10.1}$$

Here, a_T is the ratio of mechanical or dielectric relaxation time at the temperature of observation, T, to its value at a reference temperature, T_S, and C_1 and C_2 are constants. It should be noted that several food materials, such as infant formulas and dairy powders, are often stored in large bulk containers or hermetic cans. Crystallization in such conditions releases water from the crystallizing material to surroundings. This water causes a rapid increase in local water content around the growing crystals and increase the $T - T_g$ of the remaining amorphous material, which results in a dramatic increase in the rate of crystallization. Such crystallization leads to an almost instant crystallization, as reported by Kim et al. [77] and Roos [2].

10.3.2 Frozen Food Systems

Crystallization in an unfrozen phase during storage of frozen foods may include formation of ice, recrystallization of ice, and crystallization of solutes [2], which are likely to lead to defects in product quality. For example, recrystallization of ice in frozen foods often results in the formation of large ice crystals, which can significantly alter the structure and texture of frozen foods. Crystallization of lactose in ice cream is a classical example of solute crystallization, which results in an unpleasant, sandy mouthfeel.

Ice formation, crystallization of solutes, and ice recrystallization often occur in frozen food systems. These processes are governed by solvent and solute diffusion and they may be controlled by the viscosity of the unfrozen hydrophilic phase. The rate of solvent and solute mobility is related to the glass transition of the freeze-concentrated unfrozen phase. When a food material is cooled to temperatures below 0°C, ice formation may occur to an equilibrium extent and the amount of unfrozen water in the freeze-concentrated unfrozen phase can be determined from the equilibrium melting temperature curve (e.g., Roos [2]). The amount of unfrozen water with hydrophilic solutes gives the level of water plasticization and determines an effective glass transition for the freeze-concentrated system: an increase in concentration of solutes increases the effective glass transition temperature of the unfrozen phase (e.g., Roos and Karel [72]).

Roos and Karel [72] reported that ice formation in frozen food systems above T_g of the unfrozen phase was time-dependent. The rate of crystallization decreased with increasing concentration of solids of the unfrozen phase until the concentration of solids in the maximally freeze-concentrated system, C'_g, was attained. At C'_g, ice formation was suggested to cease because of limited diffusion resulting from the high viscosity around the glass transition of the unfrozen phase [72]. Donhowe and Hartel [78] showed that the rate of recrystallization of ice during storage of ice cream above the onset temperature of ice melting within the maximally freeze-concentrated matrix, T'_m, increased with increasing storage temperature and recrystallization was enhanced by temperature fluctuations. The presence of other substances in frozen food systems may stabilize those systems. For example, Goff et al. [79] found that polysaccharide stabilizers increased the viscosity of the unfrozen phase in ice cream although stabilizers did not significantly affect the glass transition temperature of the maximally freeze-concentrated unfrozen matrix, T'_g. They suggested that ice crystal growth was controlled by kinetic properties of the unfrozen, viscoelastic phase surrounding ice crystals.

There are temperature limits for each food material that allow maximum ice formation. For example, such a temperature range is often at temperatures from −40°C to −30°C for disaccharides [65]. The glass transition temperature of the maximally freeze-concentrated unfrozen matrix, T'_g, and the onset temperature of ice melting within this maximally freeze-concentrated matrix, T'_m, are independent of the initial amount of water in the system and they often increase with increasing molecular weight (e.g., Roos [65]). The concentration of solids in the maximally freeze-concentrated system, C'_g, approaches 80% for numerous mono- and disaccharides and food solids but is lower for some polymeric systems, for example, 73% for gelatinized starch (Table 10.4).

Glass transition and ice melting temperature data for freeze-concentrated systems are extremely important for the understanding of crystallization and recrystallization phenomena in frozen food systems. These data are also useful in the control of freeze-drying properties of frozen foods.

10.3.3 Fat-Based Food Systems

In fat-based food systems, the crystal form of fat may change from a less stable form to another more stable polymorphic form during storage, which may be observed as defects in product quality (e.g., Roos [2]). Such quality defects include blooming of chocolate and mechanical properties or graininess of spreads. These changes are affected by storage temperature because recrystallization to the most stable polymorphic form of fats occurs during storage. Transformation of one polymorphic form into another more stable form in solid triglycerides can occur even without intermediate melting, as was discussed by Nawar [80].

In cocoa butter, at least six polymorphic forms have been reported (e.g., Nawar [80]). These forms exhibit differences in their stability and melting point. Hartel [3] reported that crystallization of cocoa butter must be carefully controlled during tempering of chocolate to obtain the most desired polymorph (the V form) that produces the glossy surface appearance and superior product texture. According to Nawar [80], the V form is the most stable form that can be crystallized from melt. However, melting temperature of the VI form is higher than that of the V form but the VI form can be formed only by slow transformation of the V form during storage in coincidence with blooming of chocolate (producing white or grayish spots on the surface of chocolate).

Table 10.4 Glass Transition Temperature (T'_g, Onset), Concentration (C'_g), and Onset Temperature of Melting of Ice (T'_m) of Maximally Freeze-Concentrated Matrix for Various Carbohydrates and Carbohydrate-Containing Products Determined Using DSC

Material (Reference)	T'_g (°C)	T'_m (°C)	C'_g (%)
Sugars—pentose-monosaccharides			
Arabinose [65]	−66	−53	79.3
Ribose [65]	−67	−53	81.4
Xylose [65]	−65	−53	78.9
Sugars—hexose-monosaccharides			
Fructose [65]	−57	−46	82.5
Fucose [65]	−62	−48	78.4
Galactose [65]	−56	−45	80.5
Glucose [65]	−57	−46	80.0
Mannose [65]	−58	−45	80.1
Rhamnose [65]	−60	−47	82.8
Sorbose [65]	−57	−44	81.0
Sugars—disaccharides			
Lactose [65]	−41	−30	81.3
Maltose [71]	−41	−31	82.5
Maltose [65]	−42	−32	81.6
Melibiose [65]	−42	−32	81.7
Sucrose [72]	−46	−34	80.0
Sucrose (T'_g and T'_m [62]; C'_g [65])	−46	−34	81.7
α,α-Trehalose [65]	−40	−30	81.6
Sugar-containing products			
Freeze-dried skim milk [69]	−50	−32	79.5
Freeze-dried skim milk with hydrolyzed lactose [69]	−65	−40	83.1
Sugar alcohols			
Xylitol [65]	−72	−57	57.1
Sorbitol [65]	−63	−49	81.3
Maltitol [65]	−47	−37	62.9
Polysaccharides			
Maltodextrin M365, freeze-dried [71]	−43	−28	79.4
Maltodextrin M200, freeze-dried [71]	−31	−19	79.7
Maltodextrin M100, freeze-dried [71]	−23	−13	81.1
Maltodextrin M040, freeze-dried [71]	−15	−11	82.0
Gelatinized waxy corn starch [71]		−6	72.9
Gelatinized corn starch, freeze-dried [8]	−11	−11	73.0

Source: Updated from Jouppila, K., in *Carbohydrates in Food*, 2nd edn., Eliasson, A.-C. (ed.), CRC Press, Boca Raton, FL, 2006, 41–88. With permission.

10.4 PREDICTIONS OF CRYSTALLIZATION IN FOOD SYSTEMS

Crystallization in food systems occurs if a critical value of the thermodynamic driving force, that is, the critical extent of supersaturation (solution) or supercooling (melt), is exceeded. Concentration and temperature are the main parameters that determine rates of nucleation and crystal growth. Hence, these parameters are essential in understanding and predicting rates of crystallization.

Water affects concentration of crystallizing food components and their physical state. Water plasticizes most hydrophilic, amorphous food materials and water plasticization results in decreased glass transition temperatures. An increase in water content at a constant temperature increases the temperature difference of storage temperature and glass transition temperature ($T - T_g$). A critical water content of a material may be defined as the water content that depresses the T_g of the material to its storage temperature, that is, $T - T_g$ becomes equal to 0°C [74]. Critical RH can be defined as the corresponding RH, which at equilibrium corresponds to the critical water content and water activity, a_w (=0.01 RH) [74]. Critical values for water activity (storage RH) of food components often depend on their molecular weight (see Table 10.5): for example, higher critical RH was reported for disaccharides and products containing disaccharides than for products containing monosaccharides, such as milk powder in which lactose was hydrolyzed to glucose and galactose. However, critical water contents of sugar alcohols with almost similar molecular weights to those of disaccharides were much lower than those of disaccharides. When these critical values are exceeded, crystallization and other deteriorative changes, such as stickiness, caking, collapse, and nonenzymatic browning, may occur.

Table 10.5 Critical Values for Water Content and Storage RH at Room Temperature for Various Amorphous Carbohydrates and Products Containing Amorphous Carbohydrates

Material (Reference)	Temperature (°C)	Critical Water Content (g of Water/100 g of Solids)	Critical Storage RH (%)
Sugars—monosaccharides			
Fucose [59]	25	0.1[a]	
Galactose [59]	25	0.7[a]	
Glucose [59]	25	0.8[a]	
Sugars—disaccharides			
Lactose [74]	25	7.2[a]	33
Lactose, freeze-dried [69]	24	6.8[b]	37
Lactose, spray-dried [20]	23	7.5[b]	37
Lactulose [59]	25	5.7[a]	
Maltose [59]	25	6.3[a]	
Maltose, freeze-dried [7]	25	5.0[b]	28
Melibiose [59]	25	6.2[a]	
Sucrose [74]	25	4.2[a]	23
Trehalose [59]	25	7.2[a]	
Sugar alcohols			
Lactitol, freeze-dried [7]	25	1.9[b]	12
Maltitol [59]	25	1.8[a]	
Maltitol, freeze-dried [7]	25	0.8[b]	7
Sugar-containing model systems			
Lactose–albumin (3:1), spray-dried [20]	23	6.6[b]	35
Lactose–gelatin (3:1), spray-dried [20]	23	7.2[b]	36
Lactose–Na-caseinate (3:1), spray-dried [20]	23	7.2[b]	36
Lactose–WPI (3:1), spray-dried [20]	23	6.9[b]	36
Sugar-containing products			
Freeze-dried horseradish [74]	25	3.8[a]	21
Freeze-dried skim milk [69]	24	7.6[b]	37
Freeze-dried skim milk with hydrolyzed lactose [69]	24	2.0[b]	16

Table 10.5 (continued) Critical Values for Water Content and Storage RH at Room Temperature for Various Amorphous Carbohydrates and Products Containing Amorphous Carbohydrates

Material (Reference)	Temperature (°C)	Critical Water Content (g of Water/100 g of Solids)	Critical Storage RH (%)
Freeze-dried strawberries [74]	25	1.5[a]	7
Oven-dried apple slices [81]	20	3.5[c]	18[c]
Skim milk powder [82]	20	8.4[d]	43
Spray-dried skim milk [83]	20–25	7.5–8.0[e]	
Spray-dried whole milk [83]	20–25	6.5–7.0[e]	
Spray-dried skim milk [84]	25	6.4[d]	40
Spray-dried ultrafiltration retentate skim milk [85]	20	8.5–9.5[d]	50
Polysaccharides			
Maltodextrin DE 36 [59]	25	7.8[b]	
Maltodextrin DE 20 [59]	25	11.2[b]	
Maltodextrin DE 10 [59]	25	12.1[b]	
Maltodextrin DE 4 [59]	25	13.2[b]	
Starch [59]	25	26.2[b]	

Source: Updated from Jouppila, K., in *Carbohydrates in Food*, 2nd edn., Eliasson, A.-C. (ed.), CRC Press, Boca Raton, FL, 2006, 41–88. With permission.

[a] The values were based on the determination of T_g values and calculated using the k values obtained using the equation $k = 0.0293 \times T_g + 3.61$ [65].

[b] The values were based on the determination of T_g values and calculated using the k values obtained using the Gordon–Taylor equation.

[c] The values were based on the water activity (a_w) value defined as texture acceptance limit (i.e., beginning of loss of crispness) with corresponding water content from water sorption data.

[d] The values were based on the changes in water sorption behavior during storage.

[e] The values were based on the changes in α-lactose monohydrate content during storage.

The kinetics of crystallization of amorphous food components is controlled by glass transition (e.g., Sperling [48], Slade and Levine [86], Levine and Slade [87]) among other factors. The relationship between the ratio of mechanical relaxation times and $T - T_g$ can be described using the WLF equation (Williams et al. [88]), given in Equation 10.1, in which a_T is the ratio of mechanical or dielectric relaxation time at the temperature of observation, T, to its value at a reference temperature, T_S, and C_1 and C_2 are constants. Williams et al. [88] pointed out that if the T_g is used as the reference temperature, universal constants of 17.44 for C_1 and 51.6 for C_2 allowed fitting of Equation 10.1 to experimental viscosity data for a number of amorphous inorganic and organic materials. According to Williams et al. [88], the WLF equation successfully predicted the temperature dependence of mechanical and dielectric properties in amorphous materials over the temperature range from T_g to $T_g + 100$°C. Slade and Levine [76] have emphasized the applicability of the WLF equation above glass transition to model changes in relaxation times occurring in amorphous food materials.

Crystallization kinetics at a single temperature or water content follows a sigmoid relationship for crystallinity against time. This is often modeled successfully by the Avrami relationship. Such crystallization behavior is typical of starch retrogradation in bread and crystallization of amorphous sugars in food systems [2].

REFERENCES

1. Hartel, R.W. and Shastry, A.V., Sugar crystallization in food products, *Crit. Rev. Food Sci. Nutr.*, 30, 49, 1991.
2. Roos, Y.H., *Phase Transitions in Foods*, Academic Press, San Diego, CA, 1995.

3. Hartel, R.W., *Crystallization in Foods*, Aspen Publishers, Gaithersburg, MD, 2001.
4. French, D., Organization of starch granules, in *Starch: Chemistry and Technology*, Whistler, R.L., BeMiller, J.N., and Paschall, E.F. (eds.), Academic Press, New York, 1984, pp. 183–247.
5. Haque, M.K. and Roos, Y.H., Differences in the physical state and thermal behavior of spray-dried and freeze-dried lactose and lactose/protein mixtures, *Innovat. Food Sci. Emerg. Technol.*, 7, 62, 2006.
6. Mazzobre, M.F., Aguilera, J.M., and Buera, M.P., Microscopy and calorimetry as complementary techniques to analyze sugar crystallization from amorphous systems, *Carbohydr. Res.*, 338, 541, 2003.
7. Jouppila, K., Lähdesmäki, M., Laine, P., Savolainen, M., and Talja, R.A., Comparison of water sorption and crystallization behaviour of freeze-dried lactose, lactitol, maltose and maltitol, ISOPOW 10, September 2–7, 2007, Bangkok, Thailand, Poster #PP I-14, 2007.
8. Jouppila, K. and Roos, Y.H., The physical state of amorphous corn starch and its impact on crystallization, *Carbohydr. Polym.*, 32, 95, 1997.
9. Savolainen, M., Jouppila, K., Pajamo, O., Christiansen, L., Strachan, C., Karjalainen, M., and Rantanen, J., Determination of amorphous content in the pharmaceutical process environment, *J. Pharm. Pharmacol.*, 59, 161, 2007.
10. Söderholm, S., Roos, Y.H., Meinander, N., and Hotokka, M., Raman spectra of fructose and glucose in the amorphous and crystalline states, *J. Raman Spectrosc.*, 30, 1009, 1999.
11. Jørgensen, A.C., Miroshnyk, I., Karjalainen, M., Jouppila, K., Siiriä, S., Antikainen, O., and Rantanen, J., Multivariate data analysis as a fast tool in evaluation of solid-state phenomena, *J. Pharm. Sci.*, 95, 906, 2006.
12. Lane, R.A. and Buckton, G., The novel combination of dynamic vapour sorption gravimetric analysis and near infra-red spectroscopy as a hyphenated technique, *Int. J. Pharm.*, 207, 49, 2000.
13. Ottenhof, M.-A., MacNaughtan, W., and Farhat, I.A., FTIR study of state and phase transitions of low moisture sucrose and lactose, *Carbohydr. Res.*, 338, 2195, 2003.
14. Roos, Y. and Karel, M., Crystallization of amorphous lactose, *J. Food Sci.*, 57, 775, 1992.
15. Roos, Y.H., Jouppila, K., and Zielasko, B., Non-enzymatic browning-induced water plasticization: Glass transition temperature depression and reaction kinetics determination using DSC, *J. Thermal Anal.*, 47, 1437, 1996.
16. Saleki-Gerhardt, A., Ahlneck, C., and Zografi, G., Assessment of disorder in crystalline solids, *Int. J. Pharm.*, 101, 237, 1994.
17. Buckton, G. and Darcy, P., Water mobility in amorphous lactose below and close to the glass transition temperature, *Int. J. Pharm.*, 136, 141, 1996.
18. Makower, B. and Dye, W.B., Equilibrium moisture content and crystallization of amorphous sucrose and glucose, *J. Agric. Food Chem.*, 4, 72, 1956.
19. Jouppila, K. and Roos, Y.H., Water sorption and time-dependent phenomena of milk powders, *J. Dairy Sci.*, 77, 1798, 1994.
20. Haque, M.K. and Roos, Y.H., Water plasticization and crystallization of lactose in spray-dried lactose/protein mixtures, *J. Food Sci.*, 69, FEP23, 2004.
21. Miao, S. and Roos, Y.H., Crystallization kinetics and X-ray diffraction of crystals formed in amorphous lactose, trehalose, and lactose/trehalose mixtures, *J. Food Sci.*, 70, E350, 2005.
22. Omar, A.M.E. and Roos, Y.H., Water sorption and time-dependent crystallization behaviour of freeze-dried lactose–salt mixtures, *Lebensm. Wiss. u. -Technol.*, 40, 520, 2007.
23. Al-Hadithi, D., Buckton, G., and Brocchini, S., Quantification of amorphous content in mixed systems: Amorphous trehalose with lactose, *Thermochim. Acta*, 417, 193, 2004.
24. Buma, T.J. and Wiegers, G.A., X-ray powder patterns of lactose and unit cell dimensions of β-lactose, *Neth. Milk Dairy J.*, 21, 208, 1967.
25. Fries, D.C., Rao, S.T., and Sundaralingam, M., Structural chemistry of carbohydrates. III. Crystal and molecular structure of 4-O-β-D-galactopyranosyl-α-D-glucopyranose monohydrate (α-lactose monohydrate), *Acta Crystallogr.*, B27, 994, 1971.
26. Simpson, T.D., Parrish, F.W., and Nelson, M.L., Crystalline forms of lactose produced in acidic alcoholic media, *J. Food Sci.*, 47, 1948, 1982.
27. Jouppila, K., Kansikas, J., and Roos, Y.H., Glass transition, water plasticization, and lactose crystallization in skim milk powder, *J. Dairy Sci.*, 80, 3152, 1997.

28. Jouppila, K., Kansikas, J., and Roos, Y.H., Crystallization and X-ray diffraction of crystals formed in water-plasticized amorphous lactose, *Biotechnol. Prog.*, 14, 347, 1998.
29. Earl, W.L. and Parrish, F.W., A cross-polarization-magic-angle sample spinning N.M.R. study of several crystal forms of lactose, *Carbohydr. Res.*, 115, 23, 1983.
30. Vuataz, G., Preservation of skim-milk powders: role of water activity and temperature in lactose crystallization and lysine loss, in *Food Preservation by Water Activity Control*, Seow, C.C. (ed.), Elsevier, Amsterdam, the Netherlands, 1988, pp. 73–101.
31. Buckton, G., Yonemochi, E., Hammond, J., and Moffat, A., The use of near infra-red spectroscopy to detect changes in the form of amorphous and crystalline lactose, *Int. J. Pharm.*, 168, 231, 1998.
32. Drapier-Beche, N., Fanni, J., and Parmentier, M., Physical and chemical properties of molecular compounds of lactose, *J. Dairy Sci.*, 82, 2558, 1999.
33. Kedward, C.J., MacNaughtan, W., and Mitchell, J.R., Crystallization kinetics of amorphous lactose as a function of moisture content using isothermal differential scanning calorimetry, *J. Food Sci.*, 65, 324, 2000.
34. Gombás, Á., Antal, I., Szabó-Révész, P., Marton, S., and Erõs, I., Quantitative determination of crystallinity of alpha-lactose monohydrate by Near Infrared Spectroscopy (NIRS), *Int. J. Pharm.*, 256, 25, 2003.
35. Chen, X., Bates, S., and Morris, K.R., Quantifying amorphous content of lactose using parallel beam X-ray powder diffraction and whole pattern fitting, *J. Pharm. Biomed. Anal.*, 26, 63, 2001.
36. Buckton, G. and Darcy, P., The use of gravimetric studies to assess the degree of crystallinity of predominantly crystalline powders, *Int. J. Pharm.*, 123, 265, 1995.
37. Gustafsson, C., Lennholm, H., Iversen, T., and Nyström, C., Comparison of solid-state NMR and isothermal microcalorimetry in the assessment of the amorphous component of lactose, *Int. J. Pharm.*, 174, 243, 1998.
38. Palmer, K.J., Dye, W.B., and Black, D., X-ray diffractometer and microscopic investigation of crystallization of amorphous sucrose, *J. Agric. Food Chem.*, 4, 77, 1956.
39. Haque, M.K. and Roos, Y.H., Crystallization and X-ray diffraction of spray-dried and freeze-dried amorphous lactose, *Carbohydr. Res.*, 340, 293, 2005.
40. Barham, A.S., Haque, M.K., Roos, Y.H., and Hodnett, B.K., Crystallization of spray-dried lactose/protein mixtures in humid air, *J. Crystal Growth*, 295, 231, 2006.
41. Haque, M.K. and Roos, Y.H., Crystallization and X-ray diffraction of crystals formed in water-plasticized amorphous spray-dried and freeze-dried lactose/protein mixtures, *J. Food Sci.*, 70, E359, 2005.
42. Biliaderis, C.G., Lazaridou, A., Mavropoulos, A., and Barbayiannis, N., Water plasticization effects on crystallization behavior of lactose in a co-lyophilized amorphous polysaccharide matrix and its relevance to the glass transition, *Int. J. Food Prop.*, 5, 463, 2002.
43. Kedward, C.J., MacNaughtan, W., Blanshard, J.M.V., and Mitchell, J.R., Crystallization kinetics of lactose and sucrose based on isothermal differential scanning calorimetry, *J. Food Sci.*, 63, 192, 1998.
44. Roos, Y. and Karel, M., Plasticizing effect of water on thermal behavior and crystallization of amorphous food models, *J. Food Sci.*, 56, 38, 1991.
45. Briggner, L.-E., Buckton, G., Bystrom, K., and Darcy, P., The use of isothermal microcalorimetry in the study of changes in crystallinity induced during the processing of powders, *Int. J. Pharm.*, 105, 125, 1994.
46. Sebhatu, T., Angberg, M., and Ahlneck, C., Assessment of the degree of disorder in crystalline solids by isothermal microcalorimetry, *Int. J. Pharm.*, 104, 135, 1994.
47. Lechuga-Ballesteros, D., Bakri, A., and Miller, D.P., Microcalorimetric measurement of the interactions between water vapor and amorphous pharmaceutical solids, *Pharm. Res.*, 20, 308, 2003.
48. Sperling, L.H., *Introduction to Physical Polymer Science*, John Wiley & Sons, New York, 1986.
49. Supplee, G.C., Humidity equilibria of milk powders, *J. Dairy Sci.*, 9, 50, 1926.
50. Jouppila, K., Mono- and disaccharides: Selected physicochemical and functional aspects, in *Carbohydrates in Food*, 2nd edn., Eliasson, A.-C. (ed.), CRC Press, Boca Raton, FL, 2006, pp. 41–88.
51. Bushill, J.H., Wright, W.B., Fuller, C.H.F., and Bell, A.V., The crystallisation of lactose with particular reference to its occurrence in milk powder, *J. Sci. Food Agric.*, 16, 622, 1965.
52. Sebhatu, T., Elamin, A.A., and Ahlneck, C., Effect of moisture sorption on tabletting characteristics of spray dried (15% amorphous) lactose, *Pharm. Res.*, 11, 1233, 1994.

53. Elamin, A.A., Sebhatu, T., and Ahlneck, C., The use of amorphous model substances to study mechanically activated materials in the solid state, *Int. J. Pharm.*, 119, 25, 1995.
54. Haque, M.K. and Roos, Y.H., Water sorption and plasticization behavior of spray-dried lactose/protein mixtures, *J. Food Sci.*, 69, E384, 2004.
55. Drapier-Beche, N., Fanni, J., Parmentier, M., and Vilasi, M., Evaluation of lactose crystalline forms by nondestructive analysis, *J. Dairy Sci.*, 80, 457, 1997.
56. Jouppila, K. and Roos, Y.H., unpublished data, 1997.
57. Saleki-Gerhardt, A. and Zografi, G., Non-isothermal and isothermal crystallization of sucrose from the amorphous state, *Pharm. Res.*, 11, 1166, 1994.
58. Iglesias, H.A. and Chirife, J., Delayed crystallization of amorphous sucrose in humidified freeze dried model systems, *J. Food Technol.*, 13, 137, 1978.
59. Roos, Y.H. and Jouppila, K., Plasticization effect of water on carbohydrates in relation to crystallization, in *Characterization of Cereals and Flours*, Kaletunç, G. and Breslauer, K.J. (eds.), Marcel Dekker, New York, 2003, pp. 117–149.
60. Saito, Z., Lactose crystallization in commercial whey powders and in spray-dried lactose, *Food Microstruct.*, 7, 75, 1988.
61. Kouassi, K., Jouppila, K., and Roos, Y.H., Effects of κ-carrageenan on crystallization and invertase activity in lactose-sucrose systems, *J. Food Sci.*, 67, 2190, 2002.
62. Saito, Z., Particle structure in spray-dried whole milk and in instant skim milk powder as related to lactose crystallization, *Food Microstruct.*, 4, 333, 1985.
63. Würsch, P., Rosset, J., Köllreutter, B., and Klein, A., Crystallization of β-lactose under elevated storage temperature in spray-dried milk powder, *Milchwissenschaft*, 39, 579, 1984.
64. Aguilar, C.A. and Ziegler, G.R., Physical and microscopic characterization of dry whole milk with altered lactose content. 2. Effect of lactose crystallization, *J. Dairy Sci.*, 77, 1198, 1994.
65. Roos, Y. Melting and glass transitions of low molecular weight carbohydrates, *Carbohydr. Res.*, 238, 39, 1993.
66. Orford, P.D., Parker, R., Ring, S.G., and Smith, A.C., Effect of water as a diluent on the glass transition behaviour of malto-oligosaccharides, amylose and amylopectin, *Int. J. Biol. Macromol.*, 11, 91, 1989.
67. Noel, T.R., Parker, R., and Ring, S.G., Effect of molecular structure and water content on the dielectric relaxation behaviour of amorphous low molecular weight carbohydrates above and below their glass transition, *Carbohydr. Res.*, 329, 839, 2000.
68. Roos, Y. and Karel, M., Differential scanning calorimetry study of phase transitions affecting the quality of dehydrated materials, *Biotechnol. Prog.*, 6, 159, 1990.
69. Jouppila, K. and Roos, Y.H., Glass transitions and crystallization in milk powders, *J. Dairy Sci.*, 77, 2907, 1994.
70. Lloyd, R.J., Chen, X.D., and Hargreaves, J.B., Glass transition and caking of spray-dried lactose, *Int. J. Food Sci. Technol.*, 31, 305, 1996.
71. Roos, Y. and Karel, M., Water and molecular weight effects on glass transitions in amorphous carbohydrates and carbohydrate solutions, *J. Food Sci.*, 56, 1676, 1991.
72. Roos, Y. and Karel, M., Amorphous state and delayed ice formation in sucrose solutions, *Int. J. Food Sci. Technol.*, 26, 553, 1991.
73. Cardona, S., Schebor, C., Buera, M.P., Karel, M., and Chirife, J., Thermal stability of invertase in reduced-moisture amorphous matrices in relation to glassy state and trehalose crystallization, *J. Food Sci.*, 62, 105, 1997.
74. Roos, Y.H., Water activity and physical state effects on amorphous food stability, *J. Food Process. Pres.*, 16, 433, 1993.
75. Talja, R.A. and Roos, Y.H., Phase and state transition effects on dielectric, mechanical, and thermal properties of polyols, *Thermochim. Acta*, 380, 109, 2001.
76. Slade, L. and Levine, H., Beyond water activity: Recent advances based on an alternative approach to the assessment of food quality and safety, *Crit. Rev. Food Sci. Nutr.*, 30, 115, 1991.
77. Kim, M.N., Saltmarch, M., and Labuza, T.P., Non-enzymatic browning of hygroscopic whey powders in open versus sealed pouches, *J. Food Process. Pres.*, 5, 49, 1981.
78. Donhowe, D.P. and Hartel, R.W., Recrystallization of ice during bulk storage of ice cream, *Int. Dairy J.*, 6, 1209, 1996.

79. Goff, H.D., Caldwell, K.B., Stanley, D.W., and Maurice, T.J., The influence of polysaccharides on the glass transition in frozen sucrose solutions and ice cream, *J. Dairy Sci.*, 76, 1268, 1993.
80. Nawar, W.W., Lipids, in *Food Chemistry*, 3rd edn., Fennema, O.R. (ed.), Marcel Dekker, Inc., New York, 1996, pp. 225–319.
81. Konopacka, D., Plocharski, W., and Beveridge, T., Water sorption and crispness of fat-free apple chips, *J. Food Sci.*, 67, 87, 2002.
82. Lai, H.-M. and Schmidt, S.J., Lactose crystallization in skim milk powder observed by hydrodynamic equilibria, scanning electron microscopy and 2H nuclear magnetic resonance, *J. Food Sci.*, 55, 994, 1990.
83. Choi, R.P., Tatter, C.W., and O'Malley, C.M., Lactose crystallization in dry products of milk. II. The effects of moisture and alcohol, *J. Dairy Sci.*, 34, 850, 1951.
84. Warburton, S. and Pixton, S.W., The moisture relations of spray dried skimmed milk, *J. Stored Prod. Res.*, 14, 143, 1978.
85. Ozimek, L., Switka, J., and Wolfe, F., Water sorption properties of ultrafiltration retentate skim milk powders, *Milchwissenschaft*, 47, 751, 1992.
86. Slade, L. and Levine, H., Recent advances in starch retrogradation, in *Industrial Polysaccharides: The Impact of Biotechnology and Advanced Methodologies*, Stivala, S.S., Crescenzi, V., and Dea, I.C.M. (eds.), Gordon and Breach Science Publishers, New York, 1987, pp. 387–430.
87. Levine, H. and Slade, L., Influences of the glassy and rubbery states on the thermal, mechanical, and structural properties of doughs and baked products, in *Dough Rheology and Baked Product Texture*, Faridi, H. and Faubion, J.M. (eds.), Van Nostrand Reinhold, New York, 1990, pp. 157–330.
88. Williams, M.L., Landel, R.F., and Ferry, J.D., The temperature dependence of relaxation mechanisms in amorphous polymers and other glass-forming liquids, *J. Am. Chem. Soc.*, 77, 3701, 1955.

CHAPTER 11

Sticky and Collapse Temperature: Measurements, Data, and Predictions

Benu P. Adhikari and Bhesh R. Bhandari

CONTENTS

11.1 INTRODUCTION

Discrete and free flowing powder particles coming out of dryers are what a manufacturer desires. The size of powder particles can range from 10 to 200 μm. Powders of size from 200 μm to 5 mm are called powder granules. Various methods are used to produce powders, such as static hot air drying, spray drying, fluid bed drying, crystallization, and grinding. Consumers desire free flowing powder particles that do not cake during their lifetime. However, many varieties of powder particles stick with each other when they come in contact. They adhere to surfaces in contact, such as dryer bed, roof, and wall, during handling and processing. In the case of spray drying, a large quantity of powder gets deposited at the conical section, bottom, pipes, and cyclone of the dryer. In the case of belt drying, they stick to the belt surfaces. Powder particles can agglomerate to large clumps in a fluid bed dryer. They can also adhere to containers and packages and undergo caking on storage. They cake with the slight catalyzing effect of humidity, temperature, and pressure.

The tendency of powder particles to stick to each other and to chemically dissimilar surfaces such as dryer wall, storage container, and packages is known as stickiness. Thus, the term stickiness embodies two distinct interactions, cohesive and adhesive stickiness. The sticky interaction between chemically similar surfaces is known as cohesive stickiness. Cohesive stickiness, in essence, is the tendency of the surfaces of two particles to stick to each other. Adhesive stickiness is the sticky interaction between chemically dissimilar surfaces. It is the tendency of particle surfaces to stick to the surface of equipment and packages. Adhesive stickiness is also called adhesion. In some scientific fields, such as surface sciences and polymers, cohesion and adhesion are treated separately and studied independently. In the production process and storage of food powders, cohesive and adhesive interactions are so entangled that it is quite hard, if not impossible, to distinguish the contribution of one against the other. This may be the reason why the literature on theoretical treatments and practical measurements of food stickiness lump the effect due to cohesive and adhesive interactions.

There may be instances where either cohesive or adhesive stickiness can be dominant. For example, cohesive interaction or cohesive stickiness is dominant in food powder caking. Similarly, adhesive stickiness dominates when a candy surface sticks to the packaging material. In wall depositions, adhesive stickiness is dominant only on the small layer of particles sticking to the dryer wall, while cohesive stickiness builds the subsequent layers. It is not straightforward to distinguish the contributions of cohesive and adhesive interactions in food powder stickiness.

Stickiness can have both positive and negative attributes. Granulation or controlled agglomeration is a beneficial aspect of stickiness. In this process, the cohesive property of particle surfaces is harnessed to agglomerate the particles so that the resultant granule dissolves well on reconstitution. In the vast majority of processes, though, stickiness is considered to be undesirable. Stickiness is a major problem during drying and powder making process, especially during spray drying. It causes process difficulties, excessive downtime, quality degradation, and even fire hazards. Caking renders the whole lot of manufactured powder unusable. This may be the reason why the issue of stickiness is treated in considerable detail in these two instances. There are commonly two terms used in relation to stickiness.

11.1.1 Sticky Temperature (T_{st})

At sticky temperature, the powder particles show the highest tendency to stick to equipment surfaces and packages. In terms of conventional measurement, sticky temperature is the temperature at which the bulk mass of a particulate product offers maximum resistance to a shearing motion. When powder particles, equilibrated to a relative humidity (RH), are constantly sheared while slowly ramping up their temperature, at a certain temperature they exhibit a peak shearing force. At this temperature, the flow of the powder particles gets altered and they start sticking to each other or to the contact surface. If the temperature is raised further the force required to shear the powders decreases because the sample will soften and ultimately melt. The sticky temperature can also be measured or determined through the changes in the mechanical and thermal property of the materials, for example, when the solid state is changed to rubbery state or a low heat capacity solid state transformed into high heat capacity liquid state.

11.1.2 Collapse Temperature (T_c)

Sticky temperature in the context of freeze-drying is called collapse temperature. This is particularly observed at the later stage of the freeze-drying when the majority of ice is sublimed and the temperature of the dried front is higher than the critical temperature. This is due to the fact that the highly porous solid matrix of freeze-dried material can no longer support its structure against gravity. Collapse progresses in two stages. In the first stage, the collapse is localized with concentration of collapsed pores to certain areas. In the second stage, the collapse of the entire structure the takes place. Collapse temperature is normally the temperature at which the entire structure is collapsed due to viscous flow of the solid matrix.

11.2 THEORY AND MECHANISM

From the viewpoint of material property, the sticky point temperature is simply an indicator of glass to rubber transition process taking place in a powder material. Spray and freeze-drying are the rapid and low-temperature processes, respectively. The short time or low temperature during the conversion from liquid to solid state does not allow molecular rearrangement to take place for crystallization. Thus, the powder particles produced in these processes are usually in an amorphous glassy state, which is thermodynamically metastable. Hence, there is always an underlying tendency of these powder particles to move toward their thermodynamically stable state. The physical transition of a glassy solid material to a rubbery (or liquid in the strict sense) one is called glass transition. This transition Slade and Levine (1994) is associated with distinct changes in the physical properties of material, such as specific heat capacity, optical property, specific volume, viscosity, viscoelasticity, and even the texture. Glass transition temperature (T_g) is thus a temperature range at which the glass–rubber transition initiates and completes. All the stickiness measurement techniques are based on the changes in these fundamental properties of the material.

11.3 MEASUREMENT METHODS

Both T_g and T_{st} are good indicators of stickiness phenomena. T_g is usually measured by the calorimetric method, and differential scanning calorimetry (DSC) is used for this purpose. The calorimetric test is based on the distinct change in specific heat capacity of the material during the transition from one state to another. At the measured range of T_g ($T_{g,onset}$ and $T_{g,endset}$), the alteration in flow behavior of powders may or may not take place, depending on the time of contact, interfacial distance between surfaces, and surface energetics of the contacting materials. T_{st} is more intimately associated with the stickiness phenomenon and thus, it is a perceived stickiness on the equipment/probe surface or visible clumping of particles.

The classification of the measurement methods of sticky temperature is subjective. Figure 11.1 presents one of the classifications. They can be classified into direct and indirect ones. Direct methods are those that measure the T_{st} as a function of temperature and moisture content explicitly (Tables 11.1 through 11.4). Indirect methods are those that measure some other fundamental properties such as T_g which can then be implicitly correlated to sticky temperature. The tack test or the in situ stickiness test is the only test that determines the surface stickiness of droplets or particles as the drying progresses. The test methods and the equipment involved are described in the ensuing sections. The data presented in this chapter, in the majority of cases, were obtained by reading the figures and graphs from the referenced sources.

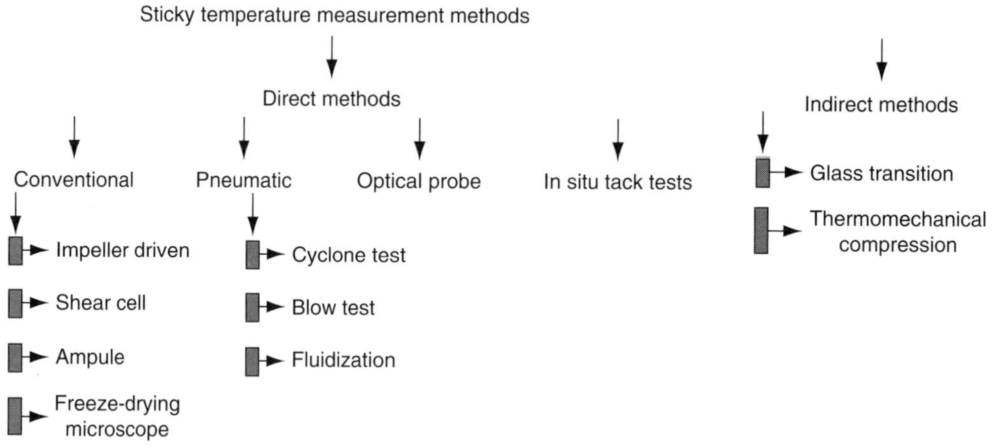

Figure 11.1 Classification of sticky temperature measurement methods.

11.3.1 Direct Methods

The direct methods measure the sticky point temperature or the degree of (adhesive or cohesive) stickiness explicitly by acquiring direct signals such as force, viscosity, flow property, and optical property of powder particles when they become sticky.

Table 11.1 Sticky Point Temperature versus Moisture Content for Spray-Dried Tomato Powder, Orange Juice Powder, and Tomato Soup and Mango Powder

Tomato Powder (Lazar et al., 1956)		Orange Juice Powder (Brennan et al., 1966)	
Moisture (%db)	T_{st} (°C)	Moisture (%db)	T_{st} (°C)
0.59	73.3	0.46	60
0.77	71.1	0.96	54
1.07	68.3	1.46	50
1.38	65.6	1.87	45
1.75	62.8	2.12	43.9
2.12	60.0	2.41	40
2.54	57.2	2.83	37.4
2.99	54.4	3.50	32
3.54	51.7	3–5[a]	52.8[a]
4.08	48.9		
4.78	46.1		
5.51	43.3		
6.25	40.6		
3–5[b]	>70[b]		

[a] Mango juice:maltodextrin:glycerol monstearate:tricalcium phosphate (35:62:1.5:1.5).
[b] Tomato soup powder.

Table 11.2 Sticky Point Temperature versus Moisture Content of Coffee Extract Powder and SMP

Coffee Extract (Wallack and King, 1988)		SMP (Hennigs et al., 2001)	
Moisture (%db)	T_{st} (°C)	Moisture (%db)	T_{st} (°C)
4.6	80.2	1.96	97.2
5.3	73.3	2.36	90.9
5.9	65.9	2.54	88.3
6.3	61.0	2.62	84.5
7.0	55.0	3.40	78.2
7.5	50.0	3.60	74.8
9.0	45.7	3.77	72.2
10.0	43.4	4.14	68.5
11.5	40.0	4.53	62.7
13.0	34.1	5.26	57.6
14.1	31.4	5.84	53.8
3–5 (Jaya and Das, 2004)	46 ± 1 (Jaya and Das, 2004)	7.54	40.9
		9.03	27.5
		10.58	23.5

Table 11.3 Sticky Point Temperature versus Moisture Content for Sucrose:Fructose and Maltodextrin:Sucrose:Fructose Composite Powders

Sucrose:Fructose (87.5:12.5) (Downton et al., 1982)		Maltodextrin:Sucrose:Fructose (50:43.7:6.3) (Wallack and King, 1988)	
Moisture (%db)	T_{st} (°C)	Moisture (%db)	T_{st} (°C)
2.02	50.8	3.5	59.0
2.87	42.3	3.8	54.2
3.26	41.5	4.0	48.1
3.62	35.0	4.4	46.0
4.19	34.5	4.5	50.18
5.80	19.8	5.0	39.9
6.61	19.5	5.8	32.13

Table 11.4 Sticky Point Temperature versus Moisture Content for WMP and SMP Using Torque Measurement Method

	Composition					T_{st} (°C)	
Sample	%Lactose	%Protein	%Fat	%Mineral	%Moisture	Torque Test	Penetration Test
WMP	37	28	27	6	2	65	60
WMP	36.5	27.6	26.6	5.9	3.3	55	—
WMP	35.5	26.8	25.9	5.7	6.1	37	—
SMP	51	37	1	8	3	57	55
SMP	50.2	36.4	1.0	7.9	4.5	—	43

11.3.1.1 Impeller-Driven Method

The impeller-driven sticky temperature measuring instrument (Figure 11.2a) was pioneered by Lazar et al. (1956) to determine the safe drying regime for tomato powder produced through spray drying. The same technique was subsequently used by Downton et al. (1982) and Wallack and King (1988). This instrument consists of a test tube in which a powder with known moisture content is placed and sealed airtight by using a rotating mercury seal. The tube is subsequently immersed in a controlled temperature bath. The temperature of the water bath is raised at reasonably low ramping rates (1°C/min–2°C/min) to equilibrate the powder particles. The powder bed within the test tube is intermittently stirred by rotating the impeller manually. The impeller is imbedded within the powder

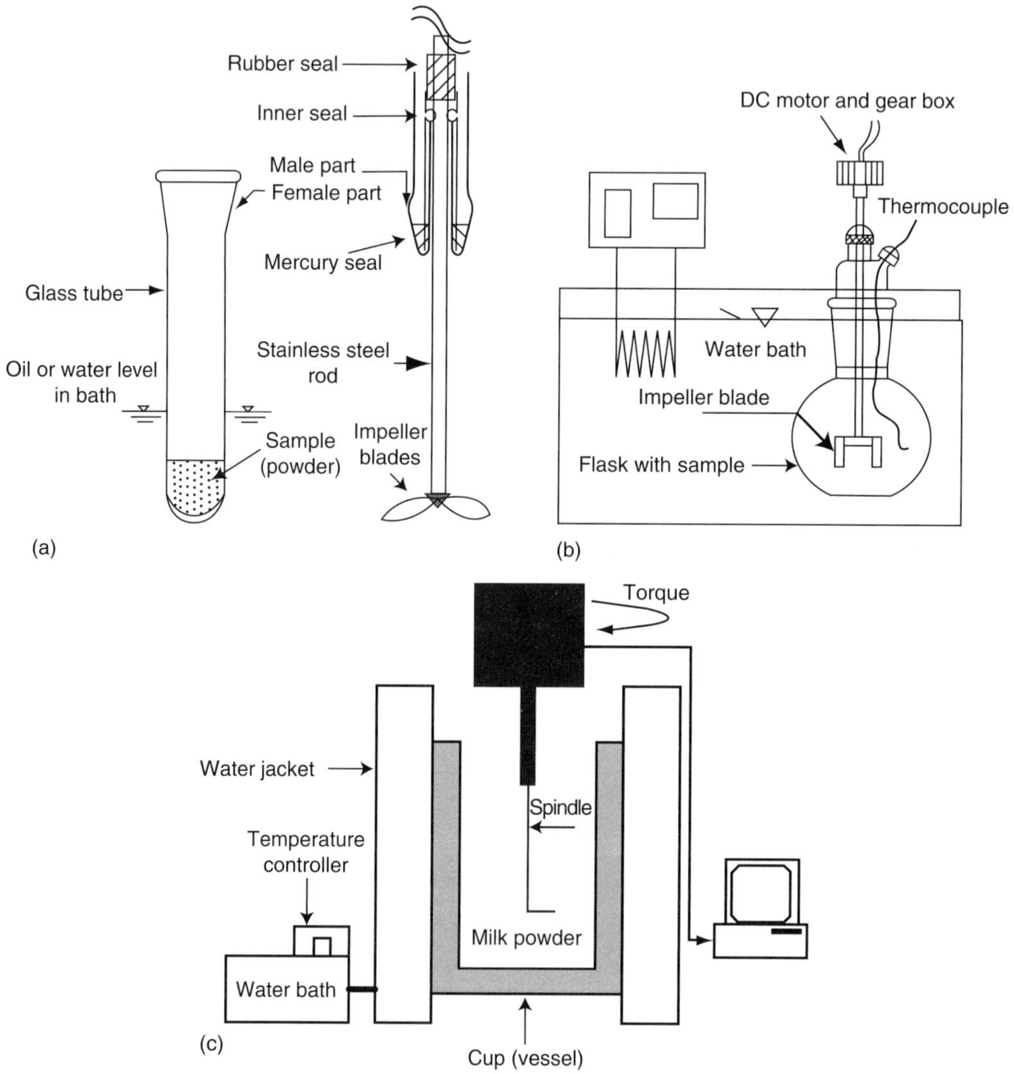

Figure 11.2 Impeller-driven sticky temperature testing instruments. (a) Hand-driven. (Adapted from Lazar et al., *Food Technol.*, 10, 129, 1956.) (b) Electrically driven. (Adapted from Hennigs, C., Kockel, T.K., and Langrish, T.A.G., *Drying Technol.*, 19, 471, 2001.) (c) Torque measurement type. (Adapted from Ozkan, N., Walisinghe, N., and Chen, X.D., *J. Food Eng.*, 55, 293, 2002.)

bed. During the ramping up, a temperature is reached at which the force required to rotate the impeller is maximum. This temperature is taken as sticky point temperature. Jaya and Das (2004) used this device to determine the sticky point temperature of instant coffee, tomato soup powder, and mango powder. Hennigs et al. (2001) further improved this instrument (Figure 11.2b) by using an electrically driven impeller. At sticky temperature, the power required to rotate the impeller increases sharply, which is recorded in a personal computer. The use of a curved-blade impeller has also been reported (Pasley et al., 1995).

Ozkan et al. (2002) introduced a system that measures the stickiness of a powder bed through the change in torque (Figure 11.2c). As shown in the figure, an "L" shaped impeller is imbedded in the sample bed. This powder bed is first equilibrated to a known moisture content and temperature. The powder is sheared by a motorized impeller and the torque is recorded. In order to maintain the physical compactness and integrity of the powder, the torque measurements obtained from the first revolution of the impeller have to be used. Furthermore, sufficient time has to be allowed so that the powder particles equilibrate to the test temperature. The authors used a 52 g sample, equilibrated for 20 min, to obtain a uniform temperature. Considerably low rotational speed (0.3 rpm) was used not to overly alter the compactness of the powder bed.

11.3.1.2 Penetration Method

Llyod et al. (1996) used penetration tests to determine the sticky point temperature of whole milk powder (WMP) and skim milk powder (SMP). For this purpose, the powder samples were compacted to form a plug with certain low consolidation load and conditioned at different temperatures ranging from room temperature to 70°C. Then, a texture analyzer was used to penetrate compacted plugs. The force required to penetrate 1 mm of the compacted powder plug was used to determine the sticky point temperature. Since it was not possible to carry out the penetration test in situ when the powder was being conditioned, all the testes were carried out immediately in an ambient condition.

Advantages: These propeller-driven sticky temperature methods provide direct measurement of T_{st}. The instruments involved are simple, cheaper, and easy to construct in-house.

Disadvantages: These tests other than that of Ozkan et al. (2002) provide only temperature values at which maximum shearing force is required. These tests are static in nature. The powder particles have to reach equilibrium in terms of RH and temperature. The resistance offered to impeller motion is mainly contributed by the particle-to-particle cohesion rather than the particle-to-equipment adhesion. These results obtained from this technique are applicable in powder processing, handling, and storage situations, but may not exactly represent the problem associated with drying because particle-to-equipment adhesion may dominate the stickiness during drying.

11.3.1.3 Ampule Method and Freeze-Drying Microscope (for Collapse Temperature)

Ampule method was originally developed to test the collapse temperature of freeze-dried materials (Tsourouflis et al., 1976). Freeze-dried powders have a highly porous structure. When they are equilibrated with a certain temperature and RH, collapse of the pores and the entire structure can take place (Levi and Karel, 1995). The collapse temperature (T_c) can be determined using a simple ampule method. The ampule is a vial made up of transparent plastic or glass. The ampules are filled with freeze-dried sample and sealed subsequently. While determining the T_c, these ampules are equilibrated with a series of temperatures subjecting to a series of water or oil baths maintained at different temperatures. After adequate time, the ampules are taken out and visually inspected for the structural collapse or their softened state. The collapse temperature is the temperature at which the pores are

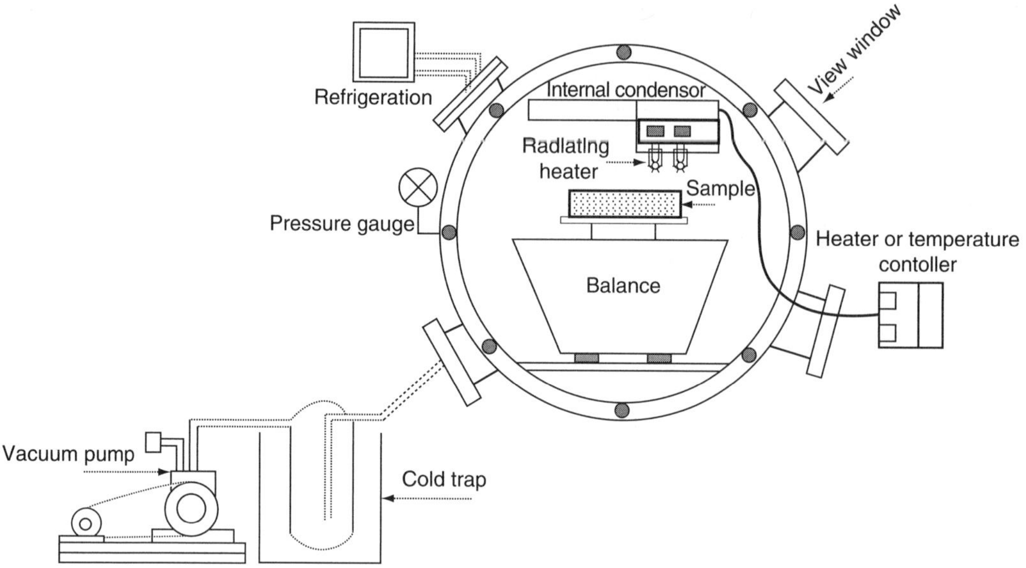

Figure 11.3 Schematic diagram of collapse temperature measuring system. (Redrawn from Bellows, R.J. and King, C.J., *AICHE Symp. Ser.*, 69, 33, 1973.)

collapsed due to viscous flow of the powder particles (Tsourouflis et al., 1976). Chuy and Labuza (1994) applied this method to determine the surface caking behavior of dairy-based food powders.

Collapse temperature and the rate of sublimation of ice were monitored by Bellows and King (1972, 1973) by using a unique experimental setup. As shown in Figure 11.3, this system consists of a vacuum chamber with a refrigeration system connected to it. Vacuum is created through the use of a vacuum pump. A balance is placed within the vacuum chamber. During the test, a prefrozen sample slab of about 5 cm in diameter and 5 mm in thickness was kept on the balance. Precalibrated thermocouples were used to monitor the temperature at which the sample collapses. The collapse of the structure was determined visually through the view window. Furthermore, the collapse temperature was also ascertained through the drying rate of the sample. This is because the collapse is associated with erratic drying rates. Heating of the sample is achieved using radiating heat on top of the sample slab. The rate of heating was precisely controlled through the use of a variable transformer.

Freeze-drying microscope was used by To and Flink (1978) to study the collapse temperature of food powders. These microscopes are increasingly being used in pharmaceutical researches to determine collapse temperature (Fonseca et al., 2004). This instrument consists of compact heating and freezing stages fitted with a microscope and a video imaging system and are available commercially from companies such as Linkam Scientific Instruments (Surrey, UK). As shown in Figure 11.4, the equipment consists of a small freeze-drying chamber fitted with a temperature-controlled stage. A vacuum pump is used to evacuate the sublimated vapor and other noncondensable gases. The drying chamber can be observed continuously through a window and attached microscope. The sample is loaded on a 0.17 mm cover slip on the top of a highly polished silver heating element for better heat transfer and temperature measurement. A platinum temperature sensor, accurate to 0.01°C, is used to obtain better and stable temperature signals. The temperature ramping is achieved within 0.1°C/min–130°C/min. The structure of the freeze-dried powder is continuously monitored through the image acquisition system until collapse of its structure takes place. The temperature at which the collapse takes place is noted.

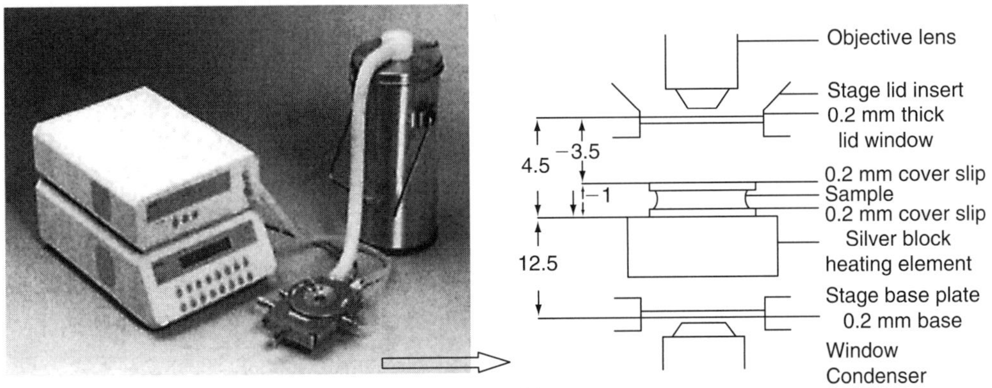

(a) (b)

Figure 11.4 Freeze drying microscope for the determination of collapse temperature of freeze-dried powders. (a) Pictorial view of the device and (b) positioning of the sample and the microscope (Model THMS600, Linkam Scientific Instruments, Surrey, U.K. With permission.).

Advantages: The ampule method is simple and inexpensive. The test can be carried out effectively anywhere. The freeze-drying microscope is fitted with state-of-the-art freeze-drying and effective imaging systems. The use of a microscope with appropriate magnification will make it possible to pinpoint the initiation of the collapse of the pores and structure quite effectively.

Disadvantages: This method is very subjective. Furthermore, it will be hard for the human eye to pinpoint the initiation and collapse of the structure of the freeze-dried products. The collapse temperature may vary depending on the amount of unfrozen water in the sample and the freeze-drying conditions used.

The collapse temperature measured by researchers for various sugars, maltodextrins, fruit juices, and other food compounds are presented in Tables 11.5 through 11.11. From Table 11.5 through 11.11, four important observations can be made. First, the T_c values of sucrose, maltose, and lactose are different and also correspond to the same trend as T_g values. Second, the T_c values of maltodextrins decrease with an increase in their dextrose equivalent (DE) values. This is an indication of molecular dependence of the T_c. Third, when materials with a higher amount of low-molecular weight sugars (e.g., orange juice) are mixed with a high-molecular weight materials (e.g., maltodextrin), plasticization occurs and the T_c of the mixture decreases. An approach to quantify this decrease and its implication on stickiness is discussed in Section 11.4.2. Fourth, when the composition is kept constant, the measured T_c depends on the thermal history of the product or the way it was frozen. Rapidly dried products tend to have higher T_c values than their slowly frozen counterparts. This effect is common to all thermocalorimetric measurements.

11.3.1.4 Optical Probe Method

A sticky point temperature measurement system that uses the optical property of powder was developed by Lockemann (1999). This method exploits the change in optical properties of powder during nonsticky to sticky transition. When the powder becomes sticky its surface reflectivity (optical property) changes accordingly. It uses an optical fiber sensor combined with an illuminator–receiver system. The sensor records the reflection of the light directed from the illuminator toward the sample through the integrated light tube. The changes in the sample optical property are monitored through the sharp rise in reflectance. The reflectance sensor records the voltage that corresponds to the quantity of light reflected by the sample. A schematic diagram of this instrument is given in Figure 11.5a. As shown in the figure, it consists of a sealed test tube containing powder

Table 11.5 Collapse Temperature (T_c) of Some of the Freeze-Dried Anhydrous Food Powders

Materials	T_c (°C)
Lactose	101.1
Maltose	96.1
Sucrose	55.6
Lactose:sucrose (50:50)	78.9
Sucrose:maltose (50:50)	73.3
Maltodextrin DE 10	248.9
Maltodextrin DE 15	232.2
Maltodextrin DE 20	232.2
Maltodextrin DE 25	204.4
Orange juice	51.7
Orange juice:maltose (100:10)	65.6
Orange juice:starch (100:2)	53.3
Sucrose:starch (100:2)	73.3
Orange juice:maltodextrin DE 10 (100:3)	75.0
Orange juice:maltodextrin DE 10 (100:5)	78.3
Orange juice:maltodextrin DE 10 (100:10)	83.9
Orange juice:maltodextrin DE 10 (100:15)	88.9
Orange juice:maltodextrin DE 10 (100:20)	100.0
Orange juice:maltodextrin DE 25 (100:2.5)	58.3
Orange juice:maltodextrin DE 25 (100:5)	62.8
Orange juice:maltodextrin DE 25 (100:10)	68.3
Orange juice:maltodextrin DE 25 (100:15)	73.3
Orange juice:maltodextrin DE 25(100:20)	83.9
Orange juice:gum arabic (100:1)	57.2
Orange juice:gum arabic (100:3)	61.1
Orange juice:gum arabic (100:6)	82.2
Orange juice:locust bean gum (100:3)	100.0
Orange juice:tragacanth gum (100:3)	98.9
Orange juice:karaya gum (100:3)	66.7
Orange juice:tapioca dextrin (100:3)	66.7
Orange juice:tapioca dextrin (100:6)	67.2
Orange juice:tapioca dextrin (100:10)	78.9

Source: From Tsourouflis, S., Flink, J.M., and Karel, M., *J. Sci. Food Agric.*, 27, 509, 1976. With permission.

sample. The test tube is immersed in a temperature-programmable oil bath at an angle of about 30°. A resistance thermometer is embedded within the powder, which records the temperature of the sample. This test tube is then mounted on a rotating drive (50 rpm). This rotation causes the particles within the tube to tumble and roll. The signals of the temperature and the reflectance sensor are shown in Figure 11.5b. The sensor signal remains almost the same as long as the powder particles remain free flowing. The signal recording system gives a sharp rise in reflectance, which is an indication of the sticky point. Some typical data obtained from this instrument are given in Table 11.12.

Advantages: This instrument is novel and can be very useful for online monitoring of stickiness in powders. So far, it has only been used to determine the sticky point temperature of solid β-carotene.

Table 11.6 Collapse Temperature (T_c) of Freeze-Dried Orange Juice Powder

Fast Freezing		Slow Freezing	
Moisture (%db)	T_c (°C)	Moisture (%db)	T_c (°C)
0.00	51.7	0.00	51.7
1.68	37.2	0.76	41.7
3.23	28.9	1.52	32.5
4.70	22.8	2.30	27.5
6.50	16.3	3.50	21.2
8.07	14.7	5.30	14.5
10.0	10.3	9.20	9.2

Table 11.7 Collapse Temperature versus Moisture Content of Freeze-Dried Maltodextrins

Moisture (%db)	T_c (°C)	Moisture (%db)	T_c (°C)
Maltodextrin DE 10		*Maltodextrin DE 15*	
0	248.9	0	232.2
2.98	161.7	1.39	73.2
5.40	122.3	4.13	132.6
7.90	89.6	9.0	69.5
11.1	52.5	11.0	52.7
13.7	37.1	14.0	30.1
Maltodextrin DE 20		*Maltodextrin DE 25*	
0	232.2	0	204.4
4.34	69.2	1.77	152.5
6.5	48.7	4.32	102.4
9.4	61.1	6.0	79.8
12.8	31.5	8.0	58.3
		13.7	20.3

Table 11.8 Collapse Temperature (T_c) of Freeze-Dried Maltose

Fast Freezing		Slow Freezing	
Moisture (%db)	T_c (°C)	Moisture (%db)	T_c (°C)
0.02	95.2	0.11	96.6
2.30	61.0	1.11	83.4
3.20	47.9	2.65	62.5
4.25	37.7	4.0	54.8
5.64	29.3	5.8	36.0
7.10	20.6	7.2	23.3
8.00	15.3	8.1	17.2

Table 11.9 Collapse Temperature (T_c) of Freeze-Dried Maltose:Sucrose (50:50) Powder

Fast Freezing		Slow Freezing	
Moisture (%db)	T_c (°C)	Moisture (%db)	T_c (°C)
0.03	79.4	0.03	74.4
1.2	72.8	0.83	63.6
1.9	62.1	1.53	56.1
3.0	48.4	3.30	39.0
4.5	35.9	5.30	19.3
5.5	27.7	7.00	8.6
6.5	19.4	8.6	3.0
8.5	5.5		

Disadvantages: This instrument is not yet tested in particulate systems that are transparent or become transparent on softening. Likewise, this method cannot provide a quantitative measure of the cohesive or adhesive force involved at the sticky state of the powders.

11.3.1.5 Pneumatic Methods

Pneumatic tests are designed to mimic the flow and no-flow scenario within a dryer, fluidized bed, and a cyclone. These tests are more realistic in the sense that both the powder properties and the process parameters such as temperature, RH, and flow pattern of air determine the stickiness of powders. Various pneumatic tests are discussed in the following sections.

11.3.1.5.1 Cyclone Stickiness Test

This is a dynamic test at which the particle surface is exposed to a controlled temperature and humid air for a short time so that the particle surface equilibrates to the air condition. Then, the sticky (cohesive) behavior of the powder particles is observed. Initially, the swirl of the air within the cyclone rotates the particles within it. When the particle surface becomes cohesive enough, the particles stick with each other and agglomerate. The agglomerated particles cannot fly along with the air-swirl. If the surface of this particle has a greater tendency to adhere to the cyclone surface, it sticks to the cyclone surface as well.

A research group at the University of Queensland, Australia, led by Bhandari and coworkers conceptualized and developed a cyclone stickiness tester for routine characterization of food powder stickiness (Figure 11.6). As shown in the figure, filtered and suitably controlled compressed air is brought to the air-heater. This air can be heated to 200°C–300°C and led to the humidifier. At the humidifier, water is sprayed through a twin-fluid nozzle. The flow rate of water and the temperature of the heating air are controlled in such a way that air with the desired temperature and RH is delivered to the cyclone. This air comes in contact with the powder particles kept at the sample holder attached to the cyclone. If the temperature of the air at a given RH exceeds the sticky point temperature, the powder particles stick to each other rapidly and agglomerate and can also stick to the cyclone wall. Finally, the agglomerated particles fail to swirl with the air. The temperature (at a given RH) of the air at which the agglomerated particles fail to swirl is taken as the sticky temperature. Some typical data obtained from cyclone stickiness tests are presented in Tables 11.13 through 11.15.

Advantages: Cyclone stickiness test has two advantages compared to the impeller-driven stickiness tests. First, it is a dynamic method that requires only the surface of the particles to come to

Table 11.10 Collapse Temperature of Maximally Freeze-Dried Samples

Solute Initial Concentration (% w/w)	Observed T_c (°C)
Sugars	
Xylose, 25%	−49
D-Fructose, 25%	−44
D-Glucose, 25%	−41.5
Maltose, 25%	−23
Raffinose, 25%	−18
Sucrose, 15%	−22.5
Sucrose, 25%	−24
Sucrose, 35%	−27
Sucrose, 45%	−29
Sucrose, 55%	−29.5
12.5% glucose + 12.5% sucrose	−33.5
Sugars with additives	
2.5% NaCl + 22.5% sucrose	−33.5
5% NaCl + 20% sucrose	−43
10% dimethyl sulfoxide + 15% sucrose	−47.5
10% glycerine + 15% sucrose	−46
12.5% sorbitol + 12.5% sucrose	−37.5
1% pectin + 24% fructose	−34
2% gelatin + 23% fructose	−27.5
Liquid foods	
Orange juice, 23%	−24
Grape fruit juice, 16%	−30.5
Lemon juice, 9%	−36.5
Apple juice, 22%	−41.5
Concord grape juice, 16%	−46
Sweetened condensed grape juice, 23%	−33.5
Pineapple juice, 10%	−41.5
Prune extract, 20%	−35.0
Coffee extract, 25%	−20

Source: Fonseca, F., Passot, S., Cunin, O., and Marin, M., *Biotechnol. Prog.*, 20, 229, 2004. With permission.
Note: The solid content indicates only the initial solution concentration.

equilibrium with the surrounding air. It has been established that particle surfaces can become sticky or nonsticky irrespective of the state of their interior and that the moisture and temperature profiles at the particle surface control the stickiness. Second, it mimics more realistically the sticky–nonsticky state of a particle during spray drying. A safe drying regime can be mapped out for any powder samples based on which the dryer outlet temperature and RH can then be adjusted in industrial operations.

Disadvantages: In drying, the droplet moves from droplet (solution) state to a dry particulate form. This instrument measures the stickiness of dry particles, which may not provide realistic results to interpret with an actual industrial drying situation.

Table 11.11 Glass Transition Temperature T'_g (of Maximally Concentrated) and Collapse Temperature (T_c) of Aqueous Solutions

Solution (%, w/w)	Molecular Weight (g/mol)	T'_g (°C)	T_c (°C)
Sucrose, 5%	342	—	−33, −34
Sucrose, 10%	342	−32 (−32.9) (Her and Neil, 1994)	−32 (−32) (McKenzie, 1975)
Maltose, 10%	342	−30 (−29.5) (Levine and Slade, 1988)	−30 (−30 to −35) (McKenzie, 1975)
Maltodextrin DE 5–8, 10%	~3000	−9, −9.5	−7 (−10) (McKenzie, 1975)
Sodium glutamate, 10%	147	−47	−46 (−50) (McKenzie, 1975)
Sodium citrate, 10%	294	−43 (−41) (Chang and Randall, 1992)	−40
Sodium ascorbate, 10%	198	−37	−37
Sorbitol, 10%	182	−45 (−46.1) (Her and Neil, 1994)	−46 (−45) (McKenzie, 1975)
Mannitol, 10%	182	−36 (−35) (Chang and Randall, 1992), (−38)	—
Sucrose 5% + maltodextrin DE 5–8, 5%	—	−24	−19 (−20) (McKenzie, 1975)
Sucrose 5% + sorbitol 5%	—	−41	−38 (−40) (McKenzie, 1975)
Sucrose 5% + mannitol 15%	—	−40	−14

Source: Fonseca, F., Passot, S., Cunin, O., and Marin, M., *Biotechnol. Prog.*, 20, 229, 2004. With permission.

11.3.1.5.2 Fluidization Test

As in the cyclone stickiness tests, fluidization allows close contact between individual particles and the fluid. In this test, the fluidization phenomenon is harnessed to characterize the sticky temperature of powder particulates (Dixon, 1999; Bloore, 2000). These authors investigated the evolution of stickiness in powder particles as a function of RH and temperature using the instrument shown in Figure 11.7. This instrument consists of an air humidification, temperature control, and temperature and RH measuring systems.

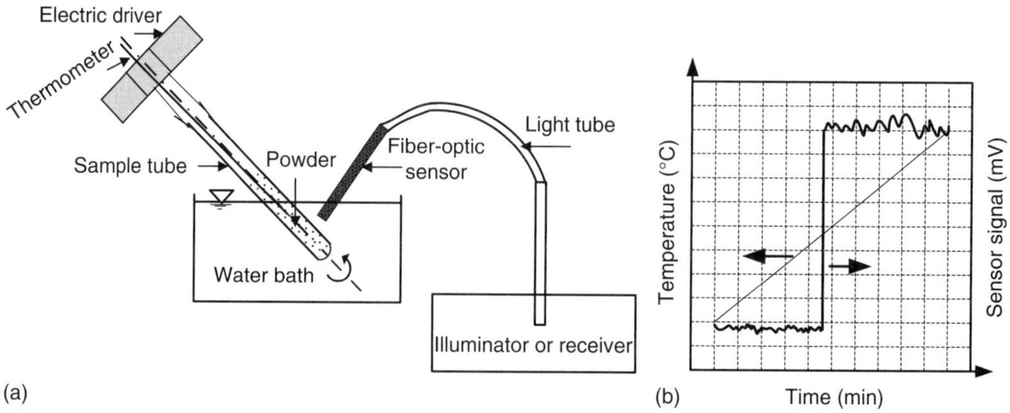

Figure 11.5 (a) Schematic diagram of the optical probe used for sticky-point determination of the free flowing particulates. (Adapted from Lockemann, C.A., *Chem. Eng. Process.*, 38, 301, 1999.) (b) Signal from the instrument (sample sticky temperature = 80°C).

Table 11.12 Sticky Temperature (T_{st}) versus Moisture Content for β-Carotene

Moisture Content (%db)	T_{st} (°C)
0.89	141.4
3.43	114.4
5.98	79.9
7.02	65.8
11.42	54.7
13.48	41.4

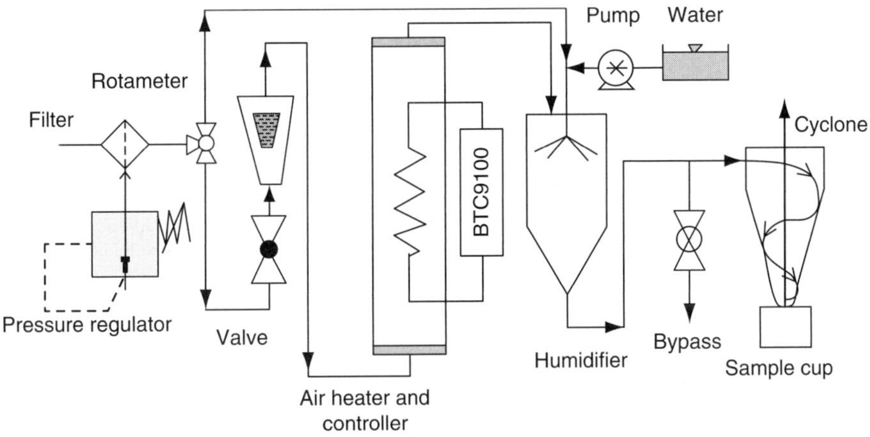

Figure 11.6 Schematic diagram of the cyclone stickiness test device.

Table 11.13 Sticky Temperature (T_{st}) and Surface Glass Transition Temperature of Apple Juice:Maltodextrin Composite Powder

Sticky Point Temperature		Surface Glass Transition Temperature	
Moisture (%db)	T_{st} (°C)	Moisture (%db)	Surface T_g (°C)
0.3	79.6	0	58.6
0.36	76.8	0.49	50.8
0.62	65.3	0.86	45.7
0.74	64.9	1.04	39.0
1.37	64.5	1.26	40.5
1.67	62.0	1.39	44.2
2.42	57.2	1.6	37.4
3.10	49.4	1.81	36.7
3.71	49.0	2.32	29.9
4.3	46.0	2.77	27.5
5.14	42.6	2.9	21.8

Source: Boonyai, P., Development of new instrumental techniques for measurement of stickiness of solid particulate food materials, PhD thesis, School of Land and Food Sciences, The University of Queensland, Australia, 2005.

Table 11.14 Sticky Temperature (T_{st}) and Surface Glass Transition Temperature of Whey Powder

Sticky Point Temperature		Surface Glass Transition Temperature	
Moisture (%db)	T_{st} (°C)	Moisture (%db)	Surface T_g (°C)
1.50	75.2	1.69	64.9
1.65	73.0	1.92	61.8
1.82	69.5	2.04	57.3
2.3	65.4	2.82	42.3
2.87	61.4	3.52	48.1
3.2	56.5	3.23	41.6
3.8	54.0	4.25	40.7
4.2	51.0	4.75	38.1
4.95	46.4	5.8	40.5
5.45	43.6	6.5	28.8
5.6	42.0	7.86	28.3
5.82	41.7	9.03	26.7

Source: Boonyai, P., Development of new instrumental techniques for measurement of stickiness of solid particulate food materials, PhD thesis, School of Land and Food Sciences, The University of Queensland, Australia, 2005.

An airstream is humidified by bubbling it through a series of containers submerged in a controlled temperature water bath. The passing of the air through these multiple containers (with water) is necessary to maintain the desired temperature and humidity of the air. The temperature and RH of the air are measured by a hygrometer before it enters a series of powder beds. At a given

Table 11.15 Sticky Temperature (T_{st}) and Surface Glass Transition Temperature of Honey Powder

Sticky Point Temperature		Surface Glass Transition Temperature	
Moisture (%db)	T_{st} (°C)	Moisture (%db)	Surface T_g (°C)
0.69	73.2	0.18	58.0
0.83	71.4	0.39	61.2
1.21	67.2	0.44	57.7
1.51	63.5	0.54	51.8
2.04	58.9	1.07	50.6
2.54	54.8	1.56	43.0
3.05	50.2	1.82	40.0
3.58	47.5	2.43	34.0
4.3	47.4	3.01	28.0
4.38	44.4	3.6	27.5
4.8	42.7	4.3	20.6
		4.75	21.2
		5.45	24.5

Source: Boonyai, P., Development of new instrumental techniques for measurement of stickiness of solid particulate food materials, PhD thesis, School of Land and Food Sciences, The University of Queensland, Australia, 2005.

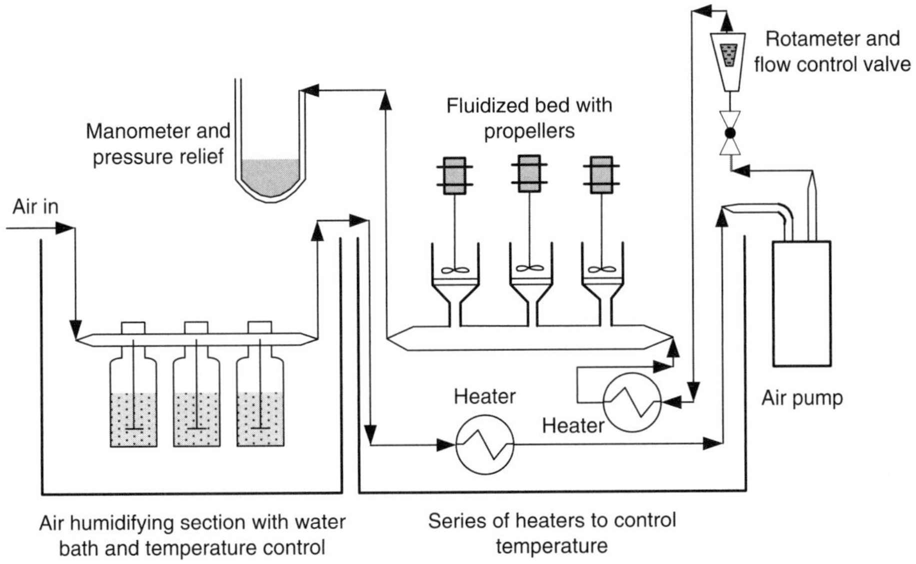

Figure 11.7 Fluidization stickiness testing instrument. (Adapted from Bloore, C., in *International Food Dehydration Conference—2000 and Beyond*, Carlton Crest Hotel, Melbourne, Australia, 2000.)

temperature, the RH of the air is slowly increased to a value at which the powder becomes sticky. Since further fluidization is impossible beyond the sticky temperature, defluidization and characteristic rat holes are observed when the powder particles stick with other. Some typical data obtained from fluidization tests are presented in Tables 11.16 through 11.20.

Advantages: Similar to cyclone stickiness test, this test allows the determination of sticky temperature as a function of RH. This can be used to construct the safe regime for any particulate products.

Disadvantages: The structure and working principle of the fluidization tester is much more complex than the cyclone stickiness tester given in Figure 11.6. This instrument will be more expensive to fabricate and require more effort and expert hand to run. In addition, it suffers from the same disadvantages as that of the cyclone stickiness tester.

11.3.1.5.3 Blow Test Method

The blow test method was developed by Paterson and coworkers (Paterson et al., 2005). The essence of this test is to harness the force of air to blow a certain size (diameter and depth) of hole in

Table 11.16 Sticky Temperature (T_{st}) of Vanilla Sustagen and SMP

Vanilla Sustagen		SMP	
RH (%)	**T_{st} (°C)**	**RH (%)**	**T_{st} (°C)**
40.5	52.5	40.1	43.9
50.2	48.6	49.8	39.8
58.2	44.4	57.7	37.3
70.3	37.5	69.7	31.9
80.2	28.6	80.0	24.4

Source: Dixon, A., Correlating food powders stickiness with composition, temperature and relative humidity, Research project, Department of Chemical Engineering, Monash University, 1999.

Table 11.17 Sticky Point Temperature of Amorphous Lactose

Chatterjee (2004)		Zuo et al. (2007)	
RH (%)	T_{st} (°C)	RH (%)	T_{st} (°C)
4.78	79.8	14.95	80.3
7.46	71.0	20.35	78.0
11.98	60.2	21.90	72.8
15.14	50.9	28.96	60.3
23.16	40.2	44.43	47.0
31.58	33.5	46.71	43.1
37.41	30.2	51.8	38.2
53.32	22.8	52.94	34.5
58.00	20.2	60.1	27.5
		64.78	22.7

Source: Chatterjee, R., Characterising stickiness of dairy powders, MTech thesis, Institute of Technology and Engineering, Massey University, Palmerston North, New Zealand, 2004; Zuo, J.Y., Paterson, A.H., Bronlund, J.E., and Chatterjee, R., *Int. Dairy J.*, 17, 268, 2007.

a preconditioned powder bed (Table 11.21). As shown in Figure 11.8, it consists of a vertical stainless steel pipe, 9 mm in diameter, with a short brass arm attached at right angles close to the bottom. A 1.3 mm diameter and 40 mm long stainless steel pipe delivers the air to the powder bed. This tube is inclined at 45° and positioned at 2 mm above the upper surface of the powder bed. Air is supplied to the tube through the vertical tube as shown in Figure 11.8a. The distributor plate or the sample bed is segmented, which helps replicate experiments by simply rotating the air tube over the segments. The vertical pipe with the air tube is positioned above the sample bed on the distributor plate and assembled in a glass enclosure as shown.

About 40 g preconditioned powder sample (at a given temperature and RH) is spread evenly onto the distributor plate and the air is blown onto the center of each segment of the plate. The airflow rate is increased until a channel appears in the sample bed. This is taken as an endpoint. The stickier powders require higher flow rates to generate a channel. Furthermore, the channels become narrower for sticky powders.

Table 11.18 Sticky Temperature (T_{st}) of WMP

Chatterjee et al. (2004)		Zuo et al. (2007)	
RH (%)	T_{st} (°C)	RH (%)	T_{st} (°C)
19.35	79.9	24.11	80.5
20.30	79.6	29.27	78.3
22.61	69.3	37.12	68.3
23.16	69.7	41.35	60.6
26.45	71.0	52.01	53.6
36.08	60.2	62.32	43.8
37.08	59.8	64.91	37.5
43.08	50.6	73.23	29.3
45.64	50.6		
50.78	40.1		

Source: Zuo, J.Y., Paterson, A.H., Bronlund, J.E., and Chatterjee, R., *Int. Dairy J.*, 17, 268, 2007; Chatterjee, R., Characterising stickiness of dairy powders, MTech thesis, Institute of Technology and Engineering, Massey University, Palmerston North, New Zealand, 2004.

Table 11.19 Sticky Temperature (T_{st}) for IWMP and AWMP

IWMP		AWMP	
RH (%)	T_{st} (°C)	RH (%)	T_{st} (°C)
20.46	79.7	28.56	79.6
22.24	78.5	31.32	79.4
29.80	70.0	49.25	70.3
30.72	69.9	50.25	69.2
48.27	61.0	60.77	60.6
49.37	60.0	67.43	50.9
55.22	50.4	68.15	50.3
67.38	40.2		
74.24	40.3		

Source: Chatterjee, R., Characterising stickiness of dairy powders, MTech thesis, Institute of Technology and Engineering, Massey University, Palmerston North, New Zealand, 2004.

IWMP, instant whole milk powder; AWMP, agglomerated whole milk powder.

Table 11.20 Sticky Temperature (T_{st}) for Milk Proteins Concentrates (MPCs)

MPC70		MPC85	
RH (%)	T_{st} (°C)	RH (%)	T_{st} (°C)
65.23	81.8	77.26	80.7
71.4	71.3	78.38	80.5
75.93	61.5	80.08	70.1
81.52	50.5	83.84	70.4
83.2	50.5	87.09	61.1
81.3	40.9	88.72	50.5
		90.26	50.7
		96.09	40.7
		97.14	40.3

Source: Chatterjee, R., Characterising stickiness of dairy powders, MTech thesis, Institute of Technology and Engineering, Massey University, Palmerston North, New Zealand, 2004.

Table 11.21 Blow Test Airflow Rate (L/min) to Create a Channel on the Powder Bed

Airflow (L/min)	Powder Characteristics
0–3	Free flowing powders
4–6	Some caking strength, when disturbed forms powders with no lumps
7–9	When disturbed, forms lumps that are fragile and easy to break into powders
10–12	Forms larger lumps with no strength
13–15	Hard to break up bed
16–18	Lumps are larger, whole segment size
19–22	Limit of tester, difficult to break up bed, forms large lumps

Source: From Paterson, A.H.J., Brooks, G.F., Bronlund, J.E., and Foster, K.D., *Int. Dairy J.*, 15, 513, 2005. With permission.

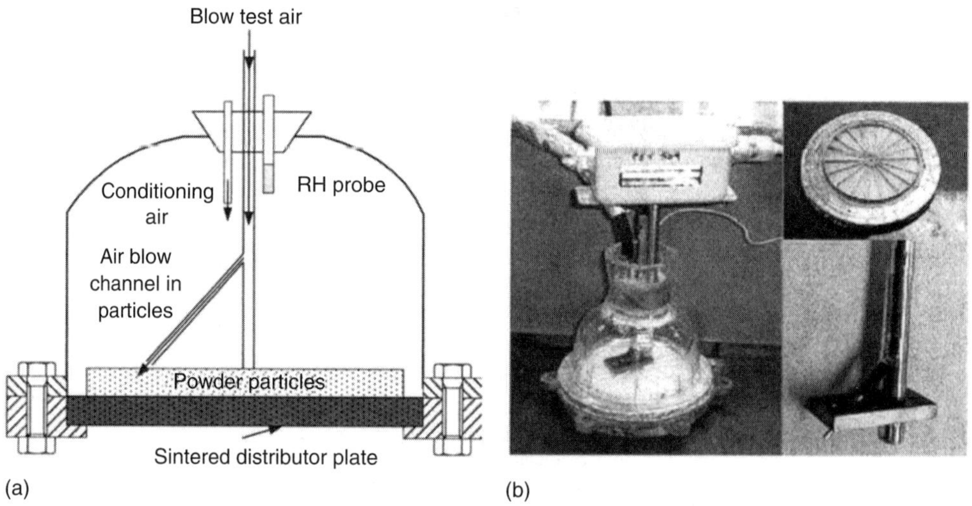

Figure 11.8 (a) Schematic diagram of a blow test instrument. (Redrawn from Paterson, A.H.J., Brooks, G.F., Bronlund, J.E., and Foster, K.D., *Int. Dairy J.*, 15, 513, 2005.) (b) Pictorial views of the blow test instrument.

Advantages: This instrument is based on simple, reliable, and easy to operate working principles. The instrument is easy to fabricate in-house and does not require highly trained personnel to run the test.

Disadvantages: The prescribed sample preparation steps are not easy to follow. Sample humidification (moisture diffusion) is a slow process and an equilibrium time of 24 h or more is needed before the test.

11.3.1.5.4 Particle-Gun Equipment

The particle-gun instrument was developed by Paterson and coworkers (Paterson et al., 2007; Zuo et al., 2007) to measure the onset of stickiness of particles traveling in air at velocities, temperatures, and RH close to that encountered in industrial spray dryers. In this test, the powder surface is allowed to come in contact with the air with known temperature and humidity. It is assumed that the surface of the particles attains the temperature and RH as that of the supplied air almost instantly and that the equilibrium time is unnecessary.

As shown in Figure 11.9, the instrument has two main sections, the first generates the controlled RH and temperature air and the second one feeds the dry particulates. The particle gun enables the particle to be fired to the target or deposition plate at a desired velocity. The first section involves pressure regulation, air–water contact (bubbling column), and primary and secondary air heaters. The RH of the airstream entering the particle gun is adjusted by increasing or decreasing the water temperature in the column. The velocity of the air at the gun tip is maintained at about 20 m/s to make it as close as possible to that in industrial spray dryers and cyclones. When this air enters into the vortex chamber, the venturi effect feeds the powder particles in from its funnel feeder. Subsequently, the powder particles are pushed to the particle-gun tube. The particles travel down the particle-gun tube and their exit trajectory is directed toward the target SS plate (diameter 75 mm) situated 16 cm apart. The air temperature and RH are measured at the exit. Twenty-five grams of powder sample is fed into the feeder funnel for each test run. The sample mass can vary depending on the degree of stickiness of the powder. Deposition of the powder particles in multiple layers on the target plate can lead to faulty results. Some powders are inherently stickier than the others and

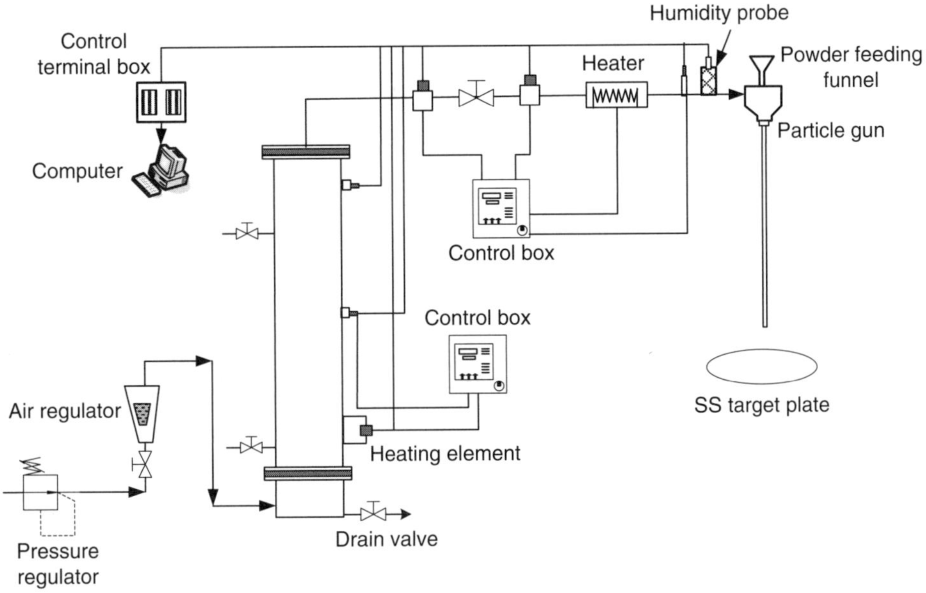

Figure 11.9 Schematic diagram of the air supply and the particle-gun apparatus. (Adapted from Zuo, J.Y., Paterson, A.H., Bronlund, J.E., and Chatterjee, R., *Int. Dairy J.*, 17, 268, 2007.)

require less sample mass. The plate is weighed before and after the experiment to determine the weight of the deposited powders. Some typical data obtained from this instrument are presented in Tables 11.22 through 11.25.

Advantages: This method is designed to mimic the possible trajectories and reported velocities of the particles in the spray dryer. This test provides insight regarding the possible enhanced sticky behavior of particles moving with a certain velocity. Furthermore, the test environment is a dynamic one in which the surface property of the particle plays a major role in stickiness.

Disadvantage: The major disadvantage of the existing experiment is that the target plate remains at ambient temperature. The temperature and RH of the air at the target plate are much different from

Table 11.22 Sticky Temperature (T_{st}) of Amorphous Lactose

Zuo et al. (2007)		Chatterjee (2004)	
RH (%)	T_{st} (°C)	RH (%)	T_{st} (°C)
7.47	70.7	44.48	56.5
10.95	60.3	46.75	42.5
14.84	50.7	51.76	37.5
23.05	40.1	52.86	34.3
31.70	33.5	60.04	26.7
38.00	30.6	58.76	17.8
53.40	22.7	64.75	22.6
54.50	22.7		

Source: Zuo, J.Y., Paterson, A.H., Bronlund, J.E., and Chatterjee, R., *Int. Dairy J.*, 17, 268, 2007; Chatterjee, R., Characterising stickiness of dairy powders, MTech thesis, Institute of Technology and Engineering, Massey University, Palmerston North, New Zealand, 2004.

Table 11.23 Sticky Temperature (T_{st}) of WMP

Zuo et al. (2007)		Chatterjee (2004) (WMP 8051)	
RH (%)	T_{st} (°C)	RH (%)	T_{st} (°C)
19.20	80.3	48.18	47.6
26.47	71.2	52.33	52.7
36.08	60.6	62.54	42.8
43.00	50.8	65.08	36.6
45.46	50.8	73.37	28.3
50.63	40.6	94.92	31.7

Source: Zuo, J.Y., Paterson, A.H., Bronlund, J.E., and Chatterjee, R., *Int. Dairy J.*, 17, 268, 2007; Chatterjee, R., Characterising stickiness of dairy powders, MTech thesis, Institute of Technology and Engineering, Massey University, Palmerston North, New Zealand, 2004.

Table 11.24 Sticky Temperature (T_{st}) of White Cheese Powder and Cheese Powder

White Cheese Powder (3190)		Cheese Powder (3139)	
RH (%)	T_{st} (°C)	RH (%)	T_{st} (°C)
37.62	48.2	37.55	48.4
41.62	41.4	41.57	41.7
42.61	37.8	42.6	38.3
49.04	33.1	49.09	33.6
50.53	28.7	50.56	28.9

Source: Chatterjee, R., Characterising stickiness of dairy powders, MTech thesis, Institute of Technology and Engineering, Massey University, Palmerston North, New Zealand, 2004.

Table 11.25 Sticky Temperature (T_{st}) of Cheese Snack Powder and Cream Powder

Cheese Snack Powder		Cream Powder CP 70	
RH (%)	T_{st} (°C)	RH (%)	T_{st} (°C)
36.62	48.0	54.19	47.6
39.54	42.7	60.65	42.9
51.67	38.0	65.16	37.5
54.63	29.5	65.14	28.6
58.95	33.8	70.10	32.7

Source: Chatterjee, R., Characterising stickiness of dairy powders, MTech thesis, Institute of Technology and Engineering, Massey University, Palmerston North, New Zealand, 2004.

the exit of the gun. Unless the target plate is enclosed within a system of the same temperature and humidity as that of the particle-gun tip, the results will not be representative of the latter. It is unlikely that the particle reaching the wall of the spray dryers have velocities to the extent of 20 m/s. Furthermore, the adherence of the particles on the target plate will not only be the function of the cohesive and adhesive forces, the force of gravity and van der Waals (in fine particles) will also facilitate the deposition. There will be blockage of the gun tube occurs if the condition is above the stickiness temperature.

11.3.1.6 *In Situ Surface Stickiness Tests*

In spray-drying operations, a droplet moves from a liquid solution to a dry particulate state. The distribution of moisture within a droplet or particle at any time is not uniform and there exists a significant moisture gradient between its interior and the surface. The sticky behavior of the droplet or particle is determined by the surface properties (moisture, temperature). Hence, it is essential that the sticky behavior of droplet or particle be determined in situ as it dries. To this end, Adhikari et al. (2003) developed a tack-based in situ surface stickiness measuring system. This instrument has been used to determine the surface stickiness of various sugars, maltodextrin, and honey.

The instrument (Figure 11.10) principally consists of a stainless steel probe (probes can be of any material) attached to the linear actuator. The linear actuator drives the probe downward at a predetermined speed (30–80 mm/min) until it makes a good contact with the droplet. The droplet sits on the top of a droplet holder. The droplet with its holder assembly is attached to a precision load cell. When the probe makes good contact (contact time 1–2 s) with the droplet or particle, it is

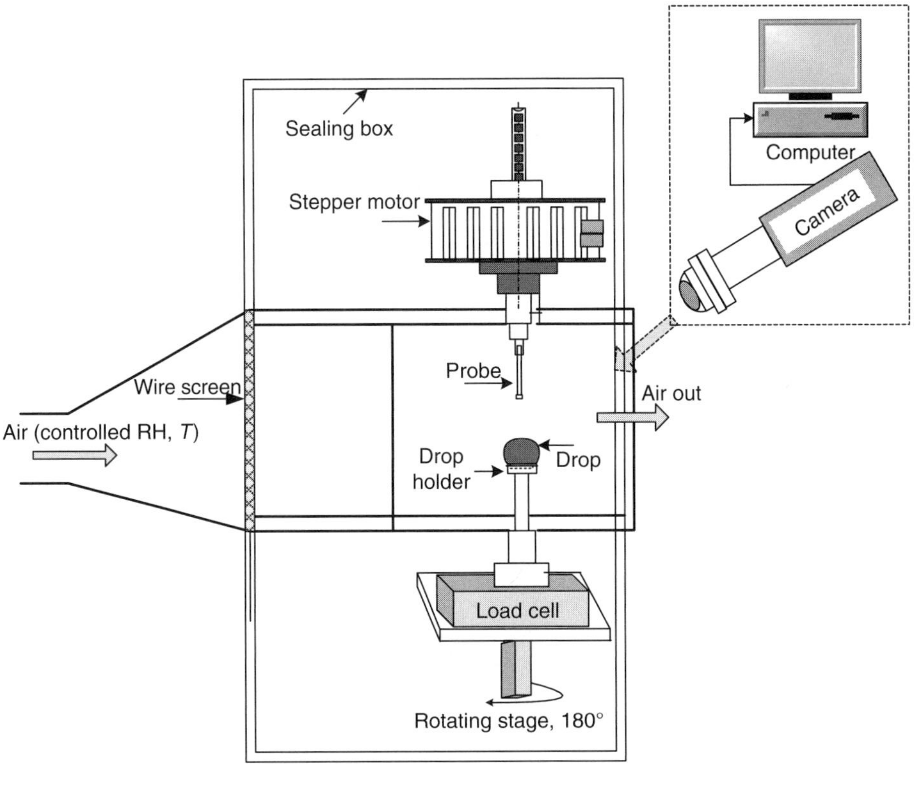

Figure 11.10 Schematic diagram of the in situ stickiness testing instrument.

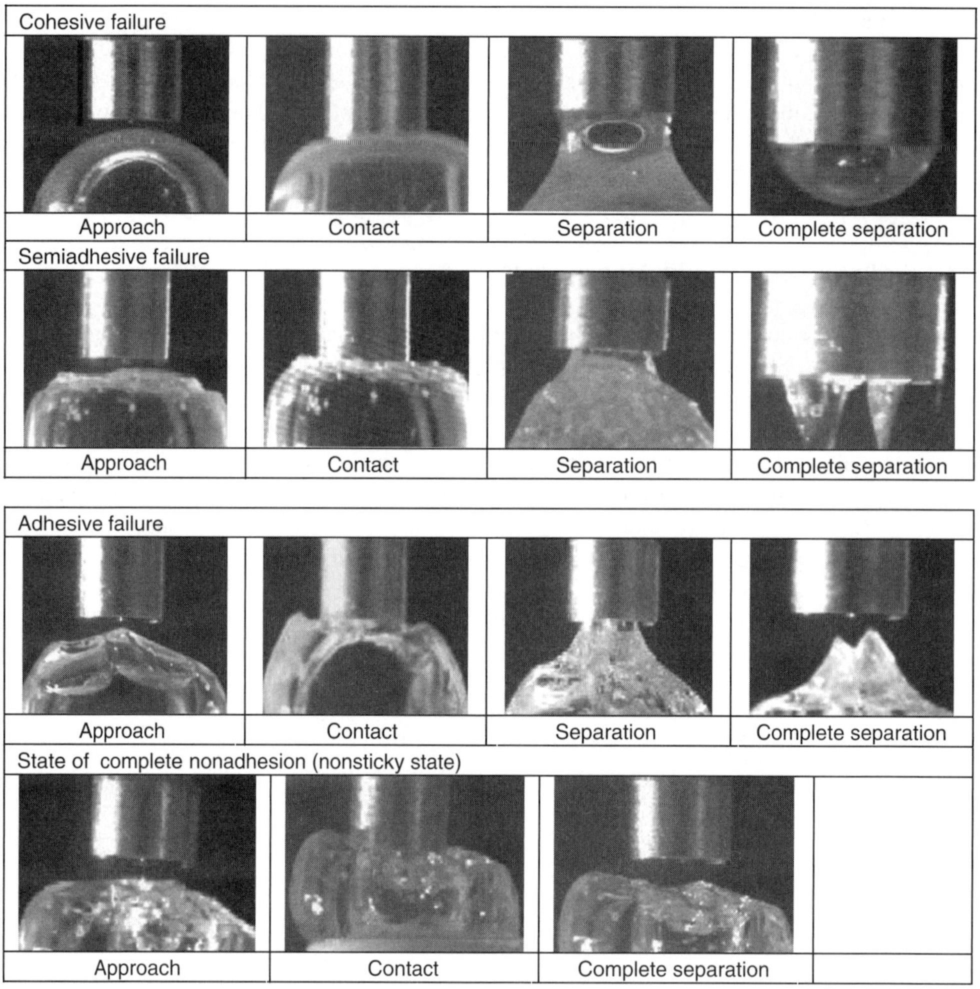

Figure 11.11 Mode of failures that determine the sticky or nonsticky state of droplet or particle surface.

withdrawn at the same speed. The force required for the probe to withdraw from the droplet or particle surface is recorded through a data acquisition system. It is essential to monitor the bonding and debonding process and the mode of failure while measurements are made. In this instrument, the images can be continuously acquired and recorded during the test (Figure 11.11).

11.3.2 Indirect Methods

The sticky temperature tests, described in Section 11.3.1, at first instance appear to be somewhat empirical in nature. However, fundamentally, they can be explained on the basis of softening of amorphous glassy powders above their T_g. At sticky point temperature liquid bridges are formed due to plasticization/softening and subsequent flow of the powder particles. This plasticization is catalyzed by the combined effect of moisture and heat (Peleg, 1993).

By determining the glass transition temperature of a system, it is possible to predict the temperature and moisture content at which the product becomes sticky. The T_g can be measured through the measurement of step changes in volume, enthalpy, and mechanical and dielectric properties. The theoretical aspects and measurement methods of glass transition have been discussed elsewhere in this handbook. T_g can be used as an indicator and subsequently correlated to determine and explain the stickiness in powders. The glass–rubber transition temperature, which specifically indicates the mechanical flow of glassy materials, is found to correlate well with stickiness and sticky temperature (Boonyai et al., 2007). The structure and working principle of this instrument are discussed in Section 11.3.2.1.

11.3.2.1 Thermomechanical Compression Test

A thermomechanical compression test (TMCT) was conceptualized by Bhandari and coworkers at the University of Queensland. It is a simple but novel method to determine the glass–rubber transition temperature ($T_{g\text{–}r}$) of food materials. This method is especially suited to investigate the glass transition behavior of high molecular weight proteins, carbohydrates, and their mixtures, which cannot be accurately measured by conventional DSC because the magnitude of change in specific heat capacity during glass transition is small.

This method involves the application of compression force at constant stress (creep test) on a thin layer (about 1 mm thickness) of powder sample. The powder sample is held in a thermally controlled sample cell. The sample cell is subsequently attached to a texture analyzer (Figure 11.12). First, the powder is compressed at a given force (say 49.05 N) for about 300 s to stabilize the sample thickness. Then, the temperature of the sample is raised at a certain rate (say 30°C/min). The change in the glassy to rubbery state is detected by the displacement of the compression probe, caused by the viscous flow of the material. The glass–rubber transition measured by this technique was found to be comparable with the T_g obtained from DSC. Typical $T_{g\text{–}r}$ values obtained from thermomechanical compression test are given in Table 11.26.

Advantages: This is a simple yet reliable technique for the determination of $T_{g\text{–}r}$ of amorphous powders. This method is specially suited to determine $T_{g\text{–}r}$ of protein and their mixtures as DSC is not effective in measuring the T_g of proteins and their mixtures. This instrument requires less capital and running costs. There is no need for specially trained personnel to run this test and the results obtained are quite repetitive. No specific sample preparation is required. One can use granules, large particles, or powders directly.

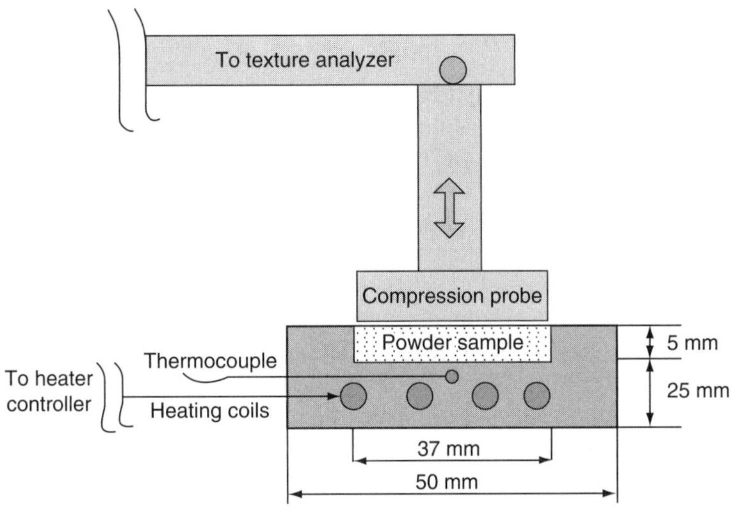

Figure 11.12 Schematic and pictorial diagram of the thermally controlled sample holder for a TMCT.

Table 11.26 **T_{g-r} Values for SMP, Honey:Maltodextrin, and Apple Juice:Maltodextrin Composite Powders**

Sample	Moisture (%db)	TMCT T_{g-r} (°C)	DSC $T_{g\text{-onset}}$ (°C)
SMP	2.01 ± 0.07	59.13 ± 1.64	68.76 ± 0.60
	3.54 ± 0.23	52.96 ± 1.34	50.91 ± 0.39
	4.83 ± 0.05	38.66 ± 0.48	39.24 ± 1.26
Honey:maltodextrin (50:50)	2.68 ± 0.36	58.5 ± 1.0	53.9 ± 1.0
Apple juice:maltodextrin(50:50)	3.38 ± 0.18	55.3 ± 1.0	50.3 ± 2.0
Whey powder	2.77 ± 0.12	69.4 ± 0.8	—
	5.84 ± 1.22	61.8 ± 0.1	—
	9.03 ± 1.41	37.7 ± 0.4	—

Source: Boonyai, P., Bhandari, B., and Howes, T., *Int. J. Food Prop.*, 9, 127, 2006; Boonyai, P., Howes, T., and Bhandari, B., *J. Food Eng.*, 78, 1333, 2007.

Disadvantages: The measurement is done at ambient condition. Special care should be taken while testing hygroscopic products. No other values such as enthalpy and specific heat can be obtained. This test is suitable for materials that are solid at room temperature because of its inability to cool down the sample.

11.4 CORRELATION AND PREDICTION OF STICKY AND COLLAPSE TEMPERATURE

All the predictive models of the sticky or collapse temperature are derived from the glass transition phenomenon. They have similarities in the fact that the T_{st} or T_c is material and moisture dependent. As the moisture and material dependence of T_g is reasonably well established, T_{st} and T_c then can be correlated by introducing an extra term (temperature offset) on the models for T_g. The various stickiness prediction models reported to date are described below.

11.4.1 Correlation Based on the Overall T_g

Various studies have shown that the stickiness and collapse (Bellows and King, 1973; Flink, 1983; Roos and Karel, 1991a; Roos and Karel, 1993) phenomena of amorphous food powders are the consequence of viscous flow of powder materials at their glass–rubber transition. These phenomena are enhanced greatly by the plasticizing effect of water, temperature, and pressure. During this transition, the viscosity of the powder surface rapidly falls below 10^8 Pa s (Downton et al., 1982) at which particles become sticky. Further plasticization leads to complete fusion of particles, one typical example of which is the collapse of the physical structure of freeze-dried powders.

Attempts have been made to develop a mathematical relationship between T_g and T_{st} for food powders. Roos and Karel (1991b) related the T_g of amorphous sucrose:fructose (7:1) mixture to the T_{st} measured by Downton et al. (1982). It was found that the T_{st} was close to the $T_{g,endset}$. Although there is no generally accepted mathematical relationship that relates the T_g with T_{st} or T_c, most of the researchers opine that T_{st} and T_c are about 20°C above the overall T_g (Roos and Karel, 1991a). Similarly, the collapse temperature of maltodextrin and their mixtures with various fruit juice powders was found to be 20°C–50°C higher than T_g (Tsourouflis et al., 1976). Since stickiness is a time-dependent phenomenon, the 20°C–50°C offset temperature reflects the time effect.

Table 11.27 Offset Temperature (x) = $T - T_g$ for Some Dairy Powders

Powder	% Fat	% Protein	% Lactose	$x = T - T_g$ (°C)
SMP A (Paterson et al., 2007)	0.62	34.27	57.84	39.1 ± 2.55
SMP B (Paterson et al., 2007)	0.79	38.19	52.98	37.9
Amorphous lactose (Paterson et al., 2007)	0	0	100	24.7
WMP A (Paterson et al., 2007)	31.09	25.91	37.82	37.5
WMP B (Paterson et al., 2007)	27.38	27.07	38.33	34.0
Milk protein concentrate	1.35	59.36	31.29	47.7
Sucrose (Foster et al., 2006)	—	—	—	23.0
Maltose (Foster et al., 2006)	—	—	—	19.0, 25.3, 29.0
Glucose:lactose (20:80) (Foster et al., 2006)	—	—	—	35.0, 37.4
Galactose:lactose (20:80) (Foster et al., 2006)	—	—	—	30.8
Fructose:lactose (10:90) (Foster et al., 2006)	—	—	—	41.3

Source: Paterson, A.H., Zuo, J.Y., Brolund, J.E., and Chatterjee, R., *Int. Dairy J.*, 17, 998, 2007; Foster, K.D., Bronlund, J.E., and Paterson, A.H.J., *J. Food Eng.*, 77, 997, 2006.

Hennigs et al. (2001) has generalized the relationship between T_{st} and T_g as shown in Equation 11.3. The sticky temperature of solid–water mixtures can be predicted using the Gordon–Taylor equation (Gordon and Taylor, 1952) with an extra offset term in it.

$$T_{st} = \frac{T_{g,s} + KmT_{g,w}}{1 + Km} + x \quad (11.1)$$

where

$T_{g,s}$ and $T_{g,w}$ are the glass transition temperatures of the solid and water, respectively (°C)
m is the moisture content on dry basis
K is the constant in Gordon–Taylor equation
x is the material-dependent offset temperature (°C)

Hennigs et al. found that the sticky point temperature of SMP could be attributed to lactose and that when $K = 7.40$ was used, the offset temperature of 23.3°C provided a good fit to experimental sticky temperature. The offset temperature data for some common diary powders provided by Paterson et al. (2007) are presented in Table 11.27).

11.4.2 Bhandari et al.'s Drying Index Method

In order to provide a simple and practical method to overcome the stickiness and achieve attainable powder recovery Bhandari et al. (1997b) developed a semiempirical approach. This approach is known as the drying index method of drying sugar and acid-rich foods. This involves finding an individual index (I_i) for each constituent of the feed matrix. The individual index of each constituent, such as fructose and glucose, depends on the ease at which it can be converted into powder form. The individual indices have to be determined empirically through experiments. The marginally successful spray-drying operation is considered to the one in which 50% of feed solids were recovered after spray drying. The drying indices of each component in the feed are obtained as follows. First, a binary mixture of each constituent with the drying additive (say maltodextrin) is prepared. Second, a mixing proportion (on dry basis) is found such that the powder recovery is close to 50%. Since the overall drying index of such a mixture is taken to be unity, the individual index is calculated using

$$\sum_{i=1}^{2} I_i x_i = 1 \tag{11.2}$$

I_i and x_i are the individual drying index and solid fraction of the binary mixture. Since the solid ratio x_i and the drying index I_i of the drying aid (say maltodextrin DE 6) are known, I_i of the remaining constituent in this binary mixture can be conveniently calculated from its fraction in the mixture. Once the individual drying indices in a multicomponent feed are known, the overall drying index (Y_I) of the feed can be calculated using

$$\sum_{i=1}^{n} I_i x_i = Y_I \tag{11.5}$$

For a marginally successful drying, the overall drying index (Y_I), should be close to unity. $Y_I < 1$ indicates unsuccessful spray drying with recovery less than 50%, whereas $Y_I > 1$ indicates successful drying with recovery greater than 50%. The higher the Y_I, the better will be the recovery of powders and vice versa. Hence, the overall drying index provides a means of feed formulation to overcome the stickiness and achieve good powder recoveries.

Table 11.28 presents the individual drying indices of some key constitutes of sugars and acid-rich foods such as fruit juices and honey. This table also provides the corresponding T_g of these constitutes at their anhydrous-glassy state. Note that these indices are somewhat arbitrary; nevertheless, they indicate the degree of ease or difficulty at which they can be converted into glassy powders. Furthermore, the table indicates that the drying indices correlate positively with the glass transition temperature. The glass transition temperature of constituents with a higher drying index is higher and vice versa.

Bhandari et al. (1997a) tested this method for honey and pineapple juice. Table 11.29 provides the overall drying indices for both honey and pineapple juice based on their composition. These overall indices indicate the extent of stickiness of these materials and also the amount of additive carriers needed for a successful spray drying. Table 11.30 provides the experimental powder recoveries in a pilot-scale spray drying trial. From this table, it can be seen that this approach is capable of not only predicting the powder recovery but also the amount of drying additive needed for a successful spray drying.

Table 11.28 Drying Indices and T_g Values of Some Sugars, Organic Acids, and Maltodextrin

Component	Drying Index	Glass Transition Temperature (T_g)
Maltodextrin DE 6	1.6	205
Lactose	1.09	101
Maltose	1.0	93
Sucrose	0.85	68.6 ± 1.0
Glucose	0.51	40.3 ± 0.7
Fructose	0.27	16.2 ± 0.8
Citric acid	−0.40	13.6 ± 0.3
Tartaric acid	—	18.5 ± 0.5
Malic acid	—	-16.1 ± 0.6

Source: Bhandari, B.R., Datta, N., Crooks, R., Howes, T., and Rigby, S., *Drying Technol.*, 15, 2509, 1997b; Truong, V., Modelling of glass transition temperature of sugar-rich foods and its relation to spray drying of such products, PhD thesis, School of Land and Food Sciences, The University of Queensland, Australia, 2003.

Table 11.29 Overall Drying Index (Y_I) for Honey and Pineapple Juice

		Honey		Pineapple Juice	
Component	Drying Index (I_i)	Solid Fraction (x_i)	$I_i x_i$	Solid Fraction (x_i)	$I_i x_i$
Fructose	0.27	0.553	0.149	0.210	0.057
Glucose	0.51	0.414	0.211	0.320	0.163
Maltose	1.00	0.034	0.034	—	—
Sucrose	0.85	0.002	0.002	0.440	0.374
Citric acid	−0.40	—	—	0.035	−0.014
Overall drying index ($Y_I = \sum I_i x_i$)		0.396	—	0.580	

Source: Bhandari, B.R., Datta, N., Crooks, R., Howes, T., and Rigby, S., *Drying Technol.*, 15, 2509, 1997b.

11.4.3 Correlations Based on the T_g of the Surface Layer

In a series of studies, Adhikari and coworkers measured the surface stickiness of droplets of sugars, maltodextrin, and their mixtures, in situ, following the drying route (Adhikari et al., 2003, 2004). They also measured the temperature and moisture histories of the droplets. It was found that the surface of the droplet or particle can become completely nonsticky much earlier than would be predicted by overall T_g values. This is because a considerable moisture gradient exists between the interior and the surface layer of droplets or particles. The formation of a glassy shell or skin takes place at the exterior of the particles even if the interior is quite viscous. This information is particularly relevant to the skin forming materials, such as maltodextrin and proteins, which become nonsticky soon after the glassy skin surrounds the viscous interior. They found that the T_g of the surface layer of the droplet or particle better correlates the stickiness scenario of their surface. They laid out the criteria for droplets or particles to enter the safe drying regime using surface layer T_g as an indicator Adhikari et al., 2005.

The drop surface remains sticky if its surface layer T_g is lower than the drop temperature (T_d). The drop surface exhibits peak tendency to stick when its surface layer T_g reaches or just crosses the drop temperature (T_d). The drop surface becomes completely nonsticky when the surface layer T_g is higher than the drop temperature (T_d) by 10°C. Hence, a safe drying regime can be defined as the regime where T_g of the surface layer is $\geq T_d + 10$°C.

They also introduced a dimensionless time ψ, defined as the ratio of the time required to enter the safe drying regime (t_{NS}) to the time needed to achieve the (desired) final average moisture content (t_{total}). ψ is an important indicator of the degree of difficulty with which a drop can enter a safe drying regime. If ψ is >1, the drop does not enter the safe drying regime. If the $\psi = 1$, it is a limit or a situation at which the drop enters the safe drying regime toward the completion of drying, that is, it is a cutoff point. If $\psi < 1$, the drop enters the safe drying regime. The smaller the ψ ratio is, the earlier the drop enters the safe drying regime. The dimensionless time (ψ) was further correlated

Table 11.30 Recoveries of Honey:Maltodextrin and Pineapple Juice:Maltodextrin Composite Powders from a Pilot-Scale Spray-Drying Trial

Honey: Maltodextrin (Solid Ratio)	Overall Index (Y_I)	Recovery (%)	Pineapple: Maltodextrin (Solid Ratio)	Overall Index (Y_I)	Recovery (%)
47:53	1.03	56.5	50:50	1.09	58.5
50:50	1.00	55.0	59:41	1.00	50.0
53:47	0.96	48.0	60:40	0.99	53.0
55:45	0.94	20.3	75:25	0.84	45.0

Source: Bhandari, B.R., Datta, N., Crooks, R., Howes, T., and Rigby, S., *Drying Technol.*, 15, 2509, 1997b.

Table 11.31 Surface Stickiness of 120 μm Initial Diameter Droplets of Fructose, Glucose, Sucrose, Citric Acid, and Maltodextrin DE 6 Dried at 65°C and 95°C, 1 m/s Slip Velocity, 2.5% ± 0.5% RH

Material	*u* Range	$(\psi) = t_{NS}/t_{total}$ 65°C	$(\psi) = t_{NS}/t_{total}$ 95°C	Remarks
Fructose	1.5–0.05	inf	inf	Sticky
Glucose	1.5–0.05	inf	inf	Sticky
Sucrose	1.5–0.05	inf	inf	Sticky
Citric acid	1.5–0.05	inf	inf	Sticky
Maltodextrin DE 6	1.5–0.05	0.007	0.05	Nonsticky after 1% of total time at 63°C, and 5% of total time at 95°C

Source: Adhikari, P., Drying kinetics and stickiness studies of single drop of sugar and acid-rich solutions, PhD thesis, School of Engineering, The University of Queensland, Australia, 2003.
t_{NS} = time to reach the nonsticky state, t_{total} = time required to reach $u = 0.05$, inf = large value.

with the degree of ease or difficulty in practical spray drying. The percent (%) powder recovery was used as a measure of spray-drying performance. Following Bhandari and coworkers' protocol (Bhandari et al., 1997a), 50% recovery was taken as a marginally successful spray drying. From the definitions of recovery and ψ, these two quantities are interrelated as follows: if $\psi = 1$, it corresponds to the 50% recovery and indicates a marginally successful spray drying; if $\psi > 1$, the recovery will be <50%. The greater the ψ value is, the lower will be the recovery. If $\psi < 1$, the recovery will be > 50%. The lower the ψ value is, the higher will be the recovery. They referred <50% recovery as unsuccessful spray drying, >50% recovery as successful spray drying, and recovery equaling 50% to be a limiting or marginally successful spray drying. The sticky-nonsticky scenario, the dimensionless time, and the surface layer T_g of sugars, acid, maltodextrin and their mixtures are presented in Tables 11.31 through 11.34.

Table 11.32 Surface Stickiness of Droplets of 120 μm Initial Diameter at 95°C, 1 m/s Slip Velocity, 2.5% ± 0.5% RH

Material	*u* Range	65°C $(\psi) = t_{NS}/t_{total}$	65°C Remarks	95°C $(\psi) = t_{NS}/t_{total}$	95°C Remarks
F:M (4:1)	1.5–0.05	inf	Sticky	inf	Sticky
G:M (4:1)	1.5–0.05	inf	Sticky	inf	Sticky
S:M (4:1)	1.5–0.05	>1	USSD	inf	Sticky
C:M (4:1)	1.5–0.05	inf	Sticky	inf	Sticky
F:M (1:1)	1.5–0.05	0.26	SSD	inf	Sticky
G:M (1:1)	1.5–0.05	0.086	VSSD	>1	USSD
S:M (3:2)	1.5–0.04	0.108	SSD	>1	USSD
S:M (1:1)	1.5–0.05	0.066	VSSD	≈1	MSSD
C:M (1:1)	1.5–0.05	>1	USSD	inf	Sticky
F:M (1:4)	1.5–0.05	0.018	VSSD	0.16	SSD
G:M (1:4)	1.5–0.05	0.016	VSSD	0.14	SSD
S:M (1:4)	1.5–0.05	0.013	VSSD	0.13	SSD
C:M (1:4)	1.5–0.05	0.022	VSSD	0.2	SSD
Maltodextrin	1.5–0.05	0.007	VSSD	0.05	VSSD

Source: Adhikari, P., Drying kinetics and stickiness studies of single drop of sugar and acid-rich solutions, PhD thesis, School of Engineering, The University of Queensland, Australia, 2003.

Note: Droplets contain fructose (F), glucose (G), sucrose (S), maltodextrin (M), and their mixtures. t_{NS} = time to enter the nonsticky regime (s), t_{total} = time required to reach $u = 0.05$ (s), inf = large value. USSD, unsuccessful spray drying; SSD, successful spray drying; VSSD, very successful spray drying; MSSD, marginally successful spray drying.

Table 11.33 Comparison of Model Predictions [Drops of S:M (3:2), 40% Initial Solids, 120 μm Drop Diameter at 2.5% ± 0.5% RH, 1 m/s Slip Velocity] with Experimental Results Obtained from a Pilot-Scale Spray Dryer [Sucrose:Maltodextrin (3:2), 50% Solids in Feed and 135°C Inlet Air Temperature]

Model Predictions Dryer Outlet Temperature (°C)	Moisture (u) Range	Final T_d (°C)	Final Surface T_g (°C)	$(\psi) = t_{NS}/t_{total}$	Remarks
85	1.5–0.028	84.9	92.8	>1	USSD
84	1.5–0.028	83.9	93.5	150/150 = 1	MSSD
76	1.5–0.034	75.9	92.4	23.2/150 = 0.145	SSD

Source: Adhikari, P., Drying kinetics and stickiness studies of single drop of sugar and acid-rich solutions, PhD thesis, School of Engineering, The University of Queensland, Australia, 2003.
T_d, drop temperature.

11.5 CONCLUDING REMARKS

Stickiness and collapse are important phenomena in air- and freeze-drying processes. Stickiness embodies cohesive and adhesive components with a characteristic complex interplay. This makes it difficult to devise a single instrument and a single interpretive model. Hence, a judicious selection of both the instrument and interpretive model has to be made when the stickiness and collapse in food are to be studied.

So far, the instruments and the interpretations of the stickiness and collapse properties have been based on bulk properties and macrolevel observations. There is a need to develop instruments that can characterize the stickiness at the molecular level. A greater understanding regarding the magnitude and interplay between the cohesive and adhesive stickiness will constitute a major future direction. It is now becoming clearer that the surface composition of food powders (produced by spray drying) is quite different from their bulk properties, particularly when the product is composed of surface active agents such as protein. The implication of this, i.e., solid segregation effect on stickiness, has to be studied. More precise surface characterization techniques such as atomic force microscopy (AFM), x-ray photoelectron spectroscope (XPS), and interfacial rheometer should be employed to study the stickiness and collapse phenomena at the molecular level.

Table 11.34 Surface Stickiness of 120 μm Diameter Drops during Spray Drying at 95°C, 63°C, Slip Velocity 1 m/s, 2.5% ± 0.5% RH

		65°C		95°C	
Material	u Range	$(\psi) = t_{NS}/t_{total}$	Remarks	$(\psi) = t_{NS}/t_{total}$	Remarks
(C + F + G + S):M (4:1)	1.5–0.05	inf	Sticky	inf	Sticky
(C + F + G + S):M (3:2)	1.5–0.05	1.00	MSSD	—	—
(C + F + G + S):M (1:1)	1.5–0.05	0.14	SSD	inf	Sticky
(C + F + G + S):M (1:4)	1.5–0.05	0.017	VSSD	0.16	SSD
(C + F + G + S):M (2:3)	1.5–0.05	—	—	1.00	MSSD
Pineapple juice: maltodextrin (3:2)[a]	—	—	53% powder recovery		

Source: Adhikari, P., Drying kinetics and stickiness studies of single drop of sugar and acid-rich solutions, PhD thesis, School of Engineering, The University of Queensland, Australia, 2003.
Note: Drops contain 40% w/w initial solids of citric acid (C) + fructose (F) + glucose (G) + sucrose (S) and maltodextrin (M) in ratios of (4:1), (1:1), and (1:4). t_{NS} = time to enter the nonsticky regime (s), t_{total} = time required to reach $u = 0.05$ (s), inf = large value.
[a] Dryer outlet temperature 65°C ± 2°C (Bhandari et al., 1997b).

REFERENCES

Adhikari, 2003. Drying kinetics and stickiness studies of single drop of sugar and acid-rich solutions. PhD Thesis, School of Engineering, The University of Queensland, Australia.

Adhikari, B., Howes, T., Bhandari, B.R., and Truong, V. 2003. In situ characterization of stickiness of sugar-rich foods using a linear actuator driven stickiness testing device. *Journal of Food Engineering*, 58(1): 11–22.

Adhikari, B., Howes, T., Bhandari, B.R., and Truong, V. 2004. Effect of addition of maltodextrin on drying kinetics and stickiness of sugar and acid-rich foods during convective drying: experiments and modelling. *Journal of Food Engineering*, 62(1): 53–68.

Adhikari, B., Howes, T., Lecomte, D., and Bhandari, B.R. 2005. A glass transition temperature approach for the prediction of the surface stickiness of a drying droplet during spray drying. *Powder Technology*, 149 (2–3): 168–179.

Bellows, R.J. and King, C.J. 1972. Freeze-drying of aqueous solutions: maximum allowable operating temperature. *Cryobiology*, 9: 559–561.

Bellows, R.J. and King, C.J. 1973. Product collapse during freeze drying of liquid foods. *AICHE Symposium Series*, 69: 33–41.

Bhandari, B.R., Datta, N., and Howes, T. 1997a. Problems associated with spray drying of sugar-rich foods. *Drying Technology*, 15(2): 671–684.

Bhandari, B.R., Datta, N., Crooks, R., Howes, T., and Rigby, S. 1997b. A semi-empirical approach to optimize the quantity of drying aids required to spray dry sugar-rich foods. *Drying Technology*, 15(10): 2509–2525.

Bloore, C. 2000. Developments in food drying technology overview. In: *International Food Dehydration Conference—2000 and Beyond*, Carlton Crest Hotel, Melbourne, Australia.

Boonyai, P. 2005. Development of new instrumental techniques for measurement of stickiness of solid particulate food materials. PhD Thesis, School of Land and Food Sciences, The University of Queensland, Australia.

Boonyai, P., Bhandari, B., and Howes, T. 2006. Applications of thermal mechanical compression tests in food powder analysis. *International Journal of Food Properties*, 9(1): 127–134.

Boonyai, P., Howes, T., and Bhandari, B. 2007. Instrumentation and testing of a thermal mechanical compression test for glass-rubber transition analysis of food powders. *Journal of Food Engineering*, 78(4): 1333–1342.

Brennan, J.G., Herrera, J., and Jowitt, R. 1971. A study of some of the factors affecting the spray drying of concentrated orange juice, on a laboratory scale. *Food Technology*, 6: 295–307.

Chang, B.S. and Randall, C.S. 1992. Use of subambient thermal analysis to optimize protein lyophilization. *Cryobiology*, 29(5): 632–656.

Chatterjee, R. 2004. Characterising stickiness of dairy powders. MTech Thesis, Institute of Technology and Engineering, Massey University, Palmerston North, New Zealand.

Chuy, L. and Labuza, T.P. 1994. Caking and stickiness of dairy based food powders as related to glass transition. *Journal of Food Science*, 59(1): 43–46.

Dixon, A. 1999. Correlating food powders stickiness with composition, temperature and relative humidity. Research project, Department of Chemical Engineering, Monash University, Clayton, Melbourne, Australia.

Downton, G.E., Floresluna, J.L., and King, C.J. 1982. Mechanism of stickiness in hygroscopic, amorphous powders. *Industrial & Engineering Chemistry Fundamentals*, 21(4): 447–451.

Flink, J.M. 1983. Structure and structure transition in dried carbohydrate materials. In: *Physical Properties of Foods AVI*, Peleg, M. and Bagley, E.B. (eds.), AVI Pub. Co., Westport, CT, pp. 473–521.

Fonseca, F., Passot, S., Cunin, O., and Marin, M. 2004. Collapse temperature of freeze-dried *Lactobacillus bulgaricus* suspensions and protective media. *Biotechnology Progress*, 20(7): 229–238.

Foster, K.D., Bronlund, J.E., and Paterson, A.H.J. 2006. Glass transition related cohesion of amorphous sugar powders. *Journal of Food Engineering*, 77(4): 997–1006.

Gordon, M. and Taylor, J.S. 1952. Ideal copolymers and the second-order transitions of synthetic rubbers. I. Noncrystalline copolymers. *Journal of Applied Chemistry*, 2(9): 493–500.

Hennigs, C., Kockel, T.K., and Langrish, T.A.G. 2001. New measurements of the sticky behavior of skim milk powder. *Drying Technology*, 19(3&4): 471–484.

Her, L.M. and Neil, S.L. 1994. Measurement of glass transition temperatures of freeze-concentrated solutes by differential scanning calorimetry. *Pharmaceutical Research*, 11(1): 54–59.

Jaya, S. and Das, H. 2004. Effect of maltodextrin, glycerol monostearate and tricalcium phosphate on vacuum dried mango powder properties. *Journal of Food Engineering*, 63(2): 125–134.

Lazar, M.E., Brown, A.H., Smith, G.S., Wong, F.F., and Lindquist, F.E. 1956. Experimental production of tomato powder by spray drying. *Food Technology*, 10: 129–134.

Levi, G. and Karel, M. 1995. Volumetric shrinkage (collapse) in freeze-dried carbohydrates above their glass transition temperature. *Food Research International*, 28(2): 145–151.

Levine, H. and Slade, L. 1988. Principles of cryostabilization technology from structure property relationships of carbohydrate water-systems—a review. *Cryo-Letters*, 9(1): 21–63.

Llyod, R.J., Chen X.D., and Hargreaves, J.B. 1996. Glass transition and caking of spray-dried lactose. *International Journal of Food Science and Technology*, 31: 305–311.

Lockemann, C.A., 1999. A new laboratory method to characterize the sticking property free flowing solids. *Chemical Engineering and Processing*, 38: 301–306.

McKenzie, A.P. 1975. Collapse during freeze-drying: Qualitative and quantitative aspects. In: *Freeze-Drying and Advanced Food Technology*, Goldblith, S.A., Rey, L., and Rothmayr, W.W. (eds.), Academic Press, New York, pp. 277–307.

Ozkan, N., Walisinghe, N., and Chen, X.D. 2002. Characterization of stickiness and cake formation in whole and skim milk powders. *Journal of Food Engineering*, 55(4): 293–303.

Pasley, H., Haloulos, P., and Ledig, S. 1995. Stickiness—a comparison of test methods and characterization parameters. *Drying Technology*, 13(5–7): 1587–1601.

Paterson, A.H.J., Brooks, G.F., Bronlund, J.E., and Foster, K.D. 2005. Development of stickiness in amorphous lactose at constant $T - T_g$ levels. *International Dairy Journal*, 15: 513–519.

Paterson, A.H., Zuo, J.Y., Brolund, J.E., and Chatterjee, R. 2007. Stickiness curves of high fat dairy powders using the particle gun. *International Dairy Journal*, 17(8): 998–1005.

Peleg, M. 1993. Glass transitions and the physical stability of food powders. In: *The Glassy States in Foods*, Blanshard, J.M.V. and Lillford, P.J. (eds.), Nottingham University Press, Nottingham, pp.435–451.

Roos, Y.H. and Karel, M. 1991a. Phase-transitions of mixtures of amorphous polysaccharides and sugars. *Biotechnology Progress*, 7(1): 49–53.

Roos, Y. and Karel, M. 1991b. Applying state diagrams to food processing and development. *Food Technology*, 45(12): 66, 68–71, 107.

Roos, Y.H. and Karel, M. 1993. Effects of glass transitions on dynamic phenomena in sugar containing food systems. In: *The Glassy State in Foods*, Blanshard, J.M.V. and Lillford, P.J. (eds.), Nottingham University Press, Nottingham, 1993, pp. 207–222.

Slade, L. and Levine, H. 1994. Water and the glass-transition—dependence of the glass-transition on composition and chemical-structure—special implications for flour functionality in cookie baking. *Journal of Food Engineering*, 22(1–4): 143–188.

To, E.C. and Flink, J.M. 1978. Collapse, a structural transition in freeze dried carbohydrates. I. Evaluation of analytical methods. *Journal of Food Technology*, 13: 551–565.

Truong, V. 2003. Modeling of glass transition temperature of sugar-rich foods and its relation to spray drying of such products. PhD Thesis, School of Land and Food Sciences, The University of Queensland, Australia.

Tsourouflis, S., Flink, J.M., and Karel, M. 1976. Loss of structure in freeze-dried carbohydrates solutions: effect of temperature, moisture content and composition. *Journal of Science of Food and Agriculture*, 27: 509–519.

Wallack, D.A. and King, C.J. 1988. Sticking and agglomeration of hygroscopic, amorphous carbohydrate and food powders. *Biotechnology Progress*, 4(1): 31–35.

Zuo, J.Y., Paterson, A.H., Bronlund, J.E., and Chatterjee, R. 2007. Using a particle-gun to measure initiation of stickiness of dairy powders. *International Dairy Journal*, 17(3): 268–273.

CHAPTER 12

State Diagrams of Foods

Didem Z. Icoz and Jozef L. Kokini

CONTENTS

12.1 STATE DIAGRAMS

In its simplest definition, state diagrams describe physical/chemical states of a material as a function of different conditions. The physical or chemical state of food materials and the thermal transitions they undergo are critical for determination of processing conditions and overall quality of the final food products, including stability, texture, and functionality. For example, the physical states of ingredients in a mixture determine the transport properties such as viscosity, density, mass and thermal diffusivity, together with reactivity of the material. Temperature, moisture content, pressure, concentration, shear, time, etc., are some of the factors that can alter the physical state of materials (Cocero and Kokini, 1991; Roos, 1995; Madeka and Kokini, 1996). Development of state diagrams of food materials enables understanding the relationship between the composition of a food product and its physical state. This knowledge in turn can be used to predict processing and storage requirements. State diagrams are critical to form a predictive basis for developing novel food products and for improving the quality of existing products.

Carbohydrates and proteins are two major components present in various food products, which can exist in an amorphous metastable state that is sensitive to moisture and temperature changes (Cocero and Kokini, 1991; Madeka and Kokini, 1994, 1996). Practically, state diagrams of carbohydrates and proteins can be used as a predictive tool to determine their behavior during baking, extrusion, sheeting, and storage (Levine and Slade, 1990; Kokini et al., 1994a). Glass transition temperature as a function of moisture content or water activity forms the simplest state diagrams for food systems and gives useful information for determination of food stability and quality.

12.2 CHARACTERIZATION OF STATE TRANSITIONS

State/phase transitions in food materials can be characterized using differential scanning calorimetry (DSC), rheological analysis, thermal expansion measurements, dielectric constant measurement, and dilatometry (Kokini et al., 1994b). As a material passes from a glassy state into a rubbery state, a discontinuity in heat capacity, thermal expansion, and dielectric constant occurs, which are identified as step changes during the experimental techniques that determine the glass transition temperature. Among the various methods of analysis, DSC and small-amplitude oscillatory rheometry measurements are the most common techniques to characterize state transitions.

12.2.1 Measurement of Glass Transition Temperature (T_g) by DSC

DSC is a common method to measure glass transitions as well as endothermic transitions (e.g., denaturation of proteins) and exothermic transitions (e.g., crystallization). DSC measures the enthalpy (H) change with respect to temperature (T), which gives the heat capacity (C_P) at constant pressure $[(\partial H/\partial T)_P = C_P]$. At T_g, an amorphous polymer goes through a step change in the magnitude of C_P with temperature. As a result, T_g is identified as the midpoint of step change in heat flow on DSC thermograms (Figure 12.1).

12.2.2 Measurement of T_g by Rheological Analysis

In a typical small oscillatory measurement, storage modulus (G') (representing the solid-like behavior) and loss modulus (G'') (representing the liquid-like behavior) are measured as a function

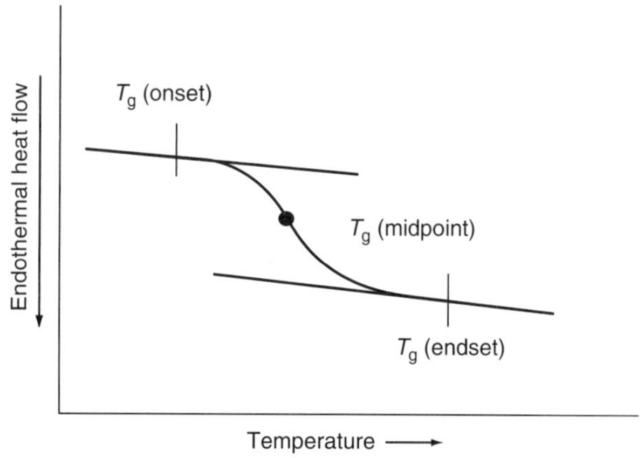

Figure 12.1 Determination of glass transition temperature from a DSC thermogram.

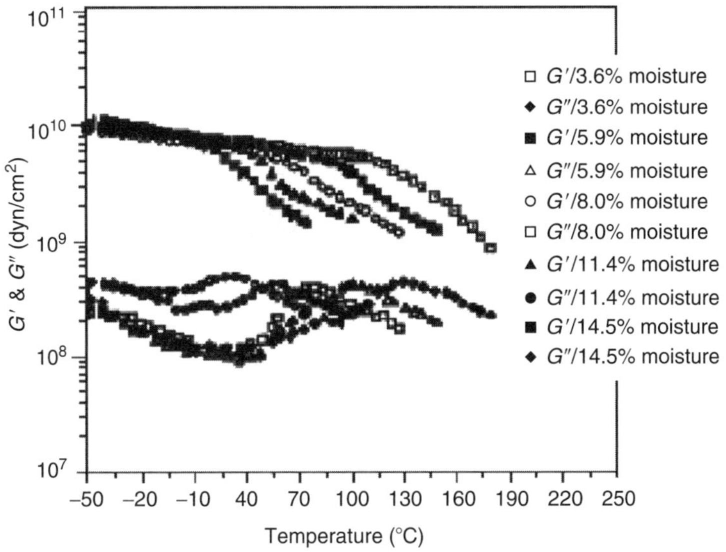

Figure 12.2 Storage modulus and loss modulus of glutenin at low moisture content as a function of temperature. (Reproduced from Kokini, J.L., Cocero, A.M., Madeka, H., and De Graaf, E., *Trends Food Sci. Tech.*, 5, 281, 1994b. With permission.)

of temperature. T_g is identified either as the onset of the drop in G' or as the peak in G'' (Figure 12.2). When the material is at the rubbery plateau region, there is little dependence of G' on the frequency at which the material is oscillated, whereas G'' shows a maximum at a specific temperature, which is identified as the T_g. The tan δ peak [tan $\delta = G''/G'$] can also be used in identification of T_g. However, in complex systems, the tan δ peak may be very broad and may not show a single maximum. Most commonly, the G'' peak is used as the marker for T_g (Cocero and Kokini, 1991; Kalichevsky and Blanshard, 1993; Kokini et al., 1994b).

12.2.3 Measurement of T_g by Dilatometry

In dilatometry, the volume change of a polymer confined in a liquid (usually mercury, since it does not have transition of its own in the temperature ranges suitable for polymers and it does not swell organic polymers) is measured as a function of temperature. T_g is determined by extrapolating the straight lines above and below the change in slope of the volume–temperature curve (Figure 12.3). Dilatometric studies also give information on the free volume that can be used in understanding the glass transition phenomena through its theoretical free volume theory (Sperling, 2001).

12.2.4 Measurement of T_g by Dielectric Thermal Analysis

Glass transition of polymers can also be characterized by measuring the changes in the dielectric constant (ε'), dielectric loss constant (ε''), and tan δ using a dielectric analyzer, where the sample is placed between parallel plate capacitors in an alternating electric field (Sperling, 2001). The response of shapes of ε', ε'', and tan δ, and the analysis as a function of temperature are similar to that in rheological analysis. Noel et al. (1992, 1996) have used dielectric methods to characterize the dielectric relaxation behavior in small carbohydrate molecules.

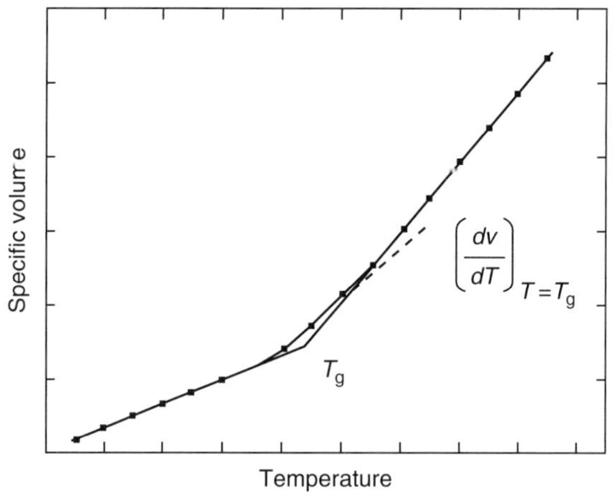

Figure 12.3 Specific volume–temperature plots showing the T_g of branched poly(vinyl acetate). (Reproduced from Sperling, L.H., *Introduction to Physical Polymer Science*, John Wiley & Sons, Inc., New York, 2001. With permission.)

12.2.5 Characterization of Chemical Reactions by Pressure Rheometry

In a polymer system, chemical reactions result from either formation of higher molecular weight products or from degradation of molecular structures. Formation of higher molecular weight products results in increase in the storage modulus (G') and loss modulus (G''). For example, Figure 12.4 shows that G' increases approximately 100 times during heating throughout the reaction zone of gliadin due to the cross-link formation that joins gliadin molecules, which results in the formation of a three-dimensional network. As the network is formed, the increase in G' is sharper

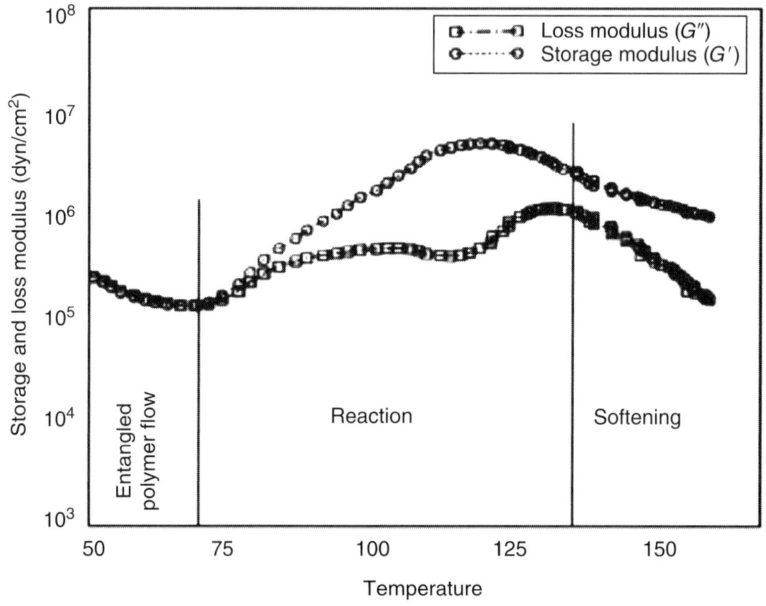

Figure 12.4 Temperature sweep of gliadin with 25% moisture. (Reproduced from Madeka, H. and Kokini, J.L., *J. Food Eng.*, 22, 241, 1994. With permission.)

than the increase in G'' (G' and G'' represent solid- and liquid-like properties, respectively). As the reaction is completed, a temperature-induced softening is observed which is identified by the decrease in G' and G'' (Figure 12.4) (Kokini et al., 1994b; Madeka and Kokini, 1994).

12.3 STATE DIAGRAMS OF PROTEINS

Figure 12.5a and b shows the physical/chemical states of soy globulins (7S and 11S soy globulins, respectively) as a function of moisture content and temperature. Soy globulins are in the glassy state at low temperatures and low moisture contents. The rubbery region occurs above the glass transition. When Figure 12.5a and b is compared, glass transition of 11S globulin occurs at higher temperatures than that of 7S globulin. The horizontal line below the rubbery state in Figure 12.5a and b indicates the glass transition temperature of maximally freeze-concentrated globulins (T_g'). At higher temperatures, entangled rubbery polymer softens and forms the entangled polymer flow region with a hypothetical transition 100°C above T_g. At this region, it is expected that the viscosity of the material decreases with increased mobility. As the temperature gets higher, the reaction region is reached, where the soy globulin molecules have enough mobility and energy to form aggregates and cross-links. As the temperature increases more, the elastomeric structure melts forming the softening regions (Figure 12.5a,b). Information on the details of how borders of each region are determined experimentally can be found in Morales-Diaz and Kokini (1998).

State diagrams for other cereal proteins such as gliadin (Kokini et al., 1994a,b; Madeka and Kokini, 1994), glutenin (Kokini et al., 1994a,b), and zein (Kokini et al., 1995; Madeka and Kokini, 1996) are given in Figures 12.6 through 12.8, respectively. Similar order–disorder transitions and chemical reactions as in Figure 12.5a and b (glassy regions, rubbery regions, entangled polymer flow region, networking reaction zone, and softening regions) are also shown in these state diagrams (Figures 12.6 through 12.8).

Gliadin, glutenin, and zein are glassy polymers at low moisture content and low temperatures (Figures 12.6 through 12.8). Water plasticizes these proteins through the hydrophilic amino acids in their structure (Kokini et al., 1994a,b). Gliadin, glutenin, and zein are determined to be in the rubbery state at room temperature at moisture contents higher than 15%, 11%, and 14%, respectively (Kokini et al., 1994a,b; Madeka and Kokini, 1994) (Figures 12.6 through 12.8). Network

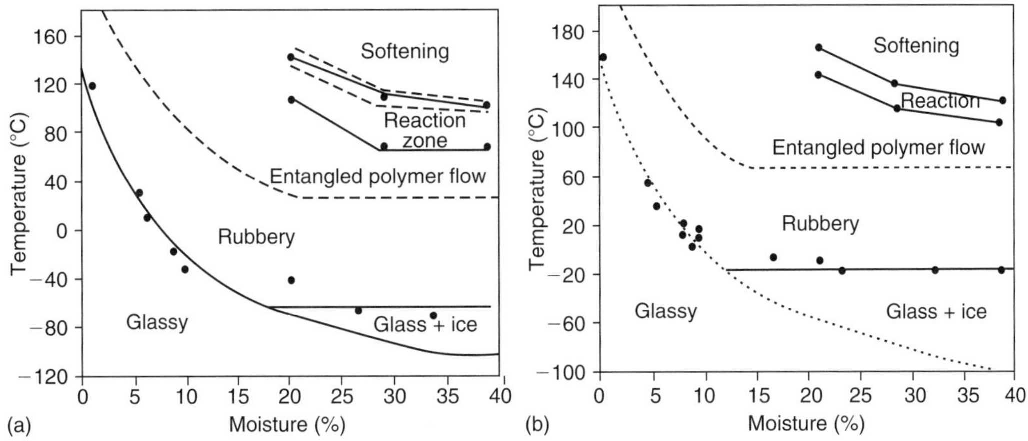

Figure 12.5 State diagram for (a) 7S soy globulin and (b) 11S soy globulin. (From Morales-Diaz, A.M. and Kokini, J.L., in *Phase/State Transitions in Foods*, Marcel Dekker, Inc., New York, 1998. With permission.)

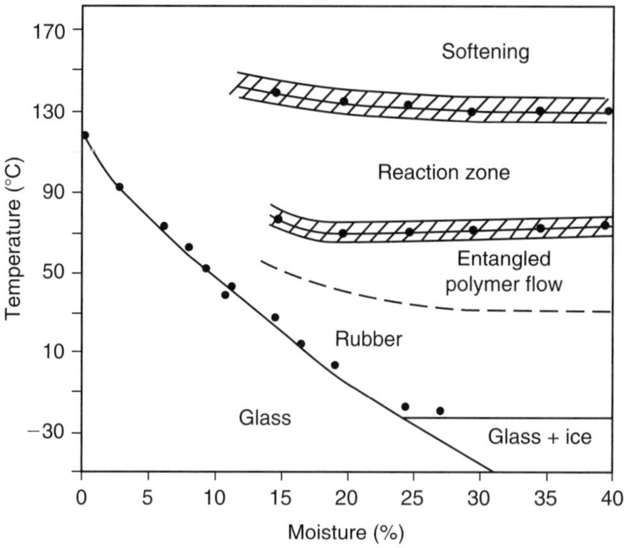

Figure 12.6 State diagram for gliadin. (From Madeka, H. and Kokini, J.L., *J. Food Eng.*, 22, 241, 1994. With permission.)

complexing reactions for gliadin, glutenin, and zein have also been shown on the state diagrams. At the end of the reaction region, the materials soften possibly due to the thermal effect of the material or due to depolymerization (Kokini et al., 1994b; Madeka and Kokini, 1994, 1996). Both reaction and softening regions for glutenin, gliadin, and zein are also strongly moisture-dependent (Figures 12.6 through 12.8).

The state diagram for gluten is given in Figure 12.9, showing glassy and rubbery regions below and above the glass transition line, respectively (Toufeili et al., 2002). Maximal freeze-concentration temperature (T'_g) is located at −20°C. The corresponding weight fraction of gluten is 76% (W'_g).

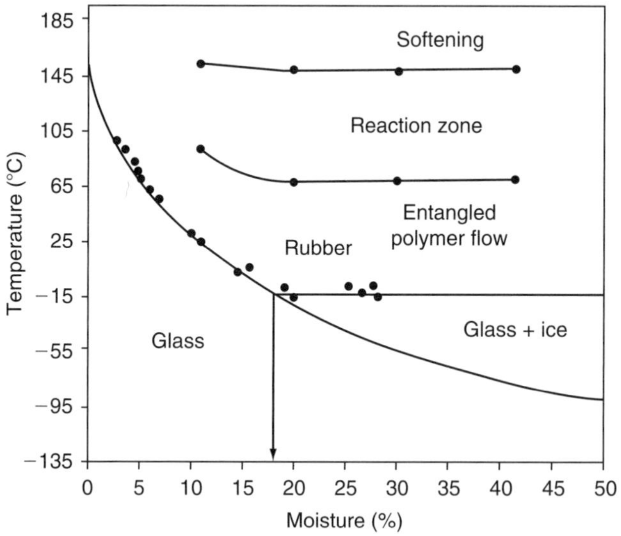

Figure 12.7 State diagram for glutenin. (From Kokini, J.L., Cocero, A.M., Madeka, H., and De Graaf, E., *Trends Food Sci. Tech.*, 5, 281, 1994b. With permission.)

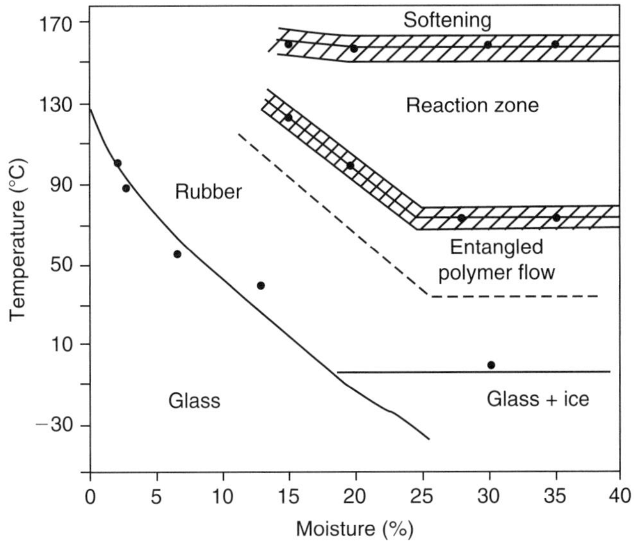

Figure 12.8 State diagram for zein. (From Kokini, J.L., Cocero, A.M., and Madeka, H., *Food Technol.*, 49, 74, 1995. With permission.)

Entangled polymer flow region occurs at 92°C–140°C and at moisture content around 10%. The network structure is formed in the reaction zone. At higher levels of moisture content, gluten forms cross-links at lower temperatures; at 120°C with 20% moisture content, and at 93°C with 30%–40% moisture levels. At moisture contents higher than 20%, softening of the cross-linked network occurs; and it occurs at lower temperatures at higher moisture contents (Figure 12.9) (Toufeili et al., 2002).

For example, during extrusion or baking, all these state diagrams would describe the specific temperature and moisture contents where the proteins would form the cross-linking reactions that in turn form the crisp texture of an extruded product or the crumb of a baked product (Kokini et al., 1994b). A hypothetical state diagram of cereal proteins during extrusion processing and storage is shown in Figure 12.10.

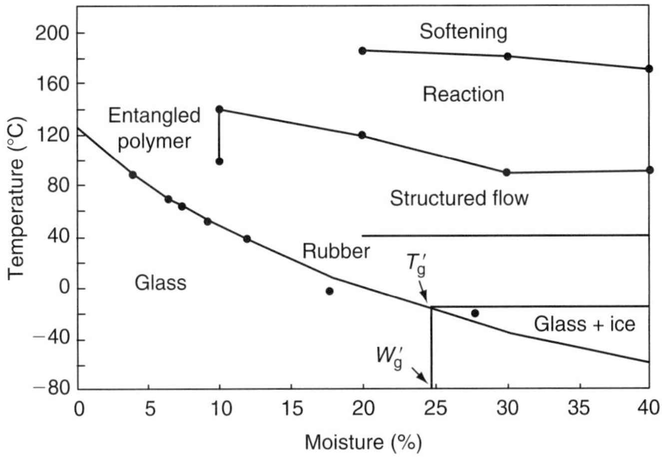

Figure 12.9 State diagram for gluten. (From Toufeili, I., Lambert, I.A., and Kokini, J.L., *Cereal Chem.*, 79, 138, 2002. With permission.)

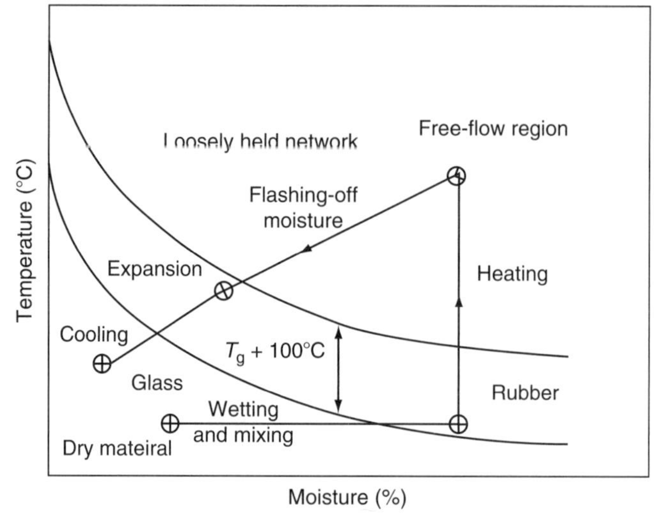

Figure 12.10 Hypothetical state diagram of cereal proteins during wetting, heating, and cooling/drying stages of extrusion cooking. (From Kokini, J.L., Cocero, A.M., Madeka, H., and De Graaf, E., *Trends Food Sci. Tech.*, 5, 281, 1994b. With permission.)

12.4 STATE DIAGRAMS OF OTHER FOOD COMPONENTS

For a binary system of a food component and water, the state diagram would show the glass transition temperature (T_g) as a function of water content, the effect of ice formation on T_g if the material is soluble in water and on ice melting temperature (T_m) at constant pressure (Figure 12.11). In other words, on this simple state diagram, stability in the glassy state and time-dependent changes in the rubbery state are represented. Figure 12.11 shows that T_g, which is the temperature

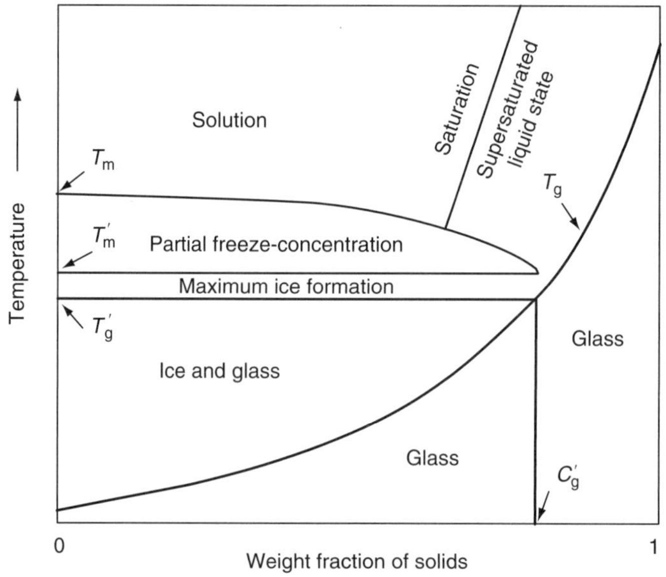

Figure 12.11 Schematic state diagram for food solutes and water. (From Roos, Y.H., *Phase Transitions in Foods*, Academic Press, Inc., San Diego, CA., 1995. With permission.)

at which the material transforms from a rigid, glassy state into a rubbery state, decreases as the water content increases because of the plasticization effect of water. $C_g^{'}$ is the solute concentration in the maximally freeze-concentrated solute and ice formation occurs at initial solute concentrations lower than $C_g^{'}$ (Figure 12.11). $T_g^{'}$ is the temperature for the onset of glass transition for the maximally freeze-concentrated solute, and $T_m^{'}$ is the temperature above which ice melting occurs. Assuming that there is no solute crystallization and the latent heat of melting ice in the food is equal to the latent heat of pure water, the difference between total water content and total amount of ice melting gives $C_g^{'}$ (Levine and Slade, 1986). According to Roos and Karel (1991a), experimental T_g versus solute concentrations provide the best $C_g^{'}$ values, determined to be around 80% (w/w) for various food components (Ablett et al., 1992; Roos, 1993). Similar state diagrams showing different regions and physical state of foods are given in Rahman (2006).

Changes in foods during dehydration, freezing, and heating/cooling can also be monitored and predicted from the state diagrams (Figure 12.12). Time-dependent and viscosity-related structural transformations, including ice formation, stickiness, collapse, and crystallization, affect these typical processes (Figure 12.13). On rapid removal of water (dehydration), the remaining solutes are transformed into an amorphous state. Structural changes, such as collapse and stickiness, can occur in food products if the temperature or relative humidity changes (Figure 12.13). These changes occur above T_g due to the tendency of the rubbery state to minimize its volume, whereas they are kinetically postponed near T_g (Roos and Karel, 1991c). Above T_g, crystallization can also occur together with the structural changes. The rate of crystallization increases as $T - T_g^{'}$ increases (Figure 12.13). During crystallization, released water is absorbed by the amorphous portions of the food resulting in lower T_g values (in other words, higher $T - T_g^{'}$ values) and rapid crystallization. If the moisture transfer between food and its environment is allowed, then the released water on crystallization is lost to the environment. The amorphous portion of the food remains at constant moisture level and the rate of crystallization is determined through constant "$T - T_g$" (Roos and Karel, 1991c). Rahman (2006) has also demonstrated a similar time-dependent structural transformation on typical state diagrams.

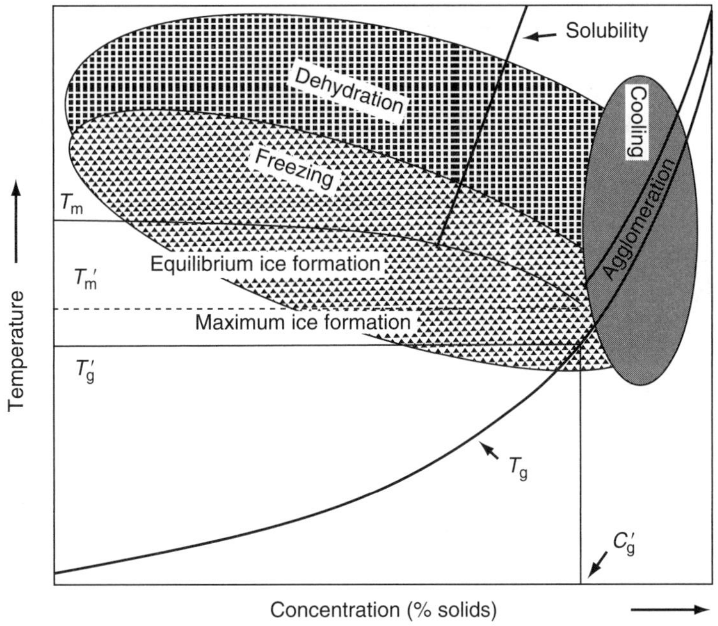

Figure 12.12 Location of various food processes including freezing, dehydration, heating, and cooling on a state diagram. (From Karel, M., Buera, M.P., and Roos, Y., in *The Glassy State in Foods*, Nottingham University Press, Loughborough, U.K., 1993, 13–34. With permission.)

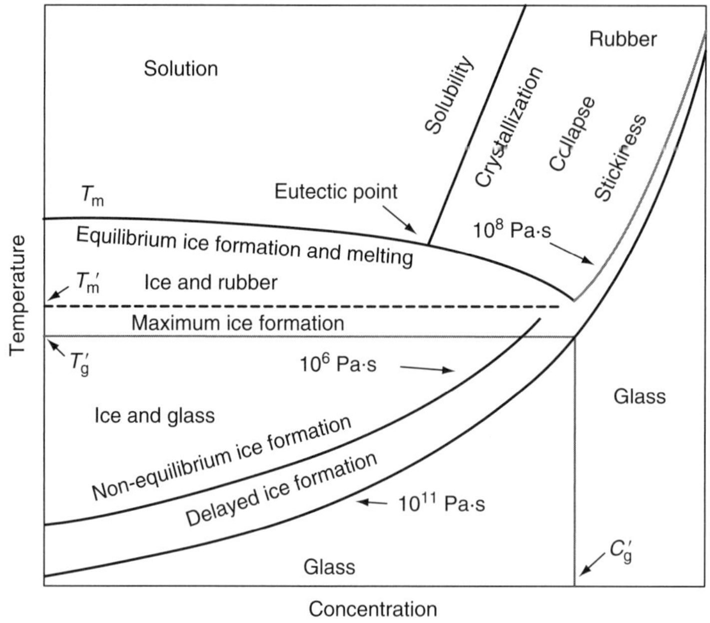

Figure 12.13 State diagram showing stability and time-dependent changes in the rubbery state with typical processes. (From Roos, Y.H. and Karel, M., *Food Technol.*, 45, 66, 1991c. With permission.)

In Figure 12.14, glass transition temperatures of different starch subfractions are shown as a function of water content (Bizot et al., 1997). Amylose has long linear chains and has the highest T_g values, whereas waxy maize starch with 98% amylopectin has relatively lower T_g values. Potato

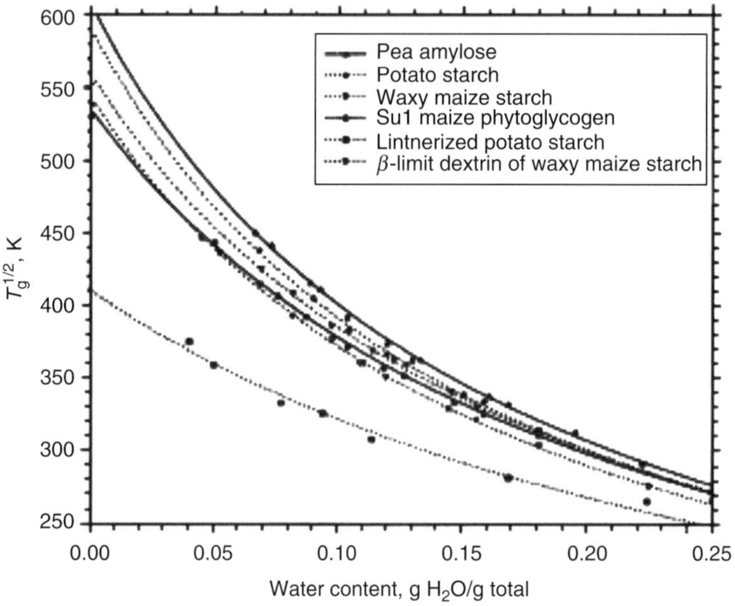

Figure 12.14 Glass transition temperature versus moisture content of starch subfractions measured by DSC with 3°C/min heating rate ($T_g^{1/2}$ represents the values at the midpoint of the transition in the second scans). (From Bizot, H., Le Bail, P., Leroux, B., Davy, J., Roger, P., and Buleon, A., *Carbohyd. Polym.*, 32, 33, 1997. With permission.)

starch with 23% amylose and 77% amylopectin content has T_g values between those of amylose and waxy maize starch. Su-1 maize phytoglycogen, which is extracted from su-1 by aqueous dispersion and alcoholic precipitation, and β-limit dextrins of waxy maize starch, have lower T_g values than those of waxy maize starch. Lintnerized potato starch, prepared by mild acid hydrolysis of potato starch and rich in 1–4 linkage, has the lowest T_g values (Figure 12.14) (Bizot et al., 1997).

For instance, the textural changes in bread after baking and before retrogradation are due to the cooling of the bread during which the rubbery state is transformed through the glassy state (Blanshard, 1986). Effective network T_g of stale bread is higher than the segmental T_g of fresh-baked bread (Levine and Slade, 1990). To obtain the soft and rubbery texture of freshly baked bread, crystalline regions should melt at the baking temperature with a required amount of moisture (Biliaderis et al., 1980; Maurice et al., 1985; Biliaderis et al., 1986; Levine and Slade, 1988). To decrease the effective T_g of stale bread and to obtain a sufficient temperature difference above T_g, a high moisture content is required for mobility and softness in the stale bread at room temperature. According to Levine and Slade (1988), a minimum of 27% of water is needed for recrystallization of wheat starch. Increasing water content in the range of 27%–50% increases the ratio of recrystallized starch due to the increased mobility of the macromolecular structure that allows the structural reorganization (Levine and Slade, 1988). Addition of sugars to bread dough prevents starch retrogradation and consequently prevents bread staling due to increasing network glass transition, also called antiplasticization (Levine and Slade, 1990). If a higher molecular weight sugar is used, a higher network glass transition will be obtained resulting in a better antistaling effect (Levine and Slade, 1988, 1990).

Similar state diagrams are also reported in the literature for various carbohydrates, including maltose and maltodextrins (Roos and Karel, 1991b); lactose (Roos and Karel, 1991c), sucrose (Slade and Levine, 1988; Roos and Karel, 1991a,c; Roos et al., 1996), glucose and fructose (Slade and Levine, 1988), lactose, sucrose, and a mixture of sucrose/fructose and sucrose/amioca (Roos and Karel, 1991d), and starch (Van de Berg, 1986; Roos and Karel, 1991b).

12.5 STATE DIAGRAMS OF FOODS

Figure 12.15 shows the approximate locations of various food products including pasta, cookies, bread, ice cream, frozen yogurt, and dried foods on a binary state diagram. Figure 12.16 shows another example of the practical use of state diagrams. The state diagram of sucrose–water binary systems is presented on which location of lean (low sugar/fat ratio) and rich (high sugar/fat ratio) cracker dough, rich cookie dough and final-baked cookie, and cracker products are positioned according to their process temperatures and typical sucrose–water content. Intersection of liquidus (nonequilibrium extension) and glass curves are represented with T_g'; intersection of liquidus and solidus curves are represented with T_e' (eutectic melting temperature) (Levine and Slade, 1993).

Lean and rich cracker dough is located above the glass curve showing that it is relatively easy to dry these products during baking. Above the vaporization temperature, the liquid water in the cracker turns into water vapor. As baking continues, the water vapor is removed more resulting in a higher concentration of dissolved sucrose. The final-baked crackers become glassy when they are cooled to room temperature (Products Box D in Figure 12.16) (Levine and Slade, 1993). Cookie dough is more complex than cracker dough because of the additional presence of crystalline sugar dissolving during mixing. The cookie dough can be either in Boxes A or B in Figure 12.16 (Levine and Slade, 1993). As the temperature increases during baking, either initial water evaporation and then sugar dissolving (path from A to H) or initial sugar dissolving and then water evaporation (path from B to H) take place. Levine and Slade (1993) described various routes for other baked products. In short, Figure 12.16 shows that the glass transition curve on the state

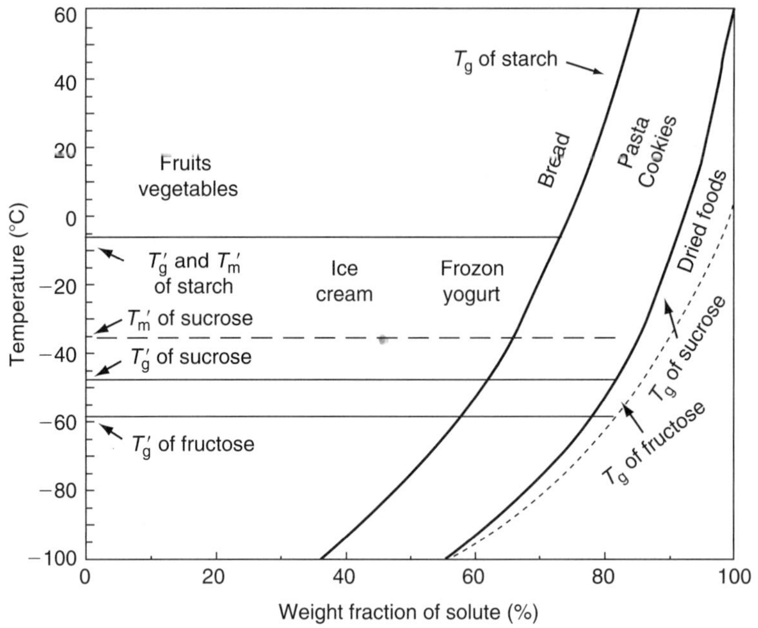

Figure 12.15 State diagram representing the position of various food products. (From Karel, M., Buera, M.P., and Roos, Y., in *The Glassy State in Foods*, Nottingham University Press, Loughborough, U.K., 1993, 13–34. With permission.)

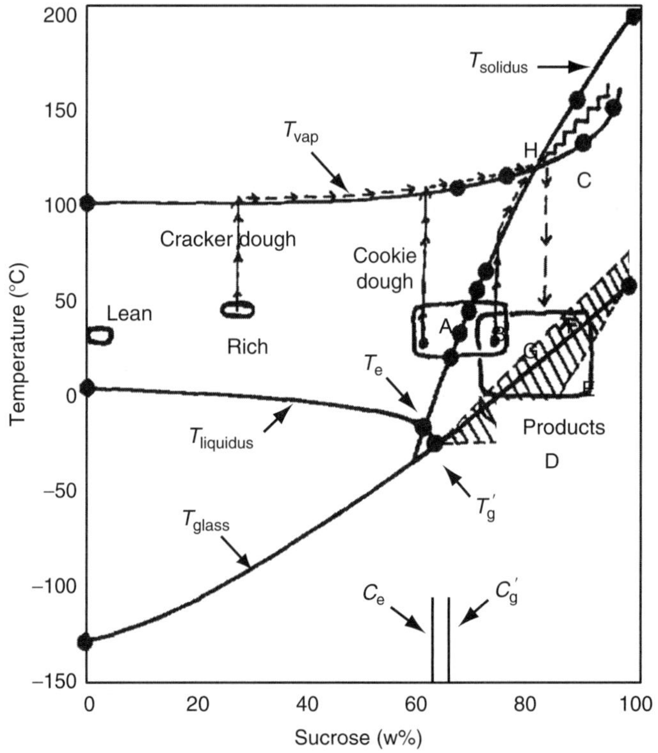

Figure 12.16 State diagram of sucrose–water system showing the location of cracker and cookie dough together with the final cracker and cookie products. (From Levine, H. and Slade, L., in *The Glassy State in Foods*, Nottingham University Press, Loughborough, U.K., 1993, 333–373. With permission.)

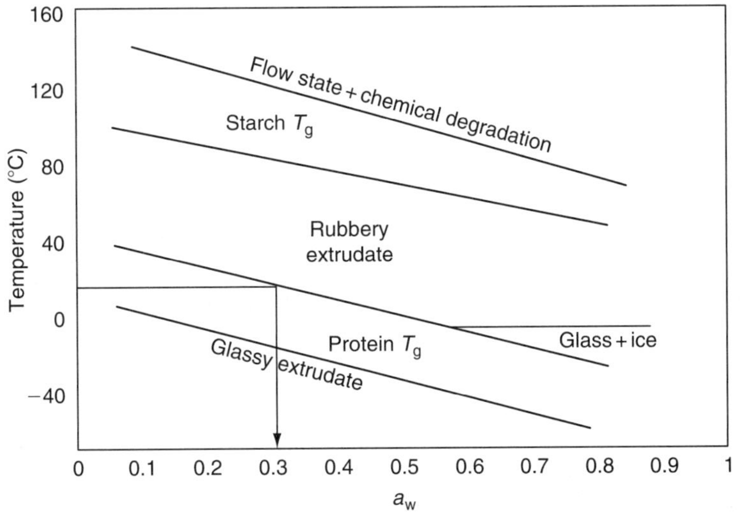

Figure 12.17 State diagram for the extruded meat–starch matrix. (The arrow indicates the water activity above which the extrudates become rubbery at room temperature.) (From Moraru, C.I., Lee, T.-C., Karwe, M.V., and Kokini, J.L., *J. Food Sci.*, 67, 3026, 2002. With permission.)

diagram of sucrose–water represents the behavior of the final-baked products. The storage stability of the products depends on the temperature, time, and possible moisture content changes during distribution and storage. High-quality products should be located in Box D in Figure 12.16 (Levine and Slade, 1993).

The state diagram for extruded meat–starch systems was developed by Moraru et al. (2002) (Figure 12.17). It has been shown that the physical state of the product is controlled by the glass transition location of the protein. The extrudates with water activities lower than 0.32 are brittle and glassy (determined through direct examination and validated by mechanical analysis), whereas samples with water activities higher than 0.57 are rubbery and pliable, as shown on the state diagram (Figure 12.17) (Moraru et al., 2002).

12.6 CONCLUSION

In this chapter, we have presented a short review of developing state diagrams of food materials, including proteins and carbohydrates and discussed how they can be used in terms of prediction of physical state of materials at different temperatures and water contents. We hope that this chapter serves as a useful starting reference to those who want to capture the key ideas behind the usage of state diagrams during food processing, distribution, and storage.

REFERENCES

Ablett, A., Izzard, M.J., and Lillford, P.J. 1992 Differential scanning calorimetric study of frozen sucrose and glycerol solutions. *J. Chem. Soc. Faraday T.* 88: 789–794.

Biliaderis, C.G., Maurice, T.J., and Vose, J.R. 1980 Starch gelatinization phenomena studied by DSC. *J. Food Sci.* 45: 1669–1680.

Biliaderis, C.G., Page, C.M., Maurice, T.J., and Juliano, B.O. 1986 Thermal characterization of rice starches: A polymeric approach to phase transitions of granular starch. *J. Agr. Food Chem.* 34: 6–14.

Bizot, H., Le Bail, P., Leroux, B., Davy, J., Roger, P., and Buleon, A. 1997 Calorimetric evaluation of the glass transition in hydrated, linear and branched polyanhydroglucose compounds. *Carbohyd. Polym.* 32(1): 33–50.

Blanshard, J.M.V. 1986 The significance of the structure and function of the starch granule in baked products. In: J.M.V. Blanshard, P.J. Frazier and T. Galliard (Eds.), *Chemistry and Physics of Baking* Royal Society of Chemistry, London, pp. 1–13.

Cocero, A.M. and Kokini, J.L. 1991 The study of the glass transition of glutenin using small amplitude oscillatory rheological measurements and differential scanning Calorimetry. *J. Rheol.* 35(2): 257–269.

Kalichevsky, M.T. and Blanshard, J.M.V. 1993 The effect of fructose and water on the glass transition of amylopectin. *Carbohydr. Polym.* 20: 107–113.

Karel, M., Buera, M.P., and Roos, Y. 1993 Effects of glass transition on processing and storage. In: J.M.V. Blanshard and P.J. Lillford (Eds.), *The Glassy State in Foods* Nottingham University Press, Loughborough, UK, pp. 13–34.

Kokini, J.L., Cocero, A.M., Madeka, H., and De Graaf, E. 1994a Order–disorder transitions and complexing reactions in cereal proteins and their effect on rheology. In: D.A. Siringer and Y.G. Yanovsky (Eds.), *Advances in Structured and Heterogenous Continua* Allerton Press, New York, pp. 215–234.

Kokini, J.L., Cocero, A.M., Madeka, H., and De Graaf, E. 1994b The development of state diagrams for cereal proteins. *Trends Food Sci. Tech.* 5: 281–288.

Kokini, J.L., Cocero, A.M., and Madeka, H. 1995 State diagrams help predict rheology of cereal proteins. *Food Technol.* 49(10): 74–81.

Levine, H. and Slade, L. 1986 A polymer physico-chemical approach to the study of commercial starch hydrolysis products (SHPs). *Carbohyd. Polym.* 6: 213–244.

Levine, H. and Slade, L. 1988 Water as a plasticizer: Physico-chemical aspects of low moisture polymeric systems. In: F. Franks (Ed.), *Water Science Reviews 3* Cambridge University Press, New York, pp. 79–186.

Levine, H. and Slade, L. 1990 Influences of the glassy and rubbery states on the thermal, mechanical, and structural properties of doughs and baked products. In: H. Faridi and J.M. Faubion (Eds.), *Dough Rheology and Baked Product Texture* Van Nostrand Reinhold, New York, pp. 157–330.

Levine, H. and Slade, L. 1993 The glassy state in applications for the food industry. In: J.M.V. Blanshard and P.J. Lillford (Eds.), *The Glassy State in Foods* Nottingham University Press, Loughborough, UK, pp. 333–373.

Madeka, H. and Kokini, J.L. 1994 Changes in rheological properties of gliadin as a function of temperature and moisture: Development of a state diagram. *J. Food Eng.* 22: 241–252.

Madeka, H. and Kokini, J.L. 1996 Effect of glass transition and aggregation on rheological properties of zein: Development of a preliminary state diagram. *Cereal Chem.* 73(4): 433–438.

Maurice, T.J., Slade, L., Page, C., and Sirett, R. 1985 Polysaccharide–water interactions—thermal behavior of starch. In: D. Simatos and J.L. Multon (Eds.), *Properties of Water in Foods* Martinus Nijhoff, Dordrecht, pp. 211–227.

Morales-Diaz, A.M. and Kokini, J.L. 1998 Understanding phase transitions and chemical complexing reactions in 7S and 11S soy protein fractions. In: M.A. Rao and R. W. Hartel (Eds.), *Phase/State Transitions in Foods* Marcel Dekker, Inc., New York, pp. 273–313.

Moraru, C.I., Lee, T.-C., Karwe, M.V., and Kokini, J.L. 2002 Phase behavior of a meat–starch extrudate illustrated on a state diagram. *J. Food Sci.* 67(8): 3026–3032.

Noel, T.R., Ring, S.R., and Whittam, M.A. 1992 Dielectric relaxations of small carbohydrate molecules in the liquid and glassy states. *J. Phys. Chem.* 96(13): 5662–5667.

Noel, T.R., Parker, R., and Ring, S.G. 1996 A comparative study of the dielectric relaxation behavior of glucose, maltose, and their mixtures with water in the liquid and glassy states. *Carbohydr. Res.* 282(2): 193–206.

Rahman, M.S. 2006 State diagram of foods: Its potential use in food processing and product stability. *Trends Food Sci. Tech.* 17: 129–141.

Roos, Y. 1993 Melting and glass transitions of low molecular weight carbohydrates. *Carbohyd. Res.* 238: 39–48.

Roos, Y.H. 1995 *Phase Transitions in Foods.* Academic Press, Inc., San Diego, CA.

Roos, Y.H. and Karel, M. 1991a Amorphous state and delayed ice formation in sucrose solutions. *Int. J. Food Sci. Tech.* 26: 553–566.

Roos, Y.H. and Karel, M. 1991b Water and molecular effects on glass transitions in amorphous carbohydrates and carbohydrate solutions. *J. Food Sci.* 56: 1676–1681.

Roos, Y.H. and Karel, M. 1991c Applying state diagrams to food processing and development. *Food Technol.* 45: 66, 68–71, 107.

Roos, Y.H. and Karel, M. 1991d Plasticizing effect of water on thermal behavior and crystallization of amorphous food models. *J. Food Sci.* 56: 38–43.

Roos, Y., Karel, M., and Kokini, J.L. 1996 Glass transitions in low moisture and frozen foods: Effect on shelf life and quality. *Food Technol.* 50(11): 95–108.

Slade, L. and Levine, H. 1988 Structural stability of intermediate moisture foods—A new understanding. In: J.M.V. Blanshard and J.R. Mitchell (Eds.), *Food Structure—Its Creation and Evaluation* Butterworths, Inc., London, pp. 115–147.

Sperling, L.H. 2001. *Introduction to Physical Polymer Science*. 3rd Edition. John Wiley & Sons, Inc. New York.

Toufeili, I., Lambert, I.A., and Kokini, J.L. 2002 Effect of glass transition and cross-linking on rheological properties of gluten: Development of a preliminary state diagram. *Cereal Chem.* 79(1): 138–142.

Van de Berg, C. 1986 Water activity. In: D. MacCarthy (Ed.), *Concentration and Drying of Foods* Elsevier Applied Science, London, pp. 11–36.

CHAPTER 13

Measurement of Density, Shrinkage, and Porosity

Panagiotis A. Michailidis, Magdalini K. Krokida, G.I. Bisharat, Dimitris Marinos-Kouris, and Mohammad Shafiur Rahman

CONTENTS

13.1 INTRODUCTION

Density, shrinkage, and porosity are included in the mass- and volume-related properties that constitute one of the main groups of mechanical properties along with rheological, morphological, acoustic, and surface properties (Rahman and McCarthy, 1999). The importance of these properties is illustrated by the following applications:

- Process design, operation, and optimization (e.g., for the prediction of the solids height in a belt dryer during dehydration)
- Product characterization—quality determination
- Estimation of other properties (e.g., thermal conductivity of food materials using the porosity of individual components or diffusion coefficient of shrinking systems using porosity and volume change)
- Handling of food materials
- Grading of fruits and vegetables (e.g., according to density)
- Separation of impurities in food materials by density differences
- Estimation of floor space during storage and transportation (using the bulk density)

13.2 DEFINITION AND TERMINOLOGY

Rahman (Rahman, 1995, 2005) presented a detailed terminology for the most common mass- and volume-related properties, which is presented in the following sections:

1. *Boundary volume*: It is the volume of a material considering the geometric boundary.
2. *Pore volume*: It is the volume of the voids or air inside a material.
3. *Density*: Density is the mass of a material per unit volume. In the case of nonhomogeneous materials, such as foods, consisting of solid, liquid, and gaseous phase(s), a number of different forms of density have been defined for the sufficient relation between mass and volume.
   a. *True density* (ρ_t): It is the density of a pure substance or a composite material calculated from its components' densities considering conservation of mass and volume.
   b. *Material density* (ρ_m): It is the density measured when a material has been thoroughly broken into pieces small enough to guarantee no closed pores remain. It is also known as substance density.
   c. *Particle density* (ρ_p): It is the density of a particle that has not been structurally modified and includes the volume of all closed pores but not the externally connected pores.
   d. *Apparent density* (ρ_a): It is the density of a substance including all pores remaining in the material.
   e. *Bulk density* (ρ_b): It is the density of a material when packed or stacked in bulk and is defined as the mass of the material per the total volume it occupies. The bulk volume of packed materials depends on the geometry, size, and surface properties of the individual particles (Lewis, 1987).
4. *Shrinkage*: Shrinkage is the volume change of a food material, which undergoes processes such as moisture loss (dehydration) during drying, ice formation during freezing, or pore formation by puffing (e.g., during extrusion cooking). Shrinkage can be defined in various ways as presented in the following section:
   a. *Apparent shrinkage* (S_a): It is the ratio of the apparent volume at a given moisture content to initial apparent volume of the material and can be expressed as

$$S_a = \frac{V_a}{V_{ai}} \tag{13.1}$$

where V_{ai} and V_a are the values of apparent volume initially and at given moisture content. It indicates the overall volume of the food material remaining at the given moisture content. The term $(1 - S_a)$ can be defined as the fraction of change from the initial apparent volume, e.g., the overall volume shrinkage of the material, while the term $S_a \times 100$ presents the percentage of the initial apparent volume that remains after the process. Shrinkage, sometimes, refers to the surface (two-dimensional [2D] shrinkage) or the length (1D shrinkage) of a food material. In the case of food materials, two types of shrinkage are usually observed.

b. *Isotropic shrinkage*: It is the uniform shrinkage in all geometric dimensions of the material. This type of shrinkage is often observed in the majority of fruits and vegetables.
c. *Anisotropic shrinkage*: It is the nonuniform shrinkage in different dimensions of the material and is observed mainly in fish and seafood that undergo significantly different shrinkage in the direction parallel to muscle fibers from that perpendicular to the fibers during air drying (Balaban and Pigott, 1986; Rahman and Potluri, 1990).

5. *Porosity*: Porosity expresses the volume fraction of void space or air in a material and is defined as the air or void volume per total volume. Porosity can be defined in various ways to characterize food products or for food process calculations.
   a. *Open pore porosity* (ε_{op}). It is the volume fraction of pores connected to the exterior boundary of the food material, expresses the volume of open pores per total material volume, and is given by

$$\varepsilon_{op} = 1 - \frac{\rho_a}{\rho_p} \tag{13.2}$$

   Open pores can be classified in two types: those that are connected to the exterior boundary only, and those that are connected to other open pores and to the exterior geometry boundary.
   b. *Closed pore porosity* (ε_{cp}): It is the volume fraction of pores closed inside the material and not connected to its exterior boundary. It expresses the volume of closed pores per total material volume and is given by the equation

$$\varepsilon_{cp} = 1 - \frac{\rho_p}{\rho_m} \tag{13.3}$$

   c. *Apparent porosity* (ε_a): It is the ratio of total air or void space to the total volume of the material enclosed inside the material boundary, expresses the volume of all pores per total material volume, and is defined as

$$\varepsilon_a = 1 - \frac{\rho_a}{\rho_m} = \varepsilon_{op} + \varepsilon_{cp} \tag{13.4}$$

   d. *Bulk porosity* (ε_b): It is the ratio of the volume of voids or air outside the boundary of individual materials to the total volume of the packed or stacked materials as bulk and is given by

$$\varepsilon_b = 1 - \frac{\rho_b}{\rho_a} \tag{13.5}$$

   e. *Bulk-particle porosity* (ε_{bp}): It is the volume fraction of the voids outside the individual particles plus the volume fraction of pores contributing to the open pore porosity when packed or stacked as bulk. It is given by the equation

$$\varepsilon_{bp} = \varepsilon_b + \varepsilon_{op} \tag{13.6}$$

   f. *Total porosity* (ε_T): It is the total volume fraction of air or void space (i.e., inside and outside of the materials) when material is packed or stacked as bulk and is expressed by the equation

$$\varepsilon_T = \varepsilon_a + \varepsilon_b = \varepsilon_{op} + \varepsilon_{cp} + \varepsilon_b \tag{13.7}$$

It should be mentioned that in the literature there are deviations from the above terminology. For example, Marousis and Saravacos (1990) used the term “particle density” to describe what is defined above as "material density." Therefore, it is important to define density before presenting or using density data in process calculations. Moreover, according to these definitions, true density refers to pure components or mixtures whose composition is known and it is estimated from the densities of the material's components. Material density characterizes a multicomponent mixture, such as foods, whose exact composition is unknown in most of the cases, and its value is found experimentally. In spite of that, the majority of literature data and models retrieved from experimental procedures for the determination of food materials' density excluded their pores and are referred to as "true density," although they express the "material density" of the foods. Chapter 12 presents a significant amount of density data and models, and the original definition as "true" has been retained. The reader should have in mind that they represent the "material density" of the foods, as they originate from experimental data.

13.3 MEASUREMENT TECHNIQUES

Table 13.1 presents the measurement techniques (methods) used for the determination of density, shrinkage, and porosity of food materials. Estimated properties refer to the structural properties' predicted values, while measured properties are those actually measured, e.g., the weight of a sample using a balance or the volume of gas displaced using a stereopycnometer. Table 13.1 also includes a few notes for each method. In the following sections, each of these techniques is presented comprehensively.

13.3.1 Density Measurement

13.3.1.1 Apparent Density

13.3.1.1.1 Geometric Dimension Method

The apparent density of a food material of a shape of regular geometry can be determined from the mass and the boundary volume calculated from the characteristic dimensions (Rahman, 1995). The equations for the volume estimation of a few well-known shapes are presented in Table 13.2. It is obvious that this method is not suitable for liquid or soft and irregularly shaped food materials, where it is not easy to measure the characteristic dimensions.

For measuring the density of frozen foods, a thick-wall cylindrical metal container can be used where the material is placed to freeze in the desired temperature. The method consists of finding the mass of the frozen sample with the known volume (that of the container). The excess frozen sample can be removed by cutting with a sharp knife (Keppeler and Boose, 1970). The container may be wrapped with electrical tape to reduce the heat gain during weighing (Rahman and Driscoll, 1994). This method is suitable for liquid and soft materials, where no void exists in the packing.

13.3.1.1.2 Buoyant Force Method

In this method, the buoyant force can be determined from sample weight in air and liquid. The apparent density can be calculated from the equation (Rahman, 1995)

$$\rho_a = \rho_l \frac{W}{G} \tag{13.8}$$

where

W and G are the weight of the sample in air and liquid, respectively

ρ_l is the density of the liquid

TABLE 13.1 Measurement Techniques for Density, Shrinkage, and Porosity Estimation

Estimated Property	Method	Measured Property	Application
Apparent density	Geometric dimension method	Mass and characteristic dimensions (for boundary volume estimation)	Regularly shaped solids
	Buoyant force method	Weight in air and buoyant force (sample must be coated)	Irregularly shaped solids
	Liquid displacement method	Mass and displacement liquid volume (sample must be coated)	Irregularly shaped coated solids
	Gas pycnometer method	Mass and displacement gas volume (sample must be coated)	Irregularly shaped coated solids
	Solid displacement method	Mass and displacement solid volume	Irregularly shaped solids
Material density	Liquid displacement method	Mass and displacement liquid volume	Grounded solids. Can be used for the measurement of density as a function of temperature
	Pycnometer method	Mass and displacement gas volume	Grounded solids
	Mercury porosimetry	Pressure versus mercury intrusion and extrusion volume	Determination of pore volume and pore size distribution of grounded solids
	Gas adsorption method	Volume adsorbed versus relative pressure	Determination of pore volume and pore size distribution of grounded solids
Particle density	Liquid displacement method	Mass and displacement liquid volume	Irregularly shaped solids
	Gas pycnometer method	Mass and displacement gas volume	Irregularly shaped solids
Bulk density	Bulk density method	Mass and container volume	Liquids or materials packed or stacked in bulk into a known volume container
Apparent shrinkage	Geometric dimension method	Characteristic dimensions (for boundary volume estimation)	Regularly shaped solids
	Liquid displacement method	Displacement liquid volume (sample must be coated)	Irregularly shaped coated solids
	Gas pycnometer method	Displacement gas volume (sample must be coated)	Irregularly shaped coated solids
	Solid displacement method	Displacement solid volume	Irregularly shaped solids
Porosity	Direct method	Bulk volume and volume of compacted material	Soft solids
	Optical microscopic method	Microscopic view of a random section of the porous material	Solids presenting sectional porosity the same as bulk porosity
	X-ray Microtomography (XMT)	Optical image of the porous material	Solids. It can be used in the visualization and measurement of microstructural features
	Density method[a]	—	Solids with known densities
	Fragile agglomerates method	Mass and optical sizing of individual agglomerates	Fragile agglomerates (instant powders)

Note: Powders are classified as irregularly shaped solids.

[a] If densities are unknown, pore volume or bulk volume should be measured using techniques for volume measurement. The volume of regularly shaped particles can also be measured with methods usually used for irregularly shaped solids but the geometric dimension method is more accurate in this case.

TABLE 13.2 Volume Calculation for Materials of Regular Geometry Shape

Regular Shape Materials	Volume
Sphere	$V = 4/3\pi r^3$
Cylinder	$V = \pi r^2 L$
Brick	$V = abc$
Cube	$V = a^3$
Prolate spheroid (Moshenin, 1986)	$V = 4/3\pi pq^2$
Oblate spheroid (Moshenin, 1986)	$V = 4/3\pi p^2 q$
Frustam right cone (Moshenin, 1986)	$V = \pi/3L(r_2^1 + r_1 r_2 + r_2^2)$

Source: Moshenin, N.N. in *Physical Properties of Plant and Animal Material*, Gordon and Breach Science Publishers, New York, 1986.

Note: V is the volume, r is the radius, L is the cylinder length or cone altitude, a, b, and c are the axis length in the 3D (length, weight, and height); p and q are the major and minor semiaxes of the ellipse of rotation, respectively; and r_1 and r_2 are the radii of base and top of the cone, respectively.

Practically, G is the buoyant force. Figures 13.1 and 13.2 schematically present two methods for weighting a sample, the analytical balance method and the top loading balance method, respectively.

Precaution should be taken during measurement as two errors may often occur in practicing this method. The first refers to mass transfer that may take place from the sample to the liquid, including any of the phases present (solid, liquid, or gas). This can be avoided by enclosing the sample in cellophane or polyethylene. It can also be coated with a thin layer of polyurethane varnish or wax; for the latter, the sample could be dipped in wax and solidified before measurement (Rahman, 2005). Loch-Bonacci et al. (1992) proposed covering of the specimen with silicon grease for making it impervious to water. It has been reported that there is no significant moisture take up when uncoated samples of fresh and dried fruits and vegetables undergo measurement in comparison to coated ones with plastic film (Lozano et al., 1980). This was due to the short time required for experimental measurement. Nevertheless, coating is the best possible option for accuracy, and care must be taken to prepare the coating (Rahman, 2005). The second source of error refers to the partial floating of the sample. In this case, a liquid of lower density than that of the sample may be used. Moshenin (1986) described a simple technique with a top-loading balance that applies to large

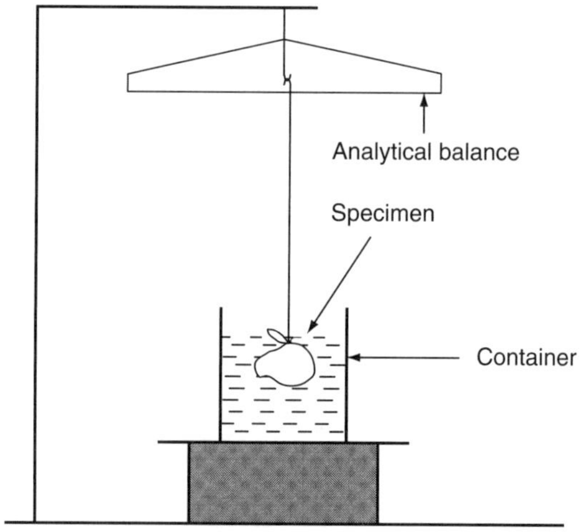

FIGURE 13.1 Analytical balance for measurement of buoyant force.

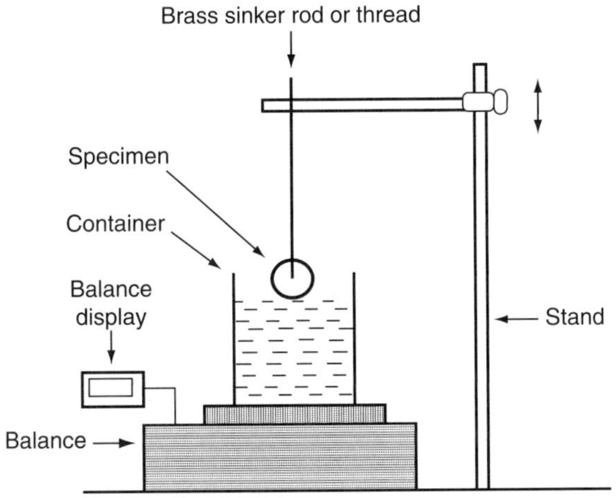

FIGURE 13.2 Top-loading balance for measurement of buoyant force for a sample lighter than liquid.

objects such as fruits and vegetables, illustrated in Figure 13.2. After the food sample is weighted on the scale in air, it is forced into the liquid (e.g., water) by means of a sinker rod. A second reading is taken for the submerged sample. The density is given by the equation

$$\rho_a = \rho_l \frac{W}{G_1 - G_2} \tag{13.9}$$

where

W is the sample weight in air (kg)
G_1 is the weight of the sample plus that of the sinker in the liquid (kg)
G_2 is the weight of the sinker submerged in the liquid (kg)

If the solid is lighter than the liquid, another solid can be attached, heavier than the liquid, to the object as a sinker. In the case of coated samples measured with this method, the following equation can be applied

$$\rho_a = \frac{W}{(G_1/\rho_l) - (G_2/\rho_l) - (W_{wax}/\rho_{wax})} \tag{13.10}$$

where

W_{wax} is the weight of the wax expressed by the difference between the weights of coated and uncoated samples (kg)
ρ_{wax} is the wax density (kg/m^3)
G_1 is the weight of the wax-coated sample plus that of the sinker in the liquid (kg)
G_2 is the weight of the sinker in the liquid (kg)

Although the wax density could be used as 912 kg/m^3, it may vary according to the source. Hence, it should be measured separately. Moshenin (1986) suggested that a solution of 3 mL wetting agent in 500 mL distilled water could reduce errors to surface tension and submergence in water. The buoyant force method was used for the measurement of the apparent density of frozen apple. The fruit, along with a sicker, was frozen at −20°C to −35°C, while the water temperature was 2°C to 3°C (Ramaswamy and Tung, 1981).

13.3.1.1.3 Volume Displacement Method

13.3.1.1.3.1 Liquid Displacement Method—The volume of a sample can be measured by direct measurement of the volume of liquid displacement. The difference between the initial volume of the liquid in the measuring apparatus and the cumulative volume of the liquid and immersed sample is the sample volume. It is better to use a nonwetting fluid such as mercury. Otherwise, coating of the samples is necessary to prevent liquid penetration in the pores. A common measuring apparatus is the specific gravity bottle with toluene (Bailey, 1912). The use of toluene as the reference liquid presents the following advantages (Moshenin, 1986):

- Low solvent action on constituents, especially fats and oils
- Smooth flow over the surface due to surface tension
- Little tendency to soak on the sample
- Stable specific gravity and viscosity when exposed to the atmosphere
- Fairly high boiling point
- Low specific gravity

Its main disadvantage is that toluene is classified as a carcinogenic material; thus experiments should be performed inside a fume chamber (Rahman et al., 2002). Moshenin (1986) proposed the following steps to measure the volume of the sample using the specific gravity bottle which gave only 2% less than the conventional air comparison pycnometer:

- Determine the exact capacity of the pycnometer by weighing it empty and filled with distilled water at 20°C.
- Determine specific gravity of the batch of toluene by comparing the weight of toluene in the bottle with the weight of distilled water at the same temperature.
- Place in the pycnometer 10 g of the solid specimen with a sufficient amount of toluene to cover the sample.
- Gradually exhaust the air from the apparatus by a vacuum pump to promote the escape of the air trapped under the surface.
- When air bubbles escape after several cycles of vacuuming and release of the vacuum, fill the apparatus with toluene and allow the temperature to reach 20°C.

In case large objects are to be measured, the classic small-neck specific gravity bottle is unsuitable and a special design apparatus should be used; such an apparatus has been practiced by Zogzas et al. (1994). Figure 13.3 schematically presents this devise for the measurement of the volume of relatively large samples of food materials. It consists of a compartment in which the sample is put and of a measuring burette. A lip can close the compartment hermetically. The apparatus is filled with a suitable portion of a liquid, and the volume displacement is measured by turning it upside down twice, once without and once with the specimen immersed. The measuring accuracy depends on the accuracy of the burette. Mercury, toluene, and *n*-heptane were tested with this apparatus giving similar satisfactory results; *n*-heptane may be adopted as it is less harmful to humans.

Rahman and Driscoll (1994) reported the use of a method for irregular and small frozen food particles such as cereals and grain. The procedures are as follows:

- Eight cylindrical glass bottles of diameter 2 cm with small necks filled three-fourths full (20 g) with the sample and the rest of the way with toluene are frozen at −40°C.
- After freezing, the bottles are immediately placed inside glass wool insulation columns of inner diameter 2 cm and outer diameter 7 cm. The temperature is recorded from one bottle by a thermocouple placed inside of its center.
- At different temperatures, the bottles are taken out, one at a time, and toluene is added to completely fill the bottle.

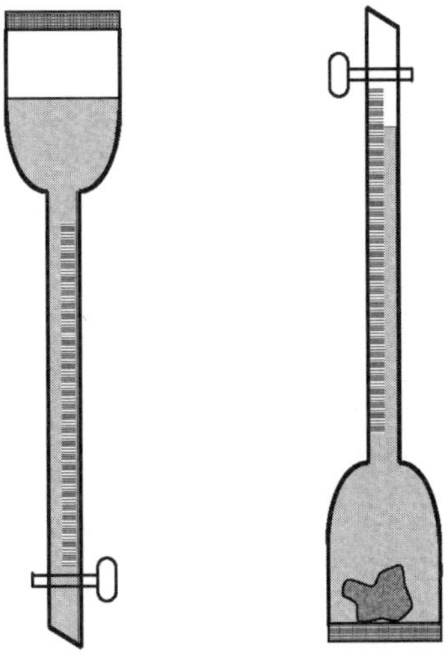

FIGURE 13.3 Experimental apparatus used for apparent volume determination.

- Bottle is closed immediately and the weight is determined. From the mass and volume of the sample, which is estimated by subtracting the volume of toluene from the volume of the bottle, the density is calculated. The volume of toluene is estimated from the mass and density at the respective temperatures. The volume of the bottle may be measured using distilled water.

The above method was used for the density measurement of frozen seafood at different temperatures and the reproducibility was found within 1% (Rahman and Driscoll, 1994). Commercial mercury porosimeters are available to measure the volume of porous and nonporous solids. The principal of mercury intrusion porosimetry is based on the fact that mercury ordinarily behaves as a nonwetting liquid (i.e., the contact angle of mercury is larger than 90°). Thus, it does not flow into the openings of porous solid bodies unless it is forced to do so by a pressure gradient (Clayton and Huang, 1984). The mercury injection method of measuring effective porosity is based on the fact that due to the surface tension and nonwetting properties of mercury, a porous sample can be immersed in mercury without the entry of mercury into the sample at atmospheric pressure. The apparent volume of the sample can be determined by displacement of mercury from a sample chamber of known volume (Rahman, 2005).

13.3.1.1.3.2 Gas Pycnometer Method—There are different commercial gas pycnometers for volume measurement, which use air, nitrogen or helium, often with no measurable difference. The technique employs Archimedes principle of fluid displacement to determine the volume. The displaced fluid is a gas, which can penetrate the finest pores to ensure maximum accuracy. From the above-mentioned gases, helium is recommended since its small atomic dimension assures penetration into crevices and pores approaching one Angstrom ($1\ \text{Å} = 10^{-10}$ m), while its behavior as an ideal gas is also desirable. Moshenin (1986) described a method to measure volume using high-pressure air. The method is schematically illustrated in Figure 13.4. According to this method, the following procedure is applied.

- Sample of the food material to be measured is placed in tank 2, and air is supplied to tank 1 when valve 2 is closed.
- When suitable manometer displacement is achieved, valve 1 is closed and equilibrium pressure P_1 is read.

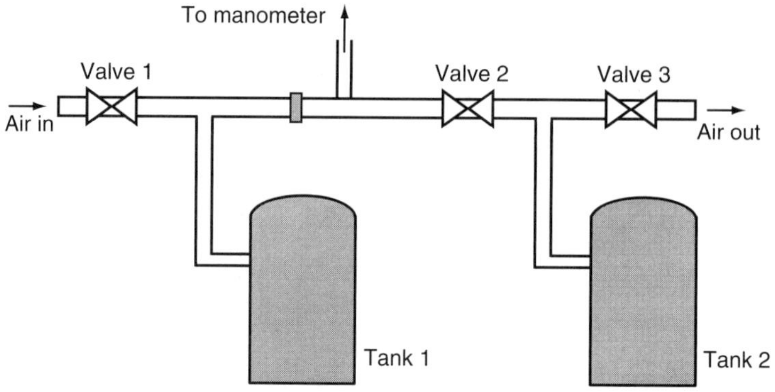

FIGURE 13.4 Air compression pycnometer.

- Valve 3 is closed, valve 2 is opened, and pressure P_3 is read. Under these conditions, the volume of the sample is measured as V_s. Specifically, the sample volume is given by the following equation, based on ideal gas law:

$$V_s = V_1\left[1 + \frac{P_3 - P_1}{P_3}\right] \tag{13.11}$$

where V_1 is the empty volume of tanks 1 and 2. The volume measurement of food samples is usually performed with different types of commercial automatic helium gas pycnometers. Figure 13.5 presents the operating principle of the Quantachrome multipycnometer. The pycnometer determines the density of solid or powder samples by measuring the pressure difference when a known quantity of helium under pressure is allowed to flow from a precisely known reference volume into a sample cell containing the material. The procedure for the volume measurement is the following:

- Correct reference volume is selected depending on the sample cup to be used.

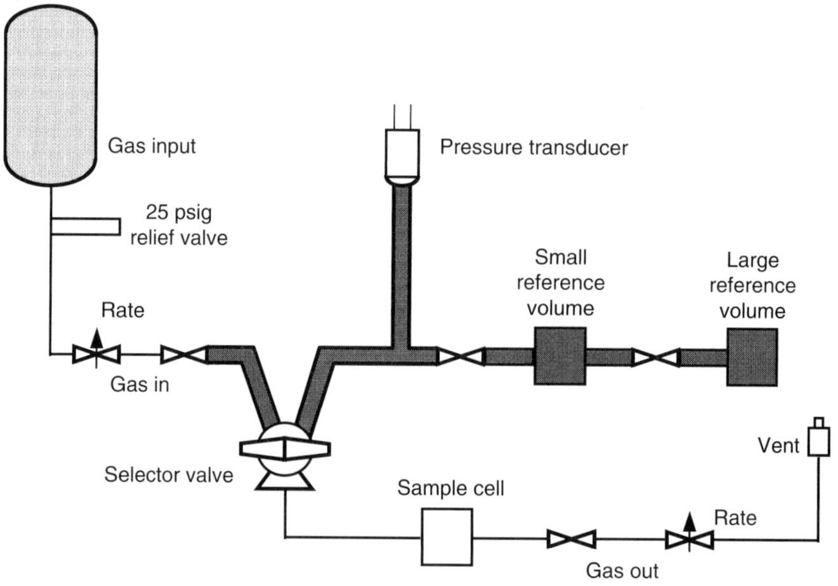

FIGURE 13.5 Flow diagram of a commercial multipycnometer.

- Sample of volume V_s is placed in the sample cup. The cup is inserted into the sample cell and the cover is replaced.
- Gas out control is opened until the zero reading stabilizes.
- Gas out valve is closed.
- Selector valve is turned to "reference."
- Gas in valve is opened and when the pressure reaches the approximate value of 17 psig the gas in valve is closed.
- When the display reading has stabilized, its value is recorded as P_1.
- Selector valve is turned to "cell."
- When the display reading has stabilized, its value is recorded as P_2.
- Gas out control is opened slowly to vent the pressure.
- Sample volume and porosity are calculated by the following equations:

$$V_s = V_c + V_R \left[\frac{P_1}{P_2} - 1 \right] \tag{13.12}$$

$$\rho_a = \frac{m_s}{V_s} \tag{13.13}$$

where V_c and V_R are constants of the device. Equation 13.12 is based on ideal gas law. Quantachrome pycnometer described above can take three sample cells with different volume each, which can be used according to the volume of the sample to be measured. The constants of Equation 13.12 depend on the cell that is used for the measurement. The calibration of the device can be performed by the use of metallic spheres of known diameter that are put in the sample cell for measurement. The experimental values of the volume obtained with these spheres in a series of measurements, where the number of spheres changes each time, are compared with the theoretical values as given in Table 13.1 and regression analysis in followed for the estimation of the constants V_c and V_R. It is important to note that for the apparent density measurement, the sample needs to be coated with wax before the procedure begins due to the helium penetration to the finest pores; otherwise, the result will be closer to the material rather than the apparent density.

13.3.1.1.3.3 Solid Displacement Method—The apparent density of an irregular solid sample can be measured by a solid (e.g., sand, mastered seeds) displacement or glass bead displacement method. The latter is more preferable as it gives reproducible results due to the uniform size and shape of the glass beads.

13.3.1.2 Material Density

13.3.1.2.1 Pycnometer Method

Before measuring the material density of a sample, it must be milled to ensure that no closed pores remain. The volume of ground material can be measured by liquid or gas displacement methods. When liquid is used, additional steps should be taken to guarantee the cover of the solid surface or pores, which include the gradual exhausting of air from the measuring apparatus by a vacuum pump to promote the escape of the air trapped under the surface, the filling of the apparatus with toluene, and the allowance for the temperature to reach 20°C, when air bubbles escape after several cycles of vacuuming and releasing the vacuum (Rahman, 2005).

Particle density can be measured by helium gas pycnometer at room temperature. However, there are negligible methods available to measure density as a function of temperature. Rahman et al. (2007) used a method to measure material density as a function of temperature by pycnometer method using a specific gravity bottle of 25 cm^3 (Rahman et al., 2007). Vegetable oil was used as a displacing liquid. Around 30 g of ground spaghetti powder or small pieces of spaghetti are placed inside a preweighed-specific gravity bottle. After weighing the bottle and sample, vegetable oil is poured into it with a dropper and filled close to its neck. In measuring material density, it is important that the displacing liquid enters into all pores in the sample (Rahman, 2005). To overcome it, the air is gradually exhausted from the bottle by applying a vacuum to promote the escape of the air trapped under the surface. When air bubbles escaped after 3–5 cycles of vacuuming and releasing the vacuum, the bottle is filled with vegetable oil and lid is placed. After wiping the excess oil on the outside surface of the bottle by a soft tissue paper, it is allowed to equilibrate at room temperature (20°C). The specific gravity bottle is filled with the sample and oil is placed in an oven and kept at a specified temperature for at least 30 min. After equilibration, the bottle is taken out from the oven and weighed after wiping out (with soft tissue paper) excess oil on the surface of the bottle. Similarly, the density of oil is measured. The bottle with the sample and oil is weighed in a balance of precision ±0.0001 g. The volume of the specific gravity bottle is determined with distilled water. The material density is estimated from the volume and mass of spaghetti assuming all pores are filled with oil.

13.3.1.2.2 Mercury Porosimetry

This method is suitable not only for measuring the pore volume, but also for the determination of pore characteristics and size distribution. It is a well-defined technique and generally accepted method for characterization of pores (Rahman, 2005; Rahman et al., 2002, 2005). Modern mercury porosimeters can achieve a pressure of more than 410 MPa. They are able to measure pore size from 0.003 to 360 nm. It is a relatively fast measurement method, as a typical analysis takes less than 1 h. To apply this method, an isotherm of pressure versus intrusion volume should be constructed. From an isotherm of this type, the total pores or the pore distribution can be derived. According to this technique, as pressure is applied, the pressure at which the mercury intrudes into the pores is measured. The smaller the pore, the higher is the pressure required to force the mercury into the pores. The amount of mercury that intrudes is monitored at a series of pressure points, representing corresponding pore sizes, as pressure is increased. Often the pressure decrease is also monitored, and a series of volumes and pore sizes are obtained as the mercury extrudes out of the pores. In this case, some amount of mercury often remains in the sample. Usually, there is a difference between the intrusion and extrusion curves due to restrictions and bottlenecks that form in the pores and this can provide valuable data regarding the pore shape (Particulars, 1998). The relation of intrusion pressure to the pore diameter is given by the following expression:

$$D_{\mathrm{u}} = \frac{4\gamma \cos\theta}{P} \tag{13.14}$$

where

P is the applied pressure (Pa)
D_{u} is the pore diameter (m)
γ is the surface tension of the fluid (N/m)
θ is the contact angle of the liquid on the capillary material

Equation 13.14 is known as Washburn equation and describes the capillary law governing the penetration of a liquid into small pores.

Mercury porosimetry is suitable for measurement of smaller open pores since it uses high pressure. An advantage of the method is that the sample need not be ground since high pressure forces mercury to penetrate into the pores, although it does not guarantee that mercury has intruded into all pores even at very high pressure (Rahman et al., 2002). Another advantage is that both apparent density and pore volume are directly measured with this method without coating the sample. When very high pressure is used, the compressibility of solids should be considered. On the other hand, under such conditions, the mercury enters the pores and compresses the trapped air in them to a negligible volume. Hence, the mercury volume injected equals the pore volume. Disadvantages of this method are

- Sample cannot be used for further tests, due to mercury contamination.
- Results may not be very precise if the volume occupied by compressed air is not determined.

13.3.1.2.3 Gas Adsorption Method

This technique is also used both for pore volume measurement and for pore characteristics and size distribution determination. It can measure pores from 4 to 5000 Å, but a measurement can last from 2 h up to 60 h (Particulars, 1998). To apply this method, an isotherm of volume adsorbed versus relative pressure should be constructed. The sample is cooled, usually to cryogenic temperatures, and is then exposed to the inert adsorptive gas, typically nitrogen, at a series of precisely controlled temperatures (Rahman, 2005). Gas molecules are adsorbed onto the sample surface during pressure increase and pressure transducers measure the volume of adsorbed gas. First, the very small micropores, less than 20 Å, are filled, followed by the free surface and finally the rest of the pores. Each pressure corresponds to a pore size and the volume of the pores is obtained from the volume of the adsorbed gas. Additional information on pore size can be obtained from the desorption isotherm derived during pressure reduction and the hysteresis observed.

13.3.1.3 Particle Density

Particle density can be measured by the volume displacement method described in apparent density measurement without coating the sample as we do not want to take into account the externally connected pores and hence gas should penetrate them. Any structural change or damage of the sample during measurement should be avoided as particle density considers all closed pores.

13.3.1.4 Bulk Density

Bulk density that takes under consideration all pores inside and outside the individual particles when a material is packed or stacked in bulk can be determined by placing a mass of particles into a container (e.g., measuring cylinder) of known volume. According to the technique followed, material is filled into the container and the excess amount on the top of the cylinder is removed by sliding a string or ruler along the top edge of the cylinder. Gentle tapping of the cylinder vertically down to a table may also be done (Rahman, 2005). The mass of the sample is measured and the bulk density is calculated by the equation

$$\rho_b = \frac{m_s}{V_C} \tag{13.15}$$

where
m_s is the measured mass (kg)
V_C is the container volume (m^3)

13.3.2 Shrinkage Measurement

The estimation of (apparent) shrinkage given by Equation 13.1 is reduced in apparent volume measurement of the material initially and at the given moisture content. Therefore, the methods described in previous section for the measurement of apparent volume, such as the geometric dimension method and the volume displacement methods, can be used for this purpose. Besides

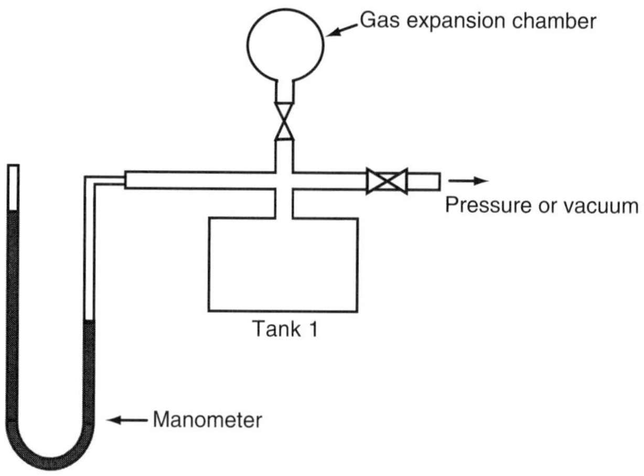

FIGURE 13.6 Gas expansion chamber for measuring volume.

the methods described for volume measurement, including the geometric dimension method and the volume displacement methods, gas expansion method is also used for the measurement of the air or gas volume contained in a pore space (Rahman, 1995). A simple flow diagram for the measuring is illustrated in Figure 13.6. The food material specimen is enclosed in container 1 under pressure or vacuum. Container 1 is connected with the evacuated container 2. Both containers are of known volume. The volume of the specimen is calculated by the equation

$$V_s = V_1 - V_2 \tag{13.16}$$

where V_1 and V_2 are the volumes of the containers 1 and 2, respectively. The above expression is derived from Boyle's law at constant temperature. To determine the solid volume of the material, it should be grounded so that no closed pores remain in the sample.

13.3.3 Porosity Measurement

13.3.3.1 Direct Method

In this method, the bulk volume of a sample of porous material is measured and then the material is compacted for the destruction of all its voids. The volume of the ground material is measured and porosity is determined from the difference of the two measured volumes. This method can be applied for very soft materials, if no repulsive or attractive force is present between the surfaces of the solid particles.

13.3.3.2 Optical Microscopic Method

Porosity can be determined from the microscopic view of a random section of the porous medium. This method is reliable if the sectional (2D) porosity is the same as the bulk (3D) porosity (Rahman, 1995).

13.3.3.3 X-Ray Microtomography

Low resolution x-ray tomography or computerized axial tomography (CAT scanning) is widely used in medicine to image body tissues in a noninvasive manner. This technology has its origins in the 1970s and conventionally has a resolution of ~100 μm, which is insufficient to explore the microstructure of many types of foam. The use of very high-energy x-rays from high-intensity synchrotrons or other sources allowed resolution of 1 μm and enabled the use of XMT (Coker et al., 1996). Two types of XMT are common: one based on absorbance of x-rays and the other based on the phase shift (or phase contrast) of incident x-rays produced by an object (Falcone et al., 2004; Snigirev et al., 1995). Noninvasive imaging technology XMT can be used in the visualization and measurement of microstructural features. It allows noninvasive imaging of sample cross-sections at various depths and facilitates accurate and hitherto impossible measurements of features like true cell size distribution (bimodal), average diameter, open wall area fraction, cell wall thickness, presence or absence of an interconnected network, and void fraction (Trater et al., 2005). All these features are impossible to ascertain using destructive, 2D imaging techniques of light microscopy and scanning electron microscopy (SEM). Another problem in SEM or optical imaging is obtaining adequate contrast between air and solid phases, for which the lighting and angle of illumination play an important role. The XMT technique was used for 3D visualizing the microstructural features of expanded extruded products (Trater et al., 2005) and bread (Falcone et al., 2004).

13.3.3.4 Density Method

Porosity is estimated from the densities of the materials using Equations 13.2 through 13.5. Density measurement methods are described above. If densities of the material are not known, porosity can be estimated from its pore volume, whose measurement can be done with liquid or gas displacement methods. Analytically, open pore, closed pore, and apparent porosity are expressed as the volume of open pores, the volume of closed pores and the volume of all pores, respectively, to the total volume of the material. Similarly, bulk porosity is given as the volume of voids outside a material's boundary per total bulk volume of stacked materials.

13.3.3.5 Fragile Agglomerates Method

Hogekamp and Pohl (2003) proposed a method for the porosity measurement of fragile agglomerates. These particles are produced by agglomeration processes (e.g., fluidized bed agglomeration or steam jet agglomeration) in the food (and pharmaceutical) industry for redispersion in liquids and are known as "instant powders." They include agglomerated skim and whole milk powder, agglomerated pectin, agglomerated flour/maltodextrin mixture, and agglomerated rice starch/maltodextrin mixtures, presenting a porosity ranging from 0.4 to 0.8 and a particle size between 0.2 and 2 mm.

According to this method, an individual particle weighing and sizing is used. A balance with 1 μg of precision allows reliable weighing of individual agglomerates as small as 300 μm diameter. The volume of the particles was determined using an optical sizing device consisting of a pivoted, semitransparent, backlit support for observation of a single particle, a charge-coupled device (CCD) camera, and a long-distance microscope giving an optical resolution of 5 μm. Three orthogonal projections of the particle in backlight can be recorded with this arrangement by rotating the support around the z-axis by 0°, 120°, and 240°. For recording and processing of the images, a computer running public domain image analysis software is used. After recording the three orthogonal projection images, the average projection area A of the particle and the volume V of a sphere of equal projection area is calculated. The imaging error involved in this method can be narrowed down for images of good illumination and contrast to ±2 pixels. Assuming that the magnification of the microscope is variable but always exactly known, the apparent diameter of a particle may be adjusted to an arbitrary value determined only by digital image resolution (a phase alternating line (PAL) image full frame, for example, would be approximately 500 × 750 pixels). Limitations of this method are

- All cavities between the closest engulfing surface and the (wider) engulfing surface were assumed by the analysis of the three projections to be part of the agglomerate. Still, this is not totally wrong, as this volume also must be filled up by liquid in the course of the agglomerate wetting process.
- It is applicable to quality control and compare purposes for particles originating from the same manufacturing process, because they may be considered to possess a similar shape.

Nevertheless, the determination of porosity for agglomerates is quite difficult and classic methods, such as mercury porosimeter, are difficult to apply because of the following reasons:

- Shape and structure of agglomerates; it is, sometimes, impossible to determine the intrusion starting point, i.e., the point where intraagglomerate voids are filled and interagglomerate pores are beginning to be intruded, as instant agglomerates possess interagglomerate pores or cavities, which are the same size as intraagglomerate voids.
- Agglomerates are very fragile and breakage is often observed.

Hogekamp and Pohl (2003) reported that, on average, the results indicate that typical absolute methodological errors of determining the porosity of agglomerates using this method are in the order of ± 0.02 for "smoothly" shaped agglomerates and in the order of ± 0.05 for "roughly" shaped ones.

13.3.4 Characterizing Pores with Mercury Porosimetry

In addition to porosity, pore characteristics and size distribution can also be determined by mercury porosimetry. Modern mercury porosimetry can achieve pressures in excess of 414 MPa, which translates into a pore size of 0.003 nm. This technique involves constructing an isotherm of pressure versus intrusion volume. Total open pores, characteristics of pores, and pore size distribution can be derived from these isotherms.

13.3.4.1 Intrusion and Extrusion Curves

The intrusion curve is to denote the volume change with increasing pressure and the extrusion curve is to indicate the volume change with decreasing pressure (Figure 13.7). Mercury intrusion and extrusion results are compared to assess capillary hysteresis, which could be used to explore characteristics of pores, such as shape and structure. Most of the porosimetry curves exhibit hysteresis. The difference in curvatures of the intrusion and extrusion curves indicates that the

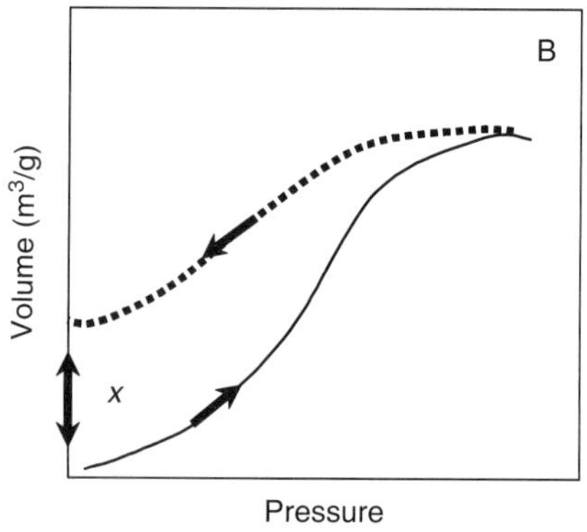

FIGURE 13.7 Intrusion and extrusion curves of mercury porosimetry showing hysteresis.

path followed by the extrusion curve is not the same as the intrusion path. At a given pressure, the volume indicated on the extrusion curve is lower than that on the intrusion curve and the difference in volume is the entrapped volume. In Figure 13.7, entrapped mercury volume is x. The intrusion–extrusion cycle does not close when the initial pressure is reached after extrusion, indicating that some mercury is permanently entrapped in the sample, thereby preventing the loop from closing. Often at the completion of an intrusion–extrusion cycle mercury slowly continues to extrude, sometimes for hours (Lowell and Shields, 1984).

13.3.4.2 Pore-Size Distribution Curves

Pore-size distribution curves are characterized by different methods. In addition to the intrusion–extrusion curve, other plots are used to characterize the pores. These are

- Derivative of the cumulative volume curve $(\mathrm{d}V/\mathrm{d}P)$ versus P or diameter
- Volume distribution $(\mathrm{d}V/\mathrm{d}r)$ versus P or diameter
- Pore-size distribution function (f_v) versus P or pore diameter

where $\mathrm{d}V$ is the volume of intruded mercury in the sample (considered exactly equal to the volume of pores) and P is the pressure (Pa). The term $\mathrm{d}V/\mathrm{d}P$ is the first derivative (slope) of the volume versus pressure or diameter data. It is used in subsequent calculations of the volume distribution function, area distribution function, and the pore number fraction. Pore-size distribution was found as the computed relation between pore radius and pore volume, assuming cylindrical pores (Lowell and Shields, 1984). Pore-size distribution function can be defined as

$$f_v = \left(\frac{P}{r}\right) \times \left(\frac{\mathrm{d}V}{\mathrm{d}P}\right) \tag{13.17}$$

where f_v is the pore-size distribution function usually expressed as $\mathrm{m}^3/\mathrm{m}/\mathrm{kg}$, and r is the pore radius (μm). The above function uses the slope times P/r, thus it becomes increasingly susceptible to minor fluctuation in volume or pressure data or to slight changes in slope. These slight variations in slope can result from actual discontinuities in the pore size distribution, that is, some pore sizes may be absent or present in a lesser volume than pores immediately adjacent. This means that pore-size distributions are usually not smooth and continuous functions (Rahman et al., 2002).

Fractal analysis is also used to characterize the pores (Rahman et al., 2005). Pfeifer and Avnir (1983) derived a scaling equation for pore-size distribution as

$$\frac{\mathrm{d}V}{\mathrm{d}D} \propto D^{2-\delta} \tag{13.18}$$

where

D is the pore diameter (m)
V is the pore volume (m^3)
δ is the fractal dimension

In terms of applied pressure, the scaling equation is

$$\frac{\mathrm{d}V}{\mathrm{d}P} \propto P^{\delta-4} \tag{13.19}$$

The fractal dimension can be estimated from the slope of the plot of log $(\mathrm{d}V/\mathrm{d}P)$ versus log P.

ACKNOWLEDGMENTS

The authors would like to thank Prof. Zacharias Maroulis and Prof. George Saravacos for their guidance and comments for the writing of this chapter.

NOMENCLATURE

a	x-axis length (m)
b	y-axis length (weight) (m)
c	z-axis length (height) (m)
D	diameter (m)
$\mathrm{d}V/\mathrm{d}P$	derivative of the cumulative volume curve
$\mathrm{d}V/\mathrm{d}r$	volume distribution
f_v	pore size distribution function ($\mathrm{m}^3/\mathrm{m\ kg}$)
G	buoyant force (kg)
L	length (m)
m	mass (kg)
p	major semiaxes of the ellipse of rotation (m)
P	pressure (Pa)
q	minor semiaxes of the ellipse of rotation (m)
r	radius (m or μm in case of pores)
S	shrinkage
V	volume (m^3)
V_c	constant
V_R	constant
W	weight (kg)
x	entrapped mercury volume (m^3/g)

Symbols

γ	surface tension (N/m)
δ	fractal dimension
ε	porosity
θ	contact angle
π	constant (= 3.14)
ρ	density ($\mathrm{kg/m}^3$)

Subscripts

α	apparent
b	bulk
bp	bulk particle
C	container
cp	closed pore
i	initial
l	liquid
m	material
op	open pore
p	particle

s	specimen (sample)
t	true
T	total
u	pore
wax	wax

REFERENCES

Bailey, C.H. 1912. A method for the determination of the specific gravity of wheat and other cereals. USDA Bureau of Plant Industry, Circular N. 99.

Balaban, M. and Pigott, G.M. 1986. Shrinkage in fish muscle during drying. *Journal of Food Science*, 51(2): 510.

Clayton, J.T. and Huang, C.T. 1984. Porosity of extruded foods. In: *Engineering and Food. Volume 2. Processing and Applications*, McKenna, B., ed. Elsevier Applied Science Publishers, Essex, pp. 611–620.

Coker, D.A., Torquato, S., and Dunsmuir, J.H. 1996. Morphology and physical properties of Fontainebleau sandstone via a tomographic analysis. *Journal of Geophysical Research*, 101: 17497–17506.

Falcone, P.M., Baiano, A., Zanini, F., Mancini, L., Tromba, G., and Montanari, F. 2004. A novel approach to the study of bread porous structure: Phase-contrast X-ray microtomography. *Journal of Food Science*, 69(1): FEP38–FEP43.

Hogekamp, S. and Pohl, M. 2003. Porosity measurement of fragile agglomerates. *Powder Technology*, 130: 385–392.

Keppeler, R.A. and Boose, J.R. 1970. Thermal properties of frozen sucrose solutions. *Transactions of the ASAE*, 13(3): 335–339.

Lewis, M.J. 1987. *Physical Properties of Foods and Food Processing Systems*. Ellis Horwwod, England and VCH Verlagsgsellschaft, FRG.

Loch-Bonacci, C.L., Wolf, E., and Gilbert, H. 1992. Quality of dehydrated cultivated mushrooms (*Agaricus bisporus*): A comparison between different drying and freeze-drying processes. *Food Science and Technology*, 25: 334–339.

Lowell, S. and Shields, J.E. 1984. *Powder Surface Area and Porosity*, 2nd edn. Chapman and Hall, London.

Lozano, J.E., Rotstein, E., and Urbicain, M.J. 1980. Total porosity and open-pore porosity in the drying of fruits. *Journal of Food Science*, 45: 1403–1407.

Marousis, S.N. and Saravacos, G.D. 1990. Density and porosity in drying starch materials. *Journal of Food Science*, 55(5): 1367.

Moshenin, N.N. 1986. *Physical Properties of Plant and Animal Material*. Gordon and Breach Science Publishers, New York.

Particulars. 1998. Newsletter of particle and surface sciences, Issue 3, Gosford, New South Wales.

Pfeifer, P. and Avnir, D. 1983. Chemistry in noninteger dimensions between two and three. I. Fractal theory of heterogenous surfaces. *Journal of Chemical Physics*, 79: 3558–3565.

Rahman, M.S. 1995. *Food Properties Handbook*. CRC Press, Boca Raton, FL.

Rahman, M.S. 2005. Mass–volume–area-related properties of foods. In: *Engineering Properties of Foods*, 3rd edn., Rao, M.A., Rizvi, S.S.H., and Datta, A.K., eds. Taylor & Francis, Boca Raton, FL.

Rahman, M.S. and Driscoll, R.H. 1994. Density of Fresh and frozen seafood. *Journal of Food Process Engineering*, 17: 121–140.

Rahman, M.S. and McCarthy, O.J. 1999. A classification of food properties. *International Journal of Food Properties*, 2(2): 93–99.

Rahman, M.S. and Potluri, P.L. 1990. Shrinkage and density of squid flesh during air drying. *Journal of Food Engineering*, 12(2): 133–143.

Rahman, M.S., Al-Amri, D., and Al-Bulushi, I.M. 2002. Pores and physicochemical characteristics of dried tuna produced by different methods of drying. *Journal of Food Engineering*, 53: 301–313.

Rahman, M.S., Al-Zakwani, I., and Guizani, N. 2005. Pore formation in apple during air-drying as a function of temperature: porosity and pore-size distribution. *Journal of the Science of Food and Agriculture*, 85(6): 979–989.

Rahman, M.S., Al-Marhubi, I.M., and Al-Mahrouqi, A. 2007. Measurement of glass transition temperature by mechanical (DMTA), thermal (DSC and MDSC), water diffusion and density methods: A comparison study. *Chemical Physics Letters*, 440: 372–377.

Ramaswamy, H.S. and Tung, M.A. 1981. Thermophysical properties of apples in relation to freezing. *Journal of Food Science*, 46: 724.

Snigirev, A., Snigireva, I., Kohn, V., Kuznetsov, S., and Schelokov, I. 1995. On the possibilities of X-ray phase contrast microimaging by coherent high-energy synchrotron radiation. *Review of Scientific Instruments*, 66(12): 5486–5492.

Trater, A.M., Alavi, S., and Rizvi, S.S.H. 2005. Use of non-invasive X-ray microtomography for characterizing microstructure of extruded biopolymer foams. *Food Research International*, 38: 709–719.

Zogzas, N.P., Maroulis, Z.B., and Marinos-Kouris, D. 1994. Densities, shrinkage and porosity of some vegetables during air drying. *Drying Technology*, 12(7): 1653–1666.

CHAPTER 14

Data and Models of Density, Shrinkage, and Porosity

Panagiotis A. Michailidis, Magdalini K. Krokida, and Mohammad Shafiur Rahman

CONTENTS

14.1 INTRODUCTION

Food structure at molecular, microscopic, and macroscopic levels can be applied not only in the study and evaluation of food texture and quality, but also to the analysis and correlation of the transport properties of foods, such as viscosity, thermal conductivity/diffusivity, and mass diffusivity. At the molecular level, Aguilera and Stanley (1999) reported that food biopolymers (proteins, carbohydrates, and lipids) of importance to transport properties are structural proteins (collagen, keratin, and elastin), storage proteins (albumins, globulins, prolamins, and glutenins), structural polysaccharides (cellulose, hemicelluloses, pectins, seaweed, and plant gums), storage polysaccharides (starch–amylose and amylopectin), and lignin (plant cell walls).

The microscopic level refers to the cells of food materials that contain several components, essential in living organisms, such as water, starch, sugars, proteins, lipids, and salts. As food materials undergo different processes, their microstructure may be preserved or destroyed for the production of useful processed products (e.g., refining of starch, oil seeds, sugars, grain, and milk), while other processes such as freezing, crystallization, milling, and emulsification cause changes in the food material structure.

At the macroscopic level, the structural (macroscopic) properties, such as density, porosity, shrinkage, particle size, and shape, can be defined as quantitative parameters of physical meaning, for the characterization of structural changes of foods during processing and storage. These properties are strongly related to the transport properties of solid and semisolid food materials. Structural generic models combining expressions for the prediction of transport properties and structural properties can be used for the calculation of the effective transport properties of food materials, such as effective thermal conductivity (Saravacos and Maroulis, 2001). The design, operation, and control of food processing operations are based on these properties, which can be measured by a variety of instruments and techniques. Food macrostructure, used in engineering and processing applications, can be related to the recent advances of food microstructure at the cellular level. Applications of food structural analysis and experimental data to the transport properties, especially mass transport, could improve food process and product development, and food product quality. Food materials are classified in two categories:

- Continuous solids, in which shrinkage and porosity develop when water is removed
- Particulate or granular materials, such as starch granules, in which porosity is a dependent variable that can be controlled, e.g., by compression or vibration

The knowledge of food physical structure is fundamental in the developing field of food material science (Krokida and Maroulis, 2000; Krokida et al., 2000; Krokida et al., 2001; Saravacos and Maroulis, 2001; Maroulis and Saravacos, 2003). In this chapter, a large number of density and porosity data for various food materials are given. Shrinkage data are also included. In addition, mathematical models for the estimation of these properties are presented as a function of a series of parameters that affect them. Generic models for the prediction of a property and models for the calculation of a property of a specific food material are proposed. Two generic semiempirical models for the prediction of density, shrinkage, and porosity as a function of properties are presented as a function of parameters, such as moisture content and temperature. The effect of drying methods, including frying, on the structural properties of food materials is illustrated with the temperature dependence of density. A few models for the calculation of structural properties of extrudate food materials are also presented. Finally, findings for the pore formation mechanisms are discussed.

14.2 DATA FOR DENSITY, SHRINKAGE, AND POROSITY

Density, shrinkage, and porosity are the most common structural properties. As they play a significant role in food texture and quality, and in food process design, extensive research has been done in resent years for the collection of data concerning these properties. The data available in the literature are presented in two forms, either tabulated or as an equation, versus one or more parameters, including moisture content (the most common one), temperature, etc. Food macrostructure has received special attention in relation to the dehydration of foods (moisture content decrease) because significant changes take place during moisture transport in solid and semisolid food materials. Tables 14.1 through 14.8 present density data for different materials. Additional tabulated data for structural properties are given by Lewis (1987) (vegetable oils, food powders, aqueous solutions of fructose, lactose, maltose, and glycerol), Keppeler and Boose (1970) (frozen sucrose solutions), and Rahman (1991) (seafood). Table 14.1 presents the value range for the density of some common materials (uranium is the heaviest element found in earth). According to the definitions given in Chapter 11, "true density" is estimated from the densities of foods' components, and "material density" is measured experimentally when the material has been thoroughly broken into pieces small enough to guarantee no closed pores remain. The majority of the literature data and models retrieved from experimental procedures use the term "true density," although they express the "material density." For the data and models presented in this chapter, the original definitions, as "true density," have been retained. It is important for the reader to have in mind that all experimental data represent "material density" of the foods.

Table 14.2 presents true density values of some of the most common food components. Most of them are in the range of 900–1670 kg/m^3, while inorganic salts are heavier. The density of

Table 14.1 Density of Some Selected Materials

Material	Density (kg/m^3)	T (°C)	Reference
Aluminum	2,640	0	Earle (1983, cited by Lewis, 1987)
Aluminum	2,707	20	Whitaker (1976)
Cast iron	7,210	0	Earle (1983, cited by Lewis, 1987)
Copper	8,900	0	Earle (1983, cited by Lewis, 1987)
Copper	8,955	20	Whitaker (1976)
Barium	3,510	25	—
Silicon	2,330	25	—
Uranium	19,100	25	—
Mild steel	7,840	18	Earle (1983, cited by Lewis, 1987)
Stainless steel	7,950	20	Earle (1983, cited by Lewis, 1987)
Concrete	2,000	20	Earle (1983, cited by Lewis, 1987)
Brick	1,760	20	Earle (1983, cited by Lewis, 1987)
Wood	200	30	Earle (1983, cited by Lewis, 1987)

Table 14.2 True Density of Some Food Components

Component	Density (kg/m^3)
Water	1000
Protein	1400
Fat	900–950
Cellulose	1270–1610
Gelatine	1270
Glucose (α)	1544
Glucose (β)	1562
Sucrose	1588
Fructose	1669
Maltose	1540
Starch	1500
Ethyl alcohol	789
Acetic acid	1049
Glycerol	1260
Citric acid	1540
NaCl	2163
KCl	1988

Sources: Adapted from Weast, R.C., *Handbook of Chemistry and Physics*, CRC Press, Cleveland, OH, 1970; Perry, R.H. and Green, D., *Perry's Chemical Engineers' Handbook*, McGraw-Hill, New York, 1997; Lewis, M.J., *Physical Properties of Foods and Food Processing Systems*, Ellis Horwood, England and VCH Verlagsgesellschaft, FRG, 1987.

1400 kg/m^3 can be considered as an average density for food components, indicating that a dry solid food material is heavier than water. Table 14.3 gives the range of variations of the density of some common food materials (Milson and Kirk, 1980). It is notable that the densities of frozen fruits and vegetables are much lower than those of their fresh counterparts. Table 14.4 gives density values of different milks at a temperature of 25°C. It can be seen that the density of milk is a little higher than that of water due to solid components in it. (X_s, X_{pr}, X_{fa}, X_{ca}, and X_{as} are the contents of total solids, proteins, fats, carbohydrates and ash, respectively, expressed as kg solid/kg sample.)

Table 14.3 Density of Some Selected Food Materials

Food	Density (kg/m^3)	Food	Density (kg/m^3)
Fresh fruit	865–1067	Fresh fish	1056
Fresh vegetables	801–1095	Frozen fish	967
Frozen fruit	625–801	Ice (0°C)	916
Frozen vegetables	561–977	Ice (−10°C)	933
Meat	1.07[a]	Ice (−20°C)	948

Sources: Adapted from Milson, A. and Kirk, D., *Principles of Design and Operation of Catering Equipment*, Ellis Horwood, Chichester, West Sussex, 1980; Lewis, M.J., *Physical Properties of Foods and Food Processing Systems*, Ellis Horwood, England and VCH Verlagsgesellschaft, FRG, 1987.

[a] Specific gravity.

Table 14.4 Density of Different Milks at 25°C

	Density (kg/m³)		
X_s (kg/kg)	Cow's Milk[a]	Soymilk (TGX)[b]	Soymilk (Doko)[c]
0.020	1000	1010	1010
0.040	1010	1020	1020
0.060	1010	1030	1030
0.080	1010	1030	1030
0.100	1020	1040	1040
0.116	1020		1050
0.124	1020		

Source: Adapted from Oguntunde, A.O. and Akintoye, O.A., *J. Food Eng.*, 13, 221, 1991.

[a] M_{pr}, 0.656; M_{fa}, 0.184; M_{ca}, 0.069; M_{as}, 0.008; M_{cf}, 0.083.
[b] M_{pr}, 0.591; M_{fa}, 0.218; M_{ca}, 0.056; M_{as}, 0.009; M_{cf}, 0.126.
[c] M_{pr}, 0.256; M_{fa}, 0.302; M_{ca}, 0.388; M_{as}, 0.054; M_{cf}, -.

Table 14.5 Bulk Density of Some Powder Food Materials

Powder	Bulk Density (kg/m³)	Reference
Oats	513	Milson and Kirk (1980)
Wheat	785	Milson and Kirk (1980)
Flour	449	Milson and Kirk (1980)
Cocoa	480	Peleg (1983)
Coffee (instant)	330	Peleg (1983)
Coffee (ground and roasted)	330	Peleg (1983)
Corn starch	560	Peleg (1983)
Milk	610	Peleg (1983)
Salt (granulated)	960	Peleg (1983)
Sugar (granulated)	800	Peleg (1983)
Sugar (powdered)	480	Peleg (1983)
Wheat flour	480	Peleg (1983)
Yeast (bakers)	520	Peleg (1983)
Egg (whole)	340	Peleg (1983)

Table 14.6 Density of Fruit Juice as a Function of Solid Content

Fruit Juice	Solid Content (%wb)	Density (kg/m³)	Reference
Apple	50.2	1227	Dickerson (1968)
Apple	12.8	1051	Dickerson (1968)
Apple	13.0	1060	Egan et al. (1981, cited by Lewis, 1987)
Bilberry	10.5	1041	Dickerson (1968)
Black currant	13.5	1055	Egan et al. (1981, cited by Lewis, 1987)
Cherry	13.3	1053	Dickerson (1968)
Grape fruit	10.4	1040	Egan et al. (1981, cited by Lewis, 1987)
Grape fruit	15.3	1062	Dickerson (1968)
Lemon	10.0	1035	Egan et al. (1981, cited by Lewis, 1987)
Lime	9.3	1035	Egan et al. (1981, cited by Lewis, 1987)
Orange	10.8	1042	Egan et al. (1981, cited by Lewis, 1987)
Orange	11.0	1043	Dickerson (1968)
Raspberry	11.5	1046	Dickerson (1968)
Strawberry	8.3	1033	Dickerson (1968)

Table 14.7 Apparent Density (kg/m^3) of Fresh (20°C) and Frozen (−30°C) Seafood

Material	X_w	X_{pr}	X_{fa}	X_{as}	Fresh	Frozen
Cuttle (mantle)	0.81	0.14	0.024	0.0097	1064	990
Squid (tail)	0.85	0.11	0.031	0.0108	1048	975
Squid (mantle)	0.82	0.13	0.033	0.0119	1061	981
Squid (mantle)	0.82	0.14	0.030	0.0131	1062	982
Prawn (abdomen)	0.75	0.19	0.029	0.0182	1088	1017
Calamari (tentacle)	0.85	0.11	0.032	0.0059	1050	972

Source: Adapted from Rahman, M.S., Thermophysical properties of seafoods, PhD Thesis, University of New South Wales, Sydney, 1991.

X, component content, wet basis (kg/kg sample).

Table 14.8 Bulk Density, True Density, and Porosity of Foods versus Moisture and Temperature (Ranges of Variation of Available Data)

	ρ_b (kg/m^3)		ρ_t (kg/m^3)		ε (−)		T (°C)		X_w (wb)	
Material	**Min**	**Max**	**Min**	**Max**	**Min**	**Max**	**Min**	**Max**	**Min**	**Max**
Baked products	161	1287	1268	1430	0.000	0.790	18	200	0.000	0.539
Bread (Zanoni et al., 1995; Sablani et al. 2002)	161	974			0.000	0.790	18	120	0.000	0.428
—	200	200					22	22	0.320	0.320
Crumb	187	974			0.000	0.790	18	100	0.418	0.428
Crust	258	895			0.000	0.710	100	120	0.000	0.000
French	161	161					22	22	0.420	0.420
Cake (Zanoni et al., 1995)	300	694							0.355	0.415
—	300	694							0.355	0.415
Cup cake batter (Baik et al., 1999; Baik and Marcotte, 2002)	236	803			0.220	0.770	25	200	0.059	0.346
—	236	803			0.220	0.770	25	200	0.059	0.346
Dough (Zanoni et al., 1995; Kawas and Moreira, 2001)	520	1287	1268	1294	0.320	0.590	28	190	0.041	0.539
Biscuit	1252	1287							0.041	0.085
Rye bread	701	820							0.459	0.539
Tortilla	520	880	1268	1294	0.320	0.590	190	190		
Wheat bread	586	750					28	28	0.420	0.451
Soy flour (Fitzpatrick et al., 2004)	600	600	1430	1430	0.580	0.580			0.062	0.062
Powder	600	600	1430	1430	0.580	0.580			0.062	0.062
Yellow cake batter (Sablani et al., 2002)	497	497					22	22	0.380	0.380
—	497	497					22	22	0.380	0.380
Cereal products	172	1130	1188	1713	0.194	0.676	20	180	0.010	0.800
Corn (Rahman, 1995; Fitzpatrick et al., 2004)	730	865	1490	1490	0.510	0.510			0.090	0.090

Table 14.8 (continued) Bulk Density, True Density, and Porosity of Foods versus Moisture and Temperature (Ranges of Variation of Available Data)

	ρ_b (kg/m^3)		ρ_t (kg/m^3)		ε (−)		T (°C)		X_w (wb)	
Material	**Min**	**Max**	**Min**	**Max**	**Min**	**Max**	**Min**	**Max**	**Min**	**Max**
Flour	730	730	1490	1490	0.510	0.510			0.090	0.090
Sweet	865	865								
Maize (Hatamipour and Mowla, 2003)	1060	1130					60	60	0.415	0.700
—	1060	1130					60	60	0.415	0.700
Millet (Baryeh, 2002)	555	689	1557	1713	0.571	0.676	20	20	0.048	0.184
—	555	689	1557	1713	0.571	0.676	20	20	0.048	0.184
Oat (Rahman, 1995)	513	513								
—	513	513								
Pearl millet (Jain and Bal, 1997)	830	866	1578	1623	0.451	0.489	22	22	0.069	0.069
Babapuri	830	830	1623	1623	0.489	0.489	22	22	0.069	0.069
Bajra 28–15	854	854	1591	1591	0.463	0.463	22	22	0.069	0.069
GHB30	866	866	1578	1578	0.451	0.451	22	22	0.069	0.069
Quinoa (Vilche et al., 2003)	667	747	1188	1188	0.194	0.438	20	20	0.044	0.205
Seeds	667	747	1188	1188	0.194	0.438	20	20	0.044	0.205
Rice (Guha et al., 1997; Kostaropoulos and Saravacos, 1997; Ramesh, 2003)	172			0	0.490	20.000	180	0	0.800	
—					0.450	0.490	20	20	0.010	0.090
Cooked							110	180	0.080	0.800
Flour	172	231					80	120		
Wheat (Rahman, 1995; Sokhansanj and Lang, 1996; Kostaropoulos and Saravacos, 1997; Teunou et al., 1999; Roman-Gutierrez et al., 2003; Fitzpatrick et al., 2004)	449	790	1332	1480	0.419	0.650	20	40	0.020	0.220
—	785	785					20	20		
Flour	449	710	1467	1480	0.520	0.650	20	20	0.020	0.132
Kenyon	686	790	1332	1374	0.419	0.485	40	40	0.080	0.220
Yeast (Rahman, 1995)	520	520								
—	520	520								
Dairy	421	1190	1133	1216	0.477	0.488	80	150	0.000	0.980
Jameed (Jumah et al., 2000)	587	623					80	150		
—	587	623					80	150		
Milk (Rahman, 1995); Teunou et al., 1999)	593	1020	1133	1133	0.477	0.477			0.046	0.980
—	1000	1020							0.876	0.980
Powder	593	610	1133	1133	0.477	0.477			0.046	0.046

(continued)

Table 14.8 (continued) Bulk Density, True Density, and Porosity of Foods versus Moisture and Temperature (Ranges of Variation of Available Data)

	ρ_b (kg/m^3)		ρ_t (kg/m^3)		ε (−)		T (°C)		X_w (wb)	
Material	**Min**	**Max**	**Min**	**Max**	**Min**	**Max**	**Min**	**Max**	**Min**	**Max**
Whey (Teunou et al., 1999)	622	622	1216	1216	0.488	0.488			0.038	0.038
Powder	622	622	1216	1216	0.488	0.488			0.038	0.038
Yogurt (Kim and Bhowmik, 1997)	421	1190							0.000	0.800
—	421	1190							0.000	0.800
Fish	317	1400	1055	1451	0.000	0.760	−40	70	0.000	0.922
Calamari (Rahman, 1995; Rahman et al., 1996)	972	1246	1074	1339	0.011	0.079	−30	70	0.000	0.845
Mantle	1000	1246	1074	1339	0.011	0.079	20	70	0.000	0.816
Tentacle	972	1050					−30	20	0.829	0.845
Wing	1047	1050					20	20	0.827	0.834
Cod (Rahman, 1995)	1100	1100								
—	1100	1100								
Cuttle (Rahman, 1995)	990	1064					−30	20	0.813	0.814
Mantle	990	1064					−30	20	0.813	0.814
Herring (Rahman, 1995)	990	990								
—	990	990								
Mussel (Rahman, 1995)	1086	1086					20	20	0.774	0.774
—	1086	1086					20	20	0.774	0.774
Octopus (Rahman, 1995)	1042	1075					20	20	0.774	0.853
Mantle	1056	1075					20	20	0.774	0.841
Tentacle	1042	1071					20	20	0.786	0.853
Perch (Balaban and Pigott, 1986; Rahman, 1995)	910	1400					−19	25	0.000	0.784
—	910	1400					−19	25	0.000	0.784
Prawn (Rahman, 1995)	1017	1093					−30	20	0.739	0.773
Abdomen	1017	1017					−30	−30		
Green	1072	1093					20	20	0.739	0.773
Tiger	1081	1081					20	20	0.757	0.757
Squid (Rahman and Potluri, 1990a; Rahman, 1995)	975	1275	1055	1451	0.000	0.125	−40	70	0.000	0.922
—	1055	1275	1055	1451	0.000	0.125	70	70	0.000	0.840
Mantle	981	1066					−40	20	0.793	0.922
Tail	975	1055					−30	20	0.813	0.847
Tentacle	1067	1067					20	20	0.786	0.786
Tuna (Rahman et al., 2002)	317	1071	1071	1319	0.000	0.760	−40	70	0.077	0.723
—	317	960	1312	1319	0.240	0.760	−40	70	0.077	0.114
Fresh	1071	1071	1071	1071	0.000	0.000	20	20	0.723	0.723

Table 14.8 (continued) Bulk Density, True Density, and Porosity of Foods versus Moisture and Temperature (Ranges of Variation of Available Data)

	ρ_b (kg/m^3)		ρ_t (kg/m^3)		ε (−)		T (°C)		X_w (wb)	
Material	**Min**	**Max**	**Min**	**Max**	**Min**	**Max**	**Min**	**Max**	**Min**	**Max**
Fruits	80	1620	1000	2029	0.000	0.950	−50	165	0.000	0.944
Ackee apple (Omobuwajo et al., 2000)	557	557			0.373	0.373			0.099	0.099
Seeds	557	557			0.373	0.373			0.099	0.099
Apple[a]	80	1466	1000	1714	0.020	0.950	−50	165	0.000	0.900
—	80	920	1000	1714	0.100	0.950	−50	70	0.000	0.900
Fresh					0.210	0.210				
Golden delicious	132	900	1038	1471	0.174	0.910	−45	165	0.000	0.894
Gortland	380	460	1390	1520	0.670	0.750				
Granny smith	420	1250	1380	1380	0.020	0.500	−35	60	0.010	0.891
Green	790	849					28	45	0.878	0.885
Jonagold	753	830			0.122	0.198				
Juice	1051	1227							0.498	0.872
Kim	669	847	1000	1000	0.137	0.310	5	40	0.360	0.880
Mutsu	765	841			0.114	0.186	60	60	0.027	0.552
Pacific rose					0.170	0.540	60	60	0.009	0.859
Packed layers	489	489					2	2		
Red	830	840	1059	1059	0.203	0.203	28	28	0.849	0.858
Sause	1232	1466							0.670	0.820
Stark delicious	180	920	1200	1200	0.130	0.800	50	50	0.000	0.850
Apricot (Rahman, 1995; Fikiin et al., 1999; Fito et al., 2001; Gezer et al., 2002)	463	1048	1003	1161	0.022	0.557	2	2	0.064	0.854
—	609	609								
Bulida	1048	1048	1057	1057	0.022	0.022			0.854	0.854
Kernel	545	559	1003	1095	0.497	0.557			0.065	0.279
Packed layers	637	637					2	2		
Pits	463	581	1053	1161	0.440	0.500			0.064	0.266
Arecanut (Kaleemullah and Gunasekar, 2002)	662	696	1139	1152	0.396	0.419	40	40	0.095	0.471
Kernel	662	696	1139	1152	0.396	0.419	40	40	0.095	0.471
Avocado (Rahman, 1995; Tsami and Katsioti, 2000; May and Perre, 2002; Medeiros et al., 2002)	1054	1060			0.090	0.734	28	70	0.012	0.750
—	1060	1060			0.090	0.734	28	70	0.012	0.750
Pulp	1054	1054								
Banana[b]	260	1620	1044	2029	0.016	0.870	−50	80	0.000	0.900
—	260	1620	1096	2029	0.050	0.870	−50	70	0.000	0.835
Macho	1027	1027	1044	1044	0.016	0.016			0.777	0.777
Puree	986	1609					10	80	0.776	0.900

(*continued*)

Table 14.8 (continued) Bulk Density, True Density, and Porosity of Foods versus Moisture and Temperature (Ranges of Variation of Available Data)

	ρ_b (kg/m^3)		ρ_t (kg/m^3)		ε (–)		T (°C)		X_w (wb)	
Material	**Min**	**Max**	**Min**	**Max**	**Min**	**Max**	**Min**	**Max**	**Min**	**Max**
Bilberry (Rahman, 1995)	1041	1041							0.895	0.895
Juice	1041	1041							0.895	0.895
Blackberry (Ratti, 2001)										
—										
Blueberry (Yang and Atallah, 1985)	190	442					−40	70	0.200	0.200
—	190	442					−40	70	0.200	0.200
Cantaloupe (Rahman, 1995)	930	930					28	28	0.928	0.928
—	930	930					28	28	0.928	0.928
Cherry (Rahman, 1995; Medeiros et al., 2002; Calisir and Aydin, 2004)	450	1071	1049	1049	0.414	0.527	20	20	0.090	0.867
—	721	721								
Juice	1053	1053							0.867	0.867
Laurel	450	615	1049	1049	0.414	0.527	20	20	0.090	0.775
Pulp	1071	1071								
Date (Sablani and Rahman, 2002)	205.5	1257	1122	1508	0.000	0.846	−45	15	0.278	0.828
Brown	1093	1257	1363	1475	0.087	0.234	−45	15	0.278	0.278
Yellow	206	1122	1122	1508	0.000	0.846	−45	15	0.828	0.828
Grape (Rahman, 1995; Ghiaus et al., 1997; Fikiin et al., 1999; Gabas et al., 1999; Azzouz et al., 2002)	368	1483	1076	1434	0.370	0.491	2	60	0.000	0.896
—	368	1483					40	60	0.100	0.810
Chasselas							60	60	0.060	0.500
Corinthian	582	762	1076	1434	0.370	0.491	26	26	0.150	0.800
Juice	1040	1062							0.847	0.896
Packed layers	557	557					2	2		
Sultanin							60	60	0.000	0.470
Grapefruit (Rahman, 1995; Chafer et al., 2003)	714	950			0.330	0.360	26	26	0.793	0.904
—	950	950					26	26	0.904	0.904
Peels	714	714			0.330	0.360			0.793	0.793
Guava (Hodgson et al., 1990; Zainal et al., 2000)	1005	1103					65	85	0.724	0.939
Juice	1005	1103					65	85	0.724	0.939
Puree	1035	1042							0.932	0.932
Guna (Aviara et al., 1999)	399	545			0.373	0.414			0.045	0.282
Seeds	399	545			0.373	0.414			0.045	0.282

Table 14.8 (continued) **Bulk Density, True Density, and Porosity of Foods versus Moisture and Temperature (Ranges of Variation of Available Data)**

	ρ_b (kg/m^3)		ρ_t (kg/m^3)		ε (−)		T (°C)		X_w (wb)	
Material	**Min**	**Max**	**Min**	**Max**	**Min**	**Max**	**Min**	**Max**	**Min**	**Max**
Hackberry (Demir et al., 2002)	536	595	1106	1106	0.310	0.540			0.153	0.504
—	536	595	1106	1106	0.310	0.540			0.153	0.504
J. drupacea (Akinci et al., 2004)	466	488			0.507	0.534	20	20	0.082	0.292
—	466	488			0.507	0.534	20	20	0.082	0.292
Karingda (Suthar and Das, 1996)	488	647	1009	1148	0.407	0.575	50	50	0.048	0.289
Kernel	501	647	1014	1133	0.429	0.506	50	50	0.048	0.262
Seeds	488	598	1009	1148	0.407	0.575	50	50	0.066	0.289
Kiwi (Fito et al., 2001)	1051	1051	1076	1076	0.007	0.007			0.815	0.815
Hayward	1051	1051	1076	1076	0.007	0.007			0.815	0.815
Lemon (Rahman, 1995; Chafer et al., 2003)	716	1035			0.370	0.420	28	28	0.804	0.918
—	768	930					28	28	0.918	0.918
Juice	1035	1035							0.900	0.900
Peels	716	716			0.370	0.420			0.804	0.804
Lime (Rahman, 1995)	1000	1035					28	28	0.899	0.907
—	1000	1000					28	28	0.899	0.899
Juice	1035	1035							0.907	0.907
Locust bean (Olajide and Ade-Omowaye, 1999; Ogunjimi et al., 2002)	559	1150	1147	1147	0.514	0.514			0.060	0.093
Seeds	559	1150	1147	1147	0.514	0.514			0.060	0.093
Longmelon (Teotia and Ramakrishna, 1989)	700	1100	1180	1600	0.068	0.563			0.040	0.085
Hull	700	700	1600	1600	0.563	0.563			0.085	0.085
Kernel	1100	1100	1180	1180	0.068	0.068			0.040	0.040
Seeds	850	850	1310	1310	0.351	0.351			0.061	0.061
Mandarin (Fito et al., 2001; Chafer et al., 2003)	849	894	1103	1103	0.240	0.330			0.724	0.750
Peels	849	894	1103	1103	0.240	0.330			0.724	0.750
Mango (Fito et al., 2001; Medeiros et al., 2002; Mujica-Paz et al., 2003)	918	1079	1082	1130	0.059	0.152			0.786	0.841
Manila	918	918	1082	1082	0.152	0.152			0.841	0.841
Pulp	1079	1079								
Tommy	1022	1022	1130	1130	0.059	0.059			0.786	0.786

(continued)

Table 14.8 (continued) Bulk Density, True Density, and Porosity of Foods versus Moisture and Temperature (Ranges of Variation of Available Data)

Material	ρ_b (kg/m^3)		ρ_t (kg/m^3)		ε (–)		T (°C)		X_w (wb)	
	Min	Max	Min	Max	Min	Max	Min	Max	Min	Max
Melon (Fito et al., 2001; Mujica-Paz et al., 2003)	899	976	1022	1043	0.060	0.133			0.910	0.944
Inodorus	976	976	1022	1022	0.060	0.060			0.910	0.910
Reticulado	899	899	1043	1043	0.133	0.133			0.944	0.944
Muskmelon (Teotia and Ramakrishna, 1989)	740	1100	1190	1660	0.076	0.554			0.042	0.091
Hull	740	740	1660	1660	0.554	0.554			0.042	0.042
Kernel	1100	1100	1190	1190	0.076	0.076			0.091	0.091
Seeds	840	840	1300	1300	0.354	0.354			0.068	0.068
Nectarine (Rahman, 1995)	990	990					28	28	0.898	0.898
—	990	990					28	28	0.898	0.898
Okra (Sahoo and Srivastava, 2002; Oyelade et al., 2003; Owolarafe and Shotonde, 2004)	190	592	1107	1107	0.394	0.465	20	20	0.075	0.467
—	450	450			0.394	0.394			0.114	0.114
Seeds	190	592	1107	1107	0.434	0.465	20	20	0.075	0.467
Orange (Rahman, 1995; Ramos and Ibarz, 1998; Fito et al., 2001; Chafer et al., 2003)	768	1294	1085	1085	0.210	0.380	0	80	0.748	0.892
—	768	1030					28	28	0.859	0.859
Juice	1042	1294					0	80	0.890	0.892
Peels	770	800	1085	1085	0.210	0.380			0.748	0.760
Papaya (Mujica-Paz et al., 2003)	968	968	1028	1028	0.058	0.058			0.928	0.928
Maradol	968	968	1028	1028	0.058	0.058			0.928	0.928
Peach (Rahman, 1995; Fikiin et al., 1999; Ramos and Ibarz, 1998; Fito et al., 2001; Mujica-Paz et al., 2003)	545	1303	1044	1070	0.046	0.091	2	70	0.820	0.885
—	608	930					28	28	0.885	0.885
Catherine	987	987	1070	1070	0.091	0.091			0.878	0.878
Criollo	996	996	1044	1044	0.046	0.046			0.849	0.849
Juice	1056	1303					5	70		
Miraflores	1038	1038	1065	1065	0.047	0.047			0.820	0.820
Packed layers	545	545					2	2		
Pear (Lozano et al., 1983; Rahman, 1995; Fito et al., 2001)	641	1140	1070	1460	0.034	0.220	28	60	0.120	0.870
—	641	1140	1090	1460	0.040	0.220	28	60	0.120	0.870
Passa crassana	1030	1030	1070	1070	0.034	0.034			0.803	0.803

Table 14.8 (continued) Bulk Density, True Density, and Porosity of Foods versus Moisture and Temperature (Ranges of Variation of Available Data)

	ρ_b (kg/m³)		ρ_t (kg/m³)		ε (−)		T (°C)		X_w (wb)	
Material	**Min**	**Max**	**Min**	**Max**	**Min**	**Max**	**Min**	**Max**	**Min**	**Max**
Pineapple (Rahman, 1995)	1010	1010					27	27	0.849	0.849
—	1010	1010					27	27	0.849	0.849
Plum (Rahman, 1995; Fito et al., 2001)	721	1130	1090	1090	0.020	0.020	26	26	0.810	0.886
—	721	1130					26	26	0.886	0.886
President	1070	1070	1090	1090	0.020	0.020			0.810	0.810
Prune (Tsami and Katsioti, 2000)					0.065	0.464	70	70	0.001	0.410
—					0.065	0.464	70	70	0.001	0.410
Raisin (Vagenas et al., 1990; Kostaropoulos and Saravacos, 1997)	582	762	1076	1434	0.370	0.650	20	20	0.075	0.800
—	582	762	1076	1434	0.370	0.490	20	20	0.140	0.800
Sultanin					0.620	0.650	20	20	0.075	0.095
Raspberry (Rahman, 1995; Fikiin et al., 1999; Ratti, 2001)	496	1046					2	50	0.885	0.885
—							20	50		
Juice	1046	1046							0.885	0.885
Packed layers	496	496					2	2		
Strawberry[c]	498	1033	1050	1050	0.064	0.471	2	70	0.055	0.920
—	900	1000			0.089	0.471	20	70	0.055	0.920
Chandler	984	984	1050	1050	0.064	0.064			0.911	0.911
Juice	1033	1033							0.917	0.917
Packed layers	498	498					2	2		
Terebinth (Aydin and Ozcan, 2002)	449	620	1031	1071	0.421	0.565	20	20	0.057	0.206
—	449	620	1031	1071	0.421	0.565	20	20	0.057	0.206
Hull	620	620	1390	1390	0.554	0.554			0.092	0.092
Kernel	1100	1100	1160	1160	0.052	0.052			0.037	0.037
Seeds	800	800	1290	1290	0.380	0.380			0.054	0.054
Legume	384	1394	1104	1641	0.026	0.515	20	70	0.048	0.246
Bean (Rahman, 1995)	384	384								
Green	384	384								
Chick pea (Konak et al., 2002)	740	799	1368	1427	0.440	0.459	20	20	0.049	0.142
Seeds	740	799	1368	1427	0.440	0.459	20	20	0.049	0.142
Faba bean (Haciseferogullari et al., 2003)	608	608	1248	1248	0.515	0.515				
—	608	608	1248	1248	0.515	0.515				
Lentil (Tang and Sokhansanj, 1993; Carman, 1996; Amin et al., 2004)	768	1394	1212	1641	0.026	0.366	20	70	0.048	0.246
Seeds	768	1394	1212	1641	0.026	0.366	20	70	0.048	0.246

(continued)

Table 14.8 (continued) **Bulk Density, True Density, and Porosity of Foods versus Moisture and Temperature (Ranges of Variation of Available Data)**

	ρ_b (kg/m^3)		ρ_t (kg/m^3)		ε (–)		T (°C)		X_w (wb)	
Material	**Min**	**Max**	**Min**	**Max**	**Min**	**Max**	**Min**	**Max**	**Min**	**Max**
Lupin (Ogut 1998)	767	868.4	1104	1104	0.114	0.305	20	20	0.077	0.161
White	767	868	1104	1104	0.114	0.305	20	20	0.077	0.161
Meat	440	1100	1100	1320	0.023	0.647	−196	75	0.535	0.789
Beef (Rahman, 1995; McDonald and Sun, 2001; Pan and Singh, 2001; McDonald et al., 2002)	920	1092	1100	1117	0.023	0.075	−20	75	0.535	0.743
Cooked	1040	1085	1100	1117	0.023	0.075	4	72	0.692	0.743
Fat	920	957					5	30		
Ground	1010	1028					30	75	0.535	0.599
Lean	1067	1077					5	30		
Muscle	984	1080					−20	30	0.740	0.740
Veal	1092	1092					30	30		
Chicken (Farkas and Singh, 1991)	440	840	1210	1320	0.331	0.647	−196	74		
—	440	840	1210	1320	0.331	0.647	−196	74		
Pork (Rahman, 1995)	1070	1090					23	23		
Boneless	1090	1090					23	23		
Ham	1070	1070								
Poultry (Rahman, 1995)	1070	1100					15	23	0.720	0.720
Muscle	1070	1100					15	23	0.720	0.720
Scallop (Chung and Merrit, 1991)	985	1061							0.789	0.789
—	985	1061							0.789	0.789
Model foods	34	1377	1196	1560	0.040	0.978	10	144	0.000	0.600
Amioca (Marousis and Saravacos, 1990; Karathanos and Saravacos, 1993)	790	1220	1263	1510	0.040	0.601	60	60	0.050	0.500
Gelatinized	900	1200	1263	1490	0.050	0.380	60	60	0.050	0.500
Gelatinized pressed					0.040	0.040			0.440	0.440
Granular	790	1220	1270	1510	0.040	0.460	60	60	0.050	0.500
Powder					0.548	0.601			0.100	0.330
Granular pressed					0.233	0.261			0.100	0.330
Guar (Carson et al., 2004)	1010	1010					10	10		
Gel	1010	1010					10	10		
Hylon 7 (Marousis and Saravacos, 1990; Karathanos and Saravacos, 1993; Shah et al., 2000)	600	1377	1196	1500	0.079	0.622	25	60	0.000	0.550
—	1364	1377	1420	1500	0.079	0.088	25	25	0.000	0.320

Table 14.8 (continued) **Bulk Density, True Density, and Porosity of Foods versus Moisture and Temperature (Ranges of Variation of Available Data)**

	ρ_b (kg/m^3)		ρ_t (kg/m^3)		ε (−)		T (°C)		X_w (wb)	
Material	**Min**	**Max**	**Min**	**Max**	**Min**	**Max**	**Min**	**Max**	**Min**	**Max**
Gelatinized	600	1100	1196	1490	0.080	0.590	60	60	0.050	0.550
Gelatinized and dried					0.342	0.342				
Granular	750	1050	1320	1500	0.210	0.483	60	60	0.000	0.430
Powder					0.622	0.622			0.100	0.100
Pectin (Tsami et al., 1999)	700	1250	1560	1560	0.200	0.540				
Gel	700	1250	1560	1560	0.200	0.540				
Potato starch (Jagannath et al., 2001)	98	458							0.033	0.075
—	98	458							0.033	0.075
Sponge (Hamdami et al., 2003)	34	34	1552	1552	0.920	0.978	103	103	0.000	0.600
—	34	34	1552	1552	0.920	0.978	103	103	0.000	0.600
Starch (Rahman, 1995; Fitzpatrick et al., 2004)	560	760	1510	1510	0.497	0.497			0.100	0.100
—	560	560								
Powder	760	760	1510	1510	0.497	0.497			0.100	0.100
Tapioca starch (Ofman et al., 2004)	93	93							0.001	0.001
—	93							0	0.001	
Nuts	57	792	1015	1470	0.339	0.644	20	30	0.027	0.350
Almond (Aydin, 2003)	475	655	1115	1115	0.339	0.529			0.028	0.200
—	525	655	1015	1115	0.355	0.529			0.028	0.200
Kernel	475	595			0.339	0.523			0.028	0.200
Cashew (Balasubramanian, 2001)	592	624	1201	1240	0.480	0.523	20	20	0.031	0.167
Raw	592	624	1201	1240	0.480	0.523	20	20	0.031	0.167
Gorgon (Jha, 1999)	57	66			0.482	0.500	30	30	0.048	0.167
Makhana	57	66			0.482	0.500	30	30	0.048	0.167
Groundnut (Baryeh, 2001)	690	792	1159	1291	0.387	0.426	20	20	0.050	0.350
Bambara	690	792	1159	1291	0.387	0.426	20	20	0.050	0.350
Hazel (Aydin, 2002; Ozdemir and Akinci, 2004)	305	586			0.431	0.644	20	20	0.027	0.167
—	305	439			0.473	0.616	20	20	0.027	0.167
Kernel	458	586			0.431	0.644	20	20	0.027	0.167
Neem (Visvanathan et al., 1996)	528	555	1020	1470	0.482	0.622			0.077	0.209
—	528	555	1020	1470	0.482	0.622			0.077	0.209
Others	330	1330	1000	1600	0.005	0.752	20	70	0.028	0.980
Bamboo (Madamba, 2003)	998	1330	1010	1600	0.005	0.170	70	70	0.100	0.920
Shoot	998	1330	1010	1600	0.005	0.170	70	70	0.100	0.920

(continued)

Table 14.8 (continued) Bulk Density, True Density, and Porosity of Foods versus Moisture and Temperature (Ranges of Variation of Available Data)

	ρ_b (kg/m^3)		ρ_t (kg/m^3)		ε (−)		T (°C)		X_w (wb)	
Material	**Min**	**Max**	**Min**	**Max**	**Min**	**Max**	**Min**	**Max**	**Min**	**Max**
Canola (Sokhansanj and Lang, 1996; Pagano and Crozza, 2002)	616	700	1050	1126	0.340	0.440	26	40	0.050	0.190
—	616	700	1050	1060	0.340	0.440	26	26	0.054	0.133
Tobin	661	678	1117	1126	0.401	0.409	40	40	0.050	0.190
Cocoa (Rahman, 1995; Bart-Plange and Baryeh, 2003; Vu et al., 2003; Fitzpatrick et al., 2004)	360	560	1440	1450	0.200	0.752			0.044	0.238
—	480	480								
Beans	505	560			0.409	0.491			0.086	0.238
Powder	360	360	1440	1450	0.200	0.752			0.044	0.044
Coffee (Rahman, 1995; Chandrasekar and Viswanathan, 1999)	330	514			0.450	0.540			0.099	0.306
Arabica	402	483			0.470	0.540			0.099	0.260
Ground	330	330								
Instant	330	330								
Robusta	434	514			0.450	0.517			0.106	0.306
Cotton (Ozarslan, 2002)	610	642	1000	1091	0.390	0.412	20	20	0.077	0.121
Seeds	610	642	1000	1091	0.390	0.412	20	20	0.077	0.121
Cumin (Singh and Goswami, 1996)	410	502	1047	1134	0.530	0.638	20	20	0.065	0.180
Seeds	410	502	1047	1134	0.530	0.638	20	20	0.065	0.180
Egg (Rahman, 1995)	340	340								
Whole	340	340								
Hemp (Sacilik et al., 2003)	512	558	1043	1043	0.427	0.465	20	20	0.079	0.173
Seeds	512	558	1043	1043	0.427	0.465	20	20	0.079	0.173
Mahaleb (Aydin et al., 2002)	566	616	1110	1250	0.490	0.507	20	20	0.028	0.093
Turkish	566	616	1110	1250	0.490	0.507	20	20	0.028	0.093
Pepper (Murthy and Bhattacharya, 1998)	513	556							0.075	0.240
Black	513	556							0.075	0.240
Salt (Rahman, 1995)	960	960								
Granular	960	960								
Soybean (Deshpante et al., 1993)	708	736	1124	1216	0.370	0.395	30	30	0.080	0.200
Grain	708	736	1124	1216	0.370	0.395	30	30	0.080	0.200

Table 14.8 (continued) Bulk Density, True Density, and Porosity of Foods versus Moisture and Temperature (Ranges of Variation of Available Data)

	ρ_b (kg/m^3)		ρ_t (kg/m^3)		ε (–)		T (°C)		X_w (wb)	
Material	**Min**	**Max**	**Min**	**Max**	**Min**	**Max**	**Min**	**Max**	**Min**	**Max**
Soymilk (Rahman, 1995)	1010	1050							0.884	0.980
—	1010	1050							0.884	0.980
Sugar (Rahman, 1995)	480	800								
Granular	800	800								
Powder	480	480								
Sunflower (Gupta and Das, 1997)	434	628	1250	1250	0.346	0.498	20	20	0.035	0.164
Kernel	574	628	1050	1250	0.453	0.498	20	20	0.035	0.164
Seeds	434	462			0.346	0.433	20	20	0.043	0.163
Tea (Teunou et al., 1999; Fitzpatrick et al., 2004)	910	913	1568	1570	0.418	0.420			0.066	0.066
Powder	910	913	1568	1570	0.418	0.420			0.066	0.066
Vegetables	75	1560	1013	2182	0.000	0.967	−50	190	0.000	0.955
Asparagus (Rahman, 1995)	577	577								
—	577	577								
Beet (Rahman, 1995)	833	1530					28	28	0.895	0.895
—	833	833								
Red	1530	1530					28	28	0.895	0.895
Beetroot (Fito et al., 2001)	964	964	1015	1015	0.043	0.043			0.800	0.800
—	964	964	1015	1015	0.043	0.043			0.800	0.800
Cabbage (Rahman, 1995; Karathanos et al., 1996)	449	449			0.050	0.920				
—	449	449			0.920	0.920				
Fresh					0.050	0.050				
Calabash (Omobuwajo et al., 2003)	429	429			0.484	0.484			0.077	0.077
Seeds	429	429			0.484	0.484			0.077	0.077
Caper (Ozcan et al., 2004)	440	440			0.369	0.369	20	20	0.453	0.453
Flower bud	440	440			0.369	0.369	20	20	0.453	0.453
Carrot[d]	110	1560	1036	1889	0.010	0.940	−50	70	0.000	0.920
—	110	1560	1050	1889	0.010	0.940	−50	70	0.000	0.920
Danver	1040	1040					28	28	0.900	0.900
Fresh					0.040	0.040				
Nantesa	975	975	1036	1036	0.137	0.137			0.890	0.890
Cauliflower (Rahman, 1995)	320	320								
—	320	320								

(*continued*)

Table 14.8 (continued) Bulk Density, True Density, and Porosity of Foods versus Moisture and Temperature (Ranges of Variation of Available Data)

	ρ_b (kg/m^3)		ρ_t (kg/m^3)		ε (–)		T (°C)		X_w (wb)	
Material	**Min**	**Max**	**Min**	**Max**	**Min**	**Max**	**Min**	**Max**	**Min**	**Max**
Celery (Karathanos et al., 1993)	100	1100			0.502	0.967	5	60	0.020	0.955
—	100	1100			0.502	0.967	5	60	0.020	0.955
Cucumber (Rahman, 1995; Fasina and Fleming, 2001)	769	959					28	28	0.954	0.954
—	769	959					28	28	0.954	0.954
Eggplant (Fito et al., 2001)	417	417			0.641	0.641			0.910	0.910
Soraya	417	417			0.641	0.641			0.910	0.910
Garlic (Lozano et al., 1983; Madamba et al., 1993; Madamba et al., 1994)	230	1280	1064	1528	0.010	0.750	60	70	0.000	0.700
—	230	1280	1064	1528	0.010	0.750	60	70	0.000	0.700
Mushroom (Loch-Bonazzi et al., 1992; Fito et al., 2001; Torringa et al., 2001)	200	1100	1250	1250	0.018	0.840	60	90	0.020	0.930
—	200	1100	1250	1250	0.018	0.840	60	90	0.020	0.928
Albidus	649	649			0.359	0.359			0.930	0.930
Fresh	830	830							0.910	0.910
Onion (Rahman, 1995; Rapusas and Driscoll, 1995; Rapusas et al., 1995; Abhayawick et al., 2002; Bahnasawy et al., 2004)	102	1229	1025	1489	0.018	0.910	28	70	0.000	0.920
—	102	1169	1416	1416	0.027	0.910	28	70	0.054	0.873
Niz	1032	1229	1063	1395	0.018	0.143	55	55	0.137	0.813
Red			1110	1110						
Spirit	1002	1149	1040	1477	0.020	0.222	55	55	0.017	0.877
Sweet vitalia	934	1046	1025	1489	0.083	0.319	55	55	0.000	0.920
White			1090	1090						
Yellow			1040	1040						
Pea (Medeiros and Sereno, 1994; Konak et al., 2002; Hatamipour and Mowla, 2003)	741	1210	1373	1429	0.440	0.460	30	65	0.049	0.750
—	1080	1210					30	65	0.100	0.700
Green	800	1030					60	60	0.375	0.750
Seeds	741	800	1373	1429	0.440	0.460			0.049	0.141
Potato[e]	180	1379	1020	2182	0.000	0.880	−50	190	0.000	0.880
—	180	1300	1020	2182	0.000	0.880	−50	190	0.000	0.880
Bintije	380	1230	1080	1350	0.010	0.700	50	50	0.000	0.770

Table 14.8 (continued) Bulk Density, True Density, and Porosity of Foods versus Moisture and Temperature (Ranges of Variation of Available Data)

	ρ_b (kg/m^3)		ρ_t (kg/m^3)		ε (−)		T (°C)		X_w (wb)	
Material	**Min**	**Max**	**Min**	**Max**	**Min**	**Max**	**Min**	**Max**	**Min**	**Max**
Blanched							60	80		
Chips							118	144	0.008	0.019
Desiree	1081	1379	1082	1755	0.001	0.278	40	70	0.091	0.804
Fresh	1054	1054	1066	1066	0.011	0.020			0.826	0.826
Katahdin	1040	1040					26	26	0.814	0.814
Kennebec	1050	1050					25	25	0.824	0.824
Monona	1040	1040					25	25	0.836	0.836
Norchip	1050	1050					26	26	0.812	0.812
Restructure	690	990	1125	1160	0.130	0.410	170	170		
Russet Burbank	1040	1040					25	25	0.829	0.829
Sweet	705	1170	1240	1480	0.140	0.320	40	40	0.100	0.750
White	769	769								
Pumpkin (Joshi et al., 1993)	75	554	1070	1533	0.515	0.951	47	47	0.040	0.289
Hull	75	75	1533	1533	0.951	0.951	47	47	0.095	0.095
Kernel	481	554	1080	1143	0.515	0.555	47	47	0.040	0.275
Seeds	404	472	1070	1179	0.555	0.657	47	47	0.055	0.289
Radish (Suzuki et al., 1976; Buhri and Singh, 1993)	942	942					40	45	0.050	0.946
—	942	942					40	45	0.050	0.946
Rutabagas (Buhri and Singh, 1993)	1008	1008					45	45	0.926	0.926
—	1008	1008					45	45	0.926	0.926
Spinach (Rahman, 1995)	224	224								
—	224	224								
Squash (Rahman, 1995; Paksoy and Aydin, 2004)	351	970			0.190	0.340	20	26	0.060	0.936
Boston marrow	970	970					23	23	0.936	0.936
Butternut	950	950					26	26	0.877	0.877
Curcubita pepo (kernel)	406	467			0.287	0.340	20	20	0.060	0.346
Curcubita pepo (seeds)	351	475			0.190	0.283	20	20	0.060	0.346
Golden delicious	900	900					23	23	0.885	0.885
Tomato (Rahman, 1995; Durance and Wang, 2002; Fitzpatrick et al., 2004)	302	1010	1490	1490	0.403	0.403	28	70	0.178	0.923
—	302	672					70	70	0.187	0.187
Cherry	1010	1010					28	28	0.923	0.923
Powder	890	890	1490	1490	0.403	0.403			0.178	0.178
Turnip (Buhri and Singh, 1993; Rahman, 1995)	952	1000					24	45	0.898	0.923
—	952	1000					24	45	0.898	0.923

(continued)

Table 14.8 (continued) Bulk Density, True Density, and Porosity of Foods versus Moisture and Temperature (Ranges of Variation of Available Data)

Material	ρ_b (kg/m^3)		ρ_t (kg/m^3)		ε (−)		T (°C)		X_w (wb)	
	Min	Max	Min	Max	Min	Max	Min	Max	Min	Max
Yam (Hsu et al., 2003)	490	930	1330	1550	0.400	0.650	60	100	0.005	0.073
Flour	490	930	1330	1550	0.400	0.650	60	100	0.005	0.073
Zucchini (Fito et al., 2001)	841	841	1013	1013	0.169	0.169			0.940	0.940
Blanco grise	841	841	1013	1013	0.169	0.169			0.940	0.940

Source: Data from Boukouvalas, Ch.J., Krokida, M.K., Maroulis, Z.B., and Marinos-Kouris, D., *Int. J. Food Prop.*, 2006b.

[a] Bai et al. (2002); Sjoholm and Gekas (1995); May and Perre (2002); Sablani and Rahman (2002); Buhri and Singh (1993); Ratti (1994); Krokida and Maroulis (1999); Rahman (1995); Nieto et al. (2004); Funebo et al. (2000); Martinez-Monzo et al. (2000); Singh et al. (1997); Fito et al. (2001); Mujica-Paz et al. (2003); Karathanos et al. (1996); Moreira et al. (2000); Krokida et al. (2000); Krokida et al. (1998); Zogzas et al. (1994); Bazhal et al. (2003); Torreggiani et al. (1995); Donsi et al. (1996); Mavroudis et al. (1998); Lozano et al. (1980); Barat et al. (1999); Mavroudis et al. (2004); Fikiin et al. (1999); Funebo et al. (2002).

[b] Wang and Chen (1999); Tsen and King (2002); Krokida and Maroulis (1997); Krokida and Maroulis (1999); Rahman (1995); Talla et al. (2004); Singh et al. (1997); Mujica-Paz et al. (2003); Krokida et al. (1998).

[c] Ratti (2001); Raghavan and Venkatachalapathy (1999); Tsami and Katsioti (2000); Ferrando and Spiess (2003); Rahman (1995); Fito et al. (2001); Fikiin et al. (1999).

[d] May and Perre (2002); Buhri and Singh (1993); Hatamipour and Mowla (2002); Ratti (1994); Krokida and Maroulis (1997); Krokida and Maroulis (1999); Rahman (1995); Lin et al. (1998); Ruiz-Lopez et al. (2004); Fito et al. (2001); Karathanos et al. (1996); Krokida et al. (1998); Zogzas et al. (1994); Lozano et al. (1983); Sanga et al. (2002); Suzuki et al. (1976).

[e] May and Perre (2002); Wang and Brennan (1995); Sablani and Rahman (2002); Buhri and Singh (1993); Gekas and Lamberg (1991); Ratti (1994); Krokida and Maroulis (1997); Krokida and Maroulis (1999); Rahman (1995); Krokida et al. (2001); Karathanos et al. (1996); McMinn et al. (1997); Pinthus et al. (1995); Krokida et al. (1998); Zogzas et al. (1994); Donsi et al. (1996); Lozano et al. (1983); Suzuki et al. (1976); Garayo and Moreira (2002).

Table 14.5 presents bulk density values for a variety of powders. It can be seen that these values are much lower than true density values of common fruits and vegetables or milk, presented here. It should be noted that the bulk density of a powder depends on how it is packed into a container. There are four basic "packing density states," which may vary by up to 30% and they are the following:

- Aerated bulk density
- Poured bulk density
- Tap density
- Compacted bulk density

Table 14.6 presents density values of fruit juice as a function of their solid content (given as percent wet basis). As in the case of milk, juices have a higher density than water. For all the examined juices, there is an increase in density as the solid content increases. For solid concentration up to about 10%, the juice density is approximately 5% higher than water, whereas for concentrated juices density is much higher and should be taken into consideration in process calculations. Table 14.7 gives apparent density values of some fresh and frozen seafood. For each material, the content of water (X_w), proteins (X_{pr}), fats (X_{fa}), and ash (X_{as}) is reported. Fresh materials have a higher density in all examined cases.

Boukouvalas et al. (2006a,b) presented a large number of data for structural properties (true density ρ_t, bulk density ρ_b' and porosity ε) of various foodstuffs, revealing the range of variation for each food material, as well as each food category, versus the corresponding ranges of material moisture content X_w and temperature T. These data have been extracted from the literature, homogenized and analyzed, and are tabulated in Table 14.8. The food materials are separated, according to their morphology, in two categories: (a) continuous materials and (b) particulate or granular materials. The experimental data introduced by Rahman (1995) were used as the initial set of data. The literature search took into account articles published since 1990, mainly in the following journals: *Journal of Food Science*, *Drying Technology*, *International Journal of Food Properties*, *Biosystems Engineering*, *Journal of Food Engineering*, *Lebensmittel-Wissenschaft und-Technologie*, *Journal of Agricultural Engineering Research*, while a few additional data retrieved by articles in other journals (*Chemical Engineering and Processing*, *Chemical Engineering Science*, *Food and Bioproducts Processing*, *Food Chemistry*, *Food Control*, *Food Research International*, *International Journal of Refrigeration*, *Powder Technology*, *Journal of Agricultural and Food Chemistry*, *International Journal of Food Science and Technology*). Most of the available data from review articles published before 1990 are also included. A total number of more than 150 papers were retrieved from the above journals and the analysis of the reported materials resulted in a collection of more than 1400 structural properties' data. These values incorporate more than 130 foodstuffs, which are classified into 11 food categories. The range of variation of bulk density, true density, and porosity for each material along with the corresponding ranges of moisture and temperature, as well as the corresponding reference(s), are presented in this table. Concisely, the range of variation of the structural properties presented is (approximate values)

- Bulk density 75–1700 kg/m^3
- True density 1000–2100 kg/m^3
- Porosity 0.00–0.98

while the ranges of temperature and moisture content are

- Temperature −50°C to 200°C
- Moisture content 0.00–0.98 kg/kg (wet basis)

It should be noted that there are cases where true density values lower than 1000 kg/m^3 are referred in the literature, but they have not been included in the following table. Structural properties of foods are affected by various factors, such as material moisture content, material morphology (continuous or granular), process method, conditions, etc. However, true density is not affected by the method or the material morphology, but it strongly depends on moisture content and temperature (Krokida et al., 2000; Saravacos and Maroulis, 2001). True density increases as the water content decreases. This should be expected since it ranges between water density and that of dry solids (Krokida et al., 2000), which is higher, according to Table 14.2.

Figures 14.1 through 14.7 analyze the data presented in Table 14.8. True density data versus moisture content and temperature are plotted in Figures 14.1 and 14.2, respectively. The data, presented in Figure 14.1, demonstrate an apparent behavior as they approach the water density when the moisture content (in wet basis) tends to unity. On the contrary, true density versus temperature presents no apparent scheme (Figure 14.2). Bulk density data versus moisture content and temperature are plotted in Figures 14.3 and 14.4, respectively. Figure 14.5 shows the data for true density versus bulk density. All the data are plotted above the diagonal, as expected since the inequation $\rho_b < \rho_t$ is valid (Saravacos and Maroulis, 2001). Porosity data versus moisture content and temperature are illustrated in Figures 14.6 and 14.7, respectively. In Figure 14.6, the porosity for granular and continuous materials is plotted separately. It can be seen that the porosity of granular materials is usually high, which is not the case for the majority of continuous materials whose porosity is low, especially at higher moisture content.

A few additional data for density, shrinkage, and porosity are presented in the following tables and graphs of this paragraph. Some of them are quite detailed but nevertheless very useful for

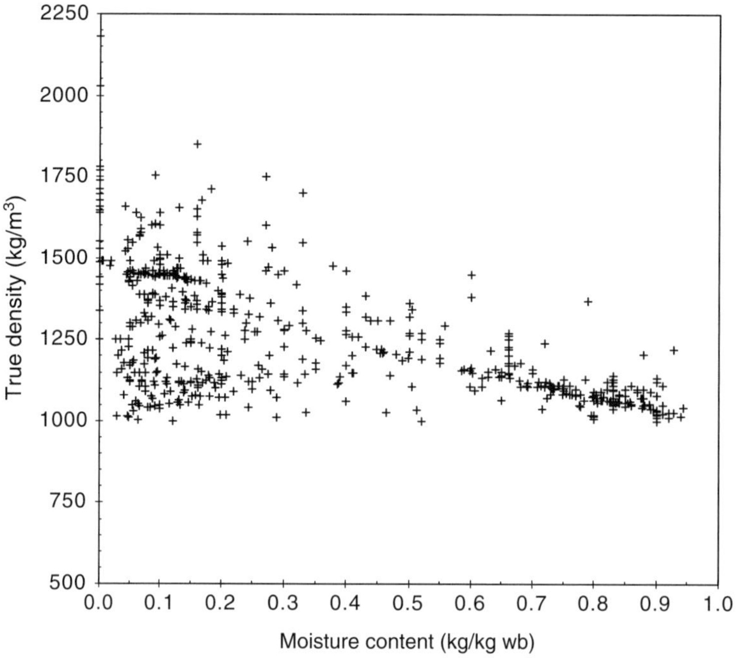

Figure 14.1 True density versus moisture content, for the foodstuffs of Table 14.8. (From Boukouvalas, Ch.J., Krokida, M.K., Maroulis, Z.B., and Marinos-Kouris, D., *Int. J. Food Prop.*, 9, 715, 2006b. With permission.)

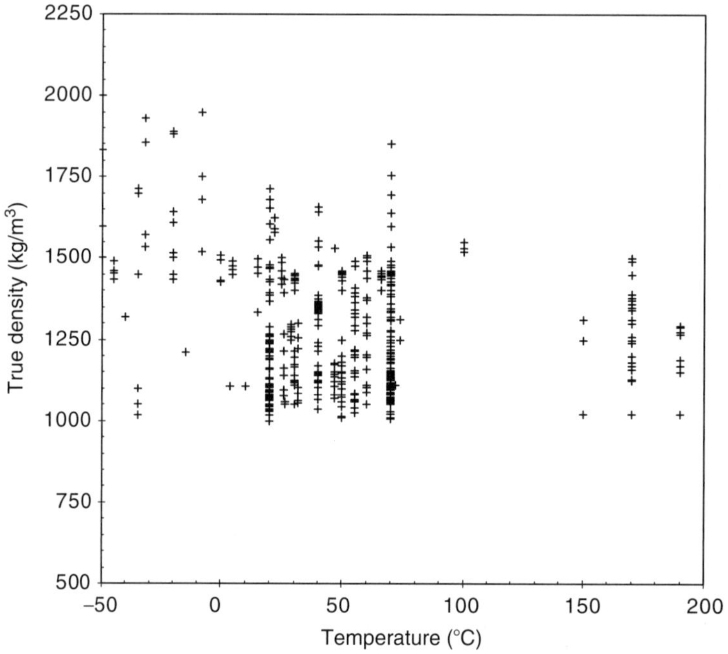

Figure 14.2 True density versus temperature, for the foodstuffs of Table 14.8. (From Boukouvalas, Ch.J., Krokida, M.K., Maroulis, Z.B., and Marinos-Kouris, D., *Int. J. Food Prop.*, 9, 715, 2006b. With permission.)

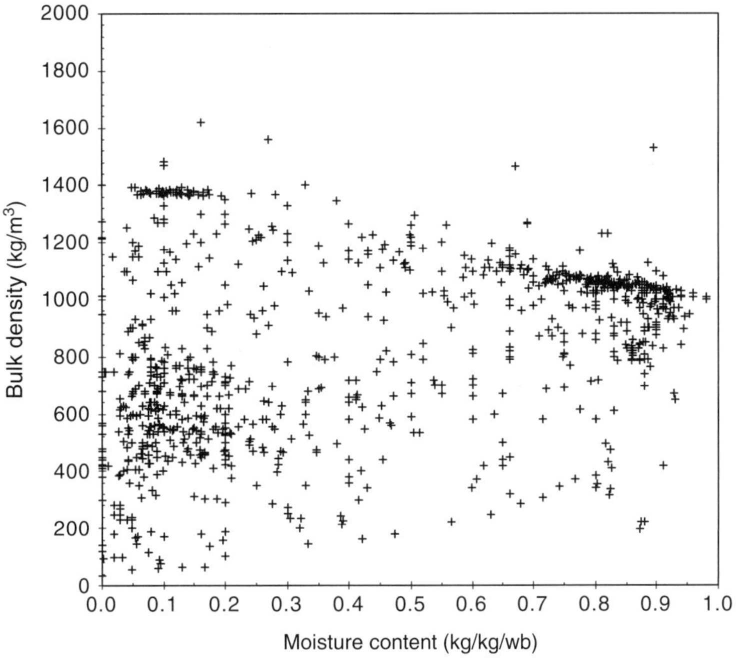

Figure 14.3 Bulk density versus moisture content, for the foodstuffs of Table 14.8. (From Boukouvalas, Ch.J., Krokida, M.K., Maroulis, Z.B., and Marinos-Kouris, D., *Int. J. Food Prop.*, 9, 715, 2006b. With permission.)

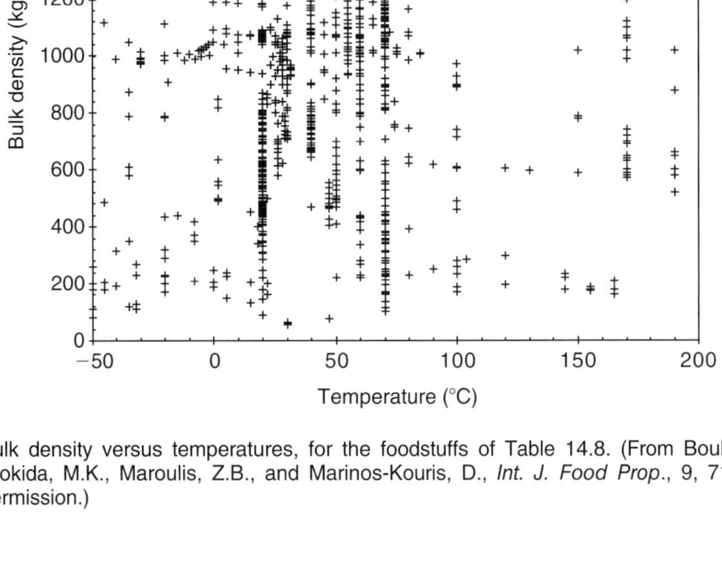

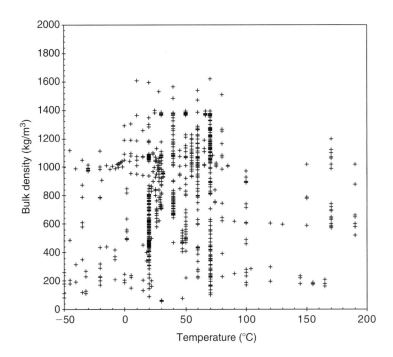

Figure 14.4 Bulk density versus temperatures, for the foodstuffs of Table 14.8. (From Boukouvalas, Ch.J., Krokida, M.K., Maroulis, Z.B., and Marinos-Kouris, D., *Int. J. Food Prop.*, 9, 715, 2006b. With permission.)

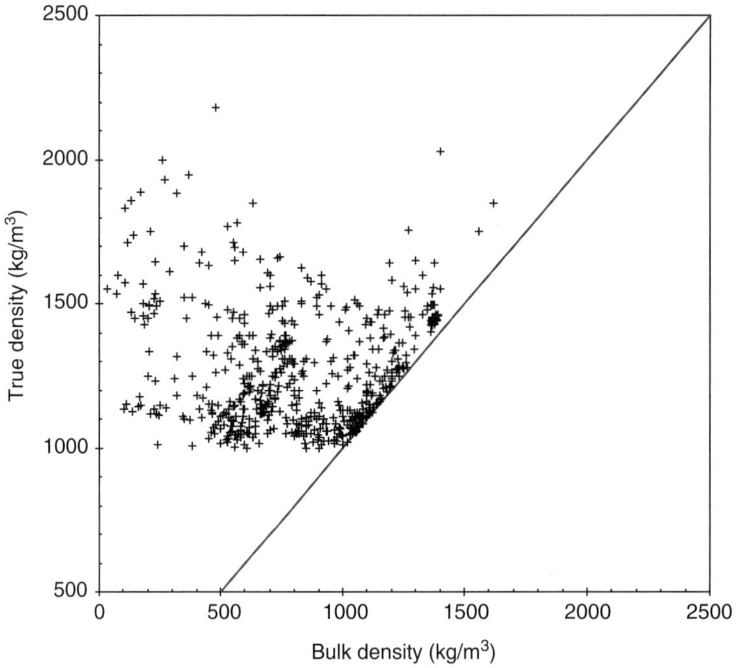

Figure 14.5 True density versus bulk density, for the foodstuffs of Table 14.8. (From Boukouvalas, Ch.J., Krokida, M.K., Maroulis, Z.B., and Marinos-Kouris, D., *Int. J. Food Prop.*, 9, 715, 2006b. With permission.)

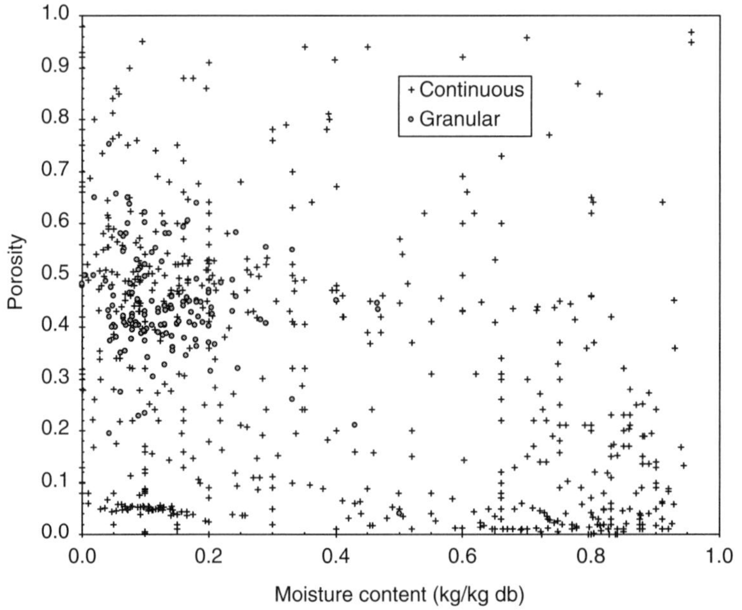

Figure 14.6 Porosity versus moisture content, for the foodstuffs of Table 14.8. (From Boukouvalas, Ch.J., Krokida, M.K., Maroulis, Z.B., and Marinos-Kouris, D., *Int. J. Food Prop.*, 9, 715, 2006b. With permission.)

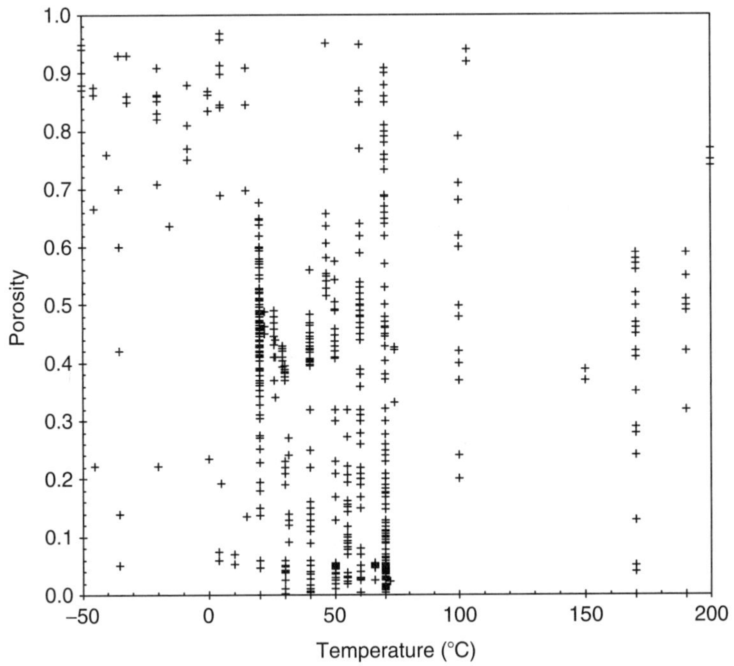

Figure 14.7 Porosity versus temperature, for the foodstuffs of Table 14.8. (From Boukouvalas, Ch.J., Krokida, M.K., Maroulis, Z.B., and Marinos-Kouris, D., *Int. J. Food Prop.*, 9, 715, 2006b. With permission.)

quality control and process design. Vagenas et al. (1990) studied the drying of raisins and the bulk and true density and porosity at various moisture contents are given in Table 14.9.

The influence of drying method on the porosity of a food material can be seen in Table 14.10, which shows apparent and true density of tuna strips dried by air-, vacuum-, and freeze-drying (Rahman et al., 2002). It is notable that true density is independent of the drying method, while apparent density is strongly influential on the drying method. Freeze-drying results in the production of highly porous materials presenting low apparent density.

Sablani and Rahman (2002) investigated the influence of one of the drying parameters in freeze-drying, the shelf temperature, and their results showed that there is significant influence on density and porosity of the dried material. Furthermore, they examined the influence of the maturity level of a food material (date) and confirmed that the structural properties of the dried product depended strongly on the quality of the fresh product that dried. Figure 14.8 illustrates these results. Jumah et al. (2000) determined the effect of inlet air temperature, air flowrate, and atomizer pressure on bulk density of spray-dried jameed (a dried fermented dairy product). Table 14.11 presents the variation of bulk density as a function of these processing parameters. Karathanos et al. (1993)

Table 14.9 Bulk Density, True Density, and Porosity of Raisins versus Moisture Content

X (kg water/kg raisin)	0.80	0.75	0.65	0.60	0.50	0.40	0.20	0.14
ρ_b (kg/m^3)	582	691	673	661	672	660	708	762
ρ_t (kg/m^3)	1076	1096	1140	1164	1214	1268	1392	1434
ε (–)	0.46	0.37	0.41	0.43	0.45	0.48	0.49	0.47

Source: Vagenas, G.K., Marinos-Kouris, D., and Saravacos, G.D., *J. Food Eng.*, 11, 147, 1990. With permission.

Table 14.10 Porosity of Dried Tuna Strips

Sample	ρ_a (kg/m^3)	ρ_t (kg/m^3)	ρ_{su} (kg/m^3)
Fresh	1098	1071	1098
Air dried	960	1312	1255
Vacuum dried	709	1319	1309
Freeze dried	317	1319	1259

Source: Rahman, M.S., Al-Amri, O.S., and Al-Bulushi, I.M., *J. Food Eng.*, 53, 301, 2002. With permission.

studied the air-drying of celery and determined the influence of air temperature on porosity development. Air-drying at low temperatures resulted in a product of greater porosity than air-drying at high temperature, as illustrated in Figure 14.9.

Table 14.12 presents data of bulk porosity for different moisture contents for Corinthian grapes, presented by Belessiotis (1995) (cited by Ghiaus et al., 1997). Farkas and Singh (1991) measured structural properties of air- and freeze-dried chicken white meat using two different measurement techniques, and true density, apparent density, porosity and micropore volume are presented in Table 14.13. In this table, true and apparent density are in kg/m^3 and micropore volume is in mL/g. Karathanos and Saravacos (1993) measured apparent porosity and pore-size distribution for granular, heat-gelatinized, and extruded starch materials, using helium stereopycnometry and mercury porosimetry and their results are presented in Tables 14.14 and 14.15.

There are foods, such as desserts and ice-cream, produced by incorporating air into a liquid for the production of foam. In these systems, air is the dispersed phase, liquid is the continuous one, and foam stabilization is due to surface-active agents collected in the interface. The density of these products presents reduced values as a result of air inclusion, and the amount of air is expressed in terms of the overrun, given by the expression

$$\text{Overrun} = \frac{\text{Volume increase}}{\text{Original volume}} \times 100 = \frac{\text{Foam volume} - \text{original liquid volume}}{\text{Liquid volume}} \times 100$$

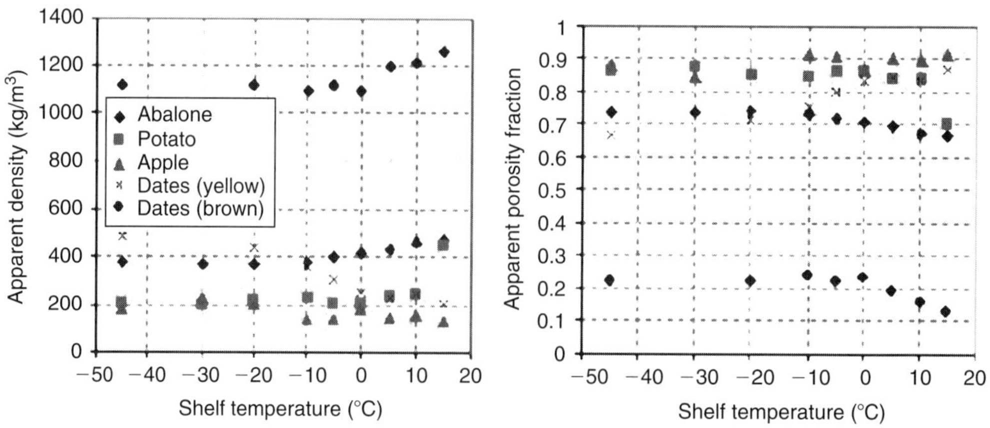

Figure 14.8 Effect of shelf temperature on apparent density and porosity of various food materials. (From Sablani, S.S. and Rahman, M.S., *Dry. Technol.*, 20, 1379, 2002. With permission.)

Table 14.11 Effect of Spray-Drying Parameters on Jameed Bulk Density

Inlet air temperature[a] (°C)	80	90	100	120	130	150
Bulk density (kg/m^3)	623	617	604	607	595	587
Air flowrate[b] (cm^3/s)	632	727	777	792		
Bulk density (kg/m^3)	629	626	617	620		
Atomizer pressure[c] (bar)	1.0	1.5	2.0	2.5		
Bulk density (kg/m^3)	609	609	619	621		

Source: Jumah, R.Y., Tashtoush, B., Shaker, R.R., and Zraiy, A.F., *Dry. Technol.*, 18, 967, 2000. With permission.

[a] Air flowrate, 792 cm^3/s; atomizer pressure, 2 bar.

[b] Inlet air temperature, 100°C; atomizer pressure, 2 bar.

[c] Inlet air temperature, 100°C; air flowrate, 632 cm^3/s.

Overrun is usually estimated as

$$\text{Overrun} = \frac{\text{Original volume weight} - \text{weight of same volume of foam}}{\text{Weight of same volume of foam}} \times 100$$

Porter (1975) and Arbuckle (1977) measured typical values for the range of runover in ice-creams and other frozen desserts and these are presented in Table 14.16.

14.3 MODELS FOR DENSITY, SHRINKAGE, AND POROSITY

Although the collection of data presented here is useful, it is often necessary as an appropriate mathematical model to be used for the calculation of a specific structural property value, e.g., for the most accurate estimation of this property under given conditions or for the incorporation of the equation in a mathematical model for the design of a process, such as drying, extrusion, etc.

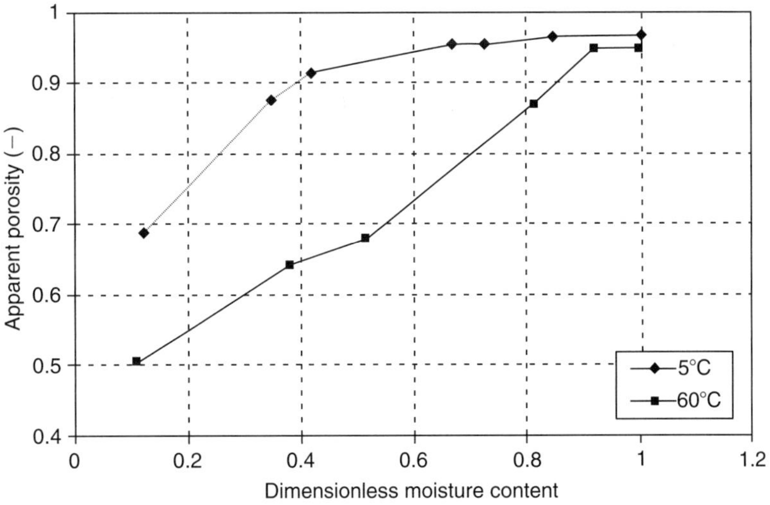

Figure 14.9 Apparent porosity of celery as a function of moisture content and drying air temperature. (From Karathanos, V.T., Kanellopoulos, N.K., and Belessiotis, V.G., *J. Food Eng.*, 29, 167, 1993. With permission.)

Table 14.12 Bulk Porosity of Corinthian Grapes

Moisture content (kg/(kg · db))	4.00	3.00	1.86	1.50	1.00	0.67	0.25	0.18
Bulk porosity (m^3/m^3)	0.46	0.37	0.41	0.43	0.45	0.47	0.49	0.47

Source: Belessiotis (1995) cited by Ghiaus, A.G., Margaris, D.P., and Papanikas, D.G., *J. Food Sci.*, 62, 1154, 1997. With permission.

Table 14.13 True Density, Apparent Density, and Porosity of Chicken White Meat

	Measurement Technique					
	Air Compression Pycnometry			Mercury Porosimenty		
Drying Method	True Density (kg/m^3)	Apparent Density (kg/m^3)	Porosity (−)	True Density (kg/m^3)	Apparent Density (kg/m^3)	Micropore Volume (ml/g)
Air-dried						
Fixed bed	1310	750	0.427	1220	580	0.3
Fluid bed	1250	840	0.331	1230	660	0.2
Frozen, fluid bed	1310	760	0.422	1230	550	0.3
Freeze-dried[a]						
Slow frozen	1210	440	0.636	1200	430	1.2
Blast frozen	1320	470	0.647	1300	430	1.4
Liquid nitrogen frozen	1260	480	0.624	1220	410	1.3

Source: Farkas, B.E. and Singh, R.P., *J. Food Sci.*, 56, 611, 1991. With permission.
[a] Slow frozen, −15°C for 24 h; blast frozen, −57°C for 30 min; liquid nitrogen frozen, immersion in liquid nitrogen (−196°C) for 1 min.

Table 14.14 Porosity and Pore-Size Distribution Data of Granular Starches

Material	Maximum Pore-Size Distribution (μm)	Apparent Porosity (−)	Mercury Porosity (−)
Granular Amioca 11% mc db, powder	3.5–1	0.601	0.497
Granular Hylon-7 11% mc db, powder	2.7–1	0.622	0.510
Granular Amioca 11% mc db, pressed	1.1–0.3	0.233	0.184
Granular Amioca 50% mc db, powder	13–2.5	0.548	0.474
Granular Amioca 50% mc db, pressed	1.7–0.7	0.261	0.248
Granular Hylon-7 70% mc db, after drying at 60°C	1.7–0.9	0.483	0.395

Source: Karathanos, V.T. and Saravacos, G.D., *J. Food Eng.*, 18, 259, 1993. With permission.
Note: mc, moisture content; db, dry basis With permission.

Table 14.15 Porosity and Pore-Size Distribution Data of Extruded and Heat-Gelatinized Starch Materials

Material	Maximum Pore-Size Distribution (μm)	Apparent Porosity (−)	Mercury Porosity (−)
Amioca extrudate, cylindrical shape (sc, 180°C, 30% mc db, 150 rpm), wet sample	40–5	0.502	0.397
Amioca extrudate, cylindrical shape (ss, 180°C, 40% mc db, 150 rpm), wet sample	60–20	0.314	0.273
Amioca extrudate, cylindrical shape (ts, 100°C, 33% mc db, 200 rpm), wet sample	10–2.5	0.168	0.167
Amioca extrudate, cylindrical shape (ts, 100°C, 33% mc db, 200 rpm), dried at 100°C for 24 h	50–10	0.407	0.366
Amioca extrudate, pressed slab (ts, 100°C, 33% mc db, 200 rpm), wet sample	3–1	0.146	0.075
Amioca extrudate, pressed and dried at 100°C for 24 h (ts, 100°C, 33% mc db, 200 rpm)	22–3[a]	0.174	0.136
Cornmeal extrudate, (ts, 140°C, 16% mc db, 300 rpm)	120–20	0.951	0.593
Gelatinized Hylon-7 dried and ground (<74 μm diameter)	20–4	0.342	0.270
Gelatinized Amioca (by heat, 80% mc db), wet sample pressed at about 100 bar	1.70.8	0.040	0.052

Source: Karathanos, V.T. and Saravacos, G.D., *J. Food Eng.*, 18, 259, 1993. With permission.
Note: ss, single screw; ts, twin screw; mc, moisture content; db, dry basis.
[a] Three maxima at 22, 12, and 5 μm.

In the following sections, a number of equations published in the literature (e.g. Rahman 1995, 2005) for the estimation of density, shrinkage, and porosity of various foodstuffs or food categories are presented. Most of them are of polynomial or equivalent type and have been derived from the fit of a simple equation to experimental data. They comprise constants with no physical meaning. In the following section, a generic semiempirical model is presented, which includes parameters of physical meaning.

Table 14.16 Overruns of Frozen Desserts

Dessert	Overrun
Ice cream (packaged)	70–80
Ice cream (bulk)	90–100
Sherbert	30–40
Ice	20–25
Soft ice cream	30–50
Ice milk	50–80
Milk shake	10–15

Source: Adapted from Arbuckle, W.S., *Ice-cream*, AVI Publishing Co., Westport, CT, 1977.

14.3.1 Prediction Models for Density

Density is one of the most basic and important variables in process design, operation, and control. Although density is a function of temperature and pressure (among others), in most applications, we assume that its value remains constant in the case of solid and liquid food materials, accepting that these materials are incompressible. On the other hand, gases and vapors are compressible and density under specific temperature and pressure is a function of these conditions.

14.3.1.1 Density of Gases and Vapors

The estimation of density for gases and vapors at low pressure (<10 atm) is based on the well-known ideal gas equation, which can be expressed as

$$PV = nRT \tag{14.1}$$

where
- P is the absolute pressure of the gas or vapor
- T is its absolute temperature
- n is the mole of the material
- V is the volume of the space, where the material is confined
- R is the ideal gas constant

In SI units, the value of constant R is 8.314 kJ/kmol · K. At high pressures (e.g., >10 atm), the pressure–volume–temperature relationship deviates from ideality, as the molecules are drawn closer together and attractive and repulsive forces between them affect their motion. In this case, the gases are characterized as real gases and a simple equation that can be applied to describe their behavior is the van der Waal's equation:

$$\left(P + \frac{n^2 a}{V^2}\right)(V - nb) = nRT \tag{14.2}$$

where a and b are the van der Waal's constants, whose values depend on the gas under estimation and can be found in most engineering handbooks.

14.3.1.2 Density of Solid Food Materials

The density of solid food materials depends, mainly, on composition and temperature. Food materials can be considered as multiphase systems, in which mass and volume conserve during the mixing process. Miles et al. (1983) and Choi and Okos (1986) proposed the following equation for the estimation of food materials' density, whose composition is known:

$$\frac{1}{\rho_t} = \sum_{j=1}^{n} \frac{x_j}{\rho_{t_j}} \tag{14.3}$$

where
- x_j are the mass fraction of food constituents
- ρ_{t_j} are their true densities

Nevertheless, this equation does not consider air phase or interaction between phases and, hence, has limited applications. In the case that volume fractions V_j of the constituents are known, food density is given by the equation:

Table 14.17 Density of Food Components as a Function of Temperature

Air	$\rho_{air} = 12.847 - 3.2358 \times 10^{-3}\ T$
Water	$\rho_w = 9.9718 \times 10^2 + 3.1439 \times 10^{-3}\ T - 3.7574 \times 10^{-3}\ T^2$
Ice	$\rho_{ic} = 9.1689 \times 10^2 - 0.1307\ T$
Protein	$\rho_{pr} = 1.3300 \times 10^3 - 0.5184\ T$
Carbohydrate	$\rho_{ca} = 1.5991 \times 10^3 - 0.31046\ T$
Fat	$\rho_{fa} = 9.2559 \times 10^2 - 0.41757\ T$
Fiber	$\rho_{fi} = 1.3115 \times 10^3 - 0.36589\ T$
Ash	$\rho_{ah} = 2.4238 \times 10^3 - 0.28063\ T$

Source: Choi, Y. and Okos, M.R., *Food Engineering and Process Applications*, Elsevier Applied Science, London, 1986. With permission.
ρ in kg/m^3 and T in °C.

$$\rho = \sum_{j=1}^{n} \rho_j V_j \tag{14.4}$$

In porous products, with porosity ε, density is described by the equation:

$$\frac{(1-\varepsilon)}{\rho} = \sum_{j=1}^{n} \frac{x_j}{\rho_j} \tag{14.5}$$

Choi and Okos (1986) presented a collection of equations for the estimation of densities of the most common food components at a temperature range of −40°C to 150°C. These equations are listed in Table 14.17 (density is expressed in kg/m^3 and T is the material temperature in °C).

The density of solid foods varies nonlinearly with the moisture content. Lozano et al. (1983) proposed the following general form of correlation to predict the density of fruits and vegetables during drying, based on data collected from drying experiments of carrot, potato, sweet potato, and garlic (whole and sliced):

$$\rho = g + hy + q\exp(-ry) \tag{14.6}$$

where g, h, q, and r are parameters depending on the examined material and $y = M_w/M_{wi}$. Different types of Equation 14.6, and values of the above parameters are presented in Table 14.18. Tables 14.19 through 14.22 summarize a few mathematical models for the prediction of the density of fruits and vegetables, nuts, meat and seafood, and some other food materials. Hsieh et al. (1977) proposed the following equation to predict the density of frozen food materials:

$$\frac{1}{\rho} = \frac{x_u}{\rho_u} + \frac{x_s}{\rho_s} + \frac{x_{ic}}{\rho_{ic}} \tag{14.7}$$

where ρ, ρ_u, ρ_s, and ρ_{ic} are the densities of the food, the unfrozen water, the dry solids, and the ice, respectively. Revision of the above equation to incorporate porosity, since it can strongly influence the density of a food, gives Equation 14.5. The density of food materials, especially fruits and vegetables, presents a sharp decrease when they freeze in comparition with the fresh products (Heldman and Singh, 1981; Heldman, 1982; Singh and Sarkar, 2005). Especially in the initial stage of freezing, there is a major dependence of density on temperature. It was reported that the density of strawberries decreases from about 1050 to 960 kg/m^3 as the product freezes, and its

Table 14.18 Parameter Values of Equation 14.6 for the Prediction of Density of Some Food Materials

Material	Range	Equation	*g*	*h*	*q*	*r*	Reference
Calamari mantle	0–0.80	$\rho_a = g + h\frac{X_w}{X_{w_i}} + q\exp\left(-r\frac{X_w}{X_{w_i}}\right)$	1.350E+03	−2.850E+02	−1.240E+02	−6.427E+00	Rahman (1991)
Calamari mantle	0–0.80	$\rho_s = g + h\frac{X_w}{X_{w_i}} + q\exp\left(-r\frac{X_w}{X_{w_i}}\right)$	8.018E+03	−1.320E+03	−6.679E+03	−1.720E−01	Rahman (1991)
Carrot	0 to turgor	$\rho_a = g + h\frac{M_w}{M_{w_i}} + q\exp\left(-r\frac{M_w}{M_{w_i}}\right)$	9.840E+02	0.000E+00	2.240E+02	1.800E+00	Lozano et al. (1983)
Carrot	0 to turgor	$\rho_s = g + h\frac{M_w}{M_{w_i}} + q\exp\left(-r\frac{M_w}{M_{w_i}}\right)$	1.497E+03	−2.940E+02	−2.530E+02	3.979E+01	Lozano et al. (1983)
Garlic (sliced)	0 to turgor	$\rho_a = g + h\frac{M_w}{M_{w_i}} + q\exp\left(-r\frac{M_w}{M_{w_i}}\right)$	1.130E+03	−5.670E+02	1.870E+02	−8.660E−01	Lozano et al. (1983)
Garlic (sliced)	0 to turgor	$\rho_s = g + h\frac{M_w}{M_{w_i}} + q\exp\left(-r\frac{M_w}{M_{w_i}}\right)$	2.694E+03	0.000E+00	−1.316E+03	−1.640E−01	Lozano et al. (1983)
Pear	0 to turgor	$\rho_a = g + h\frac{M_w}{M_{w_i}} + q\exp\left(-r\frac{M_w}{M_{w_i}}\right)$	1.251E+03	−1.530E+02	−1.070E+02	−1.330E−06	Lozano et al. (1983)
Pear	0 to turgor	$\rho_s = g + h\frac{M_w}{M_{w_i}} + q\exp\left(-r\frac{M_w}{M_{w_i}}\right)$	8.320E+02	2.200E+02	6.320E+02	2.775E+01	Lozano et al. (1983)
Potato	0 to turgor	$\rho_a = g + h\frac{M_w}{M_{w_i}} + q\exp\left(-r\frac{M_w}{M_{w_i}}\right)$	1.202E+03	−1.480E+02	2.590E+02	1.551E+01	Lozano et al. (1983)
Potato	0 to turgor	$\rho_s = g + h\frac{M_w}{M_{w_i}} + q\exp\left(-r\frac{M_w}{M_{w_i}}\right)$	1.234E+03	−1.170E+02	8.500E+01	1.904E+01	Lozano et al. (1983)
Squid mantle	0–0.80	$\rho_a = g + h\frac{X_w}{X_{w_i}} + q\exp\left(-r\frac{X_w}{X_{w_i}}\right)$	7.802E+03	−1.691E+03	−6.527E+03	−2.550E−01	Rahman (1991)
Squid mantle	0–0.80	$\rho_s = g + h\frac{X_w}{X_{w_i}} + q\exp\left(-r\frac{X_w}{X_{w_i}}\right)$	3.639E+03	−4.050E+02	−2.204E+03	−6.100E−04	Rahman (1991)
Sweet potato	0 to turgor	$\rho_a = g + h\frac{M_w}{M_{w_i}} + q\exp\left(-r\frac{M_w}{M_{w_i}}\right)$	1.266E+03	−2.190E+02	−3.190E+02	6.700E+00	Lozano et al. (1983)

Table 14.19 Density Prediction Models for Fruits and Vegetables

Apple (Lozano et al., 1980):

$$\rho_a = 684 + 68.1 \ln (M_w + 0.0054)$$
$$\rho_p = 1540 \exp(-0.051 M_w) - 1150 \exp(-2.40 M_w)$$

The above equations are valid for the whole range of moisture content above freezing.

Apple (Singh and Lund, 1984):

$$\rho_a = 852 - 462 \exp(-0.66 M_w)$$

This equation is valid up to zero moisture content above freezing.

Grapes (Ghiaus et al., 1997):

$$\rho_p = 1480.4 - 382.2 M_w + 131.8 M_w^2 - 15.48 M_w^3$$
$$\rho_b = 775.99 - 228.1 M_w + 133.6 M_w^2 - 22.19 M_w^3$$

where M_w is the water content, ranging from 0.176 to 4.0 kg/(kg · db).

Garlic (slices) (Madamba et al., 1994):

$$\rho_b = 200.6 + 280 X_w + 1.24 \times 10^4 l + 2.0 \times 10^4 X_w l$$

where

ρ_b is the bulk density (outside voids)
X_w is the water content ranging from 0.03 to 0.65
l is the slice thickness, ranging from 0.002 to 0.005 m

this equation is valid at room temperature.

temperature is lowered to −40°C (Singh and Sarkar, 2005). Figure 14.10 illustrates a typical density–temperature graph for frozen fruits.

14.3.1.3 Density of Liquid Food Materials

As in the case of solid foods, a large number of empirical equations have been proposed for density prediction of liquid foods as a function of temperature or the composition of the food. Some

Table 14.20 Density Prediction Models for Nuts

Gordon nut (large size) (Jha and Prasad, 1993):

where

$$\rho_t = 995.9 + 830 M_w - 640 M_w^2$$
$$\rho_b = 369.2 + 1290 M_w - 1020 M_w^2$$

Gordon nut (medium size) (Jha and Prasad, 1993):

$$\rho_t = 1039.7 + 690 M_w - 420 M_w^2$$
$$\rho_b = 369.7 + 1330 M_w - 1080 M_w^2$$

Gordon nut (small size) (Jha and Prasad, 1993):

$$\rho_t = 1025.7 + 930 M_w - 800 M_w^2$$
$$\rho_b = 434.0 + 1660 M_w - 8420 M_w^2$$

In the above quadratic equations for the prediction of gordon nut density (kg/m^3), the moisture content ranges between 0.15 and 0.60 kg/(kg · db) and the experiments were conducted at room temperature.

Macadamia nut (in-sheel) (Palipane et al., 1992):

$$\rho_a = 1018.6 - 44 M_w + 1300 M_w^2$$
$$\rho_b = 605.2 - 92 M_w + 800 M_w^2$$

In the above equations for the prediction of macadamia nut density (kg/m^3), the moisture content ranges between 0.02 and 0.24 kg/(kg · db) and the experiments were conducted at room temperature.

Table 14.21 Density Prediction Models for Meat and Seafood

Fresh meat products (Sanz et al., 1989):

$$\rho_a = 1053 \quad T \geq T_{fr}$$

$$\rho_a = \frac{1053}{0.9822 + 0.1131X_{w_i} + 0.2575\frac{1-X_{w_i}}{T}} \quad T < T_{fr}$$

where T_{fr} (°C) is the freezing point of the product.

Fresh seafood (Rahman and Driscoll, 1994):

$$\rho_a = 2684 - 3693X_{w_i} + 2085X_{w_i}^2$$

where X_w is the moisture content, ranging from 0.739 to 0.856 kg/kg wet basis at temperature 20°C.

Frozen seafood (at −30°C) (Rahman and Driscoll, 1994):

$$\rho_a = 1390 - 520X_{w_i} + 31.56X_{w_i}^2$$

Frozen squid mantle (Rahman and Driscoll, 1994):

$$\rho_a = 1047 + 3.603T + 0.057T^2$$

This equation is valid below the freezing point of the foodstuff up to −40°C; the moisture content is $X_{w_i} = 0.814$ kg/(kg · wb).

of them are presented in Table 14.23. Besides the development of empirical equations for the prediction of density, based on experimental data, a great effort has been made for the presentation of theoretical density prediction models. Although initial efforts have been made for the development of such models based on mass and volume conservation, it is known today that these models cannot be applied during the drying process as they do not consider phenomena, such as the

Table 14.22 Density Prediction Models for Other Food Materials

Coconut (shredded) above the freezing point (Jindal and Murakami, 1984):

$$\rho_b = 236 + 440X_w$$

Corn starch (granular and gelatinized) (Marousis and Saravacos, 1990):

$$\rho_p = 1442 + 837M_w - 3646M_w^2 + 4481M_w^3 - 1850M_w^4$$

where M_w is the moisture content of the material, ranging from 0 to 1 kg/(kg · db).

Starch materials (measured by helium stereopycnometry) (Marousis and Saravacos, 1990):

$$\rho_s = 1440 + 796M_w - 3074M_w^2 + 3320M_w^3 - 1209M_w^4 \text{ for } M_w < 0.4$$

$$\rho_s = 1500 - 291M_w \text{ for } M_w > 0.4$$

where M_w is the moisture content of the material, expressed in dry basis.

White speckled red kidney bean (Isik and Unal, 2007):

$$\rho_b = 749.02 - 8.6651M_c$$

$$\rho_t = 1078.8 + 16.906M_c$$

where M_c is the % moisture content of the material, expressed in dry basis.

Pea (during air-drying at different temperatures) (Medeiros and Sereno, 1994):

$$\rho_a = 1148 - 27.1M_w + 65.6\exp(-3.43M_w) \quad 30°C$$

$$\rho_a = 1141 - 24.2M_w + 159\exp(-3.37M_w) \quad 50°C$$

$$\rho_a = 1167 - 40.1M_w + 247\exp(-3.22M_w) \quad 65°C$$

Bamboo (Madamba, 2003):

$$\rho_a = 1338.6 + 0.99X_c - 0.051X_c^2$$

Also,

$$\rho_a = 1381.4 - 168.1\frac{X_c}{X_{c_i}} - 4.9\exp\left(3.8\frac{X_c}{X_{c_i}}\right)$$

where X_c (% wet basis) is the percent moisture content and X_{c_i} is the initial moisture content.

Source: Adapted from Rahman, M.S., Thermophysical properties of seafoods, PhD thesis, University of New South Wales, Sydney, Austrialia, 1991.

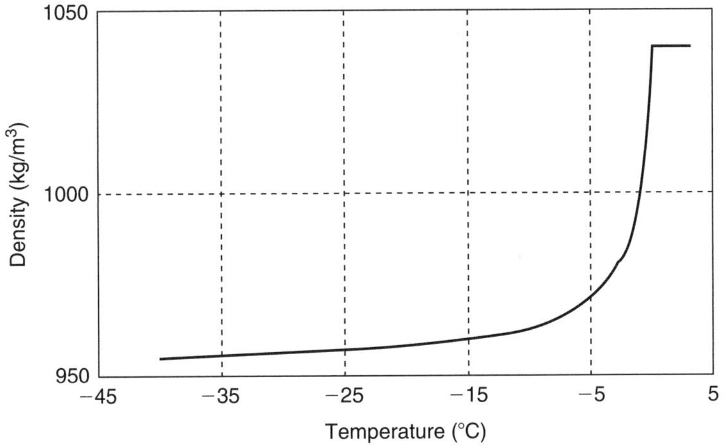

Figure 14.10 Density of frozen fruits and vegetables as a function of temperature.

Table 14.23 Density Prediction Models for Liquid Foods

Whole milk (Short, 1955):

$$\rho_a = 1035.0 - 0.358T + 0.0049T^2 - 0.00010T^3$$

Skim milk (Short, 1955):

$$\rho_a = 1036.6 - 0.146T + 0.0023T^2 - 0.00015T^3$$

where T (°C) is the temperature.

Milk cream (Phipps, 1969):

$$\rho_a = 1038.2 - 0.17T - 0.003T^2 - \left(133.7 - \tfrac{475.5}{T}\right)x_f$$

where T (°C) is the temperature and x_f is the fat content expressed as the mass fraction of fat.

Cow's milk fat (Roy et al., 1971):

$$\rho_a = 923.51 - 0.43T$$

Buffalo's milk fat (Roy et al., 1971):

$$\rho_a = 923.84 - 0.44T$$

Fruit juices (Riedel, 1949):

$$\rho_a = 16.0185 \frac{62.4}{0.206} \frac{\vartheta^2 - 1}{\vartheta^2 + 2}$$

where ϑ is the refractive index of the sugar solution.

Tomato juice (Choi and Okos, 1983):

$$\rho_a = \rho_w X_w + \rho_s X_s$$
$$\rho_w = 999.89 - 6.0334 \times 10^{-2}\,T - 3.6710 \times 10^{-3}T^2$$
$$\rho_s = 1.4693 \times 10^3 - 5.4667 \times 10^{-1}T - 6.9643 \times 10^{-3}T^2$$

where

X_w is the water mass fraction (kg/kg)
x_s is the solid mass fraction (kg/kg) of the tomato juice.

Apple juice (Bayindirli, 1992):

$$\rho_a = 830 + 350\,\exp(0.01B) - 0.567T$$

where

T (°C) is the temperature, ranging from 20°C to 80°C
B is the degrees Brix (% soluble solids in the solution) of the juice, ranging from 14° to 39° Brix.

(continued)

Table 14.23 (continued) Density Prediction Models for Liquid Foods

Peach juice (Ramos and Ibarz, 1998):

$\rho_a = 1006.6 - 0.5155T + 4.1951B + 0.0135B^2$

where

T is the temperature, ranging from 0°C to 80°C
B is the degrees Brix of the juice, ranging from 10° to 60° Brix

Orange juice (Ramos and Ibarz, 1998):

$\rho_a = 1025.4 - 0.3289T + 3.2819B + 0.0178B^2$

where

T is the temperature, ranging from 0°C to 80°C
B is the degrees Brix of the juice, ranging from 10° to 60° Brix

Pear juice (depectinized and clarified) (Ibarz and Miguelsanz, 1989):

$\rho_a = 988.8 - 0.546T + 5.13B$

where

T is the temperature, ranging from 5°C to 70°C
B is the degrees Brix of the juice, ranging from 10° to 71° Brix

Brazilian orange juice (Telis-Romero et al., 1998):

$\rho_a = 1428.5 - 0.231T - 454.9X_w$

where

T is the temperature, ranging from 0.5°C to 62°C
X_w is the water content of the juice, ranging from 0.34 to 0.73 kg water/kg juice.

formation of air phase or pores, swelling or antiswelling of the solid phases, loss of volatiles, and interaction of the constituent phases. Perez and Calvelo (1984) proposed the following model, based on the material balance equation and in the formation of an air phase, to estimate the density of beef muscle during cooking:

$$\rho_a = \rho_{a_i} \frac{1 - X_{w_i}}{(1 - X_w)\left[1 - (\rho_{a_i}/\rho_w)X_{w_i} + (\rho_{a_i}/\rho_w)[(1 - X_{w_i})/1 - X_w] + u'(X_{w_i} - X_w)^{v'}\right]} \tag{14.8}$$

where u' and v' are empirical parameters. Rahman (1991) developed a theoretical model that included the pore volume and an interaction term and can be written as

Apparent volume = Actual volume of the component phases + volume of pore or air + excess volume due to the interaction of phases.

The excess volume can be positive or negative depending on the process. In terms of density, the equation takes the form

$$\frac{1}{\rho_a} = \sum_{j=1}^{n} \frac{x_j}{\rho_{t_j}} + \varepsilon_a V_a + \varepsilon_{ex} V_a = \frac{1}{1 - \varepsilon_a - \varepsilon_{ex}} \sum_{j=1}^{n} \frac{x_j}{\rho_{t_j}} \tag{14.9}$$

if we define porosity as $\varepsilon_a = V_{air}/V_a$, excess volume fraction as $\varepsilon_{ex} = V_{ex}/V_a$, and taking into account that $V_a = 1/\rho_a$, where V_{air} is the volume of pore or air. When porosity and excess volume fraction are negligible, Equation 14.9 can be transformed to

$$\frac{1}{\rho_a} = \frac{1}{\rho_t} = \sum_{j=1}^{n} \frac{x_j}{\rho_{t_j}}$$

Table 14.24 Linear Correlation of Excess Volume (ε_{ex})

Material	Equation
Fresh seafood	$-\varepsilon_{ex} = 0.038 - 0.025X_w$
Frozen seafood (at 30°C)	$-\varepsilon_{ex} = 0.069 - 0.071X_{w_i}$
Squid mantle ($X_w = 0.814$)	$-\varepsilon_{ex} = (397 + 7.0517T) \times 10^{-4}$
Calamari mantle[a]	$-\varepsilon_{ex} = 0.043 - 0.0024\ (X_w/X_{w_i})$
Squid mantle[a]	$-\varepsilon_{ex} = 0.083 - 0.069\ (X_w/X_{w_i})$

Source: Rahman, M.S., Thermophysical properties of seafoods, PhD thesis, University of New South Wales, Sydney, Australia, 1991.

[a] Air-dried meat up to zero moisture content. T (°C), X (kg/(kg · wb)).

The above equation is the simplified density model presented as Equation 14.3. Porosity (or air volume fraction) and excess volume fraction are given by

$$\varepsilon_a = 1 - \frac{\rho_a}{\rho_s} \tag{14.10}$$

$$\varepsilon_{ex} = 1 - \frac{\rho_s}{\rho_t} \tag{14.11}$$

Rahman (1991) measured the densities of fresh, frozen, and dried seafood and used the following Ridlich-Klister (1948) equation, which calculates the excess volume of a mixture, by expanding into polynomial to predict density:

$$V_{ex} = x_1 x_2 \sum_{k=1}^{N} \overline{B}_{jk}(x_1 - x_2)^{k-1} \tag{14.12}$$

where

x_j is the mass fraction of j-component
$\overline{B}_{jk}$ is the kth coefficient of the equation

A significant improvement in density prediction was found in all cases when the above expression was used as component of the theoretical equation. The linear and quadratic correlations of ε_a and ε_{ex} for seafood are given in Tables 14.24 through 14.26 (adapted by Rahman, 1991).

Table 14.25 Quadratic Correlation of Excess Volume (ε_{ex})

Material	Equation
Fresh seafood	$-\varepsilon_{ex} = 0.787 - 1.90X_w + 1.17X_w^2$
Frozen seafood (at 30°C)	$-\varepsilon_{ex} = 0.559 - 1.29X_{w_i} + 0.768X_{w_i}^2$
Squid mantle ($X_w = 0.814$)	$-\varepsilon_{ex} = (393 + 5.877\,T - 0.0309T^2) \times 10^{-4}$
Calamari mantle[a]	$-\varepsilon_{ex} = 0.026 + 0.143\ (X_w/X_{w_i}) - 0.150\ (X_w/X_{w_i})^2$
Squid mantle[a]	$-\varepsilon_{ex} = 0.089 - 0.112\ (X_w/X_{w_i}) + 0.042\ (X_w/X_{w_i})^2$

Source: Rahman, M.S., Thermophysical properties of seafoods, PhD thesis, University of New South Wales, Sydney, Australia, 1991.

[a] Air-dried meat up to zero moisture content. T (°C), X (kg/(kg · wb)).

Table 14.26 Quadratic Correlation of Porosity (ε_a)

Material	Equation
Calamari mantle[a]	$\varepsilon_a = 0.079 - 0.164\ (X_w/X_{w_i}) + 0.099\ (X_w/X_{w_i})^2$
Squid mantle[a]	$\varepsilon_a = 0.109 - 0.219\ (X_w/X_{w_i}) + 0.099\ (X_w/X_{w_i})^2$

Source: Rahman, M.S., Thermophysical properties of seafoods, PhD thesis, University of New South Wales, Sydney, Australia, 1991.

[a] Air-dried meat up to zero moisture content. T (°C), X (kg/(kg · wb)).

14.3.1.4 Prediction Models for Shrinkage

Different models and equations have been proposed for the prediction of shrinkage in food materials. Kilpatric et al. (1955) suggested the following simple equation for apparent shrinkage:

$$S_a = \frac{V_a}{V_{a_i}} = \frac{X_w + 0.8}{X_{w_i} + 0.8} \tag{14.13}$$

where
V_a is the volume of material at mass fraction of water X_w
V_{ai} is the initial volume corresponding to initial mass fraction of water X_{wi}

The development of this expression is based on the assumption that volume change of shrinkage is equal to the volume of dehydrated water, which is valid only for the early stages of drying. Gorling (1958), Chirife (1969), and Charm (1978) used the analogy of thermal expansion to predict shrinkage with the equation:

$$S_a = 1 + \beta(X_w - X_{w_i}) \tag{14.14}$$

Suzuki et al. (1976) developed three drying models and the corresponding equations for shrinkage prediction. These models are (1) uniform drying model, where the shrinkage volume equals the water loss and the following equations are applied:

$$S_a = \frac{X_w + \Lambda}{X_{w_i} + \Lambda} \tag{14.15}$$

$$\Lambda = X_{w_e}\left(\frac{1}{\rho_{a_e}} - 1\right) + \frac{1}{\rho_{a_e}} \tag{14.16}$$

where X_{w_e} is the equilibrium mass fraction of water and ρ_{a_e} is the apparent density at equilibrium. (2) Core drying model, where the dried layer, whose density is ρ_{a_e}, can be discriminated from the wet core, whose density is equal to ρ_{a_i}. In this case, shrinkage is given by the following equations:

$$S_a = \xi \frac{X_w}{X_{w_i}} + 1 \tag{14.17}$$

$$\xi = 1 - \frac{\rho_{a_i}}{\rho_{a_e}} \frac{X_{w_e} + 1}{X_{w_i} + 1} \frac{X_{w_e}}{X_{w_i} - X_{w_e}} \tag{14.18}$$

(3) Semicore drying model, where the dried layer shrinkage is predicted by the equation

$$S_a = \theta \frac{X_w}{X_{w_i}} + \delta$$

where θ and δ are functions of ρ_{a_i}, ρ_{a_e}, X_{w_i}, X_{w_e}, and ρ_{a_e}core, and

$$\rho_d = \rho_{a_i} X_w + \rho_{a_e}(1 - X_w) \tag{14.19}$$

The above equations proposed by Suzuki et al. (1976) can predict shrinkage of vegetables (roots), such as potato, carrot, radish, and sweet potato, but only in the first stage of drying, where shrinkage changes linearly with moisture content. Lozano et al. (1983) developed a general mathematical model to predict shrinkage of fruits and vegetables during air-drying, which is given by the following equations:

$$S_a = x' + y' + z' \tag{14.20}$$

$$x' = 0.161 + 0.816 \frac{X_w}{X_{w_i}} \tag{14.21}$$

$$y' = 0.022 \exp\left(\frac{0.018}{0.025 + X_w}\right) \tag{14.22}$$

$$z' = (0.209 - S_{a_e})\left(1 - \frac{X_w}{X_{w_i}}\right) \tag{14.23}$$

Perez and Calvelo (1984) proposed the following shrinkage model for meat (beef muscle) during cooking, which takes into account the air phases besides solid and water phases:

$$S_a = 1 - \frac{\rho_{a_i}}{\rho_w} X_{w_i} + \frac{\rho_{a_i}}{\rho_w} \frac{X_w(1 - X_{w_i})}{1 - X_w} + u'(X_{w_i} - X_w)^{v'} \tag{14.24}$$

where u' and v' are empirical parameters of the model. Vacarezza (1975) proposed the following equation for the volume shrinkage of sugar beet roots:

$$S_a = \left[M_{w_i} \frac{\rho_s}{\rho_w} + 1\right]^{-1} + \frac{\rho_s/\rho_w}{\rho_s/\rho_w + 1} \frac{M_w}{M_{w_i}} \tag{14.25}$$

Mayor and Sereno (2004) presented the model (modified Perez and Calvelo):

$$S_a = \frac{1}{1 - \varepsilon}\left[1 + \frac{\rho_i}{\rho_w} \frac{M_w - M_{w_i}}{1 + M_{w_i}} - \varepsilon_i\right] \tag{14.26}$$

The above equation is applicable for apple, potato, carrot, and squid. These authors also presented a review of shrinkage prediction models classifying them into two substantially different categories regarding the approach of their development. The first one consists empirical fitted models on experimental shrinkage data as a function of moisture content, while the second approach is more fundamental, based on a physical interpretation of the food system and includes the prediction of geometrical changes based on conservation laws of mass and volume. Other much simpler equations have been proposed in the literature, like the first-order kinetics:

$$\ln S_a = -kt \tag{14.27}$$

Konanayakam and Sastry (1988) proposed for the prediction of mushroom shrinkage during blanching the following linear equation:

$$S_a = 0.17 + 0.11X_w \tag{14.28}$$

Lozano et al. (1980) developed the above equation to relate the shrinkage of apple with its moisture content. Pelegrina and Crapiste (2001) studied potato particles shrinkage during pneumatic drying and modeled the process using the following expression:

$$S_a = \frac{0.0231 + 0.2175M_w}{0.0231 + 0.2175M_{w_i}}, \; 1.89 \leq M_w \leq 4.5 \tag{14.29}$$

$$S_a = \frac{0.1866 + 0.1333M_w}{0.1866 + 0.1333M_{w_i}}, \; 0 \leq M_w \leq 1.89 \tag{14.30}$$

Hatamipour and Mowla (2002) proposed the following correlation for the shrinkage of carrot that undergoes drying in a fluidized bed dryer using an inert drying medium:

$$S_a = 0.0927M_w + 0.07752 \tag{14.31}$$

They also estimated the shrinkage across the diameter and the length of the specimen as

$$D/D_i = \sqrt{0.08741M_w + 0.13091} \tag{14.32}$$

$$L/L_i = 0.08145 \ln M_w + 0.81671 \tag{14.33}$$

Raghavan and Venkatachalapathy (1999) studied the shrinkage of strawberries with respect to the moisture ratio during microwave drying and proposed the following equations:

$$S_a = 0.0943 + 0.887M_w/M_{w_i}, \quad \text{for microwave power level 0.1 W/g} \tag{14.34}$$

$$S_a = 0.1329 + 0.8325M_w/M_{w_i}, \quad \text{for microwave power level 0.2 W/g} \tag{14.35}$$

where M_w/M_{w_i} is the moisture ratio of the food. They also present the following model for the evaluation of the equivalent diameter D_{eq} of the strawberries at various moisture ratios:

$$D_{eq}^{-1} = 0.3067 - 0.1071 \ln(M_w/M_{w_i}), \quad \text{for microwave power level 0.1 W/g} \tag{14.36}$$

$$D_{eq}^{-1} = 0.3157 - 0.0899 \ln(M_w/M_{w_i}), \quad \text{for microwave power level 0.2 W/g} \tag{14.37}$$

The equivalent diameter of an ellipsoid shape is one that corresponds to a sphere of equivalent volume V and is calculated using the relation $D_{eq} = (6V/\pi)^{1/3}$ (Saravacos and Raouzeos, 1986). Moreira et al. (2000) presented the following equation for the prediction of apple disks' shrinkage:

$$S_a = 0.2481 + 0.096M_w \tag{14.38}$$

where M_w is the moisture content (dry base). Rahman (1984) and Uddin et al. (1990) set out the following power law equation for the prediction of pineapple disk shrinkage during air-drying as a function of the slice thickness (l):

$$\frac{l_f}{l_i} = \left(\frac{M_w}{M_{w_i}}\right)^{d'} \tag{14.39}$$

where d' varies from 0.2 to 0.45 depending on the drying temperature and initial thickness.

Table 14.27 Models for the Prediction of Shrinkage during Drying

Model	Reference
$S_a = \overline{\alpha}\{1 - \exp[-\overline{b}(X_w - X_{wi})]\}$	Bala and Woods (1984)
$S_a = \alpha' + (\overline{a} + \overline{b}RH + \overline{c}T)(X_w - X_{w_i})$	Lang and Sokhansanj (1993)
$S_a = 1 + \beta(X_w - X_{w_i})$	Rahman (1995)
$S_a = 1/(\overline{a} + \overline{b}\ \exp\ X_w)$	Corrēa et al. (2004)
$S_a = \overline{a} = \overline{b}X_w$	Linear
$S_a = \overline{a}\ \exp(\overline{b}X_w)$	Exponential

Note: X_w: moisture content of the food (kg/(kg · wb)); X_{w_i}: initial moisture content (kg/(kg · wb)); β: linear shrinkage coefficient; a', $\overline{a}, \overline{b}, \overline{c}$: product-dependent parameters; T: air temperature (°C); RH: relative humidity.

The factor l_f/l_o varies from 0.40 to 0.52 at different initial thickness, and shrinkage is independent of the initial sample thickness (Rahman, 1985). Balaban and Pigott (1986) found that the shrinkage of fish muscle rectangular slab with moisture content presents significant change in different dimensions (length, width, and thickness). Rahman and Potluri (1990a) confirmed that in the case of squid mantle slab the maximum change in thickness is up to 80%, whereas the change of length and width does not exceed 20%. One different approach to derive a shrinkage model is to assume that a food material consists of solid, water, and air phases. In this case, shrinkage is given by the following equation (Rahman, 1995):

$$S_a = \frac{1}{1 - \varepsilon_a}\left(1 - \frac{\rho_{a_i}(X_{w_i} - X_w)}{\rho_w(1 - X_w)}\right) \tag{14.40}$$

Table 14.27 presents a review of a few common mathematical models used for the description of volumetric shrinkage experimental data during drying.

14.3.1.5 Prediction Models for Porosity

Porosity of food materials is usually calculated from density data or from empirical expressions of porosity versus moisture content. Practically, there is no theoretical method for porosity prediction. Kunii and Smith (1960), using theoretical concepts concerning spheres, determined two marginal values for porosity, which are 0.476 and 0.260. The first value corresponds to the loosest cubic packing of the spheres, while the second one refers to the tightest hexagonal packing. Thus, if the observed void fraction is less than 0.260, it might be considered as a result of clogging by small particles in void spaces. On the other hand, if the void fraction exceeds the value 0.476, it might be assumed that there are exceptionally large hollow spaces compared with the average void space.

Lozano et al. (1980) reported that porosity change in fruits and vegetables is the result of the shrinkage of the overall dimensions of the material, as well as the shrinkage of cells themselves, and proposed that the geometric model of Rotstein and Cornish (1978), as given by the following equations, could be used to predict porosity at full turgor:

$$\varepsilon_{a_i} = 1 - \frac{\pi\zeta^0}{6\hat{a}^3} \tag{14.41}$$

$$\zeta^0 = -2 + 4.5\hat{a} - 1.5\hat{a}^3 \tag{14.42}$$

where ζ^0 is a geometric value corresponding to the full turgor situation and $\widehat{a}$ is the cube half-side that is formed when spheres of radius r are put on a cubically truncate shape. The porosity at any moisture content is

Table 14.28 Empirical Models for Porosity Prediction of Some Foods

Open and closed pores in calamari (during air-drying up to zero moisture content) (Rahman et al., 1996):

$$\varepsilon_{\text{op}} = 0.079 - 0.164\frac{X_{\text{w}}}{X_{\text{w}_i}} + 0.099\left(\frac{X_{\text{w}}}{X_{\text{w}_i}}\right)^2$$

$$\varepsilon_{\text{cp}} = 0.068 - 0.216\frac{X_{\text{w}}}{X_{\text{w}_i}} + 0.138\left(\frac{X_{\text{w}}}{X_{\text{w}_i}}\right)^2$$

Apparent porosity of squid mantle (during air-drying up to zero moisture content) (Rahman, 1991):

$$\varepsilon_{\text{a}} = 0.109 - 0.219\frac{X_{\text{w}}}{X_{\text{w}_i}} + 0.099\left(\frac{X_{\text{w}}}{X_{\text{w}_i}}\right)^2$$

Open pore porosity of apple (during air-drying, $0.89 > X_{\text{w}} > 0$) (Lozano et al., 1980):

$$\varepsilon_{\text{op}} = 1 - \frac{852.0 - 462.0\exp(-0.66M_{\text{w}})}{1540.0\exp(-0.051M_{\text{w}}) - 1150.0\exp(-2.4M_{\text{w}})}$$

Bulk porosity of garlic slices (Madamba et al., 1994):

$$\varepsilon_{\text{B}} = 0.865 - 30\,X_{\text{w}} - 0.8\times10^3 l + 2.0\times10^3\,X_{\text{w}} l$$

where

X_{w} is the water content, ranging from 0.03 to 0.65 kg/(kg · wb)
l (m) is the slice thickness, ranging from 0.002 to 0.005 m (at room temperature).

Porosity of white speckled red kidney bean (Isik and Unal, 2007):

$$\varepsilon = 34.309 + 1.2691M_{\text{c}}$$

where M_{c} is the % moisture content of the material, expressed in dry basis.

Internal porosity of bamboo (Madamba, 2003):

$$\varepsilon = 21.4 - 0.5X_{\text{c}} + 0.003X_{\text{c}}^2$$

where X_{c} (% wet basis) is the percent moisture content.

$$\varepsilon_{\text{a}} = 1 - \frac{V_{\text{c}}V_{\text{a}_i}}{V_{\text{a}}^2}\frac{\pi\zeta^0}{6\hat{a}^3} \tag{14.43}$$

Rahman (2000) noted that the formation of pores in food materials during drying could be grouped into two generic types: one with an inversion point and another without such a point. There are cases in which during drying pores are initially collapsed, causing decrease in porosity until a critical low value is reached, while further decrease of moisture content causes the formation of pores again till the complete dehydration of the food. In other cases, the process can be introduced in the opposite direction, and a critical high porosity value is observed which is greater from the initial or final porosity of the material. On the other hand, there are many cases that present a decrease or increase in the level of pores as a function of moisture content and no inversion point is presented. Many empirical equations have been introduced in the literature for the prediction of porosity of various food materials; a few of them are summarized in Table 14.28. Hussain et al. (2002) presented the following equation for porosity prediction applicable to all foods during air-drying:

$$\varepsilon_a = 0.5X_{\text{w}}^2 - 0.8X_{\text{w}} - 0.002T^2 + 0.02T - 0.05(1 - \varepsilon_{\text{a}_i})F' \tag{14.44}$$

where F' is a factor that takes values according to the type of the product (1 for sugar-based products, 2 for starch-based products, and 3 for other products).

14.4 GENERIC STRUCTURAL PROPERTIES PREDICTION MODELS

In addition to models proposed for specific materials, effort has been made for the development of generic models of theoretical base that will be applicable for all materials using parameters of physical meaning. Two of those are presented in the following. Zogzas et al. (1994), Krokida and Maroulis (1997) and Saravacos and Maroulis (2001) developed a semiempirical model to predict structural properties, such as apparent and material density, shrinkage, and porosity. The philosophy of this model is to examine the mass and volume of each constituent phase (dry solids, water, and air) of the food material independently and also to introduce a series of empirical constants (parameters) to the model that describe the behavior of each specific material, taking into account phenomena such as air phase formation and phase interaction. The parameters involved were chosen to reflect physical meaning. The adjustment of the equations to existing experimental data for each material separately is achieved by nonlinear regression analysis.

Let us assume a moist material; it consists of three phases that are dry solids, water, and air. The mass of each of these phases is m_s, m_w, and m_{air}, respectively, while the volume is V_s, V_w, and V_{air}, respectively. Note that for continuous materials, V_{air} is the air contained in the material's internal pores (Saravacos and Maroulis, 2001), in contrast with granular materials in which the interstitial air phase is taken into account. The total mass m_T of the material sample, assuming that the mass of air is neglected due to its very low density ($\sim 1.2 \times 10^{-3}$ g/cm^3) comparing with those of water (~ 1 g/cm^3) and solid foodstuff (>1 g/cm^3), is

$$m_T = m_s + m_w \tag{14.45}$$

The total volume V_T of the sample is considered as

$$V_T = V_s + V_w + V_{air} \tag{14.46}$$

The volume of air is referred to the internal pores only. The true (particle) volume is defined as the total volume of the sample excluding the volume occupied by air inside the pores:

$$V_p = V_s + V_w \tag{14.47}$$

The apparent density of the food material ρ_a is defined as

$$\rho_a = \frac{m_T}{V_T} \tag{14.48}$$

and the material density ρ_m as

$$\rho_m = \frac{m_T}{V_p} \tag{14.49}$$

Apparent and material densities are analogous to the bulk and particle density of granular materials, respectively (Saravacos and Maroulis, 2001). The actual densities of dry solids (ρ_s) and enclosed water (ρ_w) can also be defined as

$$\rho_s = \frac{m_s}{V_s} \tag{14.50}$$

$$\rho_w = \frac{m_w}{V_w} \tag{14.51}$$

The specific volume of the sample u is defined as the total volume per unit mass of dry solids:

$$u = \frac{V_T}{m_s} \tag{14.52}$$

The material moisture content M_w on a dry basis is

$$M_w = \frac{m_w}{m_s} \tag{14.53}$$

The volume-shrinkage coefficient β' can be defined by the following equation, which represents the proportion of initial specific volume that shrinks as water is removed:

$$u = u_i - \beta' \frac{(M_{w_i} - M_w)}{\rho_w} \tag{14.54}$$

where
M_{w_i} is the initial moisture content of the moist food material
u is the specific volume at material moisture content M_w
u_i is the initial specific volume

The shrinkage coefficient β' varies between 0 (no shrinkage) and 1 (full shrinkage), as can be seen in Figure 14.11. Assuming that no volume interaction occurs between the water and the solids, combining Equations 14.49 through 14.51 and 14.53 results in

$$\rho_m = \frac{1 + M_w}{(1/\rho_s) + (M_w/\rho_w)} \tag{14.55}$$

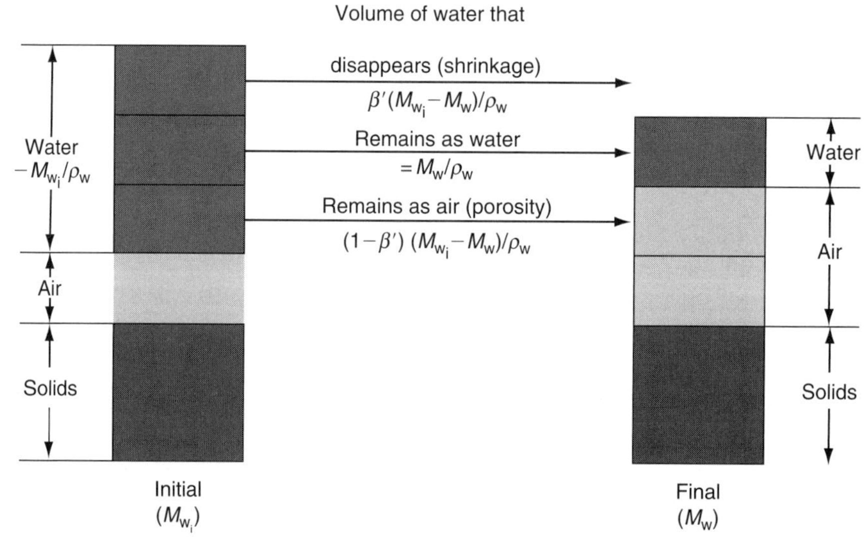

Figure 14.11 Schematic representation of shrinkage and porosity development, as part of water is removed. (From Saravacos, G.D. and Maroulis, Z.B., *Transport Properties of Foods*, Marcel Dekker, New York, 2001.)

where

ρ_s is the true density of the dry solid
ρ_w is the enclosed water density

This equation shows the dependence of moisture content on material (particle) density. Combining Equations 14.48, 14.51, 14.53, and 14.54 results in

$$\rho_a = \frac{1 + M_w}{(1 + M_{w_i}/\rho_{a_i}) + \beta'(M_w - M_{w_i}/\rho_w)} \tag{14.56}$$

where ρ_{a_i} is the initial apparent density (at $M_w = W_{w_i}$). When the zero moisture content is considered as initial moisture content ($W_{w_i} = 0$), then Equations 14.54 and 14.56 are transformed into the following expressions, respectively:

$$v = v_i + \beta' \frac{M_w}{\rho_w} \tag{14.57}$$

$$\rho_a = \frac{1 + M_w}{(1/\rho_{a_i}) + \beta'(M_w/\rho_w)} \tag{14.58}$$

where u_i and ρ_{a_i} are the specific volume and the apparent density at moisture content $M_w = 0$ (dry solid), respectively. The initial specific volume is related to the dry solid apparent density through the following equation:

$$u_i = \frac{1}{\rho_{a_i}} \tag{14.59}$$

The porosity of the food material is calculated from the following equation:

$$\varepsilon = 1 - \frac{\rho_a}{\rho_m} \tag{14.60}$$

For granular materials, this equation is equivalent to the following:

$$\varepsilon = 1 - \frac{\rho_b}{\rho_p} \tag{14.61}$$

Moreover, in fried products, when oil is considered as one more phase, Equations 14.57 and 14.58 are further transformed to equations:

$$v = v_i + \beta'\left(\frac{M_w}{\rho_w} + \frac{Y}{\rho_L}\right) \tag{14.62}$$

$$\rho_a = \frac{1 + M_w + Y}{\frac{1}{\rho_{a_i}} + \beta'\left(\frac{M_w}{\rho_w} + \frac{Y}{\rho_L}\right)} \tag{14.63}$$

where ρ_L is the oil density and Y is the oil content. The latter is given by

$$Y = \frac{m_L}{m_s} \tag{14.64}$$

where m_L is the mass of the oil in the sample of m_s dry solids. For fried products, Equation 14.55 is also transformed to equation

$$\rho_m = \frac{1 + M_w + Y}{(1/\rho_s) + (M_w/\rho_w) + (Y/\rho_L)} \quad (14.65)$$

Figure 14.11 presents the development of shrinkage and porosity during the dehydration of a food material. Depending on the material and the dehydration method used, as water is removed from the material, a part of the volume initially occupied disappears, which results in the shrinkage of the material, and the other part of its initial volume remains as air, which results in the development of porosity.

The above model was further expanded by Boukouvalas et al. (2006a,b) to incorporate the effect of temperature on the true density of food materials, given by Equation 14.55. The dry solid true density ρ_s is related to the temperature through the linear equation:

$$\rho_s = s_0 + s_1 T \quad (14.66)$$

where s_0 and s_1 are parameters. The water density is also considered temperature-dependent, and it can be calculated through a second-degree polynomial:

$$\rho_w = w_0 + w_1 T + w_2 T^2 \quad (14.67)$$

where T is the material temperature and the constants of the above equation are $w_0 = 9.97 \times 10^2$, $w_1 = 3.14 \times 10^{-3}$, $w_2 = -3.76 \times 10^{-3}$ (Maroulis and Saravacos, 2003). Rahman (2003) also developed a theoretical model to predict the porosity based on the conservation of mass and volume principle and defining a shrinkage-expansion coefficient. At first, the model for ideal condition was derived and then it was extended to nonideal condition.

14.4.1 Ideal Condition

A theoretical model (ideal condition) to predict porosity in foods during drying was developed based on conservation of mass and volume principle and assuming that the volume of pores formed is equal to the volume of water removed during drying (Rahman, 2003). In this case, there is no shrinkage (or collapse) or expansion in the initial volume of the material. If there exist no pores in the material before drying (i.e., initially), the volume fraction of pores in the material at any water level can be expressed as

$$\varepsilon_a^I = \frac{\tilde{a}}{\tilde{a} + \tilde{b}}, \quad \text{if } \varepsilon_{a_i} = 0 \quad (14.68)$$

where

$$\tilde{a} = (1 - \psi)/\rho_w, \quad \tilde{b} = \psi/\rho_m, \quad \psi = (1 - X_{w_i})/(1 - X_w)$$

where

ψ is the mass of sample at any moisture content per unit initial mass before processing (kg/kg)
$\tilde{a}$ is the volume of water removed per unit mass (m^3/kg)
$\tilde{b}$ is the volume of water remaining and solids per unit mass (m^3/kg)

The material density of a multicomponent mixture can be estimated from the following equation:

$$\frac{1}{\rho_m} = \frac{1}{\rho_t} = \frac{X_w}{\rho_w} + \sum_{j=1}^{n} \frac{x_j}{\rho_j} \quad (14.69)$$

where x_j and ρ_j are the mass fraction and density of other components except water. The mass fraction of other components except water can be estimated from the mass fraction of water:

$$x_j = x_{j_i}/\psi \tag{14.70}$$

For a two-component mixture, if the excess volume due to interaction between water and solids is zero, the material density can be expressed as

$$\frac{1}{\rho_m} = \frac{1}{\rho_t} = \frac{X_w}{\rho_w} + \frac{m_s}{\rho_s} \tag{14.71}$$

where ρ_s is the substance density for dry solids at zero moisture content (kg/m^3). In the case of porous material, the ideal porosity can be written as

$$\varepsilon_a = \varepsilon_{a_i} + (1 - \varepsilon_{a_i})\varepsilon_a^I \tag{14.72}$$

As expected, the ideal model may not be valid in many practical cases. Constant porosity and no shrinkage are often stated as key assumptions in the model for dryer design. Although the above equation does not predict the real situation, it can be used for ideal conditions. In addition, it could be used as an initial estimation when experimental results are missing in the case of nonideality and to assess processes of pore formation, which are more close to ideal conditions.

14.4.2 Nonideal Conditions

The ideal model is then extended for nonideal conditions, when there is shrinkage, collapse, or expansion, by defining a shrinkage-expansion coefficient as

$$\varepsilon_a^{II} = \frac{\phi\tilde{a}}{\phi\tilde{a} + \tilde{b}} \tag{14.73}$$

where ϕ is the shrinkage-expansion coefficient for pore formation or collapse, and the porosity can be calculated as (Rahman, 2003)

$$\varepsilon_a = \varepsilon_{a_i} + (1 - \varepsilon_{a_i})\varepsilon_a^{II} \tag{14.74}$$

The shrinkage-expansion coefficient can be estimated from the initial porosity and the actual measured porosity by the following equation:

$$\phi = \frac{\tilde{b}}{\tilde{a}}\,\frac{\varepsilon_{a_i} - \varepsilon_a}{\varepsilon_a - 1} \tag{14.75}$$

When the value of ϕ is equal to 1, the nonideal model is transformed to the ideal conditions, i.e., volume of water loss equal to the pores formed. Values of ϕ equal to zero indicate complete collapse of all pores, i.e., no pore formation occurs. Values of ϕ greater than 1 show that there is an expansion of the material's boundary during the process. Values of ϕ smaller than 1 indicate that there is shrinkage of the material by collapse of the initial air-filled pores or collapse of water-filled pores. When the moisture content of the product is very close to initial moisture content (e.g., only 2% change), this model may provide unrealistic prediction. The values of ϕ for apple (Rahman et al., 2005) during air-drying and for garlic (Sablani et al., 2007) during freeze-drying are presented in the literature. Future work needs the to be targeted to estimate the values of ϕ for a wide variation of the material's characteristics and processing conditions.

14.5 EFFECT OF DRYING METHOD ON STRUCTURAL PROPERTIES OF FOOD MATERIALS

Krokida et al. (1997, 2001) applied their model presented in the previous paragraph in the study of structural properties of fruits and vegetables and their correlation with the material moisture content. The effect of the drying method on true density, apparent density, specific volume, and porosity of banana, apple, carrot, and potato at various moisture contents is presented in Figures 14.12 through

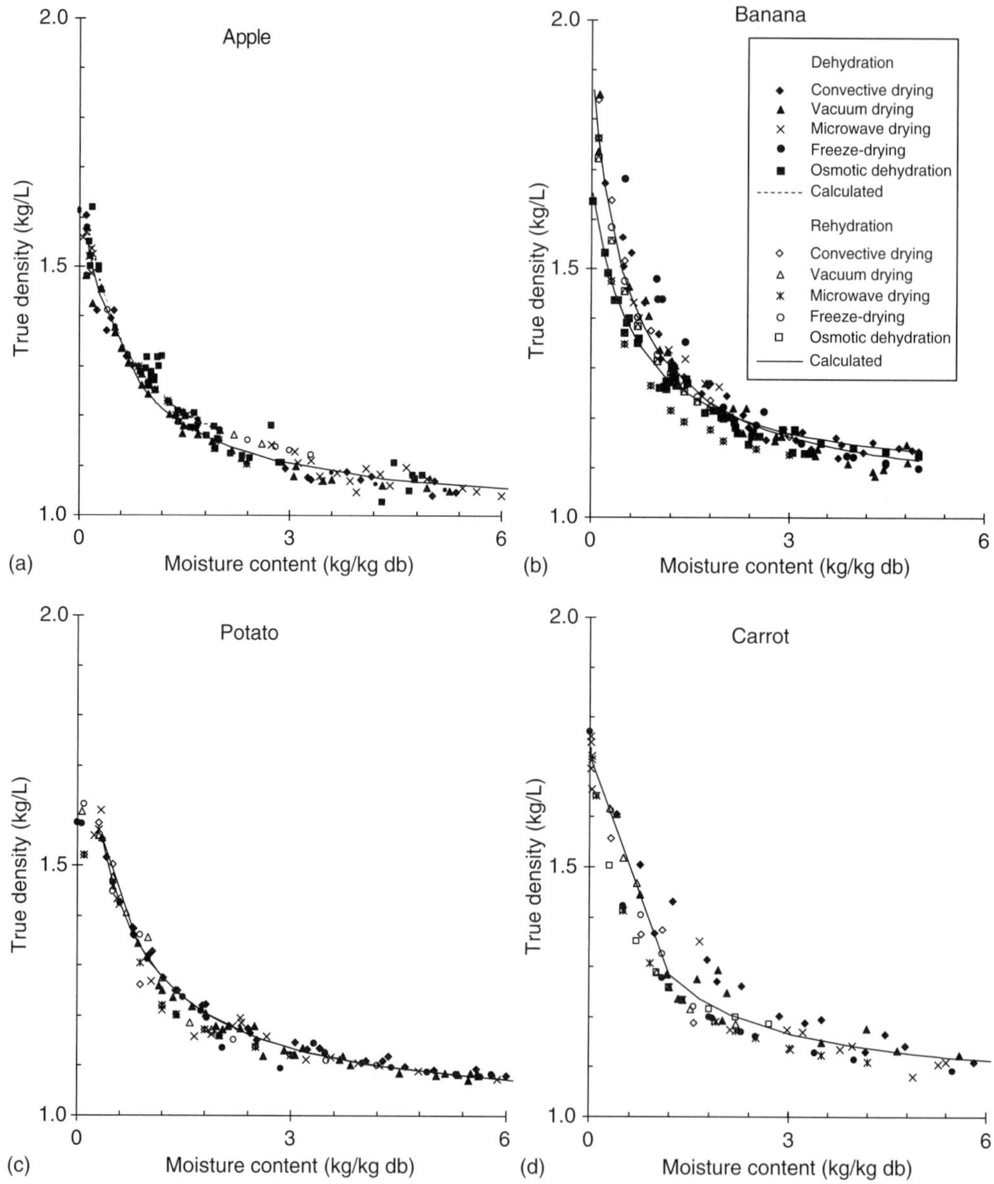

Figure 14.12 Variation of true density with material moisture content for various drying methods during dehydration and rehydration. (From Krokida, M.K. and Maroulis, Z.B., *Dry. Technol.*, 15, 2441, 1997. With permission; Krokida, M.K., Maroulis, Z.B., and Rahman, M.S., *Dry. Technol.*, 19, 2277, 2001.)

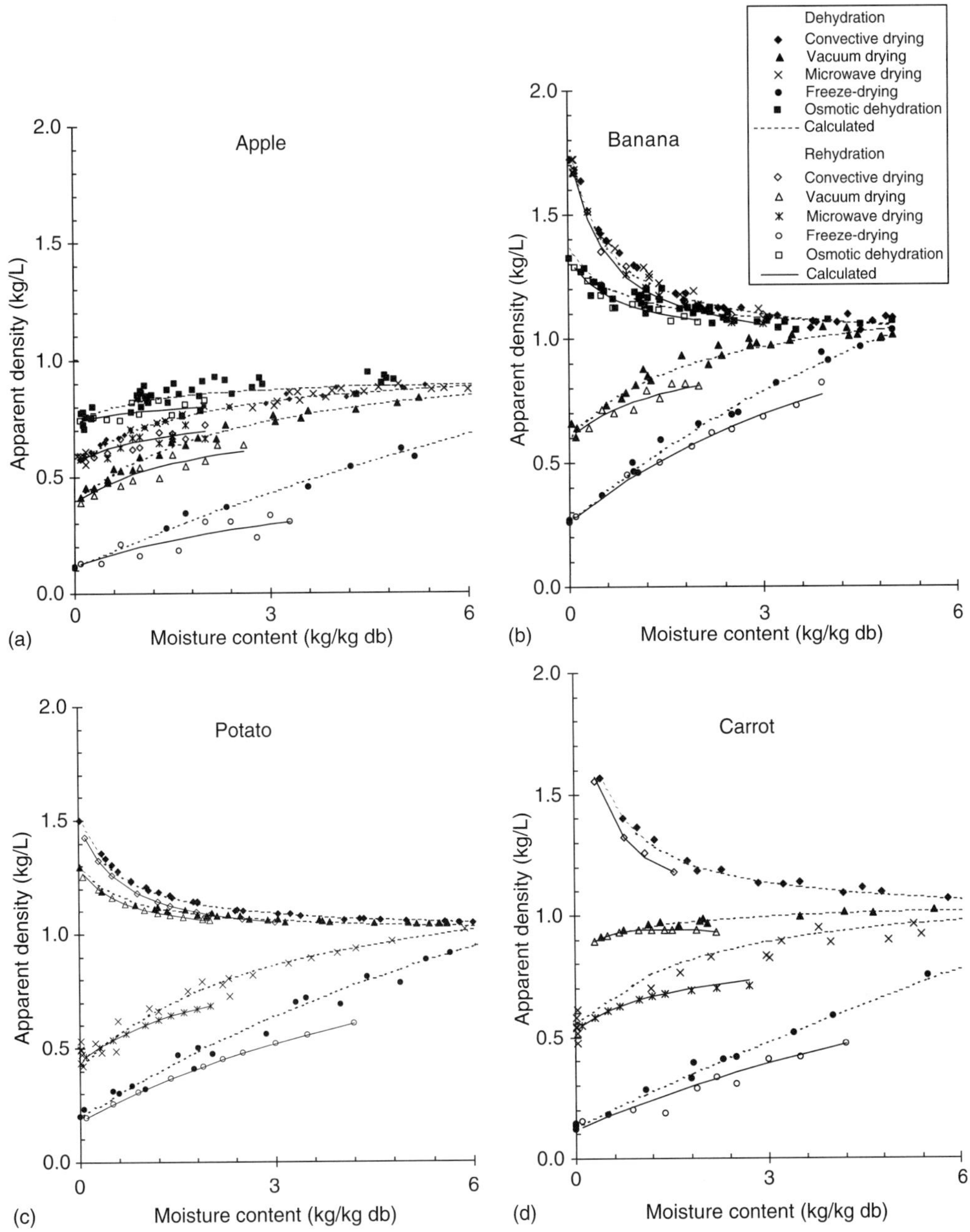

Figure 14.13 Variation of apparent density with material moisture content for various drying methods during dehydration and rehydration. (From Krokida, M.K. and Maroulis, Z.B., *Dry. Technol.*, 15, 2441, 1997. With permission; From Krokida, M.K., Maroulis, Z.B., and Rahman, M.S., *Dry. Technol.*, 19, 2277, 2001. With permission.)

14.15, using a large set of experimental measurements. Samples were dehydrated with five different drying methods: conventional, vacuum, microwave, freeze-, and osmotic-drying. The parameters of the model for each material, drying method, and process (drying and rehydration of dehydrated products) are presented in Table 14.29. The effect of the drying method on the examined properties was taken into account through its effect on the corresponding parameters. Only dry solid apparent

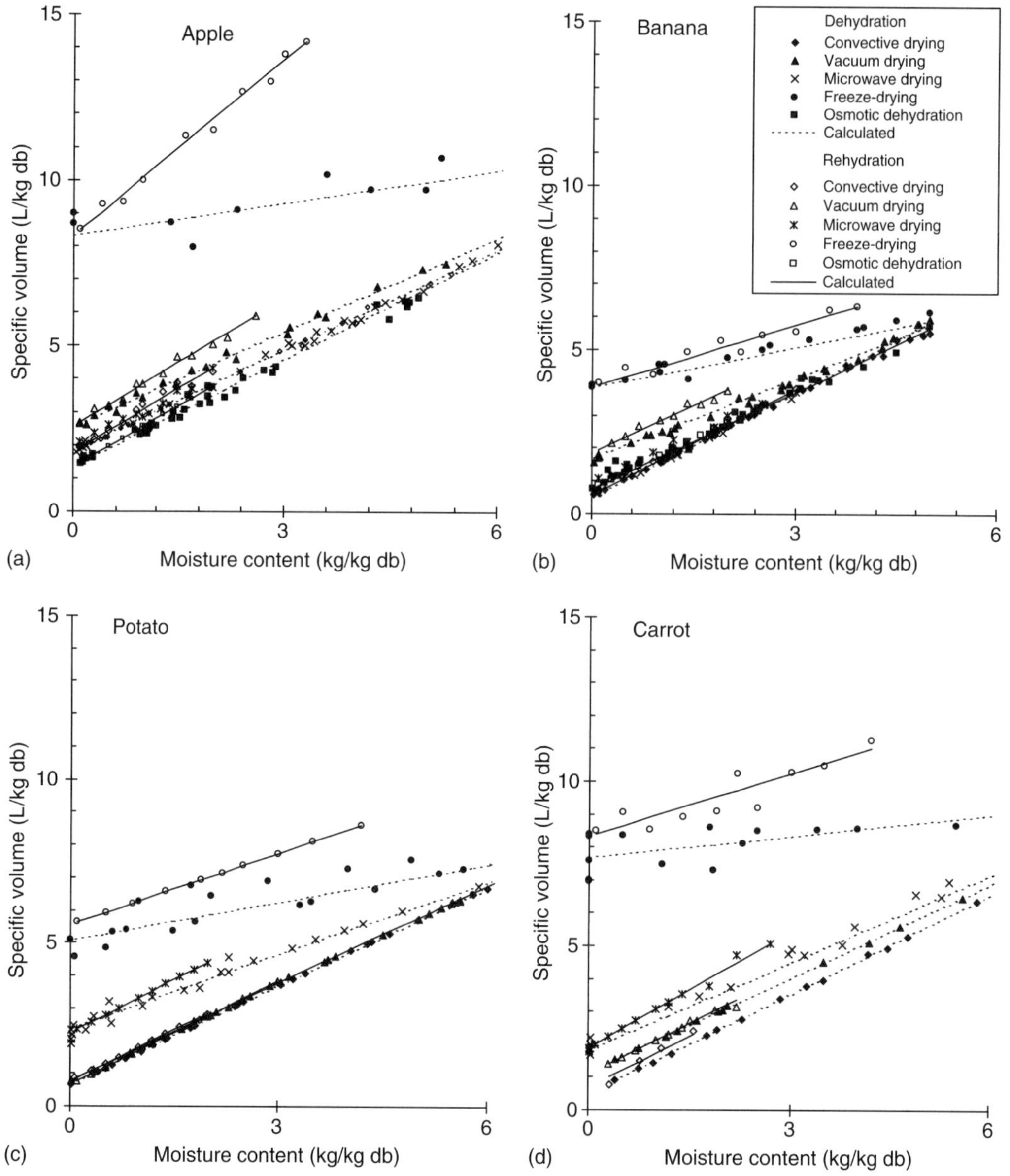

Figure 14.14 Variation of specific volume with material moisture content for various drying methods during dehydration and rehydration. (From Krokida, M.K. and Maroulis, Z.B., *Dry. Technol.*, 15, 2441, 1997. With permission; Krokida, M.K., Maroulis, Z.B., and Rahman, M.S., *Dry. Technol.*, 19, 2277, 2001.)

density was dependent on both material and drying method. Freeze-dried materials developed the highest porosity, whereas the lowest one was obtained using conventional air-drying.

It was observed that dehydrated products did not recover their structural properties after rehydration, due to structural damage that occurred during drying and the hysteresis phenomenon, which took place during rehydration. The porosity of the rehydrated products was higher during rehydration than during dehydration. The model presented was used in the rehydration process as well. Only the shrinkage coefficient, which represents volume expansion, changed on rehydration.

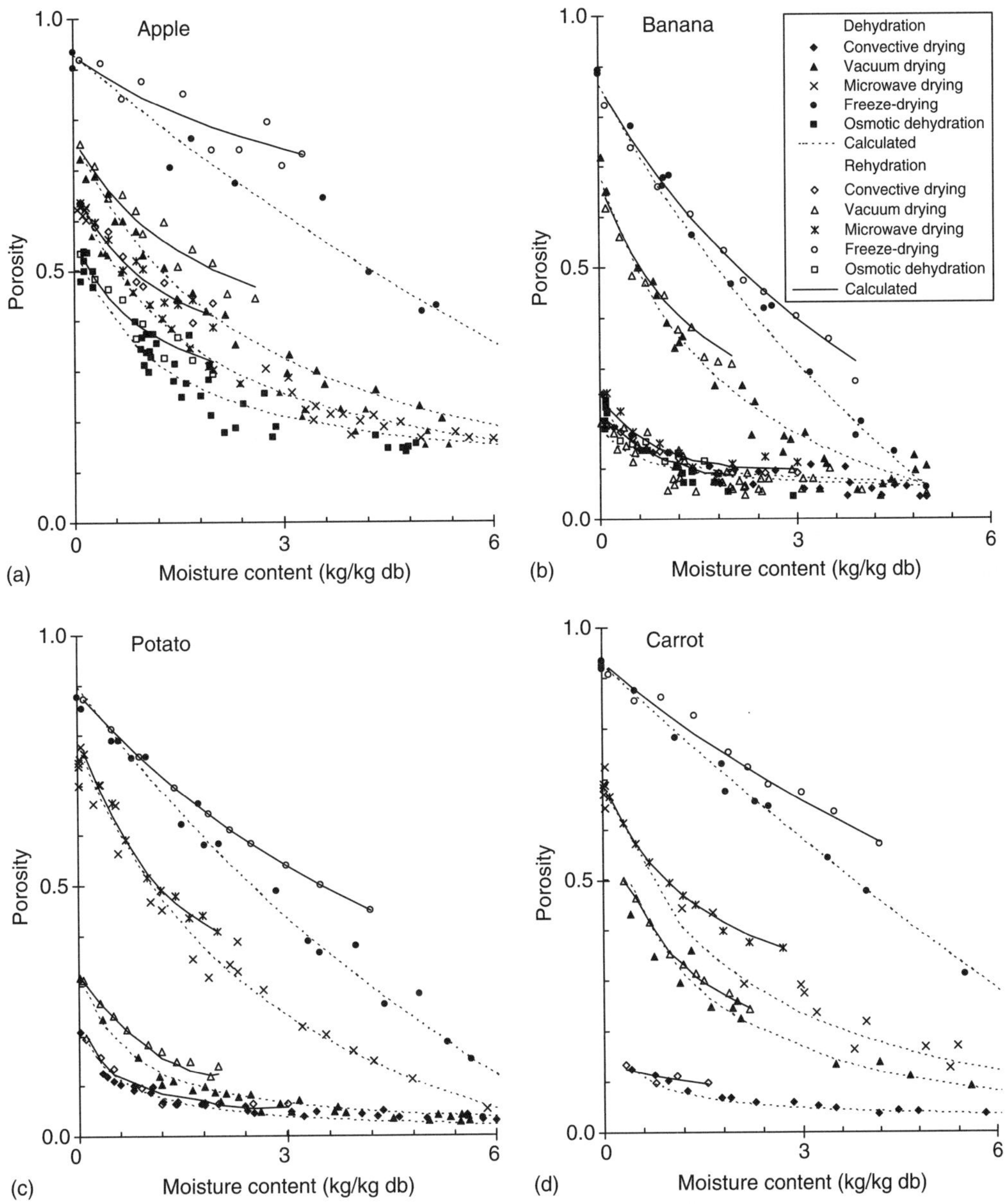

Figure 14.15 Variation of porosity with material moisture content for various drying methods during dehydration and rehydration. (From Krokida, M.K. and Maroulis, Z.B., *Dry. Technol.*, 15, 2441, 1997. With permission; Krokida, M.K., Maroulis, Z.B., and Rahman, M.S., *Dry. Technol.*, 19, 2277, 2001.)

True density, apparent density, specific volume, and porosity were investigated during deep fat frying of French fries (Krokida et al., 2000). It should be noted that frying is considered as a drying method, as food material undergoes dehydration in an oil environment. The effect of frying conditions (oil temperature, sample thickness, and oil type) on the above properties is presented in Figures 14.16 through 14.18. Moisture and oil content during deep fat frying and consequently all the examined properties are affected by frying conditions. The results showed that the porosity of French fries increases with increasing oil temperature and sample thickness, and it is higher for products fried with hydrogenated oil.

Table 14.29 Parameter Estimation of the Structural Properties Model

Material/Method	ρ_s	ρ_w	β'	ρ_{ai}
Apple				
Drying				
Convective	1.67	1.02	0.99	0.56
Vacuum			0.96	0.39
Microwave			1.01	0.56
Freeze			0.34	0.12
Osmotic			1.10	0.73
Rehydrat.				
Convective			1.30	0.56
Vacuum			1.31	0.39
Microwave			1.30	0.56
Freeze			0.81	0.12
Osmotic			1.22	0.73
Banana				
Drying				
Convective	1.90	1.02	1.04	1.81
Vacuum			0.90	0.63
Microwave			1.05	1.79
Freeze			0.43	0.26
Osmotic			1.04	1.33
Rehydrat.				
Convective			1.07	1.81
Vacuum			1.10	0.63
Microwave			1.07	1.79
Freeze			0.65	0.26
Osmotic			1.07	1.33
Carrot				
Drying				
Convective	1.75	1.02	1.02	1.60
Vacuum			0.99	0.92
Microwave			0.94	0.53
Freeze			0.30	0.14
Rehydrat.				
Convective			1.20	1.60
Vacuum			1.05	0.92
Microwave			1.20	0.53
Freeze			0.22	0.14
Potato				
Drying				
Convective	1.60	1.02	1.03	1.50
Vacuum			1.03	1.29
Microwave			0.81	0.44
Freeze			0.29	0.18

Table 14.29 (continued) Parameter Estimation of the Structural Properties Model

Material/Method	ρ_s	ρ_w	β'	ρ_{ai}
Rehydrat.				
Convective			1.07	1.50
Vacuum			1.05	1.29
Microwave			1.07	0.44
Freeze			0.74	0.18

Sources: Adapted from Krokida, M.K. and Maroulis, Z.B., *Dry. Technol.*, 15, 2441, 1997; Krokida, M.K., Maroulis, Z.B., and Rahman, M.S., *Dry. Technol.*, 19, 2277, 2001.

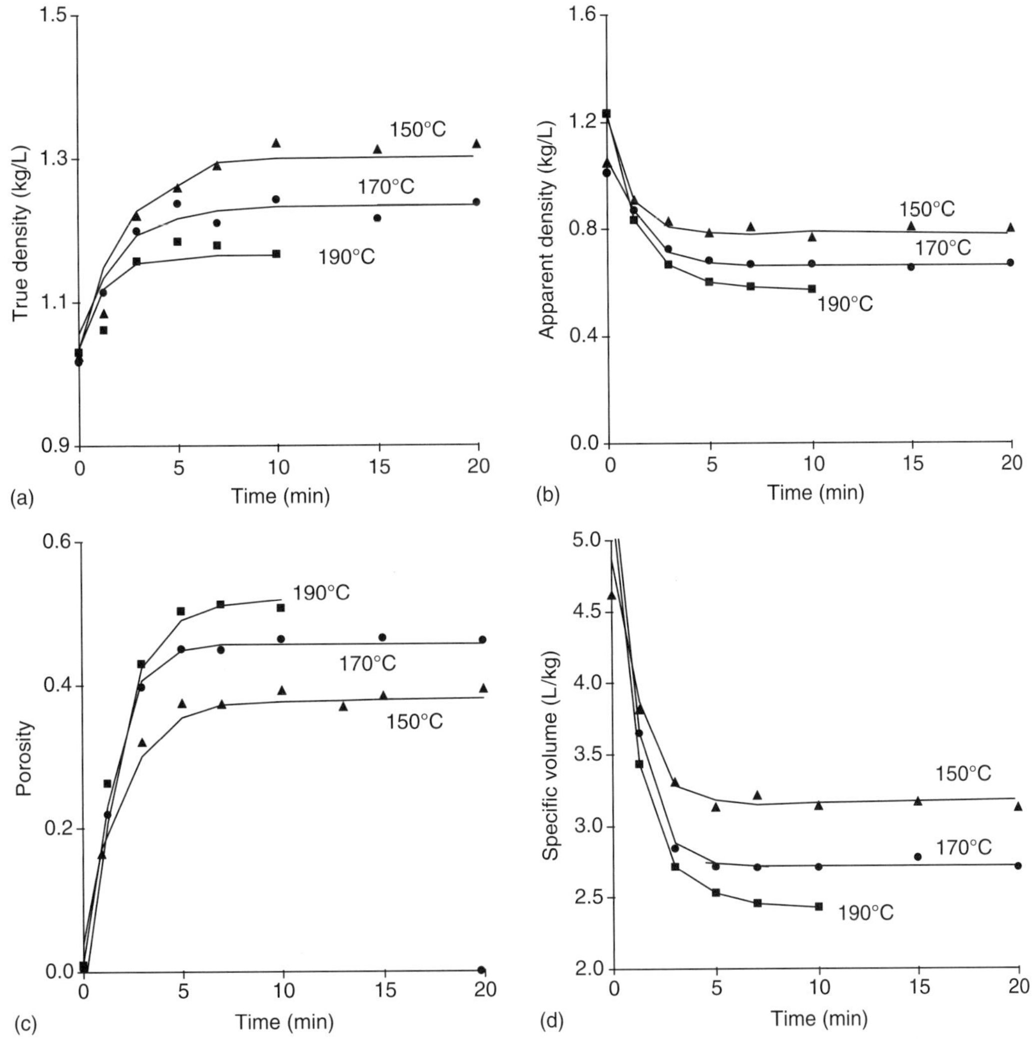

Figure 14.16 Effect of oil temperature on structural properties of French fries. (From Krokida, M.K., Kiranoudis, C.T., Maroulis, Z.B., and Marinos-Kouris, D., *Dry. Technol.*, 18, 1251, 2000. With permission.)

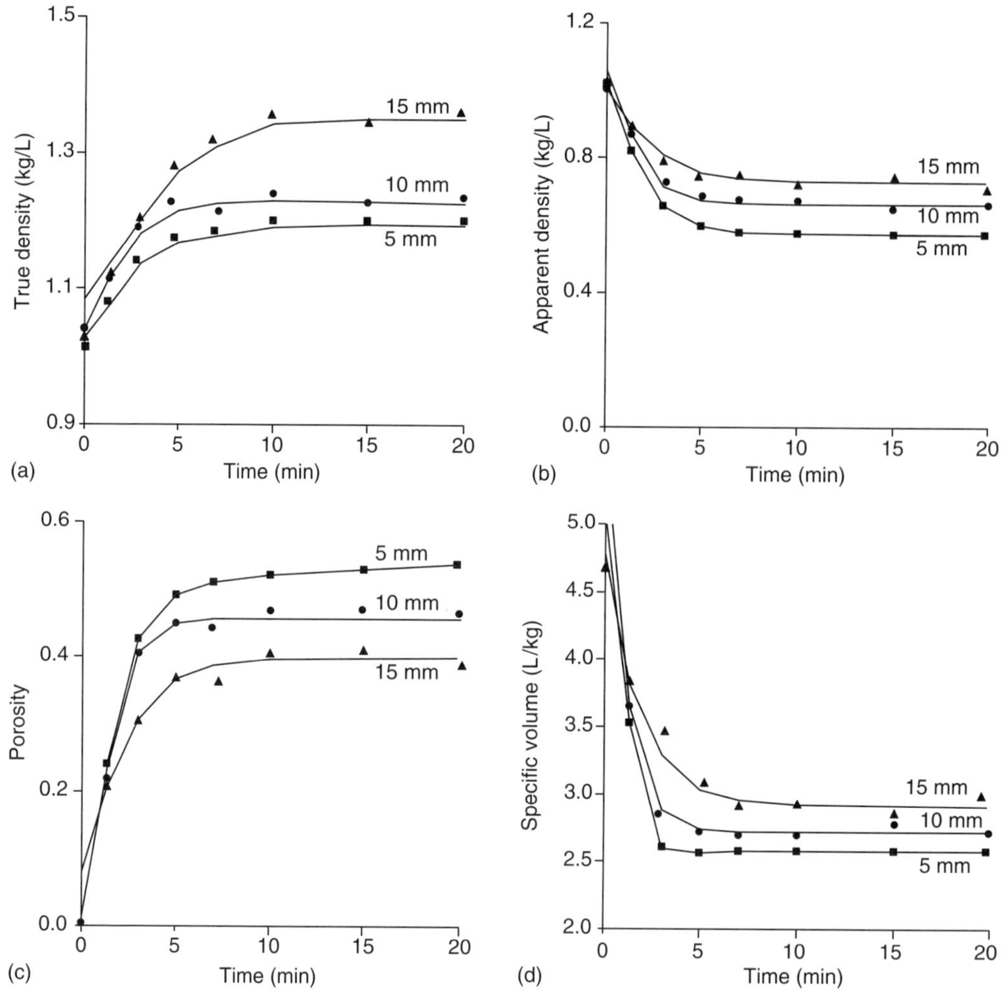

Figure 14.17 Effect of sample thickness on structural properties of French fries. (From Krokida, M.K., Kiranoudis, C.T., Maroulis, Z.B., and Marinos-Kouris, D., *Dry. Technol.*, 18, 1251, 2000. With permission.)

14.6 TEMPERATURE DEPENDENCY OF DENSITY

Boukouvalas et al. (2006a,b) collected true density data from the literature, which referred in different materials and temperature and estimated the constants of Equation 14.66 for these materials. The results are presented in Table 14.30. Due to the narrow temperature range of the experimental data of three materials (banana, carrot, and garlic), a temperature-independent approach was adopted for dry solid density ($s_2 = 0$). The assumption that the dry solid density of the materials is not strongly affected by temperature is justified by Choi and Okos (1986) who presented linear temperature-dependent correlations for pure components solid densities, but the dependency on temperature is not significant. For wheat, the temperature dependency is found to be rather significant. Although the description of the experimental data is satisfactory, the extrapolation to temperatures out of the range of the used experimental data might give significant deviations.

The ability of the developed model to predict true density versus temperature is presented in Figures 14.19 through 14.24 for all materials included in Table 14.30. Figure 14.19 presents

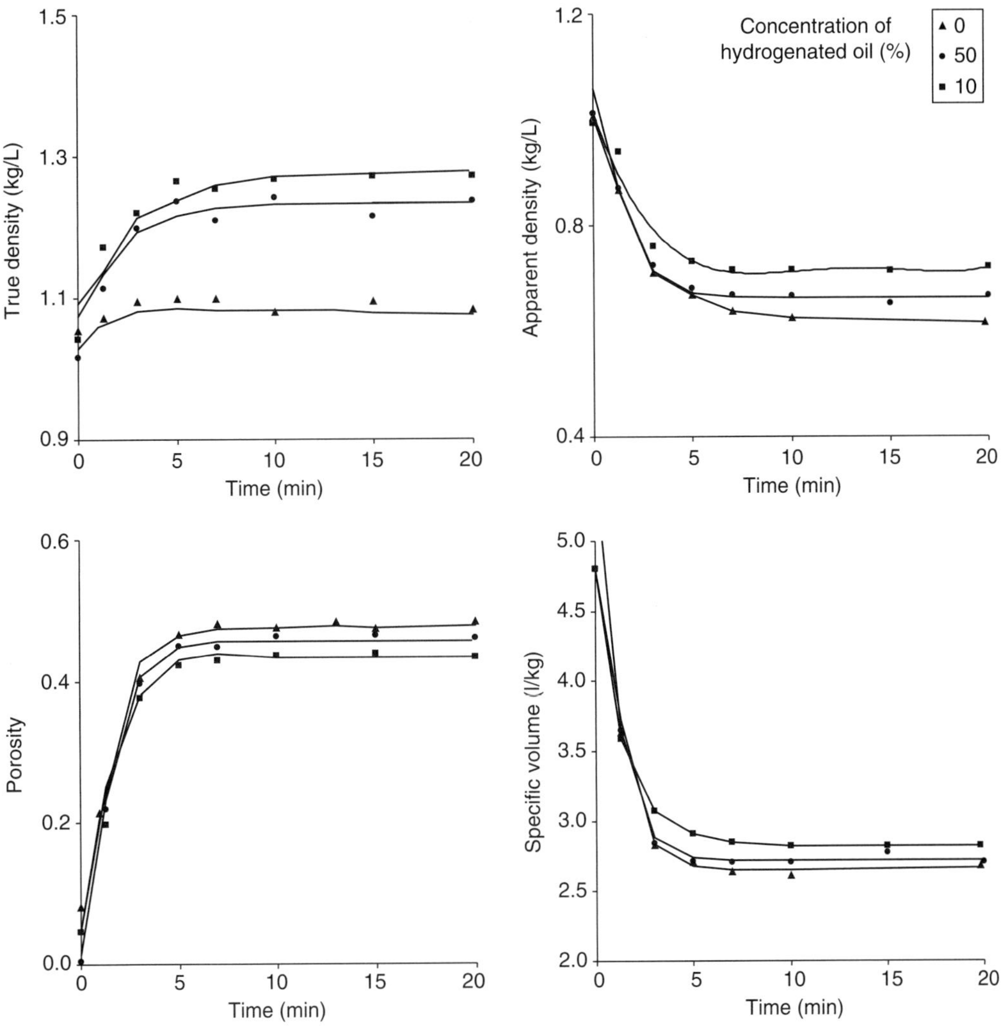

Figure 14.18 Effect of oil type on structural properties of French fries. (From Krokida, M.K., Kiranoudis, C.T., Maroulis, Z.B., and Marinos-Kouris, D., *Dry. Technol.*, 18, 1251, 2000. With permission.)

the performance of the proposed model along with the experimental data for apple, banana, potato, carrot, and onion, which are the high moisture content materials of Table 14.30. The representative temperature for all materials is 70°C, except onion presented at 55°C. Figure 14.20 shows the fit of the model for lentil, Hylon-7, garlic, and wheat, which are the low moisture content materials of Table 14.30. Lentil and garlic are presented at 70°C, while Hylon-7 at 60°C and wheat at 40°C. Figure 14.21 illustrates the performance of the model at different temperatures for apple and banana.

Figures 14.22 and 14.23 summarize the model-predicted values for various fruits and vegetables, respectively, at ambient temperature. As expected, true density ranges between the solid density (dry material) and the density of water (infinite water content) at the examined temperature. In Figure 14.24, the temperature dependency of the true density of apple and banana at $M_w = 1$ kg/(kg · db) is presented. The changes in true density are minor at near zero temperature, but they tend to be more significant at higher temperatures.

Table 14.30 Parameter Estimation of the True Density—Temperature Dependence

Material	Papers	Data (Total)	Data (Used)	SD (kg/m^3)	AAD%	*T* (°C) Min	*T* (°C) Max	S_0	S_1
Cereal products									
Wheat (Sokhansanj and Lang, 1996; Teunou et al., 1999; Roman-Gutierrez et al., 2003; Fitzpatrick et al., 2004)	4	20	20	56.6	0.9	20	40	1716	−6.91
Fruits									
Apple (Lozano et al., 1980; Zogzas et al., 1994; Donsi et al., 1996; Mavroudis et al., 1998; Krokida et al., 2000; Nieto et al., 2004)	6	45	34	234.7	4.7	20	70	1689	−0.70
Banana (Krokida and Maroulis, 1997; Krokida and Maroulis, 1999)	2	15	15	365.7	3.3	70	70	2029	0.00
Legume									
Lentils (Tang and Sokhansanj, 1993; Amin et al., 2004)	2	49	45	30.1	1.0	30	70	1551	−0.28
Model foods									
Hylon-7 (Marousis and Saravacos, 1990; Shah et al., 2000)	2	15	13	108.3	2.9	20	60	1602	−0.57
Vegetables									
Carrot (Lozano et al., 1983; Zogzas et al., 1994; Krokida and Maroulis, 1997)	3	16	14	315.1	6.3	60	70	2156	0.00
Garlic (Lozano et al., 1983; Madamba et al., 1993; Madamba et al., 1994)	3	10	9	156.1	4.9	60	70	1528	0.00
Onion (Rapusas and Driscoll, 1995; Rapusas et al., 1995; Abhayawick et al., 2002)	3	59	48	132.2	1.5	32	70	1539	−1.14
Potato (Lozano et al., 1983; Zogzas et al., 1994; Wang and Brennan, 1995; Donsi et al., 1996; Krokida et al., 2001)	7	63	58	196.8	5.9	40	190	1780	−1.05

Source: From Boukouvalas, Ch.J., Krokida, M.K., Maroulis, Z.B., and Marinos-Kouris, D., *Int. J. Food Prop.*, 9, 109, 2006a. With permission.
Note: SD and AAD% are statistical factors of the regression analysis.

14.7 PREDICTION MODELS FOR STRUCTURAL PROPERTIES OF EXTRUDATES

Extrusion cooking is a novel process for the production of food materials, such as snacks, beverage powders, pet foods, etc. It is a multivariable process and the main parameters affecting the properties of the final product include operating parameters, such as extrusion temperature, screw rotation speed, and feed rate, and material parameters, such as initial moisture content and

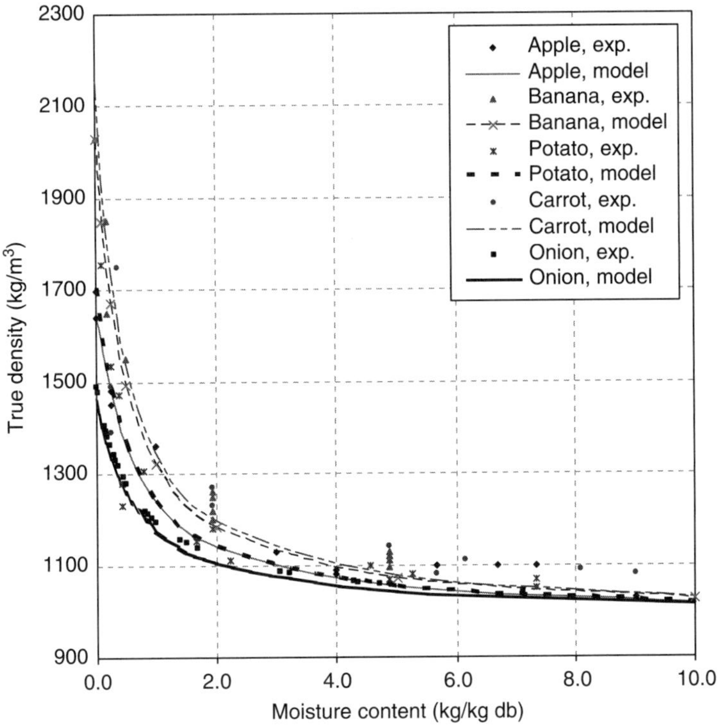

Figure 14.19 True density of apple, banana, potato, carrot, and onion at various moisture contents. (From Boukouvalas, Ch.J., Krokida, M.K., Maroulis, Z.B., and Marinos-Kouris, D., *Int. J. Food Prop.*, 9, 109, 2006a. With permission.)

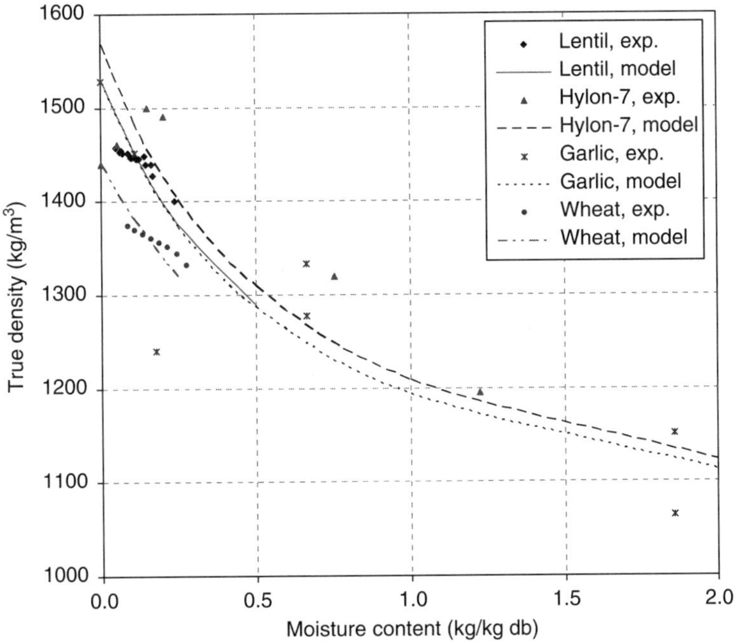

Figure 14.20 True density of lentil, Hylon-7, garlic, and wheat at various moisture contents. (From Boukouvalas, Ch.J., Krokida, M.K., Maroulis, Z.B., and Marinos-Kouris, D., *Int. J. Food Prop.*, 9, 109, 2006a. With permission.)

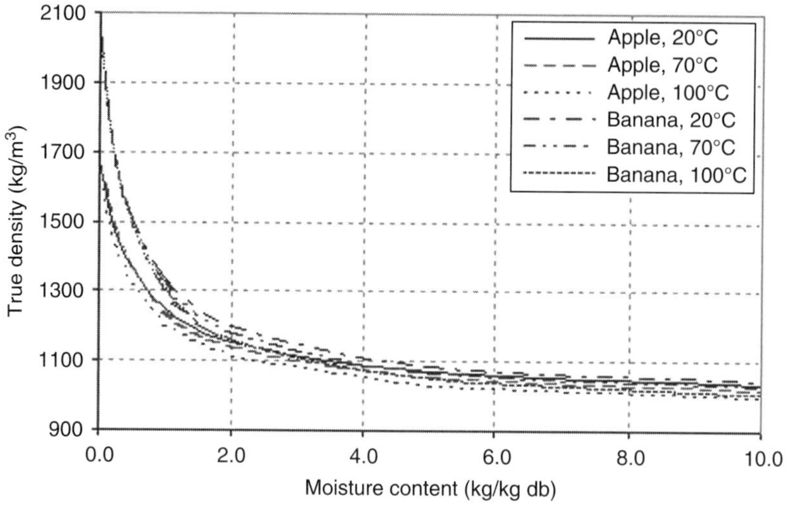

Figure 14.21 True density of apple and banana at various temperatures and moisture contents. (From Boukouvalas, Ch.J., Krokida, M.K., Maroulis, Z.B., and Marinos-Kouris, D., *Int. J. Food Prop.*, 9, 109, 2006a. With permission.)

composition of the feed. Although the moisture content of the feed decreases during extrusion (a process that lasts approximately 1 min or less) due to the vaporization of the water as the material exits from the die of the extruder, we cannot, practically, study the moisture content decrease during the process because many physical and chemical transformations take place at that time and also because the isolation of a sample from inside the extruder as it works is extremely difficult due to the conditions of high temperature, pressure, and shear rate that exist. Hence, extrusion cooking is a much different process from drying, where the study of the samples during processing is easy. For these reasons, the mathematical models for the prediction of properties, such as density and porosity

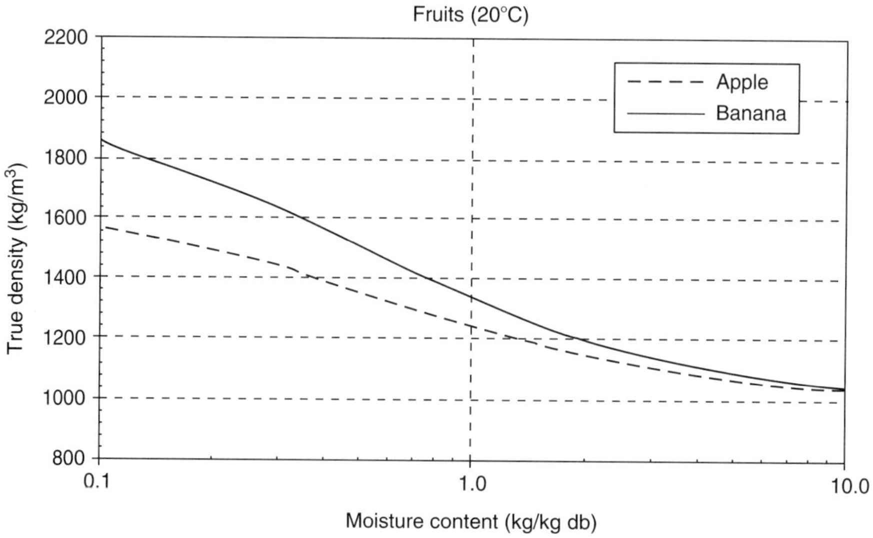

Figure 14.22 Predicted values of true density for fruits at 20°C. (From Boukouvalas, Ch.J., Krokida, M.K., Maroulis, Z.B., and Marinos-Kouris, D., *Int. J. Food Prop.*, 9, 109, 2006a. With permission.)

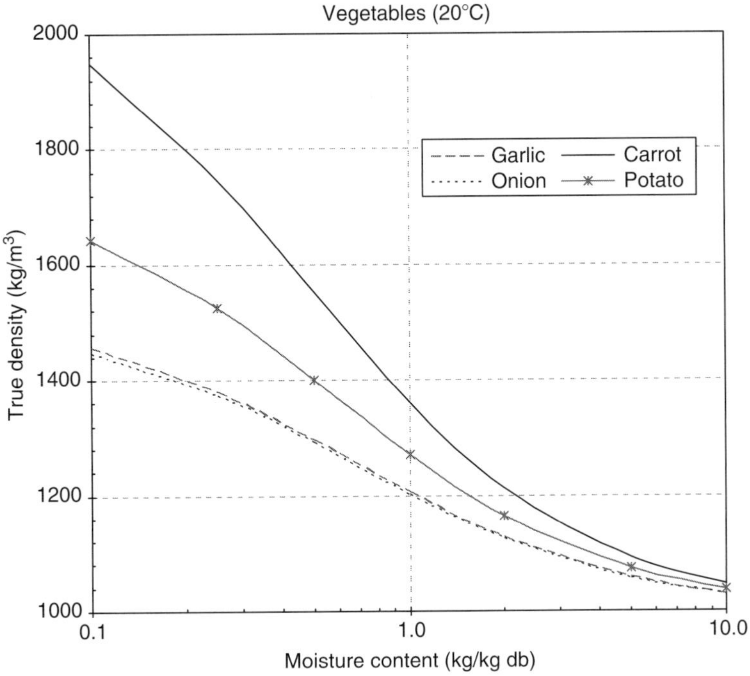

Figure 14.23 Predicted values of true density for vegetables at 20°C. (From Boukouvalas, Ch.J., Krokida, M.K., Maroulis, Z.B., and Marinos-Kouris, D., *Int. J. Food Prop.*, 9, 109, 2006a. With permission.)

of extrudates, are based on the determination of the effect of the operating parameters of the process and the feed characteristics in the final product's properties.

Thymi et al. (2005) studied the structural properties of extruded corn grits (containing 10% sugar) using a prism extruder with varied feed rate (1.16–6.44 kg/h), screw speed (150–250 rpm),

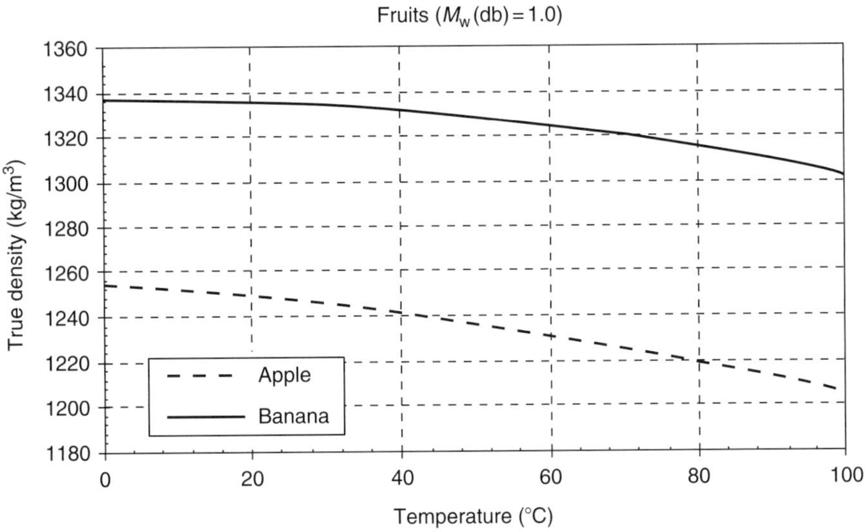

Figure 14.24 Predicted values of true density versus temperature for fruits at $M_w = 1$ kg/(kg · db). (From Boukouvalas, Ch.J., Krokida, M.K., Maroulis, Z.B., and Marinos-Kouris, D., *Int. J. Food Prop.*, 9,109, 2006a. With permission.)

extrusion temperature (100°C–260°C), and feed moisture content (0.12–0.25 kg/kg wet basis). A model consisting of the following equations was developed for the prediction of apparent and true density of the extrudates, while porosity was calculated from Equation 14.60:

$$\rho_a = \frac{m_{sa}}{V_{sa}} = n_1 \left(\frac{T}{T_r}\right)^{n_T} \left(\frac{\tau}{\tau_r}\right)^{n_\tau} \left(\frac{N}{N_r}\right)^{n_N} \left(\frac{X_c}{X_{cr}}\right)^{n_X} \quad (14.76)$$

$$\rho_p = \frac{m_{sa}}{V_p} \quad (14.77)$$

where

T is the extrusion temperature
τ is the residence time
N is the screw speed
X is the feed moisture content
m_{sa} is the mass of the sample
T_r, τ_r, N_r, and X_r are the reference values of the corresponding variables
V_{sa} is the volume of the sample, which is estimated by the formula $V_{sa} = \pi D^2 L/4$, assuming the sample is of cylindrical shape
L and D are the length and diameter of the sample, respectively
V_p is the true (particle) volume measured by a stereopycnometer
n_1, n_T, n_τ, n_N, and n_X are empirical constants

Table 14.31 presents the parameter values of the proposed model.

Lazou et al. (2007) examined the influences of process conditions and material characteristics on the porosity of corn- and legume-based extrudates. Four different types of legumes, chickpea, Mexican bean, white bean, and lentil, were used to form mixtures with corn flour in a ratio ranging from 10% to 90% corn/legume and the following simple power model was developed

$$\varepsilon = \varepsilon_0 \times \left(\frac{T}{T_r}\right)^{n_T} \times \left(\frac{\tau}{\tau_r}\right)^{n_\tau} \times \left(\frac{X_c}{X_{cr}}\right)^{n_X} \times \left(\frac{C}{C_r}\right)^{n_C} \times \left(\frac{N}{N_r}\right)^{n_N} \quad (14.78)$$

where τ is the mean residence time and C is the materials ratio (kg corn/kg legume). The influence of feed rate in the extrudates' porosity is incorporated into mean residence time, considering that mean residence time is

$$\tau = \frac{(1 - e_f) \times \rho_t \times V}{F} \quad (14.79)$$

where

e_f is the empty fraction of the extruder
ρ_t is the true density of the material (kg/m^3)
V is the free volume in the extruder barrel
F is the feed rate (kg/h)

Table 14.31 Parameter Values of Equation 14.76

Parameter	n_1	n_T	n_τ	n_N	n_X
Value	0.14	−0.40	−0.10	0.0	1.45

Source: From Lazou, A.E., Michailidis, P.A., Thymi, S., Krokida, M.K., and Bisharat, G.I., *Int. J. Food Prop.*, 10, 721, 2007. With permission.

Table 14.32 Parameter Values of Equation 14.78

Materials	ε_0	n_T	n_τ	n_X	n_C	S_R	S_E
Corn/chickpea	0.863	0.661	0.117	−0.170	0.083	0.032	0.104
Corn/Mexican bean	0.904	0.236	0.064	−0.200	0.043	0.032	0.067
Corn/white bean	0.974	0.142	0.041	−0.244	0.029	0.140	0.014
Corn/lentil	0.907	0.165	0.062	−0.214	0.052	0.132	0.022

Source: From Lazou, A.E., Michailidis, P.A., Thymi, S., Krokida, M.K., and Bisharat, G.I., *Int. J. Food Prop.*, 10, 721, 2007. With permission.

The empty fraction was correlated with extrusion conditions, and it is given by the following equation, which is valid for the extruder used (Thermoprism twin screw extruder):

$$e_f = 0.707 \times \left(\frac{N}{200}\right)^{0.141} \times \left(\frac{F}{0.001}\right)^{-0.192} \times \left(\frac{T}{200}\right)^{-0.142} \times \left(\frac{X_c}{0.100}\right)^{-0.069} \quad (14.80)$$

The parameter values of Equation 14.78 are presented in Table 14.32 for all examined mixtures. The screw rotation speed does not seem to have significant effect on porosity because it has been incorporated in the expression of mean residence time. The table also includes the standard deviation between experimental and predicted values (S_R) and the standard experimental error (S_E). Table 14.33 shows true density values for these mixtures.

In Figures 14.25 through 14.27, the porosity of extrudates is represented as a function of mean residence time, feed moisture content, extrusion temperature, and corn to legume ratio (CP, corn/chickpea; CM, corn/Mexican bean; CB, corn/white bean; CL, corn/lentil). Solid lines are used for predicted values of porosity using the mathematical model presented, and the parameter values are shown in Table 14.32.

Ali et al. (1996) studied the expansion characteristics of extruded yellow corn grit. The experiments were carried out in a single-screw extruder and the extrusion conditions were extrusion temperature 100°C–200°C, screw speed 80–200 rpm, and material moisture content 0.64 kg/(kg · wb). They described total (u_{Tp}) and open pore (u_{op}) specific volume (m^3/kg) with the following equations:

$$\begin{aligned} u_{Tp} = {} & -4.8 \times 10^{-3} + 6.7 \times 10^{-5}T + 1.97 \times 10^{-7}T^2 + 6.7 \times 10^{-5}N \\ & - 2.0 \times 10^{-7}N^2 - 7.98 \times 10^{-7}TN + 2.43 \times 10^{-9}T^2N - 7.55 \times 10^{-12}TN^2 \end{aligned} \quad (14.81)$$

$$\begin{aligned} u_{op} = {} & -5.91 \times 10^{-3} + 8.3 \times 10^{-5}T - 2.53 \times 10^{-7}T^2 + 8.5 \times 10^{-5}N \\ & - 2.68 \times 10^{-7}N^2 - 1.06 \times 10^{-7}TN + 3.43 \times 10^{-9}T^2N - 1.08 \times 10^{-11}TN^2 \end{aligned} \quad (14.82)$$

Table 14.33 True Density of Extrudates

Materials	Corn/Chickpea	Corn/Mexican Bean	Corn/White Bean	Corn/Lentil
True density ρ_t (kg/m^3)	1703 ± 387	1400 ± 107	1561 ± 134	1392 ± 61

Source: From Lazou, A.E., Michailidis, P.A., Thymi, S., Krokida, M.K., and Bisharat, G.I., *Int. J. Food Prop.*, 10, 721, 2007. With permission.

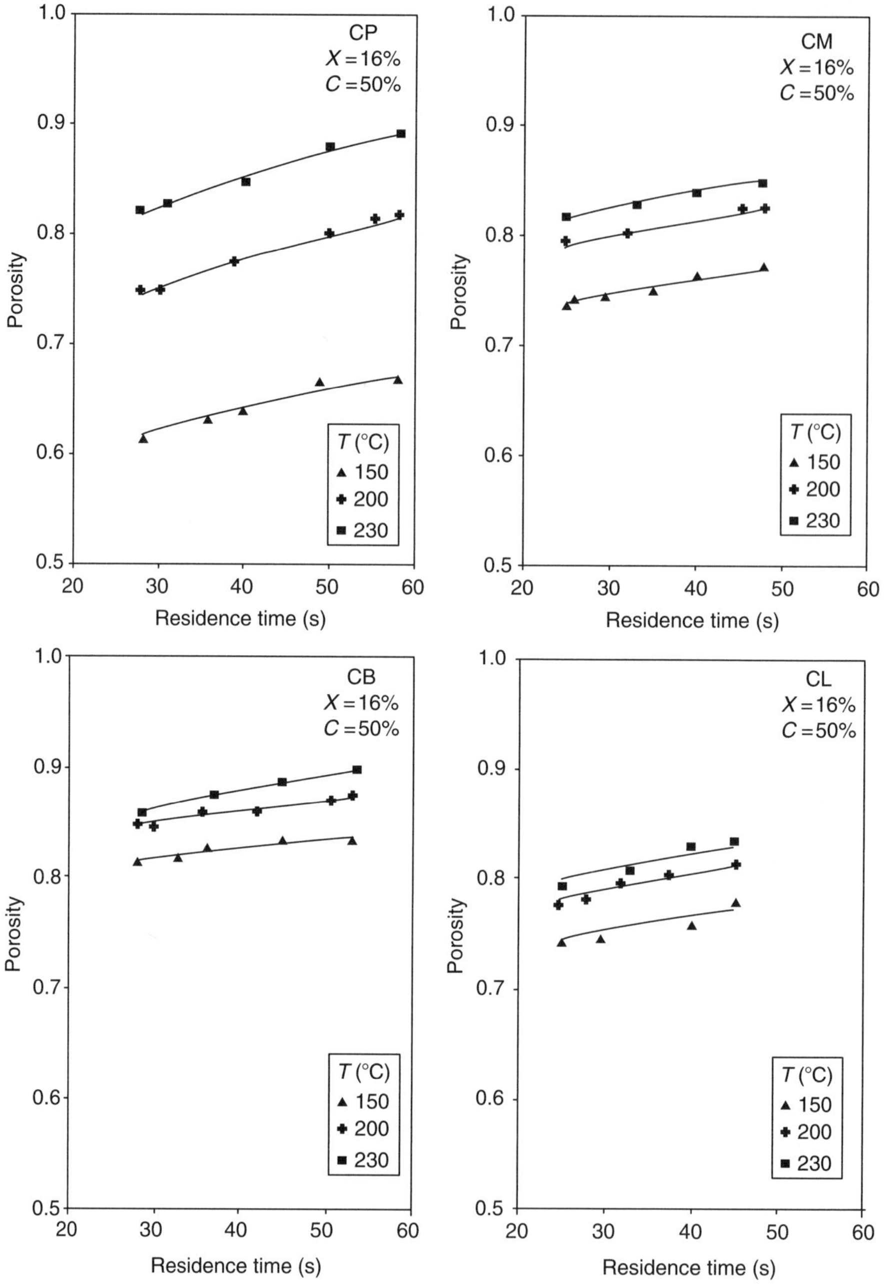

Figure 14.25 Porosity of extrudates as a function of residence time and temperature. (a) corn/chickpea mixture, (b) corn/mexican bean mixture, (c) corn/white bean mixture, and (d) corn/lentil mixture. (From Lazou, A.E., Michailidis, P.A., Thymi, S., Krokida, M.K., and Bisharat, G.I., *Int. J. Food Prop.*, 10, 721, 2007. With permission.)

where T is the extrusion (barrel) temperature (°C) and N is the screw rotation speed (rounds/min). Both open and total pore specific volume increased with the increase in temperature and screw rotation speed.

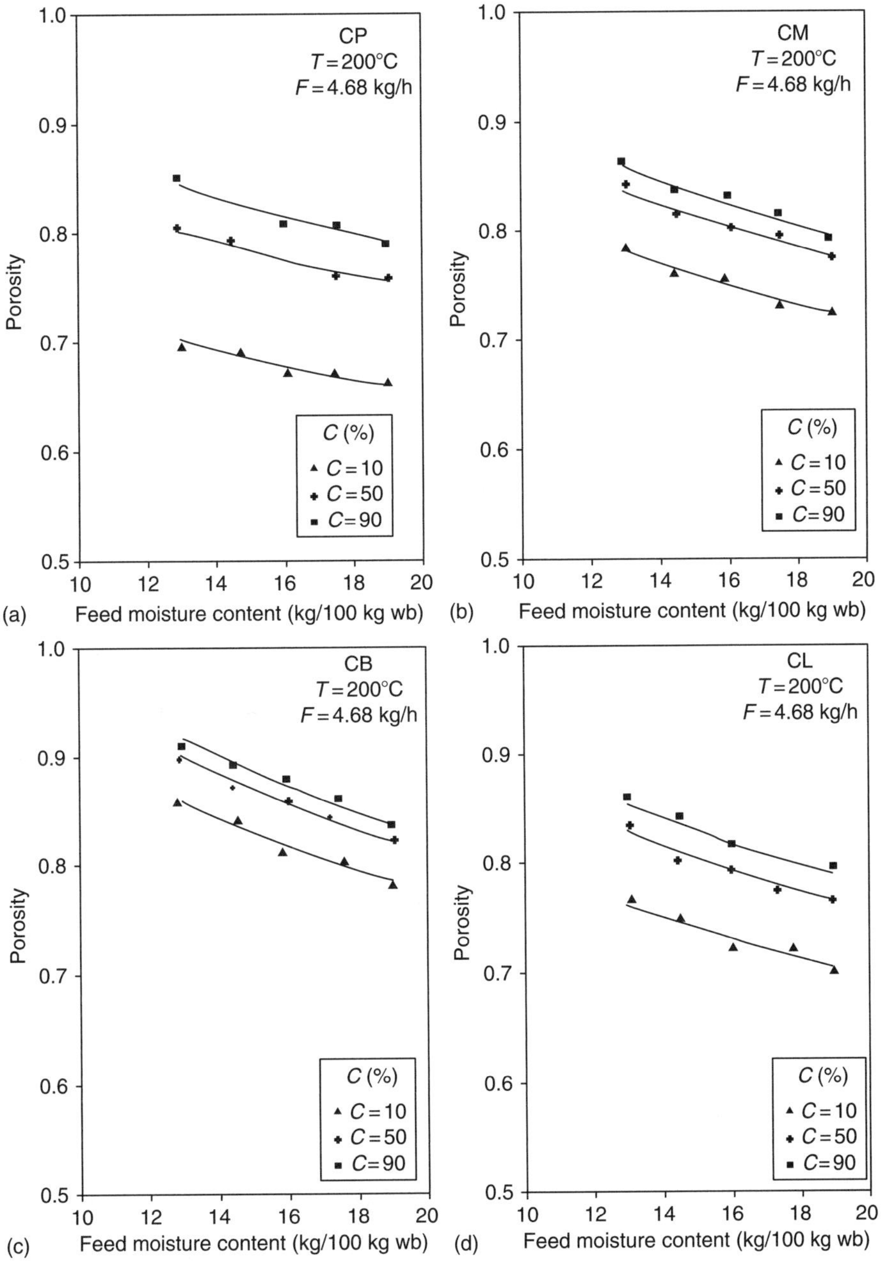

Figure 14.26 Porosity of extrudates as a function of feed moisture content and corn to legume ratio. (a) corn/chickpea mixture, (b) corn/mexican bean mixture, (c) corn/white bean mixture, and (d) corn/lentil mixture. (From Lazou, A.E., Michailidis, P.A., Thymi, S., Krokida, M.K., and Bisharat, G.I., *Int. J. Food Prop.*, 10, 721, 2007. With permission.)

14.8 MECHANISMS OF PORE FORMATION

Limited efforts have been made to explore the fundamental mechanisms responsible for the formation of pores, shrinkage, and other structural changes. In this section, some of the mechanisms are presented.

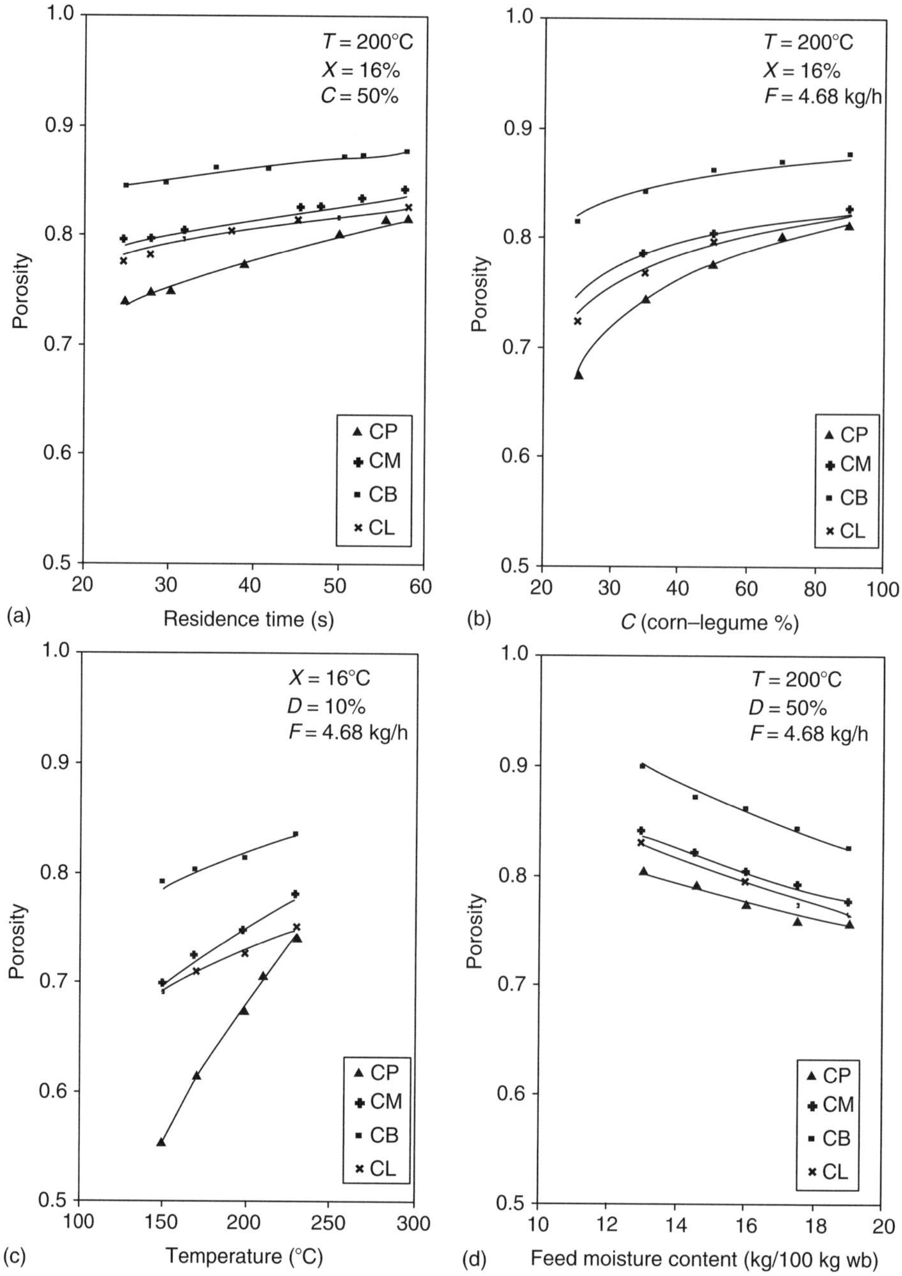

Figure 14.27 Comparative figures of porosity for all extrudates. Porosity as a function of (a) residence time, (b) corn–legume ratio, (c) temperature, and (d) feed moisture content. (From Lazou, A.E., Michailidis, P.A., Thymi, S., Krokida, M.K., and Bisharat, G.I., *Int. J. Food Prop.*, 10, 721, 2007. With permission.)

14.8.1 Mechanisms of Collapse

Collapse is a result of decrease in porosity and caused by shrinkage of the product. The factors affecting formation of pores can be grouped as intrinsic and extrinsic factors. The extrinsic factors are temperature, pressure, relative humidity, gas atmosphere, air circulation, and electromagnetic radiation applied in the process, whereas the intrinsic factors are chemical composition, inclusion of volatile components (such as alcohol and carbon dioxide), and initial structure before processing (Rahman, 2001, 2004). Genskow (1990) and Achanta and Okos (1995) mentioned several mechanisms that affect the degree of collapse or shrinkage and formation of pores. An understanding of these mechanisms would aid in achieving the desired shrinkage or collapse in the products. The following physical mechanisms play an important role in the control of shrinkage or collapse (Rahman and Perera, 1999): (1) surface tension (considers collapse in terms of the capillary suction created by a receding liquid meniscus), (2) plasticization (considers collapse in terms of the plasticizing effect of solvents on various polymer solutes), (3) electrical charge effects (considers collapse in terms of van der Waals electrostatic forces), (4) the mechanism of moisture movement in the process, and (5) gravitational effects. The rate at which shrinkage occurs is related to the viscoelastic properties of a matrix. The higher the viscosity of the mixtures caused, the lower the rate of shrinkage and vice versa (Achanta and Okos, 1996).

14.8.2 Glass Transition Concept

Levine and Slade first applied the concept of glass transition to identify or explain the physicochemical changes in foods during processing and storage (Levine and Slade, 1986, 1988; Slade and Levine, 1991). The glass transition theory is one of the concepts that has been proposed to explain the process of shrinkage, collapse, fissuring, and cracking during drying (Krokida et al., 1998; Karathanos et al., 1993, 1996; Cnossen and Siebenmorgen, 2000; Rahman, 2001). The hypothesis indicates that a significant shrinkage can be noticed during drying only if the temperature of the drying or processing is higher than the glass transition of the material at that particular moisture content (Achanta and Okos, 1996).

14.8.2.1 Support of Glass Transition Concept

The methods of freeze-drying and hot air-drying can be compared based on this theory. In freeze-drying, with the drying temperature below or close to T_g' (maximally freeze-concentrated glass transition temperature; it is independent of solid content) or T_g (glass transition as a function of solids content), the material is in the glassy state. Hence, shrinkage is negligible. As a result, the final product is highly porous. With hot air-drying, on the other hand, with the drying temperature above T_g' or T_g, the material is in the rubbery state, and substantial shrinkage occurs causing a lower level of pores. During the initial stage of freeze-drying, the composition of the freeze-concentrated phase surrounding the ice dictates the T_g'. In the initial or early stage of drying, T_g' is very relevant and the vacuum must be sufficient to ensure that sublimation occurs. At the end of the initial stage of drying, the pore size and the porosity are dictated by ice crystal size, if the collapse of the wall of the matrix that surrounded the ice crystal does not occur. The secondary stage of drying, on the other hand, refers to removal of water from the unfrozen phase. After sublimation is completed, the sample begins to warm up to the shelf temperature. At this stage, T_g of the matrix is related to the collapse and no longer to T_g'. This is because $T_g > T_g'$ (T_g increases from T_g' as the concentration of solids increases during the process of drying). Karel et al. (1994) performed freeze-drying under high vacuum (0.53 Pa) and reduced vacuum conditions (90.64 and 209.28 Pa) to obtain varying initial sample temperatures that were below (−55°C), near (−45°C), and above (−28°C). Collapse was determined by measuring apparent shrinkage before and after freeze-drying of apple, potato, and celery. Samples that dried at −55°C showed no shrinkage (more pores) whereas shrinkage increased with the increase in drying temperature, justifying the glass transition concept.

14.8.2.2 Evidence against Glass Transition Concepts

Recent experimental results dictate that the concept of glass transition is not valid for freeze-drying of all types of biological materials, indicating the need for the incorporation of other concepts (Sablani and Rahman, 2002); thus, a unified approach needs to be used. In the case of freeze-drying, pore formation in food materials showed two distinct trends when shelf temperatures were maintained at a constant level between −45°C and 15°C (Sablani and Rahman, 2002). The materials in group I (i.e., abalone, potato, and brown date) showed a decreasing trend, whereas those in group II (i.e., apple and yellow date) showed an increasing trend in pore formation. This may be due to the structural effects of the materials. However, none of the studies measured the actual temperature and moisture history of the sample passing through freeze-drying. The temperature and moisture history of the sample during freeze-drying could expose more fundamental knowledge in explaining the real process of pore formation or collapse in freeze-drying. In many cases of convection air-drying, the observations related to collapse are just the opposite of the glass transition concept (Ratti, 1994; Wang and Brennan, 1995; Del Valle et al., 1998; Rahman et al., 2005). The mechanism proposed for this was the concept of case hardening (Ratti, 1994; Achanta and Okos, 1996; Rahman et al., 2005). Bai et al. (2002) studied surface structural changes in apple rings during heat pump drying with controlled temperature and humidity. Electronic microscopy showed tissue collapse and pore formation. Case hardening occurred in the surface of the dried tissue when the apple slices were dried at 40°C–45°C and 60°C–65°C, and in the extreme case (at 60°C–65°C) cracks were formed on the surface. Low case hardening was observed in the samples dried at temperatures 20°C–25°C.

Wang and Brennan (1995) studied the structural changes in potato during air-drying (temperatures: 40°C and 70°C) by microscopy. They found that shrinkage occurs first at the surface and then gradually moves to the bottom with increase in drying time. The cell walls became elongated. The degree of shrinkage at a low drying temperature (40°C) was greater (i.e., less porosity) than that at high temperature (70°C). At a low drying rate (i.e., at low temperature), the moisture content at the center of a piece is never much higher than that at the surface, internal stresses are minimized, the material shrinks fully into a solid core, and shrinkage is uniform. At higher drying rates (i.e., higher temperature), the surface moisture decreases rapidly so that the surface becomes stiff, the outer layers of the material become rigid, and final volumes are fixed early in the drying. This causes case hardening phenomenon limiting subsequent shrinkage, and thus increases pore formation. As the drying proceeds, the tissues split and rupture internally, forming an open structure, and cracks are formed in the inner structure. When the interior finally dries and shrinks, the internal stresses pull the tissue apart. Thus, the increase in surface case hardening is increased with the increase in drying temperature, and at the extreme case cracks are formed on the surface or inside.

In the case of case hardening, the permeability and integrity of the crust play a role in maintaining the internal pressure inside the material boundary. Internal pressure always tries to puff the product by creating a force to the crust. During air-drying, stresses are formed due to nonuniform shrinkage resulting from nonuniform moisture and temperature distributions. This may lead to stress crack formation, when stresses exceed a critical level. Crack formation is a complex process influenced interactively by heat and moisture transfer, physical properties, and operational conditions (Kowalski and Rybicki, 1996; Liu et al., 1997). Liu et al. (1997) identified air relative humidity and temperature as the most influential parameters that need to be controlled to eliminate the formation of cracks. Sannino et al. (2005) used the DVS-1000 system to study drying process of lasagna pasta at controlled humidity and temperature with a sensing device to measure the electrical conductivity of pasta during the drying process. An anomalous diffusion mechanism was observed, typical of the formation of a glassy shell on the surface of the pasta slice during drying at low relative humidity, which inhibits a fast diffusion from the rubbery internal portion. Internal stresses at the interface of the glassy–rubbery surfaces are responsible for crack formation and propagation, thus causing surface breakage. An accurate control of the sample water activity and external environment humidity need to be maintained to avoid stress generation in the sample and crack formation. The glass transition concept cannot explain the effects of crust, casehardening, crack, and internal pressure. In the case of tuna meat, vacuum-drying produced a higher porosity than that in air-drying when both samples were dried at 70°C (Rahman et al., 2002). The porosity of dehydrated products is increased as vacuum pressure decreased, which means shrinkage can be prevented by controlling pressure (Krokida and Maroulis, 2000). Microwave creates a massive vaporization situation causing puffing (Pere et al., 1998). This indicates that in addition to the temperature effect, environment pressure can also affect the pore formation, and this effect cannot be explained by a single glass transition concept. Similarly in the case of extrusion, the higher the processing temperature above 100°C, the higher the porosity, which is contrary to the glass transition concept (Ali et al., 1996). This is due to the rapid vaporization of the water vapor at the exit of the die. Similarly, Rahman et al. (2005) also found a much higher pore formation at 105°C convection air-drying compared to 50°C or 80°C.

14.8.3 Rahman's Hypothesis

After analyzing experimental results from the literature, Rahman (2001) identified that the glass transition theory does not hold true for all products or processes. Other concepts, such as surface tension, pore-pressure, structure, environment pressure, and mechanisms of moisture transport also play important roles in explaining the formation of pores. Rahman (2001) hypothesized that as capillary force is the main force responsible for collapse, therefore counterbalancing this force causes the formation of pores and lower shrinkage. The counterbalancing forces are a result of generation of internal pressure due to vaporization of water or other solvents, variations in moisture transport mechanism, and pressure outside the material. Other factor could be the strength of the solid matrix (i.e., ice formation, case hardening, surface crack formation, permeability of water through crust, change in tertiary and quaternary structure of polymers, presence or absence of crystalline, amorphous, and viscoelastic nature of solids, matrix reinforcement, and residence time). Capillary force is related to the degree of water saturation of the porous matrix. In the case of apple with an initial porosity of 0.20, the final porosity of 0.56 for the dried sample could be achieved during air-drying; whereas the final porosity of dried samples is reduced to around 0.30 (nearly half) if the initial porosity of apple matrix is reduced to 0.0–0.023 by saturating with vacuum infiltration of water before air-drying (Bengtsson et al., 2003). Hussain et al. (2002) also identified that the prediction accuracy improved significantly when the initial porosity was included in the generic prediction model of porosity. This indicated that the initial air phase (related to pore or matrix pressure) has a significant effect on the subsequent pore formation in the matrix.

14.9 CHARACTERIZING PORES

14.9.1 Hysteresis in Mercury Porosimetry

Rahman et al. (2005) found the maximum hysteresis volume of 0.135 cc/g for fresh apple. This indicated the amount of mercury entrapped in the sample after a complete extrusion cycle and the complexity of the pore network. The total volume of entrapped mercury was 0.135 cc/g, which was

85% of the total mercury intruded. The shape of intrusion and extrusion curves for dried apple samples at 20 and 30 h is different from those for the fresh apple. A completely different extrusion curve was observed in the case of dried sample after 30 h. These graphs presented by Rahman et al. (2005) indicate that pore structure is changed in the course of drying in addition to porosity. Bread and cookie samples showed capillary hysteresis (Hicsasmaz and Clayton, 1992). About 95% of intruded mercury is entrapped in bread samples, while 80% mercury is entrapped in cookie samples. This indicated that both bread and cookie contain pores where narrow pore segments are followed by segments, which are larger in diameter. The fact that mercury entrapment is lower in cookies than in bread can be attributed to bread being a highly expanded product containing very large pores connected by narrow necks. This phenomenon (narrow neck pores) was also supported by measurement of the cellular structure using SEM. Another cause could, due to the filtering out of solid suspension, tend to form agglomerates in the pores during deposition.

14.9.2 Pore-Size Distribution Curve

Figures 14.28 through 14.30 show pore-size distribution curves of air-, vacuum-, and freeze-dried tuna meat (Rahman et al., 2002). In the literature, the pore-size distribution curve is usually characterized based on the number, size, and shape of peaks (Rahman and Sablani, 2003). Although pore-size distribution curves are presented in the literature, there is little information available on how this graph should be analyzed to explore the characteristics of pores. The large peak indicates that most of the pores exist in that size range, and the wider peak indicates that a relatively larger segment of pore is followed by smaller ones. The sharp peak indicates the extent of similar size pores, and the higher the height, the more pores at this size. A freeze-dried abalone sample dried at −15°C showed only one large peak at 10 μm. Increased freeze-drying temperature moves the

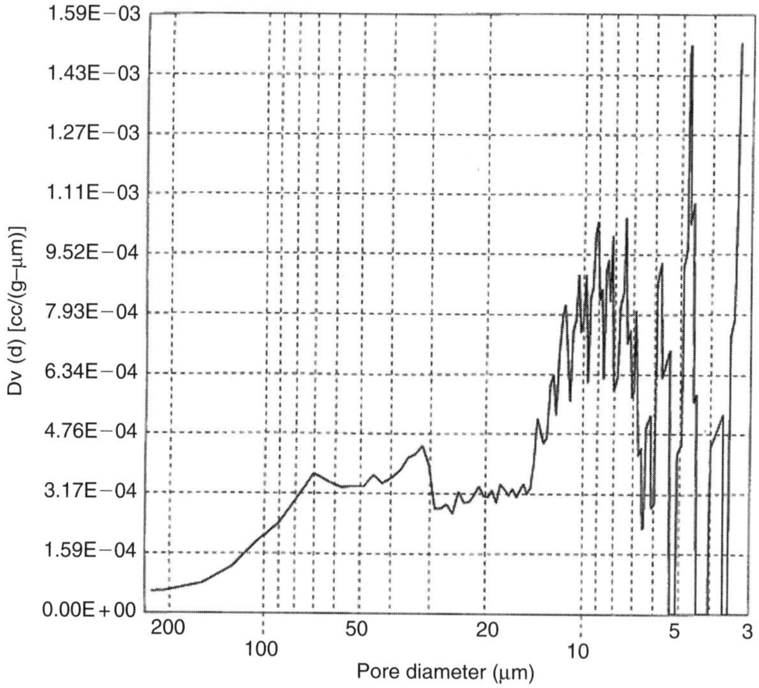

Figure 14.28 Pore-size distribution curves of air-dried tuna meat. (From Rahman, M.S., Al-Amri, O.S., and Al-Bulushi, I.M., *J. Food Eng.*, 53, 301, 2002. With permission.)

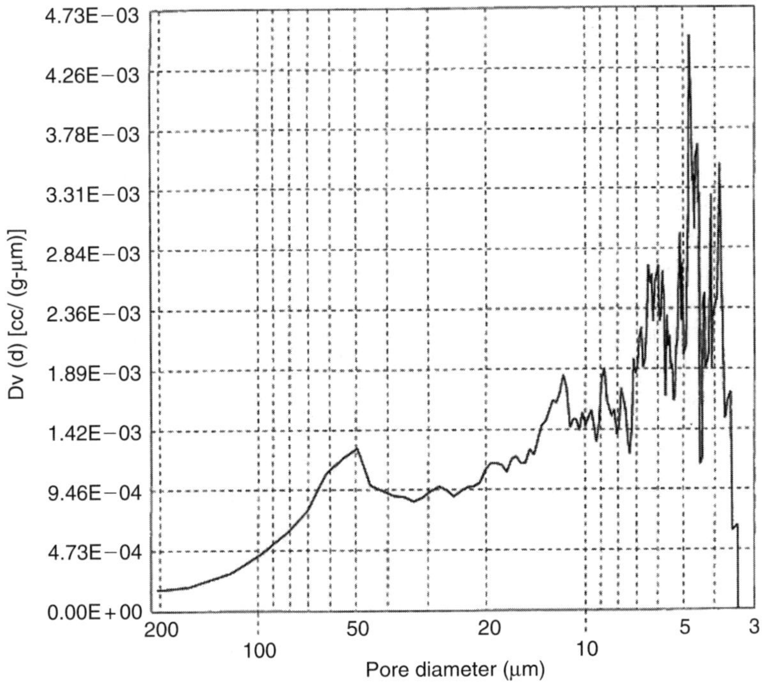

Figure 14.29 Pore-size distribution curves of vacuum-dried tuna meat. (From Rahman, M.S., Al-Amri, O.S., and Al-Bulushi, I.M., *J. Food Eng.*, 53, 301, 2002. With permission.)

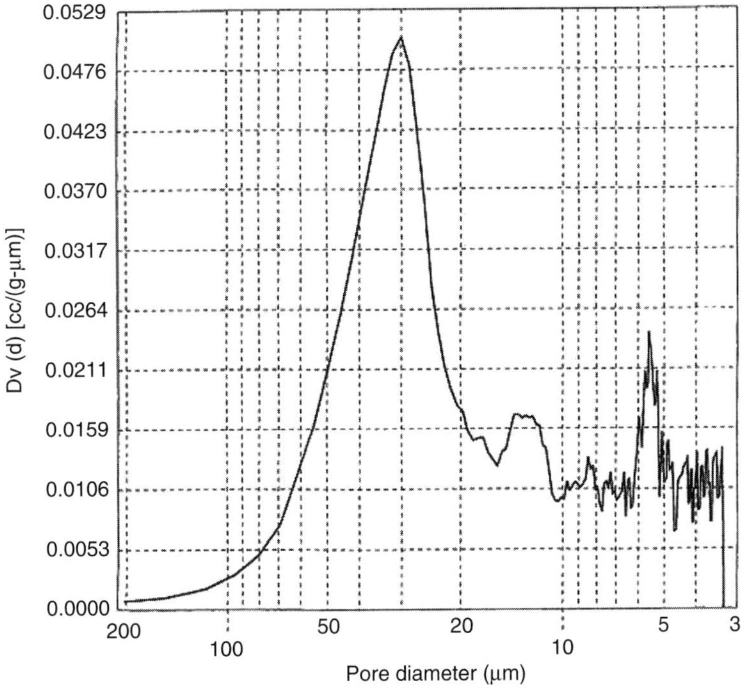

Figure 14.30 Pore-size distribution curves of freeze-dried tuna meat. (From Rahman, M.S., Al-Amri, O.S., and Al-Bulushi, I.M., *J. Food Eng.*, 53, 301, 2002. With permission.)

peaks slightly lower than 10 μm (Rahman and Sablani, 2003). Karathanos et al. (1996) found three peaks for freeze-dried carrots: two sharp peaks at 0.2 and 1.1 μm, and another shorter and wider peak at 21 μm. Two discrete peaks were found for freeze-dried potatoes, cabbage, and apples. In the case of freeze-dried garlic samples, completely different types of distribution curves were observed when samples were dried at different shelf temperature (Sablani et al., 2007). In a study by Rahman et al. (2002), the distribution graphs for tuna meat dried by different methods showed more complex and counting the peak did not characterize the entire curves. Moreover, the lines also showed significant ruggedness compared to the data for fruits and vegetables provided by Karathanos et al. (1996). The ruggedness in the case of the air-dried sample was to such an extent that it was difficult to count the number of peaks. The freeze-dried sample showed three major peaks, one at 5.5 and others at 13 and 30 μm and in addition to this there were many other small peaks making the curve rugged at the end of the graphs (at lower pore radius). Fresh apple showed two sharp peaks, one at 5.8 μm and another at 3.6 μm (Rahman et al., 2005). The second peak was so sharp that it was just a line. Apple sample air-dried for 20 h showed only one sharp peak at 3.6 μm, and the sample dried for 30 h showed three peaks at 10, 5.8, and 3.6 μm. The peak at 10 μm was much wider and shorter, which may be due to the formation of cracks or channels during longer drying. In addition, all the curves were extremely skewed to the right (toward lower pore diameter). The sharp peak indicates that most of the pores exist in that size range, and the wider peak indicates that a relatively larger segment of pores is followed by smaller ones. For Amioca starch containing 11% moisture, two peaks were found when using the low-pressure mercury porosimetry, one at the region 6–8 μm and the other at around 1–3.5 μm; another peak can be seen from the high-pressure mercury porosimetry at very small pore sizes of 3 nm (Karathanos and Saravacos, 1993). Xiong et al. (1991) studied the pore-size distribution curves of regular pasta and puffed pasta and found that regular pasta showed a horizontal line showing no peak whereas puffed pasta showed a large peak at 20 μm. Extruded regular pasta with mixtures of gluten and starch are very dense and homogeneous, whereas the puffed pasta samples were found to be porous. When gluten-starch ratio was 1, then they found two peaks one at 0.02 μm and another at 60 μm. Similarly when gluten-starch ratio was 3, two peaks were found at 0.006 and 11 μm. However, the peaks for gluten-starch ratio 3 were sharp and high. This is due to the cross-linking of protein through stronger covalent and hydrogen bonds during extrusion.

14.9.3 Fractal Analysis

Rahman et al. (2005) measured the fractal dimensions of fresh and dried apple to characterize the pores. Fractal dimensions higher than 2 and around 3 indicate the characteristics of pores in fresh and dried apples. The dimension is increased with the increasing drying time, indicating formation of micropores on the surface during air-drying. Rahman (1997) determined the fractal dimension of native, gelatinized, and ethanol deformed starch and the fractal dimension values were 3.09, 3.10, and 2.45, respectively. Lower values indicate the removal of micropores within the starch particles by ethanol modification.

14.9.4 Pore Size

Mean and median pore diameters are usually calculated. The mean pore diameter is the weighted arithmetic average of the distribution function, while the median pore diameter is the diameter at which an equal volume of mercury is introduced into larger and smaller pores than the median. Table 14.34 presents the pore diameter measured by mercury porosimetry. Pore radius in strawberry was the largest (60.8 μm), followed by apple (38.3 μm) and pear (28.7 μm) (Khalloufi and Ratti, 2003). The researchers also found that porosimetry gave lower average pore diameter compared with microscopic techniques.

Table 14.34 Pore Size (Diameter) of Different Dried Foods

Material	Drying Method	X_w (kg/(kg·wb))	ρ_a (kg/m^3)	ρ_m (kg/m^3)	ε_a	Mercury Porosimetry Range (μm)	Mercury Porosimetry Average (μm)	Microscopy Range (μm)	Microscopy Average (μm)	Reference
Abalone	Freeze-drying (PT: 15°C)	—	664	—	0.664	—	8.8 (0.2)	—	—	Rahman and Sablani (2003)
Abalone	Freeze-drying (PT: −5°C)	—	715	—	0.715	—	9.3 (0.6)	—	—	Rahman and Sablani (2003)
Abalone	Freeze-drying (PT: −20°C)	—	737	—	0.737	—	7.6 (0.4)	—	—	Rahman and Sablani (2003)
Apple	Air-drying (80°C, fresh)	0.860	849	1091	0.222	3.6–231	12.1 (2.7)	—	—	Rahman et al. (2005)
Apple	Air-drying (80°C, 20 h)	0.304	660	1352	0.512	3.6–214	12.2 (1.7)	—	—	Rahman et al. (2005)
Apple	Air-drying (80°C, 30 h)	0.103	560	1536	0.613	3.6–240	15.0 (1.0)	—	—	Rahman et al. (2005)
Apple[a]	Freeze-drying	<0.013	145	1478	0.902	—	76.6 (3.0)	46–416	146.4 (27.2)	Khalloufi and Ratti (2003)
Garlic	Freeze-drying (PT: −5°C)	0.090	469	1534	0.690	4.0–249	12.0 (2.0)	—	—	Sablani et al. (2007)
Garlic	Freeze-drying (PT: −15°C)	0.081	440	1517	0.710	4.0–227	12.0 (1.0)	—	—	Sablani et al. (2007)
Garlic	Freeze-drying (PT: −25°C)	0.061	431	1504	0.710	4.0–224	12.0 (1.0)	—	—	Sablani et al. (2007)
Pear[b]	Freeze-drying	<0.013	162	1049	0.845	—	57.4 (8.4)	44–240	96.0 (20.6)	Khalloufi and Ratti (2003)
Pear[c]	Freeze-drying	<0.013	140	1148	0.878	—	75.4 (8.4)	32–224	103.2 (10.4)	Khalloufi and Ratti (2003)
Pear[d]	Freeze-drying	<0.013	187	844	0.779	—	25.8	—	—	Khalloufi and Ratti (2003)
Strawberry[a]	Freeze-drying	<0.013	117	910	0.898	—	121.6 (4.6)	82–378	183.8 (20.4)	Khalloufi and Ratti (2003)
Strawberry[d]	Freeze-drying	<0.013	300	1476	0.797	—	121.0	—	—	Khalloufi and Ratti (2003)
Tuna meat	Air-drying (70°C)	0.114	960	1255	0.240	3.3–229	33.9 (3.1)	—	—	Rahman et al. (2002)
Tuna meat	Vacuum-drying (70°C)	0.077	709	1309	0.460	3.3–241	30.1 (4.6)	—	—	Rahman et al. (2002)
Tuna meat	Freeze-drying (PT: −20°C)	0.086	317	1259	0.760	3.3–220	27.2 (3.6)	—	—	Rahman et al. (2002)

[a] Frozen at −27°C and −17°C (hot plate temperature: 20°C–70°C).
[b] Frozen at −27°C (hot plate temperature: 20°C–70°C).
[c] Frozen at −17°C (hot plate temperature: 20°C–70°C).
[d] Collapsed sample.

ACKNOWLEDGMENT

The authors would like to thank Dr. Zacharias Maroulis for his guidance and comments for the writing of this chapter and for his contribution to tables and figures.

NOMENCLATURE

a	van der Waal's constant
a'	equation parameter
$\overline{a}$	equation parameter
$\hat{a}$	cube half-side
$\widetilde{a}$	volume of water removed per unit mass (m^3/kg)
b	van der Waal's constant
$\overline{b}$	equation parameter
$\widetilde{b}$	volume of water remains and solids per unit mass (m^3/kg)
B	degrees Brix
$\overline{B}$	equation coefficient
$\overline{c}$	equation parameter
C	materials ratio (kg corn/kg legume)
d'	equation parameter
D	diameter (m)
e_f	empty fraction of the extruder
F	feed rate (kg/h)
F'	food type factor
g	equation parameter
h	equation parameter
k	shrinkage rate constant
l	thickness (m)
L	length (m)
m	mass (kg)
M	moisture content, dry basis (kg/kg dry solid)
M_c	moisture content, % dry basis
n	mole
n_1	equation parameter
n_C	equation parameter
n_N	equation parameter
n_T	equation parameter
n_X	equation parameter
n_τ	equation parameter
N	screw rotation speed (rpm or min^{-1})
P	pressure (Pa)
q	equation parameter
r	equation parameter
R	ideal gas constant (kJ/kmol/°C)
RH	relative humidity
s	equation parameter (0, 1)
S	shrinkage
S_E	standard experimental error
S_R	standard deviation

t time (s)
T temperature (°C)
T_g glass transition temperature as a function of solids content (°C)
T_g' maximally freeze-concentrated glass transition temperature (°C)
u specific volume (m^3/kg)
u' equation parameter
v' equation parameter
V volume (m^3)
w constant (0, 1, 2)
x mass fraction (kg/kg)
x' equation parameter
X moisture content, wet basis (kg/kg sample)
X_c moisture content, % wet basis
y' equation parameter
Y oil content (kg oil/kg dry solid)
z' equation parameter

Symbols

β linear coefficient of shrinkage
β' volume-shrinkage coefficient
δ equation parameter
ε porosity
ε^{I} volume fraction of pores in ideal conditions
ε^{II} volume fraction of pores in nonideal conditions
ε_0 equation parameter
ζ^0 geometric value
θ equation parameter
ϑ refractive index
Λ equation parameter
ξ equation parameter
π constant (=3.14)
ρ density (kg/m^3)
τ residence time (s)
ϕ shrinkage-expansion coefficient
ψ sample mass at any moisture content per unit initial mass (kg/kg)

Subscripts

α apparent
air air
as ash
b bulk
c cell
ca carbohydrate
core core
cp closed pore
d dried
e equilibrium
eq equivalent

ex	excess
f	final
fa	fat
fi	fiber
fr	freezing
i	initial
ic	ice
j	index
k	index
L	oil
m	material
op	open pore
p	particle
pr	protein
r	reference
s	solid
sa	sample
su	substance
t	true
T	total
Tp	total pore
u	unfrozen water
w	water

REFERENCES

Abhayawick, L., Laguerre, J.C., Tauzin, V., and Duquenoy, A. 2002. Physical properties of three onion varieties as affected by the moisture content. *Journal of Food Engineering*, 55: 253–262.

Achanta, S. and Okos, M.R. 1995. Impact of drying on the biological product quality. In: *Food Preservation by Moisture Control, Fundamentals and Applications*, Barbosa-Canovas, G.V. and Welti-Chanes, J., eds., Technomic Publishing, Lancaster, PA, p. 637.

Achanta, S. and Okos, M.O. 1996. Predicting the quality of dehydrated foods and biopolymers: Research needs and opportunities. *Drying Technology*, 14(6): 1329–1368.

Aguilera, J.M. and Stanley, D.W. 1999. *Microstructural Principles in Food Processing and Engineering*, Aspen Publishers, Gaithersburg, MD.

Akinci, I., Ozdemir, F., Topuz, A., Kabas, O., and Canakci, M. 2004. Some physical and nutritional properties of *Juniperus drupacea* fruits. *Journal of Food Engineering*, 65(3): 325–331.

Ali, Y., Hanna, M.A., and Chinnaswamy, R. 1996. Expansion characteristics of extruded corn grits. *Food Science and Technology*, 29: 702–707.

Amin, M.N., Hossain, M.A., and Roy, K.C. 2004. Effects of moisture content on some physical properties of lentil seeds. *Journal of Food Engineering*, 65(1): 83–87.

Arbuckle, W.S. 1977. *Ice-cream*, 3rd edn., AVI Publishing Co., Westport, CT.

Aviara, N.A., Gwandzang, M.I., and Haque, M.A. 1999. Physical properties of guna seeds. *Journal of Agricultural Engineering Research*, 73: 105–111.

Aydin, C. 2002. Physical properties of hazel nuts. *Biosystems Engineering*, 82(3): 297–303.

Aydin, C. 2003. Physical properties of almond nut and kernel. *Journal of Food Engineering*, 60: 315–320.

Aydin, C. and Ozcan, M. 2002. Some physico-mechanic properties of terebinth (*Pistacia terebinthus* L.) fruits. *Journal of Food Engineering*, 53: 97–101.

Aydin, C., Ogut, H., and Konak, M. 2002. Some physical properties of *Turkish mahaleb*. *Biosystems Engineering*, 82(2): 231–234.

Azzouz, S., Guizani, A., Jomaa, W., and Belghith, A. 2002. Moisture diffusivity and drying kinetic equation of convective drying of grapes. *Journal of Food Engineering*, 55: 323–330.

Bahnasawy, A.H., El-Haddad, Z.A., El-Ansary, M.Y., and Sorour, H.M. 2004. Physical and mechanical properties of some Egyptian onion cultivars. *Journal of Food Engineering*, 62: 255–261.

Baik, O-D. and Marcotte, M. 2002. Modeling the moisture diffusivity in a baking cake. *Journal of Food Engineering*, 56: 27–36.

Baik, O.D., Sablani, S.S., Marcotte, M., and Castaigne, F. 1999. Modelling the thermal properties of a cup cake during baking. *Journal of Food Science*, 64(2): 295–299.

Bai, Y., Rahman, M.S., Perera, C.O., Smith, B., and Melton, L.D. 2002. Structural changes in apple rings during convection air-drying with controlled temperature and humidity. *Journal of Agricultural Food Chemistry*, 50: 3179–3185.

Bala, B.K. and Woods, J.L. 1984. Simulation of deep bed malt drying. *Journal of Agricultural Engineering Research, New York*, 30(3): 235–244.

Balaban, M. and Pigott, G.M. 1986. Shrinkage in fish muscle during drying. *Journal of Food Science*, 51(2): 510–511.

Balasubramanian, D. 2001. Physical properties of raw cashew nut. *Journal of Agricultural Engineering Research*, 78(3): 291–297.

Barat, J.M., Albors, A., Chiralt, A., and Fito, P. 1999. Equilibration of apple tissue in osmotic dehydration: Microstructural changes. *Drying Technology*, 17(7&8): 1375–1386.

Bart-Plange, A. and Baryeh, E.A. 2003. The physical properties of category B cocoa beans. *Journal of Food Engineering*, 60: 219–227.

Baryeh, E.A. 2001. Physical properties of bambara groundnuts. *Journal of Food Engineering*, 47: 321–326.

Baryeh, E.A. 2002. Physical properties of millet. *Journal of Food Engineering*, 51: 39–46.

Bayindirli, L. 1992. Mathematical analysis of variation of density and viscosity of apple juice with temperature and concentration. *Journal of Food Processing and Preservation*, 16: 23–28.

Bazhal, M.I., Ngadi, M.O., Raghavan, G.S.V., and Nguyen, D.H. 2003. Textural changes in apple tissue during pulsed electric field treatment. *Journal of Food Science*, 68(1): 249–253.

Bengtsson, G.B., Rahman, M.S., Stanley, R.A., and Perera, C.O. 2003. Apple rings as a model for fruit drying behavior: Effects of surfactant and reduced osmolality reveal biological mechanisms. *Journal of Food Science*, 68(2): 563–570.

Boukouvalas, Ch.J., Krokida, M.K., Maroulis, Z.B., and Marinos-Kouris, D. 2006a. Effect of material moisture content and temperature on the true density of foods. *International Journal of Food Properties*, 9(1): 109–125.

Boukouvalas, Ch.J., Krokida, M.K., Maroulis, Z.B., and Marinos-Kouris, D. 2006b. Densities and porosity: Literature data compilation for foodstuffs. *International Journal of Food Properties*, 9(4): 715–746.

Buhri, A.B. and Singh, R.P. 1993. Measurement of food thermal conductivity using differential scanning calorimetry. *Journal of Food Science*, 58(5): 1145–1147.

Calisir, S. and Aydin, C. 2004. Some physico-mechanic properties of cherry laurel (*Prunus lauracerasus* L.) fruits. *Journal of Food Engineering*, 65(1): 145–150.

Carman, K. 1996. Some physical properties of lentil seeds. *Journal of Agricultural Engineering Research*, 63: 87–92.

Carson, J.K., Lovatt, S.J., Tanner, D.J., and Cleland, A.C. 2004. Experimental measurements of the effective thermal conductivity of a pseudo-porous food analogue over a range of porosities and mean pore sizes. *Journal of Food Engineering*, 63: 87–95.

Chafer, M., Gonzalez-Martinez, G., Chiralt, A., and Fito, P. 2003. Microstructure and vacuum impregnation response of citrus peels. *Food Research International*, 36: 35–41.

Charm, S.E. 1978. *The Fundamentals of Food Engineering*, 3rd edn., AVI Publishing Co., Westport, CT.

Chung, S.L. and Merrit, J.H. 1991. Freezing time modelling for small finite cylindrical shaped foodstuff. *Journal of Food Science*, 56(4): 1072–1075.

Cnossen, A.G. and Siebenmorgen, T.J. 2000. The glass transition temperature concept in rice drying and tempering: Effect on milling quality. *Transactions of the ASAE*, 43(6): 1661–1667.

Chirife, J., 1969. El encogimiento y su influencia en la interpretacion del mecanismo de secado. *Industria Y Quimica*, 17: 145.

Choi, Y. and Okos, M.R. 1983. The thermal properties of tomato juice. *Transactions of the ASAE*, 26: 305–311.

Choi, Y. and Okos, M.R. 1986. Effects of temperature and composition on the thermal properties of foods. In: *Food Engineering and Process Applications*, Vol. 1, Transport Phenomena, Le Maguer, M. and Jelen, P., eds., Elsevier Applied Science, London.

Corrēa, P.C., Ribeiro, D.M., Resende, O., Afonso Junior, P.C., and Goneli, A.L. 2004. Mathematical modeling for representation of coffee berry volumetric shrinkage. In: *Proceeding of the 14th International Drying Symposium (IDS 2004)*, São Paulo, Brazil, pp. 742–747.

Del Valle, J.M., Cuadros, T.R.M., and Aguilera, J.M. 1998. Glass transitions and shrinkage during drying and storage of osmosed apple pieces. *Food Research International*, 31(3): 191–204.

Demir, F., Dogan, H., Ozcan, M., and Haciseferogullari, H. 2002. Nutritional and physical properties of hackberry (*Celtis australis* L.). *Journal of Food Engineering*, 54: 241–247.

Deshpante, S.D., Bal, S., and Ojha, T.P. 1993. Physical properties of soybean. *Journal of Agricultural Engineering Research*, 56: 89–98.

Dickerson, R.D. 1968. Thermal properties of foods. In: *The Freezing Preservation of Foods*, Tressler, D.K., Van Arsdel, W.B., and Copley, M.R., eds., AVI Publishing Co., Westport, CT.

Donsi, G., Ferrari, G., and Nigro, R. 1996. The effect of process conditions on the physical structure of dehydrated foods. *Food and Bioproducts Processing*, 74(2): 73–80.

Durance, T.D. and Wang, J.H. 2002. Energy consumption, density and rehydration rate of vacuum microwave and hot-air convection dehydrated tomatoes. *Journal of Food Science*, 67(6): 2212–2216.

Earle, R.L. 1983. *Unit Operations in Food Processing*, Pergamon Press, Oxford.

Farkas, B.E. and Singh, R.P. 1991. Physical properties of air-dried and freeze-dried chicken white meat. *Journal of Food Science*, 56(3): 611–615.

Fasina, O.O. and Fleming, H.P. 2001. Heat transfer characteristics of cucumbers during blanching. *Journal of Food Engineering*, 47: 203–210.

Ferrando, M. and Spiess, W.E.L. 2003. Mass transfer in strawberry tissue during osmotic treatment II: Structure–function relationships. *Journal of Food Science*, 68(4): 1356–1364.

Fikiin, A., Fikiin, K., and Triphonov, S. 1999. Equivalent thermophysical properties and surface heat transfer coefficient of fruit layers in trays during cooling. *Journal of Food Engineering*, 40: 7–13.

Fito, P., Chiralt, A., Betoret, N., Gras, M., Chafer, M., Martinez-Monzo, J., Andres, A., and Vidal, D. 2001. Vacuum impregnation and osmotic dehydration in matrix engineering. Application in functional fresh food development. *Journal of Food Engineering*, 49: 175–183.

Fitzpatrick, J.J., Barringer, S.A., and Iqbal, T. 2004. Flow property measurement of food powders and sensitivity of Jenike's Hopper design methodology to the measured values. *Journal of Food Engineering*, 61: 399–405.

Funebo, T., Ahrne, L., Kidman, S., Langton, M., and Skjoldebrand, C. 2000. Microwave heat treatment of apple before air dehydration—effects on physical properties and microstructure. *Journal of Food Engineering*, 46: 173–182.

Funebo, T., Ahrne, L., Prothon, F., Kidman, S., Langton, M., and Skjoldebrand, Ch. 2002. Microwave and convective dehydration of ethanol treated and frozen apple—physical properties and drying kinetics. *International Journal of Food Science and Technology*, 37: 603–614.

Gabas, A.L., Menegalli, F.C., and Telis-Romero, J. 1999. Effect of chemical pretreatment on the physical properties of dehydrated grapes. *Drying Technology*, 17(6): 1215–1226.

Garayo, J. and Moreira, R. 2002. Vacuum frying of potato chips. *Journal of Food Engineering*, 55: 181–191.

Gekas, V. and Lamberg, I. 1991. Determination of diffusion coefficients in volume-changing systems—application in the case of potato drying. *Journal of Food Engineering*, 14: 317–326.

Genskow, L.R. 1990. Consideration in drying consumer products. In: *Drying '89*, Mujumdar, A.S. and Roques, M., eds., Hemisphere Publishing, New York.

Gezer, I., Haciseferogullari, H., and Demir, F. 2002. Some physical properties of Hacihaliloglu apricot pit and its kernel. *Journal of Food Engineering*, 56: 49–57.

Chandrasekar, V. and Viswanathan, R. 1999. Physical and thermal properties of coffee. *Journal of Agricultural Engineering Research*, 73: 227–234.

Ghiaus, A.G., Margaris, D.P., and Papanikas, D.G. 1997. Mathematical modelling of the convective drying of fruits and vegetables. *Journal of Food Science*, 62: 1154–1157.

Gorling, P. 1958. Physical phenomena during the air drying of foodstuffs. In: *Fundamental Aspects of Dehydration of Food-Stuffs, Society of Chemical Industry*, The Macmillan Company, New York.

Guha, M., Zakiuddin Ali, S., and Bhattacharyah, S. 1997. Twin-screw extrusion of rice flour without a die: Effect of barrel temperature and screw speed on extrusion and extrudate characteristics. *Journal of Food Engineering*, 32: 251–267.

Gupta, R.K. and Das, S.K. 1997. Physical properties of sunflower seeds. *Journal of Agricultural Engineering Research*, 66: 1–8.

Haciseferogullari, H., Gezer, I., Bahtiyarca, Y., and Menges, H.O. 2003. Determination of some chemical and physical properties of sakiz faba bean (*Vicia faba* L. Var. major). *Journal of Food Engineering*, 60: 475–479.

Hamdami, N., Monteau, J-Y., and Le Bail, A. 2003. Effective thermal conductivity of a high porosity model food at above and sub-freezing temperatures. *International Journal of Refrigeration*, 26: 809–816.

Hatamipour, M.S. and Mowla, D. 2002. Shrinkage of carrots during drying in an inert medium fluidised bed. *Journal of Food Engineering*, 55(3): 247–252.

Hatamipour, M.S. and Mowla, D. 2003. Correlations for shrinkage, density and diffusivity for drying of maize and green peas in a fluidized bed with energy carrier. *Journal of Food Engineering*, 59: 221–227.

Heldman, D.R. 1982. Food properties during freezing. *Food Technology*, 36(2): 92.

Heldman, D.R. and Singh, R.P. 1981. *Food Process Engineering*, 2nd edn., AVI Publishing Co., Westport, CT.

Hicsasmaz, Z., Clayton, J.T. 1992. Characterization of the pore structure of starch based food materials. *Food Structure*, 11: 115–132.

Hodgson, A.S., Chan, H.T., Cavaletto, C.G., and Perrera, C.O. 1990. Physical–chemical characteristics of partially clarified guava juice and concentrate. *Journal of Food Science*, 55(6): 1757–1758.

Hsieh, R.C., Lerew, L.E., and Heldman, D.R. 1977. Prediction of freezing times for foods as influenced by product properties. *Journal of Food Processing Engineering*, 1: 183.

Hsu, C-L., Chen, W., Weng, Y-M., and Tsen, C-Y. 2003. Chemical composition, physical properties and antioxidant activities of yam flours as affected by different drying methods. *Food Chemistry*, 83: 85–92.

Hussain, M.A., Rahman, M.S., and Ng, C.W. 2002. Prediction of pores formation (porosity) in foods during drying: Generic models by the use of hybrid neural network. *Journal of Food Engineering*, 51: 239–248.

Ibarz, A. and Miguelsanz, R. 1989. Variation with temperature and soluble solids concentration of the density of a depectinised and clarified pear juice. *Journal of Food Engineering*, 10: 319–323.

Jain, R.K. and Bal, S. 1997. Properties of pearl millet. *Journal of Agricultural Engineering Research*, 66: 85–91.

Jha, S.N. 1999. Physical and hygroscopic properties of makhana. *Journal of Agricultural Engineering Research*, 72: 145–150.

Jha, S.N. and Prasad, S. 1993. Physical and thermal properties of gorgon nut. *Journal of Food Process Engineering*, 16(3): 237–245.

Isik, E. and Unal, H. 2007. Moisture-dependent physical properties of white speckled red kidney bean grains. *Journal of Food Engineering*, 82(2): 209–216.

Jagannath, J.H., Nanjappa, C., Das Gupta, D.K., and Arya, S.S. 2001. Crystallization kinetics of precooked potato starch under different drying conditions (methods). *Food Chemistry*, 75: 281–286.

Jindal, V.K. and Murakami, E.G. 1984. Thermal properties of shredded coconut. In: *Engineering and Food*, Vol. 1, McKennan, B.M., eds., Elsevier Applied Science Publishers, London.

Joshi, D.C., Das, S.K., and Mukherjee, R.K. 1993. Physical properties of pumpkin seeds. *Journal of Agricultural Engineering Research*, 54: 219–229.

Jumah, R.Y., Tashtoush, B., Shaker, R.R., and Zraiy, A.F. 2000. Manufacturing parameters and quality characteristics of spray dried jameed. *Drying Technology*, 18(4&5): 967–984.

Kaleemullah, S. and Gunasekar, J.J. 2002. Moisture-dependent physical properties of arecanut kernels. *Biosystems Engineering*, 82(3): 331–338.

Karel, M., Anglea, S., Buera, P., Karmas, R., Levi, G., and Roos, Y. 1994. Stability-related transitions of amorphous foods. *Thermochimica Acta*, 246: 249–269.

Karathanos, V.T. and Saravacos, G.D. 1993. Porosity and pore size distribution of starch materials. *Journal of Food Engineering*, 18: 259–280.

Karathanos, V., Anglea, S., and Karel, M. 1993. Collapse of structure during drying of celery. *Drying Technology*, 11(5): 1005–1023.

Karathanos, V.T., Kanellopoulos, N.K., and Belessiotis, V.G. 1996. Development of porous structure during air drying of agricultural plant products. *Journal of Food Engineering*, 29: 167–183.

Kawas, M.L. and Moreira, R.G. 2001. Characterization of product quality attributes of tortilla chips during the frying process. *Journal of Food Engineering*, 47: 97–107.

Keppeler, R.A. and Boose, J.R. 1970. Thermal properties of frozen sucrose solutions. *Transactions of the ASAE*, 13(3): 335–339.

Khalloufi, S. and Ratti, C. 2003. Quality determination of freeze-dried foods as explained by their glass transition temperature and internal structure. *Journal of Food Science*, 68(3): 892–903.

Kilpatric, P.W., Lowe, E., and van Arsdel, W.B. 1955. Tunnel dehydrators for fruits and vegetables. In: *Advances in Food Research*, Mrak, E.M. and Stewart, G.F., eds., Academic Press, New York, 360.

Kim, S.S. and Bhowmik, S.R. 1997. Thermophysical properties of plain yogurt as functions of moisture content. *Journal of Food Engineering*, 32: 109–124.

Konak, M., Carman, K., and Aydin, C. 2002. Physical properties of chick pea seeds. *Biosystems Engineering*, 82(1): 73–78.

Konanayakam, M. and Sastry, S.K. 1988. Kinetics of shrinkage of mushroom during blanching. *Journal of Food Science*, 53(5): 1406.

Kostaropoulos, A.E. and Saravacos, G.D. 1997. Thermal diffusivity of granular and porous foods at low moisture content. *Journal of Food Engineering*, 33: 101–109.

Kowalski, S.J. and Rybicki, A. 1996. A. Drying induced stresses and their control. In: *Drying'96, Proceedings of the 10th International Drying Symposium*, Krakow, Poland, pp. 151–158.

Krokida, M.K. and Maroulis, Z.B. 1997. Effect of drying method on shrinkage and porosity. *Drying Technology*, 15(10): 2441–2458

Krokida, M.K. and Maroulis, Z.B. 1999. Effect of microwave drying on some quality properties of dehydrated products. *Drying Technology*, 17(3): 449–466.

Krokida, M.K. and Maroulis, Z.B. 2000. Quality change during drying of food materials In: *Drying Technology in Agriculture and Food Sciences*, Mujumdar, A.S., ed., Science Publishers, Inc., Enfield (NH), USA.

Krokida, M.K., Karathanos, V.T., and Maroulis, Z.B. 1998. Effect of freeze-drying conditions on shrinkage and porosity of dehydrated agricultural products. *Journal of Food Engineering*, 35: 369–380.

Krokida, M.K., Kiranoudis, C.T., Maroulis, Z.B., and Marinos-Kouris, D. 2000. Drying related properties of apple. *Drying Technology*, 18(6): 1251–1267.

Krokida, M.K., Maroulis, Z.B., and Rahman, M.S. 2001. A Structural generic model to predict the effective thermal conductivity of granular materials. *Drying Technology*, 19(9): 2277–2290.

Kunii, D. and Smith, J.M. 1960. Heat transfer characteristics of porous rocks. *AIChE Journal*, 6(1): 71.

Lang, W. and Sokhansanj, S. 1993. Bulk volume shrinkage during drying of wheat and canola. *Journal of Food Process Engineering, Trumbull*, 16(4): 305–314.

Lazou, A.E., Michailidis, P.A., Thymi, S., Krokida, M.K., and Bisharat, G.I. 2007. Structural properties of corn-legume based extrudates as a function of processing conditions and raw material characteristics. *International Journal of Food Properties*, 10(4): 721–738.

Levine, H. and Slade, L. 1986. A polymer physico-chemical approach to the study of commercial starch hydrolysis products (SHPs). *Carbohydrate Polymer*, 6: 213–244.

Levine, H. and Slade, L. 1988. Thermomechanical properties of small-carbohydrate-water glasses and 'rubbers'. *Journal of Chemical Society—Faraday Transactions 1*, 84(8): 2619–2633.

Lewis, M.J. 1987. *Physical Properties of Foods and Food Processing Systems*, Ellis Horwood, England and VCH Verlagsgesellschaft, FRG.

Liu, H., Zhou, L., and Hayakawa, K. 1997. Sensitivity analysis for hygrostress crack formation in cylindrical food during drying. *Journal of Food Science*, 62: 447–450.

Lin, T.M., Durance, T.D., and Scaman, S.H. 1998. Characterization of vacuum microwave, air and freeze dried carrot slices. *Food Research International*, 31(2): 111–117.

Loch-Bonazzi, C., Wolff, E., and Gilbert, H. 1992. Quality of dehydrated cultivated mushrooms (*Agaricus bisporous*): A comparison between different drying and freeze-drying processes. *Lebensmittel-Wissenschaft und -Technologie*, 25: 334–339.

Lozano, J.E., Rotstein, E., and Urbicain, M.J. 1980. Total porosity and open-pore porosity in the drying of fruits. *Journal of Food Science*, 45: 1403–1407.

Lozano, J.E., Rotstein, E., and Urbicain, M.J. 1983. Shrinkage, porosity and bulk density of foodstuffs at changing moisture contents. *Journal of Food Science*, 48: 1497.

Madamba, P.S. 2003. Physical Changes in bamboo (*Bambusa phyllostachys*) shoot burning hot air drying: Shrinkage, density, and porosity. *Drying Technology*, 21(3): 555–568.

Madamba, P.S., Driscoll, R.H., and Buckle, K.A. 1993. Shrinkage, density and porosity of garlic during drying. *Journal of Food Engineering*, 23: 309–319.

Madamba, P.S., Driscoll, R.H., and Buckle, K.A. 1994. Bulk density, porosity, and resistance to airflow of garlic slices. *Drying Technology*, 12: 937–954.

Marousis, S.N. and Saravacos, G.D. 1990. Density and porosity in drying starch materials. *Journal of Food Science*, 55: 1367–1372.

Maroulis, Z.B. and Saravacos, G.D. *Food Process Design*, Marcel Dekker, New York, 2003.

Martinez-Monzo, J., Barat, J., Martinez-Gonzalez, C., and Chiralt, A. 2000. Changes in thermal properties of apple due to vacuum impregnation. *Journal of Food Engineering*, 43: 213–218.

Mavroudis, N., Gekas, V., and Sjoholm, I. 1998. Osmotic dehydration of apples. Shrinkage phenomena and the significance of initial structure on mass transfer rates. *Journal of Food Engineering*, 38: 101–123.

Mavroudis, N., Dejrnek, P., and Sjoholm, I. 2004. Studies on some raw material characteristics in different Swedish apple varieties. *Journal of Food Engineering*, 62: 121–129.

May, B.K. and Perre, P. 2002. The importance of considering exchange surface area reduction to exhibit a constant drying flux period in foodstuffs. *Journal of Food Engineering*, 54: 271–282.

Mayor, L. and Sereno, A.M. 2004. Modelling shrinkage during convective drying of food materials: A review. *Journal of Food Engineering*, 61: 373–386.

McDonald, K. and Sun, D-W. 2001. The formation of pores and their effects in a cooked beef product on the efficiency of vacuum cooling. *Journal of Food Engineering*, 47: 175–183.

McDonald, K., Sun, D-W., and Lyng, J.G. 2002. Effect of vacuum cooling on the thermophysical properties of a cooked beef product. *Journal of Food Engineering*, 52: 167–176.

McMinn, W.A.M. and Magee, T.R.A. 1997. Physical characteristics of dehydrated potatoes—Part I. *Journal of Food Engineering*, 33: 37–48.

Medeiros, G.L. and Sereno, A.M. 1994. Physical and transport properties of peas during warm air drying. *Journal of Food Engineering*, 21: 355–363.

Medeiros, M.F.D., Rocha, S.C.S., Alsina, O.L.S., Jeronimo, C.E.M., Medeiros, U.K.L., and da Mata, A.L.M.L. 2002. Drying of pulps of tropical fruits in spouted bed: Effect of composition on dryer performance. *Drying Technology*, 20(4&5): 855–881.

Miles, C.A., Beek, G.V., and Veerkamp, C.H. 1983. Calculation of thermophysical properties of foods. In: *Thermophysical Properties of Foods*, Jowitt, R., Escher, F., Hallstrom, B., Meffert, H.F.T., Spiess, W.E.L., and Vos, G., eds., Applied Science Publishers, London, pp. 269–312.

Milson, A. and Kirk, D. 1980. *Principles of Design and Operation of Catering Equipment*, Ellis Horwood, Chichester, West Sussex.

Moreira, R., Figueiredo, A., and Sereno, A. 2000. Shrinkage of apple disks during drying by warm air convection and freeze drying. *Drying Technology*, 18(1&2): 279–294.

Mujica-Paz, H., Valdez-Fragoso, A., Lopez-Malo, A., Palou, E., and Welti-Chanes, J. 2003. Impregnation properties of some fruits at vacuum pressure. *Journal of Food Engineering*, 56: 307–314.

Murthy, C.T. and Bhattacharya, S. 1998. Moisture dependant physical and uniaxial compression properties of black pepper. *Journal of Food Engineering*, 37: 193–205.

Nieto, A., Salvatori, D., Castro, M., and Alzamora, S. 2004. Structural changes in apple tissue during glucose and sucrose osmotic dehydration: Shrinkage, porosity, density and microscopic features. *Journal of Food Engineering*, 61: 269–278.

Ofman, M.H., Campos, C.A., and Gerschenson, L.N. 2004. Effect of preservatives on the functional properties of tapioca starch: Analysis of interactions. *Lebensmittel-Wissenschaft und-Technologie*, 37: 355–361.

Ogunjimi, L.A.O., Aviara, N.A., and Aregbesola, O.A. 2002. Some engineering properties of locust bean seed. *Journal of Food Engineering*, 55: 95–99.

Oguntunde, A.O. and Akintoye, O.A., 1991. Measurement and comparison of density, specific heat and viscosity of cow's milk and soymilk. *Journal of Food Engineering*, 13(3): 221.

Ogut, H. 1998. Some physical properties of white lupin. *Journal of Agricultural Engineering Research*, 69: 273–277.

Olajide, J.O. and Ade-Omowaye, B.I.O. 1999. Some physical properties of locust bean seed. *Journal of Agricultural Engineering Research*, 74: 213–215.

Omobuwajo, T., Sanni, L., and Olajide, J. 2000. Physical properties of ackee apple (*Blighia sapida*) seeds. *Journal of Food Engineering*, 45: 43–48.

Omobuwajo, T.O., Omobuwajo, O.R., and Sanni, L.A. 2003. Physical properties of calabash nutmeg (*Monodora myristica*) seeds. *Journal of Food Engineering*, 57: 375–381.

Owolarafe, O.K. and Shotonde, H.O. 2004. Some physical properties of fresh okra fruit. *Journal of Food Engineering*, 63: 299–302.

Oyelade, O.J., Ade-Omowaye, B.I.O., and Adeomi, V.F. 2003. Influence of variety on protein, fat contents and some physical characteristics of okra seeds. *Journal of Food Engineering*, 57: 111–114.

Ozarslan, C. 2002. Physical properties of cotton seed. *Biosystems Engineering*, 83(2): 169–174.

Ozcan, M., Haciseferogullari, H., and Demir, F. 2004. Some physico-mechanic and chemical properties of capers (*Capparis ovata* Desf. var. *canescens (Coss.) Heywood) flower buds. Journal of Food Engineering*, 65(1): 151–155.

Ozdemir, F. and Akinci, I. 2004. Physical and nutritional properties of four major commercial Turkish hazelnut varieties. *Journal of Food Engineering*, 63: 341–347.

Pagano, A.M. and Crozza, D.E. 2002. Near-ambient drying of canola. *Drying Technology*, 20(10): 2093–2104.

Paksoy, M. and Aydin, C. 2004. Some physical properties of edible squash (*Cucurbita pepo* L.) seeds. *Journal of Food Engineering*, 65(2): 225–231.

Palipane, K.B., Driscoll, R.H., and Sizednicki, G. 1992. Density, porosity, and composition of macadamia in shell nuts. *Food Australia*, 44(6): 276–280.

Pan, Z. and Singh, R.P. 2001. Physical and thermal properties of ground beef during cooking. *Lebensmittel-Wissenschaft und -Technologie*, 34: 437–444.

Peleg, M. 1983. Physical characteristics of food powders. In: *Physical Properties of Foods*, Peleg, M. and Bagley, E.B., eds., AVI Publishing Co., Westport, CT.

Pelegrina, A.H. and Crapiste, G.H. 2001. Modelling the pneumatic drying of food particles. *Journal of Food Engineering*, 48(4): 301–310.

Pere, C., Rodier, E., and Schwartzentruber, J. 1998. Effects of the structure of a porous material on drying kinetics in a microwave vacuum laboratory scale dryer. In: *IDS'98. 11th International Drying Symposium*, Thessaloniki-Halkidiki, Greece, pp. 1922–1929.

Perez, M.G.R. and Calvelo, 1984. A. Modelling the thermal conductivity of cooked meat. *Journal of Food Science*, 49: 152.

Perry, R.H. and Green, D. 1997. *Perry's Chemical Engineers' Handbook*, 7th edn., McGraw-Hill, New York.

Phipps, L.W. 1969. The interrelationship of viscosity, fat content, and temperature of cream between 40°C and 80°C. *Journal of Dairy Research*, 36: 417–426.

Pinthus, E.J., Weinberg, P., and Saguy, I.S. 1995. Oil uptake in deep fat frying as affected by porosity. *Journal of Food Science*, 60(4): 767–769.

Porter, J.W. 1975. *Milk and Dairy Products*, Oxford University Press, Oxford.

Raghavan, G.S.V. and Venkatachalapathy, K. 1999. Shrinkage of strawberries buring microwave drying. *Drying Technology*, 17(10): 2309–2321.

Rahman, M.S. 1984. Dehydration of pineapple and characteristics of dried pineapple powder. MSc Eng. (Chemical) thesis. Bangladesh University of Engineering and Technology, Dhaka.

Rahman, M.S. 1985. Study of osmotic and air drying of pineapple. MSc thesis. Food Engineering Thesis, University of Leeds, England.

Rahman, M.S. 1991. Thermophysical properties of seafoods. PhD thesis, University of New South Wales, Sydney, Australia.

Rahman, M.S. 1995. *Food Properties Handbook*. CRC Press, Boca Raton, FL.

Rahman, M.S. 1997. Physical meaning and interpretation of fractal dimensions of fine particles measured by different methods. *Journal of Food Engineering*, 32: 447–456.

Rahman, M.S. 2000. Mechanisms of pore formation in foods during drying: Present status. In: *Papers Presented at the Eighth International Congress on Engineering and Food (ICEF8)*, Puebla, Mexico, April 9–13, 2000.

Rahman, M.S. 2001. Toward prediction of porosity in foods during drying: a brief review. *Drying Technology*, 19(1): 1–13.

Rahman, M.S. 2003. A theoretical model to predict the formation of pores in foods during drying. *International Journal of Food Properties*, 6(1): 61–72.

Rahman, M.S. 2004. Prediction of pores in foods during processing: from regression to knowledge development from data mining (KDD) approach. In: *International Conference Engineering and Food (ICEF-9)*, Montpellier, France, 7–11 March 2004.

Rahman, M.S. 2005. Mass–volume–area-related properties of foods. In: *Engineering Properties of Foods*, 3edn., Rao, M.A., Rizvi, S.S.H., and Datta, A.K., eds., CRC, Taylor & Francis, Boca Raton, FL.

Rahman, M.S. and Potluri, P.L. 1990a. Shrinkage and density of squid flesh during air drying. *Journal of Food Engineering*, 12: 133–143.

Rahman, M.S. and Driscoll, R.H. 1994. Density of fresh and frozen seafood. *Journal of Food Process Engineering*, 17: 121–140.

Rahman, M.S. and Perera, C.O. 1999. Drying and food preservation. In: *Handbook of Food Preservation*, Rahman, M.S., ed., Marcel Dekker, New York. pp. 173–216.

Rahman, M.S. and Sablani, S.S. 2003. Structural characteristics of freeze-dried abalone: Porosimetry and puncture. *Transaction IChemE, Part C*, 81: 309–315.

Rahman, M.S., Perera, C.O., Chen, X.D., Driscoll, R.H., and Potluri, P.L. 1996. Density, shrinkage and porosity of calamari mantle meat during air drying in a cabinet dryer as a function of water content. *Journal of Food Engineering*, 30: 135–145.

Rahman, M.S., Al-Amri, O.S., and Al-Bulushi, I.M. 2002. Pores and physico-chemical characteristics of dried tuna produced by different methods of drying. *Journal of Food Engineering*, 53(4): 301–313.

Ramesh, M.N. 2003. Moisture transfer properties of cooked rice during drying. *Lebensmittel-Wissenschaft und-Technologie*, 36: 245–255.

Ramos, A.M. and Ibarz, A. 1998. Density of juice and fruit puree as a function of soluble solids content and temperature. *Journal of Food Engineering*, 35: 57–63.

Rapusas, R.S. and Driscoll, R.H. 1995. Thermophysical properties of fresh and dried white onion slices. *Journal of Food Engineering*, 24: 149–164.

Rapusas, R.S., Driscoll, R.H., and Srzednicki, G.S. 1995. Bulk density and resistance to airflow of sliced onions. *Journal of Food Engineering*, 26: 67–80.

Ratti, C. 1994. Shrinkage during drying of foodstuffs. *Journal of Food Engineering*, 23: 91–105.

Ratti, C. 2001. Hot air and freeze-drying of high-value foods: A review. *Journal of Food Engineering*, 49: 311–319.

Riedel, L. 1949. Thermal conductivity measurement on sugar solutions, fruit juices, and milk. *Chemical Engineering and Technology*, 21(17): 340–341.

Roman-Gutierrez, A., Sabathier, J., Guilbert, S., Galet, L., and Cuq, B. 2003. Characterization of the surface hydration properties of wheat flours and flour components by the measurement of contact angle. *Powder Technology*, 129: 37–45.

Rotstein, A. and Cornish, A.R.H. 1978. Prediction of the sorption equilibrium relationship for the drying of foodstuffs. *AIChE Journal*, 24: 966.

Roy, N.K., Yadav, P.L., and Dixit, R.N. 1971. Density of buffalo milk fat. II. Centrifuged fat. *Milchwissenschaft*, 26: 735–738.

Ruiz-Lopez, I.I., Cordova, A.V., Rodriguez-Jimenes, G.C., and Garcia-Alvarado, M.A. 2004. Moisture and temperature evolution during food drying: Effect of variable properties. *Journal of Food Engineering*, 63: 117–124.

Sablani, S.S. and Rahman, M.S. 2002. Pore formation in selected foods as a function of shelf temperature during freeze drying. *Drying Technology*, 20(7): 1379–1391.

Sablani, S.S., Baik, O.-D., and Marcotte, M. 2002. Neural networks for predicting thermal conductivity of bakery products. *Journal of Food Engineering*, 52: 299–304.

Sablani, S.S., Rahman, M.S., Al-Kuseibi, M.K., Al-Habsi, N.A., Al-Belushi, R.H., Al-Marhubi, I., and Al-Amri, I.S. 2007. Influence of shelf temperature on pore formation in garlic during freeze-drying. *Journal of Food Engineering*, 80: 68–79.

Sacilik, K., Ozturk, R., and Keskin, R. 2003. Some physical properties of hemp seed. *Biosystems Engineering*, 86(2): 191–198.

Sanga, E.C.M., Mujumdar, A.S., and Raghavan, G.S.V. 2002. Simulation of convection-microwave drying for a shrinking material. *Chemical Engineering and Processing*, 41: 487–499.

Sahoo, P.K. and Srivastava, A.P. 2002. Physical properties of okra seed. *Biosystems Engineering*, 83(4): 441–448.

Sannino, A., Capone, S., Siciliano, P., Ficarella, A., Vasanelli, L., and Maffezzolo, A. 2005. Monitoring the drying process of lasagna pasta through a novel sensing device-based method. *Journal of Food Engineering*, 69: 51–59.

Sanz, P.D., Domonguez, M., and Mascheroni, R.H. 1989. Equations for the prediction of thermophysical properties of meat products. *Latin American Applied Research*, 19:155.

Saravacos, G.D. and Raouzeos, G.S. 1986. Diffusivity of moisture in air drying of raisins. In: *Drying 86*, Vol. 2, Mujumdar, A.S., ed., Hemisphere Publishing, New York.

Saravacos, G.D. and Maroulis, Z.B. 2001. *Transport Properties of Foods*, Marcel Dekker, New York.

Shah, K.K., Tong, C.H., and Lund, D.B. 2000. Methodology to obtain true thermal conductivity of low porosity food powders. *Journal of Food Science*, 65(6): 962–967.

Short, A.L. 1955. The temperature coefficient of expansion of raw milk. *Journal of Dairy Research*, 22: 69.

Singh, R.K. and Lund, D.B. 1984. Mathematical modelling of heat and moisture transfer-related properties of intermediate moisture apples. *Journal of Food Processing and Preservation*, 8: 191.

Singh, K.K. and Goswami, T.K. 1996. Physical properties of cumin seed. *Journal of Agricultural Engineering Research*, 64: 93–98.

Singh, R.K and Sarkar, A. 2005. Thermal properties of frozen foods. In: *Engineering Properties of Foods*, 3rd ed. Rao, M.A., Rizvi, S.S.H., and Datta, A.K., eds., CRC, Taylor & Francis, Boca Raton, FL.

Singh, P.C., Singh, R.K., Smith, R.S., and Nelson, P.E. 1997. Evaluation of in-line sensors for selected properties measurements in continuous food processing. *Food Control*, 8(1): 45–50.

Sjoholm, I. and Gekas, V. 1995. Apple shrinkage upon drying. *Journal of Food Engineering*, 25: 123–130.

Slade, L. and Levine, H. 1991. A food polymer science approach to structure property relationships in aqueous food systems: Non-equilibrium behavior of carbohydrate-eater systems. In: *Water relationships in food*, Levine, H. and Slade, L., eds., Plenum Press, NY, pp. 29–101.

Sokhansanj, S. and Lang, W. 1996. Prediction of kernel and bulk volume of wheat and canola during adsorption and desorption. *Journal of Agricultural Engineering Research*, 63: 129–136.

Suthar, S.H. and Das, S.K. 1996. Some physical properties of karingda [*Citrullus lanatus* (Thumb) Mansf.] seeds. *Journal of Agricultural Engineering Research*, 65: 15–22.

Suzuki, K., Kubota, K., Hasegawa, T., and Hosaka, H. 1976. Shrinkage in dehydration of root vegetables. *Journal of Food Science*, 41: 1189–1193.

Talla, A., Puiggali, J-R., Jomaa, W., and Jannot, Y. 2004. Shrinkage and density evolution during drying of tropical fruits: Application to banana. *Journal of Food Engineering*, 64: 103–109.

Tang, J. and Sokhansanj, S. 1993. Geometric changes in lentil seeds caused by drying. *Journal of Agricultural Engineering Research*, 56: 313–326.

Telis-Romero, J., Telis, V.R.N., Gabas, A.L., and Yamashita, F. 1998. Thermophysical properties of Brazilian orange juice as affected by temperature and water content. *Journal of Food Engineering*, 38: 27–40.

Teotia, M.S. and Ramakrishna, P. 1989. Densities of melon seeds, kernels and hulls. *Journal of Food Engineering*, 9: 231–236.

Teunou, E., Fitzpatrick, J.J., and Synnott, E.C. 1999. Characterisation of food powder flowability. *Journal of Food Engineering*, 39: 31–37.

Thymi, S., Krokida, M.K., Pappa, A., and Maroulis, Z.B. 2005. Structural properties of extruded corn starch. *Journal of Food Engineering*, 68: 519–526.

Torreggiani, D., Toledo, R.T., and Bertolo, G. 1995. Optimization of vapor induced puffing in apple dehydration. *Journal of Food Science*, 60(1): 181–185.

Torringa, E., Esveld, E., Scheewe, I., Van den Berg, R., and Bartels, P. 2001. Osmotic dehydration as a pre-treatment before combined microwave-hot-air drying of mushrooms. *Journal of Food Engineering*, 49: 185–191.

Tsami, E. and Katsioti, M. 2000. Drying kinetics for some fruits: Predicting of porosity and color during dehydration. *Drying Technology*, 18(7): 1559–1581.

Tsami, E., Krokida, M.K., and Drouzas, A.E. 1999. Effect of drying method on the sorption characteristics of model fruit powders. *Journal of Food Engineering*, 38: 381–392.

Tsen, J-H. and King, V.A-E. 2002. Density of banana puree as a function of soluble solids concentration and temperature. *Journal of Food Engineering*, 55: 305–308.

Uddin, M.S., Hawlader, M.N.A., and Rahman, M.S. 1990. Evaluation of drying characteristics of pineapple in the production of pineapple powder. *Journal of Food Process and Preservation*, 14(5): 375.

Vacarezza, L. 1975. Cinetica y mecanismo de transporte de aqua durante la deshidratacion de la remolacha azucarera. PhD thesis, Universidad de Buenos, Aires.

Vagenas, G.K., Marinos-Kouris, D., and Saravacos, G.D. 1990. Thermal properties of raisins. *Journal of Food Engineering*, 11: 147–158.

Vilche, C., Gely, M., and Santalla, E. 2003. Physical properties of quinoa seeds. *Biosystems Engineering*, 86(1): 59–65.

Visvanathan, R., Palanisamy, P.T., Gothandapani, L., and Sreenarayanan, V.V. 1996. Physical properties of neem nut. *Journal of Agricultural Engineering Research*, 63(1): 19–26.

Vu, T.O., Galet, L., Fages, J., and Oulahna, D. 2003. Improving the dispersion kinetics of a cocoa powder by size enlargement. *Powder Technology*, 130: 400–406.

Wang, N. and Brennan, J.G. 1995. Changes in structure, density and porosity of potato during dehydration. *Journal of Food Engineering*, 24: 61–76.

Wang, Z.H. and Chen, G. 1999. Heat and mass transfer during low intensity convection drying. *Chemical Engineering Science*, 54: 3899–3908.

Weast, R.C. 1970. *Handbook of Chemistry and Physics*. CRC Press, Cleveland, OH.

Whitaker, S. 1976. *Elementary Heat Transfer Analysis*. Pergamon Press, New York.

Xiong, X., Narsimhan, G., and Okos, M.R. 1991. Effect of comparison and pore structure on binding energy and effective diffusivity of moisture in porous food. *Journal of Food Engineering*, 15: 187–208.

Yang, C.S.T. and Atallah, W.A. 1985. Effect of four drying methods on the quality of intermediate moisture lowbush blueberries. *Journal of Food Science*, 50: 1233–1237.

Zainal, B.S., Abdul Rahman, R., Ariff, A.B., Saari, B.N., and Asbi, B.A. 2000. Effects of temperature on the physical properties of pink guava juice at two different concentrations. *Journal of Food Engineering*, 43: 55–59.

Zanoni, B., Peri, C., and Gianotti, R. 1995. Determination of the thermal diffusivity of bread as a function of porosity. *Journal of Food Engineering*, 26: 497–510.

Zogzas, N.P., Maroulis, Z.B., and Marinos-Kouris, D. 1994. Densities, shrinkage and porosity of some vegetables during air drying. *Drying Technology*, 12(7): 1653–1666.

CHAPTER 15

Shape, Volume, and Surface Area

Mohammad Shafiur Rahman

CONTENTS

15.1 VOLUME

Volume is defined as the space occupied by a material. Two types of volumes are usually used by food scientists and engineers: boundary volume and pore or void volume.

15.1.1 Boundary Volume

A material's boundary volume can be measured by buoyancy force; liquid, gas, or solid displacement; gas adsorption; or it can be estimated from the material's geometric dimensions. More details of the boundary volume measurement methods are discussed in the Chapter 13. Estimation equations of the boundary volume of regular geometry are given in Table 15.1. Figure 15.1 shows a simplified shape of a pear, which considers one-half sphere and one circular cone (with

Table 15.1 Volume and Surface Area of Some Common Shapes

Sphere

$$V=\frac{4}{3}\pi r^3 \text{ and } A=4\pi r^2$$

Cylinder

$$V=\pi r^2 L \text{ and } A=2\pi r^2+2\pi rL$$

Cube

$$V=a^3 \text{ and } A=6a^2$$

Brick

$$V=abc \text{ and } A=2(ab+bc+ca)$$

Pyramid

$$V=\frac{1}{3}abL \text{ and } A=ab+aL+bL$$

Right circular cylinder

$$V=\pi r^2 h \text{ and } A=2\pi r(h+r)$$

Prolate spheroid (rotation about major axis)

$$V=\frac{4}{3}\pi ab^2 \text{ and } A=2\pi b^2+\frac{2\pi ab}{e}\sin^{-1}e$$

Prolate spheroid (rotation about minor axis)

$$V=\frac{4}{3}\pi ab^2 \text{ and } V=\frac{4}{3}\pi abc$$

Oblate spheroid (rotation about minor axis)

$$V=\frac{4}{3}\pi a^2 b \text{ and } A=2\pi a^2+\frac{\pi b^2}{e}\ln\left(\frac{1+e}{1-e}\right)$$

Right circular cone

$$V=\frac{1}{3}\pi r^2 h \text{ and } A=\pi r\sqrt{r^2+h^2}$$

Frustam right cone (truncated right circular cone)

$$V=\frac{\pi}{3}L(r_1^2+r_1r_2+r_2^2) \text{ and } A=\pi(r_1+r_2)\sqrt{L^2+(r_1-r_2)^2}$$

Notes: a and b are, respectively, major and minor semi-axes of the ellipse; e is the eccentricity given by $e=\sqrt{[1-(b/a)^2]}$.
r_1 and r_2 are the radii of base and top, respectively, and L is the altitude.

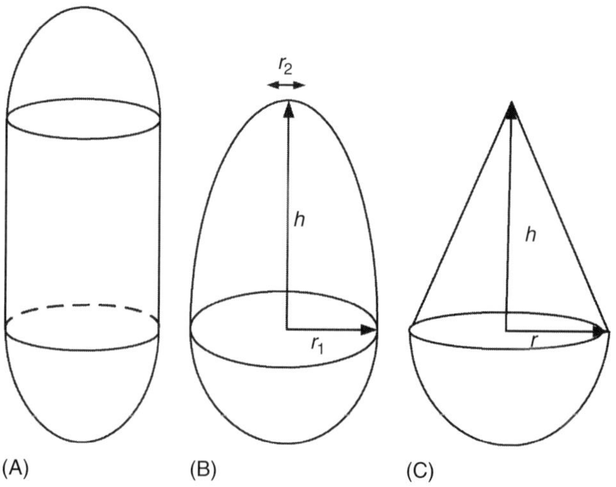

Figure 15.1 Simplified form of cucumber (A) and pear (B and C).

truncated or without truncated). Thus, the volume and surface area of pear can be estimated from the following equation (Figure 15.1C):

$$V = \frac{1}{2}\left(\frac{4}{3}\pi r^3\right) + \frac{1}{3}\pi r^2 h \tag{15.1}$$

Similarly, the surface area can also be estimated. A cucumber's volume and surface area could also be estimated considering a cylinder for the middle part and two truncated right circular cones for the two ends.

15.1.2 Pore Volume

More details of the pore volume measurement methods are discussed in Chapter 13. These methods include pycnometer, mercury intrusion, and gas adsorption.

15.2 SURFACE AREA

There are two types of surface area used in the process calculations: outer boundary surface of a particle or object and pore's surface area for a porous material. An object can be characterized as Euclidian geometry or non-Euclidian geometry. Euclidian geometry always has its characteristic dimensions and has an important common peculiarity of smoothness of surface, such as sphere, cube, or ellipsoid.

15.2.1 Measurement Techniques

Measurement techniques are different for geometric boundary area, pore surface area, and cross-sectional area.

15.2.1.1 Boundary Surface Area

Boundary surface area measurement is not as simple as the measurement of volume. Different methods used to determine boundary surface area are discussed in the following sections.

15.2.1.1.1 Profile Recordings

If an object forms a surface of revolution, like an eggshell, it is possible to perform these measurements using the profiling of the object and ascertaining the surface magnitude by linear approximation (Besch et al., 1968).

15.2.1.1.2 Image Analysis

This is mainly estimated from the geometric dimensions or measured by image analysis or contour analysis. Leaf and stalk surface area are measured by contact printing the surface on a light-sensitive paper and then the area can be estimated by a planimeter or by tracing the area on a graph paper and counting the squares or from the mass of the paper. In this method, mass–area relationship of the paper should be first developed.

15.2.1.1.3 Skin or Peeled Area Method

The surface area of fruits and vegetables can be estimated from the peeled or skin area. In this method, fruit is peeled in narrow strips and the planimeter sum of the areas of tracings of the strips are taken as the surface area. Again strips of narrow masking tape can be used to cover the surface and the surface area can be estimated from the length and width of the tape (Mohsenin, 1986).

15.2.1.1.4 Shadow-Graph Method

The simplest method of obtaining the surface area of a symmetrical convex body such as an egg is the projection method using shadow-graph or photographic enlarger. Having the profile of the egg, equally spaced parallel, perpendicular lines can be drawn from the axis of symmetry to the intersection with the profile. Then using manual computation and integration, the surface area can be obtained by summing up the surfaces of revolution for all of the divided segments (Mohsenin, 1986). An image processing based method has also been developed to measure the volume and surface area of ellipsoidal food products (Sabliov et al., 2002). The method assumes that each product has an axisymmetric geometry and is a sum of superimposed elementary frustums of right circular cones. Three steps in image processing technique are image acquisition with camera, image processing by digitizing, and volume and surface area computation. This method showed only around 1% error.

15.2.1.1.5 Simplifying to Regular Geometry

The surface area of food materials can be estimated by simplifying the geometry as mentioned earlier for pear and cucumber (Figure 15.1). The surface area of a pear can be estimated as (Figure 15.1C)

$$A = \frac{1}{2}\left(4\pi r^2\right) + \pi r\sqrt{r^2 + h^2} \qquad (15.2)$$

15.2.1.1.6 Volume–Surface Area Relationships

The surface area of a convex body can be estimated from the theoretical volume and surface area relationship. On the basis of the theories of convex bodies (Bennesen and Frenchel, 1948)

$$\frac{V^2}{A^3} \geq \frac{1}{36\pi} \qquad (15.3)$$

Polya and Szezo (1951) showed that the average projected area of a convex body is one-fourth the surface area, thus Equation 15.3 can be written as

$$A \leq \left(\frac{9\pi}{16}\right)^{1/3} V^{2/3} \qquad (15.4)$$

$$A \leq k\, V^{2/3} \qquad (15.5)$$

where k is a constant for shape factor and for a sphere, equality is achieved:

$$k = \left(\frac{9\pi}{16}\right)^{1/3} = 1.21 \qquad (15.6)$$

The value of k was found to be 1.27 for barley kernels (Bargale and Irudayaraj, 1995). However, the values of k are not widely available in the literature.

15.2.1.1.7 Surface Area Based on Heat Transfer

Another method is the use of air flow planimeter, which measures the area as a function of the surface obstructing the flow of air (Mohsenin, 1986). Transient heat transfer study can also be used to estimate the surface area.

15.2.1.2 Pore Surface Area

Pore surface area can be defined as the surface of the pores in a porous media exposed to fluid flow either per unit volume or per unit mass of the solid particles. The most definitive surface area measurements are probably those made by nitrogen adsorption using the Brunauer-Emmett-Teller (BET) theory.

15.2.1.2.1 Methods Based on Adsorption

The quantity of an inert vapor, which can be adsorbed on pore surface, is dependent on the area of the surface. The quantity of a gas or vapor adsorbed is proportional to a surface area, which inclines the tiny molecular interstices of the porous material, whereas the surface area pertinent to fluid flow does not include this portion of the surface area. Before measuring N_2 adsorption, the foods are usually degassed overnight under high vacuum with a mechanical oil pump. In many cases, intact pieces posed serious limitations on the size of the sample, which could be utilized. For whole freeze-dried foods, the mass–apparent volume ratio is quite small. This necessitates the use of sample tubes of relatively large volumes, thereby increasing what is commonly referred to as dead space error. It is, therefore, found necessary to crush or grind the food before the adsorption measurements (Berlin et al., 1966). The C values (in BET model) of 5–50 for various nitrogen adsorption isotherms suggest that nitrogen is not so strongly adsorbed on food materials (Table 15.2).

15.2.1.2.2 Methods Based on Fluid Flow

Mohsenin (1986) mentioned that Carman–Kozeny equation can be employed to measure the specific surface of the nonuniform pore space. An inert fluid needs to be used and this is more

Table 15.2 Surface Area of Freeze-Dried Foods by Nitrogen Adsorption

Food	*A* (m^2/g)	BET C
Peppers (red bell)	0.340	5
Celery (cut)	0.638	21
Shrimp (cooked deveined)	0.305	74
Fish (raw fillets)	0.581	49
Strawberries (small whole)	0.499	102
Carrots (raw diced)	0.508	31
Bananas (sliced)	0.169	136
Orange (juice)	0.121	53
Blueberries (Canadian)	0.144	—
Turkey meat (white, diced cooked)	0.285	—
Chicken meat (white, diced cooked)	0.085	—
Shrimp (raw)	0.819	29
Mushrooms (sliced)	3.830	—
Mushrooms (sliced)[a]	2.990	26

Source: Berlin, E., Kliman, P.G., and Pallansch, M.J., *J. Agric. Food Chem.*, 14, 15, 1966.

[a] Without grinding sample.

commonly used for rocks and not suitable for fragile or flexible solid as foods. More details can be found in the work of Rahman (1995). Also dead-end pore's surface could not be included in the fluid flow method and channeling could be one of the major problems.

15.2.1.2.3 Mercury Intrusion

Mercury intrusion measures the characteristics of pores. More details of the apparatus are given in Chapter 13. However, it is not as accurate as gas adsorption method since there are other assumptions involved in the calculations. Assuming the consistency of surface tension and wetting angle of mercury, cumulative surface area can be estimated as (Lowell and Shields, 1984)

$$A = 2.654 \times 10^{-5} \int_0^V P \, dV \tag{15.7}$$

It represents the surface area of all voids and pore space filled with mercury up to a given pressure. In the above equation, the average pressure in a pressure interval is multiplied by volume change in that interval. The integration term could be estimated by graphical or numerical method from the cumulative intrusion curve. Some newly developed automated porosimeters are associated with computer-aided data reduction capabilities, which perform the above integration as rapidly as data are acquired. This method gives the surface area of only open pores and thus does not provide the surface area of total pores of the entire solid.

15.2.1.3 Cross-Sectional Area

This is the area of a surface after longitudinal or transverse section of a material. It is necessary when a fluid flows over an object. This is usually measured, as discussed earlier, by image analysis, after making the material's section along its axis.

15.2.2 Prediction of Surface Area

The surface area of foods is usually correlated with characteristic dimensions or mass when it is Euclidian geometry. In case of non-Euclidian geometry mainly fractal analysis is used.

15.2.2.1 Euclidian Geometry

The boundary surface area and volume of some common shapes (Euclidian geometry) are given in Table 15.1.

15.2.2.1.1 Data on Surface Area

The boundary surface areas of different food items are compiled in Table 15.3 as a function of mass, volume, and characteristic dimensions.

15.2.2.1.2 Correlations of Surface Area

The surface area of an ellipsoid can be estimated (error less than 2%) from the length, width, and thickness from the regression equation (Igathinathane and Chattopadhyay, 2000):

$$A = \pi \times 10^{-2}\left(-1.023a^2 + 4.930\,ab + 34.36\,ac - 5.295\,bc + \frac{23.60\,bc^2}{a}\right) \tag{15.8}$$

Table 15.3 Mass, Volume, Surface Area and Characteristic Dimensions of Different Foods

Material	X_w	*m* (kg)	*V* (m^3)	*a* (m)	*b* (m)	*c* (m)	*A* (m^2)	Reference
Apple	—	—	—	—	—	—	$1.11–1.64 \times 10^{-2}$	Mohsenin (1986)
Egg	—	$5.35–6.14 \times 10^{-2}$	$5.15–5.91 \times 10^{-5}$	$2.68–2.91 \times 10^{-2}$	$2.10–2.21 \times 10^{-2}$	—	$6.95–7.61 \times 10^{-3}$	Sabliov et al. (2002)
Groundnut[a]	0.04	$0.30–0.90 \times 10^{-3}$	$0.26–0.44 \times 10^{-6}$	$0.85–1.44 \times 10^{-2}$	$3.55–9.30 \times 10^{-3}$	$5.40–8.49 \times 10^{-3}$	$1.29–3.23 \times 10^{-4}$	Olajide and Igbeka (2003)
Lemon	—	$7.27–8.46 \times 10^{-2}$	$0.80–1.00 \times 10^{-4}$	$3.30–3.72 \times 10^{-2}$	$2.39–2.57 \times 10^{-2}$	—	$0.95–1.09 \times 10^{-2}$	Sabliov et al. (2002)
Limes	—	$1.11–1.24 \times 10^{-1}$	$1.17–1.27 \times 10^{-2}$	$3.06–3.33 \times 10^{-2}$	$2.80–3.02 \times 10^{-2}$	—	$1.19–1.24 \times 10^{-2}$	Sabliov et al. (2002)
Onion (L)	—	$1.43–1.99 \times 10^{-1}$	$1.42–2.14 \times 10^{-4}$	$7.70–8.60 \times 10^{-2}$	$4.10–5.50 \times 10^{-2}$	—	$1.45–1.86 \times 10^{-2}$	Maw et al. (1996)
Onion (M)	—	$0.72–1.63 \times 10^{-1}$	$0.64–1.60 \times 10^{-4}$	$6.40–7.60 \times 10^{-2}$	$3.50–5.50 \times 10^{-2}$	—	$0.91–1.60 \times 10^{-2}$	Maw et al. (1996)
Onion (S)	—	$0.12–1.17 \times 10^{-1}$	$0.10–1.16 \times 10^{-4}$	$2.80–6.30 \times 10^{-2}$	$2.10–5.50 \times 10^{-2}$	—	$0.27–1.41 \times 10^{-2}$	Maw et al. (1996)
Paddy[b]	0.12	—	1.61×10^{-8}	7.90×10^{-3}	3.12×10^{-3}	1.96×10^{-3}	4.019×10^{-5}	Wratten et al. (1969)
Paddy[b]	0.18	—	1.97×10^{-8}	7.98×10^{-3}	3.18×10^{-3}	2.01×10^{-3}	4.245×10^{-5}	Wratten et al. (1969)
Paddy[c]	0.11	—	—	7.32×10^{-3}	3.36×10^{-3}	2.19×10^{-3}	4.750×10^{-5}	Morita and Singh (1979)
Paddy[c]	0.22	—	—	7.32×10^{-3}	3.36×10^{-3}	2.19×10^{-3}	4.750×10^{-5}	Morita and Singh (1979)
Plum	—	—	—	—	—	—	$3.50–4.52 \times 10^{-3}$	Mohsenin (1986)
Peaches	—	$1.16–1.45 \times 10^{-1}$	$1.33–1.59 \times 10^{-2}$	$2.92–3.25 \times 10^{-2}$	$2.79–3.21 \times 10^{-2}$	—	$1.30–1.46 \times 10^{-2}$	Sabliov et al. (2002)
Pear	—	—	—	—	—	—	$1.43–1.49 \times 10^{-2}$	Mohsenin (1986)
Rice[d]	0.12	—	—	7.01×10^{-3}	3.56×10^{-3}	2.39×10^{-3}	1.41×10^{-6}	Ouyang (2001)
Rice[e]	0.15	—	—	7.03×10^{-3}	3.05×10^{-3}	2.11×10^{-3}	1.23×10^{-6}	Ouyang (2001)
Tamarind (seed)	0.113	$0.57–1.18 \times 10^{-3}$	$4.3–6.9 \times 10^{-7}$	$10.8–17.3 \times 10^{-3}$	$8.2–1.26 \times 10^{-3}$	$4.5–7.5 \times 10^{-3}$	$4.41–6.85 \times 10^{-4}$	Bhattacharya et al. (1993)

Note: L, large; M, medium; and S, small.
[a] Kernel.
[b] Saturn (medium) variety.
[c] Caloro (short) variety.
[d] Tainan #9 with hull.
[e] Tainan #9 bare kernel.

Table 15.4 Values of α for Different Foods

Material	α	Reference
Apple	0.056	Frechette and Zahradnik (1965)
Eggs	0.0456–0.0507	Besch et al. (1968)

A ready reckoner table is presented for evaluating the surface area of general ellipsoids developed by numerical techniques (Igathinathane and Chattopadhyay, 1998). The lactose crystal shape was simplified to a tetrahedron with a rectangular base after viewing under a microscope. The dimensions of the base were $a{:}b{:}c = 1{:}0.6{:}1.33$. From this ratio, equations of the volume and surface area are formulated in terms of a (Hodges et al., 1993):

$$A = 2.821\,a^2 \quad \text{and} \quad V = 0.2667\,a^3 \tag{15.9}$$

The surface area can also be predicted from the empirical correlations between the surface area and mass of the food materials. Mohsenin (1986) compiled linear correlation of surface area of apple, pear, and plum. Besch et al. (1968) proposed an empirical equation as

$$A = \alpha m^{2/3} \tag{15.10}$$

where α is a constant specific to a type of food. The values of different foods are shown in Table 15.4. The surface area is also related as

$$A = E + Fm \tag{15.11}$$

The values of E and F for different foods are shown in Table 15.5. Avena-Bustillos et al. (1994) also correlated with the surface area calculated from geometric dimension for noncylindrically shaped zucchini as

$$A = (77.65 + 0.70A_c) \times 10^{-4} \tag{15.12}$$

Table 15.5 Values of *E* and *F* for Different Foods

Material	*E*	*F*	Reference
Apple (Jonathan)	4.16×10^{-3}	8.39×10^{-3}	Baten and Marshall (1943)
Apple (Jonathan)	3.43×10^{-3}	9.41×10^{-2}	Baten and Marshall (1943)
Apple (Delicious)	4.27×10^{-3}	8.32×10^{-2}	Baten and Marshall (1943)
Apple (McIntosh)	4.59×10^{-3}	8.32×10^{-2}	Baten and Marshall (1943)
Apple (Stayman)	5.43×10^{-3}	7.09×10^{-2}	Baten and Marshall (1943)
Apple (Wagner)	4.57×10^{-3}	7.74×10^{-2}	Baten and Marshall (1943)
Apple (Gravenstein)	5.32×10^{-3}	7.48×10^{-2}	Baten and Marshall (1943)
Apple (Chenango)	4.46×10^{-3}	8.26×10^{-2}	Baten and Marshall (1943)
Apple (Grimes golden)	3.80×10^{-3}	8.06×10^{-2}	Baten and Marshall (1943)
Pears (Anjou)	4.77×10^{-3}	6.26×10^{-2}	Baten and Marshall (1943)
Pears (Bosc)	4.83×10^{-3}	6.52×10^{-2}	Baten and Marshall (1943)
Apple (Bartlett)	4.89×10^{-3}	6.52×10^{-2}	Baten and Marshall (1943)
Plums (Pond)	1.61×10^{-3}	8.90×10^{-2}	Baten and Marshall (1943)
Plums (Monarch)	1.21×10^{-3}	1.03×10^{-1}	Baten and Marshall (1943)
Zucchini[a] [*m*: 0.08–0.12 kg]	1.40×10^{-2}	6.20×10^{-5}	Avena-Bustillos et al. (1994)
Zucchini[b] [*m*: 0.23–0.40 kg]	5.67×10^{-3}	8.30×10^{-5}	Avena-Bustillos et al. (1994)

[a] Cylindrical.
[b] Noncylindrical.

where A_c [$=\pi D(h+D/2)$] is the cylindrical shape for zucchini. Equation 15.12 is for noncylindrical (A_c: 290–480 m^2) and Equation 15.13 is for cylindrical (A_c: 130–1900 m^2) shape.

$$A = (37.58 + 0.61A_c) \times 10^{-4} \tag{15.13}$$

The above authors concluded that a correlation of surface area with zucchini mass produced a better fit generally than using the assumption of right cylinder formula for area estimation for noncylindrically shaped zucchini. However, for cylindrically shaped zucchini, the right cylinder assumption fitted better than using initial fruit mass. Marini (1999) developed a model to predict apple diameter as a function of fruit mass as

$$D = 17.2 + 1.36\,m - 7.82\,m^{-1/3} \tag{15.14}$$

15.2.2.2 *Non-Euclidian or Irregular Geometry*

Fractal analysis can be used in characterizing non-Euclidian geometry (Mandelbrot, 1977; Takayasu, 1990; Rahman, 1995). The non-Euclidian geometry object does not have any characteristic dimension.

15.2.2.2.1 *Methods to Estimate Fractal Dimension*

The fractal dimension can be estimated by structured walk (Richardson's plot), bulk density–particle diameter relation, sorption behavior of gases, and pore-size distribution. Richardson's method was used by Graf (1991) for fine soil particles and by Peleg and Normand (1985) for instant coffee particles to estimate fractal dimension. Yano and Nagai (1989) and Nagai and Yano (1990) used a gas adsorption method for native and deformed starch particles, and Ehrburger-Dolle et al. (1994) used a porosity method for activated carbon particles formed by different techniques or treatments. However, none of the above authors used all methods to estimate the fractal dimensions for the same material particles so that a clear physical meaning of fractal dimensions can be drawn. Nagai and Yano (1990) used two gas adsorption methods and found different fractal dimensions for the same starch particles. Thus, fractal dimension interpretation without physical understanding can be misleading or incorrectly applied. Rahman (1997) provided a better understanding of the fractal dimension by reviewing different methods available in the literature. An example is presented for starch and modified starch particles considering experimental data from the literature. His analysis can also minimize confusion and avoid misinterpretation of fractal dimensions and can provide the theoretical limitations of the fractal dimensions.

15.2.2.2.1.1 Richardson's Method—Richardson (1961) proposed a structured walk procedure for characterizing the fractal dimension of a rugged boundary. A series of polygons of side x are constructed on the perimeter using a pair of compasses (Figure 15.2). The perimeter of a fractal shape can be measured many times with many different scales of measurement. The scale of measurement is called the stride length. A larger stride length gives a lower perimeter, written as

$$p = nx + \gamma \tag{15.15}$$

In dimensionless form, the above equation can be expressed as

$$\xi = n\lambda + \frac{\gamma}{L} \tag{15.16}$$

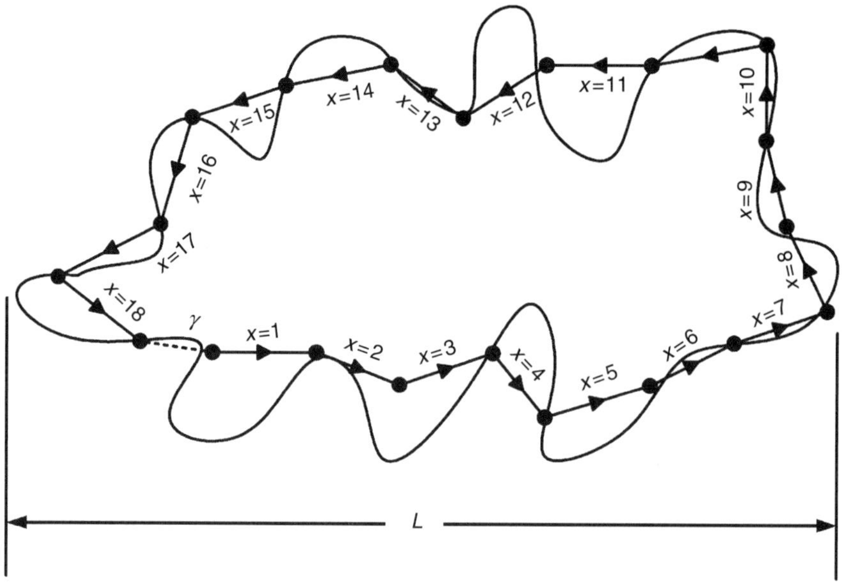

Figure 15.2 Characteristics of Richardson's method for an irregular particle. (From Rahman, M.S., *Food Properties Handbook*, CRC Press, Boca Raton, FL, 1995.)

A plot of ξ against λ on a log–log graph yields a straight line of slope s where $s = 1 - \delta_r$ or $\delta_r = 1 - s$. $\delta_r = 1.0$ indicates the smoothness and higher values of δ_r indicate the ruggedness or roughness of the boundary. The log–log plot of stride length versus perimeter is called a Richardson plot. The fractal dimension δ_r is the morphometric ruggedness of a particle boundary, which can be estimated by Richardson's method. In many cases, particles may have two fractal components with two linear segments having two different slopes (Graf, 1991). The linear segment corresponding to large stride lengths is called the structure and the linear segment corresponding to small stride lengths is called the texture (Kaye, 1989). The value of critical stride length can only be predicted by visualization of the Richardson plot. Photograph from SEM analysis can be used to construct the Richardson plot. The Richardson method provides quantitative ruggedness of the particle boundary. Fractal dimensions estimated by Richardson's method varied from 1.02 to 1.10 for native and modified starch (Rahman, 1997) and for instant coffee particles from 1.06 to 1.19 (Peleg and Normand, 1985). Higher values indicate more ruggedness or roughness of boundary, which was conclusively found from scanning electron microscopy of native and modified starch.

15.2.2.2.1.2 Gas Adsorption Method—Adsorption of a monolayer of gas molecules on the surface can be used to characterize the fractal surface of a porous structure (Figure 15.3). There is a lower cutoff and upper cutoff of molecular size between which the number of molecules vary. The scaling equation for the number of monolayer gas molecules can be expressed as (Pfeifer and Avnir, 1983)

$$N \propto A^{\delta_s/2} \tag{15.17}$$

The monolayer values of different gases can be estimated from adsorption isotherms. The fractal dimension δ_s can be estimated from the slope of the plot of log N versus log A. The fractal dimension δ_s is the ruggedness of pore surface in same size particles estimated from monolayer

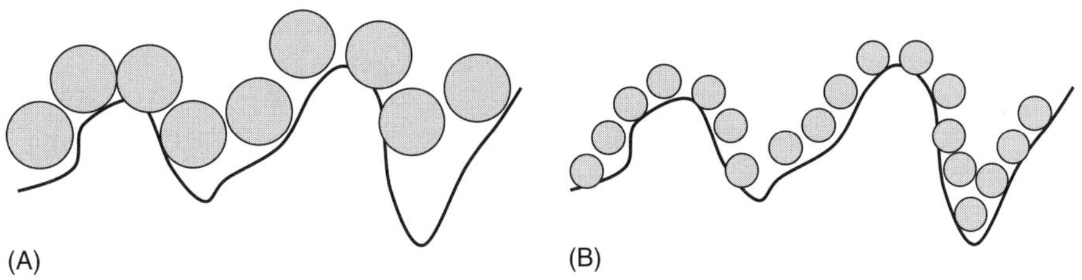

Figure 15.3 (A) Number of gas molecules is lower when molecule size is bigger; and (B) number of gas molecules is higher when molecule size is smaller. (From Rahman, M.S., *Food Properties Handbook*, CRC Press, Boca Raton, FL, 1995.)

values using a different gas within a range of molecular sizes. Pfeifer and Avnir (1983) proposed the scaling equation for a monolayer of specific adsorbate on adsorbent particles of different size as

$$N \propto (r_p)^{\delta_n - 3} \tag{15.18}$$

$$N \propto (r_p)^{\delta_w - 3} \tag{15.19}$$

where n and w indicate nitrogen and water, respectively. The above theoretical equations were developed based on the following assumptions: no adsorbate and adsorbent surface interaction, similar molecules are relaxed on the surface, and uniform adsorption on the surface. Gas adsorption method provides quantitative information of the pore surface roughness of particles. The results depend on the adsorbate–adsorbent surface interaction. Thus, the inert and interacting gases used may give different dimensions. The fractal dimension δ_n and δ_w are the ruggedness of pore surface with the increase in particle size of the same material. δ_w was 3.0 for different types of starch when sorption data were used (Table 15.6). This indicates that water sorption is not a surface phenomenon because water adsorption occurs on the water-binding sites of the starch food particles and depends on glucose residue (Nagai and Yano, 1990). Since water interacts with only polar sites of particles, δ_w is the distribution of polar sites on the surface of pores. The value of δ_n for native and gelatinized starch followed by slow cooling was 2.4, whereas in the case of gelatinized starch followed by fast

Table 15.6 Fractal Dimensions by Different Methods

Sample	δ_r	δ_s	δ_n	δ_w	δ_v	δ'_m	δ''_m
Apple (fresh)[a]	—	—	—	—	2.71	—	—
Apple (20 h dried)[a]	—	—	—	—	1.04	—	—
Apple (30 h dried)[a]	—	—	—	—	0.89	—	—
Native starch	—	2.40	2.20	3.00	—	3.00	3.00
Native starch	1.02	—	2.20	—	3.09	—	3.00
GS20	1.04	2.40	2.70	3.00	3.10	3.00	3.00
GSN	—	2.10	3.00	3.00	—	3.00	3.00
DS20	1.09	2.10	3.00	3.00	2.45	3.00	3.00
Instant coffee particle	1.05–1.20	—	—	—	—	—	—

Sources: Adapted from Rahman, M.S., *J. Food Eng.*, 32, 447, 1997; Peleg, M. and Normand, M.D., *J. Food Sci.*, 50, 829, 1985; Rahman, M.S., Al-Zakwani, I., and Guizani, N., *J. Sci. Food Agric.*, 85, 979, 2005.

Notes: GS20, gelatinized starch frozen at −20°C and freeze-dried.
GSN, gelatinized starch frozen with liquid nitrogen and freeze-dried.
DS20, deformed with ethanol, frozen at −20°C, and freeze-dried.

[a] Convection air dried at 80°C.

freezing or ethanol deformation, the value was 2.1, indicating a flattening of the pore surface. Thus, δ_n indicates more surface phenomenon since nitrogen is an inert gas.

15.2.2.2.1.3 Pore-Size Distribution Method—Pfeifer and Avnir (1983) derived a scaling equation for pore-size distribution as

$$\frac{dV_v}{dr_v} \propto (r_v)^{2-\delta_v} \tag{15.20}$$

The fractal dimension can be estimated from the slope of the plot of log (dV_v/dr_v) versus log r_v. In terms of applied pressure, the scaling equation is (Ehrburger-Dolle et al., 1994)

$$\frac{dV_v}{dP} \propto P^{\delta_v - 4} \tag{15.21}$$

The data for the above relationships could be generated from mercury porosimetry. The fractal dimension, δ_v, is the characteristic size distribution of micropores in same size particles. The pore-size distribution method provides the decreasing rate of micropores in the particle. Two or three linear portions can also be observed in the plot of log (dV_v/dr_v) versus log r_v. Initial and final segments indicate the mechanical properties of the solid and that the fractal dimension can be higher than 3. Ehrburger-Dolle et al. (1994) observed three linear sections in log (dV_v/dP) versus log P or log (dV_v/dr_v) versus log r_v plots of different active carbon particles and mentioned that only the middle linear segment gives the actual fractal dimension. The initial or final slope may lead to a fractal dimension much higher than 3 as observed by Frisen and Mikula (1988), Ehrburger-Dolle et al. (1994), and Rahman (1997), which was unrealistic. The following is the possible explanation proposed by Ehrburger-Dolle et al. (1994): below lower critical pressure (lower cutoff, i.e., initial nonlinear slope), mercury, which is not fractal in nature and above higher critical (upper cutoff) pressure, fills the intergrain voids and the result reflects the mechanical behavior of the sample. Above the higher critical pressure, the fine pores may also be formed by cracking, which is also correspondent on the chemical strength of the particles or indicates the compressibility of the particle. The fractal dimensions were measured as 2.71, 2.96, and 3.11 for fresh and air-dried samples for 20 and 3 h (drying at 80°C), respectively. The dimension was increased with the increasing drying time, indicating formation of micropores on the surface during air-drying (Rahman et al., 2005). Rahman (1997) determined the fractal dimension of native, gelatinized, and ethanol deformed starch and the fractal dimension values were 3.09, 3.10, and 2.45, respectively (Table 15.6). Lower values indicate the removal of micropores within the starch particles by ethanol modification.

15.2.2.2.1.4 Bulk Density–Particle Diameter Method—The mass of a particle increases with the increase in radius as

$$m \propto r_p^{\delta'_m} \tag{15.22}$$

The mass fractal dimension can be estimated from the above scaling equation from the slope of the plot of log m versus log r_p. Ohoud et al. (1988) derived the scaling equation for bulk density and particle size for powder as

$$\rho_B \propto r_p^{\delta'_m - \delta''_m} \tag{15.23}$$

where δ''_m is a function of the type of arrangement of the particles with respect to each other. The dimension δ'_m indicates the increase in total void or airspace by the increase in particle size and δ''_m is the packing characteristic due to the difference in particles geometry, which can be estimated from bulk density–particle diameter method. Ohoud et al. (1988) mentioned that the physical meaning of δ''_m is fundamentally different from δ'_m. The value of δ'_m is a characteristic of each individual particle, whereas δ''_m characterizes the structure of the powder at length scales above the particle size. δ''_m is 3 both for an ordered and a disordered homogeneous arrangement of spheroids, but δ''_m is 2 for ordered stacks of platelets of constant thickness and δ''_m is 1 for aligned rods of constant diameter. When bulk density of a powder is independent of r_p, then $\delta'_m = \delta''_m$. Thus, fractal dimensions depend on the method and scaling equation used. Rahman (1997) estimated the fractal dimensions by different methods for starch particles (Table 15.6). The dimensions varied with the method used for the same particle. Thus, the physical meaning of each dimension is obviously different as discussed above.

15.2.2.2.2 *Fractal Dimension Relationships*

Fractal dimension can also be estimated from the relationships of volume, surface area, characteristics dimension, and porosity.

15.2.2.2.2.1 Fractal Dimension from Porosity—Fuentes et al. (1986) proposed a theoretical equation to estimate fractal dimension from porosity as

$$\delta = 2 + 3\left[\frac{\varepsilon_V^{4/3} + (1-\varepsilon_V)^{2/3} - 1}{\varepsilon_V^{4/3}\ln\varepsilon_V^{-1} + (1-\varepsilon_V)^{2/3}\ln(1-\varepsilon_V)^{-1}}\right] \tag{15.24}$$

The above equation was developed from water conductivity through porous media.

15.2.2.2.2.2 Using Fractal Measure Relations—In general, for any nonfractal (Euclid) object, the following relation holds between its length (L), area (A), and volume (V):

$$L \propto A^{1/2} \propto V^{1/3} \tag{15.25}$$

The relation for the fractal shaped objects can be written as

$$L \propto A^{\delta/2} \propto V^{\delta/3} \tag{15.26}$$

or

$$L \propto A^{\delta/d} \propto V^{\delta/d} \tag{15.27}$$

where
- δ is the fractal dimension
- d is the Euclidian dimension

δ is the perimeter fractal dimension, $\delta + 1$ is the area fractal dimension, and $\delta + 2$ is the volume fractal dimension. In the above equation, it is assumed that length, surface, and volume fractal dimensions are equal. However, in many instances this may not be valid. The shrinkage encountered during drying of meat and potatoes is neither ideally three-dimensional nor unidimensional. Diffusion coefficient is expressed as (Gekas and Lamberg, 1991)

$$\frac{\breve{D}_e}{\breve{D}_{e,r}} = \left(\frac{V}{V_r}\right)^{2/\delta} = (S_a)^{2/\delta} \tag{15.28}$$

The fractal dimension can be estimated from the volume and thickness shrinkage coefficient as (Sjoholm and Gekas, 1995)

$$S_a = (S_c)^\delta = \left(\frac{c}{c_o}\right)^\delta \qquad (15.29)$$

The relation of S_a with the moisture content during drying and change in dimension can be used in Equation 15.29 to estimate the values of δ.

NOMENCLATURE

A	surface area (m^2)
a	largest characteristic length (m)
b	medium characteristic length (m)
C	BET parameter
c	smallest characteristic length (m)
d	Euclidian dimension or differentiation
D	diameter (m)
$\breve{D}$	diffusivity (m^2/s)
E	parameter in Equation 15.11
e	ecentricity
F	parameter in Equation 15.11
h	height (m)
k	shape factor for convex body (Equation 15.5)
L	length (m)
m	mass (kg)
N	number of monolayers
n	number of strides
P	pressure (Pa)
p	perimeter (m)
r	radius (m)
S	shrinkage
s	slope of Richardson's plot
V	volume (m^3)
X	moisture content (wet basis, kg/kg sample)
x	stride length (m)

Greek Symbols

α	parameter of Equation 15.10
γ	length of last stride (m)
λ	dimensionless stride length (Equation 15.16)
δ	fractal dimension
ξ	dimensionless perimeter (Equation 15.16)
ρ	density (kg/m^3)
ε	volume fraction (dimensionless)
π	3.1416

Subscripts

a	apparent
B	bulk
c	cylindrical shape or thickness or smallest
e	effective
e,r	effective reference
m	mass related
n	nitrogen
o	open pore or initial
p	particle or mean
r	Richardson's or reference
s	specific or surface
v	void or pore
w	water
1, 2	state or condition 1 or 2

Superscripts

$'$	state 1
$''$	state 2

REFERENCES

Avena-Bustillos, R.D.J., Krochta, J.M., Saltveit, M.E., Rojas-Villegas, R.D.J., and Sauceda-Perez, J.A. 1994. Optimization of edible coating formulations on zucchini to reduce water loss. *Journal of Food Engineering*. 21(2): 197–214.

Bargale, P.C. and Irudayaraj, J. 1995. Mechanical strength and rheological behavior of barley kernels. *International Journal of Food Science and Technology*. 30: 609–623.

Baten, W.D. and Marshall, R.E. 1943. Some methods for approximate prediction of surface area of fruits. *Journal of Agricultural Research*. 66(10): 357–373.

Bennesen, T. and Frenchel, W. 1948. *Theorie der konvexen Koerper* (Theories of convex bodies) (cited by Bargale and Irudayaraj, 1995).

Berlin, E., Kliman, P.G., and Pallansch, M.J. 1966. Surface areas and densities of freeze-dried foods. *Journal of Agricultural Food Chemistry*. 14(1): 15–17.

Besch, E.L., Sluka, S.J., and Smith, A.H. 1968. Determination of surface area using profile recordings. *Poultry Science*. 47(1): 82–85.

Bhattacharya, S., Bal, S., Mukherjee, R.K., and Bhattacharya, S. 1993. Some physical and engineering properties of tamarind (*Tamarindus indica*) seed. *Journal of Food Engineering*. 18: 77–89.

Ehrburger-Dolle, F., Lavanchy, A., and Stoeckle, F. 1994. Determination of the surface fractal dimension of active carbons by mercury porosimetry. *Journal of Colloid and Interface Science*. 166: 451–461.

Frechette, R.J. and Zahradnik, J.W. 1965. Surface area-weight relationships for McIntosh apples. *Transactions of the ASAE*. 9(4): 526.

Frisen, W.I. and Mikula, R.J. 1988. Mercury porosimetry of coals. Pore volume distribution and compressibility. *Fuel*. 67: 1516–1520.

Fuentes, C., Vauclin, M., Parlange, J., and Haverkamp, R. 1986. A note on the soil-water conductivity of a fractal soil. *Transport in Porous Media*. 23: 31–36.

Gekas, V. and Lamberg, I. 1991. Determination of diffusion coefficients in volume changing systems. Application in the case of potato. *Journal of Food Engineering*. 14: 317–326.

Graf, J.C. 1991. The importance of resolution limits to the interpretation of fractal descriptions of fine particles. *Powder Technology*. 67: 83–85.

Hodges, G.E., Lowe, E.K., and Paterson, A.H.J. 1993. A mathematical model for lactose dissolution. *The Chemical Engineering Journal*. 53: B25–B33.

Igathinathane, C. and Chattopadhyay, P.K. 1998. On the development of a ready reckoner table for evaluating surface area of general ellipsoids based on numerical techniques. *Journal of Food Engineering*. 36: 233–247.

Igathinathane, C. and Chattopadhyay, P.K. 2000. Surface area of general ellipsoid shaped food materials by simplified regression equation method. *Journal of Food Engineering*. 46: 257–266.

Kaye, B.H. 1989. *A Random Walk through Fractal Dimensions*, VCH Verlagsgellschaft mbH, Germany.

Lowell, S. and Shields, J.E. 1984. *Powder Surface Area and Porosity*. 2nd edn., Chapman and Hall, London.

Mandelbrot, B.B. 1977. *Fractals Form, Chance, and Dimension*. W.H. Freeman and Company, San Francisco, CA.

Marini, R.P. 1999. Estimating apple diameter from fruit mass measurements to time thinning sprays. *HortTechnology*. 9(1): 109–113.

Maw, B.W., Hung, Y.C., Tollner, E.W., Smittle, D.A., and Mullinix, B.G. 1996. Physical and mechanical properties of fresh and stored sweet onions. *Transactions of the ASAE*. 39(2): 633–637.

Mohsenin, N.N. 1986. *Physical Properties of Plant and Animal Materials*. Gordon and Breach Science Publishers, New York.

Morita, T. and Singh, R.P. 1979. Physical and thermal properties of short-grain rough rice. *Transactions of the ASAE*. 22: 630–636.

Nagai, T. and Yano, Y. 1990. Fractal structure of deformed potato starch and its sorption characteristics. *Journal of Food Science*. 55: 1334–1337.

Ohoud, M.B., Obrecht, F., Gatineau, L., Levitz, P., and Damme, H.V. 1988. Surface area, mass fractal dimension and apparent density of powders. *Journal of the Colloid and Interface Science*. 124: 156–161.

Olajide, J.O. and Igbeka, J.C. 2003. Some physical properties of groundnut kernels. *Journal of Food Engineering*. 58(2): 201–204.

Ouyang, Y.S. 2001. Mesomechanical characterization of in situ rice grain hulls. *Transactions of the ASAE*. 44(2): 357–367.

Peleg, M. and Normand, M.D. 1985. Characterization of the ruggedness of instant coffee particle shape by natural fractals. *Journal of Food Science*. 50: 829–831.

Pfeifer, P. and Avnir, D. 1983. Chemistry in noninteger dimensions between two and three. I. Fractal theory of heterogeneous surfaces. *Journal of Chemical Physics*. 79: 3558–3565.

Polya, G. and Szezo, G. 1951. *Isoperimetric Inequalities in Mathematical Physics* (cited by Bargole and Irudayaraj, 1995).

Rahman, M.S. 1995. *Food Properties Handbook*. CRC Press, Boca Raton, FL.

Rahman, M.S. 1997. Physical meaning and interpretation of fractal dimensions of fine particles measured by different methods. *Journal of Food Engineering*. 32: 447–456.

Rahman, M.S., Al-Zakwani, I., and Guizani, N. 2005. Pore formation in apple during air drying as a function of temperature and pore size distribution. *Journal of the Science of Food and Agriculture*. 85(6): 979–989.

Richardson, L.F. 1961. The problem of contiguity: An appendix of statistics of deadly quarrels. In: *General Systems Yearbook 6*. pp. 139–187.

Sabliov, C.M., Boldor, D., Keener, K.M., and Farkas, B.E. 2002. Image processing method to determine surface area and volume of axi-symmetric agricultural products. *International Journal of Food Properties*. 5(3): 641–653.

Sjoholm, I. and Gekas, V. 1995. Apple shrinkage upon drying. *Journal of Food Engineering*. 25: 123–130.

Takayasu, H. 1990. *Fractals in the Physical Sciences*. Manchester University Press, Manchester.

Wratten, F.T., Poole, W.D., Cheness, J.L., Bal, S., and Ramarao, V.V. 1969. Physical and thermal properties of rough rice. *Transactions of the ASAE*. 12: 801–803.

Yano, T. and Nagai, T. 1989. Fractal surface of starchy materials transformed with hydrophilic alcohols. *Journal of Food Engineering*. 10: 123–133.

CHAPTER 16

Specific Heat and Enthalpy of Foods

R. Paul Singh, Ferruh Erdoğdu, and Mohammad Shafiur Rahman

CONTENTS

16.1 INTRODUCTION

Thermophysical properties of foods are important for modeling and optimization of processes involving heating and cooling. The properties used in a mathematical model of heat transfer are usually thermal conductivity (k), specific heat (c_p), and density (ρ). An improved knowledge of these thermophysical properties of foods is essential in accurately predicting temperature changes, process duration, and energy consumption during processing (Fikiin and Fikiin, 1999). Among these properties, specific heat and density are significant in analyzing mass/energy balances; thermal conductivity is the key property in determining the rate of thermal energy transfer within a material by conduction; and the combination of these three properties, thermal diffusivity ($k/\rho c_p$), is a key property in the analysis of transient heat transfer (Heldman, 2002).

When modeling freezing and thawing processes where a phase change, from liquid to solid in freezing and solid to liquid in thawing, is included in the process, enthalpy as a thermal property is preferred since the specific heat becomes infinitely high at the freezing or thawing temperature. The overall goal of this chapter is to describe selected methods to determine the specific heat and enthalpy of foods.

16.2 SPECIFIC HEAT

Specific heat is defined as the amount of heat (J) needed to increase the temperature of unit mass (kg) of a material by unit degree (K). The unit of specific heat, therefore, becomes J/kg-K. Specific heat depends on the nature of the process of heat addition, e.g., a constant pressure or a constant volume process. The pressure dependence of specific heat can be assumed to be negligible unless extremely high pressures are applied. Since most of the food processing operations are either at or in a close range of atmospheric pressure, the specific heat of food products is usually presented at constant pressures. Specific heat is usually denoted by c_p unless it represents a constant volume process where it is referred by c_v. In addition to the dependence of specific heat on pressure and volume, it should be noted that the water content of foods greatly influences their specific heats since water has a much higher specific heat than the other major constituents (Nesvadba, 2005). Effect of temperature, on the other hand, is negligible in the unfrozen temperature range while a dramatic effect is observed in the frozen temperature range. Figure 16.1 shows the change of specific heat of a food with temperature in the frozen and unfrozen ranges, and Tables 16.1 and 16.2 show the specific heat of different foods.

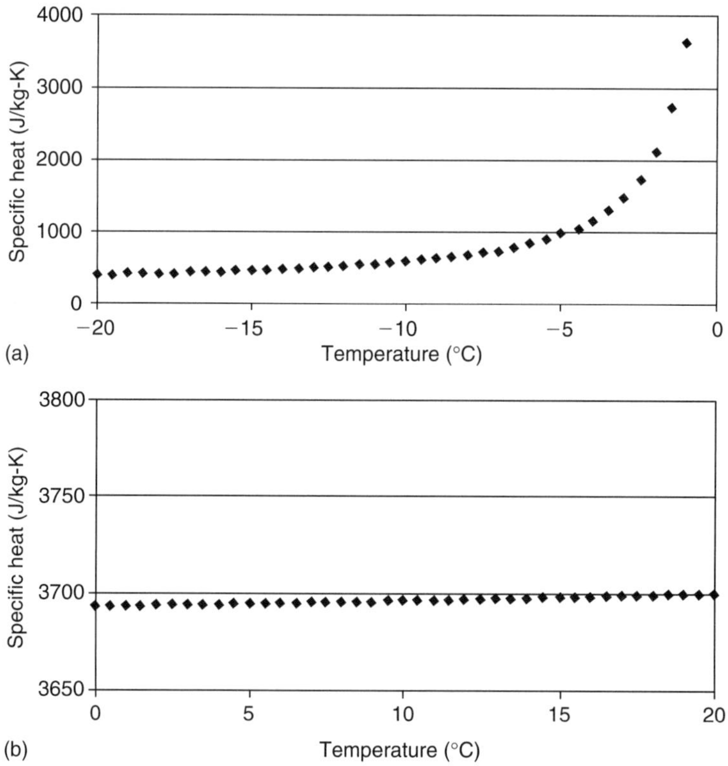

FIGURE 16.1 Typical change of specific heat of food products with temperature: (a) frozen temperature range and (b) unfrozen temperature range.

Table 16.1 Specific Heat of Food Products above Their Initial Freezing Point

Food Product	Water Content (%)	Temperature (°C)	Specific Heat (J/kg-K)	Reference
Fruits and vegetables				
Apple	75–85	—	3724–4017	Ordinanz (1946)
Apple "Golden Delicious"	87.3	20–50	3690	Ramaswamy and Tung (1981)
		−30 to −80	1950	
Apple "Granny Smith"	85.8	20–50	3580	Ramaswamy and Tung (1981)
		−30 to −80	1680	
Apple seed (Ackee—*Blighia sapida*)	9.88	80	2830	Omobuwajo et al. (2000)
Artichoke	90	—	3891	Ordinanz (1946)
Beet (dried)	4.4	27–66	2008	Stitt and Kennedy (1945)
Berry (fresh)	84–90	—	3724–4100	Ordinanz (1946)
Berry (dried)	30	—	2134	
Cabbage (dried)	5.4	27–66	2176	Stitt and Kennedy (1945)
Carrot	0.14–14.5	27–66	1439–2272	Stitt and Kennedy (1945)
Cucumber	95.23	20–95	4030	Fasina and Fleming (2001)
Fruit (fresh)	75–92	—	3347–3766	Ordinanz (1946)
Fruit (dried)	30	—	2092	
Grape "March" (fresh)[c]	—	—	3703	
Lemon "Eureka" (fresh)	—	—	3732	
Orange "Valencia" (fresh)	—	—	3515	
Orange "Navel" (fresh)	—	—	3661	
Potato	76.3–78.8	40–50	2735–3335	Rice at al. (1988)
Plum (fresh)	75–78	—	3514	Ordinanz (1946)
Plum (dried)	28–35	—	2218–2469	
Chicken and meat				
Bacon	50	—	2008	Ordinanz (1946)
Beef (lean)	72	—	3431	
Beef (mince)	—	—	3515	
Beef (boiled)	57	—	3054	Ordinanz (1946)
Bone	—	—	1674–2510	
Chicken meat (white)	—	—	3530	Siripon et al. (2007)
Chicken meat (dark)	—	—	3663	Siripon et al. (2007)
Fat (pork)	39	—	2594	Ordinanz (1946)
Goose	52	—	2929	
Pork (lean)	57	—	3054	
Sausage (fresh)	72	—	3431	
Veal	63	—	3222	
Veal (cutlet)	72	—	3431	
Veal (cutlet, fried)	58	—	3096	
Venison	70	—	3389	

(continued)

Table 16.1 (continued) Specific Heat of Food Products above Their Initial Freezing Point

Food Product	Water Content (%)	Temperature (°C)	Specific Heat (J/kg-K)	Reference
Seafood				
Calamari (tentacle)	83.09–83.91	17	3430–3690	Rahman (1993)
Calamari (mantle)	79.98–81.59	17	3350–3470	
Calamari (wings)	83.38–84.18	17	3780–3790	
Cuttle (skin)	86.99	17	3790	
Cuttle (mantle)	80.92	17	3590	
Dried fish	60	—	3012	Ordinanz (1946)
Dried–salted fish	16–20	—	1751–1841	
Fish (fresh)	80	—	3598	
King prawn (green)	75.63	17	3450	Rahman (1993)
King prawn (tiger)	76.49	17	3410	
Octopus (tentacle)	80.35	17	3290	
Shrimp (*Penaeus* spp.)	80.81	−30 to 30	2100–3630	Karunakar et al. (1998)
Squid (tentacle)	79.61–83.02	17	3500–3580	Rahman (1993)
Tuna, raw skipjack—loin meat	70.8	10–95	3192–3699	Zhang et al. (2001)
Tuna, raw skipjack—red meat	70.8	10–95	3180–3607	
Tuna, raw skipjack—viscera	70.8	10–95	3162–3624	Zhang et al. (2001)
Tuna, raw skipjack—backbone	70.8	10–95	1838–2596	
Beverages				
Cocoa	—	—	1841	Ordinanz (1946)
Cow milk	87.5	—	3849	
Cow milk (skim)	91	—	3975–4017	
Fat				
Butter	14–15.5	—	2050–2134	Ordinanz (1946)
Margarine	9–15	—	1757–2092	
Vegetable oil	—	—	1464–1883	
Soup				
Broth	—	—	3096	Ordinanz (1946)
Cabbage	—	—	3766	
Pea	—	—	4100	
Potato	88	—	3933	
Miscellaneous				
Batter (cake)	—	20	2516.8	Baik et al. (1999)
Bread (white)	44–45	—	2720–2845	Ordinanz (1946)
Bread (brown)	48.5	—	2845	
Cassava	10–68	36–51	1636–3260	Njie et al. (1998)
Castor oil	—	30.2–270.7	1937–2690	Clark et al. (1946)
Chapati (cooked)	41.16	52–62.8	2052–2031	Gupta (1990)
Chapati (puffed)	36.58	56.4–62.1	1905–1889	
Cheddar cheese	36.69–54.92	—	2340–2970	Marschoun et al. (2001)
Cheese	79.5	—	3720	Ordinanz (1946)
Coconut, shredded	1.1–51.7	25.5–47.9	2593–3234	Jindal and Murakami (1984)
Collagen	0.0–17.0	25	1558–1791	Kanagy (1971)

Table 16.1 (continued) Specific Heat of Food Products above Their Initial Freezing Point

Food Product	Water Content (%)	Temperature (°C)	Specific Heat (J/kg-K)	Reference
Cookie dough (AACC)	—	29.7–37.8	2940	Kulacki and Kennedy (1978)
Cookie dough (hard sweet)	—	29.9–38.9	2804	Ordinanz (1946)
Corn (yellow, dent)	0.9–30.2	—	1531–2460	Kazarian and Hall (1965)
Cottonseed oil	—	79.6–270.3	2176–2690	Clark et al. (1946)
Cumin seed	1.77–17.01	−70 to 50	1330–3090	Singh and Goswami (2000)
Curd (cottage cheese)	60	—	3264	Ordinanz (1946)
Dough	—	—	1883–2176	
Egg white	87	—	3849	
Egg yolk	48	—	2803	
Flour	12–13.5	—	1799–1883	
Grain	15–20	—	1883–2008	
Lentil	2.1–25.8	10–80	807–2186	Tang et al. (1991)
Linseed oil	—	29.9–271.2	2071–2748	Clark et al. (1946)
Macaroni	12.5–13.5	—	1841–1883	Ordinanz (1946)
Millet grains	10–30	—	1330–2400	Subramanian and Viswanathan (2003)
Mushroom	10.24–89.68	40–70	1715.8–3949.8	Shrivastava and Datta (1999)
	90	—	3933	Ordinanz (1946)
Mushroom (dried)	30	—	2343	
Onion slices	5.4–69.2	20	1940–3450	Rapusas and Driscoll (1995)
Pearl barley	—	—	2803–2845	Ordinanz (1946)
Perilla Oil	—	4.6–270.4	1732–2406	Clark et al. (1946)
Pistachio	5–25	—	2930–4190	Razavi and Taghizadeh (2007)
Plantain	10–57	36–51	1658–2969	Njie et al. (1998)
Porridge (buckwheat)	—	—	3222–3766	Ordinanz (1946)
Potato puree	79.5	—	3720	Nesvadba (2005)
Potato puree-sweet	77.3	5–80	3720–3770	Fasina et al. (2003)
Raisins	24.5	—	1966	Ordinanz (1946)
Raisins 'Sultana'	14–80	—	1289–1511	Vagenas et al. (1990)
Rice	10.5–13.5	—	1757–1841	Ordinanz (1946)
Salt	—	—	1130–1339	Ordinanz (1946)
Soybean	0.0–37.9	—	1576–2342	Alam and Shove (1973)
Soybean oil	—	1.2–271.3	1874–2787	Clark et al. (1946)
Soy flour, defatted	8.7–38.7	130	2128–3113	Wallapan et al. (1984)
Sugar	13.30	54.7–59.1	1298–1256	Gupta (1990)
Surimi (pacific whiting)	74–84	25–90	3530–4182	AbuDagga and Kolbe (1997)
Tofu	67.7	10	3450–3583	Baik and Mittal (2003)
		105	3556–3689	
	33.3	10	2530–2649	
		105	2663–2713	
Tortilla chips	1.3–36.1	150–190	2190–3360	Moreiara et al. (1995)
Tung oil	—	21.5–200.3	1820–2368	Clark et al. (1946)
Wheat, soft	0.7–20.3	—	1452–2184	Kazarian and Hall (1965)
Wheat flour	11.97	25.2–78.3	1720	Hwang and Hayakawa (1979)
	10.53	57–59.1	1846–1805	Gupta (1990)
Wheat dough, whole	45.37	52–62.8	2052–2031	
Yam	13–60	36–51	1766–3119	Njie et al. (1998)

Table 16.2 Specific Heat of Different Foods as a Function of Moisture Content and Temperature

Food Product	Water Content (%)	Temperature (°C)	Equation	Reference
Alfalfa pellet	—	2–110	$c_p = 941 + 10.72T$	Fasina and Sokhansanj (1996)
Apple "Golden Delicious"	87.3	−1 to 60	$c_p = 3360 + 7.5T$	Ramaswamy and Tung (1981)
Apple "Granny Smith"	84.8		$c_p = 3400 + 4.9T$	
Baked products—bread solids	—	25–85	$c_p = 980 + 4.9T$	Christenson et al. (1989)
Baked products—muffin solids			$c_p = 400 + 3.9T$	
Baked products—		30–58	$c_p = 1170 + 3.0T$	
biscuit solids		58–85	$c_p = 800 + 3.0T$	
Bean	<100	20	$c_p = 1430 + 27.57M_w$	Niesteruk (1996)
Beet root			$c_p = 1401 + 27.86M_w$	
Cabbage			$c_p = 1402 + 27.85M_w$	
Carrot			$c_p = 1373 + 28.14M_w$	
Celery			$c_p = 1401 + 27.86M_w$	
Coffee powder Mexican	—	45–140	$c_p = 921 + 7.554T$	Singh et al. (1997)
Coffee powder Colombian	—	45–110	$c_p = 1272.9 + 6.459T$	
Cucumber	<100	20	$c_p = 1385 + 28.02M_w$	Niesteruk (1996)
Lettuce			$c_p = 1342 + 28.45M_w$	
Oat	<13	—	$c_p = 1154 + 38.9M_w$	Otten and Samaan (1980)
Onion	<100	20	$c_p = 1396 + 27.91M_w$	Niesteruk (1996)
Parsley			$c_p = 1373 + 28.14\ M_w$	
Pea			$c_p = 1430 + 27.57M_w$	
Potato			$c_p = 1381 + 28.06M_w$	
Pumpkin			$c_p = 1385 + 28.02M_w$	
Rye	<13	—	$c_p = 1242 + 52M_w$	Otten and Samaan (1980)
Spinach			$c_p = 1342 + 28.45M_w$	Niesteruk (1996)
Turnip	<100	20	$c_p = 1388 + 27.99M_w$	Niesteruk (1996)
Watermelon			$c_p = 1385 + 28.02M_w$	
Wheat var. Ontario	<15.9	30–95	$c_p = 1272 + 53.1M_w$	Otten and Samaan (1980)
Wheat var. Western	<13.8		$c_p = 1377 + 44.3M_w$	
Wheat var. Frederick	<13.1		$c_p = 1311 + 48.2M_w$	
Wheat (average)	<15.9		$c_p = 1317 + 49.1M_w$	
Raisins "Sultana"	18–80	—	$c_p = 1400 + 27.82M_w$	Vagenas et al. (1990)

Note: M_w is the percent moisture content, and T is the temperature (°C).

Specific heat of foods has been investigated in detail in the literature with much emphasis on the measurement techniques. The measurement methods of specific heat are classified into the following groups: empirical equations, method of mixtures, comparison calorimeter method, adiabatic method, and differential scanning calorimetry (DSC).

16.2.1 Empirical Equations

Empirical equations to predict specific heat are based on the composition of the food materials. The earliest equation to predict the specific heat of foods is Siebel's equation (Siebel, 1892), where the data reported for specific heat of food materials are based on the following equations:

$$c_p = 837 + 3348X_{\mathrm{w}} \quad (16.1)$$

$$c_p = 837 + 1256X_{\mathrm{w}} \quad (16.2)$$

Equations 16.1 and 16.2 are reported for values above and below freezing, respectively, where X_{w} is the moisture content of the material in wet basis (fraction) and c_p is the specific heat in J/kg-K. Siebel's equation was only a function of the moisture content. Different equations were suggested by Leniger and Beverloo (1975) and by Charm (1978) incorporating the fat and nonfat solid contents of the food material. Heldman and Singh (1981) suggested the following equation to determine the specific heat of food materials using the carbohydrate, protein, fat, and ash content of a food at 20°C (Heldman, 2002):

$$c_p = 1424M_{\mathrm{c}} + 1549M_{\mathrm{p}} + 1675M_{\mathrm{f}} + 837M_{\mathrm{a}} + 4187M_{\mathrm{w}} \quad (16.3)$$

where

M_{c}, M_{p}, M_{f}, M_{a}, and M_{w} are the mass fractions of carbohydrate, protein, fat, ash, and moisture contents of the food product, respectively

c_p is the specific heat in J/kg-K

The fat content part in this equation was considered to be in the solid state. Gupta (1990) suggested the following equation to determine the specific heat of food products as a function of temperature and water content in a temperature range of 303–336 K and in a moisture content range of 0.1%–80%:

$$c_p = 2477 + 2356X_{\mathrm{w}} - 3.79T \quad (16.4)$$

where

T is the temperature (K)

X_{w} is the mass fraction of the moisture content of the product

A more comprehensive model including the effect of temperature, in addition to the composition of the food materials, was published by Choi and Okos (1986). Their model is as follows:

$$c_p = \sum_{i=1}^{n} c_{p_i} X_i \quad (16.5)$$

where

n is the number of components of a given food product

c_{p_i} is the specific heat of the ith component

X_i is the fraction of the ith component

c_p is the specific heat in J/kg-K

Table 16.3 Specific Heat of Different Food Components as a Function of Temperature

Component	Temperature Function	Standard Error	Standard % Error
Protein	$c_p = 2008.2 + 1.2089T - (1.3129 \times 10^{-3})T^2$	0.1147	5.57
Fat	$c_p = 1984.2 + 1.4733T - (1.3129 \times 10^{-3})T^2$	0.0236	1.16
Carbohydrate	$c_p = 1548.8 + 1.9625T - (5.9399 \times 10^{-3})T^2$	0.0986	5.96
Fiber	$c_p = 1845.9 + 1.8306T - (4.6509 \times 10^{-3})T^2$	0.0293	1.66
Ash	$c_p = 1092.6 + 1.8896T - (3.6817 \times 10^{-3})T^2$	0.0296	2.47
Water[a]	$c_p = 4081.7 - 5.3062T + (9.9516 \times 10^{-1})T^2$	0.0988	2.15
Water[b]	$c_p = 4176.2 - (9.0864 \times 10^{-2})T + (5.4731 \times 10^{-3})T^2$	0.0159	0.38
Ice	$c_p = 2062.3 + 6.0769T$		

Source: Adapted from Choi, Y. and Okos, M.R., in *Food Processing and Process Applications Vol. I Transport Phenomenon*, Elsevier, New York, 1986.

[a] For a temperature range of −40°C to 0°C.

[b] For a temperature range of 0°C to 150°C.

Table 16.3 gives the specific heat of the food components (protein, fat, carbohydrate, fiber, ash, ice, and water—for the temperature range of −40°C to 0°C and 0°C to 150°C) as a function of temperature. There is deviation between the experimental values and the above model due to (1) specific heat of the component phases varied with the source or origin, (2) bound water or unfrozen water has a different specific heat than bulk water, and (3) excess specific heat due to the interaction of the component phases. Rahman (1993) included the excess specific heat term in Equation 16.5 to compensate for specific heat of bound or unfrozen water and the interaction of the components. However, it is difficult to determine the excess term by simple methods.

The equations developed by Choi and Okos (1986) indicate that the specific heat increases with increasing temperature. However, the effect of temperature above and below the freezing point is quite different (Figure 16.1). Actually, for the temperature range above 0°C, the effect of temperature is quite negligible while a dramatic effect can be seen below 0°C (Figure 16.1).

Schwartzberg (1976) and Chen (1985) presented approaches to predict specific heat of frozen foods. Schwartzberg's equation to determine the specific heat of frozen foods is a function of solids, water content, and initial freezing point:

$$c_p = c_{p_u} + (bX_s - X_{w0})\Delta c_p + EX_s\left[\frac{RT_0^2}{18(T_0 - T)^2} - 0.8\Delta c_p\right] \quad (16.6)$$

where

T is the temperature of the food product (°C)
T_0 is the freezing point of water (0°C)
R is the universal gas constant
Δc_p is the difference between specific heats of water and ice (J/kg-K)
c_{p_u} is the specific heat of food above its initial freezing point
b is the amount of unfrozen water per unit weight of solids
X_s is the mass fraction of solids per unit weight of food
X_{w0} is the mass fraction of water per unit weight of food above the initial freezing point

E is the ratio of molecular weight of water (M_w) and food solids (M_s), $E = (M_w/M_s) = (18/M_s)$. Numerical value of 26,721 for $RT_0^2/18$ when temperature is in °R and an approximate value of 0.5 for Δc_p were used by Schwartzberg (1976) when the units of the specific heat were in Btu/lb-°F. In developing this equation, the high moisture food items were assumed to be ideal dilute solutions.

Chen (1985) developed the following equation for determining the specific heat of frozen foods by expanding Siebel's equation:

$$c_p = 1550 + 1260X_s + X_s\left(\frac{RT_0^2}{M_sT^2}\right) \tag{16.7}$$

where M_s is the effective molecular weight of solids. Chen (1985) provides a better description of a procedure to determine a value for the effective molecular weight.

16.2.2 Method of Mixtures

The method of mixtures is the most widely used method for measuring the specific heat of food products due to its simplicity and accuracy. In this method, the specimen of a known mass and temperature is placed in a calorimeter of a known specific heat containing water or a liquid with a known temperature and mass (Mohsenin, 1980).

The unknown specific heat of the specimen is then computed using a heat balance between the heat gained or lost by water or liquid and calorimeter and the heat lost or gained by the specimen. The heat balance equation in terms of the water equivalent of the described system is

$$c_p = \frac{c_{p_w}(m_w + W)(T_{oc} - T_{em})}{m_s(T_{os} - T_{em})} \tag{16.8}$$

where

c_p is the specific heat of the sample in the temperature range of T_{os} to T_{em} (J/kg-K)
c_{p_w} is the average specific heat of water in the temperature of T_{oc} to T_{em}
m_w is the mass of water (kg)
m_s is the mass of the sample (kg)
W is the water equivalent of the system (kg)
T_{oc} is the initial temperature of the calorimeter (°C)
T_{em} is the final equilibrium temperature of the mixture (°C)
T_{os} is the initial temperature of the sample (°C)

The water equivalent of the system is usually obtained by calibration tests. A schematic diagram of a vacuum-jacketed calorimeter is shown in Figure 16.2. This method, as indicated earlier, has been widely used by researchers due to its simplicity. An earlier work, as reported by Subramanian and Viswanathan (2003), was by Kazarian and Hall (1965) where the specific heat of soft white wheat was measured as a function of its moisture content. Hwang and Hayakawa (1979) applied this method, using a calorimeter modified from a household vacuum jar, to determine the specific heat of cookies, wheat flour, and a fresh produce. Sreenarayanan and Chattopadhay (1986) determined the specific heat of rice bran, a powdery material, by holding the samples in a capsule to prevent the materials from floating and dissolving. Recently, Subramanian and Viswanathan (2003) determined the specific heat of minor millet grains and flours while Razavi and Taghizadeh (2007) determined the specific heat of pistachio nuts as a function of moisture content, temperature, and variety. Deshpande and Bal (1999) applied this method to determine the specific heat of soybean as a function of moisture content; Madampa et al. (1995) determined the change in specific heat of garlic with its moisture content; Rapusas and Driscoll (1995) used the method of mixtures for measuring the specific heat of onion; Rahman (1993) applied this method for specific heat measurement of selected seafood including calamari, cuttlefish, prawn, and squid; Hough et al. (1986) used the method of mixtures for an Argentinian milk product and validated the results by differential scanning calorimetry; Putranon et al. (1980) applied this method for specific heat measurement of Australian paddy rice. McProud and Lund (1983) and Niekamp et al. (1984) used a modified technique of the method of mixtures. In the given procedure, the specific heat of

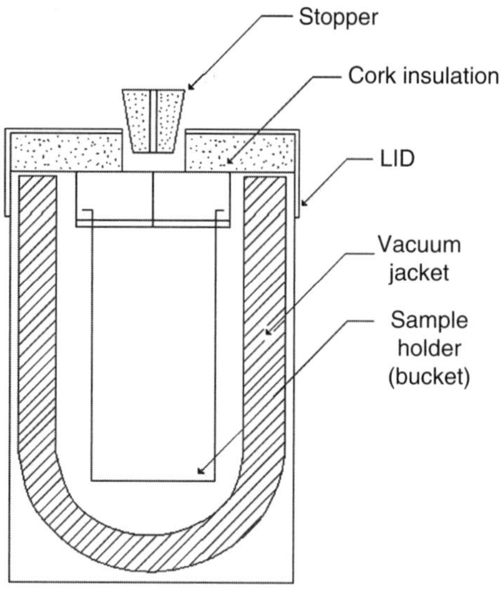

FIGURE 16.2 Vacuum-jacketed calorimeter. (From Mohsenin, N.N., *Thermal Properties of Foods and Other Agricultural Materials*, Gordon and Breach Science Publishing Group, Switzerland, 1980.)

water was determined as a correction factor for the effect of the thermos heat capacity on the specific heat of the samples where the heat losses between the system and its surroundings were neglected.

A disadvantage of the method of mixtures is that only an average specific heat value is obtained from the corresponding range. If a complete specific heat change with respect to temperature is desired, additional experiments must be run at different temperatures. It should also be borne in mind that the higher temperature differences between the sample and the calorimeter fluid would result in significant heat losses.

Due to the longer duration of this method, the common source of error can be described as thermal leakage. Possible heat generation as a result of energy added by stirring to introduce a uniform temperature distribution and a reaction between the fluid constituents and the calorimeter fluid is another source of error in the use of the method of mixtures. A third source of error can be inadequate mixing. The following is a detailed explanation of these common errors and correction methods.

16.2.2.1 Thermal Leakage

The main source of error in the method of mixtures is the thermal leakage from the calorimeter even though the heat losses are assumed to be negligible. As a result, the equilibrium temperatures measured may not give the correct values. A graphical method was proposed by Stitt and Kennedy (1945) for correction of the errors resulting from thermal leakage, so that a correction can be applied for the equilibrium temperatures measured before and after mixing. In this method, the temperature difference between two equilibrium states is taken as the length of the vertical line *DE* in Figure 16.3. This line was established such that the area of the triangle *ABD* would be equal to the area of the triangle *BCE*. This can be closely approximated by drawing parallel lines to the time axis intercepting the time temperature curve at *A* and *C*.

The water equivalent of the system obtained by calibration tests can also be used for correction. Short et al. (1942) mentioned that the transient conditions produced a difficult problem in determining the true water equivalent of the calorimeter and the different thermal conductivities of the food specimen in the calorimeter. Therefore, a single value could not be established. In the cited study, different approaches explained as follows were suggested to solve this problem:

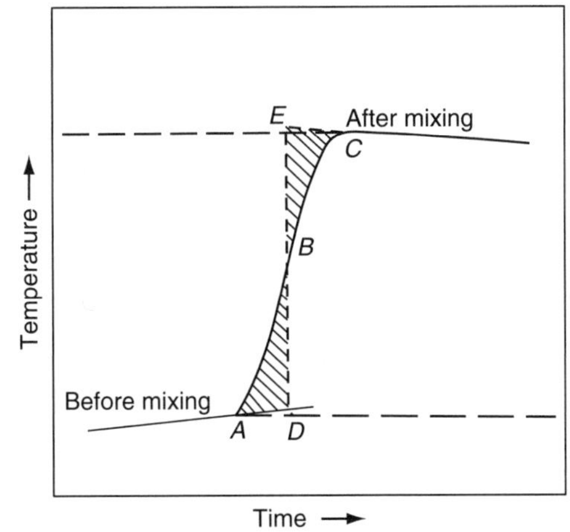

FIGURE 16.3 Graphical method for correcting heat leakage in the method of mixtures calorimetry, *D*: T_{oc} and *E*: T_{em}. (From Stitt, F. and Kennedy, E.K., *Food Res.*, 10, 426, 1945.)

- Energy was added to the food material by means of an electrical heater inside the calorimeter. However, this remedy failed to improve the precision of the results.
- Thermos jar was enclosed in a hollow jacket and the air temperature around the jar was measured at a particular point between the jacket and the jar. The air temperature around the jar varied so widely from top to bottom that a true average was not readily determined.
- Calorimeter was altered by substituting a larger container for the jacket and filling this with brine to a depth that would allow a complete immersion of the thermos jar. During the early work with the above device, it was found that a water equivalent, which was a function of temperature, had to be used and that the time between determinations of the specific heat values could not be shorter than 2–3 h because of the low thermal conductivities of the glass and thermos jar. This difficulty was avoided by allowing the brine temperature to remain approximately constant during all determinations and producing a relatively small temperature rise in the food material during the experimental determination.

In addition to the above, having the brine and food material temperatures close together at the beginning of the experiments also helped in minimizing this difficulty. These suggestions may also be improved with better insulation, mixing, and temperature measurement techniques.

16.2.2.2 Possibility of Heat Generation

Since the method of mixtures allows direct contact between the food material and the heat exchange medium, it is not easy to determine the specific heat of the hygroscopic food material as the given method does not include the evolution of heat for dissolvable chemical entities in the food material. A modified method of mixtures has been developed and used by Pealzner (1951), Li et al. (1971), and Kulacki and Kennedy (1978).

This technique consists of encapsulating the sample in a copper capsule and immersing this capsule in water in the calorimeter. Although this method completely eliminates the necessity of determining the heat of solution, it is noted to require tedious sample preparation (Hwang and Hayakawa, 1979) and possible water losses (Rodriguez et al., 1995). In addition, determination of the specific heat of the test capsule is also required. The use of capsules, however, helped the determination of specific heat of water dissolvable and hygroscopic materials (Sreenarayanan and Chattopadhay, 1986).

16.2.2.3 Mixing Problems

The density of the heat transfer medium should be lower than that of the food material so that the food can sink easily. Liquids with lower density than that of water, toluene (860 kg/m^3) and n-hexane (515 kg/m^3), have been used to provide a more suitable calorimeter solution (Mohsenin, 1980). In the case of substances where the liquid separates from fibrous material during freezing and thawing, a relatively large body of water forms in the upper portion of the container with the fibrous material at the bottom. Short et al. (1942) suggested stirring to correct this difficulty. However, stirring introduced another source of error due to addition of heat energy. The mercury-filled calorimeter, on the other hand, minimized this difficulty but apparently did not eliminate it completely (Short et al., 1942).

Parker et al. (1967) used a calorimeter based on the method of mixtures as shown in Figure 16.4 to measure the specific heat of cherries. This calorimeter consisted of two compartments made of polystyrene, and the bottom compartment was coated with a known amount of paraffin to avoid the water absorption of polystyrene. The sample was placed in the top portion of the calorimeter at room temperature and placed upside down over the bottom portion that contained water at 4.4°C in a cold room. Then, both liquids were drawn, allowing the sample to drop into the cold water and placed it again immediately to reduce heat loss. Specific heat of the sample was calculated by equating the heat loss from the sample with heat gain by water and paraffin. The heat balance required the sample, water, and paraffin, and the initial and equilibrium temperatures. Heat loss from the bottom portion was assumed negligible during the duration of the test. Turrell and Perry (1957) used ice water to measure specific heat of citrus fruits by using a similar calorimeter.

The calorimeter developed by Hwang and Hayakawa (1979) avoided the direct contact between food material and heat exchanging medium. Standard deviations of the replicated experiments were less than 2%. The calorimeter apparatus consisted of a lid and a cover, two rubber O-rings, plastic handle and jacket, vacuum bottle, plastic cup, and a copper–constantan thermocouple. Distilled water (250 g), placed in the space between the vacuum bottle and plastic cup, was used as the heat exchange medium. The use of rubber rings prevented the water loss by evaporation. A calorimeter comparably similar to the above was developed by Gupta (1990) where a copper test capsule was used to prevent the problems due to mixing. This calorimeter reported that the specific heat above 100°C can be measured by using vegetable or mineral oil as the heating medium.

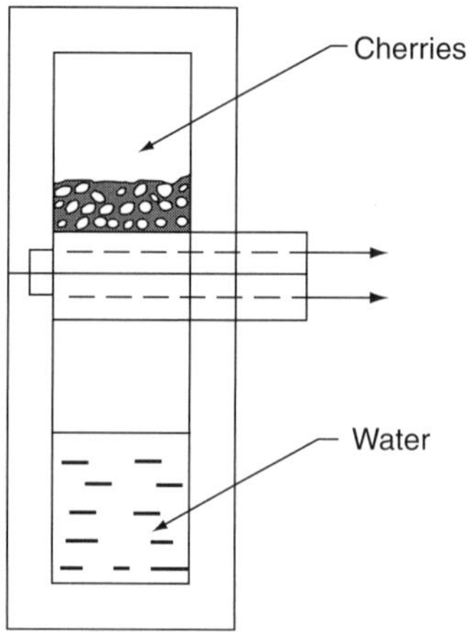

FIGURE 16.4 Two-compartment calorimeter for method of mixtures. (From Parker, R., Levin, J.H., and Monroe, G.E., *Trans. ASAE*, 10, 489, 1967.)

16.2.3 Comparison Calorimeter Method

The comparison method is used to determine the specific heat of liquid foods. A schematic diagram of a comparison calorimeter is shown in Figure 16.5a. As seen in this figure, there are two cups located in the calorimeter where cup A is filled with distilled water or another liquid with a known specific heat and cup B is filled with a liquid food of which the specific heat is to be determined. Both cups are first heated to the same temperature and then located in the calorimeter to cool. The unknown specific heat is calculated by comparing the cooling curves. Mohsenin (1980) stated that if two cups in the calorimeter are of the same size, made of same material with the same exterior finish and are of identical masses, they may be considered to be identical emitters. When two cups have the same initial temperature, then the net rate of heat losses would be equal if the temperature of the air or fluid surrounding the calorimeter is nearly constant. On the basis of this, the heat balance equation can be written as

$$\left(\frac{\Delta Q}{\Delta t}\right)_{\mathrm{A}} = \left(\frac{\Delta Q}{\Delta t}\right)_{\mathrm{B}} \tag{16.9}$$

where

ΔQ is the net rate of heat loss for cups A and B

Δt is the time that will be taken for each cup to drop through an arbitrary interval of temperature change (Figure 16.5b)

The rate of heat loss from the cups can be written as follows if the specific heats are assumed to be constant (Mohsenin, 1980):

$$\left(\frac{\Delta Q}{\Delta t}\right)_{\mathrm{A}} = (c_{p_{\mathrm{A}}} m_{\mathrm{A}} + c_{p_{\mathrm{w}}} m_{\mathrm{w}}) \frac{\Delta T}{\Delta t_{\mathrm{A}}} \tag{16.10}$$

$$\left(\frac{\Delta Q}{\Delta t}\right)_{\mathrm{B}} = (c_{p_{\mathrm{B}}} m_{\mathrm{B}} + c_{p_{\mathrm{S}}} m_{\mathrm{s}}) \frac{\Delta T}{\Delta t_{\mathrm{B}}} \tag{16.11}$$

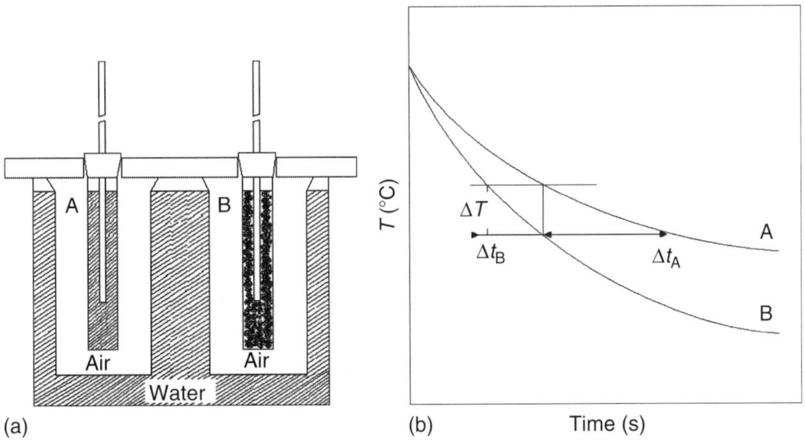

FIGURE 16.5 (a) Comparison calorimeter. (From Bowers, B.A. and Hanks, R.J., *Soil Sci.*, 94, 392, 1962.) (b) Cooling curves of the sample and reference material in a comparison calorimeter. (From Mohsenin, N.N., *Thermal Properties of Foods and Other Agricultural Materials*, Gordon and Breach Science Publishing Group, Switzerland, 1980.)

where

ΔT is an arbitrary temperature drop (Figure 16.5b)

subscripts A, B, w, and s denote cups A and B, water, and sample, respectively

Using the following equation

$$(c_{p_A} m_A + c_{p_w} m_w)\frac{\Delta T}{\Delta t_A} = (c_{p_B} m_B + c_{p_s} m_s)\frac{\Delta T}{\Delta t_B} \tag{16.12}$$

the specific heat of a sample can be found:

$$c_{p_s} = \frac{(c_{p_A} m_A + c_{p_w} m_w)\Delta t_B}{m_s \Delta t_A} - \frac{c_{p_B} m_B}{m_s} \tag{16.13}$$

In this method, temperature of the air or other medium surrounding the calorimeter is assumed to be constant due to the larger size of the heat reservoir. In addition, the heat losses due to evaporation are also accepted to be negligible. Gudzinowicz et al. (1963) used the comparison calorimeter technique to measure the specific heat of complete saturated hydrocarbons. Kovalenko (1968) also describes a calorimetric vessel in detail to determine the specific heat of liquids concluding that the suggested design of the calorimeter evaluates the specific heat of liquids with a small error of ±3.5%.

16.2.4 Adiabatic Method

The basis of the adiabatic method is the assumption that there will be no heat transfer through the chamber walls. Therefore, maintaining adiabatic conditions for this method is crucial. There are different methodologies based on adiabatic conditions.

16.2.4.1 Moline et al. (1961) Method

Moline et al. (1961) proposed a simple and rapid method to measure the specific heat of frozen foods. The apparatus is shown in Figure 16.6. This apparatus consisted of a rectangular polystyrene

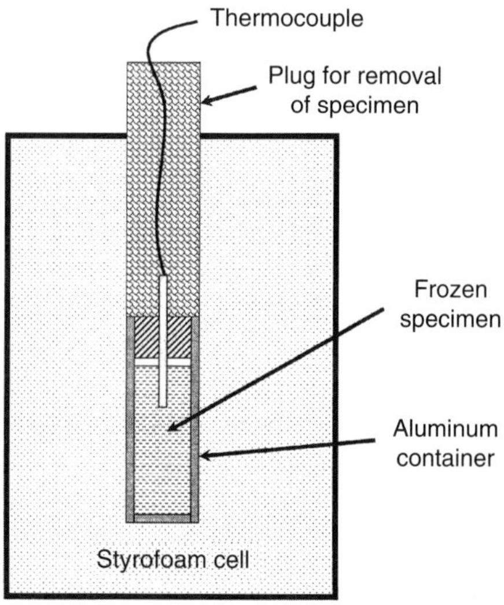

FIGURE 16.6 Specific heat apparatus for frozen sample by Moline et al. (1961) method. (From Moline, S.W., Sawdye, J.A., Short, A.J., and Rinfret, A.P., *Food Tech.*, 15, 228, 1961.)

foam ($9.5 \times 10.2 \times 25.5$ cm) with a cylindrical hole (1.4×2.3 cm) in the center where an aluminum sample container can be placed. A polystyrene plug is inserted at the top of the cell to prevent rapid heat transfer to the sample.

The specimen container is filled with the material ($\approx$20 g), weighed, and then cooled rapidly in liquid nitrogen. When the temperature equilibrium is reached in liquid nitrogen, the container is removed and quickly inserted into the polystyrene foam block maintained at room temperature. The center temperature of the specimen is measured continuously using a thermocouple and recorded over the desired range. Calibration of the apparatus with a specimen of known specific heat is required. The specific heat of the sample at a given temperature can then be calculated using the following equation:

$$c_{p_{sc}} = \frac{Q}{m_{sc}(\Delta T_c/\Delta t)} \tag{16.14}$$

where
- Q is the heat leak (J/s)
- $c_{p_{sc}}$ is the specific heat of the sample and the container (J/kg-K)
- m_{sc} is the mass of the sample and the container (kg)
- $\Delta T_c/\Delta t$ is the rate of the center temperature change (°C/s)

The heat leakage can be calculated as

$$Q = c_{p_c} m_c \frac{\Delta T_c}{\Delta t} \tag{16.15}$$

where c_{p_c} and m_c are the specific heat and mass, respectively, of the thermocouple probe used. Determination of the heat leakage of the calorimeter is an important step in this method. Sometimes, instead of using the sample itself, the use of different materials is preferred in the determination of heat leakage. Keppler and Boose (1970), for example, suggested the use of distilled water to determine the heat leakage of the calorimeter in their study to measure the thermal properties of frozen sucrose solutions.

The rate of center temperature change is determined by determining the slope of the heating curve. The specific heat of the unknown sample can then be calculated from the heat balance between the sample and the aluminum container:

$$c_{p_s} = \frac{c_{p_{sc}} m_{sc} - c_{p_{al}} m_{al}}{m_s} \tag{16.16}$$

where
- c_{p_s} and $c_{p_{al}}$ are the specific heats of the sample and the aluminum container (J/kg-K), respectively
- m_s and m_{al} are the masses of the sample and the aluminum container, respectively

16.2.4.2 Guarded Plate Method

Mohsenin (1980) presented this method to measure the specific heat of food and agricultural materials. The concept of the guarded-plate method is shown in Figure 16.7. In this method, the sample is surrounded by electrically heated thermal guards, which are to be maintained at the same temperature as that of the sample. The objective of maintaining the thermal guards and the sample at the same temperature is to prevent any possible heat loss. Then, the specific heat of the sample can

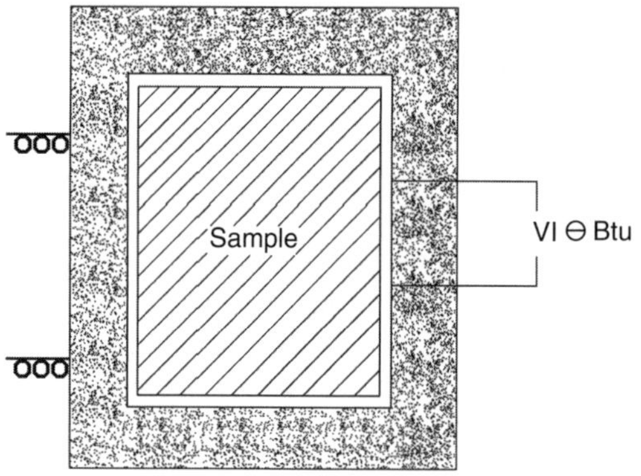

FIGURE 16.7 Concept of guarded-plate method for specific heat determination. (From Mohsenin, N.N., *Thermal Properties of Foods and Other Agricultural Materials*, Gordon and Breach Science Publishing Group, Switzerland, 1980.)

be calculated by using a heat balance where the heat gained by the sample would be equal to the heat supplied by the electric heaters. The electric heat supplied to the specimen in a given time t is set equal to the heat gained by the sample:

$$Q = m_s c_{p_s}(T_{sf} - T_{si}) = VIt$$
$$c_{p_s} = \frac{VIt}{m_s(T_{sf} - T_{si})} \quad (16.17)$$

where
- V and I are the voltage (V) and current (A) supplied to the heater during the given time interval t (s), respectively
- T_{sf} and T_{si} are the final and initial temperatures of the sample (°C), respectively
- m_s is the mass of the sample

In this method, temperature measurement of the sample and the location where the temperature is measured becomes important since the measured temperatures are used as the mass average temperature of the product. If possible, measurement at different locations and using an average would improve the results. In addition, use of advanced mathematical techniques where the objective is to determine the specific heat would be an improvement, but for this purpose knowledge of all the other parameters is necessary.

Wright and Porterfield (1970) proposed a calorimeter to measure the specific heat of a single peanut. This calorimeter consisted of a small aluminum cup surrounded by a resistance heating coil and foam insulation. Two thermocouples were used to measure the temperatures of the peanut and the calorimeter chamber. Time and temperature data for heating and cooling of the calorimeter with and without the sample were recorded after constant power was turned on for a specific time. The specific heat was calculated by heat balance based on total heat supplied and absorbed and heat of desorption of water from the sample.

Clark et al. (1946) also proposed a calorimeter to measure the specific heat of fats and oils where a constant water bath was used as calorimeter. Staph (1949) used the calorimeter approach with electrical heating to measure the specific heat of frozen foods. The sample was cut into small pieces and immersed in a kerosene bath to reduce the heat transfer resistance between the sample pieces. The temperature near the center of the sample mass was recorded when heat was supplied electrically to the food container. A stirrer in the center of the sample container was used to reduce the resistance of heat flow. The bath around the sample container was also heated to reduce heat loss from the sample wall. The specific heat can be determined from the heat input, voltage measurement, and temperature rise of the sample.

16.2.4.3 Adiabatic Chamber Method

The adiabatic chamber method is usually designed and used when no heat or moisture transfer through the chamber is required. A better and improved adiabatic condition is obtained by enclosing the test chamber in another chamber. Detailed drawing and construction for this method are presented by Mohsenin (1980). A measured quantity of heat is added by means of heating a wire buried in the bulk of the material in the container placed in the test chamber. The specific heat of the material can then be calculated from the heat balance equation:

$$Q = (mc_p\Delta T)_{\text{sample}} + (mc_p\Delta T)_{\text{container}} + (mc_p\Delta T)_{\text{chamber}} \tag{16.18}$$

With the known specific heat values of the container and chamber, the specific heat of the sample is determined from the known heat value and the measured temperatures.

16.2.5 Use of Differential Scanning Calorimetry

The DSC is a thermoanalytical technique. DSC is the most commonly used method for determining the specific heat along with the method of mixtures (Murphy et al., 1998). It is based on measuring small effects produced in thermal processes (Mohsenin, 1980). The advantages of using DSC are rapid measurements, obtaining multiple data from a single thermogram, and use of a small amount of sample yielding accurate results (Murphy et al., 1998). The thermograms show any gain or loss of heat energy at a given temperature rise over a given temperature interval. The area of the thermograms is proportional to the heat energy absorbed or released by the sample during a given process. The dynamic nature of DSC allows the determination of specific heat as a function of temperature (Singh and Goswami, 2000).

In this method, either the difference in heat gained by the sample and a reference is recorded or the temperature of the sample, increased at a constant heating rate, is monitored to determine the specific heat of the sample. Sapphire and indium is a common material used as reference in the DSC analysis. Due to the higher moisture content of the food samples, the food sample is usually placed in an encapsulator (hermetically sealed pan) during the experiment.

When the heat flow rate into a sample of a known mass is measured, the specific heat of the sample is determined using the temperature change in the sample using the following equation:

$$c_p = \frac{(\mathrm{d}Q/\mathrm{d}t)}{m_\mathrm{s}(\mathrm{d}T/\mathrm{d}t)} \tag{16.19}$$

where

$\mathrm{d}Q/\mathrm{d}t$ is the heat flow rate, W (J/s)
$\mathrm{d}T/\mathrm{d}t$ is the heating rate (K/s)
m_s is the mass of the sample (kg)

Mohsenin (1980) suggests the following equation using the parameters obtained from a DSC thermogram:

$$c_{p_\mathrm{s}} = \left(\frac{d_\mathrm{sa}}{d_\mathrm{re}}\frac{m_\mathrm{re}}{m_\mathrm{s}}\right)c_{p_\mathrm{re}} \tag{16.20}$$

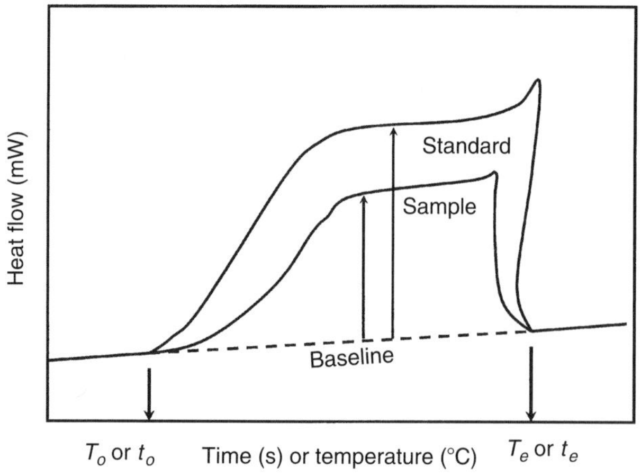

FIGURE 16.8 Typical specific heat thermogram of foods by DSC.

where

d_{sa} and d_{re} are the thermogram reflections of the sample (Figure 16.8) and the reference material from the baseline, respectively, which is obtained with a no-sample run in the DSC

m_s and m_{re} are the mass of the sample and the reference material, respectively

$c_{p_{re}}$ is the specific heat of the reference material

DSC is especially preferred for allowing the determination of specific heat as a function of temperature. Chakrabarty and Jonson (1972) determined the specific heat of tobacco by DSC, observing a 16% increase in the temperature range of 40°C–70°C. Murata et al. (1987) determined the specific heat of different grains reporting a linear change with respect to moisture content and a parabolic increase with respect to temperature. Tang et al. (1991) used DSC to measure the specific heat of lentils reporting a parabolic increase with moisture content while a linear trend was observed with temperature increase. Singh and Goswami (2000) used DSC to measure the specific heat of cumin seed. They reported that the specific heat of cumin seed increased from 1330 to 3090 J/kg-K in the temperature range of −70°C to 50°C. The factors that can affect the measured specific heat values during a DSC measurement can be defined as the sample size, heating rate, and the sample seal conditions.

16.2.5.1 Sample Size and Heating Rate

Determination of specific heat with DSC is based on the assumption that during the tests the temperature is uniform in the sample and in the sample container. Tang et al. (1991) mentioned that the thermal lag with the sample may introduce an error in the measured values of the specific heat of the biological materials due to their low thermal diffusivity. Tang et al. (1991) studied the effect of sample size and heating rate on the sample with a thermal diffusivity of 0.5×10^{-7} m^2/s. They found that the sample center temperature started to lag behind the surface temperature rise at the beginning of heating. Within less than 20 s, this difference approached a constant value varying from 0.13°C to 3.53°C for sample thicknesses of 0.8–2.4 mm and heating rates of 5–15 K/min. The specific heat values of the 2.4 mm thick samples were reported to be 250 J/kg-K, being less than that of 0.5 mm thick samples at the heating rate of 5 K/min. The measurement error of the specific heat of the 0.8 mm thick sample at 5 K/min was 30 J/kg-K. Based on this, in the given study, the scanning speed and sample thickness were chosen to be 5 K/min and less than 0.8 mm, respectively, to obtain reasonable specific heat values by reducing the errors from the sample size and heating rate.

16.2.5.2 *Sample Seal Condition*

Temperature rise during a DSC test increases the vapor pressure within the sample, and as a result, moisture leak may occur from the biological material in the form of water vapor. The latent heat absorbed in this thermal process may then introduce errors in the measurement. Thus, sample encapsulation or sample sealing must be used to get better results (Mohsenin, 1980; Tang et al., 1991).

Tang et al. (1991) studied three types of seal conditions: a complete seal where no moisture would escape from the sample pan; a poor seal where sample pan was not properly sealed (this type of sealing was reported to have no visible openings); and a no seal where the sample lid was pierced with three holes. The tests were conducted with a 10% moisture-content sample (0.8 mm thick) with a heating rate of 5 K/min. They found that the poor sealing resulted in higher values of the measured specific heat especially at elevated temperatures. The magnitude of the deviations from that of completely sealed samples depended on the degree of the sample pan seal and the amount of moisture that escaped. The weight check of the samples after each run indicated that the moisture loss from the improperly sealed samples was up to 30% of the initial moisture content of the samples. In these given cases, 20%–40% of the heat flow into the sample was reported to be used for moisture evaporation, resulting in errors of the measured specific heat values.

16.2.6 Other Methods

In addition to the processing methods, new methods to measure specific heat are reported in the literature. Zueco et al. (2004) proposed a procedure for the inverse estimation of the specific heat in a one-dimensional solid under convective and adiabatic boundary conditions where the thermal conductivity can be used to be constant or temperature dependent. They treated the given inverse problem as a function estimation problem. The numerical solution methodology for this problem is given in detail in their study where the thickness, density, thermal conductivity, and the initial temperature were the required parameters.

Nicolau et al. (2002) developed heat flux meters based on tangential temperature gradient and used them to measure specific heat and thermal conductivity of low conductivity solid materials. In this method, a plane square sample was placed between two heat flux meters while a skin heater and heat sink were placed over and under these components, respectively. Starting from a constant temperature, the whole system was heated at a constant rate moving toward a final steady-state condition. The specific heat was then obtained from the accumulated energy inside the sample during the transient regime.

Haemmerich et al. (2005) presented a method to measure the temperature-dependent specific heat of biological tissues. This methodology was based on the general heat transfer equation when a sample with a specific heat of c_p was heated by applying power with a mass-specific energy rate:

$$c_p \frac{\partial T}{\partial t} = \frac{\nabla k \nabla T}{\rho} + P \tag{16.21}$$

where

k is the thermal conductivity of the sample (W/m-K)

ρ is the density (kg/m^3)

P is the total supplied power (W/kg)

With the assumptions of a homogeneous sample, negligible heat losses to the environment obtained with a perfect insulation, uniform contact between heating elements and sample surface, and constant sample mass, Equation 16.21 leads to the following when a uniform temperature distribution is accomplished inside the sample since the thermal conduction term would be "0":

$$c_p = \frac{P}{(\partial T/\partial t)} \tag{16.22}$$

Haemmerich et al. (2005) gives a detailed explanation for this method with the required correction strategies for possible heat losses. Tanaka and Uematsu (2006) presented a new calorimeter to measure the isobaric specific heat for fluids and fluid mixtures by using the thermal relaxation technique, where the given fluid system is termed as a lumped system, and its temperature history can be considered to be a function of time only. The calorimeter, assembled in a pressure vessel to operate at different constant pressures, comprised a cylinder and metal bellows welded to the cylinder containing the fluid. The cylindrical calorimeter was surrounded by nitrogen gas to function as a poor conductor and a pressure transmitting fluid. The calculation procedures and the details of this calorimeter are given in Tanaka and Uematsu (2006).

16.3 ENTHALPY

Enthalpy is defined as the heat content in a system per unit mass (J/kg). Enthalpy can be written in terms of specific heat as

$$H = \int_{T_1}^{T_2} c_p \, \mathrm{d}T \tag{16.23}$$

If the specific heat is assumed to be constant in the given temperature range, then Equation 16.23 reduces to the following equation:

$$H = c_p(T_2 - T_1) \tag{16.24}$$

The change in the enthalpy of a food product is used to estimate the energy that must be added or removed (ASHRAE, 2006). Enthalpy properties are the required data for computations involved in process operation and equipment design for freezing and thawing foods (Chang and Tao, 1981). Table 16.4 shows the enthalpy change of different food products as a function of temperature in the freezing temperature range. Both experimental and mathematical approaches have been used to determine the enthalpy of foods. Theoretical predictions have been based on some measured properties of food products such as water and solid content and initial freezing point.

Above the freezing point, enthalpy consists of the sensible energy while it consists of both sensible and latent energy below the freezing point. For foods above their initial freezing point, the enthalpy may be determined by using the following equation in a similar way to the calculation of specific heat (Equation 16.5):

$$H = \sum_{i=1}^{n} (H_i \cdot X_i) = \sum_{i=1}^{n} \int_{T_1}^{T_2} (c_{pi} \cdot X_i) \cdot \mathrm{d}T \tag{16.25}$$

where

n is the number of components of a given food product
c_p is the specific heat of the ith component given in J/kg-K
X_i is the fraction of the ith component
T_1 and T_2 indicate the given temperature interval

Table 16.4 Enthalpy of Frozen Foods

Food	Water Content (% by Mass)		Temperature (°C) −40.0	−28.9	−23.3	−20.6	−17.8	−15.0	−12.2	−9.4	−6.7	−4.4	−2.2	−1.1
Fruits and vegetables														
Applesauce	82.8	*H* (kJ/kg)	0.0	25.6	39.5	48.8	58.2	69.8	83.7	100	130.3	165.1	265.2	337.3
		% Unfrozen water	—	5	7	9	11	14	17	20	28	41	76	100
Asparagus, peeled	92.6	*H* (kJ/kg)	0.0	18.6	32.6	37.2	44.2	51.2	60.5	69.8	86.1	102.3	146.5	234.9
		% Unfrozen water	—	—	—	4	5	6	7	9	12	20	55	—
Bilberries	85.1	*H* (kJ/kg)	0.0	23.3	34.9	41.9	51.2	58.2	69.8	86.1	104.7	130.3	202.4	346.6
		% Unfrozen water	—	—	5	6	7	9	11	14	19	27	50	100
Carrots	87.5	*H* (kJ/kg)	0.0	23.3	34.9	41.9	51.2	60.5	72.1	86.1	104.7	132.6	204.7	353.60.0
		% Unfrozen water	—	—	5	6	7	9	11	14	19	27	50	100
Cucumbers	95.4	*H* (kJ/kg)	0.0	18.6	30.2	37.2	41.9	48.8	55.8	62.8	74.4	88.4	121.0	181.4
		% Unfrozen water	—	—	—	—	—	—	—	—	7	9	18	100
Onions	85.5	*H* (kJ/kg)	0.0	23.3	37.2	46.5	55.8	65.1	79.1	93.0	121.0	153.5	244.2	346.6
		% Unfrozen water	—	5	7	8	9	12	15	18	24	35	65	100
Peaches without pits	85.1	*H* (kJ/kg)	0.0	23.3	37.2	46.5	55.8	65.1	79.1	97.7	123.3	155.8	251.2	348.9
		% Unfrozen water	—	5	7	8	10	12	15	18	26	37	69	100
Pears, Bartlett	83.8	*H* (kJ/kg)	0.0	23.3	39.5	48.8	58.2	67.5	81.4	97.7	123.3	160.5	258.2	339.6
		% Unfrozen water	—	6	8	9	10	12	15	19	27	38	72	100
Plums without pits	80.3	*H* (kJ/kg)	0.0	27.9	44.2	55.8	65.1	76.8	93.0	116.3	148.9	197.7	323.3	328
		% Unfrozen water	—	8	11	13	16	18	22	28	38	55	100	—
Raspberries	82.7	*H* (kJ/kg)	0.0	23.3	37.2	44.2	51.2	60.5	72.1	88.4	107.0	137.2	214.0	339.6
		% Unfrozen water	—	4	6	7	8	9	12	15	21	30	56	100
Spinach	90.2	*H* (kJ/kg)	0.0	18.6	32.6	37.2	44.2	51.2	60.5	67.5	81.4	97.7	137.2	216.3
		% Unfrozen water	—	—	—	—	—	—	5	7	10	14	25	50
Strawberries	89.3	*H* (kJ/kg)	0.0	20.9	34.9	41.9	48.8	58.2	67.5	79.1	95.4	118.6	179.1	295.4
		% Unfrozen water	—	—	—	5	6	7	8	10	15	21	40	79
Sweet cherries, without pits	77.0	*H* (kJ/kg)	0.0	27.9	46.5	55.8	67.5	81.4	97.7	118.6	155.8	207.0	311.7	316.3
		% Unfrozen water	—	9	12	14	17	20	25	32	43	62	100	—

(continued)

Table 16.4 (continued) Enthalpy of Frozen Foods

Food	Water Content (% by Mass)		Temperature (°C)											
			−40.0	−28.9	−23.3	−20.6	−17.8	−15.0	−12.2	−9.4	−6.7	−4.4	−2.2	−1.1
Tall peas	75.8	*H* (kJ/kg)	0.0	23.3	39.5	48.8	58.2	69.8	83.7	100.0	125.6	162.8	265.2	318.7
		% Unfrozen water	—	6	8	10	12	15	18	22	30	44	82	100
Tomato pulp	92.9	*H* (kJ/kg)	0.0	23.3	32.6	39.5	46.5	53.5	62.8	74.4	90.7	109.3	158.2	260.5
		% Unfrozen water	—	—	—	—	—	5	6	8	12	18	31	62
Fish and meat														
Cod	80.3	*H* (kJ/kg)	0.0	23.3	34.9	41.9	48.8	55.8	65.1	76.8	90.7	111.6	169.8	286.1
		% Unfrozen water	10	10	10	11	12	13	14	16	20	26	45	88
Haddock	83.6	*H* (kJ/kg)	0.0	20.9	34.9	41.9	48.8	55.8	65.1	76.8	90.7	111.6	169.8	295.4
		% Unfrozen water	8	8	9	9	10	11	12	14	17	23	42	86
Perch	79.1	*H* (kJ/kg)	0.0	20.9	32.6	39.5	46.5	53.5	62.8	74.4	88.4	107.0	158.2	272.1
		% Unfrozen water	10	10	11	11	12	13	14	16	19	24	41	83
Beef, lean, fresh[a]	74.5	*H* (kJ/kg)	0.0	20.9	34.9	41.9	48.8	55.8	62.8	74.4	88.4	111.6	172.1	276.8
		% Unfrozen water	10	10	11	12	12	13	15	18	22	28	48	92
Lean, dried	26.1	*H* (kJ/kg)	0.0	20.9	32.6	39.5	46.5	55.8	65.1	72.1	76.8	83.7	88.4	—
		% Unfrozen water	96	96	97	98	99	100	—	—	—	—	—	—
Eggs														
Yolk	50.0	*H* (kJ/kg)	0.0	20.9	32.6	37.2	44.2	51.2	58.2	67.5	76.8	88.4	109.3	151.2
		% Unfrozen water	—	—	—	—	—	—	—	—	—	23	32	66
	40.0	*H* (kJ/kg)	0.0	20.9	32.6	39.5	46.5	53.5	60.5	72.1	81.4	95.4	123.3	176.8
		% Unfrozen water	20	—	—	—	24	—	27	—	—	38	54	89
White	86.5	*H* (kJ/kg)	0.0	20.9	32.6	37.2	44.2	51.2	58.2	67.5	76.8	93.0	127.9	202.4
		% Unfrozen water	—	—	—	—	—	—	—	—	12	14	22	48
Whole, with shell[b]	66.4	*H* (kJ/kg)	0.0	20.9	30.2	34.9	41.9	46.5	53.5	62.8	72.1	86.1	114.0	169.8
Bread														
White	37.3	*H* (kJ/kg)	0.0	20.9	30.2	34.9	41.9	48.8	60.5	79.1	104.7	127.9	132.6	134.9
Whole wheat	42.4	*H* (kJ/kg)	0.0	20.9	30.2	34.9	41.9	51.2	62.8	83.7	111.6	144.2	158.2	160.5

Source: ASHRAE, in *ASHRAE Handbook 2006 Refrigeration*, American Society of Heating, Refrigerating and Air-Conditioning Engineers, Inc., Atlanta, GA, 2006. With permission.

[a] Data for chicken, veal, and venison; nearly matched data for beef of same water content (ASHRAE, 2006).

[b] Calculated for mass composition of 58% white (86.5% water) and 32% yolk (50% water).

Chen (1985) suggested the use of Equation 16.22 as a function of the mass fraction of food solids and its initial freezing temperature:

$$H = H_f + (T - T_f)\left(1 - 0.55X_s - 0.15X_s^3\right) \quad (16.26)$$

where
H is the enthalpy of the food product (J/kg)
H_f is the enthalpy of the food product at its initial freezing temperature (J/kg)
T is the temperature (°C)
T_f is the initial freezing temperature (°C)
X_s is the mass fraction of the food solids

For enthalpy values of foods below their initial freezing point, Schwartzberg (1976) derived a formula as a function of solids, water content, and initial freezing point:

$$H = (T - T_r)\left\{c_{p_u} + (bX_s - X_{w0})\Delta c_p + EX_s\left[\frac{RT_0^2}{18(T_0 - T_r)(T_0 - T)} - 0.8\Delta c_p\right]\right\} \quad (16.27)$$

where
T is the temperature of the food product (°C)
T_0 is the freezing point of water (0°C)
R is the universal gas constant
T_r is the reference temperature (−40°C) where the enthalpy is defined to be "0" (Riedel, 1957; Schwartzberg et al., 2007)
Δc_p is the difference between specific heats of water and ice (J/kg-K)
c_{p_u} is the specific heat of food above its initial freezing point
b is the amount of unfrozen water per unit weight of solids
X_s is the mass fraction of solids per unit weight of food
X_{w0} is the mass fraction of water per unit weight of food above the initial freezing point
E is the ratio of molecular mass of water (M_w) and food solids (M_s) and $E = \frac{M_w}{M_s} = \frac{18}{M_s}$

A numerical value of 26,721 for $RT_0^2/18$ when temperature is in °R, and an approximate value of 0.5 for Δc_p were used by Schwartzberg (1976) when the units of the specific heat were in Btu/lb-°F. An alternative expression for determining the enthalpy for freezing conditions can also be found in Schwartzberg (1976). By integrating Equation 16.23 between the reference temperature, T_r and the food product temperature, T, Chen (1985) also obtained a similar equation where just the value of M_s, instead of $\frac{M_w}{M_s}$, was used. Pham et al. (1994) rewrote the Schwartzberg's enthalpy model as follows:

$$H = A + c_{p_f}T + \frac{B}{T} \quad (16.28)$$

where
$c_{p_f} = c_{p_u} + (X_b - X_{w0} - 0.8EX_s)\,\Delta c_p$
$B = -\left(EX_sRT_0^2/18\right)$
A is the integration constant
T is the food product temperature

These models were obtained by integrating the given specific heat models in a certain temperature range.

As an alternative to these models obtained by integration of the given specific heat models, Chang and Tao (1981) presented correlations of enthalpies for meats, vegetables, fruits, and juices with varying moisture contents of 73%–94% in a temperature range of 230–310 K. The general form of the correlations presented by Chang and Tao (1981) was

$$H = H_f \left[y T_R + (1 - y) T_R^z \right] \tag{16.29}$$

where

H is the enthalpy of food (J/kg)
H_f is the enthalpy of the food product at its initial freezing temperature (J/kg)
T_R is the dimensionless temperature ratio, $((T - T_{ref})/(T_f - T_{ref}))$ (T_{ref} is the reference temperature with a zero enthalpy, 227.6 K, and T_f is the initial freezing temperature)
y and z are the correlation parameters

For meat group, for example, they suggested the following equations to determine the correlation parameters of y and z:

$$\begin{aligned} y &= 0.316 - 0.247(X_w - 0.73) - 0.688(X_w - 0.73)^2 \\ z &= 22.95 + 54.68(y - 0.28) - 5589.03(y - 0.28)^2 \end{aligned} \tag{16.30}$$

where X_w is the water content. The required correlations to determine the initial freezing temperature were also reported by Chang and Tao (1981). Fikiin and Fikiin (1999) also presented a set of predictive equations for determining thermal properties and enthalpy of the food materials during cooling and freezing.

In addition to the predictive equations, adiabatic calorimetry or differential scanning calorimetry techniques can also be used to measure the enthalpy of a food product. Kerr et al. (1993) reported the use of adiabatic calorimetry to measure enthalpy between −60°C and 20°C for fish, beef, egg white, egg yolk, white bread, and fruit juices. Lindsay and Lovatt (1994) used the adiabatic calorimetry technique to determine the enthalpy values of meat, fats, processed meats, butter, cheese, ice cream, and apple in the temperature range of −40°C to 40°C. Adiabatic calorimetry technique is also noted with its difficulties in measuring the enthalpy of a sample as it is removed from the freezer (Kerr et al., 1993). Kerr et al. (1993) used the differential compensated calorimetry technique to measure the enthalpy of beef, egg white, applesauce, and white bread in a temperature range of −40°C to 0°C, reporting close agreement with the previously published values.

16.4 CONCLUSION

Specific heat is a significant thermophysical property in analyzing mass/energy balances in food processing operations, and enthalpy as a thermal property is preferred in modeling freezing and thawing times of foods since the specific heat becomes infinitely high at the freezing or thawing temperature. Enthalpy change in foods during freezing and thawing should be known for computations involved in process operation and equipment design.

In this chapter, selected methods to describe specific heat and enthalpy of foods have been described. The described methods covered the use of empirical equations and experimental methods ranging from simple to sophisticated methodologies. In addition, experimental data from the literature for specific heat of different food products, enthalpy of different frozen foods with respect to temperature, and empirical equations to determine specific heat of foods as a function of temperature or moisture content were included.

As seen in Section 16.2.6, where recently developed methodologies have been described to determine the specific heat of food products, new experimental procedures have been developed to determine thermophysical properties of food products. With more accurate data on thermophysical properties of food products, better mathematical models are expected to be developed for simulation of different food-processing operations.

REFERENCES

AbuDagga, Y. and Kolbe, E., Thermophysical properties of Surimi paste at cooking temperature, *J. Food Eng.*, 32, 325, 1997.

Alam, A. and Shove, G.C., Hygroscopicity and thermal properties of Soybeans, *Trans. ASAE*, 16, 707, 1973.

ASHRAE, Thermal properties of foods, in *ASHRAE Handbook 2006 Refrigeration*, Owen, M.S. (Ed.), Ch. 9, American Society of Heating, Refrigerating and Air-Conditioning Engineers, Inc., Atlanta, GA, 2006.

Baik, O.D. and Mittal, G.S., Determination and modeling of thermal properties of Tofu, *Int. J. Food Prop.*, 6, 9, 2003.

Baik, O.D., Sablani, S.S., Marconi, M., and Castaigne, F., Modeling the thermal properties of a cup cake during baking, *J. Food Sci.*, 64, 295, 1999.

Bowers, B.A. and Hanks, R.J., Specific heat capacity of soils and minerals as determined with a radiation calorimeter, *Soil Sci.*, 94, 392, 1962.

Chakrabarty, S.M. and Jonson, W.H., Specific heat of flue cured tobacco by differential scanning calorimetry, *Trans. ASAE*, 15, 928, 1972.

Chang, H.D. and Tao, L.C., Correlations of enthalpies of food systems, *J. Food Sci.*, 46, 1493, 1981.

Charm, S.E., *The Fundamentals of Food Engineering*, 3rd edn., AVI Publishing, Westport, CT, 1978.

Chen, C.S., Thermodynamic analysis of the freezing and thawing of foods: Enthalpy and apparent specific heat, *J. Food Sci.*, 50, 1158, 1985.

Choi, Y. and Okos, M.R., Effects of temperature and composition on the thermal properties of foods, in *Food Processing and Process Applications Vol. I Transport Phenomenon*, LaMaguer, M. and Jelen, P. (Eds.), Elsevier, New York, 1986.

Christenson, M.E., Tong, C.H., and Lund, D.B., Physical properties of baked products as functions of moisture and temperature, *J. Food Proc. Preserv.*, 13, 201, 1989.

Clark, P.E., Waldenland, C.R., and Cross, R.P., Specific heats of vegetable oils from 0°C to 250°C, *Ind. Eng. Chem.*, 38, 350, 1946.

Deshpande, S.D. and Bal, S., Specific heat of soybean, *J. Food Proc. Eng.*, 22, 469, 1999.

Fasina, O.O. and Fleming, H.P., Heat transfer characteristics of cucumbers during blanching, *J. Food Eng.*, 47, 203, 2001.

Fasina, O. and Sokhansanj, S., Estimation of moisture diffusivity coefficient and thermal properties of alfalfa pellets, *J. Agric. Eng. Res.*, 63, 333, 1996.

Fasina, O.O., Farkas, B.E., and Fleming, H.P., Thermal and dielectric properties of sweet potato puree, *Int. J. Food Prop.*, 6, 461, 2003.

Fikiin, K.A. and Fikiin, A.G., Predictive equations for thermophysical properties and enthalpy during cooling and freezing of food materials, *J. Food Eng.*, 40, 1, 1999.

Gudzinowicz, B.J., Campbell, R.H., and Adams, J.S., Specific heat measurements of complete saturated hydrocarbons, *J. Chem. Eng. Data*, 8, 201, 1963.

Gupta, T.R., Specific heat of Indian unleavened flat bread (chapati) at various stages of cooking, *J. Food Proc. Eng.*, 13, 217, 1990.

Haemmerich, D., Schutt, D.J., dos Santos, I., Webster, J.G., and Mahvi, D.M., Measurement of temperature-dependent specific heat of biological tissues, *Physiol. Meas.*, 26, 59, 2005.

Heldman, D.R., Prediction models for thermophysical properties of foods, in *Food Processing Operations Modeling Design and Analysis*, Irudayaraj, J. (Ed.), Marcel Dekker Inc., New York, 2002.

Heldman, D.R. and Singh, R.P., *Food Process Engineering*. 2nd ed., Van Nostrand Reinhold. New York, 1981.

Hough, G., Moro, O., and Luna, J., Thermal conductivity and heat capacity of dulce de leche, a typical Argentine dairy product, *J. Dairy Sci.*, 69, 1518, 1986.

Hwang, M.P. and Hayakawa, K., A specific heat calorimeter for foods, *J. Food Sci.*, 44, 435, 1979.

Jindal, V.K. and Murakami, E.G., Thermal properties of shredded coconut, in *Engineering and Food*, Vol. 1, McKenna, B.M., (Ed.), Elsevier Applied Science Publishers, London, UK, 1984.

Kanagy, J.R., Specific heats of collagen and leather, *Leather Chem. Assoc.*, 5, 444, 1971.

Karunakar, B., Mishra, S.K., and Bandyopadhyay, S., Specific heat and thermal conductivity of shrimp meat, *J. Food Eng.*, 37, 345, 1998.

Kazarian, E.A. and Hall, C.W., Thermal properties of grains, *Trans. ASAE*, 8, 33, 1965.

Keppler, R.A. and Boose, J.R., Thermal properties of frozen sucrose solutions, *Trans. ASAE*, 13, 335, 1970.

Kerr, W.L., Ju, J., and Reid, D.S., Enthalpy of frozen foods determined by differential compensated calorimetry, *J. Food Sci.*, 58, 675, 1993.

Kovalenko, B.M., Calorimeter for determining the specific heat of liquids, *Meas. Tech.*, 11, 639, 1968.

Kulacki, F.A. and Kennedy, S.C., Measurement of the thermophysical properties of common cookie dough, *J. Food Sci.*, 43, 380, 1978.

Leniger, H.A. and Beverloo, W.A., *Food Process Engineering*. Springer Publishing Company, New York, 1975.

Li, K.J., Whellock, T.D., and Lanaster, E.B., Thermal properties of soft white winter wheat flour, *Chem. Eng. Symp. Series*, 67, 108, 1971.

Lindsay, D.T. and Lovatt, S.J., Further enthalpy values of foods measured by an adiabatic calorimeter, *J. Food Eng.*, 23, 609, 1994.

Madampa, P.S., Driscoll, R.H., and Buckle, K.A., Models for the specific heat and thermal conductivity of garlic, *Drying Tech.*, 13, 295, 1995.

Marschoun, L.T., Muthukumarappan, K., and Gunasekaran, S., Thermal properties of cheddar cheese: Experimental and modeling, *Int. J. Food Prop.*, 4, 383, 2001.

McProud, L.M. and Lund, D.B., Thermal properties of beef loaf produced in food service systems, *J. Food Sci.*, 48, 677, 1983.

Mohsenin, N.N., *Thermal Properties of Foods and Agricultural Materials*, Gordon and Breach Science Publishers Inc., New York, 1980.

Moline, S.W., Sawdye, J.A., Short, A.J., and Rinfret, A.P., Thermal properties of foods at low temperatures, I., *Food Tech.*, 15, 228, 1961.

Moreiara, R.G., Palau, J., Sweat, V.E., and Sun, X., Thermal and physical properties of tortilla chips as a function of drying time, *J. Food Process. Preserv.*, 19, 175, 1995.

Murata, S., Tagawa, A., and Ishibashi, S., The effect of moisture content and temperature on specific heat of cereal grains measured by DSC, *J. Jpn. Soc. Agric. Machinery*, 46, 547, 1987.

Murphy, R.Y., Marks, B.P., and Marcy, J.A., Apparent specific heat of chicken breast patties and their constituent proteins by differential scanning calorimetry, *J. Food Sci.*, 63, 88, 1998.

Nesvadba, P., Thermal properties of unfrozen foods, in *Engineering Properties of Foods*, Rao, M.A., Rizvi, S.S.H., and Datta, A.K. (Eds.), Taylor & Francis, Boca Raton, FL, 2005.

Nicolau, V.P., Güths, S., and Silva, M.G., Thermal conductivity and specific heat measurement of low conductivity materials using heat flux meters, ECPT 2002: The Sixteenth European Conference on Thermophysical Properties, September 1–4, 2002, Imperial College, London, UK.

Niekamp, A., Unklesbay, K., Unklesbay, N., and Ellersiek, M., Thermal properties of bentonite-water dispersions used for modeling foods, *J. Food Sci.*, 49, 28, 1984.

Niesteruk, R., Changes of thermal properties of fruits and vegetables during drying, *Drying Tech.*, 14, 415, 1996.

Njie, D.N., Rumsey, T.R., and Singh, R.P., Thermal properties of cassava, yam and plantain, *J. Food Eng.*, 37, 63, 1998.

Omobuwajo, T.O., Sanni, L.A., and Olajide, J.O., Physical properties of ackee apple (*Blighia sapida*) seeds, *J. Food Eng.*, 45, 43, 2000.

Ordinanz, W.O., Specific heat of foods in cooling, *Food Ind.*, 18, 101, 1946.

Otten, L. and Samaan, G., Determination of the specific heat of agricultural materials: Part II. Experimental results, *Can. Agric. Eng.*, 22, 25, 1980.

Parker, R.E., Levin, J.H., and Monroe, G.E., Thermal properties of tart cherries, *Trans. ASAE*, 10, 489, 1967.

Pealzner, P.M., The specific heat of wheat, *Can. J. Tech.*, 29, 261, 1951.

Pham, Q.T., Wee, H.K., Kemp, R.M., and Lindsay, D.T., Determination of the enthalpy of foods by an adiabatic calorimeter, *J. Food Eng.*, 21, 137, 1994.

Putranon, R., Bowrey, R.G., and Fowler, R.T., The effect of moisture content on the heat of sorption and specific heat of Australian paddy rice, *Food Tech. Aust.*, 32, 56, 1980.

Rahman, M.S., Specific heat of selected fresh seafood, *J. Food Sci.*, 58, 522, 1993.

Ramaswamy, H.S. and Tung, M.A., Thermophysical properties of apples in relation to freezing, *J. Food Sci.*, 46, 724, 1981.

Rapusas, R.S. and Driscoll, R.H., Thermophysical properties of fresh and dried white onion slices, *J. Food Eng.*, 24, 149, 1995.

Razavi, S.M.A. and Taghizadeh, M., The specific heat of pistachio nuts as affected by moisture content, temperature, and variety, *J. Food Eng.*, 79, 158, 2007.

Rodriguez, R.D.P., Rodrigo, M., and Kelly, P., A calorimetric method to determine specific heat of prepared foods, *J. Food Eng.*, 26, 81, 1995.

Rice, P., Selma, J.D., and Abdul-Rezzak, R.K., Effect of temperature on thermal properties of "Recor" potatoes, *Int. J. Food Sci. Tech.*, 23, 281, 1988.

Riedel, L., Calorimetric investigation of the meat freezing process, *Kaltechnik*, 8, 38, 1957.

Schwartzberg, H.G., Effective heat capacities for the freezing and thawing of food, *J. Food Sci.*, 41, 152, 1976.

Schwartzberg, H.G., Singh, R.P., and Sarkar, A., Freezing and thawing of foods—computation methods and thermal properties correlation, in *Heat Transfer Advances in Food Processing*, Yanniotis, S. and Sunden, B., (Eds.), WIT Press, Ashurst, UK, 2007.

Short, B.E., Woolrich, W.R., and Bartlett, L.H., Specific heat of foodstuffs, *Refrig. Eng.*, 44, 385, 1942.

Shrivastava, M. and Datta, A.K., Determination of specific heat and thermal conductivity of mushrooms (*Pleurotus florida*), *J. Food Eng.*, 39, 255, 1999.

Siebel, E., Specific heats of various products, *Ice and Refrigeration*, 2, 256, 1892.

Singh, K.K. and Goswami, T.K., Thermal properties of cumin seed, *J. Food Eng.*, 45, 181, 2000.

Singh, P.C., Singh, R.K., Bhamidipati, S.N., Singh, S.N., and Barone, P., Thermophysical properties of fresh and roasted coffee powders, *J. Food Proc. Eng.*, 20, 31, 1997.

Siripon, K., Tansakul, A., and Mittal, G.S., Heat transfer modeling of chicken cooking in hot water, *J. Food Eng.*, 40, 923, 2007.

Sreenarayanan, V.V. and Chattopadhay, P.K., Specific heat of rice bran, *Agric. Wastes*, 16, 217, 1986.

Staph, H.E., Specific heats of foodstuffs, *Refrigerating Eng.*, 57, 767, 1949.

Stitt, F. and Kennedy, E.K., Specific heats of dehydrated vegetables and egg powder, *Food Res.*, 10, 426, 1945.

Subramanian, S. and Viswanathan, R., Thermal properties of minor millet grains and flours, *Biosystems Eng.*, 84, 289, 2003.

Tanaka, K. and Uematsu, M., Calorimeter for measuring the isobaric heat capacity of fluids and fluid mixtures by the thermal relaxation method, *Rev. Sci. Inst.*, 77, 035110, 2006.

Tang, J., Sokhansanj, S., Yannacopoulos, Y., and Kasap, S.O., Specific heat capacity of lentil seeds by Differential Scanning Calorimetry, *Trans. ASAE*, 34, 517, 1991.

Turrell, F.M. and Perry, R.L., Specific heat and heat conductivity of citrus fruit, *Proc. Am. Soc. Hort. Sci.*, 70, 261, 1957.

Vagenas, G.K., Marinos-Kouris, D., and Saravacos, G.D., Thermal properties of raisins, *J. Food Eng.*, 11, 147, 1990.

Wallapan, K., Sweat, V.E., Arce, J.A., and Dahm, P.F., Thermal diffusivity and conductivity of defatted soy flour, *Trans. ASAE*, 27, 1610, 1984.

Wright, M.E. and Porterfield, J.D., Specific heat of Spanish peanuts, *Trans. ASAE*, 13, 508, 1970.

Zhang, J., Farkas, B.E., and Hale, S.A., Thermal properties of skipjack tuna (*Katsuwonus pelamis*), *Int. J. Food Prop.*, 4, 81, 2001.

Zueco, J., Alhama, F., and Fernandez, C.F.G., Inverse determination of the specific heat of foods, *J. Food Eng.*, 66, 347, 2004.

CHAPTER 17

Thermal Conductivity Measurement of Foods

Jasim Ahmed and Mohammad Shafiur Rahman

CONTENTS

17.1 DEFINITION

17.1.1 Thermal Conductivity

Transport phenomena in food processing involve momentum, heat, and mass transfer and need thermophysical properties to solve heat transfer problems. A review on the transport phenomena in food engineering is given by Welti-Chanes et al. (2005). The thermal conductivity of a food is an important thermophysical property that is commonly used in calculations involving rate of conductive heat transfer. The rate of heat flow through a material by conduction can be predicted by Fourier's law as

$$Q = -kA\frac{\partial T}{\partial x} \tag{17.1}$$

where
- Q is the rate of heat flow (J/s)
- A is the area of heat transfer normal to heat flow (m^2)
- $(\partial T/\partial x)$ is the temperature gradient along the x-direction
- k is the proportionality constant of thermal conductivity (W/m K)

Fourier's law indicates that thermal conductivity is independent of the temperature gradient but varies with temperature; however, it is not always in the same direction. Thermal conductivity can be expressed at a wider temperature range by an empirical equation:

$$k = a + bT \tag{17.2}$$

where a and b are empirical constants. Most foods contain high moisture and, therefore, thermal conductivity values are close to that of water. Water has k values in the range of 0.5 to 0.7 W/(m °C), and k increases with an increase in temperature.

17.1.2 Mechanisms of Heat Conduction

Heat conduction is the thermal transmission by molecular vibration. In the case of solids, there are two principal carriers of heat energy from molecule to molecule in conduction: (1) free electrons that are usually present in metals and semiconductors, (2) lattice waves that are always present. In metals, which have more free electrons than alloys and thermal insulating materials, heat conduction occurs mainly due to the flow of free electrons. In alloys and thermal insulating materials that have limited free electrons, heat conduction from molecule to molecule depends mainly that on lattice vibrations. This difference in the number of free electrons causes metals to have higher thermal conductivities than most alloys and insulating materials.

When heating, lattice vibrations are increased, causing the thermal conductivity of most alloys and insulating materials to increase with increasing temperature. The thermal conductivity of pure metal tends to decrease with increasing temperature. This is because the lattice vibrations impede the motion of the free electron, thus causing that component of the thermal conductivity to decrease faster than the increased component of the thermal conductivity due to greater lattice vibrations. In the case of liquids and gases, heat conduction from molecule to molecule takes place by energy interchange from molecular collisions.

Food materials are composites with various components in different states and, therefore, heat transferred by conduction may occur in several forms. The resulting effect of temperature on conductivity is not easily established by employing basic knowledge of heat transfer mechanisms in solids, gases, and liquids as described here. In addition, increasing the temperature of foodstuffs may cause several changes in their physical or chemical properties.

17.2 MEASUREMENT TECHNIQUES

Thermal conductivity may be estimated in different ways. However, mass transfer to and from solid foods differs widely from process to process, which consequently affects thermal conductivity measurement techniques. Methods of thermal conductivity measurement may be divided into three major techniques as described by Murakami and Okos (1989): (a) steady-state techniques, (b) quasi-steady-state techniques, and (c) transient techniques.

17.2.1 Steady-State Techniques

In steady-state conditions (time has no influence on the temperature distribution within an object though temperature may vary at different locations), the heat flux is measured either by using an electrical heater resistance or by an induction coil. The surface temperature is calculated by averaging the surface temperatures of the system. The heat transfer coefficient is commonly calculated by Newton's law of cooling. This technique provides the best surface heat transfer measurement, provided the test conditions are constant and there is no mass transfer. The process has limitations in measuring heat transfer of food materials since it takes a long time to reach equilibration.

17.2.1.1 Guarded Hot Plate Method

Parallel plate and guarded hot plate have found frequent use in the case of nonfood materials. Guarded hot plate is most suitable for dry homogeneous samples that can be formed into a slab. The standard guarded hot plate method is based on the steady-state longitudinal heat flow principle, which determines the thermal conductivity of the material by applying Fourier's law. According to ASTM Standards (C177-45, 1955), the experimental setup consisted of placing the sample material on each side of a hot plate. This assembly was then sandwiched between cold plates maintained at lower temperatures by liquids circulating through them. Thermal conductivity was determined by measuring the heat flux through the samples for a resulting temperature gradient. The apparatus is so called because of the edge losses from the sample being prevented by a guard ring. The unidirectional steady-state solution of the heat conduction equation for homogeneous isotropic materials with no internal heat generation can be written as

$$Q = \frac{kA(T_1 - T_2)}{l} \tag{17.3}$$

The value of k can be calculated by measuring the heat quantity Q (J/s), the inner and outer temperatures T_1 and T_2 (°C), and the sample thickness l (m). Lentz (1961) used a modified guarded hot plate apparatus in measuring the thermal conductivity of fish and poultry over a range of temperatures between 5°C and −25°C. Lentz used an insulation material of known thermal conductivity on one side of the hot plate instead of introducing a second sample to minimize the error. Mohsenin (1980) reported that with this method the measurements would take a considerable amount of time because of the several hours needed for the sample to reach a steady-state condition and because of larger sample sizes. This will eventually promote moisture migration within the sample. Therefore, it was proven more successful especially for dry homogeneous samples in the form of slabs. The method is not suitable for high-moisture materials where moisture migration takes place during steady-state conditions. The thermal conductivity of dried or frozen foods can be measured by this technique. Willix et al. (1998) used this method to generate a huge amount of data for thermal conductivity of foods above and below frozen temperatures.

17.2.1.2 Concentric Cylinder Method

This apparatus consists of two concentric cylinders, and the sample is placed between them and their ends are insulated. The heater is located at the outer cylinder whereas the coolant is circulated along the inner cylinder. The heat absorbed by the coolant is assumed to be equal to the heat that passes through the sample. Thermal conductivity can be calculated from the unidirectional radial heat transfer equation as

$$Q = A_{\text{ou}}\left[\frac{T_{\text{ou}} - T_{\text{in}}}{r_{\text{ou}}\ln(r_{\text{in}}/r_{\text{ou}})}\right] \tag{17.4}$$

Lin et al. (2003) used concentric cylinders to measure shear-dependent thermal conductivities with special design and construction. It was based on the coaxial cylinder system in which the inner cylinder is stationary and the outer cylinder can be rotated. The rest of the system consisted of a constant temperature bath, a rotating mechanism driven by a motor, a main heater, a guard heater, calibrated thermocouples (TC), power supply, and a data acquisition system.

17.2.1.3 Heat-Flux Method

A heat flow meter is a commercially available device for measuring heat flux. It measures the temperature gradient across the sample. A heat flow meter is suitable for materials with conductance (k/L) less than 11.3 $W/(m^2\,K)$ (Haas and Felselstein, 1978). Tong and Sheen (1992) proposed a technique to determine the effective thermal conductivity of a retortable multilayered plastic container. However, most available methods are not well suited for thin materials with multilayered construction and, therefore, those authors advocated a technique based on heat flux. The experimental setup is shown in Figure 17.1. A heat flux sensor was bonded to the inner surface of the plastic wall with a thin layer of high-thermal conductivity adhesive (Omegatherm 201) to assure good

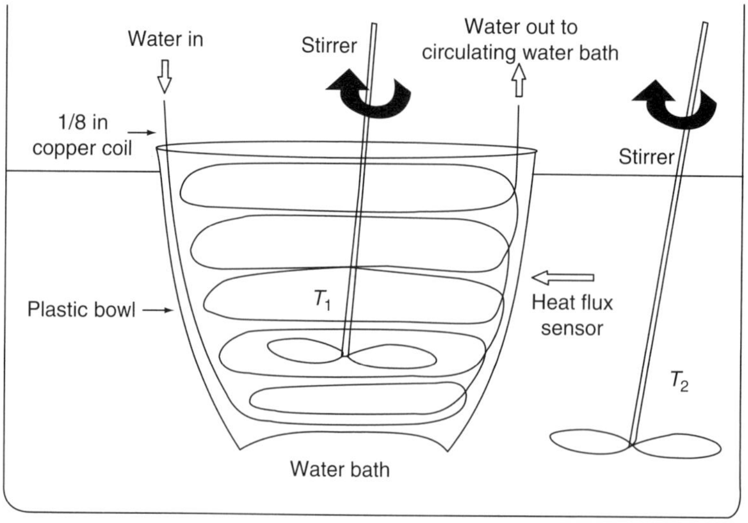

FIGURE 17.1 Thermal conductivity measurement of a multilayered plastic container by heat flux method. (From Tong, C.H. and Shen, S., *J. Food Process. Pres.*, 16, 233, 1992. With permission.)

contact between plastic wall and the heat flux sensor. Millivolt signals generated by the heat flux sensor are amplified 50 times by a DC millivolt amplifier and then recorded with a strip chart recorder. The plastic bowl filled with water is immersed up to 85% of its height in a constant temperature (T_{ou}) water bath. A copper coil (1/8 in outer diameter) connected to a constant temperature circulating water bath is immersed in the bowl to control bowl water temperature (T_{in}). A temperature difference from 5°C to 7°C was always maintained by Tong and Sheen (1992). The thermal conductivity is evaluated at $(T_{in} + T_{ou})/2$. At steady state, heat flux can be expressed as

$$W = U(T_{ou} - T_{in}) \tag{17.5}$$

where W is the heat flux (W/m^2) and U is the overall heat transfer coefficient (W/m^2 K). The overall heat transfer coefficient can be written as

$$\frac{1}{U} = \frac{1}{U_{in}} + \frac{1}{U_{ou}} + \sum_{i=1}^{N}\left(\frac{l_i}{k_i}\right) \tag{17.6}$$

where

U_{in} and U_{ou} are the internal and external heat transfer coefficient (W/m^2 K)
l_i and k_i are the thickness (m) and thermal conductivity (W/m K) of each layer
N is the number of layers

When U_{in} and U_{ou} are large, Equation 17.6 can be reduced to

$$\frac{1}{U} = \sum_{i=1}^{N}\left(\frac{l_i}{k_i}\right) \tag{17.7}$$

Since it is not possible to measure the thermal conductivity of each layer, an effective thermal conductivity can be used:

$$\sum_{i=1}^{N}\left(\frac{l_i}{k_i}\right) = \frac{l_{con}}{k_e} \tag{17.8}$$

where l_{con} and k_e are the thickness and thermal conductivity of the plastic container, respectively. Combining Equations 17.5, 17.7, and 17.8 yields

$$k_e = \frac{W l_{con}}{(T_{ou} - T_{in})} \tag{17.9}$$

Equation 17.9 can be used to calculate the effective thermal conductivity of a multilayered plastic container. The advantages of the steady-state methods are the simplicity of the mathematical equations and the high degree of control of experimental variables, which is reflected in the high-precision results (Nesvadba, 1982). The disadvantages of the method are (1) long equilibration time (up to several hours), (2) need for defined geometry of the sample, (3) moisture migration due to long equilibration time, and (4) presence of convection in case of liquid or high-moisture materials.

These well-established standard techniques work well for nonbiological materials but are not followed well by food materials (Sweat, 1986). Therefore, details of these methods are not described here.

17.2.1.4 *Differential Scanning Calorimetry*

A rapid method for measuring the thermal conductivity of a small cylindrical (disk) sample using differential scanning calorimetry (DSC) was proposed by Ladbury et al. (1990). Use of a DSC is based on a steady-state method; therefore, the thermal conductivity, k, could be calculated by the following expression:

$$k = \left(\frac{L}{\Delta Q}\right)\left(\frac{A}{\Delta T_2 - \Delta T_1}\right) \tag{17.10}$$

where
L is the sample length (m)
ΔQ is the difference in energy required to maintain pan temperature (W)
A is the sample area perpendicular to heat flow (m^2)
ΔT_2 is the final temperature difference between DSC heating pan and sample (K)
ΔT_1 is the initial temperature difference between DSC heating pan and sample (K)

The assumptions in the above equation are based on (1) good contact of sample, (2) negligible heat losses from the radial direction (i.e., thin disk-shaped sample), and (3) no thermal gradient in the heat sink.

The heat-sink temperature could be maintained by measuring and inserting a TC into the heat-sink cell screw. The diagrams of an assembled DSC head with the sample in place and DSC head cover attachment are shown in Figure 17.2a and b. Buhri and Singh (1993) experimented on the DSC method for foods and found that this method was reliable, precise, and a relatively rapid technique for determining thermal conductivity. The sample dimension was 6.5 mm internal diameter and 12.7 mm long with an aluminum bottom. The TC probe was inserted into the sample to a depth of 6.35 mm. Recently, Camirand (2004) measured the thermal conductivity of solid materials using DSC with a sample diameter less than 8.0 mm and a height less than 2.0 mm. The solid sample for which to determine the thermal conductivity is placed into the sample furnace of the calorimeter, and indium, a calibration substance, is placed on top of the sample. The reference furnace is kept empty. During melting of the calibration substance, the temperature of the calibration

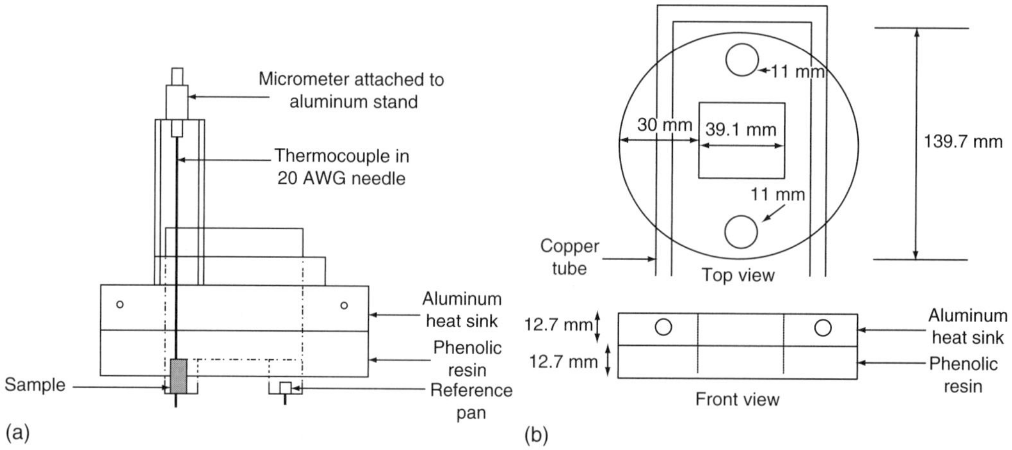

FIGURE 17.2 (a) Assembled DSC head with sample in place; (b) DSC head cover attachment for thermal conductivity measurement. (From Buhri, A.R. and Singh, R.P., *J. Food Sci.*, 58, 1145, 1993. With permission.)

substance must be constant. A scan is carried out to measure the differential power generated during the melting of the calibration substance. The resulting curve shows an increasing trend and is approximately linear during the melting and decreases exponentially after melting is completed. Measurement of the slope of the increasing part of the curve allows determination of the thermal conductivity of the sample. The analysis of heat transfer of this experiment predicts that the slope of the differential power during the transition is proportional to factor 2 and inversely proportional to the sum of the thermal resistances.

$$\text{Slope} = \frac{\mathrm{d}\Delta P}{\mathrm{d}T_\mathrm{P}} = \frac{2}{R} \tag{17.11}$$

where
$\mathrm{d}\Delta P$ is the differential power of the calorimeter obtained after the subtraction of the zero line
T_P is the programming temperature of the calorimeter
R is the total thermal resistance

$$R = R_1 + R_2 + R_\mathrm{S} \tag{17.12}$$

where
R_1 is the thermal contact resistance between the sample and the sample furnace
R_2 is the thermal contact resistance between the sample and the calibration substance
R_S is the thermal resistance of the sample

$$R_\mathrm{S} = \frac{L_\mathrm{S}}{k_\mathrm{S} A_\mathrm{S}} \tag{17.13}$$

where L_S, k_S, and A_S are the height of the sample, its thermal conductivity, and the area of a horizontal cross-section of it, respectively. Equation 17.11 states that samples with relatively small thermal conductivity will have in general a lower slope than more conductive materials. To measure the thermal conductivity of a material, it is necessary to repeat the experiment with samples having known different heights and constant cross-sectional areas.

Assuming the samples have the same thermal contact resistance $R_1 + R_2$, according to Equations 17.12 and 17.13, the total thermal resistance is a linear function of the $L_\mathrm{S}/A_\mathrm{S}$ ratio. Therefore, a plot of the total thermal resistance versus the $L_\mathrm{S}/A_\mathrm{S}$ ratio is in principle a straight line. The inverse of the slope is equal to the k_S thermal conductivity of the material and the ordinate intercept is equal to the sum of $R_1 + R_2$.

The average coefficient of variation was reported as 2.88% compared with 5.3% of the line source method. The main advantages of this method are the relatively small sample size requirement and short time requirement. However, there are some limitations of this technique. The sample length must be precise; a 10% error in length measurement would affect thermal conductivity calculation accordingly. Moreover, with long heating to reach steady state, moisture loss may occur and cause errors in readings. The major limitation is the cost of a DSC unit, which further needs a delicate modification to implement conductivity measurement.

17.2.2 Quasi-Steady-State Techniques

17.2.2.1 Fitch Method

The Fitch method is one of the most commonly used methods to measure the thermal conductivity of poor conductors. The Fitch apparatus (Fitch, 1935) consists of a heat source or sink in the form of a vessel filled with a constant-temperature liquid and a sink or source in the form of a copper

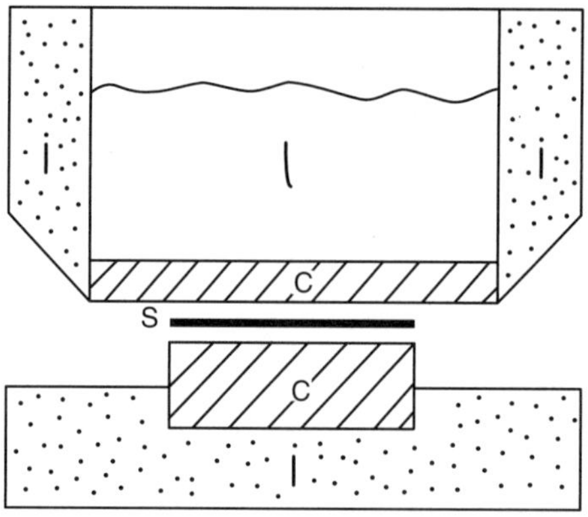

FIGURE 17.3 Fitch apparatus for measuring thermal conductivity: c, copper plug; i, insulation; l, liquid; s, sample. (Adapted from Fitch, A. L., *Am. Phys. Teacher,* 3, 135, 1935.)

plug insulated on all sides but the one facing the vessel (Figure 17.3). The sample is sandwiched between the vessel and the open face of the plug. TCs record the temperature history of the copper block, which is assumed to have a uniform temperature distribution, and the temperature of the bottom of the vessel. The model assumes a linear temperature profile, negligible heat storage in the sample, and negligible surface contact resistance. Heat transfer between the copper block and its insulation is assumed negligible. This method is the simplest because minimum equipment is necessary.

17.2.2.1.1 Theory

The quasi-steady conduction heat transfer through the sample yields the following simplified equation (Fitch, 1935; Mohsenin, 1980):

$$\frac{Ak_e(T - T_S)}{l} = m_{cp} C_{cp} \frac{\partial T}{\partial t} \tag{17.14}$$

with the initial condition

at $t = 0$, $T = T_0$

The solution of Equation 17.14 with the initial condition is

$$\ln\left[\frac{T_0 - T_S}{T - T_S}\right] = \frac{Ak_e t}{l m_{cp} C_{cp}} \tag{17.15}$$

The plot of the dimensionless temperature $[(T_0 - T_S)/(T - T_S)]$ against time (t) on semilogarithmic paper is linear with an initial curvilinear part. The thermal conductivity of the sample is calculated from the slope of the linear part of this plot with known values of heat capacity of the copper block, the thickness, and heat transfer area of the sample. However, Equation 17.15 holds true, considering some assumptions (Zuritz et al., 1989). The assumptions are (1) the contact resistance is negligible, (2) the heat storage in the sample is negligible (i.e., heat transfer in the sample is quasi-steady state), (3) the heat transfer at the edges of the sample and copper plug is negligible, (4) temperature of the heat source or sink (water tank) is constant, (5) the initial temperature of the copper plug and the sample are the same, (6) the temperature distribution in the copper plug is uniform, (7) the sample is homogeneous, and (8) the temperature history is linear on a semilog plot for the duration of the experiment. The error introduced by this assumption may be corrected by calibrating the setup using a specimen whose k is known precisely and thus obtaining a correction factor for the apparatus (Mohsenin, 1980; Rahman, 1991):

$$f = \frac{\text{true } k \text{ of specimen}}{\text{measured } k} \tag{17.16}$$

Rahman (1991) observed that the correction factor (f) varied with the thermal conductivities of the standard samples, sample thickness, and area of heat transfer, which finally led to correlate f with the sample thickness, and measured k by a quadratic equation for a defined surface area.

17.2.2.2 Cenco–Fitch Method

Bennett et al. (1962) mentioned a modification of the Fitch method known as the Cenco–Fitch method (Figure 17.4) while Kopelman (1966) rejected the use of modification on finding that a minimum error of 15% was possible. The reasons cited were (1) the unpredictable errors arising from the applied pressure, which may cause extrusion of liquid or change the thickness or the physical properties of the sample, (2) neglecting the radial heat flow, and (3) unavoidable and practically nonmeasurable heat loss from the sink or source. The above limitations have been taken care of by Zuritz et al. (1989) and Rahman (1991) to modify the Cenco–Fitch apparatus.

17.2.2.3 Zuritz et al. and Fitch's Method

Murakami et al. (1984) described a modified Fitch apparatus for measuring small disk-shaped food particles. The apparatus had a largely reduced heat transfer area compared with the regular Fitch apparatus. The heat sink and heat source were equal in diameter (6.35 mm). Later, Zuritz et al. (1989) modified the Fitch apparatus similar to that of Murakami et al. (1984) to measure the thermal conductivity of kidney beans. The method is simpler than the original Fitch apparatus (Figure 17.5). Model error analysis showed that the errors associated with sample thickness and heat transfer area had the major influence on the error in sample conductivity measurement. Ensuring good contact between sample and heat source and sink was mentioned as being critical.

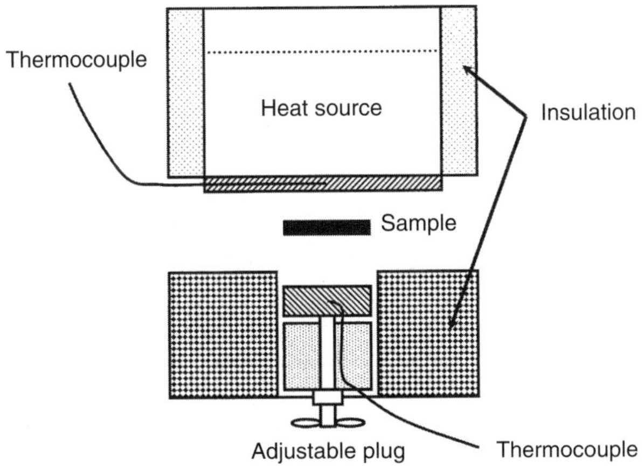

FIGURE 17.4 Cenco–Fitch apparatus for measuring thermal conductivity. (From Reidy, G.A. and Rippen, A.L., *Trans. ASAE,* 11, 248, 1971. With permission.)

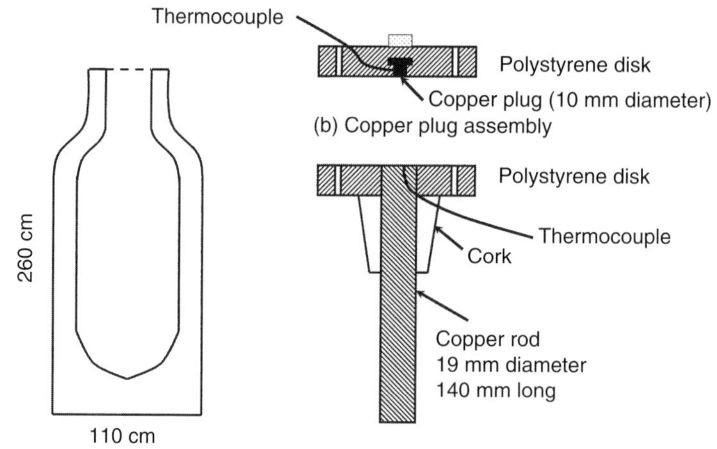

FIGURE 17.5 Schematic of Zuritz et al.–Fitch apparatus. (From Zuritz, C.A., Sastry, S.K., McCoy, S.C., Murakami, E.G., and Blaisdeh, J.L., *Trans. Am. Soc. Agric. Eng.*, 32, 711, 1989. With permission.)

Thermal conductivity measurement is difficult by probe methods while the sample does not have a regular geometrical shape, i.e., grains, beans, nuts, and sea foods. In such cases, bulk thermal conductivity, a combined property of materials and air in the pore spaces, is usually measured. Bulk conductivity may be suitable for calculating the heat transfer in processes that involve stationary materials (i.e., batch drying and cooking) but the thermal conductivity of individual particles is necessary for calculating the heat transfer when materials are either in motion or surrounded by the heating or cooling medium (Zuritz et al., 1989). Bulk thermal conductivity is not appropriate in evaluating the heat transfer in processes such as continuous flow drying and freezing, soaking, and blanching. The main advantage of this instrument is that it can measure the thermal conductivity of small food materials of 6.35 mm diameter. The instrument was calibrated by measuring the thermal conductivity of glass ($k = 1.04$ W/m K), and the observed percent error was 6.9%. The temperature history for the glass sheet and kidney bean sample measured by Zuritz et al. and Fitch's methods is shown in Figure 17.6.

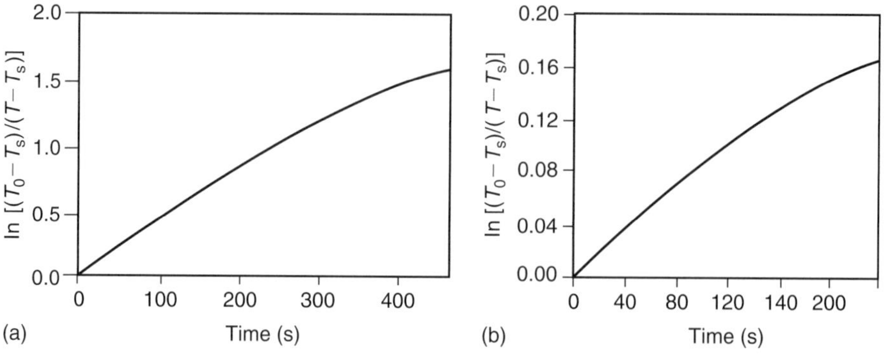

FIGURE 17.6 Temperature history of the sample measured by Zuritz et al. and Fitch's method, (a) 2.25 mm glass sheet; (b) 1.1 mm thick kidney bean. (From Zuritz, C.A., Sastry, S.K., McCoy, S.C., Murakami, E.G., and Blaisdeh, J.L., *Trans. Am. Soc. Agric. Eng.*, 32, 711, 1989. With permission.)

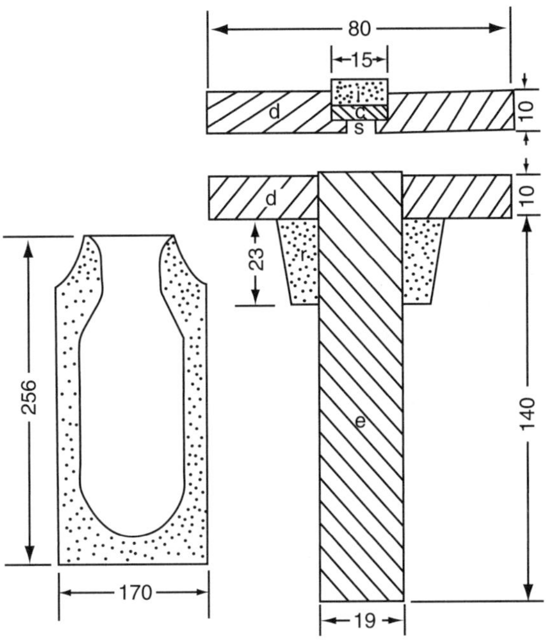

FIGURE 17.7 Schematic of Rahman–Fitch apparatus with vacuum flask, copper disk assembly, and copper rod assembly: c, copper disk; e, copper rod; i, insulation; r, rubber cork; s, sample compartment (all dimensions in mm). (From Rahman, M.S., *J. Food Eng.*, 14, 71, 1991. With permission.)

17.2.2.4 Rahman–Fitch Method

Rahman (1991a) further modified the apparatus of Zuritz et al. (1989) for measuring the thermal conductivity of fresh and frozen foods (Figure 17.7). The major advantages of this modification over other Fitch-type apparatus are (1) heat loss from the edge of the sample is minimal, and (2) all other Fitch or modified Fitch-type apparatuses permit expansion of the sample in the radial direction, causing a reduction in thickness, an increase in the heat transfer area, and extrusion of liquid from the sample. The sample compartment between the copper rod and the disk prevent the above problems.

The error of this apparatus was predicted by independent measurement of fresh apple, carrot, calamari, fat (beef), pear, and squid samples of 4 mm thickness. After calibration, this method resulted in 5% error above freezing and 9% error below freezing. However, the sensitivity of this apparatus reduced significantly for higher thermal conductivity samples, for example, frozen foods.

From a theoretical point of view, the precision and error depend on the applicability of Equation 17.15. There will be minimum error and high precision (i.e., quasi-steady state) when Equation 17.15 is satisfied. Hence, the error arises due to not satisfying quasi-steady-state conduction and depends on A, k_e, T_s, l, m_{cp}, C_{cp}, and t. The slope (S) is the response of the apparatus and can be written as

$$S = \frac{Ak_e}{lm_{cp}C_{cp}} \tag{17.17}$$

Hence, the response and precision depend on A, l, and m_{cp}. The values of A, l, and m_{cp} should be selected based on practical considerations and limitations. For example, the response increases with decreasing thickness l, which is good up to a point when l cannot be measured precisely. Again, conductivity of fresh food is slightly dependent on temperature above freezing, but below freezing conductivity changes significantly with temperature. In this case, a lower l gives a higher response, which reduces precision. Also, the highest response is limited by the heat transfer from the copper disk to the insulation. Zuritz et al. (1989) found that the heat transfer rate between the copper plug and its insulation became significant (i.e., >10%) when the temperature change in the copper plug exceeded 15°C.

In the case of the Fitch-type apparatus, there is an optimum sample thickness for best results for a given material (Mohsenin, 1980). Bennett et al. (1962) used silicon rubber ($k = 0.2465$ W/m K) for determining optimum sample thickness for their Fitch-type instrument. They determined the thermal conductivity of rubber at three different sample thicknesses and considered the optimum sample thickness as that which gave the lowest error. In that case, the optimum sample thickness was 6.35 mm. Then the authors used that sample thickness for measuring the thermal conductivity of Valencia orange. Walters and May (1963) also used a Fitch-type apparatus to determine the conductivity of chicken muscle, but they did not mention how optimum sample thickness was chosen. Zuritz et al. (1989) estimated the upper limit of optimum sample thickness by an order of magnitude analysis. They mentioned that the following relationship must hold:

$$\frac{m_{sa}C_{sa}(\delta T_s/\delta t)}{m_{cp}C_{cp}(\delta T_{cp}/\delta t)} \ll 1 \qquad (17.18)$$

Since the rate of temperature change for both the sample and the copper plug is of the same order of magnitude, inequality (13) becomes

$$\frac{m_{sa}C_{sa}}{m_{cp}C_{cp}} \ll 1 \qquad (17.19)$$

implying that a value at least one order of magnitude is lower than 1. For glass, the optimum sample thickness for Zuritz et al. (1989) apparatus was $l_{max} \ll 0.065$ m. The response (i.e., slope) of Rahman's apparatus ($m_{CD} = 8.132 \times 10^{-3}$ kg and $D_{CD} = 10 \times 10^{-3}$ m) is given in Figure 17.8a for fresh potato and apple. The slope decreased with an increase in sample thickness; however, no

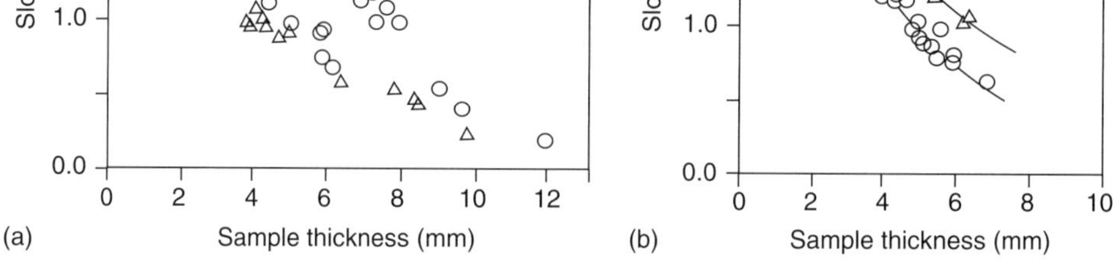

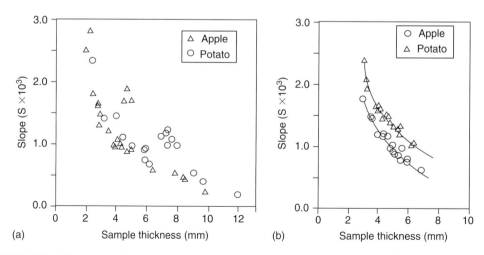

FIGURE 17.8 Effect of sample thickness on response (slope): (a) for $m_{CD} = 8.132 \times 10^{-3}$ and $D_{CD} = 10 \times 10^{-3}$; (b) for $m_{CD} = 8.648 \times 10^{-3}$ and $D_{CD} = 15 \times 10^{-3}$. (From Rahman, M.S., *J. Food Eng.*, 14, 71, 1991a. With permission.)

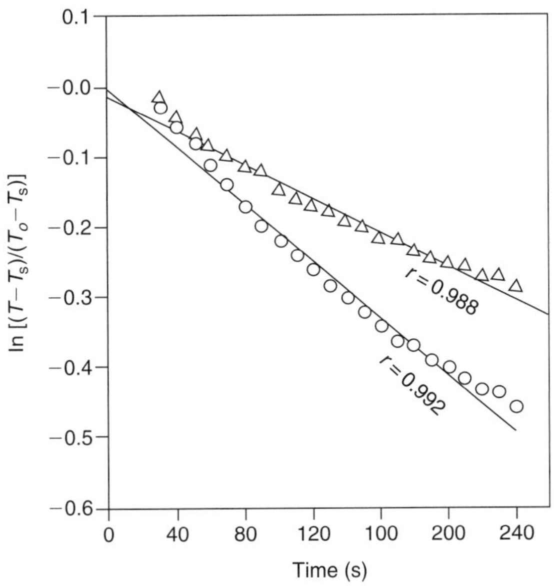

FIGURE 17.9 Plot of [ln (temperature ratio)] against cooling time, apple diameter: 14 mm; calamari mantle diameter: 15 mm. (From Rahman, M.S. and Driscoll, R.H., *Food Australia*, 43, 356, 1991.)

significant differences were observed between the response of potato and apple. This was improved by increasing the heat transfer area in the apparatus ($m_{CD} = 8.648 \times 10^{-3}$ kg and $D_{CD} = 15 \times 10^{-3}$ m), and the results are given in Figure 17.8b. Hence, the correction factor depends not only on the sample thickness, but also on the contact area. The linearity of the temperature plot of Rahman's (1991a) apparatus is given in Figure 17.9.

17.2.2.5 Temperature Profile

Two opposite surfaces of a slab are heated to an equal temperature while the other surfaces are insulated. The temperature profile of the slab is read with time, and it is fitted to a parabolic function. Thermal conductivity can be estimated from the temperature history.

17.2.3 Transient Techniques

17.2.3.1 Line Source Method

Transient (unsteady-state) heat transfer (temperature changes with location and time) is one of the most commonly used methods for food materials, where variation in the product ensues after an initial change in ambient temperature. It is also known as the line heat source method. The nonstationary technique is the most widely used for food materials due to its convenience, fast measurement, low cost, relatively small sample size, and independence of sample geometry in the case of a large sample. The method has been developed and well reviewed by various researchers, such as Stalhane and Pyk (1933), Van der Held and Van Drunen (1949), Hooper and Lepper (1950), Sweat and Haugh (1974), and Murakami et al. (1996a,b), to measure the thermal conductivity in solids and liquids. The line heat source thermal conductivity probe has been applied in different foods, such as potato salad (Dickerson, 1968), cucumber, onion, apple, peach, and applesauce (Sweat and Haugh 1974), and partly baked bread during freezing (Hamdami et al., 2004). This technique is considered a simple method because it is easy to operate with good accuracy.

17.2.3.1.1 Theory and Measurement

The details of the basic theory and mathematical derivation behind the use of the line source probe have been discussed earlier by Hooper and Lepper (1950) and Nix et al. (1967). The measurement is based on the mathematical solution of an ideal system composed by an infinite linear heat source of zero mass, immersed in an infinite solid medium. Briefly, the line heat source is embedded in the material whose thermal conductivity is to be evaluated. From a condition of thermal equilibrium, the heat source is energized and heats the medium with constant power. The temperature response of the medium is a function of its thermal properties. The thermal conductivity is found from the temperature rise measured at a known distance from the heat source. The linear range of slope of log time versus temperature plot was used for the calculation of thermal conductivity (Equation 17.20). The power input and the probe temperature rise during short contact time period are the only parameters required to measure the thermal conductivity of a given food sample.

$$T = B + \left[\frac{q}{4\pi k}\right] \ln t \quad (17.20)$$

where
- q is the heat input per unit length of the line source (W/m)
- t is the time of the experiment (s)
- B is the intercept

The following assumptions are required to validate the above equation: (1) line source is in an infinite medium, (2) temperature rise measurement occurs at a point close to the line heat source, (3) unidirectional heat flow is in the radial direction, and (4) the radius of the line source must be such that an addition of small points in the longitudinal direction can represent the line source. A plot of T versus ln t should give a straight line. A typical curve of T versus ln t is shown in Figure 17.10 (a linear portion observed at zone B).

17.2.3.1.2 Probe Size

The line source theory is based on a line heat source of infinite length and infinitesimal diameter with negligible axial heat flow. One source of error with the line heat source probe is axial flow error due to its finite length. The probe length-to-diameter ratio ($L/D = \Gamma$) indicates the validity of the

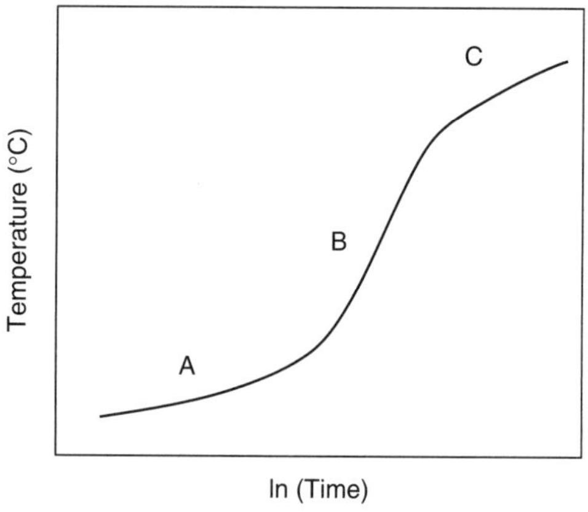

FIGURE 17.10 Typical temperature rise in the probe as a function of ln (time).

assumption. Hooper and Lepper (1950) recommended a probe length-to-diameter ratio of 100 to eliminate this error. Blackwell and Misener (1951) advocated that a value of $\Gamma = 20$ reduced this error well below their experimental error. Concerned with the large thermal mass of Hooper and Lepper's apparatus, D'Eustachio and Schreiner (1952) made a smaller probe ($\Gamma = 131$). Later, Blackwell (1956) derived an expression for calculating the upper limit to the axial flow error and found that the maximum calculated error was less than 1% for geophysical application. Sweat and Haugh (1974) observed no statistical difference ($p < 0.1$), at the 10% level, between probes with ratios varying from 93 down to 31 while testing glycerine. Sweat (1986) recommended that the ratio of the probe length to diameter should be higher than 25.

Detailed drawings of the conventional probe used are shown in Figure 17.11. Probes are commonly constructed using a heating element and a TC, which are placed inside or on the outside of the tube. Internal placement makes the probe stronger while an external placement of the heater and TC has been reported to yield data of higher accuracy and with lower standard deviation. Some recommendations for thermal conductivity probe design parameters for nonfrozen foods have been made by a group of researchers (Murakami et al., 1996a,b). The probe diameter should be as small as possible to minimize truncation error. This method also has the advantage of small size; therefore, it can be used to measure the thermal conductivity of small food samples. However, small *k* probes are difficult to make and in addition they cannot be used for a nonviscous liquid involving convection, which will develop around the heated probe. Same *k* probes can be used for different food materials only whereas the design must be based on the lowest possible thermal diffusivity to minimize the associated errors.

Generally, the *k* probe has been derived from an idealized heat transfer model and, therefore, there are obvious differences between a real and a theoretical model. These differences (finite probe diameter instead of infinitely long length; difference in thermal properties of probe compared to sample) have resulted in considerable errors in thermal conductivity measurement. The finite length of the *k* probe produces some heat due to axial flow, which could be reduced by considering Blackwell (1956) equation. During experimentation Van der Held and Van Drunen (1949) found a nonlinearity in the time–temperature plot of *k* probe data at the beginning and the authors introduced the time correction factor, which they subtracted from each time observation to correct for the effect

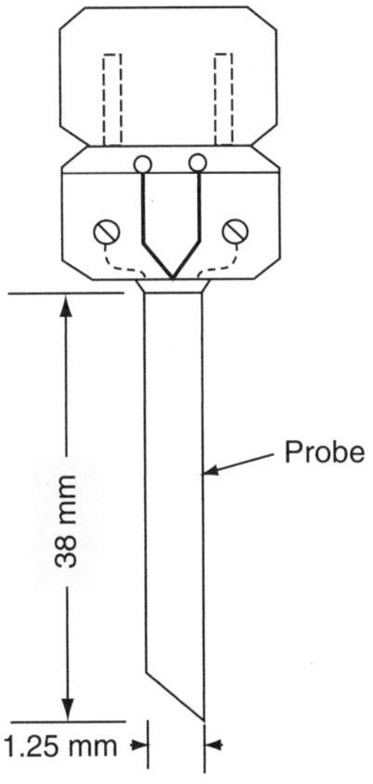

FIGURE 17.11 Thermal conductivity probe. (From Rahman, M.S., Thermophysical properties of seafoods. PhD thesis, Department of Food Science and Technology, University of New South Wales, Australia, 1991b.)

of finite heater diameter and any resistance to heat transfer between the heat source and samples. Underwood and McTaggart (1960) also proposed time correction factors, using a different method that was less time consuming than the above. Hooper and Lepper (1950) found that the above time correction factor remained constant for a specific apparatus. D'Eustachio and Schreiner (1952) stated that probably the correction factor can be eliminated for most instruments by maintaining a small probe size. Lasansky and Bilanski (1973, cited by Mohsenin, 1980), who employed a probe for measurement of conductivity of soybeans, found that even when the maximun value of an experimentally determined factor was included for conductivity calculations, the resulting conductivity was only 4% higher than that calculated by the noncorrected equation. For this reason, they did not allow for any corrections for the soybean conductivity reported. No correction factors were required with the probes when measuring the conductivity of meat except for tests at temperatures below freezing when correction factors between 0.2 and 1 were required (Sweat, 1972, cited by Sweat and Haugh, 1974). It was reasoned that a time correction factor was not required for samples at temperatures above freezing because the heat capacity of the probe was about the same as that of the food sample. At temperatures below freezing, the heat capacity of a food sample drops, changing the heat capacity ratio between the probe and the sample. Sweat (1986) concluded that the correction factor was negligible for probes with diameters as small as 0.66 mm.

Another commonly used correction factor is the calibration factor. It is recommended that the thermal conductivity probe should be calibrated with materials with known thermal conductivity values. The ratio of the published and measured k values of the calibration material is considered as the calibration factor. Simply multiplying the factor with the measured k values minimizes the error. Various researchers (Mann and Forsyth, 1956; Murakami et al., 1996b; Denys and Hendrickx, 1999) have recommended considering the calibration factor to enhance the accuracy of k measurement. Murakami et al. (1993) advocated that the time correction factor has no theoretical basis and similar results can be achieved with the optimum selection of initial time. The recent interest in high-pressure processing of foods has led to the measurement of k values at higher pressure levels. Bridgman (1923) reported that the thermal conductivity of liquids is increased by only a few percent under a pressure of about 100 MPa. Denys and Hendrickx (1999) measured thermal conductivity of food materials under high-hydrostatic pressure (0.1–400 MPa) and recommended that a probe-specific pressure and temperature- dependant calibration factor was needed. The main difference between measurements at atmospheric conditions and high pressure is that more calibration effort is demanded because high pressure has an effect on calibration factors. For successful application of the probe method at high pressure, the pressure relation of the calibration factor must be determined.

17.2.3.1.3 Sample Size

The sample size can cause errors during thermal conductivity measurement if the sample boundaries experience a temperature change. Knowledge of thermal diffusivity and the duration of the measurement affect the sample diameter. Vos (1955) advocated that when the sample size is kept within $4\alpha t/d^2 < 0.6$, the error due to the edge effect is negligible, with α being the thermal diffusivity of the sample, t time after heater is on and d the shortest distance between the probe and sample's outer edge. Therefore, it is convenient to increase the sample diameter to be able to achieve the linear section within a reasonable test time, and data should be taken before temperature changes at the sample boundaries become significant.

Nix et al. (1967) showed that an infinite solid may also be simulated by a cylinder of finite length and diameter with a length-to-diameter ratio of the sample greater than four. Sweat and Haugh (1972, cited by Mohsenin, 1980) reported the measured thermal conductivity of water containing 0.4% agar using a brass tube with an inside diameter of 1.9 cm. Agar was used to form a gel to minimize convection effects. A TC was placed adjacent to the inside wall of the tube to measure the temperature at the sample boundary. For a heater power of 1.54 W/ft length of the heater probe, which was inserted into the test sample, the outer boundaries of the sample did not experience a temperature rise greater than 0.1°C until 80–100 s had elapsed from the time the heater was energized. This indicated that a sample diameter of at least 1.9 cm was adequate for food samples that have thermal diffusivities of the same order of magnitude as that of water. Murakami et al. (1996b) and Shariaty-Niassar et al. (2000) also recommended that the specimen diameter should be more than 1 cm to exclude edge effect. Hence, to minimize this error, the measurement time can be shortened and the sample diameter can be increased. Wang and Hayakawa (1993) used O-rings in all contacting surfaces of the sample holder to prevent moisture loss at high temperatures.

17.2.3.1.4 Construction of Probe

Measurement of thermal conductivity by transient hot wire (THW) method can lead to unavoidable errors; however, these errors can be significantly reduced and sometimes compensated for by careful design and proper selection of construction materials for probes. Murakami et al. (1996a) have made some guidelines for designing and fabricating *k* probes for nonfrozen foods to minimize errors derived from the probe designing point of view. It is important to consider the location of the TC, presence of air inside the needle, material of the construction of TC, and selection of heater wire and obviously optimization of these parameters could enhance the accuracy of measurement.

The earliest version of *k* probe was designed by van der Held and Van Drunen (1949) placing heating wires and TC wires in an enclosed glass capillary tube. Later on Hooper and Lepper designed a portable *k* probe using aluminum tubing, which can be placed into the ground during measurement. D'Eustachio and Schreiner (1952) made a small probe using stainless steel tubing. Sweat and Haugh (1974) and Sweat (1974) designed a small probe from hypodermic needles. Mitchell and Kao (1978) used a ceramic tube instead of stainless steel for the probe needle while measuring the thermal resistivity of soil. The thermal resistivity of the sample was very close to the resistivity of the needle itself. Baghe-Khandan et al. (1981) developed an improved line heat source thermal conductivity probe to simplify the construction and extended the life of thermal conductivity probes, setting the probe needle with heater wire and TC on the outside of a solid sewing needle instead of inside of a hypodermic needle. The probe was an improvement compared with the previous probe, since there was no air gap between the wires and the sample. Closer contact between the wires and the sample causes a 57% lower temperature rise in the probe during the power application time. A temperature rise of 3°C was enough with the improved probe compared with a temperature rise of at least 7°C with a conventional probe. Better linearity between T and $\ln t$ was also found. However, the stuck wires introduced a distortion on the surface smoothness and exhibited asymmetric heating, which made it difficult to insert the needle into the sample.

Murakami et al. (1993) proposed a design that consists of a hypodermic stainless steel tubing, with which an insulated chromel–constantan TC wire (0.0508 mm diameter) and an insulated constantan heating wire (0.0762 mm) are introduced. Later on, Elustondo et al. (2001) modified the Murakami et al. (1993) design. The insulated constantan TC was placed inside of the hypodermic needle, entering through the bottom side and with the TC junction in the center of the heated length. The needle was placed upside down of the cylindrical container's axis through a metallic ring embedded in the isolated bottom which made one of the electrical contacts for the heating current. The needle tip was mounted into a similar ring placed on the removable lid and closed onto the electrical circuit. The authors claimed that the design allowed to measure *k* of liquid, powdered foods and paste.

Nix et al. (1967) recommended that $\beta < 0.16$ must be satisfied to achieve an accuracy of more than 1% in the equation derived for the line source technique (where $\hat{\beta} = \gamma/(\alpha t)^{0.5}$, where γ is the distance between the TC and the line heat source, α is the thermal diffusivity, t is time from the initiation of the line heat source). The critical distance γ was calculated to be <1.8 mm for beef perpendicular to the fibers, with 78.5% moisture and low fat content (Higgs and Swift, 1975). To improve the accuracy of the *k* probe by replacing air due to its low thermal conductivity, various researchers have worked to fill the voids in the *k* probe using various fillers to reduce response time. Moreley (1966) used epoxy resin as *k* probe filler for fat and meat applications. Karwe and Tong (1992) studied three filling materials: air, mercury, and omegatherm epoxy. The results indicated that large temperature gradients exist in the filling materials that have low thermal conductivity, which may violate some of the theoretical assumptions. However, by selecting a proper time interval (typically between 30 and 120 s) over which linear regression is carried out, the value of *k* can be calculated precisely. In addition, specific heat of the filling material should be as small as possible to obtain an accurate estimate of unknown *k*. Positioning of the TC in the probe has very little effect on the calculated thermal conductivity values for all these filling materials. Murakami et al. (1993) found that the thermal mass ratio of sample to probe affects the reliability of the probe data. Wang and Hayakawa (1993) used a high thermal conductivity paste to fill the void space and used a swage lock fitting to reduce heat loss from the tubing to the sample holder to eliminate metal-to metal-contact.

Gratzek and Toledo (1993) developed various probes and intended to use these at higher temperatures up to 130°C for a quick response and rapid attainment of linearity in temperature rise against ln time. The probes consisted of four sections: (1) miniature connectors, (2) Teflon support, (3) extension tubing, and (4) active probe section. Stainless steel active section (6.9 cm long 18 gage and 1.27 mm outer diameter) with Teflon insulated copper-constant TC wire (0.076 mm diameter and 0.076 mm PFA insulation) and Teflon insulated constantan heater wire (0.076 mm diameter and 0.076 mm PFA insulation) was used. The noninsulated TC junction was 3.0 cm from the probe tip, which was cemented by high-temperature epoxy and the active section was filled with silicon oil and sealed at the top with epoxy. The active section's outer diameter fit snugly into the inside of the extension tubing (1.59 mm outer diameter and 1.32 mm inside diameter and 3.56 cm long) by being epoxied together such that the active section had a length–diameter ratio of 48. The extension tubing was fitted with a ferrule and compression nut to allow placement of the active section into a high-pressure environment while the probe leads remained at ambient conditions and both the active section and half of the extension tubing were inserted into the sample. Gratzek and Toledo (1993) proposed that the experimental interval for regression was 10–25 s having $r^2 \geq 0.996$.

Kontani (1990) used a thin ribbon hot wire replacing hypodermic needle (Figure 17.12). A TC 0.08 mm diameter was spot-welded at the center of the hot ribbon to measure its temperature. The electric insulation between the hot wire and the specimen was arranged by the following three

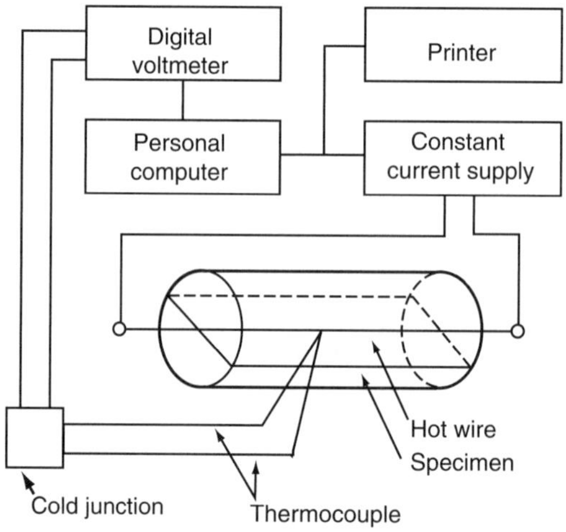

FIGURE 17.12 Schematic diagram of thin ribbon hot wire method. (Adapted from Kontani, T., *Heat Transfer Japn. Res.*, 19, 366, 1990.)

methods: (1) a Teflon sheet 20 μm thick was inserted, (2) lacquer layer approximately 10 μm thick was sprayed on the specimen, (3) after lacquer had been sprayed, the hot wire was glued with an adhesive material. If the hot wire were glued on the specimen, the scatter in data was decreased because of the stabilization of the thermal resistance between the specimen and the hot wire. The average values of the thermal conductivity did not vary significantly with different electric insulation methods. Thin ribbon type line source has not been used in the case of food materials. Hence, this ribbon type probe can be used in food materials due to its advantage of simplicity in construction versus the needle type probe. Shariaty-Niassar et al. (2000) developed a probe using a poly-(phenylene sulfide) disk with 38 mm diameter and 10 mm height as the substrate of the probe (reference specimen) and a nickel ribbon hot wire (50 μm width and 30 mm length) to measure the thermal conductivity of gelatinized potato starch under higher temperature and pressure (Figure 17.13). The pattern of the ribbon hot wire was first drawn on the surface of the substrate using a photo-lithography technique followed by nickel deposition and electroplating. Copper lead wires were joined at both ends of the hot wire to allow electric current to flow through the wire and heat it. Two additional copper lead wires were connected with nickel wires to measure the electrical resistance of the hot wire during the current flow. The authors claimed that the developed probe has successfully measured thermal conductivity of water at low, medium, high temperature and at atmospheric pressure, and the obtained results were in close agreement with values in the literature. The discrepancy at higher temperature has been attributed to some experimental errors.

Voudouris and Hayakawa (1995) studied the influence of the probe length-to-diameter ratio and filling material on the accuracy of thermal conductivity measurement. They used high thermal conductivity paste or air as filling material of the probe. Statistical analysis showed the highly significant influence of the length–diameter ratio and significant influence of the filling material.

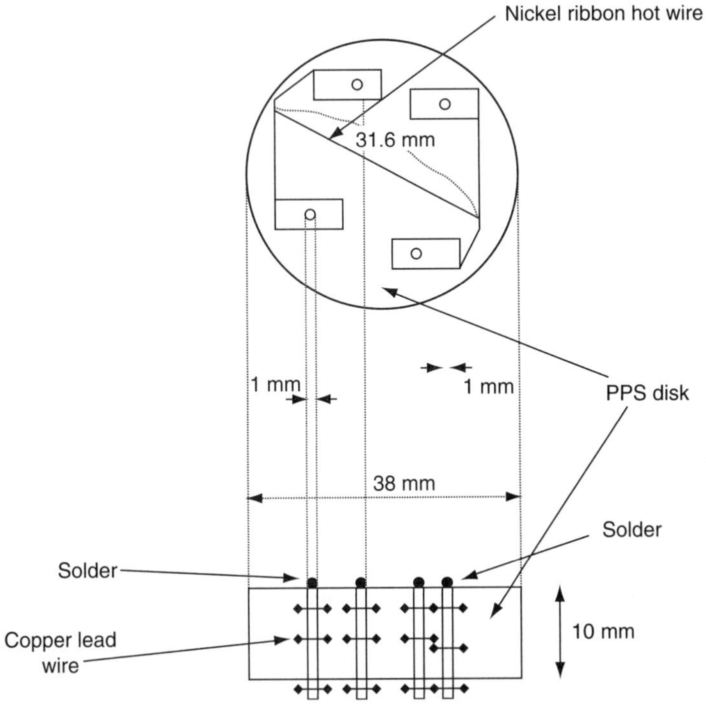

FIGURE 17.13 Schematic diagram of probe using poly-phenylene sulfide as reference specimen for thermal conductivity measurement of food at higher temperature and pressure. (From Shariaty-Niassar, M., Hozawa, M., and Tsukada, T., *J. Food Eng.*, 43, 133, 2000. With permission.)

Noncalibrated k values of the smallest probe (length–diameter ratio, 4:1) were 93% above the published value, whereas those of the largest probe (length–diameter ratio, 54:3) were identical to the published value. The use of a short air-filled probe was not recommended when the maximum slope method was applied.

Sweat (1986) observed that a line heat source device that utilizes a bare heater wire exposed directly to the sample would have calibration factors closer to 1 compared with devices where the heater wire was placed inside the metal tubing. A shorter probe was constructed with smaller diameters (1.9 mm × 0.5 mm) by wrapping insulated TC wire outside a hypodermic needle or sewing needle. These can be used for smaller samples, but are found to be fragile compared with the conventional probe. Thus, the conventional probe has more acceptability for its simplicity and ease of construction. Constantan is commonly used as a heater wire because its electrical resistance does not significantly change with temperature (Benedict, 1984; Sweat, 1986). Chromel–constantan TCs produced high-voltage outputs per degree temperature change (Sweat, 1986). Also, chromel was less likely to break than copper, according to Sweat (1986).

17.2.3.1.5 Test Duration

Test duration plays an important role in thermal conductivity measurement. Both short- and long-time durations are sources of significant errors. The method should be applied within a time period extending from a minimum initial time where truncation and nonzero radius errors are minimum to a maximum final time where noninfinite probe length and sample dimension errors start to be maximum. Longer test durations create error, due to convection and moisture migration followed by an increase in temperature at the sample boundaries. Convection in the liquid within the samples happens due to a density gradient. Nusselt (1915, cited by Van der Held and Van Drunen, 1949) measured convection by concentric cylinders. It will be possible to calculate apparent thermal conductivity from the measured heat transfer while both surfaces of a concentric cylinder are constant at variable temperature. Van der Held and Van Drunen (1949) reported that convection becomes important above $Pr{\cdot}Gr$ greater than 1000 to 1200. Those authors observed values of $Pr{\cdot}Gr$ at the time after which a deviation of the straight line (T vs. ln t) started for eight liquids of known thermophysical properties. The observed value of $Pr{\cdot}Gr$ (1070) was in close agreement with the critical point of convection as reported by Kraussold (1934) (cited by Van der Held and Van Drunen, 1949). Van der Held and Van Drunen (1949) derived an equation for the end point of the experiment at a $Pr{\cdot}Gr$ value of 1070. For water, the end point was 16 s to reduce convection. Sweat (1986) recommended that the test duration should be about 3 s for liquids and 10–12 s for most solid foods. The plots of temperature rise T versus ln t are also piecewise linear (Figure 17.11). Karwe and Tong (1992) advocated that correct thermal conductivity could be obtained only when an appropriate time interval was selected for the calculation. The time interval also depended on the type of filling material of the probe and the heat capacity of the sample.

A water bath is usually used to equilibrate the sample at a specific temperature. Gratzek and Toledo (1993) used a circulated and closed system to measure the thermal conductivity of carrot. There was no advantage in using a circulating system for thermal conductivity measurement while a closed system with high-equilibration time can break down the food components and may affect k. Gratzek and Toledo (1993) have tested a pressure vessel by applying air pressure to the reservoir to prevent boiling of water at above 100°C. These authors also used a closed ethylene glycol constant temperature bath, which had the limitation of reaching target temperature by slow heating.

17.2.3.1.6 Power or Current Supply

Power supply also affects the accuracy of the thermal conductivity measurement. This factor is interrelated to the factors discussed earlier. High-power input creates a high-temperature gradient, which may also cause moisture migration and heat convection. Sweat (1986) mentioned that the power level should be selected on the basis of the temperature rise that is necessary to get a high correlation between T and ln t. This author recommended power levels from 5 to 30 W/m.

17.2.3.1.7 Data Acquisition and Calibration

A mainframe computer-based data acquisition system can be used to control the probe heater and facilitate online data analysis. This system has been reported by various researchers. However, this system is expensive, complicated to install and maintain, and computer bus dependent (Murakami and Okos, 1989). Presently, a PC can be used for the purpose by using inexpensive hardware and software. The schematic diagram of the data acquisition system is shown in Figure 17.14. The data acquisition component collects data on temperature and electric current and simultaneously evaluates thermal conductivity. The apparatus controls the probe heater by turning it off/on at any desired time and it has several features, such as probe calibration routine, power input adjustment, calculation of time correction factors, and data and results plotting and tabulation (Murakami and Okos, 1989).

The probe is commonly calibrated with glycerine (0.286 W/m K) and 0.4% water–agar solution (0.615 W/m K) at 30°C. Thermophysical properties of ethylene glycol and glycerine are given in Table 17.1 for equipment calibration purposes. Agar is used to avoid the convection during probe heating. These calibrants are adequate in the temperature above freezing but not suitable for frozen materials due to the several times increase in thermal conductivity (Tong et al., 1993). Typical thermal conductivity values for frozen foods usually varied from 0.8 to 2.0 W/(m K) whereas metals have much higher values than these. Tong et al. (1993) mentioned that ice may be used to calibrate probes for frozen foods, but there are a few practical limitations: (1) difficulty of probe insertion, (2) variability of thermal conductivity due to the presence of air bubbles, (3) polycrystalline ice structure thermal conductivity varied by the freezing rate and number of nuclei. Tong et al. (1993) developed a standardized homogeneous bentonite paste with pure copper powder to calibrate a thermal conductivity probe for frozen foods (0.9–1.9 W/m K). The experimental values of different methods of measurement are given in Table 17.2.

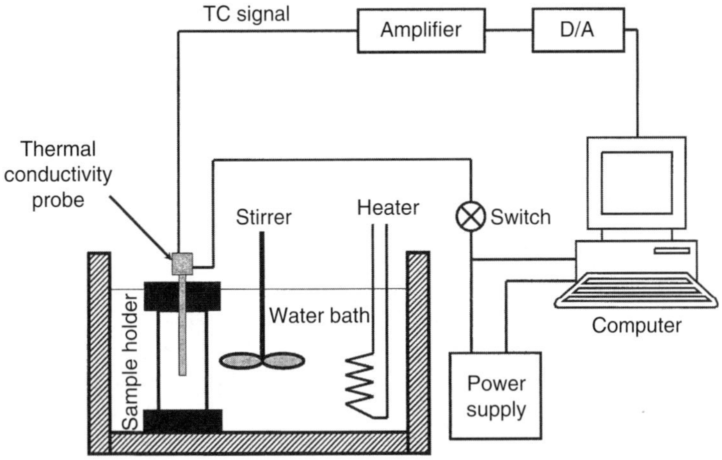

FIGURE 17.14 Probe, TC, heater, and control circuit. (From Wang, N. and Brenan, J.G., *J. Food Eng.*, 17, 153, 1992. With permission.)

TABLE 17.1 Thermophysical Properties of Ethylene Glycol [$C_2H_4(OH)_2$] and Glycerine [$C_3H_5(OH)_3$] as a Function of Temperature

T (K)	ρ (kg/m³)	C (kJ/kg K)	k (W/m K) × 10^{-1}	α (m²/s) × 10^{-8}	Pr	μ (Pa·s) × 10^{-3}	β (1/K) × 10^{-4}
Ethylene glycol							
273	1130.8	2.294	2.42	9.33	617.0	65.10	6.5
280	1125.8	2.323	2.44	9.33	400.0	42.00	6.5
290	1118.8	2.368	2.48	9.36	236.0	24.70	6.5
300	1114.4	2.415	2.52	9.39	151.0	15.70	6.5
310	1103.7	2.460	2.55	9.39	103.0	10.70	6.5
320	1096.2	2.505	2.58	9.40	73.5	7.57	6.5
330	1089.5	2.549	2.60	9.36	55.0	5.61	6.5
340	1083.8	2.592	2.61	9.29	42.8	4.31	6.5
350	1079.0	2.637	2.61	9.17	34.6	3.42	6.5
360	1074.0	2.682	2.61	9.06	28.6	2.78	6.5
370	1066.7	2.728	2.62	9.00	23.7	2.28	6.5
373	1058.5	2.742	2.63	9.06	22.4	2.15	6.5
Glycerine							
273	1276.0	2.261	2.82	9.77	85,000	10,600	4.7
280	1271.9	2.298	2.84	9.72	43,200	5340	4.7
290	1265.8	2.367	2.86	9.55	15,300	1850	4.8
300	1259.9	2.427	2.86	9.35	6780	799	4.8
310	1253.9	2.490	2.86	9.16	3060	352	4.9
320	1247.2	2.564	2.87	8.97	1870	210	5.0

Source: Incropera, F.P. and DeWitt, D.P., *Fundamentals of Heat and Mass Transfer*, John Wiley & Sons, New York, 1990. With permission.

17.2.3.2 Dual-Needle Heat-Pulse Sensor

There is an increasing need for accurate and rapid thermal measurement of food products. Recently, a method has been developed that allows measurement of all three thermal properties in a single measurement. The dual-needle heat-pulse (DNHP) probe was first introduced for the

TABLE 17.2 Experimental Thermal Conductivities of Copper Powder, Bentonite Powder, Bentonite Paste, and Copper-Filled Bentonite Paste

Sample	ρ (kg/m³)	C (kJ/kg K)	α (m²/s) × 10^{-8}	k (W/m K) × 10^{-1} Steady State	Unsteady State	Probe[a,b]
Bentonite	2752	0.95	—	—	—	—
Copper[c]	8931	0.38	—	—	—	—
Bentonite gel (8%)	1051	3.90	—	—	—	—
Bentonite gel (10%)	1065	3.80	—	—	—	—
30/10[d]	1499	2.69	23.24	9.39	9.37	9.86
40/10[d]	1751	2.30	28.42	11.56	11.45	11.94
50/10[d]	1968	2.00	44.75	16.60	17.62	17.41
55/08[d]	2134	1.87	47.68	18.88	19.03	18.14

Source: Tong, C.H., Sheen, S., Shah, K.K., Huang, V.T., and Lund, D.B., *J. Food Sci.*, 58, 186, 1993. With permission.

[a] $k = 0.657 - 0.099X_{cp} + 1.79X_{cp}^2 + 4.71X_{cp}^3$.

[b] $T = 30°C$.

[c] Three hundred mesh pure copper powder.

[d] Mass of copper powder/mass of 10% bentonite paste.

measurement of the thermal properties of soil (Campbell et al., 1991). Later, Fontana et al. (1999) applied this technique to measure thermal conductivity of foods. The theory of this technique is based on transient heat transfer and similar to the single-needle probe as discussed in an earlier section. The probe consists of two needles: one contains a heated wire and TCs while the other contains only TCs (Figure 17.15). During heat pulses, the temperature is monitored by TCs within both needles. Using the temperatures together with the known distance between needles, thermal conductivity can be calculated (Kluitrenberg et al., 1995). Fontana et al. (1999) reported that the accuracy and resolution of the dual-needle probe were 5% and 0.01 W/(m K) for a high-moisture specimen. The advantages of this technique are that it is relatively easy to use and requires a short measuring time (within 10 min). Moreover, this technique also measures other thermal properties simultaneously, e.g., thermal conductivity, heat capacity, and volumetric diffusivity. However, it gives thermal conductivity with less resolution than that of the line heat source technique. Other disadvantages are similar to those of the line heat source technique.

The advantage of using the dual-probe method together with the appropriate heat-pulse theory rather than the single probe is that all three soil thermal properties, thermal diffusivity, volumetric heat capacity, and thermal conductivity, can be determined from a single heat-pulse measurement. Instantaneous heat-pulse theory can be used with the dual-probe method to determine heat capacity from short duration heat-pulse data, but it should not be used to determine thermal diffusivity and thermal conductivity.

In designing the DHHP probe, the length, diameter, and spacing between the needles are important. The analysis of error showed that assuming an infinite length for a heat source of finite length caused errors <2% and assuming the cylindrically shaped heater to be a line heat source caused errors of <0.6% in the measurement of thermal properties (Kluitrenberg et al., 1995). As this spacing between the needles increases, the maximum temperature change at the TC decreases, the time required to reach the maximum temperature increases, and the accuracy in determination of the maximum temperature decreases. In addition, the heat pulse duration and intensity must be considered to emulate instantaneous heating. Thus, the infinite line source theory is appropriate for use in the DNHP probe.

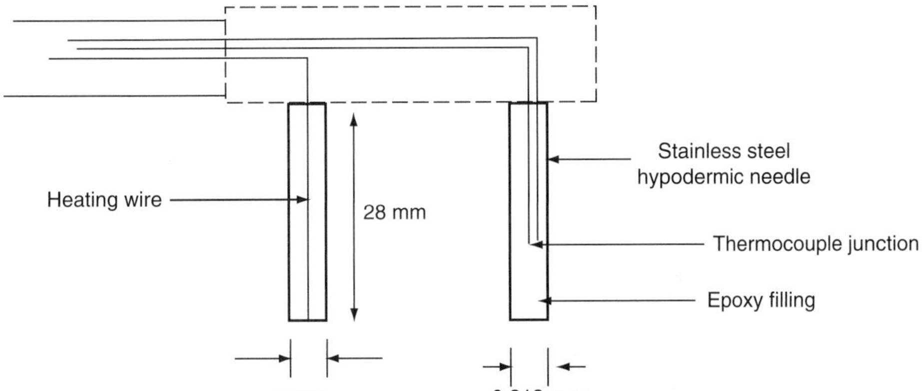

FIGURE 17.15 Schematic diagram of the DNHP sensor. (Adapted from Fontana, A.J., Varith, J., Ikediala, J., Reyes, J., and Wacker, B. Thermal properties of selected food using a dual-needle heat-pulse sensor. ASAE Meeting Presentation Paper No. 996063, 1999. With permission.)

17.2.3.3 Thermal Comparator Method

The thermal comparator, ball bearings developed first by Powell (1957), consisted of two matched 0.95 cm diameter phosphor-bronze ball bearings mounted at the base of a block of thermal insulating material (balsa wood) so that one ball was flush with the surface and the other protruded slightly to contact the sample (Figure 17.16). Two metal studs in the base of the block completed a three-point contact and ensured stability and reproducibility in the contact of the measuring ball with the sample. First, the thermal comparator is equilibrated to a higher or lower temperature than that of the sample. Then the comparator is placed to touch the sample and the measuring ball changes temperature more rapidly than the reference ball. This causes an increase in electromotive force (EMF), which can be related to the conductivity of the sample. Powell (1957) found that the EMF generated was proportional to $k^{1/2}$. Calibration is necessary with a number of materials of known thermal conductivities due to its comparative basis and nonlinear variation of k versus EMF.

Clark and Powell (1962) presented another version called direct reading thermal comparator. In this case, two TCs are attached to the sample block, one to measure the temperature at the point of contact and the other sufficiently far from the contact point. This modification avoids the heat transfer from the measuring junction through the air to the sample. In this case, the comparator was equilibrated in an oven or refrigerator. Rousan (1982) used water circulation at a constant temperature through a container containing comparator TC. This minimized the error due to inconsistent temperature differences between the sample and the comparator. The other sources of errors are as follows: (1) the reference voltage was generated by a TC that was exposed to ambient air; thus, air temperature and draft may change the voltage generated and (2) the sample temperature was not specifically controlled.

Ziegler and Rizvi (1985) modified Rousan's (1982) instrument to avoid those errors (Figure 17.17) by replacing the reference junction of Rousan's model with a constant voltage source to eliminate fluctuation in voltage reading due to variation in ambient air temperature and drafts. The reference voltage source was adjusted so that the differential EMF could be set to zero when the measuring junction was exposed to air. A jacketed stainless steel sample holder was used so that the sample could be maintained at the test temperature. Rubber insulation was used to reduce radiation heat transfer from comparator to sample. For a liquid sample, Ziegler and Rizvi (1985) used an enameled metal disk above the comparator TC to minimize surface evaporation around the TC. Silicon caulking was also used to fill the corners at the base of the measuring TC to avoid air

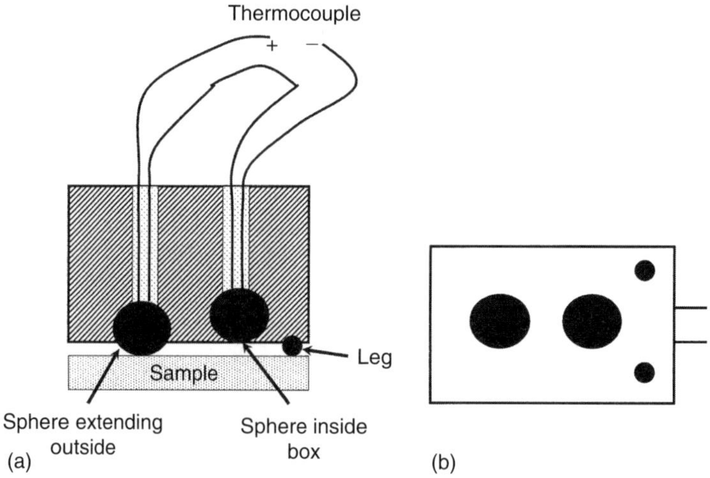

FIGURE 17.16 Two-ball thermal comparator as introduced by Powell (1957): (a) side view; (b) bottom view.

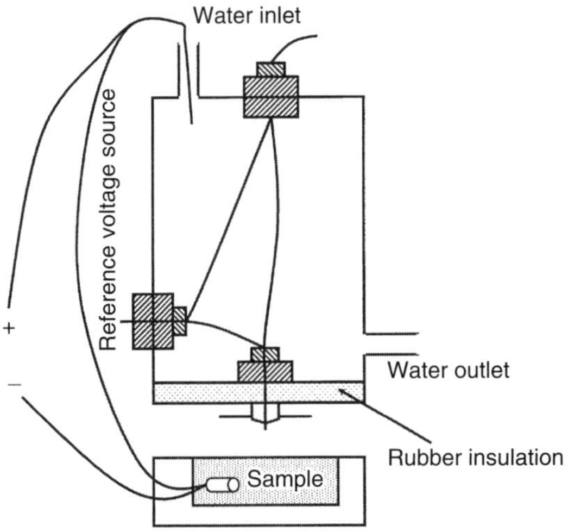

FIGURE 17.17 Schematic of the modified thermal comparator (Adapted from Ziegler, G.R. and Rizvi, S.S.H., *J. Food Sci.*, 50, 1458, 1985. With permission.)

entrapment by viscous fluids. The deviation of measurement was less than 7%. The instrument was calibrated by measuring thermal conductivities of liquids from 0.284 to 0.600 W/(m K) at 20°C and, thereafter, the calibration curve was drawn by a quadratic relation between k and EMF (volt). Therefore, calibration should be carried out before any measurement.

Recently, Carson et al. (2004) developed a comparative method that produced an uncertainty of 2%–5%. They used analytical solution for the center temperature of a sphere being cooled with the convection boundary condition. The method involved the cooling of two spheres side by side in a well-stirred ice/water bath. One sphere contained the test material and the other contained a reference material (guar gel) of known thermal conductivity. The thermal conductivity of the test sample was calculated relative to the control sample as

$$\frac{k_e}{k_c} = \left(\frac{\sigma_e}{\sigma_c}\right)\left(\frac{\rho_e}{\rho_c}\right)\left(\frac{C_e}{C_c}\right)\left(\frac{R_e^2}{R_c^2}\right)\left(\frac{\beta_{1c}^2}{\beta_{1e}^2}\right) \tag{17.21}$$

where

ρ is the density (kg/m^3)
C is the specific heat (kJ/kg °C)
R is the radius of the container (m)
β is the root of the equation

$$\beta_1 \cot\beta_1 + (Bi - 1) = 0 \tag{17.22}$$

where Bi is the Biot number (hR/k)

$$\ln\left(\frac{T - T_b}{T_0 - T_b}\right) = B - \sigma t \tag{17.23}$$

where

σ is the slope of the linear potion of the cooling curve
T_0 and T_b are the initial and bulk fluid temperature (°C)

Thermal conductivity was estimated from Equation 17.24 by minimizing function φ by changing the values of β_{1e} and β_{1c} subject to $0 \leq \beta_{1e}, \beta_{1c} \leq \pi$ (since $\beta_1 \to 0$ as $Bi \to 0$, and $\beta_1 \to \pi$ as $Bi \to \infty$):

$$\varphi = [\beta_{1e} \cot \beta_{1e} + Bi_e\{k_e(\beta_{1e})\} - 1]^2 + [\beta_{1c} \cot \beta_{1c} + Bi_c\{k_c(\beta_{1c})\} - 1]^2 \qquad (17.24)$$

Other physical properties need to be measured. Carson et al. (2004) used this method to measure the thermal conductivity of a simulated porous sample.

17.2.3.4 Thermistor-Based Method

A thermistor is a temperature-sensitive resistor. Thermistors are generally composed of ceramic semiconductor materials, such as oxides of manganese, nickel, cobalt, copper, iron, and titanium. Being resistive devices, a thermistor will heat when a current is passed through it. Those used for thermal property measurement have a negative temperature coefficient of resistance, indicating that their resistance will decrease when their temperature is increased. The negative temperature coefficient can be as large as several percent per degree Celsius, allowing the thermistor circuit to detect minute changes in temperature. Some of the commonly available thermistors are bead type, disk type, wafer and chip type, flake type, and rod type thermistors (Thermometrics, 1980). The relationship between temperature and resistance is nonlinear and is commonly described by the empirical Steinhart–Hart equation (Thermometrics, 1987):

$$T = A' + B' \ln R + C'(\ln R)^3 \qquad (17.25)$$

where

T is the temperature (K)
R is the resistance of the thermistor
A', B', and C' are curve-fitting constants

The method was primarily developed in the field of biomedical science for measuring the thermal conductivity and diffusivity of human tissue (Chato, 1968; Balasubramaniam and Bowman, 1974; Valvano, 1981). The thermistor-based method uses a small thermistor probe whose active region approximates a sphere. Because of the small diameter of this region, beads with diameters of 2.54 mm and smaller are commonly used; hence, it is applicable to small samples. Limited reports are available for measurement of thermal properties of food using thermistors. Kravets (1988) and van Gelder (1998) used thermistor-based thermal conductivity measurement of liquid milk products and moist foods (potato and lean beef) at higher temperatures. Dougherty (1987) tested glass and Teflon encapsulated probes with a variety of materials and concluded that the thermistor method is most suited for viscous liquids of higher thermal conductivity, such as castor oil and glycerol. Kravets (1988) was the first to use self-heated thermistors at elevated temperatures, and he used glass encapsulated probes to measure the thermal conductivity of milk and cream over the range of 25°C–125°C. A sample diameter of 5 mm was found to satisfy this assumption for a probe of 1.5 mm diameter (Kravets, 1988). van Gelder (1998) also used glass-coated probes (from the P60-series of thermometrics) for their suitability to measure the thermal conductivity and diffusivity of moist food materials. The thermistors used have a nominal bead diameter of 1.52 mm and an overall length of 12.7 mm. The leads were 0.203 mm in diameter. Thermistor probes were customized in-house and calibrated in three materials of known thermal conductivity and diffusivity (water, glycerol, and HTF 500, a heat transfer fluid). In most of the cases, the calibrated probe estimated thermal properties with an error of less than 5%, over the range of thermal properties spanned by those of the calibration media.

Chato (1968) solved the heat conduction equation in the medium for the case in which starting from an equilibrium situation, the surface temperature of the thermistor is suddenly elevated with a predetermined increment and held constant at the elevated temperature. The temperature solution for the medium is given by

$$\frac{T - T_0}{T_a - T_0} = \frac{a}{r}\operatorname{erfc}\left(\frac{r - a}{2\sqrt{\alpha t}}\right) \tag{17.26}$$

and the heat dissipation by the medium is given by the following equation:

$$q = -k4\pi a^2\left(\frac{\partial T}{\partial r}\right)_{r=a} = 4\pi ak(T_a - T_0) + 4a^2\sqrt{\pi k\rho c_p}\,(T_a - T_0)t^{-1/2} \tag{17.27}$$

where
- a is the radius of the thermistor bead (m)
- c_p is the specific heat of the medium (J/kg °C)
- k is the thermal conductivity of the medium (W/m °C)
- q is the heat dissipation function (W)
- r is the radial coordinate (m)
- t is the time (s)
- T is the temperature in the medium (°C)
- T_a is the surface temperature of the bead for $t > 0$ (°C)
- T_0 is the initial temperature (°C)
- α is the thermal diffusivity of the medium (m^2/s)
- r is the density of the medium (kg/m^3)

The heat dissipation by the thermistor is found to be linear in the inverse square root of time. Linear regression of measured data can be used to calculate the thermal conductivity (from the intercept of regression) and the thermal inertia (from the slope of regression). This method involved calibration with a medium of known thermal conductivity to find the effective bead radius. Beads were calibrated with agar-gelled water (1.5% and 1.75%) and paraffin. Chato (1968) reported an accuracy of 20% during measurement of k. The method was rapid and thermal properties could be measured within 2 min.

Balasubramaniam (1975) and Balasubramaniam and Bowman (1974, 1977) modified Chato's heat transfer model by considering thermal gradients in the thermistor bead: both thermistor and medium were considered as distributed thermal masses. The thermistor was modeled as a sphere embedded in an infinite, homogeneous, and isotropic medium with no surface contact thermal resistance. This coupled heat transfer problem was solved for the case where starting from a thermal equilibrium, the temperature of the thermistor was raised instantaneously to, and maintained at, an elevated temperature. The required heat generation was assumed to take place uniformly throughout the bead. This heat generation was experimentally determined to be of the form $\Gamma + \beta\ t_{-1/2}$. The temperature distributions of thermistor bead and surrounding medium were presented as indefinite integrals, which had to be solved numerically. The mathematics involved in solving this heat transfer problem can be found in Valvano (1981).

From an error analysis of the measurement of thermal conductivity, Kravets concluded that (1) error in estimation of medium thermal conductivity decreases with a higher thermal conductivity of the probe; (2) a minimum temperature ramp of 2.5°C has to be used to minimize errors in medium thermal conductivity; (3) more precise thermal conductivity reference materials are required to minimize the errors in estimated bead parameters. Of the methods discussed, a self-heated thermistor promises to be the best for measurement of thermal conductivity and thermal diffusivity. The line source probe in combination with an additional temperature sensor is capable of measuring thermal diffusivity but requires a much larger sample volume than a thermistor. A disadvantage of the thermistor-based method is that it is not absolute. Calibration with at least two reference materials is necessary. The calibration is required at every measurement temperature and demands thermal standards for both thermal conductivity and thermal diffusivity.

The composition and rheological characteristics of foods considerably change at elevated temperatures and these food characteristics may affect the adequacy of thermal contact between probe and sample, which can ultimately determine the suitability of the thermistor-based method for thermal conductivity measurement. Kravets (1988) found that thermal conductivity measurements of milk products correlated well with the physicochemical changes at higher temperatures. van Gelder (1998) reported that thermal contact between the thermistor and food samples (potato and beef) was found to be adequate for the measurements of k at selected temperatures (25°C, 50°C, and 100°C). A comparison between the bead thermistor probe and a reference line heat source probe exhibited close agreement between both methods for most of the cases.

During measurement of k by the thermistor-based method, natural convection occurred in samples of aqueous solution of thickening agents (thickened water) during the sample interval of 30 s. Statistical analysis of measurements in samples of thickened water revealed convection in samples with a viscosity of 5 cp or lower while measured with a temperature ramp of 1.5°C and 2.5°C, and in samples with a viscosity of 25 cp and lower, when measured with a temperature ramp of 5°C. A Rayleigh number was used to predict the notion of a critical Rayleigh number at the onset of convection during measurement of k. Samples with Rayleigh number in the range of 43–84 exhibited convection. The author found that when the Rayleigh number was below 43, convection could not be demonstrated. However, a glass encapsulated thermistor used by the researcher was too fragile for routine work.

17.2.3.5 Temperature History

Keppeler and Boose (1970) reported a procedure that would yield thermal conductivity, thermal diffusivity, and specific heat in a single test. This technique was based on the temperature history of a cylinder undergoing thawing. Thermal conductivity and diffusivity were evaluated during transient state while specific heat was calculated during steady state. A 5.08 cm diameter and 15 cm long insulated aluminum tube are filled with sucrose solution and then frozen. The frozen set is then moved into an ambient condition where it is allowed to thaw. The center temperature and the wall temperature of the tube with time were recorded. The thermal conductivity of the sample was calculated as

$$k_{sa} = \frac{1}{2}\left[\frac{Q}{A_{sa}} - \frac{C_{AL}}{A_{sa}} m_{AL} \frac{\Delta T_{CO}}{\Delta t}\right] \frac{r}{\Delta T_{CS}} \tag{17.28}$$

where

k_{sa} is the thermal conductivity (W/m K)
Q is the heat leak (J/s)
A_{sa} is the surface area of sample (m^2)
C_{AL} is the specific heat of the aluminum tube (J/kg K)
m_{AL} is the mass of the aluminum tube (kg)
r is the radius of the sample (m)
$(\Delta T_{CO}/\Delta t)$ is the rate of change of core temperature (°C/s)
ΔT_{CS} is the temperature difference between the core and the outside surface of the sample (°C)

Distilled water was used to determine the heat leak (Q) through the insulators and aluminum tubes. The heat leak can be calculated as

$$Q = C_{\mathrm{I}} m_{\mathrm{I}} \frac{\Delta T_{\mathrm{CO}}}{\Delta t} \tag{17.29}$$

where

C_{I} is the specific heat of ice (J/kg)
m_{I} is the mass of ice (kg)
$\Delta T_{\mathrm{CO}}/\Delta t$ is the rate of change of the core temperature (°C/s)

Transient analytical solution and temperature history of a well-defined geometry are also used to determine the thermal conductivity. Mendonca et al. (2005) determined thermal conductivity and volumetric thermal heat capacity of a spherical object from the analytical solution and temperature history of heating in hot air. The optimization technique was used to fit the temperature history data with the analytical solution.

17.2.3.6 Hot Disk Sensor

Recently, the hot disk technique that represents a transient plane source (TPS) method has gained popularity as a tool for rapid and accurate thermal conductivity measurement (Gustafsson, 1991; He, 2005; Al-Ajlan, 2006; Gustavsson and Gustafsson, 2006). The major advantages of the hot disk technique are wide thermal conductivity range (from 0.005 to 500 W/m K); wide range of materials types (from liquid, gel to solid); easy sample preparation; smaller sample size; nondestructive; relatively short measurement time (10 s to 10 min) and more importantly, high accuracy.

A precursor to this technique was first developed by Gustafsson in 1967 (cited by He, 2005) as a nonsteady-state method of measuring thermal conductivity of transparent liquids and sometimes the probe is referred to as G-probe. In the original experimental design, a thin rectangular metallic foil suspended in the liquid was heated by a constant electric current. A sensor was sandwiched between two identical thin slab samples with thickness h. The outer surfaces of the two slabs were covered with isolation material so that no heat can transfer out of these surfaces during measurement, as illustrated in Figure 17.18. The temperature distribution around the foil was measured optically as a function of time. From this result, both thermal diffusivity and thermal conductivity of the liquid could be determined.

17.2.3.6.1 Theory

The heat conduction equation can be solved by considering the hot disk that consists of a fixed number of concentric ring heat sources located in an infinitely large sample. While a constant electric current is supplied to the sensor at time $t = 0$, the temperature change in the sensor is

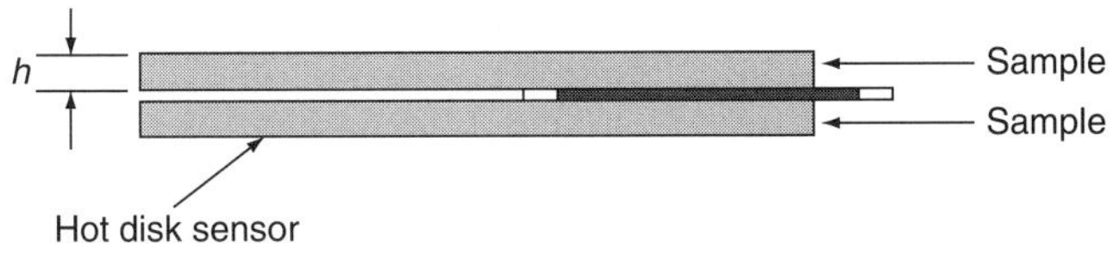

FIGURE 17.18 To measure thermal conductivity of a thin slab sample, the hot disk sensor is sandwiched between two specimens with thickness *h*. (From Rahman, M.S., *J. Food Eng.*, 14, 71, 1991a. With permission.)

recorded as a function of time. The average temperature increase across the hot disk sensor area can be measured by monitoring the total resistance of the hot disk sensor (He, 2005):

$$R(t) = R_0[1 + \Omega\{\Delta T_i + \Delta T_{ave}(\tau)\}] \quad (17.30)$$

where

R_0 is the resistance of the disk before to heating (at time $t = 0$)

Ω is the temperature coefficient of the resistivity (TCR)

ΔT_i is the constant temperature difference that develops almost momentarily over the thin insulating layers that cover the two sides of the hot disk sensor material (nickel) and that make the hot disk a convenient sensor

$\Delta T_{ave}(\tau)$ is the temperature increase in the sample surface on the other side of the insulating layer and facing the hot disk sensor (double spiral)

Rearranging the equation for the temperature increase recorded by the sensor,

$$\Delta T_{ave}(\tau) + \Delta T_i = [\{R(t)/R_0\} - 1]/\Omega \quad (17.31)$$

where ΔT_i represents the thermal contact between the sensor and the sample surface, with $\Delta T_i = 0$ indicating perfect thermal contact; ΔT_i becomes constant after a short time Δt_i, which can be calculated as

$$\Delta t_i = \delta^2/\alpha \quad (17.32)$$

where δ is the thickness of the insulating layer and α is the thermal diffusivity of the layer material.

The time-dependent temperature increase is given by theory as

$$\Delta T_{ave}(\tau) = [P_0/(\pi^{3/2}lk)]D(\tau) \quad (17.33)$$

where

P_0 is the total power output from the sensor

l is the overall length of the disk or sensor radius

k is the thermal conductivity of the sample that is being measured

$D(\tau)$ is a dimensionless time-dependent function

$D(\tau)$ is estimated as

$$\tau = [(t - t_c)/\theta]^{0.5} \quad (17.34)$$

In this equation, t is the time measured from the start of the transient recording; t_c is a time correction and θ is the characteristic time and defined as

$$\theta = l^2/\alpha \quad (17.35)$$

Equation (17.33) allows us to accurately determine ΔT_i as a function of time. If the relationship between t and τ is known, a plot ΔT_i versus $D(\tau)$ should yield a straight line. The slope of the line is $P_0/\pi^{3/2}lk$, from which thermal conductivity k can be calculated. However, the exact value of τ is generally unknown since $\tau = \sqrt{\alpha t}/l$ and the thermal diffusivity α are unknown. To calculate the thermal conductivity correctly, it is generally required to perform a series of computational plots of ΔT_i versus $D(\tau)$ for a range of α values. The correct value of α will yield a straight line for the

ΔT_i versus $D(\tau)$ plot. This optimization process can be done by an iteration process using software until an optimized value of α is found. By this technique, the thermal conductivity and the thermal diffusivity of the studied sample can be determined from one single transient recording.

17.2.3.6.2 Measurement and Construction Materials

Gustavsson and Gustafsson (2006) reported that although the operating principle of the hot disk technique is similar to traditional THW method there are a few major differences between the two techniques. The hot disk has comparatively a large heating area of sensor, which together with the sample geometry and interfacial geometry between the sample and sensor governs the heating of the sample and the resulting temperature response that has an edge over THW. The authors believe that a large contacting area provides advantages in terms of a higher sensitivity in the temperature recording and improved model fittings. Secondly, a number of sensor configurations are possible (Gustavsson et al., 1997; Gustavsson et al., 2000), which, however, must be chosen with care to ensure geometrical agreement with the physical model used in the analysis of data. The fundamental relation between sensor size, sample size, and measurement time is chosen with respect to the so-called characteristic time (θ) $[\theta = l^2/\alpha]$ of the measurement. The time regions where the sensitivity coefficients indicate maximum sensitivity in the estimation of the different thermophysical properties can be stated in terms of the dimensionless time t/θ, where t is the measurement time (Bohac et al., 2000).

The hot disk technique is based on using a thin metal strip or a metal disk as a continuous plane heat source. The metal disk or strip is sealed between two thin electrically insulating sheets (polyimide films). The hot disk uses a sensor element (Figure 17.19) in the shape of a double

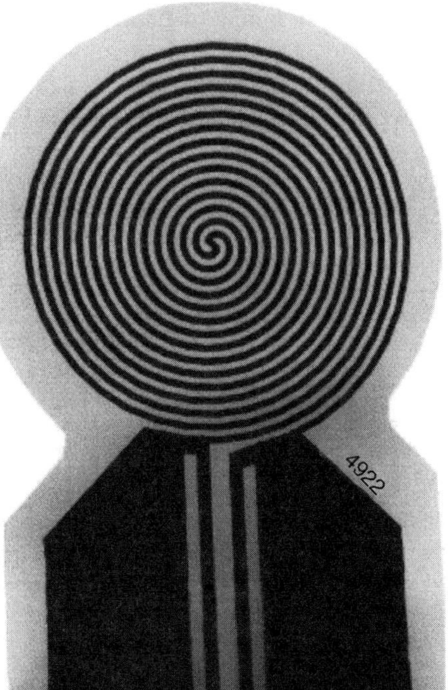

FIGURE 17.19 Hot disk sensor. The sensor insulation is Kapton of 25 μm thickness on each side. The sensing and heating pattern is made of 12 μm thick Ni. (From Gustavsson, M. and Gustafsson, S.E. *Thermochimica Acta*, 442, 1, 2006.)

spiral, which acts both as a heat source for increasing the temperature of the sample and a resistance thermometer for recording the time-dependent temperature increase in the heat source itself. Recently, Al-Ajlan (2006) studied various thermal insulation materials using the hot disk technique. The author has used a sensor element made of a 10 μm thick nickel-metal double spiral, which is supported by a material to protect its particular shape, give it mechanical strength, and keep it electrically insulated. The polyamide (Kapton) and mica are used as construction materials. The encapsulated Ni-spiral sensor is sandwiched thereafter between two halves of the sample (solid samples) or embedded in the sample (powders and liquids). The temperature increase is highly dependent on the thermal transport properties of the material surrounding the sensor. A moderately smooth sample surface is desired for this method. The thermal contact resistance between the sensor surface and sample surface causes a constant temperature difference to build up across the thin interface (gas or vacuum). This constant temperature step does not affect the measured sample properties since its influence is discarded in the calculations. By monitoring this temperature increase over a short period after the start of the experiment, it is possible to obtain precise information on the thermal transport properties of the surrounding material.

Small probes have the advantages of the ability to reduce the influence from potential convection. However, the sensitivity for small and very small probes has not yet been developed to comparable levels as can be routinely noticed while using larger sensors, especially for sensor dimensions $l \geq 10$ mm measured with TPS or hot strip techniques (Gustavsson and Gustafsson, 2006).

The theoretical background of thermal conductivity measurement using the hot disk sensor has been well described by He (2005). With proper corrections, the hot disk technique provides an excellent tool for rapid and accurate measurement of both thermal conductivity and thermal diffusivity of a wide range of materials. This technique is capable of measuring the thermal conductivity over a wide range with high accuracy, and the typical measurement time is 2.5–5 s. The hot disk technique is a valuable tool for material inspection and selection. However, the technology has not yet been exploited in the thermal conductivity measurement of food products. Recently, Gustavsson and Gustafsson (2006) have used a hot disk sensor to measure the thermal conductivity of milk (Figure 17.19) and the authors believed that it is possible to design a thermal conductivity probe, capable of recording fat content in milk with a sensitivity of better than 0.1%, within a total measurement time of 1 s.

NOMENCLATURE

A	heat transfer area (m^2)
a	empirical constant in Equation 17.2
A'	parameter in equation
B'	parameter in equation
Bi	Biot number
b	empirical constant in Equation 17.2
C	specific heat at constant pressure (J/kg K)
C'	parameter in equation
D	diameter (m)
d	parameter in equation
f	correction factor for Fitch type apparatus
g	parameter in equation
h	parameter in equation
i	parameter in equation
j	parameter in equation
k	thermal conductivity (W/m K)
l	sample thickness (m), length of hot disk

M mass fraction (dry basis)
m mass (kg)
n total number of phases
N number of layers
p parameter in equation
P pressure, power output
Pr Prandlt number
Q heat flow rate (J/s)
q heat input per unit length of line source (W/m)
r radius of sample (m)
r^2 regression coefficient
R resistance (ohm)
S response of the Fitch type apparatus
T temperature (°C or K)
t time (s)
U overall heat transfer coefficient (W/m^2 K)
W heat flux (W/m^2)
X mass fraction (wet basis)
x x-direction
Z coordination number

Greek Letter

α thermal diffusivity (m^2/s)
$\hat{\beta}$ $\gamma/(\alpha t)^{0.5}$
B thermal expansion ($1/T$)
δ differentiation
τ time-dependent function and parameter in Equation 17.33
θ characteristic time (s)
σ slope of the linear potion of the curve in Equation 17.23
μ viscosity ($Pa \cdot s$)
ρ density (kg/m^3)
ζ x-direction
Ω TCR
Γ probe length-to-diameter ratio
$\mathit{\Gamma}$ distance between the TC and the line heat source
π constant

REFERENCES

Al-Ajlan, S.A. Measurements of thermal properties of insulation materials by using transient plane source technique. *Applied Thermal Engineering* 2006, 26, 2184–2191.

Baghe-Khandan, M.S., Choi, Y., and Okos, M.R. Improved line heat source thermal conductivity probe. *Journal of Food Science* 1981, 46, 1430–1432.

Balasubramaniam, T.A. Thermal conductivity and thermal diffusivity of biomaterials: A simultaneous measurement technique, PhD dissertation, Northeastern University, Boston, MA, 1975.

Balasubramaniam, T.A. and Bowman, H.F. Temperature field due to a time dependent heat source of spherical geometry in an infinite medium. *Journal of Heat Transfer: Transactions ASME* 1974, 99(3), Series 3, 296–299.

Balasubramaniam, T.A. and Bowman, H.F. Thermal conductivity and thermal diffusivity of biomaterials: a simultaneous measurement technique. *Journal Biomechanical Engineering, Transactions ASME* 1977, 99, 148–154.

Blackwell, E.L. The axial flow error in the thermal conductivity probe. *Canadian Journal of Physics* 1956, 34(4), 412–417.

Blackwell, J.H. and Misener, A.D. Approximate solution of a transient heat flow problem. *Proceedings of Physicist Society* 1951, A64, 1132.

Bennett, A.H., Chace, W.G., and Cubbedge, R.H. Estimating thermal conductivity of fruit and vegetable components—the Fitch method. *American Society of Heating, Refrigeration and Air-condition Engineering Journal* 1962, 4(9), 8.

Benedict, R.P. *Fundamental of Temperature, Pressure and Flow Measurements*, John Willey & Sons, New York, 1984.

Bohac, V., Gustavsson, M., Kubicar, L., and Gustafsson, S.E. Parameter estimations for measurements of thermal transport properties with the hot disk thermal constants analyzer. *Review of Scientific Instruments* 2000, 71, 2452–2455.

Bridgman, P.W. The thermal conductivities of liquids. *Physics* 1923, 59, 141.

Buhri, A.R. and Singh, R.P. Measurement of food thermal conductivity using differential scanning calorimetry. *Journal of Food Science* 1993, 58, 1145–1148.

Camirand, C.P. Measurement of thermal conductivity by differential scanning calorimetry. *Thermochimica Acta* 2004, 417, 1–4.

Campbell, G.S., Calissendorff, C., and Williams, J.H. Probe for measuring soil specific heat using a heat-pulse method. *Soil Science Society of American Journal* 1991, 55, 291–293.

Carson, J.K., Lovatt, S.J., Tanner, D.J., and Cleland, A.C. Experimental measurements of the effective thermal conductivity of a pseudo-porous food analogue over a range of porosities and mean pore sizes. *Journal of Food Engineering* 2004, 63, 87–95.

Chato, J.C. A method for the measurement of the thermal properties of biological materials. *Symposium on Thermal Problems in Biotechnology*, ASME, New York, 1968, pp. 16–25.

Clark, W.T. and Powell, R.W. Measurement of thermal conduction by thermal comparator. *Journal of Science & Instrumentation* 1962, 39, 545.

Dickerson, R.W. Jr. Thermal properties of food. In: *The Freezing Preservation of Food*, Eds. Tressler, O.K., Van Arsdel, W.B., and Copley, M.J. AVI Publishing Co., Westport, 1968, 4th edn., Vol. 2, Chapter 2.

D'Eustachio, D. and Schreiner, R.E. A study of transient method for measuring thermal conductivity. *ASHVE Transactions* 1952, 58, 3312.

Denys, S. and Hendrickx, M.E. Measurement of the thermal conductivity of foods at high pressure. *Journal of Food Science* 1999, 64, 709–711.

Dougherty, B.P. An automated probe for thermal conductivity measurements. MSc thesis in Mechanical Engineering, Virginia Polytechnic Institute and State University, 1987.

Elustondo, D., Elustondo, M.P., and Urbicain, M.J. New thermal conductivity probe design based on the analysis of error sources. *Journal of Food Engineering* 2001, 48, 325–333.

Fitch, A.L. A new thermal conductivity apparatus. *American Physics Teacher* 1935, 3, 135.

Fontana, A.J., Varith, J., Ikediala, J., Reyes, J., and Wacker, B. Thermal properties of selected food suing a dual needle heat-pulse sensor. ASAE Meeting Presentation Paper No. 996063, 1999.

Gratzek, J. and Toledo, R.T. Solid food thermal conductivity determination at high temperature. *Journal of Food Science* 1993, 58, 908–913.

Gustafsson, S.E. Transient plane source technique for thermal conductivity and thermal diffusivity measurements of solid materials. *Review of Science & Instrumenttion* 1991, 62, 797–804.

Gustavsson, M. and Gustafsson, S.E. Thermal conductivity as an indicator of fat content in milk. *Thermochimica Acta* 2006, 442, 1–5.

Gustavsson, J.S., Gustavsson, M., and Gustafsson, S.E. On the use of the hot disk thermal constants analyzer for measuring the thermal conductivity of thin samples of electrically insulating materials. In: *Thermal Conductivity*, Eds. Gaal, P.S. and Apostelescu, D.E., Technomic Publishing Co. Inc., Lancaster, 1997, pp. 116–122.

Gustavsson, M., Gustavsson, J.S., Gustafsson, S.E., and Hälldahl, L. Recent developments and applications of the hot disk thermal constants analyzer for measuring thermal trasnport properties of solids. *High Temperature and High Pressure* 2000, 32, 47–51.

Haas, E. and Felselstein, G. Methods used to the thermal properties of fruits and vegetables. Special Publication No. 103, Division of Scientific Publications, The Volcani Center, Bet Dagan, Isreal, 1978.

Hamdami, N., Monteau, J., and Le Bail, A. Thermophysical properties evolution of French partly baked bread during freezing. *Food Research International* 2004, 37, 703–713.

He, Y. Rapid thermal conductivity measurement with a hot disk sensor Part 1. Theoretical considerations. *Thermochimica Acta* 2005, 436, 122–129.

Higgs, S.J. and Swift, S.P. Investigation into the thermal conductivity of beef using the line source technique. *Process Biochemistry* 1975, 11, 43–45.

Hooper, F.G. and Lepper, F.R. Transient heat flow apparatus for the determination of thermal conductivities. *ASHVE Transactions* 1950, 56, 309.

Incropera, F.P. and DeWitt, D.P. *Fundamentals of Heat and Mass Transfer*. John Wiley and Sons, New York, 1990.

Karwe, M.V. and Tong, C.H. Effect of filling material on the temperature distribution in a thermal conductivity probe and thermal conductivity measurement: A numerical study. *Journal of Food Safety* 1992, 12, 339.

Keppeler, R.A. and Boose, J.R. Thermal properties of frozen sucrose solutions. *Transaction of the ASAE* 1970, 13(3), 335.

Kluitrenberg, G.J., Bristow, K.L., and Das, B.S. Error analysis of heat pulse method for measuring soil heat capacity, diffusivity and conductivity. *Soil Science Society of American Journal* 1995, 59, 719–726.

Kontani, T. Thermal conductivity measurements of high conductivity materials by a transient hot wire method. *Heat Transfer Japanese Research* 1990, 19, 366–372.

Kopelman, I.J. Transient heat transfer and thermal properties in food systems. Ph.D. Thesis, Michigan State University, East Lansing, 1966.

Kravets, R.R. Determination of thermal conductivity of food materials using a bead thermistor. Ph.D. Thesis in Food Science and Technology, Virginia Polytechnic Institute and State University, 1988.

Ladbury, J.E.S.D., Currel, B.R., Horder, J.R., Parsonage, J.R., and Vidgeon, E.A. Application of DSC for the measurement of the thermal conductivity of elastomeric materials. *Thermochimica Acta* 1990, 169, 39.

Lentz, C.P. Thermal conductivity of meats, fats, gelatin gels, and ice. *Food Technology* 1961, 15, 243.

Lin, S.X.Q., Chen, X.D., Chen, Z.D., and Bandopadhayay, P. Shear rate dependent thermal conductivity measurement of two fruit juice concentrates. *Journal of Food Engineering* 2003, 57, 217–224.

Mann, G. and Forsyth, F.G. Measurement of thermal conductivities of samples of thermal insulating materials and insulation in situ by the heated probe method. *Modern Refrigeration* 1956, 188–191.

Mendonca, S.L.R., Filho, C.R.B., and De Silva, Z.E. Transient conduction in spherical fruits: method estimates the thermal conductivity and volumetric thermal capacity. *Journal of Food Engineering* 2005, 67, 261–266.

Mitchell, J.K. and Kao, T.C. Measurement of soil thermal resistivity. *Journal of the Geotechnical Engineering Division*, ASAE, 1978, 104, 1307–1320.

Mohsenin, N.N. *Thermal Properties of Foods and Agricultural Materials,* Gordon and Breach Science Publisher, New York, 1980.

Murakami, E.G. and Okos, M.R. Measurement and prediction of thermal properties of foods. In: *Food Properties and Computer-Aided Engineering of Food Processing Systems*, Eds. Singh, R.P. and Medina, A.G., Kluwer Academic Publishers, New York, 1989, pp. 3–48.

Murakami, E.G., Sastry, S.K., and Zuritz, C.A. A modified Fitch apparatus for measuring thermal conductivity of small food particles. ASAE Paper No. 84-6508, ASAE, St. Joseph, MI 49085, 1984.

Murakami, E.G., Sweat, V.E., Sastry, S.K., and Kolbe, E. Analysis of various design and operating parameters of the thermal conductivity probe 193. Report to the NC-136 Ad Hoc Committee on the Thermal Conductivity Probe, 1993.

Murakami, E.G., Sweat, V.E., Sastry, S.K., and Kolbe, E. Analysis of various design and operating parameters of the thermal conductivity probe. *Journal of Food Engineering* 1996b, 30, 209.

Murakami, E.G., Sweat, V.E., Sastry, S.K., Kolbe, E., Hayakawa, K., and Datta, A. Recommended design parameters for thermal conductivity probes for nonfrozen food materials. *Journal of Food Engineering* 1996a, 27, 109–123.

Moreley, M.J. Thermal conductivities of muscles, fats and bones. *Journal of Food Technology* 1966, 1, 303.

Nesvadba, P. Methods for the measurement of thermal conductivity and diffusivity of foodstuffs. *Journal of Food Engineering* 1982, 1, 93–102.

Nix, G.H., Lowery, G.W., Vachon, R.I., and Tanger, G.E. Direct determination of thermal diffusivity and conductivity with a refined line source technique. In: *Progress in Aeronautics and Astronautics: Thermophysics of Spacecraft and Planetary Bodies*, Ed. Heller, G., Academic Press, New York, 1967, pp. 865–878.

Powell, R.W. Experiments using a simple thermal comparator for measurement of thermal conductivity, surface roughness and thickness of foils or surface deposit. *Journal of Scientific & Instrumentation* 1957, 34, 485.

Rahman, M.S. Evaluation of the precision of the modified Fitch method for thermal conductivity measurement of foods. *Journal of Food Engineering* 1991a, 14, 71.

Rahman, M.S. Thermophysical properties of seafoods. PhD Thesis, Department of Food Science and Technology, University of New South Wales, Australia, 1991b.

Rahman, M.S. and Driscoll, *Food Australia* 1991, 43(88), 356–360.

Reidy, G.A. and Rippen, A.L. *Transaction of American Society of Agriculture Engineering* 1971, Vol. 11, 248–254.

Rousan, A.A. A methodology for thermal characterization of cementitious materials. Ph.D. Thesis. The Penn State University, 1982.

Stalhane, B. and Pyk, S. New method for determining the coefficients of thermal conductivity. *Teknisk Tidskift* 1933, 61(28), 389.

Sweat, V.E. Experimental values of thermal conductivities of selected fruits and vegetables. *Journal of Food Science* 1974, 39, 1081–1083.

Sweat, V.E. Thermal properties of foods. In: *Engineering Properties of Foods*, Eds. Rao, M.A. and Rizvi, S.S.H., Marcel Dekker, New York, 1986.

Sweat, V.E. and Haugh, C.G. A thermal conductivity probe for small food samples. *Transactions of ASAE* 1974, 17, 56–63.

Shariaty-Niassar, M., Hozawa, M., and Tsukada, T. Development of probe for thermal conductivity measurement of food materials under heated and pressurized conditions. *Journal of Food Engineering* 2000, 43, 133–139.

Thermometrics. *Thermistor Sensor Handbook*, Thermometrics Inc., NJ, 1980.

Thermometrics. *Thermistor Sensor Handbook*, Thermometrics Inc., NJ, 1987.

Tong, C.H. and Sheen, S. Heat flux sensors to measure effective thermal conductivity of multilayered plastic containers. *J. Food Processing & Preservation* 1992, 16, 233.

Tong, C.H., Sheen, S., Shah, K.K., Huang, V.T., and Lund, D.B. Reference materials for calibrating probes used for measuring thermal conductivity of frozen foods. *Journal of Food Science* 1993, 58, 186.

Underwood, W.M. and McTaggart, R.B. The thermal conductivity of several plastics measured by an unsteady state method. *Chemical Engineering Progress Symposium Series* 1960, 56(30), 261.

Valvano, J.W. The use of thermal diffusivity to quantify tissue perfusion. Ph.D. Thesis in Medical Engineering, Harvard University, MIT Division of Health Sciences and Technology, 1981.

Van der Held, E.F.M. and Van Drunen, F.G. A method of measuring the thermal conductivity of liquids. *Physica* 1949, 15, 865–881.

van Gelder, M.F. A Thermistor based method for measurement of thermal conductivity and thermal diffusivity of moist food materials at high temperatures. Ph.D. Thesis, Virginia Polytechnic Institute and State University, 1998.

Vos, B.H. Measurements of thermal conductivity by a non-steady state method. *Applied Scientific Research* 1955, A5, 425–438.

Voudouris, N. and Hayakawa, K. Probe length and filling material effects on thermal conductivity determined by a maximum slope data reduction method. *Journal of Food Science* 1995, 60, 456–460.

Walters, R.E. and May, K.N. Thermal conductivity and density of chicken breast muscle and skin. *Food Technology* 1963, 17, 130–133.

Wang, N. and Brennan, J.G. Thermal conductivity of potato as a function of moisture content. *Journal of Food Engineering* 1992, 17, 153–160.

Wang, J. and Hayakawa, K. Thermal conductivity of starch gels at high temperatures influenced by moisture. *Journal of Food Science* 1993, 58(4), 884–887.

Welti-Chanes, J., Vergara-Balderas, F., and Bermudez-Aguirre, D. Transport phenomena in food engineering: basic concepts and advances. *Journal of Food Engineering* 2005, 67, 113–128.

Willix, J., Lovatt, S.J., and Amos, N.D. Additional thermal conductivity values of foods measured by a guarded hot plate. *Journal of Food Engineering* 1998, 37, 159–174.

Ziegler, G.R. and Rizvi, S.S.H. Thermal conductivity of liquid foods by the thermal comparator method. *Journal of Food Science* 1985, 50, 1458–1462.

Zuritz, C.A., Sastry, S.K., McCoy, S.C., Murakami, E.G., and Blaisdeh, J.L. A modified Fitch device for measuring the thermal conductivity of small food particles. *Transaction of American Society of Agriculture Engineering* 1989, 32(2), 711–718.

CHAPTER 18

Thermal Conductivity Data of Foods

Jasim Ahmed and Mohammad Shafiur Rahman

CONTENTS

18.1 INTRODUCTION

Huge numbers of thermal conductivity data of food materials are available in the literature. In this chapter, thermal conductivity data are presented in the form of tables. The variation of thermal conductivity of food materials as a function of moisture, temperature, density, porosity, and pressure are given in various tables (Tables 18.1 through 18.63). The tabular form rather than graphical form is used to present the data to make them easier to use. The thermal conductivity data are presented in four subgroups for ease of use: (1) fruits and vegetables; (2) meat, fish, and poultry products; (3) milk and milk products; and (4) cereal and cereal-based products.

18.2 THERMAL CONDUCTIVITY OF FRUITS AND VEGETABLES

18.2.1 Fruits

Table 18.1 Thermal Conductivity of Apple as a Function of Temperature and Moisture Content

		k (W/m K)				
M_w	X_w	5°C	15°C	25°C	35°C	45°C
0.25	0.20	0.077	0.080	0.086	0.089	0.095
0.45	0.31	0.088	0.095	0.100	0.105	0.107
0.68	0.40	0.105	0.110	0.115	0.120	0.127
1.25	0.56	0.123	0.132	0.138	0.147	0.156
1.50	0.60	0.154	0.163	0.176	0.186	0.200

Source: Singh, R.K. and Lund, D.B., *J. Food Process Preserv.*, 8, 191, 1984.

Table 18.2 Thermal Conductivity, Bulk Density, and Porosity of Raisin at Selected Moisture Content

T (°C)	X_w	ρ_B (kg/m^3)	ρ_{ap}	ε_B	k_B (W/m K)
44.5	0.14	762	1434	0.47	0.137
	0.20	708	1392	0.49	0.152
	0.40	660	1268	0.48	0.230
	0.50	672	1214	0.45	0.230
	0.60	661	1164	0.43	0.300
	0.65	673	1140	0.41	0.358
	0.75	691	1096	0.37	0.383
	0.80	582	1076	0.46	0.399

Source: Vagenas, G.K., Marinos-Kowis, D., and Saravacos, G.D., *J. Food Eng.*, 11, 147, 1990. With permission.
Note: Temperature range is 39.2°C–51.8°C.

Table 18.3 Thermal Conductivity of Selected Fruits

Material	Method of Measurement	X_w	ρ_{ap} (kg/m^3)	T (°C)	k (W/m K)	Reference
Apple (green)	Probe method	0.885	790	27	0.481	Sweat (1974)
Apple (red)	Probe method	0.849	840	28	0.422	Sweat (1974)
Apple	Probe method		803	10	0.371	Liang et al. (1999)
Avocado	Probe method	0.647	1060	28	0.429	Sweat (1974)
Banana	Probe method	0.757	980	28	0.462	Sweat (1974)
Banana	Probe method		977	10	0.475	Liang et al. (1999)
Pineapple	Probe method	0.849	1010	27	0.549	Sweat (1974)
Cantaloupe	Probe method	0.928	930	28	0.571	Sweat (1974)
Pear	Probe method	0.868	1000	28	0.595	Sweat (1974)
Pear	Probe method		993	10	0.543	Liang et al. (1999)
Peach	Probe method	0.885	930	28	0.581	Sweat (1974)

Table 18.3 (continued) Thermal Conductivity of Selected Fruits

Material	Method of Measurement	X_w	ρ_{ap} (kg/m^3)	T (°C)	k (W/m K)	Reference
Peach	Probe method	0.860	1012	—	0.580	Phomkong et al. (2006)
Plum (blue)	Probe method	0.886	1130	26	0.551	Sweat (1974)
Plum	Probe method	0.675	856	—	0.540	Phomkong et al. (2006)
Nectarine	Probe method	0.898	990	28	0.585	Sweat (1974)
Strawberry	Probe method	0.888	900	28	0.462	Sweat (1974)
Strawberry	Probe method		530	20	0.520	Delgado et al. (1997)
Strawberry (frozen)	Probe method			−15	0.935	Delgado et al. (1997)
Orange (peeled)	Probe method	0.859	1030	28	0.580	Sweat (1974)
Orange	Probe method		1012	10	0.554	Liang et al. (1999)
Lime (peeled)	Probe method	0.899	1000	28	0.490	Sweat (1974)
Lemon (peeled)	Probe method	0.918	930	28	0.525	Sweat (1974)
Grapefruit (peeled)	Probe method	0.904	950	26	0.549	Sweat (1974)
Papaya	Probe method	0.877		20	0.575	Kurozawa et al. (2005)

Sources: Sweat, V.E., *J. Food Sci.*, 39, 1081, 1974; Liang, X., Zhang, Y., and Gek, X., *Meas. Sci. Technol.*, 10, 82, 1999; Phomkong, W., Srzednicki, G., and Driscoll, R.H., *Dry. Technol.*, 24, 195, 2006; Delgado, A.E., Gallo, A., De Piante, D., and Rubiolo, A., *J. Food Eng.*, 31, 137, 1997; Kurozawa, L.E., El-Aouar, A.A., Simoes, M.R., Azoubel, P.M., and Murr, F.E.X., *2nd Mercosur Congress on Chemical Engineering Proceedings*, Costa Verda, Brazil, 2005, pp. 1–9.

Table 18.4 Thermal Conductivity of Selected Fruit Juice as a Function of Total Soluble Solids and Density

	ρ_{ap} (kg/m^3)				*k* (W/m K)			
TSS (°Brix)	**0.4°C**	**22.1°C**	**42.3°C**	**66.3°C**	**0.4°C**	**22.1°C**	**42.3°C**	**66.3°C**
Yellow mombin (Asis et al., 2006)								
8.8	1003	1028	985	1005	0.556	0.593	0.574	0.610
22	1059	1055	1058	1051	0.473	0.496	0.497	0.526
32.9	1108	1110	1105	1088	0.416	0.442	0.453	0.468
44.7	1005	1051	1088	1169	0.373	0.373	0.391	0.411
	30°C	50°C	70°C	80°C	30°C	50°C	70°C	80°C
Guava (Shamsudin et al., 2005)								
10	1040	1037	1010	1005	0.56	0.60	0.59	0.59
15	1070	1064	1044	1021	0.56	0.46	0.70	0.78
20	1090	1056	1037	1023	0.54	0.54	0.52	0.54
25	1120	1105	1080	1048	0.49	0.52	0.57	0.56
30	1135	1112	1073	1039	0.51	0.54	0.52	0.61
35	1155	1145	1125	1105	0.48	0.52	0.54	0.55
40	1180	1152	1118	1020	0.48	0.48	0.49	0.49

Sources: Assis, M.M.M., Lannes, S.C. da Silva, Tadini, C.C., Telis, V.R.N., and Telis-Romero, J., *Eur. Food Res. Technol.*, 223, 585, 2006. With permission; Shamsudin, R., Mohamed, I.O., and Yaman, N.K.M., *J. Food Eng.*, 66, 395, 2005. With permission.

Table 18.5 Thermal Conductivity of Orange Juice

T (°C)	X_w	*k* (W/m K)
20	0.961	0.562
	0.936	0.452
	0.915	0.429
	0.873	0.334
	0.829	0.320
	0.758	0.315
	0.690	0.312
	0.613	0.301

Source: Ziegler, G.R. and Rizvi, S.S.H., *J. Food Sci.*, 50, 1458, 1985. With permission.

Table 18.6 Thermal Conductivity of Fruit Juices as a Function of Temperature at Selected Concentration

	Thermal Conductivity (W/m K)										
TSS (°Brix)	20°C	30°C	40°C	50°C	60°C	70°C	80°C	90°C	100°C	110°C	120°C
Peach											
14.8	0.545	0.553	0.562	0.570	0.576	0.582	0.588	0.595	0.600	0.605	0.610
20	0.524	0.534	0.541	0.548	0.555	0.562	0.568	0.575	0.580	0.585	0.590
30	0.490	0.497	0.507	0.514	0.521	0.528	0.535	0.541	0.547	0.553	0.556
40	0.453	0.464	0.473	0.482	0.490	0.496	0.502	0.506	0.510	0.513	0.517
50	0.420	0.432	0.444	0.454	0.461	0.467	0.475	0.480	0.484	0.488	0.490
60	0.390	0.402	0.413	0.423	0.432	0.440	0.447	0.453	0.457	0.460	0.462
Plum											
15.1	0.535	0.546	0.555	0.565	0.573	0.583	0.590	0.597	0.605	0.611	0.617
20	0.511	0.524	0.533	0.543	0.550	0.559	0.566	0.575	0.582	0.588	0.593
30	0.469	0.481	0.492	0.503	0.510	0.517	0.526	0.532	0.540	0.545	0.550
40	0.434	0.446	0.457	0.467	0.474	0.481	0.489	0.497	0.504	0.510	0.515
50	0.405	0.417	0.427	0.437	0.445	0.452	0.462	0.468	0.475	0.481	0.486
60	0.380	0.389	0.401	0.411	0.421	0.430	0.437	0.445	0.453	0.459	0.465
Raspberry											
9.8	0.555	0.564	0.574	0.585	0.595	0.604	0.615	0.620	0.629	0.634	0.640
15	0.529	0.541	0.553	0.563	0.574	0.584	0.591	0.599	0.605	0.611	0.615
20	0.502	0.515	0.525	0.537	0.547	0.556	0.563	0.570	0.577	0.582	0.587
30	0.462	0.472	0.483	0.492	0.501	0.510	0.518	0.526	0.533	0.539	0.545
40	0.425	0.437	0.449	0.459	0.468	0.476	0.483	0.490	0.495	0.498	0.502
50	0.396	0.407	0.417	0.427	0.435	0.443	0.450	0.456	0.461	0.466	0.469

Table 18.6 (continued) Thermal Conductivity of Fruit Juices as a Function of Temperature at Selected Concentration

	Thermal Conductivity (W/m K)										
TSS (°Brix)	**20°C**	**30°C**	**40°C**	**50°C**	**60°C**	**70°C**	**80°C**	**90°C**	**100°C**	**110°C**	**120°C**
Cherry											
15.1	0.521	0.532	0.544	0.554	0.565	0.574	0.583	0.592	0.600	0.608	0.615
20	0.502	0.513	0.524	0.535	0.544	0.553	0.562	0.571	0.579	0.587	0.595
30	0.468	0.478	0.489	0.500	0.510	0.520	0.528	0.537	0.544	0.551	0.558
40	0.436	0.447	0.458	0.468	0.477	0.487	0.495	0.503	0.510	0.517	0.523
50	0.406	0.418	0.430	0.440	0.450	0.458	0.467	0.474	0.481	0.487	0.492
60	0.380	0.390	0.399	0.408	0.417	0.425	0.432	0.439	0.446	0.451	0.456
Pear											
14	0.545	0.553	0.561	0.568	0.575	0.581	0.587	0.593	0.598	0.603	0.607
19	0.526	0.535	0.543	0.551	0.558	0.565	0.571	0.576	0.581	0.585	0.589
25	0.503	0.511	0.519	0.526	0.534	0.540	0.547	0.553	0.559	0.564	0.568
30	0.485	0.493	0.500	0.508	0.514	0.521	0.527	0.532	0.538	0.542	0.547
40	0.450	0.458	0.466	0.473	0.480	0.487	0.493	0.499	0.504	0.508	0.513
50	0.408	0.417	0.425	0.433	0.441	0.448	0.456	0.462	0.469	0.476	0.481
Apricot											
18.5	0.525	0.536	0.546	0.556	0.565	0.573	0.581	0.588	0.595	0.601	0.606
25	0.498	0.509	0.519	0.529	0.538	0.546	0.554	0.562	0.568	0.574	0.580
30	0.480	0.490	0.499	0.508	0.517	0.525	0.533	0.540	0.547	0.554	0.560
35	0.461	0.471	0.480	0.489	0.497	0.505	0.513	0.520	0.527	0.534	0.540
40	0.444	0.453	0.462	0.470	0.478	0.486	0.494	0.501	0.508	0.515	0.521
50	0.415	0.424	0.432	0.440	0.448	0.455	0.462	0.469	0.476	0.481	0.487
Sweet cherry											
14.3	0.525	0.538	0.549	0.560	0.570	0.579	0.587	0.594	0.601	0.607	0.612
20	0.505	0.516	0.527	0.537	0.546	0.555	0.563	0.569	0.576	0.582	0.586
27	0.480	0.491	0.501	0.511	0.521	0.529	0.537	0.545	0.552	0.558	0.564
35	0.454	0.464	0.473	0.482	0.491	0.498	0.507	0.514	0.522	0.528	0.534
45	0.415	0.426	0.436	0.445	0.454	0.462	0.471	0.478	0.485	0.492	0.498
50	0.401	0.410	0.420	0.429	0.438	0.447	0.455	0.463	0.470	0.478	0.484
Cherry plum											
12.2	0.543	0.552	0.560	0.568	0.577	0.584	0.593	0.600	0.606	0.613	0.618
18	0.518	0.527	0.536	0.545	0.553	0.560	0.568	0.573	0.580	0.585	0.590
24	0.495	0.504	0.512	0.520	0.528	0.535	0.542	0.548	0.554	0.560	0.565
30	0.472	0.481	0.489	0.497	0.504	0.511	0.518	0.524	0.530	0.535	0.540
40	0.434	0.443	0.451	0.460	0.467	0.474	0.481	0.488	0.594	0.500	0.505
50	0.398	0.407	0.416	0.424	0.432	0.440	0.447	0.455	0.462	0.468	0.474

Sources: From Magerramov, M.A., Abdulagatov, A.I., Azizov, N.D., and Abdulagatov, I.M., *J. Food Sci.*, 71, E238, 2006a. With permission; Magerramov, M.A., Abdulagatov, A.I., Abdulagatov, I.M., and Azizov, N.D., *J. Food Process Eng.*, 29, 304, 2006b. With permission.

Table 18.7 Effect of Shear Rate on Thermal Conductivity of Fruit Juice Concentrate

Juice Type	*T* (°C)[a]	γ (s^{-1})	*k* (W/m K)	Juice Type	*T* (°C)[a]	γ (s^{-1})	*k* (W/m K)
Orange	32.5	0	0.448	Mango	31.9	0	0.499
		89	0.479			83	0.635
		228	0.509			166	0.697
		357	0.547			249	0.755
		521	0.608			353	0.811
		789	0.689			461	0.911
		1100	0.775			544	0.957
	42.4	0	0.489			789	1.118
		87	0.505			1100	1.328
		224	0.534		42	0	0.546
		357	0.581			79	0.711
		534	0.667			166	0.813
		789	0.755			299	0.914
		1100	0.814			467	1.068
	52.6	0	0.528			550	1.169
		108	0.553			793	1.440
		249	0.586			1100	1.701
		419	0.672		51.7	0	0.589
		556	0.717			100	0.823
						197	0.928
						347	1.063
						513	1.302
						789	1.625
						1100	1.863

Source: Lin, S.X.Q., Chen, X.D., Chen, Z.D., and Bandopadhayay, P., *J. Food Eng.*, 57, 217, 2003. With permission.

[a] Average temperature.

Table 18.8 Thermal Conductivity of Shredded Coconut[a] as a Function of Moisture Content, Temperature, and Bulk Density

T (°C)	X_w	ρ_B (kg/m^3)	k_B (W/m K)	*T* (°C)	X_w	ρ_B (kg/m^3)	k_B (W/m K)
26.4	0.024	264	0.092	47.2	0.156	292	0.122
26.4	0.024	264	0.096	49.4	0.156	292	0.124
26.4	0.156	292	0.113	24.0	0.270	329	0.132
26.4	0.156	292	0.117	24.6	0.270	329	0.136
26.4	0.270	329	0.131	33.2	0.270	329	0.144
26.4	0.270	329	0.135	43.6	0.270	329	0.155
26.4	0.512	462	0.156	45.0	0.270	329	0.140
26.4	0.512	462	0.166	53.1	0.270	329	0.157
48.7	0.024	264	0.095	29.3	0.512	462	0.157
48.7	0.024	264	0.099	29.3	0.512	462	0.169
48.7	0.156	292	0.120	40.8	0.512	462	0.171
48.7	0.156	292	0.124	42.5	0.512	462	0.193
48.7	0.270	329	0.155	45.9	0.512	462	0.203

Table 18.8 (continued) Thermal Conductivity of Shredded Coconut[a] as a Function of Moisture Content, Temperature, and Bulk Density

T (°C)	X_w	ρ_B (kg/m³)	k_B (W/m K)	*T* (°C)	X_w	ρ_B (kg/m³)	k_B (W/m K)
48.7	0.270	329	0.158	47.6	0.512	462	0.213
48.7	0.512	462	0.214	49.4	0.512	462	0.217
48.7	0.512	462	0.228	52.4	0.512	462	0.210
26.9	0.024	264	0.092	27.0	0.010	254	0.085
27.0	0.024	264	0.097	27.0	0.010	254	0.086
40.8	0.024	264	0.094	27.0	0.010	293	0.090
41.4	0.024	264	0.099	27.0	0.010	293	0.091
47.2	0.024	264	0.095	27.0	0.010	323	0.095
49.4	0.024	264	0.097	27.0	0.010	323	0.098
24.0	0.156	292	0.114	27.0	0.010	338	0.097
24.6	0.156	292	0.116	27.0	0.010	338	0.101
38.3	0.156	292	0.119				
39.1	0.156	292	0.120				

Source: Jindal, V.K. and Murakami, E.G., in *Engineering and Food*, Vol. 1, *Engineering Sciences in the Food Industry,* McKenna, B.M. (ed.), *Proceedings of the 3rd International Congress on Engineering and Food,* September 26–28, 1983, Dublin, Ireland, Elsevier Applied Science, New York, 1984, 558.

[a] Fresh coconut kernel ($X_w = 0.047$ and $X_{FA} = 0.35$–0.50); average particle size, 1.67 ± 1.04 mm ($X_w = 0.024$).

Table 18.9 Thermal Conductivity of Freeze-Dried Orange Pulp at Selected Air Pressures

	Bulk Thermal Conductivity, k_B (W/m K)								
P (mm Hg)	a	b	c	d	e	f	g	h	i
760.000	0.0364	0.0373	0.0374	0.0381	0.0388	0.0390	0.0390	0.0441	0.0413
100.000	0.0359	0.0372	0.0371	0.0379	0.0381	0.0390	0.0389	0.0441	0.0410
70.000	0.0357	—	0.0365	—	—	—	—	—	0.0412
50.000	0.0361	0.0369	0.0367	0.0374	0.0379	—	—	0.0397	0.0412
25.000	—	0.0364	0.0367	0.0370	0.0376	0.0385	0.0386	0.0393	0.0402
20.000	0.0349	0.0361	0.0356	—	—	—	0.0380	—	—
15.000	—	0.0358	0.0349	0.0358	0.0369	0.0371	0.0376	—	—
10.000	0.0340	0.0343	0.0330	0.0353	0.0358	—	0.0377	0.0375	0.0386
8.000	0.0336	—	—	0.0330	—	0.0347	0.0374	—	0.0379
5.000	0.0317	0.0303	0.0308	0.0311	0.0332	—	0.0331	0.0340	0.0348
3.000	—	0.0271	—	0.0289	0.0291	0.0310	0.0316	—	—
2.000	0.0263	0.0263	—	0.0267	—	—	—	—	0.0300
1.000	0.0238	—	0.0242	—	0.0250	—	0.0274	0.0262	0.0276
0.500	—	—	0.0225	0.0232	0.0231	0.0253	0.0254	0.0259	0.0271
0.250	0.0206	0.0206	0.0221	0.0216	0.0228	0.0232	0.0248	0.0253	0.0269
0.100	—	0.0199	0.0205	0.0206	—	0.0211	0.0240	0.0247	0.0228
0.001	0.0187	0.0191	0.0194	0.0199	0.0205	0.0211	0.0215	0.0227	0.0227

Source: Fito, P.J., Piiraga, F., and Aranda, V., *J. Food Eng.*, 3, 375, 1984. With permission.

Notes: a, $\varepsilon_B = 0.874$ and $\rho_B = 166.0$; b, $\varepsilon_B = 0.866$ and $\rho_B = 176.5$; c, $\varepsilon_B = 0.861$ and $\rho_B = 183.5$; d, $\varepsilon_B = 0.850$ and $\rho_B = 198.0$; e, $\varepsilon_B = 0.845$ and $\rho_B = 204.0$; f, $\varepsilon_B = 0.839$ and $\rho_B = 213.0$; g, $\varepsilon_B = 0.828$ and $\rho_B = 226.5$; h, $\varepsilon_B = 0.821$ and $\rho_B = 236.0$; i, $\varepsilon_B = 0.814$ and $\rho_B = 245.0$; Valencia-Late variety orange; $T = 45$°C.

Table 18.10 Thermal Conductivity[a] of High Pressure Treated Commercial Apple Pulp

Pressure Level (MPa)	k (W/m K) 30°C	65°C
0.01	0.54	0.61
100	0.58	0.64
150	0.59	0.67
200	0.62	0.69
250	0.64	0.71
300	0.66	0.72
350	0.66	0.72
400	0.69	0.75

Source: Denys, S. and Hendrickx, M.E., *J. Food Sci.*, 64, 709, 1999. With permission.

[a] Measured by line heat source method.

Table 18.11 Thermal Conductivity of Sucrose Solution as a Function of Temperature

	Thermal Conductivity (W/m K) Temperature (°C)												
X_{wo}	−39	−36	−33	−30	−28	−25	−22	−19	−16	−14	−11	−8	−5
0.98	2.174	2.092	2.085	2.061	2.030	1.925	1.908	1.813	1.789	1.713	1.589	1.540	1.439
0.96	2.204	2.100	2.073	2.028	1.998	1.877	1.856	1.750	1.712	1.628	1.513	1.435	1.529
0.94	2.135	2.048	2.045	2.016	1.999	1.902	1.895	1.801	1.790	1.721	1.630	1.596	1.529
0.92	2.111	2.049	2.037	1.999	1.974	1.863	1.843	1.755	1.745	1.677	1.571	1.525	1.431
0.90	2.111	2.006	1.984	1.944	1.913	1.805	1.783	1.683	1.654	1.574	1.460	1.406	1.431
0.88	1.958	1.890	1.861	1.814	1.783	1.690	1.674	1.577	1.552	1.476	1.368	1.315	1.221
0.86	1.983	1.895	1.849	1.788	1.775	1.675	1.644	1.552	1.523	1.443	1.343	1.276	1.182
0.84	1.922	1.846	1.799	1.738	1.694	1.595	1.564	1.469	1.430	1.336	1.211	1.139	1.036
0.82	1.827	1.770	1.713	1.646	1.594	1.511	0.963	1.379	1.330	1.238	1.125	1.053	0.957
0.80	1.774	1.701	1.658	1.585	1.539	1.444	1.407	1.306	1.265	1.174	1.068	0.999	0.895
0.75	1.674	1.596	1.565	1.491	1.435	1.335	1.299	1.212	1.169	1.092	0.985	0.908	0.796
0.70	1.487	1.432	1.404	1.328	1.270	1.177	1.146	1.067	1.029	0.946	0.850	0.770	0.658
0.65	1.306	1.258	1.239	1.165	1.106	1.020	0.989	0.920	0.881	0.807	0.719	0.645	0.551

Source: Adapted from Keppeler, R.A. and Boose, J.R., *Trans. ASAE*, 13, 335, 1970.

Table 18.12 Thermal Conductivity[a] of Fructose Solution as a Function of Concentration and Temperature

	Thermal Conductivity (W/m K)					
T (°C)	$X_s = 0.10$	$X_s = 0.20$	$X_s = 0.30$	$X_s = 0.40$	$X_s = 0.50$	$X_s = 0.60$
10	0.55	0.52	0.48	0.45	0.42	0.38
20	0.56	0.54	0.50	0.46	0.43	0.39
30	0.58	0.55	0.51	0.48	0.44	0.40
40	0.59	0.56	0.52	0.49	0.45	0.41
50	0.60	0.57	0.53	0.50	0.46	0.41

Source: Muramatsu, Y., Tagawa, A., and Kasai, T., *Food Sci. Technol. Res.*, 11, 288, 2005.

[a] Measurement was carried out using transient heat flow method.

18.2.2 Thermal Conductivity of Vegetables

Table 18.13 Thermal Conductivity of Selected Vegetables at Selected Moisture and Temperature

Type of Vegetable	*T* (°C)	X_w	*k* (W/m K)	Type of Vegetable	*T* (°C)	X_w	*k* (W/m K)
Eggplant (Ali et al., 2002)	5	0.93	0.236	Pepper (Ali et al., 2002)	5	0.94	0.468
	10	0.92	0.326		10	0.94	0.481
	20	0.93	0.356		20	0.94	0.509
	30	0.93	0.389		30	0.94	0.522
	14	0.80	0.155		7	0.79	0.434
	25	0.80	0.215		29	0.79	0.462
	13	0.23	0.077		36	0.79	0.496
	26	0.23	0.083		12	0.64	0.397
Cassava (Ali et al., 2002)	6	0.80	0.495		28	0.64	0.434
	10	0.80	0.490	Ginger (Ali et al., 2002)	6	0.91	0.552
	20	0.80	0.478		10	0.90	0.550
	30	0.80	0.514		21	0.89	0.573
	5	0.58	0.462		31	0.87	0.545
	11	0.58	0.436		10	0.87	0.510
	20	0.58	0.461		21	0.87	0.531
	30	0.58	0.491		31	0.87	0.549
	6	0.25	0.347		16	0.77	0.491
	11	0.25	0.337		21	0.77	0.503
	20	0.25	0.342		30	0.77	0.523
	30	0.25	0.352		16	0.62	0.438
Radish (Ali et al., 2002)	5	0.94	0.554		21	0.62	0.444
	10	0.94	0.569		30	0.62	0.477
	20	0.94	0.574	Zucchini (Ali et al., 2002)	5	0.95	0.405
	30	0.94	0.582		10	0.94	0.431
	17	0.70	0.378		19	0.95	0.506
	27	0.70	0.410		30	0.95	0.583
	30	0.54	0.365		13	0.85	0.461
	39	0.54	0.375		20	0.85	0.518
	31	0.40	0.345		7	0.27	0.137
	40	0.40	0.335		17	0.27	0.158
Yam (Njie et al., 1998)	25	0.79	0.600		26	0.27	0.185
	25	0.54	0.570	Plantain (Njie et al., 1998)	25	0.57	0.450
	25	0.45	0.480		25	0.45	0.430
	25	0.38	0.410		25	0.32	0.290
	25	0.22	0.190		25	0.27	0.190
	25	0.16	0.160		25	0.14	0.130

Sources: Ali, S.D., Ramaswamy, H.S., and Awuah, G.B., *J. Food Process Eng.*, 25, 417, 2002; Njie, D.N., Rumsey, T.R., and Singh, R.P., *J. Food Eng.*, 37, 63, 1998.

Table 18.14 Thermal Conductivity of Vegetables

Material	MT[a]	X_w	ρ_{ap} (kg/m^3)	T (°C)	k (W/m K)	Reference
Potato						
Katahdin	A	0.814	1040	25.5	0.533	Rao et al. (1975)
Russet Burbank	A	0.829	1040	24.8	0.571	Rao et al. (1975)
Kennebec	A	0.824	1050	25.0	0.549	Rao et al. (1975)
Monona	A	0.836	1040	24.6	0.547	Rao et al. (1975)
Norchip	A	0.812	1050	25.9	0.533	Rao et al. (1975)
Potato (recorded)	B	0.763	1127	40	0.410	Rice et al. (1988)
	B	0.763	1117	50	0.470	Rice et al. (1988)
	B	0.763	1121	60	0.460	Rice et al. (1988)
	B	0.763	1105	70	0.510	Rice et al. (1988)
	B	0.763	1107	80	0.560	Rice et al. (1988)
	B	0.763	1103	90	0.560	Rice et al. (1988)
	B	0.763	1166	50	0.510	Rice et al. (1988)
	B	0.763	1115	—	0.470	Rice et al. (1988)
	B	0.754	1139	—	0.410	Rice et al. (1988)
	B	0.723	1088	—	0.360	Rice et al. (1988)
Potato	A	0.835	—	25.0	0.563	Gratzek and Toledo (1993)
Potato	A	0.835	—	75.0	0.622	Gratzek and Toledo (1993)
Potato	A	0.835	—	105.0	0.639	Gratzek and Toledo (1993)
Potato	A	0.835	—	130.0	0.641	Gratzek and Toledo (1993)
Potato	A		1071	10	0.553	Liang et al. (1999)
Potato	A	0.778	1089		0.563	Murakami (1997)
Potato blanched (10 min)	A	0.802	1088		0.551	Murakami (1997)
Potato (boiled 2.5 h)	A	0.851	1064		0.567	Murakami (1997)
Baked potato (2 h)	A	0.818			0.556	Murakami (1997)
Blanched and cooked potato	A	0.814	1086		0.557	Murakami (1997)
Cooked potato	A	0.822	1089		0.568	
Asparagus lettuce	A		1041	10	0.573	Liang et al. (1999)
Carrot	A		950	10	0.530	Liang et al. (1999)
Squash						
Butternut	A	0.877	950	26.1	0.500	Rao et al. (1975)
Golden delicious	A	0.885	900	22.9	0.471	Rao et al. 1975
Boston marrow	A	0.936	970	23.4	0.533	Rao et al. (1975)
Cucumber (Burpee)	A	0.954	950	28.0	0.598	Gratzek and Toledo (1993)
Cucumber	A		994	10	0.568	Liang et al. (1999)
Carrot	A	0.900	1040	28	0.605	Gratzek and Toledo (1993)
Carrot	A	0.923	—	25	0.571	Liang et al. (1999)
Carrot	A	0.923	—	70	0.620	Liang et al. (1999)
Carrot	A	0.923	—	105	0.649	Liang et al. (1999)
Carrot	A	0.923	—	130	0.664	Liang et al. (1999)
Beet (red, Detroit)	A	0.895	1530	28	0.601	Gratzek and Toledo (1993)
Onion	A	0.873	970	28	0.574	Gratzek and Toledo (1993)
Tomato	A		901	10	0.4–0.50	Liang et al. (1999)
Cherry tomato[b]	A	0.923	1010	28	0.571	Gratzek and Toledo (1993)
Cherry tomato[c]	A	0.923	1010	28	0.462	Gratzek and Toledo (1993)

Table 18.14 (continued) Thermal Conductivity of Vegetables

Material	MT	X_w	ρ_{ap} (kg/m^3)	T (°C)	k (W/m K)	Reference
Turnip	A	0.898	1000	24	0.563	Gratzek and Toledo (1993)
Spinach (fresh)[d]	A	0.931	524	21	0.347	Delgado et al. (1997)
Spinach frozen (without blanching)	A			−10	0.366	Delgado et al. (1997)
				−20	0.391	
Blanched spinach[c]	A	0.893	589	16	0.356	Delgado et al. (1997)
Spinach blanched and frozen	A			−10	0.399	Delgado et al. (1997)
				−20	0.434	
Sugar beets	A	0.724	1284	−11	1.038	Tabil et al. (2001)
	A	0.723	1185	15	0.584	Tabil et al. (2001)
	A	0.730	1198	30	0.528	Tabil et al. (2001)
Green radish[b]	A	—	994	10	0.537	Liang et al. (1999)
Green radish[c]	A	—	850	10	0.337	Liang et al. (1999)

Sources: Rao, M.A., Barnard, J., and Kenny, J., *Trans. ASAE*, 18, 1188, 1975; Rice, P., Selman, J.D., and Abdul-Rezzak, R.K., *Int. J. Food Sci. Technol.*, 23, 281, 1988; Gratzek, J.P. and Toledo, R.T., *J. Food Sci.*, 58, 908, 1993; Liang, X., Zhang, Y., and Gek, X., *Meas. Sci. Technol.*, 10, 82, 1999; Murakami, E.G., *J. Food Process Eng.*, 20, 415, 1997; Sweat *J. Food Sci.*, 39, 1081, 1974; Delgado, A.E., Gallo, A., De Piante, D., and Rubiolo, A., *J. Food Eng.*, 31, 137, 1997; Tabil, L.G., Eliason, M.V., and Qi, H., *ASAE Annual International Meeting*, Sacramento, CA, July 30–August 1, 2001, Paper No. 016141, 2001.

Note: A, probe method; B, from thermal diffusivity, specific heat, and density.

[a] Method of measurement.

[b] Probe inserted in side, perpendicular to core axis.

[c] Probe inserted in side, perpendicular to core axis.

[d] 70% leaves and 30% stems.

Table 18.15 Thermal Conductivity of Cumin Seed as a Function of Temperature and Moisture Content

	k (W/m K)					
X_w	−50°C	−30°C	−10°C	10°C	30°C	50°C
0.642	0.043	0.067	0.090	0.111	0.129	0.145
0.750	0.047	0.072	0.096	0.117	0.136	0.152
0.857	0.058	0.084	0.109	0.131	0.150	0.168
0.923	0.075	0.104	0.130	0.154	0.176	0.196
0.953	0.090	0.121	0.150	0.177	0.202	0.225

Source: From Singh, K.K. and Goswami, T.K., *J. Food Eng.*, 45, 181, 2000. With permission.

Note: Measurement was carried out using line heat source method.

Table 18.16 Thermal Conductivity of Radish and Alfalfa Seeds as a Function of Moisture Content, Temperature, and Bulk Density

Seed	X_w	T (°C)	ρ_B (kg/m^3)	k (W/m K)	Seed	X_w	T (°C)	ρ_B (kg/m^3)	k (W/m K)
Radish	0.046	30	690	0.052	Alfalfa	0.005	30	834	0.075
		45	689	0.061			45	833	0.080
		60	687	0.071			60	831	0.092
		80	684	0.083			80	826	0.105
	0.055	30	684	0.057		0.065	30	826	0.079
		45	683	0.067			45	823	0.083
		60	681	0.076			60	820	0.095
		80	676	0.089			80	813	0.108
	0.065	30	681	0.061		0.073	30	819	0.081
		45	679	0.071			45	816	0.089
		60	677	0.080			60	812	0.103
		80	671	0.093			80	805	0.115

Source: Yang, J. and Zhao, Y., *J. Food Process Eng.*, 24, 291, 2001.

Table 18.17 Thermal Conductivity[a] of High Pressure Treated Commercial Canned Tomato Paste

Pressure Level (MPa)	k (W/m K)	
	30°C	65°C
0.01	0.52	0.59
100	0.56	0.62
150	0.58	0.65
200	0.60	0.66
250	0.62	0.67
300	0.63	0.70
350	0.63	0.69
400	0.66	0.72

Source: Denys, S. and Hendrickx, M.E., *J. Food Sci.*, 64, 709, 1999. With permission.

[a] Measured by line heat source method.

Table 18.18 Thermal Conductivity of Mushrooms as a Function of Moisture Content, Temperature, and Bulk Density

X_w	Temperature (°C)	ρ_B (kg/m^3)	k (W/m °C)
0.102	40	111	0.208
0.102	55	111	0.212
0.102	70	111	0.215
0.102	40	383	0.218
0.102	55	383	0.221
0.102	70	383	0.226
0.102	40	656	0.230
0.102	55	656	0.233
0.102	70	656	0.235
0.500	40	111	0.342

Table 18.18 (continued) Thermal Conductivity of Mushrooms as a Function of Moisture Content, Temperature, and Bulk Density

X_w	Temperature (°C)	ρ_B (kg/m^3)	k (W/m °C)
0.500	55	111	0.344
0.500	70	111	0.349
0.500	40	383	0.353
0.500	55	383	0.357
0.500	70	383	0.361
0.500	40	656	0.366
0.500	55	656	0.369
0.500	70	656	0.372
0.900	40	111	0.504
0.900	55	111	0.508
0.900	70	111	0.510
0.900	40	383	0.514
0.900	55	383	0.517
0.900	70	383	0.520
0.900	40	656	0.522
0.900	55	656	0.528
0.900	70	656	0.531

Source: Shrivastava, M. and Datta, A.K., *J. Food Eng.*, 39, 255, 1999. With permission.

Table 18.19 Thermal Conductivity of Onion as a Function of Moisture Content and Bulk Density at 25°C

X_w	ρ_B (kg/m^3)	ε_B	k (W/m K)
Spirit[a]			
0.877	1040	0.037	0.50
0.755	1086	0.020	0.42
0.585	1156	0.069	0.27
0.499	1196	0.079	0.34
0.449	1219	0.088	0.29
0.308	1293	0.153	0.25
0.221	1343	0.162	0.18
0.156	1383	0.194	0.15
0.017	1480	0.222	0.12
Niz[a]			
0.813	1063	0.018	0.47
0.721	—	—	0.46
0.628	1138	0.023	0.40
0.480	1205	0.031	0.37
0.248	—	—	0.24
0.245	1329	0.094	0.30
0.186	1364	0.099	0.24
0.137	1395	0.143	0.24
0.000	—	—	0.14

(*continued*)

Table 18.19 (continued) Thermal Conductivity of Onion as a Function of Moisture Content and Bulk Density at 25°C

X_w	ρ_B (kg/m^3)	ε_B	k (W/m K)
Sweet Vidalia[a]			
0.920	1025	0.083	0.54
0.858	—	—	0.55
0.582	—	—	0.28
0.462	—	—	0.27
0.449	—	—	0.29
0.336	1278	0.194	0.23
0.329	—	—	0.28
0.261	1319	0.207	0.14
0.217	—	—	0.12
0.136	—	—	0.18
0.059	—	—	0.13

Source: Abhayakwick, L., Laguerre, J.C., Tauzin, V., and Duquenoy, A., *J. Food Eng.*, 55, 253, 2002. With permission.

[a] Composition of fresh samples. Spirit: X_w, 0.880; X_{pr}, 0.007; X_{FA}, 0.006; X_{as}, 0.004; X_{CA}, 0.062. Niz: X_w, 0.830; X_{pr}, 0.017; X_{FA}, 0.004; X_{as}, 0.006; X_{CA}, 0.054. Sweet Vidalia: X_w, 0.920; X_{pr}, 0.010; X_{FA}, 0.004; X_{as}, 0.005; X_{CA}, 0.044.

Table 18.20 Thermal Conductivity of Potato as a Function of Temperature and Moisture Content

Lamberg and Hallstorm (1986)					Califano and Calvelo (1991)				
X_w	T (°C)	k (W/m K)	T (°C)	k (W/m K)	X_w	T (°C)	k (W/m K)	T (°C)	k (W/m K)
0.798 (Lamberg and Hallstrom, 1986)[a]	22.5	0.620	53.0	0.665	0.80 (Califano and Calvelo, 1991)[b]	50.0	0.545	80.2	0.710
	24.0	0.635	55.0	0.665		60.0	0.555	80.3	0.720
	21.4	0.640	55.0	0.690		66.4	0.590	83.0	0.730
	29.0	0.680	59.0	0.690		66.5	0.600	83.0	0.750
	36.5	0.630	65.0	0.720		70.7	0.610	86.4	0.770
	38.5	0.630	85.0	0.690		70.7	0.620	86.4	0.850
	40.0	0.645	86.0	0.710		76.1	0.620	94.6	0.895
	47.5	0.705				76.1	0.630	94.6	0.915
0.760 (Yamada (1970) cited by Lamberg and Hallstrom, 1986)	10	0.485				76.1	0.650	100.0	0.940
0.760 (Tschubik and Maslow (1973) cited by Lamberg and Hallstrom, 1986)	75	0.556				80.2	0.670	100.0	0.960

[a] Bintje variety: ρ_{ap}, 1085 kg/m^3; T_{GP}, 66°C.
[b] Kennebec variety: ρ_{ap}, 1070 kg/m^3; ε_v, 0.017.

Table 18.21 Thermal Conductivity of Tomato Paste

		k (W/m K)		
TSS (°B)	X_w	30°C	40°C	50°C
27	0.708	0.595	0.630	0.660
30	0.678	0.560	0.590	0.620
32	0.657	0.550	0.575	0.600
36	0.618	0.525	0.546	0.557
39	0.588	0.485	0.505	0.512
44	0.538	0.460	0.478	0.490

Source: Drusas, A.E. and Saravcos, G.D., *J. Food Eng.*, 4, 157, 1985. With permission.

18.3 THERMAL CONDUCTIVITY OF FISH, MEAT, AND POULTRY

18.3.1 Egg Products

Table 18.22 Thermal Conductivity of Liquid Egg Products as a Function of Moisture Content and Temperature

	k (W/m K)										
X_w	0°C	2°C	5°C	8°C	12°C	18°C	22°C	25°C	28°C	34°C	38°C
0.518	0.407	0.403	0.403	0.400	0.399	0.397	0.396	0.393	0.393	0.390	0.389
0.591	0.433	0.435	0.433	0.430	0.429	0.428	0.428	0.424	0.426	0.421	0.421
0.664	0.481	0.464	0.467	0.459	0.459	0.458	0.461	0.453	0.459	0.452	0.453
0.736	0.493	0.498	0.493	0.494	0.491	0.490	0.493	0.486	0.490	0.485	0.486
0.809	0.541	0.526	0.529	0.522	0.522	0.522	0.525	0.518	0.522	0.517	0.518
0.882	0.554	0.560	0.555	0.552	0.551	0.552	0.558	0.547	0.555	0.547	0.550

Source: Coimbra, J.S.R., Gabas, A.L., Minim, L.A., Rojas, E.E.G., Telis, V.R.N., and Telis-Romero, J., *J. Food Eng.*, 74, 186, 2006. With permission.

Table 18.23 Thermal Conductivity of Liquid Egg Yolk[a] as a Function of Temperature

T (°C)	ρ (kg/m³)	k (W/m K)	T (°C)	ρ (kg/m³)	k (W/m K)
0.4	1133.2	0.390	38	1130.5	0.375
10	1132.5	0.387	48	1129.8	0.370
20	1132.0	0.384	60.8	1129.5	0.366
30	1131.5	0.379			

Source: Gut, J.A.W., Pinto, J.M., Gabas, A.L., and Telis-Romero, J., *J. Food Process Eng.*, 28, 181, 2005.

[a] Initial moisture content of yolk was 0.540 and mean pH 6.4.

18.3.2 Fish and Fish Products

Table 18.24 Thermal Conductivity of Fresh and Frozen Fish as a Function of Temperature and Composition

	Composition			k (W/m K)						
Fish Type	X_w	X_{FA}	X_{pr}	−20°C	−10°C	0°C	10°C	20°C	30°C	40°C
Saman	0.7730	0.0046	0.2020	1.428	1.334	0.409	0.419	0.422	0.431	0.454
Sol	0.7760	0.0046	0.2046	1.365	1.294	0.398	0.434	0.476	0.452	0.459
Black bhitki	0.7770	0.6000	0.1945	1.563	1.487	0.416	0.425	0.436	0.454	0.464
Black pomfret	0.7500	0.0476	0.1791	1.315	1.244	0.412	0.418	0.417	0.442	0.452
Mackerel	0.7735	0.0058	0.1867	1.237	1.088	0.416	0.423	0.438	0.459	—
Red bhitki	0.7953	0.0042	0.1689	1.358	1.207	0.413	0.435	0.440	0.449	0.466
Singara	0.7787	0.0057	0.2012	1.284	1.254	0.420	0.417	0.418	0.450	0.461
Hilsa	0.7470	0.0513	0.1722	1.328	1.234	0.409	0.416	0.430	0.450	0.463
Surama	0.7800	0.0077	0.1882	1.327	1.067	0.415	0.433	0.433	0.446	0.468
White pomfret	0.7483	0.0410	0.1856	1.395	1.118	0.390	0.400	0.435	0.453	0.475
Malli	0.7820	0.0045	0.1935	—	1.356	0.411	0.424	0.460	0.453	—
Rohu	0.7520	0.0057	0.2148	1.535	1.284	0.421	0.446	0.442	0.461	0.463

Source: Kumbhar, B.K., Agarwal, R.S., and Das, K., *Int. J. Refrigeration*, 4, 143, 1981.

Table 18.25 Thermal Conductivity of Fresh Seafood

	Proximate Composition					k (W/m K)		
Type	X_w	X_{pr}	X_{FA}	X_{as}	ρ_{ap} (kg/m^3)	K_T[a]	K_L[b]	**Reference**
Calamari (mantle)	0.7970	0.1570	0.0220	0.0130	1063	0.510	0.530	Rahman and Driscoli (1991)[c]
Calamari (tentacle)	0.8290	0.1070	0.0210	0.0080	1049	0.520	0.530	Rahman and Driscoli (1991)[c]
Octopus (mantle)	0.7740	0.1810	0.0190	0.0220	1074	0.530	0.520	Rahman and Driscoli (1991)[c]
Octopus (tentacle)	0.7860	0.1710	0.0230	0.0190	1071	0.530	0.530	Rahman and Driscoli (1991)[c]
Prawn (abdomen)	0.7530	0.1820	0.0120	0.0180	1082	0.490	0.510	Rahman and Driscoli (1991)[c]
King prawn (green)	0.7533	0.1822	0.0123	0.0184	1082	0.481	—	Rahman et al. (1991)[c]
King prawn (tiger)	0.7569	0.1895	0.0340	0.0157	1081	0.485	—	Rahman et al. (1991)[c]
Octopus (tentacle)	0.7754	0.1666	0.0319	0.0247	1072	0.499	—	Rahman et al. (1991)[c]
Squid (tentacle)	0.7859	0.1659	0.0344	0.0120	1067	0.475	—	Rahman et al. (1991)[c]
Octopus (mantle)	0.7918	0.1654	0.0270	0.0168	1075	0.497	—	Rahman et al. (1991)[c]
Squid (mantle)	0.7928	0.1675	0.0234	0.0134	1066	0.483	—	Rahman et al. (1991)[c]
Calamari (mantle)	0.8015	0.1599	0.0238	0.0134	1063	0.508	—	Rahman et al. (1991)[c]

Table 18.25 (continued) Thermal Conductivity of Fresh Seafood

Type	Proximate Composition				ρ_{ap} (kg/m³)	k (W/m K)		Reference
	X_w	X_{pr}	X_{FA}	X_{as}		K_T[a]	K_L[b]	
Octopus (tentacle)	0.8071	0.1358	0.0330	0.0230	1067	0.505	—	Rahman et al. (1991)[c]
Squid (tail)	0.8127	0.1346	0.0403	0.0109	1055	0.500	—	Rahman et al. (1991)[c]
Cuttle (mantle)	0.8131	0.1571	0.0130	0.0107	1059	0.485	—	Rahman et al. (1991)[c]
Calamari (wing)	0.8340	0.1139	0.0387	0.0096	1047	0.522	—	Rahman et al. (1991)[c]
Bluefish	0.7810	—	0.0370	—	—	0.489[d]	—	Radhakrishnan (1997)
Croaker	0.7960	—	0.0090	—	—	0.487[d]	—	Radhakrishnan (1997)
Salmon	0.7250	—	0.0390	—	—	0.471[d]	—	Radhakrishnan (1997)
Seabass	0.7980	—	0.0020	—	—	0.486[d]	—	Radhakrishnan (1997)[c]
Shrimp	0.8470	—	0.0000	—	—	0.543[d]	—	Radhakrishnan (1997)[c]
Mackerel	0.7310	—	0.0910	—	—	0.425[d]	—	Radhakrishnan (1997)
Spot	0.6380	—	0.1590	—	—	0.401[d]	—	Radhakrishnan (1997)
Tilapia	0.7770	—	0.0070	—	—	0.496[d]	—	Radhakrishnan (1997)
Trout	0.7910	—	0.0290	—	—	0.519[d]	—	Radhakrishnan (1997)
Tuna	0.7310	—	0.0000	—	—	0.469[d]	—	Radhakrishnan (1997)

[a] k_T is the heat transfer through the transverse direction.
[b] k_L is the heat transfer through the longitudinal direction.
[c] Measurement carried out at 20°C.
[d] Average values measured at about 5°C, 10°C, 15°C, 20°C, 25°C, and 30°C.

Table 18.26 Effect of Freezing and Thawing on Thermal Conductivity of Seafood

Type	X_w	k (W/m K)	
		Fresh	Thawed
Calamari[a]	0.831	0.554	0.550
Calamari[b]	0.799	0.544	0.527
King prawn	0.771	0.541	0.545

Source: Rahman, M.S. and Driscoli, R.H., *Food Aust.*, 43, 356, 1991.
[a] Batch 1.
[b] Batch 2.

Table 18.27 Thermal Conductivity of Shrimp and Shucked Oysters as a Function of Temperature

Type	X_w	ρ_B (kg/m^3)	T (°C)	k (W/m K)	Type	X_w	T (°C)	k (W/m K)
Shucked oysters (Hu and Mallikarjunan, 2005)		1038	10	0.590	Shrimp meat (Karunakar et al., 1998)[a]	0.780	30	0.51
		1037	20	0.604			20	0.49
		1035	30	0.618			10	0.49
		1032	40	0.632			−1	0.50
		1029	50	0.647			−2	0.96
Raw whole shrimp (Ngadi et al., 2000)	0.800	1060	25	0.560			−5	1.27
Raw ground shrimp (Ngadi et al., 2000)	0.800	1070	25	0.460			−10	1.40
Skipjack tuna loin muscle (Zhang et al., 2001)	0.708		33.2	0.560			−15	1.49
			45.9	0.570			−20	1.55
			52.7	0.570			−25	1.58
			63.9	0.540			−30	1.60
			74.9	0.590				
			84.9	0.620				
			91.6	0.660				
Albacore (raw) (Perez-Martin et al., 1989)[b,c]	0.673	1080	15–20	0.493				
Albacore (raw) (Perez-Martin et al., 1989)[b,d]	0.673	1080	15–20	0.563				
Albacore (dry) (Perez-Martin et al., 1989)[b]			20–50	0.106				

Source: Hu, X. and Mallikarjunan, P., *Lebensmittel-wissenschaft und-technologie* 38, 489, 2005.

[a] X_{pr}, 0.152; X_{FA}, 0.027; X_{as}, 0.006; X_{CH}, 0.007.

[b] X_{pr}, 0.238; X_{FA}, 0.073; X_{as}, 0.023; X_{CH}, 0.003.

[c] Heat flux perpendicular.

[d] Heat flux parallel.

Table 18.28 Effect of Heat Processing on Thermal Conductivity of Shrimp and Scallops

Process for Shrimp	X_w	ρ_B (kg/m^3)	k (W/m K)	Process for Scallops	X_w	ρ_B (kg/m^3)	k (W/m K)
Fresh black tiger shrimp	0.762	1081	0.515	Fresh sea scallops	0.863	1044	0.550
Blanched in water (4 min)	0.710	1080	0.513	Blanched in water (5 min)	0.849	1058	0.544
Blanched in 10% saline water (4 min)	0.705	1112	0.494	Blanched in 10% saline water (4 min)	0.709	1108	0.515
Cooked in CPS[a] (30 min)	0.708	1102	0.480	Canned CPS (F_o = 21.3)	0.777	1102	0.533
Canned (F_o = 3.5)	0.741	1081	0.498	Canned CPS (F_o = 3.5)	0.755	1072	0.535
Cream of potato soup	—	1042	0.577				

Source: Murakami, E.G., *J. Food Sci.*, 59, 237, 1994. With permission.
[a] CPS, cream of potato soup.

Table 18.29 Thermal Conductivity of Surimi[a] as a Function of Temperature

X_w	T (°C)	ρ (kg/m^3)	k (W/m K)	Type	T (°C)	ρ (kg/m^3)	k (W/m K)
0.740	30	1080	0.524	0.800	30	1044	0.536
	40	1065	0.525		40	1033	0.570
	50	1054	0.550		50	1021	0.571
	60	1053	0.601		60	1010	0.587
	70	1030	0.600		70	998	0.610
	80	1019	0.630		80	986	0.683
0.780	30	1055	0.534	0.840	30	1023	0.567
	40	1044	0.539		40	1011	0.577
	50	1032	0.570		50	1000	0.577
	60	1020	0.565		60	988	0.613
	70	1009	0.610		70	976	0.632
	80	997	0.680		80	965	0.708

Source: AbuDagga, Y. and Kolbe, E., *J. Food Eng.*, 32, 315, 1997. With permission.
[a] Pacific whiting surimi: 4% sucrose, 4% sorbitol, and 0.3% sodium tripolyphosphate were used.

Table 18.30 Thermal Conductivity of Dried Seafood Powder at 20°C

Type of Food	X_w	ρ_P (kg/m^3)	ε_B	k_B (W/m K)
Squid[a]	0.0460	501	0.626	0.077
		559	0.582	0.079
		606	0.549	0.080
		662	0.505	0.096
		704	0.474	0.097
		777	0.419	0.115
		805	0.398	0.124
		821	0.386	0.122

(continued)

Table 18.30 (continued) Thermal Conductivity of Dried Seafood Powder at 20°C

Type of Food	X_w	ρ_P (kg/m^3)	ε_B	k_B (W/m K)
Calamari[b]	0.0394	366	0.730	0.073
		424	0.687	0.075
		469	0.654	0.076
		511	0.623	0.082
		538	0.605	0.089
		640	0.528	0.104
		653	0.518	0.104
		685	0.495	0.106
		722	0.468	0.112
King prawn (green)[c]	0.0345	487	0.637	0.076
		531	0.604	0.079
		574	0.572	0.075
		675	0.497	0.081
		787	0.414	0.109
		828	0.383	0.107

Source: Rahman, M.S. and Driscoli, R.H., *Food Aust.*, 43, 356, 1991.

[a] X_w, 0.0394; X_{pr}, 0.7417; X_{FA}, 0.1128; X_{as}, 0.0578.

[b] X_w, 0.0345; X_{pr}, 0.6950; X_{FA}, 0.0650; X_{as}, 0.1030.

[c] X_w, 0.0460; X_{pr}, 0.7884; X_{FA}, 0.0575; X_{as}, 0.059.

18.3.3 Meat Products

Table 18.31 Effect of Blending on Thermal Conductivity of Fresh Seafood at 20°C

			Thermal Conductivity (W/m K)		
Seafood	X_w	ρ_{ap} (kg/m^3)	k_T	k_L	k_P
Calamari (mantle)	0.794	1066	0.526	0.539	0.546
Calamari (mantle)	0.816	1062	0.532	0.528	0.544
Octopus (mantle)	0.806	1069	0.521	0.524	0.548
Octopus (tentacle)	0.853	1042	0.547	0.532	0.547
Octopus (tentacle)	0.803	1065	0.519	0.513	0.542
Prawn	0.765	1077	0.501	0.528	0.544

Source: Rahman, M.S. and Driscoli, R.H., *Food Aust.*, 43, 356, 1991.

Table 18.32 Thermal Conductivity of Marinated[a] Shrimp and Catfish as a Function of Moisture Content and Temperature

Type	T (°C)	X_w	k (W/m K)	Type	T (°C)	X_w	k (W/m K)
Shrimp (nonmarinated)	10	0.762	0.41	Catfish (nonmarinated)	10	—	—
	20	0.749	0.44		20	0.786	0.48
	30	0.760	0.48		30	0.781	0.53
	40	0.759	0.48		40	0.760	0.52
	50	0.746	0.48		50	0.758	0.54
	60	0.742	0.49		60	0.748	0.54
	70	0.737	0.50		70	0.697	0.51
Shrimp (marinated)	10	0.756	0.38	Catfish (marinated)	10	—	—
	20	0.757	0.46		20	0.780	0.49
	30	0.756	0.45		30	0.785	0.52
	40	0.752	0.47		40	0.773	0.56
	50	0.756	0.50		50	0.771	0.52
	60	0.750	0.50		60	0.765	0.54
	70	0.738	0.50		70	0.708	0.53

Source: Zheng, M., Huang, Y.W., Nelson, S.O., Bartley, P.G., and Gates, K.W., *J. Food Sci.*, 4, 668, 1998. With permission.

[a] Peeled shrimp and catfish fillets were mixed with commercial lemon pepper marinade (Formula 159-J., A.C. Legg. Packing Company, Inc., Birmingham, AL). Marination formula: 11.35 kg shrimp or catfish fillet, 0.79 kg water, and 0.23 kg marinade. The marinade formula: salt, dextrose, sodium phosphates (20.69%), black pepper spice extractives and lemon oil.

Table 18.33 Thermal Conductivity of Dried Squid Meat Slab at 20°C

Sample	X_w	ρ_{ap} (kg/m^3)	ρ_B (W/m K)	k_B (W/m K)	Sample	X_w	ρ_{ap} (kg/m^3)	ρ_B (W/m K)	k_B (W/m K)
Fresh[a]	0.826	1060	1050	0.49	Fresh[c]	0.838	1060	1050	0.50
Dried	0.809	1070	1040	0.52	Dried	0.692	1120	980	0.47
	0.796	1070	1000	0.49		0.551	1170	980	0.35
	0.794	1070	1020	0.48		0.541	1170	1000	0.35
	0.791	1070	980	0.50		0.466	1200	970	0.34
	0.782	1070	1050	0.49		0.418	1210	1010	0.33
	0.755	1090	1020	0.51		0.199	1260	1020	0.17
	0.741	1090	1050	0.44		0.197	1260	990	0.11
	0.627	1140	1080	0.46		0.171	1260	920	0.11
	0.581	1160	1060	0.32		0.139	1270	710	0.04
	0.451	1200	1060	0.33		0.112	1270	700	0.05
	0.144	1260	920	0.13		0.089	1270	570	0.04
Fresh[b]	0.839	1060	1040	0.50					
Dried	0.777	1080	980	0.52					
	0.728	1120	960	0.44					
	0.693	1120	980	0.40					
	0.650	1140	960	0.46					
	0.629	1140	1070	0.40					
	0.587	1160	910	0.40					
	0.572	1160	1020	0.36					

(*continued*)

Table 18.33 (continued) Thermal Conductivity of Dried Squid Meat Slab at 20°C

Sample	X_w	ρ_{ap} (kg/m³)	ρ_B (W/m K)	k_B (W/m K)	Sample	X_w	ρ_{ap} (kg/m³)	ρ_B (W/m K)	k_B (W/m K)
	0.566	1170	930	0.40					
	0.537	1180	930	0.35					
	0.534	1180	990	0.36					
	0.527	1180	1040	0.33					
	0.434	1210	970	0.38					
	0.421	1210	1140	0.30					
	0.357	1230	960	0.25					
	0.305	1240	1190	0.32					
	0.270	1250	1100	0.15					
	0.264	1250	1030	0.20					
	0.209	1260	690	0.05					

Source: Rahman, M.S. and Potluri, P.L., *J. Food Eng.*, 12, 133, 1990. With permission.
[a] X_{pr}, 0.1546; X_{FA}, 0.0018; X_{as}, 0.0106.
[b] X_{pr}, 0.1308; X_{FA}, 0.0084; X_{as}, 0.0147.
[c] X_{pr}, 0.1407; X_{FA}, 0.0026; X_{as}, 0.0119.

Table 18.34 Thermal Conductivity of Fresh Beef Chuck Steak

X_w	X_{FA}	T (°C)	k (W/m K)	X_w	X_{FA}	T (°C)	k (W/m K)
0.614[a]	0.058	14.3	0.465	0.634	0.026	38.6	0.427
		21.7	0.442	0.634	0.026	48.8	0.449
		36.8	0.427	0.634	0.026	50.9	0.450
		45.5	0.411	0.634	0.026	52.9	0.456
		66.8	0.533	0.634	0.026	55.0	0.452
0.472	0.118	80	0.557	0.634	0.026	57.4	0.466
				0.634	0.026	59.9	0.470
				0.634	0.026	61.5	0.475
				0.634	0.026	64.4	0.487
				0.634	0.026	68.6	0.505

Source: Higgs, S.J. and Swift, S.P., *Process Biochem.*, 43, 1975.
[a] A gradual decrease in the moisture content due to drip loss. Temperature initial or bath temperature.

Table 18.35 Thermal Conductivity[a] of Fresh Beef as a Function of Composition and Temperature

Sample Type	T (°C)	X_w	X_{pr}	X_{FA}	ρ_{ap} (kg/m³)	k (W/m K)
Ground round	30	0.7460	0.2230	0.0310	1057	0.452
	50	0.7400	0.2220	0.0317	1054	0.484
	60	0.7265	0.2401	0.0333	1050	0.492
	70	0.7030	0.2608	0.0352	1015	0.522
	80	0.6821	0.2790	0.0387	1009	0.560
	90	0.6650	0.2950	0.0400	988	0.590
Whole round	30	0.7460	0.2230	0.0310	1010	0.475
	50	0.7360	0.2319	0.0322	1003	0.492
	60	0.7174	0.2480	0.0344	997	0.504
	70	0.6856	0.2760	0.0380	988	0.494
	80	0.6501	0.3071	0.0426	985	0.507
	90	0.5994	0.3517	0.0490	980	0.495

Table 18.35 (continued) Thermal Conductivity[a] of Fresh Beef as a Function of Composition and Temperature

Sample Type	T (°C)	X_w	X_{pr}	X_{FA}	ρ_{ap} (kg/m^3)	k (W/m K)
Ground shank	30	0.7450	0.2183	0.0367	1044	0.442
	50	0.7430	0.2199	0.0369	1043	0.490
	60	0.7390	0.2262	0.0351	1038	0.494
	70	0.7100	0.2527	0.0375	1030	0.514
	80	0.6597	0.2998	0.0405	1024	0.547
	90	0.7365	0.3040	0.0391	1023	0.598
Ground brisket	30	0.7225	0.1925	0.0635	1045	0.447
	50	0.6913	0.1983	0.0652	1039	0.462
	60	0.6680	0.2090	0.0685	1036	0.472
	70	0.6470	0.2324	0.0763	1029	0.484
	80	0.7530	0.2490	0.0830	1025	0.490
	90	0.7400	0.2650	0.0880	1021	0.526
Whole brisket	30	0.7530	0.1925	0.0635	916	0.436
	50	0.7400	0.1960	0.0640	914	0.460
	60	0.7310	0.2025	0.0665	911	0.458
	70	0.7000	0.2260	0.0740	906	0.469
	80	0.6280	0.2880	0.0840	901	0.460
	90	0.5770	0.3200	0.0930	893	0.440
Whole rib steak	30	0.6990	0.2262	0.0748	1029	0.459
	50	0.6929	0.2307	0.0763	1027	0.493
	60	0.6798	0.2417	0.0784	1018	0.499
	70	0.6277	0.2827	0.0895	1009	0.515
	80	0.5276	0.3625	0.1098	997	0.540
	90	0.5177	0.3714	0.1008	998	0.552
Ground sirloin tip	30	0.7450	0.2080	0.0470	1045	0.460
	50	0.7384	0.2138	0.0482	1040	0.472
	60	0.7157	0.2318	0.0524	1041	0.481
	70	0.6870	0.2550	0.0570	1035	0.503
	80	0.6502	0.2853	0.0644	1029	0.510
	90	0.6165	0.3127	0.0708	1022	0.518
Whole sirloin tip	30	0.7450	0.2080	0.0470	990	0.467
	50	0.7368	0.2140	0.0480	988	0.479
	60	0.7214	0.2274	0.0514	984	0.472
	70	0.6510	0.2810	0.0670	979	0.465
	80	0.6034	0.3246	0.0720	974	0.485
	90	0.5660	0.3580	0.0760	969	0.494
Ground rib	30	0.5780	0.1700	0.2520	1020	0.368
	50	0.5620	0.1761	0.2611	1021	0.376
	60	0.5420	0.1880	0.2690	1014	0.403
	70	0.4920	0.2280	0.2790	1008	0.419
	80	0.4790	0.2700	0.2800	1005	0.438
	90	0.4390	0.2800	0.2810	977	0.450
Ground Swiss steak	30	0.7750	0.1890	0.0360	1042	0.467
	50	0.7668	0.1958	0.0373	1040	0.438
	60	0.7549	0.2058	0.0392	1036	0.492
	70	0.7337	0.2236	0.0426	1029	0.503

(*continued*)

Table 18.35 (continued) Thermal Conductivity[a] of Fresh Beef as a Function of Composition and Temperature

Sample Type	T (°C)	X_w	X_{pr}	X_{FA}	ρ_{ap} (kg/m^3)	k (W/m K)
	80	0.7075	0.2458	0.0467	1022	0.540
	90	0.6838	0.2638	0.0500	1020	0.575
Whole Swiss steak	30	0.7750	0.1890	0.0360	983	0.467
	50	0.7680	0.1944	0.0380	980	0.482
	60	0.7602	0.2015	0.0380	979	0.499
	70	0.7389	0.2193	0.0417	975	0.509
	80	0.7003	0.2516	0.0470	971	0.501
	90	0.6476	0.2960	0.0560	966	0.508

Source: Baghe-Khandan, M.S., Okos, M.R., and Sweat, V.E., *Trans. ASAE*, 4, 1118, 1982. With permission.

[a] Measured by line source probe method.

Table 18.36 Effect of Various Cooling Techniques on Thermal Conductivity of Cooked Beef

Cooling Technique	X_w	X_{pr}	X_{FA}	X_{CH}	X_{as}	X_{salt}	K_k[a] (W/m K)	k[b] (W/m K)
Cooked only	0.743	0.187	0.036	0.003	0.032	0.024	0.464	0.442
Air blast	0.723	0.204	0.040	0.002	0.031	0.023	0.454	0.433
Vacuum	0.692	0.230	0.041	0.007	0.032	0.025	0.431	0.414
Immersion	0.735	0.196	0.037	0.001	0.032	0.023	0.462	0.441
Cold storage	0.729	0.200	0.040	0.001	0.031	0.023	0.458	0.440

Source: McDonald, K., Sun, D.W., and Lyng, J.G., *J. Food Eng.*, 52, 167, 2002. With permission.

Notes: Sample cooked to 72°C at the geometric core and allowed to cool at ambient temperature. Cooled to 4°C at the geometric core.

[a] Measured parallel to the direction of muscle fiber.

[b] Measured perpendicular to the direction of muscle fiber.

Table 18.37 Thermal Conductivity of Lamb Meat, Ground Beef, and Chicken as a Function of Temperature

Sample Type	X_w	T (°C)	X_{FA}	k (W/m K)	Sample Type	X_w	T (°C)	X_{FA}	k (W/m K)
Boneless lamb meat (Tocci et al., 1997)	0.734–0.753	−40		1.60	Ground beef (Pan and Singh, 2001)	0.600	5	0.238	0.40
		−30		1.51			27		0.40
		−20		1.22			35		0.38
		−10		1.20			50		0.35
		10		0.50			60		0.37
		20		0.51			70		0.40
		30		0.48	Ground chicken breast (Murphy and Marks, 1999)	0.729	20	0.292	0.502
Chicken drum muscle (Ngadi and Ikedialis, 1998)[a]	0.750	20	0.15–0.18	0.366		0.682	30		0.485
		40		0.413		0.658	40		0.482
		60		0.460		0.648	50		0.474
		80		0.507		0.632	60		0.464
	0.450	20		0.217		0.602	80		0.460
		40		0.264					
		60		0.311					
		80		0.358					

[a] ρ_B ranged between 762 and 1192 kg/m^3.

18.4 THERMAL CONDUCTIVITY OF DAIRY PRODUCTS

Table 18.38 Thermal Conductivity of Spray-Dried Whole and Skim Milk Powder

Whole Milk (MacCarthy, 1985)[a]				Skim Milk (MacCarthy, 1983)[b]					
512 kg/m³		605 kg/m³		(292 kg/m³)[c]		(577 kg/m³)[d]		(724 kg/m³)[e]	
T(°C)	k_B (W/m K)	T(°C)	k_B (W/m K)	T(°C)	k_B (W/m K)	T(°C)	k_B (W/m K)	T(°C)	k_B (W/m K)
11.8	0.0365	16.6	0.0583	11.9	0.0360	14.8	0.0636	18.0	0.0861
21.8	0.0540	17.7	0.0678	18.11	0.0419	15.5	0.0599	21.0	0.0845
28.3	0.0630	22.2	0.0711	19.3	0.0456	29.6	0.0661	27.9	0.0956
38.3	0.0798	26.2	0.0857	27.3	0.0463	35.8	0.0720	35.3	0.1061
42.5	0.0785	28.4	0.0862	29.7	0.0506	49.5	0.0819	40.8	0.1035
43.2	0.0771	34.3	0.0869	36.4	0.0550	49.7	0.0813	41.3	0.1003
		42.8	0.0929	41.0	0.0599			42.7	0.0994
								49.3	0.1096

[a] Method, steady-state guarded hot plate; X_{FA}, 0.261; X_w, 0.031; acidity, 0.157% as lactic; solubility, 0.75; size, 98.1% smaller than 710 μm.
[b] Method, steady-state guarded hot plate.
[c] X_{FA}, 0.15–0.30; X_w, 0.014; acidity, 0.25% as lactic.
[d] X_{FA}, 0.10–0.33; X_w, 0.042; acidity, 0.22% as lactic.
[e] X_{FA}, 0.10–0.45; X_w, 0.040; acidity, 0.18% as lactic.

Table 18.39 Effective Thermal Conductivity for Whole Milk Powder at Selected Temperature

		k (W/m °C)				
ρ_B (kg/m³)	X_w^a	10°C	20°C	30°C	40°C	50°C
550	0.014	0.058	0.060	0.062	0.064	0.066
	0.038	0.066	0.069	0.072	0.076	0.079
600	0.014	0.066	0.068	0.070	0.071	0.073
	0.038	0.073	0.077	0.079	0.082	0.085
650	0.014	0.074	0.076	0.077	0.079	0.081
	0.038	0.079	0.082	0.085	0.089	0.092
700	0.014	0.080	0.082	0.084	0.086	0.087
	0.038	0.085	0.089	0.091	0.094	0.097

Source: Muramatsu, Y., Tagawa, A., and Kasai, T., *Food Sci. Technol. Res.*, 11, 288, 2005.
[a] Protein, fat, carbohydrate, and ash mass fractions were 0.259, 0.266, 0.399, and 0.061 and 0.253, 0.260, 0.39, 0.060 for 0.014 and 0.038 water mass fraction, respectively.

Table 18.40 Effective Thermal Conductivity for Skim Milk Powder at Selected Temperature

		k (W/m °C)				
		Temperature (°C)				
ρ_B (kg/m^3)	X_w^a	10	20	30	40	50
700	0.023	0.073	0.075	0.077	0.079	0.082
	0.045	0.078	0.081	0.084	0.088	0.090
	0.076	0.082	0.087	0.091	0.097	0.102
750	0.023	0.080	0.082	0.084	0.086	0.088
	0.045	0.084	0.087	0.090	0.094	0.095
	0.076	0.089	0.094	0.100	0.105	0.111
800	0.023	0.087	0.089	0.091	0.093	0.095
	0.045	0.091	0.094	0.097	0.102	0.106
	0.076	0.096	0.100	0.105	0.111	0.122
850	0.023	0.095	0.096	0.098	0.100	0.101
	0.045	0.098	0.100	0.105	0.110	0.113
	0.076*	—	—	—	—	—

Source: Muramatsu, Y., Tagawa, A., and Kasai, T., *Food Sci. Technol. Res.*, 11, 288, 2005.

[a] Protein, fat, carbohydrate, and ash mass fractions were 0.345, 0.010, 0.541, and 0.080; 0.328, 0.010, 0.529, and 0.078; and 0.327, 0.512, 0.010, and 0.076 for 0.023, 0.045, and 0.076 water mass fraction, respectively.

Table 18.41 Thermal Conductivity of Freeze-Dried Milk[a] at Selected Vacuum

		k_B (W/m K)			
T (°C)	P (mm Hg)	ε_B: 0.831, ρ_B: 207 kg/m^3	ε_B: 0.831, ρ_B: 207 kg/m^3	ε_B: 0.831, ρ_B: 207 kg/m^3	ε_B: 0.831, ρ_B: 207 kg/m^3
45	760.000	0.0453	0.0508	0.0534	0.0595
	100.000	—	0.0506	0.0530	—
	50.000	0.0452	0.0498	0.0525	—
	25.000	0.0447	0.0495	—	—
	20.000	0.0428	0.0493	0.0511	0.0565
	15.000	0.0421	0.0466	0.0504	0.0556
	10.000	0.0387	0.0456	0.0438	—
	8.000	0.0389	0.0445	—	0.0535
	5.000	0.0356	0.0406	—	0.0526
	3.000	—	0.0387	0.0444	0.0504
	2.000	0.0307	—	—	—
	1.000	—	—	0.0383	0.0454
	0.500	—	0.0331	0.0353	0.0393
	0.250	—	—	0.0348	0.0361
	0.100	0.0254	—	0.0318	—
	0.001	0.0239	0.0277	0.0305	0.0335

Source: Fito, P.J., Piiraga, F., and Aranda, V., *J. Food Eng.*, 3, 375, 1984. With permission.

[a] X_w, 0.035; X_{FA}, 0.267; X_{su}, 0.38; X_{pr}, 0.258; X_{as}, 0.06.

Table 18.42 Thermal Conductivity of Selected Dairy Products

Type of Product	X_w	X_{FA}	ρ_{ap}[a] (kg/m^3)	k (W/m K) 0°C	15°C	20°C	30°C	40°C	Reference
Butter	0.151	0.836	942		0.227		0.233		Tavman and Tavman (1999)[b]
Butter	0.165	0.806	910	0.20		0.21			Sweat and Parmelee (1978)[c]
Pudding (DelMonte, chocolate)	0.724	0.236	1070	0.50		0.53		0.54	Sweat and Parmelee (1978)[c]
Sour cream	0.750	0.172	1000	0.47		0.46		0.49	Tavman and Tavman (1999)[b]
Filled evaporated milk	0.770	0.078	1080	0.46		0.47		0.46	Sweat and Parmelee (1978)[c]
Sweetened condensed milk	0.301	0.110	1310	0.33		0.33		0.32	Sweat and Parmelee (1978)[c]
Evaporated milk	0.770	0.077	1070	0.46		0.46		0.45	Sweat and Parmelee (1978)[c]
Half and half	0.822	0.093	1010	0.43		0.47		—	Sweat and Parmelee (1978)[c]
Cream	0.604	0.167	990	0.33		0.36		—	Sweat and Parmelee (1978)[c]
Ice milk mix	0.686	0.056	1090	0.46		0.49		0.47	Sweat and Parmelee (1978)[c]
Labne	0.691	0.209	1085		0.486		0.463		Tavman and Tavman (1999)[b]
Low fat labne	0.747	0.103	1085		0.548		0.542		Tavman and Tavman (1999)[b]
Strained yogurt	0.742	0.074	972		0.540		0.539		Tavman and Tavman (1999)[b]
Light yogurt	0.820	0.016	1033		0.571		0.583		Tavman and Tavman (1999)[b]
Pasteurized yogurt	0.825	0.041	1035		0.571		0.593		Tavman and Tavman (1999)[b]
Extra light yogurt	0.868	0.002	1025		0.584		0.596		Tavman and Tavman (1999)[b]
Yogurt	0.862			0.525[d]	0.546[e]	0.570	0.588	0.603	Kent et al. (1984)
Margarine	0.160	0.817	960	0.20		0.19		—	Sweat and Parmelee (1978)[c]
Whipped margarine	0.162	0.816	650	0.15		0.17		—	Sweat and Parmelee (1978)[c]
Diet margarine	0.567	0.401	1070	0.34		0.36		—	Sweat and Parmelee (1978)[c]

Sources: Sweat, V.E. and Parmelee, C.E., *J. Food Process Eng.*, 2, 187, 1978; Tavman, I.H. and Tavman, S., *J. Food Eng.*, 41, 109, 1999; Kent, M., Christiansen, K., van Haneghem, I.A., Holtz, E., Morley, M.J., Nesvadba, P., and Poulsen, K.P., *J. Food Eng.*, 3, 117, 1984.

[a] Bulk density for Tavman and Tavman (1999).
[b] Modified hot wire method.
[c] Line source probe method.
[d] 1°C.
[e] 10°C.

Table 18.43 Thermal Conductivity of Cheddar Cheese as a Function of Composition

X_w	X_{FA}	X_{pr}	k (W/m K)
0.337	0.349	0.261	0.354
0.348	0.352	0.261	0.356
0.442	0.234	0.283	0.391
0.474	0.172	0.298	0.423
0.500	0.170	0.303	0.432
0.549	0.094	0.326	0.472
0.379	0.323	0.267	0.388
0.456	0.194	0.280	0.428
0.502	0.102	0.310	0.465
0.454	0.217	0.303	0.425
0.412	0.240	0.285	0.422
0.394	0.258	0.281	0.410
0.471	0.172	0.301	0.448

Source: Marschoun, L.T., Muthukumarappan, K., and Gunasekaran, S., *Int. J. Food Prop.*, 4, 383, 2001.

Table 18.44 Thermal Conductivity of Various Cheeses as a Function of Temperature and Bulk Density

Type of Product	X_w	X_{FA}	ρ_{ap} (kg/m^3)	k (W/m K) 0°C	15°C	20°C	30°C	40°C	Reference
Colby	0.373	0.320	1050	0.32		0.31		0.29	Sweat and Parmelee (1978)[a]
Colby (sharp)	0.378	0.300	1090	0.31		0.28		0.27	Sweat and Parmelee (1978)[a]
Cheddar	0.372	0.320	1090	0.32		0.31		0.30	Sweat and Parmelee (1978)[a]
Cheddar	0.360	0.320	1102		0.345		0.351		Tavman and Tavman (1999)[b]
Muenster	0.443	0.315	1070	0.34		0.34		0.31	Sweat and Parmelee (1978)[a]
Port salute	0.473	0.245	1060	0.36		0.32		0.34	Sweat and Parmelee (1978)[a]
Cream	0.554	0.320	1010	0.34		0.38		0.36	Sweat and Parmelee (1978)[a]
Cream (low fat)	0.645	0.229	1000	0.42		0.37		0.40	Sweat and Parmelee (1978)[a]
Fresh cream	0.563	0.235	1014		0.433		0.434		Tavman and Tavman (1999)[b]
Swiss (without eyes)	0.352	0.320	1110	0.33		0.29		0.34	Sweat and Parmelee (1978)[a]
Monterey jack	0.395	0.330	1090	0.33		0.32		0.32	Sweat and Parmelee (1978)[a]
Gjetost	0.191	0.246	1230	0.32		0.33		0.30	Sweat and Parmelee (1978)[a]
Romano	0.313	0.273	1210	0.29		0.30		0.29	Sweat and Parmelee (1978)[a]
Brick	0.435	0.299	1090	0.32		0.30		0.30	Sweat and Parmelee (1978)[a]
Mozarella	0.455	0.170	1140	0.34		0.37		0.38	Sweat and Parmelee (1978)[a]
Mozeralla	0.444	0.239	1062		0.383		0.380		Tavman and Tavman (1999)[b]
Spread	0.478	0.195	1130	0.36		0.37		0.38	Sweat and Parmelee (1978)[a]
Spread	0.606	0.163	824		0.476		0.494		Tavman and Tavman (1999)[b]
Neufchatel	0.652	0.220	1080	0.42		0.40		0.40	Sweat and Parmelee (1978)[a]
Smoked	0.451	0.259	1130	0.32		0.32		0.33	Sweat and Parmelee (1978)[a]
Hamburger	0.410	0.248	1114		0.381		0.398		Tavman and Tavman (1999)[b]
Fresh kashkaval	0.438	0.228	1182		0.433		0.434		Tavman and Tavman (1999)[b]
Old kashkaval	0.410	0.266	1117		0.368		0.384		Tavman and Tavman (1999)[b]
Buffet kashkaval	0.498	0.143	961		0.406		0.409		Tavman and Tavman (1999)[b]
Tulum	0.410	0.289	1110		0.379		0.377		Tavman and Tavman (1999)[b]

Sources: Sweat, V.E. and Parmelee, C.E., *J. Food Process Eng.*, 2, 187, 1978; Tavman, I.H. and Tavman, S., *J. Food Eng.*, 41, 109, 1999.

[a] Line source probe method.

[b] Modified hot wire method.

Table 18.45 Thermal Conductivity of Overrun Ice Cream[a] as a Function of Porosity and Temperature

% Porosity	k (W/m K) −30°C	−25°C	−20°C	−15°C	0°C	10°C
0	1.15	1.10	1.05	1.00	0.41	0.40
13	0.92	0.90	0.85	0.83	0.37	0.40
23	0.80	—	0.72	0.70	—	0.30
33	—	0.62	0.60	—	—	—
41	0.55	—	0.53	—	—	0.22
46	0.51	0.46	0.44	0.42	—	—
60	0.33	0.32	0.29	0.30	0.17	0.17
67	0.28	0.25	0.23	0.22	0.1	—

Source: Cogne, C., Andrieu, J., Laurent, P., Besson, A., and Nocquet, J., *J. Food Eng.*, 58, 331, 2003. With permission.

[a] Ice cream composition: $X_w = 0.589$; $X_{FA} = 0.093$; $X_{em} = 0.003$; $X_{pr} = 0.058$; $X_{CH} = 0.26$.

18.5 THERMAL CONDUCTIVITY OF CEREAL-BASED PRODUCTS

18.5.1 Rice-Based Products

Table 18.46 Effective Thermal Conductivity of Paddy and Its By-Products as a Function of Moisture, Temperature, and Bulk Density

Material Type	T_m (°C)	X_w	ρ_B (kg/m^3)	k (W/m K)
Kalimpong paddy	53.5	0.120	493	0.622
	59.5	0.120	493	0.633
	66.9	0.120	493	0.680
	49.8	0.120	555	0.697
	61.2	0.120	555	0.732
	66.5	0.120	555	0.790
	51.4	0.138	555	0.767
	61.2	0.138	555	0.802
	70.1	0.138	555	0.837
Brown rice	51.5	0.120	815	0.819
	62.5	0.120	815	0.835
	69.8	0.120	815	0.854
	54.8	0.120	885	0.854
	64.5	0.120	885	0.874
	68.5	0.120	885	0.900
White rice	52.5	0.120	785	0.918
	58.3	0.120	785	0.935
	69.5	0.120	785	0.947
	55.0	0.148	785	0.982
	60.0	0.148	785	1.01
	70.0	0.148	785	1.02
Paddy husk	56.7	0.110	185	0.708
	64.04	0.110	185	0.726
	75.03	0.110	185	0.753

Source: Dua, K.K. and Ojha, T.P., *J. Agric. Eng. Res.*, 14, 11, 1969.

Table 18.47 Thermal Conductivity of Bengal Rice as a Function of Moisture Content and Temperature

X_w					
0.092		0.121		0.170	
T (°C)	*k* (W/m K)	*T* (°C)	*k* (W/m K)	*T* (°C)	*k* (W/m K)
6	0.084	3	0.085	6	0.093
17	0.098	10	0.102	17	0.107
24	0.101	24	0103	24	0.102
38	0.103	38	0.110	38	0.116
46	0.097	46	0.106	46	0.099
61	0.112	57	0.107	57	0.104
66	0.118	61	0.111	61	0.126
69	0.126	66	0.118	66	0.130
				69	

Source: Yang, W., Siebenmorgen, T.J., Thielen, T.P.H., and Cnossen, A.G., *Biosyst. Eng.*, 84, 193, 2003. With permission.
Note: Average of replicates.

Table 18.48 Effective Thermal Conductivity for Rice Flour at Selected Temperature

		k (W/m °C)				
		Temperature (°C)				
ρ_B (kg/m^3)	X_w	10	20	30	40	50
650	0.088	0.070	0.074	0.078	0.082	0.090
	0.121	0.076	0.080	0.086	0.091	0.097
	0.154	0.079	0.086	0.092	0.099	0.106
	0.181	0.084	0.092	0.100	0.107	0.120
700	0.088	0.080	0.084	0.088	0.092	0.097
	0.121	0.083	0.087	0.091	0.098	0.106
	0.154	0.089	0.094	0.102	0.109	0.118
	0.181	0.092	0.100	0.110	0.119	0.127
750	0.088	0.090	0.094	0.097	0.102	0.106
	0.121	0.093	0.099	0.106	0.112	0.117
	0.154	0.100	0.103	0.110	0.116	0.130
	0.181	0.105	0.111	0.121	0.131	0.141
800	0.088	0.100	0.103	0.109	0.113	0.122
	0.121	0.103	0.110	0.115	0.122	0.129
	0.154	0.107	0.117	0.124	0.133	0.146
	0.181	0.112	0.119	0.130	0.140	0.159

Source: Muramatsu, Y., Tagawa, A., and Kasai, T., *Food Sci. Technol. Res.*, 11, 288, 2005.

18.5.2 Wheat Products

Table 18.49 Thermal Conductivity of Raw, Parboiled, Debranned Parboiled, and Bulgur Wheat at Different Moisture Contents and Bulk Densities

	Raw		Parboiled		Debran Parboiled		Bulgar	
X_w	ρ_B (kg/m³)	k (W/m K)	ρ_B (kg/m³)	k (W/m K)	ρ_B (kg/m³)	k (W/m K)	ρ_B (kg/m³)	k (W/m K)
0.090	841	0.096	764	0.114	761	0.092	709	0.090
0.120	818	0.100	754	0.119	746	0.096	695	0.093
0.150	795	0.105	744	0.125	731	0.099	682	0.096
0.180	772	0.110	734	0.130	716	0.102	669	0.099
0.210	749	0.114	724	0.136	701	0.106	655	0.102

Source: Shyamal, D.K., Chakraverty, I.A., and Banerjee, H.D., *Energy Conserv. Manage.*, 35, 801, 1994.
Note: Temperature 25°C–30°C.

Table 18.50 Thermal Conductivity of Wheat as a Function of Moisture Content and Bulk Density

Wheat Type	X_w	ρ_B (kg/m³)	k (W/m K)
Eregli	0.093	827	0.159
	0.379	675	0.182
Saruhan	0.102	798	0.142
	0.387	698	0.201
Bulgur	0.092	765	0.164

Source: Tavman, S. and Tavman, I.H., *Int. Commun. Heat Mass Transfer*, 25, 733, 1998. With permission.

Table 18.51 Thermal Conductivity of Wheat Flour Milling Coproducts at Selected Moisture and Bulk Density

		ρ_B (kg/m³)				k (W/m K)			
Sample Type	X_w	Bran	Germ	Shorts	Red Dog	Bran	Germ	Shorts	Red Dog
HRW blend[a]									
	0.08	171	324	275	380	0.051	0.060	0.062	0.067
	0.12	164	304	265	354	0.061	0.067	0.053	0.070
	0.16	154	273	261	345	0.062	0.080	0.072	0.079
HRW blend[b]									
	0.08	155	359	272	398	0.052	0.064	0.058	0.069
	0.12	153	343	263	385	0.055	0.074	0.062	0.077
	0.16	147	314	263	360	0.056	0.080	0.070	0.081
HRS blend[c]									
	0.08	199	328	—	361	0.050	0.055	—	0.067
	0.12	187	300	—	343	0.059	0.073	—	0.077
	0.16	183	269	—	339	0.074	0.076	—	0.077

(*continued*)

Table 18.51 (continued) Thermal Conductivity of Wheat Flour Milling Coproducts at Selected Moisture and Bulk Density

		ρ_B (kg/m³)				k (W/m K)			
Sample Type	X_w	**Bran**	**Germ**	**Shorts**	**Red Dog**	**Bran**	**Germ**	**Shorts**	**Red Dog**
HRS blend[d]									
	0.08	180	333	285	372	0.055	0.060	0.062	0.062
	0.12	178	287	274	361	0.057	0.066	0.066	0.069
	0.16	169	269	270	351	0.061	0.069	0.071	0.072

Source: Kim, Y.S., Flores, R.A., Chung, O.K., and Bechtel, D.B., *J. Food Process Eng.*, 26, 469, 2003.
[a] Hard red winter wheat pilot plant blend: ($X_{pr}=0.147$; $X_{FA}=0.010$; $X_{as}=0.062$; $X_{Fib}=0.110$).
[b] Hard red winter wheat commercial blend: ($X_{pr}=0.131$; $X_{FA}=0.018$; $X_{as}=0.057$; $X_{Fib}=0.117$).
[c] Hard red spring wheat commercial blend 1: ($X_{pr}=0.163$; $X_{FA}=0.028$; $X_{as}=0.057$; $X_{Fib}=0.088$).
[d] Hard red spring wheat commercial blend 2: ($X_{pr}=0.158$; $X_{FA}=0.027$; $X_{as}=0.057$; $X_{Fib}=0.090$).

Table 18.52 Comparative Thermal Conductivities for Gluten and Glutenin as a Function of Moisture Content

	k (W/m K)			
	Gluten[a]		Glutenin[b]	
T (°C)	$X_w=0.25$	$X_w=0.30$	$X_w=0.25$	$X_w=0.30$
60	0.33	0.35	0.29	0.33
90	0.31	0.32	0.36	0.37
140	0.28	0.31	0.40	0.44
175	0.26	0.29	0.43	0.49

Source: Strecker, T.D. Cavalieri, R.P., and Pomeranz, Y., *J. Food Sci.*, 59, 1244, 1994. With permission.
[a] Density of gluten at X_w of 0.25 and 0.30 is 1150 and 1238 kg/m³.
[b] Density of glutenin at X_w of 0.25 and 0.30 is 1165 and 1211 kg/m³.

Table 18.53 Thermal Conductivity of Dough and Bread as a Function of Moisture Content, Temperature, and Bulk Density

Product Type	X_w	T (°C)	ρ_B (kg/m³)	k (W/m K)	Reference
Dough	0.435	−43.5	1100	0.920	Lind (1988)
	0.435	−28.5	1100	—	Lind (1988)
	0.435	−22.0	1100	0.880	Lind (1988)
	0.435	16.5	1100	—	Lind (1988)
	0.435	23.0	1100	0.460	Lind (1988)
	0.461	−38	1100	1.030	Lind (1988)
	0.461	−28	1100	—	Lind (1988)
	0.461	−16	1100	0.980	Lind (1988)
	0.461	19	1100	0.500	Lind (1988)
	0.461	21	1100	—	Lind (1988)
Bread					Tschubik and Maslow (1973)
Loaf	0.425	—	545	0.248	Tschubik and Maslow (1973)
Tin loaf	0.450	—	500	0.232	Tschubik and Maslow (1973)
Crust					Tschubik and Maslow (1973)

Table 18.53 (continued) Thermal Conductivity of Dough and Bread as a Function of Moisture Content, Temperature, and Bulk Density

Product Type	X_w	T (°C)	ρ_B (kg/m^3)	k (W/m K)	Reference
Loaf	0.0	—	420	0.055	Tschubik and Maslow (1973)
Tin loaf	0.0	—	300	0.041	Tschubik and Maslow (1973)
Bread (wheat)		35		0.201	Reyes et al. (2006)
				0.081	
				0..084	
Bread (rye)		35		0.174	Reyes et al. (2006)
				0.180	
Bread (corn)		35		0.098	Reyes et al. (2006)
Angel cake		35		0.083	Reyes et al. (2006)
Sponge cake		35		0.058	Reyes et al. (2006)
Muffin bakery		35		0.170	Reyes et al. (2006)
Muffin chocolate		35		0.104	Reyes et al. (2006)
Bun hamburger		35		0.101	Reyes et al. (2006)
Yellow cake Batter	0.415	—	694	0.223	Sweat (1973)

Sources: Lind, I., in *Proceedings from the International Symposium on Progress in Food Preservation Processes,* Vol. 1. CERIA, Brussels, Belgium, 1988, 101–111; Tschubik, I.A. and Maslow, A.M., in *Warmephysikalische Konstanten von Lebensmittlen und Halfabrikalen*, Fachbuchverlag, Leipzig, 1973; Reyes, C., Barringer, S.A., Uchummal-Chemminian, R., and Kaletunc, G., *J. Food Process. Preserv*., 30, 381, 2006; Sweat, V.E., Haugh, C.G., and Stadelman, W.J., *J. Food Sci.* 38, 158, 1973.

18.5.3 Soybean Products

Table 18.54 Bulk Thermal Conductivity[a,b] of Defatted Soy Flour[c] at Different Moisture Contents and Bulk Densities

Moisture Content (X_w)							
0.045		0.081		0.130		0.181	
ρ_B (kg/m^3)	k_B (W/m K)	ρ_B (kg/m^3)	k_B (W/m K)	ρ_B (kg/m^3)	k_B (W/m K)	ρ_B (kg/m^3)	k_B (W/m K)
555	0.034	517	0.059	435	0.052	331	0.055
603	0.037	570	0.061	454	0.042	367	0.045
620	0.037	603	0.061	467	0.048	386	0.050
637	0.040	624	0.072	482	0.053	398	0.054
647	0.045	644	0.086	493	0.064	405	0.063
660	0.039	655	0.078	503	0.050	414	0.051
670	0.044	664	0.079	512	0.057	420	0.055
671	0.044	662	0.082	517	0.058	432	0.060
706	0.055	691	0.097	536	0.068	476	0.074
701	0.047	703	0.086	540	0.052	475	0.058
706	0.048	707	0.088	542	0.065	486	0.065
711	0.054	709	0.087	547	0.061	492	0.063
0.253		0.276		0.284		0.320	
348	0.045	397	0.056	519	0.058	0.514[d]	0.032
387	0.063	431	0.048	535	0.045	0.739[e]	0.281

(continued)

Table 18.54 (continued) Bulk Thermal Conductivity[a,b] of Defatted Soy Flour[c] at Different Moisture Contents and Bulk Densities

Moisture Content (X_w)							
0.045		0.081		0.130		0.181	
ρ_B (kg/m^3)	k_B (W/m K)	ρ_B (kg/m^3)	k_B (W/m K)	ρ_B (kg/m^3)	k_B (W/m K)	ρ_B (kg/m^3)	k_B (W/m K)
407	0.048	452	0.055	535	0.052		
421	0.053	471	0.059	552	0.053		
439	0.069	483	0.068	556	0.063		
454	0.049	501	0.055	562	0.048		
457	0.059	499	0.060	565	0.054		
471	0.058	527	0.067	571	0.054		
550	0.089	592	0.091	625	0.098		
543	0.061	587	0.068	645	0.061		
552	0.072	600	0.079	644	0.066		
571	0.074	616	0.082	647	0.072		

Source: Wallapapan, K. and Sweat, V.E., *Trans. ASAE*, 25, 1440, 1982. With permission.
[a] Average of 16 samples.
[b] $T = 20°C–22°C$.
[c] $X_w = 0.0754$; $X_{pr} = 0.5791$; $X_{ca} = 0.2463$; $X_{oi} = 0.0121$; $X_{as} = 0.0577$; $X_{Fib} = 0.294$.
[d] Loosely packed.
[e] Tightly packed by forcing the packing bar.

Table 18.55 Bulk Thermal Conductivity of Defatted Soy Flour at Different Moisture Contents and Bulk Densities

	Thermal Conductivity, k_B (W/m K)						
	Moisture Content, X_w						
ρ_B (kg/m^3)	0.10	0.15	0.20	0.25	0.30	0.35	0.40
1000	0.144	0.200	0.251	0.290	0.339	0.374	0.409
1100	0.171	0.225	0.272	0.325	0.365	0.398	0.430
1200	0.197	0.248	0.300	0.343	0.390	0.425	0.456
1300	0.219	0.276	0.303	0.375	0.415	0.445	0.482

Source: Wallapapan, K., Sweat, V.E., Arce, J.A., and Dahm, P.F., *Trans. ASAE*, 27, 1610, 1984. With permission.

Table 18.56 Thermal Conductivity of Defatted Soybean Curd as a Function of Temperature and Moisture Content[a]

	Thermal Conductivity, k_B (W/m K)					Thermal Conductivity, k_B (W/m K)				
ε_w	Temperature (°C)				ε_w	Temperature (°C)				
	−20	−15	−10	−5		1.8	5.0	10	15	20
0.84	1.520	1.486	1.467	1.429	0.78	0.464	0.473	0.481	0.487	0.492
0.91	1.796	1.774	1.727	1.689	0.83	0.478	0.488	0.494	0.500	0.511
0.94	1.942	1.913	1.838	1.814	0.87	0.500	0.506	0.514	0.524	0.534
					0.90	0.514	0.520	0.527	0.537	0.550
					0.93	0.535	0.533	0.544	0.549	0.564

Source: Yano, T., Kong, J., Miyawaki, O., and Nakamura, K., *J. Food Sci.*, 46, 1357, 1981. With permission.
[a] $\rho_{FA} = 930$ kg/m^3 at RT; $\rho_{FA} = 900$ kg/m^3 at −10°C to −30°C; $\rho_{pr} = 1340$ kg/m^3 at RT.

18.5.4 Other Cereal and Starch Products

Table 18.57 Thermal Conductivity of Maize Grits as a Function of Temperature for Different Moisture Contents

	k (W/m K)					
X_w	30°C	45°C	60°C	75°C	90°C	100°C
0.13	0.09	0.10	0.11	0.15	0.20	0.21
0.18	0.10	0.12	0.16	0.19	0.28	0.35
0.22	0.11	0.13	0.17	0.21	0.32	0.41
0.30	0.20	0.37	0.41	0.39	0.41	—
0.38	0.40	0.41	0.42	0.41	0.52	—

Source: Halliday, P.J., Parker, R., Smith, A.C., and Steer, D.C., *J. Food Eng.*, 26, 273, 1995. With permission.

Table 18.58 Thermal Conductivity of Corn Starch[a] Gel

ρ_{ap}	X_w	T (°C)	k (W/m K)	X_w	T (°C)	k (W/m K)
	0.94	20	0.569			
	0.94	30	0.596			
	0.94	40	0.639			
1028	0.90	20	0.554	0.90	20	0.559
1028	0.90	30	0.577	0.90	25	0.564
1028	0.90	40	0.594	0.90	30	0.589
1050	0.85	20	0.547	0.85	20	0.540
1050	0.85	30	0.567	0.85	25	0.544
1050	0.85	40	0.581	0.85	30	0.529
1076	0.80	20	0.523	0.80	20	0.526
1076	0.80	30	0.538	0.80	25	0.529
1076	0.80	40	0.557	0.80	30	0.540

Source: Drusas, A., Tassopoulos, M., and Saravacos, G.D., in *Food Engineering and Process Applications*, Vol. 1, Elsevier Applied Science, New York, 1986, 141–149.

Note: Measurement was carried out using guarded hot plate method.

[a] Corn starch powder at 6%–20% solid, heating to 80°C.

Table 18.59 Thermal Conductivity of Gelatinized Starch Gel

	k (W/m K)				
	Temperature (°C)				
X_w	80	90	100	110	120
A					
0.3960	0.4363	0.4396	0.4416	0.4414	0.4416
0.4630	0.4496	0.4506	0.4530	0.4526	0.4550
0.5034	0.4587	0.4624	0.4594	0.4628	0.4633
0.5557	0.4682	0.4677	0.4691	0.4702	0.4725
0.5902	0.4807	0.4819	0.4832	0.4835	0.4854
0.6353	0.5011	0.5007	0.5009	0.5033	0.5035
0.6779	0.5170	0.5183	0.5201	0.5215	0.5227
0.7130	0.5319	0.5324	0.5362	0.5366	0.5356

(*continued*)

Table 18.59 (continued) Thermal Conductivity of Gelatinized Starch Gel

	k (W/m K)				
	Temperature (°C)				
X_w	80	90	100	110	120
B					
0.4918	0.4770	0.4785	0.4786	0.4802	0.4826
0.5212	0.4814	0.4855	0.4844	0.4881	0.4879
0.5423	0.4879	0.4888	0.4917	0.4907	0.4947
0.6044	0.4999	0.5044	0.5062	0.5039	0.5074
0.6997	0.5406	0.5413	0.5420	0.5406	0.5423
0.7497	0.5667	0.5667	0.5650	0.5627	0.5656

Source: Wang, J. and Hayakawa, K., *J. Food Sci.*, 58, 884, 1993. With permission.
Notes: A, gelatinized starch gel (amicoa; 98% amylopectin); gelatinized at 120°C for 40 min. B, gelatinized starch gel dissolved with sucrose (sucrose was dissolved in water first). Starch to sucrose weight ratio 4:1 for all moisture content.

Table 18.60 Thermal Conductivity of Freeze-Dried Starch Gel (Bone Dry) in Various Gases at 1 atm and 41°C

	k (W/m K)	
Gas	Gas-Filled Gel	Pure Gas
Air	0.0393	0.0723
Nitrogen	0.0391	0.0271
Helium	0.0149	0.0154
Vacuum	0.0091	—

Source: Saravacos, G.D. and Pilsworth, M.N., *J. Food Sci.*, 30, 773, 1965. With permission.

Table 18.61 Thermal Conductivity of Freeze-Dried Model Food Gels at 41°C

			Atmospheric Pressure		Vacuum
Food Type	X_{so}^{a}	ρ_{ap} (kg/m^3)	(W/m K)[b]	(W/m K)[c]	(W/m K)[d]
Potato starch	0.10	1090	0.0410	0.0393	0.0091
Gelatin	0.10	830	0.0413	0.0385	0.0147
Cellulose gum	0.10	1500	0.0627	0.0559	0.0195
Egg albumen	0.10	770	0.0418	0.0392	0.0129
Pectin	0.10	660	0.0474	0.0438	0.0174
Pectin	0.05	390	0.0391	0.0383	0.0120
Pectin/glucose (1:1)	0.10	1100	0.0500	0.0481	0.0179

Source: Saravacos, G.D. and Pilsworth, M.N., *J. Food Sci.*, 30, 773, 1965. With permission.
[a] X_{so} in gel before drying.
[b] At 52% relative humidity and 1 atm.
[c] Bone dried gel at 1 atm.
[d] Bone dried gel in vacuum.

Table 18.62 Thermal Conductivity of Freeze-Dried Gels as a Function of Pore Structure When Samples Are Equilibrated at 52% Relative Humidity at 41°C

Gel	X_{so}^{a}	Porosity	Pore Size Range (μm)	k (W/m K)
Starch	0.10	0.93	2–10	0.0410
Gelatin	0.10	0.93	3–15	0.0413
Egg albumin	0.10	0.97	3–15	0.0418
Cellulose gum	0.10	0.89	2–10	0.0627
Pectin	0.10	0.97	1–10	0.0474
Pectin	0.05	0.98	1–10	0.0391
Pectin:glucose (1:1)	0.10	0.96	1–8	0.0500

Source: Saravacos, G.D. and Pilsworth, M.N., *J. Food Sci.*, 30, 773, 1965. With permission.

[a] X_{so} in gel before drying.

Table 18.63 Thermal Conductivity[a] of Granular and Gelatinized Starch as a Function of Moisture Content and Temperature

	k (W/m K)					
	Temperature (°C)			Temperature (°C)		
X_w	5	25	45	5	25	45
0.550	0.492	0.490	0.512	0.475	0.486	0.495
0.600	0.493	0.508	0.543	0.489	0.504	0.538
0.650	0.514	0.529	0.541	0.500	0.520	0.545
0.700	0.536	0.537	0.563	0.524	0.534	0.558

Source: Hsu, C.-L. and Heldman, D.R., *Int. J. Food Sci. Technol.*, 39, 737, 2004.

[a] Duplicate samples with each replicate represent three repeat measurements.

NOMENCLATURE

k thermal conductivity (W/m K)
M mass fraction (dry basis)
m mass (kg)
P pressure
T temperature (°C or K)
TSS total soluble solids (°Brix)
X mass fraction (wet basis)

Greek Letters

ε volume fraction
ρ density (kg/m^3)
$\dot{\gamma}$ shear rate (s^{-1})

Subscripts

ap apparent
as ash
B bulk
CH carbohydrate
e effective
FA fat
FI fiber
L longitudinal direction
o initial
pr protein
so solid
T transverse direction
w water

REFERENCES

Abhayakwick, L., Laguerre, J.C., Tauzin, V., and Duquenoy, A. Physical properties of three onion varieties as affected by the moisture content. *Journal of Food Engineering* 2002, 55, 253–262.

AbuDagga, Y. and Kolbe, E. Thermophysical properties of surimi paste at cooking temperature. *Journal of Food Engineering* 1997, 32, 315–337.

Ali, S.D., Ramaswamy, H.S., and Awuah, G.B. Thermo-physical properties of selected vegetables as influenced by temperature and moisture content. *Journal of Food Process Engineering* 2002, 25, 417–433.

Assis, M.M.M., Lannes, S.C. da Silva, Tadini, C.C., Telis, V.R.N., and Telis-Romero, J. Influence of temperature and concentration on thermo-physical properties of yellow mombin (*Spondias mombin*, L.). *European Food Research and Technology* 2006, 223, 585–593.

Baghe-Khandan, M.S., Okos, M.R., and Sweat, V.E. The thermal conductivity of beef as affected by temperature and composition. *Transactions of the ASAE*, 1982, 4, 1118–1122.

Califano, A.N. and Calvelo, A. Thermal conductivity of potato between 50°C and 100°C. *Journal of Food Science* 1991, 56, 586–587, 589.

Cogne, C., Andrieu, J., Laurent, P., Besson, A., and Nocquet, J. Experimental data and modelling of thermal properties of ice creams. *Journal of Food Engineering* 2003, 58, 331–341.

Coimbra, J.S.R., Gabas, A.L., Minim, L.A., Rojas, E.E.G., Telis, V.R.N., Telis-Romero, J. Density, heat capacity and thermal conductivity of liquid egg products, *Journal of Food Engineering* 2006, 74, 186–190.

Delgado, A.E., Gallo, A., De Piante, D., and Rubiolo, A. Thermal conductivity of unfrozen and frozen strawberry and spinach. *Journal of Food Engineering* 1997, 31, 137–146.

Denys, S. and Hendrickx, M.E. Measurement of the thermal conductivity of foods at high pressure. *Journal of Food Science* 1999, 64, 709–713.

Drusas, A.E. and Saravcos, G.D. Thermal conductivity of tomato paste. *Journal of Food Engineering* 1985, 4, 157.

Drusas, A., Tassopoulos, M., and Saravacos, G.D. Thermal conductivity of starch gels. In: *Food Engineering and Process Applications*, Vol. 1, Eds. LeMaguer, M. and Jelen, P., Elsevier Applied Science, New York, 1986. pp. 141–149.

Dua, K.K. and Ojha, T.P. Thermal conductivity of paddy grains and its by products. *Journal of Agricultural Engineering Research* 1969, 14, 11–13.

Fito, P.J., Piiraga, F., and Aranda, V. Thermal conductivity of porous bodies at low pressure: Part I. *Journal of Food Engineering* 1984, 3, 375–388.

Gratzek, J.P. and Toledo, R.T. Solid food thermal conductivity determination at high temperatures. *Journal of Food Science* 1993, 58, 908–913.

Gut, J.A.W., Pinto, J.M., Gabas, A.L., and Telis-Romero, J. Continuous pasteurization of egg yolk: Thermophysical properties and process simulation. *Journal of Food Process Engineering* 2005, 28, 181–203.

Halliday, P.J., Parker, R., Smith, A.C., and Steer, D.C. The thermal conductivity of maize grits and potato granules. *Journal of Food Engineering* 1995, 26, 273–288.

Higgs, S.J. and Swift, S.P. Investigation into the thermal conductivity of beef using the line source technique. *Process Biochemistry*, 1975, 11, 43–45.

Hsu, C.-L. and Heldman, D.R. Prediction of the thermal conductivity of aqueous starch. *International Journal of Food Science and Technology* 2004, 39, 737–743.

Hu, X. and Mallikarjunan, P. Thermal and dielectric properties of shucked oysters. *Lebensmittel-Wissenschaft und-Technologie* 2005, 38, 489–494.

Jindal, V.K. and Murakami, E.G. Thermal properties of shredded coconut. In: *Engineering and Food*, Vol. 1, *Engineering Sciences in the Food Industry*, Ed. McKenna, B.M. *Proceedings of the 3rd International Congress on Engineering and Food*, September 26–28, 1983, Dublin, Ireland. Elsevier Applied Science New York, 1984, p. 558.

Karunakar, B., Mishra, S.K., and Bandyopadhyay, S. Specific heat and thermal conductivity of shrimp meat. *Journal of Food Engineering* 1998, 37, 345–351.

Kent, M., Christiansen, K., van Haneghem, I.A., Holtz, E., Morley, M.J., Nesvadba, P., and Poulsen, K.P. COST collaborative measurements of thermal properties of foods. *Journal of Food Engineering* 1984, 3, 117–150.

Keppeler, R.A. and Boose, J.R. Thermal properties of frozen sucrose solutions. *Transactions in ASAE* 1970, 13, 335.

Kim, Y.S., Flores, R.A., Chung, O.K., and Bechtel, D.B. Physical, chemical, and thermal characterization of wheat flour milling co-products. *Journal of Food Process Engineering* 2003, 26, 469–488.

Kumbhar, B.K., Agarwal, R.S., and Das, K. Thermal properties of fresh and frozen fish. *International Journal of Refrigeration* 1981, 4, 143–146.

Kurozawa, L.E., El-Aouar, A.A., Simoes, M.R., Azoubel, P.M., and Murr, F.E.X. Determination of thermal conductivity and thermal diffusivity of papaya (*Carica papaya* L.) as a function of temperature, *2nd Mercosur Congress on Chemical Engineering Proceedings*, Costa Verde, Brazil, 2005, pp. 1–9.

Lamberg, I. and Hallstrom, B. Thermal properties of potatoes and a computer simulation model of a blanching process. *Food Technology* 1986, 21, 577.

Liang, X., Zhang, Y., and Gek, X. The measurement of thermal conductivities of solid fruits and vegetables. *Measurement Science and Technology* 1999, 10, 82–86.

Lin, S.X.Q., Chen, X.D., Chen, Z.D., and Bandopadhayay, P. Shear rate dependent thermal conductivity measurement of two fruit juice concentrates. *Journal of Food Engineering* 2003, 57, 217–224.

Lind, I. Thawing of minced meat and dough: Thermal data and mathematical modelling. *Proceedings from the International Symposium on Progress in Food Preservation Processes,* Vol. 1. CERIA, Brussels, Belgium, 1988, pp. 101–111.

MacCarthy, D. The effective thermal conductivity of skim milk powder. In: *Engineering and Food*, Ed. McKenna, B.M., Elsevier, London, 1983.

MacCarthy, D.A. Effect of temperature and bulk density on thermal conductivity of spray-dried whole milk powder. *Journal of Food Engineering* 1985, 4, 249–263.

Magerramov, M.A., Abdulagatov, A.I., Azizov, N.D., and Abdulagatov, I.M. Thermal Conductivity of pear, sweet-cherry, apricot, and cherry-plum juices as a function of temperature and concentration, *Journal of Food Science* 2006a, 71, E238–E244.

Magerramov, M.A., Abdulagatov, A.I., Abdulagatov, I.M., and Azizov, N.D. Thermal conductivity of peach, raspberry, cherry and plum juices as a function of temperature and concentration, *Journal of Food Process Engineering* 2006b, 29, 304–326.

Marschoun, L.T., Muthukumarappan, K., and Gunasekaran, S. Thermal properties of cheddar cheese: Experimental and modeling. *International Journal of Food Properties* 2001, 4, 383–403.

McDonald, K., Sun, D.W., and Lyng, J.G. Effect of vacuum cooling on the thermophysical properties of a cooked beef product, *Journal of Food Engineering* 2002, 52, 167–176.

Murakami, E.G. Thermal processing affects properties of commercial *shrimp* and scallops. *Journal of Food Science* 1994, 59, 237–241.

Murakami, E.G. The thermal properties of potatoes and carrots as affected by thermal processing. *Journal of Food Process Engineering* 1997, 20, 415–432.

Muramatsu, Y., Tagawa, A., and Kasai, T. Thermal conductivity of several liquid foods. *Food Science and Technology Research* 2005, 11, 288–294.

Murphy, R.Y. and Marks, B.P. Apparent thermal conductivity, water content, density and porosity of thermally processed ground chicken patties, *Journal of Food Process Engineering* 1999, 22, 129–140.

Ngadi, M.O. and Ikedialis, J. Thermal properties of chicken drum muscle. *Journal of Science Food and Agriculture* 1998, 78, 12–18.

Ngadi, M.O., Mallikarjunan, P., Chinnan, M.S., Radhakrishnan, S., and Hung, Y.C. Thermal properties of shrimps, French toasts and breading. *Journal of Food Process Engineering* 2000, 23, 73–87.

Njie, D.N., Rumsey, T.R., and Singh, R.P. Thermal properties of cassava, yam and plantain. *Journal of Food Engineering* 1998, 37, 63–76.

Pan, Z. and Singh, R.P. Physical and thermal properties of ground beef during cooking. *Lebensmittel-Wissenschaft und-Technologie* 2001, 34, 437–444.

Perez-Martin, J.M.G., Banga, J.R., and Casares, J. Determination of thermal conductivity, specific heat and thermal diffusivity of albacore (*Thunnus alalunga*) Ricardo I. *Zeitschrift für Lebensmitteluntersuchung und -Forschung A*1989, 189, 525–529.

Phomkong, W., Srzednicki, G., and Driscoll R.H. Thermo-physical properties of stone fruit. *Drying Technology* 2006, 24, 195–200.

Radhakrishnan, S. Measurement of thermal properties of seafoods. MS Thesis, Virginia Polytechnic and State University, Blacksburg, VA, 1997.

Rahman, M.S. and Driscoli, R.H. Thermal conductivity of seafoods: Calamari, octopus and prawn, *Food Australia* 1991, 43, 356.

Rahman, M.S. and Potluri, P.L. Shrinkage and density of squid flesh during air drying. *Journal of Food Engineering* 1990, 12, 133.

Rahman, M.S. and Potluri, P.L. Thermal conductivity of fresh and dried squid meat by line source thermal conductivity probe. *Journal of Food Science* 1991, 56, 582–583.

Rahman, M.S., Potluri, P.L., and Varamit, A. Thermal conductivities of fresh and dried seafood powders. *Transactions of the ASAE* 1991, 34, 217–220.

Rao, M.A., Barnard, J., and Kenny, J. Thermal conductivity and thermal diffusivity of process variety squash and white potatoes. *Transactions of the ASAE* 1975, 18, 1188–1192.

Reyes, C., Barringer, S.A., Uchummal-Chemminian, R., Kaletunc, G. Thermal conductivity models for porous baked foods. *Journal of Food Processing and Preservation* 2006, 30, 381–392.

Rice, P., Selman, J.D., and Abdul-Rezzak, R.K. Effect of temperature on thermal properties of 'record' potatoes. *International Journal of Food Science and Technology* 1988, 23, 281.

Saravacos, G.D. and Pilsworth, M.N. Thermal conductivity of freeze-dried model food gels. *Journal of Food Science* 1965, 30, 773–778.

Shamsudin, R., Mohamed, I.O., and Yaman, N.K.M. Thermophysical properties of Thai seedless guava juice as affected by temperature and concentration. *Journal of Food Engineering* 2005, 66, 395–399.

Shrivastava, M. and Datta, A.K. Determination of specific heat and thermal conductivity of mushrooms (*Pleurotus florida*). *Journal of Food Engineering* 1999, 39, 255–260.

Shyamal, D.K., Chakraverty, I.A., and Banerjee, H.D. Thermal properties of raw parboiled and debranned parboiled wheat and wheat bulgar. *Energy Conservation and Management* 1994, 35, 801–804.

Singh, K.K. and Goswami, T.K. Thermal properties of cumin seed. *Journal of Food Engineering* 2000, 45, 181–187.

Singh, R.K. and Lund, D.B. Mathematical modeling of heat and moisture transfer related properties of intermediate moisture apples. *Journal of Food Processing and Preservation* 1984, 8, 191–210.

Strecker, T.D., Cavalieri, R.P., and Pomeranz, Y. Wheat Gluten and glutenin thermal conductivity and diffusivity at extruder temperatures. *Journal of Food Science* 1994, 59, 1244–1247.

Sweat, V.E. Experimental values of thermal conductivities of selected fruits and vegetables. *Journal of Food Science* 1974, 39, 1081–1083.

Sweat, V.E. and Parmelee, C.E. Measurement of thermal conductivity of dairy products and margarines. *Journal of Food Process Engineering* 1978, 2, 187–197.

Sweat, V.E., Haugh, C.G., and Stadelman, W.J., Thermal conductivity of chicken meat at temperatures between −75°C and 20°C, *Journal of Food Science*, 1973, 38, 158–160.

Tabil, L.G., Eliason, M.V., and Qi, H. Thermal properties of sugar beets, ASAE Annual International Meeting, Sacramento, CA, July 30–August 1, 2001 Paper No. 016141, 2001.

Tavman, I.H. and Tavman, S. Measurement of thermal conductivity of dairy products, *Journal of Food Engineering* 1999, 41, 109–114.

Tavman, S. and Tavman, I.H., Measurement of effective thermal conductivity of wheat as a function of moisture content. *International Communication in Heat & Mass Transfer* 1998, 25, 733–741.

Tocci, A.M., Flores, E.S.E., and Mascheroni, R.H. Enthalpy, heat Capacity and thermal conductivity of boneless mutton between −40°C and + 40°C. *Lebensmittel-Wissenschaft und-Technologie* 1997, 30, 184–191.

Tschubik, I.A. and Maslow, A.M. *Warmephysikalische Konstanten von Lebensmittlen und Halfabrikalen*, Fachbuchverlag, Leipzig, 1973.

Vagenas, G.K., Marinos-Kowis, D., and Saravacos, G.D. Thermal properties of raisins. *Journal of Food Engineering* 1990, 11, 147–158.

Wallapapan, K. and Sweat, V.E. Thermal conductivity of defatted soy flour. *Transactions of the ASAE* 1982, 25, 1440–1444.

Wallapapan, K., Sweat, V.E., Arce, J.A., and Dahm, P.F. Thermal diffusivity and conductivity of defatted soy flour. *Transactions of the ASAE* 1984, 27, 1610.

Wang, J. and Hayakawa, K. Thermal conductivity of starch gels at high temperature influenced by moisture. *Journal of Food Science* 1993, 58, 884.

Yamada, T. The thermal properties of potato (in Japanese). *Journal of Agricultural Chemical Society of Japan* 1970, 44, 587.

Yang, J. and Zhao, Y. Thermal properties of radish and alfalfa seeds. *Journal of Food Process Engineering* 2001, 24, 291–313.

Yang, W., Siebenmorgen, T.J., Thielen, T.P.H., and Cnossen, A.G. Effect of glass transition on thermal conductivity of rough rice. *Biosystems Engineering* 2003, 84, 193–200.

Yano, T., Kong, J., Miyawaki, O., and Nakamura, K. The intrinsic thermal conductivity of wet soy protein and its use in predicting the effective thermal conductivity of soybean curd. *Journal of Food Science* 1981, 46, 1357.

Yano, T., Kong, J.Y., and Miyawaki, O. Thermal conductivity of heterogeneous foodstuffs. In: *Food Process Engineering*, Vol. 1, *Food Processing Systems*, Eds. Linko, P., Malkki, Y., Olkku, J., and Larinkari, J., Applied Science, London, 1980, pp. 518–522.

Zhang, J., Farkas, B.E., and Hale, S.A. Thermal properties of skipjack tuna (*Katsuwonus pelamis*). *International Journal of Food Properties* 2001, 4, 81–90.

Zheng, M., Huang, Y.W., Nelson, S.O., Bartley, P.G., and Gates, K.W. Dielectric properties and thermal conductivity of marinated shrimp and channel catfish. *Journal of Food Science* 1998, 4, 668–672.

Ziegler, G.R. and Rizvi, S.S.H. Thermal conductivity of liquid foods by the thermal comparator method, *Journal of Food Science* 1985, 50, 1458–1462.

CHAPTER 19

Thermal Conductivity Prediction of Foods

Mohammad Shafiur Rahman and Ghalib Said Al-Saidi

CONTENTS

19.1 INTRODUCTION

The thermal conductivity of foods is affected by three factors: composition, structure, and processing conditions. Water content plays a significant role due to the relative magnitude of conductivities of water in foods. The nonaqueous part of food such as fats and oils also plays an important role in fatty foods. The structural factors consist of porosity, pore size, shape and distribution, and arrangement or distribution of different phases, such as air, water, ice, and solids. The processing factors consist of temperature, pressure, and mode of heat or energy transfer (Sablani and Rahman, 2003). Thermal conductivity values of food materials could vary up to two orders of magnitude; thus, prediction is highly complex in a generic approach. Thermal conductivity prediction models can be grouped as (1) theoretical-based models, (2) models based on distribution factors, (3) percolation theory, (4) effective medium theory, and (5) empirical-based models.

19.2 THEORETICAL-BASED MODELS

Different types of theoretical models are developed assuming the distribution of component phases and applied heat transfer fundamentals by conduction, convection, and radiation. The following sections present the most important and widely used theoretical-based models.

19.2.1 Series Model

Wiener [1912, cited in Bergman (1990)], Deissler et al. (1958), Woodside and Messmer (1961), and Brailsford and Major (1964) defined the maximum and minimum limits of thermal conductivity of a two-phase system. These are series and parallel distributions as shown in Figure 19.1. In the series distribution, two phases are thermally in series with respect to the direction of heat flow. The series distribution, which results in a minimum for k, corresponds to the weighted harmonic mean of the continuous and discontinuous phases' conductivities. Extending these for a multiphase system in series, the effective thermal conductivity can be written as

$$\frac{1}{k_{\text{se}}} = \sum_{i=1}^{n} \frac{\varepsilon_i}{k_i} \tag{19.1}$$

where

ε_i is the volume fraction of the ith component phase
k_i is the conductivity of the ith phase
k_{se} is the effective thermal conductivity of the composite medium predicted by the series model

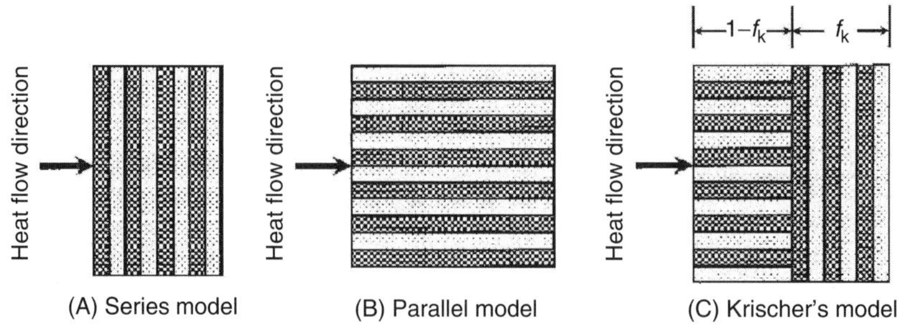

FIGURE 19.1 Graphical presentation of the (A) series, (B) parallel, and (C) Krischer's models.

19.2.2 Parallel Model

In parallel distribution, the phases are considered as thermally in parallel with respect to the direction of heat flow. The parallel distribution (maximum k) corresponds to a weighted arithmetic mean of the conductivities of the component phases and can be written as

$$k_{pa} = \sum_{i=1}^{n} \varepsilon_i k_i \tag{19.2}$$

where k_{pa} is the effective conductivity of the composite medium predicted by the parallel model. The series and parallel models are based on the electrical circuit analogy of Ohm's law considering that the current flow is related to the resistance of the wire and applied voltage. Choi and Okos (1986) proposed this model for liquid foods based on food composition. Chen (1969) proposed the model for predicting the thermal conductivity of powder foods in a packed bed considering heat transfer through solids and gaseous phases and predicted the thermal conductivity to be within 5%. This error arises due to the heat transfer through the contact point of particles. Rahman et al. (1991) found a 14%–38% error when they tested the parallel model in the case of seafood powders having a bulk density from 424 to 828 kg/m^3. Murakami and Okos (1989) considered thermal conductivities of 15 different food powders from the literature and observed that all data points fall within the range defined by the predicted values of the parallel and perpendicular models. The upper limit was the predicted value of the parallel model and the lower limit was that of the perpendicular model. These upper and lower limits are valid if there is no interaction between the component phases and no other mode of heat transfer except conduction; evaporation–condensation does not include in these effective thermal conductivity.

19.2.3 Random Model

Woodside and Messmer (1961) proposed an intermediate model, which is the weighted geometric mean of the component phases and can be written as

$$k_{ra} = k_1^{\varepsilon_1} k_2^{\varepsilon_2} k_3^{\varepsilon_3}, \ldots, k_n^{\varepsilon_n} \tag{19.3}$$

where k_{ra} is the effective thermal conductivity of the composite medium predicted by the random model. Rahman and Driscoll (1991) tested series, parallel, and random models to predict the thermal conductivity of fresh seafoods and recommended the parallel or random model because the series model was significantly different from the experimental data.

19.2.4 Murakami–Okos (1989) Model

Murakami and Okos (1989) proposed a model that was better than the Krischer's model for nonporous foods. They assumed the solid particles were parallel to the direction of heat flow and the effective solid phase was in series with the fluid phase. The resulting model was

$$\frac{1}{k_e} = \frac{1 - \varepsilon_w}{k_{so}} + \frac{\varepsilon_w}{k_w} \tag{19.4}$$

where

$$k_{so} = \left[\sum_{i=1}^{n} \varepsilon_i k_i\right] \times \left[\sum_{i=1}^{n} \varepsilon_i\right]^{-1} \tag{19.5}$$

19.2.5 Maxwell (1904) Model

Maxwell (1904) pioneered the study of thermal and electrical conductivity of two-phase mixtures. Maxwell's equation was derived on the basis of potential theory and can be written as

$$k_m = k_c\left[\frac{k_d + 2k_c - 2\varepsilon_d(k_c - k_d)}{k_d + 2k_c - \varepsilon_d(k_c - k_d)}\right] \tag{19.6}$$

where k_m is the effective conductivity predicted by the Maxwell model. The above equation was derived on the basis of randomly distributed discontinuous spheres in a continuous medium and assumes that the discontinuous spheres are far enough apart and that they do not interact. Hence, this equation is strictly applicable only when the volume fraction of the discontinuous phase is very low.

19.2.6 Maxwell–Eucken (1932) Model

Eucken (1932) modified the Maxwell model and proposed the following equation:

$$k_e = k_c\left(\frac{1 - 2\nu\varepsilon_d}{1 + 2\nu\varepsilon_d}\right) \tag{19.7}$$

where

$$\nu = \frac{k_c - k_d}{2k_c + k_d}$$

19.2.7 Kopelman (1966) Model

Thermal conductivity depends on the structure or physical arrangement and the chemical composition. The composition and structure of food systems are varied. Structurally, food systems can be divided into two groups: single phase and meta-phase. Liquid foods can be considered as single phase and emulsion, and solid or semisolid foods can be considered as meta-phase due to the presence of solid, liquid, and gaseous phases. Thus, structural models are necessary to predict the thermal conductivity of homogeneous, fibrous, or layered food systems. Kopelman (1966) proposed three structural models to predict the thermal conductivity of food materials.

19.2.7.1 Homogeneous Model

In the homogeneous systems, all the phases are dispersed randomly. When the thermal conductivity of liquid phase is much higher than the solid phase, the effective thermal conductivity can be described as

$$k_e = k_L\left[\frac{1 - \varepsilon_{so}^{2/3}}{1 - \varepsilon_{so}^{2/3}(1 - \varepsilon_{so})}\right] \tag{19.8}$$

When the thermal conductivity of liquid and solid phases are similar, the following equation can be used:

$$k_e = k_L\left[\frac{1 - \xi}{1 - \xi\left(1 - \varepsilon_{so}^{1/3}\right)}\right] \tag{19.9}$$

where

$$\xi = \varepsilon_{\mathrm{so}}^{2/3}\left(1 - \frac{k_{\mathrm{so}}}{k_{\mathrm{L}}}\right) \tag{19.10}$$

When structure does not influence thermal conductivity, then Equation 19.8 can be applied to most food systems since thermal conductivity of water is much higher than that of the other food components. Equation 19.9 is limited to very low moisture foods when water is not the continuous phase within the system (Heldman, 1992).

19.2.7.2 Anisotropic Model

If the food product contains fibers, then thermal conductivity parallel to the fiber can be expressed as

$$k_{\mathrm{pa}} = k_{\mathrm{L}}\left[1 - \varepsilon_{\mathrm{so}}\left(1 - \frac{k_{\mathrm{so}}}{k_{\mathrm{L}}}\right)\right] \tag{19.11}$$

Thermal conductivity perpendicular to the fiber is

$$k_{\mathrm{pa}} = k_i\left[\frac{1 - \varphi}{1 - \varphi\left(1 - \varepsilon_{\mathrm{so}}^{1/2}\right)}\right] \tag{19.12}$$

where

$$\varphi = \frac{\varepsilon_{\mathrm{so}}^{1/2}}{1 - (k_{\mathrm{so}}/k_{\mathrm{L}})} \tag{19.13}$$

19.2.7.3 One-Dimensional Layered Model

In this model, the system contains two distinctly different components with a clearly defined layer structure. In this case, thermal conductivity in parallel and perpendicular directions to the layers can be calculated by the following equations:

$$k_{\mathrm{pa}} = k_{\mathrm{L}}\left[1 - \varepsilon_{\mathrm{so}}\left(1 - \frac{k_{\mathrm{so}}}{k_{\mathrm{L}}}\right)\right] \tag{19.14}$$

$$k_{\mathrm{se}} = k_{\mathrm{L}}\left[\frac{k_{\mathrm{so}}}{\varepsilon_{\mathrm{so}}k_{\mathrm{L}} + k_{\mathrm{so}}(1 - \varepsilon_{\mathrm{so}})}\right] \tag{19.15}$$

In Equations 19.14 and 19.15, solids are considered as a discontinuous phase in the product. Heldman (1975) mentioned that the Kopelman (1966) model has not been applied to food systems to a great extent. Therefore, the validity of the proposed expression may be limited until more application is attempted, but the model should have a considerable value in systems containing fibers or layers (Heldman, 1975).

19.3 MODELS BASED ON DISTRIBUTION FACTORS

When distribution of different component phases are not known or difficult to assume, models based on empirical distribution factors are recommended. However, in this case, the value of distribution factors needs to be known for specific products and processing conditions.

19.3.1 Sugawara–Yoshizawa (1961) Model

Sugawara and Yoshizawa (1961) stated that those equations for predicting the thermal conductivity of porous material that took into account only the constituent's thermal conductivity and porosity are not sufficient to express the actual thermal conductivity of a composite medium. Thus, from experimental evidence of water and air-saturated porous material, the above authors derived an equation for predicting the thermal conductivity of porous and water-saturated materials. This equation can be written as

$$k_e = (1 - y)k_{so} + yk_p \tag{19.16}$$

where

$$y = \frac{2^q}{2^q - 1}\left[1 - \frac{1}{(1 - \varepsilon_p)^q}\right] \tag{19.17}$$

k_{so} and k_p are the thermal conductivities of solid and fluid phase (W/m K), respectively
ε_p is the volume fraction of the fluid phase

The appropriate value of q for any particular porous materials can be empirically determined by the best fits of the experimental data.

19.3.2 Chaurasia–Chaudhary–Bhandari (1978) Model

Chaurasia et al. (1978) developed an expression for predicting the effective thermal conductivity of two-phase systems, which takes into account not only the thermal conductivities and volume concentration but also the geometrical distribution of the constituent phases present in the system. They successfully applied this expression to the prediction of the effective thermal conductivity of suspensions, emulsions, solid–solid mixes, and porous material systems. The term porous material systems refers to a system consisting of a solid phase in the form of small particles and a saturating fluid phase filling in the pore space. This was a weighted geometric mean of k_{max} and k_{min} and was expressed as

$$k_e = (ik_c + jk_d)^f \left[\frac{k_c k_d}{gk_d + hk_c}\right]^{1-f} \tag{19.18}$$

where
k_e and k_d are the thermal conductivities of continuous and discontinuous phases (W/m K), respectively
g, h, i, and j are the model parameters
f is the weighting factor for structure

Chaurasia et al. (1978) provided relationships among these elements as follows:

$$g + h + i + j = 1 \tag{19.19}$$

$$i + j = \varepsilon_p \tag{19.20}$$

$$g + h = f \tag{19.21}$$

In solving Equation 19.18, the values of model elements *g*, *h*, *i*, and *j* and the weighting factor *f* are needed. The values can be evaluated through the following equations:

$$i = \varepsilon_{\mathrm{p}}^{1.3} \tag{19.22}$$

The element *j* can be obtained through the expression

$$j = \frac{1}{\psi}\left[\frac{1-\varepsilon_p}{1+\varepsilon_p}\right]^3 \tag{19.23}$$

where ψ is the sphericity of the dispersed phase. The value of ψ is equal to 1 for spherical shape and is equal to 0.82 for random shape. Therefore, where k_{c}, k_{d}, ε_{p}, and ψ values are known, the effective thermal conductivity can be obtained through Equation 19.18.

19.3.3 Chaudhary and Bhandari (1968) Model

Chaudhary and Bhandari (1968) attempted to develop models for predicting thermal conductivities of three-phase systems, extending series, parallel, and random models for two-phase systems. They considered all solid phases as a single phase and all liquid phases as water and air phases. They checked the validity of the random model with experimental data and found that the random model was representative only near the saturation region. The condition of $\varepsilon_{\mathrm{a}} = 0$ means water saturation and $\varepsilon_{\mathrm{w}} = 0$ means air saturation. The disagreement below the saturation region may be attributed to the failure to take into account the structure and distribution of the phases in the medium. Therefore, they proposed a semiempirical formula by inserting maximum and minimum values of the thermal conductivity of the mixture. A weighting factor was used to account for the distribution of the various phases in the mixture. If the *f*th fraction of the whole mixture is regarded as parallel distributed, then the $(1-f)$th fraction is distributed in series. The resulting equation was

$$k_{\mathrm{e}} = (k_{\mathrm{pa}})^f (k_{\mathrm{se}})^{(1-f)} \tag{19.24}$$

The numerical value of *f* needs to be determined experimentally for different compositions and conditions.

19.3.4 Herminge (1961) Model

Herminge (1961) proposed a model for predicting the thermal conductivity of moist porous materials. The thermal conductivity of the total system was calculated as

$$\frac{1}{k_{\mathrm{se}}} = \frac{\gamma}{k_{\mathrm{pa}}} + \frac{\eta}{k_{\mathrm{se}}} \tag{19.25}$$

The terms γ and η measure the relative rates of heat transfer in series and parallel systems and may be interpreted as descriptive of the geometry of the material. For any particular material, the values of γ and η may be determined from experimental data for zero and unity saturation. Herminge calculates k_{pa} and k_{se} by considering heat transfer through solid, air, and water phases with another term due to evaporation–diffusion–condensation in moist pores. However, it is not easy to determine the fraction of moist pores inside the food materials and the contribution of evaporation–diffusion–condensation in the total effective thermal conductivity.

19.3.5 Krischer's Model

Krischer (cited in Keey, 1972) proposed another model by combining parallel and series models and using the distribution factor in much the same manner as Chaudhary and Bhandari (1968):

$$\frac{1}{k_e} = \frac{1 - f_k}{k_{pa}} + \frac{f_k}{k_{se}} \tag{19.26}$$

The above equation was developed for dry porous bodies and the distribution factor is likely to be dependent only on structure. Temperature, moisture, and porosity are the important factors in the thermal conductivity of foods. Among these factors, Murakami and Okos (1989) modeled Krischer's distribution factor (f_k) as a function of moisture content and porosity by regression analysis using surface response theory. Murakami and Okos (1989) found that f_k became zero for powdered food materials at $X_w > 0.30$, which indicated that the system developed a parallel structure similar to liquid. Maroulis et al. (2002) also developed a structural generic model based on Equation 19.26 to predict the thermal conductivity of fruits and vegetables during drying. In their model, porosity was predicted from the shrinkage coefficient during drying. Similarly, this type of structural model was also used by Krokida et al. (2001) for granular materials. Carson et al. (2001) determined the effective thermal conductivity of a two-component food system by finite element methods. They found that the sizes of the individual voids were found to have a minor influence on the effective thermal conductivity, whereas more significant influence was on the choice of the structural model (i.e., void shape). Chen et al. (1998) determined Kirscher's distribution factor for apple for the temperature range of 5°C–45°C and X_w from 0.20 to 0.83. They found that the distribution factor decreased with the increase in moisture content, while it increased with the increase in temperature.

19.3.6 Renaud (1990) Model

Renaud (1990, cited in Renaud et al., 1992) proposed another model similar to Krischer's model:

$$k_e = (1 - f_{re})k_{se} + f_{re}k_{pa} \tag{19.27}$$

The values of f_{re} are given by Renaud et al. (1992) for different gels at different concentrations and temperatures below and above freezing.

19.3.7 Rahman–Potluri–Varamit (1991) Model

Rahman et al. (1991) also proposed a similar model:

$$k_e = f_{rv}\left(k_{pa} + k_{ra}\right) \tag{19.28}$$

When the above model was tested with seafood powders, f_{rv} was found to be 0.5 for all the cases studied.

19.3.8 Carson (2002) Model

Carson (2002) proposed a modified version of Maxwell's model based on a weighting parameter:

$$k_e = k_c\left[\frac{\left(f_{cn}^2/1 - f_{cn}^2\right)k_c + k_a - \left(f_{cn}^2/1 - f_{cn}^2\right)(k_c - k_a)\varepsilon_a}{\left(f_{cn}^2/1 - f_{cn}^2\right)k_c + k_a + (k_c - k_a)\varepsilon_a}\right] \tag{19.29}$$

The above equation is reduced to the series model when f_{cn} is zero and to the parallel model when f_{cn} is unity. Based on the proposed bounds for isotropic porous materials, a foam or a sponge (internal porosity) would be expected to have a f_{cn} value between 0.68 and 0.82, and a particular material (external porosity) with $k_{so}/k_a = 10$ would be expected to have an f_{cn} value between 0.4 and 0.7 (Carson et al., 2006). The author also recommended to use Krischer's model to calculate the effective thermal conductivity if Krischer's distribution factor is known.

19.3.9 Torquanto–Sen (1990) Model

A model based on the second-order bound was also proposed by Torquato and Sen (1990) and it was applied successfully by van der Sman (2008) in the case of frozen meat and fish. Torquato and Sen (1990) generalized the second-order bounds for composites consisting of aligned, anisotropic inclusions (spheroids). The lower and upper second-order bounds are

$$\frac{k_{se}}{k_c} = \frac{1 + (\varepsilon_d + \varepsilon_c f_{on})\delta}{1 + \varepsilon_c f_{on}\delta} \tag{19.30}$$

and

$$\frac{k_{pa}}{k_c} = \frac{1 + [\varepsilon_d + \varepsilon_c(1 - 2)f_{on}]\delta}{1 + \varepsilon_c(1 - 2f_{on})\delta} \tag{19.31}$$

where

$\delta = (k_d - k_c)/k_c$ is the thermal conductivity contrast between the dispersed and continuous phase

f_{on} is a factor dependent on the geometry of the inclusions and is in the range $0 < f_{on} < 1/2$

For long fibers $f_{on} = 1/2$, for spherical inclusions $f_{on} = 1/3$, and for disk-shaped platelets $f_{on} = 0$. van der Sman (2008) also presented values of f_{on} as a function of the aspect ratio of the spheroid. For thin disks the aspect ratio tends to zero and for long fibers it tends to infinity. Note that the case $f_{on} = 1/3$ coincides with the Maxwell model and $f_{on} = 0$ or $1/2$ to the series and parallel models' bounds. van der Sman (2008) applied this model in fresh and frozen meat considering ice as dispersed phase and water/fiber mixture as continuous phase. It was based on the fact that in fresh meat most of the water is contained in the meat fibers and in frozen meat most of the ice is formed in the extracellular space. For the mixture of the continuous phase, they applied the Torquato and Sen (1990) model taking the muscle fibers as the dispersed phase and the unfrozen aqueous solution (including solutes) as the continuous phase. In order to predict the ice content, they used the Schwartzberg (1976) model. van der Sman (2008) found accuracy less than 10% for the prediction of thermal conductivity for frozen meat and fish.

19.3.10 Evaporation–Condensation Mode in Pores

The above distribution concept is limited when heat and mass transport inside the product is controlled not only by a mechanism of conduction and diffusion. In the case of pores there could be a mechanism of water evaporation–condensation inside the pores of the product (De Vries et al., 1989). If a temperature gradient is applied to a porous food material, moisture migration as vapor occurs in the pore space. Water evaporates at the high temperature side, diffuses in the pore space according to the vapor pressure gradient caused by the temperature gradient, and condensates at the low temperature side. Thus, latent heat is transported through the pores (Hamdami et al., 2003).

TABLE 19.1 Values of Resistance Factor and Structural Factor for Experiments

	Moisture Content (kg Water/kg Sample)			
Parameter	**0**	**0.35**	**0.45**	**0.60**
Porosity	0.95	0.94	0.94	0.92
f_{ec}	0.00	0.24	0.14	0.19
f_k	0.93	0.80	0.58	0.45

Source: Hamdami, N., Monteau, J.Y., and Le Bail, A., *Int. J. Refrig.*, 26, 809, 2003. With permission.

By considering the effect of latent heat transport, the effective thermal conductivity in the pores was given by the following equation:

$$k_p = k_a + f_{ec}k_{ec} \tag{19.32}$$

where f_{ec} is the resistance factor against vapor diffusion. It is a weighting factor of the evaporation–condensation importance in thermal conductivity in the gaseous phase. The k_{ec} is the equivalent thermal conductivity due to the latent heat transport (evaporation–condensation) (Sakiyama et al., 1999):

$$k_{ec} = La_w\left(\frac{D}{RT}\right)\left(\frac{P}{P - a_wP_{sat}}\right)\left(\frac{dP_{sat}}{dT}\right) \tag{19.33}$$

The thermal conductivity of the porous bodies was calculated using Krischer's model considering air, solids, water, and ice (Hamdami et al., 2003). The resistance factor and structural factor for model food systems as a function of water content and porosity are given in Table 19.1. Goedeken et al. (1998) studied the effect of vapor phase moisture migration on the thermal conductivity of bread samples ($X_w = 0.05$–0.44 and $\varepsilon_a = 0.18$–0.90) sealed in a stainless steel vessel using the probe method from 25°C to 100°C. Porosity showed less enhancement for the vapor phase migration; however, the enhancement increased significantly as temperature and moisture contents were increased. The maximum enhancement could be up to six times higher at 100°C and a moisture content of 44%. They also showed that this enhancement was completely suppressed at an applied air pressure of 18 atm in the sample chamber.

19.4 PERCOLATION THEORY

The percolation theory and effective medium theory (EMT) are widely used to predict transport properties such as electrical and thermal conductivity, optical properties, and dielectric constants of heterogeneous composite materials. Negligible application of the theory for predicting the thermal conductivity of food materials has been reported. However, Mattea et al. (1986) successfully used EMT to predict the thermal conductivity of vegetables during drying. Percolation theory, introduced by Broadbent and Hammersley (1957), has been especially important in the theoretical development of the conductivity of random mixtures of conducting and nonconducting materials (i.e., highly conducting and poorly conducting materials). Kirkpatrick (1973) has reviewed the application of percolation theory to conduction problems. The prediction of percolation theory is that for a conducting substance randomly distributed in an insulating substance there exists a critical volume fraction $\varepsilon > 0$, below which the overall conductivity of a sufficiently large system is zero. In terms of a binary composite with $k_2 \gg k_1$, the prediction of percolation theory is that for a sufficiently large system $k_e = 0$ (k_2) if $\varepsilon > \varepsilon_{cr}$, and $k_e = 0$ (k_d if $\varepsilon < \varepsilon_{cr}$). In this case, ε_{cr} is referred to as the percolation threshold or simply the threshold volume fraction.

Gurland (1966) studied compacted mixtures of silver balls and kelite powder and found a conductivity threshold $\varepsilon_{cr} = 0.3$. Malliars and Turner (1971) showed that the threshold value of ε_{cr} for a compact of metal particles (nickel) and insulator (polyvinyl chloride) depended on the size and shape of metal particles. Shante and Kirkpatrick (1971) found $\varepsilon_{cr} = 0.29$ for a continuum percolation model, in which the conducting material consisted of identical spheres that were permitted to overlap and whose centers were randomly distributed. By computer studies on a network of conductors randomly placed on a single lattice, Kirkpatrick (1973) determined that the threshold fraction was 0.25. Finally, EMT (described below) yields $\varepsilon_{cr} = 1/3$. Thus, the threshold volume theoretically found varied from 1/4 to 1/3.

19.5 EFFECTIVE MEDIUM THEORY

Landauer (1952) successfully described the conductivity of several binary metallic mixtures, and Davis et al. (1975), by different reasoning, arrived at the same expression for a medium with n constituents:

$$\sum_{i=1}^{n} \varepsilon_i \frac{k_e - k_i}{k_i + 2k_e} = 0 \tag{19.34}$$

where k_e is the effective thermal conductivity of the medium (W/m^2 K). Kirkpatrick (1973) extended EMT to regular networks of resistors distributed at random, obtaining a similar equation:

$$\sum_{i=1}^{n} \frac{k_e - k_i}{k_i + ((z/2) - 1)k_e} = 0 \tag{19.35}$$

where z is the coordination number of the system (i.e., number of conductors with a site in common). Kirkpatrick (1973) assumed that a continuous medium can be represented by a cubic arrangement, for which $z = 6$. For a system of two components, with k_1 and k_2 not equal to zero, Equation 19.33 can be written as (Mattea et al., 1986)

$$\frac{k_e}{k_2} = \omega + \left(\omega^2 + \frac{2p}{z-2}\right) \tag{19.36}$$

where

$$p = \frac{k_1}{k_2}$$

$$\omega = \frac{(z\varepsilon_2/2) - 1 + p[(z/2)(1 - \varepsilon_2) - 1]}{z - 2}$$

Carson et al. (2001) used finite element numerical models to examine how the effective thermal conductivity of a material containing voids is affected by its structure. Simulated results were fitted by two commonly used effective thermal conductivity models with adjustable parameters: Krischer's model and the EMT. The sizes of the individual voids were found to have a minor influence on the effective thermal conductivity, whereas a more significant influence on thermal conductivity was the void shape.

There are many other theoretical models available in the literature (Yagi and Kunii, 1957; Kunii and Smith, 1960; Tsao, 1961; Cheng and Vachon, 1969; Chen and Heldman, 1972; Okazaki et al., 1977). The above theoretical models were developed assuming a simple structure and heat transfer mechanisms were involved. This theoretical model has limited applications in food materials due to its complex structure or arrangement of the phases. The parameters in theoretical models vary not only with different foods but also with the variety, harvesting location and time, and also the measurement location in the food materials. Due to this limitation of the theoretical models, semiempirical and empirical models are popular and widely used for food materials. The semi-empirical models, which include the weighting structure factor, have prospects in food applications due to their theoretical basis and only a single parameter is necessary to represent the structure. Theoretically, the structure factor varied from 0 to 1 if component phase thermal conductivity does not change after mixing or heating. In the case of food materials at a higher temperature, chemical and physical structure changes of the component phases may occur due to the interaction of the phases. Hence, in some cases the factor may be higher than one. For example, in the case of potato, the experimental values were much higher than the maximum limit predicted by the parallel model at a higher temperature ($t > 70$°C). Califano and Calvelo (1991) proposed two reasons for higher thermal conductivities: (1) intercellular air may be considered as air trapped in a capillary porous body, where vapor was transferred through the capillary due to temperature difference between its wall, and (2) the structural changes produced by starch gelatinization in the samples might contribute to increase in thermal conductivity at a higher temperature. A similar increase in thermal conductivity was detected by Drouzas and Saravacos (1988) for corn starch at water contents above 11.1% and temperature above 50°C.

19.6 EMPIRICAL MODELS

19.6.1 Fitting of Data for a Specific System

Thermal conductivity varies with chemical composition, distribution of the phases (i.e., geometric factor), density or porosity, temperature, and pressure. For solids and liquids, pressure has negligible effect on thermal conductivity. Thermal conductivity prediction equations of the major food components are given in Table 19.2. Water content was found to be the most important factor in determining thermal conductivity while the nonaqueous part of the food was of lesser importance. This may be due to the relative magnitude of the conductivities of water and other food constituents (Cuevas and Cheryan, 1978). Thermal conductivity of foods decreases with a decrease in moisture content. It is common to find a linear relation between thermal conductivity and moisture content or temperature. Mohsenin (1980), Miles et al. (1983), and Sweat (1986) reviewed this type of correlation for different types of food materials (Tables 19.3 and 19.4). A number of authors used the linear multiple regression equation to relate the thermal conductivity (Table 19.5).

The linear correlation of thermal conductivity with moisture content is limited to small changes in moisture and the constants of the correlation vary not only with the types of food materials but also with the variety and growing or harvesting location and time. Hence, a nonlinear correlation is necessary to cover the whole range of moisture contents. Quadratic and multiple form correlations are also used for food materials (Table 19.5). Baghe-khandan et al. (1982) presented correlations to predict the thermal conductivity of fresh whole and ground beef meat as

$$k_e = \left[303.7 - 454X_{fa} - 219X_w + 306(t)^{-0.039}\right] \times 10^{-3}(\text{whole}) \qquad (19.37)$$

TABLE 19.2 Thermal Conductivity of Major Food Components as a Function of Temperature

Material	Equation	Reference
Air (dry air)	$k=0.0184+(1.225\times10^{-4})t$	Luikov (1964)
Air (moist air: 20°C–60°C)	$k=0.0076+(7.85\times10^{-4})t+0.0156\phi$	Luikov (1964)
Air	$k=3.24\times10^{-3}+(5.31\times10^{-4})t$	Luikov (1964)
Air (0°C–1000°C)	$k=2.43\times10^{-2}+(7.98\times10^{-5})t-(1.79\times10^{-8})t^2-(8.57\times10^{-12})t^3$	Maroulis et al. (2002)
Air (*P*: mm Hg, $P\leq2$ mm Hg)	$k=0.0042P+0.01$	Fito et al. (1984)
Air (*P*: mm Hg, $P\geq2$ mm Hg)	$k_{760}/k=1+1.436\ (1/P)$	Fito et al. (1984)
Protein ($t=-40$°C to 150°C)	$k=1.79\times10^{-1}+(1.20\times10^{-3})t-(2.72\times10^{-6})t^2$	Choi and Okos (1986)
Gelatin	$k=3.03\times10^{-1}+(1.20\times10^{-3})t-(2.72\times10^{-6})t^2$	Renaud et al. (1992)
Ovalbumin	$k=2.68\times10^{-1}+(2.50\times10^{-3})t$	Renaud et al. (1992)
Carbohydrate ($t=-40$°C to 150°C)	$k=2.01\times10^{-1}+(1.39\times10^{-3})t-(4.33\times10^{-6})t^2$	Choi and Okos (1986)
Starch	$k=4.78\times10^{-1}+(6.90\times10^{-3})t$	Renaud et al. (1992)
Gelatinized starch	$k=2.10\times10^{-1}+(0.41\times10^{-3})t$	Maroulis et al. (1991)
Sucrose	$k=3.04\times10^{-1}+(9.93\times10^{-3})t$	Renaud et al. (1992)
Fat ($t=-40$°C to 150°C)	$k=1.81\times10^{-1}-(2.76\times10^{-3})t-(1.77\times10^{-7})t^2$	Choi and Okos (1986)
Fiber ($t=-40$°C to 150°C)	$k=1.83\times10^{-1}+(1.25\times10^{-3})t-(3.17\times10^{-6})t^2$	Choi and Okos (1986)
Ash ($t=-40$°C to 150°C)	$k=3.30\times10^{-1}+(1.40\times10^{-3})t-(2.91\times10^{-6})t^2$	Choi and Okos (1986)
Water ($t=-40$°C to 150°C)	$k=5.71\times10^{-1}+(1.76\times10^{-3})t-(6.70\times10^{-6})t^2$	Choi and Okos (1986)
Water	$k=5.87\times10^{-1}+[2.80\times10^{-3}(t-20)]$	Renaud et al. (1992)
Water	$k=5.62\times10^{-1}+(2.01\times10^{-3})t-(8.49\times10^{-6})t^2$	Nagaska and Nagashima (1980, Cited in Hori, 1983)
Water (0°C–350°C)	$k=5.70\times10^{-1}+(1.78\times10^{-3})t-(6.94\times10^{-6})t^2+(2.20\times10^{-9})t^3$	Maroulis et al. (2002)
Ice (*t*: −40°C to 150°C)	$k=2.22-(6.25\times10^{-3})t+(1.02\times10^{-4})t^2$	Choi and Okos (1986)

$$k_e = [668.4 - 7.4X_{fa} - 305X_w + 1.35t] \times 10^{-3}(\text{ground}) \quad (19.38)$$

The limits were $0.50 < X_w < 0.78$ and $30°C < t < 90°C$. Wang and Hayakawa (1993) developed a correlation for thermal conductivity of gelatinized starch gel ($X_w = 0.396$–0.713; $t = 80°C$–$120°C$):

$$k = 4.1676 \times 10^{-1} + (6.6375 \times 10^{-4})t + (3.6 \times 10^{-7})X_w^3 - (9.5 \times 10^{-6})t X_w \quad (19.39)$$

Wang and Hayakawa (1993) developed a correlation for thermal conductivity of gelatinized starch gel with dissolved sucrose ($X_w = 0.492$–0.750; $t = 80°C$–$120°C$):

$$k_e = 4.1830 \times 10^{-1} + (1.1312 \times 10^{-4})t + (1.009 \times 10^{-5})X_w^3 - (4.3 \times 10^{-7})t X_w \quad (19.40)$$

Mattea et al. (1989) and Lozano et al. (1979) used a nonlinear correlation, which was in exponential form, for correlating their thermal conductivity data of apple and pears during drying. The correlation was as follows:

$$k_e = 0.490 - 0.443[\exp(-0.206M_w)](\text{apple}) \quad (19.41)$$

$$k_e = 0.513 - 0.301[\exp(-1.107M_w](\text{pear}) \quad (19.42)$$

TABLE 19.3 **Numerical Values for the Coefficients *A* and *B* in the Equations for the Thermal Conductivity of an Unfrozen Food ($k_e = A + BX_w$)**

Material	X_w Range	*t* Range (°C)	*A*	*B*	Reference
Fish	—	—	0.032	0.329	Miles et al. (1983)
Mince meat	—	—	0.096	0.340	Miles et al. (1983)
Sorghum	—	—	0.564	0.086	Miles et al. (1983)
Fruit juice	—	—	0.140	0.420	Miles et al. (1983)
Fresh meat	0.600–0.800	0 to 60	0.080	0.520	Sweat (1975)
Spring wheat (hard red)	0.044–0.225	−27 to 20	0.139	0.119	Chandra and Muir (1971)
Spring wheat (hard red)	0.044–0.225	20	0.139	0.141	Chandra and Muir (1971)
Spring wheat (hard red)	0.044–0.225	5	0.144	0.095	Chandra and Muir (1971)
Spring wheat (hard red)	0.044–0.225	1	0.136	0.136	Chandra and Muir (1971)
Spring wheat (hard red)	0.044–0.225	−6	0.133	0.154	Chandra and Muir (1971)
Spring wheat (hard red)	0.044–0.225	−17	0.141	0.094	Chandra and Muir (1971)
Spring wheat (hard red)	0.044–0.225	−27	0.144	0.095	Chandra and Muir (1971)
Spring wheat (hard red)	0.014–0.148	32–60	0.129	0.274	Moote (1953, cited in Chandra and Muir, 1971)
Soft wheat (white)	0.000–0.203	21–44	0.117	0.113	Kazarian and Hall (1965, Cited in Chandra and Muir, 1971)
Bulk corn (four varieties)	0.000–0.400	40	0.108	0.180	Kustermann (1984)
Tomato paste	0.538–0.708	30	0.029	0.793	Drusas and Saravacos (1985)
Tomato paste	0.538–0.708	40	−0.066	0.978	Drusas and Saravacos (1985
Tomato paste	0.538–0.708	50	−0.079	1.035	Drusas and Saravacos (1985
Bulk rice (rough)	0.090–0.190	—	0.050	0.077	Mohsenin (1980)
Dairy product	0.160–0.822	20	0.156	0.404	Sweat and Parmelee (1978)
Dairy product	0.160–0.822	0	0.166	0.381	Sweat and Parmelee (1978)
Dairy product	0.167–0.822	40	0.169	0.395	Sweat and Parmelee (1978)

TABLE 19.4 **Numerical Values for the Coefficients c and *d* in the Equations for the Thermal Conductivity of a Food ($k_e = A + Bt$)**

Material	X_w	*t* Range (°C)	*A*	*B*	Reference
Mince meat	—	—	0.295	2.00×10^{-4}	Sorenfors (1973)
Chicken meat (dark)	0.763	0 to 20	0.481	8.65×10^{-4}	Sweat et al. (1973)
Chicken meat (white)	0.744	0 to 20	0.476	6.05×10^{-4}	Sweat et al. (1973)
Chicken meat (dark)	0.763	0 to 20	0.481	8.65×10^{-4}	Sweat et al. (1973)
Milk powder (skim) ($\rho_b = 293$ kg/m^3)	0.014	10 to 40	0.028	7.51×10^{-4}	MacCarthy (1984)
Milk powder (skim) ($\rho_b = 577$ kg/m^3)	0.042	15 to 50	0.052	5.80×10^{-4}	MacCarthy (1984)
Milk powder (skim) ($\rho_b = 724$ kg/m^3)	0.040	15 to 50	0.073	7.28×10^{-4}	MacCarthy (1984)
Apple "Golden Delicious"	0.873	0 to 25	0.394	2.12×10^{-3}	Ramaswamy and Tung (1984)
Apple "Granny Smith"	0.858	0 to 25	0.367	2.50×10^{-3}	Ramaswamy and Tung (1984)
Apple "Golden Delicious"	0.873	−25 to 0	1.290	-9.50×10^{-3}	Ramaswamy and Tung (1984)
Apple "Granny Smith"	0.858	−25 to 0	1.070	-1.11×10^{-2}	Ramaswamy and Tung (1984)
Potato	0.798	20 to 85	0.624	1.19×10^{-3}	Lamberg and Hallstrom (1986)
Tomato paste	0.618	30 to 50	0.482	1.50×10^{-3}	Drusas and Saravacos (1985)
Soy flour (defatted)	—	—	0.087	9.36×10^{-4}	Maroulis et al. (1990)

TABLE 19.5 Multiple and Quadratic Form of Thermal Conductivity Prediction Equations for Food Materials ($k_e = a + bX_w + ct$)

Material	X_w	t Range (°C)	Equation	References
Mince meat	0.40–0.80	25–50	$k_e = 0.049 + 0.340X_w + (2.40 \times 10^{-3})t$	Sorenfors (1973)
Coconut	0.02–0.51	25–50	$k_e = 0.058 + 0.181X_w + (0.90 \times 10^{-3})t$	Jindal and Murakami (1984)
Fresh meat	0.65–0.85	−40 to −5	$k_e = 0.280 + 1.900X_w - (9.20 \times 10^{-3})t$	Sweat (1975)
Apple	0.2–1.6 (M_w)	5–45	$k_e = 0.051 + 0.072M_w + (0.67 \times 10^{-3})t$	Singh and Lund (1984)
Dough wheat	—	20–90	$k_e = 0.310 - 0.820X_w + (1.30 \times 10^{-3})t$	Metel et al. (1986)
Chicken meat (dark)	0.763	−75 to −10	$k_e = 1.140 - 1.460t - (9.86 \times 10^{-5})t^2$	Sweat et al. (1973)
Chicken meat (white)	0.744	−75 to −10	$k_e = 1.070 - 1.490t - (10.4 \times 10^{-5})t^2$	Sweat et al. (1973)
Milk powder ($\rho_b = 512$ kg/m^3)	0.031	12 to 43	$k_e = 0.0076 + 0.268t - (2.30 \times 10^{-5})t^2$	MacCarthy (1985)
Milk powder ($\rho_b = 605$ kg/m^3)	0.031	17–43	$k_e = 0.0003 + 0.646t - (9.58 \times 10^{-5})t^2$	MacCarthy (1985)
Potato	0.800	50–100	$k_e = 1.05 - (1.960 \times 10^{-2})t + (19.00 \times 10^{-5})t^2$	Califano and Calvelo (1991)

Mattea et al. (1986) used other correlations for potato and pear:

$$k_e = 0.4875 - \frac{0.1931}{M_w} + 0.0227 \ln(M_w)(\text{potato}) \tag{19.43}$$

$$k_e = 0.5963 - \frac{0.1931}{M_w} + \frac{0.0301}{M_w^2}(\text{pear}) \tag{19.44}$$

Rahman and Potluri (1990) presented a more general form of correlation than Mattea et al. (1989) and Lozano et al. (1979) for correlating their data on squid meat during air-drying by normalizing both sides of the correlation with initial water content before drying or cooking. The correlation was

$$\frac{k_e}{k_e^o} = A - B\exp\left(-C\frac{X_w}{X_w^o}\right) \tag{19.45}$$

The parameters are given in Table 19.6 for different food materials. The above form of correlation can predict the thermal conductivity with less than 10% error (Rahman, 1992).

There was always a tendency to make general correlations to predict any property of food materials for use in process design equations. For example, Baroncini et al. (1980) compiled a number of general correlations to predict the thermal conductivity of a liquid. They then suggested an improved correlation for calculation of liquid thermal conductivity. Again, Kubaitis et al. (1990) studied the possibility of investigating saturation processes using a generalized expression of process rate constant. For fruits and vegetables, Lozano et al. (1983) developed a general correlation to predict the apparent shrinkage during a drying process. Miles et al. (1983) compiled the thermophysical properties prediction models of foods and gave

$$k_e = 0.344X_w - 0.0644X_{pr} - 0.133X_{fa} + 0.0008t \quad (t > t_F) \tag{19.46}$$

Riedel (1949, cited in Ramaswamy and Tung, 1981) proposed a correlation to predict the thermal conductivity of fruit juices, sugar solutions, and milk:

$$k_e = \left[0.566 + (1.799 \times 10^{-3})t - (5.882 \times 10^{-6})t^2\right] \times \left[(7.598 \times 10^{-4}) + 9.342X_w\right] \tag{19.47}$$

It was estimated that between 0°C and 180°C there was an error of 1% when this model (Equation 19.45) was used (Singh, 1992), but it is very limited in use, since it can be applied only to the indicated products. Salvadori and Mascheroni (1991) proposed a general correlation for meat products. When heat transfer is parallel to fiber,

$$k_e = 0.1075 + 0.501X_w + (5.052 \times 10^{-4})X_w t \quad (t \geq t_F) \tag{19.48}$$

TABLE 19.6 Parameters of Equation for Different Foods

Material	Process	*n*	*A*	*B*	*C*
Apple	Air-drying	15	0.155	−0.021	−3.713
Beef	Cooking	41	1.832	1.737	0.814
Pear	Air-drying	15	1.120	1.166	2.368
Potato	Air-drying	10	1.245	1.279	1.654
Squid	Air-drying	39	1.200	1.350	1.750

Source: Rahman, M.S., *J. Food Eng.*, 15, 261, 1992. With permission.

$$k_e = 0.398 + 1.448X_w^o + \frac{0.985}{t} \quad (t < t_F) \tag{19.49}$$

When heat transfer is perpendicular to fiber,

$$k_e = 0.0866 + (0.052 \times 10^{-4})X_w t \quad (t \geq t_F) \tag{19.50}$$

$$k_e = 0.378 + 1.376X_w^o + \frac{0.930}{t} \quad (t < t_F) \tag{19.51}$$

Vagenas et al. (1990) correlated the thermal conductivity of raisins ($t = t_F$ to 100°C; $X_w = 0$–1.0; and $\varepsilon_b = 0.37$–0.49) as

$$k_b = -0.118 + 0.422X_w + 0.002t + 0.120\varepsilon_b \tag{19.52}$$

Morita and Singh (1979) and Jindal and Murakami (1984) also used this form of equation. In the case of food materials, thermal conductivity empirical correlation is limited to specific materials and varieties.

19.6.2 Data Mining Approach

There has been consistent effort spent in developing generalized correlations to predict properties of food materials for use in process design and optimization (Rahman, 1992). Sweat (1986) developed a correlation using a varied data set of about 430 points for liquid and solid foods:

$$k_e = 0.58X_w + 0.155X_{pr} + 0.25X_{ca} + 0.135X_{as} + 0.16X_{fa} \tag{19.53}$$

Again the drawback of the above general correlation is that it does not include temperature, porosity terms, and geometric factors. Hence, it is not applicable in porous solid foods. Rahman et al. (1991) tested Sweat's (1986) model to predict the thermal conductivities of fresh seafood and found that the error varied from 2.5% to 7.5%. However, Sweat (1974) presented a linear model for predicting the thermal conductivity of fresh fruits and vegetables giving predictions within ±15% of most experimental values. The model was

$$k_e = 0.148 + 0.493X_w^o \tag{19.54}$$

The above model was limited to $X_w^o > 0.60$ and does not include temperature. According to Sweat, there was a strong relation between water content and thermal conductivity of all fruits and vegetables tested except for apples, which were highly porous. Hence, the general correlation should include a porosity term. For nonporous materials, the conductivity can be correlated with only water or proximate composition. In the case of porous materials, the porosity term must be included in the correlation because air has a thermal conductivity much lower than food components. A more general correlation was developed by Rahman (1992) using data of five foods (apple, beef, pear, potato, and squid). The correlation was

$$\left(\frac{k_e}{k_e^o}\right)\left(\frac{1}{1-\varepsilon_a}\right) = 1.82 - 1.66\left[\exp\left(-0.85\frac{X_w}{X_w^o}\right)\right] \tag{19.55}$$

Squid had a maximum error of 12% and potato had a minimum error of 4%. The above correlation was developed by varying the water content from 5% to 88% (wet basis), porosity from 0 to 0.5, and temperature from 20°C to 25°C. The plot of experimental data and predicted line is given in Figure 19.2D. The above correlation had a lower error than Sweat's (1974) correlation (Equation 19.54), was not limited to fruit or vegetables, and had a higher moisture range. When $\varepsilon_a = 1$, it was later identified by Rahman and Chen (1995) that the left-hand side of the above correlation becomes

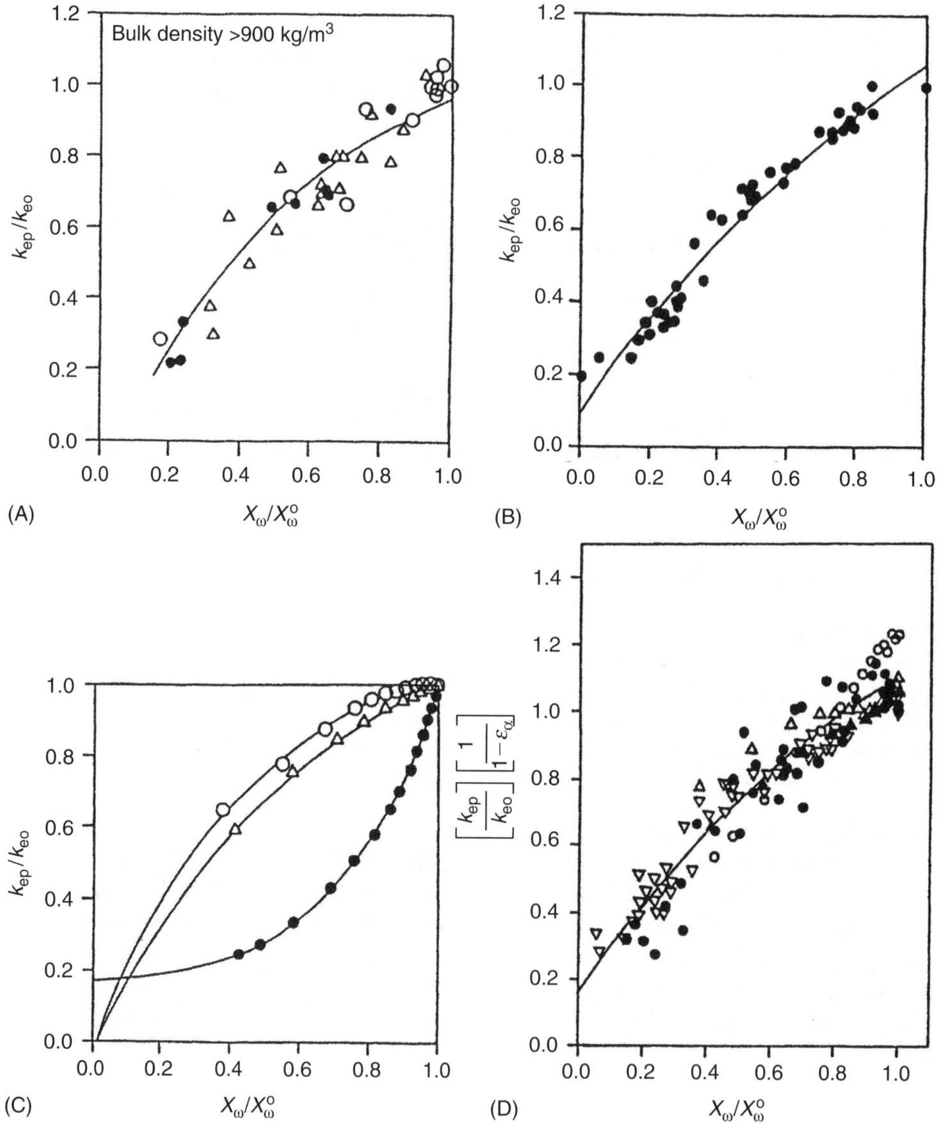

FIGURE 19.2 Dimensionless thermal conductivity as a function of dimensionless water content. (A) Squid meat (o, test 1; Δ, test 2; •, test 3). (From Rahman, M.S. and Potluri, P.L., *J. Food Sci.*, 56, 582, 1990. With permission.) (B) Beef. (From Rahman, M.S., *J. Food Eng.*, 15, 261, 1992. With permission.) (C) Fruits and vegetables (•, apple; Δ, potato; o, pear). (From Rahman, M.S., *J. Food Eng.*, 15, 261, 1992. With permission.) (D) general correlation (o, apple; Δ, pear; ∇, beef; ▼, potato; •, squid meat). (From Rahman, M.S., *J. Food Eng.*, 15, 261, 1992. With permission.)

infinity, which is physically incorrect. The other disadvantages of the above correlation are that it does not include any temperature effects on thermal conductivity and it also needs the conductivity values of the fresh foods (i.e., before processing). Thus, a more general equation should be developed using the temperature term and more experimental data points. Rahman et al. (1997) developed another improved model for fruits and vegetables during drying based on structural factors:

$$k_e = \varepsilon_a k_a + f_r[(1 - \varepsilon_a - \varepsilon_w)k_{so} + \varepsilon_w k_w] \tag{19.56}$$

A general correlation of structural factors for fruits and vegetables was developed based on 164 data points:

$$f_r = 0.996\left[1 - \varepsilon_a + \frac{k_a}{(k_w)_r}\right] \times \left(\frac{T}{T_r}\right)^{0.731} \times X_w^{0.285} \tag{19.57}$$

where

T_r indicates the reference temperature 273 K
T is the absolute temperature in K
X_w is the mass fraction of water (wet basis)

The above model was developed for X_w varying from 0.14 to 0.94, temperature from 5°C to 100°C, and apparent porosity from 0.0 to 0.56. The prediction error varied from 7% to 15% based on the type of product tested. The model, however, did not consider temperatures below 0°C and above 100°C.

Considering the wide variations in materials and processing conditions makes it difficult to develop an analytical model for the prediction of thermal conductivity. Thus, Hussain and Rahman (1999) used the artificial neural network technique for predicting thermal conductivity using a data set of 164 points. Artificial neural networks are optimization algorithms, which attempt to mathematically model the learning process. The model is a simple approximation of such a complex process, but it utilizes the basic foundations and concepts inherent in the learning processes of humans and animals. One of the major advantages of an artificial neural network (ANN) is its efficient handling of highly nonlinear relationships in data. Neural network modeling has generated increasing acceptance and is an interesting method in the estimation and prediction of food properties and process-related parameters. Similarly, ANN modeling has been used in the prediction of the thermal conductivity of bakery products (Sablani et al., 2002).

An ANN model was presented for the prediction of the thermal conductivity of foods as a function of moisture content, temperature, and apparent porosity (Sablani and Rahman, 2003). The food products considered in the study by Sablani and Rahman (2003) were apple, pear, cornstarch, raisin, potato, ovalbumin, sucrose, starch, carrot, and rice (676 data points). Thermal conductivity data of food products (0.012–2.350 W/m K) were obtained from the literature for a wide range of moisture contents ($X_w = 0.04$–0.98), temperature (−42°C–130°C), and apparent porosity (0.0–0.70). Several configurations were evaluated while developing the optimal ANN model. The optimal ANN model consists of two hidden layers with four neurons in each hidden layer. This model was able to predict thermal conductivity with a mean relative error of 13% and a mean absolute error of 0.081 W/m K. Neural network-based equations for estimation of effective thermal conductivity, k_e (W/m K) for known moisture content (X_w), temperature ratio (T/T_r, T in K), and apparent porosity (ε_a) are

$$
\begin{aligned}
X2 &= X_w(2.13) + (-1.09) \\
X3 &= (T/T_r)(3.18) + (-3.69) \\
X4 &= \varepsilon_a(2.86) + (-1) \\
X5 &= \tanh[(-0.66) + (-0.99)X2 + (-3.28)X3 + (-0.38)X4] \\
X6 &= \tanh[(1.42) + (-0.82)X2 + (-0.54)X3 + (2.11)X4] \\
X7 &= \tanh[(-3.49) + (2.11)X2 + (-3.66)X3 + (1.00)X4] \\
X8 &= \tanh[(-0.18) + (0.92)X2 + (0.36)X3 + (0.042)X4] \\
X9 &= \tanh[(0.21) + (-0.62)X5 + (-0.72)X6 + (-0.01)X7 + (0.73)X8] \\
X10 &= \tanh[(0.025) + (0.058)X5 + (-0.043)X6 + (0.17)X7 + (-0.189)X8] \\
X11 &= \tanh[(0.54) + (0.55)X5 + (0.69)X6 + (-1.13)X7 + (0.19)X8] \\
X12 &= \tanh[(-0.15) + (-2.76)X5 + (0.53)X6 + (-2.92)X7 + (0.021)X8] \\
X13 &= \tanh[(0.078) + (0.14)X9 + (-0.032)X10 + (-0.31)X11 + (-0.29)X12] \\
k_e &= X13(1.95) + (1.18)
\end{aligned}
\tag{19.58}
$$

Although Equations 19.58 are difficult to use in a normal calculator, it could be easily used in computer programming for process design and optimization in a generic approach considering the main factors of moisture content, porosity, and temperature.

19.7 CONCLUSION

Thermal conductivity can be predicted from empirical correlations if they are available for the specific product at the desired processing conditions. Alternatively, if structures or types of distribution in the component phases are known, then parallel, series, and Kopelman models could be used. If the distribution factor is known, models based on the distribution concepts, such as Krischer's models, should be used. If the above approach is difficult to use, then complex neural network-based models could be used. However, there is still a need to develop a more generic user-friendly model to predict the thermal conductivity of foods in the future.

NOMENCLATURE

A	model parameter (Equation 19.45)
a	activity
B	model parameter (Equation 19.45)
C	model parameter (Equation 19.45)
D	diffusivity (m^2/s)
d	differentiation
f	distribution factor
g	model parameter (Equation 19.18)
h	model parameter (Equation 19.18)
i	model parameter (Equation 19.18)
j	model parameter (Equation 19.18)
k	thermal conductivity (W/m K)
L	latent heat of vaporization (J/kg)
M	mass fraction (dry basis, kg/kg solids)

n	number of components in a mixture
P	pressure (Pa)
p	thermal conductivity ratio (Equation 19.36)
q	parameter of Equation 19.17
R	ideal gas constant
T	temperature (K)
t	temperature (°C)
X	mass fraction (wet basis, kg/kg sample)
X1–X13	functions in Equation 19.58
y	*p*arameter in Equation 19.16
z	coordination number

Greek Symbols

ε	volume fraction
δ	thermal conductivity contrast $[(k_d - k_c)/k_c]$
ν	parameter in Equation 19.7
ξ	model parameter of Equation 19.9
γ	distribution factor in Equation 19.25
η	distribution factor in Equation 19.25
ω	model parameter of (EMT) Equation 19.36
φ	parameter in Equation 19.12
ψ	sphericity

Subscripts

a	air or apparent
as	ash
b	bulk
c	continuous phase
ca	carbohydrate
cn	Carson's model
cr	threshold or critical
d	dispersed
e	effective
ec	evaporation–condensation
F	freezing point
fa	fat
i	*i*th component
k	Krischer model
L	liquid phase
m	Maxwell
max	maximum
min	minimum
n	*n*th term
on	Torquato–Sen
p	pore
pa	parallel
pr	protein
r	Rahman et al.'s model or reference

ra	random
re	Renaud model
rv	Rahman–Potluri–Varamit model
sat	saturation
se	series model
so	solids
w	water
1, 2, 3	component 1, 2, 3

Superscript

o	initial or before processing (treatment)

REFERENCES

Baghe-khandan, M.S., Okos, M.R., and Sweat, V.E., The thermal conductivity of beef as affected by temperature and composition, *Transactions of the ASAE*, 25, 1118–1122, 1982.

Baroncini, J.R., Filippo, P.D., Latini, G., and Pacetti, M., An improved correlation for the calculation of liquid thermal conductivity, *International Journal of Thermophysics*, 1, 159–175, 1980.

Bergman, D.J., Bulk effective moduli: Their calculation and usage for describing physical properties of composite media, in *Physical Phenomena in Granular Materials*, Cody, G.D., Geballe, T.H., and Sheng, P., Eds., Taylor & Francis, Philadelphia, PA, 1990.

Brailsford, A.D. and Major, K.G., Thermal conductivity of aggregates of several phases, including porous materials, *British Journal of Applied Physics*, 15, 313–319, 1964.

Broadbent, S.R. and Hammersley, J.M., Percolation process: I, *Proceedings of the Cambridge Philosophical Society*, 53, 629, 1957.

Califano, A.N. and Calvelo, A., Thermal conductivity of potato between 50°C and 100°C, *Journal of Food Science*, 56(2), 586–587 and 589, 1991.

Carson, J.K., Prediction of the thermal conductivity of porous foods, PhD thesis, Massey University, New Zealand, 2002.

Carson, J.K., Tanner, D.J., Lovatt, S.J., and Cleland, A.C., Analysis of the effective thermal conductivity of a two-component model food system using finite element methods, *Acta Horticulture*, 566, 391–396, 2001.

Carson, J.K., Lovatt, S.J., Tanner, D.J., and Cleland, A.C., Predicting the effective thermal conductivity of unfrozen, porous foods, *Journal of Food Engineering*, 75, 297–307, 2006.

Chandra, S. and Muir, W.E., Thermal conductivity of spring wheat at low temperatures, *Transactions of the ASAE*, 14, 644–646, 1971.

Chaudhary, D.R. and Bhandari, R.C., Heat transfer through a three phase porous media, *British Journal of Applied Physics Series*, 2(1), 815, 1968.

Chaurasia, P.B.L., Chaudhary, D.R., and Bhandari, R.C., Effective thermal conductivity of two-phase system, *Indian Journal of Pure and Applied Physics*, 16, 963–967, 1978.

Chen, A.C., Mechanisms of heat transfer through organic powder in a packed bed, PhD thesis, Michigan State University, East Lansing, MI, 1969.

Chen, A.C. and Heldman, D.R., An analysis of the thermal properties of dry food powder in a packed bed, *Tractions of the ASAE*, 15(5), 951, 1972.

Chen, X.D., Xie, G.Z., and Rahman, M.S., Application of the distribution factor concept in correlating thermal conductivity data for fruits and vegetables, *International Journal of Food Properties*, 1(1), 35–44, 1998.

Cheng, S.C. and Vachon, R.I., The prediction of the thermal conductivity of two and three phase solid heterogeneous mixtures, *International Journal of Heat and Mass Transfer*, 12, 249–264, 1969.

Choi, Y. and Okos, M.R., Effects of temperature and composition on the thermal properties of foods, in *Food Engineering and Process Applications*, Vol. 1, *Transport Phenomena*, Maguer, M. and Jelen, P., Eds., Elsevier Applied Science, London, 1986.

Cuevas, R. and Cheryan, M., Thermal conductivity of liquid foods—A review, *Journal of Food Process Engineering*, 2, 283, 1978.

Davis, H.T., Valencourt, L.R., and Johnson, C.E., Transport processes in composite media, *Journal of American Ceramic Society*, 58(9–10), 446–452, 1975.

De Vries, U., Shimer, P., and Blocksma, A.H., A quantitative model for heat transport in dough and crumb during baking, in *Cereal Science and Technology in Sweden*, Lund University, Yastad, Sweden, 1989, pp. 174–188.

Deissler, R.G., Boegli, J.S., and Ohio, C., An investigation of effective thermal conductivities of powders in various gases, *Transactions of the ASME*, October, 1417, 1958.

Drouzas, A.E. and Saravacos, G.D., Effective thermal conductivity of granular starch materials, *Journal of Food Science*, 53(1), 1795–1799, 1988.

Drusas, A.E. and Saravacos, G.D., Thermal conductivity of tomato paste, *Journal of Food Engineering*, 4, 157–168, 1985.

Eucken, A., B3. Forschungshaft no. 353, 1932.

Fito, P.J., Pinaga, F., and Aranda, V., Thermal conductivity of porous bodies at low pressure: Part I, *Journal of Food Engineering*, 3(1), 75–88, 1984.

Goedeken, D.L., Shah, K.K., and Tong, C.H., True thermal conductivity determination of moist porous food materials at elevated temperatures, *Journal of Food Science*, 63(6), 1062–1066, 1998.

Gurland, J., An estimate of contact and continuity of dispersions in opaque samples, *Transactions of Metallurgical Society of the AIME*, 236(5), 642, 1966.

Hamdami, N., Monteau, J.-Y., and Le Bail, A., Effective thermal conductivity of a high porosity model food at above and sub-freezing temperatures, *International Journal of Refrigeration*, 26, 809–816, 2003.

Heldman, D.R., *Food Process Engineering*, AVI Publishing Company Inc., Westport, CT, 1975.

Heldman, D.R., Food freezing, in *Handbook of Food Engineering*, Heldman, R. and Lund, D.B., Eds., Marcel Dekker Inc., New York, 1992.

Herminge, L., Heat transfer in porous bodies at various temperatures and moisture contents, *Tappi*, 44(8), 570, 1961.

Hori, T., Effects of rennet treatment and water content on thermal conductivity of skim milk, *Journal of Food Science*, 48, 1492–1496, 1983.

Hussain, M.A. and Rahman, M.S., Thermal conductivity prediction of fruits and vegetables using neural networks, *International Journal of Food Properties*, 2, 121–138, 1999.

Jindal, V.K. and Murakami, E.G., Thermal properties of shredded coconut, in *Engineering and Food*, Vol. 1, McKennan, B.M., Ed., Elsevier Applied Science Publishers, London, 1984.

Keey, R.B., *Drying Principles and Practice*, Pergamon Press, New York, 1972.

Kirkpatrick, S., Percolation and conduction, *Reviews of Modern Physics*, 45(4), 574, 1973.

Kopelman, I.J., Transient heat transfer and thermal properties in food systems, PhD thesis, Michigan State University, East Lansing, 1966.

Krokida, M.K., Maroulis, Z.B., and Rahman, M.S., A structural generic model to predict the effective thermal conductivity of granular materials, *Drying Technology*, 19(9), 2277–2290, 2001.

Kubaitis, Z., Jurevichiute, A., and Chalykh, A.E., The possibility of investigating saturation process using a generalized expression of the process rate constant, *European Polymer Journal*, 26, 495, 1990.

Kunii, D. and Smith, J.M., Heat transfer characteristics of porous rocks, *AIChE Journal*, 6(1), 71, 1960.

Kustermann, M., The effect of physical properties of grain on processing, in *Engineering and Food*, Vol. 1, *Engineering Sciences in the Food Industry*, McKenna, M., Ed. Elsevier Applied Science Publishers, Essex, pp. 597–601, 1984.

Landauer, R., Electrical resistance of binary metallic mixture, *Journal of Applied Physics*, 23(7), 779, 1952.

Lamberg, I. and Hallstrom, B., Thermal properties of potatoes and a computer simulation model of a blanching process, *Journal of Food Technology*, 21, 577–585, 1986.

Lozano, J.E., Urbicain, M.J., and Rotstein, E., Thermal conductivity of apples as a function of moisture content, *Journal of Food Science*, 44(1), 198, 1979.

Lozano, J.E., Rotstein, E., and Urbicain, M.J., Shrinkage, porosity and bulk density of foodstuffs at changing moisture contents, *Journal of Food Science*, 48, 1497, 1983.

Luikov, A.V., Heat and mass transfer in capillary porous bodies, *Advances in Heat Transfer*, 1, 234, 1964.

MacCarthy, D., The effective thermal conductivity of skim milk powder, in *Engineering and Food*, Vol. 1, McKennan B.M., Ed., Elsevier Applied Science Publishers, London, 1984.
MacCarthy, D.A., Effect of temperature and bulk density on thermal conductivity of spray-dried whole milk powder, *Journal of Food Engineering*, 249–263, 1985.
Malliars, A. and Turner, D.T., Influence of particle size on the electrical resistivity of compacted mixtures of polymeric metallic powders, *Journal of Applied Physics*, 42(2), 614, 1971.
Maroulis, Z.B., Drouzas, A.E., and Saravacos, G.D., Modeling of thermal conductivity of granular starches, *Journal of Food Engineering*, 11(4), 255–271, 1990.
Maroulis, Z.B., Shah, K.K., and Saravacos, G.D., Thermal conductivity of gelatinized starches, *Journal of Food Science*, 56(3), 773–776, 1991.
Maroulis, Z.B., Krokida, M.K., and Rahman, M.S., A structural generic model to predict the effective thermal conductivity of fruits and vegetables during drying, *Journal of Food Engineering*, 52, 47–52, 2002.
Mattea, M., Urbicain, M.J., and Rotstein, E., Prediction of thermal conductivity of vegetable foods by the effective medium theory, *Journal of Food Science*, 51(1), 113, 1986.
Mattea, M., Urbicain, M.J., and Rotstein, E., Effective thermal conductivity of cellular tissues during drying: Prediction by a computer assisted model, *Journal of Food Science*, 54(1), 194, 1989.
Maxwell, J.C., *Treatise on Electricity and Magnetism*, 3rd edn., Clarendon Press, Oxford, England, 1904.
Metel, S.N., Mikrukov, V.V., and Kasparov, M.N., Thermophysical characteristics of yeast dough, *Izvestiga Vysshikh Uchebnykh Zavedenii, Pishcshevaya Teknologiya*, 4, 107, 1986.
Miles, C.A., Beek, G.V., and Veerkamp, C.H., Calculation of thermophysical properties of foods, in *Physical Properties of Foods*, Jowitt, R., Escher, F., Hallstrom, B., Meffert, H.F.T., Spiess, W.E.L., and Vos, G., Eds., Applied Science Publishers, New York, 1983.
Mohsenin, N.N., *Thermal Properties of Foods and Agricultural Materials*, Gordon and Breach Science Publishers, New York, 1980.
Morita, T. and Singh, R.P., Physical and thermal properties of short-grain rough rice, *Transactions of the ASAE*, 22, 630, 1979.
Murakami, E.G. and Okos, M.R., Measurement and prediction of thermal properties of foods, in *Food Properties and Computer Aided Engineering of Food Processing Systems*, Singh, R.P. and Medina, A.G., Eds., Kluwer Academic Publishers, New York, 1989.
Okazaki, M., Ito, I., and Toei, R., Effective thermal conductivities of wet granular materials, *AIChE Symposium Series*, 73, 164, 1977.
Rahman, M.S., Thermal conductivity of four food materials as a single function of porosity and water content, *Journal of Food Engineering*, 15, 261–268, 1992.
Rahman, M.S., and Chen, X.D., A general form of thermal conductivity equation as applied to an apple: Effects of moisture, temperature and porosity, *Drying Technology*, 13, 1–18, 1995.
Rahman, M.S. and Driscoll, R.H., Thermal conductivity of seafoods: Calamari, octopus and prawn, *Food Australia*, 43(8), 356, 1991.
Rahman, M.S. and Potluri, P.L., Thermal conductivity of fresh and dried squid meat by line source thermal conductivity probe, *Journal of Food Science*, 56(2), 582, 1990.
Rahman, M.S., Potluri, P.L., and Varamit, A., Thermal conductivities of fresh and dried seafood powders, *Transactions of the ASAE*, 34(1), 217, 1991.
Rahman, M.S., Chen, X.D., and Perera, C.O., An improved thermal conductivity prediction model for fruits and vegetables as a function of temperature, water content and porosity, *Journal of Food Engineering*, 31, 163–170, 1997.
Ramaswamy, H.S. and Tung, M.A., Thermophysical properties of apples in relation to freezing, *Journal of Food Science*, 46, 724, 1981.
Ramaswamy, H.S. and Tung, M.A., A review on predicting freezing times of foods, *Journal of Food Process Engineering*, 7, 169–203, 1984.
Renaud, T., Briery, P., Andrieu, J., and Laurent, M., Thermal properties of model foods in the frozen state, *Journal of Food Engineering*, 15(2), 83–97, 1992.
Sablani, S.S. and Rahman, M.S., Using neural networks to predict thermal conductivity of food as a function of moisture content, temperature and apparent porosity, *Food Research International*, 36(6), 617–623, 2003.

Sablani, S.S., Baik, O.D., and Marcotte, M., Neural networks for predicting thermal conductivity of bakery products, *Journal of Food Engineering*, 52, 299–304, 2002.

Sakiyama, T., Akutsu, M., Miyawaki, O., and Yano, T., Effective thermal diffusivity of food gels impregnated with air bubbles, *Journal of Food Engineering*, 39, 323–328, 1999.

Salvadori, V.O. and Mascheroni, R.H., Prediction of freezing and thawing times of foods by means of a simplified analytical method, *Journal of Food Engineering*, 13, 67, 1991.

Schwartzberg, H.G., Effective heat capacities for the freezing and thawing of food, *Journal of Food Science*, 41, 53, 1976.

Shante, V.K.S. and Kirkpatrick, S., An introduction to percolation theory, *Advances in Physics*, 20, 325, 1971.

Singh, R.P., Heating and cooling processes for foods, in *Handbook of Food Engineering*, Heldman, R. and Lund, D.B., Eds., Marcel Dekker Inc., New York, 1992, Chap. 5.

Singh, R.K. and Lund, D.B., Mathematical modeling of heat and moisture transfer-related properties of intermediate moisture apples, *Journal of Food Processing and Preservation*, 8, 91–210, 1984.

Sorenfors, P., Determination of the thermal conductivity of minced meat, *Food Science and Technology*, 7(7), 236–238, 1973.

Sugawara, A. and Yoshizawa, Y., An experimental investigation on the thermal conductivity of consolidated porous materials and its application to porous rock, *Journal of Physics*, 14(4), 469, 1961.

Sweat, V.E., Experimental values of thermal conductivity of selected fruits and vegetables, *Journal of Food Science*, 39, 1080, 1974.

Sweat, V.E., Modeling the thermal conductivity of meats, *Transactions of the ASAE*, 18(3), 564, 1975.

Sweat, V.E., Thermal properties of foods, in *Engineering Properties of Foods*, Rao, M.A. and Rizvi, S.S.H., Eds., Marcel Dekker, Inc., New York, 1986.

Sweat, V.E. and Parmelee, C.E., Measurement of thermal conductivity of dairy products and margarines, *Journal of Food Process Engineering*, 2, 187–197, 1978.

Sweat, V.E., Haugh, C.G., and Stadelman, W.J., Thermal conductivity of chicken meat at temperatures between −75°C and 20°C, *Journal of Food Science*, 38, 158–160, 1973.

Torquato, S. and Sen, A.K., Conductivity tensor of anisotropic composite media from microstructure, *Journal of Applied Physics*, 67(3), 1145–1155, 1990.

Tsao, G.T., Thermal conductivity of two-phase materials, *Industrial and Engineering Chemistry*, 53(5), 395, 1961.

Vagenas, G.K., Marinos-Kouris, D., and Saravacos, G.D., Thermal properties of raisins, *Journal of Food Engineering*, 11, 147, 1990.

van der Sman, R.G.M., Prediction of enthalpy and thermal conductivity of frozen meat and fish products from composition data, *Journal of Food Engineering*, 84, 400–412, 2008.

Wang, J. and Hayakawa, K., Thermal conductivities of starch gels at high temperatures influenced by moisture, *Journal of Food Science*, 58(4), 884, 1993.

Woodside, W. and Messmer, J.H., Thermal conductivity of porous media. I. Unconsolidated sands, *Journal of Applied Physics*, 32(9), 1688, 1961.

Yagi, S. and Kunii, D., Studies on effective thermal conductivities in packed beds, *AIChE Journal*, 3(3), 373, 1957.

CHAPTER 20

Thermal Diffusivity of Foods: Measurement, Data, and Prediction

Mohammad Shafiur Rahman and Ghalib Said Al-Saidi

CONTENTS

20.1 DEFINITION

Thermal diffusivity indicates how fast heat propagates through a sample while heating or cooling. Thermal diffusivity is a lumped parameter used in the heat transfer calculation by conduction. The rate at which heat diffuses by conduction through a material depends on the thermal diffusivity and can be defined as

$$\alpha = \frac{k}{\rho C_p} \tag{20.1}$$

where
α is the thermal diffusivity (m^2/s)
ρ is the density (kg/m^3)
C_p is the specific heat (J/kg K) at constant pressure
k is the thermal conductivity (W/m·K)

20.2 MEASUREMENT TECHNIQUES

Thermal diffusivity can be determined either by direct experiment or estimated from the thermal conductivity, specific heat, and density data. The estimation of thermal diffusivity can be roughly divided into two groups: (a) direct measurement and (b) indirect prediction.

20.2.1 Direct Measurement

Direct measurement of thermal diffusivity is usually determined from the solution of one-dimensional unsteady-state heat transport equation with the appropriate boundary conditions for finite or infinite bodies by analytical method and for irregular bodies by numerical technique. The analytical solution of unsteady-state heat conduction depends on the relative importance of internal and external heat transfer resistance. The internal and external heat transfer resistance depends on the dimensionless Biot number and can be defined as

$$Bi = \frac{\text{Internal resistance}}{\text{External resistance}} = \frac{hR}{k} \tag{20.2}$$

High Biot number greater than 40 indicates that external resistance is negligible and Biot number less than 0.2 indicates that internal resistance is negligible. Between Biot number 0.2 and 40, both internal and external resistances to heat transfer are important (Singh, 1992). The mathematical expressions for temperature ratio are summarized below for different conditions (Mohsenin, 1980; Singh, 1992):

Negligible internal resistance to heat transfer

$$\frac{T - T_\mathrm{m}}{T_0 - T_\mathrm{m}} = \exp\left(-\frac{hA_\mathrm{s}}{\rho CV}\right)t \tag{20.3}$$

Negligible surface resistance to heat transfer

For infinite slab

$$\frac{T - T_\mathrm{m}}{T_0 - T_\mathrm{m}} = \sum_{n=1}^{\infty} \frac{2}{\beta_n}(-1)^{n+1}\cos\left(\frac{\beta_n r}{R}\right)\exp\left[-\beta_n^2\frac{\alpha t}{R^2}\right] \tag{20.4}$$

where R is the half thickness of a slab and r is the variable distance from the center axis and

$$\beta_n = (2n - 1)\frac{\pi}{2} \tag{20.5}$$

For infinite cylinder,

$$\frac{T - T_\mathrm{m}}{T_0 - T_\mathrm{m}} = \sum_{n=1}^{\infty} \frac{2}{\beta_n J_1(\beta_n)} J_0\left(\frac{\beta_n r}{R}\right)\exp\left[-\beta_n^2\frac{\alpha t}{R^2}\right] \tag{20.6}$$

where R is the radius of a cylinder and r is the variable distance from the center axis and

$$J_0(\beta_n) = 0 \tag{20.7}$$

For infinite sphere

$$\frac{T - T_{\mathrm{m}}}{T_0 - T_{\mathrm{m}}} = \sum_{n=1}^{\infty} 2(-1)^{n+1} \left(\frac{\beta_n r}{R}\right) \sin\left(\frac{\beta_n r}{R}\right) \exp\left[-\beta_n^2 \frac{\alpha t}{R^2}\right] \tag{20.8}$$

where R is the radius of a sphere and r is the variable distance from the center axis.

$$\beta_n = n\pi \tag{20.9}$$

Finite surface and internal resistance to heat transfer

For infinite slab

$$\frac{T - T_{\mathrm{m}}}{T_0 - T_{\mathrm{m}}} = \sum_{n=1}^{\infty} \frac{2 \sin \beta_n}{\beta_n + \sin \beta_n \cos \beta_n} \cos\left(\frac{\beta_n r}{R}\right) \exp\left[-\beta_n^2 \frac{\alpha t}{R^2}\right] \tag{20.10}$$

where R is the half thickness of a slab and r is the variable distance from the center axis. β_n is the root of the equation below

$$\beta_n \tan \beta_n = \frac{hR}{k} \tag{20.11}$$

For infinite cylinder

$$\frac{T - T_{\mathrm{m}}}{T_0 - T_{\mathrm{m}}} = \sum_{n=1}^{\infty} \frac{2(\sin \beta_n - \beta_n \cos \beta_n)}{\beta_n - \sin \beta_n \cos \beta_n} J_0\left(\frac{\beta_n r}{R}\right) \exp\left[-\beta_n^2 \frac{\alpha t}{R^2}\right] \tag{20.12}$$

where R is the radius of a cylinder and r is the variable distance from the center axis.

$$J_0(\beta_n) = 0 \tag{20.13}$$

For infinite sphere

$$\frac{T - T_{\mathrm{m}}}{T_0 - T_{\mathrm{m}}} = \sum_{n=1}^{\infty} \frac{2(\sin \beta_n - \beta_n \cos \beta_n)}{\beta_n - \sin \beta_n \cos \beta_n} \left(\frac{\beta_n r}{R}\right) \exp\left[-\beta_n^2 \frac{\alpha t}{R^2}\right] \tag{20.14}$$

where R is the radius of a sphere and r is the variable distance from the center axis.

β_n is the root of the equation below:

$$\beta_n \cot \beta_n = 1 - \frac{hR}{k} \tag{20.15}$$

The temperature history charts are also used to analyze the data due to the complexity of the series solutions. These charts are presented in Figures 20.1 through 20.3 for infinite slab, cylinder, and sphere.

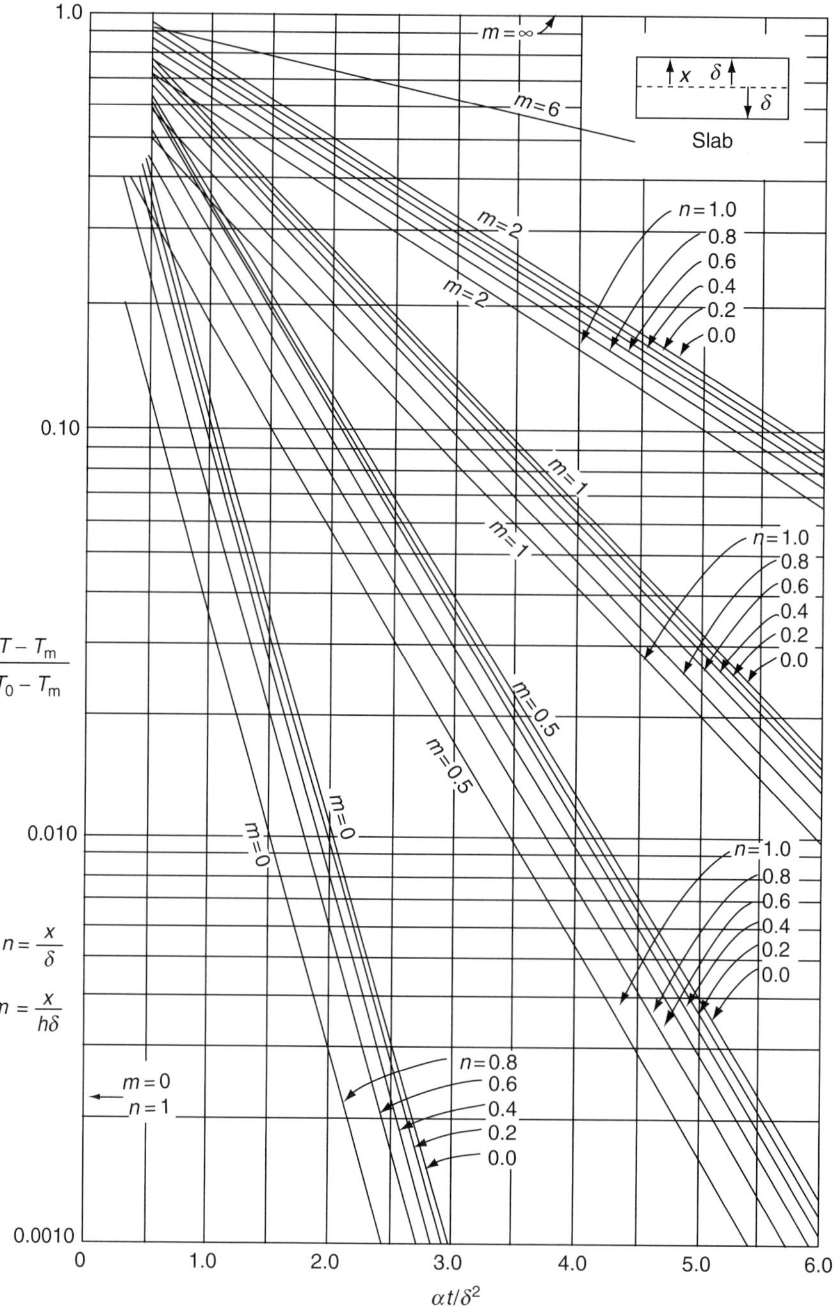

FIGURE 20.1 Unsteady-state temperature distribution in an infinite slab. (From Foust, A.S., Wenzel, L.A., Clump, C.W., Maus, L., and Andersen, L.B., *Principles of Unit Operations*, John Wiley & Sons, New York, 1960.)

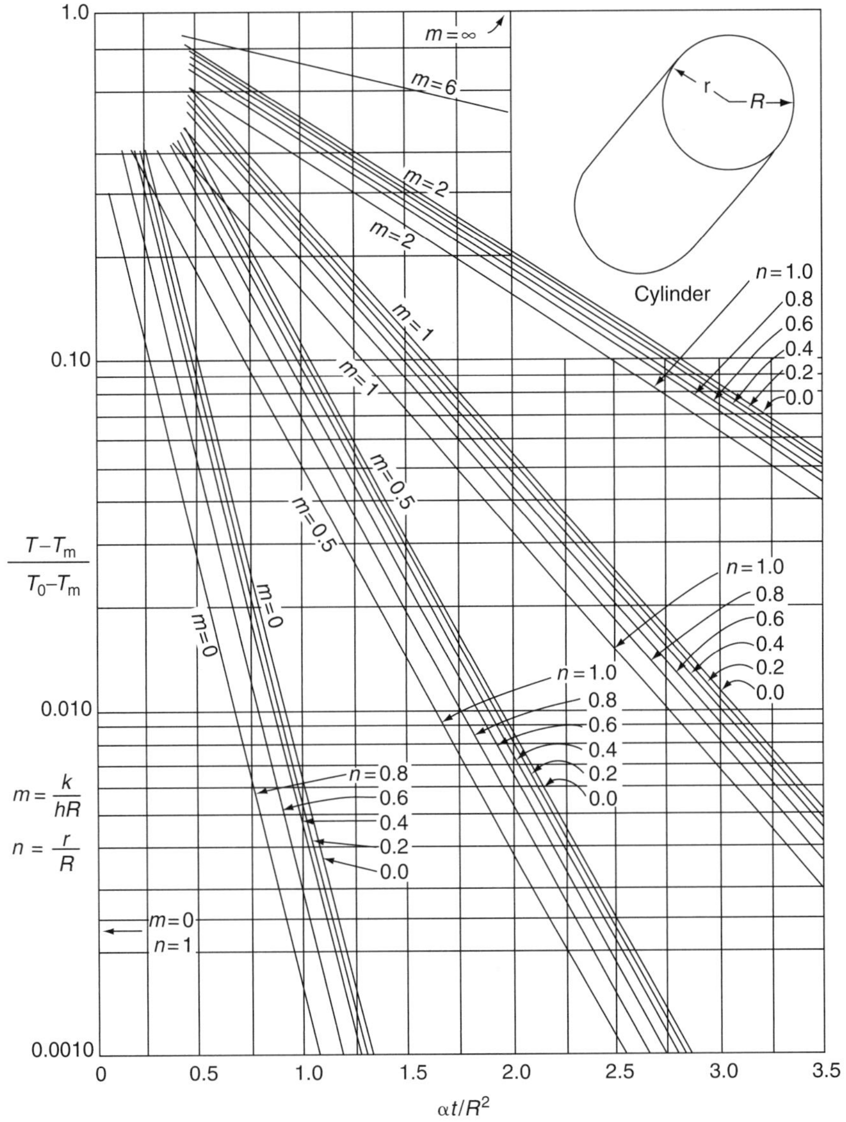

FIGURE 20.2 Unsteady-state temperature distribution in an infinite cylinder. (From Foust, A.S., Wenzel, L.A., Clump, C.W., Maus, L., and Andersen, L.B., *Principles of Unit Operations*, John Wiley & Sons, New York, 1960.)

20.2.1.1 Methods Based on Analytical Solution of Heat Conduction Equation

20.2.1.1.1 Dickerson (1965) Method

Dickerson (1965) described an apparatus to measure thermal diffusivity based on transient heat transfer. The heat transport equation can be written as

$$\frac{\Omega}{\alpha} = \frac{\delta^2 T}{\delta r^2} + \frac{1}{r}\frac{\delta T}{\delta r} \tag{20.16}$$

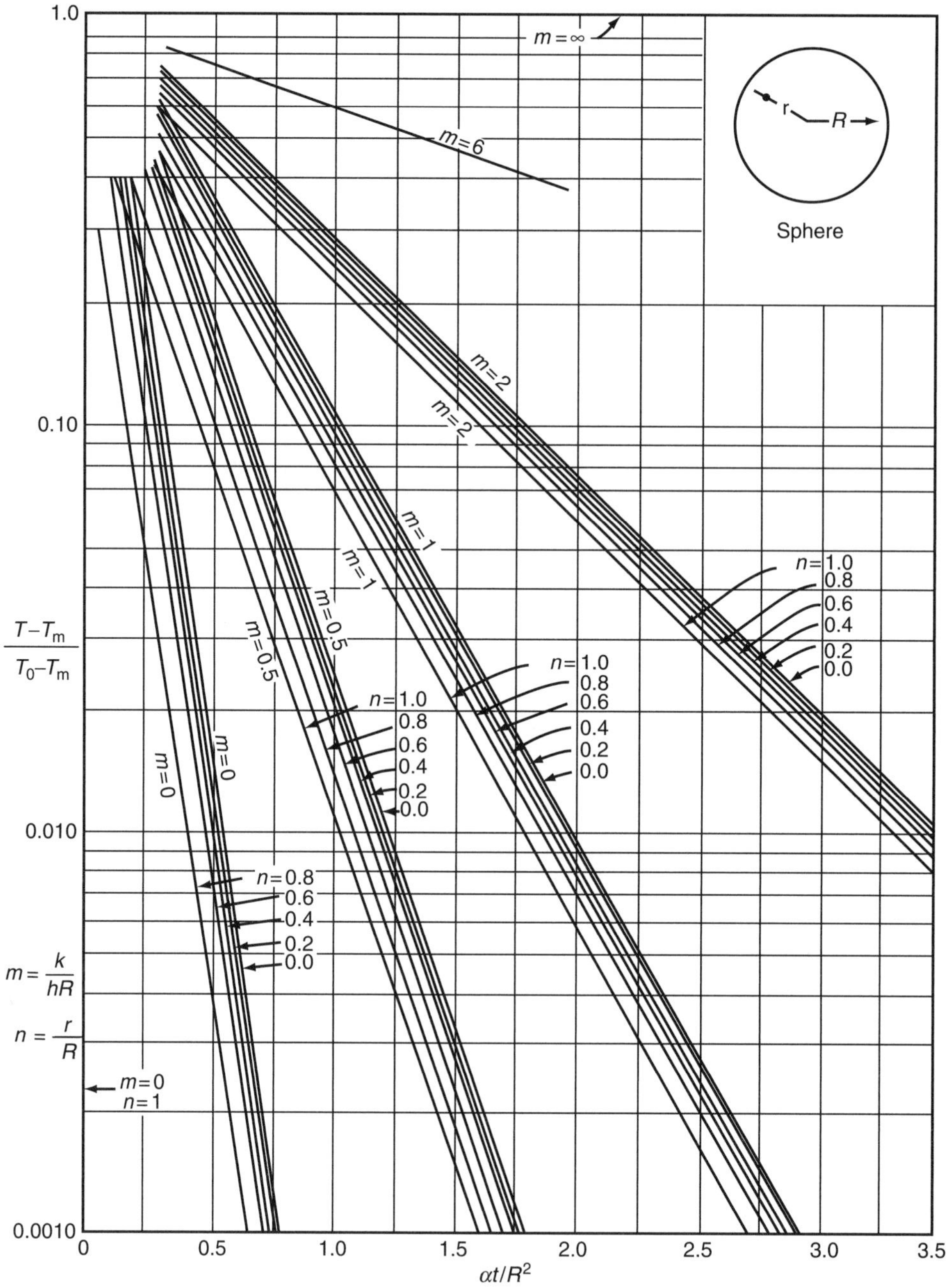

FIGURE 20.3 Unsteady-state temperature distribution in a sphere. (From Foust, A.S., Wenzel, L.A., Clump, C.W., Maus, L., and Andersen, L.B., *Principles of Unit Operations*, John Wiley & Sons, New York, 1960.)

where Ω is the constant rate of temperature rise at all points in the cylinder (°C/s). Equation 20.16 can be solved when the temperature gradient is no longer time dependent as

$$T = \frac{\Omega r^2}{4\alpha} + \omega \ln r + \varphi \tag{20.17}$$

Boundary conditions are

$$T + \Omega t = T_s, \quad t > 0, \, r = R_{sa} \tag{20.18}$$

Boundary conditions are

$$\frac{dT}{dr} = 0, \quad t > 0, \, r = 0 \tag{20.19}$$

The solution of Equation 20.17 with the above boundary conditions is

$$(T_s - T) = \frac{\Omega}{4\alpha}\left(R_{sa}^2 - r^2\right) \tag{20.20}$$

At $r = 0$ and $T = T_c$, thermal diffusivity can be expressed as

$$\alpha = \frac{\Omega R_{sa}^2}{4(T_s - T_c)} \tag{20.21}$$

A plot of sample temperature versus time is shown in Figure 20.4. Dickerson (1965) showed that 95% or more of the maximum temperature difference $(T_s - T_c)$ or the establishment of steady state takes place when $(\alpha t/R_{sa}^2) > 0.55$. Thus, the appropriate radius for the sample cylinder can be determined from the above inequality. Olson and Schultz (1942) have shown that a simulation of the mathematical model of infinite cylinder is possible if the length to diameter ratio (L/D) is greater than 4. The length of the sample holder cylinder must be four times greater than the diameter.

The Dickerson (1965) apparatus consisted of an agitated water bath and a sample holder metal cylinder (Figure 20.5). One thermocouple is soldered to the outside surface of the cylinder to monitor the surface temperature of the sample and another thermocouple probe is inserted at the center of the sample. The sample holder contains two caps at the bottom and a top made of Teflon.

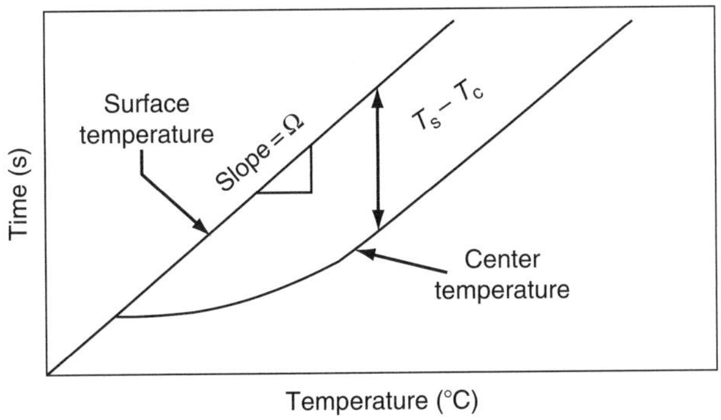

FIGURE 20.4 Temperature versus time plot of Dickerson (1965) method.

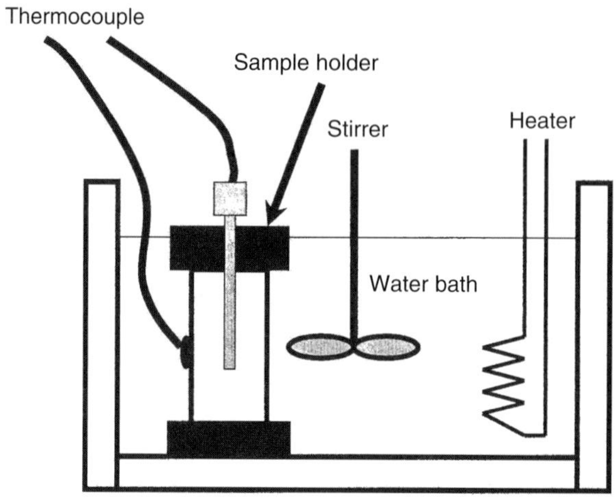

FIGURE 20.5 Thermal diffusivity measurement apparatus as used by Dickerson (1965) method.

The cylinder is then placed in the agitated water bath, and the time and temperature are recorded until a constant rate of temperature rise is obtained for both inner and outer thermocouples (Figure 20.4).

20.2.1.1.2 Hayakawa (1973) Method

If one surface of an infinite slab is insulated and the other surface is subjected to a step change in its surface temperature at the zero time of heating or cooling, then the temperature at any location can be calculated by the following equation:

$$T = 1 - \frac{4}{\pi}\sum_{n=0}^{\infty}\frac{(-1)^n}{(2n+1)}\exp\left[-\frac{\alpha(2n+1)^2\pi^2 t}{4l^2}\right]\left[\cos\frac{(2n+1)\pi\chi}{2l}\right] \tag{20.22}$$

where l is the thickness of the sample. If the temperature of the uninsulated surface T_s changes with time, Hayakawa (1973) derived the following equation for the temperature distribution in the sample using Duhamel theorem:

$$T = T_s + \frac{16l^2}{\alpha\pi^3}\sum_{m=1}^{g} a_m\left[\sum_{m=1}^{\infty} Q_{gn(m-1)} - \sum_{m-1}^{\infty} Q_{gnm}\right] \tag{20.23}$$

where

$$Q_{gnm} = (-1)^{n+1}\left[\frac{\cos(2n-1)\pi x}{2l}\right]\left[\frac{1}{(2n-1)^3}\right]\exp[-Y_n(g-m)\Delta t] \tag{20.24}$$

$$Y_n = \frac{\alpha(2n-1)^2\pi^2}{2l^2} \tag{20.25}$$

where

α_m is the slope of mth line segments (°C/s)

g is the number of line segments with which a temperature history curve on the uninsulated surface of sample material is approximated

Temperatures of the sample material are monitored experimentally at its internal locations and on its exposed surface. These temperatures are recorded at uniform time intervals ($m\Delta t$: $m = 1,2, \ldots, g$). Equation 20.23 is used to calculate theoretical temperatures at the same time intervals to compare them with recorded temperatures. For this calculation, experimental temperatures on the exposed surface are used with various assumed thermal diffusivity values. For each value assumed, the sum of squares of differences between experimental and theoretical temperatures is calculated. The diffusivity value of the sample material is determined by calculating a value at which the sum of squares of differences becomes minimal. The advantage of the procedure is that a food sample can be exposed to any time-variable heating or cooling temperatures during the experimentation.

20.2.1.1.3 Nordon and Bainbridge (1979) Method

When a cylindrical sample is transferred from a bath at temperature (T_0) to another temperature (T_{m}), the temperature at the center of the cylinder changes according to the equation (Nordon and Bainbridge, 1979):

$$\frac{T - T_0}{T_{\mathrm{m}} - T_0} = 1 - \frac{8}{\pi} \sum_{n=0}^{\infty} \sum_{m=1}^{\infty} \left[\frac{(-1)^2}{2n+1} \right] \frac{\beta_m^2 + \left[\frac{\pi(2n+1)R}{L} \right]^2}{\beta_m J_1(\beta_m)} \exp\left[\frac{\alpha t}{R^2} \right] \tag{20.26}$$

Where β_m is the root of the following equation:

$$J_0(\beta_m) = 0 \tag{20.27}$$

Theoretical values of temperature ratio can be plotted as a dimensional variable $\alpha t/R^2$. Nordon and Bainbridge (1979) mentioned that when the shape of the theoretical and experimental temperature ratio curves is identical, then a single point from the theoretical curve can be considered to calculate the thermal diffusivity. A typical theoretical plot is shown in Figure 20.6. At temperature ratio 0.5, the value of $\alpha t/R^2$ is known. When the experimental cooling time reaches the temperature ratio at 0.5, then thermal diffusivity can be calculated from $\alpha t/R^2$ by knowing the time, t.

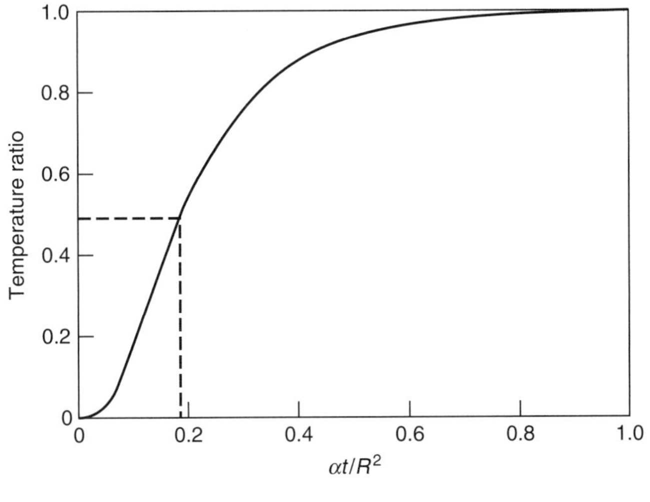

FIGURE 20.6 Computer-generated plot of normalized temperature against cd/R2. (From Jones, J.C., Wootton, M., and Green,S., *Food Austr.*, 44, 501, 1992.)

20.2.1.1.4 Olivares, Guzman, and Solar (1986) Method

Olivares et al. (1986) used the heat conduction equation for a finite cylinder. The series equation is rapidly convergent for a Fourier number greater than 0.2 (Singh, 1982), and it is possible to consider only the first two terms with accurate results (Yanez, 1983) after some time has elapsed. The resultant equation is

$$\frac{T_s - T}{T_s - T_0} = \sum_{m=1}^{2} \sum_{n=1}^{2} \frac{2(-1)^{m+1}}{\beta_m} \cos\left(\beta_m \frac{2z}{L}\right) \frac{2J_0(\beta_n r/R)}{\beta_n J_1(\beta_n)} \exp\left[-\left(\frac{\beta_n^2}{R^2} + \frac{4\beta_m^2}{L^2}\right)\alpha t\right] \quad (20.28)$$

where β_n and β_m are the roots of the cosine and Bessel functions. The values of T and T_s can be obtained experimentally by placing thermocouples at the center and outside of the can. To obtain an average thermal diffusivity, the experimental values and predicted ones for T are compared for some arbitrarily selected value of thermal diffusivity. Using an iterative technique, the values of thermal diffusivity is changed until the sum of the difference is minimized.

20.2.1.1.5 Gordon and Thorne (1990) Method

The differential equation for heat conduction in a sphere suddenly transferred to a different constant temperature surrounding is

$$\frac{\delta^2\theta}{\delta\eta^2} + \frac{2}{\eta}\frac{\delta\theta}{\delta\eta} = \frac{\delta\theta}{\delta F_0} \quad (20.29)$$

$$\theta = \frac{T - T_m}{T_0 - T_m}; \quad \eta = r/R; \quad F_0 = (\alpha t/R^2)$$

For $0 \leq \eta \leq 1$ and $F_0 \geq 0$, the solution is

$$\theta = \sum_{n=1}^{\infty} \left[\frac{2Bi \sin(\beta_n \eta)}{\eta(\beta_n^2 + Bi) \sin \beta_n} \exp(-\beta_n^2 F_0)\right] \quad (20.30)$$

where Bi is the Biot number (hR/k) and β_n is the nth root of the transcendental equation as below:

$$\beta_n \cot \beta_n = 1 - Bi \quad (20.31)$$

The center temperature ($\eta = 0$) can be written as

$$\theta_c = \sum_{n=1}^{\infty} \left[\frac{2Bi\, \beta_n}{(\beta_n^2 + Bi^2 - Bi) \sin \beta_n} \exp(-\beta_n^2 F_0)\right] \quad (20.32)$$

Gordon and Thorne (1990) provided two methods for estimating the thermal diffusivity: (i) slope method and (ii) lag method.

20.2.1.1.5.1 Slope Method—Considering a single term, Equation 20.32 can be transformed to

$$\ln \theta_c = \ln I - \frac{\beta_1^2 \alpha t}{R^2} \quad (20.33)$$

where

$$I = \frac{2Bi\beta_1}{(\beta_1^2 + Bi^2 - Bi)\sin\beta_1} \tag{20.34}$$

Thus, the logarithm of θ_c is a linear function of time and thermal diffusivity can be determined from the slope of the Equation 20.33 by knowing the first root of Equation 20.31 from the intercept. When $\theta \leq 0.4$, the first root was found to be the only significant contributing root of Equation 20.31 and use of a single root is justified. From the plot of $\ln \theta_c$ versus t, the portion of the curve satisfying the above criterion is used to determine the intercept I. Then, the value of β_1 can be determined from Equation 20.34 by binary chop iteration.

20.2.1.1.5.2 Lag Method—If a sphere at uniform temperature is transferred to a different, lower constant temperature, and temperatures at two points, the center ($\eta = 0$) and half way from the center to the surface ($\eta = 0.5$), are recorded, the temperature difference between the two points would increase with time from an initial value of zero to a maximum and then tend to zero again as the temperature within the solid tended toward that of the cooling medium. Figure 20.7 shows typical temperature profiles for the two points. The time t_{max} corresponding to the maximum temperature difference (Figure 20.7) is that period when the tangents to the two curves are parallel. The time lag (t_L) shown in Figure 20.7 is inversely related to thermal diffusivity. If α is large, heat will pass rapidly through the food and t_L will be small, and vice versa. A regression correlation was developed using random data (Gordon and Thorne, 1990):

$$t_L = \frac{3.89 \times 10^{-5}}{\alpha} - 5.3471 \tag{20.35}$$

A vigorously stirred water bath of 22 L capacity at 4°C ± 1°C was used by Gordon and Thorne (1990) to effect a high heat transfer rate ($h = 300\ W/m^2\ K$) (Figure 20.8). The ratio of water bath volume to sphere volume (200:1) satisfies the boundary conditions of a constant ambient temperature (±0.2°C). A jig can be used to locate thermocouples within the samples, which are suspended in a water bath while temperature and time are recorded. To submerge the fruits and vegetables, a piece of glass capillary tubing is inserted through the base and a piece of nylon thread tied to a lead weight is passed through this (Figure 20.8). The experimental conditions are $h = 300\ W/(m^2\ K)$, $T_0 = 20$°C, and $T_m = 40$°C. For values of thermal diffusivity between 0.85×10^{-7} and $3.23 \times 10^{-7}\ m^2/s$, a range that includes almost

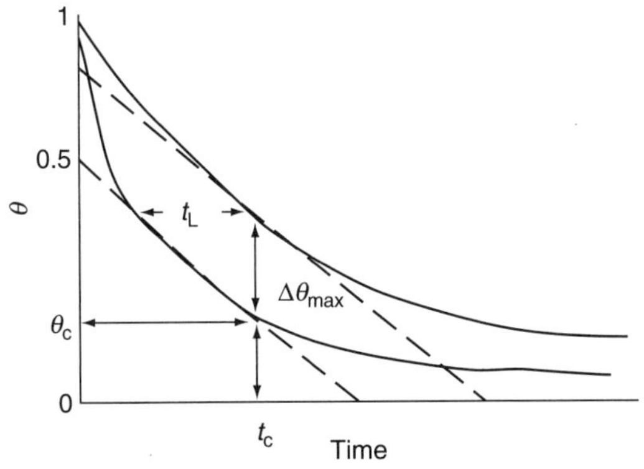

FIGURE 20.7 Estimation ohime lag from θ versus time plot. (From Gordon, C. and Thorne, S., *J. Food Eng.*, 11, 133, 1990. With permission.)

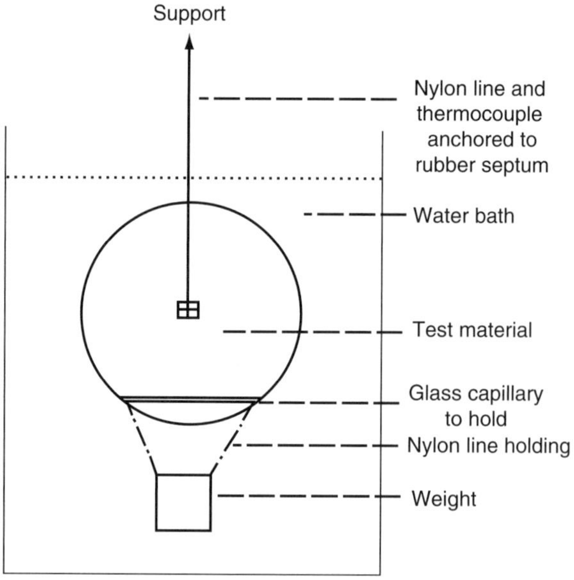

FIGURE 20.8 Sample arrangement to measure thermal diffusivity as used by Gordon and Thorne (1990). (From Gordon, C. and Thorne, S., *J. Food Eng.*, 11, 133, 1990. With permission.)

all foodstuffs, the mean and maximum errors were 1.3% and 4% by the slope method and 1.2% and 5.3% by the lag method. The slope method offers the advantage that it requires only temperature measurements at a single point at the center. For typical foods, an error of 3 mm at the center results in a maximum error of 0.8% in temperature measurement. For a thermocouple located at $\eta = 0.5$, the effect of location errors are much greater, particularly if the location error is toward the center; for example, the error in temperature measurement is 5.4% if the thermocouple is inserted 2 mm too far in.

20.2.1.1.6 Moore and Bilanski (1992) Method

The time–temperature history of a sample was measured using the plane heat source method as described by Mohsenin (1980). The construction of the device is shown in Figure 20.9. In this

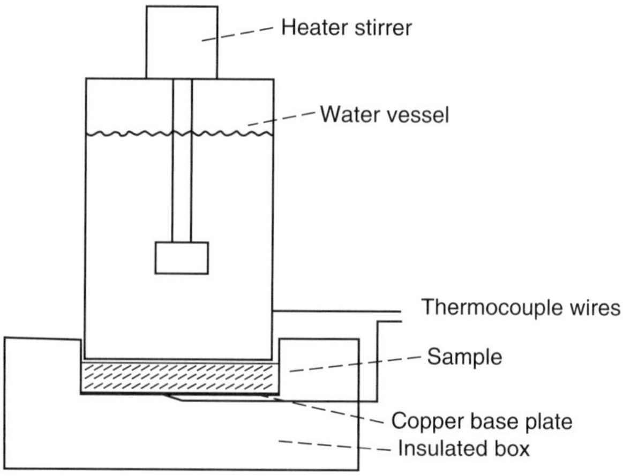

FIGURE 20.9 Thermal diffusivity measurement apparatus as used by Moore and Bilanski (1992). (From Moore, G.A. and Bilanski, W.K., *Appl. Eng. Agric.*, 8, 61, 1992.)

method, a slab of material is insulated on one plane and is subjected to an isothermal heat source on a parallel plane. To simulate a semi-infinite slab, the edges of the slab are insulated. One thermocouple is attached centrally to the underside of the water vessel so that it is in direct contact with the heat source and the sample. A second thermocouple is soldered to the underside of the 0.2 mm thick copper base plate. The heat source is maintained at approximately 80°C while the sample is placed in the cell at ambient temperature (approximately 20°C). Bulk density is controlled by placing a known mass evenly in the cell and maintaining the heat source at a known distance above the base of the cell.

The solution of the heat conduction equation to give the temperature at any time and position within the material undergoing heating or cooling depends on the boundary conditions. When a slab of material initially with uniform temperature T_0 and the heat source are placed on the face of the sample, the temperature in the slab can be written as (considering only the first term of the series solution)

$$\frac{T_m - T}{T_m - T_0} + \frac{4}{\pi}\left[\sin\left(\frac{\pi x}{2l}\right)\right] \times \left[\exp\left(-\frac{\pi^2 \alpha t}{4l^2}\right)\right] \tag{20.36}$$

The assumptions made in the derivation of this equation are (1) the material is homogeneous and of constant thermal properties, (2) there is no surface thermal resistance, i.e., an infinite heat transfer coefficient at the surface. An error of 0.5 mm in distance would lead to an error in calculation of thermal diffusivity of about 12% at a thickness of sample of 8 mm and 5% when the sample is 20 mm thick. The temperature rise was recorded until the base of the sample reached 60°C. For times greater than 100 s, the error occurred by ignoring higher terms less than 1% (Moore and Bilanski, 1992). In this case, thermal diffusivity can be calculated from the slope of the plot $\ln(T_m - T)/(T_m - T_0)$ versus time.

20.2.1.2 Method Based on Analytical Solution, *j* and *f* Factors

20.2.1.2.1 Bhowmik and Hayakawa (1979) Method

The heat conduction equation for an infinite cylinder whose surface is subject to convective heat exchange is obtained from an equation given by Carslaw and Jaeger (1959) as follows:

$$\frac{T_m - T}{T_m - T_0} = 2Bi\sum_{n=1}^{\infty}\left[\exp\left(-\alpha\beta_n^2 t/R^2\right)\right] \times \left[\frac{J_0(\beta_n r/R)}{(Bi^2 + \beta_n^2)J_1(\beta_n)}\right] \tag{20.37}$$

where β_n is the nth positive root of the equation.

$$Bi\ J_0(\beta_n) - \beta_n J_1(\beta_n) = 0 \tag{20.38}$$

The summation series may be approximated with the first term as

$$\frac{T_m - T}{T_m - T_0} = 2Bi\sum_{n=1}^{\infty}\left[\exp\left(-\alpha\beta_1^2 t/R^2\right)\right] \times \left[\frac{J_0(\beta_n r/R)}{(Bi^2 + \beta_1^2)J_1(\beta_n)}\right] \tag{20.39}$$

The temperatures at the center and surface of the cylinder can be estimated from the following equations:

$$\frac{T_m - T_c}{T_m - T_0} = 2Bi \sum_{n=1}^{\infty} \left[\exp\left(-\alpha\beta_1^2 t/R^2\right)\right] \times \left[\frac{1}{\left(Bi^2 + \beta_1^2\right)J_1(\beta_n)}\right] \tag{20.40}$$

$$\frac{T_m - T_s}{T_m - T_0} = 2Bi \sum_{n=1}^{\infty} \left[\exp\left(-\alpha\beta_1^2 t/R^2\right)\right] \times \left[\frac{J_0(\beta_1)}{\left(Bi^2 + \beta_1^2\right)J_1(\beta_n)}\right] \tag{20.41}$$

By taking the ratio of Equations 20.40 and 20.41, one can express

$$\Gamma_{cs} = \frac{T_m - T_c}{T_m - T_s} = \frac{1}{J_0(\beta_1)} \tag{20.42}$$

The constant value of Equation 20.42 can be obtained by plotting $(T_m - T_c)/(T_m - T_s)$ against the heating or cooling time of food on a rectangular graph paper (Figure 20.10). The values of β_1 may be easily determined from this constant value. According to Pflug et al. (1965), there exists a relationship between β_1 and thermal diffusivity α given by

$$\frac{f\alpha}{R^2} = \frac{2.303}{\beta_1^2} \tag{20.43}$$

where f is the slope index of the heating or cooling curve of the sample food in seconds. The values of f can be obtained from the semilogarithmic plotting of $(T_0 - T_c)$ versus t. A typical plot is shown in Figure 20.11 from experimental heat penetration data. Then α can be determined using Equation 20.43 provided f value is given. Bhowmik and Hayakawa (1979) used this method to measure the thermal diffusivity of different materials and found that the error varied from 1% to 7%.

Later Hayakawa and Succar (1983) used this technique for spherical food materials. When F_0 is moderately large, the analytical solution for transient state heat conduction in a sphere is (Carslaw and Jaeger, 1959)

$$\frac{T_m - T}{T_m - T_0} = \frac{2Bi}{r/R}\left[\left(-\beta_1^2 F_0\right)\right]\frac{\beta_1^2 + (\beta_1 - 1)^2}{\beta_1^2\left[\beta_1^2 + Bi(Bi - 1)\right]}(\sin\beta_1)\left[\sin\left(\frac{\beta_1 r}{R}\right)\right] \tag{20.44}$$

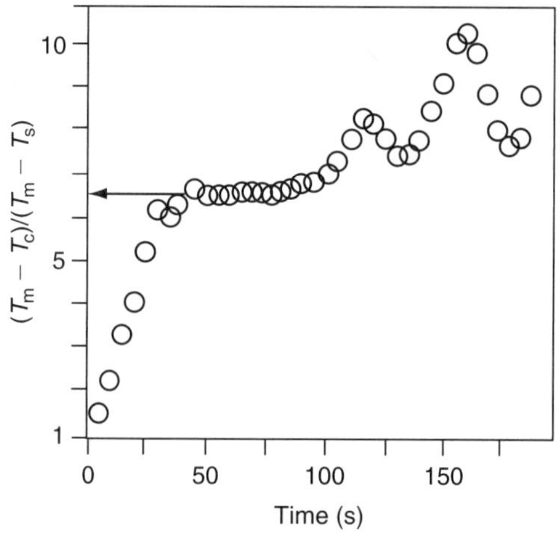

FIGURE 20.10 Graph of $(T_m - T_c)/(T_m - T_s)$ versus time for cherry tomato pulp. (From Bhowmik, S.R. and Hayakawa, K., *J. Food Sci.*, 44, 469, 1979.)

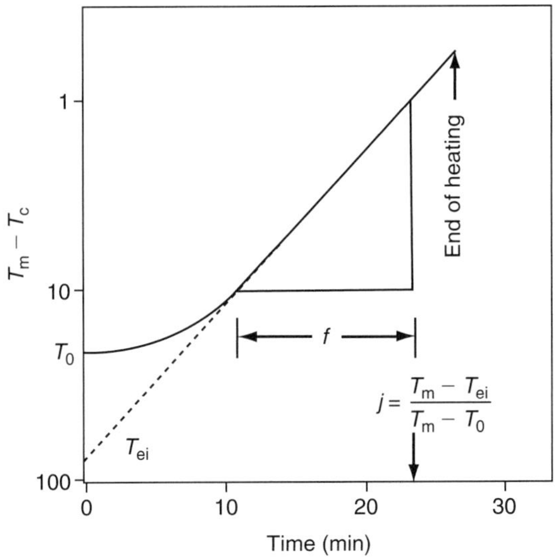

FIGURE 20.11 Typical heat penetration curve describing *f* and *j*.

where β_1 is the first positive root of the following equation:

$$\beta_1 \cot \beta_1 = 1 - Bi \tag{20.45}$$

The values of $(T_m - T)/(T_m - T_0)$ can be related to Ball's *f* and *j* values as (Ball and Olson, 1957)

$$\frac{T_m - T}{T_m - T_0} = j[10]^{t/f}, \quad t > t_1 \tag{20.46}$$

Combining Equations 20.44 and 20.46, one can express

$$\alpha = \frac{R^2 \ln 10}{\beta_1^2 f} \tag{20.47}$$

The ratio of temperatures at two locations in the sample can be written from Equation 20.44 as

$$\Gamma_{12} = \frac{T_m - T_1}{T_m - T_2} = \frac{(r_2/R)\sin(\beta_1 r_1/R)}{(r_1/R)\sin(\beta_2 r_2/R)} \tag{20.48}$$

The constant value of *r* can be determined from the plot of Γ versus *t* (Figure 20.12). The values of β_1 can be estimated from the Equation 20.48. When β_1 and *f* are known, thermal diffusivity can be calculated from Equation 20.47.

20.2.1.2.2 Uno and Hayakawa (1980) Method

Uno and Hayakawa (1980) used an analytical solution of heat conduction in a finite cylinder for the experimental determination of thermal diffusivity. This solution was derived by assuming that surface heat transfer conductances at its top, bottom, and side surfaces were all finite and different from each other. A first-term approximation of the series solution and an empirical *f* value were used

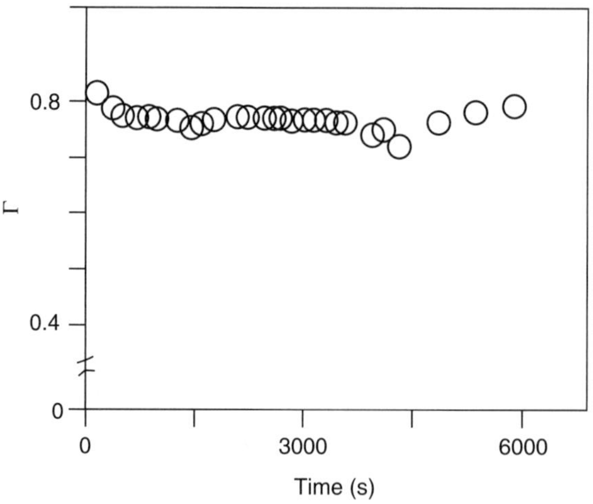

FIGURE 20.12 Plot of temperature ratio against time for a 26.7°C fresh tomato heated in a 42.9°C chamber. (From Hayakawa, K. and Succar, J., *Food Sci. Technol.*, 16, 373, 1983.)

to develop this method. Temperature distributions in a cylindrical can of heat conduction food may be estimated by the following equation (Carslaw and Jaeger, 1959):

$$\theta = \frac{T - T_{\mathrm{m}}}{T - T_{\mathrm{m}}} = \sum_{m=1}^{\infty} \sum_{n=1}^{\infty} G_m H_n \exp\left(B_{mn}\tau\right) \tag{20.49}$$

where

$$G_m = \frac{2Bi_{\mathrm{si}} J_0(p_m \eta)}{\left(Bi_{\mathrm{si}}^2 + p_m^2\right) J_0(p_m)} \tag{20.50}$$

$$H_n = \frac{2\left(q_n^2 + Bi_{\mathrm{b}}\right)\left[\sin q_n = \left(Bi_{\mathrm{t}}(\cos q_n)/q_n\right) + \left(Bi_{\mathrm{t}}/q_n\right)\right]}{q_n \cos\left(q_n \xi\right) + Bi_{\mathrm{b}} \sin\left(q_n \xi\right)} \tag{20.51}$$

$$B_{mn} = -\left(p_m^2 S^2 + q_n^2\right) \tag{20.52}$$

where

τ t/L^2, L is the height of the cylinder (m)
Bi_{si} is the side Biot number ($h_{\mathrm{si}}R/k$)
Bi_{t} is the top Biot number ($h_{\mathrm{t}}L/k$)
Bi_{b} is the bottom Biot number ($h_{\mathrm{b}}L/k$)
η is r/R
ξ is z/L
p_m and q_n are the positive roots of the following characteristic equations for an infinite cylinder and an infinite slab

$$Bi_{\mathrm{si}} J_0(p_1) = p_1 J_1(p_1) \tag{20.53}$$

$$\left(q_1^2 - Bi_{\mathrm{t}} Bi_{\mathrm{b}}\right) \sin q_1 = \left(Bi_{\mathrm{t}} + Bi_{\mathrm{b}}\right) q_1 \cos q_1 \tag{20.54}$$

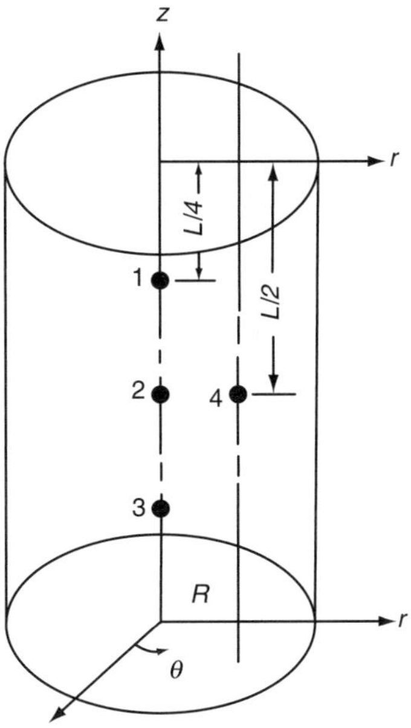

FIGURE 20.13 Locations of the thermocouples in the can for Uno and Hayakawa (1980) method. Points 1, 2, 3, and 4 are the locations of the thermocouples.

For a moderately long heat exchange time, Equation 20.49 may be approximated by the first term of the infinite series. Temperatures at four locations in the can are measured (Figure 20.13): (1) T_1 along the central axis of the cylinder and also are apart about one-quarter of the total height of the can content, (2) T_2 at the center of the cylinder, (3) T_3 below T_2 along the central axis, and (4) T_4 is positioned away from the central axis of the cylinder and its axial distance measured from the top surface of the cylinder is identical to that of location T_2. The ratios of dimensionless temperatures at two locations may be given by the following expressions after a moderately long heat exchange time:

$$\Gamma_{12} = \frac{T_m - T_1}{T_m - T_2} = \frac{q_1 \cos(q_1\xi_1) + Bi_t \sin(q_1\xi_1) J_0(p_1\eta_1)}{q_1 \cos(q_1\xi_2) + Bi_t \sin(q_1\xi_2) J_0(p_1\eta_2)} \quad (20.55)$$

$$\Gamma_{23} = \frac{T_m - T_2}{T_m - T_3} = \frac{q_1 \cos(q_1\xi_2) + Bi_t \sin(q_1\xi_2) J_0(p_1\eta_2)}{q_1 \cos(q_1\xi_3) + Bi_t \sin(q_1\xi_3) J_0(p_1\eta_3)} \quad (20.56)$$

$$\Gamma_{34} = \frac{T_m - T_2}{T_m - T_4} = \frac{J_0(p_1\eta_2)}{J_0(p_1\eta_4)} \quad (20.57)$$

where p_1 and q_1 are the first positive roots of Equations 20.53 and 20.54, respectively. By plotting Γ_{12}, Γ_{23}, and Γ_{34} against time, reasonably stable constant values of Γ_{12}, Γ_{23}, and Γ_{34} can be obtained (Figure 20.14). The values of p_1 may be obtained by trial and error by substituting Γ_{34} into Equation 20.57. By substituting this p_1 value into Equation 20.53, Bi_{si} can be obtained. From Equations 20.55 and 20.56, an equation is derived by eliminating Bi_t. This equation is used to determine a q_1 value through iterative calculations. Experimental parameters f and j are used in temperature estimation after a relatively long exposure time:

$$\frac{T_m - T}{T_m - T_0} = j[10]^{t/f} \quad (20.58)$$

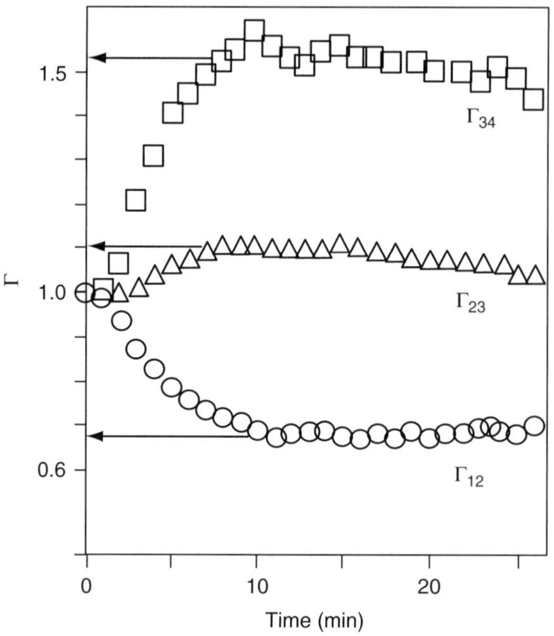

FIGURE 20.14 Temperature ratios as a function of time in a can. (From Uno, J. and Hayakawa, K., *J. Food Sci.*, 45, 692, 1980.)

Combining Equation 20.58 and first-term approximation of Equation 20.49, the value of thermal diffusivity may be obtained by

$$\alpha = \left[\frac{L^2}{f}\right] \times \left[\frac{\ln 10}{p_1^2 S^2 + q_1^2}\right] \tag{20.59}$$

where S is the shape factor of the can (L/R).

20.2.1.2.3 Singh (1982) Method

For a finite cylinder at a uniform initial temperature, exposed to a constant temperature environment and having negligible surface convective resistance, the solution of the heat conduction equation is as follows (Carslaw and Jaeger, 1959):

$$\frac{T_s - T}{T_s - T_0} = \sum_{m=1}^{\infty} \sum_{n=1}^{\infty} \frac{2(-1)^{m+1}}{\beta_m} \cos\left(\beta_m 2z/L\right) \frac{2J_0(\beta_n r/R)}{\beta_n J_1(\beta_n)} \exp\left[-\left(\frac{\beta_n^2}{R^2} + \frac{4\beta_m^2}{L^2}\right)\alpha t\right] \tag{20.60}$$

where

J_0 and h are the Bessel function of the first kind of order zero and one
R and L are the radius and length of the cylinder (m), respectively
T_s and T_0 are the surrounding and initial temperature (°C), respectively
β_m and β_n are the root of cosine and Bessel functions, respectively
z is the axis from the center along the length of the cylinder

For a long time, the first term of the series solution alone may be sufficient. For a finite cylindrical object, the solution expressed by Equation 20.60 can be simplified to retain only the terms with $m=n=1$; i.e., $\beta_m=\pi/2$, $\beta_n=2.4048$, and $J_1\,(2.4048)=0.5191$ (Appendix C). At the center of the cylindrical object, $z=0$, $r=0$, and $J_0(0)=1.0$. Thus, the approximation for long times can be expressed as

$$\frac{T_s - T}{T_s - T_0} = 2.0396\exp\left[-\left(\frac{(2.4048)^2}{R^2}+\frac{\pi^2}{L^2}\right)\alpha t\right] \tag{20.61}$$

Ball and Olson (1957) plotted experimental heat penetration curves on a semilogarithmic expression:

$$t = f\log\left(j\frac{T_s - T_0}{T_s - T}\right) \tag{20.62}$$

where

$$j = \frac{T_s - T_{ei}}{T_s - T_0} \tag{20.63}$$

where T_{ei} is the extrapolated initial temperature obtained by linearizing the entire heating curve (°C). Combining Equations 20.61 and 20.62, thermal diffusivity can be written as

$$\alpha = \frac{2.303}{\left[\left((2.4048)^2/R^2\right)+\left(\pi^2/L^2\right)\right]f} \tag{20.64}$$

Equation 20.64 allows the determination of thermal diffusivity if the heating rate parameter f (slope index of heating or cooling curve) is obtained from a heat penetration study (Figure 20.11).

20.2.1.2.4 Poulsen (1982) Method

Ball and Olson (1957) developed the relation between temperature, time, thermal diffusivity, and geometry of the bodies during heating or cooling. Time–temperature relation for an infinite cylinder with high surface heat transfer coefficient can be written as

$$t = 0.398\frac{R^2}{\alpha}\log\left[1.6\frac{T_m - T_0}{T_m - T}\right] \tag{20.65}$$

Thermal diffusivity can be calculated from Equations 20.62 and 20.65 as

$$\alpha = \frac{0.398R^2}{f} \tag{20.66}$$

In the graphical solutions of the heat equation by Gurney and Lurie (1923), straight lines give the relation between $\log[(T_m - T)/(T_m - T_0)]$ and $\alpha t/R^2$. The straight lines lie approximately between the following range:

$$0 < \frac{T_m - T}{T_m - T_0} < 0.4 \tag{20.67}$$

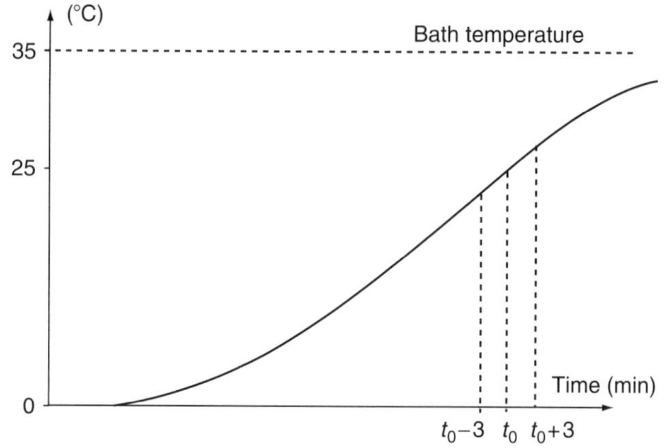

FIGURE 20.15 Curve representing temperature versus time. (From Poulsen, K.P., *J. Food Eng.*, 1, 115, 1982. With permission.)

Poulsen (1982) mentioned that temperature data can be used when 60% of the temperature interval is plotted. An example of a heating curve is presented in Figure 20.15. In this case, the thermal diffusivity was required at 25°C. The starting temperature of the cylinder was 0°C, and the temperature of the bath was 35°C. The values of f can be calculated from Equation 20.68 by taking time–temperature values T_1 and T_2 about 3 min before and after the temperature of 25°C is reached:

$$f = \frac{\log(T_{\mathrm{m}} - T_{\mathrm{c1}}) - \log(T_{\mathrm{m}} - T_{\mathrm{c2}})}{t_2 - t_1} \tag{20.68}$$

Poulsen (1982) used a thermal diffusivity cell that consists of a brass tube with a height of 20 cm and a diameter of 4.8 cm to simulate the infinite cylinder. The end caps are made of Teflon, which has low thermal diffusivity. Two thermocouples are used, one for measuring the temperature at the center of the diffusivity tube T_{c} and one for measuring the temperature of the water bath T_{m}. When f is known, thermal diffusivity can be estimated using Equation 20.66.

20.2.1.3 Nesvadba (1982) Method

One-dimensional heat conduction equation in a slab is

$$\rho(T)C(T)\frac{\delta T}{\delta t} = \frac{\delta}{\delta x}\left[k(T)\frac{\delta T}{\delta x}\right] \tag{20.69}$$

Usually, the above equation is solved with the boundary and initial conditions assuming that thermal properties do not change with temperature. However, Nesvadba (1982) considered the dependence of thermal properties on temperature in his method. The right-hand side of Equation 20.69 can be broken into

$$\frac{\delta}{\delta x}k(T)\frac{\delta T}{\delta x} = \frac{\mathrm{d}k}{\mathrm{d}T}\left(\frac{\delta T}{\delta x}\right)^2 + k(T)\frac{\delta^2 T}{\delta x^2} \tag{20.70}$$

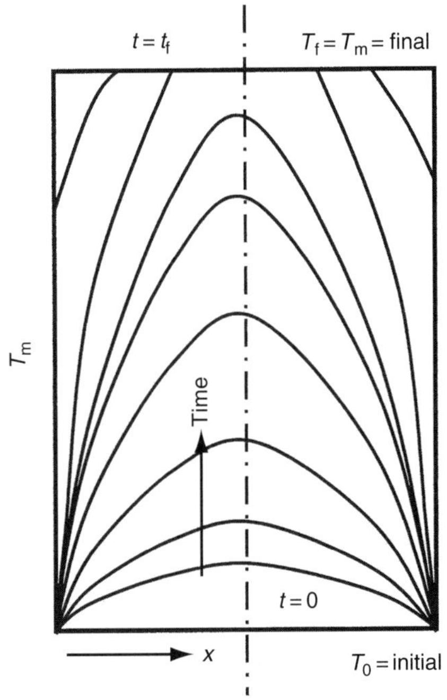

FIGURE 20.16 Variation of slab temperature as a function of time and location. (From Nesvadba, P., *J. Phys. D: Appl. Phys.*, 15, 725, 1982.)

One-dimensional transient temperature distribution $T(x, t)$ generated in the sample is measured at 20 positions at discrete times t_k. Temperature distribution in the slab has a maximum point at the slab center. A typical plot is shown in Figure 20.16. At the maximum point, the thermal diffusivity can be calculated from the equation as follows:

$$\alpha(T_{\max}) = \left(\frac{\delta T/\delta t}{\delta^2 T/\delta x^2}\right)_{x\,\max,\ t\,\max} \tag{20.71}$$

Using least squares cubic spline approximations to $T(x, t)$, the derivatives at the extrema can be evaluated. The advantages of the methods are (1) the approximating assumption of constancy of thermal properties over a small temperature is not made, (2) it enables data to be gathered in a single experiment of short duration and thus is considerably faster than the conventional methods, and (3) it uses thermal regimes that are considerably less restrictive than those of conventional methods (Nesvadba, 1982).

20.2.1.4 Numerical Method

Irregular shapes of food materials are common. Analytical techniques can be used only in the case of regular bodies. Thus, numerical technique must be used to solve the heat conduction equation with appropriate boundary conditions. Matthews and Hall (1968) used the finite difference method to determine the thermal diffusivity of potatoes during heating.

20.2.1.5 Pulse Method

Andrieu et al. (1986) used the pulse method to measure the thermal diffusivity of food materials. The three main advantages of the pulse method are (1) short experimental times, (2) low thermal perturbation, and (3) large range of application (temperature and composition). The unidirectional unsteady-state heat conduction Equation 20.69 for homogeneous materials, when thermal properties are independent of temperature, can be solved by appropriate boundary conditions. If radiant and convective heat losses on the front and on the rear face are characterized, respectively, by the equivalent heat transfer coefficients h_1 and h_2, the initial and boundary conditions can be written as follows.

Initial conditions:

$$t = 0, \quad 0 < x < \phi, \quad T = T_0 = \frac{e}{C_p \rho \phi} \tag{20.72}$$

$$\phi < x < l, \quad T = T_0 \tag{20.73}$$

Boundary conditions:

$$x = 0, \quad t > 0, \quad k\frac{dT}{dx} = h_1 T \tag{20.74}$$

$$x = l, \quad t > 0, \quad k\frac{dT}{dx} = h_2 T \tag{20.75}$$

where

l and ϕ are the sample thickness (m) and penetration depth, respectively, of pulse energy in the front face at t

e is the pulse energy by unit surface area (J/m^2)

The analytical solutions for different conditions are given by Andrieu et al. (1986):

1. $Bi_1 = Bi_2 = 0$ (i.e., no heat losses). The temperature evolution on the back face ($x = l$) is given by

$$T(l, T) = T_a\left[1 + \sum_{n=1}^{\infty} \cos(n\pi) \exp(-n^2\pi^2 t^*)\right] \tag{20.76}$$

where $T_a = e/(C_p \rho l)$ is the adiabatic temperature, $t^* = t/\mu$ is the reduced time, and μ is the time constant (l^2/α) in seconds.

2. $Bi_1 = Bi_2 = Bi$ (i.e., equal losses on each face). The solution is now given by the relation

$$T(l, T) = 2T_a \sum_{n=1}^{\infty} \left[\frac{\beta_n \cos\beta_n + Bi \sin\beta_n}{(\beta_n + Bi + 2)(Bi\beta_n)}\right] \exp(\beta_n^2 t^*) \tag{20.77}$$

where β_n are the roots of the transcendental equation:

$$2Bi - (\beta_n^2 - Bi^2)\frac{t\beta_n}{\beta_n} = 0 \tag{20.78}$$

3. $Bi_1 \neq Bi_2$ (i.e., different losses on each face). The thermogram equation is

$$T(l, T) = 2T_a\left[\sum_{n=1}^{\infty} (\beta_n \cos\beta_n + Bi_1 \sin\beta_n)\right] E_n \exp(-\beta_n^2 t^*) \tag{20.79}$$

where

$$E_n = \frac{\beta_n}{(\beta_n^2 + Bi_1^2)\left(1 + (Bi_2/\beta_n^2 + Bi_2^2)\right) + Bi_1} \tag{20.80}$$

where β_n are the roots of the transcendental equation:

$$\left(\beta_n^2 - Bi_1 Bi_2\right) t\beta_n = \beta_n (Bi_1 + Bi_2) \tag{20.81}$$

The above analytical solutions are difficult to use due to calculation or estimation of heat losses (Bi_1 or Bi_2) because the maximum value and shape of the thermogram are sensitive to these parameters. A typical thermogram is shown in Figure 20.17. Degiovanni (1977) proposed a satisfactory analysis in the case of a cylindrical sample. This method is based on the use of experimental and theoretical normalized temperatures. Theoretical reduced temperature T^* is defined by the relation

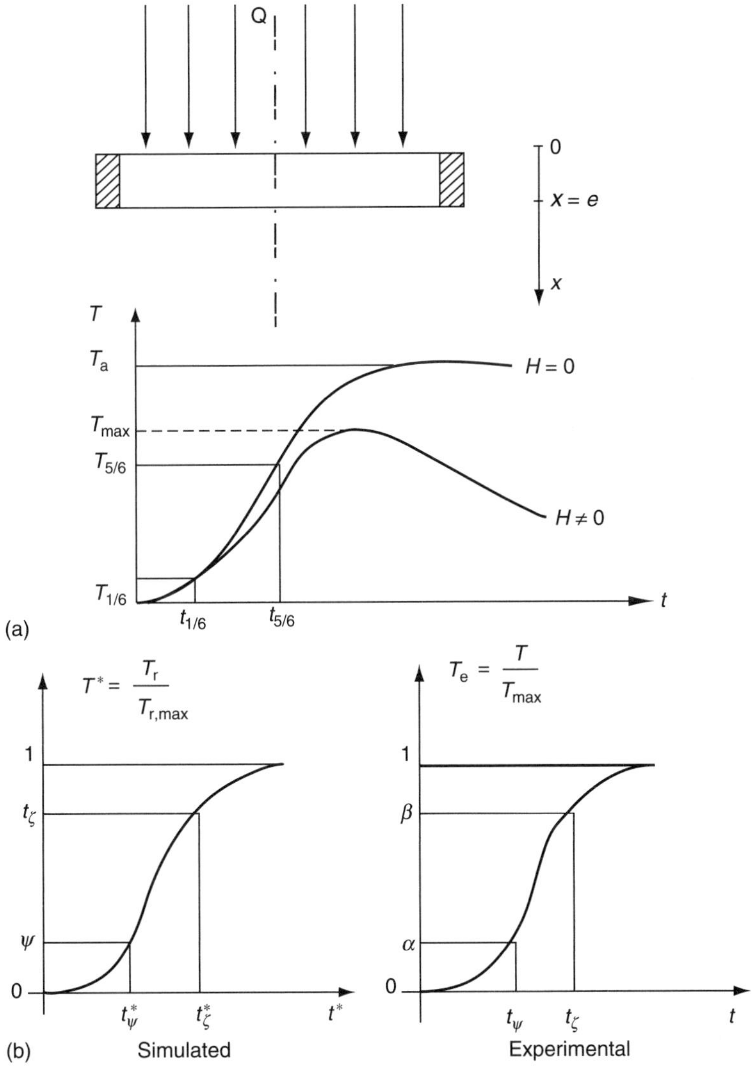

FIGURE 20.17 Schematic of reduced thermograms by pulse method. (a) Sample heating with pulse energy, shape, and peak in thermogram, (b) reduced temperature thermogram. (From Andrieu, J., Gonnet, E., and Laurent, M., *Food Engineering and Process Applications*, Le Maguer, M. and Jelen, P. (eds.), Elsevier Applied Science, London, 1986.)

$$T^* = \frac{T_r}{T_{r,\max}} \quad \text{and} \quad T_r = \frac{T(l,t)}{T_a} \tag{20.82}$$

The experimental reduced temperature T_{er} is given by the relation

$$T_{er} = \frac{T}{T_{\max}} \tag{20.83}$$

where $T_{\max}$ is the maximum temperature (Figure 20.17). After determination of the points (ψ, t_ψ) and (ζ, t_ζ) on the experimental reduced thermograms (Figure 20.17), Andrieu et al. (1986) calculated the diffusivity with the relation

$$\alpha = \frac{l^2}{t_\varsigma} t_\zeta^* \tag{20.84}$$

where the reduced time t_ζ, characteristic of heat losses, is calculated by the following empirical relation:

$$t_{5/6}^* = 0.954 - 1.58 - 1.58\left(\frac{t_{1/2}}{t_{5/6}}\right) + 0.558\left(\frac{t_{1/2}}{t_{5/6}}\right)^2 \tag{20.85}$$

where ψ and ζ are $1/2$ and $5/6$. The main disadvantage of this method is that it uses few points to calculate thermal diffusivity. The moment theory is also usually used to estimate the parameter. Andrieu et al. (1986) provided details of the moment method to estimate the thermal diffusivity. The *f*th order partial moments are defined as

$$w_f = \int_{t_\psi}^{t_\zeta} t^f T(l,t)\,\mathrm{d}t \tag{20.86}$$

If $f=0$, W_0 the zeroth moment, and if $f=-1$, then W_{-1} minus one moment (independent of timescale) (Figure 20.18). Andrieu et al. (1986) derived an equation to calculate the thermal diffusivity from partial moment:

$$\alpha = \frac{l^2 W_{-1}^*}{W_0} \tag{20.87}$$

$$W_{-1}^* = \int_{t_{1/6}^*}^{t_{5/6}^*} \frac{T^*}{t^*} \tag{20.88}$$

A schematic diagram of the equipment is shown in Figure 20.19. A pulse generator delivers a radiant energy pulse of about 800 J during 10^{-2} s. Figure 20.19 shows an apparatus for pulse method with a semiconductor thermocouple (bismuth telluride, Bi_2Te_3) with high thermoelectric power (360 μV/°C at 20°C) and short delay time $<10^{-4}$s and a potentiometric recorder (0.4 mV full scale) are used. The measurement cell, placed inside a thermoregulated chamber, has a disk shape (diameter $=$ 30 mm,

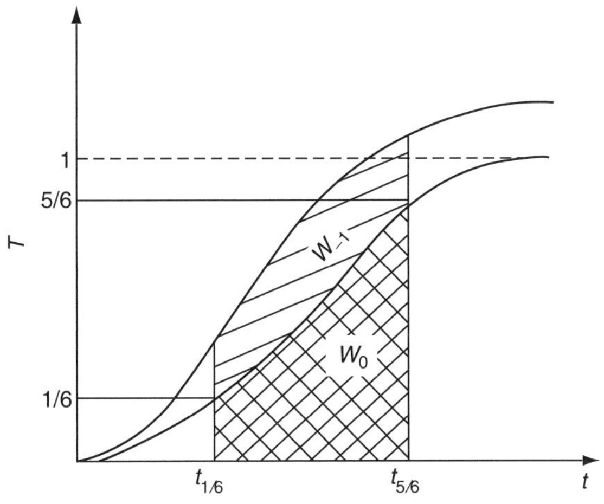

FIGURE 20.18 Schematic diagram of partial experimental moments for pulse method. (From Andrieu, J., Gonnet, E., and Laurent, M., *Food Engineering and Process Applications*, Le Maguer, M. and Jelen, P. (eds.), Elsevier Applied Science, London, 1986.)

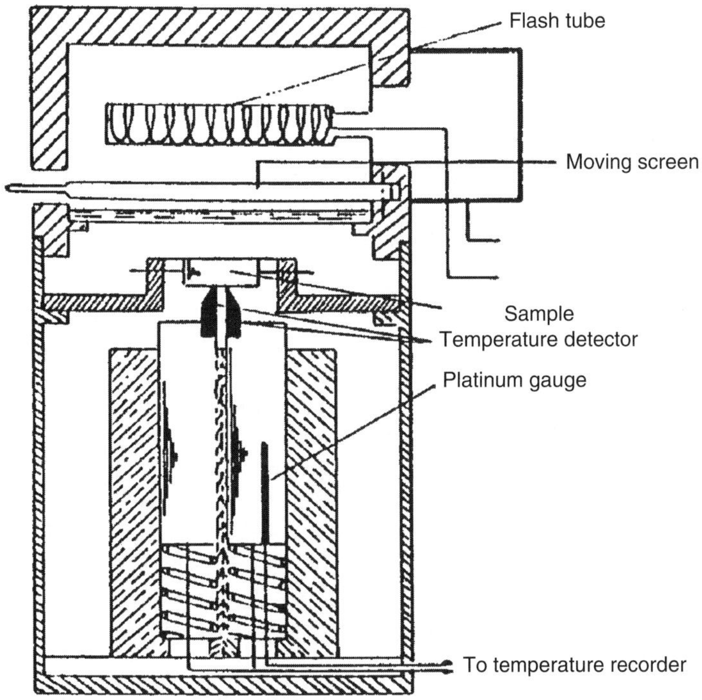

FIGURE 20.19 Thermal diffusivity measurement apparatus for pulse method. (From Andrieu, J., Gonnet, E., and Laurent, M., *Food Engineering and Process Applications*, Le Maguer, M. and Jelen, P. (eds.), Elsevier Applied Science, London, 1986.)

height = 6 mm); its bottom, made up of a thin copper sheet (1/10 mm), has to ensure good mechanical rigidity (uniform thickness) without introducing significant thermal resistance. The device allows accurate measurement of temperature less than 0.5°C. The total experimental time is about 2–3 min.

20.2.1.6 Probe Method

McCurry (1968) and Nix et al. (1967, 1969) mentioned that a thermal conductivity probe can be used for direct determination of diffusivity if an additional sensor is placed in the sample at a known distance such that $0.16 < \gamma < 3.1$. The time–temperature relationship is given by Carslaw and Jaeger (1947) as

$$T = \frac{\lambda}{2\pi k} \int_{\gamma}^{\infty} \frac{\exp(-\gamma^2)}{\gamma} \mathrm{d}\gamma \tag{20.89}$$

where γ is the dimensionless parameter and can be defined as

$$\gamma = \frac{r}{2\sqrt{\alpha t}} \tag{20.90}$$

Nix et al. (1967) gave the following series expression to evaluate the above integral as

$$T = \frac{\lambda}{2\pi k}\left[-\frac{0.58}{2} - \ln\gamma + \frac{\gamma^2}{2.1!} - \frac{\gamma^4}{4.2!} + \cdots\right] \tag{20.91}$$

Nix et al. (1967) found that the first 40 terms of Equation 20.91 need to be evaluated to ensure convergence. The values of γ can be calculated by trial-and-error method from Equation 20.91. Thermal diffusivity can be calculated from the values of γ by Equation 20.90. The advantages of this technique are the short duration of the experiment (about 5 min) and the small temperature rise in the sample (Mohsenin, 1980). An accuracy of 5% has been reported if $0.2 < \gamma < 0.5$ (Suter et al., 1975).

20.2.2 Indirect Prediction

Thermal diffusivity of food material can be calculated from experimentally determined values of thermal conductivity, specific heat, and density. This approach needs considerable time and different instrumentation. Drouzas et al. (1991) used direct (probe method) and indirect determination of thermal diffusivity of granular starch and found that the indirect method yielded more accurate values than the direct measurement as concluded from *F*-test at the level of 5% significance.

20.3 THERMAL DIFFUSIVITY DATA OF FOODS

A compilation of thermal diffusivity data of the food materials is presented in this section. Experimental data by direct measurement are presented in tabular and graphical form. Thermal diffusivity of foods as a function of water content and temperature is presented in Figures 20.20 and 20.21 and Tables 20.1 through 20.18.

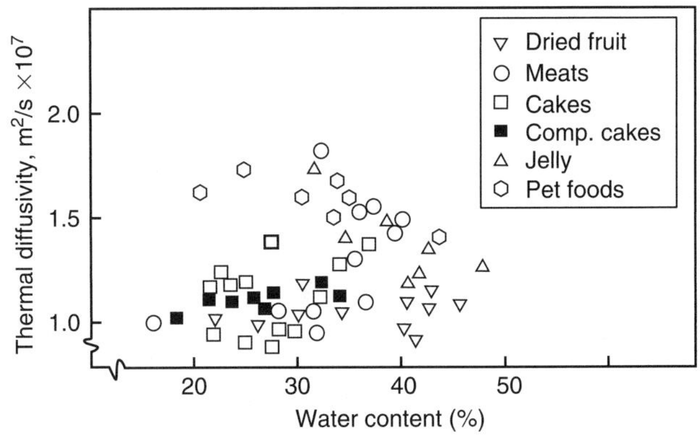

FIGURE 20.20 Thermal diffusivity as a function of water content of low and intermediate moisture foods. (From Sweat, V.E., *ASHRAE J.*, February, 44, 1987.)

20.4 THERMAL DIFFUSIVITY PREDICTION OF FOODS

Thermal diffusivity of a food material is affected by water content and temperature, as well as composition and porosity (Singh, 1982). Moisture contents in food and temperature change considerably during most of the food processing operations. Thus, prediction of thermal diffusivity by moisture content and temperature can be used in thermal analysis of food processing operations. Again the proximate composition of food varies with the type of food product. Therefore, the effect of proximate composition and porosity on thermal diffusivity should be included in the prediction models. Moreover, the nonhomogeneity of the food materials may cause the diffusivity to vary at different locations in the food materials.

Several empirical models useful in predicting thermal diffusivity of foods have appeared in the literature. Most of these models are specific to the products studied. Riedel (1969) developed a model to predict thermal diffusivity as a function of water content using a wide range of food products as

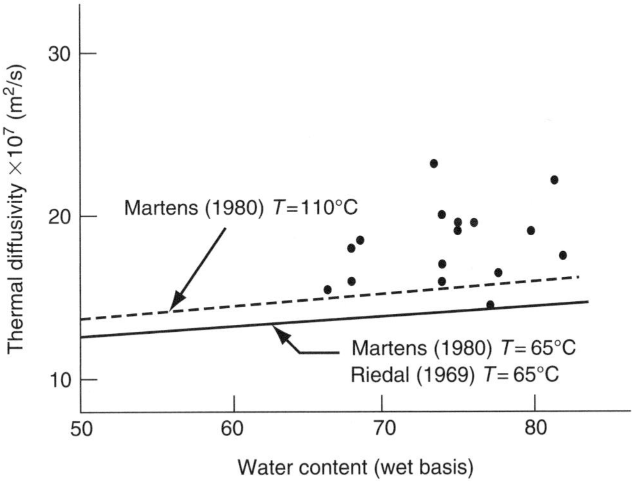

FIGURE 20.21 Thermal diffusivity as a function of water content of foods during sterilization (numbers correspond to Table 20.1). (From Olivares, M., Guzman, J.A., and Solar, I., *J. Food Process. Preservation*, 10, 57, 1986.)

Table 20.1 Thermal Diffusivity of Different Foods during Sterilization in Can

Food	X_w	T (°C)	Thermal Diffusivity (m^2/s)
Meat sauce	0.773	60–112	1.46×10^{-7}
Meat croquette	0.740	59–115	1.98×10^{-7}
Cooked seaweed	0.759	56–110	1.90×10^{-7}
Cooked chickpeas with pork sausages	0.814	71–114	2.24×10^{-7}
Frankfurters	0.734	58–109	2.36×10^{-7}
Cooked lentil with rice	0.778	64–114	1.64×10^{-7}
Cooked abalone	0.685	56–108	1.85×10^{-7}
Cooked clam	0.751	51–108	1.96×10^{-7}
Cooked clams with rice	0.801	60–114	1.98×10^{-7}
Chicken with rice	0.751	65–113	1.93×10^{-7}
Chicken with pea and diced potato and carrot	0.676	63–109	1.65×10^{-7}
Chicken with diced potato and carrot	0.737	72–109	1.70×10^{-7}
Cooked bean with corn and diced squash	0.683	56–113	1.83×10^{-7}
Cooked bovine blood	0.740	58–104	1.62×10^{-7}
Chopped meat with tomato and diced potato	0.664	65–106	1.57×10^{-7}
Chopped meat with diced potato and carrot	0.821	56–113	1.77×10^{-7}

Source: Adapted from Olivares, M., Guzman, J.A., and Solar, I., *J. Food Process. Preservation*, 10, 57, 1986.

Table 20.2 Thermal Diffusivity of Pork/Soy Hull Mixtures at 25°C

Soy Hull					Thermal Properties			
Type	Particle Size	Mixture No.	X_w	X_{SH}	ρ_{ap} (kg/m^3)	C (kJ/kg K)	K (W/m·K)	α (m^2/s)
Unprocessed	Fine (0.84 mm)	1	0.7372	0.0000	1050	4.050	0.508	1.19×10^{-7}
		2	0.7425	0.0250	1050	3.580	0.486	1.29×10^{-7}
		3	0.7552	0.0500	1040	3.050	0.488	1.54×10^{-7}
		4	0.7678	0.0750	1060	3.030	0.529	1.65×10^{-7}
		5	0.7804	0.1000	1050	3.080	0.521	1.61×10^{-7}
	Coarse (2.36 mm)	1	0.7372	0.0000	1050	4.050	0.508	1.19×10^{-7}
		6	0.7474	0.0225	1030	3.580	0.480	1.30×10^{-7}
		7	0.7649	0.0450	1040	3.440	0.505	1.41×10^{-7}
		8	0.7760	0.0675	1060	3.070	0.514	1.58×10^{-7}
		9	0.7999	0.0900	1030	3.710	0.534	1.40×10^{-7}
Processed	Fine (0.84 mm)	1	0.7372	0.0000	1050	4.050	0.508	1.19×10^{-7}
		10	0.7547	0.0200	1050	4.080	0.490	1.14×10^{-7}
		11	0.7796	0.0400	1050	3.760	0.546	1.38×10^{-7}
		12	0.8044	0.0600	1040	4.180	0.536	1.23×10^{-7}
		13	0.8292	0.0800	1040	4.100	0.554	1.30×10^{-7}
	Coarse (2.36 mm)	1	0.7372	0.0000	1050	4.050	0.508	1.19×10^{-7}
		14	0.7596	0.0175	1020	3.210	0.494	1.51×10^{-7}
		15	0.7893	0.0350	1040	3.080	0.530	1.65×10^{-7}
		16	0.8191	0.0525	1030	2.750	0.570	2.01×10^{-7}
		17	0.8488	0.0700	1040	3.000	0.572	1.83×10^{-7}

Source: Muzilla, M., Unklesbay, N., Unkles bay, K., and Helsel, Z., *J. Food Sci.*, 55, 1491, 1990.

Table 20.3 Thermal Diffusivity of Fish and Meat Products

Material	Sample	X_w	Temperature (°C)	Thermal Diffusivity (m^2/s)	Reference
Cod fish	—	0.81	5	1.22×10^{-7}	Singh (1982)
		0.81	65	1.42×10^{-7}	Singh (1982)
Cod mince[a]	Fillet	0.81	22–50	1.52×10^{-7}	Poulsen (1982)
Corned beef	—	0.65	5	1.32×10^{-7}	Singh (1982)
	—	0.65	65	1.18×10^{-7}	Singh (1982)
Beef	Chuck[b]	0.66	40–65	1.23×10^{-7}	Singh (1982)
Beef	Round[b]	0.71	40–65	1.33×10^{-7}	Singh (1982)
Beef	Tongue[b]	0.68	40–65	1.32×10^{-7}	Singh (1982)
Beef	Ground	—	−35 to −24	10.04×10^{-7}	Hayakawa (1973)
Beef	Ground	—	−17 to −27	6.08×10^{-7}	Hayakawa (1973)
Beef	Ground	—	−22 to −7	4.43×10^{-7}	Hayakawa (1973)
Halibut	—	0.76	40–65	1.47×10^{-7}	Singh (1982)
Ham	Smoked	0.64	5	1.18×10^{-7}	Singh (1982)
Ham	Smoked	0.64	40–65	1.38×10^{-7}	Singh (1982)
Lard	Purified	0.00	0–25	0.61×10^{-7}	Poulsen (1982)

[a] Density (884 kg/m^3).
[b] Data are applicable only where juices that exuded during heating remain in the food samples.

$$\alpha = 0.88 \times 10^{-7} + \left[\alpha_w - 0.88 \times 10^{-7}\right] X_w \tag{20.92}$$

The above equation indicates that thermal diffusivity of water rich foods is the sum of those of water and water-free substances such as protein, carbohydrate, and lipid in low content. Riedel (1969) suggested that the above equation is valid for liquid and solid foods containing at least 40% water and in the temperature range of 0°C–80°C. Dickerson and Read (1975) used the above model and found good agreement with experimental values of thermal diffusivity for a variety of meats (Singh, 1982). Using statistical analysis, Martens (1980) found that temperature and water content are the

Table 20.4 Thermal Diffusivity of Sucrose and Gelatin as a Function of Water Content

	Thermal Diffusivity $\times 10^{-7}$ (m^2/s)	
X_w	Sucrose[a]	Gelatin[b]
0.40	1.220	—
0.60	1.350	1.370
0.65	—	1.384
0.70	1.385	1.405
0.75	—	1.148
0.80	1.428	1.435
0.85	—	1.445
0.90	1.470	1.468
0.99	1.513	—

Source: Adapted from Andrieu, J., Gonnet, E., and Laurent, M., in *Food Engineering and Process Applications*, Le Maguer, M. and Jelen, P. (eds.), Elsevier Applied Science, London, 1986.
[a] At 33°C (Bhowmik and Hayakawa, 1979).
[b] At 20°C by pulse method.

Table 20.5 Bulk Thermal Diffusivity of Cereal Products

Cereal	T (0°C)	X_w	ρ_B (kg/m^3)	α_B (m^2/s)	Reference
White rice 1	0–30	0.129	851	1.00×10^{-7}	Jones et al. (1992)
White rice 2	0–30	0.133	823	1.00×10^{-7}	Jones et al. (1992)
White rice 3	0–30	0.132	839	1.01×10^{-7}	Jones et al. (1992)
Paddy rice	0–30	0.089	639	1.07×10^{-7}	Jones et al. (1992)
Wholemeal	0–30	0.093	542	1.02×10^{-7}	Jones et al. (1992)
Bran	0–30	0.36	235	1.72×10^{-7}	Jones et al. (1992)
Wheat	0–30	0.091	827	0.91×10^{-7}	Jones et al. (1992)
Rough rice[a]	26	0.100	632[b]	1.42×10^{-7}	Morita and Singh (1979)
Rough rice[a]	26	0.130	642[c]	1.35×10^{-7}	Morita and Singh (1979)
Rough rice[a]	26	0.160	656[d]	1.28×10^{-7}	Morita and Singh (1979)
Rough rice[a]	26	0.190	664[e]	1.22×10^{-7}	Morita and Singh (1979)
Rough rice[a]	26	0.220	632[e]	1.17×10^{-7}	—

[a] Short grain.
[b] At X_w: 0.1124.
[c] At X_w: 0.1382.
[d] At X_w: 0.1658.
[e] At X_w: 0.1945.

Table 20.6 Thermal Diffusivity of Meat Emulsions at 15°C–75°C

Emulsion	X_w	Thermal Diffusivity (m^2/s)	Emulsion	X_w	Thermal Diffusivity (m^2/s)
Base 1 model (60% beef, 20% pork, and 20% pork fat)	0.591	1.168×10^{-7}	20% vegetable oil + 7.5% binder + base 2	0.558	1.232×10^{-7}
	0.603	1.180×10^{-7}		0.558	1.210×10^{-7}
	0.621	1.200×10^{-7}		0.574	1.252×10^{-7}
	0.641	1.237×10^{-7}		0.591	1.217×10^{-7}
	0.654	1.210×10^{-7}		0.603	1.238×10^{-7}
	0.659	1.205×10^{-7}		0.629	1.242×10^{-7}
	0.672	1.242×10^{-7}		0.639	1.278×10^{-7}
	0.682	1.245×10^{-7}			
			Base 2 + 10% trip	0.610	1.178×10^{-7}
Base 1 + 7.5% binder	0.575	1.170×10^{-7}		0.617	1.188×10^{-7}
	0.581	1.165×10^{-7}		0.624	1.260×10^{-7}
	0.598	1.180×10^{-7}		0.647	1.235×10^{-7}
	0.600	1.197×10^{-7}		0.659	1.245×10^{-7}
	0.636	1.228×10^{-7}		0.676	1.220×10^{-7}
	0.651	1.232×10^{-7}		0.680	1.258×10^{-7}
Base 2 model (60% beef, 20% pork, and 20% vegetable oil)	0.581	1.188×10^{-7}	Base 2 + 15% trip	0.627	1.193×10^{-7}
	0.603	1.187×10^{-7}		0.650	1.248×10^{-7}
	0.614	1.215×10^{-7}		0.666	1.217×10^{-7}
	0.629	1.228×10^{-7}		0.674	1.280×10^{-7}
	0.632	1.243×10^{-7}		0.693	1.265×10^{-7}
	0.645	1.240×10^{-7}		0.693	1.275×10^{-7}
	0.664	1.265×0^{-7}		0.711	1.313×10^{-7}

Table 20.6 (continued) Thermal Diffusivity of Meat Emulsions at 15°C–75°C

Emulsion	X_w	Thermal Diffusivity (m^2/s)	Emulsion	X_w	Thermal Diffusivity (m^2/s)
30% vegetable oil + base 2	0.546	1.168×10^{-7}	Base 2 + 20% trip	0.619	1.233×10^{-7}
	0.546	1.187×10^{-7}		0.620	1.208×10^{-7}
	0.565	1.193×10^{-7}		0.640	1.263×10^{-7}
	0.600	1.250×10^{-7}		0.644	1.223×10^{-7}
	0.611	1.272×10^{-7}		0.659	1.238×10^{-7}
	0.660	1.283×10^{-7}		0.668	1.325×10^{-7}
				0.672	1.235×10^{-7}
				0.703	1.280×10^{-7}
				0.708	1.245×10^{-7}

Source: Adapted from Timbers, G.E., Randall, C.J., and Raymond, D.P., *Can. Inst. Food Sci. Technol. J.*, 15, 191, 1982.

Table 20.7 Thermal Diffusivity of Fruits and Vegetables

Material	Variety	Sample	X_w	Temperature[a] (°C)	Thermal Diffusivity (m^2/s)	Reference
Apple	McIntosh	Whole	—	32 to 0	0.77×10^{-7}	Gaffney et al. (1980)
Apple	Crab	—	—	16 to 0	1.30×10^{-7}	Gaffney et al. (1980)
Apple	Cox	Whole	—	20 to 4	1.34×10^{-7}	Gordon and Thorne (1990)
Apple	Jonathan	Whole	—	27 to 4	1.20×10^{-7}	Gaffney et al. (1980)
Apple	RD	Whole	—	29 to 1	1.37×10^{-7}	Gaffney et al. (1980)
Apple	RD	Flesh	—		1.11×10^{-7}	Gaffney et al. (1980)
Apple	GD	Pulp	—	4 to 26	1.50×10^{-7}	Gaffney et al. (1980)
Apple	GD	Whole	—	20 to 4	1.46×10^{-7}	Gordon and Thorne (1990)
Apple	—	—	—	27 to −18	1.50×10^{-7}	Gaffney et al. (1980)
Apple	RD	Whole	0.850	0 to −30	1.37×10^{-7}	Singh (1982)
Applesauce	—	—	0.370	5	1.05×10^{-7}	Singh (1982)
	—	—	0.370	65	1.12×10^{-7}	Singh (1982)
	—	—	0.757	21 to 50	1.49×10^{-7}	Poulsen (1982)
	—	—	0.800	5	1.22×10^{-7}	Singh (1982)
	—	—	0.800	65	1.40×10^{-7}	Singh (1982)
Avocado	Lula	Flesh	—	24 to 3	1.05×10^{-7}	Gaffney et al. (1980)
Avocado	Lula	Seed	—	24 to 3	1.10×10^{-7}	Gaffney et al. (1980)
Avocado	—	Whole	—	41 to 3	1.54×10^{-7}	Gaffney et al. (1980)
Avocado[b]	—	Flesh	—	20	1.27×10^{-7}	Andrieu et al. (1986)
Avocado[c]	—	Flesh	—	20	1.32×10^{-7}	Andrieu et al. (1986)
Banana[b]	—	Flesh	—	20	1.46×10^{-7}	Andrieu et al. (1986)
Banana[c]	—	Flesh	—	20	1.37×10^{-7}	Andrieu et al. (1986)
Banana	—	Flesh	0.760	5	1.18×10^{-7}	Gaffney et al. (1980)
Banana	—	Flesh	0.760	65	1.42×10^{-7}	Singh (1982)
Bean	—	Baked	—	4 to 122	1.68×10^{-7}	Singh (1982)
Blackberry	—	—	—	27 to −18	1.27×10^{-7}	Gaffney et al. (1980)
Carrot[b]	—	—	—	20	1.40×10^{-7}	Andrieu et al. (1986)
Carrot[c]	—	—	—	20	1.55×10^{-7}	Andrieu et al. (1986)

(continued)

Table 20.7 (continued) Thermal Diffusivity of Fruits and Vegetables

Material	Variety	Sample	X_w	Temperature[a] (°C)	Thermal Diffusivity (m^2/s)	Reference
Cherries	Tart	Flesh	—	30 to 0	1.32×10^{-7}	Gaffney et al. (1980)
Cucumber[b]	—	Flesh	—	20	1.39×10^{-7}	Andrieu et al. (1986)
Cucumber[c]	—	Flesh	—	20	1.41×10^{-7}	Andrieu et al. (1986)
Grapefruit	—	—	—	16 to 0	1.20×10^{-7}	Gaffney et al. (1980)
Grapefruit	Marsh	Whole	—	40 to 0	0.92×10^{-7}	Gaffney et al. (1980)
Grapefruit	Marsh	Whole	—	46 to 0	0.94×10^{-7}	Gaffney et al. (1980)
Grapefruit	—	Whole	—	36 to 0	0.84×10^{-7}	Gaffney et al. (1980)
Grapefruit	Marsh	Whole	—	27 to 2	0.92×10^{-7}	Gaffney et al. (1980)
Grapefruit	Marsh	Rind	—	27	1.20×10^{-7}	Gaffney et al. (1980)
Grapefruit	Marsh	Flesh	—	31	1.23×10^{-7}	Gaffney et al. (1980)
Lemon	Eureka	Whole	—	40 to 0	1.07×10^{-7}	Gaffney et al. (1980)
Lima bean	—	—	—	27 to −18	1.24×10^{-7}	Gaffney et al. (1980)
Onion	—	Whole	—	20 to 4	1.41×10^{-7}	Gordon and Thorne (1990)
Orange	—	—	—	16 to 0	1.30×10^{-7}	Gaffney et al. (1980)
Orange	Valencia	Whole	—	40 to 0	0.94×10^{-7}	Gaffney et al. (1980)
Orange	Navel	Whole	—	40 to 0	1.07×10^{-7}	Gaffney et al. (1980)
Peach	—	Flesh	—	27 to −18	1.20×10^{-7}	Gaffney et al. (1980)
Peach	—	—	—		1.11×10^{-7}	Gaffney et al. (1980)
Peach	Several	Whole		27–4	1.39×10^{-7}	Gaffney et al. (1980)
Pea	English			27 to −18	1.24×10^{-7}	Gaffney et al. (1980)
Potato[d]	—	Whole		26 to −6	1.77×10^{-7}	Hayakawa and Succar (1983)
Potato[b]	—	Flesh		20	1.48×10^{-7}	Andrieu et al. (1986)
Potato[c]	—	Flesh		20	1.53×10^{-7}	Andrieu et al. (1986)
Potato	Excel	Flesh		24–91	1.17×10^{-7}	Gaffney et al. (1980)
Potato	Irish	Flesh		27 to −18	1.23×10^{-7}	Gaffney et al. (1980)
Potato	Pungo	Whole			1.31×10^{-7}	Gaffney et al. (1980)
Potato	Several	Flesh		25	1.70×10^{-7}	Gaffney et al. (1980)
Rutabaga	—	Whole		48–0	1.34×10^{-7}	Gaffney et al. (1980)
Squash	Green	Whole		47–0	1.71×10^{-7}	Gaffney et al. (1980)
Squash	—	Whole		16–0	1.10×10^{-7}	Gaffney et al. (1980)
Squash	Several	Flesh		25	1.56×10^{-7}	Gaffney et al. (1980)
Strawberry	—	Flesh	0.920	5	1.27×10^{-7}	Gaffney et al. (1980)
Strawberry	—			27 to −18	1.47×10^{-7}	Gaffney et al. (1980)
Sugar beet	—	Flesh		14–60	1.26×10^{-7}	Gaffney et al. (1980)
Sweet potato	—	Whole		42–0	1.20×10^{-7}	Gaffney et al. (1980)
Sweet potato	Centennial	Flesh		116	1.32×10^{-7}	Gaffney et al. (1980)
Sweet potato	Goldrush	Whole		35	1.06×10^{-7}	Gaffney et al. (1980)
Sweet potato	Goldrush	Whole		55	1.39×10^{-7}	Gaffney et al. (1980)
Sweet potato	Goldrush	Whole		70	1.91×10^{-7}	Gaffney et al. (1980)
Swede	—	Whole		20–4	1.38×10^{-7}	Gordon and Thorne (1990)
Tomato	Cherry	Pulp		4–26	1.48×10^{-7}	Gaffney et al. (1980)
Tomato[d]	—	Whole		7–23	1.51×10^{-7}	Hayakawa and Succar (1983)

[a] Where two temperatures are given, the first is the initial temperature of the sample and second is that of the surroundings.
[b] Hot wire probe.
[c] Pulse method.
[d] Constant temperature ratio method.

Table 20.8 Thermal Properties of Different Food Materials

Material	Density (kg/m^3)	X_w	Temperature (°C)	Thermal Diffusivity (m^2/s) $\times 10^{-7}$	Reference
Lactose	736	0.000	0 to 50	1.64	Poulsen (1982)
Fructose[a]	782	0.000	0 to 50	1.25	Poulsen (1982)
Glucose	633	0.089	0 to 50	1.55	Poulsen (1982)
Wheat flour	713	0.110	23 to 50	1.01	Poulsen (1982)
Egg white	1065	0.875	0 to 50	1.46	Poulsen (1982)
Mayonnaise	905	0.180	0 to 50	1.07	Poulsen (1982)
Carrageenan[b]	—	0.980	0 to 50	1.57	Poulsen (1982)
Meat[c]	—	0.740	0 to 50	1.46	Poulsen (1982)
Sucrose[d]	—	0.600	50 to 90	1.39	Poulsen (1982)
	—	—	30 to 65	1.35	Poulsen (1982)
	—	—	0 to 35	1.14	Poulsen (1982)
Sucrose	—	0.500	−33.9 to −28.1	0.30	Hayakawa (1973)
Grape juice[e]	—	—	−37.8 to −19.4	0.85	Hayakawa (1973)
Grape juice[e]	—	—	−40.0 to −27.4	0.99	Hayakawa (1973)

Sources: Adapted from Poulsen, K.P., *J. Food Eng.*, 1, 115, 1982; Hayakawa, K., *J. Food Sci.*, 38, 623, 1973.
[a] Monohydrate.
[b] Gel.
[c] Imitation (1% NaCl and 25% binder).
[d] 22.5% sucrose and 17.5% binder.
[e] 40°Brix.

major factors affecting thermal diffusivity when investigating the influence of water, fat, protein, carbohydrate, and temperature (Singh, 1982). Variation in the solid fraction of fat, protein, and carbohydrate had a small influence on thermal diffusivity. Han and Loncin (1985) studied the effect of lipid on thermal diffusivity of model fish. The thermal diffusivities at 120°C and at constant water content of 65% decreased slightly with increasing lipid content. The evaluated mean value was 1.433×10^{-7} m^2/s. Between lipid contents of 1% and 10%, the deviations of all the evaluated values were less than 0.3% from the mean value and the value predicted by Riedel (1969) equation. Therefore, Han and Loncin (1985) concluded that the influence of lipid on the thermal diffusivity could be neglected without any significant errors in the practical industrial sterilization. Timbers et al. (1982) measured the thermal diffusivity of meat emulsions made of different type of fats and oils and found that thermal diffusivity was influenced by the type of materials as follows:

$$\text{Vegetable oil (30\%)} > \text{vegetable oil (20\%)} > \text{pork fat} \tag{20.93}$$

Table 20.9 Thermal Diffusivity of Potato as a Function of Temperature[a]

$T_{max,c}$ (°C)	$T_{max,s}$ (°C)	Apparent Density (kg/m^3)	Thermal Diffusivity (m^2/s)	
			Raw	Heated
72.2	57.2	1038	1.332×10^{-7}	1.371×10^{-7}
81.7	63.3	1051	1.411×10^{-7}	1.488×10^{-7}
90.6	68.3	1048	0.975×10^{-7}	1.063×10^{-7}
90.6	72.8	1049	0.961×10^{-7}	1.044×10^{-7}
100.0	61.1	1052	1.251×10^{-7}	1.335×10^{-7}
98.9	73.9	1046	1.239×10^{-7}	1.444×10^{-7}
100.0	77.2	1048	1.007×10^{-7}	1.043×10^{-7}

Source: Matthews, F.V. and Hall, C.W., *Trans. ASAE*, 11, 558, 1968.
[a] $T_{max,c}$, temperature at the center of the sample; $T_{max,s}$, temperature at the surface of the sample.

Table 20.10 Thermal Diffusivity of Sucrose Solution

	Thermal Diffusivity × 10^{-7} (m^2/s)												
	T (°C)												
X_{wo}	−39	−36	−33	−30	−28	−25	−22	−19	−16	−14	−11	−8	−5
0.98	12.97	12.50	12.03	11.49	11.27	10.67	10.38	9.72	9.25	8.68	7.62	6.83	5.31
0.96	12.95	12.50	12.03	10.93	10.68	9.74	9.69	8.69	8.37	7.59	6.48	5.41	3.68
0.94	12.51	11.74	11.08	10.73	10.28	9.61	9.55	8.81	7.98	7.39	6.30	5.22	3.38
0.92	12.34	11.47	10.83	10.07	9.80	9.05	8.69	7.90	7.34	6.58	5.41	4.26	2.52
0.90	11.46	10.83	10.83	9.27	9.09	8.23	7.98	7.04	6.44	5.69	4.54	3.43	1.94
0.88	10.92	10.18	9.26	8.34	8.17	7.44	6.99	6.22	5.64	4.82	3.80	2.76	1.52
0.86	10.58	10.18	8.81	8.05	7.83	6.99	6.67	5.8	5.15	4.39	3.41	2.41	1.29
0.84	10.49	9.76	8.54	7.67	7.17	6.58	6.06	5.26	4.69	3.81	2.94	2.03	1.03
0.82	9.84	9.21	7.75	7.03	6.57	6.03	5.58	4.79	4.19	3.43	2.60	1.79	0.86
0.80	9.86	9.11	7.64	6.85	6.34	5.75	5.29	4.59	3.94	3.26	2.42	1.58	0.76
0.75	8.9	8.28	7.05	6.11	5.62	4.87	4.49	3.81	3.28	2.6	1.88	1.23	0.57
0.70	7.81	7.32	6.03	5.15	4.64	4.02	3.73	3.11	2.53	2.01	1.41	0.87	0.40
0.65	6.65	6.32	4.94	4.05	3.75	3.16	2.95	2.42	2.01	1.53	1.05	0.63	0.29

Source: Keppler, R.A. and Boose, J.R., *Trans. ASAE*, 13, 335, 1970.

Martens (1980) performed multiple regression analysis on 246 published values on thermal diffusivity of a variety of food products and obtained the following regression equation:

$$\alpha = [0.057363X_w + 0.000288(T + 273)] \times 10^{-6} \quad (20.94)$$

The standard error of estimate was 0.14×10^{-7} m^2/s. Metel et al. (1986) provided the thermal diffusivity of dough wheat as a function of moisture content, density, and temperature as (20°C < T < 90°C)

Table 20.11 Thermal Diffusivity of Fish at Different Temperatures

	Thermal Diffusivity × 10^{-7} (m^2/s)									
	T (°C)									
Fish	X_{wo}	X_{PA}	X_p	−20	−10	0	10	20	30	40
Saman	0.773	0.0046	0.202	8.330	5.094	1.670	1.678	1.969	2.426	2.61
Sol	0.776	0.0046	0.205	7.988	5.263	1.904	2.153	2.372	2.650	3.175
Black bhitki	0.777	0.0060	0.195	9.763	5.706	1.872	2.113	2.459	2.556	2.696
Black pomphret	0.750	0.0476	0.179	8.726	4.381	1.804	1.966	2.038	2.329	2.542
Mackerel	0.774	0.0058	0.187	6.984	4.122	1.408	1.519	1.793	2.070	—
Red bhitki	0.795	0.0042	0.169	7.232	5.274	1.634	1.843	2.084	2.430	2.779
Singara	0.779	0.0057	0.201	6.836	3.827	1.130	1.508	1.717	1.894	2.131
Hilsa	0.747	0.0513	0.172	8.460	5.569	2.156	2.131	2.75	2.768	2.866
Surama	0.780	0.0077	0.188	6.689	3.920	1.469	1.609	1.994	2.092	2.304
White Pomphret	0.748	0.0410	0.186	9.846	4.385	1.753	2.034	2.434	2.815	2.938
Malli	0.782	0.0045	0.194		4.262	1.595	1.775	2.016	2.498	—
Rohu	0.752	0.0057	0.215	7.038	5.083	1.760	1.904	2.016	2.318	2.491

Source: Kumbhar, B.K., Agarwal, R.S., and Das, K., *Int. J. Refrig.*, 4, 1981.

Table 20.12 Thermal Properties of Dough and Bread, Compiled by Rask (1989)

Product	T (°C)	X_w^a	ρ_{ap} (kg/m^3)	C_p (kJ/kg k)	k (W/m·K)	$\alpha \times 10^{-7}$ (m^2/s)	Reference
Dough	—	—	—	—	0.386	—	Bakshi and Yoon (1984)
Bread solids	—	—	—	1.13	0.309[b]	—	Bakshi and Yoon (1984)
Bread (5 min)	—	0.334[b]	202[b]	2.151c	0.085[b]	—	Bakshi and Yoon (1984)
Bread (10 min)	—	0.269[b]	181[b]	1.952c	0.093[b]	—	Bakshi and Yoon (1984)
Bread crumb	—	0.340[a]	—	—	—	4.07	Johnsson and Skjoldebrand (1984)
Bread crumb	30.0	0.41	—	2.560[b]	—	—	Johnsson and Skjoldebrand (1984)
Bread crumb	100.0	0.41	—	2.626[b]	—	—	Johnsson and Skjoldebrand (1984)
Bread crust	100.0	—	—	—	—	0.367	Johnsson and Skjoldebrand (1984)
Bread crust	150.0	0	—	1.656[b]	—	—	Johnsson and Skjoldebrand (1984)
Dough	35.0	—	—	1.700[c,d]	0.550[c]	—	Kafiev et al. (1987)
Bread (rye, 10 min)	—	—	—	—	—	0.24	Kriems and Reinhold (1980)
Bread (rye, 40 min)	—	—	430[c]	—	—	0.52	Kriems and Reinhold (1980)
Bread (wheat, 10 min)	—	—	—	—	—	0.43	Kriems and Reinhold (1980)
Bread (wheat, 40 min)	—	—	290[c]	—	—	1.38	Kriems and Reinhold (1980)
Dough	−43.5	0.435	1100	1.76	0.92	4.78	Lind (1988)
Dough	−28.5	0.435	1100	1.94	—	3.95	Lind (1988)
Dough	−22.0	0.435	1100	—	0.88	—	Lind (1988)
Dough	16.5	0.435	1100	2.76	—	—	Lind (1988)
Dough	23.0	0.435	1100	—	0.46	1.45	Lind (1988)
Dough	−38.0	0.461	1100	1.760	1.03	5.3	Lind (1988)
Dough	−28.0	0.461	1100	1.88	—	—	Lind (1988)
Dough	−16.0	0.461	1100	—	0.98	4.35	Lind (1988)
Dough	19.0	0.461	1100	—	0.5	1.63	Lind (1988)
Dough	21.0	0.461	1100	2.81	—	—	Lind (1988)
Dough (wheat)	28.0	0.420	623	2.883	0.414	1.770	Lind (1988)
Bread crumb (loaf)	18.0	0.418	402	3.19	0.298	1.915	Makljukow and Makljukow (1983)
Bread crumb (tin loaf)	18.0	0.428	340	2.975	0.244	2.42	Makljukow and Makljukow (1983)
Bread crumb (crust)	140.0	0	300	1.575	0.066–0.43	0.268	Makljukow and Makljukow (1983)
Bread solids	12.5	0.366	—	—	—	—	Makljukow and Makljukow (1983)
Bread solids	<0	0.363	—	0.657	—	—	Mannheim et al. (1957)
Dough (wheat)	—	0.444[e]	750	2.8	0.5	2.375	Mannheim et al. (1957)
Dough (rye)	—	0.459[e]	820	3	0.6	2.434	Nebelung (1979)
Crust (wheat)	—	0	417	1.68	0.055	0.785	Nebelung (1979)
Crust (rye)	—	0	443	1.68	0.055	0.739	Nebelung (1979)
Crumb (wheat)	—	0.440[e]	450	2.8	0.28	2.22	Nebelung (1979)
Crumb (rye)	—	0.459[e]	500	3	0.37	2.470	Nebelung (1979)
Dough	—	0.417	—	1.600[b,c]	0.600[c]	—	Nebelung (1979)

(continued)

Table 20.12 (continued) Thermal Properties of Dough and Bread, Compiled by Rask (1989)

Product	T (°C)	X_w^a	ρ_{ap} (kg/m^3)	C_p (kJ/kg k)	k (W/m·K)	$\alpha \times 10^{-7}$ (m^2/s)	Reference
Dough	—		—	2.303	—	—	Neznanova et al. (1978)
Bread (white)	0–100	0.445	—	2.785	—	—	Ordinanz (1946)
Bread (brown)	0–100	0.485	—	2.85	—	—	Ordinanz (1946)
Dough (dry)	30	0	—	1.260[b]	—	—	Ordinanz (1946)
Dough (wet)	0–30	0.444[e]	—	2.516[b]	—	—	Polak (1984)
Bread solids	3	—	—	—	0.361	—	Polak (1984)
Bread solids	24	—	—	—	0.378	—	Tadano (1987)
Bread (rye/wheat)	—	0.467[e]	—	—	—	5.580	Tichy (1974)
Dough (wheat)	—	0.448	586	2.801	0.314	1.916	Tschubik and Maslow (1973)
Dough (wheat)	—	0.451	629	2.805	0.327	1.916	Tschubik and Maslow (1973)
Dough (rye)	—	0.536	718	3.023	0.407	1.875	Tschubik and Maslow (1973)
Dough (rye	—	0.539	701	3.027	0.396	1.875	Tschubik and Maslow (1973)
Bread (loaf)	—	0.425	545	2.742	0.248	1.666	Tschubik and Maslow (1973)
Bread (tin loaf)	—	0.450	500	2.805	0.232	—	Tschubik and Maslow (1973)
Crust (loaf)	—	0.000	420	1.675	0.055	0.800	Tschubik and Maslow (1973)
Crust (tin loaf)	—	0.000	300	1.675	0.041	—	Tschubik and Maslow (1973)
Bread white, 8 min	—	—	307	—	0.720	—	Unklesbay et al. (1981)
Bread white, 16 min	—	—	285	—	0.670	—	Unklesbay et al. (1981)
Bread white, 24 min	—	—	275	—	0.660	—	Unklesbay et al. (1981)
Bread white, 32 min	—		264	—	0.640	—	Unklesbay et al. (1981)

Source: Adapted from Rask, C., *J. Food Eng.*, 9, 167, 1989.
[a] Wet basis.
[b] Calculated.
[c] Estimated from diagram.
[d] Volumetric specific heat (opp.).
[e] Calculated from moisture content on dry basis.

Table 20.13 Thermal Properties of Bakery Products Other than Bread Compiled by Rask (1989)

Product	T (°C)	X_w	ρ_{ap} (kg/m^3)	C_p (kJ/kg·k)	k (W/m·K)	$\alpha \times 10^{-7}$ (m^2/s)	Reference
Biscuit	—	0.032	—	1.876	—	—	Hwang and Hayakawa (1979)
Biscuit	—	0.035	—	1.943	—	—	Hwang and Hayakawa (1979)
Biscuit	—	0.039	—	1.934	—	—	Hwang and Hayakawa (1979)
Cracker	—	0.027	—	1.595	—	—	Hwang and Hayakawa (1979)
Cracker	—	0.026	—	1.57	—	—	Hwang and Hayakawa (1979)
Biscuit dough (Rask, 1989)	—	0.041	1253	2.94	0.405	0.800–1.200	Kulacki and Kennedy (1978)

Table 20.13 (continued) Thermal Properties of Bakery Products Other than Bread Compiled by Rask (1989)

Product	T (°C)	X_w	ρ_{ap} (kg/m^3)	C_p (kJ/kg·k)	k (W/m·K)	$\alpha \times 10^{-7}$ (m^2/s)	Reference
Biscuit dough (hard sweet)	—	0.085	1287	2.804	0.39	0.800–1.200	Kulacki and Kennedy (1978)
Biscuit	—	—	—	—	0.07	—	Standing (1974)
Biscuit	—	—	—	—	0.16	—	Standing (1974)
Yellow cake	—	—	—	—	—	—	
Batter	—	0.415	694	2.95	0.223	1.09	Sweat (1973)
Edge 1/4 done	—	0.400	815	—	0.239	0.86	Sweat (1973)
Center 1/4 done	—	0.400	815	—	0.228	0.86	Sweat (1973)
Edge 1/2 done	—	0.390	360	—	0.147	2.14	Sweat (1973)
Center 1/2 done	—	0.390	290	—	0.195	1.61	Sweat (1973)
Edge 3/4 done	—	0.365	265	—	0.132	1.85	Sweat (1973)
Center 3/4 done	—	0.375	265	—	0.135	1.69	Sweat (1973)
Edge done	—	0.340	285	—	0.119	1.5	Sweat (1973)
Center done	—	0.355	300	2.8	0.121	1.43	Sweat (1973)
Tortilla dough	55–75	—	1102–1173	2.980–3.170	0.037–0.108	1.050–3.080	Griffith (1985)

Table 20.14 Thermal Diffusivity of Fish Jelly, Fish Sausage, and Fish Ham at Different Temperatures

X_w	X_{FM}	X_{so}	X_{SP}	X_{ws}	Temperature (°C)	Thermal Diffusivity (m^2/s)
0.5	0.263	0.079	0.079	0.079	100	1.265×10^{-7}
					100	1.286×10^{-7}
					110	1.281×10^{-7}
					110	1.305×10^{-7}
					120	1.313×10^{-7}
					120	1.295×10^{-7}
0.574	0.224	0.067	0.067	0.067	100	1.335×10^{-7}
					110	1.344×10^{-7}
					110	1.356×10^{-7}
					120	1.365×10^{-7}
					120	1.379×10^{-7}
0.651	0.184	0.055	0.055	0.055	100	1.393×10^{-7}
					110	1.412×10^{-7}
					110	1.414×10^{-7}
					120	1.423×10^{-7}
					120	1.442×10^{-7}
0.726	0.144	0.043	0.043	0.043	100	1.442×10^{-7}
					100	1.458×10^{-7}
					110	1.470×10^{-7}
					120	1.480×10^{-7}

(continued)

Table 20.14 (continued) Thermal Diffusivity of Fish Jelly, Fish Sausage, and Fish Ham at Different Temperatures

X_w	X_{FM}	X_{so}	X_{SP}	X_{ws}	Temperature (°C)	Thermal Diffusivity (m^2/s)
					120	1.500×10^{-7}
					120	1.514×10^{-7}
0.763	0.125	0.037	0.037	0.037	100	1.479×10^{-7}
					110	1.509×10^{-7}
					120	1.533×10^{-7}
					100	1.286×10^{-7}
					110	1.281×10^{-7}
					110	1.305×10^{-7}
					120	1.313×10^{-7}
					120	1.295×10^{-7}
0.574	0.224	0.067	0.067	0.067	100	1.335×10^{-7}
					110	1.344×10^{-7}
					110	1.356×10^{-7}
					120	1.365×10^{-7}

Source: Han, B.H. and Loncin, M., *Food Sci. Technol.*, 18, 159, 1985.

$$\alpha = 1.541 \times 10^{-7} + 1.34 \times 10^{-10} T \tag{20.95}$$

$$\alpha = \left[17.65 - 0.113 \times 10^{-1} X_w + 0.126 \times 10^{-1} T\right] \times 10^{-8} \tag{20.96}$$

$$\alpha = \left[17.65 - 0.84 \times 10^{-3} \rho - 0.49 \times 10^{-2} X_w + 0.89 \times 10^{-2} T\right] \times 10^{-8} \tag{20.97}$$

Based on the composition of liquid foods, Choi and Okos (1986) suggested the following model:

$$\alpha = \sum_{i=1}^{n} \alpha_i \varepsilon_i \tag{20.98}$$

Thermal properties of food components as a function of temperature are given in Table 20.19. Above freezing point, the linear model is usually used, and below freezing thermal diffusivity varies nonlinearly. Nesvadba and Eunson (1984) proposed linear and nonlinear models as

$$\alpha = \hat{a} + \hat{b} X_w \tag{20.99}$$

Table 20.15 Thermal Diffusivity of Fish Jelly, Fish Sausage, and Fish Ham as a Function of Lipid Content at 120°C

X_w	X_{LI}	X_p	X_{ca}	X_{as}	Thermal Diffusivity (m^2/s)
0.650	0.016	0.259	0.015	0.060	1.465×10^{-7}
0.650	0.021	0.242	0.013	0.070	1.460×10^{-7}
0.650	0.043	0.210	0.039	0.058	1.431×10^{-7}
0.650	0.051	0.220	0.012	0.068	1.423×10^{-7}
0.650	0.066	0.177	0.065	0.041	1.437×10^{-7}
0.650	0.097	0.186	0.010	0.057	1.416×10^{-7}

Source: Han, B.H. and Loncin, M., *Food Sci. Technol.*, 18, 159, 1985.

Table 20.16 Thermal Diffusivity of Frozen and Unfrozen Model Foods

Material	X_w	X_{ag}	X_{ST}	X_{su}	T (°C)	Thermal Diffusivity (m^2/s)	T (°C)	Thermal Diffusivity (m^2/s)
Composite	0.971	0	0.029	0	15	1.908×10^{-7}	−10	7.992×10^{-7}
Composite	0.936	0.028	0.046	0	15	1.800×10^{-7}	−10	7.668×10^{-7}
Composite	0.885	0.027	0.088	0	15	1.728×10^{-7}	−10	7.128×10^{-7}
Composite	0.971	0.029	0	0	15	1.908×10^{-7}	−10	7.992×10^{-7}
Composite	0.847	0.025	0	0.127	14	1.764×10^{-7}	−13	4.860×10^{-7}
Composite	0.699	0.021	0	0.28	12	1.692×10^{-7}	−15	2.196×10^{-7}
Composite	0.595	0.018	0	0.387	12	1.620×10^{-7}	−15	2.124×10^{-7}
Composite	0.847	0.025	0	0.127	14	1.764×10^{-7}	−13	4.896×10^{-7}
Composite	0.837	0.038	0	0.126	14	1.656×10^{-7}	−13	2.700×10^{-7}
Composite	0.826	0.025	0	0.124	14	1.620×10^{-7}	−13	2.592×10^{-7}
Gel[a]	0.63	0.02	0	0.2	20–4	1.430×10^{-7}	—	—
Potato[b]	0.85	—	—	—	15	1.476×10^{-7}	−10	7.056×10^{-7}
	0.8	—	—	—	15	1.332×10^{-7}	−10	6.660×10^{-7}
	0.75	—	—	—	15	1.296×10^{-7}	−10	4.716×10^{-7}
	0.7	—	—	—	15	1.080×10^{-7}	−10	3.852×10^{-7}
Shrimp[c]	—	—	—	—	10	1.800×10^{-7}	−10	11.484×10^{-7}
Shrimp[d]	—	—	—	—	10	1.584×10^{-7}	−10	7.776×10^{-7}

Source: Albin, F.V., Badari, N.K., Srinivasa, M.S., and Krishna, M.M.V., *J. Food Technol.*, 14, 361, 1979.
[a] Adapted from Gordon and Thorne (1990) (15% sorbitol).
[b] Mashed: X_{st}, starch; X_{su}, sucrose.
[c] Peeled (grade 200/300).
[d] Peeled (grade 300/500).

Table 20.17 Thermal Diffusivities of Different Foods at Selected Temperatures

	Thermal Diffusivity × 10^7 (m^2/s)							
	Temperature (°C)							
Material	**−40**	**−25**	**−15**	**−10**	**−7**	**5**	**25**	**45**
Ice (measured; Nesvadba, 1982)	17.60	16.50	15.30	13.60	12.50	—	—	—
Ice (calculated; Nesvadba, 1982)	15.94	14.27	12.94	12.40	12.14	—	—	—
Agar food model[a]	—	12.50	9.00	6.50	4.50	1.35	1.40	—
Agar food model (Albin et al., 1979)	—	—	—	5.60	—	1.40	—	—
Sugar food model[b]	—	4.61	2.00	0.85	0.17	1.24	1.22	1.32
Cod fillet (Nesvadba, 1982)	—	6.00	3.80	2.50	1.60	1.20	1.40	—
Cod (Reidel, 1969)	—	—	—	—	1.21	1.31	1.34	—
Cod (calculated; Sanders, 1979)	10.20	7.77	5.49	3.81	2.48	1.50	1.50	—
Mackerel (Nesvadba, 1982)	5.50	5.50	4.00	2.00	1.20	1.20	1.30	—

Sources: Adapted from Nesvadba, P., *J. Phys. D: Appl. Phys.*, 15, 725, 1982; Albin, F.V., Badari, N.K., Srinivasa, M.S., and Krishna, M.M.V., *J. Food Technol.*, 14, 361, 1979; Riedel, L., *Kaltetechnik*, 21, 315, 1969; Sanders, H.R., *Advances in Fish Science and Technology*, Connell, J.J. (ed.), Fishing News Books Ltd., England, 1979.
[a] X_{ag}, 0.0283; X_{su}, 0.0283; X_w, 0.9434 (Nesvadba, 1982).
[b] X_{su}, 0.225; X_w, 0.600; X_{BA}, 0.175 (Nesvadba, 1982).

Table 20.18 Thermal Diffusivity of Yogurt, Whole Milk Powder, Apple Pulp, Meat Paste, and Fish Paste

		Thermal Diffusivity[a] $(m^2/s) \times 10^{-7}$							
Material	ρ_B (kg/m^3)	**1°C**	**10°C**	**20°C**	**25°C**	**30°C**	**40°C**	**50°C**	**75°C**
Yogurt	—	1.28	1.32	1.4	1.37	1.46	1.33	1.43	—
Milk powder	627	—	0.906	0.904	0.952	0.88	1.11	1.049	—
Apple pulp	—	—	1.345	1.35	1.43	—	1.505	1.422	—
Meat paste	—	—	1.177	1.272	1.382	—	1.325	1.412	2.053
Fish paste	—	—	1.22	1.18	1.337	—	1.353	1.36	1.93

	Composition				
Material	X_w	X_p	X_{PA}	X_{as}	X_{ca}
Yogurt	0.862	0.042	0.011	0.010	0.075
Milk powder	0.022	0.283	0.157	0.068	0.470
Apple pulp	0.886	0.002	—	0.002	0.110
Meat paste	0.692	0.129	0.124	0.020	0.035
Fish paste	0.718	0.161	0.047	0.031	0.043

Source: Kent, M., Christiansen, K., Van Haneghem, I.A., Holtz, E., Morley, M.J., Nesvadba, P., and Poulsen, K.P., *J. Food Eng.*, 3, 117, 1984. With permission.

[a] Values are average of five methods.

$$\alpha = \hat{c} + \hat{d}T \quad (20.100)$$

$$\log \alpha = \hat{e} + \hat{f}X_{w0} \quad (20.101)$$

The model parameters are given in Tables 20.20 through 20.22. Suter et al. (1975) provided the thermal diffusivity of whole peanut kernels, ground hulls, and kernels as a linear function of temperature as follows (7°C < T < 38°C):

$$\text{Whole kernels } (M_w = 0.07)\alpha = 0.746 \times 10^{-7} - 5.226 \times 10^{-10}T \quad (20.102)$$

$$\text{Ground kernels } (M_w = 0.08)\alpha = 1.490 \times 10^{-7} - 1.580 \times 10^{-9}T \quad (20.103)$$

$$\text{Ground hulls } (M_w = 0.05)\alpha = 1.151 \times 10^{-7} - 9.290 \times 10^{-10}T \quad (20.104)$$

Table 20.19 Thermal Diffusivity of Food Components as a Function of Temperature[a]

Component	**Equation**
Water	$\alpha = 1.3168 \times 10^{-1} + 6.2477 \times 10^{-4}T - 2.4022 \times 10^{-6}T^2$
Ice	$\alpha = 1.1756 - 6.0833 \times 10^{-3}T + 9.5037 \times 10^{-5}T^2$
Protein	$\alpha = 6.8714 \times 10^{-2} + 4.7578 \times 10^{-4}T - 1.4646 \times 10^{-6}T^2$
Fat	$\alpha = 9.8777 \times 10^{-2} - 1.2569 \times 10^{-4}T - 3.8286 \times 10^{-8}T^2$
Carbohydrate	$\alpha = 8.0842 + 5.3052 \times 10^{-4}T - 2.3218 \times 10^{-6}T^2$
Fiber	$\alpha = 7.3976 \times 10^{-2} + 5.1902 \times 10^{-4}T - 2.2202 \times 10^{-6}T^2$
Ash	$\alpha = 1.2461 \times 10^{-1} + 3.7321 \times 10^{-4}T - 1.2244 \times 10^{-6}T^2$

Source: Choi, Y. and Okos, M.R., in *Food Engineering and Process Applications*, Maguer, M. and Jelen, P. (eds.), Elsevier Applied Science, London, 1986.

[a] T in °C and varied from −40°C to 150°C.

Table 20.20 Values of *a* and *b* for Different Food Materials

Food	X_w	*T* (°C)	$a \times 10^7$	$b \times 10^7$	Reference
Fish jelly	0.500–0.763	100	0.868	0.809	Han and Loncin (1985)
Fish jelly	0.500–0.763	110	0.872	0.835	Han and Loncin (1985)
Fish jelly	0.500–0.763	120	0.874	0.861	Han and Loncin (1985)
Cod mince	0.337–0.904	10	0.7	0.683	Nesvadba and Eunson (1984)
Cod mince	0.337–0.904	25	1.01	0.385	Nesvadba and Eunson (1984)
Cod mince	0.337–0.904	35	1.19	0.236	Nesvadba and Eunson (1984)
Meat analog	0.650–0.850	82	0.53	1.233	Rizvi et al. (1980)
Meat analog	0.650–0.850	93	0.657	1.05	Rizvi et al. (1980)
Meat analog	0.650–0.850	116	0.205	1.583	Rizvi et al. (1980)
Rough rice	0.100–0.220	26	1.253	−0.016	Morita and Singh (1979)
Meat emulsion[a]	0.591–0.683	15–75	0.715	0.772	Timbers et al. (1982)
Dough wheat	—	20–90	2.658	−0.016	Metel et al. (1986)

[a] 60% beef, 20% pork, and 20% pork fat.

Table 20.21 Values of c and *d* in the Equation for Thermal Diffusivity of Fish[a]: $\alpha = c + dT$

Food	X_w	*T* (°C)	C	*D*
Fish				
Saman (Kumbhar et al., 1981)	0.773	0 to 40	1.194×10^{-7}	1.997×10^{-9}
Sol	0.776	0 to 40	1.421×10^{-7}	2.343×10^{-9}
Black bhitki (Kumbhar et al., 1981)	0.770	0 to 40	1.487×10^{-7}	1.569×10^{-9}
Black pomphret (Kumbhar et al., 1981)	0.750	0 to 40	1.364×10^{-7}	1.406×10^{-9}
Mackerel (Kumbhar et al., 1981)	0.774	0 to 40	1.070×10^{-7}	1.942×10^{-9}
Red bhitki (Kumbhar et al., 1981)	0.795	0 to 40	1.223×10^{-7}	2.186×10^{-9}
Singra (Kumbhar et al., 1981)	0.779	0 to 40	0.922×10^{-7}	1.820×10^{-9}
Hilsa (Kumbhar et al., 1981)	0.747	0 to 40	1.638×10^{-7}	1.572×10^{-9}
Surama (Kumbhar et al., 1981)	0.780	0 to 40	1.115×10^{-7}	1.701×10^{-9}
White pomphret (Kumbhar et al., 1981)	0.748	0 to 40	1.365×10^{-7}	2.346×10^{-9}
Malli (Kumbhar et al., 1981)	0.782	0 to 40	1.172×10^{-7}	2.282×10^{-9}
Rohu (Kumbhar et al., 1981)	0.752	0 to 40	1.325×10^{-7}	1.421×10^{-9}
Dough wheat (Rask, 1989)		20 to 90	1.541×10^{-7}	0.134×10^{-9}

Source: Adapted from Kumbhar, B.K., Agarwal, R.S., and Das, K., *Int. J. Refrig.*, 4, 143, 1981; Rask, C., *J. Food Eng.*, 9, 167, 1989.

[a] *T* (°C).

Table 20.22 Values of e and *f* in the Equation for Thermal Diffusivity of Cod Mince in the Freezing Range: $\log_{10} \alpha = e + f X_{wo}$

Material	X_w Range	Temperature (°C)	*E*	*f*
Cod mince	0.337–0.904	−40	-0.475×10^{-7}	1.73×10^{-7}
		−30	-0.594×10^{-7}	1.83×10^{-7}
		−20	-0.755×10^{-7}	1.88×10^{-7}
		−10	-1.341×10^{-7}	2.28×10^{-7}

Source: Nesvadba, P. and Eunson, C., *J. Food Technol.*, 19, 585, 1984.

Matthews and Hall (1968) correlated thermal diffusivity of white potato as a function of temperature and found a maximum in the temperature range in which white potato starch gelatinizes (Equation 20.105)

$$\alpha = -1.962 \times 10^{-2} + 2.617 \times 10^{-4}T - 8.500 \times 10^{-7}T^2 \tag{20.105}$$

where α is in ft^2/h and T is in °F. Wadsworth and Spadaro (1969) used a fourth-degree polynomial to correlate thermal diffusivity of sweet potato as a function of temperature (30°C < T < 90°C):

$$\alpha = 0.774 \times 10^{-7} + 2.581 \times 10^{-10}T + 1.290 \times 10^{-12}T^3 - 1.419 \times 10^{-14}T^4 \tag{20.106}$$

NOMENCLATURE

A	area (m^2)
a	parameter of Equation 20.99 or slope of Equation 20.23
b	parameter of Equation 20.99
B	parameter of Equation 20.49
Bi	Biot number (Rh/k or lh/k)
C	specific heat (kJ/kg K)
c	parameter of Equation 20.100
D	diameter (m)
d	parameter of Equation 20.100 or differentiation
E	parameter of Equation 20.80
e	parameter of Equation 20.101 or pulse energy input (J/m^2)
F_0	Fourier number ($\alpha t/R^2$)
f	parameter of Equation 20.101 or slope index of heating or cooling curve
f	order of partial moment
G	parameter of Equation 20.49
g	number of line segments of time–temperature curve
H	parameter of Equation 20.49
h	heat transfer coefficient (W/m^2 K)
I	intercept of Equation 20.33
J	Bessel function
j	j-factor $(T_m - T_{ei})/(T_m - T_0)$
k	thermal conductivity (W/m·K)
L	height of cylinder (m)
l	sample thickness (m)
M	mass fraction (dry basis)
m	indicates number of terms in series
n	indicates number of terms in series
p	root of Equation 20.53
Q	parameter of Equation 20.23
q	root of Equation 20.53
R	radius (m)
r	radial distance from the center (m)
S	shape factor (L/R)
T	temperature (°C)
t	time (s)
V	volume (m^3)

W	moments
X	mass fraction (wet basis)
x	x-axis
z	z-axis

Greek Symbol

Δ	difference
λ	heat input per unit length of line heat source (W/m)
δ	partial differentiation
θ	temperature ratio $(T - T_m)/(T_0 - T_m)$
η	r/R
ψ	location in a thermocouple
ς	location in a thermocouple
ρ	density (kg/m^3)
ε	volume fraction
β	root of transcendental equation
α	thermal diffusivity (m^2/s)
γ	$r/2(\alpha t)^{0.5}$
ξ	z/L
μ	time constant (l^2/α)
τ	$\alpha t/L^2$
ϕ	penetration depth of pulse energy (m)
ω	constant of Equation 20.17
Ω	constant rate of temperature rise (°C/s)
Υ	parameter of Equation 20.24
φ	constant of Equation 20.17
Γ	temperature ratio $(T_m - T_1)/(T_m - T_2)$
ψ	location of thermogram
ς	location of thermogram
π	3.142

Subscript

a	adiabatic
ag	agar
ap	apparent
as	ash
B	bulk
b	bottom
BA	binding agent
c	center
ca	carbohydrate
cs	center-surface
er	experimental reduced temperature
ei	extrapolated initial temperature
f	order of moment
F	freezing
FA	fat
FI	fiber

FM	fish meal
$gn(m-1)$	$gn(m-1)$th parameter of Q
gnm	gnmth parameter of Q
I	ice
i	ith component
j	jth portion of line
k	kth times
LI	lipid
L	lag
m	medium or mth root
max	maximum
max, c	maximum in center
max, s	maximum in surface
n	Nth root or number of terms in series
0	initial or 0th order
p	protein
r	reduced
r, max	reduced maximum
sa	sample
SO	soybean oil
SP	soybean protein
s	surface
SH	soy hull
si	side
ST	starch
su	sucrose
t	top
WS	wheat starch
w	water
wo	water before freezing or processing
1,2,3	1,2,3,4 conditions or location
1,2	1,2 conditions or locations
2,3	2,3 conditions or location

Superscript

*	theoretical reduced

REFERENCES

Albin, F.V., Badari, N.K., Srinivasa, M.S., and Krishna, M.M.V., Thermal diffusivities of some unfrozen and frozen food models, *Journal of Food Technology*, 14, 361, 1979.

Andrieu, J., Gonnet, E., and Laurent, M., Pulse method applied to foodstuffs: Thermal diffusivity determination, in *Food Engineering and Process Applications*, Vol. 1. *Transport Phenomena*, Le Maguer, M. and Jelen, P., Eds., Elsevier Applied Science, London, 1986.

Bakshi, A.S. and Yoon, J., Thermophysical properties of bread rolls during baking, *Food Science and Technology*, 17, 90–93, 1984.

Ball, C.O. and Olson, F.C.W., *Sterilization in Food Technology—Theory, Practice and Calculations*, McGraw-Hill, New York, 1957.

Bhowmik, S.R. and Hayakawa, K., A new method for determining the apparent thermal diffusivity of thermally conductive food, *Journal of Food Science*, 44(2), 469, 1979.

Carslaw, H.S. and Jaeger, J.C., *Conduction of Heat in Solids*, Oxford University Press, England, 1947.

Carslaw, H.S. and Jaeger, J.C., *Conduction of Heat in Solids*, Oxford University Press, Oxford, England, 1959.

Choi, Y. and Okos, M.R., Effects of temperature and composition on the thermal properties of foods, in *Food Engineering and Process Applications*, Vol. 1, Transport Phenomena, Maguer, M. and Jelen, P., Eds., Elsevier Applied Science, London, 1986.

Degiovanni, A., Diffusivite et methode flash, *Revue generale de thermique*, 185, 417, 1977.

Dickerson, R.W., An apparatus for measurement of thermal diffusivity of foods, *Food Technology*, 19(5), 198, 1965.

Dickerson, R.W. and Read, R.B., Thermal diffusivity of meats, *ASHRAE Transaction*, 81(1), 356, 1975.

Drouzas, A.E., Maroulis, Z.B., Karathanos, V.T., and Saravacos, G.D., Direct and indirect determination of the effective thermal diffusivity of granular starch, *Journal of Food Engineering*, 13(2), 91, 1991.

Foust, A.S., Wenzel, L.A., Clump, C.W., Maus, L., and Andersen, L.B., *Principles of Unit Operations*, 2nd edn., John Wiley and Sons, New York, 1960.

Gaffney, J.J., Baird, C.D., and Eshleman, W.D., Review and analysis of transient method for determining thermal diffusivity of fruits and vegetables, *ASHRAE Transaction*, 86(2), 261, 1980.

Gordon, C. and Thorne, S., Determination of the thermal diffusivity of foods from temperature measurements during cooling, *Journal of Food Engineering*, 11, 133, 1990.

Griffith, C.L., Specific heat, thermal conductivity, density and thermal diffusivity of Mexican tortillas dough, *Journal of Food Science*, 50(5), 1333, 1985.

Gurney, H.P. and Lurie, J., Charts for estimating temperature distributions in heating or cooling solid shapes, *Industrial and Engineering Chemistry*, 15, 1170, 1923.

Han, B.H. and Loncin, M., Thermal diffusivities of fish products, *Food Science and Technology*, 18, 159, 1985.

Hayakawa, K., New computational procedure for determining the apparent thermal diffusivity of a solid body approximated with an infinite slab, *Journal of Food Science*, 38, 623–629, 1973.

Hayakawa, K. and Succar, J., A method for determining the apparent thermal diffusivity of spherical foods, *Food Science and Technology*, 16, 373, 1983.

Hwang, M.P. and Hayakawa, K., A specific heat calorimeter for foods, *Journal of Food Science*, 44(2), 435–438, 448, 1979.

Johnsson, C. and Skjoldebrand, C., Thermal properties of bread during baking, in *Engineering and Food*, Vol. 1, McKenna, B.M., Ed., Elsevier Applied Science Publishers, London, 1984.

Jones, J.C., Wootton, M., and Green, S., Measured thermal diffusivities of cereal products, *Food Australia*, 44(11), 501, 1992.

Kafiev, N.M., Lekhter, A.E., Leites, R.Y., Klokacheva, O.A., Panin, A.S., Ribakov, A.A., and Didenko, I.A., Thermophysical characteristics of certain bread dough improvers, *Maslozhirovaya Promyshlennost*, 1, 20, 1987.

Kent, M., Christiansen, K., Van Haneghem, I.A., Holtz, E., Morley, M.J., Nesvadba, P., and Poulsen, K.P., Cost 90 collaborative measurements of thermal properties of foods, *Journal of Food Engineering*, 3, 117, 1984.

Keppler, R.A. and Boose, J.R., Thermal properties of frozen sucrose solutions, *Transactions of the ASAE*, 13(3), 335, 1970.

Kriems, P. and Reinhold, M., Dsa backen von mischbrot (VI)-Schlussfolgerungen zurverbesserung des back-effektes und der brotqualitat zusammenfassung, Backer und Konditor, 34912, 356, 1980.

Kulacki, F.A. and Kennedy, S.C., Measurement of the thermophysical properties of common cookie dough, *Journal of Food Science*, 43, 380, 1978.

Kumbhar, B.K., Agarwal, R.S., and Das, K., Thermal properties of fresh and frozen fish, *International Journal of Refrigeration*, 4(3), 143–146, 1981.

Lind, I., Thawing of minced meat and dough: Thermal data and mathematical modeling, in *Progress in Food Preservation Processes*, Vol. 1, CERIA, Brussels, 1988.

Makljukow, I.I. and Makljukow, W.I., Thermophysikalische charakteristika fur das backstuck, in *Industrieofen der Backwarenproduktion*, Verlag Leicht und Lebensmittelindustrie, Moskau, 1983.

Mannheim, H.C., Steinberg, M.P., Nelson, A.I., and Kendall, T.W., The heat content of bread, *Food Technology*, 7, 384, 1957.

Martens, T., Mathematical model of heat processing on flat containers, PhD thesis, Katholeike University, Leuven, Belgium, 1980.

Matthews, F.V. and Hall, C.W., Method of finite differences used to relate changes in thermal and physical properties of potatoes, *Transactions of the ASAE*, 11(4), 558–562, 1968.

McCurry, T.A., The development of a numerical technique for determining thermal diffusivity utilizing data obtained through the use of a refined line-source method, Advanced Project Report, Auburn University, 1968.

Metel, S.N., Mikrukov, V.V., and Kasparov, M.N., Thermophysical characteristics of yeast dough (Izvestiga Vysshikh Uchebnykh Zavedenii), *Pishcshevaya Teknologiya*, 4, 107, 1986.

Mohsenin, N.N., *Thermal Properties of Foods and Agricultural Materials*, Gordon and Breach Science Publishers, New York, 1980.

Moore, G.A. and Bilanski, W.K., Thermal properties of high moisture content alfalfa, *Applied Engineering in Agriculture*, 8(1), 61, 1992.

Morita, T. and Singh, R.P., Physical and thermal properties of short-grain rough rice, *Transactions of the ASAE*, 22, 630, 1979.

Muzilla, M., Unklesbay, N., Unklesbay, K., and Helsel, Z., Effect of moisture content on density, heat capacity and conductivity of restructured pork/soy hull mixtures, *Journal of Food Science*, 55(6), 1491, 1990.

Nebelung, M., Model der warmeubertragung in kontinuerlich und diskontinuerlich arbeitenden kammerofen under berucksichtigung von warmeleitung, Stofftransport und reaction in einem feuchten kapillarporosen korper, PhD thesis, TU Dresden, DDR, 1979.

Nesvadba, P., A new transient method for the measurement of temperature dependent of thermal diffusivity, *Journal of Physics D Applied Physics*, 15, 725, 1982.

Nesvadba, P. and Eunson, C., Moisture and temperature dependence of thermal diffusivity of cod minces, *Journal of Food Technology*, 19, 585–592, 1984.

Neznanova, N.A., Panin, A.C., Puchova, L.L., and Skverchak, V.D., Thermophysical properties of dough and bread from grade I wheat flour during baking, *Khlebopekarnaya i Konditerskaya Promyshlennost*, 8, 13, 1978.

Nix, G.H., Lowery, G.W., Vachon, R.I., and Tanger, G.E., Direct determination of thermal diffusivity and conductivity with a refined line-source technique, *Progress Astronautics Aeronautics*, 20, 865, 1967.

Nix, G.H., Vachon, R.I., Lowery, G.W., and McCurry, T.A., The line source method. Procedure and iteration scheme for combined determination of conductivity and diffusivity. Thermal conductivity. *Proceedings of Eighth Conference, 4*, Plenum Press, New York, 1969.

Nordon, P. and Bainbridge, N.W., Some properties of char affecting the self-heating reaction in bulk, *Fuel*, 58, 450, 1979.

Olivares, M., Guzman, J.A., and Solar, I., Thermal diffusivity of nonhomogeneous food, *Journal of Food Processing and Preservation*, 10, 57, 1986.

Olson, F.C.W. and Schultz, O.T., Temperature in solids during heating and cooling, *Industrial and Engineering Chemistry*, 34, 874, 1942.

Ordinanz, W.O., Specific heat of foods in cooling, *Food Industries*, 18(12), 101, 1946.

Pflug, I.J., Blaisdell, J.L., and Kopelman, I.J., Developing temperature–time curves for objects that can be approximated by a sphere, infinite plate or infinite cylinder, *ASHRAE Transaction*, 71, 238, 1965.

Polak, M., Specific heat capacity of dough, *Mlynsko Pekarensky Prumysl*, 30(2), 42, 1984.

Poulsen, K.P., Thermal diffusivity of foods measured by simple equipment, *Journal of Food Engineering*, 1, 115, 1982.

Rask, C., Thermal properties of dough and bakery products: A review of published data, *Journal of Food Engineering*, 9, 167, 1989.

Riedel, L., Measurements of thermal diffusivity on foodstuffs rich in water, *Kaltetechnik*, 21, 315, 1969.

Rizvi, S.S.H., Blaisdell, J.L., and Harper, W.J., Thermal diffusivity of model meat analog systems, *Journal of Food Science*, 45, 1727, 1980.

Sanders, H.R., *Advances in Fish Science and Technology*, Connell, J.J., Ed., Fishing News Books Ltd., England, 1979.

Singh, R.P., Thermal diffusivity in food processing, *Food Technol.*, 36(2), 87, 1982.

Singh, R.P., Heating and cooling processes for foods, in *Handbook of Food Engineering*, Heldman, R. and Lund, D.B., Eds., Marcel Dekker Inc., New York, 1992, Chap. 5.

Standing, C.N., Individual heat transfer modes in band oven biscuit baking, *Journal of Food Science*, 39, 267, 1974.

Suter, D.A., Agrawal, K.K., and Clary, B.L., Thermal properties of peanut pods, hulls and kernels, *Transactions of the ASAE*, 18(2), 370, 1975.

Sweat, V.E., Thermal properties of low and intermediate moisture foods, *ASHRAE Journal*, February, 44, 1987.

Sweat, V.E., Experimental measurement of the thermal conductivity of a yellow cake, *Proceedings of the XIIIth International Conference on Thermal Conductivity*, University of Missouri, Rolla, 1973.

Tadano, T., Thermal conductivity of white bread, *Bulletin of the College of Agricultural and Veterinary Medicine*, 44, 18, 1987.

Tichy, O., Matematick'y model sdileni tepla a hmoty pri peceni, *Potravinarska a Chladici Technika*, 5(1), 20, 1974.

Timbers, G.E., Randall, C.J., and Raymond, D.P., Effect of composition on thermal properties of meat emulsions, *Canadian Institute of Food Science and Technology Journal*, 15(3), 191–195, 1982.

Tschubik, I.A. and Maslow, A.M., *Warmephysikalische Konstanten von Lebensmitteln und Halfabrikalen*, Fachbuchverlag, Leipzig, 1973.

Unklesbay, N., Unklesbay, K., Nahaisi, M., and Krause, G., Thermal conductivity of white bread during convective heat processing, *Journal of Food Science*, 47, 249, 1981.

Uno, J. and Hayakawa, K., A method for estimating thermal diffusivity of heat conduction food in cylindrical can, *Journal of Food Science*, 45, 692, 1980.

Wadsworth, J.I. and Spadaro, J.J., Transient temperature distribution in whole sweetpotato roots during immersion heating, *Food Technology*, 23, 219, 1969.

Yanez, S.M., Modelo general de transferencia de calor en conservas, Thesis de prado Depto. Ingenieria Quimica U. Catolica, Santiago-Chile, 1983.

CHAPTER 21

Measurement of Surface Heat Transfer Coefficient

Shyam S. Sablani

CONTENTS

21.1 INTRODUCTION

The surface heat transfer coefficient is used to quantify the rate of convective heat transfer to or from the surface of an object. Convective heat transfer is the mode of energy transfer between a solid surface and the adjacent fluid that is in motion, and it involves the combined effects of conduction and fluid motion. The faster the fluid motion, the greater the convection heat transfer. In the absence of any bulk fluid motion, heat transfer between a solid surface and the adjacent fluid is by pure conduction. The rate of convection heat transfer is observed to be proportional to the temperature difference and is conveniently expressed by Newton's law of cooling as

$$Q = hA_s(T_s - T_\infty) \tag{21.1}$$

where
- Q is the net energy added to the system (w)
- h is the surface or convective heat transfer coefficient (W/m^2 K)
- A_s is the surface area (m^2) through which convection heat transfer takes place
- T_s is the surface temperature (K)
- T_∞ is the temperature of the fluid (K) sufficiently far from the surface

The surface heat transfer coefficient is not a property of a fluid or the material. It is an experimentally determined parameter whose value depends on all the variables influencing convection such as the surface geometry, the nature of fluid motion, the properties of the fluid, and the bulk fluid velocity (Cengel, 2003). In food applications, it is needed to quantify heat transfer during heating or cooling of food materials in fluids. The surface heat transfer coefficient is an important parameter needed in the design and control of food processing equipment where fluids, such as air, nitrogen, steam, water, oil, etc., are used as heating, cooling, frying, freezing, or cooking media (Rahman, 1995).

21.2 MEASUREMENT METHODS

Experimental methods to determine surface heat transfer coefficient can be classified into three categories: (1) steady-state methods (2) quasi-steady-state methods, and (3) transient methods. A brief description of these methods is presented in this section.

21.2.1 Steady-State Method

21.2.1.1 Interference Method

Eckert and Soelnghen (1951) used a method of interfectometer to estimate surface heat transfer coefficient. They used the interference method to experimentally determine the temperature distribution in the film of a fluid near the surface. This method was based on the principle that the index of refraction of a fluid varies with its density and fluid density is a function of temperature. The rate of heat flow from the surface can be calculated as

$$Q = \frac{k_b}{\delta} A_s(T_s - T_\infty) \tag{21.2}$$

where
- k_b is the thermal conductivity of fluid film (W/mK)
- δ is the film thickness (m)

The assumptions are heat transfer resistance only in the fluid film and uniform or linear temperature gradient through the film thickness (within boundary layer). The heat transfer coefficient can be calculated from the following relationship:

$$h = \frac{k_b}{\delta} \tag{21.3}$$

21.2.1.2 *Constant Heating Method*

At steady-state condition, the surface and fluid temperatures remain constant and can be calculated from the equation

$$Q = hA_s(T_s - T_\infty) \tag{21.4}$$

where
A_s is the surface area (m^2)
T_s is the surface temperature (°C)
T_∞ is the temperature of the bulk fluid

Clary and Nelson (1969, 1970) and Feldmann (1976) used an electrical heater at the center of a body. An induction coil may also be used when materials are packed in bulk (Baumeister and Bennett, 1958). Although this method seems to be quite simple and accurate, the placement of the temperature sensors is critical (Arce and Sweat, 1980). It is difficult to accurately measure the surface heat conductance of agricultural materials with the steady-state method because the source of heat will change their physical properties and measuring surface temperatures may alter the boundary characteristics of the product (Arce and Sweat, 1980). Some investigators measured temperature at two or three points along the line perpendicular to the surface and surface temperature can be estimated through extrapolation (Feldmann, 1976). Heat transfer coefficient for a constant heating system by heating element can be calculated as

$$h = \frac{\text{(volt)(current)}}{A_s(T_s - T_\infty)} \tag{21.5}$$

In many cases, surface or bulk fluid temperature is not constant, such as heating of fluid in a tubular heat exchanger where inlet bulk fluid temperature is lower than the outlet temperature; even surface temperature is maintained as constant. In this case, the mean temperature gradient is used to calculate the heat transfer coefficient:

$$h = \frac{Q}{A_s(\Delta T)_{av}} \tag{21.6}$$

Arithmetic mean and logarithmic mean are most commonly used to estimate the mean temperature gradient.

$$\text{Log mean: } (\Delta T)_{av} = \frac{\Delta T_1 - \Delta T_2}{\ln(\Delta T_1/\Delta T_2)} \tag{21.7}$$

$$\text{Arithmetic mean: } (\Delta T)_{av} = \frac{\Delta T_1 + \Delta T_2}{2} \tag{21.8}$$

21.2.2 Quasi-Steady State

When the temperature inside the solid is uniform, then the heat balance for an initial step change in ambient temperature can be written as

$$hA_s(T_s - T_\infty) = \rho V C_p \frac{dT}{dt} \tag{21.9}$$

where
V is volume of solid
t is time
T_s is surface temperature of solid
ρ is density of solid
C_p is specific heat of solid

The above differential equation can be solved with initial condition as

$$\frac{T - T_\infty}{T_i - T_\infty} = \exp\left[-\frac{hA_s t}{\rho V C_p}\right] \tag{21.10}$$

where T_i is initial temperature of solid.

The values of h can be calculated from the slope of the line $\ln[(T - T_\infty)/(T_i - T_\infty)]$ versus t. The temperature inside the solid can be considered negligible in case of low Biot ($Bi = ha/k$), i.e., smaller than 0.1. Some authors used high thermal conductivity materials to keep the Biot number small while others used liquids where natural convection currents will keep the temperature constant (Leichter et al., 1976). Arce and Sweat (1980) mentioned that the characteristic length can be approximated by V/A_s (volume/surface area) in the case of irregularly shaped objects. The overall heat transfer coefficient for perfect mixing convection heat transfer mode in a can as mentioned by Deniston et al. (1992) can be calculated as

$$h = 2.303 \frac{V \rho C_p}{f A_s} \tag{21.11}$$

where f is the heating or cooling rate index.

21.2.3 Transient Method

In the cases where Biot number is larger than 0.1, there will be a temperature gradient within the solid and the quasi-steady state is no longer valid. Then the heat conduction partial differential equation must be solved by analytical or numerical method.

21.2.3.1 Analytical Solution

The heat transfer equation for a slab, sphere, or cylinder with the boundary conditions can be analytically solved to estimate heat transfer coefficient. The heat conduction equation for a slab in dimensionless form is

$$\frac{\partial^2 \theta}{\partial x^2} = \frac{1}{\alpha} \frac{\partial \theta}{\partial t} \tag{21.12}$$

where x is nondimensional coordinate [$x/1$, where x is axial coordinate system (m) and 1 is thickness (m)], a is the thermal diffusivity and θ is the dimensionless temperature. The boundary conditions are

$$\theta = 1 \quad \text{at} \quad t = 0 \quad \text{for} \quad -1 \leq x \leq +1 \tag{21.13}$$

$$\frac{\partial\theta}{\partial x} = 0 \quad \text{at} \quad t > 0 \quad \text{and} \quad x = 0 \tag{21.14}$$

$$\frac{\partial\theta}{\partial x} - h\theta = 0 \quad \text{at} \quad t > 0 \quad \text{and} \quad x = \pm 1 \tag{21.15}$$

The solution from Carslaw and Jaeger (1959) is

$$\theta = \sum_{i=1}^{\infty} \frac{2Bi\left[\cos(\eta_i x/l)\right](\eta_i)}{Bi(Bi+1)+\eta_i^2} \exp\left[-\eta_i^2 F_0\right] \tag{21.16}$$

where η_i is the ith root of the following equation:

$$\eta_i \cot \eta_i = Bi \tag{21.17}$$

$$\text{where} \quad F_0 = \frac{\alpha t}{a^2}; \quad Bi = \frac{hl}{k} \tag{21.18}$$

where l is half of the slab thickness (m). The solutions for other simple geometries also follow a similar form. For other geometries, the solutions are available in Carslaw and Jaeger (1959). The use of digital computers is necessary to evaluate h from the above equation. But one term approximation of the infinite series solution can also be used assuming the higher terms in the summation equation are negligible, especially after some time has elapsed. In this case, Equation 21.16 reduces to

$$\theta = \frac{2Bi[\cos(\eta/l)\sec(\eta)}{Bi(Bi+1)+\eta^2} \exp\left[-\eta^2 F_0\right] \tag{21.19}$$

The values of η can be determined from the slope of the line ln θ versus F_0. Then h can be calculated from Equation 7.17 by knowing the value of Biot number (Bi). This method unfortunately only uses a part of the data obtained in cooling/heating curves, where the temperature of the body changes slowly, thus magnifying the measurement error (Arce and Sweat, 1980). If the body is nonisotropic or nonhomogeneous with regard to their thermal properties, this will introduce another source of error. The above authors mentioned that this scheme to calculate h gave reasonable accuracy if properly used. This method is also not suitable for irregular materials.

Chang and Toledo (1990) used the first four terms of the exact series solution for an infinite slab to calculate the heat transfer coefficient of cubes immersed in an isothermal fluid. The multiplicative technique was employed in simplifying the infinite series solution for the finite cube. They calculated the time–temperature data iteratively at three locations until the h value that gave the best fit between experimental and predicted data was found.

21.2.3.2 Numerical Solution

The numerical techniques such as finite differences or finite elements are also used to calculate the heat transfer coefficient heat conductance. In the case of heterogeneous complex geometries or variable boundary conditions, only the numerical method is available to solve the heat conduction equations. This technique is the most accurate if the numerical solution uses small space and time increments to calculate the solution (Arce and Sweat, 1980).

21.2.3.3 Plank's Method

Plank's (1932) equation can also be used to predict the heat transfer coefficient during the freezing process. The equation is

$$t = \frac{\lambda}{\Delta T}\left[\frac{l}{2h} + \frac{l^2}{8k}\right] \tag{21.20}$$

where
- t is the freezing time (s)
- l is the thickness of the slab (m)
- k *is* the thermal conductivity of the frozen layer (W/mK)
- λ is the latent heat effusion (J/kg)
- ΔT is the temperature difference between the ambient and the slab (°C)

21.2.3.4 Surface Heat Flux Method

Federov et al. (1972) used heat flux pickup to measure the heat flux and the surface temperature as a function of time. Arce and Sweat (1980) mentioned that this method gave approximate values due to the presence of a device at the surface of the body.

21.3 MEASUREMENT OF HEAT TRANSFER COEFFICIENT IN DIFFERENT FOOD PROCESSES

The surface heat transfer coefficient is influenced by the composition of the fluid, the nature and geometry of particle surface, and the hydrodynamics of the fluid moving past the surface. Hence, it is important to measure the surface heat transfer coefficient under the conditions simulating the actual process. The estimation of the heat transfer coefficient falls under the category of an inverse heat conduction problem. This approach requires experimental measurement of the transient temperatures inside a body of known geometry at a specified location and estimation of transient temperatures, at the same location, by solving the governing heat conduction equations with an assumed convective boundary condition (i.e. the Biot number, *Bi*). In doing so, *Bi* is varied systematically to produce computed temperature–time histories closely matching the experimentally measured temperature histories. The following section describes the experimental protocols used for measurement of surface heat transfer coefficient under various food processing situations.

21.3.1 Canning

Accurate prediction of particle center temperature through mathematical models depends on how well overall heat transfer coefficient (U) and surface heat transfer coefficient (h) data are estimated under simulated process conditions, where both system and product parameters are expected to be influencing factors. Traditionally, U and h are determined from the measurement of the temperature responses of the particle and liquid under well-characterized initial and boundary conditions (Maesmans et al., 1992). Particle and liquid temperatures are generally measured using thermocouples. In real processing, situations involving rotational retorts, the particle motions are expected to be influenced by centrifugal, gravitational, drag, and buoyancy forces and hence, the associated U and h. The particle, attached to a rigid thermocouple, will not simulate the particle motion during agitation processing, which in turn causes some deviations in the measured heat transfer coefficients. Recently, attempts have been made to monitor the temperature of the particle

Table 21.1 Methods Employed for Determining *U* and *h*, with Canned Liquid/Particle Mixtures, Subjected to Agitation Processing

References	Motion of Test Particle	Location of Particle Temperature Measurement	Mathematical Procedure
Lenz and Lund (1978)	Restricted	Center	Lumped capacity, Duhamel's theorem
Hassan (1984)	Restricted	Surface	Integration of overall energy balance equation
Lekwauwa and Hayakawa (1986)	Restricted	Center	Empirical formulae containing heating rate index and factor, Duhamel's theorem
Deniston et al. (1987)	Restricted	Surface	Integration of overall energy balance equation
Fernandez et al. (1988)	Restricted	Center	Lumped capacity, Duhamel's theorem
Stoforos and Merson (1991)	Free	Only liquid temperature needed	Analytical solution in the Laplace transform plane
Stoforos and Merson (1991)	Free	Surface (liquid crystal)	Integration of overall energy balance equation
Weng et al. (1992)	Free	Indirect, accumulated process lethality at center	Numerical solution of heat conduction equation with convective boundary conditions
Sablani and Ramaswamy (1995)	Free	Center	Integration of overall energy balance equation and numerical solution of heat conduction equation with convective boundary conditions

without inhibiting particle motion during agitation processing. Apart from difficulties in measuring the temperature of the moving particle, the mathematical solution, of the governing equation of the energy balance on the can (containing liquid and particles), is also complex due to the time variant temperature of the can liquid. Table 21.1 summarizes published methods to determine U and h in particulate liquids in cans subjected to agitation processing together with a description of the experimental procedure and the mathematical solution. The procedures used for the determination of U and h are classified into two groups, based on the motion of the experimental particle whose temperatures are monitored, (1) fixed particle and (2) moving particle, during agitation processing.

21.3.1.1 Mathematical Procedures to Determine U *and* h

The surface heat transfer coefficient is determined using an inverse heat transfer approach, in which the boundary condition is determined using measured transient temperatures. The governing partial differential equations have to be solved to describe the conduction heat flow inside the particle with appropriate initial and boundary conditions. Experimental data needed for this analysis are the transient temperatures of both liquid and particle and thermophysical properties of the particle. The liquid bulk temperatures are measured using a needle-type copper-constantan thermocouple, with its tip located at an appropriate location of the can, and a thermocouple, embedded into a predefined location in the particle, is used to monitor particle transient temperatures.

The overall thermal energy balance on a particulate–liquid food system is used to calculate the associated convective heat transfer coefficients. The governing equation for heat transfer in such systems can be written as (all symbols are detailed in nomenclature)

$$UA_c(T_R - T_f) = m_f C_{pf}\frac{dT_f}{dt} + m_p C_{pp}\frac{d < T_p >}{dt} \tag{21.21}$$

where

A_c is surface area of can
T_f is fluid temperature
C_{pf} is specific of heat of fluid
$<T_p>$ is average particle temperature
m_f is mass of fluid
m_p is mass of particles
C_{pp} is specific heat of particle
T_R is retort temperature
U is overall heat transfer coefficient

The following are assumed in the solution of Equation 21.21: uniform initial temperature for the particle, uniform initial and transient temperatures for the liquid, constant heat transfer coefficients, constant physical and thermal properties for both liquid and particles, and no energy accumulation in the can wall.

The second term on the right side of Equation 21.21 is equal to the heat transferred to particles from the liquid through the particle surface:

$$m_p C_{pp} \frac{d < T_p >}{dt} = hA_s (T_f - T_{ps}) \tag{21.22}$$

where T_{ps} is particle surface temperature.

It is also assumed that the particle receives heat only from the liquid and not from the can wall when it impacts, i.e., heat is transferred first from the can wall to the liquid and then to the particle. For example, the heat flow in a spherical particle immersed in liquid can be described by the following partial differential equation:

$$\frac{\partial T}{\partial t} = \alpha_p \left(\frac{\partial^2 T}{\partial r^2} + \frac{2}{r} \frac{\partial T}{\partial r} \right) \tag{21.23}$$

where

T is temperature
r is radial coordinate system (m)
α_p is particle thermal diffusivity

The initial and boundary conditions are

$$T(r, 0) = T_i, \quad \text{at} \quad t = 0 \tag{21.24}$$

where T_i is initial temperature

$$\frac{dT(0, t)}{dr} = 0, \quad \text{at} \quad t > 0 \tag{21.25}$$

$$k_p \frac{dT}{dr} = h(T_f - T_{ps}) \tag{21.26}$$

where K_p is particle thermal conductivity

The solution of Equations 21.23 through 21.26 is complex because of time-dependent liquid temperatures.

21.3.1.2 Experimental Procedure

21.3.1.2.1 Restricted Particle Motion

Lenz and Lund (1978) developed a numerical solution using the fourth-order Runge–Kutta method and Duhamel's theorem to determine U and h, for the low Biot number ($Bi < 0.1$) situation. They measured the transient temperatures of a lead particle fixed at the geometric center of the can with liquid moving around it and verified that the lead particle quickly approached to one temperature at all the points. Assuming that the retort temperature instantly reached its operating condition, they proposed the following solution of Equation 21.27 for the temperature at any position in the particle:

$$\left[\frac{T_R - T}{T_R - T_i}\right] = \frac{2Bi}{r/a}\sum\left(\frac{\beta_n^2 + (Bi-1)}{\beta_n^2 + Bi(Bi-1)}\right)\frac{\sin\beta_n}{\beta_n^2}\sin\beta_n(r/a) \times \left[\exp(-\tau_p t + \left(\frac{\tau_p}{\tau_p - \tau_f}\right)\left[\exp(-\tau_f t) - \exp(-\tau_p t)\right]\right] \tag{21.27}$$

where

T_R is retort temperature

τ_p and τ_f are time constants for particle are fluid (τ is $\alpha\beta^2/a_p$; where α is thermal diffusivity, β is root of the equation: $\beta\cot(\beta) + Bi - 1 = 0$, Bi is Biot number: ha_p/k; h is heat transfer coefficient (W/m^2k^2), a_p is charateristic dimension of a particle, k is thermal conductivity).

They estimated h by minimizing the sum of the squared deviations between measured and predicted particle temperature profile. They also obtained the equation for the particle average temperature from Equation 21.27 and used it in Equation 21.21 to calculate the overall heat transfer coefficient.

Hassan (1984) derived the following equations for U and h by integrating Equations 21.21 and 21.22, respectively, allowing the heating time to approach infinity ($<T_p(\infty)> = T_f(\infty) = T_R$):

$$U = \frac{m_f C_{pf}}{A_c}\frac{T_R - T_{fi}}{\int_0^\infty (T_R - T_f)\,dt} + \frac{m_p C_{pp}}{A_c}\frac{(T_R - T_{pi})}{\int_0^\infty (T_R - T_f)\,dt} \tag{21.28}$$

where T_{fi} and T_{pi} are initial temperatures of fluid and particle.

$$h = \frac{m_p C_{pp}\left(<T_p>_{\text{final}} - <T_p>_{\text{initial}}\right)}{A_p\int_0^\infty (T_f - T_{ps})\,dt} \tag{21.29}$$

where $<T_p>$ and $<T_f>$ are average temperature of particle and fluid.

For cans subjected to end-over-end rotation, Lekwauwa and Hayakawa (1986) developed a model using an overall heat balance equation in combination with an equation for transient heat conduction in a particle. They considered the probability function representing the statistical particle volume distribution. The temperature distribution for individual particles was obtained using Duhamel's theorem and empirical formulae describing the heat transfer to spherical, cylindrical, or oblate spheroid-shaped particles in a constant temperature liquid. In their numerical solution, they assumed that within each time step the liquid temperature was a linear function of time and the coefficients of these functions were determined iteratively such that the resulting particle and liquid temperatures satisfied the overall heat balance equation.

Deniston et al. (1987) used Equations 21.28 and 21.29 to determine U and h in axially rotating cans. In their experiments, the heating time was sufficiently long to allow liquid and particle average temperatures to reach the heating media temperature to satisfy infinite time limits in the above equations. They measured the transient temperature at the particle surface using rigid-type thermocouples. They recognized the difficulties and errors associated with the measurement of surface temperatures.

Fernandez et al. (1988) determined the convective heat transfer coefficients for bean-shaped particles, in cans processed in an agitated retort. They preferred a high conductivity material such as aluminum to give a low Biot number ($Bi < 0.01$) condition and used the lumped capacity method for U and h evaluation. They measured time–temperature data for both the liquid and particle using rigid-type thermocouples and used the scheme developed by Lenz and Lund (1978) to solve the heat balance equations.

Recently, Stoforos and Merson (1995) proposed a solution to the differential equations governing heat transfer to axially rotating particulate liquids in cans. They used an analytical solution, Duhamel's theorem for particle temperature, and a numerical solution based on the fourth-order Runge–Kutta scheme for the liquid temperature. The solution avoided the need for empirical formulae or a constant heating medium within short time intervals. Their comparison between predicted values and experimental data from Hassan (1984) showed good correlation for liquid temperature; however, it deviated for particle surface temperatures.

21.3.1.2.2 Allowing Particle Motion

From liquid temperature only: Stoforos and Merson (1990) used a mathematical procedure requiring only the measurement of liquid temperature to estimate U and h in axially rotating cans. They solved an overall energy balance equation (Equation 21.21) for a can and the differential equation for a spherical particle with appropriate initial and boundary conditions (Equations 21.23 through 21.26). Since the can liquid temperature depends on both U and h, by systematically varying these coefficients and minimizing the error between experimental and predicted liquid temperatures (in Laplace plane), they estimated the U and h. This method allows for free movement of particles. The authors reported that calculated h values did not always coincide with those determined from particle surface temperature measurements.

Liquid crystal: Stoforos and Merson (1991) extended the solution procedure for the overall energy balance of the can for finite heating time, previously used by Hassan (1984) to determine U and h. They used a liquid crystal, which changes color with temperature, as a sensor to monitor the particle surface temperature. The method involved coating the particle surface with an aqueous solution of liquid crystals, videotaping of the color changes on the particle surface as a function of temperature, and comparing it with standard color charts after calibration. This method does not impose any restriction on particle motion, while monitoring the particle temperature during can rotation. They measured the liquid temperatures using rigid thermocouples for finite heating time and exponentially extrapolated them to obtain liquid temperature data for "long time" approximation of heating time in the solution equation (Equation 21.27). They used an iterative procedure to calculate the average particle temperature, required in the solution equation for hip calculation. By using the first term approximation to the series solution, for average particle temperature ($<T_{\mathrm{p}}>$) as a function of particle surface temperature and Biot number (Stoforos, 1988), they obtained an analytical expression. Initially, as a first approximation, h was estimated by assuming $<T_{\mathrm{p}}(t)> = T_{\mathrm{ps}}(t)$ in Equation 21.28; the particle average temperature was obtained from analytical expression. This value for average particle temperature was substituted into Equation 21.28 to find an improved value for h. The authors reported that one or two iterations were usually enough for convergence. However, they cautioned that the first term approximation in the expression of particle average temperature was limited to tow particle thermal diffusivity materials and suggested that for a high thermal diffusivity particle $\langle T_p(t) \rangle = T(t)$ may be a good assumption. The experiments were carried out between the temperature range of 20°C to 50°C.

Time–temperature integrator (TTI): The combined use of a TTI has been proposed in the form of microorganism, chemicals or enzymes, and a mathematical model to determine the convective heat transfer coefficient. In this approach, a particle loaded with an indicator at the center can be processed without affecting its motion in real processing conditions. The process lethality received by the particle during processing can be calculated from TTI's initial (N_0) and final (N) status:

$$F_{TTI} = D_{ref} \log\left(\frac{N_0}{N}\right) \tag{21.30}$$

F_{TTI} is integrated process lethality recorded by time temperature indicator.

By using heating liquid temperature and assuming a constant h, a time–temperature profile at the particle center can be generated using a mathematical model and F_{model} could be calculated:

$$F_{model} = \int_0^t 10^{(T - T_{ref}/z)} \mathrm{d}t \tag{21.31}$$

The value of h is estimated by minimizing the difference between F_{TTI} and F_{model}. Hunter (1972) and Heppel (1985) were the first to use this approach and used microorganisms suspended in beads to back calculate the convective heat transfer coefficients, during continuous sterilization. Weng et al. (1992) used a TTI in the form of immobilized peroxidase, and determined the heat transfer coefficients in cans at pasteurization temperatures. A polyacetal sphere loaded with the indicator at the center was hooked onto a thermocouple and placed at the geometric center of a can. They calculated the time–temperature history and associated lethality from the equation of the heat conduction with assumed h and known thermophysical properties, using an explicit finite difference method. The h was modified and lethality recalculated until the difference between the predicted and actual lethalities fell within a tolerance limit. The authors named this approach as the least absolute lethality difference (LALD) method. They also gathered the transient temperature data for liquid and particle during heating and cooling in the same experiments and estimated h by minimizing the sum of the square of the temperature difference (LSTD) between measured particle center temperature and predicted center temperature, using the mathematical model. Maesmans et al. (1994) studied the feasibility of this method and the factors that can affect the choice of this methodology. Since these factors can influence the measurement of h, care is necessary in the design of experiments to obtain accurate results with this method (Maesmans et al., 1994).

Moving thermocouple method: A methodology was developed to measure the heat transfer coefficients, while allowing the movement of the particle inside the can, by attaching it to a flexible fine wire thermocouple (Sablani and Ramaswamy, 1995, 1996, 1997; Sablani, 1996) (Figure 21.1). U was calculated from the thermal energy balance on the can (Equation 21.21) and h was evaluated using a finite difference computer program, by matching the particle time–temperature data with those obtained by solving the governing partial differential Equations 21.23 through 21.26 of conduction heat transfer in different geometries, with appropriate initial and boundary conditions.

21.3.2 Aseptic Processing

Evaluating the surface heat transfer coefficient under aseptic processing conditions is an important step in the process design of particulate liquid. The conditions that can influence the heat transfer coefficient are particle/fluid relative velocities, the shape and size of particles, pipe diameter, temperature, and concentration of carrier fluid (Ramaswamy et al., 1997). In general, h is

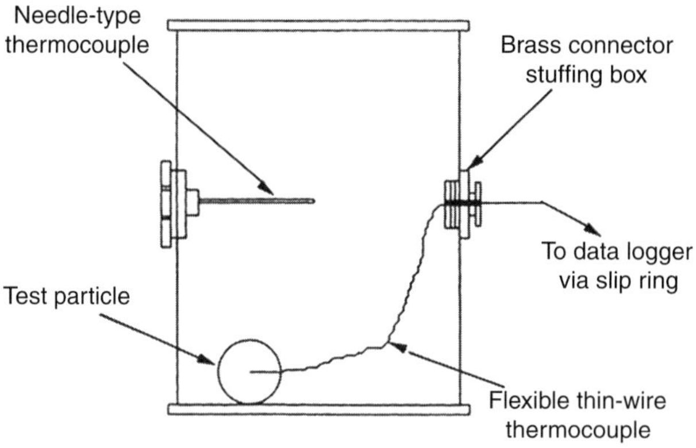

Figure 21.1 A schematic showing thermocouple-equipped particle mounted inside the can using a brass connector.

determined from the physical measurement of temperature history within a particle or particle surface under well-defined experimental conditions and by solving the transient heat transfer equation for particle geometry along with appropriate initial and boundary conditions (Table 21.2).

Table 21.2 Methods Employed for Determining *h* in Aseptic Processing Situation

References	Method	Particle Temperature/ Lethality Measurement	Mathematical Procedure
Hunter (1972) and Heppel (1985)	TTI	Microbial population reduction	Numerical solution of heat conduction equation with convective boundary conditions
Weng et al. (1992)	Free TTI	Peroxidase activity	Numerical solution of heat conduction equation with convective boundary conditions
Sastry et al. (1989)	Moving thermocouple	Center	Lumped capacity
Mwangi et al. (1993)	Melting point	Surface	Numerical solution of heat conduction equation with convective boundary conditions
Stoforos and Merson (1990) and Moffat (1990)	Liquid crystal	Surface	Numerical solution of heat conduction equation with convective boundary conditions
Whitaker (1972)	Relative velocity	Measure relative velocity	Using correlation of Ranz and Marshall (1952)
Balasubramaniam and Sastry (1994a,b)	Transmitter method	Center	Numerical solution of heat conduction equation with convective boundary conditions
Chang and Toledo (1989); Alhamdan and Sastry (1990); Zuritz et al. (1990); Chandarana et al. (1990); Awuah et al. (1993); Ramaswamy et al. (1996a,b); Awuah et al. (1996)	Restricted	Center	Numerical/analytical solution of heat conduction equation with convective boundary conditions

Ramaswamy et al. (1997) presented an excellent review of techniques used for the measurement of heat transfer coefficients in aseptic processing situations.

21.3.2.1 Time–Temperature Integrator Method

In the time–temperature integrator method, instead of comparing the experimentally measured and estimated transient temperatures, the cumulated effect of time–temperature is compared for estimation of heat transfer coefficient. Hunter (1972) and Heppel (1985) used the microbial population reduction approach by heating spore-impregnated alginate beads to estimate heat transfer coefficient. This approach led to the use of a chemical indicator. Weng et al. (1992) used a time–temperature integrator in the form of immobilized peroxidase to calculate heat transfer coefficients at pasteurization temperatures. In this method, a particle of regular geometry is loaded with the indicator (known microbial population or enzyme of known activity) at the center whose kinetic data are known. After the heating/cooling, the particle is recovered and actual destruction of microbes or inactivation of enzymes is measured for estimation of actual lethality. Assumed heat transfer coefficient and known thermophysical properties are used to compute transient temperature profile using any numerical method. The estimated lethality is then calculated from predicted time–temperature profile and compared with actual/experimental lethality. The heat transfer coefficient is then modified and lethality is recalculated until the difference between the predicted and actual lethalities is minimized within a tolerable limit. This approach of estimation of heat transfer coefficient was termed as LALD. In this approach, the weight of lethal temperatures is more pronounced than the weight of lower temperatures. However, in the approach where the least sum of squared temperature differences (LSTD) is used, each measured temperature is considered in minimizing the difference between experimental and predicted temperatures.

In this method, loading of enzymes/microbes may alter the thermophysical properties due to materials used in holding the indicator in place. Although this method is noninvasive in relation to particle trajectory as well as suitable to high-temperature applications, sample-to-sample variation in microbial population for instance can cause major problems and discrepancies in replicated data (Ramaswamy et al., 1997).

21.3.2.2 Moving Thermocouple Method

In aseptic processing, the particles move in the holding tube along with the carrier fluid. To simulate particle motion during measurement of heat transfer coefficient, Sastry et al. (1989) developed the moving thermocouple technique that involved a thermocouple hooked particle moving in a tube. The heat transfer coefficient was calculated from the experimentally measured time–temperature data using the lumped capacity approach. Sastry et al. (1990) modified the above setup by using a motor-driven setup at the downstream end of the tube to withdraw the thermocouple at the predetermined velocity and using a magnetic flowmeter to measure the fluid flow rate (Figure 21.2). Sastry (1992) indicated that the moving thermocouple method predicts conservative values. The main advantage of this technique was the measurement of transient temperatures of the moving particle. However, this approach cannot be adopted for high temperatures and pressures owing to difficulties in equipment design. Also, the fluid motion and profiles are constrained when particles are withdrawn from the tube (Ramaswamy et al., 1997).

21.3.2.3 Melting Point Method

The melting point method uses polymers that change color at specific temperatures. The polymer with calibrated color/temperature is transplanted as an indicator at the center of transparent particles. Mwangi et al. (1993) used transparent polymethylmetacrylate spheres with diameters

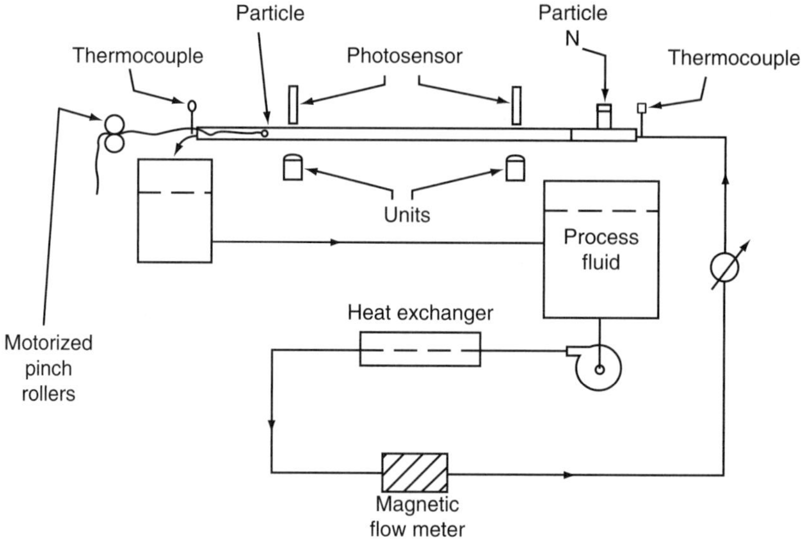

Figure 21.2 Schematic diagram of experimental setup for particle to fluid heat transfer coefficient measurement. (From Sastry, S.K., Lima, M., Bruin, J., Brunn, T., and Heskitt, B.F., *J. Food Process Eng.*, 13, 239, 1990.)

ranging from 8 to 12.7 mm. The particles were introduced through a venture into a transparent holding tube containing glycerin/water mixtures. The time at which the color change occurred was recorded in addition to time temperature data of the fluid. The temperature at the surface of the indicator within the particle was predicted with a finite difference program using an assumed value of heat transfer coefficient. An iterative procedure for minimizing the difference between time needed to reach the melting temperature was used to calculate the heat transfer coefficient. The method is noninvasive but limited to transparent tubes, and particles are not usable since the color change is irreversible (Mwangi et al., 1993).

21.3.2.4 Liquid Crystal Method

Liquid crystal method involves coating of the surface of a particle with a liquid crystal that changes color with a change in temperature. The change in color is videotaped as crystal-coated particles travel through the transparent tube. The surface transient temperatures are obtained using a color temperature-calibrated chart. This method is noninvasive and rapid. The accuracy of measured temperature thus estimated by heat transfer coefficient is limited by (1) the range of temperatures over which color changes and (2) the resolution of the video image. The method is not suitable for high temperature/pressure applications involving opaque carrier liquids.

21.3.2.5 Relative Velocity Method

Relative velocity method requires estimation of relative velocity from experimentally measured velocities of a liquid and particle. The velocities of the particle and liquid are measured using a video camera. The carrier liquid velocity is essentially measured by introducing a tracer particle into the fluid stream. Ramaswamy et al. (1992) presented an apparatus for particle fluid relative velocity measurement in tube flow at various temperatures under nonpressurized flow conditions (Figure 21.3). The relative velocity along with flow properties and particle properties is used to calculate the heat transfer coefficient from established Nusselt number correlations presented by Kramers (1960),

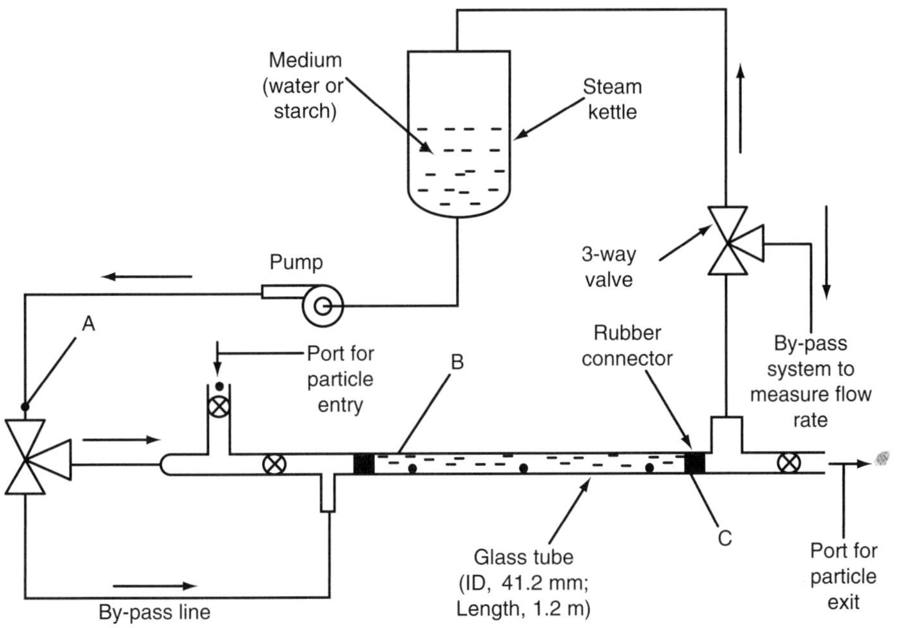

Figure 21.3 Equipment setup for the particle to fluid relative velocity measurement. (From Ramaswamy, H.S., Pannu, K., Simpson, B.K., and Smith, J.P., *Food Res. Int.*, 25, 277, 1992. With permission.)

Ranz and Marshall (1952), and Whitaker (1972). The accuracy of the heat transfer coefficient estimated using this method will depend on the reliability of the adopted correlation (Ramaswamy et al., 1997). However, the technique can be useful in characterizing the radial variations in the heat transfer coefficient (Sastry, 1992).

21.3.2.6 Transmitter Method

The transmitter technique involves placement of a miniature sensor in the particle that can transmit the signals to an external receiver. The purpose is to gather time–temperature data from a particle in motion with no obstruction such as in the case of the moving thermocouple method. Balasubramaniam and Sastry (1994a,b) used a hollow cylindrical capsule made of boron nitride with a transmitter. The magnetic signals transmitted were converted to temperature reading using a calibrated scale and an iterative procedure was used to back calculate the heat transfer coefficient. Ramaswamy et al. (1997) noted that this method has all the potentials for aseptic studies but little information has been reported in the literature. One of the reasons may be that the placement of the transmitter may cause considerable variation in the density of the particle in which it is embedded.

21.3.2.7 Stationary Particle Method

In this method, a particle is placed in a flowing stream and transient temperatures are measured within the particle at a known location and fluid. An iterative procedure is used to estimate the heat transfer coefficient by solving appropriate governing equations of transient heat transfer with convective boundary condition using the analytical or numerical method. This method may not reflect conditions in real situations since both translational rotational motions are restricted; thus it can lead to deviation in the estimated value of the heat transfer coefficient. However, several researchers (Chang and Toledo, 1989; Alhamdan and Sastry, 1990; Chandarana et al., 1990;

Zuritz et al., 1990; Awuah et al., 1993, 1996; Ramaswamy et al., 1996a,b) have used this method to study the influence of various processing parameters on the heat transfer coefficient under low- or high-temperature processing conditions. This method has been useful in providing considerable insight on the influence of various parameters on the heat transfer coefficient (Ramaswamy et al., 1997).

21.3.3 Baking

Energy balance during baking in an oven will require data on surface heat transfer coefficient on baked products in addition to thermal properties of product, construction, size of the oven and operating conditions such as the velocity, temperature, and humidity of the hot air. The direct-fired oven is the most common type for commercial baking. However, heat transfer studies in such ovens are still limited. In particular, a nonuniform temperature distribution between heating elements and product surface is an important factor as well as variable temperature profiles along the continuous oven, which makes it difficult to estimate surface heat transfer coefficients (Baik et al., 1999). There have been some studies reported on the measurement of heat transfer coefficients of the dough and the internal oven environment. In most such studies, the heat flux or heat transfer coefficient was measured directly in the oven using a heat flux sensor (typically a metal cylinder or sphere) or a commercially available h-monitor (Krist-Spit and Sluimer, 1987; Sato et al., 1987; Huang and Mittal, 1995; Li and Walker, 1996; Baik et al., 1999). These studies assumed that heat transfer between the oven and dough is the same as that between the oven and the model food (heat flux sensor).

Baik et al. (1999) described an experimental and mathematical procedure to estimate the surface heat transfer coefficient on cakes baked in a tunnel type industrial oven using h-monitor (Figure 21.4). They represented convective flux and radiative flux from wall and heating element in the form of convective and radiative heat transfer coefficients:

$$\text{Convective heat flux } Q_c = h_c(T_a - T_s) \tag{21.32}$$

$$\text{Radiative heat flux from wall } Q_{rw} = h_c(T_a - T_s) \tag{21.33}$$

$$\text{Radiative heat flux from heating element } Q_{ew} = h_c(T_a - T_s) \tag{21.34}$$

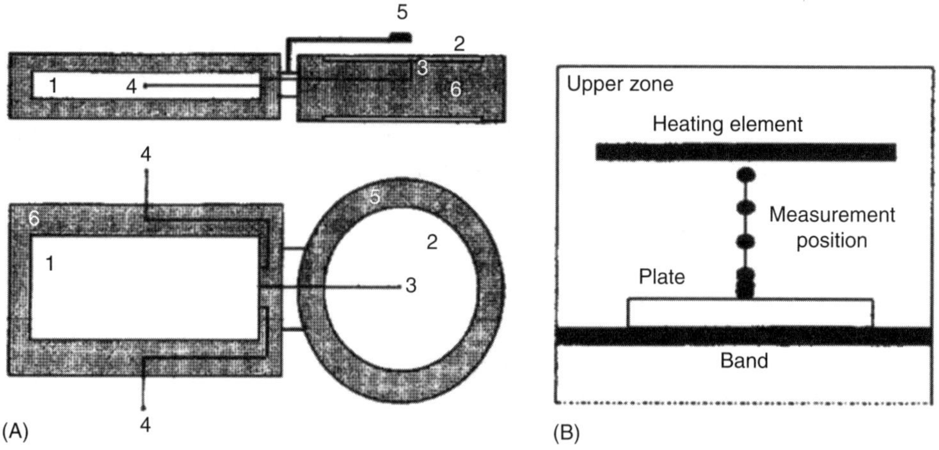

Figure 21.4 (A) Basic description of h-monitor and (B) measurement position inside the baking chamber of the tunnel type multizone industrial oven. (1) SMOLE, (2) aluminum plate, (3) thermocouple for plate temperature, (4) thermocouple position for air–mass temperature, (5) insulation, (6) thermocouple holder for air temperature measurement.

This method is useful when direct measurement of radiation heat transfer is not available. These radiant heat transfer coefficients can be treated similarly to the convective heat transfer coefficient (Kreith, 1973); in that case the total surface heat flux, Q_t, can be expressed as

$$Q_t = (h_c + h_{rw} + h_{re})(T_a - T_s) = h_t(T_a - T_s) \tag{21.35}$$

Hence h_t is defined as effective surface heat transfer coefficient. The total heat flux was measured by a commercial h-monitor and a moving temperature recorder, super multichannel occurrent logger evaluator (SMOLE) equipped with K-type thermocouples. The aluminum plate of h-monitor had emissivity similar to that of the cake (0.90–0.95). Other properties such as mass (m), specific heat (C_p), and the surface area (A) of the aluminum plate were known. By measuring the temperature for both the air and the plate at selected intervals during the baking process, the total heat flux and effective surface heat transfer coefficient were determined as

$$h_t = \frac{mC_p\Delta T}{A(T_a - T_s)\Delta t} \tag{21.36}$$

The Bi of aluminum plate of the h-monitor was <0.02, and the recorded temperature at the plate center was assumed as plate surface temperature. The temperature of the target plate (T) and the air (T_a) changed during the time interval, (Δt); thus h-values were measured as a function of time. The manufacturer recommended that for better resolution, the time period (Δt) should be kept long enough to achieve a target plate temperature change (ΔT) of $\geq$11.1°C.

21.4 IMPORTANCE

The heat transfer coefficient is used to calculate the rate of convective heat transfer to or from the surface of an object. In most food applications, it is needed to quantify the heat transfer during heating or cooling of food materials in fluids. The surface heat transfer coefficient is an important parameter needed in the design and control of food processing operations such as heating, cooling, baking, frying, freezing, or cooking. The heat transfer coefficient is not a material or fluid property but is influenced by fluid motion, geometry, and surface of the food material.

REFERENCES

Alhamdan, A. and Sastry, S.K. 1990. Natural convection heat transfer coefficient between non-Newtonian fluid and an irregular shaped particle, *Journal of Food Process Engineering*, 13: 113.

Arce, J. and Sweat, V.E. 1980. Survey of published heat transfer coefficients encountered in food refrigeration processes, *ASHRAE Transaction*, 86: 228.

Awuah, G.B., Ramaswamy, H.S., and Simpson, B.K. 1993. Surface heat transfer coefficients associated with heating of food particles in CMC solutions, *Journal of Food Process Engineering*, 16: 39–57.

Awuah, G.B., Ramaswamy, H.S., Simpson, B.K., and Smith, J.P. 1996. Fluid to particle heat transfer coefficient as evaluated in a pilot scale aseptic processing holding tube simulator, *Journal of Food Process Engineering*, 19: 241–267.

Baik, O.D., Grabowski, S., Trigui, M., Marcotte, M., and Castaigne, F. 1999. Heat transfer coefficients on cakes baked in a tunnel type industrial oven, *Journal of Food Science*, 64(4): 688–694.

Baumesiter, E.B. and Bennet, C.O. 1958. Fluid to particle heat transfer in packed beds, *AIChE Journal*, 4: 69–74.

Balasubramaniam, V.M. and Sastry, S.K. 1994a. Non-invasive estimation of heat transfer between fluid and particle in continuous tube flow using a remote temperature sensor. Paper presented at the annual meeting of the Institute of Food Technologists, Atlanta, GA, June 25–29.

Balasubramaniam, V.M. and Sastry, S.K. 1994b. Liquid to fluid convective heat transfer in non-Newtonian carrier medium during continuous tube flow, *Journal of Food Engineering*, 23: 169.

Carslaw, H.S. and Jaeger, J.C. 1959. *Conduction of Heat in Solids*, Oxford University Press, England.

Cengel, Y.A. 2003. *Heat Transfer: A Practical Approach*, 2nd edition, International Edition, McGraw-Hill, New York.

Chandrana, D.I., Gavin, A., and Wheaton, F.W. 1990. Particle/fluid interface heat transfer under UHT conditions at low particle/fluid relative velocities, *Journal of Food Process Engineering*, 13: 191–206.

Chang, S.Y. and Toledo, R.T. 1989. Heat transfer and simulated sterilization of particulates solids in a continuous flowing systems, *Journal of Food Science*, 54: 1017.

Chang, S.Y. and Toledo, R.T. 1990. Simultaneous determination of thermal diffusivity and heat transfer coefficient during sterilization of carrot dices in a packed bed, *Journal of Food Science*, 55(1): 199.

Clary, B.L. and Nelson, G.L. 1970. Determining convective heat transfer coefficients from ellipsoidal shapes, *Transaction of ASAE*, 13: 309–314.

Deniston, M.F., Hassan, B.H., and Merson, R.L. 1987. Heat transfer coefficients to liquids with food particles in axially rotating cans, *Journal of Food Science*, 52: 962–966, 979.

Deniston, M.F., Kimball, R.N., Stoforos, N.G., and Parkinson, K.S. 1992. Effect of steam/air mixtures on thermal processing of an induced convection-heating products (tomato concentrate) in a steritort, *Journal of Food Process Engineering*, 15: 49.

Eckert, E.R. and Soelnghen, E. 1951. *Proceedings of the General Discussion on Heat Transfer*, Institute of Mechanical Engineers, New York.

Federov, V.G., Ilinskiy, D.N., Gerashchenko, O.A., and Andreyeva, L.D. 1972. Heat transfer accompanying the cooling and freezing of meat carcasses, *Heat Transfer Soviet Research*, 4(4): 55.

Feldmann, C. 1976. Part 1: Transfert de chaleur et de masse entre une sphere et de L'Air a basse temperature, Conservatoire National des Art et Metiers, Paris, Memoire No. 6. March.

Fernandez, C.L., Rao, M.A., Rajeavasireddi, S.P., and Sastry, S.K. 1988. Particulate heat transfer to canned snap beans in Steritort, *Journal of Food Process Engineering*, 10: 183–198.

Hassan, B.H. 1984. Heat transfer coefficients for particles in liquid in axially rotating cans, PhD thesis, Department of Agricultural Engineering, University of California, Davis, CA.

Heppel, N.J. 1985. Measurement of the liquid–solid heat transfer coefficient during continuous sterilization of foodstuffs containing particles. Presented at the 4th Congress on Engineering and Foods, Edmonton, AB, Canada, July 7–10.

Huang, E. and Mittal, G.S. 1995. Meatball cooking-modeling and simulation, *Journal of Food Engineering*, 24: 87–100.

Hunter, G.M. 1972. Continuous sterilization of liquid media containing suspended particles, *Food Technology Australia*, 24: 158–165.

Leichter, S., Mizrahi, S., and Kopelman, I.J. 1976. Effect of vapor condensation on rate of refrigerated products exposed to humid atmosphere: Application to the prediction of fluid milk shelf life, *Journal of Food Science*, 41: 1214.

Lekwauwa, A.N. and Hayakawa, K.I. 1986. Computerized model for the prediction of thermal responses of packaged solid–liquid food mixture undergoing thermal processes, *Journal of Food Science*, 51: 1042–1049, 1056.

Lenz, M.K. and Lund, D.B. 1978. The lethality-Fourier number method. Heating rate variations and lethality confidence intervals for forced-convection heated foods in containers, *Journal of Food Process Engineering*, 2: 227–271.

Li, A. and Walker, C.E. 1996. Cake baking in conventional impingement and hybrid ovens, *Journal of Food Science*, 61: 188–191.

Kramers, H. 1960. Heat transfer from spheres to flowing media, in *Fluidization and Fluid Particle Systems*, Zenz, F.A. and Othmer, D.F., Eds., Reinhold, New York, p. 1946.

Kreith, F. 1973. *Principles of Heat Transfer*, 3rd edition, Intext Educational Publishers, New York/London.

Krist-Spit, C.E. and Sluimer, P. 1987. Heat transfer in ovens during the baking of bread, in *Cereals in a European Context*, 1st European Conference on Food Science and Technology, Morton, I.D., Ed., Ellis Horwood Publishers, Chichester, UK, pp. 346–354.

Maesmans, G., Hendrickx, M., DeCordt, S., and Fransis, A. 1992. Fluid to particle heat transfer coefficient determination of heterogeneous foods: A review, *Journal of Food Processing and Preservation*, 16: 29–69.

Maesmans, G., Hendrickx, M., DeCordt, S., and Tobback, P. 1994. Feasibility of the use of a time temperature integrator and a mathematical model to determine fluid-to-particle heat transfer coefficients, *Food Research International*, 27: 39–51.

Moffat, R.J. 1990. Experimental heat transfer, in *Proceedings of the 9th Heat Transfer Conference*, Jerusalem, Israel, pp. 187–205.

Mwangi, J.M., Rizvi, S.S.H., and Datta, A.K. 1993. Heat transfer to particles in shear flow: Application in aseptic processing, *Journal of Food Engineering*, 19: 55.

Plank, R. 1932. Ueber die Gefrierzeit von eis ind. Wasserhallegen lebensraitteln, *Z. Ges. Kalte Ind.*, 39(4): 56.

Rahman, M.S. 1995. *Food Properties Handbook*, CRC Press, Boca Raton, FL.

Ramaswamy, H.S., Pannu, K., Simpson, B.K., and Smith, J.P. 1992. An apparatus for particle to fluid relative velocity measurement in tube flow at various temperatures under nonpressurized flow conditions, *Food Research International*, 25(4): 277.

Ramaswamy, H.S., Awuah, G.B., and Simpson, B.K. 1996a. Biological verification of fluid-to-particle interfacial heat transfer coefficients in a pilot scale holding tube simulator, *Biotechnology Progress*, 12 (4): 527–532.

Ramaswamy, H.S., Awuah, G.B., Simpson, B.K., and Smith, J.P. 1996b. Influence of particle characteristics on fluid to particle heat transfer coefficient in a pilot scale holding tube simulator, *Food Research International*, 29(3): 291–300.

Ramaswamy, H.S., Awuah, G.B., and Simpson, B.K. 1997. Heat transfer and lethality considerations in aseptic processing of liquid/particle mixtures: A review, *Critical Reviews in Food Science and Nutrition*, 37(3): 253–286.

Ranz, W.E. and Marshall, W.R. Jr. 1952. Evaporation from drop, *Chemical Engineering Progress*, 48: 141–146.

Sablani, S.S. 1996. Heat transfer studies of particle/liquid mixtures in cans subjected to end-over-end processing, PhD thesis, McGill University, Montreal.

Sablani, S.S. and Ramaswamy, H.S. 1995. Fluid to particle heat transfer coefficients in cans during end-over-end rotation, *Food Science and Technology* (*lwt*), 28: 56–61.

Sablani, S.S. and Ramaswamy, H.S. 1996. Particle heat transfer coefficients under various retort conditions with end-over-end rotation, *Journal of Food Process Engineering*, 19: 403–424.

Sablani, S.S. and Ramaswamy, H.S. 1997. Heat transfer to particles in cans with end-over-end rotation: Influence of particle size and particle concentration (vv), *Journal of Food Process Engineering*, 20: 265–283.

Sastry, S.K. 1992. Liquid to particle heat transfer coefficient in aseptic processing, in *Advances in Aseptic Processing Technologies*, Nelson, P.E. and Singh, R.K., Eds., Elsevier Applied Science, London/ New York, p. 63.

Sastry, S.K., Heskitt, B.F., and Blaisdell, J.L. 1989. Experimental and modeling studies on convective heat transfer at particle–liquid interface in aseptic processing systems, *Food Technology*, 43(3): 132.

Sastry, S.K., Lima, M., Bruin, J., Brunn, T., and Heskitt, B.F. 1990. Liquid to particle heat transfer during continuous tube flow: Influence of flow rate and particle to tube diameter ratio, *Journal of Food Process Engineering*, 13(3): 239.

Sato, H., Matsumura, T., and Shibukawa, S. 1987. Apparent heat transfer in a forced convection oven and properties of baked food, *Journal of Food Science*, 52: 185–188, 193.

Stoforos, N.G. 1988. Heat transfer in axially rotating canned liquid/particulate food systems, PhD thesis, Department of Agricultural Engineering, University of California, Davies, CA.

Stoforos, N.G. and Merson, R.L. 1990. Estimating heat transfer coefficients in liquid/particulate canned foods using only liquid temperature data, *Journal of Food Science*, 55: 478–483.

Stoforos, N.G. and Merson, R.L. 1991. Measurement of heat transfer coefficients in rotating liquid/particle systems, *Biotechnology Progress*, 7: 267–271.

Stoforos, N.G. and Merson, R.L. 1995. A solution to the equations governing heat transfer in agitating liquid/particulate canned foods, *Journal of Food Process Engineering*, 18: 165–185.

Weng, Z., Hendrickx, M., Maesmans, G., and Tobback, P. 1992. The use of a time–temperature-integrator in conjunction with mathematical modeling for determining liquid/particle heat transfer coefficients, *Journal of Food Engineering*, 16: 197–214.

Whitaker, S. 1972. Forced convection heat transfer calculations for flow in pipes, past flat plates, single cylinder, single spheres, and for flow in packed beds and tube bundles, *Journal of AIChE*, 18: 361.
Zuritz, C.A., McCoy, S.C., and Sastry, S.K. 1990. Convection heat transfer coefficients for irregular particles immersed in non-Newtonian fluids during tube flow, *Journal of Food Engineering*, 11: 159.

CHAPTER 22

Surface Heat Transfer Coefficients with and without Phase Change

Liyun Zheng, Adriana Delgado, and Da-Wen Sun

CONTENTS

22.1 INTRODUCTION

In food processing, heat transfer is involved almost in every process. The transfer of heat occurs in three different modes: conduction, convection, and radiation. In these modes, heat always transfers from a high-temperature body to a low-temperature one.

Convection is the combined effects of heat conduction and radiation and fluid flow. It transfers energy between a solid surface and the adjacent liquid or gas that is in motion. Depending on the flow velocity, convection can be free (or natural) convection or forced convection. In free convection, the fluid flow is caused by buoyancy forces induced by density differences due to the variation of temperature in the fluid. In contrast, in forced convection, the fluid is forced to flow by external means such as a pump or fan. Convection also occurs during phase changes. Newton's law of cooling is used to determine the rate of convection heat transfer:

$$Q = hA(T_s - T_\infty) \tag{22.1}$$

where

h is an experimentally determined parameter defined as the convection heat transfer coefficient with its unit as W/m^2 K, and its value depends on many relevant factors such as the nature of fluid flow, properties of the fluid, and surface geometry
A is the surface area over which the fluid flows (m^2)
Q is the heat transfer rate (W)
T_s is the solid surface temperature (K)
T_∞ is the average or bulk temperature of the fluid flowing by (K)

On the basis of Equation 22.1, the convection heat transfer coefficient can be defined as follows:

$$h = \frac{q}{T_s - T_\infty} = \frac{q}{\Delta T} \tag{22.2}$$

where

q is defined as the convective heat flux (W/m^2)
ΔT is the temperature difference causing the heat flow

To calculate h from Equation 22.2, the heat flux q and the temperature difference ΔT must be known. For accurate determination of h, ΔT must be carefully defined and measured, where T_s can be obtained at the hottest location or some average values between the hottest and the coolest temperatures or over the surface; the determination of T_∞ is more difficult due to the arbitrary choice by investigators.

The convection heat transfer coefficient is sometimes referred to as the film coefficient or film conductance and represents the thermal resistance of a relatively stagnant layer of fluid between the surface and the fluid. Equation 22.1 indicates that the analysis of convective heat transfer problems significantly depends on the accurate knowledge of h. For classical flow problems such as flow over a flat plate or through a duct, h values and their determination methods have been exhaustively examined and the parameters that impact h are well categorized and quantified. For natural convection problems, the temperature distribution is the dominant parameter, while for forced convection, the fluid velocity is the governing one. Furthermore, the geometry of the problem always plays a significant role in determining the value of h. In addition, there are many practical situations (e.g., air cooling of unpackaged foods, freezing, thawing, cooking, drying, etc.) in which heat transfer to or from the product surface is combined with a certain degree of mass transfer, which in turn modifies the heat transfer. Therefore, predicting h is still a challenging issue in investigating convective heat transfer.

In the determination of h, traditionally, a dimensionless functional relationship, derived using dimension analysis, between h and the relevant physical properties and kinetic parameters of the flow situation is used. In other words, h in Equation 22.2 can be expressed in the following function:

$$h = f(\rho, V, L, \mu, c_p\Delta T, k, \beta, g) \tag{22.3}$$

where

ρ is the density of the fluid (kg/m^3)
V is the average fluid velocity (m/s)
L is a characteristic dimension (m)
μ is the dynamic viscosity of the fluid (N s/m^2)
c_p is the specific heat of the fluid (J/kg s)
k is the thermal conductivity of the fluid (W/m K)
g is the gravitational acceleration (m/s^2)

In Equation 22.3, β is the volumetric thermal expansion coefficient, which can be defined in terms of volume or density as $-1/\nu(\partial\nu/\partial T)|_P$ or $-1/\rho(\partial\rho/\partial T)|_P$. For an ideal gas, $\beta = 1/T$.

By performing a dimensional analysis on Equation 22.3, the following equation can be derived:

$$\frac{hL}{k} = f\left(\frac{\rho VL}{\mu g}, \frac{c_p\mu g}{k}, \frac{\rho^2 g\beta L^3\Delta T}{\mu^2 g^2}, \frac{V^2}{c_p g\Delta T}\right) \tag{22.4}$$

The dimensionless groups in the above equation are defined by the following specific names:

Nusselt number: $Nu = hL/k$ for expressing the ratio of convection heat transfer to fluid conduction heat transfer under the same conditions;

Reynolds number: $Re = \rho VL/\mu g$ for representing the ratio of the inertia force to the viscous force in the fluid;

Prandtl number: $Pr = (c_p\mu g/k) = (c_p\rho/k)(\mu g/\rho) = (\upsilon/\alpha)$ for showing the ratio of the fluid viscosity to the thermal conductivity of a substance with a low number indicating high convection;

Grashof number: $Gr = (\rho^2 g\beta L^3\Delta T/\mu^2 g^2) = (g\beta L^3\Delta T/\upsilon^2)$ for giving the ratio of the buoyancy forces to the viscous forces in the free convection flow system

Eckert number: $Ec = (V^2/c_p g\Delta T)$ for describing the ratio of the kinetic energy to the enthalpy of the fluid.

where v ($=\mu/\rho$) is the kinematic viscosity of the fluid (m^2/s) and α [$=k/(\rho c_p)$] is the thermal diffusivity of the fluid (m^2/s).

The *Gr* number that arises in natural convection situations, where the effect of β is significant, plays a role similar to that of the Reynolds number in forced convection, and it is also the primary variable used as a criterion for transition from laminar to turbulent boundary-layer flow. In the event of forced convection, the effect of Gr on Nu can be omitted. Equation 22.4 can then be rewritten as

$$Nu = f(Re, Pr, Gr, Ec) \tag{22.5}$$

Therefore, research on convection heat transfer mainly focuses on the finding of the above correlations, so that the surface heat transfer coefficient can be determined by the *Nu* number. However, it should be remembered that these convection correlations provide only an estimate of what is actually occurring, and uncertainties of 10%–25% are typical for this parameter.

22.2 NATURAL CONVECTION CORRELATIONS

In cooling or heating of foods, if the cooling or heating medium flows through a food product without the aid of any external means such as a fan or a pump, the heat transfers from the product surface to the medium by natural convection or free convection. In other words, in natural convection, the motion of the fluid is caused by the density difference due to the temperature difference, as a fluid with high temperature is less dense than the one with low temperature. This change in density induces a change in the gravitational force or buoyancy force, which causes the fluid to move by itself. Gravity is not the only type of force field that can produce the free convection currents; a fluid enclosed in a rotation machine, for example, is acted upon by a centrifugal force field, and free convection currents could arise if one or more of the surfaces in contact with the fluid are heated (Holman, 1986). These buoyancy forces are called body forces.

In natural convection, many correlations contain the product of the Grashof number and the Prandtl number; for convenience, the product of the two is defined as the Rayleigh number (*Ra*):

$$Ra = (Gr)(Pr) = \frac{g\beta L^3 \Delta T}{v^2}\frac{v}{\alpha} = \frac{g\beta L^3 \Delta T}{v\alpha}$$

Also, the Stanton number (*St*), based on the Nusselt number, Reynolds number, and Prandtl number and indicating the total heat transferred to total heat capacity, is sometimes used and is given below:

$$St = \frac{Nu}{(Re)(Pr)} = \frac{hL}{k}\frac{v}{VL}\frac{\alpha}{v} = \frac{h}{\rho c_p V}$$

22.2.1 Horizontal Plates

As the motion of fluid in free convection is caused by buoyancy force, for horizontal plates (either from the plate or toward the plate), two situations should be distinguished: (1) heat transfer occurs in the direction of gravitational force, and (2) heat transfers against the direction of gravitational force. Therefore, correlations are developed based on these two cases. Unless stated, the following correlations are used for estimating the average heat transfer coefficient.

22.2.1.1 Heat Transfer in the Direction of Gravitational Force

Fujii and Imura (1972) proposed the following correlation for constant heat flux and for the horizontal plate facing downward, with the length of the heated plate as the characteristic dimension L,

$$Nu = 0.58Ra^{1/5}, \quad 10^6 < Ra < 10^{11} \tag{22.6}$$

Goldstein et al. (1973) recommended the following equations with the characteristic dimension being calculated by the ratio of the surface area over the perimeter of the surface:

$$Nu = 0.96Ra^{1/6}, \quad Ra < 200 \tag{22.7a}$$

$$Nu = 0.59Ra^{1/4}, \quad Ra > 200 \tag{22.7b}$$

Other relations for heated surface facing upward (or cooled surfaces facing downward) are (Gebhart, 1971)

$$Nu = 0.54Ra^{1/4}, \quad 10^5 < Ra < 10^7 \tag{22.7c}$$

$$Nu = 0.14Ra^{1/3}, \quad 10^7 < Ra < 3 \times 10^{10} \tag{22.7d}$$

22.2.1.2 Heat Transfer against the Direction of Gravitational Force

Fujii and Imura (1972) also proposed the following correlation for heat transfer to or from horizontal plates in the opposite direction of gravitational force:

$$Nu = 0.16Ra^{1/3}, \quad Ra < 2 \times 10^8 \tag{22.8a}$$

$$Nu = 0.13Ra^{1/3}, \quad 5 \times 10^8 < Ra \tag{22.8b}$$

In Equations 22.6 and 22.8 β is evaluated at $T_\infty + 0.25(T_s - T_\infty)$ and all other properties at $T_s - 0.25(T_s - T_\infty)$, and T_s is the average wall temperature related to the heat flux (Holman, 1986; Suryanarayana, 1995). Alternatively, the following relation is recommended by McAdams (1954) (Gebhart, 1971):

$$Nu = 0.27Ra^{1/4}, \quad 3 \times 10^5 < Ra < 3 \times 10^{10} \tag{22.9}$$

For circular plates of diameter D in the stable horizontal configurations, the data of Kadambi and Drake (1959) suggest that (Lienhard and Lienhard, 2004)

$$Nu = 0.82Ra^{1/5}Pr^{0.034} \tag{22.10}$$

22.2.2 Vertical Plates

Churchill and Chu (1975a) proposed the following equation for the mean value of Nu number for vertical plates:

$$Nu = \left\{ 0.825 + \frac{0.387 Ra^{1/6}}{\left[1 + (0.492/Pr)^{9/16}\right]^{8/27}} \right\}^2 \tag{22.11a}$$

where the characteristic length L is the height of the plate. β is evaluated at T_∞ if the fluid is a gas and all other properties at the film temperature $T_f = (T_s + T_\infty)/2$.

Equation 22.11a provides a good representation for the mean heat transfer over a complete range of Ra and Pr from 0 to ∞ even though it fails to indicate a discrete transition from laminar to turbulent flow (Churchill and Chu, 1975a). For $Ra < 10^9$, the following equation gives more accurate prediction of h:

$$Nu = 0.68 + \frac{0.67 Ra^{1/4}}{\left[1 + (0.492/Pr)^{9/16}\right]^{4/9}} \tag{22.11b}$$

Equations 22.11a and b are suitable for vertical plates with uniform wall temperature. However, if uniform heat flux is applied to the vertical plates, a small adjustment to Equations 22.11a and b is needed by replacing the constant 0.492 in the denominator by 0.437. In such a case, the fluid properties are evaluated at the temperature at the mid-height of the plate, i.e., $T_f = (T_{L/2} + T_\infty)/2$ and $T_{L/2} - T_\infty$ is used to form the Raleigh number.

Alternatively, for vertical plates with uniform heat flux, the following correlations proposed by Sparrow and Gregg (1956) and Vliet and Liu (1969) can be used to calculate the local Nusselt number as a function of the length from the plate edge (Suryanarayana, 1995):

$$Nu_x = 0.6(Gr_x^* Pr)^{0.2}, \quad 10^5 < Gr_x^* Pr < 10^{13} \tag{22.12a}$$

$$Nu_x = 0.568(Gr_x^* Pr)^{0.22}, \quad 10^{13} < Gr_x^* Pr < 10^{16} \tag{22.12b}$$

where all the properties are evaluated at $T_f = (T_s + T_\infty)/2$ and x is the coordinate from the leading edge along the plate. Free convection with uniform heating is often correlated in terms of the modified Raleigh number ($Ra^* = Gr_x^* Pr$), based on the local heat flux q, to avoid explicit inclusion of the surface temperature (Churchill and Chu, 1975b). The modified Grashof number in Equation 22.12 and the modified Ra number are defined as

$$Gr_x^* = \frac{g\beta\rho^2 q x^4}{\mu^2 k}$$

$$Ra^* = \frac{g\beta q x^4}{\upsilon\alpha k}$$

where $q = h_x(T_s - T_\infty)$

22.2.3 Inclined Plates

For heat transfer in the general direction of gravitational force either from the plate or to the plate, the above correlations may be used for inclined plates, if the component of the gravity vector along the surface of the plate is used in the calculation of the Grashof number. In other words, for downward-facing heated inclined plates or upward-facing cooled inclined plates,

Equation 22.11 for vertical plates can be used by taking a modified Rayleigh number (Ra_θ) where g is replaced by $g\cos\theta$:

$$Ra_\theta = \frac{g\cos\theta\beta L^3\Delta T}{\upsilon\alpha}$$

and θ is the angle of inclination of the plate to the vertical. However if the angle is greater than 88°, it is suggested to use the correlations for horizontal plates. If the inclined plates are supplied with a uniform heat flux, Fussey and Warneford (1978) suggested the following correlations for the local heat transfer coefficients:

$$Nu_x = 0.592(Ra_x^*\cos\theta)^{0.2}, \quad Ra_x^* < \left(6.31\times10^{12}\mathrm{e}^{-0.07050\theta}\right), \quad 0 < \theta < -86.5^\circ \tag{22.13a}$$

$$Nu_x = 0.889(Ra_x^*\cos\theta)^{0.205}, \quad Ra_x^* > \left(6.31\times10^{12}\mathrm{e}^{-0.07050\theta}\right), \quad 0 < \theta < -31^\circ \tag{22.13b}$$

All the fluid properties are evaluated at the average temperature $(T_s + T_\infty)/2$. For heat transfer against the general direction of gravitational force, i.e., for an upward-facing heated inclined plate, the situation is more complicated, and the reader is advised to consult a book on specialized convection heat transfer for relevant correlations.

22.2.4 Long Horizontal Cylinders

Churchill and Chu (1975b) proposed a simple empirical expression for the mean value of the Nusselt number over the cylinder of diameter D for uniform surface temperature for all Pr and Ra values:

$$Nu = \left(0.60 + 0.387\left\{\frac{Ra}{\left[1 + (0.559/Pr)^{9/16}\right]^{16/9}}\right\}^{1/6}\right)^2 \tag{22.14a}$$

For the entire laminar regime and uniform surface temperature and all Pr, a slightly better prediction of h is given by the following expression (Churchill and Chu, 1975b):

$$Nu = 0.36 + 0.518\left\{\frac{Ra}{\left[1 + (0.559/Pr)^{9/16}\right]^{16/9}}\right\}^{1/4}, \quad 10^{-6} < Ra < 10^9 \tag{22.14b}$$

Equation 22.14a is probably a good approximation also for uniform heat flux; however, Equation 22.14c provides a possibly better correlation than Equations 22.14a and b for uniform heat flux in the laminar regime for small Pr:

$$Nu = 0.36 + 0.521\left\{\frac{Ra}{\left[1 + (0.442/Pr)^{9/16}\right]^{16/9}}\right\}^{1/4} \tag{22.14c}$$

For large temperature differences such that the variation of physical properties is significant, the properties may be evaluated at the average of the bulk and surface temperatures as a first approximation; β at $T_f = (T_s + T_\infty)/2$ for liquids, and $\beta = 1/T_\infty$ for gases. For a vertical cylinder, the same equations can be used as those for a vertical plate if the following can be satisfied (Gebhart, 1971):

$$\frac{D}{L} \geq \frac{35}{Gr^{1/4}}$$

where the Grashof number is based on L. In general, the average natural convection heat transfer coefficients for isothermal surfaces can be expressed by the following general equation for a variety of circumstances:

$$Nu = cRa^m \quad (22.14d)$$

where c and m are constants and are given in Table 22.1 for different geometries. All the physical properties are evaluated at the film temperature, $T_f = (T_s + T_\infty)/2$. Simplified equations for heat

Table 22.1 Constants for Equation 22.14d for Natural Convection

Physical Geometry	*Ra*	c	*m*
Vertical planes and cylinders (vertical height $L < 1$ m)[a]	$<10^4$	1.36	1/5[a]
	10^4–10^9	0.59	1/4[a]
	$>10^9$	0.13	1/3[a]
	10^9–10^{13}	0.021	2/5[b]
Horizontal cylinders (diameter D used for L and $D < 0.20$ m)[b]	$<10^{-5}$	0.49	0[a]
	10^{-5}–10^{-3}	0.71	1/25[a]
	10^{-3}–1	1.09	1/10[a]
	1–10^4	1.09	1/5[a]
	10^4–10^9	0.53	1/4[a]
	$>10^9$	0.13	1/3[a]
	10^{-10}–10^{-2}	0.675	0.058[b]
	10^{-2}–10^2	1.02	0.148[b]
	10^2–10^4	0.85	0.188[b]
	10^4–10^7	0.48	1/4[b]
	10^7–10^{12}	0.125	1/3[b]
Horizontal plates (upper surface of heated plates or lower surface of cooled plates)	10^5–(2×10^7)	0.54	1/4[a]
	(2×10^7)–(3×10^{10})	0.14	1/3[a]
Horizontal plates (lower surface of heated plates or upper surface of cooled plates)	10^5–10^{11}	0.58	1/5[a]
	10^5–10^{11}	0.27	1/4[b]
Vertical cylinder (height = diameter, characteristic length = D)	10^4–10^6	0.775	0.21[b]
Irregular solids (characteristic length = distance fluid particle travels in boundary layer)	10^4–10^9	0.52	1/4[b]

Sources: [a]Geankoplis, C.J., in *Transport Processes and Unit Operations*, Prentice-Hall International Inc, New Jersey, 1993; [b]Holman, J.P., in *Heat Transfer*, 6th edn., McGraw-Hill, New York, 1986.

transfer coefficients from various surfaces to air and water at atmospheric pressure are given in Rahman (1995), Geankoplis (1993), and Holman (1986). The relations for air may be extended to higher or lower pressures by multiplying by the following factors (Holman, 1986):

$$\left(\frac{p}{101.32}\right)^{1/2}, \quad \text{for laminar cases}$$

$$\left(\frac{p}{101.32}\right)^{2/3}, \quad \text{for turbulent cases}$$

where p is the pressure in kPa.

22.2.5 Spheres

Yuge (1960) recommends the following empirical expression for natural convection from spheres to air (Holman, 1986):

$$Nu = 2 + 0.392Gr^{1/4}, \quad 1 < Gr < 10^5 \tag{22.15a}$$

Equation 22.15a can be rearranged by the introduction of the Prandtl number to give

$$Nu = 2 + 0.43Ra^{1/4} \tag{22.15b}$$

It is expected that this relation would be applicable for free convection in gases and Prandtl numbers in the vicinity of 1. However, in the absence of specific information it may also be used for liquids. Properties are evaluated at the film temperature $T_f = (T_s + T_\infty)/2$, and β at T_∞.

A more complex expression (Raithby and Hollands, 1998) encompasses other Prandtl numbers (Lienhard and Lienhard, 2004):

$$Nu = 2 + \frac{0.589Ra^{0.25}}{\left[1 + (0.492/Pr)^{9/16}\right]^{4/9}}, \quad Ra < 10^{12} \tag{22.15c}$$

Equation 22.15c has an estimated uncertainty of 5% for air and a root-mean-square error of approximately 10% at higher Prandtl numbers. For higher ranges of the Rayleigh number the experiments of Amato and Tien (1972) with water suggest the following correlation (Holman, 1986):

$$Nu = 2 + 0.50Ra^{0.25}, \quad 3 \times 10^5 < Ra < 8 \times 10^8, \quad 10 \leq Nu \leq 90 \tag{22.15d}$$

22.2.6 Natural Convection in Enclosed Spaces

Free convection inside enclosed spaces occurs in a number of processing applications. The flow phenomena inside spaces are examples of complex fluid systems, which are important in energy conservation processes in buildings (e.g., in multiply glazed windows, uninsulated walls, attics), in crystal growth and solidification processes, and in hot or cold liquid storage systems (Lienhard and Lienhard, 2004). Different types of flow patterns can occur inside enclosed spaces; at low Grashof numbers heat is transferred mainly by conduction across the fluid layer, while different flow regimes are encountered when the Grashof number increases (Geankoplis, 1993). Empirical correlations and review papers on natural convection in enclosures can be found in Geankoplis (1993), Holman (1986), Yang (1987), Raithby and Hollands (1998), and Catton (1978).

22.3 FORCED CONVECTION CORRELATIONS

In the food industry, most of the cooling or heating processes occur under forced convection. Forced convection is defined as that because the heat transfer from or to the foods is due to the motion of the heating or cooling fluid caused by external means such as a fan or a pump. Therefore unlike natural convection, the movement of the fluid is not dependent on the temperature difference, and fluid properties and flow velocity play an important part in the convective heat transfer coefficients. Hence Equation 22.5 can be simplified as

$$Nu = f(Re, Pr) \tag{22.16}$$

Therefore, finding the above relation is the main task in convection heat transfer research. Depending on the fluid flow position, studies on the forced convective heat transfer can be classified into two cases: external flow and internal flow.

22.3.1 External Flow

According to the shape of the solid object, a fluid can flow externally over a plate, a cylinder, or a sphere among other shapes. Heat transfer correlations have therefore been developed accordingly. In the flow over the external surface of an object, there exists a boundary layer. Within the boundary layer, the fluid velocity changes from zero at the surface to a uniform velocity V_∞ (the free stream velocity), and the flow can be laminar or turbulent. If the flow is laminar, the layer is defined as the laminar boundary layer. Likewise if turbulent flow exists in the layer, the layer is the turbulent boundary layer. As a result, different correlations are developed for predicting convective heat transfer coefficients under different boundary layer conditions. In order to identify the boundary layer conditions, for simplicity, abrupt transition from laminar to turbulent boundary is normally assumed. Therefore, a critical Reynolds number Re_{cr} is used to indicate the abrupt point, which is commonly defined as $Re_{cr} = 5 \times 10^5$.

22.3.1.1 Flat Plates

22.3.1.1.1 Laminar Boundary Layers ($Re < Re_{cr}$)

For simple flow problems where a fluid flows parallel to the plate with uniform temperature, if all the properties are evaluated at the film temperature, $T_f = (T_s + T_\infty)/2$, i.e., average of wall and free stream temperatures, and the Reynolds number is determined at the local position, i.e., $Re_x = \rho V_\infty x/\mu$, the local Nusselt number can be correlated by the following equations (Suryanarayana, 2000; Lienhard and Lienhard, 2004):

$$Nu_x = \frac{h_x x}{k} = 0.332 Re_x^{1/2} Pr^{1/3}, \quad 0.6 \le Pr \le 50 \tag{22.17a}$$

$$Nu_x = 0.564 Re_x^{1/2} Pr^{1/2}, \quad Re_x Pr \ge 100 \quad \text{and} \quad Pr \le 0.01 \text{ or } Re_x \ge 10^4 \tag{22.17b}$$

$$Nu_x = 0.339 Re_x^{1/2} Pr^{1/3}, \quad Pr \to \infty \tag{22.17c}$$

For all Prandtl numbers, Churchill and Ozoe (1973a) and Rose (1979) recommended the following empirical correlations (Holman, 1986; Suryanarayana, 2000):

$$Nu_x = \frac{0.3387 Pr^{1/3} Re_x^{1/2}}{\left[1 + (0.0468/Pr)^{2/3}\right]^{1/4}}, \quad Re_x Pr > 100 \tag{22.18a}$$

$$Nu_x = \frac{Re_x^{1/2} Pr^{1/2}}{\left(27.8 + 75.9 Pr^{0.306} + 657 Pr\right)^{1/6}} \tag{22.18b}$$

In the range of $0.001 < Pr < 2000$, Equation 22.18a is within 1.4% and Equation 22.18b is within 0.4% of the exact numerical solution to the boundary layer energy equation. If the fluid is flowing parallel to a flat plate and heat transfer is occurring between the whole plate of length L (m) and the fluid, the average convective heat transfer coefficient $\bar{h}$ (either $\bar{q}/\Delta T$ in the uniform wall temperature problem or $q/\overline{\Delta T}$ in the uniform heat flux problem) and thus the Nu number ($Nu = 2Nu_{x=L}$) are as follows (Geankoplis, 1993):

$$Nu = 0.664 Re^{1/2} Pr^{1/3}, \quad Pr > 0.7 \tag{22.19}$$

Likewise for liquid metal flows (Lienhard and Lienhard, 2004),

$$Nu = 1.13 Re^{1/2} Pr^{1/2}, \quad Pr \ll 1 \tag{22.20}$$

Alternatively for all Prandtl numbers (Churchill, 1976),

$$Nu = \frac{0.6774 Pr^{1/3} Re^{1/2}}{\left[1 + (0.0468/Pr)^{2/3}\right]^{1/4}}, \quad Re < Re_{cr} \tag{22.21}$$

Correspondingly, if the plate is supplied with a uniform heat flux, the following correlations should be used instead (Lienhard and Lienhard, 2004; Suryanarayana, 1995):

$$Nu_x = 0.453 Re_x^{1/2} Pr^{1/3}, \quad Pr > 0.1 \tag{22.22}$$

$$Nu_x = 0.886 Re_x^{1/2} Pr^{1/2}, \quad Pr < 0.05 \tag{22.23}$$

For all Pr numbers and uniform heat flux Churchill and Ozoe (1973b) recommended the following single correlation:

$$Nu_x = \frac{0.4637 Pr^{1/3} Re_x^{1/2}}{\left[1 + (0.02052/Pr)^{2/3}\right]^{1/4}}, \quad Re_x Pr > 100 \tag{22.24}$$

22.3.1.1.2 Turbulent Boundary Layer ($Re > Re_{cr}$)

When the flow in the boundary becomes turbulent ($Re > Re_{cr}$), Equations 22.17 through 22.24 are no longer valid. For a flat plate with uniform surface temperature, the local convective heat transfer coefficients can be predicted by the following equations (Suryanarayana, 1995, 2000):

$$Nu_x = \frac{h_x x}{k} = 0.0296 Re_x^{4/5} Pr^{1/3}, \quad 0.6 \le Pr \le 60, \quad Re_{cr} < Re_x < 10^7 \tag{22.25a}$$

$$Nu_x = 1.596 Re_x (\ln Re_x)^{-2.584} Pr^{1/3}, \quad 0.6 \le Pr \le 60, \quad 10^7 < Re_x < 10^9 \tag{22.25b}$$

$$Nu_x = 0.0296Re_x^{4/5}Pr^{0.43}\left(\frac{\mu_\infty}{\mu_s}\right)^{1/4}, \quad 0.6 \le Pr \le 60, \quad Re_x > Re_{cr} \tag{22.25c}$$

For average convective heat transfer coefficient (Janna, 2000),

$$Nu = \left[0.664Re_{cr}^{1/2} + 0.0359\left(Re^{4/5} - Re_{cr}^{4/5}\right)\right]Pr^{1/3}, \quad 0.6 < Pr < 60 \tag{22.26}$$

If $Re_{cr} = 5 \times 10^5$ is inserted, Equation 22.26 can be reduced to

$$Nu = (0.0359Re^{4/5} - 830)Pr^{1/3}, \quad 0.6 < Pr < 60, \quad Re_{cr} < Re < 10^8 \tag{22.27}$$

If $Re > 10^7$ and $Re_{cr} = 5 \times 10^5$ (Suryanarayana, 2000),

$$Nu = \left[1.963Re(\ln Re)^{-2.584} - 871\right]Pr^{1/3}, \quad 0.7 < Pr < 60, \quad 10^7 < Re < 10^9 \tag{22.28}$$

An alternative expression to Equation 22.27 is the Zhukauskas–Whitaker equation, which accounts for the temperature dependence of the fluid viscosity (Whitaker, 1972):

$$Nu = 0.036(Re^{4/5} - 9200)Pr^{0.43}(\mu/\mu_s)^{1/4} \tag{22.29}$$

where μ_s is evaluated at the uniform surface temperature T_s, while all other properties are obtained based on the free stream temperature T_∞. In addition, the equation is valid only for the following conditions: $0.7 < Pr < 380$, $10^5 < Re < 5.5 \times 10^6$, and $0.26 < (\mu/\mu_s) < 3.5$. If Equation 22.29 is used to predict heat transfer to a gaseous flow, the viscosity-ratio correction term should not be used and properties should be evaluated at the film temperature (Lienhard and Lienhard, 2004).

A problem with Equation 22.29 is that it does not deal with the question of heat transfer in the rather lengthy transition region, since it is based on the assumption that the flow abruptly passes from laminar to turbulent at a critical value of x (Lienhard and Lienhard, 2004). Churchill (1976) suggested the following equations to predict heat transfer for laminar, transitional, and turbulent flow (Lienhard and Lienhard, 2004):

$$Nu_x = 0.45 + (0.3387\phi^{1/2})\left\{1 + \frac{(\phi/2{,}600)^{3/5}}{[1 + (\phi_u/\phi)^{7/2}]^{2/5}}\right\}^{1/2} \tag{22.30a}$$

where

$$\phi = Re_x Pr^{2/3}\left[1 + \left(\frac{0.0468}{Pr}\right)^{2/3}\right]^{-1/2} \tag{22.30b}$$

and ϕ_u is a number between 10^5 and 10^7. If the Reynolds number at the end of the turbulent transition region is Re_u, an estimate is $\phi_u \approx \phi(Re_x = Re_u)$. For the average heat transfer coefficient, Churchill (1976) also proposed (Lienhard and Lienhard, 2004)

$$Nu = 0.45 + (0.6774\phi^{1/2})\left\{1 + \frac{(\phi/12{,}500)^{3/5}}{[1 + (\phi_{um}/\phi)^{7/2}]^{2/5}}\right\}^{1/2} \tag{22.30c}$$

where ϕ is defined as in Equation 22.30b using Re over the length L of the plate in place of Re_x, and $\phi_{um} \approx \phi(Re = Re_u)$. This equation may be used for either uniform heat flux or uniform surface temperature. The following equation can also be used for turbulent boundary layer along the whole plate and constant wall temperature (Petukhov and Popov 1963; Schlichting 1979):

$$Nu = \frac{0.037Re^{0.8}Pr}{1 + 2.443Re^{-0.1}(Pr^{2/3} - 1)} \tag{22.31}$$

Another simplified empirical relationship for estimating the average heat transfer coefficient is as follows (Geankoplis, 1993):

$$Nu = 0.0366Re^{0.8}Pr^{1/3}, \quad Pr > 0.7, \quad Re > Re_{cr} \tag{22.32}$$

If the plate is supplied with a uniform heat flux, the local heat transfer coefficient is higher. Kays and Crawford (1980) suggested the following:

$$Nu_x = 0.03Re_x^{0.8}Pr^{0.6} \tag{22.33}$$

Thomas and Al-Sharifi (1981) also recommended the following alternatives:

$$Nu_x = \frac{\sqrt{c_{fx}/2}Re_x Pr}{2.21\ln\left(Re_x\sqrt{c_{fx}/2}\right) - 0.232\ln Pr + 14.9Pr^{0.623} - 15.6}, \quad 0.5 < Pr < 10 \tag{22.34a}$$

$$Nu_x = \frac{\sqrt{c_{fx}/2}Re_x Pr}{2.21\ln\left(Re_x\sqrt{c_{fx}/2}\right) - 0.232\ln Pr + 10Pr^{0.741} - 6.21}, \quad 10 < Pr < 500 \tag{22.34b}$$

where the local friction factor is $c_{fx} \approx 0.0592Re_x^{-0.2}$. Equation 22.30 is valid for uniform surface temperature, but it may be used for uniform heat flow if the constants 0.3387 and 0.0468 are replaced by 0.4637 and 0.02052, respectively. In laminar boundary layers, the convective heat transfer coefficient with uniform heat flux is approximately 36% higher than that with uniform surface temperature. With turbulent boundary layers, the difference is small and the correlations for the local heat transfer coefficient can be used for both uniform surface temperature and uniform heat flux (Suryanarayana, 2000).

22.3.1.2 Cylinders

Many interesting practical problems on forced convection heat transfer deal with the bodies of curvature complex shape such as cylinders or spheres. The special features of heat transfer on a curved surface manifest themselves in a boundary layer, which can no longer be simply classified as laminar or turbulent boundary layer in the same way as flows over a flat plate. The boundary layer for flows over a cylinder can be laminar or partly laminar and partly turbulent, and then followed by a turbulent wake. Therefore, correlations are not developed based on the condition of the boundary layer, but rather based on a range of Reynolds numbers.

For flows over a cylinder, the characteristic dimension is the diameter of the cylinder, Reynolds number is evaluated by using the uniform velocity V_∞, i.e., the free stream velocity, and all the properties of the fluid are calculated at $T_f = (T_s + T_\infty)/2$; the correlation is as follows:

$$Nu = \frac{hD}{k} = cRe^m Pr^n \tag{22.35}$$

Table 22.2 Constants for Equation 22.35 for Air at 0.5 < Pr < 10

Re	$cPr^{1/3}$ (Pr = 0.7)	c	*m*
4–35	0.795	0.895	0.384
35–5,000	0.583	0.657	0.471
5,000–50,000	0.148	0.167	0.633
50,000–230,000	0.0208	0.0234	0.814

The values of c, m, and n in Equation 22.35 are given in Table 22.2 (Hilpert, 1933; Morgan, 1975). For $Re\ Pr > 0.2$, Churchill and Bernstein (1977) recommend the following correlations (Suryanarayana, 2000):

$$Nu = 0.3 + \frac{0.62Re^{1/2}Pr^{1/3}}{\left[1 + (0.4/Pr)^{2/3}\right]^{1/4}}\left[1 + \left(\frac{Re}{282{,}000}\right)^{5/8}\right]^{4/5}, \quad Re > 400{,}000 \tag{22.36a}$$

$$Nu = 0.3 + \frac{0.62Re^{1/2}Pr^{1/3}}{\left[1 + (0.4/Pr)^{2/3}\right]^{1/4}}\left[1 + \left(\frac{Re}{282{,}000}\right)^{1/2}\right], \quad 100{,}00 < Re < 400{,}000 \tag{22.36b}$$

$$Nu = 0.3 + \frac{0.62Re^{1/2}Pr^{1/3}}{\left[1 + (0.4/Pr)^{2/3}\right]^{1/4}}, \quad Re < 10{,}000 \tag{22.36c}$$

For flow of liquid metals, Ishiguro et al. (1979) suggested the following correlation (Suryanarayana, 2000):

$$Nu = 1.125(RePr)^{0.413}, \quad 1 < RePr < 100 \tag{22.37}$$

Another correlation equation is given by Whitaker (1972, cited by Holman, 1986) as

$$Nu = \left(0.4Re^{0.5} + 0.06Re^{2/3}\right)Pr^{0.4}\left(\frac{\mu_\infty}{\mu_s}\right)^{0.25} \tag{22.38}$$

Equation 22.38 is valid for $40 < Re < 10^5$, $0.65 < Pr < 300$, and $0.25 < \mu_\infty/\mu_s < 5.2$; all properties are evaluated at the free stream temperature except that μ_s is at the wall temperature. For $RePr < 0.2$, Nakai and Okazaki (1975) present the following relation (Holman, 1986):

$$Nu = \left[0.8237 - \ln\left(Re^{1/2}Pr^{1/2}\right)\right]^{-1} \tag{22.39}$$

Properties in Equations 22.36, 22.37, and 22.39 are evaluated at the film temperature.

A distinction was made earlier between constant wall temperature and constant surface flux problems. However, in the flow past a cylinder, the equations apply to either case, and the distinction is not so significant (Janna, 2000).

22.3.1.3 Spheres

For flows over spheres, if the characteristic dimension is the diameter of the spheres, all properties are evaluated at T_∞, except μ_s at T_s, and if $3.5 < Re < 76{,}000$, $0.71 < Pr < 380$, and

$1 < \mu_\infty/\mu_s < 3.2$, where μ_∞ is the uniform viscosity, i.e., the free stream viscosity, Whitaker (1972) suggests the following equation for gases and liquids flowing past spheres:

$$Nu = 2.0 + (0.4Re^{1/2} + 0.06Re^{2/3})Pr^{2/5}\left(\frac{\mu_\infty}{\mu_s}\right)^{0.25} \quad (22.40)$$

Achenbach (1978) proposed the following equations for $Pr = 0.71$, where properties are evaluated at $(T_s + T_\infty)/2$ (Suryanarayana, 2000):

$$Nu = 2 + (0.25Re + 3 \times 10^{-4}Re^{1.6})^{1/2}, \quad 100 < Re < 2 \times 10^5 \quad (22.41a)$$

$$Nu = 430 + 5 \times 10^{-3}Re + 0.25 \times 10^{-9}Re^2 - 3.1 \times 10^{-17}Re^3, \quad 4 \times 10^5 < Re < 5 \times 10^6 \quad (22.41b)$$

Witte (1968a) from experimental results with liquid sodium recommends the following for liquid metals (Suryanarayana, 2000):

$$Nu = 2 + 0.386(RePr)^{1/2}, \quad 3.6 \times 10^4 < Re < 1.5 \times 10^5 \quad (22.42)$$

22.3.2 Internal Flows

Internal flows refer to flows inside tubes or ducts. In such a flow, a boundary layer also forms near the inside surface of a tube. In the boundary layer, the flow can be laminar or turbulent. However, unlike the boundary layer in external flows, the development of the boundary layer is restricted by the radius of the tube. Furthermore, in the entrance region of the tube, the thickness of the boundary layer increases rapidly, which also causes significant changes in the local convective heat transfer coefficients. When the thickness of the boundary layer becomes the maximum, the thickness then remains unchanged for the rest of the tube. Therefore, the entrance region where the flow velocity profile varies is known as the hydrodynamically developing region, while the remaining region where the velocity profile is invariant is defined as the hydrodynamically fully developed region. Correlations relating the Nusselt number, Reynolds number, and Prandtl number are therefore developed in these two regions, respectively. In these correlations, the inside diameter is taken as the characteristic dimension.

22.3.2.1 Laminar Flows

For flows inside circular pipes, if the pipes are in uniform temperature, and all fluid properties (except μ_s at T_s) are evaluated at the bulk temperature $T_b = (T_i + T_e)/2$, i.e., the average temperature between the inlet and exit; Sieder and Tate (1936) proposed the following equation for the average heat transfer coefficient over a length L of the tube in the entrance region:

$$Nu = 1.86\left(\frac{D}{L}RePr\right)^{1/3}\left(\frac{\mu}{\mu_s}\right)^{0.14} \quad (22.43)$$

Equation 22.43 is valid for the following conditions: $0.48 < Pr < 16{,}700$, $0.0044 < \mu/\mu_s < 9.75$, $(L/D) < (8/RePr)(\mu_s/\mu)^{0.42}$, and $Gz > 100$, where Gz is the Graetz number $= RePr\,D/L$. Equation 22.43 is satisfactory for small diameters and temperature differences. A more general expression covering all diameters and temperature differences is obtained by including an additional factor 0.87 $(1+0.015Gr^{1/3})$ on the right-hand side of Equation 22.43 (Knudsen et al., 1999). For $0.1 < Gz < 10^4$, the following equation is recommended (Knudsen et al., 1999):

$$Nu = 3.66 + \frac{0.19Gz^{0.8}}{1+0.117Gz^{0.467}}\left(\frac{\mu}{\mu_s}\right)^{0.14} \tag{22.44}$$

Stephan (1961, 1962) also proposes the following correlations for the entrance region for uniform surface temperature and uniform heat flux, respectively:

$$Nu = 3.657 + \frac{0.0677((D/L)RePr)^{1.33}}{1+0.1Pr((D/L)Re)^{0.83}} \tag{22.45a}$$

$$Nu = 4.364 + \frac{0.086((D/L)RePr)^{1.33}}{1+0.1Pr((D/L)Re)^{0.83}} \tag{22.45b}$$

which are valid over the range $0.7 < Pr < 7$ or if $RePr\,D/L < 33$ also for $Pr > 7$. For fully developed flow, the convective heat transfer coefficient remains a constant as given below (Janna, 2000):

For uniform surface temperature

$$Nu = 3.658 \tag{22.45c}$$

For uniform heat flux

$$Nu = 4.364 \tag{22.45d}$$

For flows inside concentric annular ducts, the annular duct is formed by two concentric tubes, and each tube has a finite wall thickness, and the annular flow area is bounded by the outer diameter of the inner tube D_i and the inner diameter of the outer tube D_o. Approximate heat transfer coefficients for laminar flow in annuli may be predicted by the following equation (Knudsen et al., 1999):

$$Nu = 1.02Re^{0.45}Pr^{0.5}\left(\frac{D_h}{L}\right)^{0.4}\left(\frac{D_o}{D_i}\right)^{0.8}\left(\frac{\mu}{\mu_i}\right)^{0.14}Gr^{0.05} \tag{22.46}$$

where

μ_i is the viscosity at the inner wall of annulus

D_h is the hydraulic mean diameter defined as the ratio of four times the area of cross section of the flow perpendicular to the direction of the flow A_c to the wetted perimeter of the duct P_w, that is:

$$D_h = 4\frac{A_c}{P_w}$$

22.3.2.2 Turbulent Flows

If flows in the pipes are turbulent, for both uniform surface temperature and uniform heat flux, the following equation can be used, which is valid for $0.7 \leq Pr \leq 16{,}700$, $Re \geq 10{,}000$, and $L/D \geq 60$ (Sieder and Tate, 1936):

$$Nu = 0.027 Re^{4/5} Pr^{1/3} \left(\frac{\mu}{\mu_s}\right)^{0.14} \tag{22.47}$$

For better predicted results, Dittus and Boelter (1930) recommend the following (Janna, 2000):

$$Nu = 0.023 Re^{4/5} Pr^n, \quad 0.7 \leq Pr \leq 160,\ Re \geq 10{,}000,\ L/D \geq 60 \tag{22.48}$$

where $n = 0.4$ for heating ($T_s > T_b$) and $n = 0.3$ for cooling ($T_s < T_b$); properties are evaluated at the fluid bulk temperature. Equation 22.48 is useful when the wall temperature is unknown. For even better prediction accuracy, Gnielinski (1976, 1990) proposes the following equations for $2300 < Re < 10^6$ and $0 < D/L < 1$ to cover both the entrance region and the fully developed region:

$$Nu = 0.0214(Re^{4/5} - 100)Pr^{2/5}\left[1 + \left(\frac{D}{L}\right)^{2/3}\right], \quad 0.5 < Pr < 1.5 \tag{22.49a}$$

$$Nu = 0.012(Re^{0.87} - 280)Pr^{2/5}\left[1 + \left(\frac{D}{L}\right)^{2/3}\right], \quad 1.5 < Pr < 500 \tag{22.49b}$$

For describing the fully developed region, ratio D/L should be set to zero in Equation 22.49. If there exists a large variation in fluid properties due to temperature, Gnielinsky (1990) suggests multiplying Equation 22.49 by $(T_b/T_s)^{0.45}$ for gases and by $(Pr/Pr_s)^{0.11}$ for liquids, where Pr_s is evaluated at the surface temperature T_s.

Equations 22.47 through 22.49 are developed for uniform surface temperature of the tube, but they can also be used for uniform heat flux in turbulent flows. If heat transfers with uniform heat flux, Petukhov (1970) recommends the following correlation for better accuracy with the conditions of $10^4 < Re < 5 \times 10^6$ and $0.08 < \mu/\mu_s < 40$:

$$Nu = \frac{(f/8)RePr}{1.07 + 12.7(f/8)^{1/2}(Pr^{2/3} - 1)}\left(\frac{\mu}{\mu_s}\right)^n \tag{22.50a}$$

where

$f = [0.79 \ln(Re) - 1.64]^{-2}$
$n = 0.11$ for liquids heating
$n = 0.25$ for liquids cooling
$n = 0$ for constant heat flux or for gases (Holman, 1986)

Alternative to Equation 22.50a, for uniform heat flux, the following Gnielinski correlation is suggested for $2300 < Re < 5 \times 10^6$ (Suryanarayana, 1995):

$$Nu = \frac{f/8(Re - 1000)Pr}{1 + 12.7\sqrt{f/8}(Pr^{2/3} - 1)}\left[1 + \left(\frac{D}{L}\right)^{2/3}\right] \tag{22.50b}$$

The Petukhov correlation (Equation 22.50a) is similar to the Gnielinksi correlation (Equation 22.50b) and is generally accurate at $10^4 < Re < 5 \times 10^6$, but Equation 22.50a generally provides slightly higher heat transfer coefficients. For uniform surface temperature, Equation 22.50 can be also used with negligible error for fluids with $Pr > 0.7$. For turbulent flows in noncircular ducts,

Equations 22.47 through 22.50 are also valid except that the Reynolds number and Nusselt number should be calculated based on the hydraulic mean diameter as the characteristic dimension.

For turbulent flows ($Re > 2300$), and for flows inside concentric annular ducts, Petukhov and Roizen (1964) suggest the following:

For heat transfer at the inner tube with outer tube insulated

$$\frac{Nu}{Nu_{\text{tube}}} = 0.86(D_i/D_o)^{-0.16} \tag{22.51a}$$

For heat transfer at the outer tube with inner tube insulated

$$\frac{Nu}{Nu_{\text{tube}}} = 1 - 0.14(D_i/D_o)^{0.6} \tag{22.51b}$$

where the characteristic dimension $D_h = D_o - D_i$ is used to determine the Reynolds number and Nusselt number, and all properties are evaluated at the fluid bulk mean temperature (arithmetic mean of inlet and outlet temperatures). If heat transfer occurs at both tubes, which have the same wall temperatures, Stephan (1962) recommends the use of the following correlation (Knudsen et al., 1999):

$$\frac{Nu}{Nu_{\text{tube}}} = \frac{0.86(D_i/D_o)^{-0.16} + \left[1 - 0.14(D_i/D_o)^{0.6}\right]}{1 + (D_i/D_o)} \tag{22.52}$$

For diameter ratios $D_i/D_O > 0.2$, Monrad and Pelton's (1942) equation is recommended for either or both the inner and outer tubes (Knudsen et al., 1999):

$$Nu = 0.020 Re^{0.8} Pr^{1/3} \left(\frac{D_o}{D_i}\right)^{0.53} \tag{22.53}$$

22.3.3 Combined Forced and Natural Convection

In any forced convection, natural convection always plays a part, and therefore the heat transfer is actually a combination of forced convection and natural convection. However, in most of the forced convections, the fluid velocity is high enough so that the contribution of the natural convection is negligible in formulating the correlations. On the other hand, if the velocity is not sufficiently high in forced convection, the contribution from the natural convection becomes significant. In this case, contributions from both the forced convection and natural convection should be considered and consequently different correlations are developed. The magnitude of the dimensionless group Gr/Re^2 describes the ratio of buoyant to inertia forces and thus governs the relative importance of free to forced convection (Awuah and Ramaswamy, 1996). The combined free and forced convection regime is generally the one for which $(Gr/Re^2) \approx 1$ (Incropera and DeWitt, 1996).

For flows in horizontal tubes, if the tubes have uniform surface temperature, Depew and August (1971) propose the following correlation to account for similar contributions from both natural and forced convections (Suryanarayana, 1995):

$$Nu = 1.75\left[Gz + 0.12(Gz Gr^{1/3} Pr^{0.36})^{0.88}\right]^{1/3} \left(\frac{\mu}{\mu_s}\right)^{0.14}, \quad \frac{L}{D} < 28.4 \tag{22.54}$$

where the Graetz number $Gz = mc_p/(kL)$, and all the properties are evaluated at the average bulk temperature, except μ_s at T_s. If the tubes are supplied with a uniform heat flux, Morcos and Bergles (1975) suggest using the following equation to evaluate the average convective heat transfer coefficient for the following conditions: $3 \times 10^4 < Ra < 10^6$, $4 < Pr < 175$, and $2 < hD^2/(k_w t) < 66$ (Suryanarayana, 1995):

$$Nu = \left\{ 4.36^2 + \left[0.145 \left(\frac{Gr * Pr^{1.35}}{P^{0.25}} \right)^{0.265} \right]^2 \right\}^{0.5} \tag{22.55}$$

where
$P = (kD/k_w t)$
t is the tube wall thickness
k_w is the tube wall thermal conductivity

All the fluid properties are evaluated at $(T_w + T_b)/2$ with T_w and T_b being the inside wall temperature and bulk temperature, respectively.

22.4 CORRELATIONS FOR CONVECTION WITH PHASE CHANGE

Heat transfer with phase change occurs regularly in food processing. Examples of phase change include boiling, condensation, melting, freezing, and drying processes. Heat transfer to a boiling liquid is very important in different kinds of chemical and biological processing, such as control of the temperature of chemical reactions, evaporation of liquid foods, etc.

22.4.1 Boiling

Boiling takes place on a solid–liquid interface when the temperature at the solid surface is higher than the saturation temperature of the liquid. During boiling, vapor is generated at the interface. Depending on the state of the fluid, boiling can be divided into pool boiling if the fluid is stationary and forced convection boiling if the fluid is in motion.

22.4.1.1 Pool Boiling

In pool boiling, there exist several distinct regimes. In nucleate boiling regime, bubbles are formed in microcavities adjacent to the solid surface. They then grow until they reach some critical size, at which point they separate from the surface and enter the fluid stream. In this regime, high heat flux is achieved at low values of temperature difference between the surface and the saturation temperature of the liquid, i.e., $\Delta T = T_s - T_{sat}$. The maximum heat flux achieved in this regime is defined as the critical heat flux (CHF), i.e., q_{max}. In the film boiling regime, as the temperature at the solid surface is much higher than the saturation temperature of the liquid, the layer of liquid closest to the surface turns to vapor; as a result, the surface is completely blanketed by the vapor. Due to the mass exchange occurring at the vapor liquid interface, bubbles of vapor periodically form and migrate upward. There also exists a transition boiling regime where both nucleate boiling and film boiling take place. In this transition regime, with the increase in the temperature at the solid surface, the area of the surface covered by film boiling increases while that with nucleate boiling decreases, with a net decrease in the average heat flux.

In the above three regimes, food processing equipment are normally designed to operate in the nucleate boiling regime below CHF to achieve a high heat transfer rate at low ΔT values. Therefore, most of the correlations developed are focused on this regime and the prediction of CHF.

It is worth noting that pool boiling correlations are generally regarded as being valid for both subcooled (the liquid temperature is below the saturation temperature) and saturated nucleate boiling (the liquid temperature is equal to the saturation temperature); and that at moderate-to-high heat flux levels, a pool boiling heat transfer correlation developed for one heated surface geometry in one specific orientation often works reasonably well for other geometries and other orientations (Carey, 2000).

22.4.1.1.1 Nucleate Boiling

Rohsenow (1952) developed the following correlation for predicting the heat transfer rate in the nucleate boiling regime (for horizontal wires):

$$q = h(T_s - T_{sat}) = \mu L_v \left(\frac{g\Delta\rho}{\sigma}\right)^{1/2} \left(\frac{c_p \Delta T}{C_s L_v Pr^n}\right)^3 \qquad (22.56)$$

where all properties are those of liquid except ρ_v, which is the density of the vapor, $\Delta\rho = \rho - \rho_v$, i.e., the density difference from the liquid and the vapor, σ is the surface tension of the liquid–vapor interface, L_v is the enthalpy of vaporization of the liquid. Constant n generally equals 1.7, but for water it is equal to 1, while constant C_s is dependent on the combination of the liquid and the material of the surface and its surface finish, which can cause significant deviations from the values predicted by Equation 22.56. Table 22.3 lists the values of the constant C_s for water ($n = 1$) as suggested by Rohsenow (1952) and Vachon et al. (1968).

Equation 22.56 may be used for geometries other than horizontal wires, since it is found that heat transfer for pool boiling is primarily dependent on bubble formation and agitation, which is dependent on surface area and not surface shape (Holman, 1986). For the critical heat flux, Kutateladze (1963) recommends the following for boiling from an infinite horizontal plate:

$$\text{CHF} = q_{max} = C\rho_v^{0.5} L_v (\sigma g \Delta\rho)^{0.25} \qquad (22.57)$$

where the constant C is between 0.12 and 0.18; e.g., Zuber (1958) theoretically estimated $C = \pi/24$, Kutateladze (1963) correlated data for $C = 0.13$, and Lienhard and Dhir (1973) and Lienhard et al. (1973) correlated data for $C = 0.15$. Stephan and Preußer (1979) also suggest the following equation for nucleate boiling regime:

Table 22.3 Constants C_s and n for Equation 22.56

Surface	C_s
Brass, nickel	0.006
Copper, polished	0.0128
Copper, lapped	0.0147
Copper, scored	0.0068
Stainless steel, ground and polished	0.008
Stainless steel, Teflon pitted	0.0058
Stainless steel, chemically etched	0.0133
Stainless steel, mechanically polished	0.0132
Platinum	0.013

$$Nu = \frac{hd_A}{k} = 0.0871\left(\frac{qd_A}{kT_s}\right)^{0.674}\left(\frac{\rho_v}{\rho}\right)^{0.156}\left(\frac{L_v d_A^2}{\alpha^2}\right)^{0.371}\left(\frac{\alpha^2 \rho}{\sigma d_A}\right)^{0.350}(Pr)^{-0.162} \quad (22.58)$$

where d_A is the bubble departure diameter $= 0.851\beta_o\sqrt{(2\sigma/g\Delta\rho)}$ with $\beta_o = \pi/4$ rad for water; 0.0175 rad for low-boiling liquids; and 0.611 rad for other liquids. Simplified empirical equations to estimate heat transfer coefficients for water boiling on the outside of submerged surfaces at 1.0 atm absolute pressure are developed (Geankoplis, 1993):

For a horizontal surface

$$h = 1043\Delta T^{1/3}, \quad \frac{q}{A} < 16 \quad (22.59a)$$

$$h = 5.56\Delta T^3, \quad 16 < \frac{q}{A} < 240 \quad (22.59b)$$

For a vertical surface

$$h = 537\Delta T^{1/7}, \quad \frac{q}{A} < 3 \quad (22.59c)$$

$$h = 7.95\Delta T^3, \quad 3 < \frac{q}{A} < 63 \quad (22.59d)$$

where q/A is in kW/m^2 and ΔT in K. If the pressure p is in atm absolute, the values of h at 1 atm obtained in Equation 22.59 may be modified to take into account the influence of pressure by multiplying by $(p/1)^{0.4}$.

22.4.1.1.2 Film Boiling

Once the critical heat flux is exceeded, film boiling occurs, in which the solid surface is fully covered by a continuous vapor film. Therefore, radiation heat transfer at very high surface temperatures through the vapor film must be considered in formulating the heat transfer coefficient. Bromley (1950) developed the following equation for horizontal cylinders:

$$Nu_v = \frac{h_b D}{k_v} = c\left[\frac{\rho_v(L_v + 0.4c_{pv}\Delta T)g\Delta\rho D^3}{k_v\mu_v\Delta T}\right]^{1/4} \quad (22.60a)$$

where subscript v refers to properties of vapor. The liquid properties and the enthalpy of vaporization are determined at T_{sat} and the vapor properties at $(T_s + T_{sat})/2$. Bromley (1950) suggests constant c of 0.62 for cylinders. Dhir and Lienhard (1971) recommend $c = 0.67$ for spheres.

If a modified Rayleigh number Ra' is introduced,

$$Ra' = \frac{\rho_v(L_v + 0.4c_{pv}\Delta T)g\Delta\rho D^3}{k_v\mu_v\Delta T}$$

Equation 22.60a can be simplified to

$$Nu_v = c(Ra')^{1/4}$$

For film boiling on spheres with cryogenic fluids, Frederking and Clark (1963) recommend

$$Nu_v = 0.14(Ra')^{1/3} \quad (22.60b)$$

For large or infinite horizontal surfaces, Berenson (1961) proposes the following equation:

$$\frac{hL'}{k_v} = 0.425\left[\frac{\rho_v(L_v + 0.4c_{pv}\Delta T)g\Delta\rho L'^3}{k_v\mu_v\Delta T}\right]^{1/4} \tag{22.60c}$$

where

$$L' = \left(\frac{\sigma}{g\Delta\rho}\right)^{1/2}$$

For a vertical flat surface, the average coefficient over the region up to a distance L is given by

$$\frac{hL}{k_v} = 0.943\left[\frac{\rho_v(L_v + 0.4c_{pv}\Delta T)g\Delta\rho L^3}{k_v\mu_v\Delta T}\right]^{1/4} \tag{22.60d}$$

The heat transfer coefficient h_b in Equation 22.60a considers only the conduction through the film but does not include the effects of radiation. The total heat transfer coefficient h may be calculated from the following empirical relation, which requires an iterative solution (Holman, 1986):

$$h = h_b\left(\frac{h_b}{h}\right)^{1/3} + h_r \tag{22.61}$$

where h_r is the radiation heat transfer coefficient and is estimated by assuming an emissivity ε equal to 1 for the liquid, that is

$$h_r = \frac{\sigma\varepsilon(T_s^4 - T_{sat}^4)}{T_s - T_{sat}} \tag{22.62}$$

where σ is the Stefan–Boltzmann constant. Once film boiling occurs, the continuous vapor film blanketed the solid surface and can sustain even if the heat flux is reduced to below the CHF. If the heat flux is further decreased to a point where the vapor formed can no longer support the liquid above the vapor layer, this heat flux is defined as the minimum heat flux (MHF). Zuber (1959) developed the expression for the minimum film boiling heat flux from a horizontal plate as follows:

$$\text{MHF} = q_{min} = c'\rho_v(L_v + 0.4c_{pv}\Delta T)\left[\frac{\sigma g\Delta\rho}{(\rho + \rho_v)^2}\right]^{1/4} \tag{22.63}$$

where for infinite horizontal surface, $c' = \pi/24$ (Zuber, 1959) and for horizontal surfaces, $c' = 0.09$ (Berenson, 1961).

22.4.1.2 Forced Convection Boiling

When a liquid is forced through a channel or over a surface maintained at a temperature higher than the saturation temperature of the liquid, forced convection boiling may result, and the effect is generally to improve heat transfer everywhere, though flow is particularly effective in raising q_{max} (CHF) (Holman, 1986; Lienhard and Lienhard, 2004). For forced convection boiling in smooth tubes, Rohsenow and Griffith (1955) recommend the addition of the effect of forced convection to the boiling heat flux as follows (Holman, 1986):

$$\left(\frac{q}{A}\right)_{\text{total}} = \left(\frac{q}{A}\right)_{\text{boiling}} + \left(\frac{q}{A}\right)_{\text{forced convection}} \tag{22.64a}$$

where $(q/A)_{\text{boiling}}$ is computed by using Equation 22.56, and the Dittus–Boelter relation (Equation 22.48) with the coefficient 0.019 instead of 0.023 is used to compute $(q/A)_{\text{forced convection}}$. The temperature difference between wall and liquid bulk temperature is used to compute the forced convection effect. Another expression to predict the heat flux during nucleate flow boiling is given by Bergles and Rohsenow (1964) (Lienhard and Lienhard, 2004):

$$q = q_{\text{fc}}\sqrt{1 + \left[\frac{q_{\text{b}}}{q_{\text{fc}}}\left(1 - \frac{q_{\text{i}}}{q_{\text{b}}}\right)\right]^2} \tag{22.64b}$$

where

- q_{fc} is the single-phase forced convection heat transfer calculated by using the correlations for forced convection given earlier
- q_{b} is the pool boiling heat flux computed by using Equation 22.56
- q_{i} is the heat flux from the pool boiling curve evaluated at the value of $(T_{\text{s}} - T_{\text{sat}})$ where boiling begins during flow boiling

Equation 22.64 is applicable to subcooled flows or other situations in which vapor generation is not too great. When saturated boiling conditions are reached, a fully developed nucleate boiling phenomenon is encountered that is independent of the flow velocity. For the fully developed boiling state, McAdams et al. (1949) suggested the following equation for low-pressure boiling water (Holman, 1986):

$$\frac{q}{A} = 2.253\Delta T^{3.96} \quad 0.2 < p < 0.7 \tag{22.65a}$$

For higher pressures (Levy, 1959)

$$\frac{q}{A} = 283.2p^{4/3}\Delta T^3 \quad 0.7 < p < 14 \tag{22.65b}$$

where

- q/A is in W/m^2
- ΔT is in °C
- p is in MPa

For the critical heat flux in flow boiling, a technique similar to the one employed in Equation 22.64 is suggested (Gambill, 1962), that is, to calculate the CHF in flow boiling by a superposition of the forced convection effect and the critical heat flux for pool boiling. For film boiling heat flux during forced flow normal to a cylinder, Bromley et al. (1953) recommend (Lienhard and Lienhard, 2004)

$$q = \text{constant}\left\{\frac{k_{\text{v}}\rho_{\text{v}}L_{\text{v}}\left[1 + (0.968 - 0.163/Pr_{\text{v}})c_{\text{pv}}(T_{\text{s}} - T_{\text{sat}})/L_{\text{v}}\right]\Delta T V_{\infty}}{D}\right\}^{1/2} \tag{22.66}$$

Equation 22.66 is valid for $V_{\infty}^2/(gD) \geq 4$ and $Pr_{\text{v}} \geq 0.6$. Bromley et al.'s (1953) data fixed the constant at 2.70, while Witte (1968b) recommended a value of 2.98 for the constant for flow over a sphere.

Flow boiling in tubes is a complex convection process that is encountered in the air-conditioning, heating, and refrigeration industries. The models to describe flow boiling generally differentiate between nucleate-controlled heat transfer and convective boiling heat transfer, where relations to predict the heat transfer coefficient are typically formulated to impose a gradual suppression of nucleate boiling and gradual increase in liquid film evaporation as the vapor fraction increases (Carey, 2000). One method of this type is developed by Kandlikar (1990), Kandlikar and Nariai (1999), and Kandlikar et al. (1999) for both horizontal and vertical tubes. Kandlikar's method consists of calculating the ratio of the flow boiling heat transfer coefficient (h_{fb}) over the liquid-only heat transfer coefficient (h_{lo}) from each of the following two correlations and choosing the bigger value (Lienhard and Lienhard, 2004):

$$\left.\frac{h_{fb}}{h_{lo}}\right|_{nbd} = (1-x)^{0.8}\left[0.6683Co^{-0.2}f_0 + 1058Bo^{0.7}F\right] \tag{22.67a}$$

$$\left.\frac{h_{fb}}{h_{lo}}\right|_{cbd} = (1-x)^{0.8}\left[1.136Co^{-0.9}f_0 + 667.2Bo^{0.7}F\right] \tag{22.67b}$$

where subscripts nbd mean nucleate boiling dominant and cbd mean convective boiling dominant. The convection number Co and boiling number Bo are defined, respectively, as

$$Co = \left(\frac{1-x}{x}\right)^{0.8}\left(\frac{\rho_v}{\rho}\right)^{0.5}$$

$$Bo = \frac{q}{GL_v}$$

where

x is the vapor fraction or quality
G is the superficial mass flux through the pipe in $\text{kg/m}^2\text{s}$

F in Equation 22.67 is a fluid-dependent parameter whose value is tabulated for a variety of fluids, and f_0 is an orientation factor. For vertical tubes f_0 is set to 1, while for horizontal tubes

$$f_0 = 1, \quad \text{for } Fr_{lo} \geq 0.04$$

$$f_0 = (25Fr_{lo})^{0.3}, \quad \text{for } Fr_{lo} < 0.04$$

and Fr is the Froude number defined as $G^2/\rho gD$. The following simplified relation can be used for forced convection boiling inside vertical tubes (Geankoplis, 1993):

$$h = 2.55(\Delta T)^3 e^{p/1551} \tag{22.68}$$

where ΔT is in K and p in kPa.

22.4.2 Condensation

Condensation is also a common phenomenon in the food industry, which is a process of converting vapor back into liquid, and is the reverse process of evaporation. Condensation occurs when the surface temperature drops below the saturation temperature corresponding to the vapor pressure. This temperature is also known as the dew point. Similar to the two modes of boiling, i.e., nucleate and film boiling, condensation also has two modes: dropwise and filmwise. However unlike boiling, the mode of condensation depends on the surface characteristics rather than on the temperature difference. Dropwise condensation takes place when the rate of condensation is low (e.g., the presence of a noncondensable gas) or when the liquid does not wet the surface; however, such condensation cannot be sustained and will eventually become filmwise. Film condensation occurs when the liquid wets the surface or when a drop of the liquid on the surface tends to spread to the surface. Therefore in filmwise condensation, the cooled surface is completely covered with a layer of the condensate. In most applications, film condensation is expected.

22.4.2.1 Laminar Film Condensation

When the Reynolds number in the flow of the condensate film is smaller than 450, the film condensation is laminar. The earliest correlation for laminar film condensation was derived by Nusselt (1916), who proposed the following local heat transfer coefficient for vertical surfaces:

$$Nu_x = \frac{hx}{k} = 0.707\left(\frac{\rho L_v g \Delta\rho x^3}{\Delta T_c \mu k}\right)^{1/4} \tag{22.69a}$$

where $\Delta T_c = T_{sat} - T_s$, and all properties without subscript are for condensate evaluated at the arithmetic mean temperature $(T_{sat} + T_s)/2$. Equation 22.69a assumes that all the condensate is at the saturation temperature. However in reality, the condensate should be a subcooled liquid at the surface temperature. Rohsenow (1956) considered the heat transfer due to the cooling of the condensate below the saturation temperature and proposed the following modification to Equation 22.69a:

$$Nu_x = \frac{hx}{k} = 0.707\left[\frac{\rho(L_v + 0.68c_p\Delta T_c)g\Delta\rho x^3}{\Delta T_c \mu k}\right]^{1/4} \tag{22.69b}$$

Therefore, the average heat transfer coefficient over a length L can be derived as

$$Nu = \frac{hL}{k} = 0.943\left[\frac{\rho(L_v + 0.68c_p\Delta T_c)g\Delta\rho L^3}{\Delta T_c \mu k}\right]^{1/4} \tag{22.70}$$

An alternative correlation in terms of the Reynolds number Re and the condensate number Co can be obtained as

$$Co = \frac{h}{k}\left(\frac{\mu^2}{\rho g \Delta\rho}\right)^{1/3} = 0.924Re^{-1/3} \tag{22.71}$$

Equations 22.69 through 22.71 can be also used for inclined surfaces by replacing g by $g\cos(\theta)$ where θ is the angle of inclination of the plate to the vertical. Equations 22.69 through 22.71 are generally for $Re < 10$. For a higher Reynolds number $(10 < Re < 450)$, White (1988) suggests the following correlation:

$$Co = \frac{h}{k}\left(\frac{\mu^2}{\rho g \Delta\rho}\right)^{1/3} = \frac{Re}{1.47Re^{1.22} - 1.3} \tag{22.72}$$

The average condensation heat transfer coefficient on a horizontal cylinder at a uniform surface temperature is correlated by the following relation (Suryanarayana, 1995):

$$Nu = \frac{hD}{k} = 0.729\left(\frac{(L_v + 0.68c_p\Delta T_c)\rho g\Delta\rho D^3}{\Delta T_c\mu k}\right)^{1/4} \tag{22.73}$$

For film condensation on spheres at a uniform temperature, Dhir and Lienhard (1971) showed that the constant in Equation 22.73 should be replaced by 0.815.

22.4.2.2 Turbulent Film Condensation

Turbulent film condensation occurs when the Reynolds number based on the film thickness is greater than 450 or 1800 depending on the definition of the Reynolds number, that is $Re = \dot{m}/P_w\mu$ or $Re = 4\dot{m}/P_w\mu$, respectively. For vertical surfaces, Labuntsov (1957) suggests

$$Co = \frac{h}{k}\left(\frac{\mu^2}{\rho g\Delta\rho}\right)^{1/3} = \frac{Re}{2188 + 41\left(Re^{0.75} - 89.5\right)Pr^{-0.5}} \tag{22.74a}$$

Alternatively, the following expression may also be used (McAdams, 1954; Geankoplis, 1993):

$$Nu = \frac{hL}{k} = 0.0077\left(\frac{g\rho^2 L^3}{\mu^2}\right)^{1/3} Re^{0.4} \tag{22.74b}$$

22.5 HEAT TRANSFER COEFFICIENTS AS APPLIED TO THE FOOD INDUSTRY

Rahman (1995) presented a comprehensive review and discussion of the empirical equations for predicting heat transfer coefficients along with the techniques available for measuring h. A further and extensive update of literature data of heat transfer coefficients in food processing was carried out by Zogzas et al. (2002) and Krokida et al. (2002), where predictive correlations for h are given for different foods and processing operations (baking, blanching, cooling, drying, freezing, storage, and sterilization). Goldstein et al. (2005) reviewed 1886 papers published in 2001 on heat transfer and also mentioned important conferences and meetings on heat transfer and related fields, as well as books on heat transfer published during 2002. The information given below attempts to complement the correlations given in the previous sections, and to report recent published literature on heat transfer coefficients as applied to food processing.

22.5.1 Air Impingement Applications

Use of air impingement is a promising development in rapid thermal processing of foods. Applications of these systems, which involve arrays of jets that impinge air on the surface of the product, include operations such as drying, baking, freezing, and thawing (Sarkar and Singh, 2003). Sarkar and Singh (2003) developed the following correlation for impingement heat transfer coefficients under freeze-thaw conditions:

$$Nu = fRe^b \tag{22.75}$$

Table 22.4 Empirical Constants in Equations 22.75 and 22.76 for Slot and Circular Jet

	Parameter Values					
Jet Type	***H* (cm)**	$a \times 10^2$	$b \times 10^{-1}$	$c \times 10^{-2}$	d	R^2
Circular jet	13.1	2.5793	7.5586	2.3918	1.0757	0.951
	7.6	0.6784	8.9247	2.1709	1.0284	0.938
Slot jet	13.1	5.9814	6.9852	3.8006	0.8253	0.9552
	7.6	8.6696	6.5204	4.4658	0.6699	0.9525

Source: Sarkar, A. and Singh, R.P., *J. Food Sci.*, 68, 910, 2003.

where
- the Nusselt and Reynolds numbers are defined with respect to nozzle diameter D (hydraulic diameter in m)
- b is an empirical constant
- f is a function of the H/D ratio and r/D ratio given below as

$$f = a^{\left[1+c(r/D)^d\right]} \tag{22.76}$$

where
- H is the distance of nozzle to plate in m
- r is the radial distance from stagnation point in m

Table 22.4 shows the parameters a, b, c, and d and the corresponding regression coefficient R^2 for H/D ratios of 6.62 and 3.84 corresponding to H of 0.131 m and 0.076 m, respectively. These parameters are valid for temperature ranges between −50°C and 0°C.

22.5.2 Forced Air Cooling and Freezing

Becker and Fricke (2004) reviewed heat transfer data for the cooling and freezing of foods and developed nine Nusselt–Reynolds–Prandtl correlations shown in Table 22.5. It is well known that the knowledge of heat transfer coefficients is necessary either for the design of refrigeration, storage,

Table 22.5 Nusselt–Reynolds–Prandtl Correlations for Cooling and Freezing of Selected Food Items

Food Type	*Re* Range	*Nu–Re–Pr* Correlation
Beef patties (unpackaged)	$2{,}000 < Re < 7{,}500$	$Nu = 1.37Re^{0.282}Pr^{0.3}$
Cake (packaged and unpackaged)[a]	$4{,}000 < Re < 80{,}000$	$Nu = 0.00156Re^{0.960}Pr^{0.3}$
Cheese (packaged and unpackaged)[b]	$6{,}000 < Re < 30{,}000$	$Nu = 0.0987Re^{0.560}Pr^{0.3}$
Chicken breast (unpackaged)	$1{,}000 < Re < 11{,}000$	$Nu = 0.0378Re^{0.837}Pr^{0.3}$
Fish fillets (packaged and unpackaged)[c]	$1{,}000 < Re < 25{,}000$	$Nu = 0.0154Re^{0.818}Pr^{0.3}$
Fried potato patties (unpackaged)	$1{,}000 < Re < 6{,}000$	$Nu = 0.00313Re^{1.06}Pr^{0.3}$
Pizza (packaged and unpackaged)[d]	$3{,}000 < Re < 12{,}000$	$Nu = 0.00517Re^{0.981}Pr^{0.3}$
Sausage (unpackaged)	$4{,}500 < Re < 25{,}000$	$Nu = 7.14Re^{0.170}Pr^{0.3}$
Trayed entrees (packaged)[e]	$5{,}000 < Re < 20{,}000$	$Nu = 1.31Re^{0.280}Pr^{0.3}$

Source: Becker, B.R. and Fricke, B.A., *Int. J. Refrig.*, 27, 540, 2004.
[a] Packaging consists of either an aluminum tray or aluminum foil cover and a paper tray.
[b] Packaging consists of a pouch, a paper tray with plastic lid, or a plastic box with lid.
[c] Packaging consists of either a hard plastic tray or a plastic pouch.
[d] Packaging consists of either a cardboard backing or a cardboard backing and shrink wrap.
[e] Packaging consists of an aluminum or plastic tray, an aluminum or plastic tray cover with paper lid, a plastic tray with film lid, or a paper tray with paper or plastic lid.

and freezing equipment or for adapting or modifying operating conditions of existing units. The equations in Table 22.5 were obtained from the cooling curves and surface heat transfer data for various food items provided by members from different sectors of the food refrigeration industry (food refrigeration equipment manufacturers, designers of food refrigeration plants, and food processors). In addition, Becker and Fricke (2004) also collected surface heat transfer coefficients from the literature. The *Nu–Re–Pr* correlations given for cake, chicken breast, fish fillets, fried potato patties, and pizza were found to be more representative of the data than the correlations given for beef, patties, cheese, sausage, and trayed entrees.

For chilling of carrots, Chuntranuluck et al. (1998) obtained the following correlations:

For peeled carrots

$$Nu = 0.267Re^{0.597}, \quad R^2 = 0.913 \tag{22.77a}$$

For unpeeled carrots

$$Nu = 0.704Re^{0.466}, \quad R^2 = 0.870 \tag{22.77b}$$

Experimental trials corresponding to Equation 22.77 were conducted across a range of practical conditions such as air velocities between 0.5 and 3 m/s, relative humidity values of 0.70–0.90, air temperatures between 0°C and 10°C, and product initial temperatures of 20°C–30°C.

Cleland and Valentas (1997) stated that the following correlation was found and worked well for predicting cooling times of objects with curved surfaces, such as beef carcasses (Willix et al., 2006):

$$h = 12.5V^{0.6} \tag{22.78}$$

where V is the velocity of free airstream in m/s. If the heat transfer medium is air, the turbulence intensity (Tu), which characterizes the velocity variation around its means over a short period, is a parameter that was rarely considered in food engineering studies with the exception of a few studies (Kondjoyan et al., 1993; Kondjoyan and Daudin, 1995; Hu and Sun, 2001; Verboven et al., 2001). Willix et al. (2006) measured heat transfer coefficients at 11 different positions over the surface of a fiberglass model of a side of beef and considered the influence of the turbulence intensity of the free airstream on the heat transfer coefficient. Since the cooling rate at the deep leg or deep butt position is used to represent the cooling rate of the overall carcass, the correlations given below correspond to the average value of the heat transfer coefficient measured at the inside leg, outside leg, and rump positions:

$$h = 8.3V^{0.77}, \quad Tu = 2.5\% \tag{22.79a}$$

$$h = 21V, \quad Tu > 20\% \tag{22.79b}$$

An approach for obtaining h values during air blast cooling, which considers the mass transfer due to moisture evaporation, was developed by Ansari and Khan (1999), who first developed a correlation, which predicted the effective heat transfer coefficient by including the effects of moisture evaporation. Then, they regressed the effective h with the produce temperature and used the linear regression obtained to solve the transient heat conduction equation with pure convection surface boundary condition equation. For a tomato sample they obtained

$$h = 5.2138T + 33.3025 \tag{22.80}$$

where T is in °C.

22.5.3 Heat Transfer Coefficients in Heat Exchangers

22.5.3.1 *Scraped Surface Heat Exchangers*

Heat exchange for sticky and viscous foods such as heavy salad dressings, margarine, chocolate, peanut butter, fondant, ice cream, and shortenings is possible only by using scraped surface heat exchangers (SSHEs) (Rao and Hartel, 2006). Early efforts to estimate heat transfer in SSHEs were based on the application of the penetration theory. Latinen (1958) applied the penetration theory to a two-bladed votator-type SSHE and arrived at an empirical relation for the Nusselt number (Rao and Hartel, 2006):

$$Nu = 1.6Re_{\mathrm{r}}^{0.5}Pr^{0.5} \tag{22.81a}$$

where Re_{r} is the rotational Reynolds number. Limitations of Equation 22.81a included the independence of heat transfer on viscosity and velocity of liquid through the exchanger. Skelland et al. (1962) introduced in the analysis parameters such as effects of rotation, number of blades, fluid physical characteristics, etc., and gave the following equation to predict the inside heat transfer coefficient for a votator-type SSHE (Geankoplis, 1993):

$$Nu = \frac{hD}{k} = \alpha Pr^{\beta}\left[\frac{(D-D_{\mathrm{s}})\rho V}{\mu}\right]^{1.0}\left(\frac{DN}{V}\right)^{0.62}\left(\frac{D_{\mathrm{s}}}{D}\right)^{0.55}(n_{\mathrm{B}})^{0.53} \tag{22.81b}$$

with

$\alpha = 0.014$, $\beta = 0.96$, for viscous liquids
$\alpha = 0.039$, $\beta = 0.76$, for nonviscous liquids

where

D is the diameter of the vessel in m
D_{s} is the diameter of rotating shaft in m
V is the axial flow velocity of liquid in m/s
N is the agitator speed in rev/s
n_{B} is the number of blades on agitator

Equation 22.81b is valid for

$$0.15 < \left[\frac{(D-D_{\mathrm{s}})\rho V}{\mu}\right]^{1.0} < 5.0,\ 0.79 < Re_{\mathrm{r}} < 194, \text{ and } 1000 < Pr < 4000.$$

Trommelen (1967) and Trommelen and Beek (1971) proposed the following equation resulting from penetration theory modified by a correction factor that was determined experimentally (Rao and Hartel, 2006):

$$Nu = 1.13(RePrn_{\mathrm{B}})^{0.5}(1-f) \quad \begin{array}{l} 300 < Re_{\mathrm{r}} < 3600 \\ 119 < Pr < 2650 \\ 700 < Pe < 8640 \end{array} \tag{22.81c}$$

where $f = 2.78(Pe + 200)^{-0.18}$ and the Peclet number (Pe) is defined as the product of Re and Pr numbers. Cuevas and Cheryan (1982) studied heat transfer coefficients of a Contherm-type SSHE for water and soybean extracts, and, respectively, obtained the following (Rao and Hartel, 2006):

$$h = 1709V^{0.42}V_\theta^{0.43}\frac{D_0}{D} \tag{22.82a}$$

$$h = 905.5V^{0.22}V_\theta^{0.33}S^{-0.16}\frac{D_0}{D} \tag{22.82b}$$

where

V and V_θ are the axial and rotational velocity, respectively, in ft./s
D_0 is the diameter of the tube in ft.
S is the solid content of feed in % w/w
h is in Btu/h ft.^{2} °F

Typical overall heat transfer coefficients in food applications are 1700 W/m^2 K for cooling margarine with NH_3, 2270 W/m^2 K for heating apple sauce with steam, 1420 W/m^2 K for chilling shortening with NH_3, and 2270 W/m^2 K for cooling cream with water (Geankoplis, 1993).

22.5.3.2 Plate Heat Exchanger

A correlation for the determination of convective heat transfer coefficients for stirred yogurt during the cooling process in a plate heat exchanger was obtained by Alonso et al. (2003). Generalized Reynolds and Prandtl dimensionless numbers were used to describe the non-Newtonian behavior of stirred yogurt, which are defined as

$$Re_g = \frac{V\rho D_e}{\mu_{app}}$$

$$Pr_g = \frac{c_p\mu_{app}}{k}$$

where

D_e is the hydraulic diameter in m defined in terms of plate distance (b) and is equal to $D_e = 2b$
μ_{app} is the apparent viscosity in (Pa s), which is defined to consider the viscosity dependency on the shear rate (γ) and temperature as

$$\mu_{app} = \left[\mu_{app}(\gamma)\right]e^{E/RT}$$

where

$\mu_{app}(\gamma)$ is the term that takes into account the viscosity dependence on the shear rate, and E/RT is a term of the Arrhenius type to include the effects of temperature on the flow behavior
E (J/mol) is the activation energy
R is the constant of ideal gas ($R = 8.31451$ J/mol K)
T is the absolute temperature

The corresponding correlation is as follows:

$$Nu = 1.759Re_g^{0.455}Pr_g^{0.3} \tag{22.83}$$

The generalized *Re* number of Equation 22.83 varies from 0.51 to 14.27 corresponding to shear stresses between 24 and 59 Pa.

22.5.3.3 *Tubular Heat Exchanger*

Ditchfield et al. (2005) obtained heat transfer coefficients for banana puree in a tubular heat exchanger with two heating sections, each 3.048 m long with an internal diameter of 1.22×10^{-2} m. Three volumetric flow rates (2.5×10^{-5}, 3.7×10^{-5}, and 4.7×10^{-5} m^3/s), three steam temperatures (110.0°C, 121.1°C, and 132.2°C), and two length/diameter ratios (250 and 500) were studied. The empirical correlation as a function of the flow rate ($\dot{m}$), length to diameter ratio (L/D), and steam temperature (T) is the following:

$$Nu = 9.66 + 3.64\frac{L}{D} - 236502\dot{m} + 0.011T - 52432\frac{L}{D}\dot{m} - 0.021\frac{L}{D}T + 2722\dot{m}T \tag{22.84}$$

Equation 22.84 is valid for $2.4 \times 10^{-5} < \dot{m} < 4.8 \times 10^{-5}$, $110°C < T < 132.2°C$, and $L/D = 250$ and $L/D = 500$. An empirical correlation was also proposed for relating the Nusselt number to the flow behavior index (n), Graetz number, and viscosity (Ditchfield et al., 2005):

$$Nu = 4.00\left(\frac{3n+1}{4n}\right)^{-2.79} Gz^{0.37}\left(\frac{\mu}{\mu_s}\right)^{0.077} \tag{22.85}$$

22.5.3.4 *Falling Film Evaporator*

A falling film evaporator is essentially a shell and tube heat exchanger that is widely used for concentrating liquid food such as fruit juices. In order to extrapolate them to a multiple effect unit, Prost et al. (2006) obtained experimental values of the inner liquid film coefficient in a single effect evaporator under different operating conditions. The model fluid studied was a solution of sucrose in water, which is alike to fruit juice from the thermohydraulic behavior point of view. The obtained values were correlated in terms of the dimensionless heat transfer coefficient (h^+) defined as

$$h^+ = h_i\left(\frac{\mu^2}{\rho^2 k^3 g}\right)^{1/3} \tag{22.86}$$

where h_i is the inner or liquid side heat transfer coefficient. The resulting equation was

$$h^+ = 1.6636Re^{-0.2648}Pr^{0.1592} \qquad \begin{array}{l} 15 < Re < 3000 \\ 2.5 < Pr < 200 \\ R^2 = 0.988 \end{array} \tag{22.87}$$

22.5.4 Deep-Fat Frying

A heat transfer correlation equation between average *Nu* and *Ra* numbers was derived for chicken drum shaped bodies during natural convective heat transfer using aluminum chicken drum shaped models (Ngadi and Ikediala, 2005). The following linear relationship was obtained:

$$\log Nu = 0.637471 + 0.11247 \log Ra \tag{22.88}$$

Equation 22.88 is based on the characteristic diameter defined as the average diameter of three diameter measurements at three positions of the model chicken drums. Heat transfer coefficient values obtained for different combinations of temperatures (60°C, 90°C, and 120°C), sizes (small, medium, and large), and oil viscosities (0.0014, 0.017, and 0.048 Ns/m^2) ranged from 67 to 163 W/m^2 K. Equation 22.88 can be used to estimate heat transfer coefficients of chicken drum during deep-fat frying under certain conditions.

22.5.5 Drying

A correlation of the type $Nu = cRa^m$ was obtained by Rahman and Kumar (2006), which considers the effect of shrinkage on the heat transfer coefficient during drying. Potato cylinders of 0.05 m length and 0.01 m diameter were dried under natural convection using 40°C, 50°C, and 60°C air temperatures. The following correlations for Nusselt number are proposed:

$$Nu = 0.28(Ra)^{0.42}, \quad R^2 = 0.88, \quad T_{\text{air}} = 40°\text{C} \qquad (22.89\text{a})$$

$$Nu = 1.96(Ra)^{0.17}, \quad R^2 = 0.78, \quad T_{\text{air}} = 50°\text{C} \qquad (22.89\text{b})$$

$$Nu = 3.83(Ra)^{0.10}, \quad R^2 = 0.76, \quad T_{\text{air}} = 60°\text{C} \qquad (22.89\text{c})$$

The uncertainty in the average value of the heat transfer coefficient is in the range of 2%–8%. It was observed that the heat transfer coefficient with shrinkage was higher than that obtained without shrinkage and became almost double at the end of the drying period. An empirical correlation of the ratio of the heat transfer coefficient with shrinkage, h_{sh}, and without shrinkage, h, as a function of drying time (t in min) and temperature of drying air (T in °C) was also proposed:

$$\frac{h_{\text{sh}}}{h} = -2.0863 + 0.049t^{0.517} + T^{0.2729} \qquad (22.90)$$

Anwar and Tiwari (2001a) developed correlations for some crops (green chillies, green peas, white gram or Kabuli chana, onion flakes, potato slices, and cauliflowers) under open sun drying conditions. The values of the constants c and m of the equation of the type $Nu = cRa^m$ along with the heat transfer coefficients are given in Table 22.6. The values shown in Table 22.6 are valid for $2.5 \times 10^5 < Gr < 4.0 \times 10^5$.

The knowledge of heat transfer coefficients under natural convection may find immediate application in the design of solar energy assisted natural convection air dryers for tropical areas, especially in rural and remote regions where electricity is either not available or intermittently available (Rahman and Kumar, 2006). Anwar and Tiwari (2001b) also obtained correlations for the convective heat transfer coefficient under forced convection for the crops listed in Table 22.6. The equations estimated h under a simulated condition of forced mode in indoor open (case I) and closed

Table 22.6 Values of c, m, and h for $Nu = cRa^m$ for Some Crops under Open Sun Drying

Crop	c	m	h (W/m^2 K)
Green chillies	1.00	0.23	3.71
Green peas	1.18	0.28	8.22
Kabuli chana	1.13	0.28	8.45
Onion flakes	1.00	0.31	14.03
Potato slices	1.75	0.32	25.98
Cauliflower	1.75	0.27	9.99

Source: Anwar, S.I. and Tiwari, G.N., *Energy Conv. Manag.*, 42, 627, 2001a.

Table 22.7 Values of c, m, and h for $Nu = cRa^m$ for Indoor Open (Case I) and Closed (Case II) Simulation under Forced Mode

Crop	c	m	h (W/m² K)
Case I			
Green chillies	1.00	0.39	1.31
Green peas	0.95	0.88	3.65
Kabuli chana	0.96	0.88	3.95
Onion flakes	0.99	0.75	4.75
Potato slices	1.00	0.72	5.40
Cauliflower	1.00	0.57	12.80
Case II			
Green chillies	0.99	0.38	1.25
Green peas	0.94	0.90	3.37
Kabuli chana	0.86	0.95	3.86
Onion flakes	0.99	0.59	4.85
Potato slices	0.99	0.69	5.82
Cauliflower	0.99	0.58	10.94

Source: Anwar, S.I. and Tiwari, G.N., *Energy Conv. Manag.*, 42, 1687, 2001b.

(case II) conditions. Hot air was blown over the crops at 80°C and 0.4 m/s. The results are shown in Table 22.7.

The experimental errors in terms of percent uncertainty ranged between 7% and 24% and between 6% and 20% in cases I and II, respectively, and the different values of *c*, *m*, and *h* were found to be within the range of the percent experimental error. Akpinar (2004) determined the heat transfer coefficients for fruits and vegetables, namely, mulberry, strawberry, apple, garlic, potato, pumpkin, eggplant, and onion, under forced convection drying. A heat convector was used to blow hot air at 80°C and 0.4 m/s over the product surface, which was placed on a wire mesh tray of 0.17×0.30 m². Values of constants *c*, *m*, and *h* are given in Table 22.8. The products in Tables 22.6 through 22.8 received some treatments before the experiments such as peeling, size reduction, and slicing. As can be seen in Table 22.8, *h* values ranged between 0.644 and 7.121 W/m² K. The experimental error ranged between 3.1% and 12.7%.

Table 22.8 Values of c, m, and h for $Nu = cRa^m$ in Indoor Open Simulation under Forced Mode

Product	c	m	h (W/m² K)
Mulberry	0.982	0.204	2.007
Strawberry	0.987	0.291	3.615
Apple	0.993	0.256	2.874
Garlic	0.982	0.036	0.644
Potato	0.984	0.336	4.897
Pumpkin	0.989	0.287	3.530
Eggplant	0.968	0.310	4.026
Onion	0.987	0.391	7.121

Source: Akpinar, E.K., *Int. Commun. Heat Mass Transfer*, 31, 585, 2004.

22.5.6 Heat Transfer in Two-Phase Solid–Liquid Food Flows

The fluid particle and the wall-fluid heat transfer coefficients, h_{fp} and h_w respectively, are crucial design parameters in aseptic processes. h_{fp} in particular plays a key role in establishing proper processing schedules from both quality and safety points of view (Awuah and Ramaswamy, 1996). Many researchers have developed dimensionless correlations for predicting h_{fp} in the form of $Nu = a + bRe^c Pr^d$. Barigou et al. (1998) reviewed the measurement techniques and mathematical models for estimating convective heat transfer coefficients for particles in continuous flow and provided an extensive list of correlations. On the other hand, the wall-fluid heat transfer coefficient h_w has been much less studied though it is a crucial parameter for designing the length of the heating tube in a heat-hold-cool system, and also the length of the holding tube (Barigou et al., 1998). It seems that experimental and theoretical study relevant to real solid–liquid flow situations is still needed, since much of the work has not been targeted in realistic food flow regimes.

22.5.7 Heat Transfer in Agitated Vessels

Liquid foods are often heated or cooled in agitated vessels. This is usually done by heat transfer surfaces, which may be in the form of cooling or heating jackets in the wall of the vessel or coils of pipe immersed in the liquid (Geankoplis, 1993).

22.5.7.1 Vessel with Heating Jacket

Correlations for the heat transfer coefficient from the agitated Newtonian liquid inside the vessel to the jacket walls of the vessel have the following form (Geankoplis, 1993):

$$\frac{hD_t}{k} = a\left(\frac{D_a^2 N\rho}{\mu}\right)^b \left(\frac{c_p\mu}{k}\right)^{1/3} \left(\frac{\mu}{\mu_s}\right)^m \qquad (22.91)$$

where

- D_t is the inside diameter of the tank in m
- D_a is the diameter of the agitator in m
- N is the rotational speed in revolutions per second

All the liquid physical properties are evaluated at the bulk liquid temperature except μ_s, which is evaluated at the wall temperature. Table 22.9 shows the values of the constants in Equation 22.91 and the Reynolds number range for different types of agitators used in heating and mixing applications.

22.5.7.2 Vessel with Heating Coils

Correlations for the heat transfer coefficient to the outside surface of the coils are given below (Geankoplis, 1993):

For a paddle agitator with no baffles

$$\frac{hD_t}{k} = 0.87\left(\frac{D_a^2 N\rho}{\mu}\right)^{0.62} \left(\frac{c_p\mu}{k}\right)^{1/3} \left(\frac{\mu}{\mu_s}\right)^{0.14}, \quad 300 < Re < 4\times 10^5 \qquad (22.92)$$

Table 22.9 Values of Constants *a*, *b*, and *m* of Equation 22.91

Type of Agitator	*a*	*b*	*m*	*Re*
Paddle agitator with no baffles[a]	0.36	2/3	0.21	$300 < Re < 3 \times 10^5$
Paddle, baffled/unbaffled jacketed vessels[b]	0.36	2/3	0.14	$Re > 4000$
	0.415	2/3	0.24	$20 < Re < 4000$
Flat-blade turbine agitator with no baffles[a]	0.54	2/3	0.14	$30 < Re < 3 \times 10^5$
Flat-blade turbine agitator with baffles[a]	0.74	2/3	0.14	$500 < Re < 3 \times 10^5$
Retreating-blade turbine[b]	0.68	2/3	0.14	
Anchor agitator with no baffles[a]	1.0	1/2	0.18	$10 < Re < 300$
	0.36	2/3	0.18	$300 < Re < 4 \times 10^4$
Anchor[b] (anchor to wall clearance >14 cm)	0.55	2/3	0.14	$4{,}000 < Re < 37{,}000$
Helical ribbon agitator with no baffles[a]	0.633	1/2	0.18	$8 < Re < 10^5$

Sources: [a]Geankoplis, C.J., in *Transport Processes and Unit Operations*, Prentice-Hall International Inc., New Jersey, 1993; [b]Singh, R.P., In *Handbook of Food Engineering*, Marcel Dekker, Inc., New York, 1992, 247–276.

If the heating or cooling coil is in the form of vertical tube baffles with a flat-blade turbine, the following correlation can be used:

$$\frac{hD_0}{k} = 0.09\left(\frac{D_a^2 N\rho}{\mu}\right)^{0.65}\left(\frac{c_p\mu}{k}\right)^{1/3}\left(\frac{D_a}{D_t}\right)^{1/3}\left(\frac{2}{n_b}\right)^{0.2}\left(\frac{\mu}{\mu_f}\right)^{0.4} \tag{22.93}$$

where

D_0 is the outside diameter of the coil tube in m
n_b is the number of vertical baffle tubes
μ_f is the viscosity at the mean film temperature

22.6 SOFTWARE AVAILABLE FOR FOOD THERMAL PROCESSING

The advances in computer-aided simulation have allowed the use of computational fluid dynamics (CFD) to solve problems of fluid flow, heat transfer, and other areas of science and engineering. There are several types of general purpose and well-established packages available, such as ABAQUS, ADINA, ANSYS-FLUENT, ANSYS CFX, MARC, MSC/NASTRAN, NISA II, PHOENICS, FEMLAB, EFD Lab, STAR-CD to cite a few, which in most of the cases are a type of finite element analysis (FEA). Research centers have also developed programs aimed at improving the efficiency of processing, for example, the Food Product Modeller (Version 2) from the Mirinz Centre, AgResearch Ltd. (New Zealand), which can be used to determine chilling, freezing, thawing, or heating process requirements for a great variety of foods. Other examples are the SURFHEAT and HEATSOLV software from the Aberdeen University (UK) for predicting heat transfer coefficients in various food processing geometries and heat flows and temperatures in foodstuffs, respectively. A summary of various CFD models developed recently for analyzing food thermal processes is provided by Wang and Sun (2006).

22.7 CONCLUSION

Apart from engineering properties (e.g., thermal conductivity, specific heat, and density), heat transfer coefficients play a key role in the proper design of food industrial plants, processing equipment, and in establishing adequate operating conditions from both quality and food safety standpoints. However in many instances, the heat transfer coefficient is inherently a local phenomenon, and a global single value is assumed to be representative of the convection process. Since only under simple flow situations the convective heat transfer coefficients can be obtained by solving the boundary layer equations, empirical equations of the type $Nu = cRe^m Pr^n$ are widely used in practice. It can be seen that with c, m, and n being constants, it is difficult to describe to certain accuracy the change in Nusselt number with Reynolds and Prandtl numbers over a wide range of these parameters. It is therefore very important to always provide information regarding the specific range and conditions from which the correlations were developed, since they are only applicable for the range of test data. In addition, because many situations involve coupled transport processes, information on whether an effective or true convective heat transfer coefficient should also be provided.

Considering the great diversity of food products of different shapes, sizes and composition, and food processes, data on heat transfer coefficients are still needed. Further developments in instrumentation (e.g., heat flux sensors, remote temperature sensors, etc.) will help to obtain important information on the heat transfer parameters. Therefore, research on measuring or estimating heat transfer coefficients is still necessary and a challenging issue.

NOMENCLATURE

A	surface area (m^2)
a	constant
b	constant
c	constant
c_{fx}	local friction factor
c_p	specific heat (J/kg K)
C_s	constant
d	constant
D	diameter (m)
D_h	hydraulic diameter (m)
f	darcy friction factor, function in Equations 22.75 and 22.81c
f_O	orientation factor in Equation 22.67
F	parameter in Equation 22.67
g	gravitational acceleration (9.807 m/s^2)
G	superficial mass flux (kg/m^2s)
H	height (m)
h	heat transfer coefficient (W/m^2 K or W/m^2 °C)
h_x	local heat transfer coefficient (W/m^2 K or W/m^2 °C)
k	thermal conductivity (W/m K or W/m °C)
k_w	wall thermal conductivity (W/m K or W/m °C)
L	length (m)
L_v	enthalpy of vaporization (J/kg)
m	constant
$\dot{m}$	flow rate (m^3/s, kg/s)
n	constant, flow behavior index
n_b	number of vertical baffle tubes
n_B	number of blades on agitator
N	speed (rev/s)

p	pressure (kPa, MPa)
P	parameter $= kD/k_w t$
P_w	perimeter (m)
q	heat flux (W/m^2)
Q	rate of energy (W)
r	radial distance (m)
S	solid content (% w/w)
t_w	tube wall thickness (m)
T	temperature (K, °C)
T_b	bulk temperature (K, °C)
T_e	exit temperature (K, °C)
T_f	film temperature (K, °C)
T_i	inlet temperature (K, °C)
T_s	surface temperature (K, °C)
T_{sat}	saturation temperature (K, °C)
T_w	inside wall temperature (K, °C)
T_∞	free stream temperature (K, °C)
V	velocity (m/s)
v	specific volume (m^3/kg)
x	distance (m), vapor fraction or quality

Greek Letters

α	thermal diffusivity (m^2/s)
β	volumetric thermal expansion coefficient (K^{-1})
ε	emissivity
ϕ	function in Equation 22.30
γ	shear rate (s^{-1})
μ	dynamic viscosity (Ns/m^2)
ν	kinematic viscosity $= \mu/\rho$ (m^2/s)
ρ	density (kg/m^3)
σ	surface tension (N/m), Stefan–Boltzmann constant (5.67×10^{-8} $W/m^2\ K^4$)

Dimensionless Parameters

Bo	boiling number, q/GL_v
Co	convection or Condensate number, $[(1-x)/x]^{0.8}(\rho_v/\rho)^{0.5}$, $h/k(\mu^2/\rho g\Delta\rho)^{1/3}$
Ec	Eckert number, $V^2/c_p(T_s - T_\infty)$
Fr	Froude number, $G^2/\rho g D$
Gr	Grashof number, $(g\beta\rho^2\Delta T L^3)/\mu^2$
Nu	Nusselt number, hL/k, hx/k, hD/k
Nu_x	local Nusselt number
Pr	Prandtl number, $c_p\mu/k$, ν/α
Ra	Rayleigh number, $GrPr$
Re	Reynolds number, $\rho VL/\mu$, $\rho Vx/\mu$, $\rho VD/\mu$
Gz	Graetz number, $RePr\ D/L$, $mc_p/(kL)$
Pe	Peclet number, $RePr$
St	Stanton number, $Nu/(RePr)$

REFERENCES

Achenbach, E. 1978. Heat transfer from spheres up to $Re = 6 \times 10^6$. In *Proceedings of the 6th International Heat Transfer Conference*, Vol. 5, Hemisphere Publishing, Washington, DC, pp. 341–346.

Akpinar, E.K. 2004. Experimental determination of convective heat transfer coefficient of some agricultural products in forced convection drying. *International Communications in Heat and Mass Transfer*, 31(4), 585–595.

Alonso, I.M., Hes, L., Maia, J.M., and Melo, L.F. 2003. Heat transfer and rheology of stirred yoghurt during cooling in plate heat exchangers. *Journal of Food Engineering*, 57, 179–187.

Amato, W.S. and Tien, C. 1972. Free convection heat transfer from isothermal spheres in water. *International Journal of Heat and Mass Transfer*, 15(2), 327–339.

Ansari, F.A. and Khan, S.Y. 1999. Application concept of variable effective surface film conductance for simultaneous heat and mass transfer analysis during air blast cooling of food. *Energy Conversion and Management*, 40, 567–574.

Anwar, S.I. and Tiwari, G.N. 2001a. Evaluation of convective heat transfer coefficient in crop drying under open sun drying conditions. *Energy Conversion and Management*, 42, 627–637.

Anwar, S.I. and Tiwari, G.N. 2001b. Convective heat transfer coefficient of crops in forced convection drying—an experimental study. *Energy Conversion and Management*, 42, 1687–1698.

Awuah, G.B. and Ramaswamy, H.S. 1996. Dimensionless correlations for mixed and forced convection heat transfer to spherical and finite cylindrical particles in an aseptic processing holding tube simulator. *Journal of Food Process Engineering*, 19, 269–287.

Barigou, M., Mankad, S., and Fryer, P. 1998. Heat transfer in two-phase solid-liquid food flows: A review. *Transactions of the Institution of Chemical Engineers*, 76, Part C, 3–29.

Becker, B.R. and Fricke, B.A. 2004. Heat transfer coefficients for forced-air cooling and freezing of selected foods. *International Journal of Refrigeration*, 27, 540–551.

Berenson, P.J. 1961. Film boiling heat transfer from a horizontal surface. *Journal of Heat Transfer, Transactions of the ASME*, 83C, 351–356.

Bergles, A.E. and Rohsenow, W.M. 1964. The determination of forced convection surface-boiling heat transfer. *Journal of Heat Transfer, Transactions of the ASME, Series C*, 86(3), 365–372.

Bromley, L.A. 1950. Heat transfer in stable film boiling. *Chemical Engineering Progress*, 46, 221–227.

Bromley, L.A., LeRoy, N.R., and Robbers, J.A. 1953. Heat transfer in forced convection film boiling. *Industrial & Engineering Chemistry*, 45(12), 2639–2646.

Catton, I. 1978. Natural convection in enclosures. In *Proceedings of the 6th International Heat Transfer Conference*, August 7–11, Toronto, pp. 13–31.

Carey, V.P. 2000. Heat and mass transfer. In F. Kreith (Ed.), *The CRC Handbook of Thermal Engineering*, CRC Press, Boca Raton, FL, pp. 91–107.

Chuntranuluck, S., Wells, C.M., and Cleland, A.C. 1998. Prediction of chilling times of foods in situations where evaporative cooling is significant—Part 3. Applications. *Journal of Food Engineering*, 37, 143–157.

Churchill, S.W. 1976. A comprehensive correlation equation for forced convection from flat plate. *AIChE Journal*, 22(2), 264–268.

Churchill, S.W. and Bernstein, M. 1977. A correlating equation for forced convection from gases and liquids to a circular cylinder in crossflow. *Journal of Heat Transfer*, 99, 300–306.

Churchill, S.W. and Chu, H.H.S. 1975a. Correlating equations for laminar and turbulent free convection from a vertical plate. *International Journal of Heat and Mass Transfer*, 18, 1323–1329.

Churchill, S.W. and Chu, H.H.S. 1975b. Correlating equations for laminar and turbulent free convection from a horizontal cylinder. *International Journal of Heat and Mass Transfer*, 18, 1049–1053.

Churchill, S.W. and Ozoe, H. 1973a. Correlations for laminar forced convection in flow over an isothermal flat plate and in developing and fully developed flow in isothermal tube. *Journal of Heat Transfer, Transactions of the ASME, Series C*, 95, 416–419.

Churchill, S.W. and Ozoe, H. 1973b. Correlations for laminar forced convection with uniform heating in flow over a plate and in developing and fully developed flow in a tube. *Journal of Heat Transfer, Transactions of the ASME, Series C*, 95, 78–84.

Cleland, D.J. and Valentas, K.J. 1997. Prediction of freezing times and design of food freezers. In K.J. Valentas, E. Rotstein, and R.P. Singh (Eds.), *Handbook of Food Engineering Practice*, CRC Press, Boca Raton, FL, pp. 71–124.

Cuevas, R. and Cheryan, M. 1982. Heat transfer in a vertical, liquid-full scraped-surface heat exchanger. Application of the penetration theory and Wilson plots models. *Journal of Food Process Engineering*, 5, 1–21.

Depew, C.A. and August, S.E. 1971. Heat transfer due to combined free and forced convection in a horizontal and isothermal tube. *Journal of Heat Transfer, Transactions of the ASME*, 93C, 380–384.

Dhir, V.K. and Lienhard, J.H. 1971. Laminar film condensation on plane and axi-symmetric bodies in non-uniform gravity. *Journal of Heat Transfer, Transactions of the ASME*, 93C, 97–100.

Ditchfield, C., Tadini, C.C., Singh, R.K., and Toledo, R.T. 2005. Experimental determination of heat transfer coefficients in banana puree. In *Proceedings of the Eurotherm 77 Heat and Mass Transfer in Food Processing*, Pisa, ETS, 1, pp. 165–168.

Dittus, F.W. and Boelter, L.M.K. 1930. Heat transfer in automobile radiators of the tubular type. *University of California (Berkeley) Publication in Engineering*, 2, 443–461.

Frederking, T.H.K. and Clark, J.A. 1963. Natural convection film boiling on a sphere. In K.D. Timmerhaus (Ed.), *Advances in Cryogenic Engineering*, Vol. 8, Plenum Press, New York, pp. 501–506.

Fujii, T. and Imura, H. 1972. Natural-convection heat transfer from a plate with arbitrary inclination. *International Journal of Heat and Mass transfer*, 15(4), 755–764.

Fussey, D.E. and Warneford, I.P. 1978. Free convection from a downward facing inclined plate. *International Journal of Heat and Mass Transfer*, 21, 119–126.

Gambill, W.R. 1962. Generalized prediction of burnout heat flux for flowing, subcooled, wetting liquids. In *AIChE Preprint 17, 5th National Heat Transfer Conference*, Houston.

Geankoplis, C.J. 1993. *Transport Processes and Unit Operations*. Prentice-Hall International, Inc., New Jersey.

Gebhart, B. 1971. Convection with body forces. In *Heat Transfer*, 2nd edn., McGraw-Hill, New York, pp. 316–388.

Gnielinski, V. 1976. New equation for heat and mass transfer in turbulent pipe and channel flow. *International Chemical Engineering*, 16(2), 359–368.

Gnielinski, V. 1990. Forced convection in ducts. In G.F. Hewitt (Ed.), *Handbook of Heat Exchanger Design*, Hemisphere Publishing, New York.

Goldstein, R.J., Sparrow, E.M., and Jones, D.C. 1973. Natural convection mass transfer adjacent to horizontal plates. *International Journal of Heat and Mass Transfer*, 27, 466–468.

Goldstein, R.J., Eckert, E.R.G., Ibele, W.E, Patankar, S.V., Simon, T.V., Kuehn, T.H., Strykowski, P.J., Tamma, K.K., Bar-Cohen, A., Heberlein, J.V.R., Davidson, J.H., Bischof, J., Kulacki, F.A., Kortshagen, U., Garrick, S., and Srinivasan, V. 2005. Heat transfer—a review of 2002 literature. *International Journal of Heat and Mass Transfer*, 48, 819–927.

Holman, J.P. 1986. *Heat Transfer*, 6th edn. McGraw-Hill, New York.

Hilpert, R. 1933. Warmeabgabe von geheizen Drahten und Rohren. *Forschung auf dem Gebiet des Ingenieurwesen*, 4, 220.

Hu, Z. and Sun, D.-W. 2001. Predicting local surface heat transfer coefficients during air-blast chilling by different turbulent k-ε models. *International Journal of Refrigeration*, 24(7), 702–717.

Incropera, F.P. and DeWitt, D.P. 1996. *Fundamentals of Heat and Mass Transfer*, 4th edn., John Wiley & Sons, New York.

Ishiguro, R., Sugiyama, K., and Kumada, T. 1979. Heat transfer around a circular cylinder in a liquid-sodium cross flow. *International Journal of Heat and Mass Transfer*, 22, 1041–1048.

Janna, W.S. 2000. *Engineering Heat Transfer*, 2nd edn., CRC Press, Boca Raton, FL.

Kadambi, V. and Drake, R.M., Jr., 1959. Free convection heat transfer from horizontal surfaces for prescribed variations in surface temperature and mass flow through the surface. Technical Report in Mechanical Engineering HT-1, Princeton University, Princeton, June 30.

Kandlikar, S.G. 1990. A general correlation for saturated two-phase flow boiling heat transfer inside horizontal and vertical tubes. *Journal of Heat Transfer*, 112(1), 219–228.

Kandlikar, S.G. and Nariai, H. 1999. Flow boiling in circular tubes. In S.G. Kandlikar, M. Shoji, and V.K. Dhir (Eds.), *Handbook of Phase Change: Boiling and Condensation*, Taylor & Francis, Philadelphia, pp. 367–402.

Kandlikar, S.G., Tian, S.T., Yu, J., and Koyama, S. 1999. Further assessment of pool and flow boiling heat transfer with binary mixtures. In G.P. Celata, P. Di Marco, and R.K. Shah (Eds.), *Two-Phase Flow Modeling and Experimentation*, Edizione ETS, Pisa.

Kays, W.M. and Crawford, M.E. 1980. *Convective Heat and Mass Transport*. McGraw-Hill, New York.

Knudsen, J.G., Hottel, H.C., Sarofim, A.F., Wankat, P.C., and Knaebel, K.S. 1999. Heat and mass transfer. In R.H. Perry, D.W. Green, and J.O. Maloney (Eds.), *Perry's Chemical Engineers' Handbook*, 7th edn., McGraw-Hill, New York.

Kondjoyan, A. and Daudin, J.D. 1995. Effects of free stream turbulence intensity on heat and mass transfers at the surface of a circular cylinder and an elliptical cylinder, axis ratio 4. *International Journal of Heat and Mass Transfer*, 38(10), 1735–1749.

Kondjoyan, J.D., Daudin, J.D., and Bimbenet, J.J. 1993. Heat and mass transfer coefficients at the surface of elliptical cylinders placed in a turbulent air flow. *Journal of Food Engineering*, 20, 339–367.

Krokida, M.K., Zogzas, N.P., and Maroulis, Z.B. 2002. Heat transfer coefficient in food processing: Compilation of literature data. *International Journal of Food Properties*, 5(2), 435–450.

Kutateladze, S.S. 1963. *Fundamentals of Heat Transfer*, Academic Press, New York.

Labuntsov, D.A. 1957. Heat transfer in film condensation of pure steam on vertical surfaces and horizontal tubes. *Teploenergetika*, 4, 72–80.

Latinen, G.A. 1958. Discussion of the paper correlation of scraped film heat transfer in the votator. *Chemical Engineering Science*, 9, 263–266.

Levy, S. 1959. Generalized correlation of boiling heat transfer. *Journal of Heat Transfer*, 81C, 37–42.

Lienhard, J.H. and Dhir, V.K. 1973. Hydrodynamic prediction of peak pool-boiling heat fluxes from finite bodies. *Journal of Heat Transfer, Transactions of the ASME*, 95C, 152–158.

Lienhard IV, J.H. and Lienhard V., J.H. 2004. *A Heat Transfer Textbook*, 3rd edn., Phlogiston Press, Cambridge, MA.

Lienhard, J.H., Dhir, V.K., and Riherd, D.M. 1973. Peak pool boiling heat-flux measurements on finite horizontal flat plates. *Journal of Heat Transfer, Transactions of the ASME*, 95C, 477–482.

McAdams, W.H. 1954. *Heat Transmission*, 3rd edn., McGraw Hill Book Company, New York.

McAdams, W.H., Minden, C.S., Carl, R., Picornell, P.M., and Dew, J.E. 1949. Heat transfer at high rates to water with surface boiling. *Industrial & Engineering Chemistry*, 41, 1945–1953.

Monrad, C.C. and Pelton, J.F. 1942. Heat transfer by convection in annular spaces. *AIChE Journal*, 38, 593–611.

Morcos, S.M. and Bergles, A.E. 1975. Experimental investigation of combined forced and free laminar convection in horizontal tubes. *Journal of Heat Transfer Transactions of the ASME*, 97C, 212–219.

Morgan, V.T. 1975. The overall convective heat transfer from smooth circular cylinders. In T.F. Irvine and J.P. Hartnett (Eds.), *Advances in Heat Transfer*, Vol. 11, Academic Press, New York.

Nakai, S. and Okazaki, T. 1975. Heat transfer from a horizontal circular wire at small Reynolds and Grashof numbers—1 Pure Convection. *International Journal of Heat and Mass Transfer*, 18, 387–396.

Ngadi, M. and Ikediala, J.N. 2005. Natural heat transfer coefficients of chicken drum shaped bodies. *International Journal of Food Engineering*, 1(3), Article 4.

Nusselt, W. 1916. Die Oberflächenkondensation des Wasserdampfes. *Zeitschrift Verein Deutscher Ingenieure*, 60, 541–546 and 569–575.

Petukhov, B.S. 1970. Heat transfer and friction in turbulent pipe flow with variable physical properties. In J.P. Hartnett and T.F. Irvine (Eds.), *Advances in Heat Transfer*, Vol. 6, Academic Press, New York, pp. 503–565,.

Petukhov, B.S. and Popov, V.N. 1963. Theoretical calculation for heat exchange and frictional resistance in turbulent flow in tubes of an incompressible fluid with variable physical properties. *Journal of High Temperature*, 1, 69–83.

Petukhov, B.S. and Roizen, L.I. 1964. Generalized relationships for heat transfer in turbulent flow of gas in tubes of annular section. *Journal of High Temperature*, 2, 65–68.

Prost, J.S., Gonzalez, M.T., and Urbicain, M.J. 2006. Determination and correlation of heat transfer coefficients in a falling film evaporator. *Journal of Food Engineering*, 73, 320–326.

Rahman, S. 1995. Heat transfer coefficients in food processing. In S. Rahman (Ed.), *Food Properties Handbook*, CRC Press, Boca Raton, FL, pp. 393–456.

Rahman, N. and Kumar, S. 2006. Evaluation of convective heat transfer coefficient during drying of shrinking bodies. *Energy Conversion and Management*, 47, 2591–2601.
Raithby, G.D. and Hollands, K.G.T. 1998. Natural convection. In W.M. Rohsenow, J.P. Hartnett, and Y.I. Cho (Eds.), *Handbook of Heat Transfer*, 3rd edn., McGraw-Hill, New York, pp. 4-1–4-87.
Rao, C.S. and Hartel, R.W. 2006. Scraped surface heat exchangers. *Critical Reviews in Food Science and Nutrition*, 46, 207–219.
Rohsenow, W.M. 1952. A method of correlating heat transfer data for surface boiling of liquids. *Transactions of the ASME*, 74, 969–975.
Rohsenow, W.M. 1956. Heat transfer and temperature distribution in laminar film condensation. *Transactions of the ASME*, 78, 1645–1648.
Rohsenow, W.M. and Griffith, P. 1955. Correlation of maximum heat flux data for boiling of saturated liquids. *AIChE-ASME Heat Transfer Symposium*, Louisville, KY.
Rose, J.W. 1979. Boundary layer flow on a flat plate. *International Journal of Heat and Mass Transfer*, 22, 969.
Sarkar, A. and Singh, R.P. 2003. Spatial variation of convective heat transfer coefficient in air impingement applications. *Journal of Food Science*, 68(3), 910–916.
Schlichting, H. 1979. *Boundary Layer Theory*, 7th edn., McGraw-Hill Book Company, New York.
Sieder, E.N. and Tate, C.E. 1936. Heat transfer and pressure drop of liquids in tubes. *Industrial & Engineering Chemistry*, 28, 1429–1435.
Singh, R.P. 1992. Heating and cooling processes for foods. In D.R. Heldman and D.B. Lund (Eds.), *Handbook of Food Engineering*, Marcel Dekker, New York, pp. 247–276.
Skelland, A.H., Oliver, D.R., and Tooke, S. 1962. Heat transfer in a water-cooled scraped-surface heat exchanger. *British Chemical Engineering*, 7, 346–353.
Sparrow, E.M. and Gregg, J.L. 1956. Laminar free convection from a vertical plate with uniform surface heat flux. *Transactions of the ASME*, 78, 435–440.
Stephan, K. 1961. Gleichungen für den Wärmeübergang laminar strömender Stoffe in ringförmigen Querschnit. *Chemie Ingenieur Technik*, 33, 338–343.
Stephan, K. 1962. Wärmeübergang bei turbulenter und bei laminarer Strömung in Ringspalten. *Chemie Ingenieur Technik*, 34, 207–212.
Stephan, K. and Preußer P. 1979. Wärmeübergang und maximale Wärmestrondichte beim Behältersieden binärer und ternärer Flüssigkeitsgemische. *Chemie Ingenieur Technik*, 51, (synopse MS 649/ 97).
Suryanarayana, N.V. 1995. *Engineering Heat Transfer*, West Publishing Company, St. Paul, MN.
Suryanarayana, N.V. 2000. Forced convection–external flows. In: F. Kreith (Ed.), *The CRC Handbook of Thermal Engineering*, CRC Press LLC, Boca Raton, FL, pp. 26–46.
Thomas, L.C. and Al-Sharifi, M.M. 1981. An integral analysis for heat transfer to turbulent incompressible boundary layer flow. *Journal of Heat Transfer*, 103, 772–777.
Trommelen, A.M. 1967. Heat transfer in a scraped-surface heat exchanger. *Transactions of the IChemE*, 45, 176–178.
Trommelen, A.M. and Beek, W.J. 1971. Flow phenomena in a scraped-surface heat exchanger (votator-type). *Chemical Engineering Science*, 26, 1933–1942.
Vachon, R.I., Nix, G.H., and Tanger, G.E. 1968. Evaluation of constants for the Rohsenow pool-boiling correlation. *Journal of Heat Transfer*, 90, 239–247.
Verboven, P., Scheerlinck, N., De Baerdemaeker, J., and Nicolaï, B.M. 2001. Sensitivity of the food centre temperature with respect to the air velocity and the turbulence kinetic energy. *Journal of Food Engineering*, 48, 53–60.
Vliet, G.C. and Liu, C.K. 1969. An experimental study of turbulent natural convection boundary layers. *Journal of Heat Transfer*, 91C, 517–531.
Wang, L. and Sun, D.-W. 2006. Heat and mass transfer in thermal food processing. In D.-W. Sun (Ed.), *Thermal Food Processing, New Technologies and Quality Issues*, CRC Press/Taylor & Francis, Boca Raton, FL, pp. 35–71.
Whitaker, S. 1972. Forced convection heat transfer correlation for flow in pipes past flat plates, single cylinders, single spheres, and for flow in packed beds and tube bundles. *AIChE Journal*, 18, 361–371.
White, F.M. 1988. *Heat and Mass Transfer*. Addison-Wesley, Reading, MA.

Willix, J., Harris, M.B., and Carson, J.K. 2006. Local surface heat transfer coefficients on a model beef side. *Journal of Food Engineering*, 74, 561–567.

Witte, L.C. 1968a. An experimental study of forced-convection heat transfer from a sphere to liquid sodium. *Journal of Heat Transfer*, 90, 9–14.

Witte, L.C. 1968b. Film boiling from a sphere. *Industrial & Engineering Chemistry Fundamentals*, 7(3), 517–518.

Yang, K.T. 1987. Natural convection in enclosures. In S. Kakaç, R.K. Shah, and W. Aung (Eds.), *Handbook of Single-Phase Convective Heat Transfer*, Wiley-Interscience, New York, Chapter 13.

Yuge, T. 1960. Experiments on heat transfer from spheres including combined forced and natural convection. *Journal of Heat Transfer, Transactions of the ASME, Series C*, 82(1), 214–220.

Zogzas, N.P., Krokida, M.K., Michailidis, P.A., and Maroulis, Z.B. 2002. Literature data of heat transfer coefficients in food processing. *International Journal of Food Properties*, 5(2), 391–417.

Zuber, N. 1958. On the stability of boiling heat transfer. *Transactions of the ASME*, 80, 711–720.

Zuber, N. 1959. Hydrodynamic aspects of boiling heat transfer. AEC Report AECU-4439, Physics and Mathematics.

CHAPTER 23

Surface Heat Transfer Coefficient in Food Processing

Panagiotis A. Michailidis, Magdalini K. Krokida, and Mohammad Shafiur Rahman

CONTENTS

23.1 INTRODUCTION

Heat transfer coefficients are used in the design, operation, optimization, and control of several food processing operations and equipment. They are related to the basic heat transport properties of foods (thermal conductivity), and they depend strongly on the characteristics of fluid flow (velocity and turbulence intensity) and solid (shape, dimensions, surface roughness, surface temperature, and outgoing fluxes), as well as the heat transfer equipment (food–equipment interface) and the thermophysical properties of the system (density, specific heat, and thermal conductivity). Table 23.1 presents some important heat transfer operations, which are used in food processing. In all of these operations, heat must be supplied to or removed from the food material with an external heating or cooling medium, through the interface of some type of processing equipment. Some operations, such as evaporation, involve mass transfer, but the controlling transfer mechanism is heat transfer (Heldman and Lund, 1992; Valentas et al., 1997). Heat transfer is also very important for some mass transfer operations (in which the controlling transfer mechanism is mass transfer or both heat and mass transfer, acting simultaneously), such as crystallization and drying used for the purification of components and food preservation, respectively. Heat transfer coefficients are empirical transfer constants that characterize a given operation from theoretical principles, but

Table 23.1 Heat Transfer Operations in Food Processing

Operations	Objective
Blanching	Enzyme inactivation
Evaporation	Concentration of liquid foods
Freezing	Food preservation
Frying	Food preparation
Pasteurization	Microorganisms and enzymes inactivation
Refrigeration	Preservation of fresh foods
Sterilization	Microorganisms inactivation

they are either obtained experimentally or correlated in empirical equations applicable to particular transfer operations and equipment (Saravacos and Maroulis, 2001).

In this chapter, after the initial definitions, the range of variation in heat transfer coefficient for different processes used in food industry and experimental data for these processes are reported. Other sections present a comprehensive review of correlations for the estimation of heat transfer coefficient, classified in 14 different food processes, and a revised approach for the calculation of heat transfer coefficient based on the heat transfer factor. Finally, mass transfer data that can be used in heat transfer processes by applying the Chilton–Colburn analogy are presented.

23.2 DEFINITION

The heat transfer coefficient h at a solid–fluid interface is defined in Newton's law of heat transfer as

$$\dot{q} = hA\Delta T \tag{23.1}$$

where

$\dot{q}$ is the heat flux (W)
A is the heat transfer surface area (m^2)
ΔT is the temperature difference (K)

A similar definition is applicable to liquid–fluid interfaces. Heat transfer coefficient depends on the local rheological and thermophysical properties of the fluid film in contact with the solid surface and on the geometry of the system. Heat transfer is considered to take place by heat conduction through a film of thickness δ and thermal conductivity k. Thus, the heat transfer coefficient is equivalent to

$$h = k/\delta \tag{23.2}$$

However, the above equation is difficult to apply because the film thickness δ cannot be determined accurately as it varies with the conditions of flow at the interface. The overall heat transfer coefficient U between two fluids separated by a conducting wall is given by the equation

$$\dot{q} = UA\Delta T \tag{23.3}$$

where ΔT is the overall temperature difference. The coefficient U is related to the heat transfer coefficients h_1 and h_2 of the two sides of the wall and the wall heat conduction x/k_w by the equation

$$\frac{1}{U} = \frac{1}{h_1} + \frac{x}{k_w} + \frac{1}{h_2} + \frac{1}{h_{d_1}} + \frac{1}{h_{d_2}} \tag{23.4}$$

where
- x is the wall thickness
- k_w is the wall thermal conductivity
- h_{d_1} and h_{d_2} are the fouling coefficients of the wall–liquid interfaces of the two sides of the wall, respectively

These coefficients are a result of the fouling of the heat transfer surface, caused by particles and soluble solids deposit on it. In most practical cases, the values of these coefficients increase with the operating time of the equipment, which has to be cleaned to restore its efficient performance.

23.3 DATA FOR HEAT TRANSFER COEFFICIENT

The surface and the overall heat transfer coefficients are specific for each processing equipment and fluid system and are related to the physical structure of the food materials. In most cases, they are determined from the experimental measurements in pilot plants or industrial equipment. Measurement techniques for the estimation of heat transfer coefficients are classified into the following categories (Rahman, 1995):

1. Steady-state methods
   a. Interference method
   b. Constant heating method
2. Quasi-steady-state method
3. Transient method
   a. Analytical solution
   b. Numerical solution
   c. Plank's method
   d. Evaporation rate method
4. Surface heat flux method

It should be noted that all the above methods are applicable when particles or solids are in rest and fluid is in motion or rest. Special methods are necessary when both solid and fluid are moving; such methods have been reposted by Sastry et al. (1989), Ramaswamy et al. (1992), and Mwangi et al. (1993). Typical values of heat transfer coefficients are shown in Table 23.2 (Hallstrom et al., 1988; Perry and Green, 1997; Rahman, 1995; Saravacos, 1995; Saravacos and Maroulis, 2001).

Figure 23.1 schematically presents the ranges of variation in heat transfer coefficient in processes commonly used in the food industry. It is evident that when phase change takes place in a process, heat transfer flux is high for a given surface area or, alternatively, the area required for the transfer of a certain amount of heat is limited compared with that necessary under turbulent flow conditions. As expected, air processes and heat transfer under natural convection present very low values of heat transfer coefficient. Generally, these data are useful in the preliminary design as they contribute to a rough estimation of the process variables' values.

23.4 CORRELATIONS FOR HEAT TRANSFER COEFFICIENT

Although heat transfer coefficient data are useful for the initial estimation of the heat transfer equipment and the parameter values of the process in which heat transfer occurs, the usage of correlations for the heat transfer coefficient is necessary for the detailed design of the process, the development of its simulation models, and the examination of sensitivity analysis situations depending on the factors and the response variables selected. Correlations of heat transfer data

Table 23.2 Typical Heat Transfer Coefficient *h* and Overall Coefficients *U* in Food Processing Operations

Heat Transfer System	h (W/m$^2\cdot$K)
Air/process equipment, natural convention	5–20
Baking ovens	20–80
Air-drying, constant rate period	30–200
Air-drying, falling rate period	20–60
Vacuum self-dryer	5–6
Rotary vacuum dryer	28–284
Indirect rotary dryer	11–57
Jacketed trough dryer	11–85
Agitated tray dryer	28–340
Drum dryer	1,135–1,700
Water, turbulent flow	1,000–3,000
Boiling water	5,000–10,000
Film condensation of water vapor	5,000–6,700
Dropwise condensation of water vapor	40,000–50,000
Individual liquid nitrogen droplets on contacting food material	600–800
Refrigeration, air-cooling	20–200
Cryogenic freezing	90–200
Canned foods, retorts	150–500
Aseptic processing, particles	500–3,000
Freezing, air/refrigerants	20–500
Single-screw extruder	227
Twin-screw extruder	312–891
Co-rotating twin-screw extruder (melt conveying section)	500
Co-rotating twin-screw extruder (solid feed conveying section)	28.4
Frying, oil/solids	250–1,000
Contact frying (double-sided pan at constant pressure)	210–310
Contact frying (double-sided pan at increased pressure)	392–458
Heat exchangers (tubular/plate)	500–3,500 (overall *U*)
Scraped surface heat exchanger	1,276–3,123 (overall *U*)
Evaporators	500–3,000 (overall *U*)

are used for estimating the heat transfer coefficient h in various processing equipment and operating conditions. These correlations contain, in general, dimensionless numbers, characteristic of the heat transfer mechanism, the flow conditions, and the thermophysical and transport properties of the fluids. Rahman (1995) presented a comprehensive review of heat transfer correlations that apply in most of the food processes. Tables 23.3 through 23.17 summarize a large number of these correlations classified into the following processes/equipment:

- Natural convection around different shapes and surfaces
- Forced convection upon flat surfaces and around different shapes
- Flow in tubes and channels
- Heat exchangers
- Extruders
- Freezing
- Sterilization—retort canning
- Sterilization—aseptic canning
- Flow through packed beds

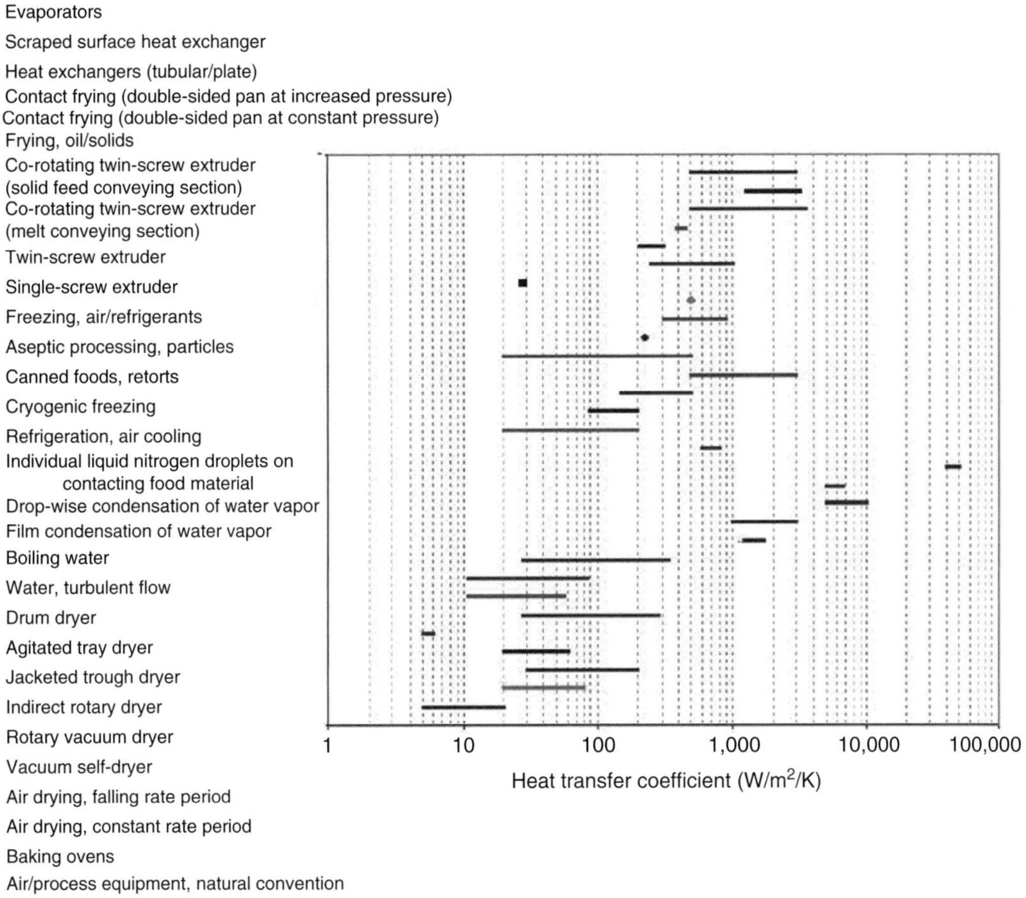

FIGURE 23.1 Range of variation in heat transfer coefficient in some common food processes.

- Flow through fluidized beds
- Mixed convection
- Convection with phase change
- Evaporation
- Dryers

Saravacos and Moyer (1967) proposed the following empirical correlation for the prediction of heat transfer in agitated vessels:

$$Nu = \bar{C}Re^{0.66}Pr^{1/3}(\eta_l/\eta_w)^{0.14} \tag{23.5}$$

$$Re = \frac{D^2 N\rho}{\eta_l}, \quad \eta_l = K\dot{\gamma}^{n-1} \tag{23.6}$$

where

D is the diameter of the impeller

η_l is the apparent viscosity estimated at the agitation speed N

K and n are the rheological constants of the fluid at the mean temperature (flow consistency coefficient and flow behavior index, respectively)

coefficient $\bar{C}$ is 0.55 for Newtonian and 1.474 for non-Newtonian fluids

Table 23.3 Heat Transfer Correlations for Natural Convection around Different Shapes and Surfaces

1. Analytical solution of the energy equation for a sphere at rest in a motionless fluid (Rahman, 1995):

$$Nu_b = 2$$

2. Natural convection (Schlichting, 1968):

$$Nu_b = e' \sqrt[4]{Gr_b Pr_b}$$

$$e' = -0.00035(\ln Pr_b)^2 + 0.0446(\ln Pr_b) + 0.5283$$

3. Outer surface of a horizontal cylinder (Whitaker, 1976):

$$Nu_a = 0.52 \sqrt[4]{Gr_a Pr_a}$$

$$Gr_a = \frac{g\beta\rho_a^2 D_T^3 \Delta T}{\eta_a^2}, \quad Nu_a = \frac{hD_T}{k_a}, \quad Pr_a = \frac{C_a \eta_a}{k_a}$$

$$10^3 < Gr_a Pr_a < 10^9$$

where
- g is the acceleration due to gravity
- β is the volumetric thermal expansion coefficient of the fluid
- D_T is the cylinder diameter, h is the heat transfer coefficient
- ρ_α is the average between surface and bulk fluid density
- k_α is the average between surface and bulk thermal conductivity of the fluid
- C is the average between surface and bulk-specific heat of the fluid at constant pressure
- ΔT is the temperature difference between the surface and the bulk
- η_α is the average between surface and bulk viscosity of the fluid

4. Horizontal flat surface (Hallstrom et al., 1988):

$$Nu_b = 0.59(Gr_b Pr_b)^{0.25}, \quad 10^4 < Gr_b Pr_b < 10^9$$

$$Nu_b = 0.13(Gr_b Pr_b)^{0.33}, \quad Gr_b Pr_b > 10^9$$

$$Gr_b = \frac{g\beta\rho_b^2 L^3 \Delta T}{\eta_b^2}, \quad Nu_b = \frac{hL}{k_b}, \quad Pr_b = \frac{C_b \eta_b}{k_b}$$

5. Long cylinder in an infinite fluid (McAdams, 1954):

$$Nu_b = 0.525 \sqrt[4]{Gr_b Pr_b}$$

$$Pr_b > 0.6, \; Gr_b Pr_b < 10^4$$

6. Cylinder (Fishenden and Saunders, 1950):

$$Nu_b = 0.47 \sqrt[4]{Gr_b Pr_b}$$

$$Gr_b = \frac{g\beta\rho_b^2 D_T^3 \Delta T}{\eta_b^2}, \quad Nu_b = \frac{hD_T}{k_b}, \quad Pr_b = \frac{C_b \eta_b}{k_b}$$

7. Liquids between two horizontal surfaces (Jakob and Cupta, 1954):

$$Nu_{av} = 0.300 \sqrt[4]{Gr_{av} Pr_{av}}, \quad \text{(laminar range, } 2 \times 10^3 < Gr_{av} Pr_{av} < 150 \times 10^3\text{)}$$

$$Nu_{av} = 0.1255 \sqrt[3]{Gr_{av} Pr_{av}}, \quad \text{(turbulent range, } 200 \times 10^3 < Gr_{av} Pr_{av} < 3 \times 10^9\text{)}$$

$$Nu_{av} = 0.3615 Gr_{av}^{1/4} \left(\frac{\delta}{L}\right)^{0.058}, \quad \text{(laminar range)}$$

where
- δ is the layer thickness (m), ranging from 3.05 to 15.24 mm
- L is the horizontal width of the heating surface (m)
- ΔT is the temperature difference of the two surfaces; heated surface temperature varied from 20.4°C to 74.08°C

Table 23.3 (continued) Heat Transfer Correlations for Natural Convection around Different Shapes and Surfaces

8. Vertical plate (Whitaker, 1976):

$$Nu_a = 0.58\sqrt[4]{Gr_a Pr_a}, \quad Gr_a Pr_a < 10^9$$

$$Nu_a = 0.021\sqrt[3]{(Gr_a Pr_a)}^2, \quad Gr_a Pr_a > 10^9$$

$$Gr_a = \frac{g\beta\rho_a^2 H^3 \Delta T}{\eta_a^2}, \quad Nu_a = \frac{hH}{k_a}$$

9. Single sphere in a large body of fluid (Ranz and Marshall, 1952; Whitaker, 1976):

$$Nu_a = 2 + 0.60\sqrt[4]{Gr_a}\sqrt[3]{Pr_a}$$

$$Gr_a = \frac{g\beta\rho_a^2 D_{pa}^3 (T_s - T_b)}{\eta_b}, \quad Nu_a = \frac{hD_{pa}}{k_a}, \quad Pr_a = \frac{C_a \eta_a}{k_a}$$

$$Gr_a^{1/4} Pr_a^{1/3} < 200$$

where
- D_{pa} is the sphere diameter
- T_s is the surface temperature
- T_b is the temperature far from the surface
- $\beta = 1/T_{av}$ for ideal gases
- T_{av} is the average temperature between surface and the main body of the fluid

10. Mushroom-shaped particles (aluminum cast) in water (Alhamdan et al., 1990):

$$Nu_b = 5.53(Gr_b Pr_b)^{0.21}, \quad \text{for heating}$$

$$Nu_b = 0.08(Gr_b Pr_b)^{0.27}, \quad \text{for cooling}$$

$$Gr_b = \frac{g\beta\rho_b^2 D_{ca}^3 (T_{sc} - T_{eg})}{\eta_b^2}, \quad Nu_b = \frac{hD_{ca}}{k_b}, \quad Pr_b = \frac{C_{so}\eta_b}{k_b}$$

$$2 \times 10^6 < Gr_b Pr_b < 3 \times 10^9$$

where
- D_{ca} is the diameter of the cap
- T_{sc} is the temperature in the center of the solid
- T_{eg} is the equilibrium temperature at the end of the heating or cooling process
- C_{so} is the specific heat of the solid; particle oriented with stem and upward, water as a fluid, temperature 20°C–80°C

11. Mushroom-shaped particles in aqueous solutions of sodium carboxymethylcellulose[a] (Alhamdan and Sastry, 1990):

$$Nu_b = 1.88 \times 10^{-3}(Gr_b Pr_b)^{0.388}, \quad \text{for heating}$$

$$Nu_b = 0.1250(Gr_b Pr_b)^{1.113}, \quad \text{for cooling}$$

$$Gr_b = \frac{g\beta\rho_b^2 D_{ca}^3 (T_{sc} - T_{eg})}{E}, \quad Nu_b = \frac{hD_{ca}}{k_b}, \quad Pr_b = \frac{C_{so}E}{k_b}, \quad E = \frac{K[(3n+1)/n]^n 2^n D_{ca}^{n-1}}{4u_s^{1-n}}, \quad 4.0 < Gr_b Pr_b < 8.0$$

where
- K is the consistency coefficient of the fluid
- n is the flow behavior index of the fluid
- u_s is the fluid velocity around the particle surface; carbomethyl cellulose (CMC) concentration 0.5%–1.2% (w/w), temperature 20°C–80°C

[a] Power law fluid (non-Newtonian).

Table 23.4 Heat Transfer Correlations for Forced Convection upon Flat Surfaces and around Different Shapes

1. Flow past a thin flat (flow direction is parallel to the flat surface) (Eckert and Drewitz, 1943; Howarth, 1957):

$$Nu_b = 0.644 Re_b^{1/2} Pr_b^{1/3}, \quad 0.6 \le Pr_b \le 10$$

$$Nu_b = 0.678 Re_b^{1/2} Pr_b^{1/3}, \quad Pr_b \to \infty$$

$$Nu_b = 1.03 Re_b^{1/2} Pr_b^{1/3}, \quad Pr_b \to \infty$$

$Re_b < 2 \times 10^5$ for all expressions (laminar flow).

$$Re_b = \frac{L u_{ap} \rho_b}{\eta_b}, \quad Nu_b = \frac{hL}{k_b}, \quad Pr_b = \frac{C_b \eta_b}{k_b}$$

where L is the length of the plate in the flow direction.

2. Non-Newtonian (starch solution) fluid flowing past specially constructed silicon cube (Chandarana et al., 1990):

$$Nu_b = 2.0 + 0.0282 Re_b^{1.6} Pr_b^{0.89}$$

$$Re_b = \frac{D_{pa}^n u_{av}^{2-n} \rho_b}{K 8^{n-1} \Omega^n}, \quad Nu_b = \frac{h_{pa} D_{pa}}{k_b}, \quad Pr_b = \frac{C_b K \Omega^n 2^{n-3}}{k_b (D_{pa}/u_{av})^{n-1}}, \quad \Omega = \frac{3n+1}{n}$$

$1.23 < Re_b < 27.38$ (laminar flow), $9.47 < \mathrm{Prb} < 376.18$

for $T_b = 115.6°C$, $T_s = 129.4°C$, starch solution 2% w/w, u_s: 0.43–1.21 cm/s, K: 0.204–0.00420, n: 0.78–1.0.

3. Water flowing past specially constructed silicon cube (Chandarana, 1990):

$$Nu_b = 2.0 + 0.0333 Re_b^{1.08}$$

$$Re_b = \frac{D_{pa} u_{av} \rho_b}{\eta_b}, \quad Nu_b = \frac{h_{pa} D_{pa}}{k_b}$$

$287.3 < Reb < 880.76$ (laminar flow)

for $T_b = 115.6°C$, $T_s = 129.4°C$, u_s: 0.43–1.21 cm/s

4. Fluid over a flat plate: flow parallel to the surface of plate (Whitaker, 1976):

$$Nu_b = 0.036 (Re_b^{0.8} Pr_b^{0.43} - 17400) + 298 Pr_b^{1/3}$$

$Re_b > 2 \times 10^5$ (turbulent flow)

5. Flow normal to a single cylinder: parallel to cylinder diameter (Whitaker, 1976):

$$Nu_b = \left(0.4 Re_b^{1/2} + 0.06 Re_b^{2/3}\right) Pr_b^{0.4} (\eta_b/\eta_w)^{1/4}$$

$$Re_b = \frac{D_T u_{ap} \rho_b}{\eta_b}, \quad Nu_b = \frac{h D_T}{k_b}$$

$1.0 < Re_b < 1.0 \times 10^5$, $0.67 < Pr_b < 300$, $0.25 < \eta_b/\eta_w < 5.2$

6. Air or water flow past a single sphere (Whitaker, 1976):

$$Nu_b = 2 + \left(0.4 Re_b^{1/2} + 0.06 Re_b^{2/3}\right) Pr_b^{0.4} (\eta_b/\eta_w)^{1/4}$$

$$Re_b = \frac{D_{pa} u_{ap} \rho_b}{\eta_b}, \quad Nu_b = \frac{h D_{pa}}{k_b}$$

$3.5 \times 10^4 < Re_b < 7.6 \times 10^4$, $0.71 < Pr_b < 380$, $1.0 < \eta_b/\eta_w < 3.2$

7. Flow past a single sphere (Hallstrom, 1988):

$$Nu_b = 0.92 + 0.99 (Re_b Pr_b)^{1/3}$$

$$Nu_b = 2 + 0.6 Re_b^{0.5} Pr_b^{0.33}$$

These equations can also be used for irregular shape particles by defining a characteristic length of the particle as

$$l = \frac{A_s}{2\pi r_{pr}}$$

Table 23.4 (continued) Heat Transfer Correlations for Forced Convection upon Flat Surfaces and around Different Shapes

where
A_s is the heat transfer area (m^2)
r_{pr} is the radius of the projected area of the particle in the direction of flow (m)

8. Ellipsoidal surface of axis ratio 4:1 (Ko and Sogin, 1958):

$$\frac{h}{G_{ai}C_b}Pr_b^{2/3}\left(\frac{G_{ai}D_{er}}{\eta_b}\right)^{1/2} = 0.52$$

where
D_{er} is the diameter of the ellipsoid of revolution (m)
G_{ai} is the air mass velocity ($kg/m^2 \cdot s$)

For axis ratio 3:1 the right side constant is 0.60.

9. Ellipsoidal sphere: air flow perpendicular to minor axis (Clary and Nelson, 1970):

$$Nu_b = 0.489Re_b^{0.557}Pr_b^{1/3}\left(\frac{D_{ma}}{D_{mi}}\right)^{-0.070}\left(\frac{D_{me}}{D_{mi}}\right)^{-0.440}$$

$$Re_b = \frac{D_{mi}u_{ap}\rho_b}{\eta_b}, \quad Nu_b = \frac{hD_{mi}}{k_b}, \quad Pr_b = \frac{C_b\eta_b}{k_b}$$

$$3 \times 10^4 < Re_b < 1.5 \times 10^5, \ 1.33 < D_{ma}/D_{mi} < 3.00, \ 1.00 < D_{me}/D_{mi} < 2.50$$

where D_{ma}, D_{mi}, and D_{me} are the major-, minor-, and semiaxis of ellipsoid.

10. Elliptical cylinder: air flow parallel to major axis (Kondjoyan et al., 1993):

$$Nu_b = 0.00523Re_b + 17.90, \quad \gamma = 1 \text{ and } 2$$

$$Nu_b = 0.00433Re_b + 19.51, \quad \gamma = 3, 4, \text{ and } 5$$

$$Re_b = \frac{D_{ma}u_{ap}\rho_b}{\eta_b}, \quad Nu_b = \frac{hD_{mi}}{k_b}$$

where γ is the major to minor axis ratio; major axis: 0.10–0.20 m, u_a (fluid velocity at the main stream): 0.5–2.0 m/s, Tu (turbulence intensity of air in the main stream direction): 17%–60%.

11. Irregular particle (mushroom shaped in rest) immersed in non-Newtonian (power law pseudoplastic) fluid during tube flow (fluid flow parallel to longitudinal axis on the cap) (Zuritz et al., 1990):

$$Nu_b = 2 + 28.37Re_b^{0.33}Pr_b^{0.143}\frac{D_{ca}}{D_T}$$

$$Re_b = \frac{\rho_b u_{av}^{2-n}l^n}{2^{n-3}K\left(\frac{3n+1}{n}\right)^n}, \quad Nu_b = \frac{hD_{ca}}{k_b}, \quad Pr_b = \frac{2^{n-3}K\left(\frac{3n+1}{n}\right)^n C_b\left(\frac{L}{u_{av}}\right)^{1-n}}{k_b}$$

where
D_{ca} is the diameter of the projected cap perimeter of the mushroom-shaped particle
D_T is the tube diameter
C_b is the bulk fluid-specific heat; fluid is CMC, fluid flow rate: 0.06–0.29 kg/s, particle volume: 5×10^{-6}–10×10^{-6} m^3, particle surface area: 13.37×10^{-4} to 22.88×10^{-4} m^2 D_{ca}: 2.065–2.817 cm, D_{ca}/D_{pi}: 0.4130–0.5634, K: 2.3–30 Pa·s^n, n: 0.71–0.95, $T = 71°C$

12. Rotating cylinder in heated air (with and without cross flow) (Kays and Bjorklund, 1958):

$$Nu = 0.135\left[0.5\left(Re_{co}^2 + Re_{cs}^2 + Gr_{co}\right)Pr_{cs}\right]^{1.3}$$

$$Re_{cs} = \frac{D_c u_s \rho_{ai}}{\eta_{ai}}, \quad Gr_{co} = \frac{g\beta_{ai}D_c^3\rho_{ai}^2}{\eta_{ai}^2}, \quad Nu = \frac{h_{co}D_c}{k_{ai}}, \quad Pr_{co} = \frac{C_{ai}\eta_{ai}}{k_{ai}}$$

where
u_s is the cross flow gas velocity
h_{co} is the external heat transfer, and subscript ai refers to air

13. Finite cylindrical wall to air (Flores and Mascheroni, 1988):

$$Nu_b = 0.148Re_b^{0.633} 5 \times 10^3 < Re_b \le 5 \times 10^4$$

$$Nu_b = 0.0208Re_b^{0.814} 5 \times 10^4 < Re_b \le 2.3 \times 10^5$$

Table 23.5 Heat Transfer Correlations for Flow in Tubes and Channels

1. Newtonian fluid flowing through channel (Graetz, 1883, 1885; Leveque, 1928):

$$Nu_b = 1.75 Gz_b^{1/3}, \quad \text{laminar flow}$$

2. Newtonian fluid flowing through channel (Sieder and Tate, 1936):

$$Nu_b = 1.75 Gz_b^{1/3} (\eta_b/\eta_w)^{0.14}, \quad \text{laminar flow}$$

3. Newtonian fluid flowing in a circular tube (Sieder and Tate, 1936)

$$Nu_b = 1.86 (Re_b Pr_b)^{1/3} (D_T/L)^{1/3} (\eta_b/\eta_w)^{0.14}$$

$$Re_b = \frac{D_T u_{av} \rho_b}{\eta_b}, \quad Nu_b = \frac{h D_T}{k_b}, \quad Pr_b = \frac{C_b \eta_b}{k_b}$$

$$13 < Re_b < 2{,}030,\ 0.48 < Pr_b < 16{,}700,\ 0.0044 < \eta_b/\eta_w < 9.75$$

$$(Re_b Pr_b)^{1/3} (L/D_T)^{-1/3} (\eta_b/\eta_w)^{0.14} > 2$$

where
L is the length of the tube
η_w is the mean wall temperature; heat transfer coefficient is based on log mean temperature along the pipe length; all thermophysical properties refer to the mean bulk fluid temperature

4. Non-Newtonian fluid flowing through channel (Pigford, 1955):

$$Nu_b = 1.75 \Omega^{1/3} Gz_b^{1/3}$$

$$Gz_b = \frac{m C_b}{k_b L}, \quad Nu_b = \frac{h D_T}{k_b}$$

$$\Omega = \frac{3n+1}{4n}, \quad \text{for pseudoplastic fluids}$$

$$\Omega = \frac{1 - (\hat{\tau}_y/\hat{\tau}_w)}{1 - ((4/3)(\hat{\tau}_y/\hat{\tau}_w)) + ((1/3)(\hat{\tau}_y/\hat{\tau}_w))^4}, \quad \text{for Bingham plastics}$$

5. Non-Newtonian fluids flowing through tubes in laminar flow (Metzner et al., 1957; Christiansen et al., 1966):

$$Nu_b = 1.75 \Lambda^{1/3} Gz_b^{1/3} (K_b/K_w)^{0.14}$$

$$\Lambda^{1/3} = \Omega^{1/3} = \left(\frac{3n+1}{4n}\right)^{1/3}, \quad \text{for } n > 0.4 \text{ and any } Gz_b \text{ number}$$

$$\Lambda^{1/3} = 1.18 - 0.24n, \quad \text{for } 0 < n < 0.4 \text{ and } Gz_b = 5$$

$$\Lambda^{1/3} = 1.30 - 0.60n, \quad \text{for } 0 < n < 0.4 \text{ and } Gz_b = 10$$

$$\Lambda^{1/3} = 1.40 - 0.72n, \quad \text{for } 0 < n < 0.4 \text{ and } Gz_b = 15$$

$$\Lambda^{1/3} = 1.57 - 0.35n, \quad \text{for } 0 < n < 0.4 \text{ and } Gz_b = 25$$

$$\Lambda^{1/3} = 1.57 - 0.35n, \quad \text{for } 0 < n < 0.4 \text{ and } Gz_b > 25$$

where Λ is the non-Newtonian to Newtonian heat transfer coefficient ratio.

6. Power law fluids flowing through tubes in laminar flow (Leninger and Beverloo, 1975):

$$Nu = 0.683 \frac{n+1}{n} Gz^{1/3}, \quad \text{for constant wall temperature}$$

$$Nu = 0.770 \frac{n+1}{n} Gz^{1/3}, \quad \text{for constant heat flux}$$

Table 23.5 (continued) Heat Transfer Correlations for Flow in Tubes and Channels

7. Non-Newtonian fluids flowing through a horizontal tube (Filkova et al., 1986):

$$Nu_x = 1.229\ \mathrm{M}^{-0.36}$$

$$Nu_x = 1.972\ Gz_x^{0.274}\left[\frac{K}{K_w}\frac{3n+1}{2(3n-1)}\right]^{0.14}$$

$$Nu_x = \frac{h_x D_T}{k_a},\quad Gz_x = \frac{mC_a}{k_a\bar{x}},\quad Re_a = \frac{D_T^n u_{av}^{2-n}\rho_a}{K}$$

$$M = \frac{\bar{x}}{Pe_a D_T}\frac{4n}{3n+1},\quad Pe_a = \frac{D_T u_{av} C_a \rho_a}{k_a}$$

$$31 < Re_a < 1{,}228,\ 55 < Gz_x < 376,\ 13{,}158 < Pe_a < 90{,}476,$$

$$10^{-5} < M < 10^{-2},\ Gr_a Pr_a < 1867$$

where
$\bar{x}$ is the axial distance
m is the mass flow rate (kg/s); this equation has been tested for CMC (0.9%–1.8%) and tomato juice, T_w: 40°C–90°C

8. Turbulent flow in pipes (Whitaker, 1976):

$$Nu_b = 0.015 Re_b^{0.83} Pr_b^{0.42}(\eta_b/\eta_w)^{0.14}$$

$$Re_b = \frac{D_T u_{av}\rho_b}{\eta_b},\quad Nu_b = \frac{hD_T}{k_b},\quad Pr_b = \frac{C_b\eta_b}{k_b}$$

$$2300 < Re_b < 10^5,\ 0.48 < Pr_b < 592,\ 0.44 < \eta_b/n_w < 2.5,\ 96 < L/D_T < 196$$

where h is the heat transfer coefficient based on log mean temperature along the pipe length.

$$Nu_b = 0.023 Re_b^{0.8} Pr_b^{0.4},\quad \text{for heating}$$

$$Nu_b = 0.023 Re_b^{0.8} Pr_b^{0.3},\quad \text{for cooling}$$

$$2000 < Re_b < 10^7,\ 0.5 < Pr_b < 120$$

9. Turbulent flow in pipes (Ditus and Boelter, cited by Metzner, 1956):

$$Nu_b = 0.023 Re_b^{0.8} Pr_b^{0.4},\quad \text{for heating}$$

$$Nu_b = 0.017 Re_b^{0.8} Pr_b^{0.3},\quad \text{for cooling}$$

10. Turbulent flow in pipes for liquids (e.g., oil) whose viscosity rapidly changes with temperature (Sieder-Tate):

$$Nu_b = 0.0243 Re_b^{1/3}(\eta_b/\eta_s)^{0.14}$$

11. Turbulent flow in noncircular duct (Whitaker, 1976):

$$Nu_b = 0.015\left(\frac{D_h u_{av}\rho_b}{\eta_b}\right) Pr_b^{0.43}(\eta_b/\eta_w)^{0.14}$$

where D_h is the hydraulic diameter, which is defined as

$$D_h = 4\frac{\text{Cross-sectional area}}{\text{Wetted perimeter}}$$

12. Turbulent flow in noncircular duct (Hallstrom, 1988):

$$Nu_b = 0.027\left[1 + \left(\frac{D_T}{L}\right)^{0.66}\right] Re_b^{0.8} Pr_b^{0.37}$$

(continued)

Table 23.5 (continued) Heat Transfer Correlations for Flow in Tubes and Channels

13. Turbulent flow in tube when fouling takes place inside the tube (Gnielinski, 1975):

$$Nu = \frac{(f/8)(Re - 1{,}000)Pr}{1 + 12.7(f/8)^{0.5}(Pr^{0.66} - 1)}\hat{F}$$

$$f = (1.82\lg Re - 1.64)^{-2}$$

$$\hat{F} = 1 + (D_{eq}/L)^{2/3}$$

$$Nu = 0.032Re_b^{0.8}Pr_b^{0.37}(D_T/L)^{0.054}, \quad \text{for } L > 60D_T$$

14. Non-Newtonian (power law) fluids in turbulent flow (Wilkinson, 1960):

$$Nu_b = 0.023Re_b^{0.8}Pr_b^{0.4}(\eta_b/\eta_w)^{0.14}$$

$$Re_b = \frac{D_T^n u_{av}^{2-n}\rho_b}{K}, \quad \eta = K\left(\frac{u_{av}}{D_T}\right)^{n-1}, \quad Pr_b = \frac{C_p K u_{av}^{n-1}}{D_T^{n-1}}, \quad K = \hat{Y}\exp\left(-\Delta\hat{E}/T\right)$$

where $\hat{Y}$ is a parameter of the equation.

15. Suspension through tube (Orr and Valle, 1954):

$$Nu_b = 0.027\left(\frac{D_T u_{av}\rho_{su}}{\eta_{su}}\right)^{0.8}\left(\frac{C_p\eta_{su}}{k_{su}}\right)^{1/3}\left(\frac{\eta_i}{\eta_{iw}}\right)^{0.14}$$

$$\eta_{su} = \frac{\eta_i}{(1 - \varepsilon_{sb}/\varepsilon_{se})^{1.8}}$$

$$10^4 < Re < 3 \times 10^5$$

where ε_{sb} and ε_{se} are the volume fraction of solids in the suspension and in the sedimented suspension.

16. Slurries in turbulent flow (Salamone and Newman, 1955):

$$\frac{hD_T}{k_l} = 0.0131\left(\frac{D_T u_{av}\rho_{su}}{\eta_{su}}\right)^{0.62}\left(\frac{C_l\eta_{su}}{k_l}\right)^{0.72}\left(\frac{k_{pa}}{k_l}\right)^{0.05}\left(\frac{D_T}{D_{pa}}\right)^{0.05}\left(\frac{C_{pa}}{C_l}\right)^{0.35}$$

where subscript l refers to liquid properties.

17. Particles in bundle (Schlunder, 1972):

$$Nu_{bundle} = \Xi Nu_{single\ particle}$$

$$\Xi = 1 + 1.5(1 - \varepsilon_{FL})$$

where ε_{FL} is the volume fraction of fluid ranging between 0.4 and 1.

18. Fluids flowing past a staggered tube bank (Whitaker, 1976):

$$Nu_b = \left(0.4Re_b^{0.5} + 0.2Re_b^{2/3}\right)Pr_b^{0.4}(\eta_b/\eta_w)^{0.14}$$

$$80 < Re_b < 200,\ 0.42 < \varepsilon_{FL} < 0.65$$

This equation has been applied for air and oil; it can also be applied in packed beds for variable velocity near the surface and bulk.

19. Falling films on the exterior of a vertical tube without evaporation, turbulent regime (Wilke, 1962):

$$h = \left(\frac{\nu_l^2}{gk_l^3}\right)^{1/3} = 8.7 \times 10^{-3}\left(\frac{4\Gamma}{\eta_l}\right)^{0.4}\left(\frac{\nu_l}{a_l}\right)^{0.344}$$

This equation has been applied for water and mixtures of water–ethylene glycol.

Table 23.6 Heat Transfer Correlations for Heat Exchangers

1. Plate heat exchanger (Auth and Loiano, 1978, cited by Jackson and Lamb, 1981):

$$Nu_b = \hat{a} Re_b^{\hat{b}} Pr_b^{\hat{c}} (\eta_b/\eta_s)^{\hat{d}}$$

$$0.15 < \hat{a} < 0.4,\ 0.65 < \hat{b} < 0.85,\ 0.30 < \hat{c} < 0.45,\ 0.05 < \hat{d} < 0.2$$

$$Nu_b = \dot{a}\left[Re_b Pr_b (D_{eq}/L)\right]^{0.33} (\eta_b/\eta_s)^{0.14}, \quad \text{for laminar flow}$$

$$1.86 < \dot{a} < 4.5$$

where L is the plate length.

2. Plate heat exchanger (Dutta and Chanda, 1991):

$$Nu_b Pr_b^{-0.33} \left(\frac{\eta_b}{\eta_w}\right)^{0.14} = 0.244 \left(\frac{A_D}{A_{pr}}\right)^{1.32} \left(1 + \frac{l_{ri}}{l_{ga}}\right)^{1.216} \left(1 + \frac{\hat{\theta}}{180}\right)^{1.55} Re_b^{0.30}$$

$$Re_b = \frac{2m}{l\eta_b}, \quad Nu_b = \frac{hD_{eq}}{k_b}, \quad Pr_b = \frac{C_b \eta_b}{k_b}$$

$$120 < Re_b < 200$$

where
- A_D is the developed area
- A_{pr} is the projected area
- l_{ri} is the rise of ribs above the plate surface
- l_{ga} is the gap between plates
- $\hat{\theta}$ is the angle of ribs with the flow direction (degree)
- l is the plate width
- D_{eq} is the equivalent diameter of the channel between plates

These equations have been developed in plate exchangers of length 0.69 m, width 0.21 m, thickness 0.001–0.0023, projected plate area 0.1349 m^2, developed plate area 0.1349–0.1415 m^2, gap between adjacent plates 0.00176 m, equivalent diameter 0.0024–0.0035 m, ribs to flow direction angle 0°–180°, and rise of ribs 0–0.0027 m for water and water–glycerin solution

3. Scraped surface heat exchanger

Product side (Skelland et al., 1962):

$$\frac{h_p D_T}{k_p} = Y \frac{(D_T - D_{sa}) u_{av} \rho_p}{\eta_p} \left(\frac{C_p \eta_p}{k_p}\right)^{\kappa} \left(\frac{D_T N}{u_{av}}\right)^{0.62} \left(\frac{D_{sa}}{D_T}\right)^{0.55} \nu^{0.53}$$

where
- Y and κ are 0.014 and 0.96, respectively, for cooling viscous liquids; 0.039 and 0.70, respectively, for cooling thin mobile liquids
- D_{sa} is the diameter of the shaft; and υ is the number of rows of scraper blades

Trommelen (1970):

$$h_p = 1.13 \left(k_p \rho_p C_p N \upsilon\right)^{0.5} \hat{f}$$

where $\hat{f}$ is an empirical correction factor (values are given by Van Boxtel and De Fielliettaz Goethart for different food systems, cited by Rahman, 1995).

4. Jacket side (water cooled system) (Van Boxtel and De Fielliettaz Goethart, 1983, 1984):

$$Nu_b = 18.2 + 0.0158 Re_b^{0.8} Pr_b^{0.40}$$

This equation has been developed for flow rate 0.171–0.417 kg/s and jacket side heat transfer coefficient 3220–5572 W/m^2 °C.

Table 23.7 Heat Transfer Correlations for Extruders

1. Single-screw extruder (fed with hard wheat flour dough) (Levine and Rockwood, 1986):

$$Nu_b = 2.2Br_b^{0.79}$$

$$Nu_b = \frac{h_{ou}H}{k_b}, \quad Pe_b = \frac{\rho_b C_b \pi N D_{sr} H}{L}, \quad Br_b = \frac{K(\pi N D_{sr})^{n+1}}{k_b(T_i - T_s)H^{n-1}}$$

$$15 < Nu_b < 50, \ 15 < Br_b < 58$$

where
H is the tread or screw channel depth
D_{sr} is the diameter of the screw
N is the screw rotational rate
T_i is the initial temperature
T_s is the barrel temperature
L is the length of the screw. This correlation characterizes the specific extruder

2. Twin-screw extruder, corotating (Todd, 1988):

$$Nu_b = 0.94\left(\frac{D_{sr}^2 N \rho_b}{\eta_b}\right)^{0.28}\left(\frac{\eta_b C_b}{k_b}\right)^{0.33}\left(\frac{n_b}{\eta_w}\right)^{0.14}$$

$$Nu_b = \frac{h_{in}D_{sr}}{k_b}$$

3. Twin-screw extruder (Mohamed and Ofoli, 1989):

$$Nu_b = 0.0042Gz_b^{1.406}Br_b^{0.851}$$

$$Nu_b = \frac{h_{in}D_{sr}}{k_b}, \quad Gz_b = \frac{mC_b}{k_b}, \quad Br_b = \frac{K_i L^2 \dot{\gamma}_a^{n+1}}{k_b(T_i - T_w)}\exp\left(-\Delta\hat{E}/RT_i\right)$$

where
K_i is the reference consistency coefficient
T_i is the reference temperature
$\dot{\gamma}$ is the average shear rate
$\Delta\hat{E}$ is the activation energy

Table 23.8 Heat Transfer Correlations for Freezing

1. Flat plate surface during air freezing (Heldman, 1980):

$$Nu_x = 0.665Re_x^{0.571}, \quad \text{for acrylic}$$

$$Nu_x = 0.579Re_x^{0.582}, \quad \text{for ground beef}$$

$$1 \times 10^4 < Re_x < 5 \times 10^5$$

air speed: 0–14 m/s, temperature: −28.4°C to −17.8°C, plate thickness: 0.945 and 1.89 cm.

2. Belt freezer (air to disk face heat transfer coefficient) (Flores and Mascheroni, 1988):

$$Nu_b = 10.998Re_b^{0.277}$$

Parallel to belt airflow, material: copper, both surfaces

$$Nu_b = 7.891Re_b^{0.328}$$

Parallel to belt airflow, material: unpacked hamburger, both surfaces

$$Nu_b = 4.293Re_b^{0.395}$$

Parallel to belt airflow, material: packed hamburger, both surfaces

$$Nu_b = 0.412Re_b^{0.578}$$

Transverse to belt airflow (downward), material: copper, both surfaces

$$Nu_b = 0.326Re_b^{0.640}$$

Table 23.8 (continued) Heat Transfer Correlations for Freezing

Transverse to belt airflow (downward), material: hamburger, both surfaces

$$Nu_b = 4.190 Re_b^{0.355}$$

Transverse to belt airflow (upward), material: hamburger, upper surface only

$$Nu_b = 1.312 Re_b^{0.496}$$

Transverse to belt airflow (upward), material: hamburger, lower surface only

$$Re_b = \frac{D u_{av} \rho_b}{\eta_b}, \quad Nu_b = \frac{hD}{k_b}$$

$$20 \times 10^{-3} < Re_b < 70 \times 10^{-3}$$

where D is the disk diameter.

3. Heat transfer due to temperature gradient without vapor condensation in the case of refrigerated food products (Leichter et al., 1976):

$$Nu_b = 0.860 Gr_b^{0.25}$$

$$Nu_b = \frac{h_{woc} H_T}{k_b}, \quad Gr_b = \frac{g \beta \rho_b^2 H^3 \Delta T}{\eta_b^2}$$

where H_T is the height of the cylinder and the subscript woc indicates the process without condensation of water vapor.

The effect of temperature and relative humidity of atmosphere in the heat transfer coefficient on a cold surface (at 0°C) is given by the expressions

$$h_{wic}/h_{woc} = 0.584 + 0.0113 \cdot \%\text{RH}, \quad T_b = 15^\circ\text{C}$$

$$h_{wic}/h_{woc} = 0.760 + 0.0104 \cdot \%\text{RH}, \quad T_b = 20^\circ\text{C}$$

$$h_{wic}/h_{woc} = 0.813 + 0.0112 \cdot \%\text{RH}, \quad T_b = 25^\circ\text{C}$$

$$h_{wic}/h_{woc} = 0.879 + 0.0123 \cdot \%\text{RH}, \quad T_b = 30^\circ\text{C}$$

where %RH is the percent relative humidity, ranging from 40% to 100%, and the subscript wic indicates the process with condensation of water vapor.

Table 23.9 Heat Transfer Correlations for Sterilization—Retort Canning

1. Retort canning (sterilization) (Merson et al., 1980)

$$Nu_b = 0.17 Re_b^{0.52} Pr_b^{1/3} (H_c/e)^{1/3}$$

$$Re_b = \frac{D_r \pi D_c N \rho_b}{\eta_b}, \quad Nu_b = \frac{h_{ov} D_c}{k_b}, \quad Pr_b = \frac{C_b \eta_b}{k_b}$$

where
- h_{ov} is the overall heat transfer coefficient
- N_c is the can rotational speed
- H_c is the can height
- D_c is the can diameter
- e is the head space height with can vertical; bulk fluid properties are evaluated at the average bulk temperature between the start and the end of the process.

2. Steam heated steritort (container–fluid interface for fluids without particles) (Lenz and Lund, 1978):

$$Nu_b = 115 + 15 Re_b^{0.3} Pr_b^{0.08}$$

$$Re_b = \frac{(D_r/2)^2 N \rho_b}{\eta_b}, \quad Nu_b = \frac{h_{in} D_r}{2k_b}, \quad Pr_b = \frac{C_b \eta_b}{k_b}$$

This equation has been applied for reel speed: 3.5–8 rpm, ε_l: 0.32–0.45, steam temperature 121°C, D_p: 0.97–3.81 cm, container size: 303 × 406 and 608 × 700.

(continued)

Table 23.9 (continued) Heat Transfer Correlations for Sterilization—Retort Canning

3. Steam heated steritort, container–fluid interface for fluids with spherical particles (Lenz and Lund, 1978):

$$Nu_b = -33 + 53Re_b^{0.28}Pr_b^{0.14}\left[\frac{D_p}{D_r/2(1-\varepsilon_l)}\right]$$

$$Re_b = \frac{(D_r/2)^2 N\rho_p}{\eta_b}, \quad Nu_b = \frac{hD_r}{2k_p}, \; Pr_b = \frac{C_p\eta_b}{k_p}$$

This equation has been applied in conditions similar to the previous case.

4. Steam heated retort and axially rotated cans (Soule and Merson, 1985):

$$Nu_w = 0.434Re_w^{0.571}Pr_w^{0.278}\left(\frac{H_c}{D_c}\right)^{0.356}\left(\frac{\eta_b}{\eta_w}\right)^{0.154}$$

$$h_{in} = 1.07h_{ov}$$

$$Re_w = \frac{D_c^2 N_c\rho_w}{\eta_w}, \quad Nu_w = \frac{h_{ov}D_c}{k_w}, \quad Pr_w = \frac{C_w\eta_w}{k_w}$$

$12 < Re_w < 4.4 \times 10^4$, $2.2 < Pr_w < 2{,}300$, $1.11 < H_c/D_c < 1.61$, $1.22 < \eta_b/\eta_w < 1.79$

where

h_{in} and h_{ov} are the internal film and overall coefficient, respectively

N_c is the can rotational speed; wall fluid properties are evaluated at the arithmetic average of the initial and final temperatures of the can wall; head space 1.0 cm, vacuum 20 in Hg, N_c: 0–2.5 rps

5. Steam heated, axially rotating cans (Deniston et al., 1987):

$$Nu_b = 1.87 \times 10^{-4}Re_p^{1.69}\left(\frac{a_p}{\omega D_p^2}\right)^{0.126}B^{0.53}F^{-0.171}$$

$$B = \frac{\rho_p - \rho_b}{6\dot{c}\rho_b}\,\frac{\omega^2 D_{cx} + 2g}{\omega^2 D_{cx}}\,\frac{D_p}{D_{cx}}$$

$$F = (1-\varepsilon_p)\frac{H_{ce}}{D_{ci}}\,\frac{\omega D_{cx}^2}{a_b}$$

$$Re_p = \frac{D_p u_p \rho_b}{\eta_b}, \quad Nu_b = \frac{h_{ov}D_{cx}}{k_b}, \quad \dot{c} = \frac{Ar_b}{4Re_p^2}, \quad N_c = \frac{N_r D_r}{D_{cx}}$$

$$Ar_b = \frac{4D_p^3\rho_b(\rho_p - \rho_b)\left[g + \omega^2(D_{cx}/2)\right]}{3\eta_b^2}$$

where

$\dot{c}$ is the drag coefficient

Ar_b is the Archimedes number

H_{ce} is the effective can length (internal can length: head space)

D_{ci} and D_{cx} are the internal and external can diameter, respectively

ω is the can angular velocity ($2\pi N_c$)

N_c is the can rotational speed

N_r is the reel rotation

6. Steritort (concentrated tomato 7.2°Brix) (Deniston et al., 1992):

$$Nu_b = 0.703 + 0.417Re_b^{1.103}Pr_b^{0.324}\left(\frac{H_{cx}}{D_{cx}}\right)^{0.069}\left(\frac{P_{ss}}{P_{TO}}\right)^{0.806}$$

$$Re_b = \frac{D_{cx}^2 N_c\rho_b}{\eta_p}, \quad Nu_b = \frac{h_{ov}D_{cx}}{k_p}, \quad Pr_b = \frac{C_p\eta_p}{k_p}$$

$1.99 < Re_b < 5.38$, $23{,}500 < Pr_b < 34{,}800$, $1.12 < H_{cx}/D_{cx} < 1.51$, $0.75 < P_{ss}/P_{TO} < 1.00$

where

H_{cx} is the can height

D_{cx} is the external can diameter

P_{ss} is the absolute steam pressure

P_{TO} is the absolute total pressure

h_{ov} is the overall heat transfer coefficient based on total external surface (heating medium [can wall] internal fluid); reel rotation speed: 4–10 rpm, air: 0%–25%, can size: 211 × 300, 300 × 407, 307 × 503, $T_{re} = 121.1$°C, Φ: 60.9–84.9

Table 23.9 (continued) Heat Transfer Correlations for Sterilization—Retort Canning

7. Steritort (particle–fluid heat transfer coefficient for canned snap beans) (Fernandez et al., 1988):

$$Nu_b = 2.69 \times 10^4 Re_b^{0.294} Pr_b^{0.33} \Psi^{6.98}$$

$$\Psi = \frac{\pi\left(\frac{6V_{pa}}{\pi}\right)^{2/3}}{A_{pa}}$$

$$Re_b = \frac{D_c N_c \rho_b}{\eta_b}, \quad Nu_b = \frac{h_{pa} D_{pa}}{k_b}, \quad Pr_b = \frac{C_b \eta_b}{k_b}$$

Beans mass: 164–244 g, reel speed: 2–8 rpm, steam temperature 115.6°C, can size: 303 × 406, head space: 0.64 cm.

8. Steritort (cans containing non-Newtonian liquids) (Rao et al., 1985):

$$Nu_b = 2.60(Gr_b Pr_b)^{0.205} + 7.15 \times 10^{-7}[Re_b Pr_b (D_c/H_c)]^{1.837}$$

$$Re_b = \frac{D_c^2 N_c^{2-n} \rho_b}{8^{n-1} K\left(\frac{3n+1}{4n}\right)^n}, \quad Nu_b = \frac{h_{ov} D_c}{k_b}, \quad Pr_b = \frac{C_b \eta_b}{k_b}, \quad Gr_b = \frac{g D_c^3 \rho_b^2 \beta \Delta T}{\eta_b}, \quad \eta_b = \frac{K 8^{n-1}}{N_c^{1-n}}\left(\frac{3n+1}{4n}\right)^n$$

$$0.4 < Re_b < 96, \; 205 < Pr_b < 5100, \; 24 < Nu_b < 160, \; 100 < Gr_b < 5.2 \times 10^5, \; 1.13 < H_c/D_c < 1.37$$

This model has been applied for aqueous guar gum solutions (i. 0.3%, $K = 0.041$, $n = 0.58$, ii. 0.4%, $K = 0.077$, $n = 0.69$, iii. 0.5%, $K = 0.178$, $n = 0.63$, iv. 0.75%, $K = 0.922$, $n = 0.58$) and glycerin ($K = 0.082$, $n = 1.00$); T_{steam}: 110°C–115.6°C, $e = 6.35$ mm, steam heated retort, axially rotated can.

9. Steretort (cans containing Newtonian liquids) (Anantheswaran and Rao, 1985a):

$$Nu_b = 2.9 Re_b^{0.436} Pr_b^{0.287}$$

$$Re_b = \frac{D_r^2 N_r \rho_b}{\eta_b}, \quad Nu_b = \frac{h_{in} D_r}{k_b}, \quad Pr_b = \frac{C_b \eta_b}{k_b}$$

$$11 < Re_b < 2.1 \times 10^5, \; 2.8 < Pr_b < 498, \; 0.73 < H_{cx}/D_{cx} < 1.37$$

where

N_r is the speed of rotation

D_r is the diameter of rotation or reel diameter; end-over-end rotation in steam at atmospheric pressure steretort, can size: 303 × 406, N_r: 0–38.6 rpm, D_r: 0–29.8 m, headspace volume: 3%–9%.

10. Steretort, cans containing non-Newtonian liquids (Anantheswaran and Rao, 1985b):

$$Nu_b = 1.41 Re_b^{0.482} Pr_b^{0.355}$$

$$Re_b = \frac{D_r^2 N_r^{2-n} \rho_b}{8^{n-1} K \Omega^n}, \quad Nu_b = \frac{h_{in} D_r}{k_b}, \quad Pr_b = \frac{C_b K 8^{n-1} \Omega^n}{k_b N_r^{1-n}}, \quad 0.73 < H_{cx}/D_{cx} < 1.37$$

$70 < Re_b < 1.2 \times 10^4$, $Pr_b = 48$, for aqueous guar gum solution 0.3%

$4 < Re_b < 1.17 \times 10^3$, $508 < Pr_b < 953$, for aqueous guar gum solution 0.4%

$1 < Re_b < 478$, $1250 < Pr_b < 2800$, for aqueous guar gum solution 0.5%

$0.1 < Re_b < 46.2$, $1.66 \times 10^4 < Pr_b < 5.7 \times 10^4$, for aqueous guar gum solution 0.75%

end-over-end rotation in steam retort, can size: 303 × 406, N_r: 0–38.6 rpm, D_r: 0–29.8 m

11. Retort-canned gelatinized starch (Ramaswamy et al., 1993):

$$\ln h_{ov} = 5.99 - 2.72 \times 10^{-3} T_{re} + 3.76 \times 10^{-2}(N_r/60) - 3.55\eta_{ini}$$

$$\ln Nu = 4.604 + 152.26 Fr - 3.29 \times 10^{-4} \eta_{insp}$$

$$Nu = \frac{h_{ov} D_r}{k_b}, \quad Fr = \frac{D_r N_r^2}{g}, \quad \eta_{insp} = \frac{(K 8^{n-1} \Omega^n)/N_r^{1-n}}{\eta_{water}}$$

where

Fr is the rotational Froude number

η_{ini} is the initial apparent viscosity of the product

η_{insp} is the specific apparent viscosity; end-over-end rotation of can, product concentration: 3%–4%, reel rotation speed: 10–20 rpm, can headspace: 6.4–12.8 mm, T_{re}: 110°C–130°C, heating medium: steam (75%)–air (25%) mixture

(*continued*)

Table 23.9 (continued) Heat Transfer Correlations for Sterilization—Retort Canning

12. Flame sterilization (flame heating of CMC solutions) (Teixeira Neto, 1982):

$$Nu = 0.433 Re_{ci}^{0.56} Pr_{ci}^{0.60} Re_{co}^{-0.68}$$

$$Re_{ci} = \frac{D_r \pi D_c N \rho_{ci}}{\eta_{ci}}, \quad Re_{co} = \frac{D_c \bar{u}_c \rho_{ai}}{\eta_{ai}}, \quad Nu = \frac{h_{ov} D_c}{k_{ci}}, \quad Pr_{ci} = \frac{C_{ci} \eta_{ci}}{k_{ci}}$$

where
- $\bar{u}_c$ is the cylinder surface peripheral velocity
- ρ_{ai} is the air density, η_{ai} is the air viscosity
- D_c is the can diameter
- D_r is the reel diameter, and subscripts ci and co refer to internal and outside of the can, respectively

13. Liquids heating in a can with different modes of outer heating (Duquenoy, 1980):

$$Nu = 17.0 \times 10^{-5} Re^{1.449} Pr^{1.190} We^{-0.551} Fr^{0.932} \varepsilon_p^{0.628}$$

$$Re = \frac{\omega D_c}{2\eta_p} \frac{2 D_c H_c}{D_c/2 + H_c}, \quad Nu = \frac{h_{in} D_c}{2k_p}, \quad Pr = \frac{C_p \eta_p}{k_p}, \quad We = \frac{\pi \omega^2 H^2 D_c}{\sigma_p}, \quad Fr = \frac{D_r N_r^2}{g}, \quad \omega = 2\pi N$$

$2.07 < Re < 9.91$, $0.8 < Pr < 8.06$, $2.73 < Nu < 5.05$

where
- h_{in} is the inside heat transfer coefficient
- ω is the speed of rotation (rad/s), ε_p is the product volume fraction compared to total volume of can, We is the Weber number, and σ_p is the surface tension of liquid product (N/m); T_i: 20°C–60°C, ω:

0.21–4.19 rad/s, D_c: 13.75–24.25 mm, H: 45.4–55.7 mm, k_p: 0.118–0.670 W/m, ρ_p: 792–1150 kg/m^3, C_p: 1448–4332 J/kg, V_p: 0.75–0.99.

Table 23.10 Heat Transfer Correlations for Sterilization—Aseptic Canning

1. Aseptic canning based on slip velocity (Kramers, 1946):

$$Nu_b = 2.0 + 1.3 Pr_b^{0.15} + 0.66 Re_s^{0.5} Pr_b^{0.31}$$

2. Aseptic canning based on slip velocity (Ranz and Marshall, 1952):

$$Nu_b = 2.0 + 0.6 Re_s^{0.5} Pr_b^{0.33}$$

3. Aseptic canning based on slip velocity (Whitaker, 1972):

$$Nu_b = 2.0 + \left(0.4 Re_s^{0.5} + 0.06 Re_s^{2/3}\right) Pr_b^{0.4} (\eta_b/\eta_s)^{0.25}$$

$$Re_s = \frac{|u_{pa} - u_l| D_{pa} \rho_l}{\eta_l}$$

where u_{pa} and u_l are the particle and liquid fluid velocity, respectively.

4. Aseptic canning (Sastry et al., 1990):

$$Nu = 26.81 + 0.00455 Re_{pa} (D_{pa}/D_T)$$

$$Nu = 6.023 \times 10^{-6} Re_{pa}^{1.79} (D_{pa}/D_T)^{1.71} Fr_{pa}^{-0.64}$$

$$Nu = 0.0046 Re_{pa} + 41.54 (D_{pa}/D_T) - 35.65 Fr_{pa} - 5.24$$

$$Re_{pa} = \frac{D_{pa} u_{av} \rho_b}{\eta_b}, \quad Nu = \frac{h_{pa} D_{pa}}{k_b}, \quad Fr_{pa} = \frac{u_{av}^2}{g D_{pa}\left[\left(\rho_p/\rho_b\right) - 1\right]}$$

$3600 < Re_{pa} < 27300$, $0.39 < Fr_{pa} < 14.83$, $0.2618 < D_{pa}/D_T < 0.6273$

D_{pa}: 1.33–2.39 cm, flow rate: 1.33×10^{-4}–7.98×10^{-4} m^3/s.

Table 23.10 (continued) Heat Transfer Correlations for Sterilization—Aseptic Canning

5. Aseptic canning (Chandarana et al., 1990):

$$h_{pa} = 1.14 \times 10^{-4} Re_{pa}^{0.07} \Phi^{1.94}$$

where Φ is the surface area to volume ratio of the particle.

6. Aseptic processing (shear flow) (Mwangi et al., 1993):

$Nu_b = 0.184 Re_b^{0.58}$, laminar flow, solid fraction: 2.12%

$Nu_b = 0.195 Re_b^{0.74}$, laminar flow, solid fraction: 3.22%

$Nu_b = 0.0078 Re_b^{0.95}$, turbulent flow, solid fraction: 1.22%

$Nu_b = 0.034 Re_b^{0.80}$, turbulent flow, solid fraction: 3.22%

$Nu_b = 0.100 Re_b^{0.58} Pr_b^{0.33}$, laminar flow, $Re_b < 563$, $3.68 < Pr_b < 5.52$

$Nu_b = 0.0336 Re_b^{0.80} Pr_b^{0.32}$, turbulent flow, $563 < Re_b < 6000$, $3.68 < Pr_b < 5.52$

D_{pa}: 8.0–12.7 mm, D_{pa}/D_T: 0.156–0.25

Table 23.11 Heat Transfer Correlations for Flow through Packed Beds

1. Forced convection through packed bed (Yoshida et al., 1962):

$$j = 0.91 Re_b^{-0.51} \Psi \quad Re_b < 50$$

$$j = 0.61 Re_b^{-0.41} \Psi \quad Re_b > 50$$

$$j = \frac{h}{C_b G}\left(\frac{C_b \eta_b}{k_b}\right)^{2/3}, \quad Re_{av} = \frac{G}{A_{sp} \eta_{av} \Psi}$$

where
G is the superficial mass velocity ($kg/m^2/s$)
Ψ is the particle shape factor (Gamson presented the following values of this factor for packed bed correlations: 1.00 for sphere, 0.91 for cylinder, 0.86 for flake, 0.80 for berl saddle, 0.79 for rasching ring, 0.67 for partition ring)

Properties' values are calculated in temperature, which is the arithmetic mean of inlet T_0 and outlet T_e temperature.

It should be noted that the heat transfer rate from particle to fluid in a packed bed can be calculated by the following equation:

$$\dot{q} = h A_{sp} A_{cs} H (T_0 - T_e)$$

where
A_{cs} is the sectional area of the bed (m^2)
A_{sp} is the total particle surface area per unit volume of the bed (m^2/m^3)
H is the bed height (m)

2. Packed bed of sphere and liquid (Rohsenow and Choi, 1961):

$$Nu_b = 0.8 Re_b^{0.7} Pr_b^{0.33}$$

3. Packed bed during drying (Hallstrom et al., 1988):

$$Nu_{pa} = 0.3 Re_{pa}^{1.3}$$

$$Re_{pa} = \frac{D_{pa} u_{ap} \rho_b}{\eta_b}, \quad Nu_{pa} = \frac{h_{pa} D_{pa}}{\psi}$$

where ψ is the mass transfer coefficient.

4. Packed bed (particle-to-fluid heat transfer coefficient) (Whitaker, 1976):

$$Nu_b = \left(0.4 Re_b^{0.5} + 0.2 Re_b^{2/3}\right) Pr_b^{0.4}$$

$$Re_b = \frac{D_{pa} G}{\eta_b (1 - \varepsilon_{FL})}, \quad Nu_b = \frac{h D_{pa}}{k_b} \frac{\varepsilon_{FL}}{1 - \varepsilon_{FL}}, \quad D_{pa} = \frac{6 V_{pa}}{A_s}$$

$$3.7 < Re_b < 8000, \; Pr_b = 0.7, \; \eta_b/\eta_w = 10.34 < \varepsilon_{FL} < 0.74$$

where A_s is the particle surface.

Table 23.12 Heat Transfer Correlations for Flow through Fluidized Beds

1. Fluidized bed (bed-to-wall heat transfer) (Levenspiel and Walton, 1954):

$$\frac{h_w D_{ec}}{k_{ai}} = 0.0018 \frac{D_{pa} G_{ai}}{\eta_{ai}} \left(\frac{D_{pa}}{D_T}\right)^{-1.16}$$

$$\frac{h_w}{C_{ai} G_{ai}} = 10 \left(\frac{D_{ec} G_{ec}}{\eta_{ai}}\right)^{-0.65}$$

$$\frac{h_w}{C_{ai} G_{ai}} = 0.6 \left(\frac{D_{pa} G_{ai}}{\eta_{ai}}\right)^{0.7}$$

$$D_{ec} = \frac{\pi D_{pa}}{6(1 - \varepsilon_{ai})}, \quad G_{ec} = \frac{G_{ai}}{\varepsilon_{ai}}$$

where ε_{ai} is the air volume fraction and the subscript ec refers to effective; D_p: 0.35–4.33 mm.

2. Fluidized bed, bed-to-cylindrical surface heat transfer coefficient during freezing (Sheen and Whitney, 1990):

$$Nu_b = 37.47 Re_b^{0.222} (H_i/H_s)^{0.436} (D_p/H_s)$$

$$Re_b = \frac{H_s u_{ap} \rho_b}{\eta_b}, \quad Nu_b = \frac{h_w H_s}{k_b}$$

$$32{,}000 < Re_b < 83{,}000, \; 0.46 < H_i/H_s < 0.92, \; 0.05 < D_{pa}/H_s < 0.07$$

where
u_{ap} is the superficial air velocity
H_s is the height of the cylindrical heat transfer surface
H_i is the initial height of the bed and the physical properties are those of air

3. Fluidized bed (fluid-to-particle heat transfer coefficient) (Wen and Chang, 1967):

$$j = 0.097 Re_{pa}^{-0.502} Ar^{0.198}$$

$$j = \frac{h_{pa}}{C_g G_g} \left(\frac{C_g \eta_g}{k_g}\right)^{2/3}, \quad Re_{pa} = \frac{D_{pa} G_g}{\eta_g} \text{ at minimum fluidization}, \quad Ar = \frac{g D_{so}^3 \left(\rho_{so} - \rho_g\right)^2}{\eta_g^2}$$

where subscripts so and g refer to solid and gas, respectively.

4. Fluidized bed drying, fluid-to-particle heat transfer coefficient in the case of rotary stream (Shu-De and Fang-Zhen, 1993):

$$Nu = -0.937 + 0.133 Re_T^{0.622} - 0.42 Re_T^{-1.285} \hat{j}^{-0.790}$$

$$Re_T = \frac{D_T u_{ap} \rho_g}{\eta_g}, \quad Nu = \frac{h_{pa} D_{pa}}{k_g}$$

$$12 < Re_T < 189, \; 0.0863 < \hat{j} < 0.17$$

where
$\hat{j}$ is the mass flow ratio of solid to gas
D_T is the inside diameter of the drying chamber
u_{ap} is the gas velocity; inlet gas temperature: 100°C, D_T: 15.0 cm, chamber height: 1.5 m

5. Fluidized bed drying, air-to-particle heat transfer coefficient (Ramirez et al., 1981):

$$Nu = 4.948 \times 10^{-3} Re_{pa}^{1.14} \left(\frac{D_{pa}}{H}\right)^{0.71} \left(\frac{D_{pa}}{D_T}\right)^{-0.94} Pr^{0.33}$$

$$Re_{pa} = \frac{D_{pa} u_{av} \rho_b}{\eta_b}, \quad Nu = \frac{h D_{pa}}{k_b}$$

$$0.28 < Re_{pa} < 3.0, \; 0.007 < D_{pa}/D_T < 0.012, \; 0.0016 < D_{pa}/H < 0.0152, \; 0.39 < H/D_T < 1.49$$

D_{pa}: 0.0125–0.035 cm, minimum fluidization velocity: 0.4113–1.67 cm/s, H: 2–8 cm, airflow rate: 5.3–20.29 L/min.

Table 23.12 (continued) Heat Transfer Correlations for Flow through Fluidized Beds

6. Fluidized bed, total heat transfer coefficient, particle-to-fluid and particle-to-particle (Wen and Chang, 1967):

$$Nu = \frac{0.0529 Re_{pa}^{0.552} Ar^{0.038} Cr^{0.236}}{\varepsilon_g}$$

$$Re_{pa} = \frac{D_{pa} G_g}{\eta_g}, \quad Nu = \frac{h_{pa} D_{pa}}{2k_{so}}, \quad Cr = \frac{C_{so}\rho_{so}}{C_g \rho_g}$$

$$300 < Re_{pa} < 5{,}000,\ 5.86 \times 10^6 < Ar < 2.55 \times 10^7,\ 30.6 < Nu < 89.4$$

where
ρ_g and ρ_{so} are the apparent density of air and grain, respectively
u_a is the superficial air velocity
G_g is the mass flow rate of air
ε_g is the porosity of fluidized bed; D_{pa}: 4.76–6.35 mm, ε_g: 0.425–0.898, G_g: 1.557–16.637 kg/(m$^2 \cdot$ s), air temperature: 52°C–150°C

7. Fluidized bed: total heat transfer coefficient (Shirai et al., 1965):

$$h = 7.4\left(\frac{u}{u_{min}}\right)^{0.36} D_{pa}^{-0.06} D_T^{0.27}$$

where
h is the particle-to-fluid and particle-to-particle heat transfer coefficient
u_{min} is the minimum fluidizing velocity; D_{pa}: 4.50–8.20 mm, D_T: 7.99–28.01 cm, u_{min}: 7.99×10^{-3}–1.05×10^{-1} m/s, u/u_{min}: 2–8

Table 23.13 Heat Transfer Correlations for Mixed Convection

1. Mixed convection from an immersed solid (fluid-to-cylindrical particles heat transfer coefficient during heating in non-Newtonian fluid) (Awuah et al., 1993):

$$Nu_b = 2.45(Gr_b Pr_b)^{0.108}, \quad \text{for carrot}$$

$$Nu_b = 2.02(Gr_b Pr_b)^{0.113}, \quad \text{for potato}$$

$$Nu_b = \frac{hD_{pa}}{k_b}, \quad Re_b = \frac{D_{pa}^n u_{av}^{2-n} \rho_b}{8^{n-1} K\Omega^n}, \quad Pr_b = \frac{C_b K \Omega^n 2^{n-3}}{k_b (D_{pa}/u_{av})^{n-1}}, \quad Gr_b = \frac{g\beta\rho_b^2 D_{pa}^3 (T_b - T_i)}{\left(\frac{K\Omega^n 2^{n-1}}{4u_{av}^{1-n} D^{n-1}}\right)^2}$$

$$3 \times 10^3 < Gr_b Pr_b < 6 \times 10^4$$

where
T_b is the temperature of the bulk fluid
T_i is the initial temperature of the solid; T_b: 50°C–80°C, fluid: CMC (0%–1%), fluid flow parallel to the length of the cylinder (upward and downward), u_{av}: 0.2×10^{-3}–0.7×10^{-3}, D_{pa}: 0.016–0.023 m

2. Mixed convection in a horizontal tube (Colburn, 1933):

$$Nu_b = 1.75 Gz_b \left(1 + 0.015 Gr_a^{1/3}\right)\left(\frac{\eta_b}{\eta_w}\right)^{0.14}$$

$$3.7 \times 10^3 < Gr_a < 3.0 \times 10^8,\ 0.76 < Pr_a < 160$$

This equation has been applied for water, air, and light oil; L/D_T: 24–400, Gr number is calculated for the average conditions between surface and bulk.

3. Flow in horizontal tubes where natural convection is significant (Eubank and Proctor, 1951; McAdams, 1954):

$$Nu_b = 1.75\left\{Gz_b + 12.6\left[Pr_w Gr_w \left(\frac{D_T}{L}\right)^{0.4}\right]^{1/3}\right\}\left(\frac{\eta_b}{\eta_w}\right)^{0.14}$$

$$3.3 \times 10^5 < Gr_w Pr_w < 8.6 \times 10^8,\ 12 < Gz_b < 4{,}900,\ 140 < Pr_w < 15{,}200$$

where subscript w refers to wall; L/D_T: 61–235. This equation has been applied in petroleum oil.

(continued)

Table 23.13 (continued) Heat Transfer Correlations for Mixed Convection

4. Mixed convection in pseudoplastic fluids in horizontal tubes of circular cross section (Metzner and Gluck, 1960):

$$Nu_b = 1.75\left\{ Gz_b + 12.6\left[Pr_w Gr_w\left(\frac{D_T}{L}\right)^{0.4}\right]^{1/3}\right\}\Omega^{1/3}\left(\frac{K_b}{K_w}\right)^{0.14}$$

$$0.32 < n < 0.75,\ Gz_b < 1{,}000$$

This equation has been applied for applesauce, banana puree, ammonium alginate, and carbopol.

5. Mixed convection for non-Newtonian fluid in a horizontal tube (Oliver and Jenson, 1964: Rodriguez-Luna et al., 1987):

$$Nu_b = 1.75\left[Gz_b + 0.0083(Pr_w Gr_w)^{0.75}\right]^{1/3}\left(\frac{K_b}{K_w}\right)^{0.14}$$

6. Mixed convection for Newtonian fluid in a horizontal tube (Jackson et al., 1961; Oliver, 1962):

$$Nu_b = 2.67\left[Gz_b^2 + 0.0087^2(Pr_w Gr_w)^{1.5}\right]^{1/6}$$

based on logarithmic mean temperature between wall and bulk

$$Nu_b = 1.75\left[Gz_b + 0.0083(Pr_w Gr_w)^{0.75}\right]^{1/3}\left(\frac{\eta_b}{\eta_w}\right)^{0.14}$$

based on arithmetic mean temperature between wall and bulk

$$1.57\times10^6 < Gr_w < 3.14\times10^6,\ 33 < Gz_b < 1{,}300,\ Pr_w < 0.71,\ L/D_T = 31$$

7. Mixed convection in horizontal tube (combined heat transfer) (Oliver, 1962):

$$Nu_b = 1.75\left[Gz_b + 5.6\times10^{-4}(Pr_b Gr_b L/D_T)^{0.70}\right]^{1/3}\left(\frac{\eta_b}{\eta_w}\right)^{0.14}$$

based on arithmetic temperature difference between fluid and wall

$1.0\times10^4 < Gr_w < 1.1\times10^5$, $7 < Gz_b < 11$, $1.9 < Pr_w < 3.7$ (water)

$4.9\times10^4 < Gr_w < 1.6\times10^5$, $24 < Gz_b < 187$, $4.8 < Pr_w < 7.0$ (ethyl alcohol)

$29 < Gr_w < 64$, $20 < Gz_b < 176$, $62 < Pr_w < 326$ (glycerol–water)

Fluid properties are calculated at the average inlet and outlet bulk temperature.

8. Mixed convection in a horizontal tube (Brown and Tomas, 1965):

$$Nu_b = 1.75\left[Gz_b + 0.012(Gz_b Gr_b)^{1/3}\right]^{1/3}$$

based on arithmetic mean temperature

$$7.1\times10^4 < Gr_b < 8.9\times10^5,\ 19 < Gz_b < 112,\ 3.5 < Pr_b < 7.4,\ 36 < L/D_T < 108$$

This equation has been applied for water.

9. Mixed convection in a horizontal isothermal tube (Depew and August, 1971):

$$Nu_b = 1.75\left[Gz_b + 0.12\left(Gz_b Gr_b^{1/3} Pr_b^{0.36}\right)^{0.88}\right]^{1/3}\left(\frac{\eta_b}{\eta_w}\right)^{0.14}$$

based on arithmetic mean temperature

$0.7\times10^5 < Gr_w < 5.8\times10^5$, $25 < Gz_b < 338$, $5.7 < Pr_w < 8.0$ (water)

$2.7\times10^5 < Gr_w < 9.9\times10^5$, $36 < Gz_b < 712$, $14.2 < Pr_w < 16.1$ (ethyl alcohol)

$510 < Gr_w < 900$, $53 < Gz_b < 188$, $328 < Pr_w < 391$ (glycerol–water)

$L/D_T = 28.4$

10. Mixed convection for air flowing in a horizontal isothermal tube (Yousef and Tarasuk, 1982):

$$\text{Region I: } Nu_b = 1.75\left[Gz_b + 0.245\left(Gz_b^{1.5} Gr_b^{1/3}\right)^{0.882}\right]^{1/3}\left(\frac{\eta_b}{\eta_w}\right)^{0.14}$$

$$1\times10^4 < Gr_b < 8.7\times10^4,\ 20 < Gz_b < 110,\ 0.0073 < \xi < 0.040$$

Table 23.13 (continued) Heat Transfer Correlations for Mixed Convection

$$\text{Region II: } Nu_b = 0.969 Gz_b^{0.82}\left(\frac{\eta_b}{\eta_w}\right)^{0.14}$$

$$0.8\times10^4 < Gr_b < 4\times10^4,\ 3.2 < Gz_b < 20,\ 0.04 < \xi < 0.25$$

$$\text{Region III: } Nu_b = 2.0$$

$$0.8\times10^4 < Gr_b < 4\times10^4,\ Gz_b < 3.2,\ \xi < 0.25$$

where
$\xi = x/(Re_b Pr_b D_T)$ and x is the distance in x-axis
Region I corresponds to the area near the tube inlet
Region II is further downstream
Region III is the tube area far from its inlet

Heat transfer coefficient and Grashof number are calculated on logarithmic mean temperature difference.

11. Laminar mixed convection for non-Newtonian foods in a horizontal tube (Rodriguez-Luna et al., 1987):

$$Nu_b = 1.75\left[Gz_b + 0.0083(Pr_w Gr_w)^{0.75}\right]^{1/3}\Omega^{1/3}\left(\frac{K_b}{K_w}\right)^{0.14}$$

$$Nu_b = \frac{hD_T}{k_b},\quad Nu_b = \frac{hD_T}{k_b},\quad Pr_w = \frac{C_w\eta_w}{k_w},\quad Gr_w = \frac{g\beta\rho_w^2 D_T^3\Delta T}{\eta_w}$$

$$0.8\times10^4 < Gr_b < 4\times10^4,\ 500 < Gz_b < 7000,\ 500 < Gz_b < 7{,}000,\ 0.18 < n < 0.63,\ T_w = 70°C,$$

$$1.01 < (K_b/K_w)^{0.14} < 1.08,\ 1.06 < \Omega^{1/3} = \left(\frac{3n+1}{4n}\right)^{1/3} < 1.21,\ 18.8 < \frac{L}{D_T} < 126.6$$

where L is the tube length and subscripts b and w refer to bulk and wall conditions, respectively.

12. Combined forced and natural convection in a horizontal cylinder (water flow parallel) (Kitamura et al., 1992):

$$\frac{Nu_m}{Nu_f} = 1.18\left(\frac{Gr_m}{Nu_m Re^2}\right)^{0.14} \text{ and } \frac{Nu_m}{Nu_n} = 1.62\left(\frac{Gr_m}{Nu_m Re^2}\right)^{-0.07}$$

$$Nu_n = 0.586 Ra_m^{0.2},\quad Nu_f = 0.364 Re_a^{0.55} Pr_a^{0.37}$$

$$Nu = \frac{hD_T}{k_b},\quad Gr_m = \frac{g\beta q_s\rho_a^2 D_T^4}{k_a\eta_a^2},\quad Pr_a = \frac{C_a\eta_a}{k_a},\quad Ra_m = Pr_a Gr_m$$

$$10^5 < Ra_m < 10^9,\ 100 < Re_a < 2{,}000,\ 0.3 < \frac{Gr_m}{Nu_m Re_a^2} < 1{,}000$$

13. Mixed convection from a rectangular box to a cold fluid (laminar flow range) (Hanzawa et al., 1991):

$$Nu_b = 9.2 Re_b^{-0.088} Gr_b^{0.14}\left(\frac{H_{bo}}{l}\right)^{-0.60}$$

$$Re_b = \frac{l u_{av}\rho_b}{\eta_b},\quad Nu_b = \frac{hl}{k_b},\quad Gr_b = \frac{g\beta\rho_b^2 l^3(T_s - T_0)}{\eta_b^2}$$

$$100 < Re_b < 500,\ 4\times10^5 < Gr_b < 2\times10^7$$

where
l is a characteristic length (i.e., space between heated surface and box)
H_{bo} is the box height; T_s: 40°C–80°C, T_i: 13°C–21°C, H_{bo}/l: 6.25–15.0, $\varphi = 180°$

Table 23.14 Heat Transfer Correlations for Convection with Phase Change

1. Film condensation outside a horizontal tube (Bird et al., 1960):

$$h = 0.954\left(\frac{gk_{cv}^3\rho_{cv}^2 L}{m\eta_{cv}}\right)^{1/3} \text{ or } h = 0.725\left(\frac{gk_{cv}\lambda_v\rho_{cv}^2}{\eta_{cv}D_T(T_d - T_w)}\right)^{1/4}$$

$$Re = \frac{m}{L\eta_{cv}} < 1{,}000$$

where
T_d is the dew point
m/L is the mass rate of condensation per unit length of tube
λ_v is the latent heat of vaporization, and the subscript cv refers to condensing vapor. Physical properties are calculated at the average film temperature of the condensed vapor

(continued)

Table 23.14 (continued) Heat Transfer Correlations for Convection with Phase Change

2. Film condensation outside a vertical tube or wall (Bird et al., 1960):

$$h = \frac{4}{3}\left(\frac{gk_{cv}^3\rho_{cv}^2}{3\Gamma\eta_{cv}}\right)^{1/3} \quad \text{or} \quad h = \frac{2\sqrt{2}}{3}\left(\frac{gk_{cv}^3\lambda_v\rho_{cv}^2}{\eta_{cv}L(T_d - T_w)}\right)^{1/4}$$

$$\Gamma = \frac{m}{\pi D_T}$$

where Γ is the total rate of condensate flow from the bottom of the condensing surface per unit width of the surface.

This equation is applied for relatively short tubes (up to 15 cm long).

3. Film condensation outside a vertical tube or wall (turbulent flow) (Bird et al., 1960):

$$h = 0.033\left(\frac{gk_{cv}^3\rho_{cv}H(T_d - T_w)}{\lambda_v\eta_{cv}^3}\right)^{1/2} \quad \text{or} \quad h = 0.021\left(\frac{gk_{cv}^3\rho_{cv}^2\Gamma}{\eta_{cv}^3}\right)^{1/3}, \text{ for small } \Delta T$$

4. Nucleate boiling of liquid from a hot surface (case of pool boiling) (Rohsenow, 1951):

$$\frac{C_l(T_s - T_{sv})}{\lambda_v} = \Theta\left(\frac{1}{\eta_l\lambda_v}\frac{\dot{q}}{A_s}\sqrt{\frac{\dot{q}}{A_s}\frac{\sigma}{g(\rho_l - \rho_v)}}\right)^{0.33}\left(\frac{C_l\eta_l}{k_l}\right)^{1.7}$$

where

$\dot{q}$ is the heat flow rate
A_s is the surface area
T_s is the surface temperature
T_{sv} is the saturation temperature of the liquid
σ is the interfacial tension between liquid and vapor
Θ is the Rohsenow constant dependent on fluid and heating surface and is given in Table 24.17. Fluid properties are evaluated at the saturation temperature corresponding to local pressure

It should be noted that the heat flow rate is related to heat transfer coefficient through the equation

$$\dot{q} = hA_s(T_s - T_{sv})$$

Also, the maximum flux in nucleate boiling is given by the equation (Zuber, 1958)

$$\left(\frac{\dot{q}}{A_s}\right)_{max} = \frac{\pi}{24}\lambda_v\rho_v^{1/2}[g\sigma(\rho_l - \rho_v)]^{1/4}\left(1 + \frac{\rho_v}{\rho_l}\right)^{1/2}$$

or by the equation (Rohsenow and Griffith, 1956)

$$\left(\frac{\dot{q}}{A_s}\right)_{max} = 0.0121\rho_v\lambda_v\left(\frac{\hat{g}}{g}\right)^{0.25}\left(\frac{\rho_l - \rho_v}{\rho_v}\right)^{0.6}$$

where $\hat{g}$ is the induced gravitational field such as in the case of centrifuge.

5. Nucleate boiling of liquid from a hot surface (case of pool boiling) (Kutateladze, cited by Jackson and Lamb, 1981):

$$\frac{h}{k_l}\left(\frac{\sigma}{g(\rho_l - \rho_v)}\right)^{1/2} = 0.0007\left(\frac{\dot{q}}{a_l\lambda_v\rho_v}\frac{P}{\sigma}\frac{\sigma}{g(\rho_l - \rho_v)}\right)^{0.7}\left(\frac{C_l\eta_l}{k_l}\right)^{-0.35}$$

where P is the absolute pressure.

6. Nucleate boiling both inside and outside of tubes (Gilmour, 1959):

$$\frac{h}{C_l\hat{G}}\left(\frac{C_l\eta_l}{k_l}\right)^{0.6}\left(\frac{g\sigma\rho_l}{P^2}\right)^{0.425} = 0.001\left(\frac{D_T\hat{G}}{\eta_l}\right)^{-0.3}$$

$$\hat{G} = \frac{m_v\rho_l}{A_s\rho_v}$$

where m_v is the vapor mass flow rate.

7. Nucleate boiling of liquid from a hot surface (Palen and Taborek, 1962):

$$\left(\frac{hD_T}{k_l}\right) = 0.225\left(\frac{C_l\eta_l}{k_l}\right)^{0.69}\left(\frac{\dot{q}D_T}{\eta_l\lambda_v}\right)^{0.69}\left(\frac{PD_T}{\sigma}\right)^{0.31}\left(\frac{\rho_l - \rho_v}{\rho_v}\right)^{0.31}$$

Table 23.14 (continued) Heat Transfer Correlations for Convection with Phase Change

8. Stable film boiling on a submerged horizontal tube (Breen and Westwater, 1962):

$$h\left(\frac{\chi\eta_v\Delta T}{g\vartheta\rho_v k_v^3(\rho_l-\rho_v)}\right)^{1/4}=0.59+0.069\frac{\chi}{D_T}$$

$$\chi=2\pi\left[\frac{\sigma}{g(\rho_l-\rho_v)}\right]^{1/2},\quad \vartheta=\lambda_v\left(1+0.34\frac{C_v}{\lambda_v}\Delta T_v\right)^2$$

where
ΔT_v is the temperature drop across vapor film
χ is the wavelength of the smallest wave growth on a flat horizontal surface

9. Film boiling of liquid nitrogen droplet on the food surface (cryogenic freezing) (Baumeister et al., 1966):

$$h_{av}=1.2\dot{K}V_i^{-1/12}\ \text{for small spherical drops}$$

$$h_{av}=\frac{1.29\ddot{K}^{1/4}V_i^{5/6}-3.89\times10^{-8}\ddot{K}^{1/4}+1.2\dot{K}V_i^{11/12}}{V_i}\ \text{for large drops}$$

$$\dot{K}=\left(\frac{g\rho_l\rho_v\lambda*k_v^3}{\eta_v\Delta T}\right)^{1/4},\quad \ddot{K}=\left(\frac{g^{1/2}\rho_l^{1/2}\rho_v\sigma_l^{1/2}\lambda*k_v^3}{\eta_v\Delta T}\right),\quad \lambda^*=\lambda_v\left(1+\frac{7}{20}\frac{C_v}{\lambda_v}\Delta T\right)$$

where
ΔT is the mean temperature difference between droplet and food surface
D_i is the initial average diameter (mean of major and minor axes of spheroid) of droplet
V_i is the initial volume
h_{av} is the average heat transfer coefficient between droplet and food surface

10. Single nitrogen droplet film boiling on a food surface (Awonorin and Lamb, 1988):

$$h_{av}=13.5\frac{k_v}{D_i}\left(\frac{\eta_v^2C_v\lambda_v}{k_v^2\Delta T}\right)^{0.17}\left(\frac{g\rho_l\rho_vC_vD_i^3}{\eta_vk_v}\right)^{0.25}\left(\frac{\rho_v}{\rho_l}\right)^{0.33}\left(\frac{\eta_vC_v}{k_v}\right)^{1.75}$$

$$0.1<D_i<2.5\ \text{mm},\ 360<\rho_l/\rho_v<668,\ 0.9<\left(\frac{k_v^2\Delta T}{\eta_v^2C_v\lambda_v}\right)<2.2$$

Table 23.15 Heat Transfer Correlations for Evaporation

1. Evaporating liquid films from a surface of water falling films flowing along the outside surface of a vertical tube (Chun and Seban, 1971):

$$h=\left(\frac{4}{3}\right)^{1/3}\left(\frac{gk_l^3}{\nu_l^2}\right)^{1/3}\left(\frac{4\Gamma}{\eta_l}\right)^{-1/3},\quad \text{laminar region}$$

$$h=0.606\left(\frac{gk_l^3}{\nu_l^2}\right)^{1/3}\left(\frac{\Gamma}{\eta_l}\right)^{-0.22},\quad \text{laminar region when surface ripples exist}$$

$$h=\left(\frac{\nu_l^2}{gk_l^3}\right)^{1/3}=3.8\times10^{-3}\left(\frac{4\Gamma}{\eta_l}\right)^{0.4}\left(\frac{\nu_l}{a_l}\right)^{0.65},\quad \text{turbulent regime}$$

$$h=11.4\times10^{-3}\left(\frac{\Gamma}{\eta_l}\right)^{0.4},\quad \text{for Prandtl number of 5 and a high}$$

Reynolds number

$$\Gamma=\frac{g\rho_l}{\nu_l}\frac{\delta^3}{3}$$

where
δ is the film thickness
ν_l is the kinematic viscosity of the liquid
α_l is the thermal diffusivity of the liquid

These properties are given by the equations

(continued)

Table 23.15 (continued) Heat Transfer Correlations for Evaporation

$$\nu_l = \eta_l/\rho_l, \quad a_l = k_l/\rho_l C_l$$

Γ expresses the mass flow rate per unit width of the wall.

Laminar-layer surface presents capillary waves when (Kapitza, cited by Dukler, 1960)

$$\left(\frac{4\Gamma}{\eta_l}\right) = 2.43\left(\frac{g\eta_l^4}{\sigma^3\rho_l}\right)^{-1/11}$$

The separation of the laminar and turbulent regime starts at the wavy laminar regime at Reynolds number

$$\left(\frac{4\Gamma}{\eta_l}\right) = 5800\left(\frac{\nu_l}{a_l}\right)^{-1.06} = 0.215\left(\frac{g\eta_l^4}{\sigma^3\rho_l}\right)^{-1/3}$$

2. Climbing film evaporator (Jackson and Lamb, 1981):

$$\frac{h_{mix}}{h_l} = \ddot{c}\left(\frac{1}{X}\right)^{\dot{n}}$$

$$X = \left(\frac{\rho_v}{\rho_l}\right)^{0.5}\left(\frac{\eta_l}{\eta_v}\right)^{0.1}\left(\frac{1-x'}{x'}\right)^{0.90}$$

$$2.17 < \ddot{c} < 7.55, \; 0.328 < \dot{n} < 0.75$$

where
- h_l is the heat transfer coefficient of a single liquid flow at the same condition (given by Dittus–Boelter equation for single flow of liquid)
- $\dot{c}$ and $\dot{n}$ are constants
- X is the Lockhart–Martinnelli parameter, and x' is the mass fraction of vapor in the mixture

3. Climbing film evaporator (Bourgois and Le Maguer, 1983a,b, 1987):

$$Nu_z = \frac{Pe_i}{4}(\hat{B} + 2\hat{C}Z), \quad \text{liquid zone}$$

$$Nu_z = 8.5Re_a^{0.2}Pr_a^{1/3}S^{2/3}, \quad \text{boiling zone}$$

$$h_z = 615.5 - 112.0Z + 41.55Z^2, \quad \text{boiling zone}$$

$$\hat{B} = -0.458 + \frac{9{,}445}{Pe_i} + \frac{5.22 \times 10^6}{Pe_i^2}$$

$$\hat{C} = 0.238 + \frac{755}{Pe_i} + \frac{1.350 \times 10^9}{Pe_i^2}$$

$$Z = z/H_T$$

$$Nu_z = \frac{h_z H_T}{k_i}, \quad Pe_i = Re_i Pr_i, \quad Re_i = \frac{4\Gamma}{\pi\eta_i D_T}, \quad Pr_i = \frac{\eta_i C_i}{k_i}, \quad \text{liquid zone}$$

$$Nu_z = \frac{h_z z}{k_l}, \quad Re_a = \frac{z u_l \rho_l}{\eta_l}, \quad Pr_a = \frac{\eta_l C_l}{k_l}, \quad S = \frac{u_v}{u_l}, \quad \text{boiling zone}$$

where
- H_T is the height and D_T is the diameter of the tube
- z is the local distance in the tube
- S is the slip ratio, and subscripts i and l refer to inlet conditions of liquid and liquid, respectively

System at atmospheric pressure, steam at 1.357×10^5 Pa absolute, liquid flow rate ranging from 0.692×10^{-2} to 1.30×10^{-2} m/s. All properties of the liquid are calculated at the average of bulk and wall temperatures at the local position z.

Nu_z expression for the boiling zone is valid when

$$u_v^* = \frac{u_v\sqrt{\rho_v}}{\sqrt{gD_T(\rho_l - \rho_v)}} < 2.5$$

where
- u_v is the volumetric vapor flux
- u_v^* is the modified volumetric vapor flux

These equations have been applied for sucrose solutions and tomato juice.

Table 23.16 Heat Transfer Correlations for Dryers

1. Thin layer air-drying (McCabe and Smith, 1976):

$$h = 0.0204G_{ai}^{0.8}, \quad \text{parallel flow}$$

$$h = 1.17G_{ai}^{0.37}, \quad \text{perpendicular flow}$$

   where G_{ai} is the air mass velocity.
2. Drum drying (evaporating vapor film heat transfer coefficient) (Hougen, 1940):

$$h_v = 8.484 \times 10^{-8}u_{dr}^{0.8}\lambda_v\Delta P/\Delta T$$

   where
   u_{dr} is the drum speed
   ΔP is the difference between saturation vapor pressure and vapor pressure of air
   ΔT is the vapor film temperature difference
3. Spray drying (air film heat transfer coefficient of the droplet in hot air) (Ranz and Marshall, 1952):

$$Nu = 1.6\left(1 + 0.3Pr^{1/3}Re^{1/2}\right)$$

$$Re = \frac{D_l u_{re}\rho_{ai}}{\eta_{ai}}, \quad Nu = \frac{h_{ai}D_l}{k_{ai}}, \quad Pr = \frac{C_{ai}\eta_{ai}}{k_{ai}}$$

   where
   D_l is the droplet diameter
   u_{re} is the relative velocity of droplet and air
4. Tunnel-based convection oven (Mureau and Barreteau, 1981):

$$Nu = 4.5Re^{0.27}, \quad Re < 40$$

$$Nu = 0.70Re^{0.61}, \quad Re > 40$$

$$Re = \frac{Lu_{av}\rho_b}{\eta_b}, \quad Nu = \frac{hL}{k_b}$$

   where
   L is the width of the tunnel
   u_{av} is the average air velocity

Table 23.17 Values of Rohsenow Constant

Liquid	Surface	Θ	Reference
Water	Platinum wire	0.0130	Rohsenow (1952)
Carbon tetrachloride	Emery-polished copper	0.0070	Vachon et al. (1968)
Water	Brass tube	0.0060	Rohsenow (1952)
Water	Paraffin-treated copper	0.0147	Vachon et al. (1968)
Water	Emery-polished copper	0.0128	Vachon et al. (1968)
Water	Scored copper	0.0068	Vachon et al. (1968)
Water	Ground and polished stainless steel	0.0080	Vachon et al. (1968)
Water	Teflon-pitted stainless steel	0.0058	Vachon et al. (1968)
Water	Chemically etched stainless steel	0.0133	Vachon et al. (1968)
Water	Mechanically polished stainless steel	0.0132	Vachon et al. (1968)
Ethyl alcohol	Polished-plated chromium	0.0027	Rohsenow (1952)
Benzene	Polished-plated chromium	0.0100	Rohsenow (1952)
n-Pentane	Lapped copper	0.0049	Vachon et al. (1968)
n-Pentane	Emery rubber copper	0.0074	Vachon et al. (1968)
n-Pentane	Polished-plated chromium	0.0150	Rohsenow (1952)
n-Pentane	Emery-polished copper	0.0154	Vachon et al. (1968)
n-Pentane	Emery-polished nickel	0.0127	Vachon et al. (1968)

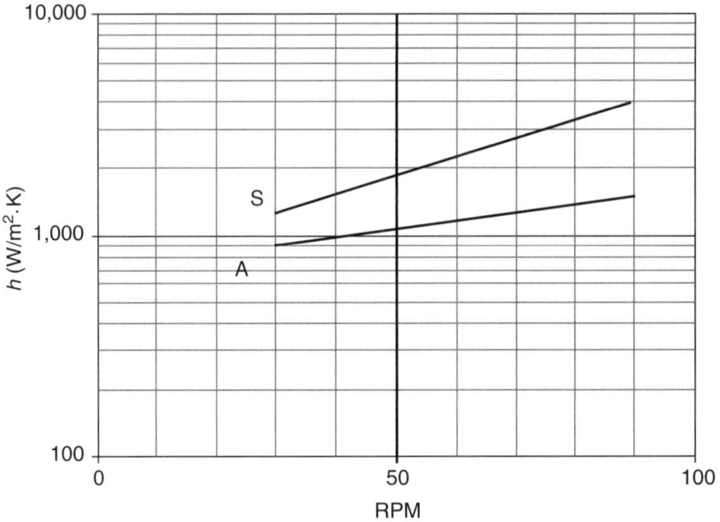

FIGURE 23.2 Heat transfer coefficients in agitated kettle. S, sucrose solution 40°Brix; A, applesauce; rpm, 1/min. (From Saravacos, G.D. and Maroulis, Z.B., *Transport Properties of Foods*, Marcel Dekker, New York, 2001.)

The shear rate $\dot{\gamma}$ for the pilot-scale-agitated kettle, described in this reference (0.40 m diameter, anchor agitator), was calculated from the empirical relation $\dot{\gamma} = 13N$. The heat transfer coefficients h at the internal interface of the vessel for a sugar solution and for applesauce increased linearly with the speed of agitation (rpm), as shown in Figure 23.2.

Besides the correlations presented above, there exist a few simplified dimensional equations, applicable to specific equipment geometries and system conditions, for the estimation of heat transfer coefficient of air and water in some important operations (Perry and Green, 1984; Geankoplis, 1993). These are presented in Table 23.18. In Table 23.18, ΔT is the temperature difference (°C), D_0 is the outside diameter (m), L is the length (m), G is the mass flow rate (kg/m^2·s), Γ is the irrigation flow rate of the films (kg/m·s), and N is the number of horizontal tubes in a vertical plane.

In the case of evaporation of fluid foods, heat transfer controls the evaporation rate and high heat transfer coefficients are essential in the various types of equipment. Prediction of the heat transfer coefficients in evaporators is difficult, and experimental values of the overall heat transfer coefficient U are used in practical applications. Equation 23.4 can be written as the following expression:

$$\frac{1}{U} = \frac{1}{h_i} + \frac{x}{k} + \frac{1}{h_0} + \mathrm{FR} \tag{23.7}$$

Table 23.18 Simplified Heat Transfer Coefficient Equations for Air and Water

Natural convection of air:
Horizontal tubes, $h = 1.42\ (\Delta T/D_0)^{1/4}$
Vertical tubes, $h = 1.42\ (\Delta T/L)^{1/4}$
Falling films of water:
$h = 9150\ \Gamma^{1/3}$
Condensing water vapors:
Horizontal tubes, $h = 10{,}800/[(N_b D_0)^{1/4}(\Delta T)^{1/3}]$
Vertical tubes, $h = 13{,}900/[L^{1/4}\ (\Delta T)^{1/4}]$

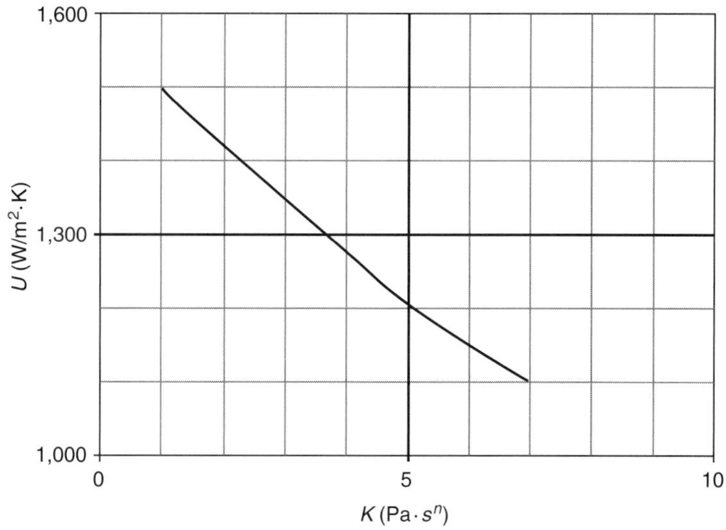

FIGURE 23.3 Overall heat transfer coefficient U of fruit purees in agitated kettle. (From Saravacos, G.D. and Maroulis, Z.B., *Transport Properties of Foods*, Marcel Dekker, New York, 2001.)

where FR is the fouling resistance that becomes important in the evaporation of liquid foods containing colloids and suspensions, which tend to deposit on the evaporator walls, reducing significantly the heat transfer rate. Falling film evaporators are used extensively in the concentration of fruit juices and other liquid foods because they are simple in construction and they achieve high heat transfer coefficients. Figure 23.3 shows the overall heat transfer coefficient for an agitated vessel. Figure 23.4 illustrates overall heat transfer coefficients U for apple juices in a pilot plant falling film evaporator, 3 m long and 5 cm diameter tube (Saravacos and Moyer, 1970).

Higher U values were obtained in the evaporation of depectinized (clarified) apple juice (1200–2000 $W/m^2 \cdot K$) than the unfiltered (cloudy) juice, which tended to foul the heat transfer surface as the concentration was increased. The U value for water, under the same conditions, was

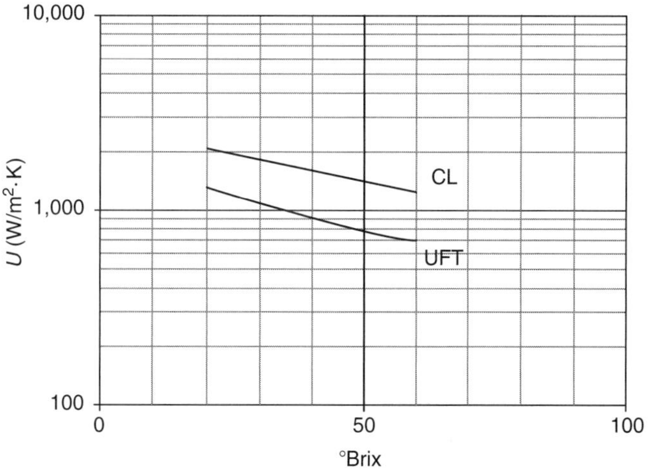

FIGURE 23.4 Overall heat transfer coefficients U in evaporation of clarified (CL) and unfiltered (UFT) apple juice at 55°C. (From Saravacos, G.D. and Maroulis, Z.B., *Transport Properties of Foods*, Marcel Dekker, New York, 2001.)

higher as expected and equal to 2300 $W/(m^2 \cdot K)$. Jet impingement ovens and freezers operate at high heat transfer rates due to the high air velocities at the air–food interface. Heat transfer coefficients of 250–350 W/m^2 K can be obtained in ovens, baking cookies, crackers, and cereals. Ultrasounds can substantially improve the air-drying rate of porous foods like apples (acoustically assisted drying). Ultrasound of 155–163 db increased the moisture diffusivity at 60°C from 7×10^{-10} to 14×10^{-10} m^2/s (Mulet et al., 1999).

23.5 HEAT TRANSFER FACTOR IN FOOD PROCESSES

Saravacos and Maroulis (2001) and Krokida et al. (2001) retrieved a significant number of recently reported heat transfer coefficient data in food processing from the literature (including the following journals: *Drying Technology*, *Journal of Food Science*, *International Journal for Food Science and Technology*, *Journal of Food Engineering*, *Transactions of the ASAE*, and *International Journal of Food Properties*).

The collected data refer to seven different processes presented in Table 23.19 and include about 40 food materials in Table 23.20. Most of the data were available in the form of empirical equations using dimensionless numbers. All available empirical equations were transformed in the form of heat transfer factor versus Reynolds number given by the equation

$$j_H = aRe^{n'} \tag{23.8}$$

The thermophysical properties of the materials and characteristics of the systems were taken into account to achieve this transformation (Krokida et al., 2002b). This equation was also fitted to all data for each process and the resulting equations characterize the process, since they are based on the data from all available materials. The results are classified based on the process and material and are

Table 23.19 Number of Available Equations for Each Food Process

Process	No. of Equations
Baking	
Forced convection	1
Blanching	
Steam	1
Cooling	
Forced convection	9
Drying	
Convective	16
Fluidized bed	1
Rotary	4
Freezing	
Forced convection	6
Storage	
Forced convection	4
Sterilization	
Aseptic	9
Retort	3
Total no. of equations	54

Table 23.20 Number of Available Equations for Each Food Material

Material	No. of Equations
Apples	1
Apricots	1
Barley	2
Beef	1
Cakes	1
Calcium alginate gel	1
Canola seeds	1
Carrot	1
Corn	2
Corn starch	1
Figs	1
Fish	1
Grapes	3
Green beans	1
Hamburger	2
Maize	1
Malt	1
Meat carcass	1
Model food	4
Newtonian liquids	1
Nonfood material	3
Particulate liquid foods	3
Peaches	1
Potatoes	2
Raspberries	1
Rice	1
Soya	2
Soybean	1
Strawberries	1
Sugar	1
Wheat	3
Spherical particles	1
Tomatoes	1
Corn cream	1
Rapeseed	1
Meatballs	1
Total no. of equations	54

presented in Table 23.21. Heat transfer coefficient values for process design can be obtained easily from the proposed equations and graphs. The range of variation in this uncertain coefficient can also be obtained to carry out valuable process sensitivity analysis. Estimations for materials, not included in the data, can also be made using similar materials or average values. It is expected that the resulting equations are more representative and predict the heat transfer coefficients more accurately. The results of fitting the equation to all data, for each process, are summarized in Table 23.22.

Table 23.21 Parameters of the Equation $j_H = aRe^{n'}$ for Each Process and Each Material

Process/Product (Reference)	A	n'	Min *Re*	Max *Re*
Baking				
Cakes (Baik et al., 1999)	0.801	−0.390	40	3,000
Blanching				
Green beans (Zhang and Cavalieri, 1991)	0.00850	−0.443	150	1,500
Cooling				
Apples (Fikiin et al., 1999)	0.0304	−0.286	4,000	48,000
Apricots (Fikiin et al., 1999)	0.114	−0.440	2,000	25,000
Figs (Dincer, 1995)	8.39	−0.492	3,500	9,000
Grapes (Fikiin et al., 1999)	0.472	−0.516	1,300	17,000
Model food (Alvarez and Flick, 1999)	2.93	−0.569	2,000	12,000
Peaches (Fikiin et al., 1999)	0.186	−0.500	3,700	43,000
Raspberries (Fikiin et al., 1999)	0.0293	−0.320	1,300	16,000
Strawberries (Fikiin et al., 1999)	0.136	−0.440	1,900	25,000
Tomatoes (Dincer, 1997)	0.267	−0.550	1,000	24,000
Drying				
Convective				
Barley (Sokhansanj, 1987)	3.26	−0.650	20	1,000
Canola seeds (Lang et al., 1996)	0.458	−0.241	30	50
Carrot (Mulet et al., 1989)	0.692	−0.486	500	5,000
Corn (Fortes and Okos, 1981)	1.06	−0.566	400	1,100
Corn (Torrez et al., 1998)	4.12	−0.650	20	1,000
Grapes (Ghiaus et al., 1997)	0.665	−0.500	8	50
Grapes (Vagenas et al., 1990)	0.741	−0.430	1,000	3,000
Maize (Mourad et al., 1997)	11.9	−0.901	150	1,500
Malt (Lopez et al., 1997)	0.196	−0.185	60	80
Potatoes (Wang and Brennan, 1995)	0.224	−0.200	2,000	11,000
Rice (Torrez et al., 1998)	4.12	−0.650	20	1,000
Soybean (Taranto et al., 1997)	2.48	−0.523	200	1,500
Wheat (Lang et al., 1996)	149	−0.340	50	100
Wheat (Sokhansanj, 1987)	3.26	−0.650	20	1,000
Fluidized bed				
Corn starch (Shu-De et al., 1993)	0.101	−0.355	3,200	13,000
Rotary				
Fish (Shene et al., 1996)	0.00160	−0.258	80	300
Soya (Alvarez and Shene, 1994)	0.00960	−0.587	10	100
Soya (Shene et al., 1996)	0.000300	−0.258	20	80
Sugar (Wang et al., 1993)	0.805	−0.528	1,500	17,000
Freezing				
Beef (Heldman, 1980)	0.650	−0.418	80	25,000
Calcium alginate gel (Sheng, 1994)	48.6	−0.535	300	600
Hamburger (Flores et al., 1988)	8.87	−0.672	7,500	150,000
Hamburger (Tocci and Mascheroni, 1995)	4.67	−0.645	9,000	73,000
Meat carcass (Mallikarjunan and Mittal, 1994)	0.228	−0.269	1,800	20,000
Meatballs (Tocci et al., 1995)	0.536	−0.485	3,400	28,000

Table 23.21 (continued) Parameters of the Equation $j_H = aRe^{n'}$ for Each Process and Each Material

Process/Product (Reference)	*A*	n′	Min *Re*	Max *Re*
Storage				
Potatoes (Xu and Burfoot, 1999)	0.658	−0.425	70	90
Wheat (Chang et al., 1993)	0.0136	−0.196	1,500	10,000
Sterilization				
Aseptic				
Model food (Balasubramaniam and Sastry, 1994)	0.500	−0.507	5,000	20,000
Model food (Sastry et al., 1990)	0.448	−0.519	2,400	45,000
Model food (Zuritz et al., 1990)	3.42	−0.687	2,000	11,000
Non-food material (Kramers, 1946)	0.748	−0.512	3,000	85,000
Non-food material (Ranz et al., 1952)	0.662	−0.508	3,000	85,000
Non-food material (Whitaker, 1972)	0.517	−0.441	3,000	85,000
Particulate liquid foods (Mankad et al., 1997)	0.225	−0.400	140	1,500
Particulate liquid foods (Sannervik et al., 1996)	0.0493	−0.199	1,800	5,200
Spherical particles (Astrom and Bark, 1994)	2.26	−0.474	4,300	13,000
Retort				
Newtonian liquids (Anantheswaran et al., 1985a)	2.74	−0.562	11,000	400,000
Particulate liquid foods (Sablani et al., 1997)	0.564	−0.403	30	1,600
Corn cream (Zaman et al., 1991)	0.108	−0.343	130,000	1,100,000

Source: Adapted from Saravacos, G.D. and Maroulis, Z.B., *Transport Properties of Foods*, Marcel Dekker, New York, 2001.

All the equations referred to in Table 23.19 are presented in Figure 23.5 to define the range of variation in the heat transfer factor j_H versus Reynolds number (*Re*). The range of variation by process is sketched in Figure 23.6. The data of Tables 23.21 and 23.22 demonstrate the importance of the flow conditions (Reynolds number, *Re*) and the type of food process and product on the heat transfer characteristics (heat transfer factor, j_H). As expected from theoretical considerations and experience in other fields, the heat transfer factor, j_H, decreases with a negative exponent of about −0.5 of the *Re*. The highest heat transfer factor values are obtained in drying and baking operations,

Table 23.22 Parameters of the Equation $j_H = aRe^{n'}$ for Each Process

Process	*A*	*n′*	Min *Re*	Max *Re*
Baking	0.80	−0.390	40	3,000
Blanching	0.0085	−0.443	150	1,500
Cooling	0.143	−0.455	1,000	48,000
Drying/convective	1.04	−0.455	8	11,000
Drying/fluidized bed	0.10	−0.354	3,200	13,000
Drying/rotary	0.001	−0.161	10	300
Freezing	1.00	−0.486	80	150,000
Storage	0.259	−0.387	70	10,000
Sterilization/aseptic	0.357	−0.450	140	45,000
Sterilization/retort	1.034	−0.499	30	110,000

Source: Saravacos, G.D. and Maroulis, Z.B., *Transport Properties of Foods*, Marcel Dekker, New York, 2001.

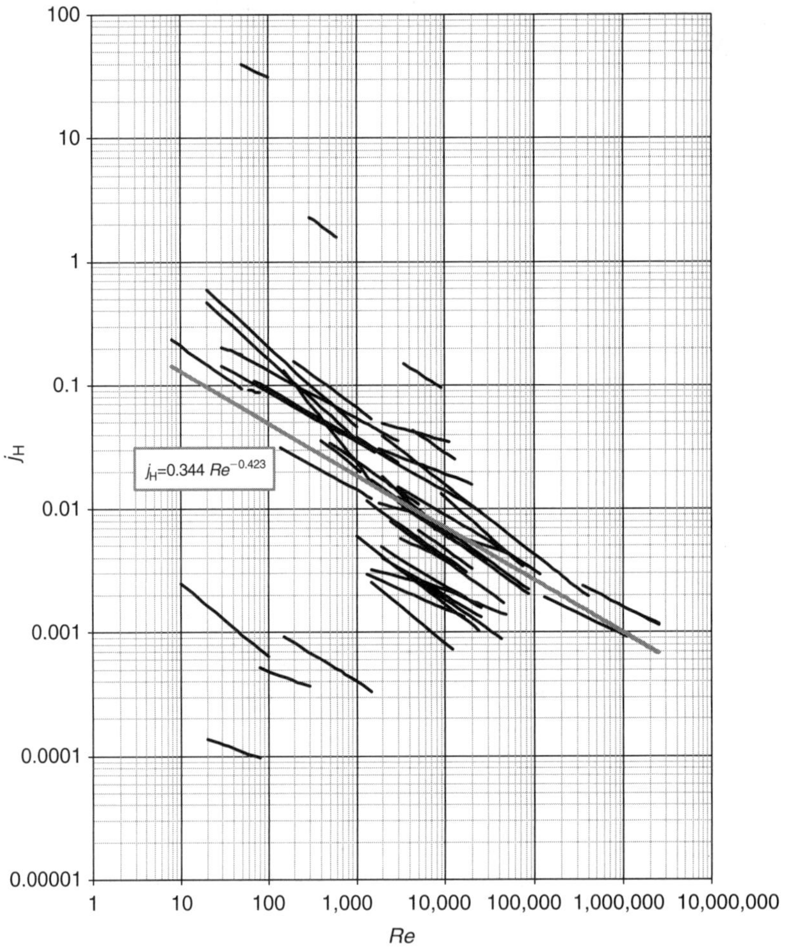

FIGURE 23.5 Heat transfer factor versus Reynolds number for all the examined processes and materials. (From Saravacos, G.D. and Maroulis, Z.B., *Transport Properties of Foods*, Marcel Dekker, New York, 2001.)

while the lowest values are in cooling and blanching. Granular food materials, such as corn and wheat, appear to have better heat transfer characteristics than large fruits, e.g., apples (Table 23.21).

In drying, heat transfer takes place in a variety of mechanisms, and the heat transfer factor has been calculated for different dryer types, separately (Krokida et al., 2002a). The general presentation of Figure 23.6 for convective drying has been analyzed for the following drying methods:

- Flat plate drying
- Fluidized bed drying
- Packed bed drying
- Rotary drying

Figures 23.7 through 23.10 illustrate the heat transfer factor for each of the above drying methods. The effect of food material is obvious in these diagrams. The range of variation per dryer type is sketched in Figure 23.11.

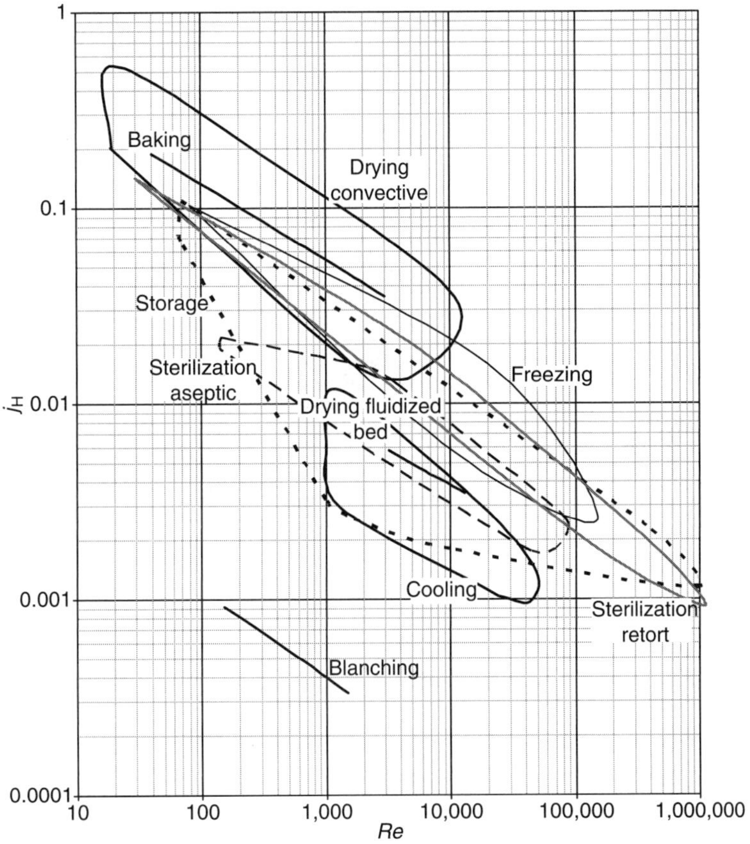

FIGURE 23.6 Ranges of variation in the heat transfer factor versus Reynolds number for all the examined processes. (From Saravacos, G.D. and Maroulis, Z.B., *Transport Properties of Foods*, Marcel Dekker, New York, 2001.)

23.6 PREDICTION OF HEAT TRANSFER COEFFICIENT USING MASS TRANSFER DATA

One of the fundamental theories in transport phenomena is the analogy in momentum, heat, and mass transfer. Due to this analogy, data of any type of transfer are equally useful in the design and operation of thermal processes, as they may be used for the evaluation of the heat/mass transfer mechanisms and the estimation of heat transfer coefficient (or vice versa). Chilton and Colburn quantified this analogy proposing the well-established Chilton–Colburn (or Colburn) analogy, expressed by the following equation:

$$j_H = j_M = \frac{f}{2} \tag{23.9}$$

where

j_H is the heat transfer factor
j_M is the mass transfer factor
f is the Fanning friction factor

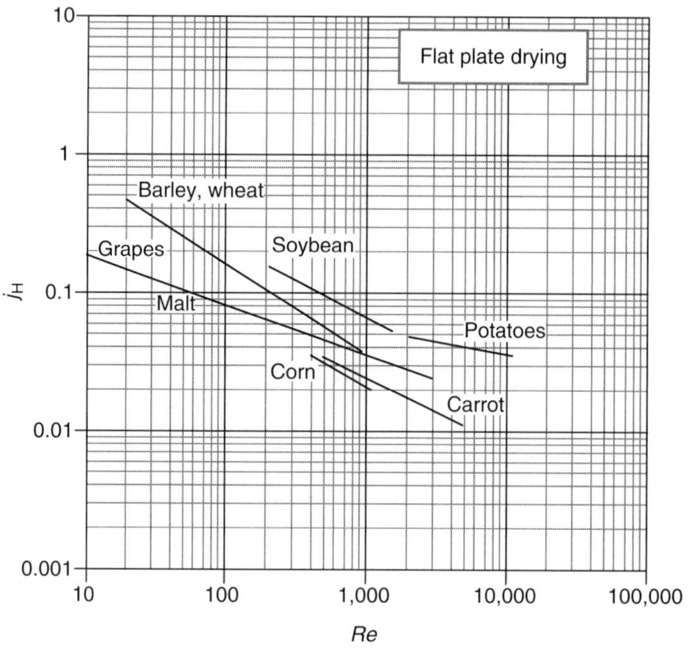

FIGURE 23.7 Heat transfer factor versus Reynolds number for flat plate drying and various materials.

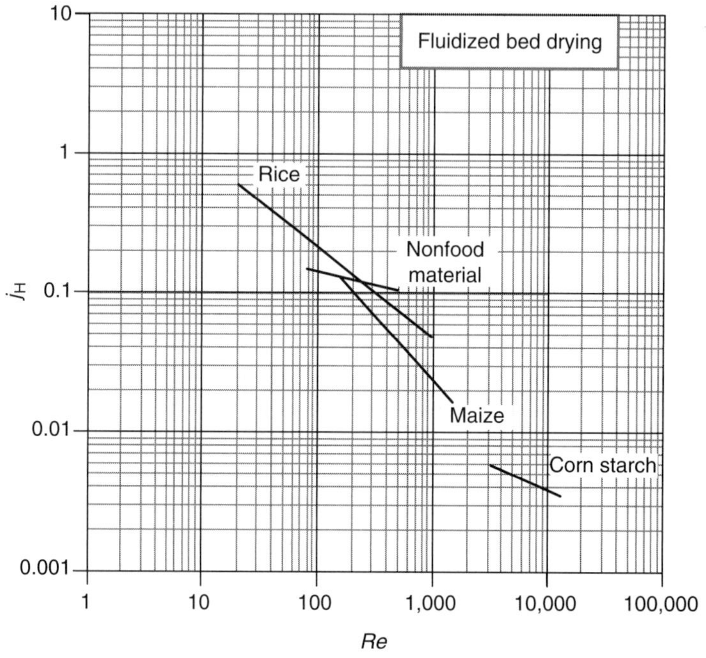

FIGURE 23.8 Heat transfer factor versus Reynolds number for fluidized bed drying and various materials.

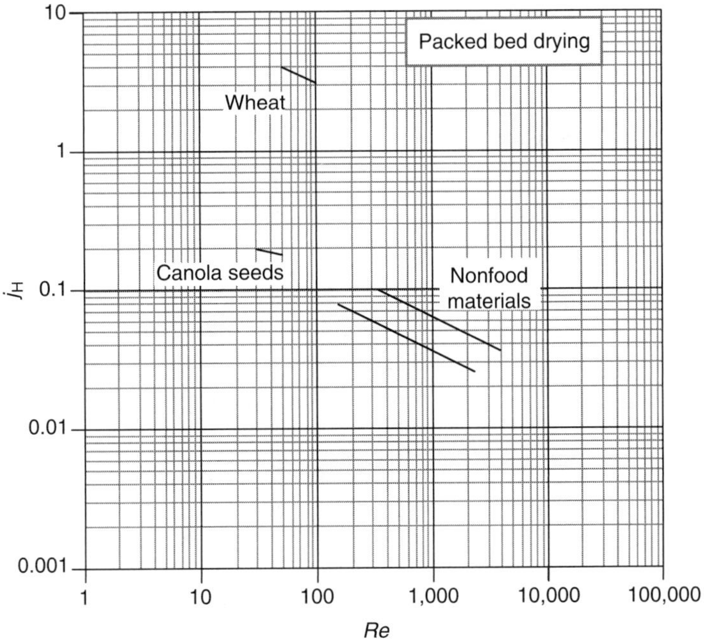

FIGURE 23.9 Heat transfer factor versus Reynolds number for packed bed drying and various materials.

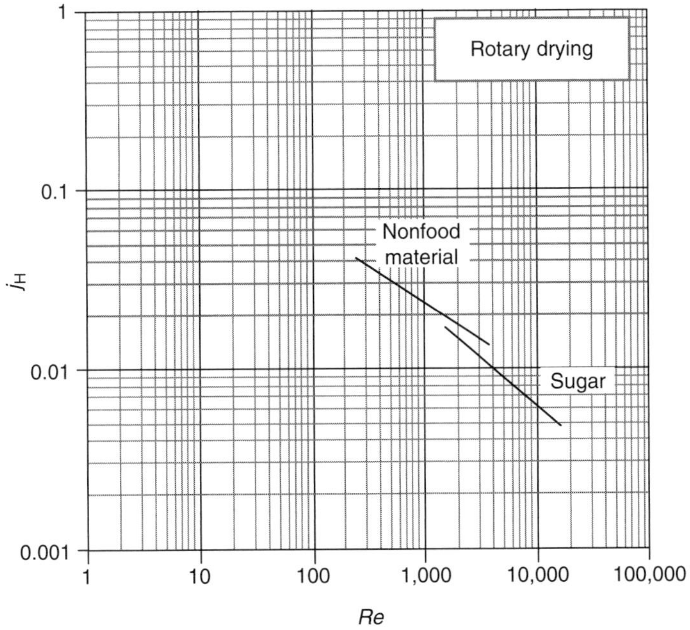

FIGURE 23.10 Heat transfer factor versus Reynolds number for rotary drying and various materials.

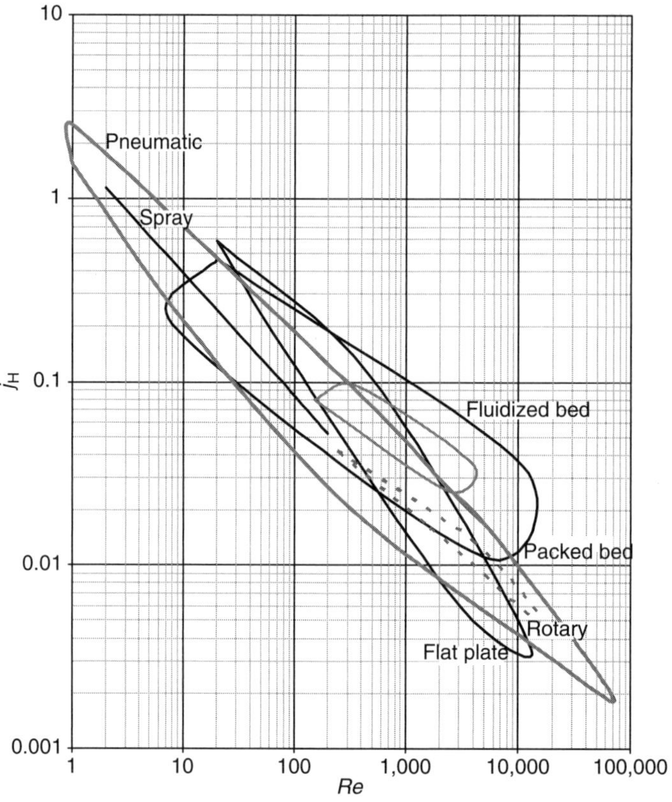

FIGURE 23.11 Ranges of variation in the heat transfer factor versus Reynolds number for all the examined drying methods.

Analytically, heat and mass transfer factors are given by the equations

$$j_{\mathrm{H}} = St_{\mathrm{H}} Pr^{2/3} \tag{23.10}$$

$$j_{\mathrm{M}} = St_{\mathrm{M}} Sc^{2/3} \tag{23.11}$$

$$St_{\mathrm{H}} = \frac{h}{u\rho C} \tag{23.12}$$

$$St_{\mathrm{M}} = \frac{h_{\mathrm{M}}}{u\rho} = \frac{k_{\mathrm{C}}}{u} \tag{23.13}$$

$$Pr = \frac{\eta C}{k} \tag{23.14}$$

where

h is the heat transfer coefficient
h_{M} is the mass transfer coefficient
k_{C} is the mass transfer coefficient expressed in concentration units
u is the average (bulk) fluid velocity
ρ is the density of the fluid
η is the dynamic viscosity of the fluid
C is the specific heat of the fluid
k is the thermal conductivity of the fluid

All fluid properties refer to bulk temperature. The Chilton–Colburn analogy in air–water mixtures (applications in drying and air conditioning) is simplified when the Pr and Sc are approximately equal ($Pr \cong Sc \cong 0.8$). Therefore, Equation 23.9, by substituting Equations 23.10 and 23.11, becomes

$$St_{\mathrm{H}} = St_{\mathrm{M}} \quad \text{or} \quad \frac{h}{u\rho C} = \frac{k_{\mathrm{C}}}{u} = \frac{h_{\mathrm{M}}}{u\rho} \quad \text{or} \quad \frac{h}{\rho C} = k_{\mathrm{C}} \tag{23.15}$$

In terms of the mass transfer coefficient h_{M}, the above relationship gives

$$h_{\mathrm{M}} = \frac{h}{C} \tag{23.16}$$

The specific heat of atmospheric air at ambient conditions is approximately $C = 1000\ \mathrm{J/(kg \cdot K)}$. Therefore, Equation 16 yields $h = 1000\ h_{\mathrm{M}}$, where h is in $\mathrm{W/(m^2 \cdot K)}$ and h_{M} in $\mathrm{kg/(m^2 \cdot s)}$. If the units of h_{M} are taken as $\mathrm{g/(m^2 \cdot s)}$, the above relationship is written as (Saravacos, 1997)

$$h(\mathrm{W/m^2\ K}) \cong h_{\mathrm{M}}(\mathrm{g/m^2\ s}), \quad \text{for atmospheric air}$$

A similar relationship is obtained between the coefficients h and k_{C}

$$h(\mathrm{W/m^2\ K}) \cong k_{\mathrm{C}}\ (\mathrm{mm/s}), \quad \text{for atmospheric air}$$

Krokida et al. (2001) retrieved recently reported mass transfer coefficient data in food processing from the literature (which include materials such as corn, grapes, maize, meat, model food, potatoes, rice, carrots, and milk) and, following the same procedure as for heat transfer coefficient data, transformed them in the form of mass transfer factor versus Reynolds number given by the equation

$$j_{\mathrm{M}} = aRe^{n'} \tag{23.17}$$

The results are classified based on the process and material, which are presented in Tables 23.23 and 23.24. All the equations used are presented in Figure 23.12 to define the range of variation in

Table 23.23 Parameters of the Equation $j_{\mathrm{M}} = aRe^{n'}$ for Each Process and Each Material

Process/Product (Reference)	*A*	*n′*	Min *Re*	Max *Re*
Drying				
Convective				
Corn (Torrez et al., 1998)	5.15	−0.575	20	1,000
Grapes (Ghiaus et al., 1997)	0.004	−0.462	10	40
Grapes (Vagenas et al., 1990)	0.741	−0.430	900	3,000
Maize (Mourad et al., 1997)	34.6	−1.000	5	15
Rice (Torrez et al., 1998)	5.15	−0.575	20	1,000
Carrot (Mulet et al., 1987)	0.69	−0.486	500	5,000
Spray				
Milk (Straatsma et al., 1999)	2.947	−0.890	1	2
Freezing				
Meat (Tocci et al., 1995)	2.496	−0.495	2500	70,000
Storage				
Potatoes (Xu and Burfoot, 1999)	0.667	−0.428	50	55
Sterilization				
Model food (Fu et al., 1998)	11.220	−1.039	6500	26,000

Source: Saravacos, G.D. and Maroulis, Z.B., *Transport Properties of Foods*, Marcel Dekker, New York, 2001.

Table 23.24 Parameters of the Equation $j_M = aRe^{n'}$ for Each Process

Process	*A*	*n'*	Min *Re*	Max *Re*
Drying/convective	23.5	−0.882	5	5,000
Drying/spray	2.95	−0.889	1	2
Freezing	0.10	−0.268	2,500	70,000
Storage	0.67	−0.427	50	55
Sterilization	11.2	−1.039	6,500	26,000

Source: Saravacos, G.D. and Maroulis, Z.B., *Transport Properties of Foods*, Marcel Dekker, New York, 2001.

the mass transfer factor j_M versus Reynolds number, *Re*. The range of variation by process is sketched in Figure 23.13. Especially for drying processes, the available mass transfer data have been classified according to the drying method and the corresponding equipment (Krokida et al., 2002a,b). Figure 23.14 presents the mass transfer factor versus Reynolds number for all the examined drying types and materials, while Figure 23.15 illustrates the ranges of variation in the mass transfer factor versus Reynolds number for all the examined drying types. Figure 23.16 shows the effect of Reynolds number to mass transfer factor in the case of fluidized bed drying for all the examined materials.

Regression analysis of published mass transfer data shows the similarity between the heat transfer factor j_H and the mass transfer factor j_M. Mass transfer coefficient values for process design can be obtained from the proposed equations and graphs. The range of variation in both heat and mass transfer coefficients, which generally presents a degree of uncertainty, can also be obtained from the presented data to carry out valuable process sensitivity analysis. Estimations for materials not included in the data can also be made using similar materials or average values.

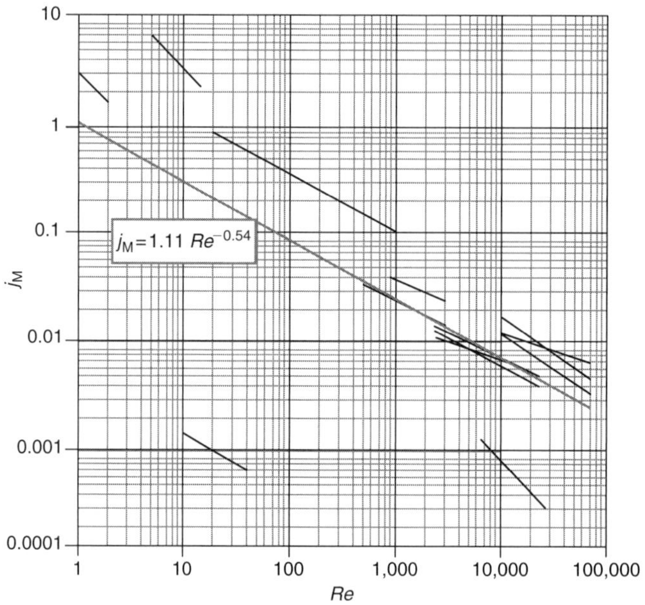

FIGURE 23.12 Mass transfer factor versus Reynolds number for all the examined processes and materials. (From Saravacos, G.D. and Maroulis, Z.B., *Transport Properties of Foods*, Marcel Dekker, New York, 2001.)

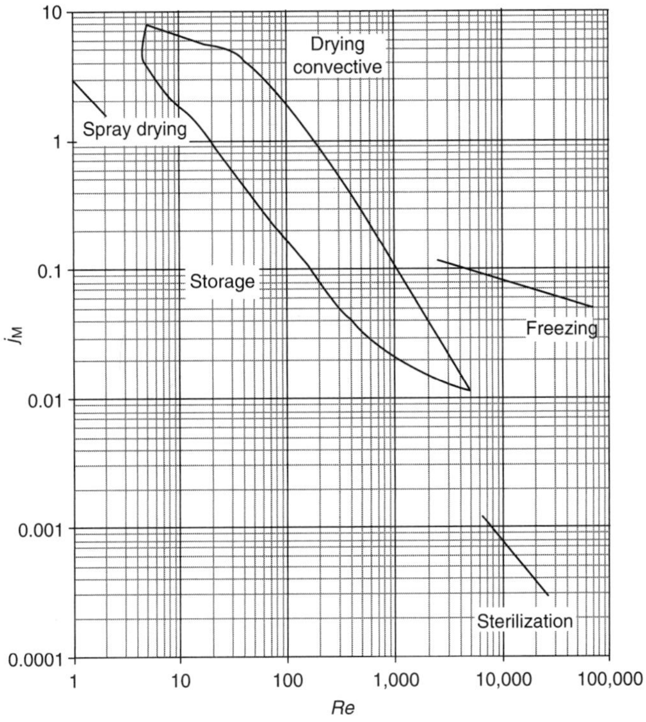

FIGURE 23.13 Ranges of variation in the mass transfer factor versus Reynolds number for all the examined processes. (From Saravacos, G.D. and Maroulis, Z.B., *Transport Properties of Foods*, Marcel Dekker, New York, 2001.)

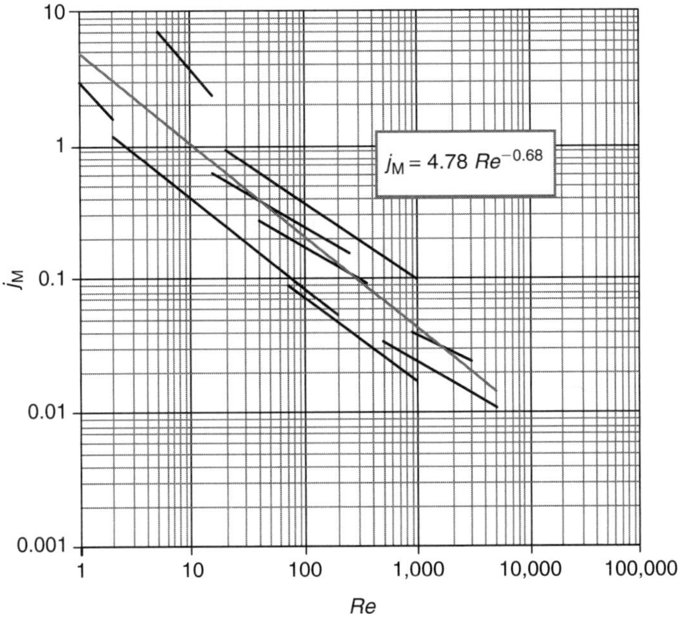

FIGURE 23.14 Mass transfer factor versus Reynolds number for all the examined drying types and materials.

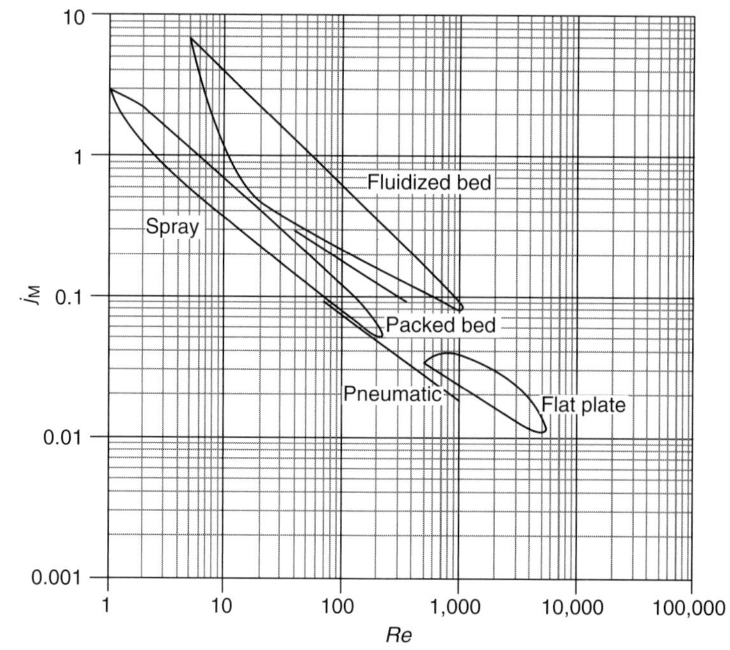

FIGURE 23.15 Ranges of variation in the mass transfer factor versus Reynolds number for all the examined drying types.

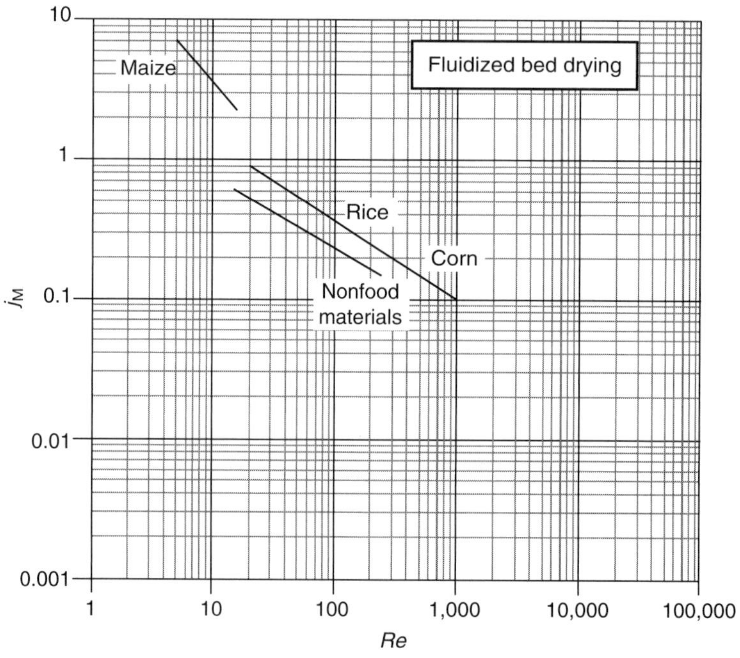

FIGURE 23.16 Mass transfer factor versus Reynolds number for fluidized bed drying and various materials.

ACKNOWLEDGMENT

The authors would like to thank Dr. Zacharias Maroulis for his guidance and comments for the writing of this chapter and for his contribution in preparing the tables and figures.

NOMENCLATURE

α	equation parameter
$\dot{\alpha}$	equation parameter
$\hat{a}$	equation parameter
A	heat transfer surface area (m^2)
Ar	Archimedes number
$\hat{b}$	equation parameter
B	equation parameter
$\hat{B}$	equation parameter
Br	Brinkman number
$\dot{c}$	drag coefficient
$\ddot{c}$	equation parameter
$\hat{c}$	equation parameter
C	specific heat at constant pressure (kJ/kg/°C)
$\bar{C}$	equation parameter
$\hat{C}$	equation parameter
Cr	equation parameter (dimensionless)
D	diameter (m)
$\hat{d}$	equation parameter
e	can headspace height (m)
e'	equation parameter
E	material, fluid characteristic
$\hat{E}$	activation energy (J)
f	friction factor
$\hat{f}$	correction factor
F	equation parameter
$\hat{F}$	correction function
Fr	Froude number (dimensionless)
g	acceleration due to gravity (m/s^2)
$\hat{g}$	induced gravitational field
G	air mass velocity (kg/m^2/s)
$\hat{G}$	equation parameter
Gr	Grashof number (dimensionless)
Gz	Graetz number (dimensionless)
h	heat transfer coefficient (W/m^2/°C)
H	height or screw channel depth (m)
h_d	fouling coefficient of the wall/liquid interface (W/m^2/°C)
h_M	mass transfer coefficient
j	Colburn transfer factor
$\hat{j}$	mass flow ratio of solid to gas
k	thermal conductivity (W/m/K)
K	consistency coefficient of the fluid (Pa · s^n)
$\dot{K}$	equation parameter

$\ddot{K}$	equation parameter
k_C	mass transfer coefficient expressed in concentration units
l	characteristic length (m)
L	film thickness or length (m)
m	mass flow rate (kg/s)
M	equation parameter
n	flow behavior index of the fluid
n'	equation parameter
$\dot{n}$	equation parameter
N	rotational or agitation speed (s^{-1} or rps)
N_b	number of horizontal tubes in a vertical plane
Nu	Nusselt number (dimensionless)
P	pressure (Pa)
Pe	Peclet number (dimensionless)
Pr	Prandtl number (dimensionless)
$\dot{q}$	heat flux (kW)
q_s	parameter equation
r	radius (m)
Ra	Rayleigh number (dimensionless)
Re	Reynolds number (dimensionless)
S	vapor velocity to liquid film velocity ratio
Sc	Schmidt number (dimensionless)
St	Stanton number (dimensionless)
T	temperature (°C)
Tu	turbulence intensity
u	fluid velocity (m/s)
u_v	volumetric vapor flux (m^3/s)
u_v^*	modified volumetric vapor flux (dimensionless)
$\bar{u}_c$	cylinder surface peripheral velocity (m/s)
U	overall heat transfer coefficient ($W/m^2/°C$)
V	volume (m^3)
We	Weber number
x	wall thickness (m)
x'	mass fraction of vapor in mixture
$\bar{x}$	distance in x-axis
X	Lockhart–Martinnelli parameter
$\hat{Y}$	equation parameter
Z	equation parameter

Greek Symbols

α	thermal diffusivity (m^2/s)
β	volumetric thermal expansion coefficient of the fluid
γ	major to minor axis ratio
Γ	condensation rate per unit length of perimeter (kg/s/m)
$\dot{\gamma}$	shear rate (s^{-1})
δ	film or layer thickness (m)
Δ	difference
ε	volume fraction
η	viscosity ($Pa \cdot s$)

ϑ	equation parameter
$\hat{\theta}$	orientation of ribs in plate heat exchanger
Θ	Rohsenow constant
κ	equation parameter
Λ	non-Newtonian to Newtonian heat transfer coefficient ratio
ν	kinematic viscosity (m^2/s)
λ	latent heat of vaporization (kJ/kg)
λ^*	modified latent heat of vaporization (kJ/kg)
ξ	equation parameter
Ξ	correction function
π	constant (= 3.14)
ρ	density (kg/m^3)
σ	surface or interfacial tension (N/m)
$\hat{\tau}$	shear stress
υ	number of rows of scraper blades
Υ	equation parameter
Φ	surface area to volume ratio
χ	wavelength
ψ	mass transfer coefficient
Ψ	particle shape factor
ω	can angular velocity
Ω	non-Newtonian to Newtonian heat transfer rates
%RH	percent relative humidity

Subscripts

ai	air
ap	approach
av	average
b	bulk fluid
bo	box
c	can
ca	mushroom cup
ce	effective can length
ci	can internal
co	can outside
cs	can surface or cross sectional
cv	condensing vapor
cx	can external
d	due point
D	developed area
dr	drum
e	exit or outlet
ec	effective
eg	equilibrium at the end of heating or cooling process
eq	equivalent
er	ellipsoid of revolution
f	forced convection
FL	fluid
ga	gap between plates

h	hydraulic
H	heat
i	inlet or initial
ini	inside initial
insp	inside specific
iw	liquid at wall
l	liquid
m	mixed convection
M	mass
ma	major axis
me	semiaxis
mi	minor axis
min	minimum
mix	mixture
n	natural convection
ou	outside
ov	overall
p	product
pa	particle (or sphere)
pr	projected
r	reel
re	retort or relative
ri	rise of ribs
s	surface
sa	shaft
sb	bulk suspension
sc	center of solid
se	sedimented suspension
so	solid
sp	specific
sr	screw
ss	absolute steam pressure
su	suspension
sv	saturation
T	tube or cylinder
TO	total
v	vapor
w	wall
woc	without condensation of water vapor
wic	with condensation of water vapor
x	x-axis
y	y-axis
z	local distance in z-direction

REFERENCES

Alhamdan, A. and Sastry, S.K. 1990. Natural convection heat transfer between non-Newtonian fluids and an irregular shaped particle. *Journal of Food Process Engineering*, 13(2): 113.

Alhamdan, A., Sastry, S.K., and Blaisdell, J.L. 1990. Natural convection heat transfer between water and an irregular-shaped particle. *Transactions of the ASAE*, 33(2): 620.

Alvarez, G. and Flick, D. 1999. Analysis of heterogeneous cooling of agricultural products inside bins. *Journal of Food Engineering*, 39(3): 239–245.
Alvarez, P. and Shene, C. 1994. Experimental determination of volumetric heat transfer coefficient in a rotary dryer. *Drying Technology*, 12(7): 1605–1627.
Anantheswaran, R.C. and Rao, M.A. 1985a. Heat transfer to model Newtonian liquid foods in cans during end-over-end rotation. *Journal of Food Engineering*, 4: 1.
Anantheswaran, R.C. and Rao, M.A. 1985b. Heat transfer to model non-Newtonian liquid foods in cans during end-over-end rotation. *Journal of Food Engineering*, 4: 21.
Astrom, A. and Bark, G. 1994. Heat transfer between fluid and particles in aseptic processing. *Journal of Food Engineering*, 21(1): 97–125.
Awonorin, S.O. and Lamb, J. 1988. Heat transfer coefficient for nitrogen droplets film-boiling on a food surface. *International Journal of Food Science and Technology*, 23: 391.
Awuah, G.B., Ramaswamy, H.S., and Simpson, B.K. 1993. Surface heat transfer coefficients associated with heating of food particles in CMC solutions. *Journal of Food Process Engineering*, 16(1): 39.
Baik, O.D., Grabowski, S., Trigui, M., Marcotte, M., and Castaigne, F. 1999. Heat transfer coefficients on cakes baked in a tunnel type industrial oven. *Journal of Food Science*, 64(4): 688–694.
Balasubramaniam, V.M. and Sastry, S.K. 1994. Convective heat transfer at particle–liquid interface in continuous tube flow at elevated fluid temperatures. *Journal of Food Science*, 59(3): 675–681.
Baumeister, K.J., Hamill, T.D., and Schoessow, G.J. 1966. A generalized correlation of vaporisation times of drops of film boiling on a flat plate. *Proceedings of the 3rd International Heat Transfer Conference*, Chicago.
Bird, R.B., Stewart, W.E., and Lightfoot, E.N. 1960. *Transport Phenomena*, John Wiley and Sons, Inc., New York and London.
Bourgois, J. and Le Maguer, M. 1983a. Modelling of heat transfer in a climbing-film evaporator: Part I. *Journal of Food Engineering*, 2: 63.
Bourgois, J. and Le Maguer, M. 1983b. Modelling of heat transfer in a climbing-film evaporator: Part II. *Journal of Food Engineering*, 2: 225.
Bourgois, J. and Le Maguer, M. 1987. Heat transfer correlation for upward liquid film heat transfer with phase change: Application in the optimization and design of evaporators. *Journal of Food Engineering*, 6(4): 291.
Breen, B.P. and Westwater, J.W. 1962. Effect of diameter of horizontal tubes on film boiling heat transfer. *Chemical Engineering Progress*, 58(7): 67.
Brown, A.R. and Tomas, M.A. 1965. Combined free and forced convection heat transfer for laminar flow in horizontal tubes. *Journal of Mechanical Engineering Science*, 7(4): 440.
Chandarana, D.I., Gavin, A., and Wheaton, F.W. 1990. Particle/fluid interface heat transfer under UHT conditions at low particle/fluid relative velocities. *Journal of Food Process Engineering*, 13: 191.
Chang, C., Converse, H., and Steele, J. 1993. Modelling of temperature of grain during storage with aeration. *Transactions of the ASAE*, 36(2): 509–519.
Christiansen, E.B., Jensen, G.E., and Tao, F.S. 1966. Laminar flow heat transfer. *AIChE Journal*, 12: 1196.
Chun, K.R. and Seban, R.A. 1971. Heat transfer to evaporating liquid films. *Journal of Heat Transfer*, 93(4): 391.
Clary, B.L. and Nelson, G.L. 1970. Determining convective heat transfer coefficients from ellipsoidal shapes. *Transactions of the ASAE*, 3: 309–314.
Colburn, A.P. 1933. A method of correlating forced convection heat transfer data and comparison with fluid friction. *Transaction of the AIChE*, 29: 174.
Deniston, M.F., Hassan, B.H., and Merson, R.L. 1987. Heat transfer coefficients to liquids with food particles in axially rotating cans. *Journal of Food Science*, 52(4): 962–966.
Deniston, M.F., Kimball, R.N., Stoforos, N.G., and Parkinson, K.S. 1992. Effect of steam/air mixtures on thermal processing of an induced convection-heating product (tomato concentrate) in a steritort. *Journal of Food Process Engineering*, 15: 49.
Depew, C.A. and August, S.F. 1971. Heat transfer due to combined free and forced convection in a horizontal and isothermal tube. *Journal of Heat Transfer-Transaction of the ASME*, 93: 380.
Dincer, I. 1995. Transient heat transfer analysis in air cooling of individual spherical products. *Journal of Food Engineering*, 26(4): 453–467.

Dincer, I. 1997. New effective Nusselt–Reynolds correlations for food-cooling applications. *Journal of Food Engineering*, 31(1): 59–67.

Dukler, A.E. 1960. Fluid mechanics and heat transfer in vertical falling film systems. *Chemical Engineering Progress Symposium Series*, 56(30): 1.

Duquenoy, A. 1980. Heat transfer to canned liquids. In: *Food Process Engineering, Vol. 1. Food Process Systems*, Linko, P. et al. eds., Applied Science Publishers Ltd., London.

Dutta, T.K. and Chanda, B.C. 1991. The effect of surface designs of plates on heat transfer in a plate heat exchanger. *Indian Chemical Engineer*, 33(4): 42.

Eckert, E. and Drewitz, O. 1943. *The Heat Transfer to a Plate at High Speed*. NASA Technical Memorandum 1045.

Eubank, C.L. and Proctor, W.S. 1951. M.Sc. Thesis in Chemical Engineering. MIT.

Fernandez, C.L., Rao, M.A., and Rajavasireddi, S.P. 1988. Particulate heat transfer to canned snap beans in a steritort. *Journal of Food Process Engineering*, 10: 183.

Fikiin, A.G., Fikiin, K.A., and Triphonov, S.D. 1999. Equivalent thermophysical properties and surface heat transfer coefficient of fruit layers in trays during cooling. *Journal of Food Engineering*, 40(1–2): 7–13.

Filkova, I., Koziskova, B., and Filka, P. 1986. Heat transfer to a power law fluid in tube flow: An experimental study. In: *Food Engineering and Process Applications, vol. 1. Transport phenomena*, Le Maguer, M. and Jelen, P. eds., Elsevier Applied Science Publishers Ltd., London.

Fishenden, M. and Saunders, O.A. 1950. *An Introduction to Heat Transfer*, Oxford University Press, Oxford.

Flores, E.S. and Mascheroni, R.H. 1988. Determination of heat transfer coefficients for continuous belt freezers. *Journal of Food Science*, 53(6): 1872.

Fortes, M. and Okos, M. 1981. Non-equilibrium thermodynamics approach to heat and mass transfer in corn kernels. *Transactions of the ASAE*, 761–769.

Fu, W.R., Sue, Y.C., and Chang, K.L.B. 1998. Distribution of liquid–solid heat transfer coefficient among suspended particles in vertical holding tubes of an aseptic processing system. *Journal of Food Science*, 63(2): 189–191.

Geankoplis, C.J. 1993. *Transport Processes and Unit Operations*, 3rd ed. Prentice Hall, New York.

Ghiaus, A.G., Margaris, D.P., and Papanikas, D.G. 1997. Mathematical modeling of the convective drying of fruits and vegetables. *Journal of Food Science*, 62(6): 1154–1157.

Gilmour, C.H. 1959. Performance of vaporizers: Heat transfer analysis of plant data. *Chemical Engineering Progress Symposium Series*, 55(29): 67.

Gnielinski, V. 1975. Neue Gleichungen fur den Warme-und Stoffubergang in Turbulent Durchstromten Rohren und Kanalen. *Forschung im Ingenieu rwesen*, 41(1): 8.

Graetz, L. 1883. Uber die Warmeleitungsfahigkeit von flussigkeiten. *Annalen Der Physik und Chemie*, 18: 79.

Graetz, L. 1885. Uber die Warmeleitungsfahigkeit von flussigkeiten. *Annalen Der Physik*, 25: 337.

Hallstrom, B., Skjoldebrand, C., and Tragardh, C. 1988. *Heat Transfer and Food Products*, Elsevier Applied Science Publishers Ltd., London.

Hanzawa, T., Yu, H.T., Hsiao, Y., and Sakai, N. 1991. Characteristics of heat transfer from package in refrigerating room with laminar downflow. *Journal of Chemical Engineering Japan*, 24(6): 726.

Heldman, D.R. 1980. Predicting of food product freezing rates. In: *Food Process Engineering, vol. 1*. Linko, P., Malkki, Y., Olkku, J. and Larinkari, J. eds. Applied Science Publishers Ltd., London pp. 40–45.

Heldman, D.R. and Lund, D.B. eds. 1992. *Handbook of Food Engineering*. Marcel Dekker, New York.

Hougen, O.A. 1940. Typical dryer calculations. *Chem. Met. Eng.*, 47(1): 15.

Howarth, L. 1957. *On the solution of the laminar boundary layer equations*, *Proceedings of the Royal Society of London, Series A*, 164: 547.

Jackson, A.T. and Lamb, J. 1981. *Calculations in Food and Chemical Engineering*, The Macmillan Press Ltd., London.

Jackson, T.W., Spurlock, J.M., and Purdy, K.R. 1961. Combined free and forced convection in a constant temperature horizontal tube. *AIChE Journal*, 7: 38.

Jakob, M. and Cupta, P.C. 1954. Heat transfer by free convection through liquids between two horizontal surfaces. *AIChE Symposium Series, No. 9*, 50: 15.

Kays, W.M. and Bjorklund, I.S. 1958. Heat transfer from a rotating cylinder with and without crossflow. *Transactions of the ASME, series C*, 80: 70.

Kitamura, K., Honma, M., and Kashiwagi, S. 1992. Heat transfer by combined forced and natural convection from a horizontal cylinder. *Heat Transfer Japanese Research*, 21(1): 63.

Ko, S. and Sogin, H.H. 1958. Laminar mass and heat transfer from ellipsoidal surfaces of fineness ratio 4 in axisymmetrical flow. *Transaction of the ASME*, 80: 387.

Kondjoyan, A., Daudin, J.D., and Bimbenet, J.J. 1993. Heat and mass transfer coefficients at the surface of elliptical cylinders placed in a turbulent air flow. *Journal of Food Engineering*, 20: 339.

Kramers, H. 1946. Heat transfer from spheres to flowing media. *Physica*, 12: 61.

Krokida, M.K., Zogzas, N.P., and Maroulis, Z.B. 2001. Mass transfer coefficient in food processing: Compilation of literature data. *International Journal of Food Properties*, 4(3): 373–382.

Krokida, M.K., Maroulis, Z.B., and Marinos-Kouris, D. 2002a. Heat and mass transfer coefficients in drying: Compilation of literature data. *Drying Technology*, 20(1): 1–18.

Krokida, M.K., Zogzas, N.P., and Maroulis, Z.B. 2002b. Heat transfer coefficient in food processing: Compilation of literature data. *International Journal of Food Properties*, 5(2): 435–450.

Lang, W., Sokhansanj, S., and Rohani, S. 1996. Dynamic shrinkage and variable parameters in Bakker-Arkema's mathematical simulation of wheat and canola drying. *Drying Technology*, 12(7): 1687–1708.

Leichter, S., Mizrahi, S., and Kopelman, I.J. 1976. Effect of vapor condensation on rate of refrigerated products exposed to humid atmosphere: Application to the prediction of fluid milk shelf life. *Journal of Food Science*, 41: 1214.

Leninger, H.A. and Beverloo, W.A. 1975. *Food Process Engineering*. Riedel, Dordrecht.

Lenz, M.K. and Lund, D.B. 1978. The lethality-Fourier number method. Heating rate variations and lethality confidence intervals for forced-convection heated foods in containers. *Journal of Food Processing and Preservation*, 2: 227.

Levenspiel, O. and Walton, J.S. 1954. Bed-wall heat transfer in fluidized systems, *Chemical Engineering Progress Symposium Series No 9*, 50: 1.

Lévéque, A. 1928. Les lois de la transmission de chaleur par convection. *Annales des Mines ou Recueil de Mémoires sur l'Exploitation des Mines et sur les Sciences et les Arts qui s'y Rattachent*, Mémoires, Tome XIII(13): 201–239.

Levine, L. and Rockwood, J. 1986. A correlation for heat transfer coefficients in food extruders. *Biotechnology Progress*, 2(2): 105.

Lopez, A., Virseda, P., Martinez, G., and Llorka, M. 1997. Deep layer malt drying modelling. *Drying Technology*, 15(5): 1499–1526.

Mallikarjunan, P. and Mittal, G.S. 1994. Heat and mass transfer during beef carcass chilling—modelling and simulation. *Journal of Food Engineering*, 23(3): 277–292.

Mankad, S., Nixon, K.M., and Fryer, P.J. 1997. Measurements of particle–liquid heat transfer in systems of varied solids fraction. *Journal of Food Engineering*, 31(1): 9–33.

McAdams, W.H. 1954. *Heat Transmission*, McGraw-Hill, New York.

McCabe, W.L. and Smith, J.C. 1976. *Unit Operations of Chemical Engineering*, 3rd ed. McGraw-Hill, New York.

Merson, R.L., Leonard, S.J., Mejia, E., and Heil, J. 1980. Temperature distributions and liquid-side heat transfer coefficients in model liquid foods in cans undergoing flame sterilization heating. *Journal of Food Processing and Preservation*, 4: 85.

Metzner, A.B. 1956. Non-Newtonian technology: Fluid mechanics, mixing and heat transfer. In: *Advances in Chemical Engineering*, Vol. 1, Academic Press, New York.

Metzner, A.B. and Gluck, D.F. 1960. Heat transfer to non-Newtonian fluids under laminar-flow conditions. *Chemical Engineering Science*, 12: 185.

Metzner, A.B., Vaughn, R.D., and Houghton, G.L. 1957. Heat transfer to non-Newtonian fluids. *AIChE Journal*, 3(1): 92.

Mohamed, I.O. and Ofoli, R.Y. 1989. Average heat transfer coefficients in twin screw extruders. *Biotechnology Progress*, 5(4): 158.

Mourad, M., Hemati, M., Steinmetz, D., and Laguerie, C. 1997. How to use fluidization to obtain drying kinetics coupled with quality evolution. *Drying Technology*, 15(9): 2195–2209.

Mulet, A., Berna, A., and Rosselo, C. 1989. Drying of carrots. I. Drying Models. *Drying Technology*, 7(3): 537–557.

Mulet, A., Berna, A., Borras, M., and Pinaga, F. 1987. Effect of air flow rate on carrot drying. *Drying Technology*, 5(2): 245–258.

Mulet, A., Carcel, J., Rosselo, C., and Simal, S. 1999. Ultrasound mass transfer enhancement in food processing. *Proceedings of the 6th Conference of Food Engineering CoFE'99*, G.V. Barbosa-Canovas and S.P. Lombardo, eds., AIChE, New York, pp. 74–82.

Mureau, P. and Barreteau, D. 1984. *Modélisation d'un four à rouleaux Document interne Gaz de France*-DETN, Paris.

Mwangi, J.M., Rizvi, S.S.H., and Datta, A.K. 1993. Heat transfer to particles in shear flow: Applications in aseptic processing. *Journal of Food Engineering*, 19: 55.

Oliver, D.R. 1962. The effect of natural convection on viscous flow heat transfer in horizontal tubes. *Chemical Engineering Science*, 17: 335.

Oliver, D.R. and Jenson, V.G. 1964. Heat transfer to pseudoplastic fluids in laminar flow in horizontal tubes. *Chemical Engineering Science*, 19: 115.

Orr, C. and Valle, D. 1954. Heat transfer properties of liquid–solid suspensions. *Chemical Engineering Progress Symposium series No. 9*, 50: 29.

Palen, J.W. and Taborek, J.J. 1962. Refinery kettle reboilers: Proposed method for design and optimization. *Chemical Engineering Progress*, 58(7): 37.

Perry, R.H. and Green, D. 1984. *Perry's Chemical Engineers' Handbook,* 5th ed. McGraw-Hill, New York.

Perry, R.H. and Green, D. 1997. *Perry's Chemical Engineers' Handbook*, 7th ed. McGraw-Hill, New York.

Pigford, R.L. 1955. Nonisothermal flow and heat transfer inside vertical tube. *Chemical Engineering Progress Symposium Series*, 51, 79.

Rahman, S. 1995. *Food Properties Handbook*, CRC Press, Boca Raton, FL.

Ramaswamy, H.S., Pannu, K., Simpson, B.K., and Smith, J.P. 1992. An apparatus for particle-to-fluid relative velocity measurement in tube flow at varius temperatures under nonpressurized flow conditions. *Food Research International*, 25(4): 277.

Ramaswamy, H.S., Abbatemarco, C., and Sablani, S.S. 1993. Heat transfer rates in a canned model food as influenced by processing in an end-over-end rotary steam/air retort. *Journal of Food Processing and Preservation*, 17(4): 269.

Ramirez, J., Ayora, M., and Vizcarra, M. 1981. Study of the behaviour of heat and mass transfer coefficients in gas-solid fluidized bed systems at low Reynolds numbers. In: *Chemical Reactors*, Fogler, H.S. ed., Am. Chem. Soc.

Ranz, W.E. and Marshall, W.R. 1952. Evaporation for drops. I. *Chemical Engineering Progress*, 48: 141.

Rao, M.A., Cooley, H.J., Anantheswaran, R.C., and Ennis, R.W. 1985. Convective heat transfer to canned liquid foods in steritort. *Journal of Food Science*, 50: 150.

Rodriguez-Luna, G., Segurajauregui, J.S., Torres, J., and Brito, E. 1987. Heat transfer to non-Newtonian fluid foods under laminar flow conditions in horizontal tubes. *Journal of Food Science*, 52(4): 975.

Rohsenow, W.M. 1952. A method of correlating heat-transfer data for surface boiling of liquids. *Transaction of the ASME*, 75: 969–976.

Rohsenow, W.M. and Choi, H. 1961. *Heat, Mass, and Momentum Transfer*, Prentice Hall, Englewood Cliffs, NJ.

Rohsenow, W.M. and Griffith, P. 1956. Correlation of maximum heat flux data for boiling of saturated liquids. *Chemical Engineering Progress Symposium Series*, 52(18): 47.

Salamone, J.J. and Newman, M. 1955. Water suspensions of solids. *Industrial and Engineering Chemistry*, 47: 283.

Sablani, S.S., Ramaswamy, H.S., and Mujumdar, A.S. 1997. Dimensionless correlations for convective heat transfer to liquid and particles in cans subjected to end-over-end rotation. *Journal of Food Engineering*, 34(4): 453–472.

Sannervik, J., Bolmstedt, U., and Tragardh, C. 1996. Heat transfer in tubular heat exchangers for particulate containing liquid foods. *Journal of Food Engineering*, 29(1): 63–74.

Saravacos, G.D. 1995. Mass transfer properties of foods. In: *Engineering Properties of Foods*, 2nd ed. M.A. Rao and S.S.H. Rizvi, eds., Marcel Dekker, New York.

Saravacos, G.D. 1997. Moisture transport properties of foods. In: *Advances in Food Engineering*. Narasimham, G., Okos, M.R. and Lombardo, S. eds. Purdue University, West Lafayette, IN. pp. 53–57.

Saravacos, G.D. and Maroulis, Z.B. 2001. *Transport Properties of Foods*, Marcel Dekker, New York.

Saravacos, G.D. and Moyer, J.C. 1967. Heating rates of fruit products in an agitated kettle. *Food Technology*, 21: 54–58.

Saravacos, G.D. and Moyer, J.C. 1970. Concentration of liquid foods in a falling film evaporator. *New York State Agricultural Experiment Station Bulletin No. 4*, Cornell University, Geneva, New York.

Sastry, S.K., Heskitt, B.F., and Blaisdell, J.L. 1989. Experimental and modeling studies on convective heat transfer at the particle–liquid interface in aseptic processing systems. *Food Technology*, 43(3): 132.

Sastry, S.K., Lima, M., Brim, J., Brunn, T., and Heskitt, B.F. 1990. Liquid-to-particle heat transfer during continuous tube flow: Influence of flow rate and particle to tube diameter ratio. *Journal of Food Process Engineering*, 13(3): 239.

Schlichting, H. 1968. *Boundary Layer Theory*, 6th ed. McGraw-Hill Book Co. Inc., New York.

Schlunder, E.U. 1972. *Einfuhrung in die Warmeubertrangung*, Friedr. Vieweg and sons, Braunschweig, Wiesbaden.

Sheen, S. and Whitney, L.F. 1990. Modelling heat transfer in fluidized beds of large particles and its applications in the freezing of large food items. *Journal of Food Engineering*, 12(4): 249.

Shene, C., Cubillos, F., Perez, R., and Alvarez, P. 1996. Modelling and simulation of a direct contact rotary dryer. *Drying Technology*, 14(10): 2419–2433.

Sheng, H.L. 1994. Mathematical model for freezing of calcium alginate balls. *Journal of Food Engineering*, 21(3): 305–313.

Shirai, T., Yoshitome, H., Shoji, Y., Tanaka, S., Hojo, K., and Yoshida, S. 1965. Heat and mass transfer on the surface of solid spheres fixed within fluidized. *Chemical Engineering*, Japan, 29(11): 880–884.

Shu-De, Q. and Fang-Zhen, G. 1993. The experimental study of rotary-stream fluidized bed drying. *Drying Technology*, 11(1): 209–219.

Sieder, E.N. and Tate, G.E. 1936. Heat transfer and pressure drop of liquids in tubes. *Industrial and Engineering Chemistry*, 28: 1429.

Skelland, A.H.P., Oliver, D.R., and Tooke, S. 1962. Heat transfer in a water-cooled scraped-surface heat exchanger. *British Chemical Engineering*, 7: 346.

Sokhansanj, S. 1987. Improved heat and mass transfer models to predict grain quality. *Drying Technology*, 5(4): 511–525.

Soule, C.L. and Merson, R.L. 1985. Heat transfer coefficients to Newtonian liquids in axially rotated cans. *Journal of Food Process Engineering*, 8: 33.

Straatsma, J., van Houwelingen, G., Steenbergen, A.E., and De Jong, P. 1999. Spray drying of food products: 1. Simulation model. *Journal of Food Engineering*, 42(2): 67–72.

Taranto, O., Rocha, S., and Raghavan, G. 1997. Convective heat transfer during coating of particles in two-dimensional spouted beds. *Drying Technology*, 15(6–8): 1909–1918.

Teixeira Neto, R.O. 1982. Heat transfer rates to liquid foods during flame sterilization. *Journal of Food Science*, 47(2): 476.

Tocci, A.M. and Mascheroni, R.H. 1995. Heat and mass transfer coefficients during the refrigeration, freezing and storage of meats, meat products and analogues. *Journal of Food Engineering*, 26(2): 147–160.

Todd, D. 1988. Heat transfer in twin screw extrusion. *ANTEC'88 Conference Proceedings of Plastics Engineers*, 46th Annual Technical Conference and Exhibit. Atlanta, Georgia, USA, pp. 54–57.

Torrez, N., Gustafsson, M., Schreil, A., and Martinez, J. 1998. Modeling and simulation of crossflow moving bed grain dryers. *Drying Technology*, 16(9&10): 1999–2015.

Trommelen, A.M. 1970. Physical aspects of scraped-surface heat exchangers. PhD thesis, University of Delft, the Netherlands.

Vachon, R.I., Nix, G.H., and Tanger, G.E. 1968. Evaluation of constants for Rohsenow pool-boiling correlation. *Journal of Heat Transfer*, 90: 239.

Vagenas, G., Marinos-Kouris, D., and Saravacos, G. 1990. An analysis of mass transfer in air-drying of foods. *Drying Technology*, 8(2): 323–342.

Valentas, K.J., Rotstein, E., and Singh, R.P. 1997. *Handbook of Food Engineering Practice*. CRC Press, New York.

Van Boxtel, L.B.J. and De Fielliettaz Goethart, R.L. 1983. Heat transfer to water and some highly viscous food systems in a water-cooled scraped surface heat exchanger. *Journal of Food Process Engineering*, 7: 17.

Van Boxtel, L.B.J. and De Fielliettaz Goethart, R.L. 1984. Heat transfer in a scraped surface heat exchanger during cooling of some highly viscous food products. In: *Thermal Processing and Quality of Foods*, Zeuthen, P. et al. eds., Elsevier Applied Science Publishers Ltd., England.

Wang, N. and Brennan, J.G. 1995. Mathematical model of simultaneous heat and moisture transfer during drying of potato. *Journal of Food Engineering*, 24(1): 47–60.

Wang, F., Cameron, I., Lister, J., and Douglas, P. 1993. A distributed parameter approach to the dynamics of rotary drying processes. *Drying Technology*, 11(7): 1641–1656.

Wen, C.Y. and Chang, T.M. 1967. Particle-to-particle heat transfer in air-fluidized beds. *Proceedings of the International Symposium on Fluidization*, June 6–7, Eindhoven, the Netherlands.

Whitaker, S. 1972. Forced convection heat transfer calculations for flow in pipes, past flat plates, single cylinders, single spheres, and for flow in packed beds and tube bundles. *AIChE Journal*, 18: 361.

Whitaker, S. 1976. *Elementary Heat Transfer Analysis*, Pergamon Press Inc., New York.

Wilke, W. 1962. Warmeubergang an Reisefilme. Verein Deutscher Ingenieure—Forschungsheft, 490.

Wilkinson, W.L. 1960. *Non-Newtonian Fluids*, Pergamon Press, New York.

Xu, Y. and Burfoot, D. 1999. Simulating the bulk storage of foodstuffs. *Journal of Food Engineering*, 39(1): 23–29.

Yoshida, F., Ramaswami, D., and Hougen, O.A. 1962. Temperatures and partial pressures at the surfaces of catalyst particles. *AIChE Journal*, 8(1): 5.

Yousef, W.W. and Tarasuk, J.D. 1982. Free convection effects on laminar forced convection heat transfer in a horizontal isothermal tube. *Journal of Heat Transfer*, 104: 145.

Zaman, S., Rotstein, E., and Valentas, K.J. 1991. Can material influence on the performance of rotating cookers. *Journal of Food Science*, 56(6): 1718–1724.

Zhang, Q. and Cavalieri, R. 1991. Thermal model for steam blanching of green beans and determination of surface heat transfer coefficient. *Transactions of the ASAE*, 34(1): 182–186.

Zuber, N. 1958. On the stability of boiling heat transfer. *Transaction of the ASME*, 80: 711.

Zuritz, C.A., McCoy, S.C., and Sastry, S.K. 1990. Convective heat transfer coefficients for irregular particles immersed in non-Newtonian fluid during tube flow. *Journal of Food Engineering*, 11: 159.

CHAPTER 24

Acoustic Properties of Foods

Piotr P. Lewicki, Agata Marzec, and Zbigniew Ranachowski

CONTENTS

24.1 INTRODUCTION

Acoustic emission (AE) generated during biting and mastication of food is an important attribute of food texture. Acoustic properties of snack-type food as well as fresh fruits and vegetables are important from the quality point of view. Crunchiness and crispness are the sensory attributes preferred by consumers and are regarded as signs of freshness and propriety of processing. Lack of these attributes suggests low quality and is at least partly responsible for the unacceptance of the product by consumers.

Crunchiness/crispness is a subjective assessment of quality and until now, there are divergences in the description and definition of this sensory attribute. From the point of research, analyzing the crunchiness/crispness sound percept by a human ear is considered as the main physical index influencing the consumers' assessment of quality.

Elastic waves generated during the disintegration of food can be used to identify the acoustic properties of the product. Moreover, analysis of the emitted sound can be used to follow changes in the texture of food, caused during storage or by modifications of recipes or processing parameters.

And lastly, the acoustic properties of food can be used to numerically express one of its important quality attributes, namely the crunchiness/crispness.

24.2 BASICS OF ACOUSTIC EMISSION

An initial inhomogeneity of internal energy distribution can appear in any solid, gaseous, or liquid body. In actual materials, inhomogeneity characteristics are due to structural defects, impurities, internal stresses, and other factors and can be micro- or macroscopic [1]. The following structure elements are examples of body inhomogeneities: (1) surface between different phases of the material, (2) cellular walls and intercellular objects, and (3) gaseous pores in starch polymers, forming the cereal food products.

Inhomogeneities have no influence on the state of global mechanical equilibrium of that body. Introduction of external stimuli, such as external force, chemical reaction, or thermal excitation, disturbs the equilibrium state and causes the onset of vibrations of particles in the body volume. The vibration can remain active and propagate through the medium even after decay of the stimuli. The reason for the extended reaction of the medium can be explained by the elastic and inertia properties of the majority of bodies. The elastic properties are responsible for the reaction against the disturbance of the equilibrium. The inertia properties, however, are the source of forces opposing to any change in state of the matter, i.e., opposing to the decrease in pending kinetic activity even when the local equilibrium is achieved. Therefore, the movement of body particles results in a series of transformations, from the kinetic energy supported by inertial forces to the potential energy supported by elastic forces. This process can continue infinitely even after the decay of the stimulus, but is not another property of the real matter—exhibiting the ability to attenuate the internal vibrations. However, the kinetic energy of the movement gradually gets converted into heat and changes the microstructure of the body (dislocation movements, creation of micro- and macro-cracks). In this way, the attenuated vibrations gradually fade.

The disturbance of the equilibrium that has appeared in time t_0, in a certain spot of a body, would cover the entire body after a sufficient period of time. This effect occurs because every particle of the body interacts with its neighbors and passes a part of its kinetic energy to them. This phenomenon is called wave propagation [2,3]. Two basic modes of wave propagation can occur: (1) a longitudinal wave, propagation of which causes the elements of the medium to move in the direction of the wave propagation; (2) a transverse wave, with the vibrating motion perpendicular to the direction of the wave propagation (Figure 24.1).

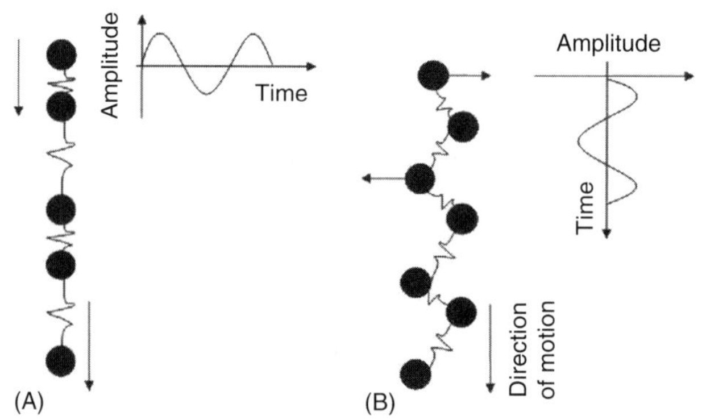

Figure 24.1 Vibrations of an element in (A) longitudinal wave and (B) and transverse wave. (From Sliwinski, A., *Ultrasound and Its Application*, 2nd ed, WNT, Warszawa, 2001 (in Polish); Malecki, I., *Physical Foundations of Technical Acoustics*, Pergamon Press, Oxford, 1969. With permission.)

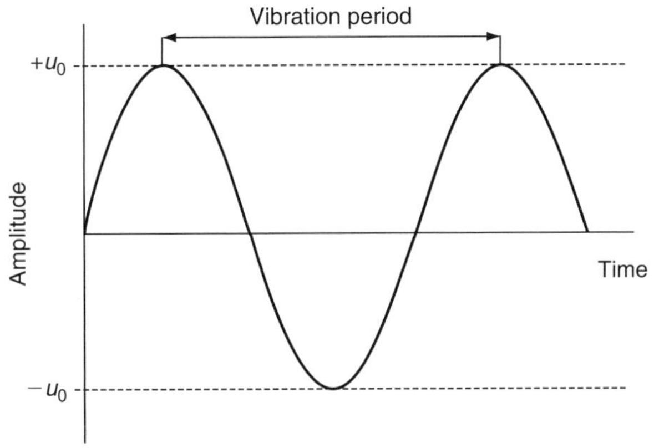

Figure 24.2 Elastic wave.

In liquids and gases, only longitudinal wave can be excited. Two other modes of wave propagation, namely Rayleigh–Lamb and Love waves, occur near the surface of the solids. The number of times a complete motion cycle takes place in every point of the body during a period of 1 s is called the frequency and is measured in hertz (Hz) [4]. The distance between the equilibrium location and the maximum displacement during the motion cycle is the amplitude [5] (Figure 24.2).

Generally, disturbance of mechanical and thermodynamic equilibrium in solids, liquids, and gases leads to the release of local deformation processes and invokes the dissipation of energy. The major part of this emerging energy gets converted into the work done by stresses, heat, and a relatively small portion radiates in the form of elastic waves. The resulting waves propagate through the material and reach its outer surface. When a sensor of elastic waves is placed on the body surface, it can detect the onset of the described phenomenon [6]. The particles of the air surrounding the vibrating body can also transmit the waves to the microphone situated at a certain distance from the body.

The explanation of the terminology used has its roots in the past—the elastic waves arising in loaded mechanical elements were situated in the audible range of frequencies (16–16000 Hz), and hence, scientists who first observed these phenomena introduced the terms noise, ultrasonic emission, or AE. It can be stated that when a sudden release of internal energy in the loaded matter takes place, and if an elastic wave's propagation pattern is recorded, then the research is inclined toward the measurement of AE.

An AE source is defined as a certain location or element of the structure that emits the AE signal. Usually, an electric waveform $u(t)$ recorded on the output of the sensor prepared for storage or processing is considered as the AE signal. The AE sources can be divided into groups with respect to the processes involved in their activities [7]. These include (1) crack formation and propagation, (2) relocation of defects in crystalline lattice, (3) phase transitions and chemical reactions, (4) creation of transitions between the energy states of the atoms, (5) body movements with the accompanying friction mechanisms, and (6) biological processes running in live tissues and organisms.

The highest AE amplitude levels are detected during the crack formations and destruction processes in solids under mechanical stresses, and the lowest are generated when transitions between the energy states of the atoms occur. The AE signal arising in the processes stated earlier varies with respect to its amplitude and frequency. The nature of waveforms of different frequencies is identical. However, different effects related to the interactions between the propagating elastic wave and surrounding material can occur. The microscopic structure of a solid medium is mostly inhomogeneous (i.e., polycrystalline or composite-like) and anisotropic, and this decisively influences the mechanism of energy absorption of the wave. This phenomenon depends on the mutual relation between the dimensions of the material particles and the length of the propagating wave [1]. The relation between the wave frequency f (s^{-1}), wavelength λ (m), and wave velocity c ($m\ s^{-1}$) in a body is as follows:

$$\lambda = \frac{c}{f} \tag{24.1}$$

A typical AE source is characterized by small dimensions with respect to the element or the volume of its occurrence. Therefore, it can be treated as a source of a spherical wave. The AE propagation mechanism in air can be explained as the spread of harmonic changes in pressure. In solids and liquids, the oscillations of particles are also transferred in all directions from the source. However, the propagation of acoustic waves in vacuum is not possible. The direction of the propagation pattern may vary from its symmetrical form when structural inhomogeneities, inclusions, or particle migrations occur. An acoustic wave traveling through the boundary between two bodies undergoes reflection, refraction, and scattering, which cause problems in tracing the origin of the AE activity. Furthermore, according to Doppler's effect, a moving AE source is subject to variations in produced wave frequency [8].

The deformation processes can occur in a micro- and macro scale and, therefore, cause generation of pulse trains of AE signals (bursts) differing in amplitude and intensity. Time dependence of the AE signals reflects the type of the process, i.e., the source of emission and kinetics of the process. The American Standard E 1316 91b [9] deals with the basic term definitions for AE, listing two types of AE: continuous and discrete. The AE signal is identified as discrete when the time delays between pulses having relatively large amplitudes are greater than or equal to the duration of the pulses itself. This condition is related to the phenomena of micro- and macro crack formation and their growth. Referring to the above-mentioned standard, continuous AE is caused by "rapidly occurring" AE sources, since the signals they generate are mixed. A good example of this kind of AE source is a steel ball sliding across the smooth surface, causing abrasive wear of it in the bearings [10].

The relaxation processes in the investigated body cause the origin of the AE signal. By assuming an isotropic solid with loss-less structure and analyzing where a local change of stress field was induced, the propagating elastic wave can be described by the following equation [2]:

$$\rho \frac{\partial^2 U}{\partial t^2} = (\delta + \eta)\text{grad div } U + \delta\text{div grad } U \tag{24.2}$$

where
- ρ is the medium density ($kg\ m^{-3}$)
- U is the displacement vector (m)
- δ, η are Lame constants, reflecting the mechanical properties of the investigated medium ($N\ m^{-2}$)
- t is the time (s)

The analytical solution of Equation 24.2 for a particular body is extremely complex, owing to the body's geometrical configuration, along with the reflection, refraction, inhomogeneities, or scattering phenomena occurring at its boundaries. Generally, it can be stated that the AE signal registered on the tested body surface is a product of AE-source parameters and properties of a body in which AE wave propagates.

To construct a mathematical model of a system comprising an AE source, its environment, and wave sensor, some simplifications are applied that result in imperfections in the description of the real effects. The main problems are related to the negligence of the source dimensions and wave transformations in the medium. The recorded AE signal is also distorted by the receiving sensor, because this device is capable of detecting the limited frequency bandwidth. Owing to these problems, usually the determination of dependence between a registered AE signal and its origin is impossible [11]. Researchers investigating on AE signal have avoided these limitations by constructing certain signal parameters, called AE signal descriptors, that can be measured or calculated in practice.

24.3 MEASUREMENT OF ACOUSTIC EMISSION

The AE method has now found wide application in testing the mechanical properties, phase transitions, and in identifying the physical and chemical processes in different materials. As explained before, the AE signal propagates from its source, across the material under test to the sensors placed on the surface of that material [12]. To gain the maximum benefit from the application of the AE method, the use of the proper sensor is important. There are different contact and noncontact sensors available in the market. Noncontact sensors are made as laser Doppler interferometers. Generally, they can be divided into groups with respect to the operating frequency band, and the typical parameters are listed below:

0.1 Hz–15 kHz
1 Hz–40 kHz
1 kHz–150 kHz
100 kHz–1 MHz

It is important to achieve proper contact between the AE sensor and the body under test. Any air slot or inclusion of a material resistant to elastic waves may cause AE signal attenuation and reflection, introducing severe errors to the measuring procedure. The AE signal amplifiers and registering devices must operate at a suitable speed and the lowest internal noise level to process the data properly. It is also important to suppress the noises from the environment (including the machine loading the tested object) [13]. The typical arrangement to measure AE signals consists of (1) AE sensor, (2) low-noise preamplifier, matched to the signal level produced by the AE sensor, (3) filter to cut off the undesired frequency bands, (4) main amplifier to establish the desired signal voltage for further processing, (5) high-speed analog-to-digital converter with an interface to the computer, and (6) computer for data storing and processing (Figure 24.3).

Traditionally, acoustic measurements of food were carried out with microphones as the sensors of AE signal; however, for this study, the contact method was preferred, i.e., using the sensor that remains in mechanical contact with the product sample. In recent years, many food technology

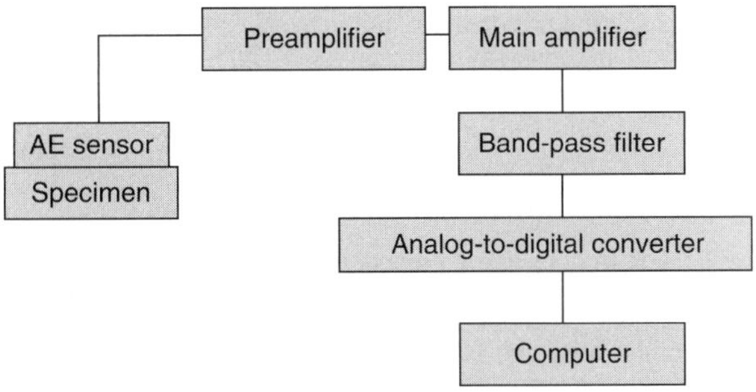

Figure 24.3 Arrangement of a stand for AE measurement and analysis.

laboratories are carrying out the process of registration of AE signal during the crushing of food products. The instrumentation for such tests consists of a texturometer equipped with a microphone probe, acoustic signal amplifier, and a computer to control the loading process and to register the experimental results, as presented in Figure 24.4A through C. The acoustic waves travel across the body under test, then across the air, and finally excite the thin membrane placed in the microphone probe. The microphone converts the vibrations into electric waveforms ready for further computer processing. The sensitivity of the microphone probe is proportional to the membrane area and the spatial direction of the microphone probe. The place and direction of the probe with respect to the specimen also has a significant influence on the registered signal level. It should be stated that the application of sensitive microphone probes requires accurate calibrations of the instrumentation [14]. The method presented in Figure 24.4A through C cannot be assumed as an effective method when applied under industrial conditions. The sounds generated by crushed-food specimen are relatively weak, and the microphone records the unwanted sounds coming from the environment, echoes, etc. Elastic wave traveling across the bulk of the specimen is subject to high reflection phenomenon on the body–air interface owing to the significant difference in the specific density of both the media. Therefore, the way the signal is propagated differs from actual conditions of human sensing, where the bone sound-transmission plays the most significant role. However, the results of the application of a microphone probe for a sound registration in testing of food products can be found in many research papers [15–20].

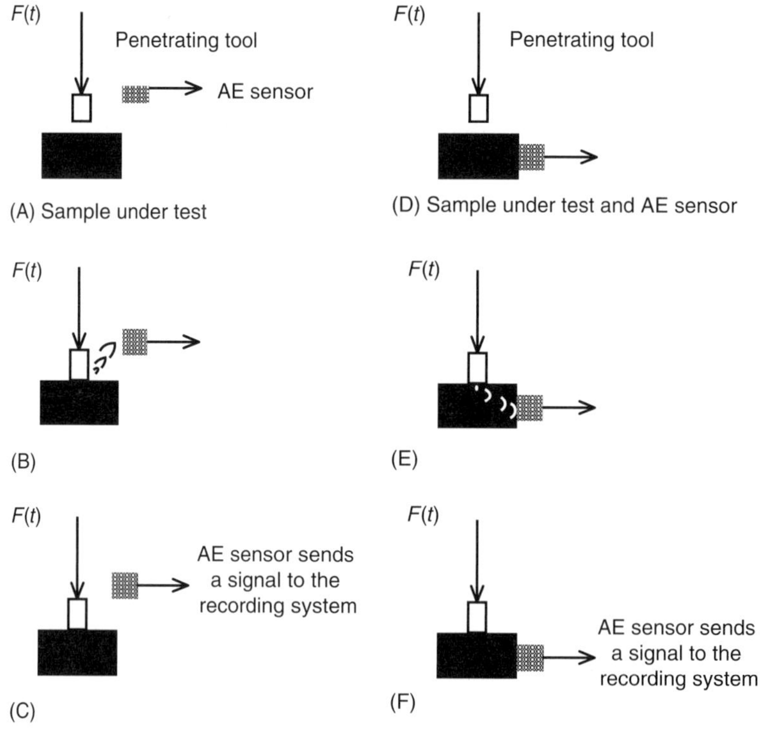

Figure 24.4 The schematic diagram of testing a product sample with the application of loading machine and a sensor of AE. (A and D) penetrating tool moves toward a product; (B) penetrating tool induces a mechanical stress in the product volume; the released stress wave travels across the product body, and then a part of it propagates across the air toward a microphone; (C) a microphone sends AE signal as an electric waveform to the recording system; (E) penetrating tool induces a mechanical stress in the product volume; the released stress wave travels across the product body; (F) contact AE sensor sends AE signal as an electric waveform to the recording system.

A contact method of registering AE signals of food specimens, in good agreement with the idea presented in Figure 24.4D through F, has been designed [21]. A contact sensor capable of measuring an acceleration evoked in the investigated product was mounted between the lower end of the moving traverse and penetrating tool of the loading machine [22]. The sensor was placed on a conical adapter to prevent it from the damage caused by the static load produced by the loading machine. Its sensitivity within the operating frequency band (1–15,000 Hz) was set to 8 mV per 1 m s^{-2}. It is possible to determine the amplitude of the measured vibration (in micrometer) by double integration of the current–voltage registered during the investigation. Since the accelerometer receives the vibration directly from the penetrating tool through a metal joint, its mechanism of sensing the processes accompanying the deformation of the food specimen remains in close relation to the conditions existing in the human mouth. The electric signal from the sensor is fed to the low-noise amplifier, where it is magnified hundred times (40 dB). A 12-bit analog-to-digital converter of type ADLINK 9112 was applied to convert the current values of the AE signal and to convert the stress level into digital form, which can be stored and processed in a computer. The converter operates at a rate of 44,100 sampling processes per second. The custom-designed software helps to produce two types of records as a result of the performed test. The AE signal is stored in popular Windows multimedia .wav format, which makes it possible to play and edit the signal with the application of standard software. The signal, proportional to the force loading the specimen, is stored in the text format compatible with the requirements of popular worksheet programs.

A similar instrumentation was applied to investigate the AE signal generated during the wear processes in bearings and to investigate the mechanical durability of wood and fiberglass specimens. Alchakra et al. [23] applied both an accelerometer operating at 0.2–20 kHz range and a microphone probe operating at 0–40 kHz range in tests of destruction of concrete, graphite, and polystyrene. They have also observed that the loading machine used in the investigation had not emitted measurable background noise within the used frequency range.

From the point of sensory assessment, the acoustic investigation of food should be made in the frequency range from 1 to 16 kHz, because the waves of those frequencies evoke audible sensations in humans. This effect was described by Dacremont [16], who had registered the bone transmission of sound from the mouth to the microphone placed in the mastoid. The effectiveness of bone sound-transmission was compared with the air propagation during biting of crisp-bread specimens. Dacremont [16] stated that the sounds of frequency 5–10 kHz propagate through a bone tract, and the sounds of frequency 1–2 kHz propagate through air tract. There are numerous literatures confirming that the frequency range stated here is carefully chosen for the sensory investigation of food [24–27]. Lee et al. [28] has stated that the proper frequency range should reach a minimum of 12 kHz for crispy products. The AE frequencies generated by typical crispy products are in the 5–12.8 kHz range, while those generated during chewing of potato chips are in 3–6 kHz range, and when specimens of chips are compressed in the loading machine, the AE lies in the 1.9–3.3 kHz range [29].

24.4 ANALYSIS OF RECORDED SIGNAL OF ACOUSTIC EMISSION

A typical record of the time dependence of AE signal is presented in Figure 24.5. A section of the signal where measurable oscillations are detected is called the AE event. Within a time of an event, the consecutive AE signal amplitudes exceed the preset threshold, called the AE discrimination level [30]. Modern computerized AE processing equipment is capable of producing different AE descriptors during the investigation. The most commonly used descriptors are listed in Table 24.1 [31] and are discussed in the later sections.

To determine the time-domain derivatives listed in Table 24.1, some assumptions were made with respect to the shape of the AE waveform. The basic assumption made after practical observations is the approximation of the registered AE signal produced by the impulse AE source to the

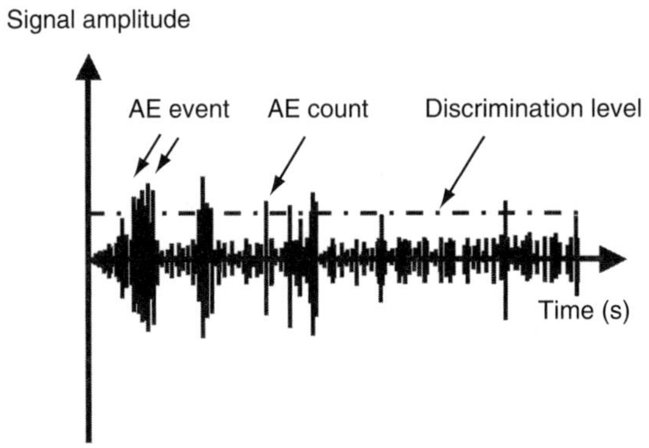

Figure 24.5 Typical record of AE signal. The AE event is the duration of an effect, when the AE signal crosses a certain preset signal threshold called the discrimination level. The AE count is registered at every moment when the current AE amplitude crosses that threshold.

shape of a damped sinusoid. It is possible to preset a certain signal threshold and register (in other words, count) every moment when current amplitude of the AE signal would cross that threshold. This strategy of tracing the activity of AE source is called AE count processing. The time when the consecutive AE signal amplitudes, belonging to the single AE impulse, exceed the preset threshold is called AE event (Figure 24.5). The AE waveform of the single event can then be approximated using the following formula:

$$\nu(t) = A_{\mathrm{m}} \exp(\alpha_1 t) \sin(2\pi f_0 t) \tag{24.3}$$

Table 24.1 Most Commonly Used AE Signal Descriptors

Time-domain derivatives
Counts number processing
Events number processing
Zero crossing
Peak amplitude
Average amplitude
Area of waveform above mean
"Half life," i.e., time to reach half area (see the above point)
Energy derivatives
Peak root-mean-square (rms)
Average rms
Sum of rms
Energy of single count of event
Frequency-domain derivatives
Frequency in signal spectrum with maximum intensity
Median frequency, dividing signal spectrum for parts with equal energies
Maximum of the signal spectrum or local maximum
Average frequency of registered waveform train
Frequency band of AE signal crossing the preset energy level
Signal power or area of signal spectrum above the preset level in defined frequency bands

where

$\nu(t)$ is the transient value of amplitude (V)
A_m is the maximum signal amplitude registered in the event (V)
f_0 is the frequency in the signal spectrum with maximum intensity (s^{-1})
α_1 is the decrement of the AE signal damping in the system, including the tested body and the AE sensor (s^{-1})
t is the time (s)

Several AE analyzers have the ability of registering the number of counts or number of events in the preset time unit. This method of tracing the activity of the AE sources is called AE count/event number measuring. It is possible to determine the number of counts, N_c, within a single AE event when a certain signal threshold, A_t, is preset using the following formula [7]:

$$N_c = \frac{f_0}{\alpha_1} \ln\left(\frac{A_m}{A_t}\right) \tag{24.4}$$

The registered AE signal, presented in Figure 24.5, with amplitude–time coordinates can be characterized by global quantitative descriptor E, the area of waveform above the mean. This parameter is often colloquially called AE signal energy, because it is to some extent proportional to the exact value of the energy. Assuming that N signal samples were recorded during the investigation, the AE signal energy can be calculated (in volts) from the following equation:

$$E = \sum_{m=1}^{N} \nu(m \cdot T_1) \tag{24.5}$$

where

m is an index of a stored-signal sample
T_1 is an inverse of a sampling frequency (s)

The third part in Table 24.1 lists the AE descriptors as the frequency-domain derivatives. They deal with differences existing in the AE signal, by analyzing its spectral characteristics (Figure 24.6). This method, contrary to the procedure of the AE signal energy determination, does not depend on a specimen volume under the test. If the AE signal, $\nu(t)$, is absolutely integrable, it can be associated

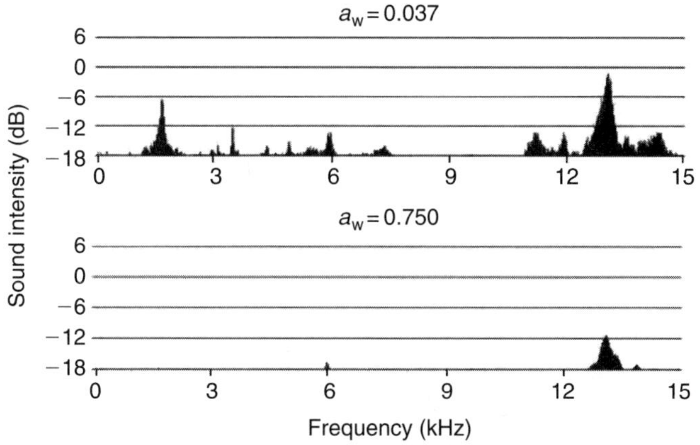

Figure 24.6 AE signal spectral characteristics of the rye-extruded crisp bread as a function of water activity.

with its spectral density function, $A(\omega)$, where ω is a linear analogue of frequency f, $\omega = 2\pi f$, using a Fourier transform:

$$\nu(t) = \frac{1}{\pi} \int_0^{\infty} A(\omega) \sin[\omega t + \varphi(\omega)] d\omega \tag{24.6}$$

where φ is an argument representing a phase of transformed signal. There are numerous computer procedures to derive a discrete image of $A(\omega)$ that consists of a set of coefficients, c_n. In this study, the procedure was designed to carry out this task by organizing the recorded AE signal sample in time windows at a length of 0.25 s. To reject the influence of background noise, one dominant AE burst was detected (if any present) in each section. All the bursts were processed to obtain their power-spectrum function, keeping the same phase of each burst at the transformation process. This algorithm is sometimes called event filtering and enables to suppress the random noise accompanying the recorded signal. As a result, for each time window, the procedure produces a series of coefficients, c_n, and each of them represents AE signal power in a frequency of about 11 Hz. The whole series of c_n cover the desired spectral range of 11–15,000 Hz. The recorded time-dependent AE signal, $\nu(t)$, of each recording is converted into a vector of digital samples. The algorithm performing the $\nu(mT_1) \rightarrow c_n$ transform is based on the standard approximated formula:

$$c_n \approx \frac{1}{N} \sum_{m=0}^{N-1} v(m\,T_1) \bmod \left(e^{\frac{jn2\pi m}{N}} \right) \tag{24.7}$$

where

j denotes square root of -1

mod denotes the modulus of a complex number

In the analysis of acoustic properties of foods, descriptors such as loudness, pitch, energy of the emitted sound, and the number of peaks are mostly used. Much less attention is paid to the analysis of frequency of emitted sound [32].

24.5 MEASUREMENT OF ACOUSTIC EMISSION IN FOOD

During the early decades of acoustic testing of food, sounds emitted by the products were registered with the use of microphones operating in the acoustic frequency band (16–16,000 Hz). After a desired amplification, the collected data were recorded on a tape recorder [33]. In 1980s, a computer equipped with a soundcard or other analog-to-digital converter was introduced in these investigations. An important part of food sounds registration procedure is a kind of applied method of product crushing. Numerous investigators [33,34] examined the sounds during food mastication, biting with their incisors, or breaking by hand. From the beginning of 1970s, machines for the mechanical loading of the products have been applied for tests. The machines (of universal application like Instron or specialized texturometers) were equipped with the operating tool for mechanical loading of the sample, moving with a controlled speed, and an arrangement enabling the registration of applied force [17]. The experiments carried out by Drake [35] revealed that the sounds emitted during the process of food crushing differ from each other with respect to amplitude, frequencies found in signal spectrum representing dominant intensity, and in characteristics of intensity changes in the registered signal amplitude.

In 1976, Vickers and Bourne [33] concluded that biting crisp foods generate specific sounds of sharp, short, and noisy type. Spectral analysis on these sounds revealed that the investigated signal included the components within 0–10 kHz frequency band. These researchers hypothesized that the perceived crispness level is proportional to the sound amplitude of the signal of the product registered by the instrumentation.

In 1979, Vickers and Wasserman [36] attempted to verify the hypothesis that the sound of crushed product includes the pattern specific to the structure of the tested sample. The probability of right product recognition by listening to the prepared records was determined, and about 18 products were examined. Two of them were recognized successfully at a rate greater than 50%, while the success of other product recognitions varied from 0% to 44% of the cases.

The typical process in food technology, in which raw material is processed into a final product, usually produces a complex and multiphase structure. The thermodynamic state of the ingredients and phases in the products containing large amounts of water reaches uniformity in short time, and therefore, the physicochemical equilibrium is established. This is caused mostly by the existing possibilities of molecular movements and molecule mobility in water solution. These movements are remarkably reduced or blocked in low-moisture products, resulting in an internal stress. The magnitude of this stress depends on the parameters of the technological process. In other words, the stresses formed in the product by a technological process cannot relax. The example of blocked relaxation is the process of drying. In extruded products, the internal stresses are created during kneading and flow through a high-temperature nozzle, and later, they are preserved in the material owing to immediate evaporation of water.

Generation of AE signal in dry food products is caused by the sudden release of accumulated elastic energy. This effect can be invoked by crack formation and propagation, as well as by the destruction processes due to application of external stress, for example, during biting. Crack formation and propagation are accompanied by high-level AE, especially in more crunchy, i.e., less deformable and more resistive-to-stress materials. The nature of this phenomenon is not entirely recognized. Its understanding is complicated by the AE signal transformation, where the signal travels across its path from the crack, through the matter under test, toward the AE sensor. Dry food products often have a cellular structure that highly influences the nature of the emitted signal. The interior of the cells is filled with air, and the AE source is formed by cracking the cell walls [29]. The destruction process, including the greater thickness of dry food products causes the generation of a series of sharp electron-acoustic (EA) waveforms, and each AE pulse is invoked by destroying the single cell wall. When the large destruction processes run simultaneously, a specific sound is emitted. The sound amplitude generated by the product of that kind is proportional to that produced by a single cell break and to the number of cell walls per dimension unit of the investigated product [16]. To explain the anomalies in acoustic properties of such two-phase material, complicated formulas and models are proposed, such as the approach described by Malecki [2]. The consequence of two different phases existing in a medium is that the two bands of wavelengths have privileged conditions to propagate, one band in the first phase and the other in the second phase.

Spectral characteristics are valuable sources of parameters of the generated AE signal. Signal differences presented in the averaged spectral characteristics of crisp bread baked by the traditional method, biscuits, crackers, and cornflakes are presented in Figure 24.7. Each product shown in Figure 24.7 emits a signal with unique spectral characteristics. Both low- and high-frequency components are present in these characteristics. The chemical composition influences fewer characteristics than the way the product is produced.

Regions in the frequency domain where the high level of power-spectrum function is observed for different food products are as follows [37]:

- Rye-crisp-bread baked by traditional method: 2–8 kHz and 13–14 kHz
- Crackers: 2–3 kHz and 14 kHz
- Low-fat biscuits: 2–6 kHz and 12–14 kHz
- Rich-in-fat biscuits: 2–4 kHz and 13 kHz

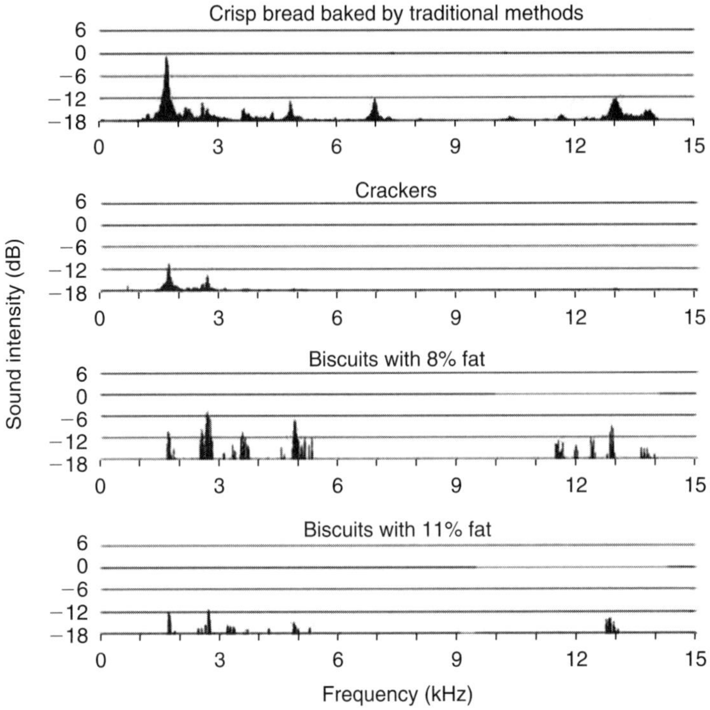

Figure 24.7 Spectral characteristics of different food products recorded during a bending test.

Food technologists attribute great importance to water content and its migration when testing the texture changes of food. Interesting experimental results were obtained by Tesch et al. [17] who attempted to equilibrate the water activities (a_w) of cheese balls and croutons stored in desiccators. In their study, food samples with water activity ranging from 0.11 to 0.75 were investigated. During specimens' compression process using the Instron machine, acoustic signatures of the crushing process were recorded using a microphone probe. The following AE signal descriptors were used for product characterization: mean (averaged) value of the time signal, peak amplitude and amplitude standard deviation of time signal, and mean magnitude of its Fourier spectrum. Although the scatter of all four descriptors was "fairly large," the researchers concluded that all the descriptors clearly illustrated the fact that at a_w above 0.65, the AE signal practically vanished, owing to the plasticizing effect of water. Maximum product crispness was observed at $a_w = 0.23$. For higher water activity levels, AE intensity loss was observed.

Other investigations [16,25], made with different texturometers for crushing the products with different moisture levels, generally agreed with those obtained by Tesch et al. [17]. However, no distinct maximum limit of the AE signal intensity descriptors was observed in some products, but AE descriptors measured in all of them demonstrated a remarkable variation of the AE intensity in the range of a_w from 0.4 to 0.6. This was confirmed by sensory tests for crispness done by the trained assessors.

There are several studies on the sound generated by extruded crisp bread [16,24,25] that describe the results obtained using a microphone probe. In the experiment done by Daceremont [16], a sound transmitted by a bone tract was registered. To perform this, the probe was placed on the human mastoid. The other probe was capable of registering a sound transmitted by an air tract. The conclusion of the investigation was that the sounds in the frequency band of 5–10 kHz propagate by both the air and the bone tract, while those in the frequency band of 1–2 kHz propagate only by air.

The number of AE counts in biscuits was also determined by Chen et al. [20] to determine their crunchiness. Registered with a microphone probe, the AE signal was processed in the acoustic detector of the acoustic emission detector (AED) type. The procedure was designed to analyze the sound intensity, and on that basis, to recognize the acoustic events of $\sim$250 ms duration. The instrumentation was capable of altering the event duration gate to fit the different food products: chips and products with crunchiness lower than that of chips. They concluded that the method used in their experiment was capable of detecting the variations of the crunchiness in the materials they examined.

The above-presented data show that the AE of food can be analyzed in multiple ways and expressed by a number of physical indices. Descriptors used in the research of acoustic properties of food are listed in Table 24.2, which shows the complexity of the AE analysis.

Volume stress can exist not just in dry products. Its presence was also confirmed by Konstankiewicz and Zdunek [13] in hydrated bodies under conditions where they contain liquid in the cell interior, which exerts a pressure onto the surrounding cell wall. In a tissue that is under turgor pressure, cell walls are stressed and are perceived as firmness by the consumers. Fruits and vegetable tissues have a very complex structure, in which the basic elements are cell walls, responsible for the mechanical strength. This strength varies among cell walls, and hence, the natural dispersion of internal stress and stored energy can be observed. A part of that energy remains in stressed cell membranes (i.e., in wall and plasmalemma) and in the intercellular layers of the lamellae. Application of the AE method to investigate moist plant tissues requires the identification of possible AE sources. Changes in the tissue structure, evoked by external forces or internal stress due to moisture variations, are one of the possible origins of AE [38]. Sound emitted by fresh fruits and vegetables is caused by a rapid loss of intercellular pressure [33]. Zdunek and Konstankiewicz [39] presented a description of the processes, believed to be the possible reasons for the generation of AE signal.

Deformation of the plant tissue by external force causes changes in both the state of internal stress and internal energy. If the external force reaches a critical level, cellular walls would break and a new configuration of stresses would be formed. Evolution of the destruction process of such a structure would depend on the structural parameters of the investigated material, cell turgidity level, and properties of external factors (e.g., load increase speed) [13,40]. The phenomenon described earlier, with respect to the crack formation and propagation in plant tissue, becomes a source of AE signal (Figure 24.8).

The AE method enables the investigation of the influence of different factors on the plant tissue-destruction processes. The analysis of the dependence of AE descriptors on the mechanical factors (e.g., stress level) helps in determining the mechanical conditions leading to the onset of the plant-tissue breaking processes. Moreover, the assessment of the amount of the breaking processes is also possible [39]. The effectiveness of the AE method in recognizing and investigating the plant-tissue breaking processes was observed to be high [40]. It was concluded that there is a possibility of applying the described method in the investigation of potato tubers, and the frequency spectrum of the registered AE signal has specific maxima at 60, 75, 115, and 135 kHz. The other result described was that the initial AE activity occurs at product elongation, equaled to approximately 65% of the critical one.

24.6 CALCULATION OF ACOUSTIC EMISSION DESCRIPTORS IN FOOD

The acoustic activity of food can be expressed by several descriptors. These include AE signal energy, number of AE events, energy of the AE event, partition-power spectrum slope, and acoustograms. The AE signal energy is calculated based on the recorded time-dependent AE signal. The calculation is carried out according to Equation 24.5.

Table 24.2 Descriptors of AE Measured Instrumentally

Acoustic Parameters	Products	Analyzed Frequency Range	References
Amplitude	Carrot, almond, extruded bread, dry biscuit	10 Hz–10 kHz, 20 Hz–20 kHz	[16]
	Snack food	1 Hz–10 kHz	[59]
	Dry apple, dry carrot	1 Hz–15 kHz	[61]
	Cereal flakes	0–10 kHz	[19]
	Biscuits	1 Hz–12 kHz	[20]
	Crisp bread	1 Hz–15 kHz	[38,44,48]
	Cereal flakes	1 Hz–15 kHz	[63]
	Apple, potato	100 Hz–1 MHz	[57]
	Cornflakes	1 Hz–15 kHz	[62]
Number of sound bursts	Celery fresh, turnips, wafer		[60]
Number of acoustic events	Biscuits	1 Hz–12 kHz	[20]
	Crisp bread	1 Hz–15 kHz	[44,48]
	Cereal flakes	1 Hz–15 kHz	[63]
	Apple, potato	100 Hz–1 MHz	[57]
	Cornflakes	1 Hz–15 kHz	[62]
Energy	Sponge fingers, wafer biscuit, ice cream wafers	25 Hz–20 kHz	[15]
	Crisp bread	1 Hz–15 kHz	[21,45]
	Bread, crackers, cake	1 Hz–15 kHz	[43,48]
	Cornflakes	1 Hz–15 kHz	[63]
	Cornflakes	6.3 Hz–20 kHz	[62]
Mean sound pressure level (N/m^2)	Sponge fingers, wafer biscuit, ice cream wafers	0.5–3.3 kHz	[64]
	Extruded bread, dry biscuits, toasted rust roll	6.3 Hz–20 kHz	[58]
Mean sound pressure level (dB)	Biscuits, apples	25 Hz–20 kHz	[15]
Acoustic intensity (W/m^2)	One food among several	0.5–3.3 kHz	[64]
Partition-power spectrum	Crisp bread, crackers	1 Hz–15 kHz	[44,48]
	Cereal flakes	1 Hz–15 kHz	[63]
	Dry apple, dry carrot	1 Hz–15 kHz	[61]
Fast Fourier analysis	One food among several	0.5–3.3 kHz	[64]
	Apples	25 Hz–20 kHz	[15]
	Potato chips	0–20 kHz	[28]
	Carrot, almond, extruded bread, dry biscuit	10 Hz–10 kHz, 20 Hz–20 kHz	[16]
	Crisp bread, crackers	1 Hz–15 kHz	[43,48]
	Cereal flakes	1 Hz–15 kHz	[63]
	Dry apple, dry carrot	1 Hz–15 kHz	[61]
	Extruded bread, dry biscuits, toasted rusk roll	6.3 Hz–20 kHz	[58]

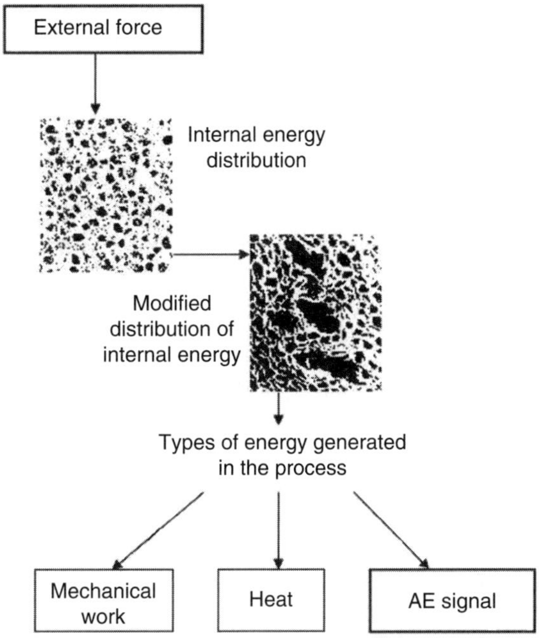

Figure 24.8 The scheme of AE signal creation in plant tissue. (From Konstankiewicz, K. and Zdunek, A., *Acta Agrophis*., 24, 87, 1999 (in Polish). With permission.)

Registered AE presented in the amplitude–time coordinates exhibited a number of visible AE bursts caused by the matrix-break processes. More precise signal investigation revealed a typical pattern of bursts among all the records. Within that pattern, a series of AE impulses of similar amplitude and duration occurred. A characteristic noisy and crispy sound of food products was formed by these impulses. Therefore, they can be treated as the integral elements of the analyzed signal and are called AE events.

A procedure to calculate the parameters of a single AE event, designed by Ranachowski [41], is capable of analyzing the records of 10 s of AE signal converted into a set of digital samples. The recognition of the occurrence of the AE event is done in the following way: AE event duration is measured as a time when the signal energy, calculated using Equation 24.5, exceeds a preset level. Both the start and the end events were determined on the basis of the energetic criterion mentioned earlier. Under practical conditions, the AE signal energy does not fade to reach a zero level of energy, owing to the accompanying background noise. Because of this, the end of the event is recognized when the energy of the three signal samples is lower than half of the preset level applied to detect the start of the event. When the end of the event is reached, the next two signal samples are omitted to prevent the procedure from the false recognition of the event already processed. Practical examination of the procedure determined the optimal detection threshold for cornflakes to be 2 V and for crisp bread as 3 V. Application of higher thresholds could result in losing of the weaker AE events, while low threshold causes false recognition due to background noise. A software procedure described earlier enables determination of the following parameters of the detected AE events: event duration and peak amplitude of the AE signal within the event, number of events per second, and energy of a single event.

The elastic wave propagates in a different way with respect to the frequency. In the 1 kHz and 10 kHz regions, acoustic wavelength, according to Equation 24.1, differs 10 times. Since food texture presents a resonance character to the propagating elastic wave, its wavelength plays a significant role in sound generation. Moreover, low-frequency waves propagate mostly across the air cavities of a product, while the high-frequency components prefer solid-state phases.

The shape of the acoustic-spectral characteristics (Figure 24.3) reveals that, in most of the investigated foods, there are two regions in the frequency domain, where the high level of power-spectrum function is observed. Taking into account this observation, a dimensionless AE descriptor, independent of the sample volume, was proposed. This coefficient is called partition-power spectrum slope (β). According to Equation 24.8, it is calculated as a ratio of AE signal power spectrum registered in the high-frequency range, labeled as P_{high}, and AE signal power registered in the low-frequency range, labeled as P_{low}.

$$P_{\text{high}} = \sum_{n\mapsto \text{high 1, kHz}}^{n\mapsto \text{high 2, kHz}} c_n, \quad P_{\text{low}} = \sum_{n\mapsto \text{low 1, kHz}}^{n\mapsto \text{low 2, kHz}} c_n, \quad \beta = \frac{P_{\text{high}}}{P_{\text{low}}} \tag{24.8}$$

The best measuring conditions for measuring β coefficient exist during the compression test. In this test, elastic waves propagate from the destructed region of the product toward the compressing plate across the larger volume than during the bending process. The researchers of this study suggest that determination of a single AE signal descriptor to characterize complex food products of multilayered composition results in poor effects. Acoustic investigation of an inhomogeneous material can be carried out with the application of an experimental results presentation technique called acoustogram. An acoustogram presents collected data of the process of a product deformation and is a three-dimensional graph. The horizontal axis represents time, the vertical axis indicates the spectral characteristic of the measured signal, and the signal intensity is visualized using color coding. A single graph "slice" is performed by applying Equation 24.7. The acoustograms presented here are constructed with the horizontal resolution of 0.25 s and vertical resolution 22 Hz, in the frequency band of 1–15 kHz. The construction of color scale refers to the reference signal level A_0 (equal to 10 mV rms in 22 Hz band), by assuming that signals weaker than −16 dB in relation to A_0 would be presented in dark blue. The other signal intensities are presented in different colors.

The software designed to create acoustograms also produces a digital image of the data, which enables precise comparison of the product properties. Figures 24.9 and 24.10 demonstrate the application of the described procedure to present experimental data of bending of four samples of flat rye-bread, differing in water activity level and the method of their production. The bands of high level of power-spectrum function are clearly visible in the figure. The effect of signal-energy loss caused by the water sorption process can also be observed. In the latter case, high-frequency components increase with respect to the low-frequency components. The acoustograms were also applied for testing the bread staling process [42].

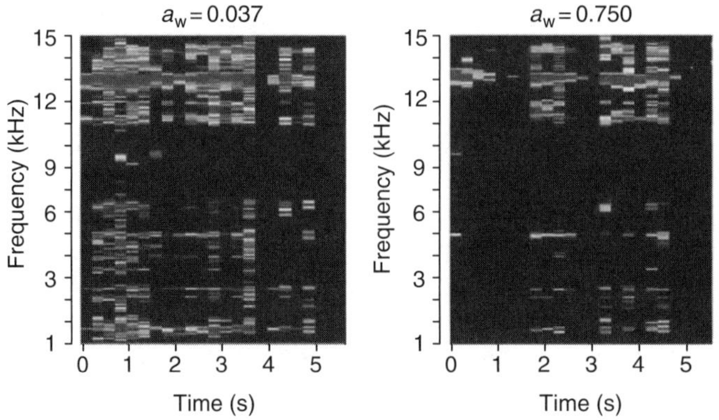

Figure 24.9 Acoustograms of bending of rye-extruded crisp bread as affected by water activity.

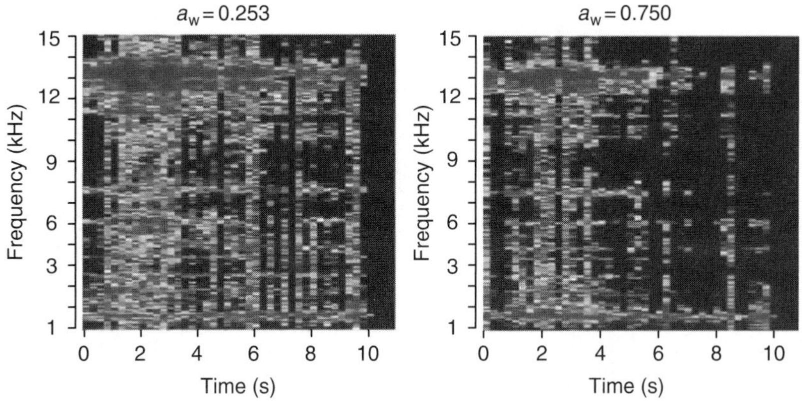

Figure 24.10 Acoustograms of bending of rye-crisp-bread baked by the traditional method, affected by water activity.

The AE method is capable of delivering a useful tool to determine the existing product crispness level. It was proposed by the authors of this chapter to introduce the crispness index χ, combining both mechanical and acoustic properties of a tested product. The coefficient can be determined as a ratio of the registered number of AE events and the mechanical work applied to destruct the product under normalized conditions specific to the product type. Hence,

$$\chi = \frac{\text{AE number of events}}{W} \tag{24.9}$$

where

$W = u \int_0^t F(t)\mathrm{d}t$ is the work of mechanical destruction of the product (mJ)
$F(t)$ is the current level of the force registered by a loading machine during a test (N)
u is the velocity of loading cross-head (m s^{-1})
t is the test duration (s)

24.7 FACTORS AFFECTING ACOUSTIC EMISSION IN FOOD

Investigations have been made to determine the influence of the mechanical-loading speed and the kind of product manufacturing method on the parameters of the generated AE signal [43]. The influence of the mechanical-loading method at the identical loading speed of 20 mm/min on the AE signal energy is presented in Figure 24.11. The results showed that the generated AE signal energy has a significantly greater level when compression is applied in comparison with bending. This effect is caused by the varied size of the contact of loading tool-tested product areas. In addition, the process of product destruction differed during both the tests, since during bending, the contact occupies a rather small specimen area that evokes only a single process of product cell-breaking. The other difference is that at the upper specimen surface, the loading tool evokes compression stresses, while at the lower surface stretching stresses exist. When compression as a method of loading was used, the area of loading tool-tested product was large and caused higher AE energy level. Contrasting effects were observed during the investigation of a product containing more fat, i.e., crackers [44]. It is worth mentioning that the spectral characteristics derived for both the destruction methods, presented in Figures 24.12 and 24.13, are different. Both the destruction methods evoke AE signal in low- and high-frequency regions. The amount of low-frequency band energy dominates during bending, while high-frequency band dominates during compression test.

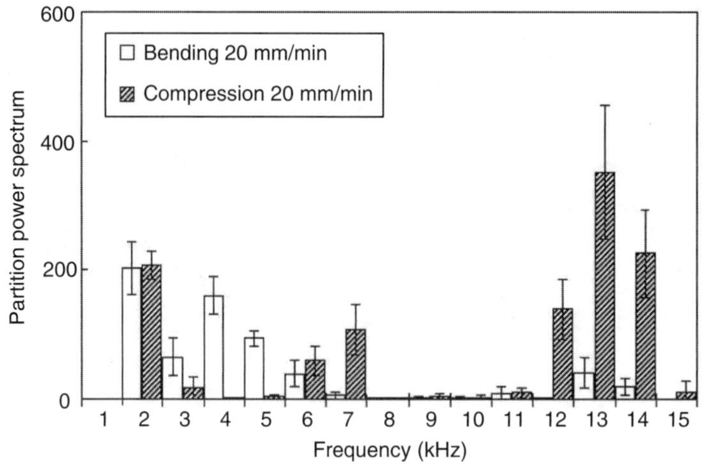

Figure 24.11 Effect of two mechanical-loading methods on the spectral characteristics (presented in 1 kHz bands) of the generated AE signal in low-fat cakes.

When loading speeds of bending processes were set to 20 and 50 mm min^{-1}, the spectral characteristics of AE signal presented the activity in the 13 kHz region. The same speeds applied during the compression test resulted in the shift of the significant activity level to the band of 1–7 kHz and 12–14 kHz (Figures 24.12 through 24.15). Increase in mechanical-loading speed led to the increase in AE signal energy, regardless of the kind of loading method applied (Figures 24.11 and 24.14). The reason of that effect can be that when lower speed of mechanical loading is used, the process of energy dissipation in the product becomes more stable, and low number of micro-cracks arising evokes low level of the AE signal. On the contrary, high rate of the release of accumulated energy causes high intensity of AE.

The analysis of the spectral characteristics, presented in Figures 24.12 and 24.13, leads to the conclusion that the compression test should be a recommended method of mechanical testing of the cereal products. Application of the compression test enables increased contact between the

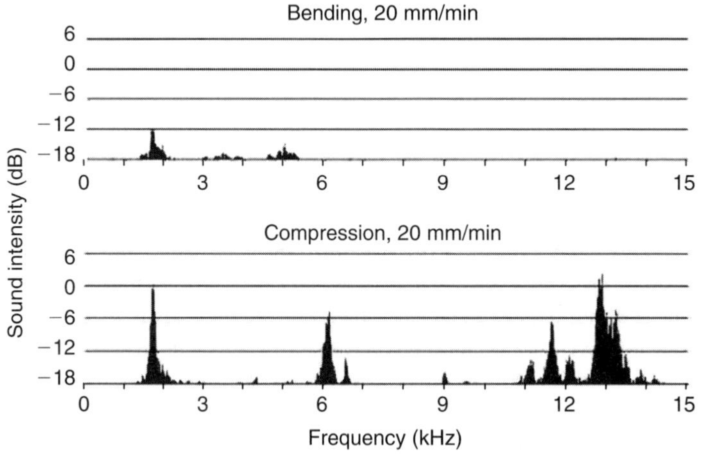

Figure 24.12 Comparison of the two mechanical-loading methods on the spectral characteristics of the generated AE signal in low-fat cakes. The upper graph illustrates the bending process and the lower graph illustrates the compression process at 20 mm/min.

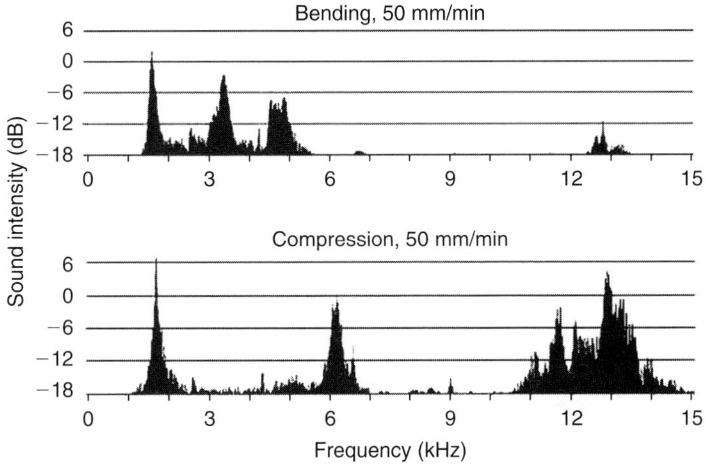

Figure 24.13 Comparison of the two mechanical-loading methods on the spectral characteristics of the generated AE signal in low-fat cakes. The upper graph illustrates the bending process and the lower graph illustrates the compression process at 50 mm/min.

elastic waves propagating in the specimen and the flat plate of the loading tool—guiding the waves to the sensor. A similar comparison of the methods of the specimen loading was made with respect to the crisp bread. During the compression or the bending tests, the spectral characteristic of the AE signal presented the activity in 1–3 kHz and 12–15 kHz regions. Acoustic energy was generated at a higher level during the compression test than that during the bending test of the crisp bread. The dependence of the partition-power spectrum slope, β, on water activity measured in extruded rye- and wheat-crisp-bread specimens is presented in Figure 24.16. Till the level of $a_w \cong 0.4$, the β coefficient remained unchanged, whereas after crossing of this level a significant increase in β was observed. A similar trend in β changes was observed in extruded wheat-crisp-bread specimens and crisp bread baked by the traditional method [21,37,45].

The other character of spectral characteristics versus water activity was observed in products compressed in bulk, such as cornflakes [26,46]. Regions of a high level of power spectrum in this

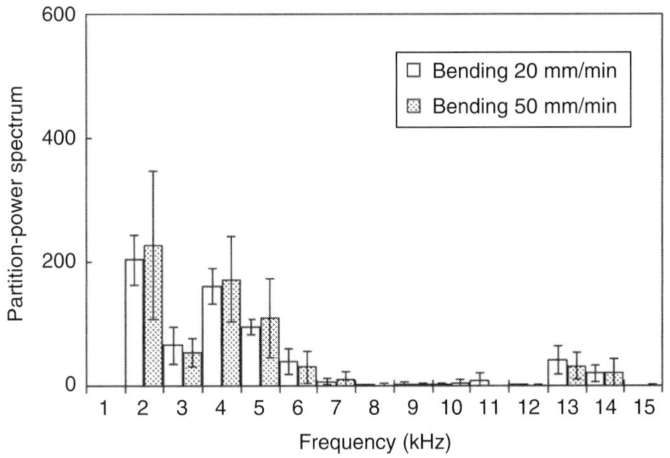

Figure 24.14 Effect of the two mechanical-loading velocities on the spectral characteristics (presented in 1 kHz bands) of the generated AE signal in low-fat cakes during bending test.

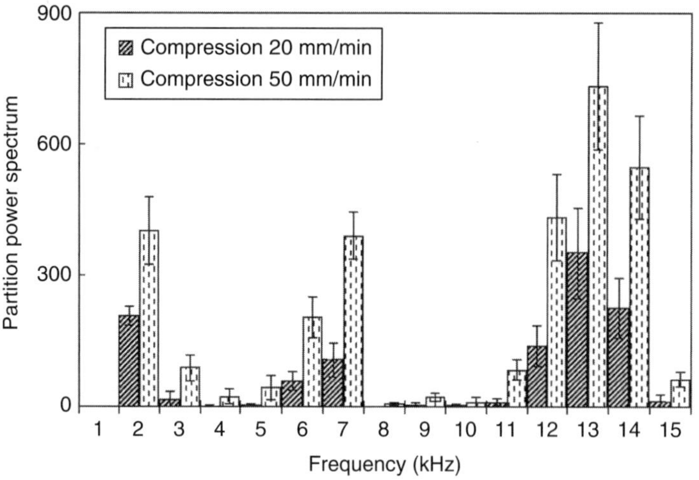

Figure 24.15 Effect of the two mechanical-loading velocities on the spectral characteristics (presented in 1 kHz bands) of the generated AE signal in low-fat cakes during compression test.

product were situated in the frequency bands of 4–9 kHz and 14–15 kHz [47]. In the range of products of water activity from 0.05 to 0.45, the partition-power spectrum slope was constant (Figure 24.17) and at higher water activities β decreased.

Application of different manufacturing technologies (backing with traditional method and extrusion) results in the formation of different product structures. The dependence of the detected number of events per second on water activity—calculated for three food products: extruded wheat, rye bread, and rye bread baked by traditional method—is presented in Figure 24.18. However, from the statistical point of view, no valid difference was found when the events rate for the extruded wheat and rye bread at the same water activity level was compared. Spectral characteristics of these products are presented in Figure 24.19. The experimental results presented in Figures 24.18 and 24.19 lead to the conclusion that the manufacturing technology affects the parameters of the emitted AE more significantly than the chemical composition of the investigated products. The influence of the applied manufacturing technology on the acoustic properties of the

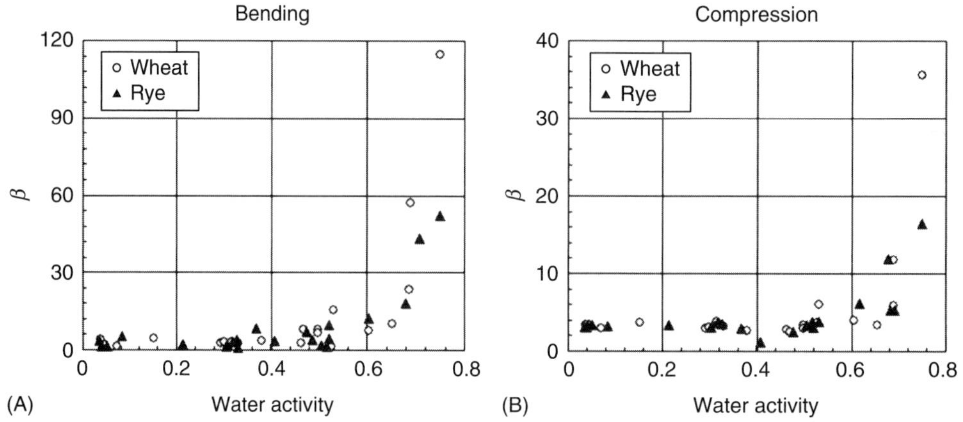

Figure 24.16 Influence of the mode of deformation and water activity of extruded rye- and wheat-crisp-bread on partition-power spectrum slope β. (A) Bending and (B) compression.

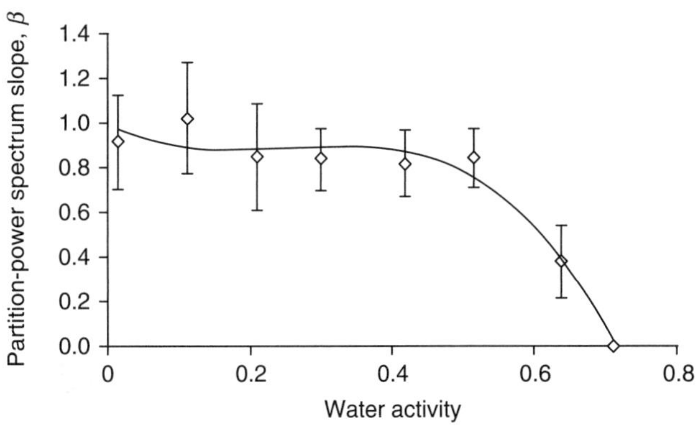

Figure 24.17 Influence of water activity on the partition-power spectrum slope β for cornflakes. (From Kuropatwa, M., Influence of water activity on mechanical and acoustic properties of corn flakes. M.Sc. thesis, Warsaw Agricultural University (SGGW), Warszawa, 2003 (in Polish).)

product can be most efficiently demonstrated by a comparison of a spectral characteristics of the product [43].

The influence of water activity on acoustic properties of dry foods is very pronounced. The resulting 3 s long records registered in the region of maximal AE signal activity for five levels of water activity are presented in Figure 24.20. The graphical resolution of Figure 24.20 enables the detection of short AE signal bursts of 5 ms duration. When a recording speed of 44,100 signal samples per second was used, the real AE signal record, stored in the memory of a computer consisting of a file of samples registered with a time delay T_1, equaled 22.7 μs, successively. The large number of discrete signal samples can be treated as a real image of the effective value (rms) of the emitted AE signal. The discrete form of the signal, stored and processed in the computer, is denoted as $\nu(m \cdot T_1)$ and using Equation 24.5, the AE signal energy can be calculated. A remarkable influence of specimen water activity on the AE signal energy is presented in Figure 24.21.

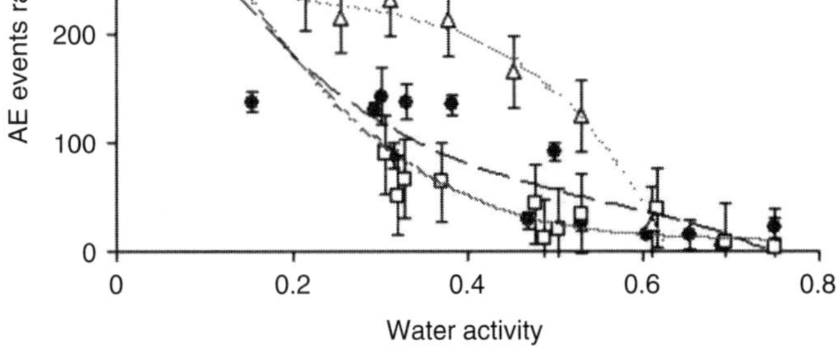

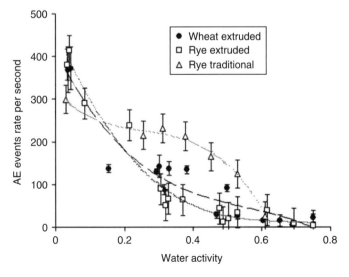

Figure 24.18 Influence of the mode of processing and water activity on the number of acoustic events generated in crisp breads.

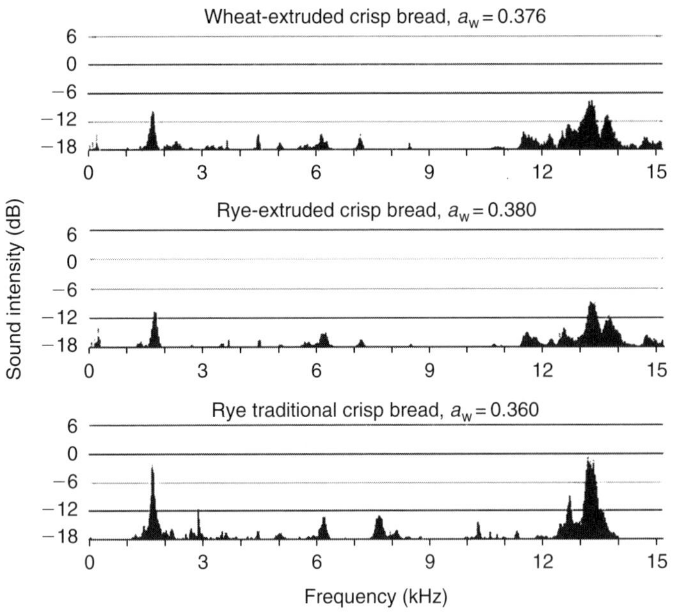

Figure 24.19 AE signal spectral characteristics of crisp breads.

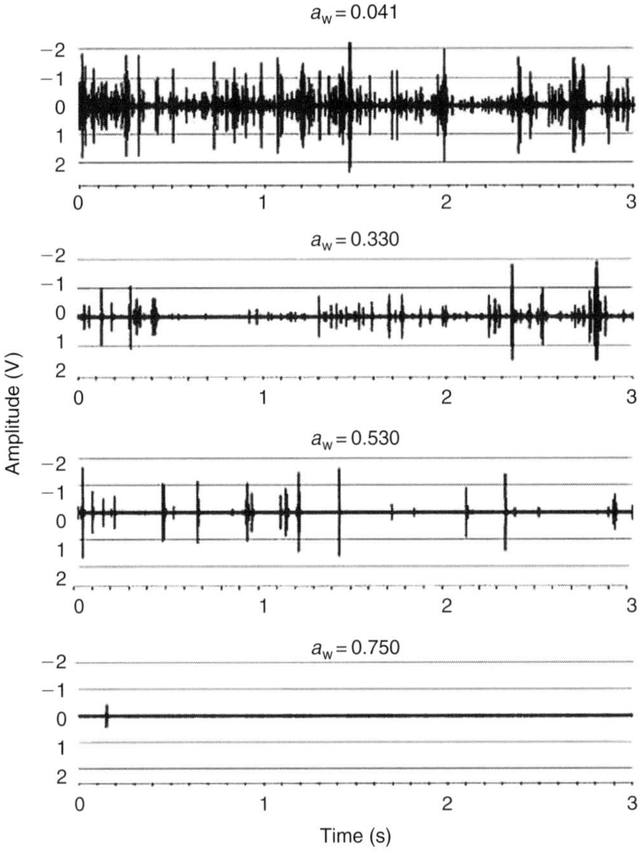

Figure 24.20 Test results of bending process of extruded wheat-crisp-bread specimens for different water activity levels. Signal amplitude is presented in volts, as is registered at the output of AE sensor with added 100 times amplification.

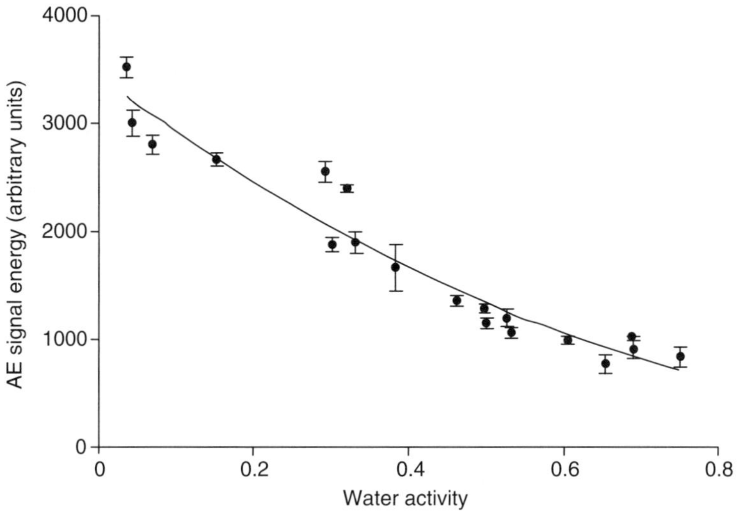

Figure 24.21 Influence of water activity on AE signal energy (Equation 24.5) in flat, extruded wheat bread.

Water activity also affects the number of acoustic events [48], which decrease with increasing water activity (Figure 24.18). This is caused by the altered stress distribution occurring in the moisturized structures [49]. The increase in water content causes dissipation of the elastic energy stored in the material and in turn, reduces the possibility of crack occurrence [50].

It is interesting to note that the energy of a single acoustic event is a little dependent on the water activity of dry foods (Figure 24.22). In crisp-extruded breads, the energy of a single acoustic event was constant in water activities range of 0.03–0.5, and thereafter, it decreased by about 15% [48]. This suggests that even at high water activities, there are domains capable of breaking and generating vibrations with such energy as that emitted in the dry material. The probability of occurrence of such domains decreases with increasing water activity; hence, the number of acoustic events is affected by the water activity and is lower when the wetness of the material is higher.

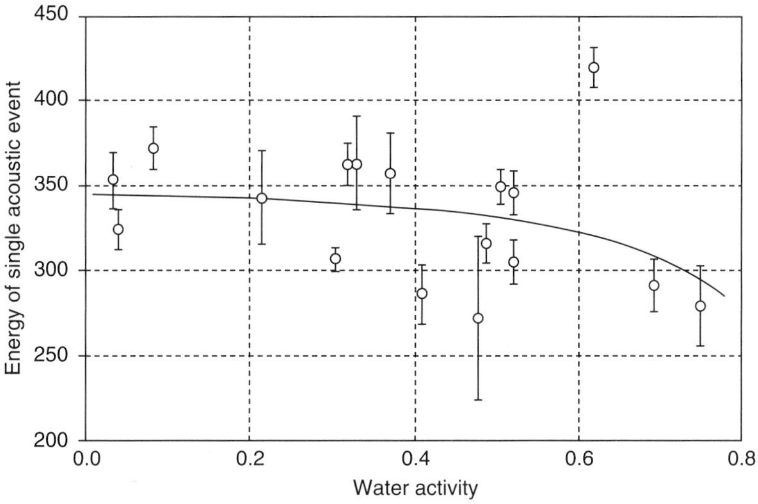

Figure 24.22 Influence of water activity on the energy of a single acoustic event.

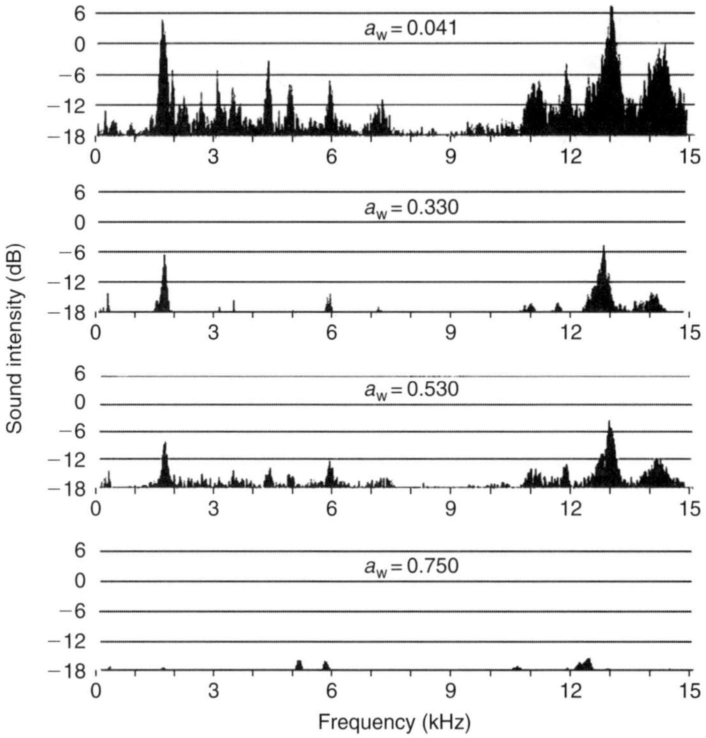

Figure 24.23 Spectral characteristics of AE signal generated during the bending process of wheat-extruded crisp bread with different water activity levels.

Spectral characteristics of the AE signal generated in the bending process of extruded wheat-crisp-bread specimens with different water activity levels are presented in Figure 24.23. It can be observed that in the investigated material, there are two regions in the frequency domain where the high level of power-spectrum function is observed. For different kinds of extruded crisp bread, these regions are 1–3 kHz and 12–15 kHz.

The AE generated during crushing of cereal products constitutes a significant factor of the features, described as crispness and crunchiness. The consumers are capable of distinguishing between crispy and crunchy sounds. A crispy sound is sharp such as that produced while walking on snow or icy ground, whereas the crunchy ones sound longer and are more firm as that produced while walking on gravel or dry leaves [32]. Recent investigations made by the authors of this work were focused on the phenomena of sound generation with respect to the perceived intensity of crispness and crunchiness. The main reason for the loss of these features is the increase in water content in cereals caused by sorption processes from the environment or its migration from the product interior [51]. The investigations done by Marzec et al. [27] and Lewicki et al. [26,37] proved that food remains crispy within a narrow range of water activity, which is a specific value for a given product.

American researchers have proposed a theory stating that crispness is a product texture feature mostly of acoustic nature [33,35]. Mohamed and Jowitt [15], and Liu and Tan [52] have stated that crispy sounds are related to the structure breaking processes and these effects appear when the energy accumulated in the product is instantaneously released in the form of elastic waves with acoustic frequencies. Sound generation is also related to crack increase and fast material destruction.

The AE signals can be used to detect the changes in the quality of food. An example is shown in Figure 24.24, which presents the influence of the staling process of bread on the AE signal [53]. It is

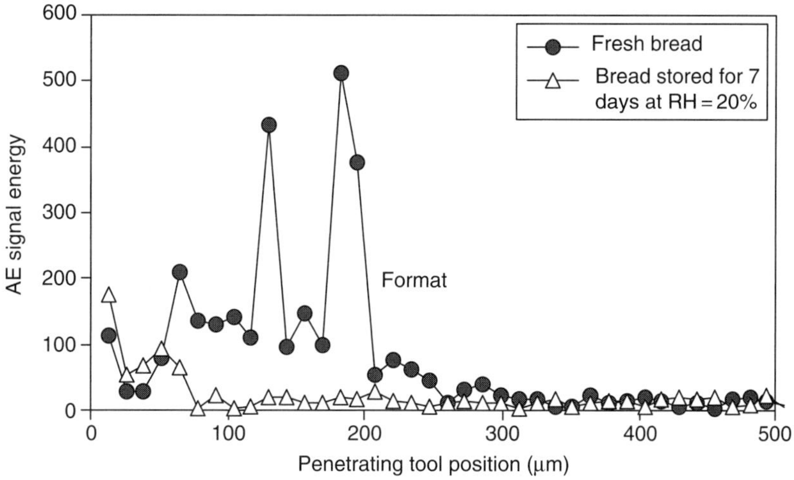

Figure 24.24 Changes in acoustic properties of bread crust during storage.

evident that fresh bread shows a crisp crust of about 200 μm thickness, which emits a strong acoustic signal during penetration. The moist crumb is acoustically inactive. During storage, the moisture moving from the interior of the loaf moistens the surface and the crust loses its attractive property. There is no difference in acoustic activity between the crust and the crumb after 7 days of storage of bread.

The AE method is capable of delivering a useful tool to determine the existing product crispness level. It is proposed to introduce the crispness index, χ, combining both mechanical and acoustic properties of the tested product. The coefficient is calculated according to Equation 24.9.

Figure 24.25 presents the influence of water activity on the crispness index, χ, measured in two kinds of crisp breads. The products were baked by the traditional method. They were bended with the speed of 20 mm min^{-1} and the system amplification was set to 20 dB. As can be observed, the χ coefficient decreases with respect to the increasing water activity, although there are no statistically measurable variations of the coefficient in the region of water activity from 0.2 to 0.4. The sensory tests have also revealed that in that region of water activity, the product is assessed as

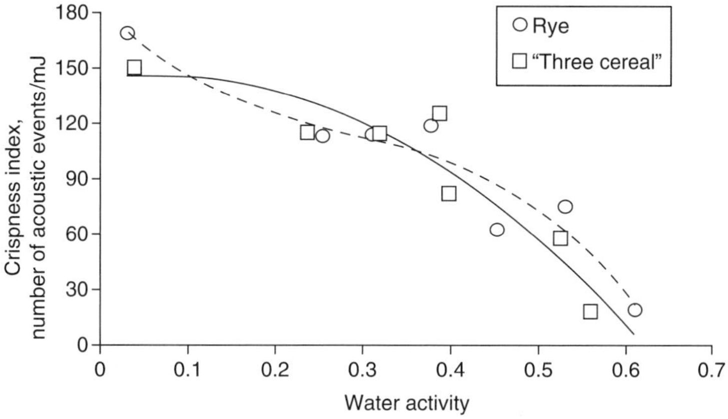

Figure 24.25 Effect of water activity on the crispness index of crisp breads.

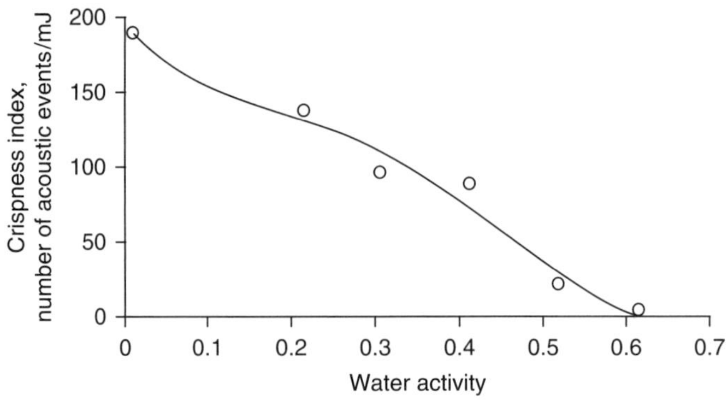

Figure 24.26 Effect of water activity on the crispness index of crackers.

crispy and acceptable [54]. Hence, the proposed coefficient can be applied as a useful tool to characterize the textural properties of cereal products [44].

The analogical trend of the dependence of crispness coefficient on a_w was determined in crackers (Figure 24.26) by performing the bending test at the speed of 60 mm/min and with the system amplification set to 40 dB. It was found that the product crispness decreases exponentially with an increase in water content. Again, in the region of water activity from 0.2 to 0.4, the statistically measurable variations of the crispness index could be neglected.

The profile assessment of the texture (quantitative descriptive analysis, QDA) and general sensory assessment of the crackers as a function of water activity were undertaken by Gondek and Marzec [54]. The researchers were focused on the assessment of mechanical (kinetics) and acoustic features of the texture, and these features were considered as optimal means to describe the textural properties of the crunchy and crispy products [29,55]. The investigation proved that the acoustic impression is a more sensitive indicator of the quality changes occurring in the product due to water sorption than mechanical impressions [54].

24.8 CORRELATIONS BETWEEN ACOUSTIC EMISSION DESCRIPTORS AND SENSORY ASSESSMENT OF QUALITY

The comparison between the measured number of AE events and the sensory assessment of the sound intensity as a function of water activity is presented in Figure 24.27. Sensory assessment of crackers with water activity from 0.225 to 0.670 was carried out by trained persons using QDA, according to Stone and Sidel [56]. The procedure used in the assessment is described in the Standard ISO 13299:2300 (E). In the preliminary tests, acoustic descriptors accounting for sensory-texture profile were chosen. The loudness of the sound was defined as the intensity of emission generated during the mastication of the material with teeth. A general texture quality was defined as the sensory impression obtained from mechanical and acoustic senses and their harmonization [54]. The investigation revealed that crackers in a_w region from 0.225 to 0.490 were acceptable by the consumers, and there were no significant changes in both the investigated parameters in that region.

Figures 24.28 and 24.29 present the relationships between acoustic properties assessed by the sensory impression and those measured instrumentally. It is evident that the general quality of the sensory impression is related to the AE energy and to the crispness index, having a logarithmic-type relationship. Although both AE energy and crispness index change with water activity in the range from 0.225 to 0.490, the total sensory quality changes insignificantly within this a_w range.

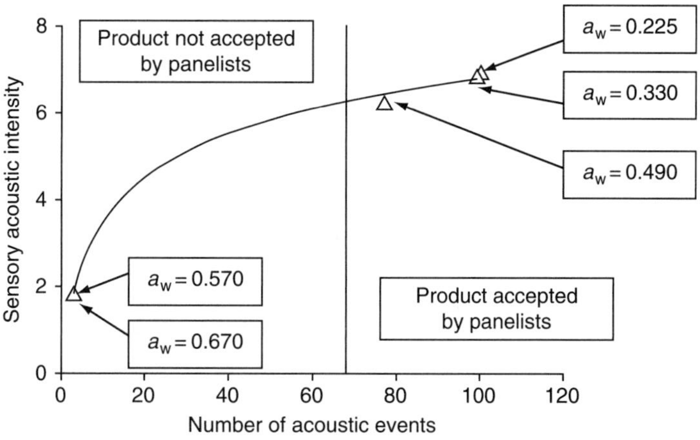

Figure 24.27 Relationship between the measured number of acoustic events and the sensory assessment of sound intensity of crackers as a function of water activity.

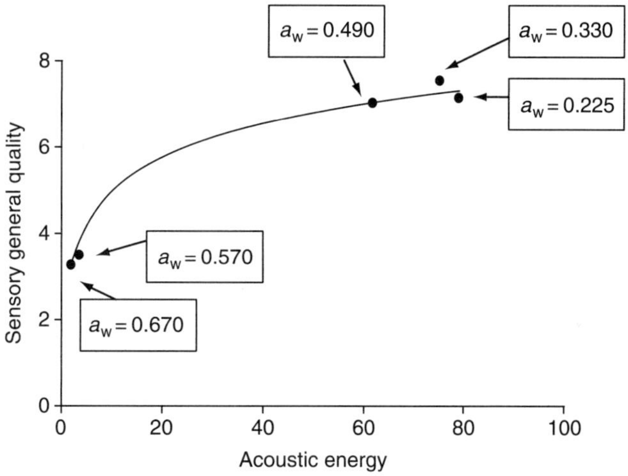

Figure 24.28 Relationship between the measured acoustic energy and the sensory assessment on the general quality of crackers as a function of water activity.

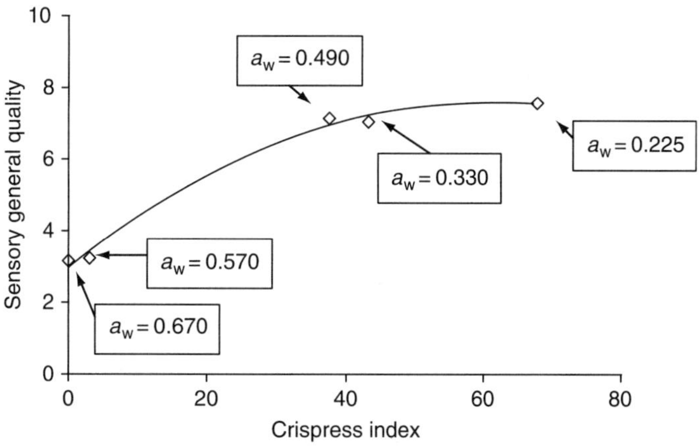

Figure 24.29 Relationship between crispness index and the general sensory quality of crackers as a function of water activity.

24.9 RECAPITULATION

The AE method can be applied to determine the quality characteristics and microstructure changes in the investigated material subjected to mechanical stress. The analysis of the emitted sound delivers the information about the nature of the destruction processes and enables detection of the onset of these phenomena that appears before the occurrence of the maximum material stress. The determined AE signal energy contains information about the scale of the destruction processes. Frequency of the emitted sound is closely related to the kind of destruction process and the type of the tested material, and therefore, it can be used as a descriptor for the quality assessment of fruits and vegetables and for the assessment of the crispness index for snacks. Examples of the AE signal processing described in this work are the evidence for the possibility of determining the AE descriptors in food products. These descriptors can be applied in quality testing and determination of crispness in cereals. To perform a proper analysis of the registered AE signal, sufficient knowledge on the influence of product texture on the image of emitted signal is needed. It is also worth mentioning that a variety of software tools to process AE signals, both in frequency and time domain, have been recently prepared to improve the research work.

ACKNOWLEDGMENT

The authors gratefully acknowledge the financial support of the State Committee for Scientific Research (Grant No. 3PO6T 040 25).

REFERENCES

1. Sliwinski, A. *Ultrasound and Its Application*, 2nd ed. WNT, Warszawa, 2001 (in Polish).
2. Malecki, I. *Physical Foundations of Technical Acoustics*. Pergamon Press, Oxford, 1969.
3. Cempel, Cz. *Applied Vibroacoustics*. PWN, Warszawa, 1978.
4. Speaks, C.E. *Introduction to Sound*. Singular Publishing, San Diego, 1999.
5. Ozimek, E. *Sond and Its Perception. Physical and Psychoacoustic Aspects*. PWN, Warszawa, 2002 (in Polish).
6. Scott, I. *Basic Acoustic Emission*. Gordon and Breach Science Publishers, New York, 1991.
7. Malecki, I. and Ranachowski, J. *Acoustic Emission. Sources, Methods and Application*. Pascal Publications, Warszawa, 1994 (in Polish).
8. Kamiński, Z. *Physics for the Candidates to Study in Universities*. Vol. 1. WNT, Warszawa, 1984 (in Polish).
9. American Standard E 1316 91b (ASTM Standards, 1991).
10. Holroyd, T. and Randall, N. Use of acoustic emission for machine condition monitoring. *British Journal of Nondestructive Testing*, 35(2), 75–78, 1993.
11. Rajewska, K. Experimental analysis of destruction of dried material, in *Problems of Deformation and Destruction of Dried Materials,* S.J. Kowalski, Ed. Wydawnictwo Politechniki Poznańskiej, Poznań, 2000, pp. 129–150 (in Polish).
12. Malecki, I. and Opilski, A. Characteristics and classification of AE signals: Basic conceptions, in *Acoustic Emission. Sources. Methods and Applications*, Malecki, I. and Ranachowski, J., Eds. Wydawnictwo Biuro Pascal, Warszawa, 1994, pp. 18–33 (in Polish).
13. Konstankiewicz, K. and Zdunek, A. Method of acoustic emission in investigation of plant tissue cracking. *Acta Agrophisica*, 24, 87–95 1999 (in Polish).
14. Bogusz, B. Generation of mechanical acoustic pulses, in *Materials of 50th Open Seminar of Acoustics*. Polish Acoustic Society, Gliwice, 2003, pp. 522–525 (in Polish).

15. Mohamed, A.A.A., Jowitt, R., and Brennan, J.G. Instrumental and sensory evaluation of crispness. I—in friable foods. *Journal of Food Engineering*, 1, 55–75, 1982.
16. Dacremont, C. Spectral composition of eating sounds generated by crispy, crunchy and crackly foods. *Journal of Texture Studies*, 26, 27–43, 1995.
17. Tesch, R., Normand, M., and Peleg, M. Comparison of the acoustic and mechanical signatures of two cellular crunchy cereal foods at various water activity levels. *Journal of Food and Agricultural Science*, 70, 347–354, 1996.
18. Roudaut, G., Dacremont, C., Valles Pamies, B., Colas, B., and Le Meste, M. Crispness: A critical review on sensory and material science approaches. *Trends in Food Science & Technology*, 13, 217–227, 2002.
19. Chaunier, L., Courcoux, P., Valle, G.D., and Lourdin, D. Physical and sensory evaluation of cornflakes crispness. *Journal of Texture Studies*, 36, 93–118, 2005.
20. Chen, J., Karlsson, C., and Povey, M. Acoustic envelope detector for crispness assessment of biscuits. *Journal of Texture Studies*, 36, 139–156, 2005.
21. Marzec, A., Lewicki, P.P., Ranachowski, Z., and Debowski, T. The influence of moisture content on spectral characteristic of acoustic signals emitted by crisp bread samples, in *Proceedings of the AMAS Course on Nondestructive Testing of Materials and Structures*, Deputat, J. and Ranachowski, Z. Eds. Centre of Excellence for Advanced Materials and Structures, Warszawa, 2002, pp. 127–135.
22. *Piezoelectric Accelometer and Vibration Preamplifier Handbook* (catalog), Brüel & Kjær, Naerum, Dennmark, 1987.
23. Alchakra, W., Allaf, K., and Ville, J.M. Acoustical emission technique applied to the characterisation of brittle materials. *Applied Acoustics*, 52(1), 53–69, 1997.
24. Roudaut, G., Dacremont, C., and Le Meste, M. Influence of water on the crispness of cereal-based foods: Acoustic, mechanical, and sensory studies. *Journal of Texture Studies*, 29, 199–213, 1998.
25. Roudaut, G. Low moisture cereal products: Texture changes versus hydration, in *Proceedings of the AMAS Course on Nondestructive Testing of Materials and Structures II*. Deputat, J. and Ranachowski, Z., Eds. Centre of Excellence for Advanced Materials and Structure, Warszawa, 2003, pp. 77–89.
26. Lewicki, P.P., Gondek, E., and Ranachowski, Z. Influence of water activity on acoustic emission of breakfast cereals, in *Proceedings of the AMAS Course on Nondestructive Testing of Materials and Structures II*. Deputat, J. and Ranachowski, Z., Eds. Centre of Excellence for Advanced Materials and Structure, Warszawa, 2003, pp. 93–109.
27. Marzec, A., Lewicki, P.P., and Ranachowski, Z. Influence of water activity on mechanical and acoustic properties of wheat crisp bread. *Inzynieria Rolnicza*, 5(38), 101–108, 2002 (in Polish).
28. Lee, W.E., Glembin, C.T., and Munday, E.G. Analysis of food crushing sounds during mastication: Frequency-time studies. *Journal Texture Studies*, 19, 27–38, 1988.
29. Duizier, L., A review of acoustic research for studying the sensory perception of crisp, crunchy and crackly textures. *Trends in Food Science & Technology*, 12, 17–24, 2001.
30. Ranachowski, Z., Methods of measuring and analyzing of the acoustic emission signal, Reports of the Institute of Fundamental Technological Research, Warszawa, 1977 (in Polish).
31. Wade, A.P. Strategies for characterization of chemical acoustic emission signals near the conventional detection limit. *Analitica Chimica Acta*, 246, 23–42, 1991.
32. Luyten, H., Plijter, J.J., and van Vliet, T. Crispy/crunchy crusts of cellular solid foods. A literature review with discussion. *Journal of Texture Studies*, 35, 445–492, 2004.
33. Vickers, Z.M. and Bourne, M.C. Crispness in foods—a review. *Journal of Food Science*, 41, 153–157, 1976.
34. Drake, B.K. Food crunching sounds. An introductory study. *Journal of Food Science*, 28, 233–241, 1963.
35. Drake, B.K. Food crunching sounds comparison of objective and subjective data. *Journal of Food Science*, 30, 556–559, 1965.
36. Vickers, Z.M. and Wasserman, S.S. Sensory qualities of food sounds based on individual perceptions. *Journal of Texture Studies*, 10, 319–332, 1979.
37. Lewicki, P.P., Marzec, A., and Ranachowski, Z. Acoustic properties of crunchy products, in *3rd International Workshop on Water in Food*. Lausanne, Switzerland, 29–30th March, 2004 (CD).

38. Kowalski, S. Deformation and fracture of dried materials in thermodynamic description. *Zeszyty Naukowe Politechniki Lodzkiej, Inzynieria Chemiczna i Procesowa*, 24, 9–28, 1999 (in Polish).
39. Zdunek, A. and Konstankiewicz, K. Acoustic emission in investigation of planet tissue micro-cracking. *Transactions of the American Society of Agricultural Engineers*, 47, 1171–1177, 2004.
40. Zdunek, A. and Konstankiewicz, K. Acoustic emission as a method for the detection of fractures in the plant tissue caused by the external forces. *International Agrophysics*, 11, 223–227, 1997.
41. Ranachowski, Z. Instrumentation designed to investigate texture parameters of cereal food, in *Structures—Waves—Human Health. Acoustical Engineering*, Pamuszka, R. Ed. Polish Acoustical Society, Division in Krakow. Krakow, 2005, pp. 137–140.
42. Ranachowski, Z., Lewicki, P.P., and Marzec, A. Investigation of staling of bread using mechanical and acoustic methods, in *Proceedings of 51st Open Seminar of Acoustics*. Polish Acoustic Society. Gdansk, Poland, 2004, pp. 429–433.
43. Marzec, A., Lewicki, P.P., and Ranachowski, Z. Analysis of selected descriptor of acoustic emission signal generated in samples of flat crisp bread, in *Proceedings of 8th International Scientific Conference. Theoretical and Applicatory Problems of Agricultural Engineering*. Wroclaw Agricultural University Press, Wroclaw, 2005, vol. 2, pp. 66–69.
44. Marzec, A., Lewicki, P.P., and Ranachowski, Z. Mechanical and acoustic properties of dry cereal products. *Inzynieria Rolnicza*, 9(69), 207–214, 2005 (in Polish).
45. Marzec, A., Lewicki, P.P., Ranachowski, Z., and Debowski, T. Cereal food texture evaluation with application of mechanical and acoustical method, in *Proceedings of the AMAS Course on Nondestructive Testing of Materials and Structures II*. Deputat, J. and Ranachowski, Z. Eds. Centre of Excellence for Advanced Materials and Structure. Warszawa, 2003, pp. 111–133.
46. Ranachowski, Z., Gondek, E., Lewicki, P.P., and Marzec, A. Investigation of acoustic properties of compressed wheat bran flakes. *Archives of Acoustics*, 30, 255–265, 2005.
47. Kuropatwa, M. Influence of water activity on mechanical and acoustic properties of corn flakes. M.Sc. Thesis, Warsaw Agricultural University (SGGW), Warszawa, 2003 (in Polish).
48. Marzec, A., Lewicki, P.P., and Ranachowski, Z. Influence of water activity on acoustic properties of flat extruded bread. *Journal of Food Engineering*, 79, 410–422, 2007.
49. Lewicki, P.P., Marzec, A., Ranachowski, Z., and Debowski, T. Spectral characteristic of acoustic signals emitted by flat rye bread samples, in *Proceedings of 49th Open Seminar of Acoustics*. Polish Acoustic Society. Warszawa, Poland, 2002, pp. 453–458.
50. Poliszko, S., Klimek, D., and Moliñski, W. Acoustic emission activity of rehydrated corn extrudates, in *Properties of Water in Foods*, Lewicki, P.P. Ed. Warsaw Agricultural University Press, Warszawa, 1995, pp. 25–30.
51. Katz, E.E. and Labuza, T.P. Effect of water on the sensory crispness and mechanical deformation of snack food products. *Journal of Food Science*, 46, 403–409, 1981.
52. Liu, X. and Tan, J. Acoustic wave analysis for food crispness evaluation. *Journal of Texture Studies*, 30, 397–408, 1999.
53. Marzec, A., Lewicki, P.P., and Ranachowski, Z. Wavelet decomposition of acoustic emission signal generated in process of thin layers breaking, in *Proceedings of 50th Open Seminar of Acoustics*. Polish Acoustic Society, Gliwice, Poland, 2003, pp. 261–264.
54. Gondek, E. and Marzec, A. Influence of water activity on sensory assessment of texture and general quality of crackers. *Inżynieria Rolnicza*, 7(82), 181–187, 2006 (in Polish).
55. Guinard, J.X. and Mazzucchelli, R. The sensory perception of texture and mouthfeel. *Trends in Food Science & Technology*, 7, 213–219, 1996.
56. Stone, H. and Seidel, J.L. *Sensory Evaluation Practices*. Academic Press, Orlando, 1985.
57. Zdunek, A. and Bednarczyk, J. Effect of mannitol treatment on ultrasound emission during texture profile analysis in potato and apple tissue. *Journal of Texture Studies*, 37, 339–359, 2006.
58. Luyten, H. and van Vliet, T. Acoustic emission, fracture behavior and morphology of dry crispy foods. A discussion article. *Journal of Texture Studies*, 37, 221–240, 2006.
59. Duzier, L.M., Campanella, O.H., and Barnes, G.R.G. Sensory, instrumental and acoustic characteristics of extruded snack food products. *Journal of Texture Studies*, 29, 397–411, 1998.

60. Vickers, Z.M. Relationships of chewing sounds to judgments of crispness, crunchiness and hardness. *Journal of Food Science*, 47, 121–124, 1981.
61. Pasik, S. Dried fruits and vegetables texture evaluation with application of acoustical method. M.Sc. Thesis. Warsaw Agricultural University (SGGW). Department of Food Engineering and Process Management. Warszawa, 2006 (in Polish).
62. Marzec, A., Lewicki, P.P., and Kuropatwa, M. Influence of water activity on texture of corn flakes. *Acta Agrophysica*, 2006 (accepted for publication).
63. Gondek, E., Lewicki, P.P., and Ranachowski, Z. Influence of water activity on the acoustic properties of breakfast cereals. *Journal of Texture Studies*, 37, 497–515, 2006.
64. Seymour, S.K. and Hamann, D.D. Crispness and crunchiness of selected low moisture foods. *Journal of Texture Studies*, 19, 51–59, 1988.

Appendix A

Table A.1 Vapor Pressure, Saturated Liquid Density, Thermal Expansion Coefficient, Compressibility Coefficient, and Refractive Index of Water

t (°C)	Vapor Pressure (kPa)	Saturated Liquid Density (kg/m^3)	Thermal Expansion Coefficient (K^{-1} $\times$ 10^{-3})	Compressibility Coefficient (atm^{-1} $\times$ 10^{-6})	Refractive Na$^{0.5893}$
0	0.6102	999.84	−0.07	50.6	1.33464
10	1.2259	999.70	0.088	48.6	1.33389
20	2.3349	998.20	0.207	47.0	1.33299
25	3.1634	997.05	0.255	46.5	1.33287
30	4.2370	995.65	0.303	46.0	1.33192
40	7.3685	992.22	0.385	45.3	1.33051
50	12.3234	988.05	0.457	45.0	1.32894
60	19.8984	983.21	0.523	45.0	1.32718
70	31.1282	977.79	0.585	45.2	1.32511
80	47.3117	971.83	0.643	45.7	1.32287
90	70.0485	965.32	0.698	46.5	1.32050
100	101.2300	958.35	0.752	48.0	1.31783

t (°C)	Heat Capacity (kJ/kg K)	Latent Heat of Vaporization (kJ/kg)	Enthalpy (kJ/kg)	Entropy (kJ/kg K)	Free Energy (kJ/kg)	Velocity of Sound (m/s)
0	4.2174	2500.5	0.000	0.0000	0.000	1402.74
10	4.1919	2476.9	42.03	0.1511	42.03	1447.59
20	4.1816	2453.4	83.86	0.2963	83.86	1482.66
25	4.1703	2441.7	104.74	0.3669	104.74	1497.00
30	4.1782	2429.9	125.61	0.4364	146.46	1509.44
40	4.1783	2406.2	167.34	0.5718	167.33	1529.18
50	4.1804	2382.2	209.11	0.7031	209.10	1542.87
60	4.1841	2357.9	250.91	0.8304	250.89	1551.30
70	4.1893	2333.1	292.78	0.9542	292.75	1555.12
80	4.1961	2307.8	334.72	1.0747	334.67	1554.81
90	4.2048	2281.9	376.75	1.1920	376.68	1550.79
100	4.2156	2255.5	418.88	1.3063	418.77	1543.41

(*continued*)

Table A.1 (continued) Vapor Pressure, Saturated Liquid Density, Thermal Expansion Coefficient, Compressibility Coefficient, and Refractive Index of Water

t (°C)	Dielectric Constant	Viscosity (Pa · s) × 10^{-3}	Thermal Conductivity (W/m · K)	Surface Tension × 10^{-2} (N/m)	Electrical Conductivity	Enthalpy of Ionization (kJ/mol)
0	87.69	1.788	0.550	75.62	1.61	62.81
10	83.82	1.305	0.576	74.20	2.85	59.64
20	80.08	1.004	0.598	72.75	4.94	57.00
25	78.25	0.903	0.608	71.96	6.34	55.84
30	76.49	0.801	0.617	71.15	8.04	54.75
40	73.02	0.653	0.633	69.55	12.53	52.75
50	69.70	0.550	0.647	67.90	18.90	50.90
60	66.51	0.470	0.658	66.17	27.58	49.13
70	63.45	0.406	0.667	64.41	38.93	47.39
80	60.54	0.355	0.675	62.60	53.03	45.64
90	57.77	0.315	0.680	60.74	69.65	43.86
100	55.15	0.282	0.683	58.84	88.10	42.05

Source: Horvath, A.L. *Handbook of Aqueous Electrolyte Solutions*, Ellis Horwood, England, 1985.

Table A.2 Physical Constants of Water

Property	Values
Molecular weight	18.01534
Melting point at 101.32 kPa (1 atm)	0°C
Boiling point at 101.32 kPa (1 atm)	100°C
Critical temperature	374.15°C
Triple point	0.0099°C and 610.4 kPa
Heat of fusion at 0°C	334.0 kJ/kg
Heat of vaporization at 100°C	2257.2 kJ/kg
Heat of sublimation at 0°C	2828.3 kJ/kg

Source: Fennema, O.R., *Food Chemistry*, 2nd ed., Fennema, O.R., ed., Marcel Dekker, New York, 1985, pp. 23–67.

Table A.3 Vapor Pressures of Water and Ice at Subfreezing Temperatures

t (°C)	Vapor Pressure (kPa)		Water Activity
	Liquid Water[a]	Ice[b]	
0	0.6104[b]	0.6104	1.000[c]
−5	0.4216[b]	0.4016	0.953
−10	0.2865[b]	0.2599	0.907
−15	0.1914[b]	0.1654	0.864
−20	0.1254[d]	0.1034	0.820
−25	0.0806[d]	0.0635	0.790
−30	0.0509[d]	0.0381	0.750
−40	0.0189[d]	0.1290	0.680
−50	0.0064[d]	0.0039	0.620

Source: Fennema, O.R., *Food Chemistry*, 2nd ed., Fennema, O.R., ed., Marcel Dekker, New York, 1985, pp. 23–67.

[a] Supercooled at all temperatures except degree centigrade.
[b] Observed data.
[c] Applies only to pure water.
[d] Calculated data.

Appendix B

Table B.1 Properties of Saturated Steam

t (°C)	Vapor Pressure (kPa)	Specific Volume (m^3/kg)		Enthalpy (kJ/kg)		Entropy (kJ/kg·K)	
		Liquid	Saturated Vapor	Liquid (H_c)	Saturated Vapor (H_v)	Saturated Liquid	Vapor
0.01	0.6113	0.0010002	206.1360	0.00	2501.4	0.0000	9.1562
3	0.7575	0.0010001	168.1320	12.57	2506.9	0.0457	9.0773
6	0.9349	0.0010001	137.7340	25.20	2512.4	0.0912	9.0003
9	1.1477	0.0010003	113.3680	37.80	2517.9	0.1362	8.9253
12	1.4022	0.0010005	93.7840	50.41	2523.4	0.1806	8.8524
15	1.7051	0.0010009	77.9260	62.99	2528.9	0.2245	8.7814
18	2.064	0.0010014	65.0380	75.58	2534.4	0.2679	8.7123
21	2.487	0.0010020	54.5140	88.14	2539.9	0.3109	8.6450
24	2.985	0.0010027	45.8830	100.70	2545.4	0.3534	8.5794
27	3.567	0.0010035	38.7740	113.25	2550.8	0.3954	8.5156
30	4.246	0.0010043	32.8940	125.79	2556.3	0.4369	8.4533
33	5.034	0.0010053	28.0110	138.33	2561.7	0.4781	8.3927
36	5.947	0.0010063	23.9400	150.86	2567.1	0.5188	8.3336
40	7.384	0.0010078	19.5230	167.57	2574.3	0.5725	8.2570
45	9.593	0.0010099	15.2580	188.45	2583.2	0.6387	8.1648
50	12.349	0.0010121	12.0320	209.33	2592.1	0.7038	8.0763
55	15.758	0.0010146	9.56800	230.23	2600.9	0.7679	7.9913
60	19.940	0.0010172	7.67100	251.13	2609.6	0.8312	7.9096
65	25.030	0.0010199	6.19700	272.06	2618.3	0.8935	7.8310
70	31.190	0.0010228	5.04200	292.98	2626.8	0.9549	7.7553
75	38.580	0.0010259	4.13100	313.93	2635.3	1.0155	7.6824
80	47.390	0.0010291	3.40700	334.91	2643.7	1.0753	7.6122
85	57.830	0.0010325	2.82800	355.90	2651.9	1.1343	7.5445
90	70.140	0.0010360	2.36100	376.92	2660.1	1.1925	7.4791
95	84.550	0.0010397	1.98190	397.96	2668.1	1.2500	7.4159
100	101.350	0.0010435	1.67290	419.04	2676.1	1.3069	7.3549
105	120.820	0.0010475	1.41940	440.15	2683.8	1.3630	7.2958
110	143.270	0.0010516	1.21020	461.30	2691.5	1.4185	7.2387
115	169.060	0.0010559	1.03660	482.48	2699.0	1.4734	7.1833
120	198.530	0.0010603	0.89190	503.71	2706.3	1.5276	7.1296
125	232.100	0.0010649	0.77060	524.99	2713.5	1.5813	7.0775
130	270.100	0.0010697	0.66850	546.31	2720.5	1.6344	7.0269
135	313.000	0.0010746	0.58220	567.69	2727.3	1.6870	6.9777
140	316.300	0.0010797	0.50890	589.13	2733.9	1.7394	6.9299
145	415.400	0.0010850	0.44630	610.63	2740.3	1.7907	6.8833

(*continued*)

Table B.1 (continued) Properties of Saturated Steam

t (°C)	Vapor Pressure (kPa)	Specific Volume (m^3/kg)		Enthalpy (kJ/kg)		Entropy (kJ/kg · K)	
		Liquid	Saturated Vapor	Liquid (H_c)	Saturated Vapor (H_v)	Saturated Liquid	Vapor
150	475.800	0.0010905	0.39280	632.20	2746.5	1.8418	6.8379
155	543.100	0.0010961	0.34680	653.84	2752.4	1.8925	6.7935
160	617.800	0.0011020	0.30710	675.55	2758.1	1.9427	6.7502
165	700.500	0.0011080	0.27270	697.34	2763.5	1.9925	6.7078
170	791.700	0.0011143	0.24280	719.21	2768.7	1.0419	6.6663
175	892.000	0.0011207	0.21680	741.17	2773.6	2.0909	6.6256
180	1002.100	0.0011274	0.19405	763.22	2778.2	2.1396	6.5857
190	1254.400	0.0011444	0.15654	807.62	2786.4	2.2359	6.5079
200	1553.800	0.0011565	0.12736	852.45	2793.2	2.3309	6.4323
225	2548.000	0.0011992	0.07849	966.78	2803.3	2.5639	6.2503
250	3973.000	0.0012512	0.05013	1085.36	2801.5	2.7927	6.0730
275	5942.000	0.0013168	0.03279	1210.07	2785.0	3.0208	5.8938
300	8581.000	0.0010436	0.02167	1344.00	2749.0	3.2534	5.7045

Source: Keenan, J.H., Keyes, F.G., Hill, P.G., and Moore, J.G., *Steam Tables: Thermodynamic Properties of Water Including Vapor, Liquid, and Solid Phases (Metric Measurements)*, Wiley, New York, 1969.

Table B.2 Specific Volume of Superheated Steam

		Specific Volume (m^3/kg)							
		Temperature (°C)							
P (kPa)	T_s (°C)	100	150	200	250	300	360	420	500
10	45.81	17.196	19.512	21.825	24.136	26.445	29.216	31.986	35.679
50	81.33	3.4180	3.8890	4.3560	4.8200	5.2840	5.8390	6.3940	7.1340
75	91.78	2.2700	2.5870	2.9000	3.2110	3.5200	3.8910	4.2620	4.7550
100	99.63	1.6958	1.9364	2.1720	2.4060	2.6390	2.9170	3.1950	3.5650
150	111.37		1.2853	1.4443	1.6012	1.7570	1.9432	2.1290	2.3760
400	143.63		0.4708	0.5342	0.5951	0.6458	0.7257	0.7960	0.8893
700	164.97			0.2999	0.3363	0.3714	0.4126	0.4533	0.5070
1000	179.91			0.2060	0.2327	0.2579	0.2873	0.3162	0.3541
1500	198.32			0.1325	0.1520	0.1697	0.1899	0.2095	0.2352
2000	212.42				0.1114	0.1255	0.1411	0.1562	0.1757
2500	223.99				0.0870	0.0989	0.1119	0.1241	0.1399
3000	233.90				0.0706	0.0811	0.0923	0.1028	0.1162

Source: Keenan, J.H., Keyes, F.G., Hill, P.G., and Moore, J.G., *Steam Tables: Thermodynamic Properties of Water Including Vapor, Liquid, and Solid Phases (Metric Measurements)*, Wiley, New York, 1969.

Table B.3 Enthalpy Volume of Superheated Steam

		Enthalpy (kJ/kg)							
		Temperature (°C)							
P (kPa)	T_s (°C)	100	150	200	250	300	360	420	500
10	45.81	2687.5	2783.0	2879.5	2977.3	3076.5	3197.6	3320.9	3489.1
50	81.33	2682.5	2780.1	2877.7	2976.0	3075.5	3196.8	3320.4	3488.7
75	91.78	2679.4	277.82	2876.5	2975.2	3074.9	3196.4	3320.0	3488.4
100	99.63	2676.2	2776.4	2875.3	2974.3	3074.3	3195.9	3319.6	3488.1
150	111.37		2772.6	2872.9	2972.7	3073.1	3195.0	3318.9	3487.6
400	143.63		2752.8	2860.5	2964.2	3066.8	3190.3	3315.3	3484.9
700	164.97			2844.8	2953.6	3059.1	3184.7	3310.9	3481.7
1000	179.91			2827.9	2942.6	3051.2	3178.9	3306.5	3478.5
1500	198.32			2796.8	2923.3	3037.6	3169.2	3299.1	3473.1
2000	212.42				2902.5	3023.5	3159.3	3291.6	3467.6
2500	223.99				2880.1	3008.8	3149.1	3284.0	3462.1
3000	233.90				2855.8	2993.5	3138.7	3276.3	3456.5

Source: Keenan, J.H., Keyes, F.G., Hill, P.G., and Moore, J.G., *Steam Tables: Thermodynamic Properties of Water Including Vapor, Liquid, and Solid Phases (Metric Measurements)*, Wiley, New York, 1969.

Appendix C

Table C.1 Bessel Functions of First Kind

x	$J_0(x)$	$J_1(x)$	x	$J_0(x)$	$J_1(x)$
0.0	1.000	0.000	2.50	−0.048	0.497
0.1	0.988	0.050	2.60	−0.097	0.471
0.2	0.990	0.100	2.70	−0.142	0.442
0.3	0.978	0.148	2.80	−0.185	0.410
0.4	0.960	0.196	2.90	−0.224	0.375
0.5	0.938	0.242	3.00	−0.260	0.339
0.6	0.912	0.287	3.10	−0.292	0.301
0.7	0.881	0.329	3.20	−0.320	0.261
0.8	0.846	0.369	3.30	−0.344	0.221
0.9	0.808	0.406	3.40	−0.364	0.179
1.0	0.765	0.440	3.50	−0.380	0.137
1.1	0.720	0.471	3.60	−0.392	0.095
1.2	0.671	0.498	3.70	−0.399	0.054
1.3	0.646	0.511	3.80	−0.402	0.013
1.4	0.567	0.542	3.90	−0.402	−0.027
1.5	0.512	0.558	4.00	−0.397	−0.066
1.6	0.455	0.570	4.10	−0.389	−0.103
1.7	0.398	0.578	4.20	−0.376	−0.139
1.8	0.340	0.582	4.30	−0.361	−0.172
1.9	0.282	0.581	4.40	−0.342	−0.203
2.0	0.224	0.577	4.50	−0.320	−0.231
2.1	0.167	0.568	4.60	−0.296	−0.256
2.2	0.110	0.556	4.70	−0.269	−0.279
2.3	0.055	0.540	4.80	−0.240	−0.298
2.4	0.002	0.520	4.90	−0.210	−0.315
			5.00	−0.178	−0.328

Source: Tabulated values are available in the *Handbook of Mathematical Functions*, U.S. Department of Commerce, Applied Mathematics Series No. 55 (cited by Whitaker, S., *Elementary Heat Transfer Analysis*, Pergamon Press, New York, 1976).

Index

A

B

C

D

E

F

G

H

Q

R

S

T

U

V

W